Aquatic Parasitology: Ecological and Environmental Concepts and Implications of Marine and Freshwater Parasites

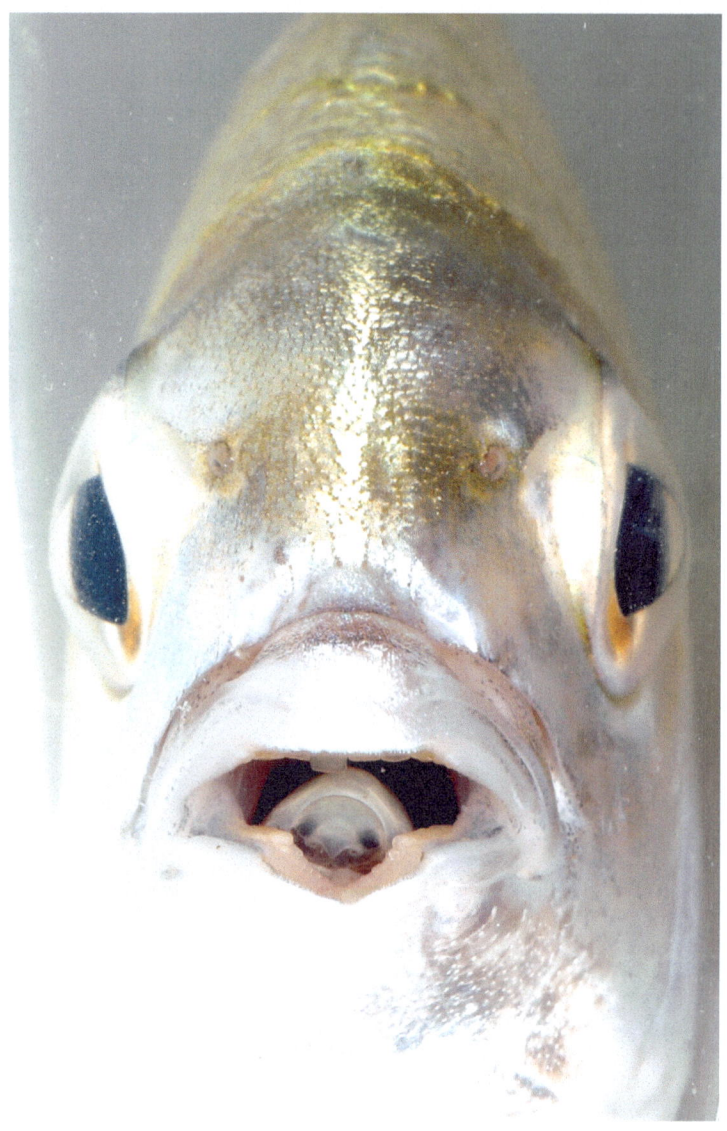

"The famous tongue replacement isopod, *Ceratothoa famosa*, in the mouth of the Cape seabream, *Diplodus capensis*"

Nico J. Smit • Bernd Sures
Editors

Aquatic Parasitology: Ecological and Environmental Concepts and Implications of Marine and Freshwater Parasites

Springer

Editors
Nico J. Smit
Water Research Group
Unit for Environmental Sciences and
Management
North-West University
Potchefstroom, South Africa

Bernd Sures
Department of Aquatic Ecology and
Centre for Water and Environmental
Research (ZWU)
University of Duisburg-Essen
Essen, Germany

ISBN 978-3-031-83902-3 ISBN 978-3-031-83903-0 (eBook)
https://doi.org/10.1007/978-3-031-83903-0

© The Editor(s) (if applicable) and The Author(s) 2025. This book is an open access publication.

Open Access This book is licensed under the terms of the Creative Commons Attribution-NonCommercial-NoDerivatives 4.0 International License (http://creativecommons.org/licenses/by-nc-nd/4.0/), which permits any noncommercial use, sharing, distribution and reproduction in any medium or format, as long as you give appropriate credit to the original author(s) and the source, provide a link to the Creative Commons license and indicate if you modified the licensed material. You do not have permission under this license to share adapted material derived from this book or parts of it.

The images or other third party material in this book are included in the book's Creative Commons license, unless indicated otherwise in a credit line to the material. If material is not included in the book's Creative Commons license and your intended use is not permitted by statutory regulation or exceeds the permitted use, you will need to obtain permission directly from the copyright holder.

This work is subject to copyright. All commercial rights are reserved by the author(s), whether the whole or part of the material is concerned, specifically the rights of translation, reprinting, reuse of illustrations, recitation, broadcasting, reproduction on microfilms or in any other physical way, and transmission or information storage and retrieval, electronic adaptation, computer software, or by similar or dissimilar methodology now known or hereafter developed. Regarding these commercial rights a non-exclusive license has been granted to the publisher.

The use of general descriptive names, registered names, trademarks, service marks, etc. in this publication does not imply, even in the absence of a specific statement, that such names are exempt from the relevant protective laws and regulations and therefore free for general use.

The publisher, the authors and the editors are safe to assume that the advice and information in this book are believed to be true and accurate at the date of publication. Neither the publisher nor the authors or the editors give a warranty, expressed or implied, with respect to the material contained herein or for any errors or omissions that may have been made. The publisher remains neutral with regard to jurisdictional claims in published maps and institutional affiliations.

This Springer imprint is published by the registered company Springer Nature Switzerland AG
The registered company address is: Gewerbestrasse 11, 6330 Cham, Switzerland

If disposing of this product, please recycle the paper.

Preface

What could be more fascinating than the lives of parasites? These organisms must always coexist with others. At first glance, this coexistence seems negative for one partner—the host—who suffers from various effects caused by the parasite. But is that the whole story? Certainly not. Parasites can also have positive impacts: they structure communities, act as ecosystem engineers, drive evolutionary processes, and may even help reduce pollutant concentrations in their hosts' tissues. These effects challenge the simplistic view that parasites are merely harmful organisms intent on damaging their hosts.

Of course, there are fatal consequences as well, such as when humans are infected with malaria pathogens, where the disease effects can only be managed—at least partially—with medication. Overall, parasites exhibit a wide range of behaviours, from adventurous life cycles to pathogenic effects, and even surprising benefits to their hosts. How could one not be captivated by such creatures?

We, Nico Smit and Bernd Sures, have shared this fascination for parasites for many years. Naturally, it made sense for us to combine our passion into a book. Both of us are long-time parasitologists with complementary expertise and significant overlap in our scientific interests. The idea for a joint book titled *Aquatic Parasitology: Ecological and Environmental Concepts and Implications of Marine and Freshwater Parasites* was quickly formed and discussed with Springer. Through collaboration with colleagues worldwide, we were able to present a global perspective on aquatic parasitology, ensuring that both the Southern and Northern Hemispheres were equally represented. We also secured funding to make this book freely available for download by anyone interested in science, thus sharing our fascination for parasites and reducing barriers to scientific knowledge.

The writing and editing process was both productive and enjoyable. During this time, we visited each other in Germany and South Africa and worked intensively on the texts for several weeks, which deepened our friendship—a bond that now extends to our families. This is undoubtedly another unexpected, yet positive outcome brought about by our shared passion for parasites.

We hope all readers enjoy this book and that many will come to share our enthusiasm for parasites.

Potchefstroom, South Africa Nico J. Smit
Essen, Germany Bernd Sures

Contents

Part I Introduction and Lifecycle Strategies of Aquatic Parasites

1. Introduction to Aquatic Parasitology: Ecological and Environmental Concepts and Implications of Marine and Freshwater Parasites.................................. 3
   Nico J. Smit and Bernd Sures

2. The Protists: Small But Mighty.......................... 9
   Sonja Rückert

3. Biology and Life Cycles of Microsporidia and Myxozoa....... 41
   Daniel Grabner and Ivan Fiala

4. Ecology and Life Cycles of Aquatic Pathogenic Fungi 71
   Ché Weldon

5. Biology and Life Cycle of Helminths...................... 89
   Bernd Sures, Dakeishla M. Díaz-Morales, Russell Q-Y. Yong, Anja Erasmus, and Jessica Schwelm

6. Biology and Life Cycles of Parasitic Arthropoda Infesting Aquatic Hosts 125
   Kerry A. Hadfield, Anja Erasmus, Niel L. Bruce, and Nico J. Smit

Part II Ecological Principles and Latest Research Developments in Aquatic Parasitology

7. Unveiling the Hidden Players: Exploring the Intricate Dance of Aquatic Parasites, Host Biodiversity and Ecosystem Health................................... 167
   Clarisse Louvard, Kerry A. Hadfield, Maarten P. M. Vanhove, Bernd Sures, and Nico J. Smit

8. Impact of Aquatic Parasites on Host Community Structure 199
   Kim N. Mouritsen

9 **Molecular Approaches for Investigating the Population Genetic Structure of Aquatic Parasites**.................... 229
Simonetta Mattiucci, Paolo Cipriani, Daniele Canestrelli, Giuseppe Nascetti, and Marialetizia Palomba

10 **Chemosensory Behaviour of Free-Living Stages of Aquatic Parasites** 255
Clayton Vondriska and Paul C. Sikkel

11 **Evolutionary Biology of Aquatic Parasites**................. 277
Robert Poulin and Jerusha Bennett

12 **Parasites Under Extreme Conditions** 297
Christian Selbach and Rachel A. Paterson

13 **Aquatic Parasite Conservation** 325
Marliese Truter, Bjoern C. Schaeffner, and Nico J. Smit

14 **Detecting and Assessing Aquatic Parasite Diversity Using Environmental DNA**............................. 361
Kamil Hupało, Isabel Blasco-Costa, Alejandro Trujillo-González, and Florian Leese

15 **Ecology of Marine Mammal Parasites** 383
Kristina Lehnert and Jesús S. Hernández-Orts

16 **Host–Parasite Trophic Interactions as Revealed by Stable Isotope Analyses: Determinants for Trophic and Isotopic Niches of Hosts and Their Associated Parasites**.............. 415
Milen Nachev, Philip M. Riekenberg, Maik A. Jochmann, Ana Born-Torrijos, Marcel T. J. van der Meer, Nico J. Smit, Torsten C. Schmidt, David W. Thieltges, and Bernd Sures

Part III Environmental Aquatic Parasitology and Ecological Applications: Impacts on Animal and Human Life

17 **Parasites as Biological Tags in Aquatic Hosts**............... 445
Juan T. Timi

18 **Invasion Biology in the Context of Aquatic Host–Parasite Interactions** 471
Dakeishla M. Díaz-Morales, Bernd Sures, E. Rosa Jolma, and David W. Thieltges

19 **Aquatic Foodborne Zoonoses**............................ 493
Shokoofeh Shamsi

20 **Environmental Parasitology: Interactions Between Parasites and Pollutants** 509
Bernd Sures, Milen Nachev, Daniel Grabner, Victor Wepener, and Sonja Zimmermann

21	**Parasites and Pollutants: Allometric Toxicology for Parasitologists**	535
	Andy Dobson	
22	**Climate Change and Parasitism in Aquatic Ecosystems**	547
	David J. Marcogliese	
23	**Parasites in Aquaculture**	595
	Cecilia Power, Sho Shirakashi, Nathan J. Bott, and Barbara F. Nowak	

Species List of Parasites, Hosts and Vectors of Aquatic Parasitology: Ecological and Environmental Concepts and Implications of Marine and Freshwater Parasites 621
Russell Q. -Y. Yong

Glossary of Terms Used in Aquatic Parasitology: Ecological and Environmental Concepts and Implications of Marine and Freshwater Parasites 649

About the Editors

Nico J. Smit is a Professor of Ecology in the School of Biological Sciences at North-West University, Potchefstroom Campus, South Africa. Nico's research focuses on the biodiversity, taxonomy, and ecology of marine and freshwater parasites and he has authored and co-authored more than 250 scientific papers and three edited books on these and other related topics. To date, 23 PhD and 38 MSc students have graduated under his supervision. His research and teaching excellence has internationally been recognised through visiting Professor appointments at University of Duisburg-Essen, Germany, University of Queensland, Australia, and Masaryk University, Czech Republic, and through his current appointment as Adjunct Professor of Marine Biology and Ecology at the Rosenstiel School of Marine, Atmospheric and Earth Science, University of Miami, USA. Nico has contributed to the management of national and international academic societies as president of the Parasitological Society of Southern Africa (PARSA), president of the South African Society of Aquatic Scientists (SASAqS) and committee member of the International Symposium on Fish Parasites (ISFP). He is currently rated by the South African National Research Foundation (NRF) as an internationally acclaimed scientist in the field of aquatic parasitology.

Bernd Sures is a highly regarded scientist in the fields of parasitology, ecotoxicology, and aquatic ecology. He is a Professor at the University of Duisburg-Essen and Founding Director of the Research Center One Health Ruhr of the Research Alliance Ruhr in Germany. He has established himself as a leader in research on the interactions between parasites, pollutants, and aquatic organisms. He is also Extraordinary Professor at the Unit for Environmental Sciences and Management at North-West University, Potchefstroom Campus, South Africa. His research has greatly advanced our understanding of the interactions between parasites and pollutants and their combined impact on the health of aquatic organisms. Driven by a passion for uncovering the effects of multiple stressors on aquatic environmental health, he frequently integrates ecotoxicological and parasitological approaches in his work. Bernd has published >250 scientific papers and has coordinated numerous large-scale, interdisciplinary research projects. He is also recognised for his commitment to fostering collaboration and mentorship in environmental sciences, actively supporting both national and international research initiatives and the next generation of scientists.

Part I

Introduction and Lifecycle Strategies of Aquatic Parasites

Introduction to Aquatic Parasitology: Ecological and Environmental Concepts and Implications of Marine and Freshwater Parasites

Nico J. Smit and Bernd Sures

Abstract

This book provides a comprehensive exploration of aquatic parasitology, addressing the ecological and environmental roles of marine and freshwater parasites. Spanning 23 chapters, it offers a detailed overview of parasites' life cycle strategies, their ecological implications and their impact on human and animal health. The book fills an obvious gap in the literature by focusing on the environmental aspects of aquatic parasitology, integrating traditional ecological concepts with modern molecular techniques such as environmental DNA (eDNA) and stable isotope methodologies. It emphasises the importance of understanding parasite–host dynamics in the context of global change, including climate change and habitat destruction, and advocates for the conservation of parasitic species as vital components of biodiversity. By presenting the latest research developments and future trends, this volume aims to inspire further studies in aquatic parasitology, contributing to the sustainability and health of aquatic ecosystems worldwide. The book is a valuable resource for those seeking to understand the complex interactions between aquatic parasites and their environments and the broader implications for ecology, conservation and the one health approach.

N. J. Smit (✉)
Water Research Group, Unit for Environmental Sciences and Management, North-West University, Potchefstroom, South Africa
e-mail: nico.smit@nwu.ac.za

B. Sures
Water Research Group, Unit for Environmental Sciences and Management, North-West University, Potchefstroom, South Africa

Department of Aquatic Ecology and Centre for Water and Environmental Research (ZWU), University of Duisburg-Essen, Essen, Germany

Research Center One Health Ruhr, Research Alliance Ruhr, University Duisburg-Essen, Essen, Germany
e-mail: bernd.sures@uni-due.de

1.1 Introduction

What could be more fascinating than the life of parasites? It is a perfect blend of mystery, surprise and complex interactions between organisms, often leading to unexpected outcomes. This duality highlights parasites' ambivalent nature: they are disease agents, yet they also drive evolutionary progress. Parasites and their hosts engage in a continuous evolutionary arms race, leading to numerous reciprocal adaptations. Studying parasites is valuable across ecological, medical and scientific domains. Ecologically, parasites can shape the behaviour, population dynamics and evolution of their hosts, regulating populations and sustaining biodiversity. Researching parasites allows for a deeper understanding of these ecological relationships and the equilibrium of natural systems. Medically, parasites cause significant diseases in humans and animals, prompting research into prevention, diagnosis and treatment. Understanding their life cycles and transmission is crucial for controlling and preventing outbreaks. From a 'One Health' perspective, studying parasites also sheds light on the impact of global change effects, such as climate change and habitat destruction, on disease dynamics. As climates change, the range and prevalence of parasitic infections are expected to shift, making it crucial to understand these patterns for future disease control and to incorporate parasites more effectively into ecological concepts.

Over the past two decades, several books have delved into the biology of aquatic parasites, primarily focusing on taxonomic classifications and the basic biological attributes of mostly fish parasites. Notable works include those edited by Patric Woo and colleagues on fish parasites (Woo 2006; Woo and Buchmann 2012) and Klaus Rohde on marine parasites (Rohde 2005). Additionally, comprehensive guides on their identification, diseases caused by these parasites, and methodologies for dealing with them from a research perspective have been published (Klimpel et al. 2019; Scholz et al. 2018; Piazzon et al. 2021). Specialised books have also explored specific groups, such as digenetic trematodes (Madhavi and Bray 2018), parasitic crustaceans (Smit et al. 2019) and polystomid flatworms (Du Preez et al. 2023). However, there has been a gap in the literature addressing the ecological and environmental features of marine and freshwater parasites comprehensively.

This book seeks to fill that gap by providing an extensive overview of the ecological and environmental aspects of aquatic parasitology. Spanning 23 chapters, and authored by leading experts, this book presents the current state of knowledge and future trends in the field. The book is divided into three parts to introduce readers to (1) the fundamental life cycle strategies of various aquatic parasite groups; (2) the ecological implications of these parasites that include the latest research developments in aquatic parasitology, such as molecular tools, environmental DNA (eDNA) and stable isotope methodologies; and (3) the impact and implications of aquatic parasites on animal and human life, including their role in invasions, pollution detection, aquaculture and ramifications of anthropogenic climate change. While this book does not aim to provide exhaustive detail on every aspect, it offers a concise overview of the most critical topics. It equips readers with a clear understanding of the life cycles of key parasite groups, their integration into new scientific concepts and the broader ecological and environmental implications.

1.2 Basic Life Strategies of Aquatic Parasites

The first part of the book explores the diverse life strategies of aquatic parasites, ranging from protists to arthropods (Part I; Chaps. 2–6). Aquatic parasitic protists, discussed in Chap. 2, comprise a polyphyletic group of microscopic organisms thriving in aquatic environments. These protists, including amoebae, ciliates, flagellates and sporozoans, exhibit a wide range of host relationships, from benign to pathogenic, affecting humans, animals and plants. The ecological impact of these relationships and their adaptive strategies are key themes explored throughout this chapter.

In the third chapter, the focus shifts to the life cycles and biology of Microsporidia and Myxozoa, two parasitic groups with complex life cycles involving multiple hosts. Despite their taxonomic differences, both groups share similarities in their spore-based transmission and significant ecological impacts on their hosts. This chapter also examines the unique adaptations and evolutionary trajectories that have enabled these parasites to thrive in diverse aquatic environments.

The review of aquatic pathogenic fungi in Chap. 4 highlights their role in aquatic ecosystems and their ability to adapt to water-based environments. These fungi impact a variety of aquatic hosts, including amphibians and fish, and their life cycles and ecological roles are influenced by environmental factors and human activities. This chapter underscores the importance of understanding fungal parasitism in maintaining ecosystem balance and health.

Chapter 5 covers the biology and life cycles of helminths inhabiting both freshwater and marine systems. Discussing their diverse life cycles, morphological adaptations and ecological impacts, this chapter provides a comprehensive understanding of their role in aquatic ecosystems. The complex interactions between helminths and their hosts, along with their influence on host population dynamics, are some of the key areas explored.

Parasitic arthropods, particularly crustaceans, form the main theme of Chap. 6, which details their complex life cycles and host interactions. The biology of various parasitic crustaceans, including copepods and isopods and their ecological implications for host populations and aquatic environments are thoroughly examined. This chapter further provides insights into the evolutionary pressures and ecological niches that contribute to the large diversity of parasitic arthropods found in the aquatic environment.

These five chapters of Part I lay the foundation for understanding the diverse and intricate life strategies of aquatic parasites, highlighting the evolutionary and ecological adaptations that enable them to thrive in various aquatic environments. By examining the biology and life cycles of these parasites, readers gain a foundational understanding of their roles and impacts within aquatic ecosystems, setting the stage for the more focused ecological and applied studies that follow.

1.3 Ecological Principles in Aquatic Parasitology

The second part of the book (Part II; Chaps. 7–16) addresses the broader ecological principles that govern the interactions between aquatic parasites and their environments. The first chapter in this part, Chap. 7, explores the intricate roles of aquatic parasites in ecosystems, emphasising their impact on host populations, biodiversity and ecosystem health. Key concepts within ecological parasitology such as parasite-induced direct trait modification, direct parasite competition and parasite-induced density-mediated indirect effects (DMIEs), amongst others, are explained through examples from the aquatic environment. Further key themes in this chapter are the complex interactions between aquatic parasites and hosts, including behavioural modifications and co-evolutionary dynamics. The chapter is concluded with a call for future research on parasite-induced trait-mediated indirect effects (TMIEs) and density-mediated indirect effects (DMIEs) within aquatic food webs, as well as the broader ecological consequences of parasite–host dynamics in aquatic settings.

The influence of parasites on aquatic community structure is highlighted in Chap. 8, through direct and indirect effects on host populations and biotic processes. This chapter synthesises evidence showing the context-dependent role of parasitism in maintaining biodiversity and ecological stability. It provides a detailed analysis of how parasites can drive evolutionary and ecological processes in aquatic environments.

Molecular approaches to studying the genetic population structure of aquatic parasites are discussed in Chap. 9. This chapter examines how genetic and molecular tools provide insights into host–parasite interactions and the evolutionary processes shaping these relationships. Through a case study on the population genomics of anisakid nematodes of the genus *Anisakis*, the authors highlight the advancements in molecular techniques and their applications in understanding parasite diversity and adaptation.

Chemosensory behaviour of free-living stages of aquatic parasites is reviewed in Chap. 10. This review focuses on how chemical cues influence parasite behaviour, aiding in host location and environmental navigation. In addition, this chapter also provides a detailed look at the sensory adaptations that enable parasites to successfully locate and infect their hosts in aquatic environments.

In Chap. 11, the evolutionary biology of aquatic parasites is discussed, exploring how aquatic habitats influence parasite evolution and the impact of anthropogenic changes on these processes. This chapter emphasises the role of environmental pressures in shaping the life history strategies and evolutionary trajectories of aquatic parasites.

Adaptations of aquatic parasites to extreme conditions, such as deep-sea and polar environments, are the main theme of Chap. 12. This chapter highlights the specialised lifestyles of parasites in these habitats and their ecological significance. The unique adaptations that enable survival in extreme conditions are discussed, providing insights into the resilience and diversity of aquatic parasites.

The importance of conserving aquatic parasites is advocated in Chap. 13, emphasising their contribution to our understanding of aquatic biodiversity and the need to integrate parasite conservation into broader biodiversity efforts. Here the authors also provide a quantitative analysis of the literature on aquatic parasite conservation spanning from 1990 to 2023. This chapter further discusses the ecological roles of parasites and the potential consequences of their loss on ecosystem function and stability, concluding with an in-depth case study on chondrichthyan parasites to illustrate the critical role of parasites in ecosystem health and the importance of their conservation.

The use of environmental DNA (eDNA) in detecting and assessing aquatic parasite diversity is reviewed in Chap. 14. Current applications, challenges and future potential of eDNA methodologies in parasitology are discussed, highlighting the transformative impact of this technology on parasite detection and monitoring.

One of the most often overlooked and least researched group of aquatic parasites are those of marine mammals, most probably due to the limited opportunities to sample their free-ranging and vulnerable hosts. In Chap. 15, the authors provide a complete synthesis of our knowledge on all the various groups of both protists and metazoan parasites of marine mammals by reviewing marine mammal parasite diversity, life cycles and pathogenicity. This chapter further provides information on how these parasites can impact host populations, leading to decreased fitness and mortality of marine mammals. The chapter concludes by highlighting the ecological differences between the parasites from different hemispheres, emphasising the importance of continued investigation in this field.

In Chap. 16, stable isotope analysis (SIA) is introduced as a powerful tool for examining trophic interactions and nutrient fluxes within host–parasite systems. The chapter provides an overview of stable isotope chemistry and isotope ecology, making the topic accessible to those unfamiliar with the field. It reviews studies on host–parasite interactions, discussing fractionation patterns between parasitic taxa and their hosts, and the impact of parasitism on host isotopic composition. Nutrient uptake in parasites and nutrient fluxes within host–parasite systems using compound-specific stable isotope analysis are also explored, presenting new avenues for future research in this area.

The second part of this book (Chaps. 7–16), therefore, underscores the pivotal ecological principles that shape aquatic parasitology. By integrating evolutionary, molecular and ecological perspectives, it provides a comprehensive understanding of the multifaceted roles parasites play in aquatic environments. This foundation is crucial for applying knowledge to real-world management and conservation efforts, as explored in the subsequent part.

1.4 Ecological Applications of Aquatic Parasitology

The final part (Part III; Chaps. 17–23) addresses practical applications of ecological research on aquatic parasites. The close association between parasites and their hosts provides the opportunity for researchers to gain valuable ecological information on the hosts and their populations, communities and even the ecosystem through in-depth studies of their parasites. This is specifically applicable to aquatic ecosystems and forms the main theme of Chap. 17, which focuses on the use of parasites as biological tags. This chapter compiles research on how parasites can provide valuable ecological information about their hosts, including insights into host behaviour, migration patterns and environmental interactions. The applications of using parasites as bioindicators for environmental monitoring and assessment are also discussed.

Chapter 18 delves into the significance of parasite invasion biology, by examining host–parasite interactions, as well as the stages, mechanisms and effects of biological invasions in aquatic environments from a parasitological perspective. The various invasion hypotheses are discussed along with their impacts on native aquatic ecosystems. Direct and indirect effects of invasive parasites on native species and aquatic ecosystems are illustrated through examples. The chapter concludes with management and mitigation strategies for addressing the challenges posed by aquatic parasite invasions.

With a focus on aquatic food-borne zoonoses, Chap. 19 highlights the rapid increase in global seafood consumption and the lag in seafood safety guidelines in many countries. Key information on prevalent aquatic parasites of zoonotic concern, including well-known and neglected species in freshwater and marine aquaculture, is provided. The chapter emphasises that food-borne parasitic diseases are common in both developing and developed countries and advocates for a 'One Health' approach to ensure safe seafood consumption globally.

Chapter 20 provides a comprehensive overview of Environmental Parasitology, a field that has evolved over the past 30 years. The chapter delves into the complex interactions between parasites and pollutants, emphasising their combined effects on host health. It also explores how parasites can modify pollutant toxicity and act as indicators of environmental changes. Additionally, the chapter reviews the impact of pollutants on parasite life cycles and the broader implications for populations and communities.

Following on from Chap. 20 regarding the interaction between aquatic parasites and pollution, Chap. 21 provides a unique and completely novel mathematical framework to examine these dynamics, focusing on parasitic helminths, their hosts and pollutants. Although this new model is based on examples from the terrestrial environment, the model is designed to provide general insights into these interactions and can be applied to aquatic parasites and environments. The chapter concludes by emphasising the importance of further research to understand the complex interactions between parasites, their hosts and pollutants and their implications for ecosystems.

In Chap. 22, the implications of climate change on aquatic parasitology are reviewed, showing how shifting environmental conditions are affecting parasite distribution, prevalence and host–parasite interactions. Projections on future trends are provided, along with discussions on the challenges and opportunities for managing parasitic diseases in a changing climate.

Shifting the focus to the socio-economic impacts of aquatic parasitology, the concluding chapter (Chap. 23) outlines how parasitic infections affect fisheries, aquaculture and public health, presenting economic analyses of the costs and benefits of managing parasitic diseases. Advocacy for greater investment in parasitological research and management is presented to mitigate the economic impacts of parasitic infections on aquatic resources.

Through these chapters, this part (Chaps. 17–23) illustrates the diverse and impactful applications of ecological research on aquatic parasites. It emphasises the significance of parasites as both indicators and agents in the management and conservation of aquatic ecosystems, providing a compelling case for their inclusion in broader environmental and resource management strategies.

1.5 Conclusion

This book aims to offer a comprehensive and intriguing exploration of aquatic parasitology, highlighting the ecological and environmental significance of marine and freshwater parasites. By understanding the latest research developments in this field, readers will gain valuable insights into the complex and fascinating world of aquatic parasites. The integration of traditional ecological concepts with modern molecular techniques provides a holistic view of the field, emphasising the importance of interdisciplinary approaches in understanding and managing aquatic parasitism. The book also underscores the necessity of conserving parasitic species as integral components of aquatic biodiversity, advocating for a balanced perspective that recognises their ecological importance. Through the collective efforts of leading experts, this volume aims to inspire further research and innovation in the study of aquatic parasites, ultimately contributing to the sustainability and health of aquatic ecosystems worldwide.

Acknowledgements All the chapters of this book have been peer-reviewed by experts in the various fields of aquatic parasitology covered in this book, and the editors would hereby like to thank all the reviewers for their time and constructive comments on each chapter. The editors would also like to thank Niel Bruce for language editing of chapters; our graphical illustrator Drikus Roets for producing world-class illustrations and figures; our scientific illustrator Anja Erasmus for most probably the best illustrations of different parasite and hosts species ever published; Kerry Hadfield for help with constructing the photo plates; Russell Yong for compiling the species list; and Chandra le Roux for help with the glossary. Finally, we would also like to thank Annette Klaus, editor of *Veterinary Medicine and Parasitology* at Springer, for initiating this project and the North-West University (NWU) Unit for Environmental Sciences and Management for funding the open access fees of this book, making it accessible to all researchers, students and the interested public, particularly in developing countries. This book is contribution number 972 of the NWU-Water Research Group.

References

du Preez LH, Landman WJ, Verneau O (2023) Polystomatid flatworms, Zoological monographs, vol 9. Springer, Cham

Klimpel S, Kuhn T, Münster J, Dörge DD, Klapper R, Kochmann J (2019) Parasites of marine fish and cephalopods. Springer, Cham

Madhavi R, Bray RA (2018) Digenetic trematodes of Indian marine fishes. Springer, Dordrecht

Piazzon MC, Wiegertjes G, Bron JE, Bobadilla A (eds) (2021) Fish parasites: A handbook of protocols for their isolation, culture and transmission. 5m Books, Essex

Rohde K (2005) Marine parasitology. CSIRO, Melbourne

Scholz T, Vanhove MPM, Smit NJ, Jayasundera Z, Gelnar M (eds) (2018) A guide to the parasites of African freshwater fishes, vol 18. ABC Taxa, Belgium

Smit NJ, Bruce NL, Hadfield KA (eds) (2019) Parasitic Crustacea: State of knowledge and future trends. Springer, Cham

Woo PTK (2006) Fish diseases and disorders. Vol 1: Protozoan and metazoan infections. CABI, Wallingford

Woo PTK, Buchmann K (2012) Fish parasites—pathobiology and protection. CABI, Wallingford

Open Access This chapter is licensed under the terms of the Creative Commons Attribution-NonCommercial-NoDerivatives 4.0 International License (http://creativecommons.org/licenses/by-nc-nd/4.0/), which permits any non-commercial use, sharing, distribution and reproduction in any medium or format, as long as you give appropriate credit to the original author(s) and the source, provide a link to the Creative Commons license and indicate if you modified the licensed material. You do not have permission under this license to share adapted material derived from this chapter or parts of it.

The images or other third party material in this chapter are included in the chapter's Creative Commons license, unless indicated otherwise in a credit line to the material. If material is not included in the chapter's Creative Commons license and your intended use is not permitted by statutory regulation or exceeds the permitted use, you will need to obtain permission directly from the copyright holder.

The Protists: Small But Mighty

Sonja Rückert

Abstract

Protists are a polyphyletic group of diverse, mostly microscopic, unicellular organisms that encompass all species that are not animals, plants or fungi. One common characteristic of protists is that they prefer aquatic or moist habitats and here we find marine picoplankton, the smallest free-living protists, as well as the giant kelp, which is the largest. Of the many known protist species and life forms, roughly 15,000 are parasitic. The relationship of parasitic protists with their hosts ranges from harmless to causing significant diseases in humans, animals and plants. Here we focus on parasitic protists in the aquatic realm, including amoebae, ciliates, flagellates and sporozoans.

S. Rückert (✉)
Department of Eukaryotic Microbiology, University of Duisburg-Essen, Essen, Germany

Centre for Water and Environmental Research (ZWU), University of Duisburg-Essen, Essen, Germany
e-mail: sonja.rueckert@uni-due.de

2.1 Introduction

Single-celled organisms that are characterized by a membrane-bound nucleus are eukaryotic microorganisms called protists or protozoans. These terms are often used interchangeably but differ by the range of organisms they encompass. Protists include the Protozoa, some algae and slime moulds. In this chapter the general term protist will be used for these three taxa.

Protists represent a diverse group of eukaryotic microorganisms that do not fit neatly into the categories of plants, animals or fungi. It does not therefore represent a phylogenetic grouping per se, as protists are spread across the eukaryotic tree of life (Fig. 2.1; Burki et al. 2020). They encompass a remarkable range of forms and lifestyles and play crucial roles in various ecological processes. Most of them are up to a couple of 100 μm in size, but some can be quite large, visible by the naked eye, with the largest being giant kelp, which can grow up to 30 m. They inhabit a wide range of ecosystems, including all aquatic habitats, from freshwater to hypersaline environments.

Among the vast diversity of protists (around 40,000 species are currently described), some species (~one third) have evolved a parasitic lifestyle. These parasitic protists are adept at exploiting their hosts, showcasing intricate adaptations

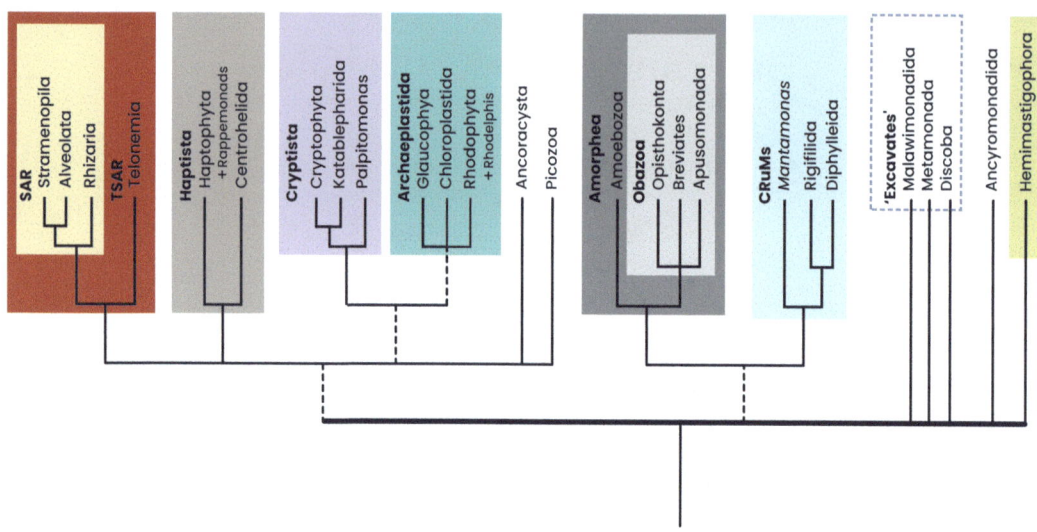

Fig. 2.1 Current 'supergroups' in the eukaryotic Tree of Life. The summary tree shows the current accepted 'supergroups' based on available phylogenomic studies (adapted from Burki et al. 2020)

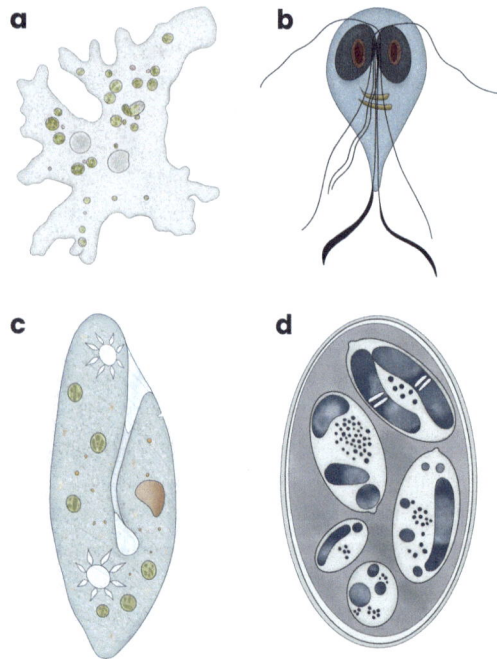

Fig. 2.2 Representative sketches of the four main groups of parasitic protists (**a**) amoebae, (**b**) flagellates, (**c**) ciliates, (**d**) sporozoans

and strategies to secure their survival and reproduction.

Protists are often categorized into four broad groupings based on their locomotion or reproduction, the amoebae, ciliates, flagellates and sporozoans. In short, Amoebae have no defined shape and the term amoeboid is used to describe an amorphous shape. They use pseudopodia, which are extensions of the cell membrane filled with cytoplasm, to move (Fig. 2.2a). Flagellates encompass a broad variety of organisms that use one flagellum or multiple flagella for their motility. Flagella are slender (whip-like) extensions from the cell membrane, with an inner centre made up of microtubules, where nine pairs of microtubules surround a central pair of single microtubules in the 2 + 9 configuration (Fig. 2.2b). They perform a propeller or whip-like movement. Ciliates are covered in rows of small hair-like structures called cilia, but can also have areas with special cilia for different purposes (e.g. feeding via the oral groove/funnel) (Fig. 2.2c). The cilium structure is similar to that of the flagellum. Cilia perform undulating movements in waves along the cells, synchronized through a connection between their basal bodies. Sporozoa are an assemblage of species that form a spore stage during their life cycle for transmission (Fig. 2.2d).

While most protists are free-living, the transmission of parasitic protists occurs in four different ways. (1) Direct transmission can occur between hosts; (2) life stages are passed into the environment within the faeces from one host and are ingested orally by the next host; (3) a vector is involved in the transmission; (4) prey organisms infested with life stages are consumed by a predator.

This chapter provides an overview of the diverse groups of parasitic protists found in aquatic environments. These include but are not limited to members of the Amoebozoa, Rhizaria (Phytomyxea, Ascetosporea), Alveolata (Dinoflagellata, Perkinsea, Ciliates, Apicomplexa), Discoba (Euglenozoa, Heterolobosea) and Metamonada. Each of these groups encompasses a unique set of characteristics, life cycles and host-specific interactions. Thus, the study of parasitic protists in aquatic environments is a field of growing interest, as their ecological interactions, host-parasite dynamics and impact on ecosystem health are increasingly recognized. The chapter follows the four groups, described above. Within these groups the described taxa follow the accepted taxonomy.

2.2 Amoebae

Amoebae are typically free-living but, in certain conditions, some can become opportunistic parasites, which can be categorized into three systematic groups: Amoebozoa, Rhizaria and Heterolobosea.

2.2.1 Amoebozoa

Amoebozoa use their pseudopodia for movement and feeding. They are found in all types of aquatic habitats, from freshwater to saltwater, and play important roles in aquatic ecosystems as predators, scavengers, and opportunistic parasites of other organisms. Amoebozoa can be divided into two main groups based on their morphology and genetic characteristics, the Lobosa and Conosa (Smirnov et al. 2011).

The Lobosa are characterized by the presence of lobed or irregular-shaped pseudopodia. Some species of Lobosa are known to be parasites of fish, other aquatic organisms and also humans. For example, *Vermamoeba vermiformis* (formerly *Hartmannella vermiformis*) belongs to the Class Tubulinea and is predominantly a free-living amoeba, but it can cause infections in other organisms such as fish and humans, but its pathogenicity is still under debate (Scheid 2019). The trophozoites are longer than broad with a slug-like morphology. They move in a monopodial manner, but become bi- or multipodal, when they change direction (Delafont et al. 2018). The life cycle consists of two stages, the trophozoite that feeds, moves and divides by binary fission, and a dormant cyst stage that enables *V. vermiformis* to survive in unfavourable conditions (Scheid 2019). It has been recorded from lakes, tap water, bottled mineral water, thermal water, and recreational waters such as swimming pools (Chelkha et al. 2020). This species can affect the gills of fishes such as Nile tilapia *Oreochromis niloticus* (see Milanez et al. 2017) and has been isolated from tissue lesions of four different freshwater fish species (Dykova et al. 2005). It has also been shown to be responsible for some cases of keratitis, causing corneal damage in humans (Delafont et al. 2018). In addition, this amoeba can harbour other microorganisms in its cytoplasm that can pose health risks to humans, these include, for example, species of *Legionella*, *Mycobacteria*, *Chlamydiae*, *Pseudomonas*, viruses and more (Milanez et al. 2017; Delafont et al. 2018). Therefore, more research is needed to understand the role of *V. vermiformis* in infections, as it appears that this generally harmless free-living amoeba has the potential to become pathogenic and carry other microbial pathogens.

The class Discosea includes free-living amoeba that can also be associated with infections in higher eukaryotes, specifically in fish, other aquatic organisms and humans. This group of amoebae is characterized by their flat morphology and disc-shaped pseudopodia. They are typically classified as parasites, but infections are often rapid and serious and therefore a name such as opportunistic pathogens should be considered as preferential. There are several genera that have been identified to be possibly involved with gill diseases in fish, *Acanthamoeba*, *Mayorella*,

Thecamoeba, *Vannella* and *Vexillifera* (see Bermingham and Mulcahy 2007).

The species *Neoparamoeba perurans* is the proven causative agent of amoebic gill disease (AGD) in Atlantic salmon *Salmo salar*, leading to gill lesions and mortality in aquaculture (Young et al. 2007; see Chap. 23). It is an ectoparasite with a direct life cycle that has no resting cyst stage (Nowak 2007). *Neoparamoeba perurans* has digitiform pseudopodia when free in the water column and mamilliform pseudopodia, when adhered to a surface (Young et al. 2007). AGD puts a huge burden on Atlantic salmon farming as mortalities can be high (up to 85%) (Padrós and Constenla 2021). The closely related *Paramoeba* species, *P. invadens*, is responsible for mass mortalities of sea urchins in Nova Scotia, Canada (Feehan et al. 2013) and it has been shown that the disease is the main agent controlling sea urchin populations in kelp beds in Nova Scotia (Feehan and Scheibling 2014). *Vibrio* species are also associated with *N. perurans*, thus *V. vermiformis* could be considered a vector for bacterial disease (MacPhail et al. 2021).

Another example of usually free-living bacteriovores, but opportunistic pathogens, are species of *Acanthamoeba* that can cause keratitis, a potentially blinding infection of the cornea, and granulomatous amoebic encephalitis, a rare but often fatal infection of the brain and central nervous system in humans (Marciano-Cabral et al. 2000). They can live in all habitats and occur in water sources such as rivers, lakes, oceans, swimming pools and tap water, to name a few (Henriquez 2017). *Acanthamoeba* spp. have a simple life cycle that includes an active trophozoite stage and an inactive cyst stage. In addition to being an opportunistic parasite itself, they can also be a vector for disease-causing bacteria such as *Legionella pneumophila*, *Staphylococcus aureus*, *Escherichia coli* and *Pseudomonas aeruginosa* (see Henriquez 2017).

The Conosa encompasses three lineages of amoeba that have cilia or flagella, or have lost them secondarily, the Archamoebae, Mycetozoa and Variosea (Smirnov et al. 2011). Morphologically this group has clearly pointed pseudopodia or branching subpseudopodia.

Within the family Archamoebae, the genus *Entamoeba* is best known for being a commensal within the human intestine, but some species, such as *Entamoeba histolytica*, are known to cause amoebic dysentery. The life cycle encompasses two life stages, the trophozoite and the cyst stage. Infection happens through the ingestion of cysts from contaminated water. The cysts are resistant to chlorination, gastric acidity, as well as desiccation, and only a small number of cysts are required to cause an infection. Excystation happens in the small bowel where eight trophozoites will move on to the large bowel. Trophozoites divide by binary fission and either form new cysts, which often stay unrecognized, or they can cause inflammations of the bowel wall. The amoeba can spread to the liver and the central nervous system, the lung or the heart (see Royer and Petri 2014).

2.2.2 Rhizaria

The Rhizaria are a supergroup of protists that are best known for their free-living species like the Foraminifera (forming beautiful shells that make up the sands of some beaches in the Indo-Pacific), but some groups are parasitic and of special interest as parasites of commercially important invertebrate species (Burki and Keeling 2014). The parasitic forms can be found within the classes Phytomyxea, infecting plants, algae and stramenopiles, and the Ascetosporea infecting invertebrates (Ward et al. 2018).

In the aquatic realm only a few parasite species of Phytomyxea are known. These species are of importance as they infect primary producers such as brown algae, diatoms and seagrass (Neuhauser et al. 2011). Flagellated free-swimming zoospores are the infective life stages, which release a small protoplast after piercing the cell with an extrusome (Garvetto et al. 2023), before developing into two different types of plasmodia (sporogenic and sporangial). The host organisms show signs of hypertrophy or galls.

Garvetto et al. (2023) have shown that the intracellular plasmodia within the host cells feed via phagocytosis.

Phytomyxea species that infect brown algae are *Maullinia ectocarpii* infecting *Ectocarpus siliculosus* and *Phagomyxa algarum* infecting, e.g., *Bachelotia antillarum* (Neuhauser et al. 2011). Maier et al. (2000) used zoospores of *M. ectocarpii* and were able to infect ten species of cultured brown algae that originated from across the globe. Besides hypertrophy of host cells, they were also able to show infection of oogonia in Giant Kelp *Macrocystis pyrifera*, which affected the host's reproduction. The infection of brown algae with *P. algarum* is mostly restricted to the thallus of the host. Infected cells change colour and quickly deteriorate (Karling 1944).

An example of a seagrass infecting Phytomyxea is *Plasmodiophora bicaudata*. Infections are visible through the formation of galls at the internodes of *Zostera* species. This species has a worldwide distribution (Den Hartog 1989) and also affects the root growth negatively.

Not much is known about the effects of infections of diatoms with Phytomyxea species like *Phagomyxa odontellae* that infects diatoms of the genus *Odontella*. This intracellular parasite feeds on the cytoplasm, and included chloroplasts, which can be found in digestive vacuoles (Schnepf et al. 2000).

Commercially important invertebrates that are infected by Ascetosporea are molluscs and crustaceans. There are five orders, Haplosporida, Paramyxida, Mikrocytida, Claustrosporida and Paradinida, of which the first three cause named diseases such as MSX (*Haplosporidium nelsoni*), bonamiosis (*Bonamia* sp.) and Aber disease (*Marteilia refringens*) in oysters and debilitating disease in crabs (*Paramikrocytos canceri*) (Ward et al. 2018).

Haplosporidium nelsoni, for example, has been reported from Atlantic oyster *Crassostrea virginica*, Pacific oyster *C. gigas* and European flat oyster *Ostrea edulis*, mainly in the northern hemisphere, causing mass mortalities, and consequently reducing biomass and harvest. There seems to be a natural temperature-related limitation of the spread of the disease. The life cycle is not well known and there are speculations about direct transmission or the utilization of an intermediate hosts (Fig. 2.3), the latter being more likely (Ford et al. 2018). MSX is rarely found where winter temperatures are consistently below 3 °C, but due to climate change this disease could spread further north (Brander 2009).

Bonamia is another haplosporidian genus with species that cause severe diseases in oysters. Infections with this microcell parasite are often lethal and infections with two described *Bonamia* species are notifiable to the World Organization of Animal Health (OIE) (Bateman et al. 2020).

The Paramyxida species *Marteilia refringens* is highly pathogenic to molluscs, especially oysters, and is also a notifiable disease to the OIE. The life cycle of this species is not yet completely described (Bateman et al. 2020), but it is expected that an intermediate host, e.g. different zooplankton species such as copepods, is involved (Arzul et al. 2014).

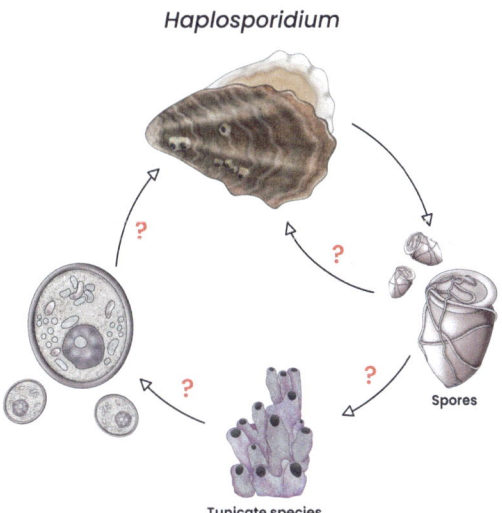

Fig. 2.3 Putative life cycle of *Haplosporidium nelsoni*. Spores are released from the oyster host (*Crassostrea virginica*) and supposed to be responsible for transmission. A tunicate like *Styela* sp. is suggested to be potential intermediate or reservoir host (adapted from Fernández Robledo et al. 2018)

Decapod crabs can be infected with *Paramikrocytos canceri*, a Mikrocytida species. Infection occurs in the antennal gland of juvenile crabs (Edwards et al. 2019), causing a potentially emerging disease, with yet undetermined pathogenicity. Parasite stages can be found in the epithelial cells of the antennal gland causing hypertrophy. Epithelial cells underlying the cuticle and nephrocytes in the gills are infested in later stages of infection (Edwards et al. 2019). The host cytoplasm is invaded by uni- and multinucleated cells (Onut-Brännström et al. 2023). A survey revealed that *P. canceri* can be found in many invertebrate species, which hints towards transmission including an intermediate host (Hartikainen et al. 2014).

Not much is known about the last two groups of Ascetosporea, the Claustrosporida and Paradinida. The genus *Claustrosporidium* comprises two species, *C. gammari* and *C. aselli* both infecting the freshwater amphipod *Rivulogammarus pulex* (Hine et al. 2020). Three life stages have so far been described for *C. gammari*, the plasmodia, sporoblasts and spores. The spores of these two species do not have an orifice, separating them from the Haplosporida. The order Paradinida comprises two genera, *Paradinium* and *Atelodinium* (Ward et al. 2018), both infecting a variety of copepods. *Paradinium* species were first described to possibly belonging to the Dinozoa (Chatton 1920), but SSU rDNA sequence data by Skovgaard and Daugbjerg (2008) revealed their phylogenetic position to be close to the Haplosporida. The life stages of *Paradinium* spp. include amoeboid cells in the host's body cavity that are interconnected by pseudopodia. A solid plasmodium forms and enters the intestine. The plasmodium is excreted through the anus where a cyst, the gonosphere, is formed, which is then attached to the urosome. Flagellated spores are formed within the gonosphere that are inferred to be the infective stages (Skovgaard and Daugbjerg 2008). The effect of the infection on the copepods is not clear, but copepods are primary consumers that are important players at the base of the aquatic (here marine) food webs. Therefore, we can speculate that there will be changes in the dynamics of the food webs when parasites are present.

2.2.3 Heterolobosea

The Heterolobosea belong to the supergroup Excavata and therein to the taxon Discoba, based on morphological characteristics such as the discoidal shape of the mitochondrial cristae (Page and Blanton 1985), but also phylogenetic analyses (Roger et al. 1996). They are also called amoeboflagellates due to the three different life stages in their life cycle, trophic amoeba, flagellates and cysts (Pánek et al. 2016). The pseudopods of the amoeba stage develop and move in an eruptive way (Fehling et al. 2007) compared to other lobose amoeba. They have a global distribution and occur in many habitats including extreme environments. Most Heterolobosea are free-living and heterotropic in the soil, marine and freshwater environments, but *Naegleria fowleri*, the most notorious species within this group, is known for its facultative parasitic lifestyle.

Naegleria fowleri, also known as the 'brain-eating amoeba', belongs to the class Schizopyrenidae, occurs in freshwater and is adapted to warmer temperatures, with an optimum of about 40 °C. They live in both naturally and artificially heated water bodies (Hausmann et al. 2003). Infections with *N. fowleri* are rare but can be devastating for humans and are often lethal as they can lead to primary amoebic meningo-encephalitis (PAM). Normally, *N. fowleri* is free-living and bacteria eating. The infective pathway is through the nasal epithelium and olfactory system when swimming. The three life stages occur in contaminated freshwater (Fig. 2.4). The infective amoeboid trophozoite can either form a resting cyst or a flagellated stage for dissemination. The trophozoite stage is the infective stage that enters the brain where mass development starts. The trophozoites are found in the cerebrospinal fluid and tissue (CDC 2022).

Fig. 2.4 Warning sign about *Naegleria fowleri* at a stream in New Zealand

2.3 Ciliates

The phylum Ciliophora, commonly known as ciliates, is another group of parasitic protists found in aquatic habitats. Ciliates are a diverse group of unicellular protists that are characterized by the presence of cilia (hair-like structures) that are used for movement and feeding, the morphology of the cortex, nuclear dualism (one or more macronuclei and one to several micronuclei) and a conjugation stage during the sexual phase of the life cycle (Hausmann et al. 2003). They are found in all aquatic habitats, from freshwater to hypersaline waters, and play important roles in aquatic ecosystems as predators, primary producers and parasites of other organisms. A form of symbiosis can be found in one third of all ciliate species in several classes (Armophorea, Heterotrichea, Litostomatea, Nassophorea, Oligohymenophorea, Plagiopylea, Phyllopharyngea, Prostomatea and Spirotrichea) (Mayén-Estrada et al. 2021). Many aquatic organisms have ciliate symbionts as epibionts, ectoparasites or endobionts. Most of these ciliate species are harmless, but some are pathogenic for a diverse range of aquatic organisms, including fish and crustaceans. A well-known example of a disease that affects a wide range of marine fish species is marine white spot disease, or marine ich, which is caused by *Cryptocaryon irritans* (Colorni and Burgess 1997), but it is often not clear to distinguish between commensalism and parasitism. Looking at invertebrates, the most specious symbiont or parasite group for copepods is the ciliates, especially the peritrichs and suctorians (Bass et al. 2021). If prevalence is high, it could lead to changes in trophic links and instability in food webs (Pérez 2009; Gómez-Gutiérrez et al. 2003, 2012). This chapter includes only examples of species that have been determined as parasites.

2.3.1 Heterotrichea

The class Heterotrichea comprises large ciliates with an obvious adoral zone made up of polykinetids or membranelles (AZM) (Lynn 2010) and has been established based on SSU rRNA phylogenetic information. Species of the family Folliculinidae are widespread, typically marine and attach to various invertebrates including bivalves, crustaceans, polychaetes tubes, hydroids and bryozoans (Mayén-Estrada et al. 2021). *Microfolliculina limnoria* is suggested to be ectoparasitic of wood-boring *Limnoria* isopods, as they suppress feeding and may hinder their dispersal (Delgery et al. 2006).

Another species *Halofolliculina corallasia*, a colonial heterotrich ciliate is a known parasite of corals. It causes Skeleton Eroding Band (SEB) syndrome and has been described affecting large, branching corals (e.g. *Acropora* spp.) from the Indo-Pacific and the Red Sea (Antonius and Lipscomb 2001; Winkler et al. 2004). It was the first described stony coral disease by a protist and eukaryote (Antonius and Lipscomb 2001). The ciliate forms a lorica that is deeply embedded in the outer layer of the coral skeleton. In this process part of the skeleton is damaged. Due to the vast numbers (up to >400 individuals/mm^2) in the outer layer of the skeleton, it is destroyed (Winkler et al. 2004). The name of the syndrome stems from the bands that can be seen on the coral due to the aggregation of the loricae. Environmental stress is thought to be the main reason for the susceptibility of corals to SEB (Winkler et al. 2004).

2.3.2 Litostomatea

There are two major groups in the class Litostomatea, the subclass Haptoria (predators) and the subclass Trichostomatia (endosymbionts). The Litostomatea are characterized by the structure of the somatic kinetids (Small and Lynn 1981), and this taxonomic grouping has been confirmed by phylogenetics. The Trichostomatia includes many endosymbiotic species and some that are inferred to be parasitic in aquatic organisms, they also include the only known human pathogenic ciliate *Balantidium coli*.

Morado and Small (1995) listed five species of the genus *Hemiophrys* (Haptoria) that infect different *Gammarus* species, a *Cyclops* sp. and two *Asellus* species on the gills, body tergites pereopods and exoskeleton. Fenchel (1965) found *H. baltica* on five species of *Gammarus* and observed that this species feeds on the blood cells and tissue of its hosts. In addition, two *Balantidium* species have been reported from the intestines of two *Orchestia* and one *Talorchestia* amphipod species (Morado and Small 1995). Even though there was no mention of any effect of these infestations on the amphipods, there were hundreds counted in each individual (Watson 1916), which suggests a possible impact on food uptake.

2.3.3 Oligohymenophorea

The class Oligohymenophorea is the most diverse group of ciliates occurring mainly in freshwater and marine habitats. Except for the peritrichs, the ciliature of the species in this class is holotrichous, but their overall morphology can vary greatly. The common characteristic is that they only have a few oral polykinetids or membranelles (Lynn 2010). Depending on the lifestyle of the species the life cycles can differ between subclasses and will be discussed in the relevant groups presented here. There are species that show an affinity to diseased or dead organisms, which is assumed to have led to different forms of symbiosis with organisms such as crustaceans, molluscs and fishes (Lynn 2010). The degree of pathogenicity varies, but infections can cause occasional mass mortalities under culture conditions, which happens rarely in nature.

For example, *Ichthyophthirius multifiliis*, a hymenostome ciliate of the order Ophryoglenida is a parasite that causes white spot disease, also known as ichthyophthiriasis (ich), in freshwater fishes, leading to high mortality rates in aquaria and aquaculture operations, where fishes are crowded (see Chap. 23). Infections depend on the conditions of the fish and occur mostly when fishes are under stress due to unfavourable environmental conditions such as high temperature (von Gersdorff Jørgensen et al. 2022). The life cycle of *I. multifiliis* has four phases (compare Coyne et al. 2011; von Gersdorff Jørgensen et al. 2022; and Fig. 2.8 on similar *Cryptocaryon irritans* life cycle): (1) The infective, free-swimming **theront** bores into the epithelium, where it develops into the trophont. (2) The **trophont** is the feeding stage that ingests body fluids, cell debris and cells to grow and mature within 3–8 days. These life stages can be seen as white spots by the naked eye. The trophont can be easily identified due to its large, horseshoe-shaped macronucleus. (3) When mature, the trophonts leave the host as **tomonts** that swim freely for a few hours before they settle and encyst on the bottom or any other substrate. The tomonts start dividing by binary fission and form hundreds of daughter cells, the **tomites**. (4) The cysts rupture and the tomites are released that develop into the infective **theront** stage.

The length of the life cycle depends on the surrounding temperature and is faster with increasing temperature. Symptoms of the disease are changes in swimming behaviour, including fishes swimming and rubbing along the wall. Mass mortalities occur when infections are severe, e.g., on the gills causing hyperplasia, fusion of gill lamellae and blood clots, or the wounds and lesions are secondarily infected by bacteria.

The subclass Hymenostomatia includes one of the best studied ciliate taxa, the genus *Tetrahymena* (see Lynn 2010), and there are reports on some parasitic species within this genus. One of these species is *Tetrahymena corlissi*, which causes white spots on the fish skin, in this case necrotic tissue, partially peeling off the skin and causing scale shift. These ciliates were also found in high numbers in the blood vessels (Hoffman et al. 1975). The life cycle so far has only been described from cells fed with oligochaete tissue (Lynn 1975). There is an interesting relationship between a predator and its free-living tetrahymenid prey, transforming into a parasite/host relationship as predator defence strategy. This relationship was described by Washburn et al. (1988) for the treehole mosquito *Aedes sierrensis* and the ciliate *Lambornella clarki*. The mosquito releases a waterborne factor that induces cell division and transformation into parasitic theronts of the ciliate. These in return infect and encyst in the larval mosquito. The ciliates move into the haemocoel and reproduce rapidly which leads to the death of their predator host. With the shift in trophic level, *L. clarki* avoids predation (Washburn et al. 1988).

Some ciliates in the subclass Apostomatia infest the haemolymph of amphipods and euphausiids, such as *Fusiforma themisticola*. This species is found in the haemolymph of the hyperiid amphipod *Themisto libellula* (Fig. 2.5), which is an important trophic link in Arctic waters. Prokopowicz et al. (2010) suggested that the ciliate infection could have an impact on trophic processes. Another species within this subclass is *Vampyrophrya pelagica* a histophagous ciliate, infecting calanoid copepods. The metamorphosis of the ciliate into a feeding stage is initiated by two different ways, the injury of the host or the ingestion of the host by a predator, in the otherwise similar life cycles (Fig. 2.6) (Grimes and Bradbury 1992).

Infections with the histophagous scuticociliate *Orchitophyra stellarum* of the subclass Scuticociliatia were first discovered in the gonads of male sea stars (Bouland and Jangoux 1988). Infected sea stars have a reduced fecundity due to the destruction of the germinal epithelium and phagocytoses of sperm cells. Leighton et al. (1991) reported increased virulence leading to mortalities in a then new host *Pisaster ochraceus* (ochre seastar), but infections have since been reported from many invertebrates, including systemic infections in blue crabs *Callinectes sapidus* (Small et al. 2013).

The subclass Peritrichia includes many species that are parasites of aquatic organisms, some

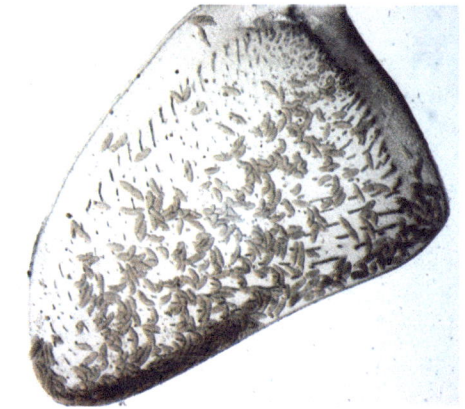

Fig. 2.5 Leg part of hyperiid amphipod *Themisto libellula* filled with *Fusiforma themisticola*, an apostomate ciliate

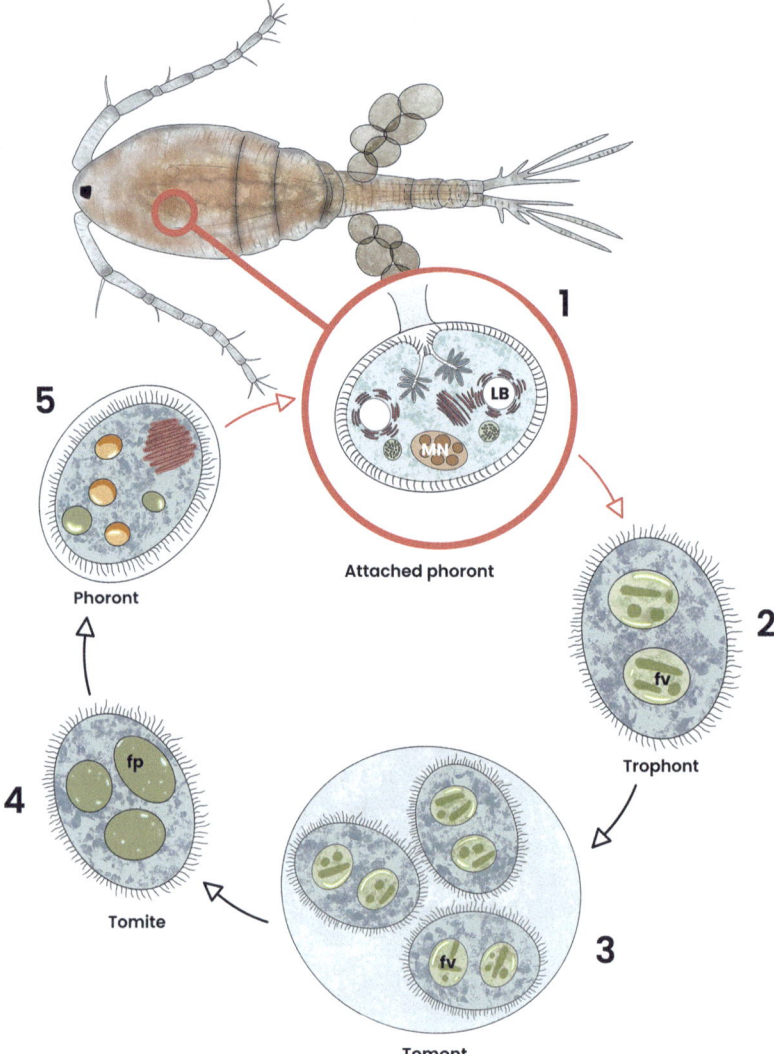

Fig. 2.6 Life cycle of histotrophic apostome ciliate, *Vampyrophyra pelagica*. (1) Phoronts are attached with a stalk to copepod exoskeleton. Excystation is triggered by injury of the host or ingestion by predator. (2) Trophonts enter host and start feeding on copepod tissue. (3) Grown trophonts metamorphose into tomonts, which produce different numbers of tomites. (4) Tomites are released and search for a new host. (5) On the new host the tomites encyst and become phoronts, which starts a new cycle. *fp* food plaquette, *fv* food vacuole, *LB* lipid body, *MN* macronucleus (adapted from Grimes and Bradbury 1992; Ohtsuka et al. 2015)

sessile (Order Sessilida) and some mobile (Order Mobilida). The Sessilida are characterized by the presence of a stalk or adhesive disc that is used for attachment to substrates or hosts. For example, *Epistylis* is a genus of peritrich ciliates that is known to attach to the skin of fishes. They are not obligate parasitic, but due to the penetration of the skin and dissolving of skin tissue in the attachment phase, it leaves inflamed lesions (red-sore disease) (Rogers 1971). The lesions leave access points for secondary bacterial infections. The life cycle of this ciliate includes a free-swimming telotroch that is infecting a new host. Another example is *Zoothamnium* sp., a peritrich ciliate that is known to cause mortalities in shrimps, especially in culture facilities. Zoothamniosis restricts breathing and movement of the shrimp and restricts the ability to moult, which can cause mass mortalities (Mahasri et al. 2019).

The most prominent mobile peritrichs are the symbiotic or parasitic trichodinids. Most species occur on the gills or the skin of fish, with a few exceptions that occur in the urinary tract and intestines. Amphibians and invertebrates can also be infected. There are ten genera within the family Trichodinidae. The most striking feature of these parasites that is used for morphological description and identification (Lom 1958) is the denticles that form a ring in the centre of the

Fig. 2.7 Micrograph of the adhesive disc of the trichodinid ciliate *Trichodina magna*, an ectoparasite of freshwater fishes. © Liesl van As

saucer-shaped ciliate (Fig. 2.7). They also have a suction cup with which they can attach to the host. Trichodinids move on the surface of their host feeding on organic material and bacteria. In low numbers they do not have any effect on their host, but when occurring in high numbers they can cause lesions on gills and skin that can lead to mortalities. Trichodinids have a direct life cycle with asexual reproduction (binary fission) and sexual reproduction (conjugation). Transmission happens on contact with an infected individual or in highly contaminated waters. There are rarely any mass outbreaks in wild fish populations, but these do occur in captivity (e.g. fish farms).

2.3.4 Phyllopharyngea

The class Phyllopharyngea encompasses free-living and symbiotic species, which show diverse body plans based on their different life forms (Lynn 2010). In general, their body is flattened ventrally with a long, arched ciliary band on the right, and a shorter band on the left (Bastos Gomes et al. 2017). At least two species of the subclass Cyrtophoria, namely *Chilodonella hexasticha* and *C. piscicola*, can infect fishes and cause hyperplasia of the gill epithelium, as well as necrosis (Bastos Gomes et al. 2017). These ciliates use their cytostome (mouth structure) to feed on skin and gill mucus and cells causing inflammatory responses. *Chilodonella* species reproduce mainly by transverse binary fission but sexual reproduction through conjugation occurs as well. Species of two rather rare genera within this subclass infect marine mammals, e.g. *Planilamina ovata* and *Kyaroikeus paracetarius*. Both named species were isolated from the blowholes of a beluga whale *Delphinapterus leucas*, in captivity, which showed signs of disease. As both species were also isolated from the healthy whale, but in much lower numbers, Jin et al. (2021) inferred the species to be parasitic. The ciliates were attached to flocs formed by epithelial cells, which the ciliates fed on. Morphological differences in the parasitic species compared to their free-living counterparts, such as large number of right kineties, dense cilia, prominent oral cavity, amongst others, suggest adaptation to their new environment (Jin et al. 2021).

Species of the subclass Rhynchodia have been described as endobionts of freshwater and marine mussels as well as other invertebrates (Hausmann et al. 2003). The species *Stegotricha enterikos* parasitizes the digestive gland of the Pacific oyster *Crassostrea gigas*. The species divides by binary fission, conjugation has not yet been observed. This species does not appear to be pathogenic (Bower and Meyer 1993), therefore its parasitic status is questionable.

Species in the subclass Suctoria are presumed to be the most widespread symbionts within the class Phyllopharyngea (Lynn 2010). They can be found as ectobionts on many crustaceans and other aquatic organisms. Suctorians have no lorica, two types of tentacles, a stalk and a ramous macronucleus. Mature forms of these species are sessile, producing swarmers through budding that infect new specimens or the current host. Even though not much is known about potential damage they can cause, the species *Ephelota gigantea* has a known negative effect on Wakame (*Undaria pinnatifida*) culture in Japan, due to an unpleasant smell they cause, when damaged with the stalks remaining on the Wakame fronds, decreasing the commercial value (Sato et al. 2015).

2.3.5 Prostomatea

The class Prostomatea includes only a small number of species compared to other classes, but they can be very abundant. They are ovoid to cylindrical, largely radially symmetrical and prostomatous (Lynn 2010). Some of the species are small <20 µm. They are holotrichous with single or multiple caudal cilia. Most species are free-living and heterotrophs. In 2002, a fish parasite was added to this group based on SSU rDNA sequence data, the ectoparasite *Cryptocaryon irritans* (Wright and Colorni 2002).

Cryptocaryon irritans is a notorious marine parasite that is especially harmful for fishes cultured in temperate and tropical waters (see Chap. 23). It causes cryptocaryonosis, which is referred to as marine white spot disease, or marine ich. It has been thought that there is a close relationship of this species to the previously mentioned freshwater species *Ichthyophthirius multifiliis*, due to several common characteristics concerning their life cycle, lifestyle and dispersal (Wright and Colorni 2002), which is remarkable considering the different habitats and is a great example of convergent evolution. *Cryptocaryon* has a special arrangement of the nuclei with four linked macronuclear segments in the phoront and young trophont stages. Growing into the protomont, these segments fuse and form an elongated, twisted macronucleus (Lynn 2010).

The life cycle of *C. irritans* encompasses four phases, similar to *I. multifiliis* (see above and compare Colorni 1987; Li et al. 2022; Watanabe et al. 2016; Fig. 2.8): (1) **theronts** are the infective stage that are free-swimming until entering the epithelium of the fish host's skin. The theronts are pear-shaped and can survive for up to 2 days, but infectivity decreases after a few hours. (2) Within the epithelium the theronts develop into the next stage, the **trophonts**. In this parasitic lifestage, they feed on body fluids, tissue debris and cells for 3–7 days, growing in size that the trophonts can be seen as white spots on the surface of the fish. (3) Mature trophonts leave the host as round **protomonts** and sink to the bottom, where they encyst, after ~20 h. (4) They then transform into the **tomonts**, which are the reproductive stages. The reproductive phase can last for 3–28 days in a whole cycle. The encysted tomonts undergo multiple binary fissions (palintomy) to produce numerous **tomites**. The cysts rupture and the tomites are released into the water, they elongate and grow into the infective theront stage, which restarts the cycle. Infections with *C. irritans* cause high mortality in fishes, due to its virulence, direct life cycle, low host specificity and a wide geographical distribution (Li et al. 2022).

2.3.6 Spirotrichea

The class Spirotrichea is a diverse group of ciliates encompassing seven subclasses. Most species are free-living and only a few symbionts have been recognized. Most of these seem to be harmless commensals (Bradbury 1994). One species that is known to cause damage to the eye of the scallop *Chlamys opercularis* belongs to the subclass Licnophoria, *Licnophora auerbachii* (Harry 1980). The body is divided into three parts, a disc for attachment, a neck region and an oral disc. The species causes damage to the eye by its suction attachment and movement of the cilia that lead to detachment of pigment cells. *Licnophora* is feeding on detached epithelial cells, which Harry (1980) recognized due to the pigmented cells in the ciliate's food vacuoles. The consequences of the damage to the eye are not clear. Species of this genus have also been reported from sea cucumbers, limpets and even cyanobacteria (Lynn 2010).

2.4 Flagellates

Flagellates are unicellular protists that move using one or more whip-like flagella. They are found in all aquatic habitats, from freshwater to saltwater, and play important roles in aquatic ecosystems as primary producers, predators and parasites of other organisms. Flagellates are not a monophyletic group like the ciliates but can be found in several supergroups, including the Discoba (Euglenozoa, Kinetoplastea), Metamonada and TSAR (Dinoflagellata, Perkinsozoa).

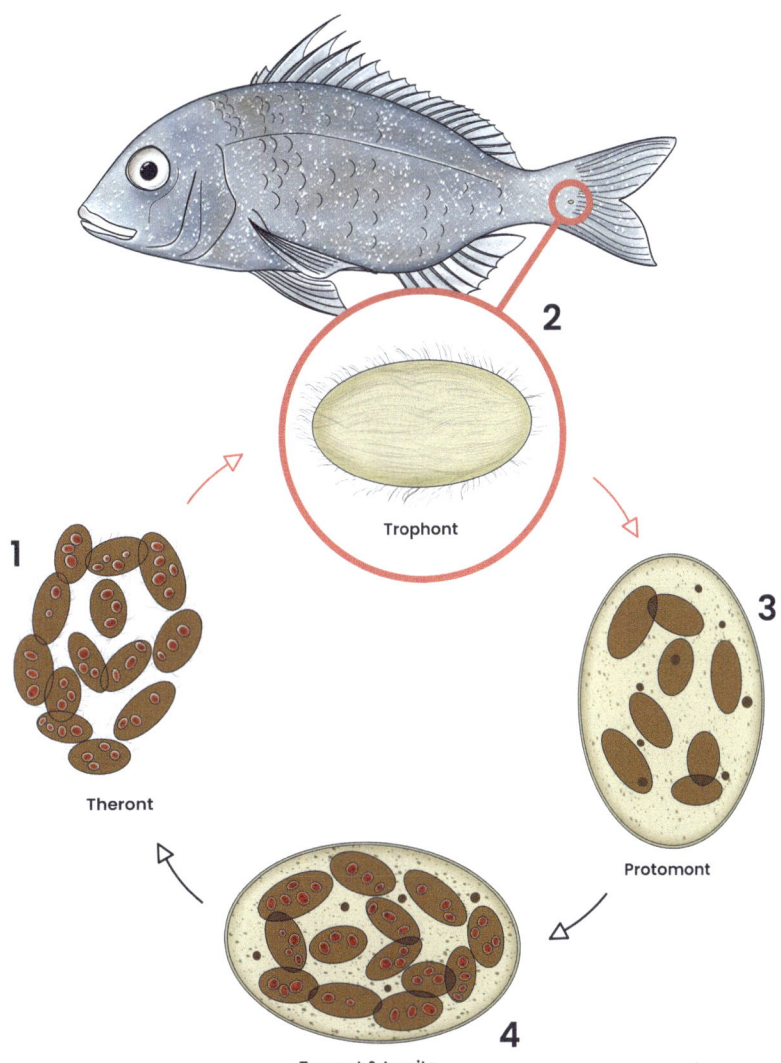

Fig. 2.8 Life cycle of *Cryptocaryon irritans*. (1) Theronts are the free-swimming infective stages of *C. irritans*. They enter the epithelium of the fish host. (2) In the epithelium they develop into trophonts which feed on fluids, tissue and cells. The trophonts grow (white spots) and leave the host when mature. (3) Mature trophonts develop into protomonts, which sink to the bottom where they encyst. (4) Here they change into the reproductive stages, the tomonts. Through palintomy they form numerous tomites. At the end of the reproductive phase the cyst ruptures and releases the tomites. In the water the tomites grow into the infective theront stages searching for a new host to restart the cycle (adapted from Watanabe et al. 2016, Reef Stable 2024)

2.4.1 Discoba (Euglenozoa)

The Euglenozoa are a diverse group of flagellates that are commonly found in freshwater habitats, but also in marine and even hypersaline environments. They are characterized by the presence of one or two flagella and a flexible proteinaceous structure called the pellicle. Some species of Euglenozoa are photosynthetic, while others are heterotrophic or mixotrophic, meaning they can photosynthesize and feed on other organisms. Within the Euglenozoa the class Kinetoplastea, characterized by a large mass of DNA in the mitochondrion, situated near the base of the flagella, encompasses free-living, commensal as well as parasitic species. Both classes Bodonida and Trypanosomatida contain parasitic species, the latter exclusively.

The best known bodonid parasite is the fish-infecting ectoparasite *Ichthyobodo necator*. This small flagellate (7–15 μm) is flattened and has two flagella, with which the organism attaches itself to the skin or gills of freshwater fishes. When free-swimming, *Ichthyobodo* shows a characteristic circular movement. Fishes have an increased mucus production, hyperplasia of gill and epithelial cells, as well as fused gill lamellae. *Ichthyobodo* is extremely pathogenic for juvenile fish, which are very susceptible to infections, causing mass mortalities. In salmonids, especially in hatcheries, *Ichthyobodo* affects osmoregulation, so that the seawater adaptation of juvenile fish is disturbed (Robertson 1985, Urawa 1996). Transmission is direct from fish to fish, or through water masses where free-swimming forms are present. These parasites reproduce asexually by longitudinal binary fission.

Internal parasites of this class belong to the genus *Cryptobia* (often synonymized with *Trypanoplasma*), which was originally described from snail reproductive systems, but now includes species that infect other invertebrates as well as vertebrates. Woo (2003) lists 52 species, five from the skin and gills, seven from the intestines and 40 from the blood. The taxonomy of these species is still under debate. An important species for salmonids on the Pacific Coast of North America is *Cryptobia salmositica,* a blood parasite. This species is elongate, with two flagellae (anterior flagellum free, recurrent flagellum attached but free when extending over the body) and a pulsating vacuole. The latter organelle allows this species to occur outside of the fish as ectoparasites. This enables the parasite to be directly transmitted between fish in contrast to the usual transmission pathway via a vector, in this case the leech *Piscicola salmositica*. It has been shown that sculpins act as reservoir hosts for the parasites. Infected fishes show several clinical signs, including exophthalmia, anaemia, oedema and anorexia. This parasite causes mortalities due to rapid multiplication by binary fission, which occurs mostly in aquaculture settings (see Woo 2003).

In contrast to the Bodonida, the Trypanosomatida possess only one smooth flagellum that is free or connected to the surface as undulating membrane. There are different morphological forms depending on the position of the kinetoplast, kinetosome and flagellar groove, which are typical for specific genera or life stages (Hausmann et al. 2003). They are notorious for causing human diseases such as African trypanosomiasis, better known as sleeping sickness in humans, but they can infect all classes of vertebrates (Simpson et al. 2006), invertebrates as vectors and even protists and plants (Hausmann et al. 2003).

Atlantic cod, *Gadus morhua*, is infected with two different trypanosome species. Both species are transmitted by a leech vector, but by different species (1) *Trypanosoma murmanensis* is transmitted by the leech *Johanssonia* sp. (Khan 1976) and (2) *Trypanosoma pleuronectidium* is transmitted by *Calliobdella nodulifera* (Karlsbakk and Nylund 2006).

Hemmingsen et al. (2005) hypothesize that the introduction of the red king crab *Paralithodes camtschaticus* to the Barents Sea is indirectly the reason for an increase in trypanosome infections in cod. This is due to the leech vector that transmits the trypanosome to cod. In their life cycles, the leech needs to lay eggs on a hard substrate, and they leave the fish host for that. The leech *Johanssonia arctica* seems to favour the carapace of the red king crab to do so. With the increased populations of red king crabs, the leech population increased, which likewise means an increase of trypanosome transmissions to the cod. *Trypanosoma murmanensis* can kill juvenile fish and can negatively affect adult fish, making them more vulnerable to predation. Through the introduction of the red king crab to the Barents Sea, the health of the local cod population has been negatively impacted.

2.4.2 Metamonada

The Metamonada are anaerobic flagellates that lack mitochondria and possess a minimum of four flagella. While the Trichomonadida and Oxymonadida are mainly found in terrestrial habitats, Trimastigida and Diplomonadida occur in aquatic habitats.

One of the most important diplomonad genera that causes gastrointestinal illness in humans and other vertebrates (mammals, amphibians, birds) worldwide is *Giardia* (see Chaps. 19 and 23). Diplomonads are diplozoic, with two identical nuclei and eight flagella. The life cycle (Fig. 2.9) includes only two stages, the trophozoites that replicate through longitudinal binary fission in

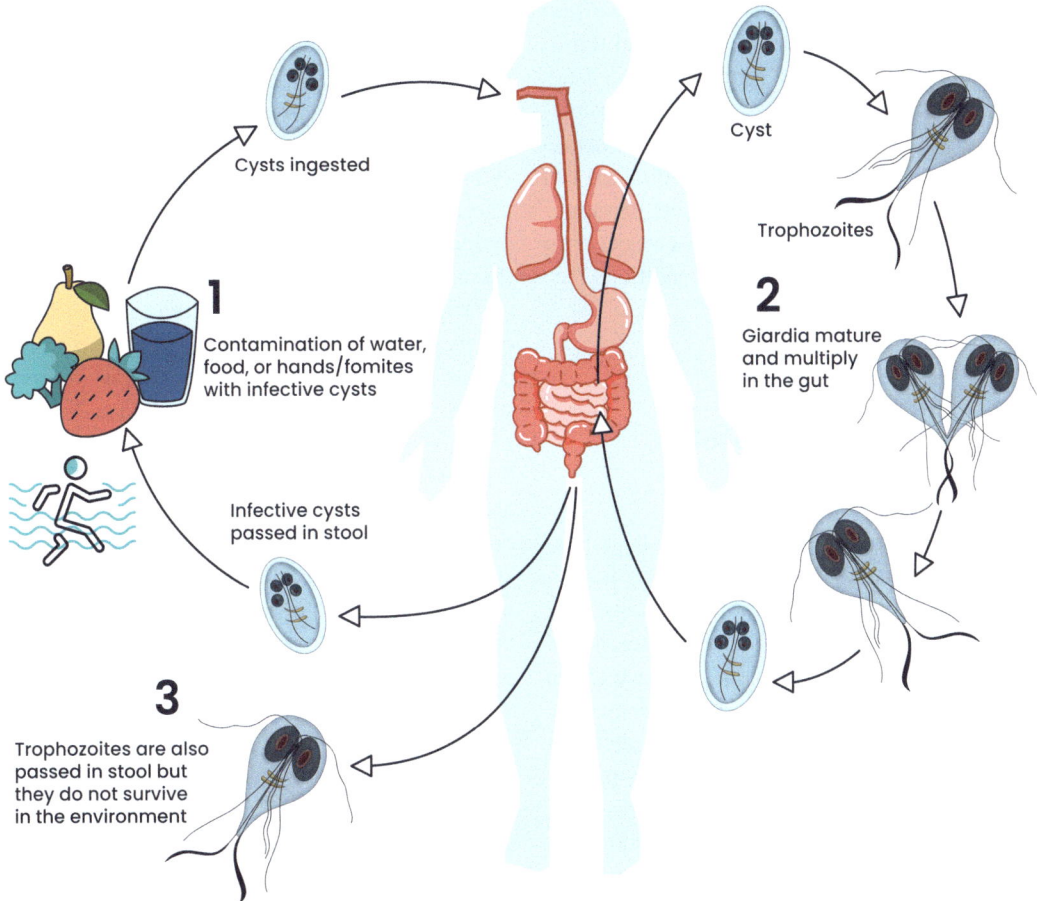

Fig. 2.9 Life cycle of a Diplomonadida (*Giardia* sp.). The life cycle of *Giardia* sp. encompasses only two life stages, the trophozoite and the cyst stage. (1) The infection with Giardia starts with the ingestion of cysts in contaminated water or food. (2) In the small intestine the trophozoites are released, which replicate through longitudinal binary fission. The trophozoites encyst in the colon and are released with the faeces. (3) Trophozoites can also be released with the faeces but are not able to survive in the environment (adapted from Esch and Petersen 2013)

the small intestine. These trophozoites then encyst in the colon during dehydration of the faeces (Hausmann et al. 2003). Cysts are released with the faeces. Transmission occurs through the ingestion of infective cysts present in contaminated water or food. Proper water treatment, personal hygiene and safe food practices are crucial for the prevention of infections (Ryan and Cacciò 2013). There are six described *Giardia* species of which only *G. duodenalis* (synonyms *G. intestinalis* and *G. lamblia*) infects humans, but is also found in many other organisms, which hints towards a zoonotic organism. The contamination of freshwater as well as marine habitats is of concern, not only for direct infective potential for humans, but because this species has also been found in shellfish and faeces of seals, sea lions and whales (Lasek-Nesselquist et al. 2008). These animals might have active infections, or just transfer these cysts passively, raising concern for rapid transmission and distribution.

Another diplomonad parasite that causes morbidity and mortality in fish, particularly salmonids in culture operations, is *Spironucleus salmonicida*, causing economic losses. Most prominent life stages are the trophozoites that show morphological similarities to *Giardia*. Longitudinal binary fission is the asexual reproduction process. No cyst forms have yet been reported so there is a debate about the transmission pathway. This parasite can cause spironucleosis, which is usually a chronic infection, but can also cause severe systemic infections in salmonids, especially in fish farms, and can occur in freshwater as well as in marine conditions (see Ástvaldsson 2019).

2.4.3 TSAR (Telonemia, Stramenopila, Alveolata, Rhizaria)

Dinoflagellata are a diverse group of flagellates within the Alveolata that are found in mainly marine but also freshwater environments. The motile stages are characterized by the presence of two flagella, one of which is in a groove called the sulcus (Hausmann et al. 2003). A general morphological feature of the dinoflagellates is the dinokaryon (their nucleus), which has permanent condensed chromosomes and lacks histones. Some species of dinoflagellates are photosynthetic, while others are heterotrophic or mixotrophic. Many species of dinoflagellates (~150) are known to be parasites of fishes, crustaceans, invertebrates and protists, including even other parasitic dinoflagellates (Coats 1999). There are endo- and ectoparasitic dinoflagellates, the latter making use of special structures to attach to and feed on the host. The distinction between predation and parasitism is not always straightforward in this group.

Species of the genus *Blastodinium*, belonging to the Dinophyceae, are endoparasites of copepods that parasitizes the host gut. This parasitic life stage can be quite large (>1 mm) and consists of hundreds of cells lacking flagella, whereas the dispersal stage has the typical dinoflagellate morphology (Skovgaard et al. 2012). Infected copepods are smaller than uninfected ones, and in females the infection can lead to castration. The life cycle starts with the oral ingestion of a dinospore, which will grow into a trophocyte in the gut. Then the already mentioned multicellular life stage develops. This stage is enveloped by a cuticle. The trophocyte divides and produces a gonocyte, which will produce sporocytes. This part of the cycle can be repeated multiple times. When the cuticle ruptures the non-motile sporocytes are released into the water with the faeces. These binucleated cells produce flagella and divide into four uni-nucleate dinospores (Skovgaard et al. 2012).

Within the Dinophyceae there is another curious parasite group, the genus *Haplozoon*. Species of this genus infect the intestines of polychaete worms, especially species of the Maldanidae, the bamboo worms. The identified parasitic stage does not look like a dinoflagellate (Fig. 2.10), is also 'multicellular' and has previously even been described as a metazoan species. The life cycle is still unknown but assumed to be complex. The trophont can be described as having three compartments: (1) the trophocyte at the anterior end with which the species attaches to the intestinal wall. It bears a suction disc and one to multiple stylets. (2) The gonocytes, which is the midsection made up of several 'cells', usually in one row. (3) The sporocytes, the most posterior compartment with one to multiple rows of more rounded cells. The latter are supposed to detach, leave the host and become dinospores, but this has not been observed. The surface is covered in the so-called thecal barbs, inferred as enlarging the surface area for surface mediated nutrition (compare Rueckert and Leander 2008).

One notable parasitic dinoflagellate in fish is the perdinian species *Amyloodinium ocellatum*, which causes amyloodiniosis in brackish and marine environments, and is especially threatening for the aquaculture industry. The economic impact of this disease is significant, as it can lead to high mortality rates (in less than 12 h) in affected fish (Beraldo and Massimo 2022). The life cycle is direct and consists of three stages (Fig. 2.11): (1) the motile, infective dinospore, (2) the parasitic trophont and (3) the reproductive encysted tomont (Bower et al. 1987). The trophont is attached to the fish (or other aquatic organisms) epithelia with the so-called rhizoid. They feed on epithelial cells and can be found mostly on gills and in the host mouth cavity, but

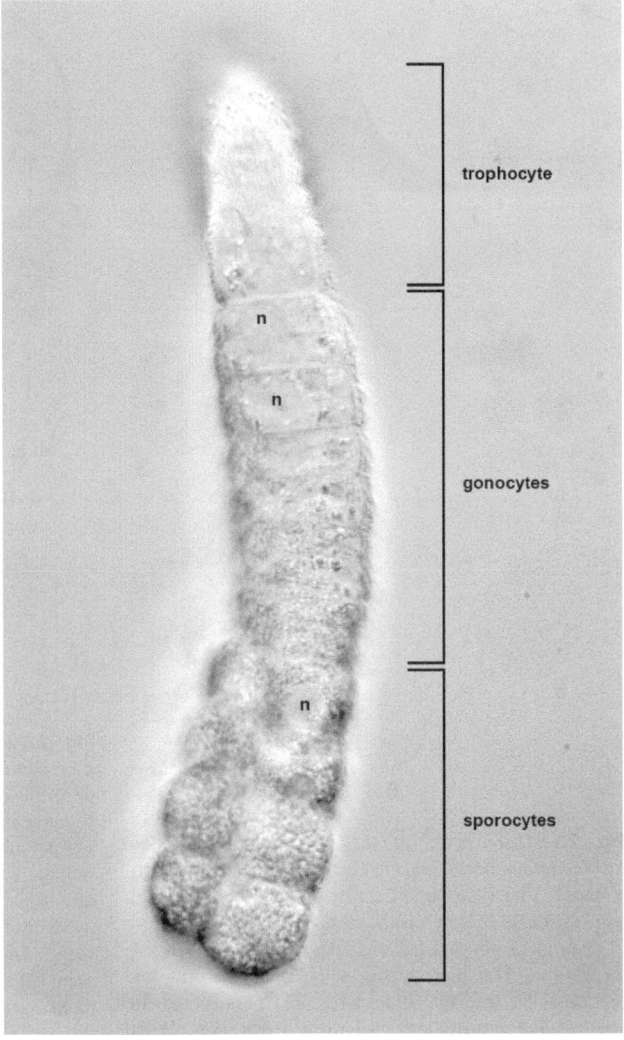

Fig. 2.10 Light micrograph of *Haplozoon praxillellae*, a parasitic dinoflagellate. The three body compartments trophocyte, gonocyte and sporocyte show different morphological characteristics. The nucleus (n) is visible in several 'cells'

in heavy infections can also affect other parts of the skin. These infections can cause different levels of gill hyperplasia, inflammation, haemorrhage and necrosis (Beraldo and Massimo 2022).

Perkinsus species are single-celled parasites that belong to the superclass Perkinsozoa in the Alveolata. Perkinsids in general combine morphological characteristics of dinoflagellates and apicomplexans (Hausmann et al. 2003). They have flagellar structures similar to those of dinoflagellates, but also possess an apical complex similar to the Apicomplexa, which makes them interesting model organisms to understand the biology of the Myzozoa (dinoflagellates and apicomplexans). Species of the genus *Perkinsus* can have massive negative impact on commercially important shellfish species such as oysters, abalone, clam and others (Einarsson et al. 2021), causing diseases and mass mortalities that can lead to significant economic losses to aquaculture or the fishing industry. Perkinsosis caused by two species *Perkinsus marinus* and *P. olseni* are listed diseases in the Aquatic Animal Health Code (https://www.woah.org/en/what-we-do/standards/codes-and-manuals/aquatic-code-online-access/). Transmission of this parasite is direct,

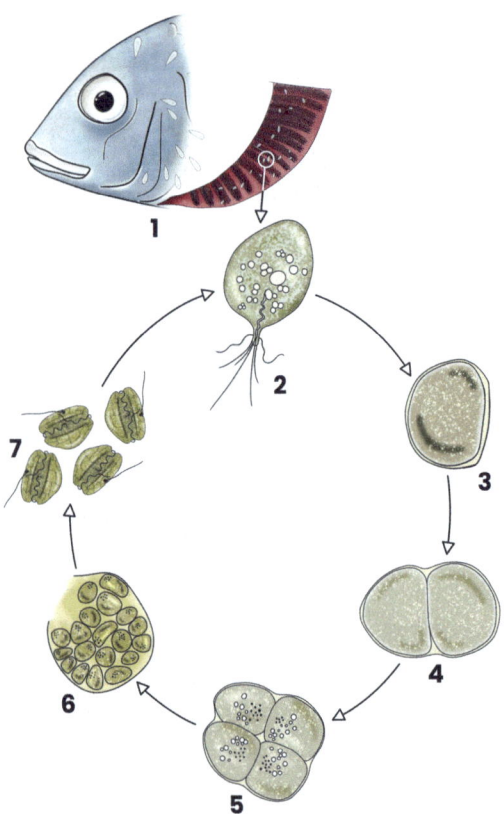

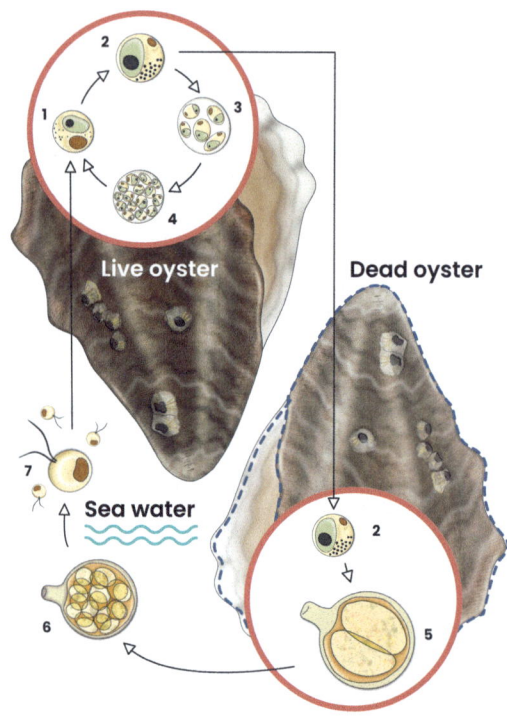

Fig. 2.11 Life cycle of the parasitic dinoflagellate *Amyloodinium ocellatum*. (1) The fish is infested with trophonts. (2) These trophonts are attached to the epithelium with rhizoids, feeding on the host cells. (3) The mature trophont detaches and becomes the tomont with encystment. (4–6) This reproductive stage starts dividing and produces the infective dinospores. (7) Dinospores are released and actively swim to infect a new host. When attached to a new host the dinospore transforms into the trophont and the cycle starts again (adapted from Bower et al. 1987, AquaSymbio 2024)

Fig. 2.12 Life cycle of *Perkinsus marinus* infecting *Crassostrea virginica*. (1–2) In the live oyster a young trophozoite develops into a mature trophozoite with a big vacuole, mostly within the haemocytes. (3–4) Propagation via palintomy results in a schizont with 4 to numerous immature trophozoites. When the schizont ruptures immature trophozoites are released and spread through the oyster with the haemolymph or can be released with the faeces. In the dead and decaying oyster, a mature trophozoite (2) enlarges and develops into a prezoosporangium or hypnospore (5) covered in a thick wall. (6) Released into the marine environment, the hypnospore becomes a zoosporangium with a tube for discharge of zoospores. The zoospores are formed by multiple fission and released into the environment (7), where they can freely swim and infect a new host (adapted from Yokoyama et al. 2015, Fernández Robledo et al. 2018)

as the filter feeding molluscs take up life stages (zoospores or trophozoites) that are phagocytosed by host haemocytes. The trophozoites undergo asexual reproduction by successive binary fission. Immature trophozoites either re-infect the mollusc tissue or are released into the water with the faeces or by the decaying host (Petty 2010). In water the cell enlarges and develops a thick wall, becoming a hypnospore. The next step is zoosporulation by multiple fission and the resulting flagellated zoospores are released via the discharge tube (Fig. 2.12) (Ben-Horin et al. 2015).

2.5 Sporozoans

When considering the sporozoans, we need to further define what organisms should be included here. Sporozoa can be broadly defined as organisms that form a spore stage for transmission. This would also include, for example, the Microsporidia and Myxozoa that are discussed in Chap. 3, but also the Ascetospora that were already introduced earlier in this chapter (Sect. 2.2.2), as they also have amoeboid stages in their life cycles. Here we use the term Sporozoa in its traditional sense following Cox (1994), including protists that are characterized by the presence of a specialized organelle called the apical complex, which is used for host cell invasion and nutrient uptake. Therefore, this chapter is focused on the phylum Apicomplexa in accordance with Adl et al. (2019).

Apicomplexans are a diverse group of unicellular organisms that has been treated as being obligate parasitic, including many important parasites of mostly humans, livestock and wildlife. Even though most people are aware of these parasites in relation to humans and terrestrial animals, recent environmental surveys on protists in soils and aquatic environments have shown that they are highly abundant and diverse in these environments. Most abundant within the apicomplexans in mainly marine environments are the gregarines (del Campo et al. 2019). There are two major groups of Apicomplexa following the taxonomy of Adl et al. (2019), the Aconoidasida (lacking the conoid in the apical complex) encompassing the Haemosporidia, Piroplasmorida, Nephromycida and the Conoidasida (with complete apical complex) containing Coccidida, Gregarinasina and the Blastogregarinea. The exact placement of the genus *Cryptosporidium* is still under debate; therefore, it is here treated separately. The general life cycles encompass merogony, gamogony and sporogony but can differ depending on the taxonomic group.

2.5.1 Haemospororida

Members of this group are not necessarily known for infections in organisms living in marine or freshwater environments, such as mammals, fish or invertebrates. Specific genera have been reported though to cause malaria-like diseases in seabirds that are often infected with ectoparasites that are vectors of, e.g., *Plasmodium* spp. transmitted by midges, mosquitos and flies, *Haemoproteus* spp. transmitted by midges and *Leucocytozoon* spp. transmitted by flies (Zagalska-Neubauer and Bensch 2015; Khan et al. 2019). Not only seabirds but also ducks, geese and swans are affected by these three genera. *Plasmodium* and *Leucocytozoon* can have severe effects on infected birds, causing anaemia, weight loss and death, especially in young birds. *Plasmodium relictum* has, for example, been reported to cause high mortalities in wild penguins (Friend and Franson 1999). The general life cycle of the Haemospororida includes the vector that bites the potentially uninfected bird and transmits the infective sporozoites via its saliva. The sporozoites enter the tissue and produce merozoites, which then penetrate red blood cells. There they mature into infective gametocytes, which are taken up by a second vector. In the midgut of the second vector sexual and asexual reproduction yields oocysts that are filled with infective sporozoites and encapsulated on the outer intestinal wall. The oocysts rupture and the sporozoites move to the salivary glands and with the next blood meal they infect another bird (Friend and Franson 1999) (Fig. 2.13).

2.5.2 Piroplasmorida

The Piroplasmorida do not possess oocysts, spores, pseudocysts or flagellated life stages. The transmission of these parasites involves ectoparasites as vectors. The parasite infects lymphocytes, erythrocytes and other blood, and blood forming cells, in the intermediate host (Hausmann et al. 2003). The life cycle is comparable with that of the Haemospororida. Of the eight genera within this group listed in Adl et al. (2019), three occur in aquatic organisms: (1) *Babesia*, blood parasites of seabirds, e.g., the African penguin *Spheniscus demersus*, suspected to be transmitted by the tick *Ornithodoros capensis* (Snyman et al. 2020). (2) *Haemohormidium*, blood parasites of different marine and freshwater fish species. So (1972) found seven fish species to be infected with two species of *Haemohormidium* in Newfoundland waters. It has been shown experimentally by Khan (1980) that vectors of these blood parasites are leeches.

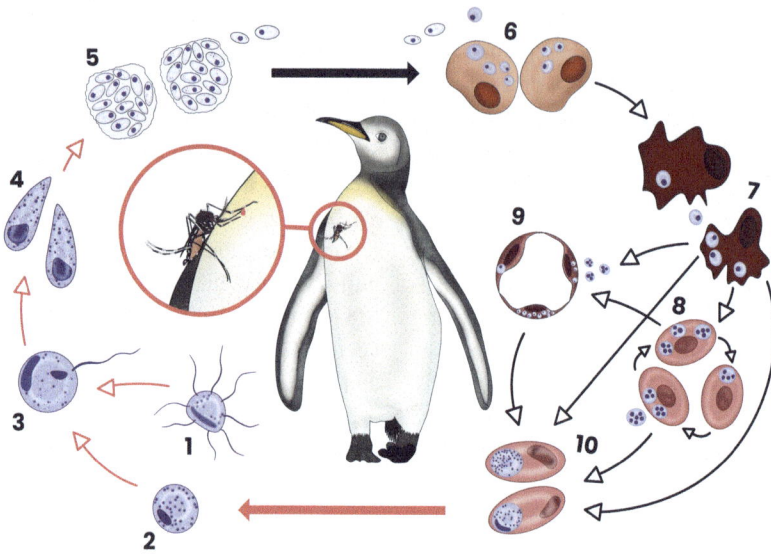

Fig. 2.13 General life cycle of the Haemosporida exemplified by penguin infecting *Plasmodium relictum*. (1–2) With the blood meal of an already infected host (e.g. penguin) the vector (e.g. mosquito) takes up macro- and microgametocytes, which develop into a macrogamete and through exflagellation eight microgametes. (3) In the midgut of the vector the macrogametes are fertilized producing ookinetes (4). The ookinete moves to the basal lamina of the epithelial where it embeds. (5) Sporogony occurs forming sporozoites within the developing oocyst. The sporozoites move towards the salivary glands and with the next bite of the vector they are transmitted to the new host. (6–7) The sporozoites first invade macrophages, monocytes and endothelial cells. They develop into the first generation of meronts outside the erythrocytes. Through asexual multiplication they produce merozoites. Once released, they are distributed with the blood stream following several rounds of merogony in different host tissues, which increases the parasite load of the host. (8) Merozoites can invade erythrocytes, growing into trophozoites that then develop into a second generation of meronts that will produce merozoites. (9) Merozoites invading endothelial cells will also develop via meronts into merozoites. (10) All merozoites can enter gametogony to produce gametocytes, which develop into macro- or microgametocytes, the final life cycle stages in the avian host. Life cycle stages in the mosquito demarked by red arrows, in the penguin by black arrows (adapted from Friend and Franson 1999; Grilo et al. 2016)

2.5.3 Nephromycida

The Nephromycida encompass two genera, *Nephromyces* and *Cardiosporidium*. Both are found within ascidian hosts. Species of *Nephromyces* are found in the renal sac, an 'organelle' of uncertain function, and are restricted to molgulid tunicates. There are several extracellular life stages found within the renal sac, such as sporozoites, spores and large filamentous trophozoites (Saffo and Nelson 1983; Muñoz-Gómez et al. 2019). It is inferred to be mutualistic or commensalistic rather than parasitic (Muñoz-Gómez et al. 2019) and is transferred horizontally via seawater. Species of *Cardiosporidium* are, in contrast, parasites in the haemocytes but occur freely in the heart cavity and pericardial body of ascidians and were first described in the vase tunicate *Ciona intestinalis* (Ciancio et al. 2008). Several life stages have been described such as plasmodia, plasmodia with merogonic stages or merozoites, merozoites, sporozoites and gamonts. The transmission process is not yet known.

2.5.4 Coccidia

Another important, and the largest group of apicomplexan parasites, are the Coccidia infecting primarily vertebrates including mammals, birds, fishes, reptiles and amphibians, but also invertebrates. They are obligate intracellular parasites and present three sequential life cycle stages, merogony, gamogony and sporogony. Coccidia are divided into Adeleorina, blood parasites of vertebrates and invertebrates and the Eimeriorina, often referred to as true coccidia that mostly infect the intestinal tract of vertebrates (Adl et al. 2019; Duszynski et al. 2018). Due to the number of Coccidia genera and species, only selected examples will be presented below to illustrate the range of possible infections in aquatic environments.

The Adeleorina have one life cycle stage that sets them apart from the Eimeriorina, the syzygy stage, which is the association between micro- and macrogamonts before the microgamont produces two to four microgametes. The genera *Haemogregarina* and *Hepatozoon* seem to be the most widespread blood parasites in aquatic environments. The first haemogregarine described from a reptile was *Haemogregarina stepanowi* infecting the European pond turtle *Emys orbicularis*, with a broad geographical distribution. Subsequently, it has been shown that there are genetically distinct lineages that are still in need of formal species description (Maričić et al. 2023). These parasites are transmitted by leeches. One of the most widespread fish-infecting Adeleorina is *H. bigemina*, which has been recorded from roughly 100 fish species (Davies et al. 2004). Trophozoites have been detected in erythrocytes, which enlarge and mature into meronts. The meronts divide by binary fission, producing the characteristic paired gamonts. An interesting finding here concerns the transmission of the parasites. While it is usually leeches, Davies and Smit (2001) discovered another haematophagous vector for this species, gnathiid isopods, in which they found developmental stages (gamonts, immature oocysts, mature oocysts and sporozoites) (Fig. 2.14).

The Eimeriorina include most coccidian genera. There is no syzygy stage, meaning the micro- and macrogametes develop independently. Microgamonts produce large numbers of microgametes (Adl et al. 2019). Coccidian infections can be found in all aquatic environments from freshwater to the marine deep sea. Fishes are mostly infected by three genera *Eimeria*, *Goussia* and *Calyptospora*. While the first two genera are transmitted directly through the faecal–oral route, species of the genus *Calyptospora* have a heteroxenous life cycle involving invertebrates such as crustaceans or annelids (Davies and Ball 1993). *Eimeria* species have been reported to cause serious health problems in freshwater catfish cultures of species of *Clarias* and *Heteroclarias*. *Eimeria* infections were observed through oocysts in the faeces, mucous material and intestine. Infected fish had decreased abdominal fat, haemorrhages in the intestine and pale gills, which resulted in poor overall body condition. Farm management had an influence on infection rates, e.g. fish fed with dead poultry or fish were more likely to be infected (Adah et al. 2022).

One zoonotic parasite in the Apicomplexa is *Toxoplasma gondii*. This species is distributed worldwide and infects a broad range of intermediate hosts. It can cause morbidity and death in humans, domestic animals, terrestrial and aquatic wildlife (Shapiro et al. 2019; see Chap. 19). The definitive host is restricted to domestic and wild cats. While *T. gondii* is primarily transmitted via the consumption of infested meat or cat faeces, oocysts have been detected in aquatic environments, soils and food, and there is also the route from mother to foetus (Tenter et al. 2000). The oocyst plays a central role in the success of this apicomplexan species. The oocysts of *T. gondii* can withstand many disinfectants and unfavourable environmental conditions, which is the reason that they can remain in any environment for several years. Originally it was thought that *T. gondii* was not waterborne, but more and more cases of *Toxoplasmosis* outbreaks have been traced to both natural waters, but also drinking water sources have put *T. gondii* into focus as an emerging waterborne disease (Shapiro et al. 2019). Several outbreaks have been reported in humans, e.g., in Brazil (Keenihan et al. 2002) and Canada (Bowie et al. 1997). Marine mammals have also been of interest (see Chap. 15) and several marine mammals have been recorded to show signs of clinical toxoplasmosis, including cetaceans, phocids, otariids, walruses, sirenians and

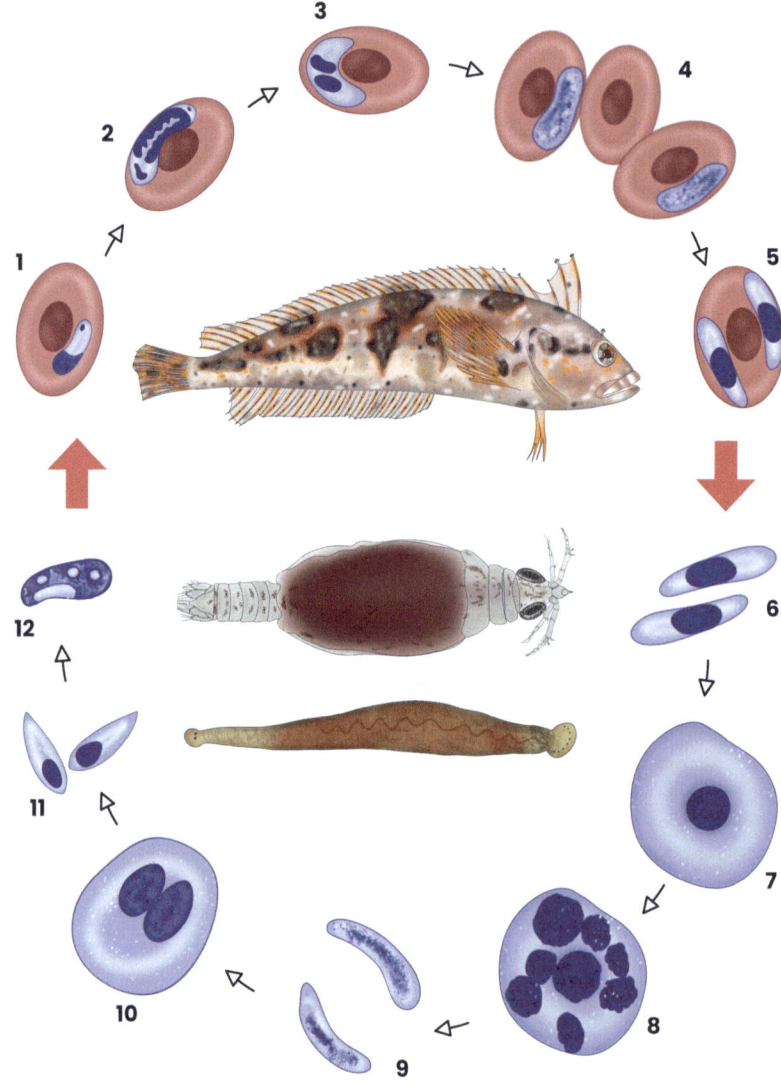

Fig. 2.14 Proposed life cycle of *Haemogregarina bigemina* involving two vectors. (1–5) Different developing stages of *H. bigemina* in the peripheral blood of a fish host. (6–12) Life cycle stages found in two different vectors that are proposed to transmit this blood parasite, a leech and a gnathiid isopod. (1) Small trophozoite in an erythrocyte grows and develops into a meront (2). (3) A stage of pregamontic binary fission. (4) Immature gamonts. (5) Maturing gamonts, pre-paired. (6) Free gamonts. (7) Immature oocyst. (8) Mature oocyst with eight nuclei. (9) Free sporozoites. (10) Meront. (11) First generation merozoites. (12) Second generation merozoite (adapted from Davies and Smit 2001; Hayes et al. 2011)

sea otters (Miller et al. 2018). Even more curious is the fact that serological exposure has been reported for all marine environments and even Antarctica, where no felids are known to be present. The oocyst presence in aquatic environments is mostly due to surface water run off that transports oocysts from faeces or manure into the adjacent water bodies. Fish and invertebrates might not serve as true intermediate hosts, but oocysts can stay viable when ingested and be further transmitted when, e.g., fish and other aquatic invertebrates are fed on, by humans or wildlife (Shapiro et al. 2019). In future it will be important to develop quick and reliant detection methods as well as avoidance and inactivation methods.

2.5.5 Gregarinasina

A group of apicomplexans that can be found in freshwater, marine and terrestrial environments are the gregarines. They are unique as they only infect invertebrates, which is one of the reasons why they are the most understudied group, as they rarely cause any economic losses (Leander 2008). Phylogenetically they are at the base of the Apicomplexa and have previously been considered to be primitive. Gregarines are typically found in the digestive tracts, coeloms and reproductive vesicles of more or less all invertebrates, including annelids, crustaceans and molluscs amongst others. Due to the number of known and potential invertebrate hosts, it is not surprising that they are extremely abundant in eDNA studies (del Campo et al. 2019). Eugregarines and archigregarines occur in freshwater and marine habitats, the former in both, the latter is exclusively marine. Gregarines are extracellular and most of them have monoxenous life cycles including only one host species. Infections occur through the faecal-oral route. The general life cycle (Fig. 2.15) exemplified here for the archigregarine *Selenidium pendula* involves both sexual and for some species asexual reproduction. The life cycle starts with the ingestion of oocysts. In the intestine sporozoites are released that partially invade the cell of the intestinal lining. The gregarine starts to grow and develops into the feeding stage, the trophozoite, which detaches and remains free in the intestinal lumen. Depending on the species asexual reproduction, the so-called schizogony, can occur at this stage. Trophozoites develop into gamonts that pair up in syzygy. A cyst (gamontocyst) forms around them and a similar number of male and female gametes are produced. The gametes fuse into zygotes with another cyst forming around them (oocyst). Through meiosis and mitosis, the sporozoites are produced (specific numbers depend on gregarine species) (see Schrével and Desportes 2016; Rueckert and Horák 2017).

While gregarines are not typically considered pathogenic to their hosts, they can have significant impacts on their growth, reproduction and survival. Mita et al. (2012) reported on an eugregarine (*Lankesteria ascidiae*) as causative agent of long faeces syndrome in an inland culture of the vase tunicate *Ciona intestinalis*. Diseased animals can die as quick as 1 week after the recognition of the disease. The parasitic lifestyle of gregarines is questionable as their effects are very variable, which places them in different positions on the spectrum of symbiosis (Rueckert et al. 2019) (Fig. 2.16). Due to their unique position, they are assumed to provide important insights into the evolution towards intracellular parasitism within the Apicomplexa.

2.5.6 Cryptogregarea

This group contains only a single genus, *Cryptosporidium*. Its systematic placement is under constant debate and dependent on methods and raw data used (Ryan et al. 2021; Salomaki et al. 2021). One of the reasons is that this genus combines characteristics of both gregarines and coccidians. The general life cycle starts with the ingestion of oocysts. Sporozoites are released and enter epithelial cells of mainly the gastrointestinal tract. *Cryptosporidium* spp. are intracellular but extracytoplasmic, within a parasitovorous vacuole. The parasite undergoes asexual reproduction (schizogony and merogony). Following that, sexual reproduction starts with gametogony producing micro- (male) and macrogametes (female). These fuse to form zygotes and develop to oocysts with either a thin or a thick wall. Sporulation of the oocysts yields the infective sporozoites. Thin-walled oocysts can auto-infect the same host, whereas thick-walled cysts are released with the faeces (see CDC 2012). *Cryptosporidium* spp. are zoonotic and cause water- and foodborne diseases. *Cryptosporidium* spp. are common in both freshwater and marine environments and can infect a wide range of hosts, including humans, livestock and wildlife (see Chap. 19). Due to the development of detection and typing tools in the past 50 years, at least 44 species are described, six avian, four piscine, one amphibian, four reptile and 29 mammalian species (Ryan et al. 2021). Oocysts of *Cryptosporidium* spp. are transmitted via the faecal-oral route, but there are several transmis-

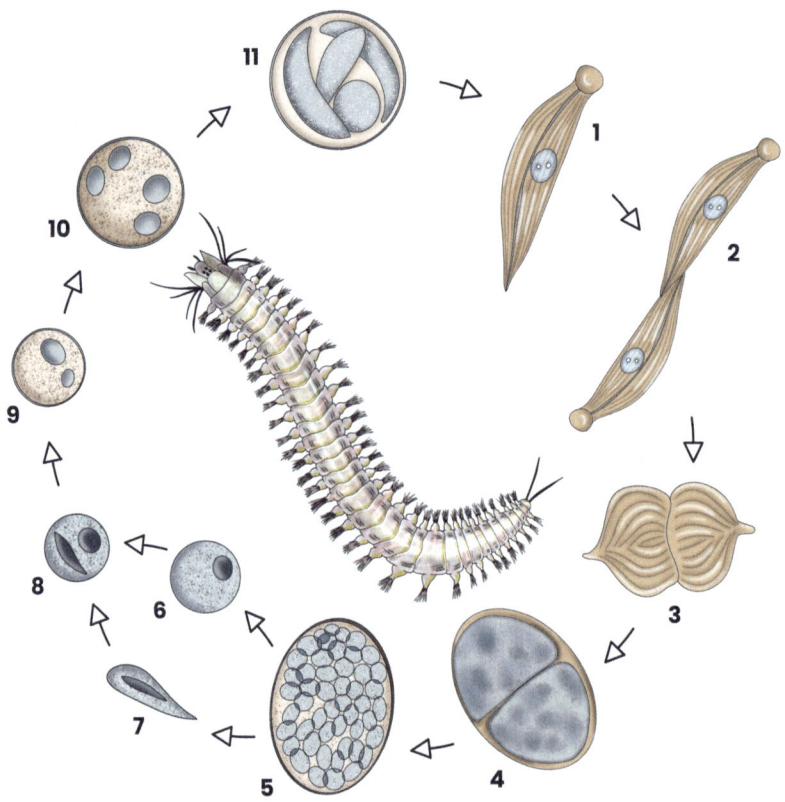

Fig. 2.15 Life cycle of gregarine apicomplexan *Selenidium pendula*. (1) Trophozoites in the intestine of polychaete *Scolelepis squamata* grow into gamonts. (2–3) Two gamonts pair up in caudal syzygy. (4) The two gamonts in syzygy encyst forming the gamontocyst. (5) In the sexual phase the gametes are produced through nuclear multiplication, resulting in equal numbers of female (6) and male (7) gametes within the gametocyst. (8) Zygotes form through fertilization. (9–10) During sporogony a cyst (oocyst) forms around the zygote and through meiosis and mitosis four sporozoites are produced. (11) The oocyst with four infective sporozoites will be ingested by a new host to restart the cycle (adapted from Schrével and Desportes 2016; Rueckert and Horák 2017)

sion pathways (Fig. 2.17), and can cause severe gastrointestinal illness, especially in children and immunocompromised individuals. According to Khalil et al. (2018) *Cryptosporidium* spp. cause a global annual loss of ~4.2 million disability-adjusted life years. Waterborne *Cryptosporidium* spp. outbreaks are the bulk (>60%) of reported waterborne outbreaks of protists due to drinking water and recreational water and in addition *Cryptosporidium* spp. are the fifth most important foodborne parasites globally (Ryan et al. 2021). Tryland (2018) lists several marine mammals including dugongs, California sea lions, ringed, bearded, harp and grey seals, bowhead and right whales, as well as harbour porpoises from which *Cryptosporidium* spp. have been reported (see also Chap. 15). Many of the waterborne transmissions are connected to rivers and estuaries, but observational data on *Cryptosporidium* spp. transmission via rivers is still scarce. Thus, Vermeulen et al. (2019) used a modelling approach to first estimate *Cryptosporidium* occurrence in rivers worldwide and second to determine hotspots. This model is also suggested to be useful to analyse future scenarios, including global change processes.

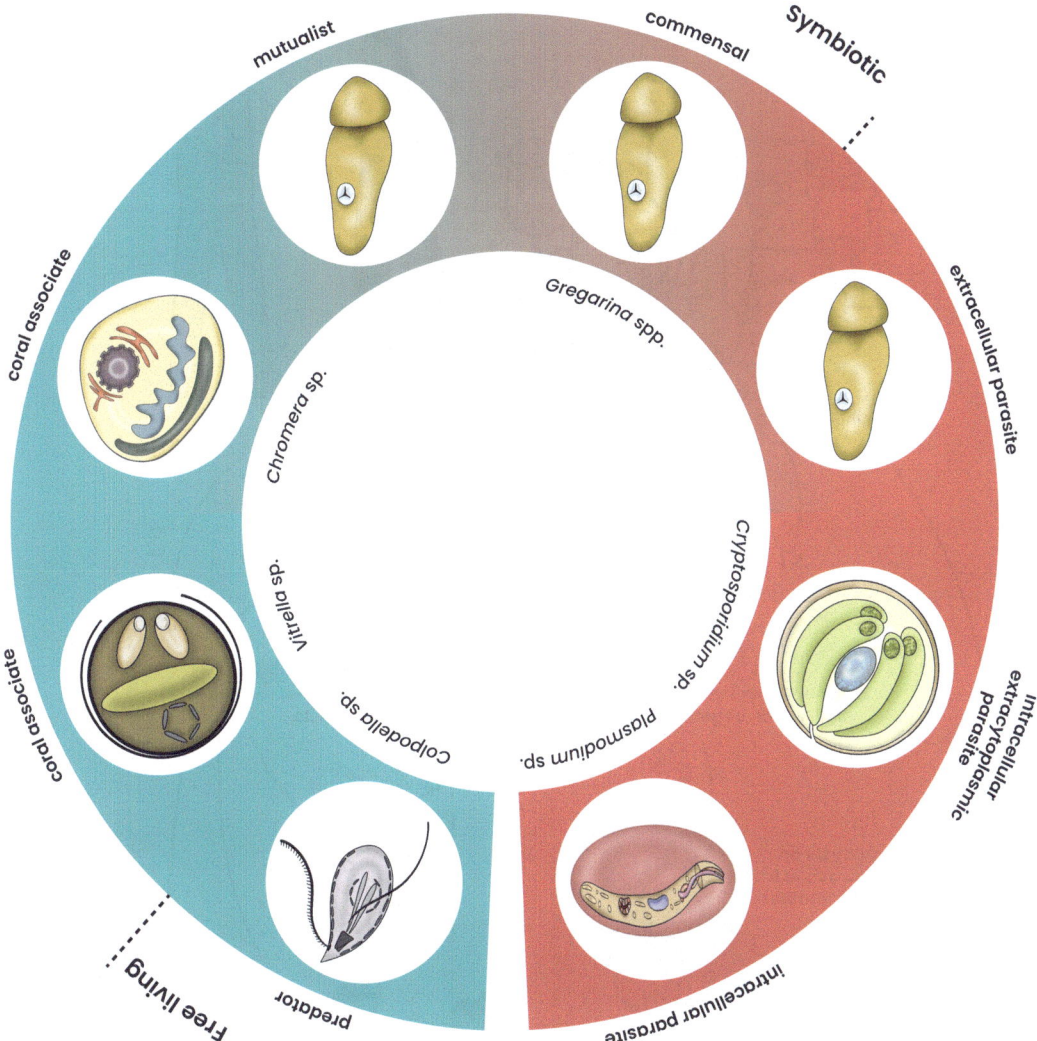

Fig. 2.16 The symbiotic spectrum of the Apicomplexa and close relatives. In this sketch apicomplexans and their close relatives are placed on a symbiosis spectrum from free-living to intracellular parasitism (green to red). *Colpodella* is a free-living, heterotrophic predator. *Chromera* (brown) and *Vitrella* (green) are phototrophic coral associates. All three are close relatives of the Apicomplexa. Gregarines play an important role in the understanding of the development of intracellular parasitism in the Apicomplexa, as they can be mutualistic, commensalistic or extracellular parasitic. *Cryptosporidium* is an intracellular, but extracytoplasmic parasite, while *Plasmodium* is an intracellular parasite (adapted from Rueckert et al. 2019, Richtová et al. 2021)

Fig. 2.17 Overview of various *Cryptosporidium* oocyst transmission pathways (adapted from Robertson et al. 2020)

2.6 Conclusion

Amoebae, ciliates, flagellates and sporozoans are diverse groups of unicellular organisms that play important roles in aquatic ecosystems, including that as parasites of other aquatic organisms. The influence of amoebae on other organisms and consecutively ecosystems is not clear for all groups, but some species cause significant mortalities in various aquaculture operations. Further research is needed to better understand the biology and ecology of these parasitic amoebae, and their interaction with other microorganisms, as well as to develop effective methods for their control and management in aquatic environments. A lot remains unknown when it comes to the relationship between the symbiotic ciliates and their hosts, but some examples show that these small protists can have huge impacts on their hosts and/or on food webs and ecological processes. The impacts of parasitic flagellates on aquatic ecosystems are significant, as they can cause diseases, reduced fitness and the death of their hosts. The different groups of apicomplexan genera, such as *Cryptosporidium*, *Eimeria* and *Toxoplasma*, can have significant impacts on both wild and cultured aquatic organisms, as well as human health. Research is needed to better understand the biology and ecology of these parasites in aquatic environments, as well as to develop effective methods for their control and management.

Understanding the biology, ecology and impacts of these diverse groups of parasitic protists in aquatic habitats is essential for comprehending the intricate web of interactions within these ecosystems.

Acknowledgements The author would like to thank Prof Dr Fiona Henriquez for reading parts of this chapter and providing some valuable thoughts and suggestions as well as Prof Liesl van As, University of the Free State, South Africa, for the photo used in Fig. 2.7.

References

Adah AD, Lawal S, Oniye SJ et al (2022) Occurrence and risk factors associated with *Eimeria* species infections in *Clarias gariepinus* and *Heteroclarias* species. Niger J Parasitol 43(1):93–101

Adl SM, Bass D, Lane CE et al (2019) Revisions to the classification, nomenclature, and diversity of eukaryotes. J Eukaryot Microbiol 66(1):4–119

Antonius AA, Lipscomb D (2001) First protozoan coral-killer in the Indo-Pacific. Atoll Res Bull 481:1–21

AquaSymbio (2024) *Amyloodinium ocellatum* [online]. Available at: http://aquasymbio.fr/en/amyloodinium-ocellatum. Accessed 18 June 2024

Arzul I, Chollet B, Boyer S et al (2014) Contribution to the understanding of the cycle of the protozoan parasite *Marteilia refringens*. Parasitology 141:227–240

Ástvaldsson A (2019) Pathogenesis and cell biology of the salmon parasite *Spironucleus salmonicida*. Dissertation, Uppsala University

Bass D, Rueckert S, Stern R et al (2021) Parasites, pathogens, and other symbionts of copepods. Trends Parasitol 37(10):875–889

Bastos Gomes R, Jerry DR, Miller TL et al (2017) Current status of parasitic ciliates *Chilodonella* spp. (Phyllopharyngea: Chilodonellidae) in freshwater fish aquaculture. J Fish Dis 40(5):703–715

Bateman KS, Feist SW, Bignell JP et al (2020) Marine pathogen diversity and disease outcomes. In: Behringer DC, Lafferty KD, Silliman BR (eds) Marine disease ecology. Oxford University Press, Oxford, pp 3–43

Ben-Horin T, Bidegain G, Huey L et al (2015) Parasite transmission through suspension feeding. J Invertebr Pathol 131:155–176

Beraldo P, Massimo M (2022) Amyloodiniosis. In: Kibenge SFB, Baldisserotto B, Chong RS-M (eds) Aquaculture pathophysiology, Volume I. Finfish diseases. Academic, London, pp 475–483

Bermingham ML, Mulcahy MF (2007) *Neoparamoeba* sp. and other protozoans on the gills of Atlantic salmon *Salmo salar* smolts in seawater. Dis Aquat Org 76:231–240

Bouland C, Jangoux M (1988) Infestation of *Asterias rubens* (Echinodermata) by the ciliate *Orchitophrya stellarum*: effect on gonads and host reaction. Dis Aquat Org 5:239–242

Bower SM, Meyer GR (1993) *Stegotricha enterikos* gen. nov., sp. nov. (class Phyllopharyngea, order Rhynchodida), a parasitic ciliate in the digestive gland of Pacific oysters (*Crassostrea gigas*), and its distribution in British Columbia. Can J Zool 71:2005–2017

Bower CE, Turner DT, Biever RC (1987) A standardized method of propagating the marine fish parasite, *Amyloodinium ocellatum*. J Parasitol 73(1):85–88

Bowie WR, King AS, Werker DH (1997) Outbreak of toxoplasmosis associated with municipal drinking water. The BC Toxoplasma Investigation Team. Lancet 350:173–177

Bradbury PC (1994) Ciliates of fish. In: Kreier JP (ed) Parasitic protozoa, vol 8, 2nd edn. Academic, San Diego, pp 81–138

Brander KM (2009) Fisheries and climate. In: Wefer G, Lamy F, Mantoura F (eds) Marine science frontiers for Europe. Springer, Berlin, pp 483–490

Burki F, Keeling PJ (2014) Rhizaria. Curr Biol 24(3):R103–R107

Burki F, Roger AJ, Brown MW et al (2020) The new tree of eukaryotes. Trends Ecol Evol 35(1):43–55

CDC (Centers for Disease Control and Prevention) (2012) *Cryptosporidium* (also known as "Crypto"). In: Parasites. Centers for Disease Control and Prevention. https://www.cdc.gov/parasites/crypto/index.html. Accessed 6 Aug 2023

CDC (Centers for Disease Control and Prevention) (2022) *Naegleria fowleri*—primary amebic meningoencephalitis (PAM)—amebic encephalitis. In: Parasites. Centers for Disease Control and Prevention. https://www.cdc.gov/parasites/naegleria/pathogen.html. Accessed 4 Aug 2023

Chatton É (1920) Les Péridiniens parasites: morphologie, reproduction, ethologie. Arch Zool Exp Gén 59:1–475

Chelkha N, Hasni I, Cherif Louazani A et al (2020) *Vermamoeba vermiformis* CDC-19 draft genome sequence reveals considerable gene trafficking including with candidate phyla radiation and giant viruses. Sci Rep 10:5928

Ciancio A, Scippa S, Finetti-Sialer M et al (2008) Redescription of *Cardiosporidium cionae* (Van Gaver and Stephan, 1907) (Apicomplexa: Piroplasmida), a plasmodial parasite of ascidian haemocytes. Eur J Protistol 44(3):181–196

Coats DW (1999) Parasitic lifestyles of marine dinoflagellates. J Eukaryot Microbiol 46(4):402–409

Colorni A (1987) Biology of *Cryptocaryon irritans* and strategies for its control. Aquaculture 67(1/2):236–237

Colorni A, Burgess P (1997) *Cryptocaryon irritans* Brown 1951, the cause of 'white spot disease' in marine fish: an update. Aquar Sci Conserv 1:217–238

Cox FEG (1994) The evolutionary expansion of the Sporozoa. Int J Parasitol 24(8):1301–1316

Coyne RS, Hannick L, Shanmugam D et al (2011) Comparative genomics of the pathogenic ciliate *Ichthyophthirius multifiliis*, its free-living relatives and a host species provide insights into adoption of a parasitic lifestyle and prospects for disease control. Genome Biol 12:R100

Davies AJ, Ball SJ (1993) The biology of fish coccidia. Adv Parasitol 32:293–366

Davies AJ, Smit NJ (2001) The life cycle of *Haemogregarina bigemina* (Adeleina: Haemogregarinidae) in South African hosts. Folia Parasitol 48:169–177

Davies AJ, Smit NJ, Hayes PM et al (2004) *Haemogregarina bigemina* (Protozoa: Apicomplexa: Adeleorina)—past, present and future. Folia Parasitol 51:99–108

Delafont V, Rodier A-H, Maisonneuve E et al (2018) *Vermamoeba vermiformis*: a free-living amoeba of interest. Microb Ecol 76:991–1001

del Campo J, Heger TJ, Rodriguez-Martinez R et al (2019) Assessing the diversity and distribution of Apicomplexans in host and free-living environments using high-throughput amplicon data and a phylogenetically informed reference framework. Front Microbiol 10:2373

Delgery CC, Cragg SM, Busch S et al (2006) Effects of the epibiotic heterotrich ciliate *Mirofolliculina limnoriae* and of moulting on faecal pellet production by the wood-boring isopods, *Limnoria tripunctata* and *Limnoria quadripunctata*. J Exp Mar Biol Ecol 334:165–173

den Hartog C (1989) Distribution of *Plasmodiophora bicaudata*, a parasitic fungus on small *Zostera* species. Dis Aquat Org 6:227–229

Duszynski DW, Kvičerová J, Seville RS (2018) The biology and identification of the Coccidia (Apicomplexa) of carnivores of the world. Academic, London

Dykova I, Pindova Z, Fiala I et al (2005) Fish-isolated strains of *Hartmannella vermiformis* Page, 1967: morphology, phylogeny and molecular diagnosis of the species in tissue lesions. Folia Parasitol 52:295–303

Edwards M, Coates CJ, Rowley AF et al (2019) Host range of the mikrocytid parasite *Paramikrocytos canceri* in decapod crustaceans. Pathogens 8:252

Einarsson E, Lassadi I, Zielinski J et al (2021) Development of the myzozoan aquatic parasite *Perkinsus marinus* as a versatile experimental genetic model organism. Protist 172:1–15

Esch KJ, Petersen CA (2013) Transmission and epidemiology of zoonotic protozoal diseases of companion animals. Clin Microbiol Rev 26:58–85

Feehan CJ, Scheibling RE (2014) Disease as a control of sea urchin populations in Nova Scotian kelp beds. Mar Ecol Prog Ser 500:149–158

Feehan CJ, Johnson-Mackinnon J, Scheibling RE et al (2013) Validating the identity of *Paramoeba invadens*, the causative agent of recurrent mass mortality of sea urchins in Nova Scotia, Canada. Dis Aquat Org 103:209–227

Fehling J, Stoeckert D, Baldauf SL et al (2007) Photosynthesis and the eukaryotic tree of life. In: Falkowski PG, Knoll AH (eds) Evolution of primary producers in the sea. Academic, London, pp 75–107

Fenchel T (1965) On the ciliate fauna associated with the marine species of the amphipod genus *Gammarus* J. G. *fabricius*. Ophelia 2(2):281–303

Fernández Robledo JA, Marquis ND, Countway PD et al (2018) Pathogens of marine bivalves in Maine (USA): A historical perspective. Aquaculture 493:9–17

Ford SE, Stokes NA, Alcox KA et al (2018) Investigating the life cycle of *Haplosporidium nelsoni* (MSX): a review. J Shellfish Res 37(4):679–693

Friend M, Franson JC (1999) Field manual of wildlife diseases. General field procedures and diseases of birds. Federal Government Series, Madison

Garvetto A, Murúa P, Kirchmair M et al (2023) Phagocytosis underpins the biotrophic lifestyle of intracellular parasites in the class Phytomyxea (Rhizaria). New Phytol 238:2130–2143

Gómez-Gutiérrez J, Peterson WT, Robertis AD et al (2003) Mass mortality of krill caused by parasitoid ciliates. Science 301:339

Gómez-Gutiérrez J, Strüder-Kypke MC, Lynn DH et al (2012) *Pseudocollinia brintoni* gen. nov., sp. nov. (Apostomatida: Colliniidae), a parasitoid ciliate infecting the euphausiid *Nyctiphanes simplex*. Dis Aquat Org 99:57–78

Grilo ML, Vanstreels RET, Wallace R (2016) Malaria in penguins—current perceptions. Avian Pathol 45:393–407

Grimes BH, Bradbury PC (1992) The biology of *Vampyrophrya pelagica* (Chatton & Lwoff, 1930), a histophagous apostomes ciliate associated with marine calanoid copepods. J Protozool 39:65–79

Harry OG (1980) Damage to the eyes of the bivalve *Chlamys opercularis* caused by the ciliate *Licnophora auerbachii*. J Invertebr Pathol 36:283–291

Hartikainen H, Stentiford GD, Bateman KS et al (2014) Mikrocytids are a broadly distributed and divergent radiation of parasites in aquatic invertebrates. Curr Biol 24:807–812

Hausmann K, Hülsmann N, Radek R (2003) Protistology, 3rd edn. E. Schweizerbart'sche Verlagsbuchhandlung (Nägele u. Obermiller), Science, Stuttgart

Hayes PM, Wertheim DF, Smit NJ et al (2011) Three-dimensional visualisation of developmental stages of an apicomplexan fish blood parasite in its invertebrate host. Parasit Vectors 4:219

Hemmingsen W, Jansen PA, MacKenzie K et al (2005) Crabs, leeches and trypanosomes: an unholy trinity? Mar Pollut Bull 50:336–339

Henriquez FL (2017) Acanthamoeba and its ocular impact. Optician. Available via opticionline. https://www.opticianonline.net/cpd-archive/4432. Accessed 13 Aug 2023

Hine PM, Morris DJ, Azevedo C et al (2020) Haplosporosomes, sporoplasmosomes and their putative taxonomic relationships in rhizarians and myxozoans. Parasitology 147(14):1614–1628

Hoffman GL, Lando M, Camper JE et al (1975) A disease of freshwater fishes caused by *Tetrahymena corlissi* Thompson, 1955, and a key for identification of holotrich ciliates of freshwater fishes. J Parasitol 61(2):217–223

Jin D, Qu Z, Wei B et al (2021) Two parasitic ciliates (Protozoa: Ciliophora: Phyllopharyngea) isolated from respiratory-mucus of an unhealthy beluga whale: characterization, phylogeny and an assessment of morphological adaptations. Zool J Linn Soc 191:941–960

Khalil IA, Troeger C, Rao PC et al (2018) Morbidity, mortality, and long-term consequences associated with diarrhoea from Cryptosporidium infection in children younger than 5 years: a meta-analyses study. Lancet Glob Health 6:e758–768

Karling JS (1944) *Phygomyxa algarum* n.gen., n.sp., an unusual parasite with plasmodiophoralean and protomyxean characteristics. Am J Bot 31:38–52

Karlsbakk E, Nylund A (2006) Trypanosomes infecting cod *Gadus morhua* L. in the North Atlantic: a resurrection of *Trypanosoma pleuronectidium* Robertson, 1906 and delimitation of *T. murmanense* Nikitin, 1927 (emend.), with a review of other trypanosomes from North Atlantic and Mediterranean teleosts. Syst Parasitol 65:175–203

Keenihan SH, Schetters T, Taverne J (2002) Toxoplasmosis in Brazil. Trends Parasitol 18:203–204

Khan RA (1976) The life cycle of *Trypanosoma murmanensis* Nikitin. Can J Zool 54:1840–1849

Khan RA (1980) The leech as a vector of a fish piroplasm. Can J Zool 58(9):1631–1637

Khan JS, Provencher JF, Forbes MR et al (2019) Parasites of seabirds: A survey of effects and ecological implications. Adv Mar Biol 82:1–50

Lasek-Nesselquist E, Bogomolni AL, Gast RJ et al (2008) Molecular characterization of *Giardia intestinalis* haplotypes in marine animals: variation and zoonotic potential. Dis Aquat Org 81:39–51

Leander BS (2008) Marine gregarines: evolutionary prelude to the apicomplexan radiation? Trends Parasitol 24:60–67

Leighton BJ, Boom JDG, Bouland C et al (1991) Castration and mortality in *Pisaster ochracaeus* parasitized by *Orchitophrya stellarum* (Ciliophora). Dis Aquat Org 10:71–73

Li Y, Jiang B, Mo Z et al (2022) *Cryptocaryon irritans* (Brown, 1951) is a serious threat to aquaculture of marine fish. Rev Aquac 14:218–236

Lom J (1958) A contribution to the systematics and morphology of endoparasitic trichodinids from amphibians, with a proposal of uniform specific characteristics. J Protozool 5:251–263

Lynn DH (1975) The life cycle of the histophagous ciliate *Tetrahymena corlissi* Thompson, 1955. J Protoyool 22(2):188–195

Lynn DH (2010) The ciliated Protozoa. Characterisation, classification, and guide to the literature, 3rd edn. Springer, New York

MacPhail DPC, Koppenstein R, Maciver SK et al (2021) Vibrio species are predominantly intracellular within cultures of *Neoparamoeba perurans*, causative agent of amoebic gill disease (AGD). Aquaculture 532:736083

Mahasri G, Rozi, Mukti AT et al (2019) The correlation between ectoparasite infestation and the total plate count of *Vibrio* sp. in pacific white shrimp (*Litopenaeus vannamei*) in ponds. IOP Conf Ser Earth Environ Sci 236:1–7

Maier I, Parodi E, Westermeier R et al (2000) *Maullinia ectocarpii* gen. et sp. nov. (Plasmodiophorea), an intracellular parasite in *Ectocarpus siliculosus* (Ectocarpales, Phaeophyceae) and other filamentous brown algae. Protist 151:225–238

Marciano-Cabral F, Puffenbarger R, Cabral GA (2000) The increasing importance of *Acanthamoeba* infections. J Eukaryot Microbiol 47(1):29–36

Maričić M, Danon G, Faria JF et al (2023) Molecular screening of haemogregarine hemoparasites (Apicomplexa: Adeleorina: Haemogregarinidae) in populations of native and introduced pond turtles in Eastern Europe. Microorganisms 11(4):1063

Mayén-Estrada R, Dias RJP, Ramírez-Ballesteros M et al (2021) Ciliates as symbionts. In: Pereira L, Gonçalves AM (eds) Plankton communities. IntechOpen, London, pp 1–21

Milanez GD, Masangkay FR, Thomas RC et al (2017) Molecular identification of *Vermamoeba vermiformis* from freshwater fish in lake Taal, Philippines. Exp Parasitol 183:201–206

Miller MA, Shapiro K, Murray M et al (2018) Protozoan parasites of marine mammals. In: Gulland FMD, Dierauf LA, Whitman KL (eds) CRC Handbook of marine mammal medicine. CRC, New York p, pp 425–470

Mita K, Kawai N, Rueckert S et al (2012) Large-scale infection of the ascidian *Ciona intestinalis* by the gregarine *Lankesteria ascidiae* in an inland culture system. Dis Aquat Org 101:185–195

Morado JF, Small EB (1995) Ciliate parasites and related diseases of crustacea: A review. Rev Fish Sci 3(4):275–354

Muñoz-Gómez SA, Durnin K, Eme L et al (2019) *Nephromyces* represents a diverse and novel lineage of the Apicomplexa that has retained apicoplasts. Genome Biol Evol 11(10):2727–2740

Neuhauser S, Kirchmair M, Gleason FH et al (2011) Ecological roles of the parasitic phytomyxids (plasmodiophorids) in marine ecosystems—a review. Mar Freshw Res 62(4):365–371

Nowak BF (2007) Parasitic diseases in marine cage culture—an example of experimental evolution of parasites? Int J Parasitol 37:581–588

Ohtsuka S, Suzaki T, Kanazawa A et al (2015) Biology of symbiotic apostome ciliates: Their diversity and importance in the aquatic ecosystems. In: Ohtsuka S, Suzaki T, Horiguchi T et al (eds) Marine protists. Springer, Tokyo, pp 441–463

Onut-Brännström I, Stairs CW, Campos A et al (2023) A mitosome with distinct metabolism in the uncultured protist parasite *Paramikrocytos canceri* (Rhizaria, Ascetosporea). Genome Biol Evol 15(3):1–16

Padrós F, Constenla M (2021) Diseases caused by Amoeba in fish: an overview. Animals 11:991

Page FC, Blanton RL (1985) The Heterolobosea (Sarcodina: Rhizopoda), a new class uniting the Schizopyrenida and the Acrasidae (Acrasida). Protistologica 21:121–132

Pánek T, Simpson AGB, Brown MW et al (2016) Heterolobosea. In: Archibald JM, Simpson AGB, Slamovits CH (eds) Handbook of the protists. Springer, Cham, pp 1–41

Pérez JM (2009) Parasites, pests, and pets in a global world: New perspectives and challenges. J Exot Pet Med 18(4):248–253

Petty D (2010) *Perkinsus* infections of bivalve molluscs. Fisheries and Aquatic Sciences Department, Florida Cooperative Extension Service, Institute of Food and Agricultural Sciences, University of Florida. Available via IFAS. https://shellfish.ifas.ufl.edu/wp-content/uploads/Perkinsus-Infections-of-Bivalve-Molluscs.pdf. Accessed 15 Oct 2023

Prokopowicz AJ, Rueckert S, Leander BS et al (2010) Parasitic infection of the hyperiid amphipod *Themisto libellula* in the Canadian Beaufort Sea (Arctic Ocean), with a description of *Ganymedes themistos* sp. n. (Apicomplexa, Eugregarinorida). Polar Biol 33:1339–1350

Reef Stable (2024) Marine Ich. [online] Available at: https://reefstable.com/solutions/illness/marine-ich/. Accessed 18 June 2024

Richtová J, Sheiner L, Gruber A et al (2021) Using diatom and apicomplexan models to study the heme pathway of *Chromera velia*. Int J Mol Sci 22:6495

Robertson DJ (1985) A review of *Ichthyobodo necator* (Henneguy, 1883) an important and damaging parasite. In: Muir JF, Roberts RJ (eds) Recent advances in aquaculture, vol 2. Croom Helm, London, pp 1–30

Robertson LJ, Johansen OH, Kifleyohannes T et al (2020) *Cryptosporidium* infections in Africa—How important is zoonotic transmission? A review of the evidence. Front Vet Sci 7:575881

Roger AJ, Smith MW, Doolittle RF et al (1996) Evidence for the Heterolobosea from phylogenetic analysis of genes encoding glyceraldehyde-3-phosphate dehydrogenase. J Eukaryot Microbiol 43(6):475–485

Rogers WA (1971) Disease in fish due to the protozoan *Epistylis* (Ciliata: Peritricha) in the Southeastern U.S. In: Proceedings of 25th annual conference Southeastern Association of Game and Fish Commissioners, vol 25, pp 493–496

Royer TL, Petri WA Jr (2014) Waterborne parasites I *Entamoeba*. In: Batt CA, Tortorello ML (eds) Encyclopedia of food microbiology, 2nd edn. Academic, Amsterdam, pp 782–786

Rueckert S, Horák A (2017) Archigregarines of the English Channel revisited: New molecular data on *Selenidium* species including early described and new species and the uncertainties of phylogenetic relationships. PLoS One 12(11):e0187430

Rueckert S, Leander BS (2008) Morphology and molecular phylogeny of *Haplozoon praxillellae* n. sp. (Dinoflagellata): a novel intestinal parasite of the maldanid polychaete *Praxillella pacifica* Berkeley. Eur J Protistol 44(4):299–307

Rueckert S, Betts EL, Tsaousis AD (2019) The symbiotic spectrum: Where do the gregarines fit? Trends Parasitol 35(9):687–694

Ryan U, Cacciò SM (2013) Zoonotic potential of *Giardia*. Int J Parasitol 43:943–956

Ryan UM, Feng Y, Fayer R et al (2021) Taxonomy and molecular epidemiology of *Cryptosporidium* and *Giardia*—a 50 year perspective (1971–2021). Int J Parasitol 51:1099–1119

Saffo MB, Nelson R (1983) The cells of *Nephromyces*: developmental stages of a single life cycle. Can J Bot 61(12):3230–3239

Salomaki ED, Terpis KX, Rueckert S et al (2021) Gregarine single-cell transcriptomics reveals differential mitochondrial remodeling and adaptation in apicomplexans. BMC Biol 19:77

Sato Y, Muto T, Endo Y et al (2015) Morphological, developmental, and ecological characteristics of the suctorian ciliate *Ephelota gigantea* (Ciliophora, Phyllopharyngea, Ephelotidae) found on cultured Wakame seaweed in Northeastern Japan. Acta Protozool 54:295–303

Scheid PL (2019) *Vermamoeba vermiformis*—a free-living Amoeba with public health and environmental health significance. Open Parasitol J 7:40–47

Schnepf E, Drebes G, Elbrächter M et al (2000) *Pirsonia guinardiae*, gen. et spec. nov.: A parasitic flagellate on the marine diatom *Guinardia flaccida* with an unusual mode of food uptake. Helgol Meeresunters 44:275–293

Schrével J, Desportes I (2016) Gregarines. In: Mehlhorn H (ed) Encyclopedia of parasitology. Springer, Berlin, pp 1142–1188

Shapiro K, Bahia-Oliveira L, Dixon B et al (2019) Environmental transmission of *Toxoplasma gondii*: Oocysts in water, soil and food. Food Waterborne Parasitol 15:e00049. https://doi.org/10.1016/j.fawpar.2019.e00049

Simpson AGB, Stevens JR, Lukeš J et al (2006) The evolution and diversity of kinetoplastid flagellates. Trends Parasitol 22(4):168–174

Skovgaard A, Daugbjerg N (2008) Identity and systematic position of *Paradinium poucheti* and other *Paradinium*-like parasites of marine copepods based on morphology and nuclear-encoded SSU rDNA. Protist 159:401–413

Skovgaard A, Karpov SA, Guillou L (2012) The parasitic dinoflagellates *Blastodinium* spp. Inhabiting the gut of marine, planktonic copepods: morphology, ecology, and unrecognized species diversity. Front Microbiol 3:1–22

Small EB, Lynn DH (1981) A new macrosystem for the phylum Ciliophora Doflein, 1901. BioSystems 14:387–401

Small HJ, Miller TL, Coffey AH et al (2013) Discovery of an opportunistic starfish pathogen, *Orchitophrya stellarum*, in captive blue crabs, *Callinectes sapidus*. J Invertebr Pathol 114(2):178–185

Smirnov AV, Chao E, Nassonova ES et al (2011) A revised classification of naked lobose amoebae (Amoebozoa: Lobosa). Protist 162:545–570

Snyman A, Vanstreels RET, Nell C et al (2020) Determinants of external and blood parasite load in African penguins (*Spheniscus demersus*) admitted for rehabilitation. Parasitology 147(5):577–583

So BKF (1972) Marine fish haematozoa from Newfoundland waters. Can J Zool 50(5):543–554

Tenter AM, Heckeroth AR, Weiss LM (2000) *Toxoplasma gondii*: from animals to humans. Int J Parasitol 30:1217–1258

Tryland M (2018) Zoonoses and public health. In: Gulland FMD, Dierauf LA, Whitman KL (eds) CRC handbook of marine mammal medicine. CRC, New York, pp 47–61

Urawa S (1996) The biopathology of ectoparasitic protozoans on hatchery-reared Pacific salmon. Sci Rep Hokkaido Salmon Hatchery 46:175–203

Vermeulen LC, van Hengel M, Kroeze, et al (2019) *Cryptosporidium* concentrations in rivers worldwide. Water Res 149:202–214

von Gersdorff Jørgensen L, Puspasari K, Insariani (2022) Infection by *Ichthyophthirius multifiliis*. In: Kibenge FSB, Baldisseretto B, Chong RS-M (eds) Aquaculture pathophysiology. Vol. 1 Finfish diseases. Academic, London, pp 493–503

Ward GM, Neuhauser S, Groben R et al (2018) Environmental sequencing fills the gap between parasitic haplosporidians and free-living giant amoebae. J Eukaryot Microbiol 65:574–586

Washburn JO, Gross ME, Mercer DR et al (1988) Predator-induced trophic shift of a free-living ciliate: Parasitism of mosquito larvae by their prey. Science 240(4856):1193–1195

Watanabe Y, Nishida S, Zenke K et al (2016) Development of the macronucleus of *Cryptocaryon irritans*, a parasitic ciliate of marine teleosts, and its ingestion and digestion of host cells. Fish Pathol 51(3):112–120

Watson ME (1916) A new infusorian parasite in sand fleas. J Parasitol 2(3):145–146

Winkler R, Antonius A, Renegar DA et al (2004) The skeleton eroding band disease on coral reefs of Aqaba, Red Sea. Mar Ecol 25(2):129–144

Woo PTK (2003) *Cryptobia* (*Trypanoplasma*) *salmositica* and salmonid cryptobiosis. J Fish Dis 26:627–646

Wright A-DG, Colorni A (2002) Taxonomic reassignment of *Cryptocaryon irritans*, a marine fish parasite. Eur J Protistol 37:375–378

Yokoyama H, Itoh N, Ogawa K (2015) Fish and shellfish diseases caused by marine protists. In: Ohtsuka S, Suzaki T, Horiguchi T et al (eds) Marine protists. Springer, Tokyo, pp 533–549

Young ND, Crosbie PBB, Adams MB et al (2007) *Neoparamoeba perurans* n. sp., an agent of amoebic gill disease of Atlantic salmon (*Salmo salar* L.). Int J Parasitol 37:1469–1481

Zagalska-Neubauer M, Bensch S (2015) High prevalence of *Leucocytozoon* parasites in freshwater breeding gulls. J Ornithol 157:525–532

Open Access This chapter is licensed under the terms of the Creative Commons Attribution-NonCommercial-NoDerivatives 4.0 International License (http://creativecommons.org/licenses/by-nc-nd/4.0/), which permits any non-commercial use, sharing, distribution and reproduction in any medium or format, as long as you give appropriate credit to the original author(s) and the source, provide a link to the Creative Commons license and indicate if you modified the licensed material. You do not have permission under this license to share adapted material derived from this chapter or parts of it.

The images or other third party material in this chapter are included in the chapter's Creative Commons license, unless indicated otherwise in a credit line to the material. If material is not included in the chapter's Creative Commons license and your intended use is not permitted by statutory regulation or exceeds the permitted use, you will need to obtain permission directly from the copyright holder.

Biology and Life Cycles of Microsporidia and Myxozoa

Daniel Grabner and Ivan Fiala

Abstract

The knowledge on life cycles and biology of the Microsporidia and the Myxozoa is summarised. These two groups are placed together due to their similarities (intracellular development, spore stages for transmission, ejectable polar "filament") and the historical fact that they were assigned to the common taxon "Cnidospora" prior to being correctly classified. We would like to emphasise here that these groups are taxonomically far apart. While the Microsporidia are related to the fungi, the Myxozoa belong to the Cnidaria and are therefore metazoans. Both are important parasites of aquatic organisms. Microsporidians can infect a wide range of hosts from protists to vertebrates. They can be directly (horizontally) transmitted via spores or vertically via the gonads to the offspring.

D. Grabner (✉)
Aquatic Ecology, University of Duisburg-Essen, Essen, Germany

Centre for Water and Environmental Research, University of Duisburg-Essen, Essen, Germany
e-mail: daniel.grabner@uni-due.de

I. Fiala
Institute of Parasitology, Biology Centre, Czech Academy of Sciences,
České Budějovice, Czech Republic
e-mail: fiala@paru.cas.cz

Life cycles can be simple (monoxenous, one host species) or complex (heteroxenous, multi-host life cycles), some including phases of both horizontal and vertical transmission. Myxozoans, on the other hand, require two obligate hosts in their life cycle, a vertebrate intermediate and an invertebrate final host. Transmission to the vertebrate host is mediated by free-floating actinospores, while the invertebrate host becomes infected by myxospores that are mostly in sediments. Both groups contain species of economic relevance and ecological impact.

3.1 Introduction

More than 120 years ago the Microsporidia and the Myxozoa were originally placed in a single taxon, the "Cnidospora" (Bütschli 1881). Since then, there has been quite a taxonomic odyssey for both groups, but it is now generally accepted that the Microsporidia belong to the kingdom Fungi (see also Chap. 4) and the Myxozoa to the phylum Cnidaria. Nevertheless, both groups share common characteristics prompting us to cover both in a joint chapter. Both groups have spores as the characteristic stage for transmission from one host to another. These spores contain a coiled polar filament that can be ejected within milliseconds. Although the function of this filament is completely different for microsporidians and myxozoans, this complex structure is an impressive example of the homologous development of a cellular component. Another similarity is the infectious cell (or group of cells) called "sporoplasm". Furthermore, in both groups, the development in the host consists of phases of multiplication followed by phases of spore formation.

There are major differences between these two groups. For example, the Microsporidia have an intracellular development while the Myxozoa develop extracellularly in the host tissue. The polar filament is an attachment structure in the Myxozoa that prevents dislocation of the spore from the host surface, while in the Microsporidia, the polar filament injects the infective sporoplasm into the host cell. Additionally, most microsporidians have single-host life cycles and there is a huge variety of host groups, ranging from unicellular eukaryotes to vertebrates, while myxozoans have obligate two-host life cycles between a vertebrate and an invertebrate host.

Both groups are highly abundant parasites and the number of currently described species is probably not even close to the true diversity, and we are also lacking knowledge on the life cycles, biology, and ecological role of most species. The present chapter summarises the current knowledge about biology, life cycles, and taxonomy of both groups.

3.2 Microsporidia

3.2.1 Introduction

The Microsporidia are a fascinating group of intracellular parasites that are found in an exceptionally wide range of hosts from protists, invertebrates to mammals from terrestrial, freshwater, and marine habitats (Smith 2009; Vávra and Lukes 2013). The infectious stage which is responsible for the transmission from host to host is the uniquely complex microsporidian spore. Approximately 1500 microsporidian species have been described thus far; however, a large number of unclassified sequence isolates, as well as recent assessments of freshwater and marine microsporidians, indicate that only a small part of the Microsporidia diversity has been assessed to date (Murareanu et al. 2021; Chauvet et al. 2023).

Some microsporidian species are of human relevance as disease agents and are considered to be emerging diseases in humans and animals, often with a water or foodborne transmission (Didier 2005; Stentiford et al. 2016, 2019; Cali et al. 2016). Furthermore, several species with terrestrial life cycles can potentially have considerable economic impact, e.g. *Nosema apis* and *Nosema ceranae* for beekeeping or *Nosema bombycis* in sericulture (Grupe and Quandt 2020; Wei et al. 2022). In addition, we know today that about half of the described microsporidian genera occur in aquatic hosts (Stentiford et al. 2013b) and more and more of these microsporidian species are being recognised as pathogens of economically relevant aquatic animals such as crustaceans and fish (Stentiford and Dunn 2014; Stentiford et al. 2013b; Schuster et al. 2022; Bojko and Stentiford 2022). Besides their negative effects, some microsporidians are also used for pest control, e.g. *Paranosema* (*Nosema*) *locustae* against grasshoppers (Cali et al. 2016), or *Amyblyospora* spp. (among others) against mosquitoes (Andreadis 2007). Recent findings also indicate that microsporidians impair the transmission of malaria by mosquitoes and might have potential as a malaria control (Herren et al. 2020).

Knowledge on the so-called "expanded microsporidians" has recently greatly increased, including several basal taxa summarised as "short-branch microsporidians" including the genera *Paramicrosporidium*, *Mitosporidium*, and *Nucleophaga* and a number of sequence isolates obtained from environmental samples (Bass et al. 2018). This book chapter is limited to the "canonical", "classical", or "long-branch" Microsporidia (lineages that were among the first to be discovered; separated in molecular phylogeny by a long branch from the basal groups), since information on the hosts and the biology of the "short-branch" Microsporidia is still limited (Doliwa et al. 2021).

3.2.2 Taxonomy

After the discovery of *Nosema bombycis*, the first described microsporidian species (Nägeli 1857), the Microsporidia were originally classified as early-branching, basal eukaryotes. Based on morphological and molecular characteristics, the Microsporidia are today placed in the taxon Opisthosporidia (Eukaryota, Opisthokonta), a clade of early diverging fungi (Karpov et al. 2014; Hirt et al. 1999; see also Chap. 4). With the fungi, the Microsporidia share a number of morphological and biochemical traits such as the presence of chitin in the spore wall, trehalose as storage substance, similar mitosis and meiosis, and *O*-mannosylation (Han and Weiss 2017). Morphological characteristics of the Microsporidia are the spore with a multi-layered wall and the polar filament apparatus. Flagella are lost in all microsporidian groups while functional mitochondria are found in the early diverging microsporidian groups (Bass et al. 2018; Wadi and Reinke 2020). Furthermore, their Golgi apparatus is highly derived, and the division of the nucleus takes place with the preservation of the nuclear envelope (Dunn and Smith 2001).

Basal among the "expanded Microsporidia" are groups belonging to the "short-branch" microsporidians, while the metchnikovellids (e.g. *Amphiamblys* sp., parasites of gregarines) are located at the basal branches of the "canonical" microsporidians that comprise most of the known microsporidian diversity (Bass et al. 2018; Wadi and Reinke 2020; Park and Poulin 2021) (Fig. 3.1).

The taxonomy of the Microsporidia is often problematic. Morphological identification is based on ultrastructure of developmental stages which may result in misleading classifications (Vávra and Lukes 2013). In addition, the morphological variability of the spore stages makes it difficult to classify some of the species (e.g. Stentiford et al. 2013a; Sokolova et al. 2016; Vavra et al. 2018; Casal et al. 2012; Scholz et al. 2017). Therefore, the taxonomy of the Microsporidia is mostly based on DNA sequences (Vossbrinck and Debrunner-Vossbrinck 2005; Vossbrinck et al. 2014). For this purpose, sequences of the ribosomal RNA gene (rRNA), the large subunit of the RNA polymerase II (RPB1), and some other protein-coding genes have been used (Vossbrinck et al. 2014; Hirt et al. 1999; Ironside 2013; Doliwa et al. 2023). However, most of the sequence entries available in the microsporidian databases are parts of the small subunit (SSU) rRNA gene that is also commonly used for molecular barcoding (Vossbrinck et al. 2014). Notwithstanding its frequent use, the SSU rRNA gene is not necessarily well suited for elucidating the more ancient divergences in the microsporidian tree and can fail to differentiate closely related species (Bojko et al. 2022). Further, variable intragenomic copies of the SSU genes might create ambiguous results for barcoding and molecular phylogenetic analyses (Ironside 2013). Today, the increasing number of sequenced microsporidian genomes provides an additional tool, particularly for deep phylogenetic analyses to resolve the relationship of major microsporidian clades. Figure 3.1 shows a simplified tree of the major microsporidian orders based on the most recent published results (Bojko et al. 2022; Wadi and Reinke 2020). Please note that there is no formal taxonomical classification available for the Microsporidia to date. The

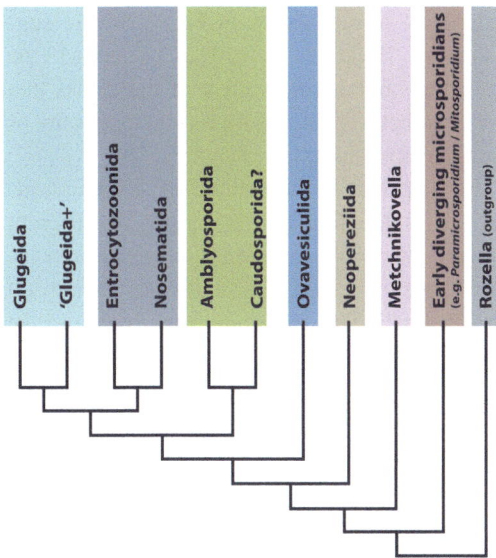

Fig. 3.1 Simplified phylogenetic tree of the major groups of the Microsporidia based on Bass et al. (2018), Bojko et al. (2022), and Wadi and Reinke (2020)

order-level names shown in Fig. 3.1 were suggested by Bojko et al. (2022) to replace the ambiguous numbering system of the clades that was used before, but still no formal taxonomy of the Microsporidia exists.

Most species of the "canonical" microsporidian clades are found in a variety of different hosts; however, the majority of those species show a preference for a particular host group and also habitat. The Glugeida mainly parasitise marine or freshwater fish and crustaceans, while the majority of Nosematida-species are found most frequently in terrestrial insect hosts (Fig. 3.1). The Amblyosporida commonly parasitise mosquitoes (Culicidae) that develop in freshwater and terrestrial habitats, but also show life cycles involving insect and crustacean hosts. Most Neopereziida parasitise freshwater fish, crustaceans, and oligochaetes but also terrestrial insects. The Enterocytozoonida are found using a wide range of hosts, but mostly crustaceans, insects, and fish. The Ovavesiculida comprise only a few known species and are found in nematode or insect hosts from terrestrial environments (Fig. 3.1). Species infecting mammals (including humans) are found in the Neopereziida, Nosematida, Glugeida, and the Enterocytozoonida (for details, see Bojko et al. 2022) (Fig. 3.1). The order basal to the "canonical" microsporidians are the Metchnikovellida that are hyperparasites of gregarinids (Wadi and Reinke 2020). The order "Caudosporida" still requires more data to substantiate that the included species form a major microsporidian clade, as well as the order "Glugeida+" that consists of several weakly supported "orphan lineages" (Bojko et al. 2022).

3.2.3 Biology and Life Cycle

Structural and Physiological Characteristics
The Microsporidia are intracellular parasites that can infect a variety of different tissue types. A few species even develop inside of the host nucleus (Stentiford et al. 2013b; Hedrick et al. 1991). They show some particular features at the molecular level compared to other eukaryotes, like 16S and 23S ribosomes instead of 18S and 28S ribosomes, the lack of 5.8S ribosomes and particularly small genomes (Cali et al. 2016). While genomes of microsporidians are generally small, their size can vary by about two orders of magnitude between species. In addition, the number of predicted protein-coding genes can vary up to threefold (Williams et al. 2022) indicating lineage-specific gene loss instead of a general trend to genome reduction in the long-branch Microsporidia (Bass et al. 2018). In contrast, the basal microsporidian groups (Metchnikovellids, Chytridiopsids) possess mitochondria and genomes that are more similar to those of fungi (Corsaro 2022). The reduction of protein-coding genes results in a severe reduction in key metabolic pathways, with some species even having lost the ability to undergo glycolysis (Akiyoshi et al. 2009). Furthermore, microsporidian mitochondria are reduced to the so-called mitosomes that have lost their function for aerobic respiration but retained pathways for iron-sulphur cluster assembly (Williams et al. 2002; Goldberg et al. 2008). Consequently, most molecules and energy (ATP, GTP, NAD$^+$) are obtained directly from the host cell (Dean et al. 2018; Corradi 2015). This highlights the strong dependence of microsporidians on host cellular metabolism, although the degree of dependency can differ significantly between species (Desjardins et al. 2015).

The Microsporidian Spore
The microsporidian spore is the only stage that is viable outside of the host. Spores are mostly ovoid but can be spherical, rod-shaped, or crescent shaped, with a spore size between about 1 and 20 μm (Han and Weiss 2017). The spore is enclosed by an outer exospore (composed of proteins and keratin), an inner endospore (containing chitin), and a membrane containing the cytoplasm and nucleus (single or diplokaryon) of the spore, which is called sporoplasm. The spore also contains the polar filament apparatus, with the polar tube connected to the anterior anchoring disk, the polaroplast at the anterior end of the spore (consisting of stacks of flattened membranes), the polar tube that is coiled around the nucleus, and the posterior vacuole (Keeling and Fast 2002) (Fig. 3.2).

During infection, the spore is first activated by an external stimulus, e.g. from the intestinal tract of the host (Keeling and Fast 2002). The spore "fires" the polar tube by an increase of osmotic pressure and subsequent influx of water that occurs within milliseconds. The resulting swelling of the polaroplast induces the explosive ejection of the polar tube that penetrates the host cell. Subsequently, the posterior vacuole presses the sporoplasm through the polar tube into the host cell cytoplasm. There it is enclosed by a membrane derived from the polaroplast, while the former plasma membrane remains inside the spore (Vávra and Lukes 2013). In some microsporidian species, the spore is taken up by the host cell via phagocytosis, and the polar tube is used to escape the phagolysosome (Franzen 2004). A single microsporidian species often forms two or more spore types that are used either for autoinfective cycles inside of the same host or the infection of new host individuals (Vávra and Lukes 2013).

Development in the Host Tissues

After oral uptake of the spores by the host, the development of microsporidians usually begins in the gastrointestinal tract. The further spread of the parasite within the host can occur by different mechanisms (Vávra and Lukes 2013). In different genera, autoinfective spores have been identified that mediate the dissemination of the parasite within the host (Iwano and Ishihara 1991). The parasites can infect mobile host cells (Becnel et al. 1989) or can be distributed to daughter cells during host cell division (Terry et al. 1999a). During their development, the parasites are in close contact with the endoplasmic reticulum and mitochondria of the host (Desportes-Livage 2000; Williams 2009; Terry et al. 1999a). The first phase of development is merogony that progresses by binary or multiple fission, resulting in multinucleate plasmodia (Han and Weiss 2017). The second phase is sporogony where spores develop via sporonts and sporoblasts (Vávra and Lukes 2013), taking place in the cytoplasm of the host cell or, in some species, inside of sporophorous vesicles (Keeling and Fast 2002) (see Fig. 3.3). In these phases, some microsporidian species have single nuclei, while some are diplokaryotic during the whole development, or only in certain phases (Desportes-Livage 2000). There can be different ways of merogonic proliferation

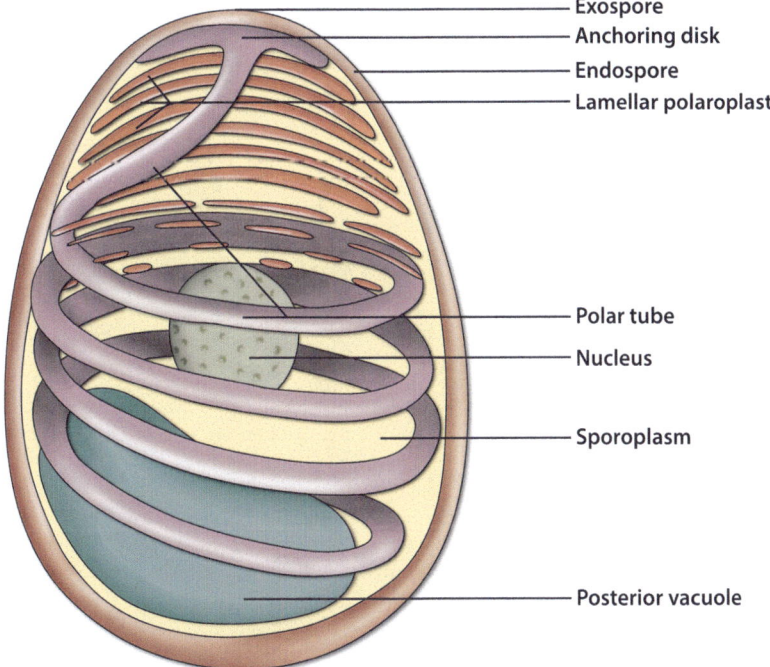

Fig. 3.2 Schematic of a microsporidian spore (based on Cali et al. 2016)

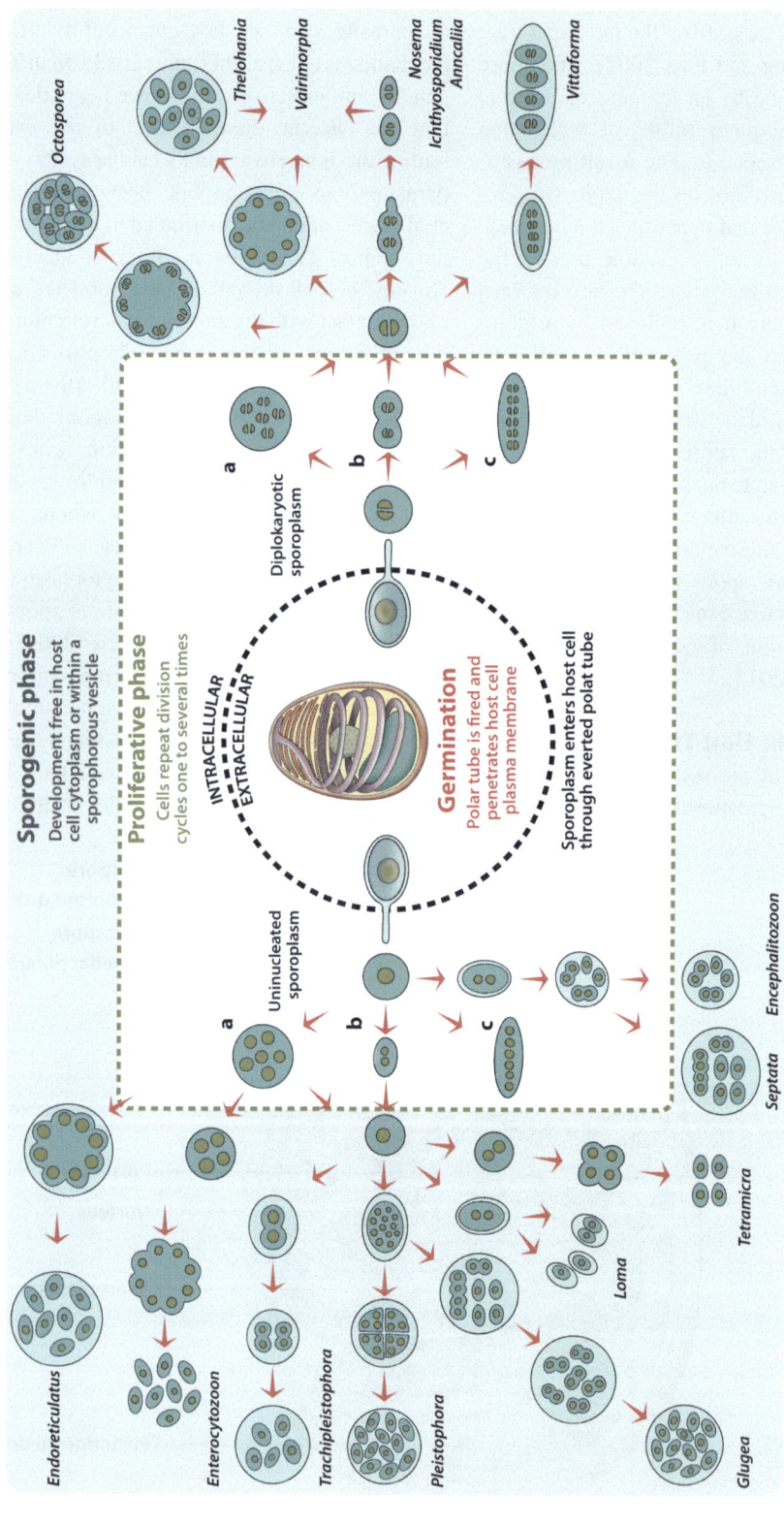

Fig. 3.3 Representation of microsporidian developmental cycles. Cells produce one of three basic developmental forms during proliferative phase: (**a**) elongated moniliform multinucleate cells that divide by multiple fission (e.g., some *Nosema* species); (**b**) cells divide immediately after karyokinesis by binary fission; (**c**) rounded plasmodial multinucleate cells that divide by plasmotomy (e.g., *Endoreticulatus*). It should be noted that in the *Thelohania* cycle and the *Thelohania*-like part of the *Vairimorpha* cycle, the diplokarya separate and continue their development as cells with isolated nuclei (from Cali and Takvorian 2014)

and sporogony in the same microsporidian species that show considerable morphological differences, e.g. production of single and multinucleated spores (Cali et al. 2016). The different spore types can be autoinfective for the same host individual, dedicated for transovarial transmission, or transmission to other host individuals or species in the case of multi-host life cycles (e.g. Amblyosporidae) (Becnel and Andreadis 2014) (Fig. 3.3).

Microsporidians have haplo-diploid or diploid-tetraploid generations (Vávra and Lukes 2013; Corradi 2015) and sexual reproduction has been demonstrated in some species, e.g. *Edhazardia aedis* (Becnel et al. 1989), while other species reproduce only asexually (Corradi 2015). Meiotic divisions occur during sporogony, finally resulting in haploid spores (Desportes-Livage 2000). There is evidence that sexual reproduction was lost independently in several microsporidian lineages (Ironside 2007; Haag et al. 2013) and it might be limited to horizontally transmitted lineages in otherwise-mostly vertically transmitted species (Wilkinson et al. 2011).

Life Cycle and Transmission

Microsporidia are transmitted from one host to the next either by spores that are released to the environment and taken up by the next host (horizontal transmission, Fig. 3.4) or directly to the offspring via the gonads of the females (vertical transmission) (Dunn and Smith 2001). In most cases, microsporidians use only one host species, but there are also a number of species that include two hosts in their life cycle (e.g. Amblyosporida) including a complex switch of different sporogonic cycles (Smith 2009).

In the case of horizontal transmission, spores can be released after death of the host (Dunn and Smith 2001), transmitted from infected to uninfected individuals by inter-host cannibalism (e.g. *Cucumispora dikerogammari*) (Ovcharenko et al. 2010), or shed continuously in the case of intestinal infections (e.g. *Nosema apis*) (de Graaf et al. 1994). Predators that consume infected hosts can further spread the spores in the host population (Wang-Peng et al. 2018). Virulence in horizontally transmitted species depends on the tissue location of the parasite and the associated path of spore release. Low virulence is found in gut-infecting species that continuously produce and release spores (Solter 2006), but high virulence is found in species that depend on host death to release spores (Dunn and Smith 2001). Horizontal transmission is found in species with a feeding strategy that favours the oral uptake of spores from the environment (Dunn et al. 2001).

In the case of vertical transmission, the microsporidians are transmitted from one host generation to the next within the same lineage, mostly via the ovary to the developing eggs (Dunn et al. 2001). Vertical transmission is found in all major microsporidian groups (sensu Bojko et al. 2022) and mostly in microsporidians infecting insects and crustaceans (Terry et al. 2004). However, there is also evidence for vertical transmission of Microsporidia infecting fish, e.g. *Ovipleistophora ovale* from golden shiner (*Notemigonus crysoleucas*) (Phelps and Goodwin 2008) or *Pseudoloma neurophilia* from the zebrafish (*Danio rerio*) (Kent and Bishop-Stewart 2003). In addition, indications for vertical transmission of microsporidians were found in snails and mussels (McClymont et al. 2005; Olivares 2005). In most cases, vertical transmission occurs in combination with phases of horizontal transmission (Dunn et al. 2001), whereas vertical transmission as the only transmission mode is rare [e.g. *Nosema granulosis* infecting the amphipod *Gammarus duebeni* (Terry et al. 1998, 1999b, 2004)]. *Nosema granulosis* is highly adapted to the vertical mode of transmission, as it targets specific cell lines in developing gonadal tissues of the developing embryo (Dunn et al. 1998, 2001).

Since the parasite transmission depends on host reproduction, the pathology for the host is usually low during phases of vertical transmission (Dunn et al. 2001). Uncontrolled multiplication of the parasite in the host embryos must be controlled to avoid the impairment of the development. This is achieved by a reduced replication of the parasite or a specific location in the embryo

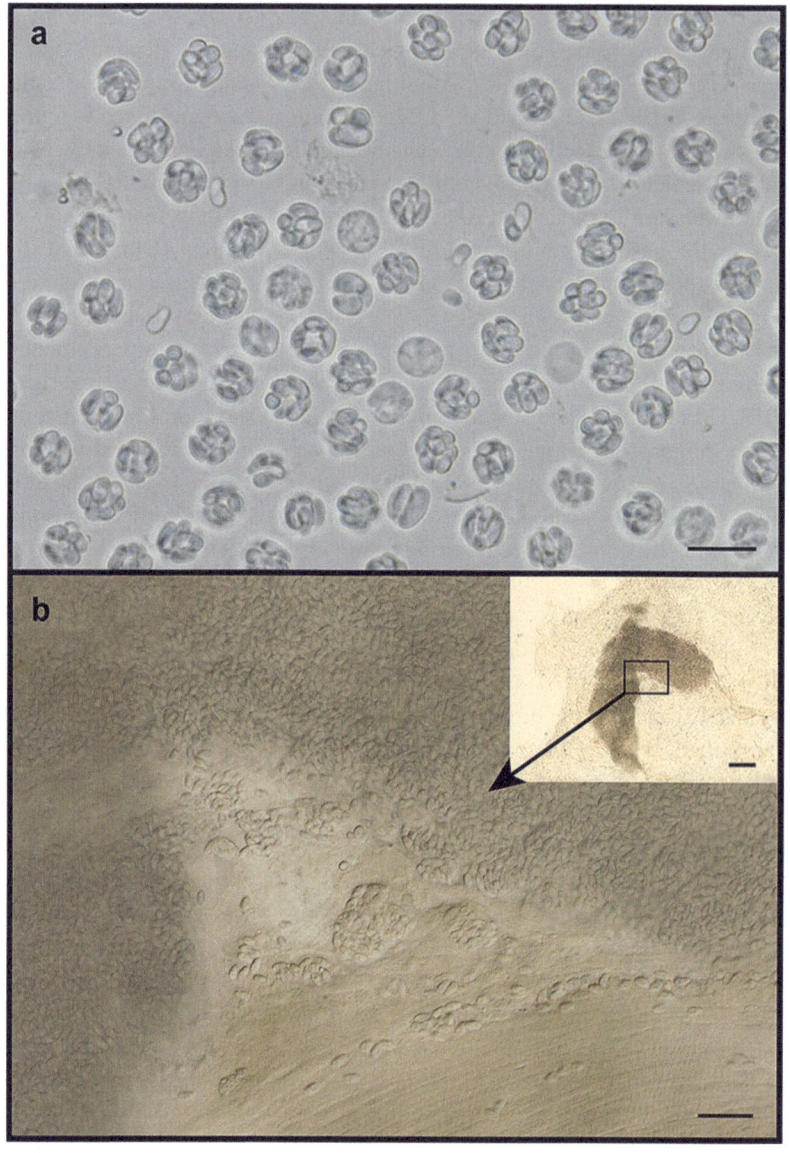

Fig. 3.4 (a) Fresh smear of *Dictyocoela* sp. octospores (scale bar 10 μm); (b) fresh smear of microsporidian spores in muscle tissue. Small inset (scale bar 200 μm) shows masses of spores (dark area) in muscle. Area of the black rectangle is shown in the larger image (scale bar 20 μm). Both isolated from *Gammarus fossarum*, Rumbach, Germany

that does not interfere with embryogenesis (Dunn et al. 2001).

Vertical transmission via eggs requires the infection of female hosts; infection of male hosts would be a dead end for the parasite. Therefore, some microsporidians distort the sex ratio of the host to increase the efficiency of transovarial transmission. In fact, microsporidians are the only known eukaryote group in which this phenomenon has been observed (Dunn et al. 2001). There are two strategies described in Microsporidia leading to sex-ratio distortion. One of them is male-killing, which is found in species using both horizontal and vertical transmission. In the latter case, the parasite shows a sex-specific development leading to differential virulence in males and females. In males, spores develop for horizontal transmission, while females transmit the parasite vertically. Consequently, infected male hosts show a higher mortality than females, which can lead to a female-biased sex ratio in the host population (Dunn et al. 2001). This mechanism of sex-ratio distortion is frequently found among the

Amblyosporida (Terry et al. 2004). The second strategy that enhances the vertical transmission is feminisation of male hosts. It has been shown for the microsporidians *Nosema granulosis* (Nosematida), *Dictyocoela duebenum* (Glugeida), *Fibrillanosema crangonycis* (Neopereziida), and an undescribed microsporidian from the group Neopereziida that are all presumed to be exclusively vertically transmitted (Terry et al. 1998, 2004; Mautner et al. 2007; Slothouber Galbreath et al. 2004; Ironside et al. 2003a, b). In addition, there is evidence that host feminisation related to vertical transmission occurs in even more amphipod-microsporidian systems and potentially in other host groups (Haine et al. 2004; Terry et al. 2004; Dunn et al. 2001). All the feminising microsporidians infect amphipod hosts which is explained by the mechanism of sex determination in this host group, which can be disturbed by the parasite through endocrine disruption (Rodgers-Gray et al. 2004).

Multi-host Life Cycles

There are a number of microsporidians with complex life cycles that include more than a single-host species. Particularly the Amblyosporida include species that alternate between horizontal and vertical transmission and include a host switch, often between mosquitoes and copepods (Andreadis 2007) (Fig. 3.5). There are also indications that several fish microsporidians use an additional crustacean host in their life cycle (Stentiford et al. 2018; Ironside et al. 2008; Nylund et al. 2010; Yanagida et al. 2022; Lovy and Friend 2017; Stratton et al. 2022). Furthermore, there is indirect evidence that non-host vectors like fish or birds might play a role in the dispersal of microsporidians, especially in aquatic ecosystems (Prati et al. 2023; Slodkowicz-Kowalska et al. 2006).

Host Specificity of Microsporidia

Some microsporidia are apparently host-specific, at least for a genus or family (see Park and Poulin 2021 and the references therein), leading to co-evolution with their hosts, as has been shown for amphipod- and mosquito-infecting species (Quiles et al. 2019, 2020; Park et al. 2020; Andreadis et al. 2012). Nevertheless, host shifts can occur during phases of horizontal transmission in otherwise-predominantly vertically transmitted species, e.g. the amphipod-infecting genus *Dictyocoela* (Wilkinson et al. 2011). On the other hand, it is particularly noteworthy that some microsporidia have an astonishingly wide range of possible hosts. Recently, Trzebny et al. (2023) showed that microsporidians infecting mosquitoes occurred in at least three different host species and species of the *Enterocytozoon*-group can be found in invertebrates, fish, birds, and mammals, including humans (Stentiford et al. 2019). One of the most striking examples of variable host use, however, is *Trachypleistophora hominis* that can alternatively infect humans or insects (Field et al. 1996; Hollister et al. 1996; Weidner et al. 1999).

Hyperparasitism

The flexibility concerning the host choice of some species probably explains the high number of hyperparasitic Microsporidia that infect both host tissues and the parasite stages developing within it. This was found, for example, for the microsporidian *Neoflabelliforma magnivora* that is infecting both the tissues of the oligochaete host and its myxozoan parasite, affecting the development of actinospores (Morris and Freeman 2010), and *Ovipleistophora diplostomuri* infecting tissues of both the bluegill sunfish, *Lepomis macrochirus* and the trematode *Posthodiplostomum minimum* (Lovy and Friend 2017).

The genus *Unikaryon* exclusively contains microsporidians parasitic in trematodes and cestodes from vertebrate and invertebrate hosts (Sokolova et al. 2021; Sene et al. 1997; Canning and Nicholas 1974; Canning et al. 1983). Several hyperparasitic species are also found in the genus *Nosema*, parasitising myxozoans (Diamant and Paperna 1985), cestodes (Dissanaike 1957), and trematodes. In the latter host, the microsporidians often impair the development of the trematode (e.g. Sprague 1964; Hussey 1971; Colley et al. 1975; Levron et al. 2005; Levron et al. 2004; Canning et al. 1974; Toguebaye et al. 2014; Cort 1960). In addition, microsporidian hyperparasitic

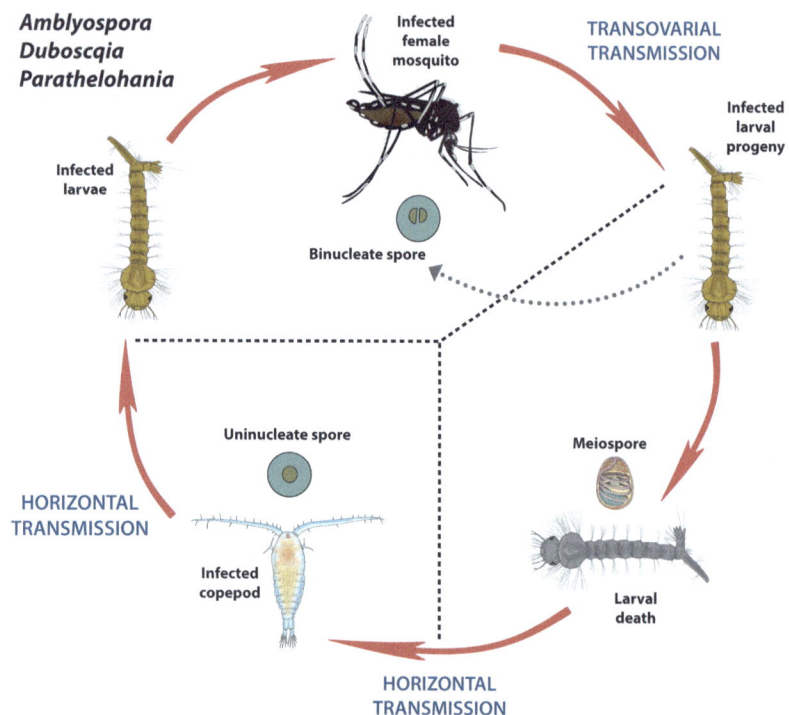

Fig. 3.5 Example of multi-host life cycle in the Amblyosporida (based on Andreadis 2007)

infections were shown in acanthocephalans, parasitic copepods, and Paramyxida (de Buron et al. 1990; Nylund et al. 2010; Freeman et al. 2003; Stentiford et al. 2017).

3.2.4 Ecological Consequences of Host–Microsporidian Interactions

A large number of microsporidians are known from a variety of aquatic hosts, where they have a number of effects (e.g. Stentiford et al. 2013b; Grabner 2017; Grabner et al. 2020, 2022). Apart from regulating host populations by increasing mortality rate (e.g. Kohler and Hoiland 2001), there are a number of sublethal effects that microsporidians can induce. In this section, examples are given for the effects of these parasites with possible consequences for the host populations. For example, the infection with *Vavraia culicis* causes early maturation in mosquito larvae resulting in reduced adult size (Stentiford et al. 2013b). Moreover, some microsporidians can alter host behaviour which will likely affect population growth and structure. Three-spined sticklebacks (*Gasterosteus aculeatus*) showed increased social and shoaling behaviour when infected with the microsporidian *Glugea anomala* (Ward et al. 2005; Petkova et al. 2018). Further, microsporidian-induced behavioural changes directly affecting the reproduction of the host were found in two-spotted gobies (*Gobiusculus flavescens*). Fish infected with microsporidians showed markedly decreased courtship behaviour compared to uninfected fish, even though overall condition of the host was not affected by the parasites (Pélabon et al. 2005). Likewise, the local occurrence of hosts can be affected by microsporidian parasites. A recent study on microsporidians in amphipods demonstrated that these parasites can influence the spatial distribution of the host population by modulating the drift behaviour of their amphipod hosts (Prati et al. 2023).

Fitness consequences of microsporidian infections for the host can be deduced based on the modulation of the host response to other stressors, which also has practical consequences for ecotoxicological studies (Grabner and Sures

2019). It has to be noted that the effects of microsporidian infections on their host are not necessarily all negative. For example, it was shown that *Nosema granulosis* can also have a positive effect on its amphipod host *Gammarus roeselii*. For this vertically transmitted microsporidian, an increased survival rate was found for amphipod offspring that were infected with this microsporidian (Haine et al. 2007).

3.2.5 Distribution of Microsporidians

The factors regulating microsporidian local diversity are poorly understood. When studying the microsporidian diversity in amphipods, Prati et al. (2022) found that neither host density nor water parameters or restoration stage were predictors of the diversity of generalist microsporidian species in anthropogenically impacted environments. Other factors might be more relevant for changes in microsporidian communities, with biological invasions being highly relevant. There are some prominent examples of biological invasions of microsporidians, e.g. *Cucumispora dikerogammari* from the "killer-shrimp" *Dikerogammarus villosus* (Bacela-Spychalska et al. 2012), *Cucumispora ornata* from the "demon-shrimp" *Dikerogammarus haemobaphes* (Bojko et al. 2015), *Fibrillanosema crangonycis* in *Crangonyx pseudogracilis* (Slothouber Galbreath et al. 2004), or *Ecytonucleospora* (=*Enterocytozoon*) *hepatopenaei* that was co-introduced with ornamental shrimp from aquarium pet trade (Schneider et al. 2022). Even though the transfer of non-indigenous microsporidians from neozoic amphipods or other host species is rarely observed, this mechanism might become relevant when introduced microsporidians start to adapt to indigenous host species, thereby posing a threat to their populations and affecting microsporidian diversity (Slothouber Galbreath et al. 2004; Bacela-Spychalska et al. 2012; Grabner et al. 2015; Schneider et al. 2022; Dewangan et al. 2023).

3.2.6 Outlook

Even though microsporidians are probably one of the most ubiquitous parasite groups, our knowledge on the biology and ecology of most species (apart from some human or economically relevant species) is still limited. Recent advances in the taxonomy of microsporidians (e.g. by using whole genomes for species comparison) help to clarify species identity and metagenomic studies focusing on microsporidians will be a tool to improve the census of microsporidian diversity. Nevertheless, we need targeted studies on the life cycles and the ecological effects of microsporidian parasites to understand their role, particularly in aquatic ecosystems.

3.3 Myxozoa

3.3.1 Introduction

Myxozoa are a significant group of aquatic microscopic parasites of cnidarian origin. During their evolution from free-living cnidarian ancestors, they adapted their morphology to the parasitic lifestyle, forming various plasmodial stages that produce spores serving for myxozoan transmission (Fig. 3.6). The only exception of their extreme morphological simplification is *Buddenbrockia*, with its worm-like stage possessing four longitudinal muscles (Okamura et al. 2002). Myxozoans infect aquatic vertebrates and especially fish, as their intermediate hosts, with annelids and bryozoans as definitive hosts. Despite their inconspicuous nature, Myxozoa have significant ecological and economic implications. Some myxozoan species are linked to emerging diseases in iconic fish species such as salmon and trout, causing diseases like whirling disease, proliferative kidney disease, and ceratomyxosis (Kent and Hedrick 1985, Bartholomew and Wilson 2002, Bartholomew 2012). These infections pose a substantial economic threat to aquaculture and fisheries, impacting the health of farmed and hatchery fish and reducing the marketability of wild fish (see also Chap. 23).

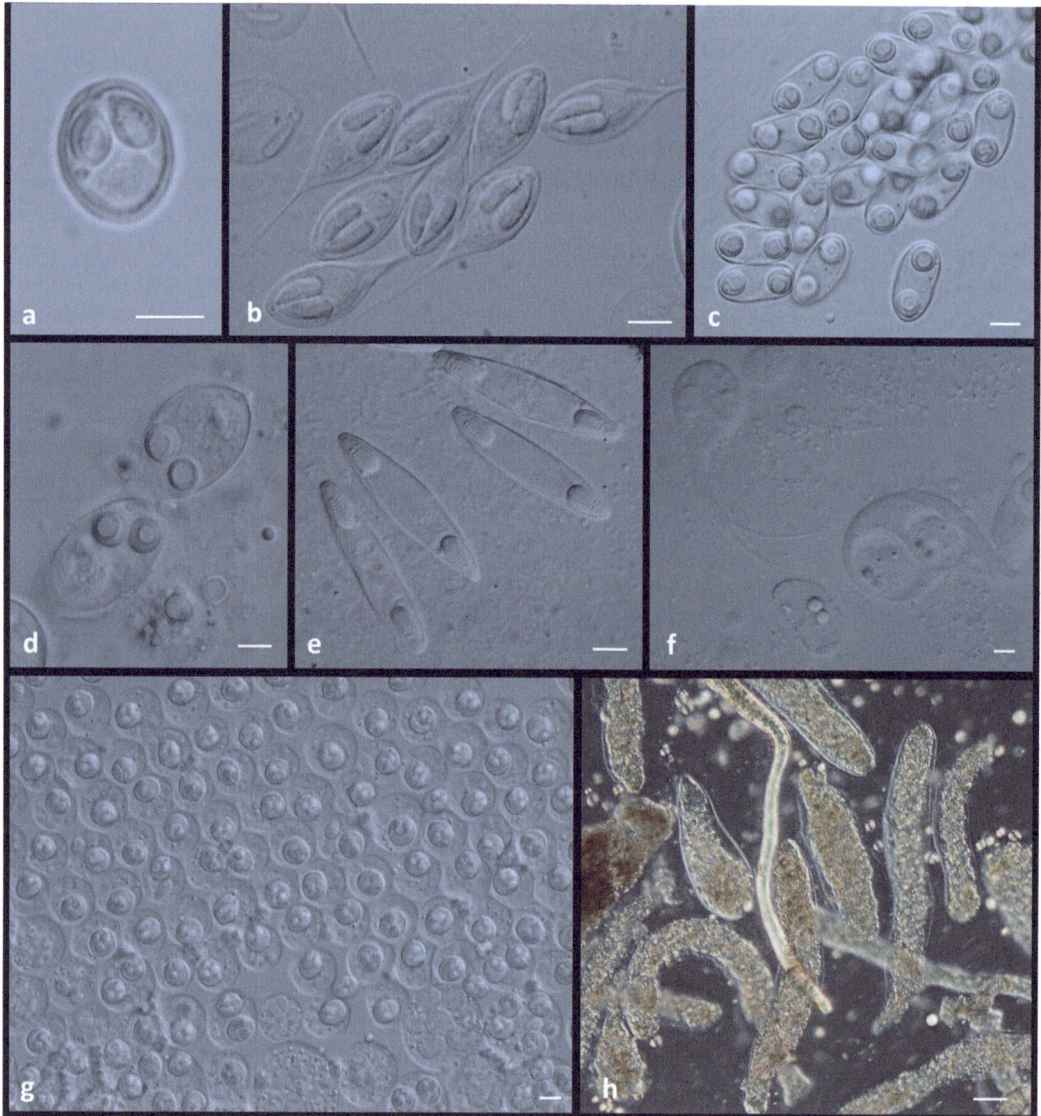

Fig. 3.6 (**a**) *Myxobolus cerebralis*, spore from cartilage of *Salmo trutta*; (**b**) *Henneguya psorospermica*, spores from gills of *Esox lucius*; (**c**) *Zschokkella nova* from the gall bladder of *Ctenopharyngodon idella*; (**d**) *Sinuolinea lophii* from the urinary bladder of *Lophius piscatorius*; (**e**) *Myxidium coryphaenoideum* from the gall bladder of *Coryphaenoides rupestris*; (**f**) *Ceratomyxa* sp., diasporic plasmodia and spore from the gall bladder of *Lophius piscatorius*; (**g**) *Chloromyxum* sp., mono-, di-, and polysporic plasmodia containing spores from the gall bladder of *Leuciscus leuciscus*; (**h**) *Myxidium lieberkuehni*, early plasmodia from the urinary bladder of *Esox lucius*. Scale bar: (**a–g**) 5 μm; (**h**) 50 μm

Myxozoans are increasingly recognised as significant components of freshwater and marine ecosystems. With approximately 2600 species described (Okamura et al. 2018), they represent a staggering 18% of all known cnidarian species. Very likely many more species are not known yet, as taxonomists continue to describe new species and reveal species complexes. This is evident in the substantial increase in the number of newly described species since Canning and Okamura's report in 2004, which documented around 1300 described species, only half of the known diversity documented 14 years later. Furthermore, modern metabarcoding methods initiated the identification of a substantial number of potential myxozoan species through the analysis of environmental DNA (Hartikainen et al. 2016, Lisnerová et al. 2023; see Chap. 16). This underscores the hidden diversity of these endoparasitic cnidarians, which may rival or even surpass the number of species of their free-living counterparts. This striking diversity highlights the importance of understanding Myxozoa and their unique ecological roles.

Myxozoa present a fascinating opportunity for biologists and researchers to explore the evolution of early metazoans. They exemplify how seemingly simple diploblastic organisms have evolved into complex endoparasites, engaging in sophisticated interactions with their hosts. Over time, myxozoans have evolved complex life cycles, including annelids and bryozoans as definitive hosts and extending their intermediate host range beyond fish to include amphibians, reptiles, waterfowl, and even small mammals. This adaptability and diversification make them a subject of significant scientific interest.

3.3.2 Taxonomy

Within Myxozoa, two subclasses are recognised: the Malacosporea and the Myxosporea. The Malacosporea comprises only five described species (Patra et al. 2017), retaining primitive features such as epithelia and muscles. They develop as inactive sacs or active myxoworms in their definitive hosts, freshwater bryozoans. In contrast, the Myxosporea have undergone substantial radiation, with more than 2600 described species. The first myxosporean species was reported by Jurine (1825) but a formal taxonomy of myxosporeans was only established in 1881, with the classification of the Myxosporida together with the Microsporidia and other spore-forming organisms within the Sporozoa by Bütschli (1881). The number of recognised myxosporean genera grew over the centuries, resulting in 64 genera in 17 families (Lom and Dyková 2006). The genera *Myxobolus* and *Henneguya* are the most species rich, together accounting for more than 1000 described species infecting freshwater hosts (Eiras et al. 2021 Rangel et al. 2023), whereas the gall bladder-infecting *Ceratomyxa* or histozoic *Kudoa* are typical myxosporeans that infect marine fish. Myxozoan classification traditionally relied heavily on spore morphology. The unique features of myxospores, such as the number and configuration of shell valves and the number and arrangement of polar capsules, were used to categorise them at various taxonomic levels, from orders to families and genera (Lom and Arthur 1989). Additional characteristics, such as the presence of ridges and striations on the spore valves, number of turns of the polar filament, and the presence or absence of a mucous envelope, were also considered. Furthermore, aspects of the host life cycle and information about the vegetative stages of the parasite were used for species-level classification. However, the spore-based classification had its limitations, as similar morphological characteristics are found in multiple genera, making it challenging to differentiate between species accurately. This inconsistency between spore-based classification and molecular phylogeny was a significant issue, as molecular markers revealed unexpected relationships among myxosporean species (Fiala 2006).

The taxonomy of Myxozoa has seen significant changes and updates. The classification now considers a cnidarian origin for the myxozoans but the history of myxozoan classification experienced a number of revolutionary changes due to the change in taxonomic understanding, including placing myxozoans together with microsporidians as protists. Early observations by Štolc (1899) suggested that myxozoans should be classified with Metazoa due to their multicellular spores. Weill (1938) noted similarities between polar capsules in myxozoans and nematocysts in cnidarians, further supporting their potential connection. Myxozoa were later recognised as a phylum within Metazoa by Grassé (1970) and were confirmed as such through molecular data by Smothers et al. (1994), with an apparently close relation to nematodes. Later, Siddall et al. (1995) proposed a closer relationship to *Polypodium hydriforme,* an enigmatic cnidarian parasite.

Despite the recognition in the 1970s that Myxozoa are most likely multicellular organisms, textbooks often continued to classify myxozoans as protists until the 2000s, contributing to confusion. Definitive proof of metazoan and cnidarian origin came in Holland's recognition of minicollagens, taxonomically restricted cnidarian genes, in the genome of *Tetracapsuloides bryosalmonae* (Holland et al. 2011).

Present-day classification acknowledges the complex life cycles, including both actinospore and myxospore phases. The Actinosporea class, initially thought to be distinct, was later suppressed, recognising that these phases represent different stages of the same species (Wolf and Markiw 1984). There have been modifications at the higher taxonomic ranks, including the class Malacosporea, which accommodates genera like *Buddenbrockia* and *Tetracapsuloides*, known to parasitise bryozoans and fish (Canning et al. 2000). Molecular data have played an essential role in unravelling the relationships among myxozoan species, revealing previously unknown lineages and hidden diversity within the group (Kodádková et al. 2015, Lisnerová et al. 2020).

3.3.3 Phylogeny of the Myxozoa

The phylogeny of the Myxozoa is now based on the use of molecular markers, particularly the small subunit ribosomal RNA gene (SSU rDNA) (e.g. Fiala 2006, Bartošová et al. 2009). High rates of change in the myxozoan SSU rDNA, especially in their variable regions, indicate the rapid evolution of myxozoans. This can lead to difficulties in sequence alignment, which is essential for molecular phylogenetic analyses. In addition to SSU rDNA, the large subunit ribosomal RNA (LSU rDNA) gene, with its conserved and variable regions, has been found to be informative for myxosporean phylogeny (Bartošová et al. 2009). However, SSU rDNA remains the primary choice when analysing myxozoan phylogeny due to the high number of records available in databases.

The phylogeny of myxozoans involves several lineages based on their genetic characteristics (Fig. 3.7). The first lineage in the myxozoan evolutionary tree leads to the extant malacosporean species, as suggested by SSU analyses and supported by their "primitive" morphological features. Following the separation of the Malacosporea from the Myxosporea, the *Sphaerospora* sensu stricto clade separated from the remaining Myxosporea. The latter underwent extensive radiation, diversifying into polychaete-infecting (mostly marine species) and oligochaete-infecting (mostly freshwater species) lineages (Holzer et al. 2018).

The Malacosporea lineage includes species with life cycles in both fish and bryozoan hosts, demonstrating worm-like and sac-like morphologies in the latter. This clade comprises several malacosporean representatives of *Buddenbrockia* and *Tetracapsuloides*, and several as yet unknown malacosporean species.

The polychaete myxosporean lineage includes species with life cycles mostly in the marine environment, with polychaete definitive hosts. It has six main clades, each with different characteristics (Fig. 3.7). These clades encompass species with the same site of infection forming clades infecting biliary tract, urinary tract, and histozoic tissues. The relationships within these clades are still being studied, and the positions of some of the clades can vary depending on the analytical method and marker used. The *Ceratomyxa* clade is the most taxon-rich group within the polychaete lineage, primarily composed of *Ceratomyxa* species infecting gall bladder of many marine fish hosts including elasmobranchs (Fiala et al. 2015). Interestingly, there is a radiation of *Ceratomyxa* species infecting freshwater fish in Amazon basin within the *Ceratomyxa* clade (Zatti et al. 2023). Histozoic myxosporeans cluster into three well-defined clades and include important fish pathogens such as *Ceratonova shasta*, *Enteromyxum leei*, and *Kudoa thyrisites*.

The oligochaete-infecting myxosporean lineage comprises myxozoans primarily found in freshwater environments that use oligochaetes as definitive hosts. Although traditionally thought to be separate from marine myxozoans, some exceptions have emerged. This lineage has

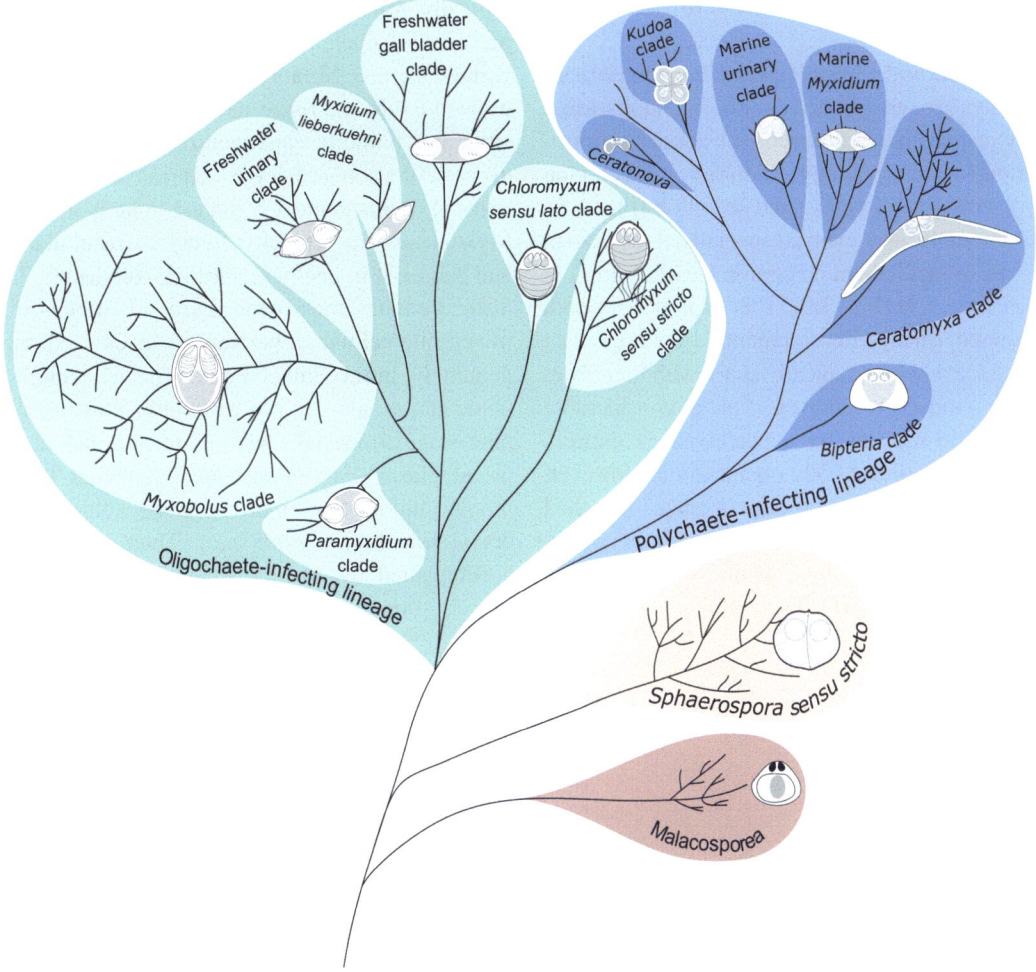

Fig. 3.7 Schematic illustration of the myxozoan phylogeny

seven clades, with the *Myxobolus* clade the largest within the myxozoan phylogenetic tree (Fig. 3.7). The *Myxobolus* clade contains mainly species of *Myxobolus* and *Henneguya* that show poly- and paraphyletic relationships based on morphological characteristics which is not well aligned with the relationships as revealed by DNA sequence data. The *Myxobolus* clade has eight clades (Liu et al. 2019), largely reflecting host phylogeny with two large clades infecting mostly cypriniform and perciform fishes, respectively.

Additionally, the *Sphaerospora* sensu stricto clade comprises myxosporeans primarily infecting the urinary tracts of marine and freshwater fishes, elasmobranchs, and amphibians. Sphaerosporids are distinctive for their exceptionally long SSU rDNA sequences, which include insertions in their variable regions (Holzer et al. 2007). The clade includes *Sphaerospora molnari*, responsible for skin and gill sphaerosporosis in common carp (*Cyprinus carpio*) in central Europe. This parasite is also noteworthy for its multicellular blood stages in the life cycle, which proliferate rapidly and exhibit unique motility, possibly aiding them in evading contact with host immune cells (Hartigan et al. 2016).

3.3.4 Biology and Life Cycle

Structural and Physiological Characteristics

As myxozoans adapted to their endoparasitic way of life, they underwent substantial simplification of morphology, assuming plasmodial forms capable of uptake and secretion (Fig. 3.8). Additionally, they evolved specialised myxozoan spores to facilitate transmission between hosts in their life cycle (Fig. 3.9). This evolutionary path obscured many developmental features that myxozoans would typically share with their free-living cnidarian relatives. For a considerable period, myxozoans were perceived as entirely lacking tissue-level body organisation. However, the discovery of malacosporean stages has shed light on the existence of tissues, such as epidermal epithelia and musculature in myxozoans (Canning et al. 2007). The cellular organisation typically follows eukaryotic patterns, with the presence of nuclei, endoplasmic reticulum, and Golgi cisternae. Notably, their mitochondria tend to feature tubular cristae, as opposed to plate-like structures found in most metazoans, and nuclear division does not involve centrioles, which is unusual in this context. Cilia are notably absent. Myxozoans are typically found extracellularly and can exist in coelozoic or histozoic forms. In histozoic states, the trophic phases are often positioned intercellularly, but several species exhibit genuinely intracellular stages. The extrasporogonic stages, also known as trophozoites or trophic stages, may take various forms, including pseudoplasmodia (uninucleate), syncytial plasmodia (multinucleate), or organised cellular layers surrounding a syncytium. Their growth is accompanied by an increase in the number of nuclei, some serving somatic functions, while

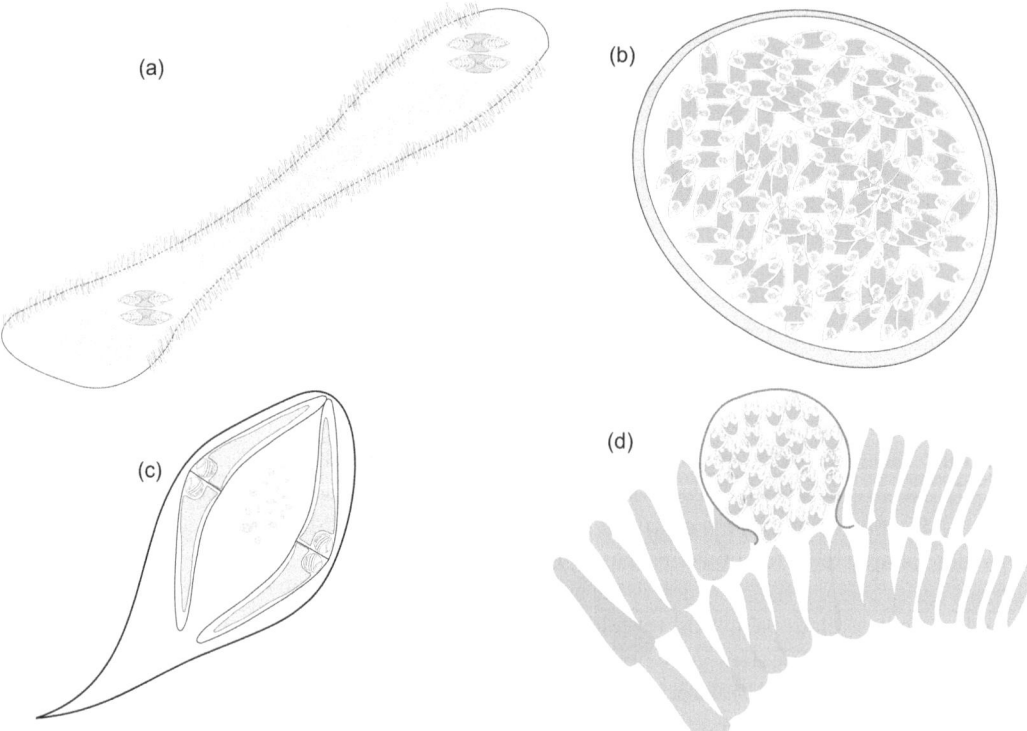

Fig. 3.8 Different types of myxozoan plasmodia. (**a**) Coelozoic polysporic plasmodium of *Myxidium lieberkuehni* with developing spores and surface villosities from the urinary bladder of *Esox lucius*; (**b**) histozoic polysporic plasmodium of *Myxidium rhodei* developing in the kidney glomerulus of *Rhodeus amarus*; (**c**) coelozoic disporic plasmodium typical for *Ceratomyxa* spp. inhabiting the gall bladder of marine fish; (**d**) histozoic polysporic plasmodium of *Myxobolus intimus* developing in the secondary gill lamella of *Rutilus rutilus*

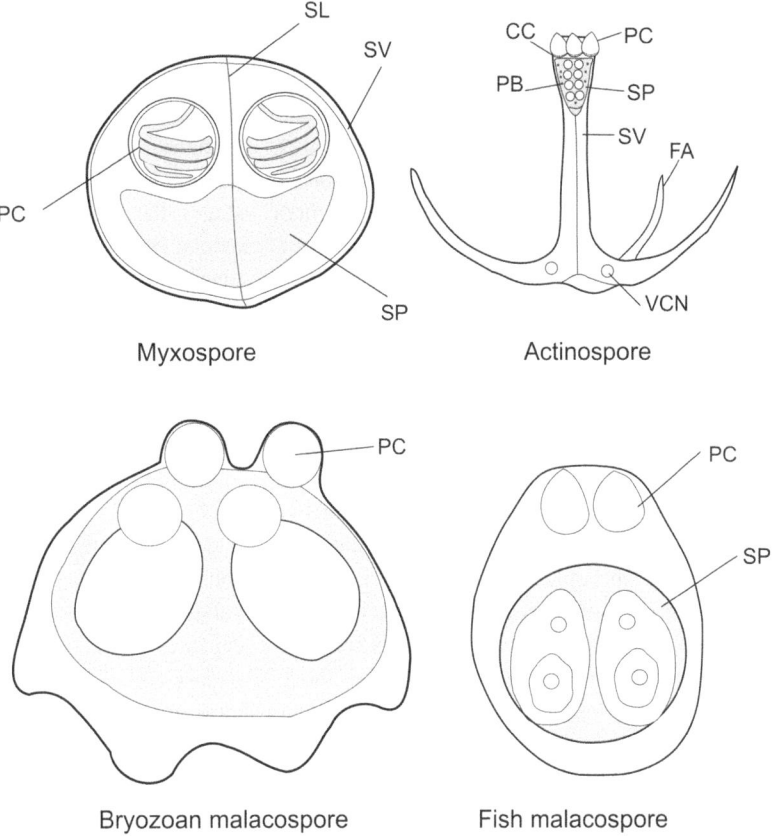

Fig. 3.9 Myxozoan spore types. *CC* capsulogenic cell, *FA* floating appendix, *PB* polar body, *PC* polar capsule, *SL* suture line, *SP* sporoplasm, *SV* shell valve, *VCN* valve cell nucleus

others are destined for spore formation. While cell division through binary fission is documented, particularly in the bryozoan phase of malacosporean myxozoans, proliferation by endogeny stands out as a highly characteristic feature of the group. In the process of endogeny, following nuclear division, one of the nuclei and its surrounding cytoplasm become enclosed by endoplasmic reticulum, giving rise to a secondary cell. This secondary cell resides within a membrane-bound vacuole within the primary cell. Endogeny can extend to tertiary or quaternary levels and is responsible for proliferation when two or more endogenous cells are generated within the cytoplasm of another cell, subsequently being released as the parent cell breaks down.

A key mechanism for Myxozoa's survival and transmission is their multicellular spores (Fig. 3.9), which enclose infectious amoeboid cells (sporoplasms) and cells with polar capsules. These polar capsules contain eversible filaments used for attaching to host surfaces and are homologous to nematocysts found in free-living cnidarians. The attachment of the spore to the host initiates the infection process, as the sporoplasm or their secondary cells invade host tissues.

The myxospore-stage develops in plasmodia or pseudoplasmodia in the vertebrate host and serves for transmission to annelid hosts. The myxospore shell is composed of shell valves, two in Bivalvulida and four and more in Multivalvulida, that adheres together forming a suture line. Myxospores contain polar capsules and sporoplasm, which is the infective germ. The number and morphology of shell valves, polar capsules, and sporoplasms vary by genus and are key deterministic characters in taxonomy (Lom and Arthur 1989, Lom and Dyková 2006). The shell valves have different surface characteristics, including projections and mucous envelopes, which may disappear after spore release. These structures help to keep spores afloat and promote wide dispersal. Polar capsules have a coiled polar filament, which can rapidly extrude and serve to attach spores to the annelid host's intestinal surface. The main morphotypes of myxospores, which represent the majority of myxozoan diversity, are those found in the species of the genera *Myxobolus*, *Henneguya*, *Sphaerospora*, *Ceratomyxa*, *Myxidium*, *Zschokkella*, *Chloromyxum,* and *Kudoa* (Fig. 3.10). The size of myxospores typically ranges between 10 and 20 μm, but several *Ceratomyxa* species extend these dimensions up to 740 μm in *C. maxima* (Eiras 2006). The equivalent to the myxospores in the Malacosporea has been named "fishmalacospores" that are characterised by having two shell valves, enclosing two polar capsules and one diploid sporoplasm (Morris and Adams 2008). The maintenance of spore infectivity varies between Malacosporea and Myxosporea, with myxosporean spores released from fish hosts remaining infectious for months to years, while fishmalacosporean spores are relatively short-lived (de Kinkelin et al. 2002).

Actinospore stages are characterised by spores with triradiate symmetry. The majority of actinospores contain three polar capsules and three shell valves, which create an opening for the tips of the polar capsules. However, in the case of sphaeractinomyxon, the polar capsules are positioned beneath the spore surface. Typically, there is a sporoplasm resembling a plasmodium with numerous nuclei and infectious cells located behind the spore structure. In most actinosporean stages, the shell valves behind the spore structure extend into elongated, hollow caudal projections that move apart from each other. These projections can inflate to their full length osmotically when the spore is discharged from the host. There are approximately 20 recognised actinospore collective groups; among these, aurantiactinomyxon stands out for its diversity, with 61 distinct types identified based on morphological features (Rocha 2023).

Malacospores developed in bryozoans are spherical and feature eight non-hardened shell valves. These spores are equipped with four polar capsules and contain two sporoplasms with secondary cells. The sporoplasms contain sporoplasmosomes—electron-dense bodies. Notably, the polar capsules located at the spore's anterior pole have their mouths covered by a dense cap, which is further overlaid by a structure composed of fibrous components.

General Life Cycle Characteristics

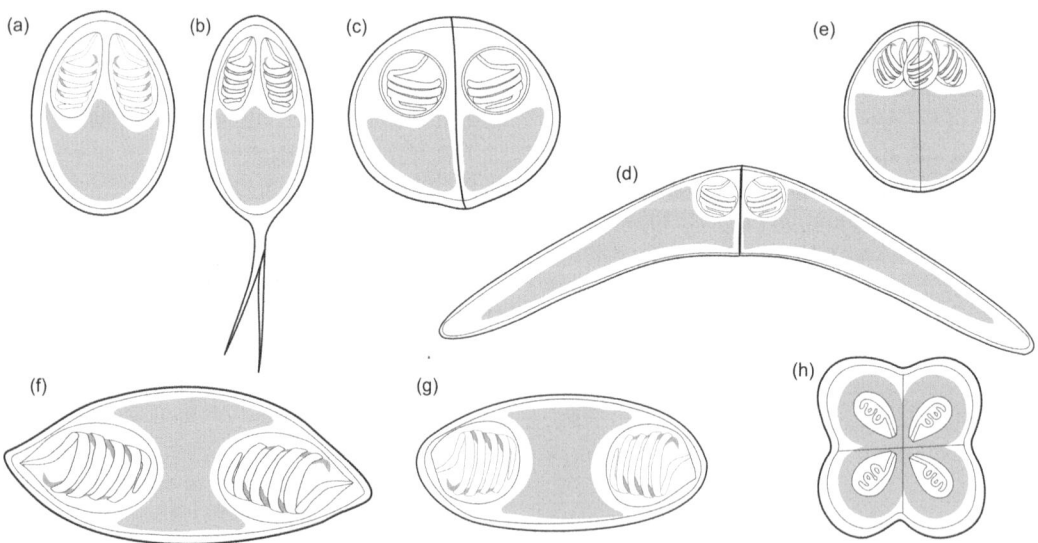

Fig. 3.10 Spore drawings of the typical myxosporean genera. (**a**) *Myxobolus*; (**b**) *Henneguya*; (**c**) *Sphaerospora*; (**d**) *Ceratomyxa*; (**e**) *Chloromyxum*; (**f**) *Myxidium*; (**g**) *Zschokkella*; (**h**) *Kudoa*

As mentioned earlier, the life cycles of myxozoans are complex and involve a vertebrate intermediate host and a definitive host, typically annelids or bryozoans (Fig. 3.11). Fish make up the largest number of known intermediate hosts, but amphibians, reptiles and, exceptionally, birds and mammals are also recorded as myxozoan hosts (Eiras 2005, Prunescu et al. 2007). There is no clear pattern of host-specificity among myxozoans. Some species are strictly host-specific, while others show low specificity and can be found in various hosts (Feist and Longshaw 2006). Myxozoans are found worldwide in all types of aquatic environments from marine and estuarine, to freshwater.

Myxosporean Life Cycle

Although it is now well known that the life cycle of myxosporeans includes two hosts, for a long time, it was believed that myxospores were transmitted directly from one fish to another. The first documented myxozoan life cycle was that of the myxosporean *Myxobolus cerebralis*, which is responsible for salmonid whirling disease. Wolf and Markiw (1984) revealed that the life cycle of *M. cerebralis* consists of two distinct spore forms, which develop alternately within vertebrate and invertebrate hosts. Since 1984, numerous life cycles have been explored; however, many of these studies identify myxozoan transmission based on DNA sequence comparisons only (Lom and Dyková 2006). The life cycle of the Myxozoa begins when actinospores in the water come into contact with a susceptible fish species. Effective host attachment necessitates precise timing of polar capsule discharge, ensuring successful connection during the transient host encounter. The Actinospores apical region must be in close proximity to the host's surface, typically within the length of the polar filament. The polar capsule discharge is likely to be stimulated by chemical and mechanical host stimuli (Kallert et al. 2005). The sequence of actinospore invasion events involves hydrodynamic forces opening the spore

Fig. 3.11 Myxozoan life cycle

valve shell, spore anchoring through the polar filament to the host surface with subsequent sporoplasm penetration. When the sporoplasm is successfully transmitted to the vertebrate host, it starts active amoeboid movement via pseudopodia facilitated by proteolytic activity (Adkison and Hedrick 2001). It takes the sporoplasm only a few minutes to penetrate the fish host tissue. Then, the primary cell of the sporoplasm degrades, releasing secondary cells that either become blood stages or invade other tissues for further development. The proliferation of secondary cells is stimulated by rapid synchronised mitosis (Daniels et al. 1976, El-Matbouli et al. 1995). This leads to the production of cell doublets through the endogenous division of secondary cells into an outer enveloping cell and an inner cell. The life cycle continues by migration of parasite stages to the location where the sporogonic stage, either a plasmodium or a pseudoplasmodium, takes form. Both plasmodia and pseudoplasmodia might extend temporary cytoplasmic projections known as pseudopodia and potentially divide through plasmotomy, a process that divides the cell into two or more daughter parts. Plasmodial stages can either be histozoic, residing within the host's tissues and often resembling "cysts", or coelozoic, inhabiting cavities within body organs, primarily the urinary tract or gall bladder. Sporogony in species with large polysporic plasmodia begins with one generative cell enveloping another (known as the pericyte), forming a pansporoblast. These cells undergo synchronous differentiation into valvogenic, capsulogenic, and sporoplasmogenic cells. Valvogenic cells surround the sporoplasm(s) and capsulogenic cells and lose their nuclei and organelles as they mature. Sporoplasmogenic cells are uni- or bi-nucleate and often contain sporoplasmosomes, structures of unknown function. Capsulogenic cells have an extensive endoplasmic reticulum, and an external tube that eventually forms the polar capsule. The polar filaments in mature cells are sealed with a plug and positioned near the sutural line. Myxospores often exhibit distinct ornamentation like ridges, bristles, pits, mucous envelopes, or various valve cell extensions, potentially influencing environmental distribution and transmission (Lom and Dyková 2006).

Upon ingestion of the myxospore by the definitive host (an annelid where the sexual phase occurs) the sporoplasm of the myxospore initiates a merogony phase. In *M. cerebralis*, the resultant uninucleate cells produce a tetranucleate stage, which consists of two sporogonic cells enveloped by a pair of pericytes, forming the future pansporocyst. During the further development of these cells, the enveloping cells house eight pairs of gametic cells (α and β cells). After meiosis, which involves the formation of synaptonemal complexes, these gametic cells fuse to produce eight zygotes. These zygotes then develop through cell division and differentiation into eight triradiate actinospores (Lom and Dyková 2006).

Malacosporean Life Cycle

Malacosporean infection typically begins with the contact of the malacospore with the fish, as in the case of Myxosporea. An infectious amoebic sporoplasm is released from the spore and enters the host through the host's epidermis and mucus cells (Grabner and El-Matbouli 2010). In *Tetracapsuloides bryosalmonae*, early stages consist of unicellular forms, likely representing secondary cells of the sporoplasm. Infection targets kidneys, the final site of infection, where multiple proliferation cycles within the kidney interstitium occur before spore formation. Later on, larger pseudoplasmodia develop in kidney tubules, containing several secondary cells and secondary/tertiary cell doublets from which the fishmalacospores are developed. Spores are released from the fish through urine to the aquatic environment where they are filtered by bryozoan definitive hosts. Here, unicellular stages initially occur in the body cavity and can remain inactive within the host. After activation, multicellular stages form clusters of cells by aggregation or mitosis. Further development leads to the formation of sacs, reaching up to 300 μm in diameter, in which spores are developed. In the case of worm-like *Buddenbrockia*, early unicellular stages are located within the extracellular matrix. These stages form clusters of cells and penetrate the peritoneum towards the coelom where they form an outer epithelial layer and an inner compact mass of cells, that later differentiates into musculature. The sporogony takes place in the internal cavity of the *Buddenbrockia* worm. Mature spores consist of two sporoplasms, four capsulogenic cells, and eight valve cells.

3.3.5 The Myxozoan Genome

The shift from a free-living cnidarian ancestor to a typical microscopic myxozoan parasite is marked by a significant reduction in both genome size and gene content (Chang et al. 2015, Alama-Bermejo and Holzer 2021). The pressures associated with parasitic lifestyles have instigated evolutionary constraints on myxozoan genomes, which are among the most compact genomes found in animals. The Myxozoan genomes encompass a wide size range from 22.5 to 188.5 Mb, while their free-living counterparts typically range from over 200 Mb to approximately 1 Gb (Adachi et al. 2017). Within myxozoans, gene reductions primarily impact processes related to development and cell differentiation (Chang et al. 2015). This is consistent with the observation that nearly all myxozoans lack a complex multicellular body. The reduction in protein-coding genes in many myxozoans is quite remarkable, with *Kudoa iwatai* possessing only 5533 genes, in contrast to the 17,440 genes found in *Polypodium hydriforme*, an enigmatic cnidarian parasite closely related to the Myxozoa (Chang et al. 2015). However, *Thelohanellus kitauei* was found to have 16,638 genes (Yang et al. 2014) similar to the free-living cnidarian *Hydra magnipapillata* with 17,918 genes (Chapman et al. 2010). Myxozoan genomes appear to lack key genes and signalling pathways that are associated with body plans, such as Hox-like, Wnt pathway, and Hedgehog pathway (Faber et al. 2021). It is noteworthy that only in *Tetracapsuloides bryosalmonae* some homeobox proteins have been identified, which are different from the classical Hox cluster. In contrast to the remarkably compact nuclear genome size of Myxozoa, *Enteromyxum leei* has demonstrated a distinctive characteristic. It possesses the largest mitochondrial genome within metazoans, comprising eight circular chromosomes, each approximately 23 kb in size (Yahalomi et al. 2017). One of the most remarkable features of myxozoans is their ability to adapt to life without relying on aerobic cellular respiration. A prime example of this adaptation is found in *Henneguya salminicola*, which is the only known metazoan species to have entirely lost its mitochondrial genome (Yahalomi et al. 2020).

3.3.6 Outlook

Myxozoa are a diverse and enigmatic group of endoparasites that are increasingly being recognised for their ecological and economic significance. They exemplify the complex nature of host–parasite interactions and present a unique opportunity to explore the evolution of early metazoans. As research continues, the hidden world of Myxozoa promises to reveal more about these intriguing parasites, their intricate life cycles, diversity, and genomics.

References

Adachi K, Miyake H, Kuramochi T et al (2017) Genome size distribution in phylum Cnidaria. Fish Sci 83:107–112. https://doi.org/10.1007/s12562-016-1050-4

Adkison MA, Hedrick RP (2001) Purification of the serine protease from the triactinomyxon stage of *Myxobolus cerebralis*. In: Proceedings of the 7th annual symposium on whirling disease, Salt Lake City, 8–9 Feb 2001

Akiyoshi DE, Morrison HG, Lei S et al (2009) Genomic survey of the non-cultivatable opportunistic human pathogen, *Enterocytozoon bieneusi*. PLoS Pathog 5(1):e1000261. https://doi.org/10.1371/journal.ppat.1000261

Alama-Bermejo G, Holzer AS (2021) Advances and discoveries in myxozoan genomics. Trends Parasitol 37(6):552–568. https://doi.org/10.1016/j.pt.2021.01.010

Andreadis TG (2007) Microsporidian parasites of mosquitoes. Am Mosq Control Assoc 23(sp2):3–29. https://doi.org/10.2987/8756-971x(2007)23[3:Mpom]2.0.Co;2

Andreadis TG, Simakova AV, Vossbrinck CR et al (2012) Ultrastructural characterisation and comparative phylogenetic analysis of new Microsporidia from Siberian mosquitoes: evidence for coevolution and host switching. J Invertebr Pathol 109(1):59–75. https://doi.org/10.1016/j.jip.2011.09.011

Bacela-Spychalska K, Wattier RA, Genton C et al (2012) Microsporidian disease of the invasive amphipod *Dikerogammarus villosus* and the potential for its transfer to local invertebrate fauna. Biol Invasions 14(9):1831–1842. https://doi.org/10.1007/s10530-012-0193-1

Bartholomew JL (2012) Salmonid Ceratomyxosis. In: Blue book: Suggested procedures for the detection and identification of certain finfish and shellfish pathogens, 4th edn. Fish Health Section, American Fisheries Society, Bethesda

Bartholomew JL, Wilson JC (eds) (2002) Whirling disease: reviews and current topics. In: American Fisheries Society symposium 29. American Fisheries Society, Bethesda, p 262

Bartošová P, Fiala I, Hypša V (2009) Concatenated SSU and LSU rDNA data confirm the main evolutionary trends within myxosporeans (Myxozoa: Myxosporea) and provide an effective tool for their molecular phylogenetics. Mol Phylogenet Evol 53:81–93. https://doi.org/10.1016/j.ympev.2009.05.018

Bass D, Czech L, Williams BAP, Berney C et al (2018) Clarifying the relationships between Microsporidia and Cryptomycota. J Eukaryot Microbiol 65(6):773–782. https://doi.org/10.1111/jeu.12519

Becnel JJ, Andreadis TG (2014) Microsporidia in insects. In: Wittner M, Weiss LM (eds) The Microsporidia and microsporidiosis. American Society for Microbiology, Washington, DC, pp 447–501. https://doi.org/10.1128/9781555818227.ch14

Becnel JJ, Sprague V, Fukuda T et al (1989) Development of *Edhazardia aedis* (Kudo, 1930) n. g., n. comb. (Microspora: Amblyosporidae) in the mosquito *Aedes aegypti* (L.) (Diptera: Culicidae). J Protozool 36(2):119–130. https://doi.org/10.1111/j.1550-7408.1989.tb01057.x

Bojko J, Stentiford GD (2022) Microsporidian pathogens of aquatic animals. In: Weiss LM, Reinke AW (eds) Microsporidia. Experientia supplementum, vol 114. Springer, Cham, pp 247–283. https://doi.org/10.1007/978-3-030-93306-7_10

Bojko J, Dunn AM, Stebbing PD et al (2015) *Cucumispora ornata* n. sp. (Fungi: Microsporidia) infecting invasive 'demon shrimp' (*Dikerogammarus haemobaphes*) in the United Kingdom. J Invertebr Pathol 128:22–30. https://doi.org/10.1016/j.jip.2015.04.005

Bojko J, Reinke AW, Stentiford GD et al (2022) Microsporidia: a new taxonomic, evolutionary, and ecological synthesis. Trends Parasitol 38(8):642–659. https://doi.org/10.1016/j.pt.2022.05.007

Bütschli O (1881) Myxosporidien. Zoologischer Jahresbericht für 1880(1):162–164

Cali A, Takvorian PM (2014) Developmental morphology and life cycles of the microsporidia. In: Wittner M, Weiss LM (eds) The microsporidia and microsporidiosis. American Society for Microbiology, Washington, DC, pp 85–128

Cali A, Becnel JJ, Takvorian PM (2016) Microsporidia. In: Archibald J, Simpson A, Slamovits C (eds) Handbook of the protists. Springer, Cham, pp 1559–1618. https://doi.org/10.1007/978-3-319-32669-6_27-1

Canning EU, Nicholas JP (1974) Light and electron microscope observations on *Unikaryon legeri* (Microsporida, Nosematidae), a parasite of the metacercaria of *Meigymnophallus minutus* in *Cardium edule*. J Invertebr Pathol 23(1):92–100. https://doi.org/10.1016/0022-2011(74)90078-0

Canning EU, Okamura B (2004) Biodiversity and evolution of the Myxozoa. Adv Parasitol 56:43–131. https://doi.org/10.1016/S0065-308X(03)56002-X

Canning EU, Foon LP, Joe LK (1974) Microsporidian parasites of trematode larvae from aquatic snails in West Malaysia. J Protozool 21(1):19–25. https://doi.org/10.1111/j.1550-7408.1974.tb03611.x

Canning E, Barker RJ, Hammond JC et al (1983) *Unikaryon slaptonleyi* sp. nov. (Microspora: Unikaryonidae) isolated from echinostome and strigeid larvae from *Lymnaea peregra*: Observations on its morphology, transmission and pathogenicity. Parasitology 87:175–184. https://doi.org/10.1017/S0031182000052549

Canning EU, Curry A, Feist SW et al (2000) A new class and order of myxozoans to accommodate parasites of bryozoans with ultrastructural observations on *Tetracapsula bryosalmonae* (PKX organism). J Eukaryot Microbiol 47:456–468. https://doi.org/10.1111/j.1550-7408.2000.tb00075.x

Canning EU, Curry A, Hill SLL et al (2007) Ultrastructure of *Buddenbrockia allmani* n. sp. (Myxozoa, Malacosporea), a parasite of *Lophopus crystallinus* (Bryozoa, Phylactolaemata). J Eukaryot Microbiol 54:247–262. https://doi.org/10.1017/S0031182001001184

Casal G, Matos E, Garcia P et al (2012) Ultrastructural and molecular studies of *Microgemma carolinus* n. sp. (Microsporidia), a parasite of the fish *Trachinotus carolinus* (Carangidae) in Southern Brazil. Parasitology 139(13):1720–1728. https://doi.org/10.1017/S0031182012001011

Chang ES, Neuhof M, Rubinstein ND et al (2015) Genomic insights into the evolutionary origin of Myxozoa within Cnidaria. Proc Natl Acad Sci USA 112:14912–14917. https://doi.org/10.1073/pnas.1511468112

Chapman J, Kirkness E, Simakov O et al (2010) The dynamic genome of *Hydra*. Nature 464:592–596. https://doi.org/10.1038/nature08830

Chauvet M, Monjot A, Lepère C (2023) High diversity of microsporidian parasites and new planktonic hosts in freshwater and marine ecosystems. Limnol Oceanogr 68(4):928–941. https://doi.org/10.1002/lno.12321

Colley FC, Joe LK, Zaman V et al (1975) Light and electron microscopical study of *Nosema eury-*

tremae. J Invertebr Pathol 26:11–20. https://doi.org/10.1016/0022-2011(75)90163-9

Corradi N (2015) Microsporidia: Eukaryotic intracellular parasites shaped by gene loss and horizontal gene transfers. Annu Rev Microbiol 69:167–183. https://doi.org/10.1146/annurev-micro-091014-104136

Corsaro D (2022) Insights into Microsporidia evolution from early diverging Microsporidia. In: Weiss LM, Reinke AW (eds) Microsporidia, Experientia supplementum. Springer, Cham, pp 71–90. https://doi.org/10.1007/978-3-030-93306-7_3

Cort WW (1960) Studies on a microsporidian hyperparasite of strigeoid trematodes. I. Prevalence and effect on the parasitised larval trematode. J Parasitol 46(3):317–326. https://doi.org/10.2307/3275494

Daniels SB, Herman RL, Burke CN (1976) Fine structure of an unidentified protozoon in the epithelium of rainbow trout exposed to water with *Myxosoma cerebralis*. J Protozool 23:402–410. https://doi.org/10.1111/j.1550-7408.1976.tb03795.x

Dean P, Sendra KM, Williams TA et al (2018) Transporter gene acquisition and innovation in the evolution of Microsporidia intracellular parasites. Nat Commun 9(1):1709. https://doi.org/10.1038/s41467-018-03923-4

de Buron I, Loubès C, Maruand J (1990) Infection and pathological alterations within the acanthocephalan *Acanthocephaloides propinquus* attributable to the microsporidian hyperparasite *Microsporidium acanthocephali*. Trans Am Microsc Soc 109(1):91–97. https://doi.org/10.2307/3226598

de Graaf DC, Raes H, Jacobs FJ (1994) Spore dimorphism in *Nosema apis* (Microsporida, Nosematidae) developmental cycle. J Invertebr Pathol 63(1):92–94. https://doi.org/10.1006/jipa.1994.1015

de Kinkelin P, Gay M, Forman S (2002) The persistence of infectivity of *Tetracapsula bryosalmonae*-infected water for rainbow trout, *Oncorhynchus mykiss* (Walbaum). J Fish Dis 25:477–482. https://doi.org/10.1046/j.1365-2761.2002.00382.x

Desjardins CA, Sanscrainte ND, Goldberg JM et al (2015) Contrasting host-pathogen interactions and genome evolution in two generalist and specialist microsporidian pathogens of mosquitoes. Nat Commun 6:7121. https://doi.org/10.1038/ncomms8121

Desportes-Livage I (2000) Biology of Microsporidia. In: Petry F (ed) Cryptosporidiosis and microsporidiosis, vol 6. Contributions to microbiology. Karger, Basel, pp 140–165. https://doi.org/10.1159/000060359

Dewangan KN, Pang J, Zhao C et al (2023) Host and transmission route of *Enterocytozoon hepatopenaei* (EHP) from dragonfly to shrimp. Aquaculture 574:739642. https://doi.org/10.1016/j.aquaculture.2023.739642

Diamant A, Paperna I (1985) The development and ultrastructure of *Nosema ceratomyxae* sp. nov., a microsporidian hyperparasite of the myxosporean *Ceratomyxa* sp. from red sea rabbitfish (Siganidae). Protistologica 21:249–258

Didier ES (2005) Microsporidiosis: An emerging and opportunistic infection in humans and animals. Acta Trop 94(1):61–76. https://doi.org/10.1016/j.actatropica.2005.01.010

Dissanaike A (1957) The morphology and life cycle of *Nosema helminthorum* Moniez, 1887. Parasitology 47(3-4):335–346. https://doi.org/10.1017/S0031182000022022

Doliwa A, Dunthorn M, Rassoshanska E et al (2021) Identifying potential hosts of short-branch Microsporidia. Microb Ecol 82(2):549–553. https://doi.org/10.1007/s00248-020-01657-9

Doliwa A, Grabner D, Sures B et al (2023) Comparing Microsporidia-targeting primers for environmental DNA sequencing. Parasite 30:52. https://doi.org/10.1051/parasite/2023056

Dunn AM, Smith JE (2001) Microsporidian life cycles and diversity: The relationship between virulence and transmission. Microbes Infect 3(5):381–388. https://doi.org/10.1016/S1286-4579(01)01394-6

Dunn AM, Terry RS, Taneyhill DE (1998) Within-host transmission strategies of transovarial, feminising parasites of *Gammarus duebeni*. Parasitology 117(Pt 1):21–30. https://doi.org/10.1017/s0031182098002753

Dunn AM, Terry RS, Smith JE (2001) Transovarial transmission in the Microsporidia. Adv Parasitol 48:57–100. https://doi.org/10.1016/s0065-308x(01)48005-5

Eiras JC (2005) An overview on the myxosporean parasites in amphibians and reptiles. Acta Parasitol 50:267–275

Eiras JC (2006) Synopsis of the species of *Ceratomyxa* Thélohan, 1892 (Myxozoa: Myxosporea: Ceratomyxidae). Syst Parasitol 65:49–71. https://doi.org/10.1007/s11230-018-9791-3

Eiras JC, Cruz CF, Saraiva A et al (2021) Synopsis of the species of *Myxobolus* (Cnidaria, Myxozoa, Myxosporea) described between 2014 and 2020. Folia Parasitol 68:012. https://doi.org/10.14411/fp.2021.012

El-Matbouli M, Hoffmann RW, Mandok C (1995) Light and electron microscopic observations on the route of the triactinomyxon sporoplasm of *Myxobolus cerebralis* from epidermis into rainbow trout (*Oncorhynchus mykiss*) cartilage. J Fish Biol 46:919–935. https://doi.org/10.1111/j.1095-8649.1995.tb01397.x

Faber M, Shaw S, Yoon S et al (2021) Comparative transcriptomics and host-specific parasite gene expression profiles inform on drivers of proliferative kidney disease. Sci Rep 11(1):2149. https://doi.org/10.1038/s41598-020-77881-7

Feist SW, Longshaw M (2006) Phylum Myxozoa. In: Woo PTK (ed) Fish diseases and disorders. CABI, Wallingford, pp 230–296

Fiala I (2006) The phylogeny of Myxosporea (Myxozoa) based on small subunit ribosomal RNA gene analysis. Int J Parasitol 36(14):1521–1534. https://doi.org/10.1016/j.ijpara.2006.06.016

Fiala I, Hlavničková M, Kodádková A et al (2015) Evolutionary origin of *Ceratonova shasta* and phylogeny of the marine myxosporean lineage. Mol

Phylogenet Evol 86:75–89. https://doi.org/10.1016/j.ympev.2015.03.004

Field AS, Marriott DJ, Milliken ST et al (1996) Myositis associated with a newly described microsporidian, *Trachipleistophora hominis*, in a patient with AIDS. J Clin Microbiol 34(11):2803–2811. https://doi.org/10.1128/Jcm.34.11.2803-2811.1996

Franzen C (2004) Microsporidia: how can they invade other cells? Trends Parasitol 20(6):275–279. https://doi.org/10.1016/j.pt.2004.04.009

Freeman MA, Bell AS, Sommerville C (2003) A hyperparasitic microsporidian infecting the salmon louse, *Lepeophtheirus salmonis*: an rDNA-based molecular phylogenetic study. J Fish Dis 26(11-12):667–676. https://doi.org/10.1046/j.1365-2761.2003.00498.x

Goldberg AV, Molik S, Tsaousis AD et al (2008) Localisation and functionality of microsporidian iron-sulphur cluster assembly proteins. Nature 452(7187):624–628. https://doi.org/10.1038/nature06606

Grabner DS (2017) Hidden diversity: parasites of stream arthropods. Freshw Biol 62(1):52–64. https://doi.org/10.1111/fwb.12848

Grabner DS, El-Matbouli M (2010) *Tetracapsuloides bryosalmonae* (Myxozoa: Malacosporea) portal of entry into the fish host. Dis Aquat Organ 90(3):197–206. https://doi.org/10.3354/dao02236

Grabner D, Sures B (2019) Amphipod parasites may bias results of ecotoxicological research. Dis Aquat Organ 136(1):123–134. https://doi.org/10.3354/dao03355

Grabner DS, Weigand AM, Leese F, Winking C, Hering D, Tollrian R, Sures B (2015) Invaders, natives and their enemies: distribution patterns of amphipods and their microsporidian parasites in the Ruhr Metropolis, Germany. Parasit Vectors 8:419. https://doi.org/10.1186/s13071-015-1036-6

Grabner D, Weber D, Weigand AM (2020) Updates to the sporadic knowledge on microsporidian infections in groundwater amphipods (Crustacea, Amphipoda, Niphargidae). Subterr Biol 33:71–85. https://doi.org/10.3897/subtbiol.33.48633

Grabner D, Doliwa A, Sworobowicz L et al (2022) Microsporidian diversity in the aquatic isopod *Asellus aquaticus*. Parasitology 149(13):1729–1736. https://doi.org/10.1017/S003118202200124X

Grassé PP (1970) Embranchement des Myxozoaires. In: Grassé PP, Poisson RR, Tuzet O (eds) Précis de Zoologie vol 1, Invertébrés. Masson et Cie, Paris

Grupe AC, Quandt CA (2020) A growing pandemic: A review of *Nosema* parasites in globally distributed domesticated and native bees. PLoS Pathog 16(6):e1008580. https://doi.org/10.1371/journal.ppat.1008580

Haag KL, Sheikh-Jabbari E, Ben-Ami F et al (2013) Microsatellite and single-nucleotide polymorphisms indicate recurrent transitions to asexuality in a microsporidian parasite. J Evol Biol 26(5):1117–1128. https://doi.org/10.1111/jeb.12125

Haine ER, Brondani E, Hume KD et al (2004) Coexistence of three Microsporidia parasites in populations of the freshwater amphipod *Gammarus roeseli*: evidence for vertical transmission and positive effect on reproduction. Int J Parasitol 34(10):1137–1146. https://doi.org/10.1016/j.ijpara.2004.06.006

Haine ER, Motreuil S, Rigaud T (2007) Infection by a vertically-transmitted microsporidian parasite is associated with a female-biased sex ratio and survival advantage in the amphipod *Gammarus roeseli*. Parasitology 134(Pt 10):1363–1367. https://doi.org/10.1017/S0031182007002715

Han B, Weiss LM (2017) Microsporidia: Obligate intracellular pathogens within the fungal kingdom. Microbiol Spectr 5(2). https://doi.org/10.1128/microbiolspec.FUNK-0018-2016

Hartigan A, Estensoro I, Vancová M et al (2016) New cell motility model observed in parasitic cnidarian *Sphaerospora molnari* (Myxozoa:Myxosporea) blood stages in fish. Sci Rep 6:39093. https://doi.org/10.1016/j.ijpara.2016.07.006

Hartikainen H, Bass D, Briscoe AG et al (2016) Assessing myxozoan presence and diversity using environmental DNA. Int J Parasitol 46:781–792. https://doi.org/10.1016/j.ijpara.2016.07.006

Hedrick RP, Groff JM, Baxa DV (1991) Experimental infections with *Nucleospora salmonis* n. g., n. sp.: An intranuclear microsporidium from chinook salmon (*Oncorhynchus tshawytscha*). Dis Aquat Organ 10:103–108

Herren JK, Mbaisi L, Mararo E et al (2020) A microsporidian impairs *Plasmodium falciparum* transmission in *Anopheles arabiensis* mosquitoes. Nat Commun 11(1):2187. https://doi.org/10.1038/s41467-020-16121-y

Hirt RP, Logsdon JM Jr, Healy B et al (1999) Microsporidia are related to Fungi: evidence from the largest subunit of RNA polymerase II and other proteins. Proc Natl Acad Sci U S A 96(2):580–585. https://doi.org/10.1073/pnas.96.2.580

Holland JW, Okamura B, Hartikainen H et al (2011) A novel minicollagen gene links cnidarians and myxozoans. Proc Roy Soc Ser B 278:546–553. https://doi.org/10.1098/rspb.2010.1301

Hollister WS, Canning EU, Weidner E et al (1996) Development and ultrastructure of *Trachipleistophora hominis* n.g., n.sp. after in vitro isolation from an AIDS patient and inoculation into athymic mice. Parasitology 112(Pt 1):143–154. https://doi.org/10.1017/s0031182000065185

Holzer AS, Wootten R, Sommerville C (2007) The secondary structure of the unusually long 18S ribosomal RNA of the myxozoan *Sphaerospora truttae* and structural evolutionary trends in the Myxozoa. Int J Parasitol 37:1281–1295. https://doi.org/10.1016/j.ijpara.2007.03.014

Holzer AS, Bartošová-Sojková P, Born-Torrijos A et al (2018) The joint evolution of the Myxozoa and their alternate hosts: A cnidarian recipe for success and vast biodiversity. Mol Ecol 27:1651–1666. https://doi.org/10.1111/mec.14558

Hussey KL (1971) A microsporidan hyperparasite of strigeoid trematodes, *Nosema strigeoideae* sp. n. J Protozool 18(4):676–679. https://doi.org/10.1111/j.1550-7408.1971.tb03396.x

Ironside JE (2007) Multiple losses of sex within a single genus of Microsporidia. BMC Evol Biol 7:48. https://doi.org/10.1186/1471-2148-7-48

Ironside JE (2013) Diversity and recombination of dispersed ribosomal DNA and protein coding genes in Microsporidia. PLoS One 8(2):e55878. https://doi.org/10.1371/journal.pone.0055878

Ironside JE, Dunn AM, Rollinson D et al (2003a) Association with host mitochondrial haplotypes suggests that feminising Microsporidia lack horizontal transmission. J Evol Biol 16(6):1077–1083. https://doi.org/10.1046/j.1420-9101.2003.00625.x

Ironside JE, Smith JE, Hatcher MJ et al (2003b) Two species of feminising microsporidian parasite coexist in populations of *Gammarus duebeni*. J Evol Biol 16(3):467–473. https://doi.org/10.1046/j.1420-9101.2003.00539.x

Ironside JE, Wilkinson TJ, Rock J (2008) Distribution and host range of the microsporidian *Pleistophora mulleri*. J Eukaryot Microbiol 55(4):355–362. https://doi.org/10.1111/j.1550-7408.2008.00338.x

Iwano H, Ishihara R (1991) Dimorphism of spores of *Nosema* spp. in cultured cell. J Invertebr Pathol 57(2):211–219. https://doi.org/10.1016/0022-2011(91)90119-B

Jurine LL (1825) Histoire des poissons du Lac Léman. Mém Soc Phys Hist Nat Genève 3

Kallert DM, El-Matbouli M, Haas W (2005) Polar filament discharge of *Myxobolus cerebralis* actinospores is triggered by combined non-specific mechanical and chemical cues. Parasitology 131:609–616. https://doi.org/10.1017/S0031182005008383

Karpov SA, Mamkaeva MA, Aleoshin VV et al (2014) Morphology, phylogeny, and ecology of the aphelids (Aphelidea, Opisthokonta) and proposal for the new superphylum Opisthosporidia. Front Microbiol 5:112. https://doi.org/10.3389/fmicb.2014.00112

Keeling PJ, Fast NM (2002) Microsporidia. biology and evolution of highly reduced intracellular parasites. Annu Rev Microbiol 56:93–116. https://doi.org/10.1146/annurev.micro.56.012302.160854

Kent ML, Bishop-Stewart JK (2003) Transmission and tissue distribution of *Pseudoloma neurophilia* (Microsporidia) of zebrafish, *Danio rerio* (Hamilton). J Fish Dis 26(7):423–426. https://doi.org/10.1046/j.1365-2761.2003.00467.x

Kent ML, Hedrick RP (1985) PKX, the causative agent of proliferative kidney disease (PKD) in Pacific salmonid fishes and its affinities with the Myxozoa. J Protozool 32:254–260. https://doi.org/10.1111/j.1550-7408.1985.tb03047.x

Kodádková A, Bartošová-Sojková P, Holzer AS et al (2015) *Bipteria vetusta* n. sp.—an old parasite in an old host: tracing the origin of myxosporean parasitism in vertebrates. Int J Parasitol 45:269–276. https://doi.org/10.1016/j.ijpara.2014.12.004

Kohler SL, Hoiland WK (2001) Population regulation in an aquatic insect: The role of disease. Ecology 82(8):2294–2305. https://doi.org/10.2307/2680232

Levron C, Ternengo S, Toguebaye BS et al (2004) Ultrastructural description of the life cycle of *Nosema diphterostomi* sp. n., a Microsporidia hyperparasite of *Diphterostomum brusinae* (Digenea: Zoogonidae), intestinal parasite of *Diplodus annularis* (Pisces: Teleostei). Acta Protozool 43:329–336. https://doi.org/10.1016/j.ejop.2005.05.001

Levron C, Ternengo S, Sikina Toguebaye B et al (2005) Ultrastructural description of the life cycle of *Nosema monorchis* n. sp. (Microspora, Nosematidae), hyperparasite of *Monorchis parvus* (Digenea, Monorchiidae), intestinal parasite of *Diplodus annularis* (Pisces, Teleostei). Eur J Protistol 41(4):251–256. https://doi.org/10.1016/j.ejop.2005.05.001

Lisnerová M, Blabolil P, Holzer AS et al (2020) Myxozoan hidden diversity: the case of *Myxobolus pseudodispar* Gorbunova, 1936. Folia Parasitol 67:019. https://doi.org/10.14411/fp.2020.019

Lisnerová M, Holzer A, Blabolil P et al (2023) Evaluation and optimisation of an eDNA metabarcoding assay for detection of freshwater myxozoan communities. Environ DNA 5:312–325. https://doi.org/10.1002/edn3.380

Liu Y, Lövy A, Gu ZM et al (2019) Phylogeny of Myxobolidae (Myxozoa) and the evolution of myxospore appendages in the *Myxobolus* clade. Int J Parasitol 49:523–530. https://doi.org/10.1016/j.ijpara.2019.02.009

Lom J, Arthur JR (1989) A guideline for the preparation of species descriptions in Myxosporea. J Fish Dis 12:151–156. https://doi.org/10.1111/j.1365-2761.1989.tb00287.x

Lom J, Dyková I (2006) Myxozoan genera: definition and notes on taxonomy, life-cycle terminology and pathogenic species. Folia Parasitol 53:1–36

Lovy J, Friend SE (2017) Phylogeny and morphology of *Ovipleistophora diplostomuri* n. sp. (Microsporidia) with a unique dual-host tropism for bluegill sunfish and the digenean parasite *Posthodiplostomum minimum* (Strigeatida). Parasitology 144(14):1898–1911. https://doi.org/10.1017/S0031182017001305

Mautner SI, Cook KA, Forbes MR et al (2007) Evidence for sex ratio distortion by a new microsporidian parasite of a *Corophiid amphipod*. Parasitology 134(Pt 11):1567–1573. https://doi.org/10.1017/S0031182007003034

McClymont EH, Dunn AM, Terry RS et al (2005) Molecular data suggest that microsporidian parasites in freshwater snails are diverse. Int J Parasitol 35(10):1071–1078. https://doi.org/10.1016/j.ijpara.2005.05.008

Morris DJ, Adams A (2008) Sporogony of *Tetracapsuloides bryosalmonae* in the brown trout *Salmo trutta* and the role of the tertiary cell during the vertebrate phase of myxozoan life cycles. Parasitology 135:1075–1092. https://doi.org/10.1017/S0031182008004605

Morris DJ, Freeman MA (2010) Hyperparasitism has wide-ranging implications for studies on the invertebrate phase of myxosporean (Myxozoa) life cycles. Int J Parasitol 40(3):357–369. https://doi.org/10.1016/j.ijpara.2009.08.014

Murareanu BM, Sukhdeo R, Qu R et al (2021) Generation of a Microsporidia species attribute database and analysis of the extensive ecological and phenotypic diversity of Microsporidia. mBio 12(3):e0149021. https://doi.org/10.1101/2021.02.21.432160

Nägeli C (1857) Über die neue Krankheit der Seidenraupe und verwandte Organismen. Botanische Zeitung 15:760–761

Nylund S, Nylund A, Watanabe K et al (2010) *Paranucleospora theridion* n. gen., n. sp. (Microsporidia, Enterocytozoonidae) with a life cycle in the salmon louse (*Lepeophtheirus salmonis*, Copepoda) and Atlantic salmon (*Salmo salar*). J Eukaryot Microbiol 57(2):95–114. https://doi.org/10.1111/j.1550-7408.2009.00451.x

Okamura B, Curry A, Wood TS et al (2002) Ultrastructure of *Buddenbrockia* identifies it as a myxozoan and verifies the bilaterian origin of the Myxozoa. Parasitology 124:215–223. https://doi.org/10.1017/S0031182001001184

Okamura B, Hartigan A, Naldoni J (2018) Extensive uncharted biodiversity: The parasite dimension. Integr Comp Biol 58:1132–1145. https://doi.org/10.1093/icb/icy039

Olivares CA (2005) Evidence of a parasite protist in *Eurhomalea lenticularis* (Sowerby, 1835) (Mollusca: Bivalvia): A case of intraoocytarian parasitism. J Nat Hist 39(23):2073–2082. https://doi.org/10.1080/00222930500060447

Ovcharenko MO, Bacela K, Wilkinson T et al (2010) *Cucumispora dikerogammari* n. gen. (Fungi: Microsporidia) infecting the invasive amphipod *Dikerogammarus villosus*: a potential emerging disease in European rivers. Parasitology 137(2):191–204. https://doi.org/10.1017/S0031182009991119

Park E, Poulin R (2021) Revisiting the phylogeny of Microsporidia. Int J Parasitol 51(10):855–864. https://doi.org/10.1016/j.ijpara.2021.02.005

Park E, Jorge F, Poulin R (2020) Shared geographic histories and dispersal contribute to congruent phylogenies between amphipods and their microsporidian parasites at regional and global scales. Mol Ecol 29(17):3330–3345. https://doi.org/10.1111/mec.15562

Patra S, Hartigan A, Morris DJ et al (2017) Description and experimental transmission of *Tetracapsuloides vermiformis* n. sp. (Cnidaria: Myxozoa) and guidelines for describing malacosporean species including reinstatement of *Buddenbrockia bryozoides* n. comb. (syn. *Tetracapsula bryozoides*). Parasitology 144:497–511. https://doi.org/10.1017/S0031182016001931

Pélabon C, Borg ÅA, Bjelvenmark J et al (2005) Do microsporidian parasites affect courtship in two-spotted gobies? Mar Biol 148(1):189–196. https://doi.org/10.1007/s00227-005-0056-8

Petkova I, Abbey-Lee RN, Lovlie H (2018) Parasite infection and host personality: *Glugea*-infected three-spined sticklebacks are more social. Behav Ecol Sociobiol 72(11):173. https://doi.org/10.1007/s00265-018-2586-3

Phelps NB, Goodwin AE (2008) Vertical transmission of *Ovipleistophora ovariae* (Microspora) within the eggs of the golden shiner. J Aquat Anim Health 20(1):45–53. https://doi.org/10.1577/H07-029.1

Prati S, Grabner DS, Pfeifer SM et al (2022) Generalist parasites persist in degraded environments: a lesson learned from microsporidian diversity in amphipods. Parasitology 149(7):973–982. https://doi.org/10.1017/S0031182022000452

Prati S, Enss J, Grabner DS et al (2023) Possible seasonal and diurnal modulation of *Gammarus pulex* (Crustacea, Amphipoda) drift by microsporidian parasites. Sci Rep 13(1):9474. https://doi.org/10.1038/s41598-023-36630-2

Prunescu CC, Prunescu P, Pucek Z et al (2007) The first finding of myxosporean development from plasmodia to spores in terrestrial mammals: *Soricimyxum fegati* gen. et sp. n. (Myxozoa) from *Sorex araneus* (Soricomorpha). Folia Parasitol 54:159–164

Quiles A, Bacela-Spychalska K, Teixeira M et al (2019) Microsporidian infections in the species complex *Gammarus roeselii* (Amphipoda) over its geographical range: evidence for both host-parasite co-diversification and recent host shifts. Parasit Vectors 12(1):327. https://doi.org/10.1186/s13071-019-3571-z

Quiles A, Wattier RA, Bacela-Spychalska K et al (2020) *Dictyocoela* Microsporidia diversity and co-diversification with their host, a gammarid species complex (Crustacea, Amphipoda) with an old history of divergence and high endemic diversity. BMC Evol Biol 20(1):149. https://doi.org/10.1186/s12862-020-01719-z

Rangel LF, Santos MJ, Rocha S (2023) Synopsis of the species of *Henneguya* Thélohan, 1892 (Cnidaria: Myxosporea: Myxobolidae) described since 2012. Syst Parasitol 100:291–305. https://doi.org/10.1007/s11230-023-10088-2

Rocha S (2023) Synopsis of the aurantiactinomyxon collective group (Cnidaria, Myxozoa), with a discussion on the validity of morphotype definition and demise of guyenotia. Syst Parasitol 100:307–323. https://doi.org/10.1007/s11230-023-10089-1

Rodgers-Gray TP, Smith JE, Ashcroft AE et al (2004) Mechanisms of parasite-induced sex reversal in *Gammarus duebeni*. Int J Parasitol 34(6):747–753. https://doi.org/10.1016/j.ijpara.2004.01.005

Schneider R, Prati S, Grabner D et al (2022) First report of microsporidians in the non-native shrimp *Neocaridina davidi* from a temperate European stream. Dis Aquat Organ 150:125–130. https://doi.org/10.3354/dao03681

Scholz F, Fringuelli E, Bolton-Warberg M et al (2017) First record of *Tetramicra brevifilum* in lumpfish (*Cyclopterus lumpus*, L.). J Fish Dis 40(6):757–771. https://doi.org/10.1111/jfd.12554

Schuster CJ, Sanders JL, Couch C et al (2022) Recent advances with fish Microsporidia. In: Weiss LM, Reinke AW (eds) Microsporidia. Experientia supplementum, vol 114. Springer, Cham, pp 285–317. https://doi.org/10.1007/978-3-030-93306-7_11

Sene A, Ba CT, Marchand B et al (1997) Ultrastructure of *Unikaryon nomimoscolexi* n. sp. (Microsporida, Unikaryonidae), a parasite of *Nomimoscolex* sp. (Cestoda, Proteocephalidea) from the gut of *Clarotes laticeps* (Pisces, Teleostei, Bagridae). Dis Aquat Organ 29:35–40. https://doi.org/10.3354/dao029035

Siddall ME, Martin DS, Bridge D et al (1995) The demise of a phylum of protists: phylogeny of Myxozoa and other parasitic Cnidaria. J Parasitol 81:961–967. https://doi.org/10.2307/3284049

Slodkowicz-Kowalska A, Graczyk TK, Tamang L et al (2006) Microsporidian species known to infect humans are present in aquatic birds: implications for transmission via water? Appl Environ Microbiol 72(7):4540–4544. https://doi.org/10.1128/AEM.02503-05

Slothouber Galbreath JG, Smith JE, Terry RS et al (2004) Invasion success of *Fibrillanosema crangonycis*, n.sp., n.g.: a novel vertically transmitted microsporidian parasite from the invasive amphipod host *Crangonyx pseudogracilis*. Int J Parasitol 34(2):235–244. https://doi.org/10.1016/j.ijpara.2003.10.009

Smith JE (2009) The ecology and evolution of microsporidian parasites. Parasitology 136(14):1901–1914. https://doi.org/10.1017/S0031182009991818

Smothers JF, van Dohlen CD, Smith LH et al (1994) Molecular evidence that the myxozoan protists are metazoans. Science 265:1719–1721. https://doi.org/10.1126/science.8085160

Sokolova YY, Senderskiy IV, Tokarev YS (2016) Microsporidia *Alfvenia sibirica* sp. n. and *Agglomerata cladocera* (Pfeiffer) 1895, from Siberian microcrustaceans and phylogenetic relationships within the "Aquatic outgroup" lineage of freshwater Microsporidia. J Invertebr Pathol 136:81–91. https://doi.org/10.1016/j.jip.2016.03.009

Sokolova YY, Overstreet RM, Heard RW et al (2021) Two new species of *Unikaryon* (Microsporidia) hyperparasitic in microphallid metacercariae (Digenea) from Florida intertidal crabs. J Invertebr Pathol 182:107582. https://doi.org/10.1016/j.jip.2021.107582

Solter LF (2006) Transmission as a predictor of ecological host specificity with a focus on vertical transmission of Microsporidia. J Invertebr Pathol 92(3):132–140. https://doi.org/10.1016/j.jip.2006.03.008

Sprague V (1964) *Nosema dollfusi* n. sp. (Microsporidia, Nosematidae), a hyperparasite of *Bucephalus cuculus* in *Crassostrea virginica*. J Protozool 11(3):381–385. https://doi.org/10.1111/j.1550-7408.1964.tb01767.x

Stentiford GD, Dunn AM (2014) Microsporidia in aquatic invertebrates. In: Weiss LM, Becnel J (eds) Microsporidia: pathogens of opportunity. Wiley, Chichester, pp 579–604. https://doi.org/10.1002/9781118395264.ch23

Stentiford GD, Bateman KS, Feist SW et al (2013a) Plastic parasites: extreme dimorphism creates a taxonomic conundrum in the phylum Microsporidia. Int J Parasitol 43(5):339–352. https://doi.org/10.1016/j.ijpara.2012.11.010

Stentiford GD, Feist SW, Stone DM et al (2013b) Microsporidia: diverse, dynamic, and emergent pathogens in aquatic systems. Trends Parasitol 29(11):567–578. https://doi.org/10.1016/j.pt.2013.08.005

Stentiford GD, Becnel J, Weiss LM et al (2016) Microsporidia—Emergent pathogens in the global food chain. Trends Parasitol 32(4):336–348. https://doi.org/10.1016/j.pt.2015.12.004

Stentiford GD, Ramilo A, Abollo E et al (2017) *Hyperspora aquatica* n.gn., n.sp. (Microsporidia), hyperparasitic in *Marteilia cochillia* (Paramyxida), is closely related to crustacean-infecting microspordian taxa. Parasitology 144(2):186–199. https://doi.org/10.1017/S0031182016001633

Stentiford GD, Ross S, Minardi D et al (2018) Evidence for trophic transfer of *Inodosporus octospora* and *Ovipleistophora arlo* n. sp. (Microsporidia) between crustacean and fish hosts. Parasitology 145(8):1105–1117. https://doi.org/10.1017/S0031182017002256

Stentiford GD, Bass D, Williams BAP (2019) Ultimate opportunists—The emergent *Enterocytozoon* group Microsporidia. PLoS Pathog 15(5):e1007668. https://doi.org/10.1371/journal.ppat.1007668

Štolc A (1899) Actinomyxidies, nouveau groupe de Mésozoaires parent des Myxosporidies. Bul Int Acad Sci Bohème 22:1–12

Stratton CE, Moler P, Allain TW et al (2022) The plot thickens: *Ovipleistophora diplostomuri* infects two additional species of Florida crayfish. J Invertebr Pathol 191:107766. https://doi.org/10.1016/j.jip.2022.107766

Terry RS, Smith JE, Dunn AM (1998) Impact of a novel, feminising microsporidium on its crustacean host. J Eukaryot Microbiol 45(5):497–501. https://doi.org/10.1111/j.1550-7408.1998.tb05106.x

Terry RS, Dunn AM, Smith JE (1999a) Segregation of a microsporidian parasite during host cell mitosis. Parasitology 118(Pt 1):43–48. https://doi.org/10.1017/s0031182098003540

Terry RS, Smith JE, Bouchon D et al (1999b) Ultrastructural characterisation and molecular taxonomic identification of *Nosema granulosis* n. sp., a transovarial transmitted feminising (TTF). Microsporidium. J Eukaryot Microbiol 46(5):492–499. https://doi.org/10.1111/j.1550-7408.1999.tb06066.x

Terry RS, Smith JE, Sharpe RG et al (2004) Widespread vertical transmission and associated host sex-ratio distortion within the eukaryotic phylum Microspora. Proc Biol Sci 271(1550):1783–1789. https://doi.org/10.1098/rspb.2004.2793

Toguebaye BS, Quilichini Y, Diagne PM et al (2014) Ultrastructure and development of *Nosema podo-*

cotyloidis n. sp. (Microsporidia), a hyperparasite of *Podocotyloides magnatestis* (Trematoda), a parasite of *Parapristipoma octolineatum* (Teleostei). Parasite 21:44. https://doi.org/10.1051/parasite/2014044

Trzebny A, Mizera J, Dabert M (2023) Microsporidians (Microsporidia) parasitic on mosquitoes (Culicidae) in central Europe are often multi-host species. J Invertebr Pathol 197:107873. https://doi.org/10.1016/j.jip.2022.107873

Vávra J, Lukes J (2013) Microsporidia and 'the art of living together'. Adv Parasitol 82:253–319. https://doi.org/10.1016/B978-0-12-407706-5.00004-6

Vavra J, Fiala I, Krylova P et al (2018) Molecular and structural assessment of Microsporidia infecting daphnids: The "obtusa-like" Microsporidia, a branch of the monophyletic Agglomeratidae clade, with the establishment of a new genus *Conglomerata*. J Invertebr Pathol 159:95–104. https://doi.org/10.1016/j.jip.2018.10.003

Vossbrinck CR, Debrunner-Vossbrinck BA (2005) Molecular phylogeny of the Microsporidia: Ecological, ultrastructural and taxonomic considerations. Folia Parasitol 52(1–2):131–142. https://doi.org/10.14411/fp.2005.017

Vossbrinck CR, Debrunner-Vossbrinck BA, Weiss LM (2014) Phylogeny of the Microsporidia. In: Weiss LM, Becnel JJ (eds) Microsporidia: pathogens of opportunity, 1st edn. Wiley-Blackwell, Hoboken, pp 203–220. https://doi.org/10.1002/9781118395264.ch6

Wadi L, Reinke AW (2020) Evolution of Microsporidia: An extremely successful group of eukaryotic intracellular parasites. PLoS Pathog 16(2):e1008276. https://doi.org/10.1371/journal.ppat.1008276

Wang-Peng S, Zheng X, Jia WT et al (2018) Horizontal transmission of *Paranosema locustae* (Microsporidia) in grasshopper populations via predatory natural enemies. Pest Manag Sci 74(11):2589–2593. https://doi.org/10.1002/ps.5047

Ward AJW, Duff AJ, Krause J et al (2005) Shoaling behaviour of sticklebacks infected with the microsporidian parasite, *Glugea anomala*. Environ Biol Fishes 72(2):155–160. https://doi.org/10.1007/s10641-004-9078-1

Wei J, Fei Z, Pan G et al (2022) Current therapy and therapeutic targets for microsporidiosis. Front Microbiol 13:835390. https://doi.org/10.3389/fmicb.2022.835390

Weidner E, Canning EU, Rutledge CR et al (1999) Mosquito (Diptera: Culicidae) host compatibility and vector competency for the human myositic parasite *Trachipleistophora hominis* (phylum Microspora). J Med Entomol 36(4):522–525. https://doi.org/10.1093/jmedent/36.4.522

Weill R (1938) L'interprétation des Cnidosporidies et la valeur taxonomique de leur cnidome. Leur cycle comparé à la phase larvaire des Narcomeduses cuninides. Trav Stat Zool Wimer 13:727–744

Wilkinson TJ, Rock J, Whiteley NM et al (2011) Genetic diversity of the feminising microsporidian parasite *Dictyocoela*: new insights into host-specificity, sex and phylogeography. Int J Parasitol 41(9):959–966. https://doi.org/10.1016/j.ijpara.2011.04.002

Williams BAP (2009) Unique physiology of host-parasite interactions in Microsporidia infections. Cell Microbiol 11(11):1551–1560. https://doi.org/10.1111/j.1462-5822.2009.01362.x

Williams BAP, Hirt RP, Lucocq JM et al (2002) A mitochondrial remnant in the microsporidian *Trachipleistophora hominis*. Nature 418(6900):865–869. https://doi.org/10.1038/nature00949

Williams BAP, Williams TA, Trew J (2022) Comparative genomics of Microsporidia. In: Weiss LMR, A. W. (eds) Microsporidia, Experientia supplementum, vol 114. Springer, Cham, pp 43–69. https://doi.org/10.1007/978-3-030-93306-7_2

Wolf K, Markiw ME (1984) Biology contravenes taxonomy in the Myxozoa: new discoveries show alternation of invertebrate and vertebrate hosts. Science 225:1449–1452. https://doi.org/10.1126/science.225.4669.1449

Yahalomi D, Haddas-Sasson M, Rubinstein ND et al (2017) The multipartite mitochondrial genome of *Enteromyxum leei* (Myxozoa): Eight fast-evolving megacircles. Mol Biol Evol 34:1551–1556. https://doi.org/10.1093/molbev/msx072

Yahalomi D, Atkinson SD, Neuhof M et al (2020) A cnidarian parasite of salmon (Myxozoa: *Henneguya*) lacks a mitochondrial genome. Proc Natl Acad Sci U S A 117:5358–5363. https://doi.org/10.1073/pnas.1909907117

Yanagida T, Asai N, Yamamoto M et al (2022) Molecular and morphological description of a novel microsporidian *Inodosporus fujiokai* n. sp. infecting both salmonid fish and freshwater prawns. Parasitology 150(1):1–14. https://doi.org/10.1017/S003118202200141X

Yang Y, Xiong J, Zhou Z et al (2014) The genome of the myxosporean *Thelohanellus kitauei* shows adaptations to nutrient acquisition within its fish host. Genome Biol Evol 6:3182–3198. https://doi.org/10.1093/gbe/evu247

Zatti SA, Araújo BL, Adriano EA et al (2023) A new freshwater *Ceratomyxa* species (Myxozoa: Ceratomyxidae) parasitising a sciaenid fish from the Amazon Basin, Brazil. Parasitol Int 97:102796. https://doi.org/10.1016/j.parint.2023.102796

Open Access This chapter is licensed under the terms of the Creative Commons Attribution-NonCommercial-NoDerivatives 4.0 International License (http://creativecommons.org/licenses/by-nc-nd/4.0/), which permits any non-commercial use, sharing, distribution and reproduction in any medium or format, as long as you give appropriate credit to the original author(s) and the source, provide a link to the Creative Commons license and indicate if you modified the licensed material. You do not have permission under this license to share adapted material derived from this chapter or parts of it.

The images or other third party material in this chapter are included in the chapter's Creative Commons license, unless indicated otherwise in a credit line to the material. If material is not included in the chapter's Creative Commons license and your intended use is not permitted by statutory regulation or exceeds the permitted use, you will need to obtain permission directly from the copyright holder.

Ecology and Life Cycles of Aquatic Pathogenic Fungi

Ché Weldon

Abstract

Aquatic pathogenic fungi play a pivotal role in aquatic ecosystems due to their diversity, abundance and impacts on the other organisms and the ecosystem. Taxonomic surveys suggest a vast, largely undiscovered fungal diversity. Various well-documented examples of disease-inducing species exist within the Ascomycota, Oomycetes, Chytridiomycetes and other fungal-like organisms, with impacts upon diverse hosts, including algae, fish, amphibians, reptiles, crustaceans, molluscs and corals. The non-photosynthetic nature of fungi underscores their ecological roles as saprophytes or parasites, often with intricate interactions among fungal species and their hosts. Aquatic pathogenic fungi are crucial components of aquatic environments, including in aquacultural systems, adapting their life cycles to overcome challenges presented by water-based ecosystems. Life cycles involve multiple stages of development that, regardless of variation, have an acute ability to disperse in water and locate their hosts. The ability to produce spores plays a vital role in the dispersal, survival and infection capabilities of these fungi. Environmental factors strongly influence the prevalence and virulence of these fungi by directly or indirectly affecting their growth, reproduction and pathogenicity in aquatic ecosystems. These factors, varying with fungal species, aquatic ecosystems and host demographics, are significantly impacted by human activities like aquaculture, habitat alteration, water pollution and introduction of non-native species, which can create conducive environments for fungal proliferation or disrupt natural ecosystems, thereby contributing to disease outbreaks.

4.1 Introduction

Life in modern human civilisation is undeniably intertwined with the benefits provided by some familiar fungal species. Button mushrooms, *Agaricus bisporus*, are an excellent source of nutrition and flavour in a variety of dishes. The fermentation process induced by brewer's yeast, *Saccharomyces cerevisiae*, is a vital component in baking and in the brewing of alcoholic beverages, while penicillin produced by *Penicillium chrysogenum* has forever changed the way modern medicine fights infectious diseases. The diversity in shape, size and structure shown by the fungal kingdom is truly astounding and one frequently must examine their cellular design to

C. Weldon (✉)
Unit for Environmental Sciences and Management, North-West University, Potchefstroom, South Africa
e-mail: che.weldon@nwu.ac.za

uncover shared morphological characteristics. Fungi are not unlike other eukaryotes in that their cells share similar kinds of organelles, but they are defined by certain morphological structures that are unique to the kingdom. These include organelles involved in the growth of hyphal tips and plugs that protect wounded colonies from haemorrhaging cytoplasm. Notwithstanding these subcellular structures, they are better known for the presence of chitin and other polymers in the cell wall, and the formation of microscopic spores (Watkinson 2016).

As a group of organisms, fungi evolved and diversified into a vast kingdom of approximately 120,000 described species. Remarkably, estimates based on modern statistical and phylogenetic interpretation of environmental sequence data place the actual number of fungi at a minimum of 2.2 million species (Hawksworth and Lücking 2017). They are a major component of the microbiome in almost every habitat on earth. Aquatic fungi do not conform to a particular morphological or phylogenetic group but are rather categorised as fungi dependent on aquatic habitats for at least part of their life cycle. These fungi are abundant in all major aquatic ecosystems including freshwater, marine and brackish environments. Six major phylogenetic groups of aquatic fungi are currently recognised and are often portrayed with their closest fungi-like organisms in representative trees (Fig. 4.1).

The following list of taxonomic groups is by no means comprehensive, but contains the major groups of fungi responsible for causing diseases in aquatic organisms:

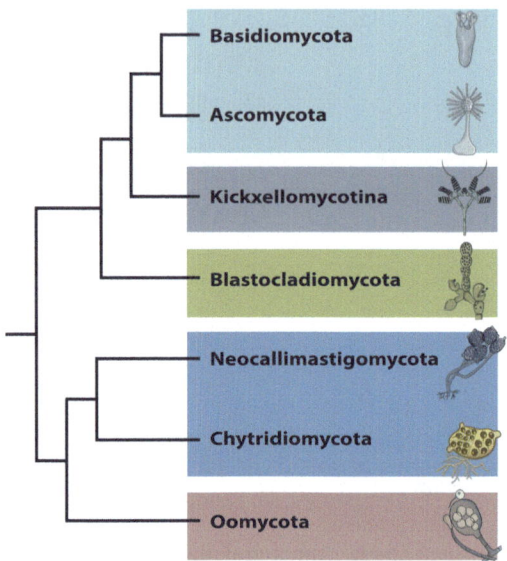

Fig. 4.1 Basic phylogenetic relationship between major phyla of aquatic fungi (adapted from Moore et al. 2020)

(i) *Oomycetes*. These are filamentous, water-dwelling fungus-like microorganisms. Despite their appearance and lifestyle, they are not true fungi but belong to the kingdom Stramenopila (Alexopoulos et al. 1996). Oomycetes include several well-known species that are significant aquatic pathogens including *Saprolegnia* spp. and *Aphanomyces* spp. *Aphanomyces invadans* causes epizootic ulcerative syndrome (EUS) in freshwater and brackish water fishes, affecting more than 100 fish species (Kamilya and Baruah 2014).

(ii) *Chytrids*. A group of fungi characterised by their simple structure and unique life cycle, including a motile, flagellated spore called a zoospore (Rasconi et al. 2012). Several chytrid species are known to be aquatic pathogens, causing diseases in various aquatic organisms, in particular algae and amphibians. The chytrid *Batrachochytrium dendrobatidis*, for example, is responsible for the lethal amphibian disease chytridiomycosis, which has been implicated in devastating amphibian population declines and extinctions worldwide (Fisher and Garner 2020).

(iii) *Ascomycota*. One of the largest and most diverse groups of fungi, with many aquatic species included. They produce non-motile sexual spores referred to as ascospores that are stored in sac-like structures called asci (singular ascus). Some aquatic pathogenic fungi of this phylum can cause disease in both marine and freshwater organisms. For instance, species belonging to the genus *Fusarium* and related genera can infect fish and other aquatic animals (Hassan et al. 2020). The significance of *Aspergillus* spp.

in the development of fish mycosis cannot be overstated, as infections in freshwater fish have been on the rise in recent years, leading to the emergence of piscine aspergillosis, a devastating disease characterised by tissue-penetrating hyphae that develop beyond the original infection site.

(iv) *Microsporidia.* Obligate intracellular fungal-like parasites known for their small size and unicellular nature (see Chap. 3). About 20 microsporidian genera infect fishes, 50 genera infect aquatic arthropods and at least 21 genera infect aquatic non-arthropod invertebrates, protists and hyperparasites of aquatic hosts (Vávra and Lukeš 2013). It is important to note that the taxonomy and classification of fungi are continuously evolving with ongoing research (Hawksworth and Lücking 2017). As our understanding of these organisms improves, new taxa may be identified, and the relationships between different groups may be revised. Nonetheless, these four major taxonomic groups represent some of the primary fungi responsible for causing diseases in aquatic ecosystems.

Fungi usually occupy ecological niches as saprophytes in dead organic matter and soil or as parasites of animals, plants and humans (Ziarati et al. 2022). It is important to consider that fungus–animal interactions can, however, range from parasitic to symbiotic and fungi in turn are a food source for various aquatic invertebrates. Aquatic pathogenic fungi are those species that can be grouped on the shared characteristic of causing diseases in various organisms that inhabit aquatic environments. These fungi have evolved to thrive and interact with aquatic ecosystems, where they can infect a wide range of hosts. They are specifically adapted to infect and parasitise aquatic plants, fish, crustaceans, molluscs and microorganisms in these ecosystems (Johnson et al. 2009; Frenken et al. 2017; Grossart et al. 2019). Similar to other fungi, aquatic pathogenic fungi produce spores as a means of reproduction and survival. Unlike terrestrial fungi, however, aquatic fungi have adapted certain life cycle strategies to overcome the unique challenges presented by water-based environments. Notably, these fungi have adapted specialised spore types that can effectively disperse in aquatic environments, allowing them to infect new hosts submerged in water or spread throughout the water body. Moreover, aquatic pathogenic fungi have evolved various mechanisms to infect their hosts, such as direct penetration of host tissues or through spore germination on the host surface. Many of these pathogens exhibit some degree of host specificity, some with high host specificity infecting only certain species or groups of organisms, while others display low host specificity by infecting a wide variety of aquatic hosts. Once inside the host, they colonise and proliferate, leading to disease development. In extreme cases these epidemics can have significant ecological impacts on aquatic ecosystems. These disease outbreaks may lead to population declines or affect food webs by disrupting the balance between predators and prey. Similarly, fungal disease outbreaks in aquaculture can lead to huge economic losses, accounting for the second most serious cause of loss in aquaculture after bacterial disease outbreaks (Gozlan et al. 2014; see also Chap. 23).

In this chapter, the life cycles of selected species from various taxonomic groups and host types are examined. This section is followed by a case study on risks involved with trade-mediated spread of the amphibian chytrid fungus. Next, an overview on the major ecological impacts that aquatic pathogenic fungi have on humans, other organisms and ecosystems is presented. The chapter is concluded with a summary of the main arguments.

4.2 Life Cycles of Selected Aquatic Pathogenic Fungi

The life cycles of aquatic fungi may vary depending on the specific fungal species and the conditions of their habitat, and the nature of their association with other organisms. Some pathogenic aquatic fungi may display complex life cycles involving multiple stages of development, while for others, the life cycle may be much simpler. However, they are all characterised by a remarkable ability to disperse in water and locate their host organisms. Spore production is a crucial aspect of their life cycle, as spores play a vital role in the dispersal, survival and infection of these fungi. The mechanisms of spore production can vary among different groups of aquatic pathogenic fungi. Oomycetes, including aquatic pathogenic taxa such as species of *Saprolegnia* and *Aphanomyces*, produce spores through a process called sporulation. The main types of spores produced by oomycetes are zoospores and oospores. Chytrids produce zoospores that are motile and possess a single whip-like flagellum. The zoospores are formed within specialised structures called zoosporangia. When the zoosporangia mature, they release the motile zoospores into the surrounding water. The zoospores actively swim and seek out suitable hosts for infection. Some other groups of aquatic pathogenic fungi, such as certain species of Ascomycota and Basidiomycota, may produce asexual spores (conidia) or sexual spores (ascospores or basidiospores) through specialised reproductive structures. Spore dispersal contributes to disease transmission and the persistence of infections in aquatic environments. Many aquatic pathogenic fungi release their spores directly into the water. These spores are then dispersed by water currents and movements, which carry them over varying distances where they can encounter and infect susceptible hosts. In the case of oomycetes and chytrids, the motile zoospores play an active role in dispersal, although water currents could transport the motile zoospores to new locations. In some cases, certain aquatic pathogenic fungi, especially those belonging to Ascomycota and Basidiomycota, produce aerial spores that can be carried by wind currents. When released from the sporulating structures, such as conidiophores, the spores are lifted into the air, but can settle on water surfaces or be inhaled by aquatic hosts and result in new infections.

Demonstrated here are the life cycles of a selection of fungal pathogens known for their ability to cause disease epidemics in their respective host species.

Aphanomyces *Aphanomyces* is a genus of aquatic oomycetes (water moulds) that includes several species that are well-known aquatic pathogens. The life cycle of *Aphanomyces* is completed when the zoospores released into the environment find and infect new hosts, germinate and produce sporangia and spores that, in turn, lead to the release of more zoospores. This life cycle allows the fungus to persist and spread in aquatic environments, contributing to its ability to cause diseases in various aquatic organisms, such as epizootic ulcerative syndrome in fish caused by *A. invadans* and crayfish plague caused by *A. astaci* (Kumar et al. 2020; Oidtmann et al. 2002).

The life cycle of *A. invadans* (Fig. 4.2) begins with the release of motile zoospores (1) from mature sporangia into the water. These zoospores are specialised structures that contain the genetic material of the fungus, protected by a thick wall (not true cell walls). These zoospores are microscopic, flagellated cells that possess two flagella. These flagella enable them to actively move within the aquatic environment. The zoospores are attracted to chemical signals released by potential hosts, a process called chemotaxis. This chemotactic behaviour guides them towards suitable hosts for infection. Once the zoospores locate a susceptible host, they attach to the host surface and encyst. The cyst (2) is a protective structure that surrounds the zoospore and helps it to anchor to the host's tissues. After encystment, the zoospore loses its flagella. The cyst germinates, and the infective hyphae (3) emerge from it. The hyphae penetrate the host's tissues and begin to grow inside the host's cells. The fungus starts to obtain nutrients from the host and proliferates, causing damage to the host's cells and tissues. As the hyphae grow and develop, they eventually form specialised structures called sporangia (4) within the host's tissues. The sporangia are sac-like structures that contain numerous spores. The mature sporangia burst (5), releasing the zoospores into the aquatic environment. These newly released motile zoospores then disperse in search of new hosts, continuing the life cycle of *Aphanomyces*.

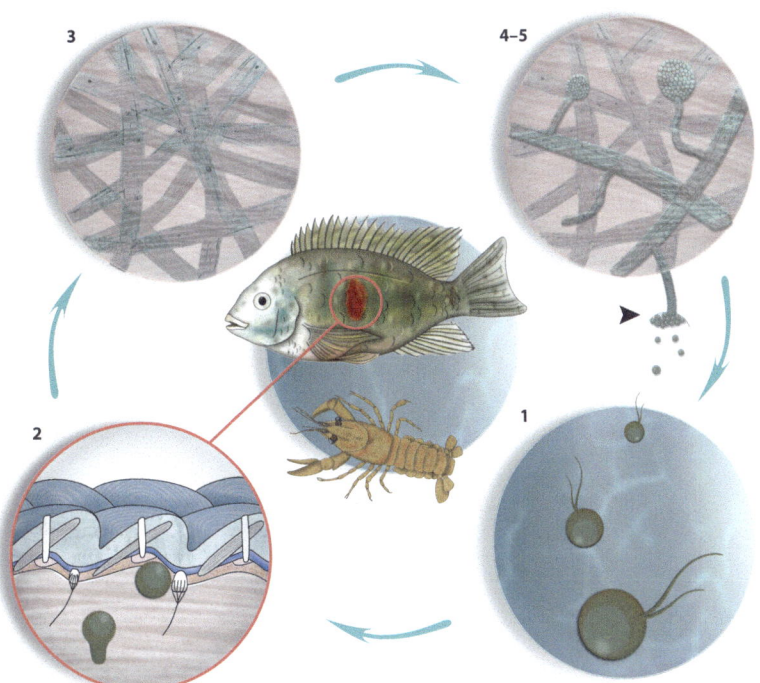

Fig. 4.2 Life cycle of *Aphanomyces*, depicting key stages in its reproductive and developmental processes in fish (caused by *Aphanomyces invadans*) and crayfish (caused by *Aphanomyces astaci*): (1) motile zoospore, (2) encysted zoospore in host tissue, (3) hyphae inside host cells, (4) sporangia containing zoospores, (5) zoospore release from mature sporangia

Aspergillus sydowii Sea fan corals (particularly species of *Gorgonia*) may become infected with *Aspergillus sydowii*, an otherwise predominantly soil-dwelling saprotroph (Alker et al. 2001). This disease is characterised by lesions, galls and a purpling of the coral tissue that can lead to death and opportunistic epidemics driven by compromised host immunity and elevated environmental temperatures.

The life cycle of *A. sydowii* (Fig. 4.3) in sea fan corals begins when spores (1) are released into the environment, likely from terrestrial sources or marine debris. Once in the marine environment, the spores encounter the sea fan corals. They can enter the coral's tissue through natural openings, such as feeding pores or injuries, or by direct penetration of the coral's surface (2). Once inside the coral, the spores germinate and start to grow thread-like structures called hyphae (3). As the fungal colony grows and spreads within the coral, it causes damage to the coral's tissue and disrupts its normal physiological processes. This can result in disease symptoms, such as tissue discolouration (4), lesions and tissue necrosis. The fungus then produces asexual spores, known as conidia, within specialised structures called conidiophores (5). These conidia (6) are then released from the infected coral into the surrounding seawater, completing the life cycle and potentially spreading the fungus to other corals.

Chytrid Fungus The two species of amphibian chytrid fungi, *Batrachochytrium dendrobatidis* and *B. salamandrivorans*, are the only members of the Chytridiomycota that infect vertebrates. Their zoospore driven infection strategy targets keratinised skin of amphibians and mouthparts of their larvae and may result in the disease chytridiomycosis. This epidermal ulcerative disease is responsible for significant declines and extinctions of amphibian populations worldwide, making batrachochytrids some the most devastating of all pathogens affecting vertebrates (Fisher and Garner 2020).

The life cycle of *Batrachochytrium* (Fig. 4.4) begins with the release of motile zoospores (1). Chytrid zoospores have a single flagellum and are highly motile in water. They can actively swim through aquatic environments, searching for amphibian hosts. Once they encounter

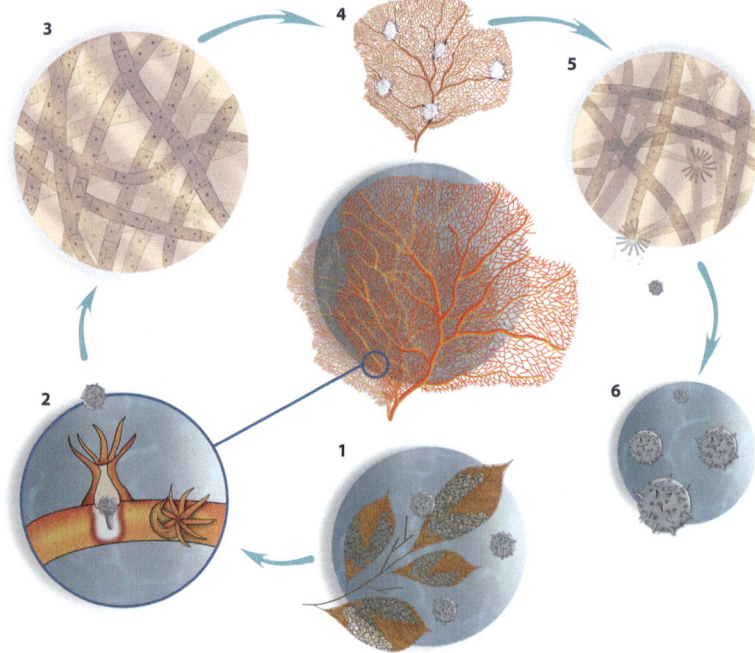

Fig. 4.3 Life cycle of *Aspergillus sydowii*, depicting key stages in its reproductive and developmental processes in sea fan corals: (1) conidiospores released from marine debris, (2) spores penetrate coral surface and germinate, (3) thread-like hyphae in host tissue, (4) tissue discolouration and necrosis in coral, (5) conidiophores containing spores, (6) mature conidia

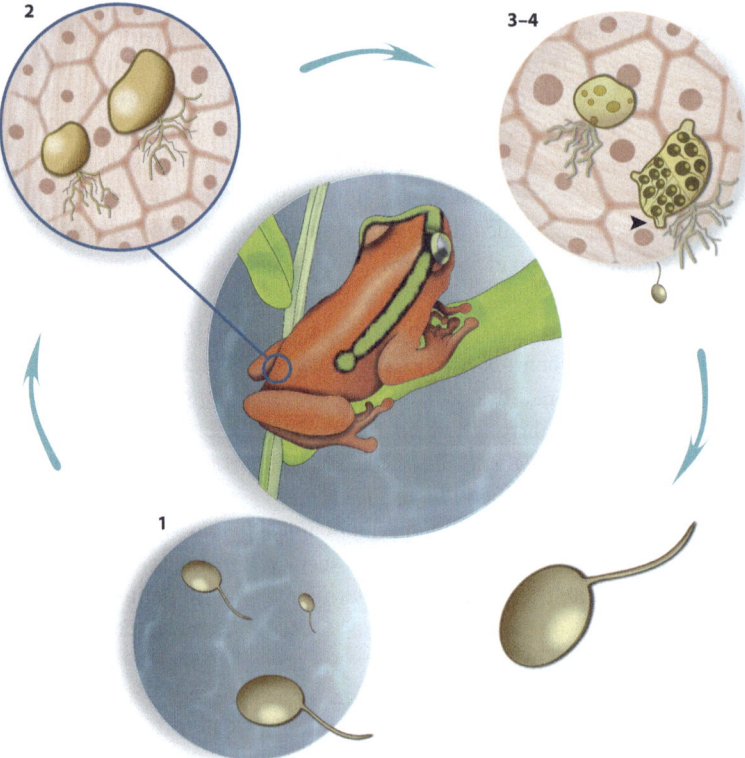

Fig. 4.4 Life cycle of *Batrachochytrium dendrobatidis*, depicting key stages in its reproductive and developmental processes in frogs: (1) motile zoospores, (2) immature sporangia in epidermal skin of host, (3) mature sporangia containing zoospores, (4) discharge papilla (arrow) where zoospores are released

amphibian skin, they attach to the surface and start to infect the epidermis of the host. The zoospores soon lose their flagella and transform into a non-motile, non-flagellated and intracellular form, the immature sporangium (2). This form of the fungus starts to grow inside the host's skin cells, develop rhizoids, and feed on the keratin-rich tissues. As the fungus grows, it leads to various pathological effects such as hyperkeratosis and hyperplasia, often fatally weakening the frog. As the infection progresses, the sporangia mature (3) containing the new generation of zoospores. Discharge papillae (4, arrow) develop through which the newly formed zoospores are released into the environment, continuing the life cycle.

Fusarium The life cycle of *Fusarium*, a genus of filamentous fungi, is complex and involves several stages. *Fusarium* species are widespread in diverse habitats, including soil, plants and water. Most *Fusarium* species are plant pathogens, causing significant damage to crops, while others can also cause infections in humans and animals. Of note are some members of the *Fusarium solani* species complex (e.g. *F. crassum, F. falciforme, F. keratoplasticum*) that are capable of infecting sea turtles as well as their eggs, causing fusariosis and leading to high mortality rates in this threatened group of reptiles (Gleason et al. 2020; Greeff-Laubscher and Jacobs 2022). They are also known to cause lethal fusariosis in sharks and stingrays (Crow et al. 1995; Fernando et al. 2015).

It has been proposed that these fungi grow on floating particles of plant tissues and other debris, which are transferred by wind, currents and turtles to the spawning sites of the turtles (Gleason et al. 2020). *Fusarium solani* (Fig. 4.5) can endure adverse environmental conditions by surviving as chlamydospores (1, germ tube indicated with an arrow), which are thick-walled resting structures. When favourable conditions arise, such as suitable moisture levels and temperature, the chlamydospores germinate to produce hyphae (2). The hyphae of *F. solani* are large and septate, and typically colonise the surface of their chondrichthyan or reptilian hosts. The fungus then spreads within the host's tissues via specialised infection structures like mycelial coils (3) and starts to feed on the host's nutrients, causing disease development. Typical symptoms include yellowing and the formation of necrotic lesions; infected eggs develop yellowish-blue infection zones that eventually become necrotic lesions and ultimately kill the developing embryos. Under suitable environmental conditions, *F. solani* may produce conidia carried on conidiophores (4) with different types of spores, including large sickle-shaped (5) macroconidia with multiple septations, and smaller unicellular (6) microconidia. These spores serve as a means of reproduction and dissemination. Once mature, the conidia are released from the conidiophores and are carried by air, water, or other means, allowing the fungus to disperse and potentially infect new hosts.

Microsporidia Microsporidia is a group of intracellular parasitic fungi that has gone through a long taxonomic odyssey, known for their small size and unicellular nature and are presented in more detail in Chap. 3. The life cycle of Microsporidia (see Fig. 3.3 in Chap. 3) infecting crustaceans typically involves direct transmission through ingestion of spores. The intracellular nature of these parasites allows them to avoid host immune responses and makes them challenging to treat or control once an infection is established. Microsporidian infections in crustaceans can lead to various pathological effects, affecting their growth, reproduction and survival, and may have significant implications for aquaculture and wild crustacean populations (Stentiford et al. 2013). For more details on the biology of Microsporidia please see Chap. 3.

4.3 Risk of Trade-Mediated Spread of Fungi: Chytridiomycosis Case Study

Concern has been raised that the worldwide trade of animals has increased the potential for the translocation of diseases, which pose serious risks to animal health and even human health in

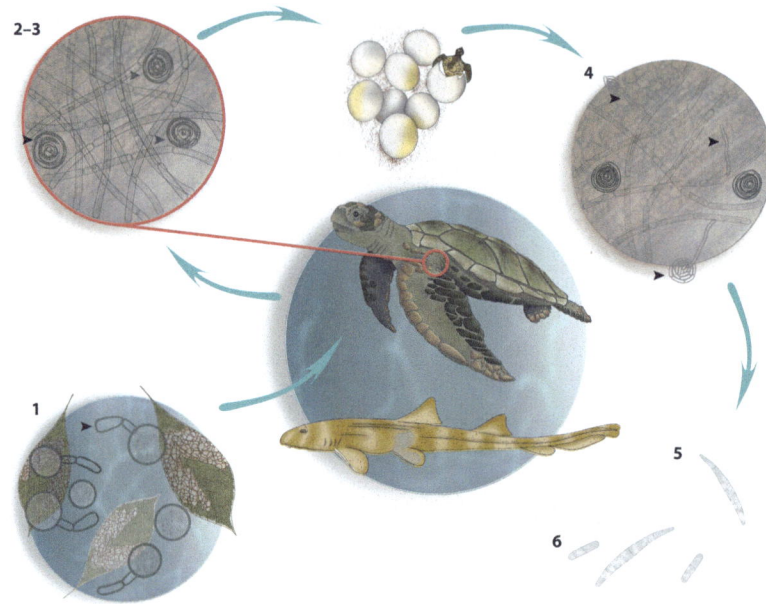

Fig. 4.5 *Fusarium solani* infects marine turtles and chondrichthyans. Life cycle of *F. solani* in marine turtles and their eggs is depicted by the following key stages in its reproductive and developmental processes: (1) clamydospores on decomposing plant material, indicating germination (arrow) under favourable conditions, (2) large, septate hyphae, (3) formation of distinct mycelial coils (arrows), (4) conidiophores, (5) macroconidia, (6) microconidia

the case of zoonotic diseases (Marano et al. 2007). For example, global exports from South Africa of the African clawed frog (*Xenopus laevis*), a known carrier of the amphibian chytrid fungus *Batrachochytrium dendrobatidis* (Fig. 4.6), reached 400,000 animals from 1940 to 1970. By 1970, *X. laevis* had become the world's most widely distributed amphibian, with South Africa supplying 48 countries with live colonies on all inhabited continents (Van Sittert and Measey 2016). There was a reduction in exports between 1998 and 2004, with a little over 10,000 frogs being exported annually; soon after, all exports of live *X. laevis* were terminated (Weldon et al. 2007). This highly adaptable amphibian established invasive populations on four continents as a consequence of its global trade (Tinsley and McCoid 1996; Fouquet and Measey 2006). Today, *B. dendrobatidis* is recognised globally for contributing to the decline of 501 amphibian species, of which 90 are now presumed extinct (Scheele et al. 2019). Import risk assessments are often used to evaluate the risks of disease introduction via international trade, and to consider consequences to animal health and the environment.

A risk analysis on the likelihood that *B. dendrobatidis* may have caused disease in amphibians acquired via the former trade in *X. laevis* identifies several major steps that pose variable risks to naïve amphibian populations in the introduced range (see also Chap. 18). Commercial suppliers of *X. laevis* visiting an infected site risked spreading *B. dendrobatidis* to multiple source sites through the re-use of unsterilised equipment, or to holding facilities (steps 1 and 2; Fig. 4.7). The next step in the trade pathway was the distribution of frogs to local and international users (step 3). Local destinations were more vulnerable to pathogen exposure if no prior quarantine and pathogen screening was conducted. Import facilities that operated as breeder-suppliers potentially spread the pathogen to end-users nationally and internationally (step 4). Customs and biosecurity practices in the respective importing countries could, at these stages, have determined whether the risk of exposure was eliminated, reduced, or remained unchanged. Finally, *B. dendrobatidis* could have spread from the end-users and importers to the environment (step 5). The risk of transmission to native species is increased when end-users lack relevant training, or due to negligence regarding biosecurity protocols and awareness. Countries that do not have animal health legislation in place to deal with infectious diseases of wildlife are more

4 Ecology and Life Cycles of Aquatic Pathogenic Fungi

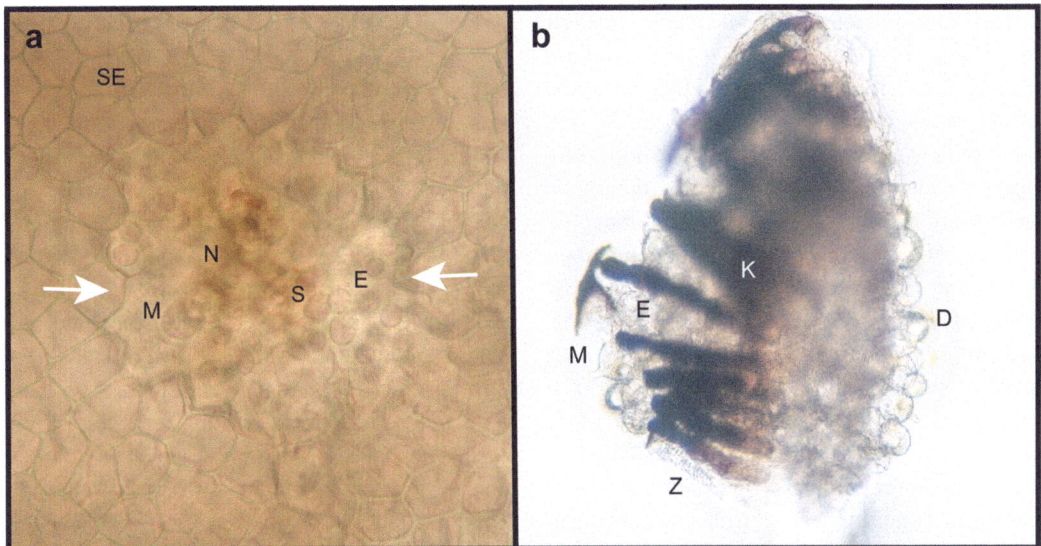

Fig. 4.6 *Batrachochytrium dendrobatidis* micrographs of (**a**) infected digit tip of *Cacosternum boettgeri*. E, empty sporangium; M, mature sporangium; N, necrotic tissue displaying hyperkeratotic region; SE, stratified squamous epithelium, arrows indicate extent of skin legion. (**b**) In vitro culture isolated from an *Amietia delalandii* tadpole. E, empty sporangium; D, developing sporangium; K, tadpole kerotodont; M, sporangium with discharge papillae; Z, zoospores

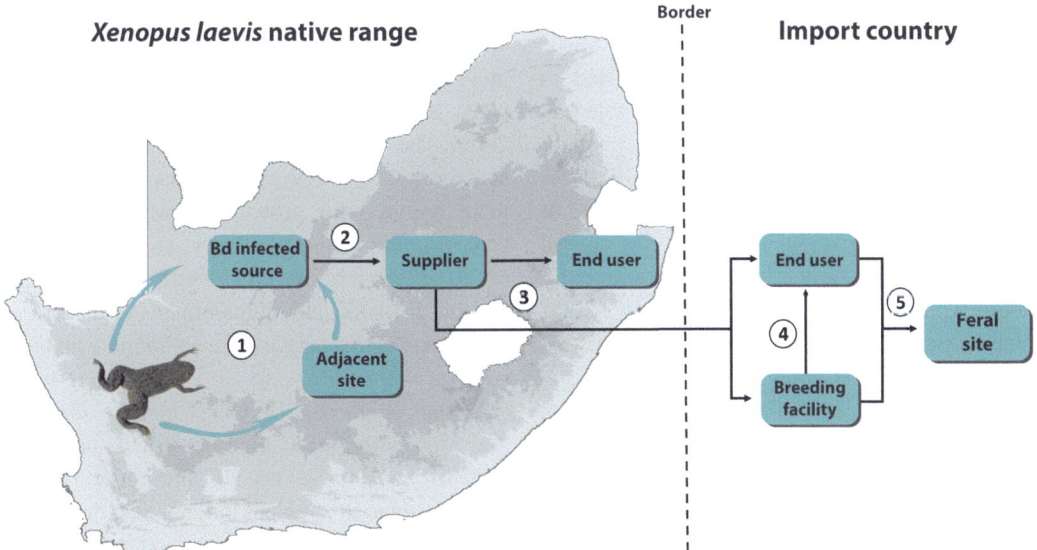

Fig. 4.7 A simplified representation of the African clawed frog trade pathway, indicating major linkages by which *Batrachochytrium dendrobatidis* can spread

likely to experience exposure from the environment (Pessier and Mendelson 2010).

Due to the epidemiological nature of a transmissible disease, the risk of an amphibian from an importing country acquiring chytridiomycosis because of the risk process (trade in vector host species) can be traced back to the initial incidence of the disease at the start of the risk process (the collection of wild hosts in their native range, in this case *X. laevis* from southern Africa). The potential risk that is present at the start of the pathway is only realised during the event of a susceptible host in the importing country becoming exposed to the pathogen. Concurrently, the level of risk when transmission is most certain to occur is not dependent on the variance of any preceding step in the dissemination pathway, but rather depends on host susceptibility, population density and other intrinsic characteristics (Murray et al. 2009). It is therefore difficult to predict the level of risk in specific case studies without considering the host–pathogen relationship and particular environmental conditions for the species in question. Thus, the consequences of exposure will occur irrespective of the variation in steps preceding exposure. A more informative measure of risk is the likelihood of transmission taking place, which is influenced by variation in the preceding steps of the release and exposure processes. The inputs that make up the release and exposure routes serve as an estimate of the risk that culminates in transmission to naïve hosts, and variation of these inputs can be used to devise a decision tree that results in variable risk outcomes.

A strict biosecurity protocol and early detection system are essential to control the spread of emerging fungal pathogens within the animal trade (Weldon and Fisher 2011). Such a protocol should function around controllable units, predicated on the steps that structure a trade pathway. The spread pathway could be interrupted if the pathogen is eradicated during a specific step, either by deliberate human intervention (e.g. early detection) or by strict biosecurity management (e.g. the application of disinfectants). The risk of spreading pathogens does not necessarily outweigh the contributions of the trade to scientific research and development (Garner et al. 2009). Instead, careful consideration of the likelihood of risks in the trade pathway can generate control measures to reduce the risks from occurring, while allowing for the advancement of knowledge and technologies through ethically sound animal science.

4.4 Ecological Role and Impact on Aquatic Ecosystems, Organisms and Humans

The ecological role of aquatic pathogenic fungi is multifaceted and an essential part of shaping aquatic ecosystems. These fungi are key components of the natural balance in aquatic environments that contribute to ecosystem dynamics in several ways. Essentially, fungal pathogens contribute to the natural balance of aquatic ecosystems by controlling the populations of certain organisms. The effects of fungal epidemics on freshwater algal populations and communities have been well documented for parasitic chytrids (Rasconi et al. 2011). These zoosporic fungi have been linked to suppression or retardation of phytoplankton blooms (Canter and Lund 1951), selective effects on species composition and successions (Donk and Ringelberg 1983) and steering the outcome of interspecific competition among subdominant species (Canter and Lund 1969). The sheer number of zoospores produced during fungal epidemics, comparable to the abundance of edible phytoplankton, implies that they are a major food source for zooplankton when such conditions prevail (Kagami et al. 2004). The cholesterol- and polyunsaturated fatty acids-rich zoospores provide essential nourishment required for growth and reproduction of heterotrophic nanoflagellates (Klein Breteler et al. 1999). Moreover, chytrids also infect diverse colonial and filamentous cyanobacteria including *Anabaena flosaquae*, *Gomphosphaeria* sp., or *Microcystis* sp. in freshwater lakes (Rasconi et al. 2009). Consequently, chytrids may modify the food web scheme in aquatic ecosystems, not only via transferring nutrients from inedible algae to zooplankton, but also due to the inherent nutritional value contained in their lipid globules (Kagami et al. 2007).

The impact of aquatic pathogenic fungi on prey species can have cascading effects on higher trophic levels. By affecting the abundance and distribution of certain organisms, these fungi can influence the structure of aquatic food webs and the interactions between predators and prey. Trophic transfer is demonstrated within a wide range of aquatic invertebrate phyla infected with microsporidians (Stentiford et al. 2013; see also Chap. 3). Evidence exists for Microsporidia-mediated host population densities and host population cycles of caddisflies (*Brachycentrus americanus*) (Kohler and Hoiland 2001). Moreover, infection of an amphipod (*Gammarus duebeni celticus*) with the microsporidian *Pleistophora mulleri* results in reduced predation on the isopod *Asellus aquaticus* and weakens its impact as a biological control agent of another invasive species, *Gammarus tigrinus* (MacNeil et al. 2003). Further up the food chain, the microsporidian *Thelohania contejeani* reduces predatory success of a crayfish species on amphipods (Hatcher and Dunn 2011). Thus, in natural ecosystems, pathogenic fungi act as regulators of population sizes and help maintain the balance of aquatic species.

Fig. 4.8 *Salamandra salamandra gallaica* in a moribund state (**a**), displaying skin haemorrhaging (arrows) and ulcerous lesions (**b**) associated with advanced *Batrachochytrium salamandrivorens* infection (photo: Jonas Virgo)

However, pathogenic fungi can also pose challenges to the health of aquatic organisms, jeopardising their survival and hindering conservation efforts, especially in cases where already-endangered aquatic species are susceptible to these pathogens. A tipping point in algae is reached when the growth of fungal parasites is enhanced and exceeds growth rates or resilience of their hosts resulting in epidemic events (Kagami et al. 2007). Aquatic pathogenic fungi can cause devastating disease outbreaks in aquatic organisms such as amphibians, fish, crustaceans, molluscs and aquatic plants (Raghukumar 2012). These outbreaks can lead to mass mortalities, affecting population dynamics and biodiversity in aquatic environments. Nowhere is this more evident than in freshwater systems of the world where amphibian populations have been devastated by the disease chytridiomycosis caused by chytrid fungi in the genus *Batrachochytrium* (see also Chap. 18). Initially the genus was thought to consist of only one species, the globally distributed *B. dendrobatidis* capable of infecting any member in the class Amphibia (Berger et al. 1998). A second species, *B. salamandrivorans*, was discovered more than a decade later, which appears to be restricted to hosts in the order Urodeles (salamanders and newts; Fig. 4.8) from Europe and Southeast Asia and is equally capable of inducing chytridiomycosis in susceptible hosts (Martel et al. 2013; Laking et al. 2017). Presently, five deeply divergent lineages of *B. dendrobatidis* with variable pathogenicity have been discovered, and evidence of recombinants with enhanced virulence has been found in Brazil and South Africa (O'Hanlon et al. 2018). Consequently, more than 500 amphibian species have experienced chytridiomycosis-associated declines and extinction of almost 90 species is linked to this pathogen (Scheele et al. 2019).

A considerable portion of the global food security is provided through fisheries, but harvesting wild stock cannot keep up with the growing demand (Kent 1997). A large part of this demand is now met by the aquaculture industry, the farming of aquatic organisms for food, recreation and conservation purposes (see also Chap. 23). Globally, aquaculture is the fastest growing protein production sector, with production measured in millions of metric tons and valued in the billions of dollars (Smith et al. 2010; Garlock et al. 2020). Pathogenic fungi can be a major concern in aquaculture, and disease outbreaks can result in substantial economic losses for the industry, even threatening the livelihood of aquaculture facilities (Spring and Fegan 2005; see also Chap. 23). Significantly, fungal diseases in aquaculture are second only to bacterial diseases in terms of economic losses, and all organisms including algae, fish, crustaceans, or molluscs are vulnerable to infection by various fungal species (Ramaiah 2006). While rare, some aquatic pathogenic fungi have zoonotic potential, thus having the ability to transmit diseases from aquatic organisms to humans. Therefore, individuals who encounter infected aquatic organisms or contaminated water may be at risk of infection. Indeed, most of the fish-derived zoonotic diseases are transmitted to humans, mainly via the consumption of improperly cooked or raw fish or fish products (Ziarati et al. 2022; also see Chap. 19). While there are relatively fewer known examples of zoonotic aquatic fungi compared to other zoonotic pathogens, a notable example is *Saprolegnia* spp., ubiquitous aquatic fungi that can infect a variety of aquatic animals, including fish, amphibians and crustaceans. Although they primarily affect aquatic organisms, there have been occasional reports of infections in humans, particularly in cases of compromised immune systems (Azizi et al. 2012).

Large-scale die-offs of aquatic organisms caused by fungal diseases can lead to the release of organic matter, nutrients and toxins into the water. This can have implications for water quality, potentially leading to eutrophication and other environmental issues (Duarte et al. 2015). When aquatic organisms succumb to fungal diseases, they become part of the nutrient cycle in the ecosystem. Decomposition of infected organisms by fungi and other decomposer organisms releases organic matter and nutrients into the water column, making them available for uptake by other organisms. This breaking down of dead organic matter is therefore essential for nutrient cycling and maintaining water quality. Per implication, the presence and prevalence of aquatic pathogenic fungi can serve as indicators of the overall health and ecological condition of aquatic ecosystems, since their growth patterns respond to the slightest increase in pollutant concentrations (Gerhardt 2002). Certain moulds including *Trichoderma* spp., *Aspergillus* spp., *Fusarium* spp., *Penicillium* spp. and *Candida albicans* have often been used as biological indicators for specific contaminants in the environment (Zaghloul et al. 2020). Increases in disease occurrence may suggest environmental stressors or changes in the ecosystem in need of conservation management or public health attention.

4.5 Environmental and Demographic Factors That Influence Disease Development

Environmental factors play a significant role in shaping the prevalence and virulence of aquatic pathogenic fungi. These factors can directly or indirectly influence the growth, reproduction and pathogenicity of these fungi in aquatic ecosystems. Moreover, these factors can vary depending on the specific species of fungi and the characteristics of the aquatic ecosystem they inhabit and the demographics of their hosts. Human activities, such as aquaculture, habitat alteration, water pollution and introduction of non-native species, can influence the prevalence and virulence of aquatic pathogenic fungi. These activities can create favourable conditions for the fungi to thrive or disrupt natural ecosystems, leading to disease outbreaks.

Most fundamental to the existence of aquatic pathogenic fungi is the requirement of a moist environment for spore germination and dispersal. Inanimate spore dispersal is facilitated by the flow of water in lotic systems or the presence of currents in lentic systems and oceans (Magyar et al. 2016). Not only do water flow and currents affect fungal spore distribution, but also they affect their ability to encounter suitable hosts. Strong currents may disperse spores over greater distances, potentially leading to widespread infections (Golan and Pringle 2017). Even in the absence of currents, other parameters that influence water quality such as pH, dissolved oxygen levels, salinity and nutrient availability can significantly affect the abundance and activity of aquatic pathogenic fungi. For example, a study on the influence of water contamination on aquatic fungal communities of the Tietê River, South America, found that diversity was affected by pH and dissolved iron, while many factors including dissolved oxygen, pH, nitrate and seasonality among others influenced community composition (Ortiz-Vera et al. 2018). The source of water quality change need not be local to affect fungal interactions, as demonstrated by increased susceptibility of crayfish to microsporidian infections following global environmental disturbances that resulted in ocean acidification (France and Graham 1985).

It is often unpredictable how a fungal species will react to variation in environmental conditions and their responses do not always result in the same outcome. Carnter and Jaworski explored the infectivity of chytrid zoospores, highlighting the zoospores' sensitivity to light and their reduced ability to attach to hosts in the absence of light. Members of the chytridiomycete fungus genus *Rhizophydium* express variable responses to light intensity; whereas zoospores of *R. planktonicum* cannot find and infect their phytoplanktonic hosts during low light conditions, *R. sphaerocarpum* zoospores are able to infect their hosts in complete darkness (Barr and Hickman 1967; Bruning 1991).

Temperature is another crucial factor that affects the growth and activity of aquatic pathogenic fungi. Different fungal species have specific thermal ranges in which they thrive and reproduce optimally. Changes in temperature can therefore influence the rate of fungal growth, spore production and the duration of infection. For some fungi, warmer temperatures may enhance their virulence, leading to more severe disease outbreaks. One example is *Batrachochytrium dendrobatidis*, whose virulence increases as temperature shifts towards the growth optimum of the fungus (Conradie et al. 2011). In aquaculture, a sudden drop in water temperature is often associated with an increase in fungal spores in the water (Sarkar et al. 2022). At the same time, fish immune systems are compromised due to suppressed mucus production in the dermal layer of their skin, increasing susceptibility to fungal infections like saprolegniosis and leading to *Aphanomyces* outbreaks (Matthews 2019).

The presence of suitable hosts, predators and competitors is another fundamental factor influencing the life cycle of aquatic pathogenic fungi. Generally, interactions between aquatic pathogenic fungi and other organisms impact the frequency of infections and the overall disease dynamics in the ecosystem. Some interactions may enhance fungal survival, while others may limit their growth. In certain species, epidemics can manifest even when algal densities are relatively low, while in numerous instances, a specific threshold density must be met (Rasconi et al. 2011). High host densities can lead to increased transmission opportunities for the fungi. Furthermore, stressed or immunocompromised hosts may be more susceptible to infection, leading to higher disease severity. In fact, outbreaks of fish infections often occur when fish become stressed due to a combination of factors related to inadequate aquaculture management, including overcrowding, transportation and poor fish storage (Sarkar et al. 2022). The specific growth rate of host and parasite can determine the nature of fungal epidemics, since epidemic proportions can be reached when algal growth is inhibited (e.g. in conditions of low nutrient levels) or when fungal growth is stimulated (e.g. when host density is high) (Kagami et al. 2007). Moreover, understanding the dynamics of the fundamental ecological processes behind fungal diseases relies on basic demographic information of both host and pathogen. This principle was clearly illustrated through a significant size-dependent amplification of the impact of aspergillosis on sea fan corals, causing a decline in reproduction among many of the largest and typically most fertile individuals, and leading to a population structure dominated by smaller, less fecund colonies (Bruno et al. 2011). A similar pattern emerged from various studies on chytrid–algae interactions, indicating a preference for larger host species and species with larger host cells (Sommer 1987; Kagami et al. 2007). Larger host size holds many advantages for parasitic chytrids, including increased growth and fecundity due to more available resources, and avoiding being grazed upon by zooplankton due to restrictions on ingestion size of infected host cells. Conversely, ex situ experiments have shown that chytrid infections of phytoplankton can be reduced by introducing grazers to aquaria (Kagami et al. 2004) and, in natural systems, cold seasonal temperatures inhibit activity of zooplankton grazers resulting in increased fungal infections (Kudoh and Takahashi 1990; Rasconi et al. 2011). The opposing impacts of zooplankton predation and chytrid infections in relation to the size and edibility of phytoplankton resources can play a pivotal role in shaping the structure and dynamics of the food web in aquatic ecosystems.

4.6 Conclusion

Fungi are a major component of the microbiome in almost every habitat on Earth. Aquatic pathogenic fungi encompass a diverse array of taxonomic groups that have adapted to thrive in aquatic environments and cause diseases in various aquatic organisms. These fungi can be found in both freshwater and marine ecosystems, infecting a wide range of aquatic organisms.

The study of aquatic pathogenic fungi is essential for understanding the dynamics of aquatic ecosystems and conservation efforts, and managing the health of aquatic organisms, particularly in aquaculture settings where disease outbreaks can lead to economic losses. Studying these fungi can help identify and implement effective disease management strategies and contribute towards exploring their role in the broader context of aquatic ecology. It is essential to recognise that aquatic pathogenic fungi are just one component of a complex and interconnected web of interactions in aquatic ecosystems (Grossart et al. 2019). Their ecological role is intertwined with that of other organisms, abiotic factors and ecological processes. Demographic influences such as the presence of multiple host species and high host densities can influence the prevalence and virulence of aquatic pathogenic fungi. Some fungi may have a broader host range, and the availability of multiple host species can enhance their ability to persist and spread in the environment.

Seasonal changes can influence the life cycle of aquatic pathogenic fungi. Temperature fluctuations, changes in host abundance and variations in water conditions can lead to seasonal patterns in the prevalence and activity of these fungi. However, sudden fluctuations in water temperature and various environmental factors resulting from climate change or other factors may influence the prevalence and distribution of aquatic pathogenic fungi (Perrone et al. 2020). As a result, disease patterns in aquatic ecosystems may be altered, impacting both wildlife and human activities reliant on these ecosystems. Furthermore, the interactions between aquatic organisms and pathogenic fungi in a dynamic environment can drive evolutionary processes. Over time, hosts may develop resistance to specific pathogens, and fungi may evolve to become more virulent or find new ways to infect hosts, leading to an ongoing co-evolutionary dynamic (Balodi et al. 2017).

Understanding the impact of environmental factors on aquatic pathogenic fungi and their biotic interactions is essential for disease management and conservation efforts. It is essential for predicting disease outbreaks and allows researchers and policymakers to implement strategies to mitigate outbreaks, protect vulnerable aquatic populations and preserve the health and balance of aquatic ecosystems that can lead to sustainable utilisation of aquatic resources.

References

Alexopoulos CJ, Mims CW, Blackwell M (1996) Phylum Myxomycota: True slime molds. In: Introductory mycology, 4th edn. Wiley, New York, pp 775–808

Alker AP, Smith GW, Kim K (2001) Characterization of *Aspergillus sydowii* (Thom et Church), a fungal pathogen of Caribbean sea fan corals. Hydrobiologia 460:105–111

Azizi IG, Fard MH, Tahmasbipour S (2012) The effect of aquatic and alcoholic extracts of *Citrullus colocynthis* on growth of the *Saprolegnia parasitica*.

WJFMS 4(3):258–262. https://doi.org/10.5829/idosi.wjfms.2012.04.03.61251

Balodi R, Ghatak LV, Bisht S, Shukla N (2017) Reproductive fitness of fungal phytopathogens: Deriving co-evolution of host–pathogen systems. In: Ghatak A, Ansar M (eds) The phytopathogen. Apple Academic, New York, pp 41–64

Barr DJS, Hickman CJ (1967) Chytrids and algae. II. Factors influencing parasitism of *Rhizophydium sphaerocarpum* on *Spirogyra*. Canad J Bot 45(4):431–440. https://doi.org/10.1139/b67-043

Berger L, Speare R, Daszak P, Green DE, Cunningham AA, Goggin CL, Slocombe R, Ragan MA, Hyatt AD, McDonald KR, Hines HB, Lips KR, Marantelli G, Parkes H (1998) Chytridiomycosis causes amphibian mortality associated with population declines in the rain forests of Australia and Central America. Proc Natl Acad Sci USA 95(15):9031–9036. https://doi.org/10.1073/pnas.95.15.9031

Bruning K (1991) Effects of temperature and light on the population dynamics of the *Asterionella-Rhizophydium* association. J Plankton Res 13(4):707–719. https://doi.org/10.1093/plankt/13.4.707

Bruno JF, Ellner SP, Vu I, Kim K, Harvell CD (2011) Impacts of aspergillosis on sea fan coral demography: modeling a moving target. Ecol Monogr 81(1):123–139. https://doi.org/10.1890/09-1178.1

Canter HM, Lund JWG (1951) Studies on plankton parasites: III. Examples of the interaction between parasitism and other factors determining the growth of diatoms. Ann Bot 15(3):359–371. https://doi.org/10.1093/oxfordjournals.aob.a083287

Canter HM, Lund JWG (1969) The parasitism of planktonic desmids by fungi. Österr Bot Z 116(1/5):351–377. https://www.jstor.org/stable/43337618

Conradie W, Weldon C, Smith KG, Du Preez LH (2011) Seasonal pattern of chytridiomycosis in common river frog (*Amietia angolensis*) tadpoles in the South African Grassland Biome. Afr Zool 46(1):95–102. https://hdl.handle.net/10520/EJC18175

Crow GL, Brock JA, Kaiser S (1995) *Fusarium solani* fungal infection of the lateral line canal system in captive scalloped hammerhead sharks (*Sphyma lewini*) in Hawaii. J Wildl Dis 31(4):562–565. https://doi.org/10.7589/0090-3558-31.4.562

Donk EV, Ringelberg J (1983) The effect of fungal parasitism on the succession of diatoms in Lake Maarsseveen I (The Netherlands). Freshw Biol 13(3):241–251. https://doi.org/10.1111/j.1365-2427.1983.tb00674.x

Duarte S, Bärlocher F, Trabulo J, Cássio F, Pascoal C (2015) Stream-dwelling fungal decomposer communities along a gradient of eutrophication unraveled by 454 pyrosequencing. Fungal Divers 70:127–148

Fernando N, Hui SW, Tsang CC, Leung SY, Ngan AH, Leung RW, Groff JM, Lau SK, Woo PC (2015) Fatal *Fusarium solani* species complex infections in elasmobranchs: the first case report for black spotted stingray (*Taeniura melanopsila*) and a literature review. Mycoses 58(7):422–431. https://doi.org/10.1111/myc.12342

Fisher MC, Garner TW (2020) Chytrid fungi and global amphibian declines. Nat Rev Microbiol 18(6):332–343

France RL, Graham L (1985) Increased microsporidian parasitism of the crayfish *Orconectes virilis* in an experimentally acidified lake. Wat Air Soil Poll 26:129–136

Frenken T, Alacid E, Berger SA, Bourne EC, Gerphagnon M, Grossart HP, Gsell AS, Ibelings BW, Kagami M, Kupper FC, Letcher PM, Loyau A, Miki T, Nejstgaard JC, Rasconi S, Rene A, Rohrlack T, Rojas-Jimenez K, Schmeller DS, Scholz B, Seto K, Sime-Ngando T, Sukenik A, Van de Waal DB, Van den Wyngaert S, Van Donk E, Wolinska J, Wurzbacher C, Agha R (2017) Integrating chytrid fungal parasites into plankton ecology: research gaps and needs. Environ Microbiol 19(10):3802–3822. https://doi.org/10.1111/1462-2920.13827

Fouquet A, Measey GJ (2006) Plotting the course of an African clawed frog invasion in Western France. Anim Biol 56(1):95–102.

Garner TW, Walker S, Bosch J, Leech S, Rowcliffe MJ, Cunningham AA, Fisher MC (2009) Life history tradeoffs influence mortality associated with the amphibian pathogen Batrachochytrium dendrobatidis. Oikos 118(5):783–791

Garlock T, Asche F, Anderson J, Bjørndal T, Kumar G, Lorenzen K, Smith MD, Tveterås R (2020) A global blue revolution: aquaculture growth across regions, species, and countries. Rev Fish Sci Aquac 28(1):107–116. https://doi.org/10.1080/23308249.2019.1678111

Gerhardt A (2002) Bioindicator species and their use in biomonitoring. Environ Monit 1:77–123

Gleason FH, Allerstorfer M, Lilje O (2020) Newly emerging diseases of marine turtles, especially sea turtle egg fusariosis (SEFT), caused by species in the *Fusarium solani* complex (FSSC). Mycol 11(3):184–194. https://doi.org/10.1080/21501203.2019.1710303

Golan JJ, Pringle A (2017) Long-distance dispersal of fungi. Microbiol Spectr 5(4):e0047. https://doi.org/10.1128/microbiolspec.funk-0047-2016

Gozlan RE, Marshall WL, Lilje O, Jessop CN, Gleason FH, Andreou D (2014) Current ecological understanding of fungal-like pathogens of fish: what lies beneath? Front Microbiol 5:e62. https://doi.org/10.3389/fmicb.2014.00062

Greeff-Laubscher M, Jacobs K (2022) *Fusarium* species isolated from post-hatchling loggerhead sea turtles (*Caretta caretta*) in South Africa. Sci Rep 12(1):e5874. https://doi.org/10.1038/s41598-022-06840-1

Grossart HP, Van den Wyngaert S, Kagami M, Wurzbacher C, Cunliffe M, Rojas-Jimenez K (2019) Fungi in aquatic ecosystems. Nat Rev Microbiol 17(6):339–354

Hassan O, Hassan A, Abd El Ghany N, El-baky A, Hanna M, Abd El Aziz M (2020) A contribution on the pathogenicity of *Fusarium oxysporum* isolated from cultured Nile tilapia (*Oreochromis niloticus*) with trials for the treatment. Egypt J Aquat Res 24(5):197–215. https://doi.org/10.21608/ejabf.2020.104139

Hatcher MJ, Dunn AM (2011) Parasites in ecological communities: from interactions to ecosystems. Cambridge University Press, Cambridge

Hawksworth DL, Lücking R (2017) Fungal diversity revisited: 2.2 to 3.8 million species. Microbiol Spectr 5(4):10–1128. https://doi.org/10.1128/microbiolspec.funk-0052-2016

Johnson PT, Ives AR, Lathrop RC, Carpenter SR (2009) Long-term disease dynamics in lakes: Causes and consequences of chytrid infections in *Daphnia* populations. Ecology 90(1):132–144. https://doi.org/10.1890/07-2071.1

Kagami M, Van Donk E, de Bruin A, Rijkeboer M, Ibelings BW (2004) *Daphnia* can protect diatoms from fungal parasitism. L&O 49(3):680–685. https://doi.org/10.4319/lo.2004.49.3.0680

Kagami M, de Bruin A, Ibelings BW, Van Donk E (2007) Parasitic chytrids: their effects on phytoplankton communities and food-web dynamics. Hydrobiologia 578:113–129. https://doi.org/10.1007/s10750-006-0438-z

Kamilya D, Baruah A (2014) Epizootic ulcerative syndrome (EUS) in fish: history and current status of understanding. Rev Fish Biol Fish 24:369–380. https://doi.org/10.1007/s11160-013-9335-5

Kent G (1997) Fisheries, food security, and the poor. Food Policy 22(5):393–404. https://doi.org/10.1016/S0306-9192(97)00030-4

Klein Breteler WCM, Schogt N, Baas M, Schouten S, Kraay GW (1999) Trophic upgrading of food quality by protozoans enhancing copepod growth: role of essential lipids. Mar Biol 135:191–198

Kohler SL, Hoiland WK (2001) Population regulation in an aquatic insect: the role of disease. Ecology 82(8):2294–2305. https://doi.org/10.1890/0012-9658(2001)082[2294:PRIAAI]2.0.CO;2

Kudoh S, Takahashi M (1990) Fungal control of population changes of the planktonic diatom *Asterionella formosa* in a shallow eutrophic lake. J Phycol 26(2):239–244. https://doi.org/10.1111/j.0022-3646.1990.00239.x

Kumar PI, Sarkar P, Raju SV, Manikandan V, Gurua A, Arshad A, Elumalaid P, Arockiaraja J (2020) Pathogenicity and pathobiology of epizootic ulcerative syndrome (EUS) causing fungus *Aphanomyces invadans* and its immunological response in fish. Rev Fish Sci 28(3):358–375. https://doi.org/10.1080/23308249.2020.1753167

Laking AE, Ngo HN, Pasmans F, Martel A, Nguyen TT (2017) *Batrachochytrium salamandrivorans* is the predominant chytrid fungus in Vietnamese salamanders. Sci Rep 7(1):e44443. https://doi.org/10.1038/srep44443

MacNeil C, Dick JT, Hatcher MJ, Terry RS, Smith JE, Dunn AM (2003) Parasite-mediated predation between native and invasive amphipods. P Roy Soc Lond B Biol Sci 270(1521):1309–1314. https://doi.org/10.1098/rspb.2003.2358

Magyar D, Vass M, Li DW (2016) Dispersal strategies of microfungi. In: Li DW (ed) Biology of microfungi, Fungal biology. Springer, Cham, pp 315–371

Martel A, Spitzen-van der Sluijs A, Blooi M, Bert W, Ducatelle R, Fisher MC, Woeltjes A, Bosman W, Chiers K, Bossuyt F, Pasmans F (2013) *Batrachochytrium salamandrivorans* sp. nov. causes lethal chytridiomycosis in amphibians. Proc Natl Acad Sci USA 110(38):15325–15329. https://doi.org/10.1073/pnas.1307356110

Marano N, Arguin PM, Pappaioanou M (2007) Impact of globalization and animal trade on infectious disease ecology. Emerg Infect Dis 13(12):1807–1809

Matthews E (2019) Environmental factors impacting *Saprolegnia* infections in wild fish stocks. Doctoral dissertation, Cardiff University. https://orca.cardiff.ac.uk/id/eprint/130012

Moore D, Robson GD, Trinci AP (2020) 21st century guidebook to fungi. Cambridge University Press, Cambridge

Murray KA, Skerratt LF, Speare R, McCallum H (2009) Impact and dynamics of disease in species threatened by the amphibian chytrid fungus, Batrachochytrium dendrobatidis. Conserv Biol 23(5):1242–1252.

O'Hanlon SJ, Rieux A, Farrer RA, Rosa GM, Waldman B, Bataille A, Kosch TA, Murray KA, Brankovics B, Fumagalli M, Martin MD, Wales N, Alvarado-Rybak M, Bates KA, Berger L, Böll S, Brookes L, Clare F, Courtois EA, Cunningham AA, Doherty-Bone TM, Ghosh P, Gower DJ, Hintz WE, Höglund J, Jenkinson TS, Lin C, Laurila A, Loyau A, Martel A, Meurling S, Miaud C, Minting P, Pasmans F, Schmeller DS, Schmidt BR, Shelton JMG, Skerratt LF, Smith F, Soto-Azat C, Spagnoletti M, Tessa G, Toledo LF, Valenzuela-Sánchez A, Verster R, Vörös J, Webb RJ, Wierzbicki C, Wombwell E, Zamudio KR, Aanensen DM, James TY, Gilbert MTP, Weldon C, Bosch J, Balloux F, Garner TWJ, Fisher MC (2018) Recent Asian origin of chytrid fungi causing global amphibian declines. Science 360(6389):621–627. https://doi.org/10.1126/science.aar1965

Oidtmann B, Heitz E, Rogers D, Hoffmann RW (2002) Transmission of crayfish plague. Dis Aquat Org 52(2):159–167. https://doi.org/10.3354/dao052159

Ortiz-Vera MP, Olchanheski LR, da Silva EG, de Lima FR, Martinez LRDPR, Sato MIZ, Jaffé R, Alves R, Ichiwaki S, Padilla G, Araújo WL (2018) Influence of water quality on diversity and composition of fungal communities in a tropical river. Sci Rep 8(1):e14799. https://doi.org/10.1038/s41598-018-33162-y

Perrone G, Ferrara M, Medina A, Pascale M, Magan N (2020) Toxigenic fungi and mycotoxins in a climate change scenario: Ecology, genomics, distribution, prediction and prevention of the risk. Microorganisms 8(10):e1496. https://doi.org/10.3390/microorganisms8101496

Pessier AP, Mendelson JR (2010) A manual for control of infectious diseases in amphibian survival assurance colonies and reintroduction programs: proceedings from a workshop: 16–18 February 2009 San Diego

Zoo. IUCN/SSC Conservation Breeding Specialist Group

Raghukumar C (ed) (2012) Biology of marine fungi, vol 53. Springer, Berlin

Rasconi S, Jobard M, Jouve L, Sime-Ngando T (2009) Use of calcofluor white for detection, identification, and quantification of phytoplanktonic fungal parasites. Appl Environ Microbiol 75(8):e2545–e2553. https://doi.org/10.1128/AEM.02211-08

Rasconi S, Jobard M, Sime-Ngando T (2011) Parasitic fungi of phytoplankton: ecological roles and implications for microbial food webs. Aquat Microb Ecol 62(2):123–137. https://doi.org/10.3354/ame01448

Rasconi S, Niquil N, Sime-Ngando T (2012) Phytoplankton chytridiomycosis: community structure and infectivity of fungal parasites in aquatic ecosystems. Environ Microbiol 14(8):2151–2170. https://doi.org/10.1111/j.1462-2920.2011.02690.x

Ramaiah N (2006) A review on fungal diseases of algae, marine fishes, shrimps and corals. Indian J Mar Sci 35(4):380–387. http://drs.nio.org/drs/handle/2264/570

Sarkar P, Raju VS, Kuppusamy G, Rahman MA, Elumalai P, Harikrishnan R, Arshad A, Arockiaraj J (2022) Pathogenic fungi affecting fishes through their virulence molecules. Aquaculture 548:737553. https://doi.org/10.1016/j.aquaculture.2021.737553

Scheele BC, Pasmans F, Skerratt LF, Berger L, Martel AN, Beukema W, Acevedo AA, Burrowes PA, Carvalho T, Catenazzi A, De La Rivam I, Fisher MC, Flechas SV, Foster CN, Frías-Álvarez P, Garner TWJ, Gratwicke B, Guayasamin JM, Hirschfeld M, Kolby JE, Kosch TA, La Marca E, Lindenmayer DB, Lips KR, Longo AV, Maneyro R, McDonald CA, Mendelson J III, Palacios-Rodriguez P, Parra-Olea G, Richards-Zawacki CL, Rödel MO, Rovito SM, Soto-Azat C, Toledo LF, Voyles J, Weldon C, Whitfield SM, Wilkinson M, Zamudio KR, Canessa S (2019) Amphibian fungal panzootic causes catastrophic and ongoing loss of biodiversity. Science 363(6434):1459–1463. https://doi.org/10.1126/science.aav0379

Smith MD, Roheim CA, Crowder LB, Halpern BS, Turnipseed M, Anderson JL, Asche F, Bourillon L, Guttormsen AG, Khan A, Liguori LA, McNevin A, O'Connor MI, Squires D, Tyedmers P, Brownstein C, Carden K, Klinger DH, Sagarin R, Selkoe KA (2010) Sustainability and global seafood. Science 327(5967):784–786. https://doi.org/10.1126/science.1185345

Sommer U (1987) Factors controlling the seasonal variation in phytoplankton species composition—a case study for a deep, nutrient rich lake. Prog Phycol Res 5:124–178. https://oceanrep.geomar.de/id/eprint/14142

Spring P, Fegan DF (2005) Mycotoxins—a rising threat to aquaculture. In: Nutritional biotechnology in the feed and food industries. Proceedings of Alltech's 21st annual symposium, Lexington, Kentucky, USA, Alltech UK, pp 323–331

Stentiford GD, Feist SW, Stone DM, Bateman KS, Dunn AM (2013) Microsporidia: diverse, dynamic, and emergent pathogens in aquatic systems. Trends Parasitol 29(11):567–578. https://doi.org/10.1016/j.pt.2013.08.005

Tinsley RC, McCoid MJ (1996) Feral populations of Xenopus outside Africa. In: Biology of Xenopus. Oxford University Press, pp 81–94

Van Sittert L, Measey GJ (2016) Historical perspectives on global exports and research of African clawed frogs (Xenopus laevis). Trans R Soc S Afr 71(2):157–166

Vávra J, Lukeš J (2013) Microsporidia and 'the art of living together'. Adv Parasitol 82:253–319. https://doi.org/10.1016/B978-0-12-407706-5.00004-6

Watkinson SC (2016) Mutualistic symbiosis between fungi and autotrophs. In: Watkinson SC, Boddy L, Money N (eds) The fungi, 3rd edn. Academic, New York, pp 205–243

Weldon C, Fisher MC (2011) The effect of trade-mediated spread of amphibian chytrid on amphibian conservation. In: Fungal diseases: An emerging challenge to human, animal, and plant health. The National Academies Press, Washington, DC, pp 355–367

Weldon C, De Villiers AL, Du Preez LH (2007) Quantification of the African clawed frog trade from South Africa, with implications for biodiversity conservation. Afr J Herpetol 56(1):77–83. https://doi.org/10.1080/21564574.2007.9635553

Zaghloul A, Saber M, Gadow S, Awad F (2020) Biological indicators for pollution detection in terrestrial and aquatic ecosystems. Bull Natl Res Cent 44(1):1–11. https://doi.org/10.1186/s42269-020-00385-x

Ziarati M, Zorriehzahra MJ, Hassantabar F, Mehrabi Z, Dhawan M, Sharun K, Bin Emran T, Dhama K, Chaicumpa W, Shamsi S (2022) Zoonotic diseases of fish and their prevention and control. Vet Q 42(1):95–118. https://doi.org/10.1080/01652176.2022.2080298

Open Access This chapter is licensed under the terms of the Creative Commons Attribution-NonCommercial-NoDerivatives 4.0 International License (http://creativecommons.org/licenses/by-nc-nd/4.0/), which permits any non-commercial use, sharing, distribution and reproduction in any medium or format, as long as you give appropriate credit to the original author(s) and the source, provide a link to the Creative Commons license and indicate if you modified the licensed material. You do not have permission under this license to share adapted material derived from this chapter or parts of it.

The images or other third party material in this chapter are included in the chapter's Creative Commons license, unless indicated otherwise in a credit line to the material. If material is not included in the chapter's Creative Commons license and your intended use is not permitted by statutory regulation or exceeds the permitted use, you will need to obtain permission directly from the copyright holder.

Biology and Life Cycle of Helminths

5

Bernd Sures ⓘ, Dakeishla M. Díaz-Morales ⓘ, Russell Q-Y. Yong ⓘ, Anja Erasmus ⓘ, and Jessica Schwelm ⓘ

Abstract

Helminths are worm-like parasites that utilise a variety of vertebrates and invertebrates as hosts. They include various phyla such as Platyhelminthes (flatworms), Acanthocephala (thorny-headed worms) and Nematoda (roundworms). The Platyhelminthes include the Monogenea, Cestoda and Trematoda. Although some species of these phyla are free-living (e.g. some flatworms such as Turbellaria and several taxa within the Nematoda), the species with parasitic lifestyles are very diverse and exhibit a wide range of life cycles and transmission strategies. They are found in almost all freshwater and marine systems, and many of them play an important role and have a significant impact on ecological interactions and ecosystem health. Often, these parasites can also bridge between different ecosystems, that is, between terrestrial and aquatic systems, due to the variety and diversity of hosts they utilise. In this chapter, we cover the biological underpinnings of these parasite groups, including their life cycles, morphology and impacts on their hosts. This forms a valuable basis and reference for the helminth examples given in the following chapters.

5.1 Introduction

There are several groups of parasitic metazoans with a 'worm-like' shape, also known as helminths. The animal taxa grouped together as hel-

B. Sures (✉) · J. Schwelm
Department of Aquatic Ecology and Centre for Water and Environmental Research (ZWU), University of Duisburg-Essen, Essen, Germany

Research Center One Health Ruhr, Research Alliance Ruhr, University Duisburg-Essen, Essen, Germany

Water Research Group, Unit for Environmental Sciences and Management, North-West University, Potchefstroom, South Africa
e-mail: bernd.sures@uni-due.de;
jessica.schwelm@uni-due.de

D. M. Díaz-Morales
Department of Aquatic Ecology and Centre for Water and Environmental Research (ZWU), University of Duisburg-Essen, Essen, Germany

Research Center One Health Ruhr, Research Alliance Ruhr, University Duisburg-Essen, Essen, Germany

School of Aquatic and Fishery Sciences, University of Washington, Seattle, WA, USA
e-mail: diazdakeishla@gmail.com

R. Q-Y. Yong · A. Erasmus
Water Research Group, Unit for Environmental Sciences and Management, North-West University, Potchefstroom, South Africa
e-mail: 49933884@mynwu.ac.za;
23599235@mynwu.ac.za

© The Authors(s) 2025
N. J. Smit, B. Sures (eds.), *Aquatic Parasitology: Ecological and Environmental Concepts and Implications of Marine and Freshwater Parasites*, https://doi.org/10.1007/978-3-031-83903-0_5

minths do not form a phylogenetically systematic unit, but contain a number of phyla, many of which are completely independent. Because they are multicellular organisms, they tend to be larger than the protists and can usually be seen with the naked eye. Parasitic worms live in (endoparasites), or occasionally on (ectoparasites) their hosts, from which they obtain their food, often for long periods of time (ranging from weeks to years). They obtain food and shelter by interfering with their host's ability to absorb nutrients. Helminths have a life cycle that includes several larval stages, some of which occur outside the host. All parasitic worms produce eggs when they reproduce. These eggs usually have a strong shell that protects them from a range of environmental conditions. The eggs of several taxa can therefore survive in the environment for many months or years. According to our current understanding, helminths include different phyla such as Platyhelminthes (flatworms), Acanthocephala (thorny-headed worms) and Nematoda (roundworms).

5.2 Platyhelminthes

Flatworms (Platyhelminthes) are a phylum with a few free-living, predatory species, a few ectoparasitic and many endoparasitic worms, including a number of human pathogenic parasites (see also Chap. 19). The Platyhelminthes are divided into one group of mostly free-living species, the Turbellaria and three groups of parasitic species: the endoparasitic Trematoda and Cestoda, and the Monogenea (Fig. 5.1), the latter being mostly ectoparasitic on fish and amphibians. While the monophyly of the latter three classes—collectively known as the Neodermata—seems to be given, the Turbellaria, whose classification is still in progress, are considered a paraphyletic group based on morphological and molecular evidence.

The systematic relationships among the four major lineages within the Neodermata are also subject to discussion and re-organisation. In contrast to Fig. 5.1, recent results suggest that the monogenean subclass Polyopisthocotylea form a group with the Trematoda, rather than being a sister group to the Monopisthocotylea, and the Cestoda might also be placed differently (Brabec et al. 2023). Irrespective of the ongoing phylogenetic discussion, we will present the morphology and life cycles of the involved parasite groups largely independent of the final phylogenetic relationships.

5.2.1 Biology and Life Cycle of Monogenea

Monogeneans are common parasites of the gills and skin of fish, amphibians, reptiles, cetaceans or cephalopods and one species, *Oculotrema hippopotami*, is known to infect hippopotamuses. Some species become endoparasitic by colonising the nose, throat, cloaca, or bladder of their respective hosts, such as the genera *Enterogyrus*, *Polystoma* and *Urogyrus*, among others (Mehlhorn 2016; Kuchta et al. 2018). The Monogenea have been divided into two subclasses, namely Monopisthocotylea and Polyopisthocotylea. The Monopisthocotylea differ from the Polyopisthocotylea with respect to the microhabitats of infection, the structure of their gut, their behaviour and their diet, and in the latter having a more complex attachment organ (Fig. 5.2). Monogeneans are found in both freshwater and marine ecosystems, with more than 5,600 described species belonging to 750 genera (Gibson 2023). Monogeneans are highly specific to their host and, in some cases, to a particular geographic location or microhabitat making them ideal as biological tags to discern between fish stocks (see Chap. 17). The diet of monogeneans includes host mucus, blood, or tissue and their size ranges from approximately 300 μm up to a few centimetres.

5.2.1.1 Monogenean Life Cycles
Monogeneans lack intermediate hosts and, therefore, have a simple (monoxenous) life cycle (Fig. 5.3). Most monogeneans infect only one host (e.g. fish) and their ontogeny involves only two developmental stages: an adult and a ciliated larval stage called an oncomiracidium. The adults

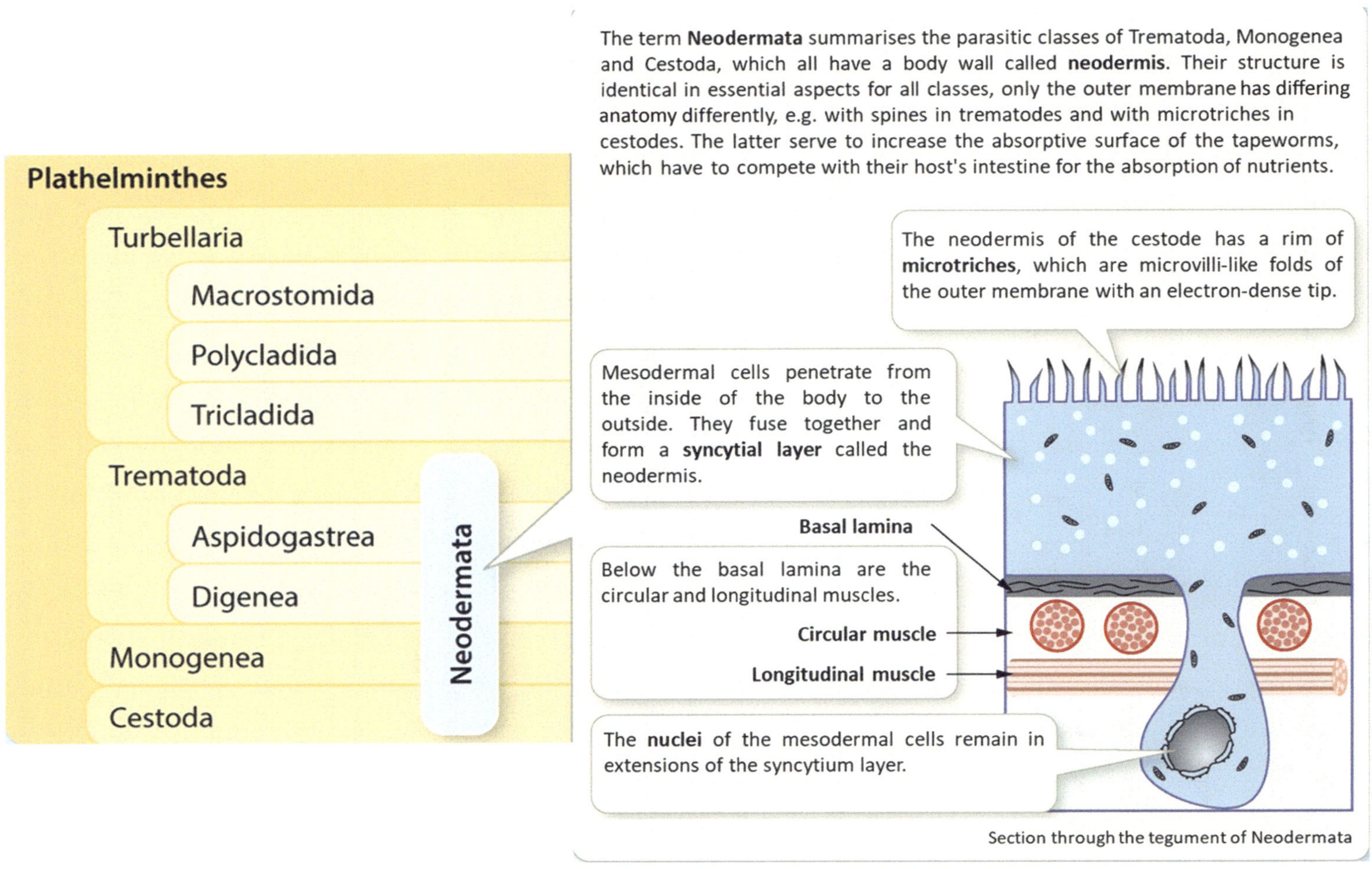

5 Biology and Life Cycle of Helminths

The term **Neodermata** summarises the parasitic classes of Trematoda, Monogenea and Cestoda, which all have a body wall called **neodermis**. Their structure is identical in essential aspects for all classes, only the outer membrane has differing anatomy differently, e.g. with spines in trematodes and with microtriches in cestodes. The latter serve to increase the absorptive surface of the tapeworms, which have to compete with their host's intestine for the absorption of nutrients.

The neodermis of the cestode has a rim of **microtriches**, which are microvilli-like folds of the outer membrane with an electron-dense tip.

Mesodermal cells penetrate from the inside of the body to the outside. They fuse together and form a **syncytial layer** called the neodermis.

Basal lamina

Below the basal lamina are the circular and longitudinal muscles.

Circular muscle

Longitudinal muscle

The **nuclei** of the mesodermal cells remain in extensions of the syncytium layer.

Section through the tegument of Neodermata

Plathelminthes
- Turbellaria
  - Macrostomida
  - Polycladida
  - Tricladida
- Trematoda
  - Aspidogastrea
  - Digenea
- Monogenea
- Cestoda

Neodermata

Fig. 5.1 Phylogenetics within the phylum Platyhelminthes and illustration of the neodermis. (Adapted from: Sures 2021)

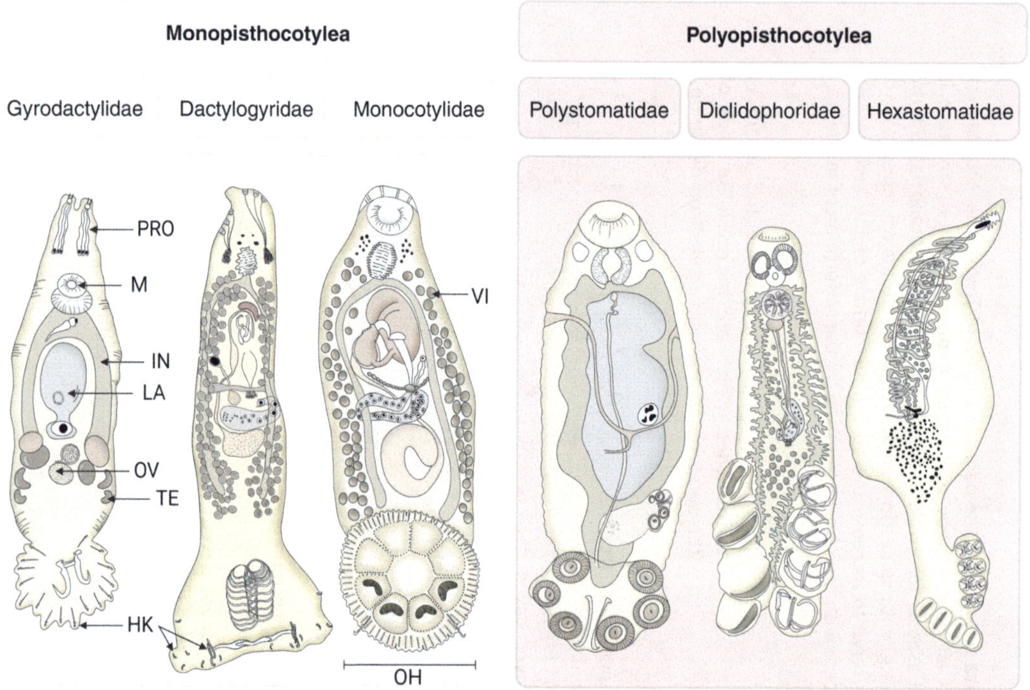

Fig. 5.2 Morphology of different families of monogeneans. *HK* hooks, *IN* intestine, *LA* larva, *M* mouth, *OH* opisthaptor, *OV* ovary, *PRO* prohaptor, *TE* testis, *VI* vitellarium. (Source: Mehlhorn 2016). Illustrations by Erasmus, A. Created in BioRender. Schwelm, J. (2025) https://BioRender.com/b44y859

are protandrous hermaphrodites and can self-fertilise, although cross-fertilisation is more common. They are either oviparous (Fig. 5.3a; e.g. *Dactylogyrus* sp. infecting the gills of cyprinids) or viviparous (Fig. 5.3b; e.g. *Macrogyrodactylus* sp. infecting the skin of siluriformes), presenting either an oncomiracidial larva or sequential embryony resembling a 'Russian doll', respectively (Kuchta et al. 2018). Fertilised eggs have a resistant shell and are often equipped with filaments. Eggs are usually laid in water and either settle to the bottom of the habitat or remain attached to the host's skin or gills by the filament. Upon development of the embryo into the oncomiracidium, this ciliated larva hatches in the water and actively searches for a suitable host. Some monogeneans continue the life cycle on the same host while most infect different individual fish. A unique version of this life cycle is that of *Diplozoon* spp., which includes an additional larval stage called a diporpa (Fig. 5.3c). As the oncomiracidium develops into a diporpa, it attaches to the fish host and looks for another diporpa with which to fuse. The two become life-long partners and may stay together for years. Larvae that do not successfully find their partner will eventually die and fail to continue the life cycle. Another remarkable case is that of *Polystoma integerrimum* (Fig. 5.3d). The adult stage of this parasite is endoparasitic in anurans and can infect the bladder and cloaca. The life cycle can take two pathways. The oncomiracidium (larval stage) can infect the outer gill branchia of tadpoles and grow into the gills as neotenic larval forms that produce eggs, from which a new generation of oncomiracidia will hatch and infect the inner gills of tadpoles (dashed arrow in Fig. 5.3d). Upon the tadpole's metamorphosis, the worm will migrate to the intestine and reach the target site, the frog's bladder. The oncomiracidium can also directly infect tadpoles and develop into endoparasitic adults without the need of an additional generation of oncomiracidia (Mehlhorn 2016).

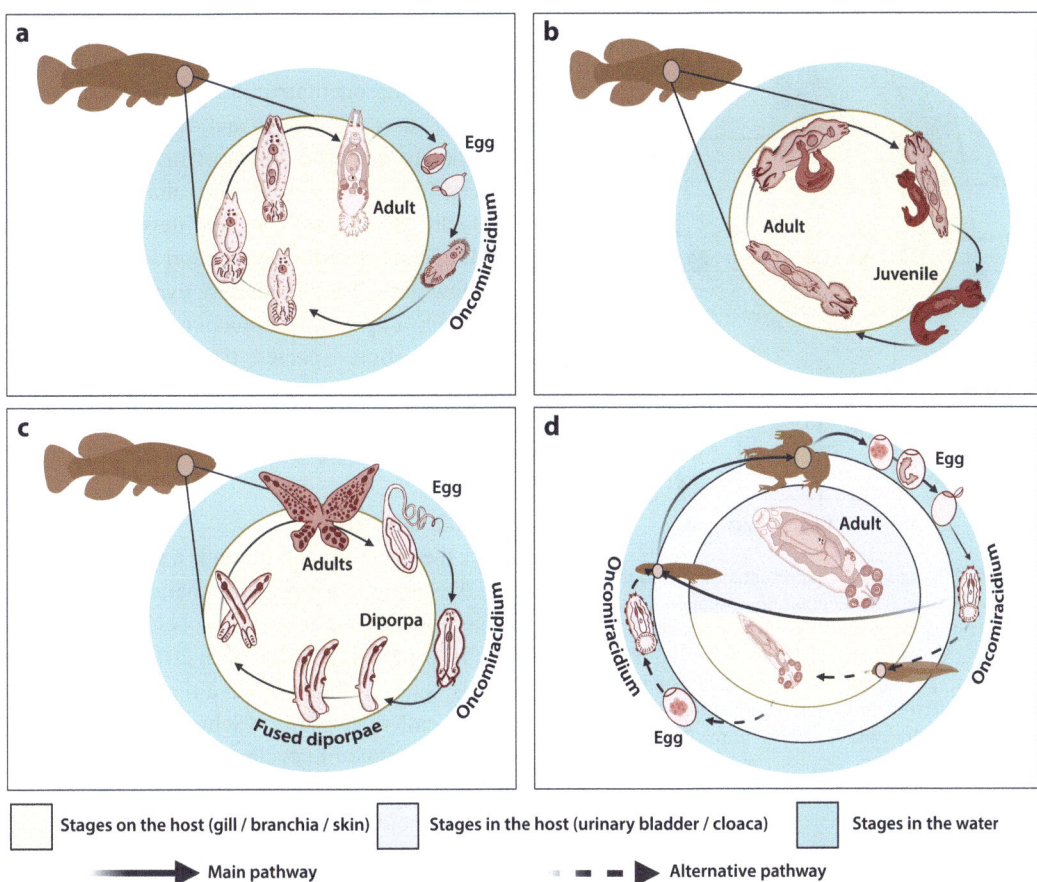

Fig. 5.3 Examples of monogeneans with an oviparous (**a**; *Dactylogyrus* sp.) or viviparous (**b**; *Macrogyrodactylus* sp.) life cycle. Some life cycles include diporpa as a larval stage (**c**; *Diplozoon* spp.). Monogeneans are found mostly in the gills of fish; however, with few exceptions, they present themselves as endoparasites (**d**; *Polystoma integerrimum*). (Drawings based on Kuchta et al. 2018 [**a**, **b**], and Mehlhorn 2016 [**c**, **d**]). Created in BioRender. Schwelm, J. (2025) https://BioRender.com/t89d703

5.2.2 Morphology of Adult and Larval Stages

5.2.2.1 Adult

Adult monogeneans are characterised by attachment organs – 'haptors' – on the anterior and the posterior parts of the worm (see Fig. 5.2). The prohaptor is located in the anterior part and usually contains suckers with sclerites (hardened structures) or clamps, and bothria. In the posterior part, the adult is equipped with an opisthaptor, which is composed of large anchors (i.e. hamuli), bars, small hooks, clamps and needles that are used as anchoring organs. The complexity of the clamps tends to be proportional to the degree of sedentarism of the worm. The anchors are paired and their structure includes inner and outer roots, points and anchor filaments. The bars connect the pair of anchors and the needles are splinter-like structures whose function is not yet elucidated. The prohaptor is used for feeding and its attachment structures are usually sclerotised. The digestive system consists of a mouth, prepharynx, pharynx and a blind intestine. The excretory system contains protonephridia, collecting ducts, paired contractile bladders and excretory pores that are located anteriorly. All monogeneans have both male and female reproductive systems (Fig. 5.4), although they remain separated and mostly engage in cross-fertilisation.

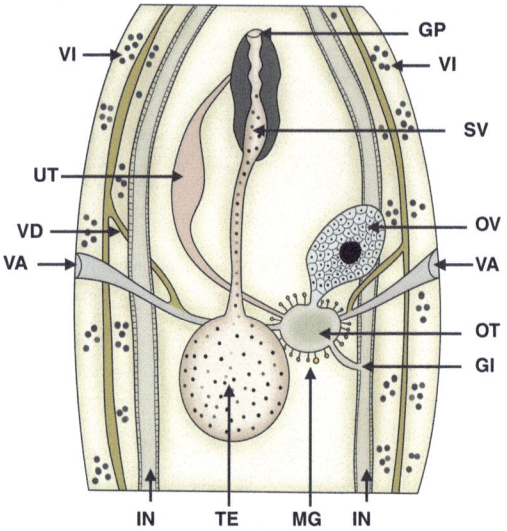

Fig. 5.4 Diagrammatic representation of monogenean reproductive systems. *GI* genitointestinal duct, *GP* genital pore, *IN* intestinal branch, *MG* Mehlis' glands, *OT* ootype, *OV* ovary, *SV* seminal vesicle, *TE* testis, *UT* uterus, *VD* vitelloduct, *VA* vagina, *VI* vitellarium. (Adapted from: Mehlhorn 2016). Illustration by Erasmus, A. Created in BioRender. Schwelm, J. (2025) https://BioRender.com/g14l497

In males, the reproductive system includes a male copulatory organ (which can be muscular, sclerotised, or an eversible cirrus), testes, an ejaculatory duct, a vas deferens, and a seminal vesicle. Female reproductive organs comprise an ovary, an oviduct, an ootype, the Mehlis' gland, a uterus, a seminal receptacle and a vagina. The structure and arrangement of the reproductive organs, particularly the male copulatory organ and the vagina, are helpful for species identification. Some species also contain photoreceptors (i.e. eye-spots) and papillae that aid with detecting mechanical stimuli (e.g. *Entobdella* spp., from Goater et al. 2014).

5.2.2.2 Egg and Oncomiracidium

Monogenean eggs are ovoid or fusiform, may have one or two filaments, and are often found in the substrate of aquatic systems or on the external surfaces of fishes (e.g. skin). Inside the egg, an embryo will develop into a ciliated oncomiracidium. When the timing or conditions are right (Whittington and Kearn 2011), the oncomiracidium hatches into the external environment. Upon hatching, the oncomiracidium will search for a suitable host to infect and, once located, it attaches to it with the help of its haptors. After successfully recognising and attaching to the host, the larva transforms and develops into an adult. Oncomiracidia have a short life span of up to a few days. They often have eye-spots that are usually lost upon development to the adult stage; however, some species retain the eye-spots. Although host recognition is still under investigation, oncomiracidia can react positively to light, host chemical cues, water current and gravity.

5.2.2.3 Diporpa

After the oncomiracidium finds a suitable host, it can develop into a diporpa, a characteristic larval stage found in monogeneans belonging to the family Diplozoidae (Mehlhorn 2016). Originally, it was believed that this larva represented a new genus called *Diporpa* (Schmidt and Roberts 1990). Once the life cycle of *Diplozoon* spp. was comprehended, it was recognised as a larval stage of diplozoids. The morphology of diporpae depends on the stage of maturation but generally it consists of two oral suckers, one ventral sucker, a pharynx, dorsal papillae, cerebral ganglia and both female and male reproductive organs (Avenant-Oldewage and Milne 2014; Mehlhorn 2016). Species such as *Paradiplozoon ichthyoxanthon* can have up to four pairs of permanent attachment clamps (Avenant-Oldewage and Milne 2014). The ventral sucker and dorsal papillae serve as attachment organs when diporpae meet. When one diporpa uses its ventral sucker to attach to the papilla of the other, fusion begins, along with further development including the maturation of gonads (Schmidt and Roberts 1990). Despite the fact that the worms contain both female and male reproductive organs, as mentioned above, cross-fertilisation occurs. This happens through the collocation of the vagina near the vas deferens of the other larva. After successful fusion, the larvae are indistinguishable from each other and, as such, the diporpa pair can mature and live in partnership for years.

5.2.3 Effects on the Host

The effects of monogeneans on the host can be severe. Attachment organs such as fish gills are mechanically damaged during attachment and reattachment, as some monogeneans are mobile and therefore can injure the host's tissue upon migration. Additionally, the constant active nutrient uptake from the host, coupled with the ease with which this parasite can reproduce and attain high infection intensities, can severely impact the host. Some of the effects that monogeneans have on their host include increased serum antibodies, anaemia, tissue damage and necrosis, as often seen in *Gyrodactylus salaris* infecting Atlantic salmon (*Salmo salar*) in Scandinavian waters (Bakke et al. 2007). This parasite's viviparous reproductive strategy, which involves the production of already-gravid neonates, makes it very easy to achieve high infection loads. Utilising proteolytic enzymes, the parasite dissolves the host's tissue for subsequent ingestion. This parasite significantly reduces salmon populations due to its high capacity for reproduction and severe damage to the host's tissue, even eradicating entire host populations (Bakke et al. 2007). A similar example is the invasive monogenean, *Neoheterobothrium affine*, lethally parasitising olive flounder, *Paralichthys olivaceus*, in Japan and Korea (see Yoshinaga et al. 2009). Despite some monogeneans being fatal to their host, severe pathological effects are most often observed in fish captivity or under intensive aquaculture conditions (see Chap. 23) and less often in the wild.

5.2.4 Biology and Life Cycle of Trematoda

Trematodes, also referred to as flukes, are a diverse class of endoparasitic flatworms that parasitise both invertebrates and vertebrates. They comprise two subclasses: the Aspidogastrea and the Digenea. Digeneans are the most diverse subclass of trematodes, with approximately 18,000 species currently known, relating to roughly 2,500–2,700 nominal genera (Kostadinova and Pérez-del Olmo 2014; Pérez-Ponce De León and Hernández-Mena 2019), whereas Aspidogastrea is thought to be an archaic class of trematodes, with only about 60 species described belonging to four families (Alves et al. 2015). Trematodes have a worldwide distribution and can be found in both freshwater and marine environments. With few exceptions (e.g. schistosomes), trematodes are hermaphroditic. During their life cycle, the different developmental stages alternate between sexual and asexual reproduction, a process referred to as metagenesis.

5.2.4.1 Trematode Life Cycles

Aspidogastrea Life Cycle

The life cycle of aspidogastreans is the simplest among trematodes (see Fig. 5.5). Aspidogastrean adults differ from digenean adults in having a ventrally located adhesive disc (Baer's disc) that aids in attachment to the host through septa and alveoli, or suckers (Fig. 5.5a). They often lack asexually produced free-living larval stages, usually having only one host, in most cases a mollusc, although some species infect vertebrates (e.g. elasmobranchs, teleosts and turtles) as facultative or compulsory hosts. In the case of aspidogastreans that have an obligate vertebrate host, molluscs serve as intermediate hosts. *Lobatostoma manteri* is an example that uses the Snubnosed Dart *Trachinotus blochii* (Carangidae) as final host and prosobranch snails as intermediate host (see Rohde 1973).

Despite the fact that aspidogastreans lack asexually produced free-living larval stages, some species, for example, *Multicotyle purvisi* and *Cotylaspis insignis* (Fig. 5.5c), have cotylocidial larvae produced through sexual reproduction (Rohde 1972; Rosen et al. 2016). Cotylocidia can be found as fully developed or developing larvae in eggs. After the eggs are laid, the larva hatches in response to a light stimulus and can live for several hours to several days. The larva is positively phototactic and, after hatching, it either swims, attaches to the substrate, or drifts in the water until it encounters a mollusc to infect. Infection occurs passively through syphon current and breathing action of the mollusc or

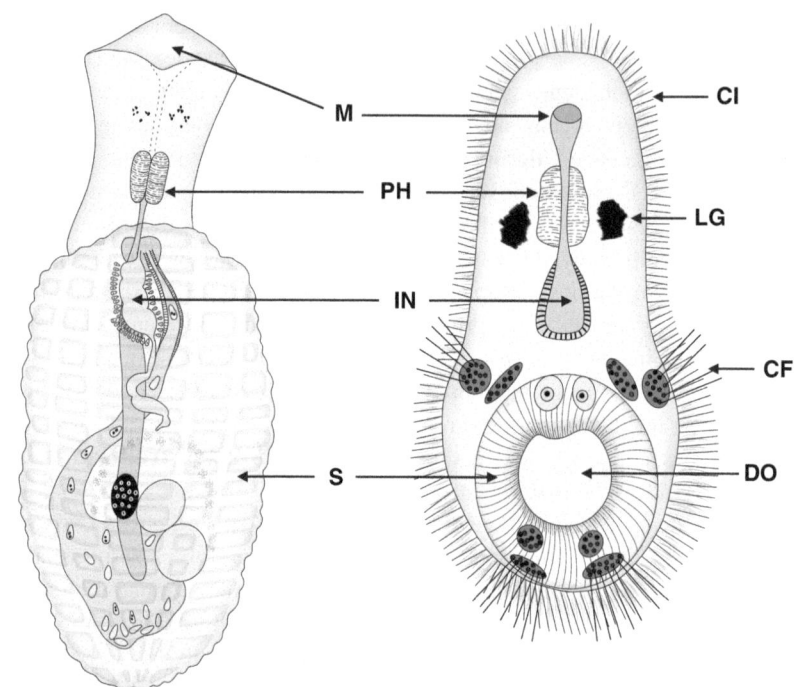

Fig. 5.5 Adult (left) and larval (right) Aspidogastrean trematode from ventral view. *CF* ciliated field, *CI* cilia, *DO* development of opisthaptor, *IN* intestine, *LG* lateral larval glands, *M* mouth, *PH* pharynx, *S* ventral sucker (Baer's disc). (Adapted from: Mehlhorn 2008). Illustrations by Erasmus, A. Created in BioRender. Schwelm, J. (2025) https://BioRender.com/b80q726

simply through ingestion. The cotylocidium generally has a high resemblance to the adult larval stage and, therefore, it does not undergo significant metamorphosis. Aspidogastreans have a low specificity towards their molluscan hosts where they reach maturity allowing the life cycle to continue.

Digenean Life Cycle

The life cycle of digeneans often consists of two or three hosts, while some species may have only one, and still others up to four hosts (see Fig. 5.6). The definitive host is usually a vertebrate. In the definitive host, sexual reproduction takes place, leading to eggs being shed into the host gut or bloodstream and ultimately released into the environment. Depending on the species, a ciliated miracidial larva will hatch in the external environment and actively search for the first intermediate host, or wait passively inside the egg until it is ingested by the first intermediate host, which is usually a mollusc. Once the gastropod is infected, the miracidium migrates to the target organ (usually the gonads) and develops into a mother sporocyst. This mother sporocyst will continue developing and produce a generation of daughter sporocysts or rediae (i.e. parthenitae). These parthenitae maintain the infection in the gastropod via asexual reproduction, massively multiplying and indefinitely producing both new generations of parthenitae and the next, free-living, larval stage, the cercariae. This stage of asexual reproduction occurs throughout the gastropod's lifetime, and it is typically only limited by the host's resources. Once developed, cercariae exit the parthenitae and emerge into the external environment. The emergence of cercariae can be triggered by several factors such as light, temperature, or circadian rhythms, synchronised with periods of high activity of the next host. Upon emergence, cercariae seek to infect the next host, which is often a second intermediate host belonging to molluscs, crustaceans, fishes, amphibians, or aquatic insects in trixenous (three-host) life cycles. This asexual cercarial production and infection of a second intermediate host contributes to the dispersal of the parasite. Due to the lecithotrophic nature of the larvae, their lifes-

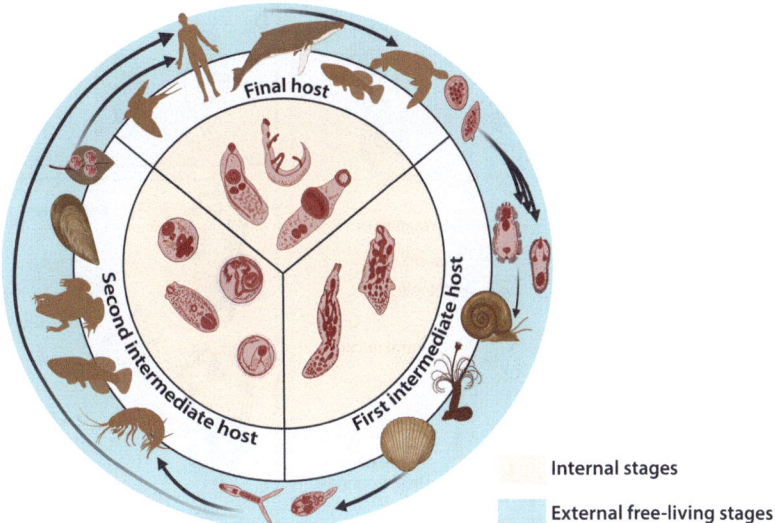

Fig. 5.6 The life cycle of digenean trematodes often consists of three hosts. In the first intermediate host, asexual reproduction occurs and parthenitae (i.e. rediae or sporocysts) will produce cercariae. Cercariae infect the second intermediate host and encyst as metacercariae, which will be trophically transmitted to the final host. In some species, cercariae encyst on substrates such as vegetation or directly infect the final host. After transmission to the final host, sexual reproduction takes place and eggs are produced. Eggs, once shed together with faeces, will hatch into miracidia, the infective larval stage to the first intermediate host, and as such the life cycle proceeds. Created in BioRender. Schwelm, J. (2025) https://BioRender.com/w57e362

pan is typically limited to a few hours; consequently, they must emerge in synchrony with the activity of the downstream host. When the cercaria successfully infects the second intermediate host, it encysts as a metacercaria and rests in a semi-dormant stage awaiting trophic transmission to the final host. During this stage, the second intermediate host can accumulate several and even hundreds of metacercarial cysts before trophic transmission, increasing the effectiveness of the life cycle through exogenous accumulation (Galaktionov and Dobrovolskij 2003). In species with dixenous (two-host) life cycles, encystment can occur on substrates such as vegetation (e.g. *Fasciola* spp. and *Fasciolopsis* spp.). Alternatively, cercariae can directly infect the definitive host via direct penetration (e.g. Schistosomatidae) or through trophic interactions without undergoing encystment (e.g. Azygiidae, Bivesiculidae) (Schell 1970). Following ingestion by a suitable host, the metacercaria will excyst and migrate to a suitable organ, where it can develop into an adult and reproduce sexually, continuing the life cycle.

5.2.5 Morphology of Digenean Adult and Larval Stages

5.2.5.1 Adult

The morphology of adult digeneans (Fig. 5.7a) is highly diverse. Oral and ventral suckers (acetabula) that serve as attachment organs can be observed in many species while some lack suckers entirely (e.g. Bivesiculidae, Cyclocoelidae). Spines on the oral sucker and tegument can also serve as attachment organs in other species (e.g. Echinostomatidae). Adult trematodes possess digestive, excretory, nervous and reproductive systems that occur in the parasite's parenchyma, a complex acellular matrix possessed in lieu of a coelom. All of these systems are surrounded by the tegument (i.e. neodermis) that serves as a barrier against external agents and as a sensory organ, while facilitating metabolic exchange to meet nutritional and excretory requirements (Goater et al. 2014). The neodermis (Fig. 5.1) consists of a glycocalyx, a living plasma membrane that provides protection against host defences, and a syncytium, which is highly meta-

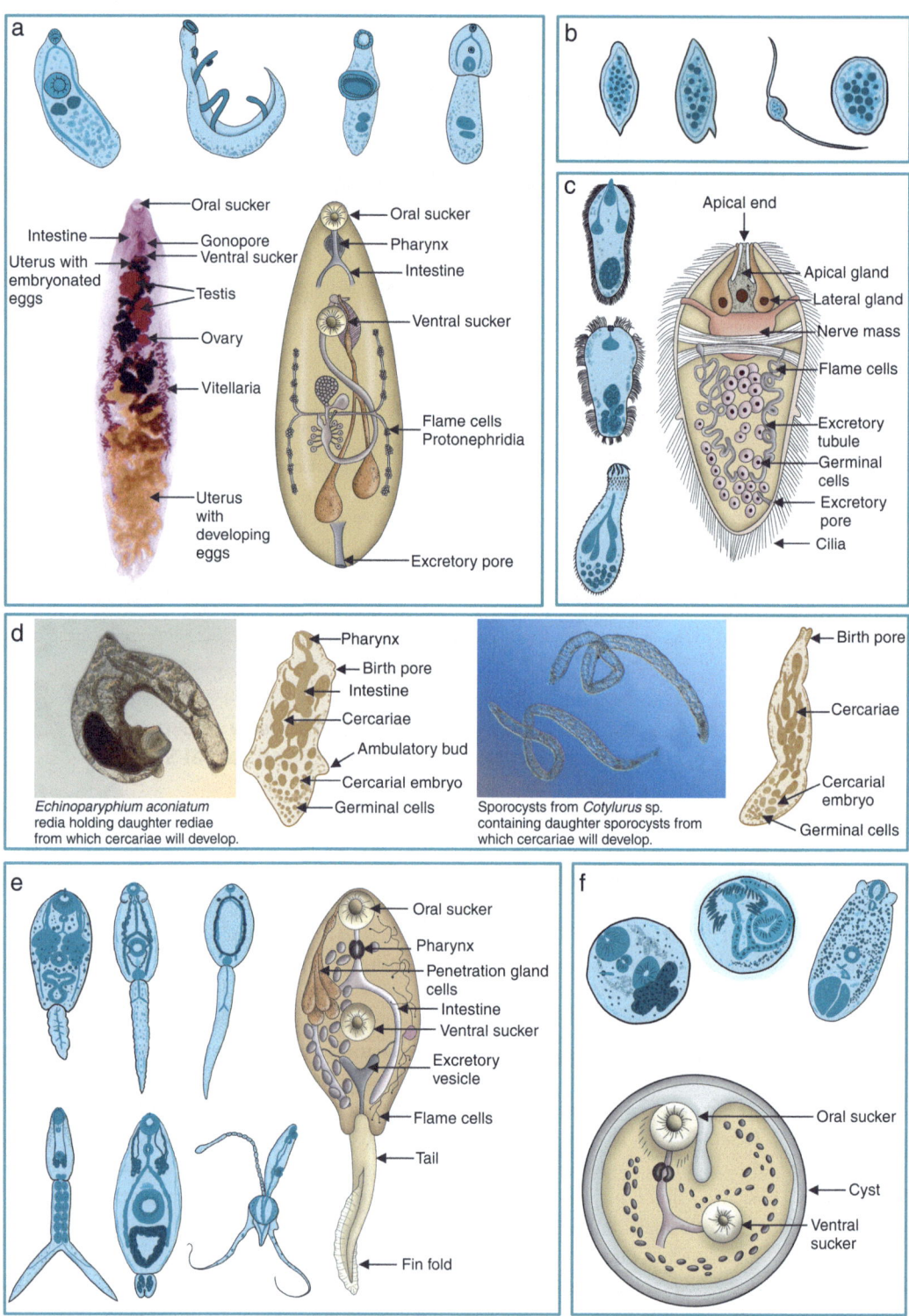

Fig. 5.7 Morphology schematics and morphological variations (blue sketches) of adult (**a**), eggs (**b**), miracidia (**c**), parthenitae (**d**), cercariae (**e**), and metacercariae (**f**) of digeneans. (Drawings were reprinted or adapted from illustrations by Sures 2021 [schematic **a** and pictures in **a**, **d**], Mehlhorn 2016 [**b**], Galaktionov and Dobrovolskij 2003 [**c**], Schell 1970 [schematics in **d** and **e**], and Goater et al. 2014 [blue sketches in **e**]). Redrawn by Erasmus, A. and Schwelm, J. Created in BioRender. Schwelm, J. (2025) https://BioRender.com/r87p220

bolically active and serves as the primary portal for nutrient exchange. The digestive system is incomplete and is usually (but not always) composed of a mouth, a pre-pharynx, a pharynx, an oesophagus and caeca (Fig. 5.7a). The number and ending position of caeca and the presence or absence of a pharynx are commonly used as an identification trait. In addition to active feeding, trematodes can also absorb nutrients through their tegument. A typical trematode will have flame cells (i.e. protonephridia), ducts and an excretory vesicle for excretion. Similarly, the nervous system in some families (e.g. Deropristidae, Cryptogonimidae and Lepocreadiidae) consists primarily of eye-spots and a transversal commissure that connects a pair of nerve cords extending from a primitive brain (Schell 1970). The nervous system includes reactions to mechanical, light and osmotic stimuli. The reproductive system is usually hermaphroditic (Fig. 5.7a) with the exception of species of the family Schistosomatidae, which have separate sexes. The male reproductive organ consists of testes, a vas deferens, a cirrus sac (including a prostate gland and a seminal vesicle) and a protrusible cirrus. The female's reproductive system comprises a gonopore, a seminal receptacle, a single ovary producing egg cells and an oviduct. The Laurer's canal connects the oviduct to the rest of the body. Fertilised eggs travel through the oviduct to an ootype surrounded by the Mehlis' gland, where they meet the vitelline duct connecting eggs to vitellaria, which supply the eggs with yolk and molecules that aid in producing the eggshell.

5.2.5.2 Egg

Trematode eggs (Fig. 5.7b) are made up of an ovum, sperm, lipoproteins and vitelline cells (Galaktionov and Dobrovolskij 2003). When fertilised, these components result in the miracidium (or cotylocidium in some aspidogastrids), the first of the infective free-living trematode larvae (Fig. 5.7c). The eggshell's features, including shape, size and colour, are often essential for identification (Schell 1970). These features highly depend on the hatching strategy, which can occur in the external environment or once ingested by a suitable intermediate host. For instance, eggs that hatch in the external environment contain an operculum through which the larva can emerge. With very few exceptions, the developed larva has ciliated epidermal plates that are used for locomotion. The larva is rich in germinal cells and contains penetration and apical glands, subepidermal cells, flame cells, excretory tubules and a central nervous system (CNS), with a few species also containing eye-spots and spines (Galaktionov and Dobrovolskij 2003).

5.2.5.3 Parthenitae: Sporocyst and Redia

Upon infection of the first intermediate host, the miracidium will undergo metamorphosis into a mother sporocyst. The mother sporocyst will produce a second generation of daughter sporocysts or daughter rediae (Fig. 5.7d). Sporocysts can appear as an ovoid mass with an excretory system and a birth pore or, in some cases, as branched sporocysts with pigmented brood sacs (e.g. *Leucochloridium paradoxum*). Because sporocysts are typically immobile and are not equipped with a mouth or intestine, they do not actively consume host tissue. Conversely, rediae have a mouth, pharynx and intestine and actively feed on host tissue or even the sporocysts of other competing trematode species. An example of the latter is the capacity of *Himasthla elongata* rediae to displace *Renicola parvicaudatus* infections in periwinkles by preying on sporocysts through the division of labour into different castes (soldier and reproductive castes), which has been observed in other trematode species (Hechinger et al. 2011; Nielsen et al. 2014). Like the sporocyst, rediae also have a birth pore and flame cells. These parthenitae (i.e. daughter rediae and sporocysts) contain a mixture of stem and somatic cells (Goater et al. 2014) and will continuously produce the next free-swimming infective larval stages, that is, cercariae.

5.2.5.4 Cercaria

The morphology of cercariae (Fig. 5.7e) often includes a tail that allows self-propelling in the water column. Depending on the species, this tail can have numerous expressions, such as bifurcation (e.g. furcocercous cercariae) and finfolds (e.g. parapleurolophocercous cercariae). Interestingly, cercariae present different adaptations depending on the behaviour and distribution of the intermediate or definitive host. For instance, cercariae from the family Opecoelidae are tail-truncated larvae that attach to the substrate and ambush intermediate hosts associated with benthic environments such as amphipods (Galaktionov and Dobrovolskij 2003). Other cercariae swim in the water column so they can be taken up by filter feeders such as bivalves, while others have a sword-like structure (i.e. stylet) and penetration glands that secrete, among other things, proteolytic enzymes, to facilitate direct penetration of the host tissue and encyst as metacercariae. Even more intriguing are hemiurid cercariae, which float in the water column and have refractile droplets attractive to zooplankton such as copepods that serve as second intermediate hosts (Galaktionov and Dobrovolskij 2003). These transmission strategies involve trade-offs between transmission rate and absolute lifespan, with less-active cercariae (e.g. hemiurids) living for days to months and active swimming cercariae (e.g. furcocercous) only living for a few hours. Cercariae also have an excretory system, which includes an excretory vesicle, flame cells and excretory ducts. The shape of the excretory vesicle, the number of flame cells, and the length and terminal location of the ducts are critical parameters for species identification. The presence of eye-spots, suckers (ventral and oral), a pharynx, the number and position of cystogenous (cyst-forming) glands, the presence, number and arrangement of ventral and dorsal spines, and the morphology of intestinal ceca are also important aspects of cercarial morphology. Finally, other essential diagnostic features include those associated with penetration, such as the presence and shape of the stylet and the presence and number of penetration glands.

5.2.5.5 Metacercaria

As with cercariae, the metacercarial larval stage comes with variations (Fig. 5.7f). Some species form metacercariae that (1) are encysted without having metabolic exchange with the host; (2) those that are encysted and remain metabolically active; and (3) those that remain unencysted (Galaktionov and Dobrovolskij 2003). Metacercariae that do not present metabolic exchange with the host usually have a thick multilayered cyst. This characteristic is common among metacercariae that do not undergo further metamorphosis. Members of the family Echinostomatidae, Monorchiidae and Renicolidae are exemples. Metacercariae that experience further metamorphic changes, including growth into progenetic cysts, have a simplified thin-walled cyst consisting of a maximum of two walls, such as members of the families Heterophyidae, Plagiorchiidae, Ochetosomatidae and Lecithodendriidae. Other species entirely lack cysts and are instead surrounded by a capsule of connective tissue produced by the host (e.g. Gymnophallidae, Strigeidae, Diplostomidae) while undergoing further metamorphosis. Other variations include metacercariae that encyst on vegetation, such as *Fasciola* spp., which produce a four-layered cyst and are therefore particularly well-protected against harmful environmental influences.

5.2.6 Effects on the Host

The effects that trematodes have on their hosts vary depending on the life stage and the location of infection. In the case of adult trematodes occurring in the gut of vertebrates, they often feed on the intestinal mucosa, and the extent of the damage is primarily limited to mechanical harm to the intestinal epithelium, including necrosis and degeneration (Dezfuli et al. 2018). However, such infections are rarely fatal. In infections of humans with schistosomes, acute symptoms such as fever, cough, myalgia, malaise, diarrhoea, abdominal pain and eosinophilia occur among others (Gryseels et al. 2006). However, in chronic infections, symptoms predominantly arise from eggs trapped in tissues, rather than adult worms. This is mostly attributed to proteolytic enzymes secreted by the eggs that induce inflammation and, ultimately, fibrosis (Gryseels et al. 2006). Lesions can be fatal, particularly when this fibrotic tissue occurs in the liver (i.e. hepatic schistosomiasis) (Gryseels et al. 2006), making it the second most devastating disease worldwide after malaria. During the parthenitic stage, infections by trematode intermediate stages can castrate the gastropod host by replacing the gonadal tissue with parasite biomass through feeding (Kuris 1974). Moreover, trematodes can hijack energy reserves from the host, such as carbohydrates, to supply the energy requirements of the parasite (Pinheiro et al. 2009). As a result of castration, some trematodes can induce gigantism in snails depending on the life history of the gastropod. Sousa (1983) hypothesised that in short-lived semelparous snails, the parasite could benefit from inducing gigantism by exploiting the available space for cercarial production over the snail's lifetime. However, the parasite has not developed this adaptation for long-lived gastropods, as the longer lifespan of the gastropod host already allows the parasite to reproduce asexually over an extended period. Finally, infections with metacercariae are thought to have lower virulence due to their semi-dormant character. All of the effects that metacercariae will have on their host will likely depend on the infection intensity, how metabolically active metacercariae are and the presence and thickness of the barrier (i.e. cyst) between the larva and host. Some trematode larvae invade the eye's vitreous humour of their fish host (e.g. *Diplostomum* spp.) and, at high infection intensities, they can induce dietary changes by impairing vision and therefore the ability to see smaller prey (Vivas Muñoz et al. 2021). In other hosts, such as bivalves, metacercarial infections can induce effects such as biochemical alterations and decreased filtration capacity in mussels (Magalhães et al. 2020). A more severe example of metacercarial effects is that of the trematode *Ribeiroia ondatrae* on its amphibian host. When cercariae encyst in tadpoles, they often result in leg malformations, and a high proportion of the frogs often do not survive the infection (Johnson et al. 2004). Since infected frogs die before reaching reproductive age, this parasite directly reduces the population's fitness. Other trematode metacercariae can cause behavioural changes such as decreased burrowing capacity due to foot damage in cockles (Mouritsen 2002; Babirat et al. 2004) and changes in microhabitat selection. In exceptional cases, non-encysting metacercariae have been observed to be highly active, and to even ingest host tissue, as is the case with accacoeliid and hemiurid metacercariae infecting *Physalia physalis* (Louvard et al. 2023). However, this is an open venue of research since the biology and life cycles of accacoeliid and hemiurid are scarcely understood.

5.2.7 Biology and Life Cycle of Cestoda

Cestodes, commonly known as tapeworms, exhibit highly specialised lifestyles, inhabiting almost exclusively the digestive tracts of a wide range of vertebrate hosts, including aquatic organisms, such as elasmobranchs (sharks, rays and chimaeras), ray-finned fishes (teleosts), reptiles and marine mammals. Cestodes that parasitise fishes (Chondrichthyes and Actinopterygii) make up approximately one-third of the total diversity of tapeworm species, which amounts to 1,670 species out of around 5,000 known cestode species (Scholz and Kuchta 2022).

Cestodes have evolved remarkable adaptations to ensure their survival and reproduction within their hosts. The systematics of cestodes is an ongoing field of research, and our understanding of their evolutionary relationships continues to evolve as new data become available. Traditionally, two subclasses of cestodes were recognised (Eucestoda and Cestodaria). However, recent studies show that while the subclass Eucestoda is supported as a monophyletic group, the monophyly of the Cestodaria is not supported (Scholz and Kuchta 2022; Caira et al. 2017a). Nevertheless, the two groups differ in some significant characteristics, such as the number of larval hooks and the morphology of adults. Species of Cestodaria possess ten hooks (decanth) in the larval stage and show unsegmented bodies without a scolex as adults. The subclass Eucestoda features larvae with six hooks (hexacanth) and includes numerous medically and economically relevant species. In this chapter, we will only very briefly discuss the Cestodaria and concentrate mainly on the Eucestoda.

5.2.7.1 Cestodaria

Although the monophyly of the group traditionally referred to as Cestodaria is not supported (Caira et al. 2017a; Scholz and Kuchta 2022), we will use the term Cestodaria here in the absence of an alternative to summarise and present the two most basal orders of tapeworms, namely the Amphilinidea and the Gyrocotylidea. Cestodaria, unlike Eucestoda, are often seen as more primitive or ancestral tapeworms. Cestodaria typically have a simpler body structure compared to Eucestoda. They lack the complex segmentation seen in Eucestoda, instead exhibiting a more continuous, unsegmented body plan. Another notable distinction is in their life cycles. Cestodaria often have more direct life cycles compared to the complex, multi-host life cycles seen in Eucestoda. While members of the Eucestoda typically require multiple intermediate hosts, Cestodaria often have simpler life cycles with fewer host species involved.

The order Amphilinidea stands out as a rather small order, comprising merely eight recognised species in six genera. These monozoic, large cestodes are characterised by a flattened leaf-like body, without any distinct organs of attachment (Fig. 5.8). As adults, they inhabit the body cavity of chondrosteans, freshwater and marine teleosts, and a single species has been identified in freshwater turtles. Most authors consider the Amphilinidea to be one of the two evolutionarily ancient groups of cestodes. However, insufficient data are available to assess the interrelationships between this small order and other cestode orders, or to decipher patterns in its host communities and geographic distribution (Scholz and Kuchta 2017; Scholz and Kuchta 2022). Similar to the Amphilinidea, the Gyrocotylidea is also a rather small order. It consists of ten recognised species, all of which belong to one genus. They are also large and monozoic with a flattened leaf-like body (Fig. 5.8). Unlike the Amphilinidea, they have a sucker-like organ at the anterior end. The body usually ends in a rosette-like adhesive organ (also called a rosette organ) in the shape of a funnel. The lateral margins of the body are usually, but not always, plicate or crenulated (Kuchta et al. 2017). They are known from all oceans and mainly inhabit the deep sea. As adults, they exclusively parasitise holocephalans and are considered highly host-specific, with each species parasitising only a single holocephalan species.

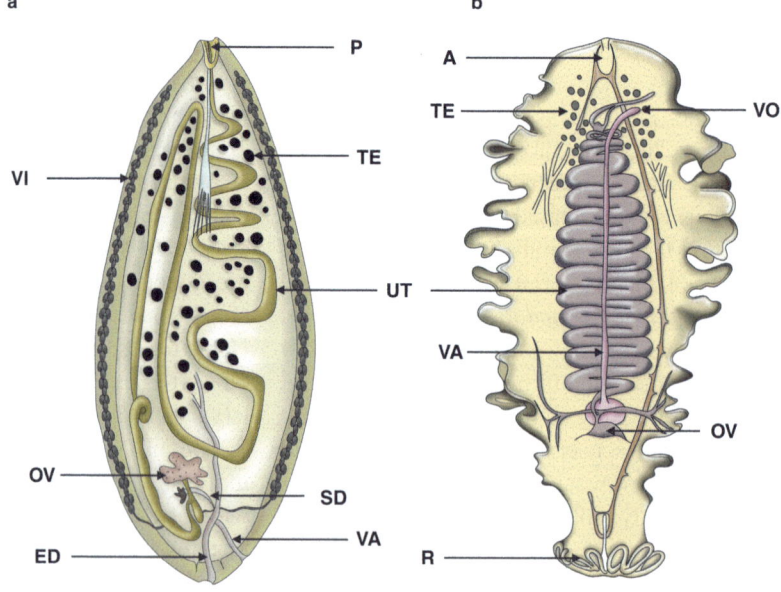

Fig. 5.8 Schematic morphology of cestodes belonging to the order (**a**) Amphilinidea and (**b**) Gyrocotylidea. *A* acetabulum, *ED* ejaculatory duct, *OV* ovary, *P* proboscis, *R* rosette, *SD* sperm duct, *TE* testis, *UT* uterus, *VA* vagina, *VI* vitellarium, *VO* vaginal opening. Illustrations by Erasmus, A. Created in BioRender. Schwelm, J. (2025) https://BioRender.com/r07v162

However, it is common for a holocephalan species to host more than one species of Gyrocotylidea (Kuchta et al. 2017).

5.2.8 Eucestoda Life Cycles

Current understanding of host specificity within the class Cestoda reveals a considerable degree of variability. Some cestodes display a strict preference for specific hosts, while others exhibit a less strict host specificity. It is important to note that the life cycles within the different orders presented here can also vary greatly from the exemplary cases shown below.

The life cycle of eucestodes is complex and usually involves two intermediate hosts (see Fig. 5.9). The eggs are passed in the faeces of the definitive host. The coracidium hatches from the egg and must be consumed by the first intermediate host. In its intestine, its ciliated epithelium dissolves and the oncosphere is released. The oncosphere penetrates the gut wall, enters the hemocoel and undergoes development, transforming into the procercoid larva. Once the infected first intermediate host is ingested by the second intermediate host, usually through predation, the procercoid larva is released, penetrates the gut wall and migrates to the species-specific infection sites, such as the body cavity, musculature or other tissues. Here, it undergoes further development and massive growth, transforming into the plerocercoid larva, which is the infective stage for the definitive host. The plerocercoid larva can remain viable in the second intermediate host for extended periods. After the second intermediate host is consumed by the definitive host, the plerocercoid larva attaches itself to the intestinal wall of the final host and eventually develops into an adult tapeworm. The adult tapeworm absorbs nutrients from the host's digestive system and produces eggs, which are

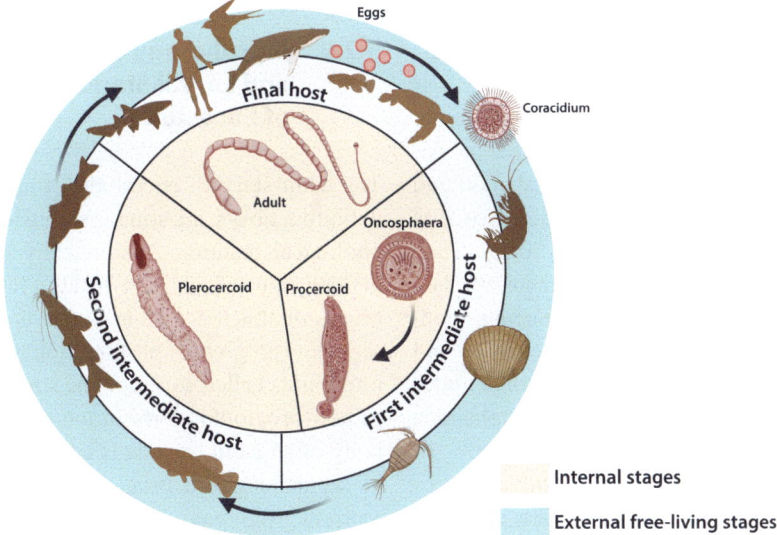

Fig. 5.9 Schematic life cycle of Eucestoda. Members of the Eucestoda usually have a complex life cycle with two intermediate hosts. Eggs are passed in the definitive host's faeces. The coracidium hatches and is consumed by the first intermediate host. In its intestine, the oncospherea is released and penetrates the gut wall, developing into the procercoid larva usually in the haemocoel. When the first intermediate host is eaten by the second intermediate host, the procercoid larva is released and migrates to specific locations, growing into the plerocercoid larva. After preying on the second intermediate host, the plerocercoid larva will be released in the intestine of the final host, where it attaches to the intestinal wall and becomes an adult tapeworm, which then can produce eggs that are shed in the host's faeces, completing the cycle. Created in BioRender. Schwelm, J. (2025) https://BioRender.com/h51m960

shed in the host's faeces, completing the life cycle. Paratenic hosts (e.g. piscivorous fishes) are often used to bridge the gap between definitive hosts and planktivorous fishes (Goater et al. 2014).

Tapeworms belonging to the family Diphyllobothriidae, commonly referred to as broad tapeworms, are predominantly large parasites that primarily infect wildlife but can also affect humans as either their natural or accidental hosts (see also Chap. 19). The life cycles of the Diphyllobothriidae are invariably associated with aquatic environments, whether freshwater or marine, due to the fact that the initial larval stage (coracidium, see Fig. 5.9) is aquatic and requires a transition to the first aquatic intermediate host, which is typically a copepod. These tapeworms also have two intermediate hosts: copepods serve as the first intermediate hosts, and vertebrates act as the second intermediate hosts (Scholz et al. 2019). The cestode order Onchoproteocephalidea can be divided into two groups considering morphology and host associations: taxa that primarily parasitise freshwater fishes, frogs, snakes and lizards (i.e. those formerly included in the Proteocephalidea and referred to as Onchoproteocephalidea I in relevant literature [de Chambrier et al. 2017]) and those that parasitise elasmobranchs (i.e. batoids and sharks) and referred to as Onchoproteocephalidea II (Caira et al. 2017b; Scholz and Kuchta 2022). They encompass the most species-rich genus of elasmobranch tapeworms, *Acanthobothrium*, with more than 200 recognised species (Scholz and Kuchta 2022). Tapeworms of the former order Proteocephalidea (now Onchoproteocephalidea I) are parasites of freshwater teleosts (60% of species) and some tetrapods (amphibians, reptiles and one in mammals). They are currently placed in a single family, Proteocephalidae (Scholz and Kuchta 2022). Adult bothriocephalidean cestodes parasitise ray-finned fishes with a few exceptions that use lungless salamanders as definitive hosts (Kuchta and Scholz 2017). The non-monophyletic group 'Tetraphyllidea' consists of ten clades, including cestodes with highly diverse morphology (Caira and Jensen 2022).

Each of the ten clades generally parasitises a relatively small subgroup of elasmobranchs, that is, sharks and rays (Caira et al. 2017c). One of the most common metazoan parasites of marine fishes are cestodes of the order Trypanorhyncha, adults of which parasitise elasmobranchs (Palm et al. 2009). They are currently known from almost all major groups of elasmobranchs worldwide and adults have been described from all eight orders of sharks (24 of 35 families) and all four orders of batoids (Goater et al. 2014; Beveridge et al. 2017). In some trypanorhynch species, procercoids develop in crustaceans and plerocercoids in planktivorous fishes. The caryophyllideans are small intestinal parasites of freshwater teleosts. As adults they are mainly associated with cyprinids, catfishes and catostomids, although a few parasitise the coelom of freshwater oligochaetes. The life cycle includes the definitive fish host and an oligochaete as second intermediate host that is castrated when infected (Goater et al. 2014).

5.2.9 Morphology of Adults and Larval Stages of Eucestoda

The adult stage of eucestodes, found within the definitive host's intestine, exhibits distinct morphological features. Anatomically, cestodes are divided into a scolex, or head, which bears the organs of attachment, a neck that is the region of segment (proglottid) proliferation and a chain of proglottids called strobila. The strobila elongates as new proglottids form in the neck region. The body of an adult cestode is composed of a chain of segments called proglottids, each containing reproductive organs. The proglottids are arranged sequentially along the anterior–posterior axis, with the youngest, or immature, proglottids located near the neck region and the mature proglottids situated towards the posterior end (Fig. 5.10). An exception among the Eucestoda are the Caryophyllidea, which have a scolex like the other Eucestoda, but are characterised by the possession of a monozoic body type, which is

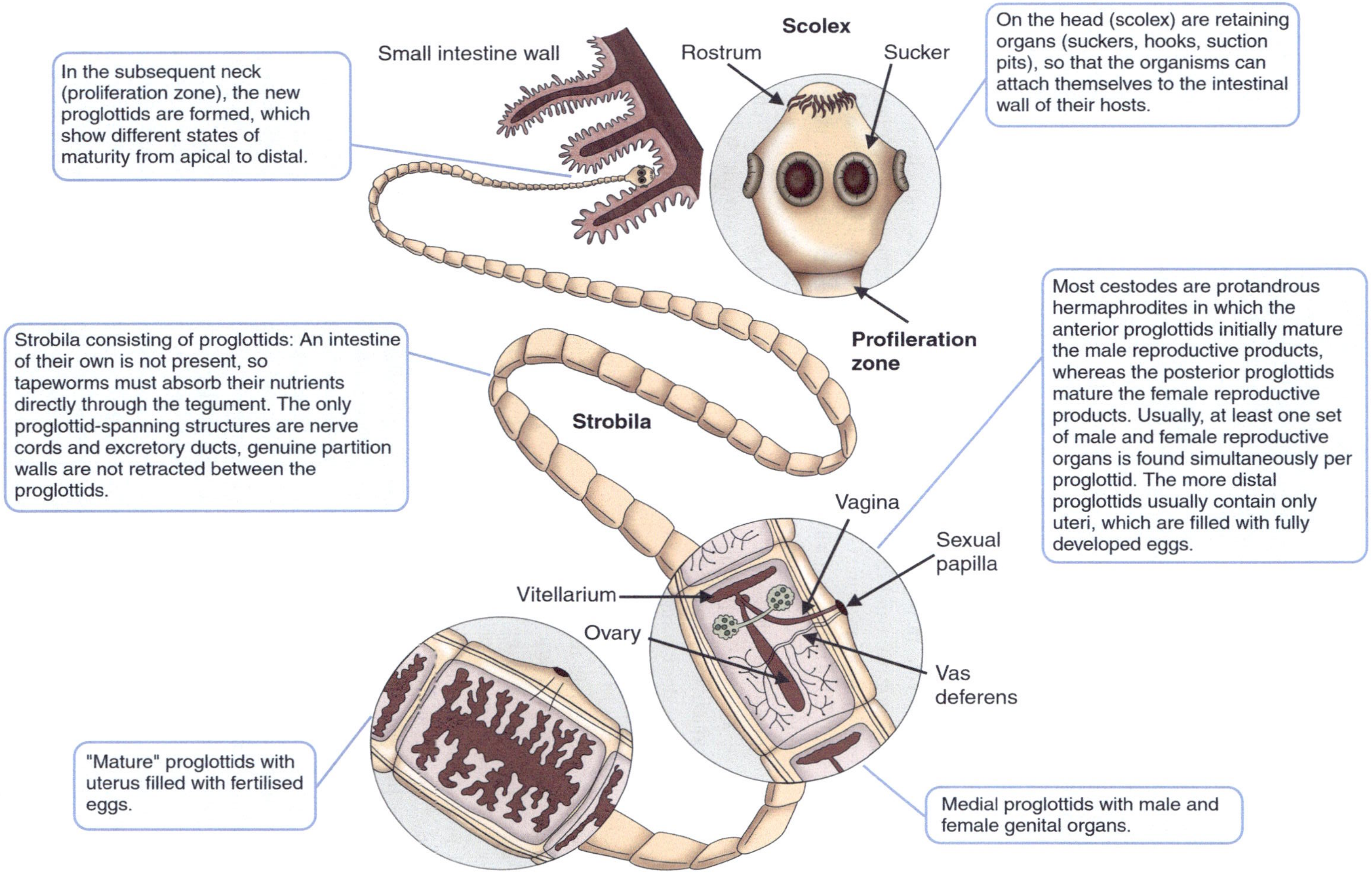

Fig. 5.10 Schematic representation of an adult tapeworm with detailed zooms of proglottids at different stages of development. (Adapted from Sures 2021). Illustrations by Erasmus, A. Created in BioRender. Schwelm, J. (2025) https://BioRender.com/631334

astrobilate and consists of only one set of male and female organs (Oros et al. 2010; Scholz and Oros 2017).

The scolex is equipped with specialised structures for attachment to the intestinal wall of the definitive host, which varies in morphology across cestode taxa but commonly features bothria, bothridia and acetabula (Goater et al. 2014). Cestodes lack an intestine; instead, all nutrients are absorbed through the body surface, the neodermis. In order to enhance the efficiency of nutrient absorption, microtriches have evolved within the neodermis (see Fig. 5.1). These microtriches increase the absorptive surface area, aiding the parasite in its competition with the host for nutrients within the intestine. The tegument of eucestodes also plays a role in immune evasion, enabling the tapeworm to avoid detection and attack by the host's immune system (Goater et al. 2014; Lucius et al. 2018). The majority of cestodes exhibit monoecious characteristics, wherein each proglottid in the adult stage possesses a reproductive system comprising both male and female reproductive organs (Fig. 5.10). The male reproductive system comprises testes responsible for sperm production, along with a vas deferens that connects to a cirrus, a muscular copulatory organ. The female reproductive system consists of one or more ovaries, oviducts and a seminal receptacle for receiving sperm during copulation. Fertilised eggs are produced within the gravid proglottids, typically situated at the posterior end of the worm, poised for release and initiation of the life cycle. While both self- and cross-fertilisation can occur, self-fertilisation is generally avoided due to protandry or protogyny (Goater et al. 2014).

The development of cestodes from eggs to sexually mature worms is a typically metamorphic process. During this development, the larva progresses through at least two stages within one or more intermediate hosts, with the final host needing to ingest the last intermediate host. Cestode larvae undergo striking morphological transformations while in the intermediate host (Fig. 5.11). The initial stage, known as the six-hook larva or oncosphere, emerges from the cestode egg in all eucestodes. This oncosphere assumes a spherical shape and is encased in a protective outer layer called the oncosphere membrane (Fig. 5.11a). This membrane often features hooks or spines that aid in tissue penetration within the intermediate host. In the context of the taxa being discussed here, the oncosphere possesses a membrane covered in cilia, which represents a specific adaptation to the aquatic environment, enabling purposeful movements in water. Because of the ring-like arrangement of these cilia, this particular form is referred to as a coracidium (Fig. 5.11b).

After oral uptake by the first intermediate host, the larva undergoes metamorphosis, forming specialised larval stages known as metacestodes. The larval stage in the first intermediate host is termed the procercoid, characterised by its elongated and cylindrical shape. It lacks a cavity and retains the six larval hooks on a bladder-like structure called the cercomer, situated at the posterior end (Fig. 5.11c). The procercoid larva undergoes profound physiological changes and experiences significant growth. It derives nutrients from the host's tissues and develops specialised structures for attachment and migration within the host. These structures can include hooks, spines, or adhesive organs that assist in anchoring to the host's tissues or within its body cavity. After the procercoid larva has completed its growth and development within the first intermediate host, it awaits ingestion by the next host in the life cycle, often a vertebrate predator. Upon being consumed by the second intermediate host, the procercoid larva continues its development, transforming into the subsequent larval stage known as the plerocercoid (Fig. 5.11d) (Mehlhorn and Piekarski 2002; Goater et al. 2014). The plerocercoid, in general, retains a solid body with an elongated shape reminiscent of the procercoid larva. However, it is distinguished by its larger size and more intricate structure, possessing a well-defined scolex and a relatively short, subtly segmented strobila. In the case of trypanorhynch plerocercoids, they are often encased in a fleshy capsule or vesicle known as the blastocyst (Goater et al. 2014). The plerocercoid stage is usually located within the tissues or body cavity of the second intermediate host, which can

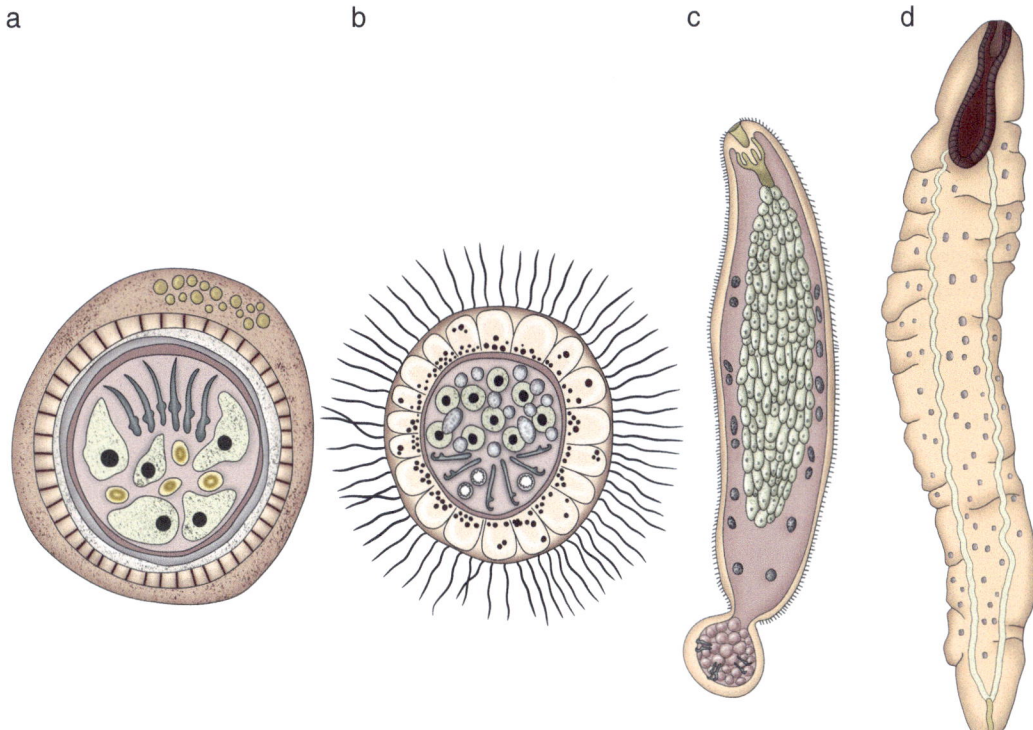

Fig. 5.11 Larval stages of eucestodes with an aquatic life cycle. Oncosphera (**a**), Coracidium (**b**), Procercoid (**c**), and Plerocercoid (**d**). (Modified from: Mehlhorn and Piekarski 2002). Illustrations by Erasmus, A. Created in BioRender. Schwelm, J. (2025) https://BioRender.com/n63a884

encompass a diverse array of organisms such as fish, crustaceans, or amphibians. The choice of this host depends on the cestode species and the dietary preferences of the definitive host (Mehlhorn and Piekarski 2002, Goater et al. 2014). Upon consumption of the second intermediate host by the definitive host, the plerocercoid larva is released and attaches itself to the intestinal wall of the definitive host. It subsequently undergoes further development, ultimately maturing into an adult tapeworm.

5.2.10 Effects on the Host

Cestodes might have distinct impacts on both intermediate and definitive hosts within their life cycles, although negative effects often only occur in cases of severe infestations. These impacts can influence the physiology, behaviour and overall fitness of the infected hosts, including humans. Obviously, adult tapeworms residing in the digestive tracts of definitive hosts can compete for nutrients, potentially leading to reduced nutrient absorption by the host. This competition for resources may result in malnutrition or reduced body condition (Dalton et al. 2004). The presence of adult tapeworms can also elicit immune responses in the definitive host. In some cases, chronic inflammation may occur in the intestinal lining, leading to gastrointestinal disturbances and potential tissue damage (Dezfuli et al. 2018). The most prominent example of an aquatic cestode affecting humans is the fish tapeworm, *Dibothriocephalus latus* (formerly known as *Diphyllobothrium latum*). It infects a wide range of fish species as intermediate hosts and can cause diphyllobothriasis in humans who consume raw or undercooked infected fish (Scholz and Kuchta 2022; see Chap. 19 for details).

Larvae of diphyllobothriideans can also cause severe harm in their intermediate hosts. For example, plerocercoids of the genus *Spirometra* can cause sparganosis in humans and other mammals, a serious condition that can affect various tissues and organs in the host, making it a significant health concern (Scholz et al. 2019).

In some cases, cestode larvae can manipulate the behaviour of their intermediate hosts to increase the likelihood of predation by the definitive host. For instance, *Schistocephalus solidus* larvae alter the behaviour of both infected copepods and fishes, making both of them more prone to predation by the downstream host, as soon as the respective stage (either procercoid or plerocercoid) is infective (Hammerschmidt et al. 2009; Barber and Scharsack 2010). For these larval stages, it is also known that they can impair the growth, reproduction, or survival in copepod first intermediate hosts (Franz and Kurtz 2002) and in its second intermediate fish host, the three-spined stickleback (*Gasterosteus aculeatus*) (Wohlleben et al. 2022). Plerocercoids continue to grow for several months, nearly reaching sexual maturity within the fish host. Once inside the definitive (fish-eating) bird host, adults of *S. solidus* have a relatively short lifespan, surviving only for a few days but producing a substantial number of eggs during this brief period (Barber and Scharsack 2010; Scholz and Kuchta 2022).

5.3 Acanthocephala

Acanthocephala, also commonly referred to as thorny-headed worms, is a group of exclusively endoparasitic species that occur as adults in the intestines of vertebrates of all classes. Of the approximately 1,300 species described so far, the vast majority are parasites of fish (Perrot-Minnot et al. 2023). A total of four classes have been morphologically classified for the Acanthocephala: the Archiacanthocephala, with four orders (Aporhynchida, Gigantorhynchida, Oligacanthorhynchida and Moniliformida), the Eoacanthocephala with two orders (Gyracanthocephala and Neoechinorhynchida), the Palaeacanthocephala with three orders (Echinorhynchida, Polymorphida and Heteramorphida) and the Polyacanthocephala with one order (Polyacanthorhynchida) (Amin 2013).

5.3.1 Biology and Life Cycle of Acanthocephala

Acanthocephalans are transmitted trophically between predators and prey (see Fig. 5.12). Vertebrates serve as definitive hosts harbouring the dioecious adult Acanthocephala in their intestine, while arthropods (mainly insect larvae and crustaceans) are used as intermediate hosts (Sures 2015; see Fig. 5.12). For acanthocephalans with aquatic life cycles, the intermediate hosts usually belong to the Crustacea such as Ostracoda, Amphipoda, Isopoda and Decapoda (Kennedy 2006). In the final host, the females release eggs into the host's intestine after mating and are excreted into the environment with the host's faeces. In the eggs, the first larval stage, the acanthor, develops. After ingestion by the intermediate host, the acanthor hatches in the intestinal lumen of the intermediate host and penetrates the intestinal wall into the hemocoel. Within the hemocoel, the larvae develop via an intermediate so-called acanthella stage into the cystacanth larva, which is infectious for the final host. If intermediate hosts with an infectious cystacanth are eaten by suitable final hosts, the cystacanths enter the intestine of the final host, protrude their proboscis, attach themselves and start growing. The key stimulus for evagination of the proboscis is bile acids, which are secreted into the intestine by the host (Sures and Siddall 1999). As soon as the worms reach sexual maturity, they reproduce. Adult worms usually live less than a year in the gut of the final host (Kennedy 2006). In addition to this basic life cycle, several acanthocephalan species also utilise paratenic hosts for the transmission of cystacanths to the definitive host. Such paratenic or transport hosts do not provide a suitable environment for the cystacanths to develop, but the larvae are at least able to survive in these hosts for some time, usually in extraintestinal positions (Sures and Siddall 2001). For

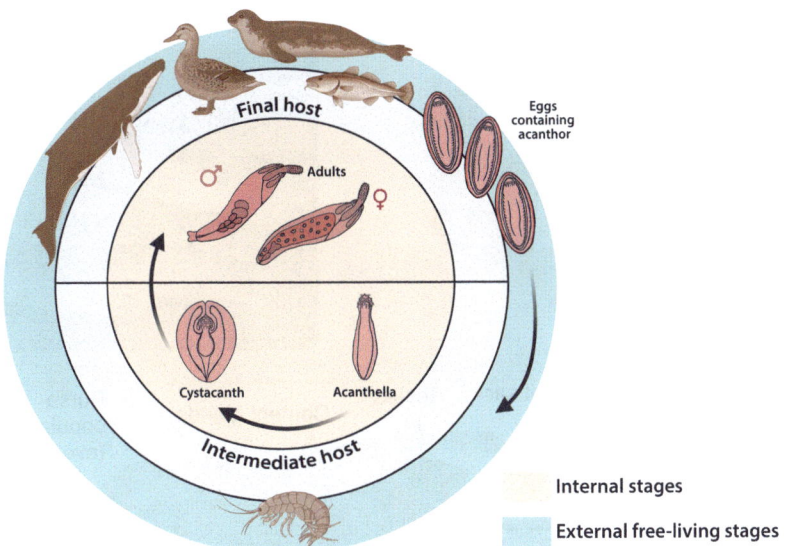

Fig. 5.12 Schematic life cycle of Acanthocephala. Acanthocephalans usually have a two-host life cycle with adults in the intestine of vertebrate hosts. Females release eggs containing the acanthor that will enter the environment and are eaten by suitable intermediate hosts. In the intermediate host, the acanthor transforms into an acanthella that further develops into a cystacanth, which is often visible by the naked eye. Cystacanths often influence the behaviour of their intermediate hosts, making them easier prey for the final hosts. After infected intermediate hosts have been eaten by a suitable definitive host, the acanthocephalan matures in the intestine and the cycle is completed. Created in BioRender. Schwelm, J. (2025) https://BioRender.com/n88k017

example, invasive goby species in the Rhine River largely exhibit high prevalences and abundances of *Pomphorhynchus* sp. cystacanths (Emde et al. 2014), which cannot develop into adults within the gobies. However, the cystacanths remain infectious in the gobies for a longer period of time, so that they settle and grow in the intestine after experimental application to chub (Nachev et al. 2024).

The role that paratenic hosts play in the completion of life cycles is controversial, as the presence of a cystacanth in a particular organism does not necessarily indicate whether this cystacanth remains infectious to the final host. On the contrary, when considering trophic relationships, paratenic hosts can serve as ideal bridges to facilitate entry into a definitive host that would not typically feed on the intermediate host of the respective acanthocephalan species (Kennedy 2006). For example, amphipods are primarily consumed by fish rather than seals. Consequently, many fish species that are typical prey for seals appear to function as paratenic hosts for *Corynosoma semerme* (Helle and Valtonen 1981; Valtonen and Crompton 1990). In addition, Acanthocephalans can be transmitted to a vertebrate host through a process known as post-cyclic transmission. This means that if a predator ingests an adult parasite residing in a definitive host, the parasite has the potential to survive and parasitise the predator, even after reaching maturity in the former vertebrate host (Nickol 2003; Kennedy 2006). The inclusion of paratenic hosts or post-cyclic transmission allows for greater flexibility in the life cycle, so that the Acanthocephala can respond to changes in their environment or to the variable availability of suitable final hosts.

5.3.2 Morphology of Adult and Larval Stages

Acanthocephala often occur in high densities in the intestines of their hosts (Fig. 5.13a). These high densities are necessary because the individuals are normally firmly attached to the gut and

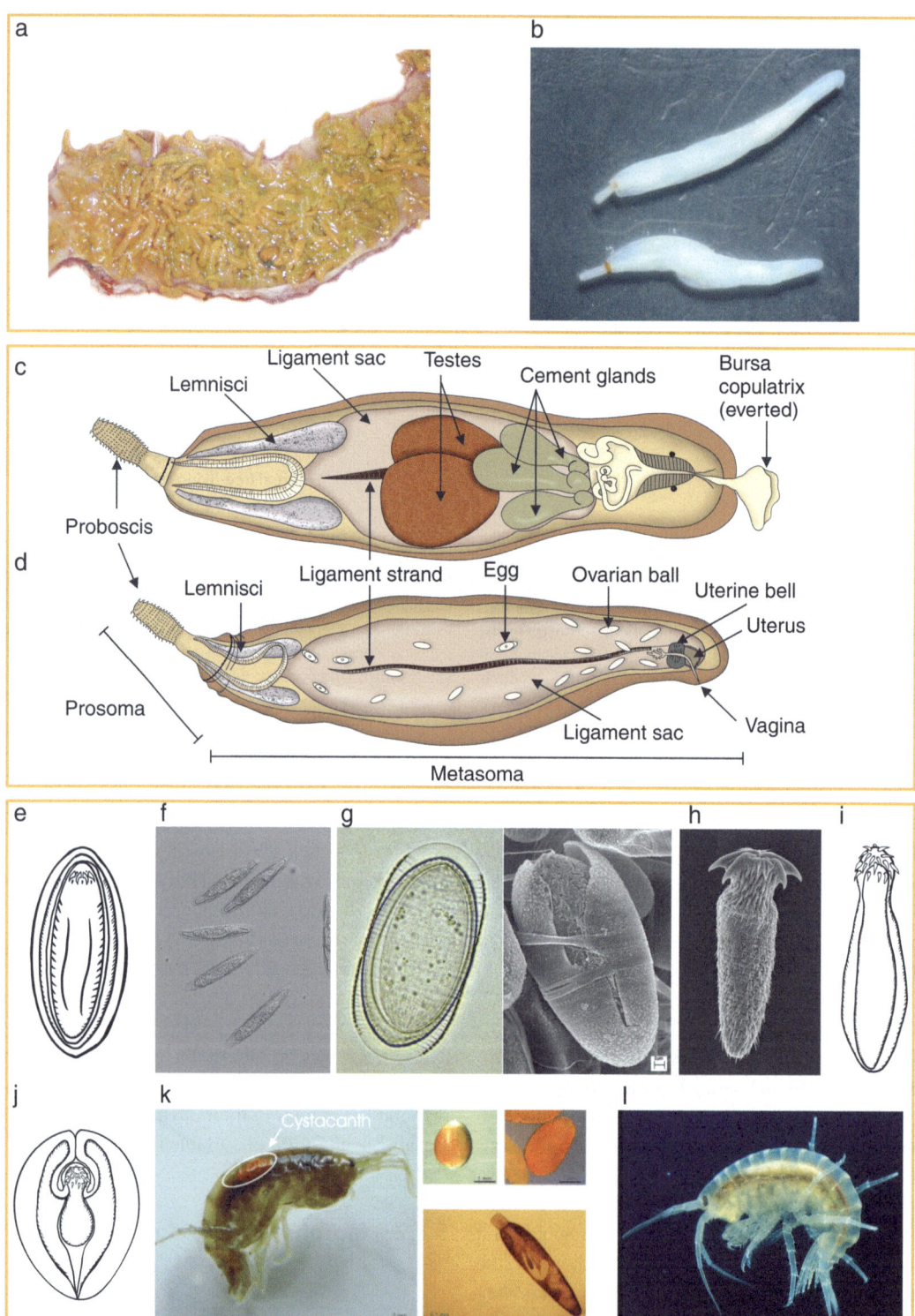

Fig. 5.13 Illustration of different stages of acanthocephalans with: opened intestine of a barbel (*Barbus barbus*) with intensive infestation of *Pomphorhynchus* sp. (**a**); individuals of *Acanthocephalus rhinensis* (**b**); schematic drawing of a male (**c**) and a female (**d**) acanthocephalan; schematic drawing of an embryonated egg including an

can therefore only mate if they are in close proximity to each other. In the gut, acanthocephalans absorb nutrients from their hosts (see also Chap. 16). Similar to tapeworms, acanthocephalans do not have their own intestines, but absorb all nutrients through their tegument. This tegument is enlarged by inwardly folding the outer membrane to effectively compete with the host's intestine for nutrient absorption (Taraschewski 2000, 2015). The body of acanthocephalans is divided into two parts, an anterior prosoma and a posterior, tube-like metasoma (Fig. 5.13b–d). The prosoma contains the proboscis (Fig. 5.13b), a retractable and extendable anterior body part with hooks with which the acanthocephalans can anchor themselves in the intestinal wall of their final hosts, and a pair of lemnisci (Fig. 5.13c), whose function is not fully understood, but might serve as a counter bearing for the in- and evagination of the proboscis. As the Acanthocephala do not have a digestive tract, the metasoma mainly contains the sexual organs. The male reproductive system (Fig. 5.13c) consists of two ovoid testes, vasa efferentia, vas deferens (plus seminal vesicle in a few species), one to eight cement glands, a cement reservoir (in Eoacanthocephala) and a penis (Taraschewski 2015). To facilitate copulation, the males also contain an extendable bursa copulatrix in the form of a funnel, which allows connection with the female's genital opening during mating. After mating, the genital pore of a female is sealed with a copulatory cap produced by the cement glands in order to prevent further mating by another individual (Taraschewski 2015). The genital complex of female acanthocephalans (Fig. 5.13d) changes fundamentally over time (Taraschewski 2015). Initially, floating ovaries are present in which the oocytes mature. After individual fertilisation of the oocytes, these ovaries are subsequently degraded, leaving only the developing eggs in the metasoma. The efferent duct of the female genital system begins with the opening of the funnel-shaped uterine bell, leading into a structure called the egg-sorting apparatus, then progressing into the uterus, and finally ending with the vagina equipped with its sphincter muscle. During development, eggs float through the body cavity of the females. Immature eggs with developing acanthors pass through an opening of the selector apparatus and remain in the body cavity, whereas only completely embryonated eggs (which differ in shape and size compared to the immature) are released by the uterus. Once these eggs (Fig. 5.13e, f) are excreted with the faeces of the host, they can be taken up orally by suitable intermediate hosts, where the acanthor hatches in the intestine (Fig. 5.13g, h) before penetrating the intestinal wall and developing via the acanthella stage (Fig. 5.13i) into the cystacanth in the hemocoel (Fig. 5.13j–l). The acanthor contains apical hooks as well as spines on its body to facilitate penetration of the intermediate host's intestine. Depending on the species, fully developed cystacanths can occur in the hemocoel either as round cysts (e.g. *Pomphorhynchus* sp., *Polymorphus* sp.; Fig. 5.13k) or as elongated stages (e.g. *Paratenuisentis ambiguus*, *Acanthocephalus* sp.; Fig. 5.13l), whose outer appearance already resembles the adult worm.

5.3.3 Effects on the Host

A well-known and interesting effect that acanthocephalans exert on their hosts is the cystacanth-induced behavioural change of infected intermediate hosts (Moore 1984; Sures 2015; Cozzarolo and Perrot-Minnot 2022). While under normal circumstances, uninfected gammarid amphipods prefer dark and shaded areas, infected conspecifics are found at a higher percentage in

Fig. 5.13 (continued) acanthor (**e**); embryonated eggs of *A. rhinensis* (**f**); acanthor of *Paratenuisentis ambiguus* surrounded by egg shells (left) and hatching (right) (**g**); hatched acanthor of *Moniliformis moniliformis* (**h**); schematic drawing of the intermediate acanthella stage (**i**); schematic drawing of a cystacanth (**j**); cystacanths of *Pomphorhynchus* sp. (**k**); and cystacanth of *Paratenuisentis ambiguus* (**l**). (Photographs by Milen Nachev [**a**]; Bernd Sures [**b**]; Daniel Grabner [**f**]; Felix Reitze [**g**, **h**]; from Sures 2015 [**k**]; and Armin Svoboda [**l**]). Illustrations (**c**, **d**) by Erasmus, A. Created in BioRender. Schwelm, J. (2025) https://BioRender.com/c97y218

open water, where they are more vulnerable to predation by putative final hosts. Behavioural manipulation by acanthocephalans is complex, with some manipulated traits, such as geotaxis and phototaxis, showing a certain degree of specificity. For example, *Polymorphus minutus* parasitising ducks reverses geotaxis of gammarids but not phototaxis, while the fish-infecting species *Pomphorhynchus laevis* and *P. tereticollis* reverse phototaxis but not geotaxis (Tain et al. 2006). Behavioural changes in intermediate hosts, which lead to greater susceptibility to predation and thus to an increase in transmission rates, have been described for many species of aquatic and terrestrial acanthocephalans. From the mechanistic perspective, amphipods infected with *P. laevis* and *P. tereticollis* have been observed to exhibit an increased brain serotonin immunoreactivity, which is proportional to the intensity of their phototaxis (Tain et al. 2006).

Adult acanthocephalans can harm their final hosts due to their anchoring in the intestinal wall. The extent of possible damage differs mainly due to the depth of penetration of the proboscis into the intestinal wall, which varies among acanthocephalan species. For instance, some species like *Pomphorhynchus* spp. can penetrate deeply and even project the anterior part of their proboscis into the peritoneal cavity. In contrast, other species, such as *Paratenuisentis ambiguus*, are only loosely attached and may even change their position within the intestine (Taraschewski 2000). The movement of the proboscis and the associated penetration into the host tissue results from muscle contraction, which increases the hydrostatic pressure within the presomal body cavity and thus leads to proboscis eversion (Hammond 1966; Herlyn and Taraschewski 2017). Accordingly, initial anchoring occurs by an alternating retraction and extension of the proboscis, probably assisted by proteolytic enzymes secreted by the anchoring individual. The presence of proteolytic enzymes (a trypsin-like collagenolytic proteinase) has been described for selected acanthocephalan species (Polzer and Taraschewski 1994), but these have not been characterised in detail. The long-term anchoring of the worms by means of their proboscis must be strong enough to withstand the pressure of the food slurry flowing through the host's digestive tract (Hernández-Orts et al. 2012). Depending on the depth of anchor penetration and frequency of the change of position in the intestine, even heavily infected hosts show no obvious signs of damage, for example, slimming or malformation, as demonstrated by barbel (*Barbus barbus*) infected with average intensities of around 100 *Pomphorhynchus* sp. (Nachev and Sures 2009).

A striking trait of adult acanthocephalans is their uptake and accumulation of enormous amounts of toxic metals in their bodies without showing any obvious adverse effects themselves (Sures et al. 2023; see also Chaps. 20 and 21). Their enormous metal accumulation capacity even leads to a reduction of pollutant concentrations in tissues of their final hosts (summarised in Sures et al. 2017, Sures and Nachev 2022). This pollutant reduction in host tissues has led to interesting considerations and discussions. Since lower concentrations of metals are generally associated with lower toxicity, infected fish are potentially less harmed by contaminants compared to their uninfected counterparts. On the other hand, the parasites' possible beneficial effects in lowering metal concentrations in the host tissue may be outweighed by the negative effects of acanthocephalan attachment in the gut or possible nutrient deprivation. Such trade-offs between potential positive and negative effects must be assessed individually for each host-parasite-pollutant combination, as we are still far from understanding antagonistic effects of parasites and pollutants (see also Chaps. 20 and 21).

5.4 Nematoda

Nematoda, often referred to as roundworms, is a remarkably diverse and abundant phylum, comprising unsegmented worm-like organisms that inhabit both terrestrial and aquatic habitats and occur as free-living or parasitic organisms. Presently, there are nearly 30,000 recognised nematode species, but conservative estimates suggest the actual number may be closer to 500,000 or even a million, with approximately half of these being parasitic (Hodda 2022a, b; Nisa et al. 2022). Reflecting their species diversity and thus diverse ecology and lifestyles, the life cycle and morphology of the parasitic nematodes also vary greatly.

5.4.1 Biology and Life Cycle of Nematoda

Nematode life cycles are highly variable, with some species having only one and others many obligate hosts. However, most nematode life cycles consist of several stages including eggs, larvae (or juveniles) and adults (Wharton 1986). Nematodes reproduce sexually, with most species presenting sexual dimorphism and oviparity, although certain taxa are ovoviviparous or viviparous. Hermaphroditism and parthenogenesis are also observed in some cases (Lucius et al. 2018; McClelland 2005). After sexual reproduction, the life cycle continues with the production and shedding of eggs, which often display an elliptical outline, symmetrically positioned poles, and can have thick or thin shells, rendering them durable and resistant (Lucius et al. 2018). Female nematodes can produce exorbitant amounts of eggs, for example, *Ascaris lumbricoides* can excrete up to 200,000 eggs per day, although not all of them are fertilised (Mehlhorn 2016). Eggs can vary from a one- to eight-cell stage or might already contain the first juvenile larval stage. Strictly speaking, the term 'larva' might not be entirely accurate, as there are no fundamental differences from the adult stage, and no metamorphosis is initiated (Lucius et al. 2018). Not all nematode life cycles include the presence of eggs. Species such as *Dracunculus medinensis* are viviparous and juveniles are released directly by female nematodes into the water column (Mehlhorn 2012). Juvenile nematodes undergo growth by moulting from one larval stage to the next, that is, L1, L2, L3, L4, preadult and finally adulthood. At each stage of the moulting process, the nematodes will produce a new cuticle and undergo ecdysis, that is, shedding of the external old cuticle. As nematode larvae advance through these stages, they typically grow in size and complexity. For aquatic species, the development of juveniles to adulthood can occur completely in a single host (e.g. *Trichinella zimbabwensis* in the Nile crocodile; Fig. 5.14), while for complex life cycle species, part of the development occurs in an additional (intermediate) host such as crustaceans as for the life cycle of *Anguillicola crassus* (Fig. 5.15) and *Anisakis* spp. (Fig. 5.16) (Mehlhorn 2016).

5.4.2 Morphology of Adult and Larval Stages

Parasitic nematodes typically appear as colourless worms, displaying bilateral symmetry and adopting an elongated cylindrical form that tapers at both ends. The size of nematodes varies, ranging from less than 1 mm to over 1 m when fully mature. In extraordinary instances, such as *Placentonema gigantissima* found in sperm whales, nematodes can even attain lengths of up to 9 m (Lucius et al. 2018). Males are noticeably smaller than females and possess species-specific copulatory structures, such as one to two spicules or a bursa copulatrix, and a coiled tail. In females, the tail is often sharper than that of males of the same species (Mehlhorn 2012). Females often contain two ovaries connected to an oviduct and uterus through which eggs pass until reaching a single vagina that opens into a sexual porus (Fig. 5.17; Mehlhorn 2016). They possess a fluid-filled pseudococlom and a complete digestive system, with the mouth located at the anterior end and the anus situated at or near the posterior end. The body is enveloped by a non-cellular cuticle, which is secreted by the underlying hypodermis (Fig. 5.17; McClelland 2005). In addition to the main component collagen, proteins and lipids are embedded in the cuticle, influencing the physical properties of the animal's outer covering. Due to the variety of habitats they inhabit, the cuticle has been adapted to the conditions of each specific environment (Paululat and Purschke 2023). With the exception of some species, nematodes are directly enclosed by this non-cellular cuticle. During growth, this layer, which also lines the ectodermal anterior

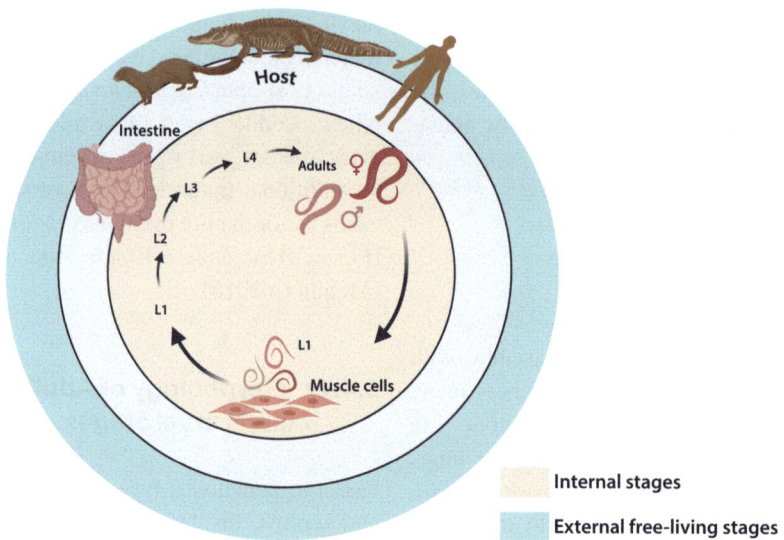

Fig. 5.14 Life cycle of *Trichinella zimbabwensis*. The host (mammals and reptiles) becomes infected through the oral consumption of muscle tissue containing the encapsulated L1. After digestion of the muscles and the capsule, the larvae in the anterior part of the small intestine bore into several adjacent cells of the mucosal epithelium, transforming it into a syncytium. Within 30 h, four moults, rapid growth into adults, and copulation occur. The female gives birth to approximately 1,500 L1 over 1–6 weeks, which then bore into the subepithelial connective tissue. From there, they are swept through the lymph or bloodstream, passing through the right atrium, lungs, and the systemic circulation to reach all organs in the body. In the striated muscles, especially those with high-mobility potential such as the eye, tongue, and limb muscles, the L1 larvae burrow into a muscle cell. Over the course of eight weeks, the larvae grow to a length of 1 mm in this environment. At this point, they become infectious for the next hosts. (Lucius et al. 2018). Created in BioRender. Schwelm, J. (2025) https://BioRender.com/x64k890

and posterior gut, as well as the vagina and the excretory pore, must be moulted. This occurs at species-specific intervals during the maturation process.

The digestive tract traverses the nematode's body as a straight tube and typically opens ventrally, usually terminating in front of the posterior end. It is composed of a mouth, an oesophagus (pharynx), a central intestine and a rectum. The oral features of parasitic nematodes are adapted to their specific feeding strategies and play a crucial role in their ability to establish themselves within their host organisms. These features can vary among different parasitic nematode species, but some common structures and adaptations include lips, hooks, teeth, stylets, or buccal cavities. Often, lips surround the mouth opening, which are covered by the cuticle and help guide the nematode's feeding process. These structures are often found in groups of three. One lip is positioned dorsally, while the other two are ventro-lateral. In some species, small teeth can be found within or at the base of the oral cavity (buccal cavity). Additionally, species-specific grooves and folds surrounding the mouth are common and serve to increase suction power. Larvae often possess mouth hooks as well. Many parasitic nematodes possess a specialised piercing organ called a stylet. The stylet is a rigid, needle-like structure that can be extended from the mouth to penetrate host tissues, allowing the nematode to feed on cellular contents, fluids, or blood. The stylet is often equipped with barbs or teeth to anchor the nematode in place while feeding (Mehlhorn 2012).

Among nematodes, a diverse array of mechanisms exists for nutrient acquisition. During their free-living juvenile stage, they consume bacteria and possibly other microorganisms. While still enclosed within the second-stage moult, third-

5 Biology and Life Cycle of Helminths

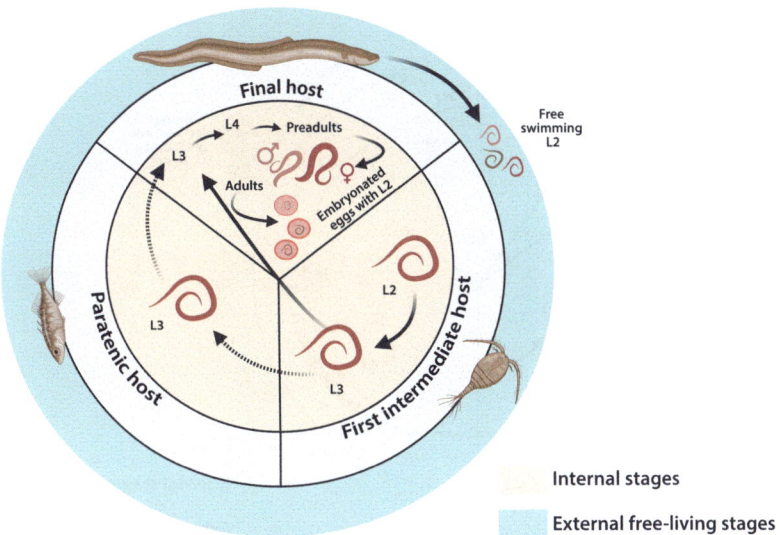

Fig. 5.15 Life cycle of *Anguillicola crassus* (adapted from: Dangel et al. 2015). Adult worms reside in the swim bladder of the final host, typically an eel. Females lay eggs within the swim bladder, from which second-stage larvae (L2) hatch and penetrate through the *ductus pneumaticus*, reaching the host's digestive tract and eventually being released into the water along with faeces. Employing wriggling body movements, these larvae probably encourage predation by copepods, acting as intermediate hosts. Upon ingestion by copepods, the larvae penetrate the wall of the digestive system, entering the haemocoel and developing into third-stage larvae. The copepod, now hosting the infective larval stage (L3), is often consumed by smaller fish, serving as paratenic hosts. Here, the L3 will usually migrate into the swim bladder wall but will not develop further (see Sures et al. 1999). Alternatively, the copepod may be ingested by the final host, the eel, during its feeding activities. Upon ingestion of the infective larvae (L3), whether through predation of the paratenic host or the first intermediate host, the larvae migrate through the intestinal wall and the body cavity of the eel to reach the swim bladder. Once in the swim bladder wall, the larvae undergo moulting to develop into the fourth-stage larvae (L4). The L4 then migrates into the swim bladder lumen to moult to preadults and eventually reaching the adult stage. Male and female worms mate within the swim bladder, and females produce eggs, completing the life cycle. (Moravec 2006). Created in BioRender. Schwelm, J. (2025) https://BioRender.com/k53m709

stage juveniles/larvae seem unable to feed, although there may be occasional exceptions among certain species. Upon reaching the adult stage, they engage in parasitism, infiltrating and exploiting almost all organs and tissues. This provides them with a wide range of nutrient resources, spanning from various tissues and tissue fluids to what is ingested by the host. The substantial variation in mouth structures among nematodes mirrors the diversity of foods they exploit. The mouth ranges from being relatively simple, as expected in grazers, to rather complex in those that attach themselves and feed on body fluids. The nematode pharynx consists of a syncytium of radial muscles. The robust musculature associated with the pharynx is essential for pumping food against the high internal pressure of the worm, facilitating the propulsion of food through the rest of the worm's gut (Goater et al. 2014). Nematodes are also capable of directly obtaining nutrients via their tegument due to the cuticle's high permeability, which permits non-electrolyte, ionic and small organic substances to pass through (Lucius et al. 2018; Mehlhorn 2016).

The excretory system of nematodes commonly consists of a glandular system (renette cells) and a channel system (H-cells) (Mehlhorn 2016). The glandular system is located ventrally, while the channel system is located laterally, eventually opening through a ventral excretory pore near the body's anterior part.

Nematodes have a relatively uniform nervous system structure. It comprises a simple central

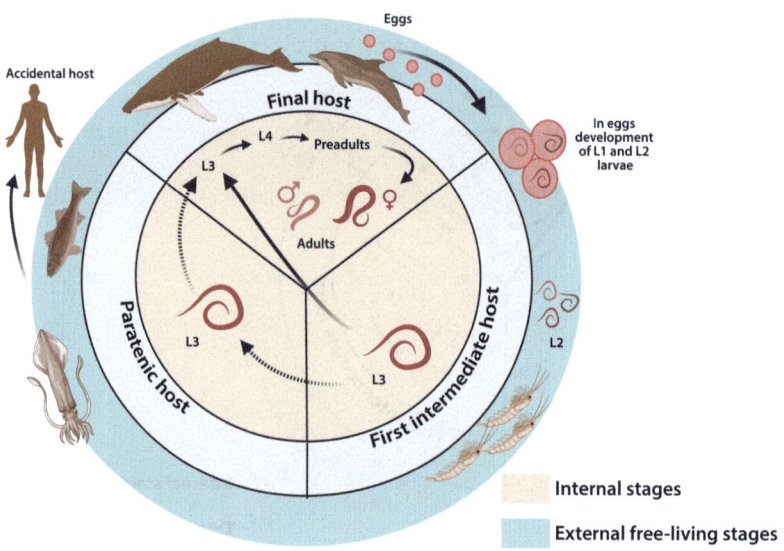

Fig. 5.16 Life cycle of *Anisakis* spp. (Adapted from: Fiorenza et al. 2020). The adults, which reside in the stomach of marine mammals, such as whales and dolphins, produce eggs. A sheathed L2 develops from the eggs they excrete, hatching in water and being ingested by various small crustaceans, such as copepods or krill, which serve as intermediate hosts. Inside these crustaceans, the larva grows into the infectious L3. Marine mammals could theoretically become infected directly through the infected crustaceans, but they usually acquire the infection by consuming paratenic hosts (fish or squid). The larvae penetrate the gut wall of the paratenic host and migrate to liver, muscles, or tissues of the body cavity, where they encyst. The life cycle is completed when a marine mammal consumes the infected fish or squid (or the infected crustacean host). The final moulting occurs in the definitive host. The released larvae mature into adult worms in the stomach of the marine mammal, and the cycle continues as the adult worms produce eggs, which are then released in the faeces. Humans can become accidental hosts in this life cycle if they consume raw or undercooked fish or squid containing the infective cysts. In humans, the worms rarely reach sexual maturity but are typically found as L4 larvae in the stomach, occasionally in the intestine, causing serious abdominal pain (Lucius et al. 2018). Created in BioRender. Schwelm, J. (2025) https://BioRender.com/o46q858

nervous system (CNS) composed of several apical ganglia and nerve fibres that encircle the pharynx. Additionally, there is a ventral nerve cord. From this cord, a dorsal nerve cord emerges, with each of its cells connecting to a cell from the ventral cord via the ring nerve system, and this connection passes through the body wall (Fig. 5.17). Notably, only a few nerve cells are formed in total. The muscle cells of nematodes are cross-striated and longitudinally placed, containing cytoplasmic protrusions that stretch to the nerve strands. These nerve strands, along with the excretion channels, separate each of the four quadrants in which the muscle cells are arranged. The sensory organs of parasitic nematodes are simple sensillae that serve as mechanoreceptors or chemoreceptors. Special clusters of sensory cells are found at both the anterior and posterior ends (Mehlhorn 2012). There are two types of sensillae, the phasmids, which can be found in pairs and surrounded by glands in the anus, and the amphids, which act as chemoreceptors (Mehlhorn 2016). The shape and arrangement of these sensillae is an essential taxonomic feature. The bacillary cells present in Trichuridae and Trichinellidae also serve as sensory cells in addition to their excretory function (Mehlhorn 2016).

5.4.3 Effects on the Host

Parasitic nematodes can inflict a wide range of damage upon their hosts, from effects on their physiology to their fitness and behaviour; in severe cases, they can be lethal (McClelland 2005; Moravec 2013). The best-known examples

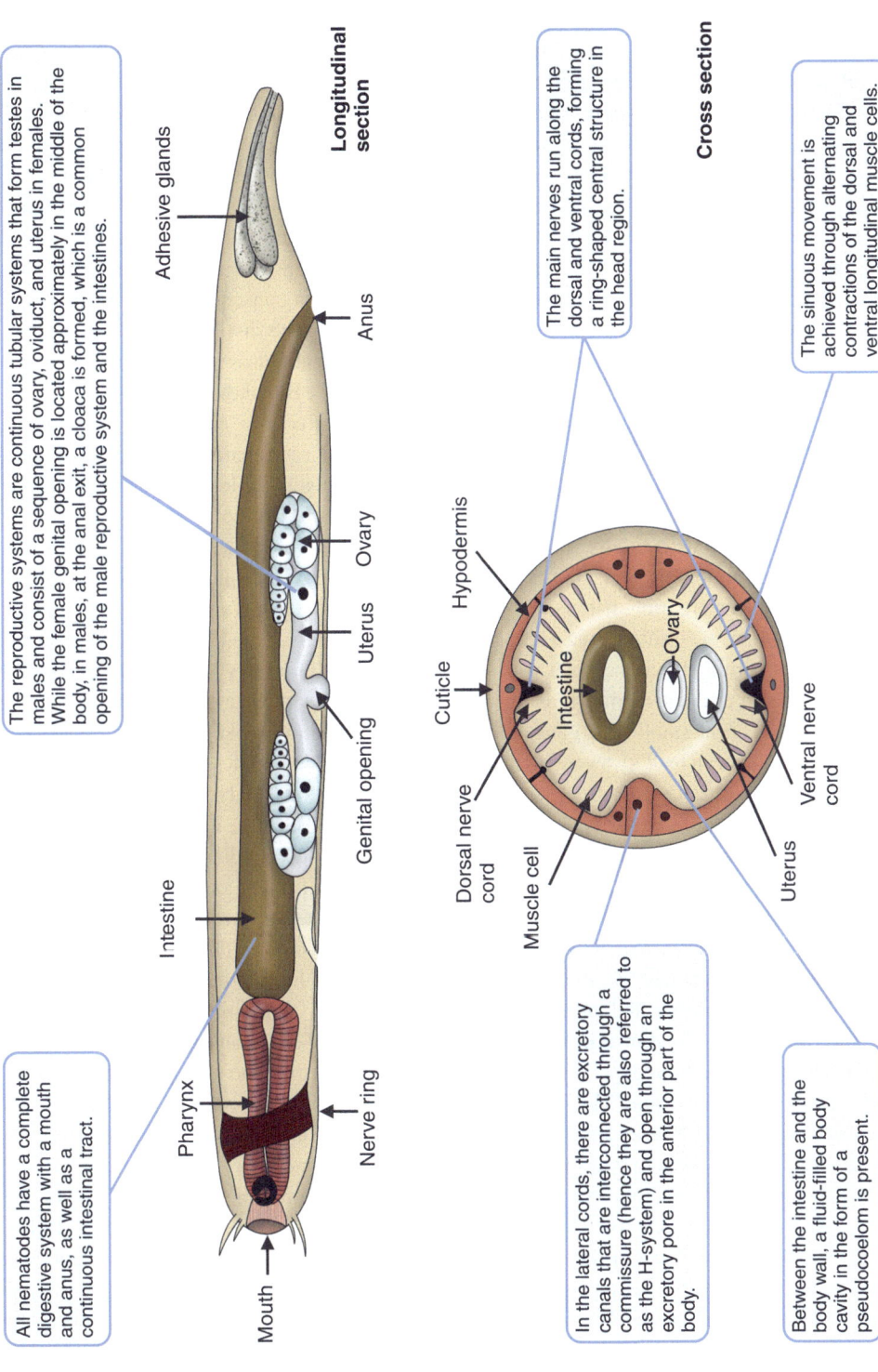

Fig. 5.17 Schematic representation of a female nematode in longitudinal section and in cross-section. (Adapted from: Sures 2021). Illustrations by Erasmus, A. Created in BioRender. Schwelm, J. (2025) https://BioRender.com/n63a884

concern their final vertebrate hosts. For instance, *Trichinella* spp. is known to induce symptoms in mammals such as diarrhoea, fever and muscle damage including myositis, pain and stiffness that can result in paralysis of the respiratory tract (Mehlhorn 2016). Another well-known parasitic disease in mammals (e.g. humans) is anisakiasis, caused by *Anisakis* sp. (Fig. 5.16). The worms seldom reach sexual maturity and are typically found as L4 larvae in the stomach or body cavity and occasionally in the intestine. Since the detection of larvae can be challenging, their presence may go unnoticed, but allergic reactions can manifest shortly after ingestion. In some cases, severe abdominal pain, referred to as 'acute abdomen', may occur, posing a life-threatening condition that requires immediate surgical removal of the parasites (Lucius et al. 2018). Clinical observations may reveal inflamed sections of the digestive tract with intense eosinophilic infiltrates or granulomas. Anisakiasis is prevalent in countries where raw fish is consumed, such as in sushi or sashimi (see Chap. 19). However, even thoroughly cooked fish with *Anisakis* infestation can cause discomfort, as the worms contain potent, heat-stable allergens. In sensitive individuals, this can lead to severe allergic reactions (Lucius et al. 2018).

Although most attention has been given to the impact of parasitic nematodes in humans, aquatic vertebrates have also been reported to experience severe repercussions from infection. For instance, marine mammals such as seals and whales are often severely affected by nematode infections (e.g. *Uncinaria* sp. hookworms and pseudaliid lungworms; see also Chap. 15) showing signs of respiratory obstruction, haemorrhagic enteritis and anaemia, which can often be fatal (McClelland 2005). Another example is *Stenurus* spp., which can infect the cranial sinus system and interfere with hearing and navigation in toothed whales (Odontocetes) increasing the rate of stranding events (McClelland 2005). Effects of this nature can also be observed in lower-order aquatic vertebrates such as fishes. The nematode *Anguillicola crassus* (Fig. 5.15), which infects the swim bladder of eels, has been observed to inflame and provoke necrosis of this organ (Würtz and Taraschewski 2000), as well as disruption of the gas glands that ensure its effective regulation (Würtz et al. 1998; Würtz and Taraschewski 2000). The physical impact on the swim bladder, combined with the immunological cost of the infection (e.g. oxygen demand), can impair the host's overall performance by jeopardising its swimming speed and endurance (Palstra et al. 2007). The effect on swimming speed and endurance may have additional ramifications for eel migration, spawning activity and survival (Kirk 2003), resulting in consequences for the eel population including the disruption of their catadromous life cycle and mass mortality events (Sures and Knopf 2004; Bychkova et al. 2022).

Nematode infections can also have negative repercussions in intermediate hosts ranging from molecular changes to lethal effects, with the degree of mortality directly correlated with the intensity of infection. In less severe cases, effects such as physiological and behavioural changes can be observed. One of the primary pathogenic mechanisms involves the larvae causing mechanical damage to internal organs as they navigate the host's tissues. Nematodes from the family Camallanidae can provoke damage to the reproductive system of female copepods resulting in castration. Simultaneously, these nematodes may release toxic substances, exacerbating the harm inflicted on the intermediate host and contributing to the overall pathogenicity of the infection (Moravec 2013). For instance, the larval sealworm (*Phocanema decipiens*) can anaesthetise the intermediate host through the production of volatile ketones and interfere with predator avoidance behaviour, thereby facilitating transmission to the final host (McClelland 2005). Similarly, *Skrjabinoclava morrisoni* (Acuariidae) larvae increase daytime surface activity of the amphipod intermediate host, increasing the chances of transmission to the final host, the sandpiper *Calidris pusilla* (McCurdy et al. 1999). Another example within the family Anisakidae is *Sulcascaris sulcata*, known for causing the weakening of the adductor muscle in scallops. The adductor muscle is crucial for the scallop's ability to open and close its shell. The weakening of this muscle can have significant consequences for the scallop's mobility and overall health, impacting its ability to feed and evade potential predators (McClelland 2005). These altered behaviours substantially increase the probability of predation by the final or paratenic host, facilitating the successful completion of the nematode life cycle (Moravec 2013).

Above, we have summarised some of the examples for effects that nematodes can have on their hosts throughout their life cycle. However, effects can be much more complicated, particularly when considering interactions with other parasite groups. Nematodes, for instance, can make their hosts more prone to secondary infections, for example, *Anguillicola crassus*, which strongly favours secondary viral or parasitic infections (Sures 2001; Sures 2006).

5.5 Conclusion

Parasitic helminths, with their complex life cycles involving multiple hosts, play a crucial role in regulating host populations and nutrient transfer within aquatic ecosystems. Despite their reputation for causing diseases in all classes of vertebrates and invertebrates, helminths also act as indicators of ecosystem health emphasising the need for a more comprehensive understanding of their ecology. Recognition of the presence and roles of aquatic helminths is essential for effective conservation and management strategies, which underscores the importance of assessing the entire biocenosis, including parasites, to better understand, protect and preserve their delicate balance and biodiversity.

Acknowledgements The following figures were created with BioRender.com: Fig. 5.2 agreement number: CN27JUT5JE (Schwelm, J. (2025) https://BioRender.com/b44y859); Fig. 5.3 agreement number: QN27JUQCKX (Schwelm, J. (2025) https://BioRender.com/t89d703); Fig. 5.4 agreement number: SZ27JUQQW8 (Schwelm, J. (2025) https://BioRender.com/g14l497); Fig. 5.5 agreement number: JL27JUR54N (Schwelm, J. (2025) https://BioRender.com/b80q726); Fig. 5.6 agreement number: LT27JUTYRF (Schwelm, J. (2025) https://BioRender.com/w57e362); Fig. 5.7 agreement number: WL27JUUTYK (Schwelm, J. (2025) https://BioRender.com/r87p220); Fig. 5.8 agreement number: DO27JUV8NZ (Schwelm, J. (2025) https://BioRender.com/r07v162); Fig. 5.9 agreement number: ZC27JUVWBC (Schwelm, J. (2025) https://BioRender.com/h51m960); Fig. 5.10 agreement number: IR27JUW952 (Schwelm, J. (2025) https://BioRender.com/63l334); Fig. 5.11 agreement number: SA27JUWIXR (Schwelm, J. (2025) https://BioRender.com/n63a884); Fig. 5.12 agreement number: LH27JUYF8X (Schwelm, J. (2025) https://BioRender.com/n88k017); Fig. 5.13 agreement number: XX27JUXJ9T (Schwelm, J. (2025) https://BioRender.com/c97y218); Fig. 5.14 agreement number: WQ27JUYUX3 (Schwelm, J. (2025) https://BioRender.com/x64k890); Fig. 5.15 agreement number: VE27JUZ6S0 (Schwelm, J. (2025) https://BioRender.com/k53m709); Fig. 5.16 agreement number: WW27JUZMHZ (Schwelm, J. (2025) https://BioRender.com/o46q858); and Fig. 5.17 agreement number: WL27JUUTYK (Schwelm, J. (2025) https://BioRender.com/n63a884).

References

Alves PV, Vieira FM, Santos CP et al (2015) A checklist of the Aspidogastrea (Platyhelminthes: Trematoda) of the world. Zootaxa 3918:339–396. https://doi.org/10.11646/zootaxa.3918.3.2

Amin OM (2013) Classification of the acanthocephala. Folia Parasitol 60:273–305. https://doi.org/10.14411/fp.2013.031

Avenant-Oldewage A, Milne SJ (2014) Aspects of the morphology of the juvenile life stages of *Paradiplozoon ichthyoxanthon* Avenant-Oldewage, 2013 (Monogenea: Diplozoidae). Acta Parasitol 59:247–254

Babirat C, Mouritsen KN, Poulin R (2004) Equal partnership: two trematode species, not one, manipulate the burrowing behaviour of the New Zealand cockle, *Austrovenus stutchburyi*. J Helminthol 78:195–199. https://doi.org/10.1079/joh2003231

Bakke TA, Cable J, Harris PD (2007) The biology of gyrodactylid monogeneans: the "Russian-Doll Killers.". Adv Parasitol 64:161–378. https://doi.org/10.1016/S0065-308X(06)64003-7

Barber I, Scharsack JP (2010) The three-spined stickleback-*Schistocephalus solidus* system: an experimental model for investigating host-parasite interactions in fish. Parasitology 137:411–424. https://doi.org/10.1017/S0031182009991466

Beveridge I, Haseli M, Ivanov VA et al (2017) Trypanorhyncha Diesing, 1863. In: Caira JN, Jensen K (eds) Planetary biodiversity inventory (2008–2017): tapeworms from vertebrate bowels of the earth. University of Kansas, Natural History Museum, Lawrence, pp 401–429

Brabec J, Salomaki ED, Kolísko M et al (2023) The evolution of endoparasitism and complex life cycles in parasitic platyhelminths. Curr Biol 33:1–7. https://doi.org/10.1016/j.cub.2023.08.064

Bychkova EI, Yakovich MM, Degtyarik SM (2022) Alien species of fish Helminths of Belarus. Russ J Biol Invasions 13:15–21. https://doi.org/10.1134/S2075111722010040

Caira JN, Jensen K (2022) Diversity and phylogenetic relationships of 'tetraphyllidean' Clade 3 (Cestoda) based on new material from orectolobiform sharks in Australia and Taiwan. Folia Parasitol 69. https://doi.org/10.14411/fp.2022.010

Caira JN, Jensen K, Georgiev BB et al (2017a) An overview of tapeworms from vertebrate bowels of the earth. In: Caira JN, Jensen K (eds) Planetary biodiversity inventory (2008–2017): tapeworms from vertebrate bowels of the earth. University of Kansas, Natural History Museum, Lawrence, pp 1–20

Caira JN, Jensen K, Ivanov VA (2017b) Onchoproteocephalidea II Caira, Jensen, Waeschenbach, Olson & Littlewood, 2014. In: Caira JN, Jensen K (eds) Planetary biodiversity inventory (2008–2017): tapeworms from vertebrate bowels of the earth. University of Kansas, Natural History Museum, Lawrence, pp 279–304

Caira JN, Kl J, Ruhnke TR (2017c) "Tetraphyllidea" van Beneden, 1850 relics. In: Caira JN, Jensen K (eds) Planetary biodiversity inventory (2008–2017): tapeworms from vertebrate bowels of the earth. University of Kansas, Natural History Museum, Lawrence, pp 371–400

Cozzarolo CS, Perrot-Minnot MJ (2022) Infection with an acanthocephalan helminth reduces anxiety-like behaviour in crustacean host. Sci Rep 12:21649. https://doi.org/10.1038/s41598-022-25484-9

Dalton JP, Skelly P, Halton DW (2004) Role of the tegument and gut in nutrient uptake by parasitic platyhelminths. Can J Zool 82:211–232. https://doi.org/10.1139/z03-213

Dangel KC, Keppel M, Le TTY et al (2015) Competing invaders: performance of two *Anguillicola* species in Lake Bracciano. Int J Parasitol Parasites Wildl 4:119–124. https://doi.org/10.1016/j.ijppaw.2014.12.010

de Chambrier A, Scholz T, Mariaux J, Kuchta R (2017) Onchoproteocephalidea I Caira, Jensen, Waeschenbach, Olson & Littlewood, 2014. In: Caira JN, Jensen K (eds) Planetary biodiversity inventory (2008–2017): tapeworms from vertebrate bowels of the earth. University of Kansas, Natural History Museum, Lawrence, pp 251–277

Dezfuli BS, Giari L, Squerzanti S et al (2018) Histological damage and inflammatory response elicited by *Monobothrium wageneri* (Cestoda) in the intestine of *Tinca tinca* (Cyprinidae). Parasit Vectors 4:1–11. https://doi.org/10.1186/1756-3305-4-225

Emde S, Rueckert S, Kochmann J et al (2014) Nematode eel parasite found inside acanthocephalan cysts—a "Trojan horse" strategy? Parasit Vectors 7:1–5. https://doi.org/10.1186/s13071-014-0504-8

Fiorenza EA, Wendt CA, Dobkowski KA et al (2020) It's a wormy world: meta-analysis reveals several decades of change in the global abundance of the parasitic nematodes Anisakis spp. and Pseudoterranova spp. in marine fishes and invertebrates. Glob Chang Biol 26:2854–2866. https://doi.org/10.1111/gcb.15048

Franz K, Kurtz J (2002) Altered host behaviour: manipulation or energy depletion in tapeworm-infected copepods? Parasitology 125:187–196. https://doi.org/10.1017/S0031182002001932

Galaktionov KV, Dobrovolskij AA (eds) (2003) The biology and evolution of trematodes: an essay on the biology, morphology, life cycles, transmissions, and evolution of digenetic trematodes. Springer-Science+Business Media, Dordrecht. https://doi.org/10.1007/978-94-017-3247-5

Gibson D (2023) World list of monogenea. In: Bánki O, Roskov Y, Döring M et al (eds) Catalogue of Life Checklist, ver 10/2023. https://doi.org/10.48580/df7lv-3cv

Goater TM, Goater CP, Esch GW (2014) Parasitism: the diversity and ecology of animal parasites, 2nd edn. Cambridge University Press

Gryseels B, Polman K, Clerinx J, Kestens L (2006) Human schistosomiasis. Lancet 368:1106–1118. https://doi.org/10.1016/S0140-6736(06)69440-3

Hammerschmidt K, Koch K, Milinski M et al (2009) When to go: optimisation of host switching in parasites with complex life cycles. Evolution 63:1976–1986. https://doi.org/10.1111/j.1558-5646.2009.00687.x

Hammond RA (1966) The proboscis mechanism of *Acanthocephalus Ranae*. J Exp Biol 45:203–213. https://doi.org/10.1242/jeb.45.2.203

Hechinger RF, Wood AC, Kuris AM (2011) Social organization in a flatworm: trematode parasites form soldier and reproductive castes. Proc R Soc B Biol Sci 278:656–665. https://doi.org/10.1098/rspb.2010.1753

Helle E, Valtonen ET (1981) Comparison between spring and autumn infection by *Corynosoma* (Acanthocephala) in the ringed seal *Pusa hispida* in the Bothnian Bay of the Baltic Sea. Parasitology 82:287–296. https://doi.org/10.1017/S0031182000057036

Herlyn H, Taraschewski H (2017) Evolutionary anatomy of the muscular apparatus involved in the anchoring of Acanthocephala to the intestinal wall of their vertebrate hosts. Parasitol Res 116:1207–1225. https://doi.org/10.1007/s00436-017-5398-x

Hernández-Orts JS, Timi JT, Raga JA et al (2012) Patterns of trunk spine growth in two congeneric species of acanthocephalan: investment in attachment may differ between sexes and species. Parasitology 139:945–955. https://doi.org/10.1017/S0031182012000078

Hodda M (2022a) Phylum Nematoda: a classification, catalogue and index of valid genera, with a census of valid species. Zootaxa 5114:1–289. https://doi.org/10.11646/zootaxa.5114.1.1

Hodda M (2022b) Phylum Nematoda: trends in species descriptions, the documentation of diversity, systematics, and the species concept. Zootaxa 5114:290–317. https://doi.org/10.11646/zootaxa.5114.1.2

Johnson PTJ, Sutherland DR, Kinsella JM, Lunde KB (2004) Review of the trematode genus *Ribeiroia* (Psilostomidae): ecology, life history and pathogenesis with special emphasis on the amphibian malformation problem. Adv Parasitol 57:191–253. https://doi.org/10.1016/S0065-308X(04)57003-3

Kennedy CR (2006) Ecology of the Acanthocephala. Cambridge University Press, Cambridge, p 249

Kirk RS (2003) The impact of *Anguillicola crassus* on European eels. Fish Manag Ecol 10:385–394. https://doi.org/10.1111/j.1365-2400.2003.00355.x

Kostadinova A, Pérez-del Olmo A (2014) The systematics of the Trematoda. In: Toledo R, Fried B (eds) Digenetic trematodes. Advances in experimental medicine and biology. Springer Science+Business Media, New York, pp 21–44

Kuchta R, Scholz T (2017) Bothriocephalidea Kuchta, Scholz, Brabec & Bray, 2008. In: Caira JN, Jensen K (eds) Planetary biodiversity inventory (2008–2017): tapeworms from vertebrate bowels of the earth. University of Kansas, Natural History Museum, Lawrence, pp 29–45

Kuchta R, Scholz T, Hanoen H (2017) Gyrocotylidea Poche, 1926. In: Caira JN, Jensen K (eds) Planetary biodiversity inventory (2008–2017): tapeworms from vertebrate bowels of the earth. University of Kansas, Natural History Museum, Lawrence, pp 191–199

Kuchta R, Basson L, Cook C et al (2018) A systematic survey of the parasites of freshwater fishes in Africa. In: Scholz T, Vanhove M, Smit N et al (eds) A guide to the parasites of African freshwater fishes. Abc Taxa; ©CEBioS, Royal Belgian Institute of Natural Sciences, pp 137–402

Kuris A (1974) Trophic interactions: similarity of parasitic castrators to parasitoids. Q Rev Biol 49:129–148. https://doi.org/10.1086/408018

Louvard C, Yong RQ, Cutmore SC, Cribb TH (2024) The oceanic pleuston community as a potentially crucial life-cycle pathway for pelagic fish-infecting parasitic worms. Int J Parasitol 54:267–278. https://doi.org/10.1016/j.ijpara.2023.11.001

Lucius R, Loos-Frank B, Lane RP (2018) Biologie von Parasiten. Springer, Berlin. https://doi.org/10.1007/978-3-662-54862-2

Magalhães L, Freitas R, de Montaudouin X (2020) How costly are metacercarial infections in a bivalve host? Effects of two trematode species on biochemical performance of cockles. J Invertebr Pathol 177. https://doi.org/10.1016/j.jip.2020.107479

McClelland G (2005) Nematoda (roundworms). In: Rohde K (ed) Marine parasitology. Csiro Publishing, pp 104–121

McCurdy DG, Forbes MR, Boates JS (1999) Evidence that the parasitic nematode *Skrjabinoclava* manipulates host *Corophium* behavior to increase transmission to the sandpiper, *Calidris pusilla*. Behav Ecol 10:351–357. https://doi.org/10.1093/beheco/10.4.351

Mehlhorn H (2008) Encyclopedia of parasitology, 3rd edn. Springer, Berlin

Mehlhorn H (2012) Die Parasiten der Tiere: Erkrankungen erkennen, bekämpfen und vorbeugen 7. Auflage. Springer, Berlin

Mehlhorn H (2016) Animal parasites: diagnosis, treatment, prevention. Springer, Cham

Mehlhorn H, Piekarski G (2002) Grundriss der Parasitenkunde 6. Auflage. G. Fischer Verlag, Stuttgart

Moore J (1984) Altered behavioral responses in intermediate hosts -- an acanthoceptalan parasite strategy. Am Nat 123:572–577. https://doi.org/10.1086/284224

Moravec F (2006) Dracunculoid and anguillicoloid nematodes parasitic in vertebrates. Academia, Praha

Moravec F (2013) Parasitic nematodes of freshwater fishes of Europe, 2nd edn. Academia, Praha

Mouritsen KN (2002) The parasite-induced surfacing behaviour in the cockle *Austrovenus stutchburyi*: a test of an alternative hypothesis and identification of potential mechanisms. Parasitology 124:521–528. https://doi.org/10.1017/S0031182002001427

Nachev M, Sures B (2009) The endohelminth fauna of barbel (*Barbus barbus*) correlates with water quality of the Danube River in Bulgaria. Parasitology 136:545–552. https://doi.org/10.1017/S003118200900571X

Nachev M, Hohenadler M, Bröckers N, Grabner D, Sures B (2024) Suitability of invasive gobies as paratenic hosts for Pomphorhynchus sp. Parasitology 151, 1522–1529. https://doi.org/10.1017/S0031182024001197

Nickol BB (2003) Is postcyclic transmission underestimated as an epizootiological factor for acanthocephalans? Helminthologia 40:93–95

Nielsen SS, Johansen M, Mouritsen KN (2014) Caste formation in larval *Himasthla elongata* (Trematoda) infecting common periwinkles *Littorina littorea*. J Mar Biol Assoc United Kingdom 94:917–923. https://doi.org/10.1017/S0025315414000241

Nisa U, Tantray A, Shah A (2022) Shift from morphological to recent advanced molecular approaches for the identification of nematodes. Genomics 114:110295. https://doi.org/10.1016/j.ygeno.2022.110295

Oros M, Scholz T, Hanzelová V, Mackiewicz JS (2010) Scolex morphology of monozoic cestodes (Caryophyllidea) from the palaearctic region: a useful tool for species identification. Folia Parasitol 57:37–46. https://doi.org/10.14411/fp.2010.006

Palm HW, Waeschenbach A, Olson PD, Littlewood DTJ (2009) Molecular phylogeny and evolution of the Trypanorhyncha Diesing, 1863 (Platyhelminthes: Cestoda). Mol Phylogenet Evol 52:351–367. https://doi.org/10.1016/j.ympev.2009.01.019

Palstra AP, Heppener DFM, van Ginneken VJT et al (2007) Swimming performance of silver eels is severely impaired by the swim-bladder parasite *Anguillicola crassus*. J Exp Mar Bio Ecol 352:244–256. https://doi.org/10.1016/j.jembe.2007.08.003

Paululat A, Purschke G (2023) Nematoda (Fadenwürmer, Rundwürmer). In: Metazoa - Morphologie und Evolution der vielzelligen Tiere. Springer, Berlin, pp 103–117. https://doi.org/10.1007/978-3-662-66184-0_7

Perrot-Minnot MJ, Cozzarolo CS, Amin O et al (2023) Hooking the scientific community on thorny-headed worms: interesting and exciting facts, knowledge gaps and perspectives for research directions on Acanthocephala. Parasite 30:23. https://doi.org/10.1051/parasite/2023026

Pérez-Ponce de León G, Hernández-Mena DI (2019) Testing the higher-level phylogenetic classification of Digenea (Platyhelminthes, Trematoda) based on nuclear rDNA sequences before entering the age of the 'next-generation' Tree of Life. J Helminthol 93, 260–276. https://doi.org/10.1017/S0022149X19000191

Pinheiro J, Maldonado Júnior A, Lanfredi RM (2009) Physiological changes in *Lymnaea columella* (Say, 1817) (Mollusca, Gastropoda) in response to *Echinostoma paraensei* Lie and Basch, 1967 (Trematoda: Echinostomatidae) infection. Parasitol Res 106:55–59. https://doi.org/10.1007/s00436-009-1630-7

Polzer M, Taraschewski H (1994) Proteolytic enzymes of *Pomphorhynchus laevis* and in three other acanthocephalan species. J Parasitol 80:45–49. https://doi.org/10.2307/3283343

Rohde K (1972) The Aspidogastrea, especially *Multicotyle purvisi*. Adv Parasitol 70:77–151. https://doi.org/10.1016/s0065-308x(08)60173-6

Rohde K (1973) Structure and development of *Lobatostoma manteri* sp.nov. (Trematoda: Aspidogastrea) from the great barrier reef, Australia. Parasitology 66:63–83. https://doi.org/10.1017/S0031182000044450

Rosen R, Berg E, Peng L et al (2016) Location and development of the cotylocidium within the egg of *Cotylaspis insignis* (Trematoda: Aspidogastridae). Comp Parasitol 83:6–10. https://doi.org/10.1654/1525-2647-83.1.6

Sayyaf Dezfuli B, Castaldelli G, Giari L (2018) Histopathological and ultrastructural assessment of two mugilid species infected with myxozoans and helminths. J Fish Dis 41:299–307. https://doi.org/10.1111/jfd.12713

Schell SC (1970) How to know the trematodes. WM. C. Brown Company Publishers

Schmidt G, Roberts L (1990) Foundations of parasitology, 8th edn. The McGraw-Hill Companies, Inc., New York, p 10020

Scholz T, Kuchta R (2017) Amphilinidea Poche, 1922. In: Caira JN, Jensen K (eds) Planetary biodiversity inventory (2008–2017): tapeworms from vertebrate bowels of the earth. University of Kansas, Natural History Museum, Lawrence, pp 21–28

Scholz T, Kuchta R (2022) Fish tapeworms (Cestoda) in the molecular era: achievements, gaps and prospects. Parasitology 149:1876–1893. https://doi.org/10.1017/S0031182022001202

Scholz T, Oros M (2017) Caryophyllidea van Beneden in Carus, 1863. In: Caira JN, Jensen K (eds) Planetary biodiversity inventory (2008–2017): tapeworms from vertebrate bowels of the earth. University of Kansas, Natural History Museum, Lawrence, pp 47–64

Scholz T, Kuchta R, Brabec J (2019) Broad tapeworms (Diphyllobothriidae), parasites of wildlife and humans: recent progress and future challenges. Int J Parasitol Parasites Wildl 9:359–369. https://doi.org/10.1016/j.ijppaw.2019.02.001

Sousa WP (1983) Host life history and the effect of parasitic castration on growth: a field study of *Cerithidea californica* and its trematode parasites. J Exp Biol 73:273–296. https://doi.org/10.1016/0022-0981(83)90051-5

Sures B (2001) The use of fish parasites as bioindicators of heavy metals in aquatic ecosystems: a review. Aquat Ecol 35:245–255. https://doi.org/10.1023/A:1011422310314

Sures B. (2006) How parasitism and pollution affect the physiological homeostasis of aquatic hosts. Journal of Helminthology 80:151–157

Sures B (2015) Ecology of the acanthocephala. In: Schmidt-Rhaesa A (ed) Gastrotricha, cycloneuralia and gnathifera. Handbook of zoology, vol 3. de Gruyter, Berlin, pp 337–345. https://doi.org/10.1515/9783110274271

Sures B (2021) Evolution und Systematik der Tiere. In: Boenigk J (ed) Biologie: Der Begleiter in und durch das Studium. Springer, Heidelberg, pp 989–1014. https://doi.org/10.1007/978-3-662-61270-5_38

Sures B, Knopf K (2004) Parasites as a threat to freshwater eels? Science 304:208–209. https://doi.org/10.1126/science.304.5668.209

Sures B, Nachev M (2022) Effects of multiple stressors in fish: how parasites and contaminants interact. Parasitology 149:1822–1828. https://doi.org/10.1017/S0031182022001172

Sures B, Siddall R (1999) *Pomphorhynchus laevis*: the intestinal acanthocephalan as a lead sink for its fish host, chub (*Leuciscus cephalus*). Exp Parasitol 93:66–72. https://doi.org/10.1006/expr.1999.4437

Sures B, Siddall R (2001) Comparison between lead accumulation of *Pomphorhynchus laevis* (Palaeacanthocephala) in the intestine of chub (*Leuciscus cephalus*) and in the body cavity of goldfish (*Carassius auratus auratus*). Int J Parasitol 31:669–673. https://doi.org/10.1016/S0020-7519(01)00173-4

Sures B, Knopf K, Taraschewski H (1999) Development of *Anguillicola crassus* (Dracunculoidea, Anguillicolidae) in experimentally infected Balearic congers *Ariosoma balearicum* (Anguilloidea, Congridae). Dis Aquat Organ 39:75–78

Sures B, Lutz I, Kloas W (2006) Effects of infection with *Anguillicola crassus* and simultaneous exposure with Cd and 3,3′,4,4′,5-pentachlorobiphenyl (PCB 126) on the levels of cortisol and glucose in European eel (*Anguilla anguilla*). Parasitology 132:281–288. https://doi.org/10.1017/S0031182005009017

Sures B, Nachev M, Selbach C, Marcogliese DJ (2017) Parasite responses to pollution: what we know and where we go in 'environmental parasitology'. Parasit Vectors 10:65. https://doi.org/10.1186/s13071-017-2001-3

Sures B, Nachev M, Schwelm J, Grabner D, Selbach C (2023) Environmental parasitology: stressor effects on aquatic parasites. Trends Parasitol 39:461–474. https://doi.org/10.1016/j.pt.2023.03.005

Tain L, Perrot-Minnot MJ, Cézilly F (2006) Altered host behaviour and brain serotonergic activity caused by acanthocephalans: evidence for specificity. Proc R Soc B Biol Sci 273:3039–3045. https://doi.org/10.1098/rspb.2006.3618

Taraschewski H (2000) Host-parasite interactions in acanthocephala: a morphological approach. Adv Parasitol 46:1–179. https://doi.org/10.1016/s0065-308x(00)46008-2

Taraschewski H (2015) Acanthocephala: functional morphology. In: Schmidt-Rhaesa A (ed) Gastrotricha, cycloneuralia and gnathifera. Handbook of zoology, vol 3. de Gruyter, Berlin, pp 301–315. https://doi.org/10.1515/9783110274271

Valtonen TE, Crompton DWT (1990) Acanthocephala in fish from the Bothnian Bay, Finland. J Zool 220:619–639. https://doi.org/10.1111/j.1469-7998.1990.tb04739.x

Vivas Muñoz JC, Feld CK, Hilt S et al (2021) Eye fluke infection changes diet composition in juvenile European perch (*Perca fluviatilis*). Sci Rep 11:1–14. https://doi.org/10.1038/s41598-021-81568-y

Wharton DA (1986) Life cycle. In: A functional biology of nematodes, Functional biology series. Springer, Boston. https://doi.org/10.1007/978-1-4615-8516-9_6

Whittington ID, Kearn GC (2011) Hatching strategies in monogenean (Platyhelminth) parasites that facilitate host infection. Integr Comp Biol 51:91–99. https://doi.org/10.1093/icb/icr003

Wohlleben AM, Steinel NC, Meyer NP et al (2022) The timing and development of infections in a fish-cestode host-parasite system. Parasitology 149:1173–1178. https://doi.org/10.1017/S0031182022000567

Würtz J, Taraschewski H (2000) Histopathological changes in the swimbladder wall of the European eel *Anguilla anguilla* due to infections with *Anguillicola crassus*. Dis Aquat Org 39:121–134. https://doi.org/10.3354/dao039121

Würtz J, Knopf K, Taraschewski H (1998) Distribution and prevalence of *Anguillicola crassus* (Nematoda) in eels *Anguilla anguilla* of the rivers. Dis Aquat Org 32:137–143. https://doi.org/10.3354/dao032137

Yoshinaga T, Tsutsumi N, Hall KA, Ogawa K (2009) Origin of the diclidophorid monogenean *Neoheterobothrium hirame* Ogawa, 1999, the causative agent of anemia in olive flounder *Paralichthys olivaceus*. Fish Sci 75:1167–1176. https://doi.org/10.1007/s12562-009-0148-3

Open Access This chapter is licensed under the terms of the Creative Commons Attribution-NonCommercial-NoDerivatives 4.0 International License (http://creativecommons.org/licenses/by-nc-nd/4.0/), which permits any non-commercial use, sharing, distribution and reproduction in any medium or format, as long as you give appropriate credit to the original author(s) and the source, provide a link to the Creative Commons license and indicate if you modified the licensed material. You do not have permission under this license to share adapted material derived from this chapter or parts of it.

The images or other third party material in this chapter are included in the chapter's Creative Commons license, unless indicated otherwise in a credit line to the material. If material is not included in the chapter's Creative Commons license and your intended use is not permitted by statutory regulation or exceeds the permitted use, you will need to obtain permission directly from the copyright holder.

6. Biology and Life Cycles of Parasitic Arthropoda Infesting Aquatic Hosts

Kerry A. Hadfield, Anja Erasmus, Niel L. Bruce, and Nico J. Smit

Abstract

Parasitic arthropods, particularly crustaceans, play significant roles in aquatic ecosystems, utilising various hosts for their life cycles. This chapter delves into the biology and life cycles of selected parasitic arthropods in aquatic ecosystems, especially highlighting crustaceans that exhibit significant diversity, often involving complex life cycles with multiple hosts. Ten major crustacean groups, in six classes, are discussed in the chapter. Life cycle notes on certain parasitic copepods from Cyclopoida, Harpacticoida, Monstrilloida and Siphonostomatoida are provided. Within the class Ichthyostraca, the Branchiura (or ectoparasitic fish lice) and the Pentastomatida (worm-like crustaceans) are highlighted. The third class, Malacostraca, includes amphipods, such as whale lice and jelly parasites, as well as the parasitic isopods from Bopyridae, Cymothoidae and Gnathiidae. The parasitic ostracods and the tiny Tantulocarida are mentioned, followed by the final aquatic crustacean class, Thecostraca. This last class includes the Ascothoracida (endoparasites within echinoderms and cnidarians), the Cirripedia (comprising the Acrothoracica, Rhizocephala and Thoracica) and what is known of the enigmatic Facetotecta. Additionally, the terrestrial arthropods within Arachnida and Hexapoda (Insecta) parasitising aquatic hosts are discussed. Throughout this chapter, the ecological implications of parasitic crustaceans are emphasised, particularly their roles in host population dynamics and potential co-evolutionary interactions. Understanding these life cycles provides valuable insights into the ecological impact of parasitic arthropods on their hosts and the broader aquatic environment.

K. A. Hadfield (✉) · A. Erasmus · N. J. Smit
Water Research Group, Unit for Environmental Sciences and Management, North-West University, Potchefstroom, South Africa
e-mail: kerry.malherbe@nwu.ac.za; 23599235@mynwu.ac.za; nico.smit@nwu.ac.za

N. L. Bruce
Water Research Group, Unit for Environmental Sciences and Management, North-West University, Potchefstroom, South Africa

Biodiversity & Geosciences Program, Queensland Museum, South Brisbane BC, QLD, Australia
e-mail: niel.bruce@qm.qld.gov.au

6.1 Introduction

Parasitic arthropods are a diverse group of organisms that include insects, arachnids and crustaceans. These parasites live on or in other organisms (hosts), often causing harm, and can be found in both terrestrial and aquatic environments. The latter group often plays significant roles in the ecosystem, directly impacting their hosts, host populations and other ecological components in the aquatic environment. They can live either on the surface (ectoparasites) or inside the host body (endoparasites). Understanding the biology and impact of these parasitic arthropods is vital for understanding and effectively managing the effects they may have on both wild and cultured aquatic species. Mites, ticks and insects, such as lice, are known to infect selected aquatic hosts, but within the aquatic environment, the parasitic crustaceans appear to be the most dominant. The Crustacea exhibit the highest diversity in parasitic forms, constituting over a quarter of all known species within the arthropods.

Crustaceans are a diverse group primarily found in aquatic environments within the phylum Arthropoda and subphylum Crustacea, known to inhabit a wide array of niches all over the world. These parasitic crustaceans have adapted to a parasitic lifestyle, dependent on a host organism for various aspects of their life cycle, including nutrition and reproduction.

While the majority of crustacean parasites exhibit simple life cycles, a few have more intricate developmental pathways. In direct life cycles, a solitary obligatory host is involved (monoxenous), whereas in indirect life cycles, a minimum of two obligatory hosts play essential roles, one acting as the secondary intermediate host, and the other as the final (or definitive) host (heteroxenous).

The life cycles of many parasitic crustaceans still remain largely underexplored, primarily due to the inherent challenges associated with monitoring these organisms in or on their aquatic hosts, especially those within marine environments. The small size of many parasites, coupled with their often inaccessible locations on hosts, presents formidable obstacles in tracking and observing morphological changes across different life stages. Consequently, it comes as no surprise that only a limited number of parasitic crustacean species have had their life histories extensively documented. Understanding the life history and ecology of parasitic arthropods is important as they can have significant implications for the health of host populations and ecosystems. Thus, this chapter aims to provide insight into the life cycles and biology of these aquatic parasites.

The subphylum Crustacea has been taxonomically recognised as having a segmented body with a hard exoskeleton, jointed appendages and typically two pairs of antennae. Traditionally, the Crustacea and Hexapoda (containing the insects) were viewed as sister taxa, forming a group called Pancrustacea. More recently, Hexapoda has been included within the Crustacea (Schwentner et al. 2017); however, this placement has not completely been resolved (Fig. 6.1). Since the phylogenetic placement of these groups does not influence their basic morphology or life history, we will, for the purpose of this chapter, first discuss the traditional six major Crustacea classes considered as parasites of aquatic hosts (listed alphabetically, then subdivided as needed to ensure clarity and comprehension), including the Copepoda, Ichthyostraca, Malacostraca, Ostracoda, Tantulocarida and Thecostraca. This will be followed by a brief discussion of the two terrestrial arthropod groups with aquatic hosts, the Arachnida (order Ixodida, including the ticks and mites) and Insecta (order Phthiraptera, comprising the lice) (Fig. 6.1).

6.2 Copepoda

Copepods, a highly diverse and prolific group of crustaceans, are abundant in aquatic environments, showcasing their adaptability as free-living, symbiotic or parasitic organisms. While these small crustaceans exhibit significant diversity in morphological characters, they often have a tapered body shape accompanied by long antennae (Fig. 6.2). Parasitic copepods differ dramatically in morphology, making generalisations on their biology quite challenging. Their host selection is equally diverse, including a broad spectrum of organisms across nearly every animal phylum, from sponges and echinoderms to fish and mammals (Boxshall 2005).

Furthermore, the attachment sites on selected hosts are far from uniform, which is unsurprising

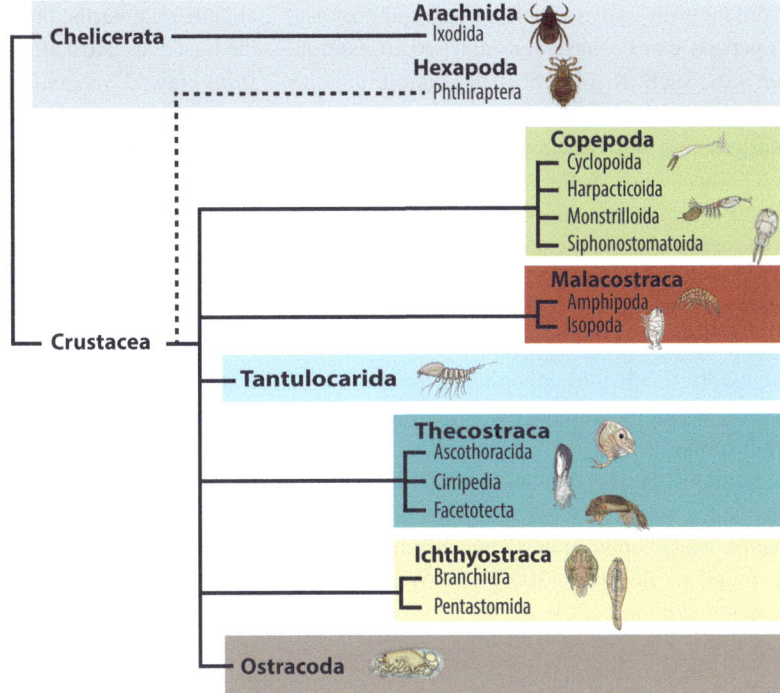

Fig. 6.1 Taxon tree showing the taxonomic relationships of selected aquatic and terrestrial parasitic arthropod groups discussed in this chapter

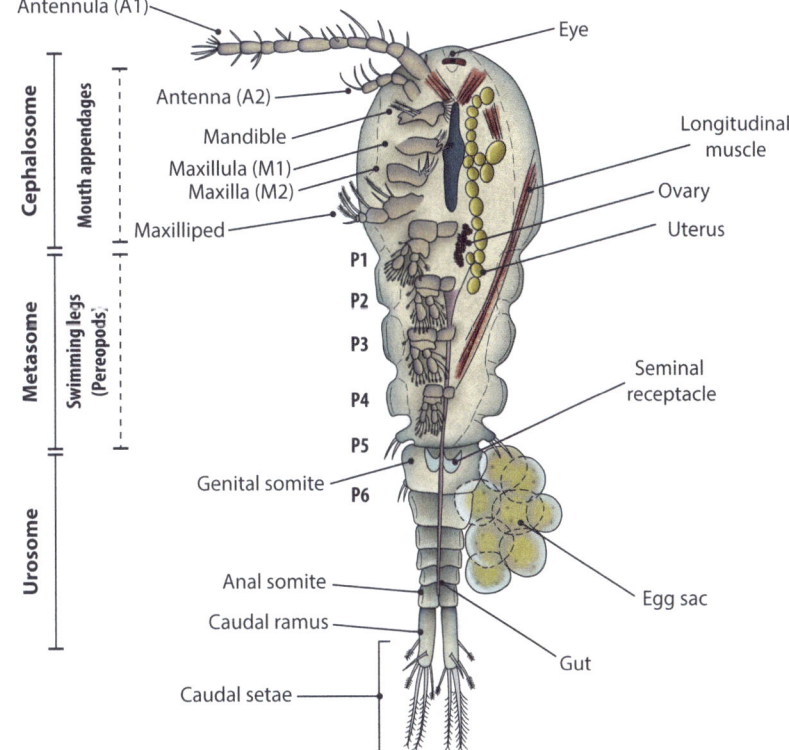

Fig. 6.2 Generalised morphology of a copepod, ventral view with external appendages shown on the left side, internal organs visible on the right side. Abbreviations: *A1* Antenna 1, *A2* Antenna 2, *M1* Maxilla 1, *M2* Maxilla 2, *P1–6* pereopods [Adapted from Santhosh et al. (2018) and The Robinson Library (2024)]

given their diversity of morphology and host use. Copepods can be observed attached to external surfaces, such as gills, nostrils, mantle cavities and genital folds, while others reside as endoparasites within the host muscles, digestive tracts and body cavities. Some of these copepods maintain a permanent attachment, whereas others possess the remarkable ability to relocate to new hosts. Overall, a typical copepod life cycle unfolds through distinct stages, commencing with nauplii (with a maximum of six stages) that eventually moult into copepodids (maximum of six stages), culminating in the development of the adult forms.

Some of the representative parasitic copepod life cycles will be discussed further. More information on copepod biology and life history can be found in Boxshall (2005) and Williams and Bunkley-Williams (2019).

6.2.1 Biology and Life Cycle of Cyclopiform Copepods (Cyclopoida)

Copepods within this order are distributed globally, occupying various ecological roles. While the majority of these copepods are free-living or predatory, a significant fraction is parasitic, targeting a wide array of hosts, including molluscs, sea anemones, sea squirts, fishes and crustaceans. These parasitic copepods distinguish themselves from their free-living counterparts by the notably short antennulae (first antennae), measuring only half the length of their bodies, as well as uniramous antennae (second antennae). Furthermore, the abdomen is distinctly narrower than the thorax. In terms of reproduction, females differ from males as the genital and first abdominal somites fuse during the final moult, forming a genital double-somite.

Cyclopiform copepods resemble the free-living copepod *Cyclops*, with well-defined body segmentation and distinct tagmosis of the prosome (comprising the cephalosome and metasome or thorax) and the urosome (constituting the abdomen), along with all appendages. These parasitic copepods attach themselves to various locations on the host body, including the gill chambers, nostrils, body surface, fins and around the host eyes. Host attachment mechanisms range from clawed antennae to modified ventral body structures that facilitate adherence to the host.

Some cyclopoid species are known to serve as intermediate hosts for cestodes and nematodes, and host numerous pathogenic parasites that affect both humans and fish. Additionally, some taxa, like the anchor worms of the genus *Lernaea*, can be problematic in aquaculture. These parasites bore into the skin of the host fish, leading to a condition called lernaeosis, characterised by haemorrhagic ulcers at the attachment site.

The typical life cycle of *Lernaea* begins with the release of nauplii larvae from eggs (Fig. 6.3). These nauplii undergo two moults, successively producing larger nauplii II and III, all of which remain free-living. The third nauplius stage subsequently transforms into the first copepodid stage, which is infective and seeks out a host. It commonly attaches to the gills but may also attach to the host skin or fins, undergoing four more moults before reaching maturity. Each of the five copepodid stages gains an additional body segment with each moult, and the sex of the adult can be discerned at the last copepodid stage.

At this point, copulation takes place, after which the male copepod detaches and typically dies within approximately 24 h. The mated female can detach from the gills and seek the final attachment site on either the same host or a different one, where it will assume a sedentary lifestyle. The mated female burrows beneath the host skin, penetrates the muscles, and transforms into a mesoparasite (Fig. 6.3). Throughout this process, the female thorax and abdomen extend, external segmentation is lost and the abdomen widens at the posterior end. Simultaneously, the anchoring holdfast's branches extend outward from the cephalothorax. The post-metamorphic female's egg sacs develop and may be released within a few days (14–28 days) or overwinter on the fish host until optimal conditions for release are met.

For other cyclopoid families, the life cycle varies slightly. For example, in the family Ergasilidae, only females are parasitic, and predominantly found on freshwater fish hosts. The life cycle of these parasites has six nauplius

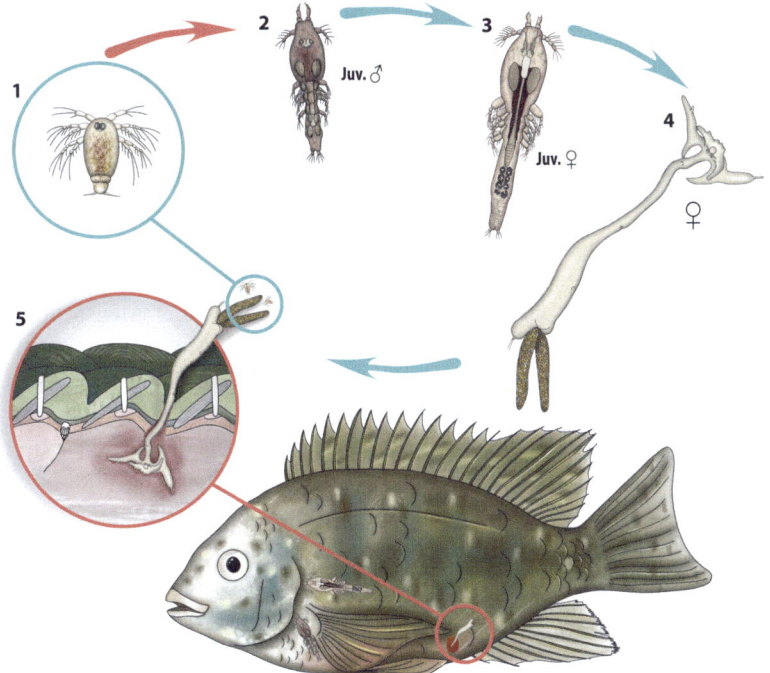

Fig. 6.3 Typical life cycle of *Lernaea* (Copepoda: Cyclopoida), parasitising freshwater fish. Nauplius larvae (*1*) are released from the egg sacs of the female. Nauplii (with approximately three instars) develop into infective copepodid stages (*2*) seeking a host. Copepodids moult four more times gaining additional segmentation with each moult. The sex can be determined at the last copepodid stage when copulation takes place. Only the female (*3*) continues to develop into a post-metamorphic stage (*4*) and assumes a sedentary lifestyle embedded in host tissue (*5*). The red arrow indicates the free-living stage. Schematic, not to scale [Adapted from Kearn (2004a)]

stages, five copepodid stages and adults. The naupliar and copepodid stages are free-living, with males dying after mating and females becoming infective as they search for a host. Eggs develop within external membranous sacs, with fertilised eggs later released into the water column, ultimately leading to the emergence of the new hatching nauplii.

There are also families of Cyclopoida with completely different life cycles. Copepods in the family Thaumatopsyllidae follow a life cycle reminiscent of monstrilloids (see Sect. 6.2.3 below), exhibiting a protelean strategy that relies on brittle stars as hosts, encompassing both an external and internal cycle. In contrast to monstrilloids, thaumatopsyllids have one infective and one or more endoparasitic naupliar stages, in addition to five or six non-feeding planktonic copepodid stages that culminate in the adult stage.

6.2.2 Biology and Life Cycle of Worm-Like Copepods (Harpacticoida)

Members of the order Harpacticoida are primarily free-living copepods, with a limited number of documented symbiotic or parasitic associations. These copepods have been found on the external surfaces of various hosts, including turtles, whales and manatees, where they can either act as epibionts or potential parasites. Additionally, they have been observed in the gills or on the external surfaces of octopuses. The precise nature of some of these possible parasitic interactions remains somewhat enigmatic.

While many of these organisms are assumed to follow the conventional copepod life cycle, López-González et al. (2000) proposed an intriguing alternative cycle for cholidyinid harpacticoids. Evidently, the free-swimming naupliar stages are similar to those of other copepods, but the copepodid stages develop *inside* the octopus host. This deduction stems from the discovery of juvenile copepods within the connective tissue of the octopus's integument. Copepodids settle onto the suitable host, penetrate the integument and embed themselves into the host tissue. Within the host, they undergo moulting and presumably complete the typical copepodid six-stage cycle. Once the copepodid reaches sexual maturity, it exits the host through a slit in the host integument.

As adults are capable of swimming and have been found on various regions of the octopus host, including the gills, arms and mantle, it is assumed that they remain in contact with the host and feed on its superficial tissues throughout the remainder of their life cycle. Mating is also assumed to take place outside of the host.

6.2.3 Biology and Life Cycle of Monstrilloid Copepods (Monstrilloida)

Monstrilloid copepods are only partially or temporarily parasitic. These crustaceans are characterised by their protelean nature, where they have parasitic post-naupliar and pre-adult stages, but the adults are non-feeding and lead a free-living existence. The order Monstrilloida consists of a single family, Monstrillidae, with endoparasitic forms known to inhabit a variety of hosts, including polychaetes, molluscs and other invertebrates.

These copepods start their life cycle as free-swimming, infective nauplius larvae, which hatch from eggs and search for a suitable host (Fig. 6.4). The nauplii rely on the yolk stores passed down from their mother (lecithotrophic) to sustain them until they attach to and burrow into the host

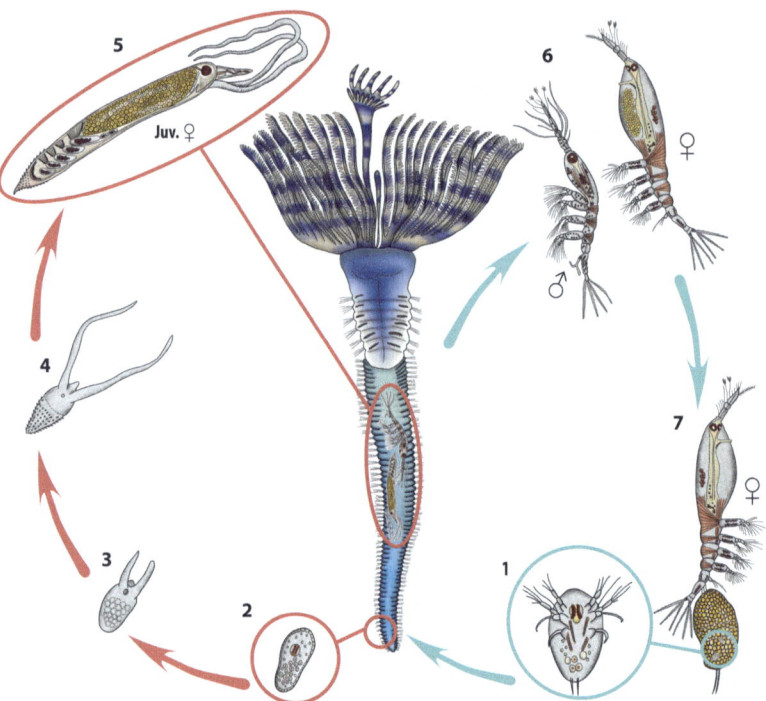

Fig. 6.4 Typical life cycle of Monstrilloida (Copepoda) infecting serpulid hosts. The infective nauplius stage (*1*), released from the female egg sac, penetrates the host tissue developing into endoparasitic naupliar stages (*2–4*) until they reach the endoparasitic copepodid stage (*5*). The copepodids then exit the host as pre-adults by rupturing the body wall. Free-swimming, reproductive adults (*6*) develop, and eggs are attached to paired ovigerous spines (*7*). Red arrows indicate the endoparasitic stages, green arrows indicate the free-living stages. Schematic, not to scale [Adapted from Huys (2014)]

tissue. Equipped with mandibles, which have a pair of terminal claws, and antennae, they are able to attach and burrow into host tissue. Once inside the host haemolymph, the nauplius undergoes a metamorphosis, transforming into a sac-like second naupliar stage with a protective sheath enveloping its body. This stage is unrecognisable as a typical crustacean, featuring two anteroventral root-like processes or feeding tubes that absorb host fluids (Suárez-Morales 2011).

Post-naupliar development unfolds within this protective sheath, with approximately three endoparasitic copepodid stages. During this time, the copepods grow in size and begin to alter the host body. Upon completing their development, the copepods exit the host by rupturing the body wall, emerging as pre-adults (Fig. 6.4). These pre-adults possess fully formed appendages and appendage armature but are structurally simple in comparison to other crustaceans. They undergo one final moult, transforming into free-living, reproductive adults, with all cephalic appendages absent, except for the antennulae. In females, egg sacs are absent and, instead, eggs are attached to paired ovigerous spines through a mucous secretion produced by the terminal part of the oviduct.

6.2.4 Biology and Life Cycle of Siphon-Mouth Copepods (Siphonostomatoida)

Most fish parasitic copepods belong to this order. These copepods are predominantly found in marine waters and are characterised by a siphon-like mouth tube or oral cone that houses stylet-like mandibles used for extracting bodily fluids, blood or tissue from the host. Species of this order all share a dorsoventrally flattened body plan, featuring an anterior cephalothorax and a post-cephalothoracic genital trunk.

The Pennellidae are renowned for being some of the largest copepods. Some pennellids can reach lengths of up to 250 mm and carry egg sacs that may exceed 350 mm. They exhibit diverse life cycles, encompassing both direct and indirect cycles, as well as monoxenous and heteroxenous variations. One example of such a life cycle is observed in *Peniculus minuticaudae*, as detailed by Ismail et al. (2013). The infective copepodid hatches from an egg and actively swims in search of a suitable host. Upon finding a host, the copepodid uses its antennae to attach (particularly on the fins) and subsequently moults into the first chalimus stage. These copepods undergo four chalimus stages of development, during which they attach to the host using a frontal filament located at the anterior end of the infective copepodid. The final moulting leads to the development of an adult. These organisms have well-developed swimming legs in pre-metamorphic adults, suggesting potential for host switching, site migration on the same host or copulation. Copulation typically involves the adult male holding a chalimus female using the antennae (A2) and maxillipeds, and occurs before the female's detachment from the host. Once fertilised, the pre-metamorphic female seeks a new settlement site and undergoes significant development into a post-metamorphic adult female. These females produce eggs and egg strings that surpass the length of their bodies, resulting in a life cycle with two free-swimming and two parasitic phases.

The family Caligidae stands as one of the most species-rich copepod groups parasitic on fish. These organisms hold particular importance as they can inflict severe damage on fish in aquaculture, especially salmonids, resulting in substantial losses. Caligids use an alternative method for attaching to hosts. Instead of relying on their antennae only, they use the suctorial capacity of their cephalothorax. The life cycle generally consists of two nauplius stages, a copepodid stage, four chalimus stages and an adult (Fig. 6.5). This cycle bears several resemblances to that of the Pennellidae, with adult caligids demonstrating a notable ability to move swiftly across a surface, enabling them to locate suitable feeding sites or mates. The nauplii stages do not involve feeding and rely on lecithotrophic nutrition.

Another intriguing family is the Nicothoidae, which primarily dwells on the body surface, gills, marsupium or egg masses of other marine crustaceans. These copepods can pose threats to com-

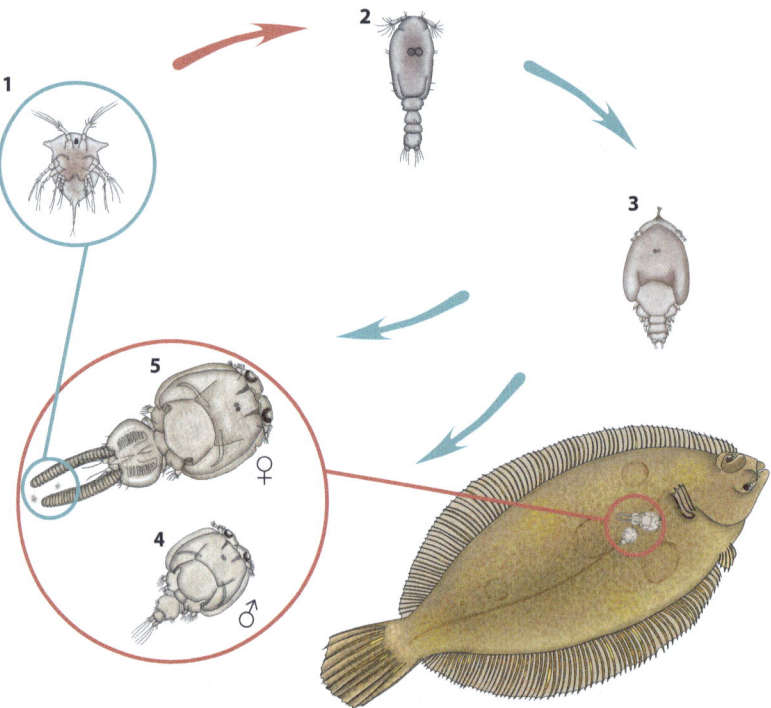

Fig. 6.5 Typical life cycle of the siphon-mouth copepod *Caligus* (Siphonostomatoida: Caligidae). The free-living nauplius larvae (*1*), consisting of two instars, are released from the uniseriate egg strings and develop into one infective copepodid instar (*2*) attaching to their host with antennae. Subsequently, four chalimus instars (*3*) of development occur attaching to their host using a frontal filament. Final moulting leads to adult male (*4*) and female (*5*) stages which rely on the suctorial attachment of their cephalothorax. The red arrow indicates the free-living stage. Schematic, not to scale [Adapted from Kearn (2004b)]

mercially significant lobsters and spider crabs. Nicothoids avoid being removed by the host by mimicking the eggs of the host. A representative species, *Choniomyzon inflatus*, exhibits three distinct stages: a nauplius, at least two copepodids and the adult stages. The life cycle unfolds with free-living nauplii hatching from the egg sac of the adult female, moulting into the first infective copepodid stage in the water and attaching themselves to a suitable host. The copepodid subsequently settles on the host gills or body surface, undergoes the next copepodid stage and moves to where the host marsupium and egg masses are located. Here it matures while attached to the host egg mass until it becomes an adult. The adults engage in mating, and the females produce eggs, completing the life cycle (Otake et al. 2013).

6.3 Ichthyostraca

6.3.1 Branchiura

Members of the subclass Branchiura are ectoparasitic crustaceans in the class Ichthyostraca. They are known to feed primarily on freshwater fishes, although records exist of their presence on alligators, marine fish, salamanders and tadpoles. They predominantly inhabit freshwater environments but can also be found infesting estuarine and nearshore marine fishes. These dorsoventrally flattened parasites are not permanently attached to their hosts and function as temporary parasites with the ability to swim freely throughout their lives. Branchiurans are characterised by their flat and oval shape, possession of two large and separate compound eyes, a single nauplius eye, four pairs of strong swimming legs and a broad shield-like carapace with well-developed lobes typically covering the lateral legs of their bodies (Fig. 6.6). The maxillulae (first maxillae) in this group are used for host attachment and can exhibit sucking capabilities. In species of three of the four genera within this group, namely *Argulus*, *Chonopeltis* and *Dipteropeltis*, the first maxillae are modified as suction cups, whereas in species of the fourth genus, *Dolops* Audouin, 1837, they are functional claws.

These parasites attach to the skin of their hosts and feed on their blood, mucus and external tissues. Their attachment is non-permanent, and these parasites can move on and off the host (except for species of *Chonopeltis*, which are poor swimmers). Of the four recognised genera in the family Argulidae, a monotypic family in the subclass Branchiura, the life cycle of *Argulus* is the most well documented. This is the most speciose genus in the family and thus the most extensively studied. Subsequent remarks will reflect more on the general life cycle of *Argulus*, with short notes on the other genera.

6.3.1.1 Biology and Life Cycle of Fish Lice (*Argulus* spp.)

The genus *Argulus* consists of obligate parasites that feed on the blood, mucus or extracellular material of fish hosts, and are often referred to as fish lice. These crustaceans disperse readily and thus infect many hosts, via their swimming abilities as well as using water currents. However, this dispersal capacity may limit their ability to find suitable mates, consequently reducing their reproductive success.

In the search for a suitable mate, adult males exhibit a greater degree of freedom from their hosts, spending more time swimming than females. Males are capable of detecting and responding to sex pheromones emitted by females attached to fish hosts. Copulation typically occurs on the external surface of the host. In the case of *Argulus japonicus*, males transfer sperm via a spermatophore, which is expelled from the male's genital aperture. Subsequently, the sperm is conveyed to a socket on the third pair of legs of the male and further transmitted to the spermathecae of the female through the spermathecal spines (Avenant-Oldewage and Everts 2010).

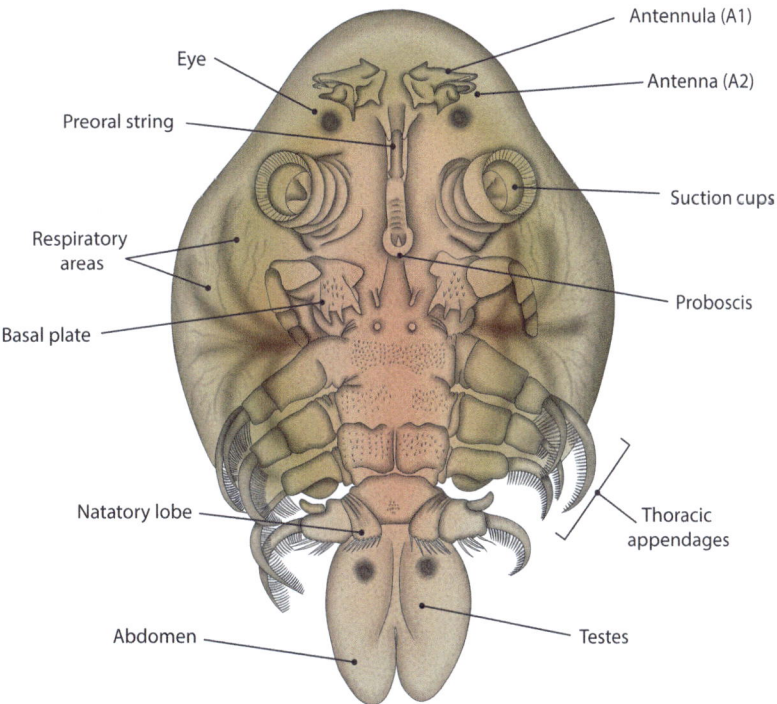

Fig. 6.6 Generalised morphology of a male fish louse from the genus *Argulus* (Ichthyostraca: Branchiura), ventral view. Abbreviations: *A1* Antenna 1, *A2* Antenna 2

Females feed continuously, grow faster than males and store energy for egg production. They only detach from the host when the need arises to lay eggs on a suitable substratum. Gravid females need to deposit their eggs in areas of high fish abundance, typically found in spawning and nursery grounds. The sites are often characterised by dark-coloured stones, wood debris or vegetation in shallow littoral zones, depending on the species and fish habitats. These locations are carefully chosen to minimise exposure to strong currents. The eggs are laid in disordered rows, forming what are known as egg strings, each containing up to 1,200 eggs, and are securely attached to the substrate using an adhesive substance. After depositing an egg clutch, the female seeks another host to feed on to restore her energy reserves before laying her next clutch. The hatching period for eggs spans from 12–80 days, contingent on the species and water temperature, with the possibility that some eggs overwinter.

In the life cycle of some *Argulus* species, the initial stage is a metanauplius-like larval phase (Fig. 6.7). These larvae hatch and must quickly locate an available host. They have a naupliar swimming apparatus (antennal exopods and mandibular palps involved in locomotion) and lack functional legs (Møller and Olesen 2014). However, they possess fully developed postmandibular appendages and differentiated first thoracopods, distinguishing them from true metanauplius larvae. The free-swimming larval stage consists of a single instar, equipped with all the essential features for adopting a parasitic lifestyle (Møller and Olesen 2014). When necessary, these larvae can transfer to a more suitable host, to which they securely attach, rarely separating until reaching adulthood (Mikheev et al. 2015).

Fig. 6.7 Typical life cycle of the fish louse, *Argulus* (Branchiura). Eggs deposited in substratum (*1*) hatch into a metanauplius-like larval stage (*2*) of a single instar, equipped with swimming apparatus to find a suitable host. Further development occurs gradually through moulting until the fifth stage (*3*) whereafter the maxillula develops from a claw into a circular sucker. Subsequent moulting results in adult male (*4*) and female (*5*) branchiuran parasites, attached to the host external surface, where copulation takes place. Females only detach to deposit eggs on substrata. The red arrow indicates the free-living stage. Schematic, not to scale

In some *Argulus* species, the initial development stage occurs later in their life cycle. These larvae hatch with four pairs of fully developed thoracopods, which they use for locomotion, rendering the naupliar swimming apparatus (which they lack) unnecessary. These parasitic juvenile-like larvae bear a resemblance to adults but lack suction discs, as well as several setae and spines. They are free-swimming and the number of larval stages varies from 8–12 development stages, all occurring on fish hosts. Moults occur in different intervals until reaching maturity. Most of the larval transformations are gradual, particularly concerning the thoracic legs and reproductive organs. However, a notable transformation occurs during the fifth larval stage, as the maxillula evolves from a slender appendage ending with a distal claw into a short, circular sucker. By the time the sixth stage is reached, a fully developed suction disc is in place.

Furthermore, some *Argulus* species may serve as hosts for other parasites. *Argulus coregoni* and *A. foliaceus*, for instance, have been reported as intermediate hosts for skrjabillanid nematodes. Additionally, *Argulus mexicanus* and *A. yucatanus* have been identified as intermediate hosts for daniconematid nematodes (Moravec et al. 2009; Neethling and Avenant-Oldewage 2016). Some *Argulus* species are also known to transmit infectious rhabdoviruses, such as Spring viraemia of carp (SVC) virus (Pfeil-Putzien 1978; Hadfield and Smit 2019).

6.3.1.2 Other Branchiura Genera

The remaining three genera within Branchiura have received comparatively less attention and are relatively understudied, with some aspects of their life cycle completely unknown. Although the available data suggest that these genera share numerous similarities with the *Argulus* life cycle, several distinctions stand out.

As mentioned above, species of *Dolops* have maxillulae that are robust hooks rather than suction cups. These branchiurans do not have metanauplius-like larval development but instead have juvenile-like larvae reminiscent of those found in *Argulus*. Notably, these larvae have maxillulae with a shape similar to the adult form, terminating in a strong, recurved hook. Sperm transfer occurs via chitinous spermatophores. A single brood can consist of up to 177 ellipsoid eggs, deposited in consecutive rows, with a gelatinous layer that protects and firmly anchors them to the substrate. Eggs are initially white and gradually transition to a brown hue over four days. Unhatched eggs turn black. Larvae emerge within a span of 16–57 days, appearing as miniature adults, and the post-hatching larval survival period ranges from 4 to 6 h.

Species of *Chonopeltis* possess a conical or funnel-shaped shield and differ from species of *Argulus* in that they lack metanauplius-like or juvenile-like larvae. There is no naupliar swimming apparatus, and thoracopods are not completely developed so how these parasites move around remains unknown. These non-swimming larvae are generally smaller in size than their counterparts in other branchiuran genera, featuring maxillulae with a robust bipartite distal hook but lacking naupliar appendages and functional thoracopods. The antennae are uniramous, with minimal major setation, displaying distinct differences from other branchiurans. Sperm transfer is achieved through a spermatophore and females of this genus tend to produce smaller broods, containing around 130 eggs. Typically, *Chonopeltis* larvae undergo approximately eight to ten developmental stages, with stage 1 associated with fish hosts.

Species of *Dipteropeltis* are distinguished by the notably elongate abdominal lobes present in adult females. Based on observations of juvenile larvae thus far, they seem to closely resemble miniature versions of the adults. While adult organisms employ suction cups for attachment that allows them to 'walk' along surfaces, they are not able to swim when detached from the host. Juveniles, however, can swim and use their legs for short bursts of movement (as observed by Gaboardi et al. in 2023). It is possible that the elongate carapace of *Dipteropeltis* spp. may also play a role in facilitating their swimming capabilities.

6.3.2 Pentastomatida

The pentastomids are a distinct and unique group within the parasitic crustaceans. These organisms are obligate parasites that mostly have a complex and indirect life cycle. They are predominately endoparasitic and do not conform to the typical body shape of other parasitic crustaceans. Pentastomids have elongate, worm-like bodies that are dorsoventrally flattened, making them appear tongue-like, hence the common name 'tongue worms'. These organisms, characterised by the lack of true segmentation and simple body plan, are covered in a protective chitinous cuticle. Moreover, they possess a unique set of five anterior appendages. Among these, one functions as the mouth, while the other four are equipped with hooks used to firmly anchor the parasite in place and facilitate the process of feeding upon the host (Fig. 6.8).

Adult pentastomids usually occur in the upper and lower respiratory tracts, including the lungs and trachea, of a wide range of vertebrates including amphibians, birds, mammals, reptiles and humans. Adult pentastomids can be readily distinguished from other parasites due to the presence of two pairs of retractable hooks located on either side of the mouth (Hadfield 2019). They feed on blood or mucous depending on the specific location within the host respiratory system.

The life cycle of pentastomids starts with the release of embryonated eggs by adult parasites (Fig. 6.9), which are then either expelled from the definitive host via nasal mucus and saliva or ingested and excreted through the digestive system, typically in faeces. The released eggs remain viable for prolonged periods of time and must be consumed by an appropriate host to continue the life cycle.

Once within the intermediate host (most often a fish, invertebrate or small herbivorous mammal), the pentastomid larva hatches and migrates through the host's inner organs via the bloodstream (Fig. 6.9). Over time, these larvae undergo multiple moults (usually from two to eight, depending on the species) inside the intermediate host, experiencing growth and maturation before forming a cyst within the host body. These cysts are usually in the visceral tissue such as the lymph nodes, liver, intestine, swim bladder and reproductive organs of the intermediate host. The larvae are then referred to as nymphs, closely resembling worms and earning them the colloquial name 'tongue worms'. The encapsulated nymphs continue to develop until they reach a stage where they become infective to the definitive host, although many may perish along the way.

The final host becomes infected with these nymphs either through the consumption of raw or uncooked host tissue, or when a mature nymph excysts and migrates to the respiratory tract of its host. Excystation appears to be triggered by the death of the intermediate host when it is consumed by the definitive host (Riley 1986). If consumed, the nymph penetrates the wall of the digestive system (oesophagus, stomach or intestine), migrates through the host body to the respiratory tract and matures into an adult pentastomid (Fig. 6.9). In most species of pentastomids, sexual dimorphism is not highly evident, with the primary distinction being the considerable size difference between females and males, with females consistently much larger.

The males fertilise the females, but their own lifespan is relatively short, with females storing sperm following insemination (Riley 1986). Females potentially mate only once, but males copulate with more than one female. Gravid females may carry millions of eggs, depending on the species, which mature as they descend from the uterus and are eventually released. In some pentastomids, females have a sieve-like uterus which permits only large, fully developed eggs to be released while retaining smaller, immature ones. Others have eggs that mature as they travel down the uterus, thus both methods ensure that only fully embryonated eggs are released by the definitive host.

There are some infrequently encountered species, of the genera *Sebekia* and *Subtriquetra*,

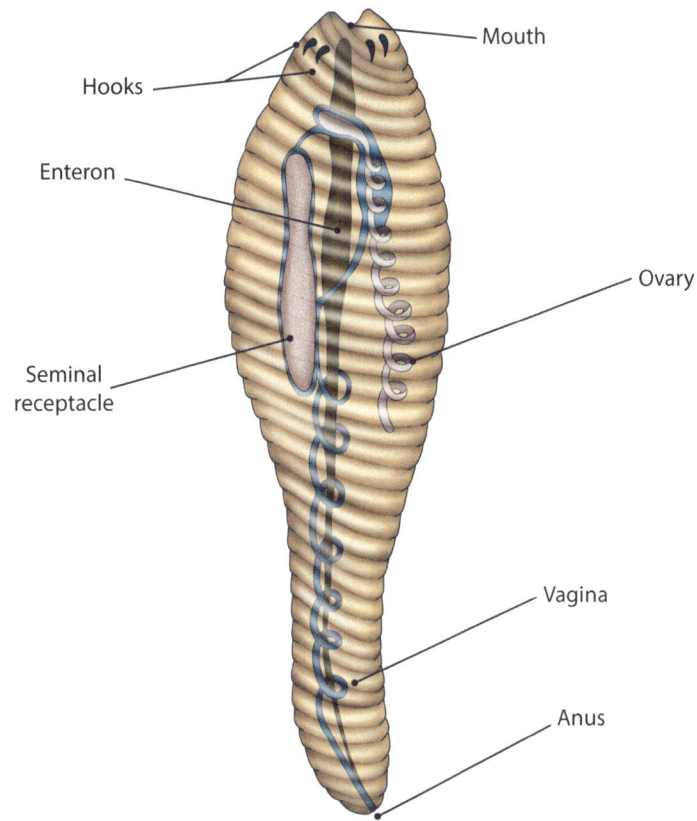

Fig. 6.8 Generalised morphology of the crustacean 'tongue worms' (Ichthyostraca: Pentastomatida), with internal organs visible [Adapted from Maiti (2022)]

which have free-swimming larvae (Riley 1986; Junker et al. 1998). These infective primary larvae make contact with a suitable intermediate host, eliminating the need for egg transfer to a new host.

Furthermore, species from the genus *Reighardia*, specifically *Reighardia sternae*, have a direct life cycle in birds. These parasites are found in the air sacs of gulls, terns and other sea birds, especially juveniles, with transmission likely occurring through the ingestion of regurgitated food (from parents to young) or faeces (Christoffersen and De Assis 2013). Due to their direct life cycle, these parasites can also induce autoinfection in the host, a rare occurrence in pentastomids. Female specimens within this group also have significantly shorter lifespans than other pentastomids, typically dying shortly after laying eggs (Paré 2008). At least two other genera, *Linguatula* and *Sambonia*, are also known to have species with direct life cycles.

Humans can become accidental hosts for pentastomids by eating the raw flesh of infected hosts or ingesting infective eggs (via water, contaminated vegetation or secretions from the infected definitive host). Domestic dogs are generally the cause of human infections in these cases (Paré 2008). The severity of the resulting disease, known as pentastomiasis, varies among individuals, but humans generally exhibit a high level of tolerance to pentastomid infections.

6.4 Malacostraca

6.4.1 Amphipoda

Amphipods (often referred to as sea fleas, sand hoppers or scuds) are a diverse and species-rich group of crustaceans characterised by their typically compressed bodies, absence of a carapace, possession of three pairs of pleopods (also known

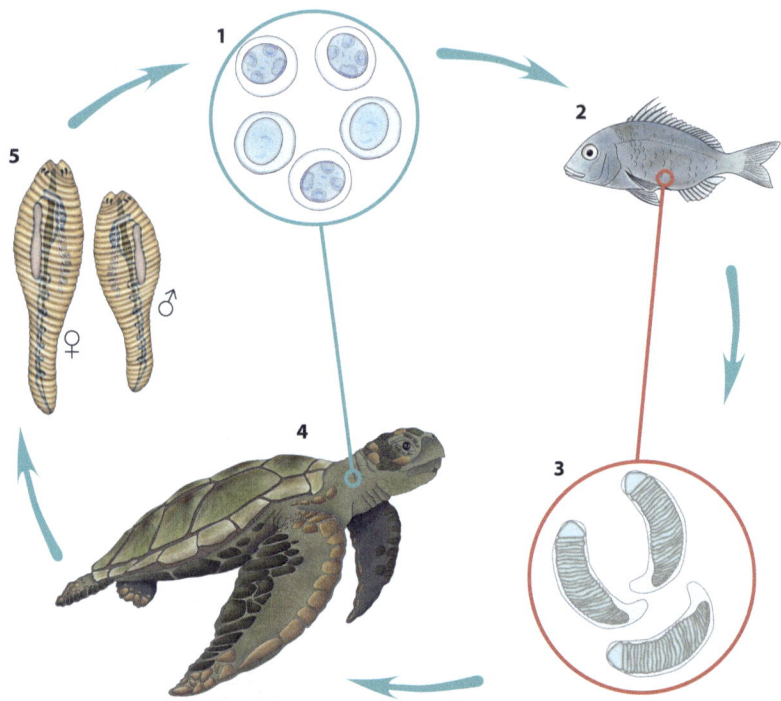

Fig. 6.9 Typical life cycle of aquatic Pentastomatida ('tongue worms') infecting fish (intermediate) and reptilian hosts (definitive). Eggs (*1*) are expelled from the definitive host (either via nasal mucus or faeces). Once the appropriate intermediate host (*2*) consumes the eggs, the larvae migrate via the bloodstream and undergo two to eight moults, eventually forming cysts. The encapsulated nymph stage (*3*) becomes infective over time, and once consumed by the definitive host, penetrates the intestinal wall and migrates to the respiratory tract of the final host (*4*) to mature into adult males and females (*5*). Schematic, not to scale [Adapted from Maiti (2022))]

as swimmerets) and three pairs of uropods. This crustacean order falls within the class Malacostraca, with the body divided into a head, a thorax (pereon) and an abdomen (pleon) (Fig. 6.10). The name Amphipoda, meaning 'different feet', alludes to the distinctly dimorphic forms of the pereopods (legs), which, among other characters, sets them apart from the closely related Isopoda (meaning 'same feet').

Amphipods thrive in a wide array of environments including the aquatic realms such as marine, freshwater and brackish waters, as well as terrestrial habitats. Some amphipods have free-living lifestyles, while others establish different types of symbiotic associations with other organisms. Only a few taxa have adopted micropredatory or parasitic habits, such as the families Cyamidae (colloquially known as 'whale lice'), Melitidae and Trischizostomatidae, as well as the suborder Hyperiidea. Further discussion will delve into two of these groups.

6.4.1.1 Biology and Life Cycle of Whale Lice (Cyamidae)

Species of Cyamidae, often referred to as whale lice, are dorsoventrally flattened crustaceans (in contrast to other amphipods) that are unable to swim. The cyamid body is posteriorly reduced and the first two and last three pairs of legs are broad, flattened prehensile appendages for secure attachment to the host skin. Cyamids attach to cetaceans such as whales, dolphins and porpoises by direct transmission from one host to another. This typically occurs during social interactions such as mating, nursing of young, and caregiving. Cyamids have a high degree of host specificity and will die if they are detached from the host. Females are generally broader and shorter than

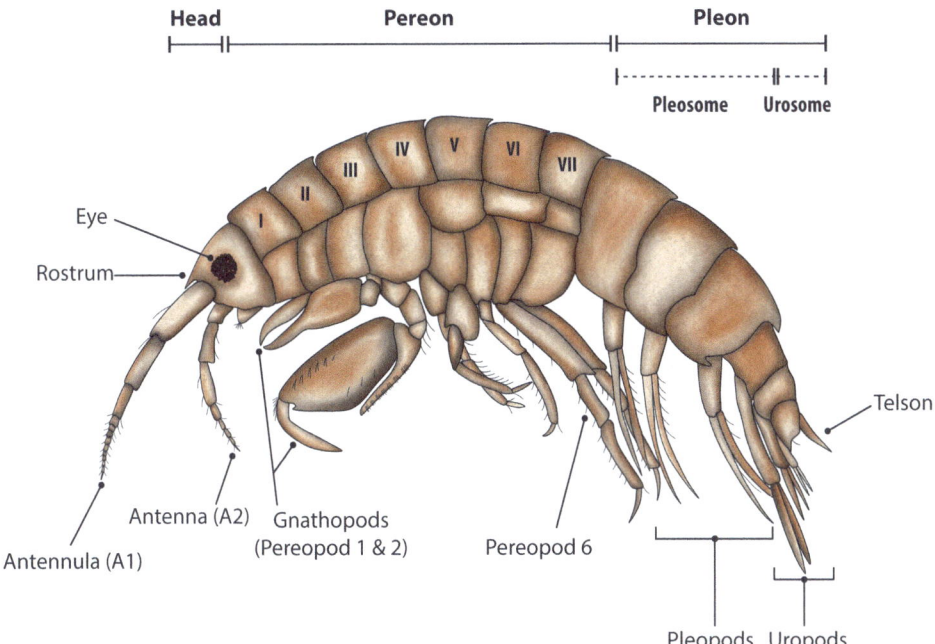

Fig. 6.10 Generalised morphology of an amphipod (Malacostraca: Amphipoda), lateral view. Abbreviations: *A1* Antenna 1, *A2* Antenna 2, *I–VII* pereon segments [Adapted from The Animal Files (2024)]

males (5–27 mm in length) and can produce a varying number of eggs, ranging from 50 to over a thousand, contingent upon the species.

Whale lice usually attach to areas on the host body that are sheltered from strong water currents (see Chap. 15, Fig. 15.12). These parasites primarily feed on the outer layer of the host skin, with some evidence suggesting potential consumption of other substances adhering to the skin, such as bacteria and algae, or plankton (Rowntree 1996). Nonetheless, stable isotope analyses reveal that the carbon and nitrogen isotopes in a whale louse are from the host (Schell et al. 2000). Certain species have been associated with skin lesions on the hosts. Cyamids have been discovered inside fresh and healing ulcerative wounds of unknown origin, with more extensive lesions displaying signs of inflammation and haemorrhaging. The extent to which whale lice may contribute to the maintenance of these lesions and hinder the host healing process remains speculative (Lehnert et al. 2021). Generally, whale lice only inflict minor damage when present in small numbers, but large infestations can cause significant harm to the cetacean hosts.

The direct life cycle of one of the largest species within this family, *Cyamus scammoni*, characterised by males reaching lengths of up to 27 mm and females up to 16 mm, has been described by Leung (1976). This species is commonly referred to as the grey whale louse as it predominantly parasitises the grey whale, *Eschrichtius robustus*. Mating occurs when the adults attain sexual maturity, with females depositing eggs into the brood pouch. A female typically carries around 980 to 1,078 eggs, although not all of them are fertilised, with approximately 60% being successful. Overlapping oostegites protect the eggs and developing larvae during the incubation period. After 2–3 months, the young are released from the brood pouch as miniature adults with fully developed hooked pereopods, allowing them to secure attachment to the host (Fig. 6.11). The number of instars is estimated to be seven or eight, based on the size and development of morphological structures (Leung 1976). The entire life cycle spans approximately 8–9 months, likely overlapping, as cyamids of various developmental stages are always present.

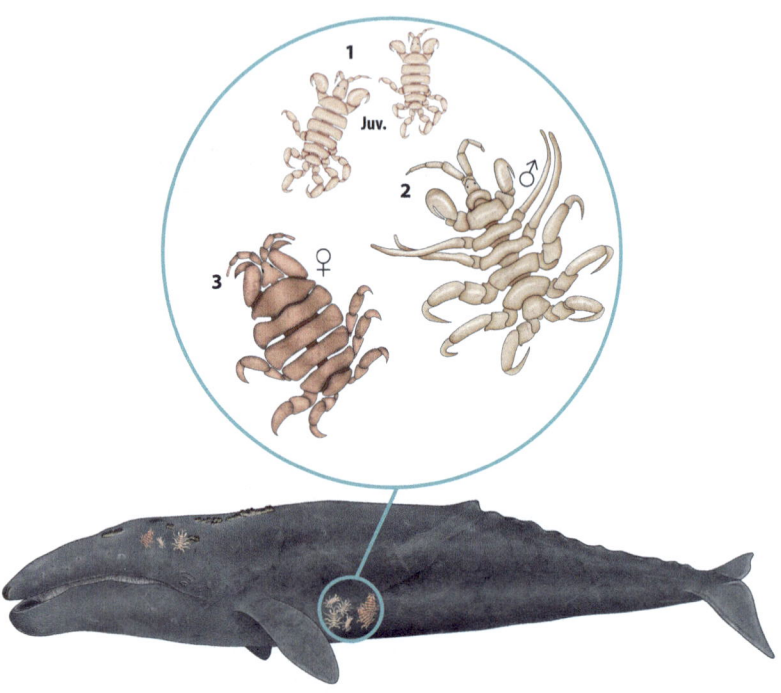

Fig. 6.11 Typical life cycle of whale lice (Cyamidae) that parasitise cetaceans (e.g. the grey whale). Juveniles (*1*) are released from the female brood pouch. These young resemble miniature adults and will usually develop into adult male (*2*) and female (*3*) cyamids over seven or eight instars. Schematic, not to scale

6.4.1.2 Biology and Life Cycle of Jelly Parasites (Hyperiidea)

Amphipods belonging to the suborder Hyperiidea are found solely within marine environments, with a preference for pelagic zones and tropical or warm temperate regions. They are recognised by their sizable cephalothorax featuring exceptionally large eyes and exhibit sizes ranging from a millimetre to well over a centimetre. These amphipods are frequently associated with other zooplankton, and their interactions range from parasitic relationships to commensalism, particularly with gelatinous organisms such as jellyfish, ctenophores, molluscs and tunicates. Certain species exhibit host specificity, while others display a more generalist approach. They are sometimes referred to as parasitoids (see Williams and Bunkley-Williams 2019), indicating that their larval stage relies on the host, and it is common to find remnants of host tissue within the amphipod stomach contents, with some ultimately killing these hosts.

Hyperiids are reliant on gelatinous zooplankton for juvenile development. These obligatory hosts provide multifunctional resources: a nurturing environment (nursery), a source of sustenance and a protective haven. As the young amphipods mature, the association becomes less prominent but remains essential as a source of food, shelter, refuge and transportation. While some species directly consume the zooplankton's tissue, others exclusively feed on the planktonic matter collected by the host (Lützen 2005).

The life cycle of hyperiids has various stages: a pantochelis stage (the marsupial stage), protopleon stages 1–5 (post-marsupial phase), followed by a juvenile and the adult phase (Fig. 6.12). Adult mating occurs on host plankton, with adult males typically maintaining a free-swimming, predatory lifestyle, while ovigerous females stay on the host brooding their eggs (approximately 50–600). Males leave the host to copulate with females on other hosts as they have superior swimming abilities (Dittrich 1987). The eggs develop into pantochelis larvae within the marsupium, displaying different and distinct features to adult hyperiids, such as an unsegmented and limbless metasome and urosome, four cheliform pereopods and an inability to swim or feed independently.

Subsequently, these larvae undergo metamorphosis into protopleon larvae, marked by seg-

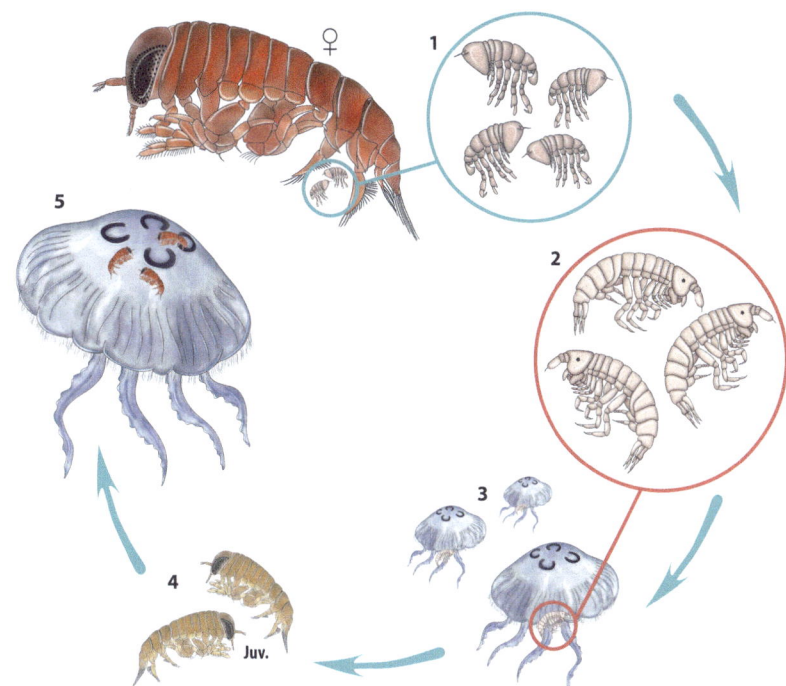

Fig. 6.12 Typical life cycle of jellyfish parasitoids (Hyperiidea). Eggs from the brood pouch develop into pantochelis larvae (*1*) in the marsupium and subsequently metamorphose into protopleon larvae (*2*) comprising approximately five instars. During demarsupiation, the larvae are deposited in the gonadal and gastric tissues of various jellyfish hosts (*3*). The protopleon larvae develop into juveniles (*4*), resembling adults (*5*). Schematic, not to scale [Adapted from Crossley et al. (2009)]

mented metasomes and rudimentary pleopods. In some species, this marks the initial larval stage, progressing through approximately five developmental stages, each characterised by the gradual acquisition of adult features. At this point, females release the larvae from their marsupium and deposit them onto a host, a process known as demarsupiation. The females swim from one host to another, depositing larvae in strategic locations, which may involve creating small openings in suitable gelatinous host tissues. This typically occurs in areas where they can readily access an ample food supply, such as in proximity to gonadal and gastric tissues. Females may even divide a host gonad with their mouthparts, inserting the larvae deep within the organ. As the larvae mature, they leave the gonad and feed on organisms captured by the host. In other cases, larvae deposited on a host enter the branchial cavity and consume the host wall tissue or ingest collected matter.

Protopleon larvae continue their development until they metamorphose into the first juvenile stage, closely resembling miniature adults (Fig. 6.12). These juveniles have a segmented pleon and a full set of differentiated pleopods. At this point, they gain the ability to swim, seek out alternative hosts and start to feed. However, the free-swimming phase is temporary as they will attach to another host organism as quickly as possible.

Certain species parasitic on salps (also known as planktonic tunicates) can exhibit the remarkable behaviour of consuming or entirely removing internal organs and meticulously cleaning the inner walls with their mouthparts and pereopods. The larvae are then released into this emptied barrel, where they cluster closely together. The female, using the salp as a nursery, provides maternal care, ensuring that no larvae escape from the barrel. She keeps the barrel in almost constant motion using her pereopods and only departs intermittently to secure prey to feed the brood (Lützen 2005).

Female hyperiids with juveniles on a host ctenophore have been observed to remain in proximity to the host oral region while traversing the body surface. The juveniles attach themselves to the inner surface of the ctenophore and can move freely about the host, where they feed on the host epidermal tissue (Mazda et al. 2019). In this situation, the female does not exert control over the ctenophore, and the host was most likely chosen as the vulnerable offspring are protected from strong currents and predators.

6.4.2 Isopoda

Parasitic isopods are a versatile group known to infest both fish and crustacean hosts, primarily in marine environments, particularly those with warm waters. Most parasitic isopods are ectoparasites that typically feed on host blood or haemolymph, using biting and piercing mouthparts. The common name, 'isopod', reflects the similarity in shape and size of all of the legs. Isopods normally have seven pairs of thoracic legs (known as pereopods) and five pairs of appendages on the abdomen, the pleopods, which serve the purpose of respiration and propulsion (Fig. 6.13). The isopod body is usually dorsoventrally flattened and lacks a carapace. Additional characteristics of the Isopoda include the presence of oostegites in females, which constitute the brood pouch or marsupium, and the occurrence of biphasic moulting.

Several parasitic isopod families are associated with fish. They can be permanent parasites on fish hosts, opportunistic feeders or temporary parasites. The latter group, often referred to as micropredators, does not usually cause mortality of their hosts and is only parasitic for a particular life stage or portion of their life cycle (e.g. only attaching when feeding). Additionally, certain isopod species live within the body cavity or hae-

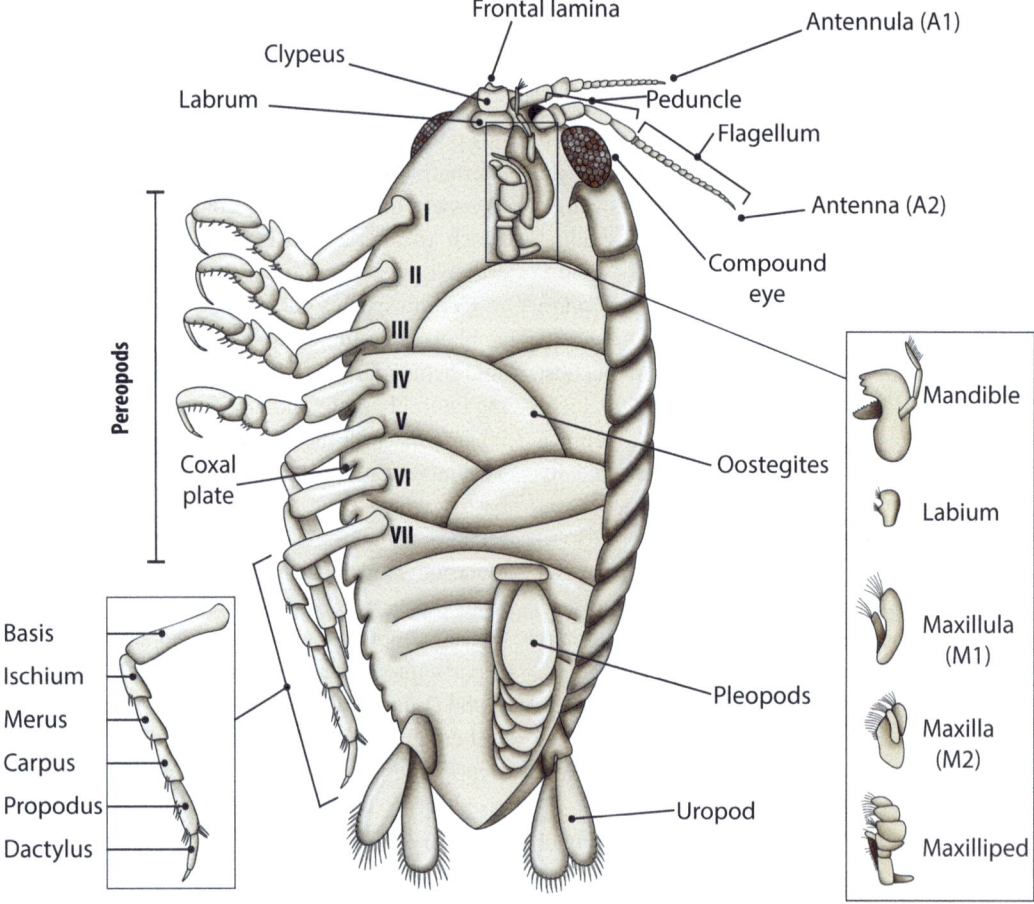

Fig. 6.13 Generalised morphology of Isopoda (Cymothooidea), female, ventral view. Abbreviations: *A1* Antenna 1, *A2* Antenna 2, *I–VII* thoracic (pereon) segments [Adapted from Kensley and Schotte (1989)]

mocoel of their hosts, with some even exhibiting hyperparasitic tendencies, targeting other crustaceans (as detailed in Hadfield 2019).

Selected families from two groups, the Cymothooidea and Epicaridea, will be elaborated on to demonstrate the variability in the life cycles within Isopoda.

6.4.2.1 Biology and Life Cycle of the Fish-Associated Isopods (Cymothooidea)

Families within the superfamily Cymothooidea, which are associated with fish, include the Aegidae, Barybrotidae, Corallanidae, Cymothoidae, Gnathiidae and Tridentellidae. Of these, only the Cymothoidae are obligate parasites. Notably, isopods from this superfamily adhere to a monoxenous lifestyle, meaning they rely exclusively on one type of host throughout the entire life cycle (no intermediate stage).

Cymothoidae

The family Cymothoidae consists of permanent ectoparasites that infest predominantly teleost fishes in marine waters. They are often found within the buccal cavity, branchial chamber or on the external surfaces of fish, although some species can also embed into the fish tissue, particularly in freshwater environments. Cymothoids are relatively large compared to most free-living species, ranging from a little under 1 cm up to 6 cm. The attachment site is often genus or species-specific. Cymothoids occupying the branchial cavity often have a twisted and asymmetrical body shape, while those inhabiting external surfaces are typically symmetrical and display countershading. All cymothoids possess characteristic adaptations for parasitism, such as hook-like pereopods that firmly grip the host tissue, and mouthparts that form a 'buccal cone' enabling blood-feeding.

Cymothoids have a protandrous hermaphroditic reproductive strategy, transitioning from a male to a female as part of their development. Adult cymothoid parasites are permanently attached to the individual host, with a free-swimming (immature) stage that locates a host (Fig. 6.14). The prevalence and life history of these parasites are believed to be influenced by factors such as seasonality and associated effects, including rainfall, salinity, ocean currents and temperature fluctuations.

Cymothoid isopods are typically encountered in pairs, comprising a larger female and a significantly smaller male. Initially, the immature isopod develops into a mature male following attachment to the host but will undergo a sex change when circumstances demand it. The sexual inversion is under the control of the Bellonci organs and androgenic glands. As the male undergoes the transformation into a female, observable changes include an enlargement of the thorax, a reduction in the appendix masculina, an increase in ovaries and a decrease in testes. The presence of a female prevents another male from undergoing a sex change, ensuring that only one transition occurs when multiple individuals are present.

Copulation takes place on the host, where the male typically moves to the posterior end of the female. The female elevates her abdomen, allowing the male to pass beneath her and rotate until the two ventral surfaces make contact for mating. After a duration of approximately 5–10 min, the isopods return to their original positions (Legrand 1952).

The female takes on the role of nurturing developing embryos within a brood pouch, referred to as a marsupium, formed by the ventral oostegites. After hatching (typically ranging from 37–1,600 offspring), the embryos progress through the first (pre-manca) and then the second (manca) larval stage (Fig. 6.14). This second stage has six pairs of hooked pereopods and is undifferentiated in terms of sexual characteristics.

Once ready, the gravid female releases the manca larvae at brief intervals, a process facilitated by a series of oostegite contractions, through an opening situated towards the posterior of the brood pouch. This methodical process unfolds gradually, with the release of a new manca larva occurring every few minutes. Under conditions of stress for the host, such as post-capture, there is a potential for a sudden burst release. During this event, the female may discharge all her larvae, regardless of the stage of development.

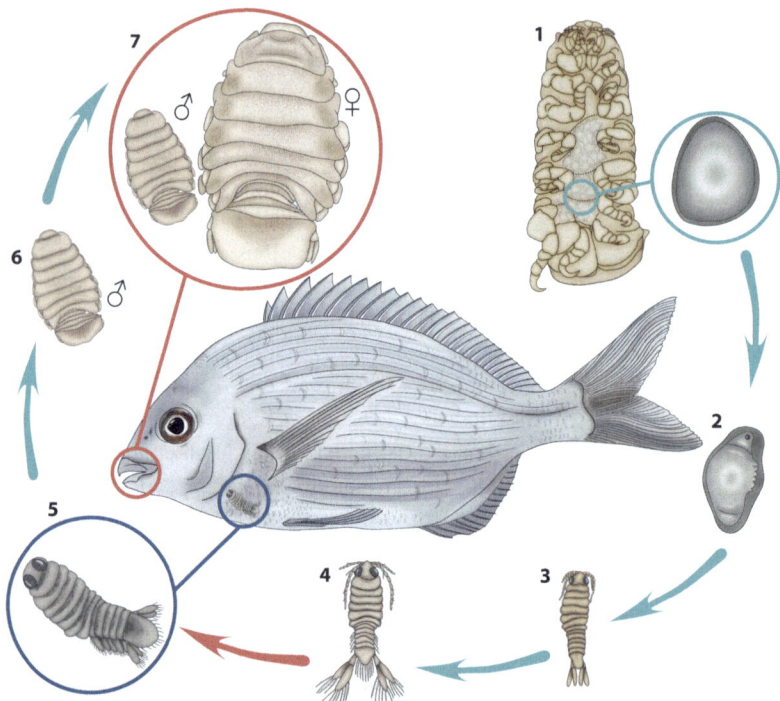

Fig. 6.14 Typical life cycle of a cymothoid isopod (Cymothooidea: Cymothoidae) infesting the mouth of a fish host. Eggs develop in a specialised brood pouch of the female (*1*), within the ventral oostegites. Subsequently, the embryos (*2*) hatch into the first larval stage (premanca) (*3*), followed by the second larval stage (manca) (*4*), all within the brood pouch. The manca is then released in intervals searching for a suitable fish host to infest. Once attached to a host, they develop the seventh pair of legs and are known as infective juvenile 'aegathoid' stages (*5*). Juveniles commence feeding and transition into males (*6*), and one will moult into a female (*7*). Adult males and females are permanently attached to an individual host. The red arrow indicates the free-living stage. Schematic, not to scale

The newly released mancae initially feed off the yolk while actively searching for a suitable host using their long setae and well-developed eyes (Fig. 6.14). Typically, these isopods do not infect adult fishes; rather, they exhibit a preference for young, mobile fish. Maximum infectivity is immediately after release, with the ability to attach to a host diminishing over time. Nevertheless, they remain motile for an extended duration, up to four days, with some individuals even persisting for as long as two months. These infective stages may resort to host-switching if they fail to find a suitable host and have not yet undergone the transition into juveniles.

When a permanent host is located, the isopods attach anywhere on the host body and then move to the preferred attachment site. Once settled, these isopods moult, shed their setae and can no longer swim effectively. They lose the capability to migrate to another host but develop a seventh pereonite and pair of pereopods. At this stage, they are identified as juveniles ('aegathoid') or pre-adults and commence feeding on the host. Up to three juvenile instars can occur on the host. After a short time, juveniles transition and mature into males (immature and then mature), undergo another transitional stage, and ultimately develop into females (immature, non-ovigerous and then ovigerous), completing the life cycle (Fig. 6.14).

Ovigerous female cymothoids do not feed, but non-ovigerous females and male cymothoids do, potentially harming the host organisms. Mature females alternate between reproductive and interreproductive stages, separated by moulting events. They may begin feeding on the host as early as three days after releasing the manca larvae and can produce a new brood as soon as 18 days after the previous release. The duration of an entire life cycle can range from a few days to years depending on the species (Smit et al. 2014).

Cymothoids can induce adverse effects on the host organisms, with the nature and severity of these effects contingent upon the specific isopod species, its location on the host and the intensity of the infection. Fish hosts of these isopods may exhibit localised damage or lesions at the infection site, experience diminished growth and condition factors, suffer from anaemia, undergo behavioural alterations and, in extreme cases, succumb to the infection and die.

Gnathiidae

Gnathiid isopods differ from other isopods in several key aspects. Unlike most isopods, which possess seven prominent pairs of legs, gnathiids have only five pairs of functional legs. Due to the fusion of the first thoracic segment with the cephalosome (head), the first appendage manifests as gnathopods in juveniles and pylopods in adults. The other appendage is lost as the last thoracic segment is reduced and lacks appendages. Gnathiids also display a polymorphic nature, with each life stage exhibiting distinct morphological characteristics, especially between adults and larvae. Furthermore, while only the larval stages are temporary parasites, the adults are free-living. This bi-phasic life cycle encompasses three ectoparasitic larval stages, which come in two distinct forms: the unfed parasitic phase known as the zuphea and the blood-filled praniza phase.

The life history of this family has been well-documented for various species, yet gaps in our knowledge persist. Gnathiids feed on the blood and tissue fluids of both teleost and elasmobranch hosts. They typically exhibit widespread points of attachment across different areas of the host bodies, including the body surface, buccal cavity, eyes, fins, gills, nares and, in the case of elasmobranchs, cloaca and claspers. The size of the praniza larva is commonly influenced by the duration of its feeding, varying considerably between the different genera and species. The life cycle also varies among species, often contingent on the water temperature, ranging from a few weeks to as long as five years to completion (Smit and Davies 2004). Warmer and temperate waters tend to yield higher growth rates than cold water and, consequently, shorter life spans.

Non-feeding adult gnathiids are often found in benthic environments, using diverse refugia such as small spaces within dead coral, barnacles, sponges or worm tubes. In some cases, a solitary male gnathiid may live in a harem consisting of numerous females and offspring (with up to 42 females observed with *Paragnathia formica*). Females are attracted to a male through the release of pheromones. When other males are present, they are either inhibited from maturation or eliminated by the dominant male. Male competition may occur and involves physical confrontations using their large mandibles.

Females attain maturity within the benthic cavity and mate with a male, fertilising the eggs. To accommodate the growing number of eggs in the brood pouch, the female internal organs are displaced. The gravid female retains the embryos within the marsupium until the first zuphea stage is ready for release. The brood size, which ranges from 20 to over 200 larvae, correlates with the size of the individual larvae produced by the females. Once ready, the first zuphea stage (Z1) is expelled from the female and swims rapidly, searching for a suitable host to initiate feeding (Fig. 6.15). These larvae use their hooked pereopods and specialised mouthparts (for sucking and piercing) to penetrate the host tissues and feed until they are engorged.

As the zuphea feeds, the anterior hindgut expands with the liquid meal, leading to a significant transformation. The elastic intersegmental membrane becomes stretched to the extent that the segmentation between pereonites 5 and 7 becomes indistinguishable, now referred to as the

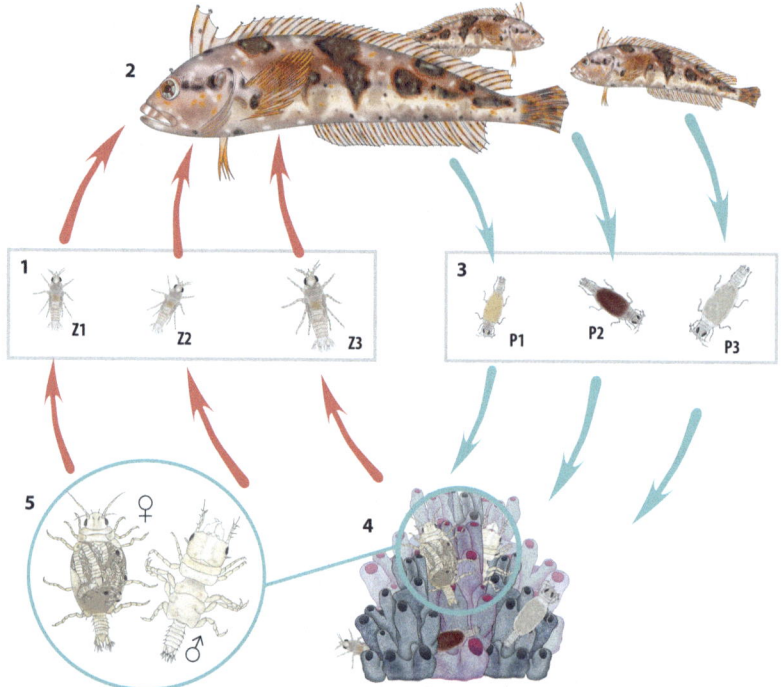

Fig. 6.15 Typical life cycle of a gnathiid isopod (Cymothooidea: Gnathiidae) infesting a teleost (e.g. *Clinus superciliosus*). The zuphea stage 1 (Z1) (*1*), released from the female marsupial, attaches to a host (*2*) to feed on blood (red arrows indicate infective stages) until they reach the fed praniza stage (P1) (*3*). The praniza then detaches and digests the bloodmeal (green arrows indicate free-living stages) and moults into the subsequent zuphea stage while hidden in suitable substrata (*4*). This process repeats twice until the third praniza (P3) emerges and moults into an adult male or female (*5*). Schematic, not to scale [Adapted from Smit et al. (2003)]

swollen praniza larva (P1) (Fig. 6.15). The duration of feeding varies by species, larval stage and host, with a zuphea usually spending only a few hours feeding on teleosts but several weeks on elasmobranchs.

Upon concluding feeding, the praniza larvae will leave the host and search for a suitable substratum, often found in the form of cavities and sponges. Here, they digest their recent meal and undergo a moult to reach the subsequent zuphea stage (Z2). This transition includes a two-phase moult, first for the posterior and then the anterior segments. The zuphea stage is distinguished by its more segmented appearance, with the intersegmental membrane now concealed beneath the fourth pereonite. A similar cycle of feeding and moulting is repeated twice more until the third-stage praniza (P3) emerges (Fig. 6.15). This final larval stage will detach and moult into a male or female adult that will need to survive on its last ingested meal. Typically, males moult before females, ensuring that they are ready to fertilise the females immediately following their moult. Females are semelparous, reproducing only once, and males are presumed to die after depletion of the nutrients taken in during the final feeding stage.

Often only the final-stage (P3) gnathiids have been found on elasmobranch hosts. Ota et al. (2012) conducted a study that established a connection between first- and second-stage larvae, which fed on teleost fishes, and a third-stage larva feeding on an elasmobranch host. This finding suggests that certain species of gnathiids may exhibit host shifts, even among hosts with very different body fluid composition.

Furthermore, gnathiids have been recognised as vectors for specific blood parasites, such as *Haemogregarina bigemina*, which is transmitted by the larvae of *Gnathia africana*. These crustaceans are also suspected of transmitting viral erythrocytic necrosis (VEN) and have associations with nematode larvae, flagellates and fungal-like structures (Hadfield and Smit 2019).

6.4.2.2 Biology and Life Cycle of the Crustacean-Associated Isopods (Epicaridea)

Within the Epicaridea, two superfamilies Bopyroidea and Cryptoniscoidea are most known to infest crustaceans. Epicarid isopods exhibit a heteroxenous lifestyle, which necessitates the involvement of at least two distinct types of hosts to complete the life cycle. Most species in this group use copepods as intermediate hosts and another crustacean (usually decapods, cirripedes or peracarids) as definitive hosts. While Cryptoniscoidea is composed of endoparasitic (and some hyperparasitic) species, members of Bopyridae are mostly ectoparasitic. Female cryptoniscoids are the most extensively modified parasitic isopods. Cryptoniscids are often unrecognisable as isopods, as they sometimes have a sac-like shape that lacks the typical pereopods and oostegites found in this order (Williams and Boyko 2012). Although female bopyrids are also slightly modified (sometimes asymmetrical, with modified segmentation and appendages), they are still mostly recognisable as isopods and will be the focus for the rest of this section.

Bopyridae

This family exhibits typical isopod characteristics, with seven pereonites, seven pairs of legs and a brood pouch formed by oostegites. However, the asymmetrical and irregular female body shape, as well as the highly modified morphology, can make them challenging to identify as isopods. These parasitic isopods are commonly found in the host branchial chamber, resulting in a conspicuous protuberance. Some genera and species attach to the host abdomen.

Adult males are considerably smaller in size than females and are typically located on the ventral side of the female, positioned between the pleopods. While females primarily feed on the host haemolymph, it is possible that males do not engage in feeding at all. In most cases, sex determination in these isopods is epigenetically regulated, where the first isopod to settle becomes female and subsequent individuals typically do not develop beyond the male stage.

Adult males usually fertilise the eggs within the female marsupium. In each brood, females produce a substantial number of eggs, often numbering in the tens of thousands. Once these eggs hatch, epicaridium larvae emerge within the marsupium (Fig. 6.16). These larvae emerge from the brood pouch of the female and swim rapidly in search of an appropriate copepod intermediate host (often a calanoid). Epicaridium larvae have six pairs of pereopods, with the last pair being notably longer, and the head extending ventrally.

Upon locating the copepod host, the larvae pierce the host exoskeleton and commence feeding on its blood, using clawed pereopods and styliform suctorial mouthparts. Within a few days, they undergo metamorphosis into microniscus larvae, which remain attached to the intermediate host (Fig. 6.16). Microniscus larvae are more elongate, with anteriorly directed heads and buds of the seventh pereopods being present. They continue to develop for several weeks on the copepod host before detaching and transforming into infective cryptoniscus larvae (Williams and Boyko 2012).

The cryptoniscus larva, elongate and rounded anteriorly while tapering posteriorly, has seven pairs of pereopods. This larva must search for a definitive decapod crustacean host. Once a suitable host is located, the cryptoniscus larva settles and metamorphoses into a bopyridium juvenile (Fig. 6.16). This bopyridium migrates to its final attachment site, with the initial juvenile becoming an adult female. Adult females primarily feed on the host haemolymph or ovarian fluids. Subsequent larvae that arrive, often attracted by pheromones released by the female, will develop into males.

Similar to some other parasitic crustaceans, bopyrids and epicarideans in general, can function as parasitic castrators of the host. While some species do not completely disrupt the host reproduction, others can severely impair the host. This may involve the cessation of egg development, feminisation of male hosts, a decrease in

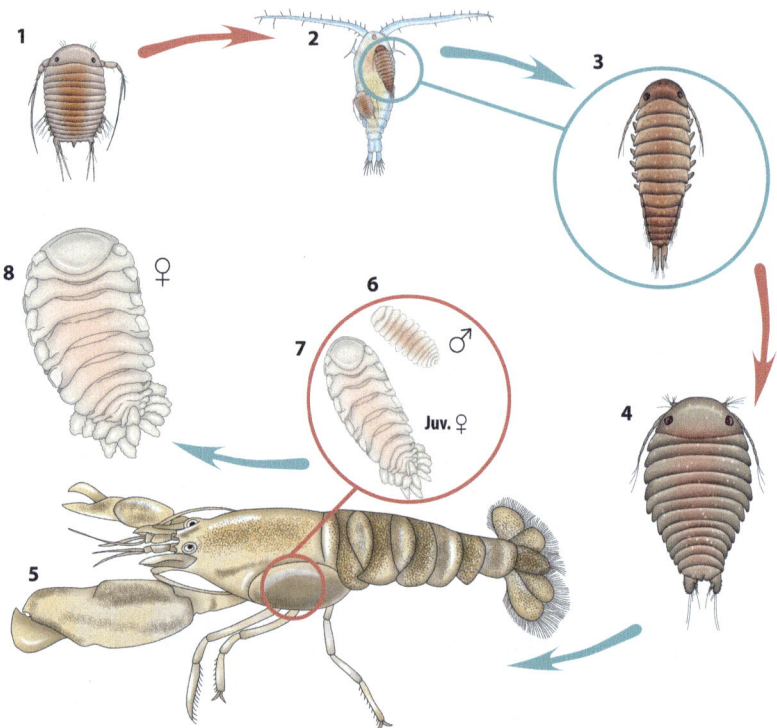

Fig. 6.16 Typical life cycle of bopyrid isopod (Epicaridea: Bopyridae) infecting a copepod (intermediate) and shrimp (definitive) host. An epicaridium larva (*1*) is released from the female brood pouch as an infective pelagic larva and infects the intermediate copepod host (*2*) to develop into microniscus larvae (*3*). The bopyrid then develops into another pelagic larval stage, cryptoniscus (*4*), until it infects the definitive shrimp host (*5*) where it develops into a male (*6*) and juvenile female (bopyridium) (*7*) until they reach sexual maturity (*8*). Red arrows indicate the free-living stages, green arrows indicate infective stages. Schematic, not to scale [Adapted from Baeza (2015)]

host growth and a reduction in the host metabolic activity. Additionally, the morphology and behaviour of the host may be affected.

6.5 Ostracoda

Ostracods are small 'bivalved' crustaceans, superficially resembling minute clams, and are commonly found in both freshwater and marine environments. The body, enclosed within a carapace, is a flattened and hinged bivalve shell. While most are free-living, predators or scavengers, a few engage in symbiotic or parasitic associations with other organisms. This parasitic behaviour extends to various hosts, such as gammaridean amphipods, groundwater isopods, polychaete worms, sea urchins and sharks, though the specifics of these relationships are not always fully known (Hart and Hart 1974).

While most free-living ostracods can reproduce sexually or asexually, parasitic species adopt gonochoric (sexual) reproduction, with females being larger in size than males. Ostracods display direct development, with all pre-adult instars closely resembling the adult stage, albeit smaller in size and with fewer appendages. Larvae and juveniles are not differentiated in these organisms. To facilitate parasitic interactions, ostracods frequently have compressed valves that lack ornamentation and are almost flat along the ventral margin. This adaptation enhances their ability to establish more effective contact with host organisms.

Generally, ostracods progress through a life cycle featuring seven to eight larval and juvenile instars, culminating in an adult stage. This is predominantly seen in the order Podocopa. Conversely, members of the order Myodocopa exhibit a sequence of four to seven larval stages before transitioning to adulthood. The first instar that emerges from the egg possesses antennulae, antennae and mandibles. Furthermore, the initial three instars lack fully functional thoracic legs and rely on their antennae (A2) for mobility and host attachment. The mandible pierces the host epidermis, thus facilitating feeding.

As each moult occurs and the ostracod matures through its various stages, the existing appendages undergo progressive development in each moult. Moreover, additional appendages are incrementally added starting with the maxilla, followed by the first, second, and third thoracic legs (Fig. 6.17), and culminating in the development of genitalia. The maxilla is an additional appendage used to embed the parasite within the host tissue. The addition of the fourth through seventh limbs is a stepwise process, with one pair being added at each successive moult. With each of these moults, the ostracod's capacity for movement and attachment increases. In certain parasitic genera, the thoracic legs are equipped with spines and terminal claws to enable secure attachment. The sexual organs begin to develop in the last two or three instars. At this stage, two distinct size classes emerge, with adult males distinguished by the presence of a copulatory organ (anlage), which females lack (Kretzler 1984).

6.6 Tantulocarida

Within the realm of micro-crustacean parasites, the Tantulocarida represent one of the most diminutive and most intriguing groups. These minute (<0.3 mm) organisms infest a wide range of crustacean hosts, spanning a wide spectrum of marine environments, from the intertidal to abys-

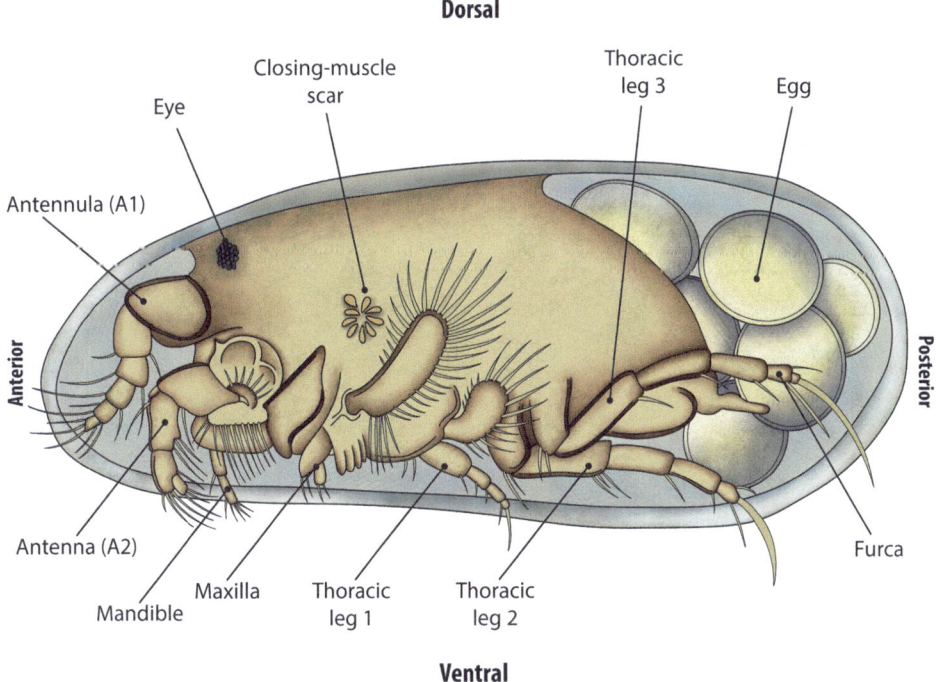

Fig. 6.17 Generalised morphology of Ostracoda, lateral view with front carapace removed. Abbreviations: *A1* Antenna 1, *A2* Antenna 2 [Adapted from British Geological Survey (2024)]

sal depths, and exist at temperatures ranging from the tropics to polar regions. The Tantulocarida follow a protelean life cycle characterised by obligatory ectoparasitic larvae and non-feeding adults (Huys et al. 2014). The parasitic larvae infest a diversity of hosts including amphipods, copepods, cumaceans, isopods, ostracods and tanaids (Boxshall and Vader 1993), with most displaying a high level of host specificity (Petrunina and Huys 2020).

Tantulocarids rank among the tiniest arthropods known to science, with the tantulus larvae recognised as being the smallest segmented larvae within the Crustacea. For example, one species, *Serratotantulus chertoprudae*, measures a mere 76 μm in total length (see Petrunina and Huys 2020). Another remarkable trait of this group is the apparent lack of a typical crustacean moulting process between successive stages in the dual life cycle.

Tantulocarids lack typical appendages found in most arthropods, such as functional legs or eyes. Instead, they have specialised, biramous thoracic feeding limbs used for grasping the host (Fig. 6.18). The tantulus larva is composed of an anterior tagma, known as the prosome. This section consists of the cephalon and a thorax containing six pedigerous segments. Additionally, there is a limbless tagma referred to as the urosome, which includes the last thoracic segment and a single-segmented abdomen featuring paired caudal rami. Furthermore, the tantulus lacks cephalic appendages and remains permanently affixed to the host through a frontal oral disc.

The double life cycle is another unique facet of the tantulocarids. These parasites undergo two

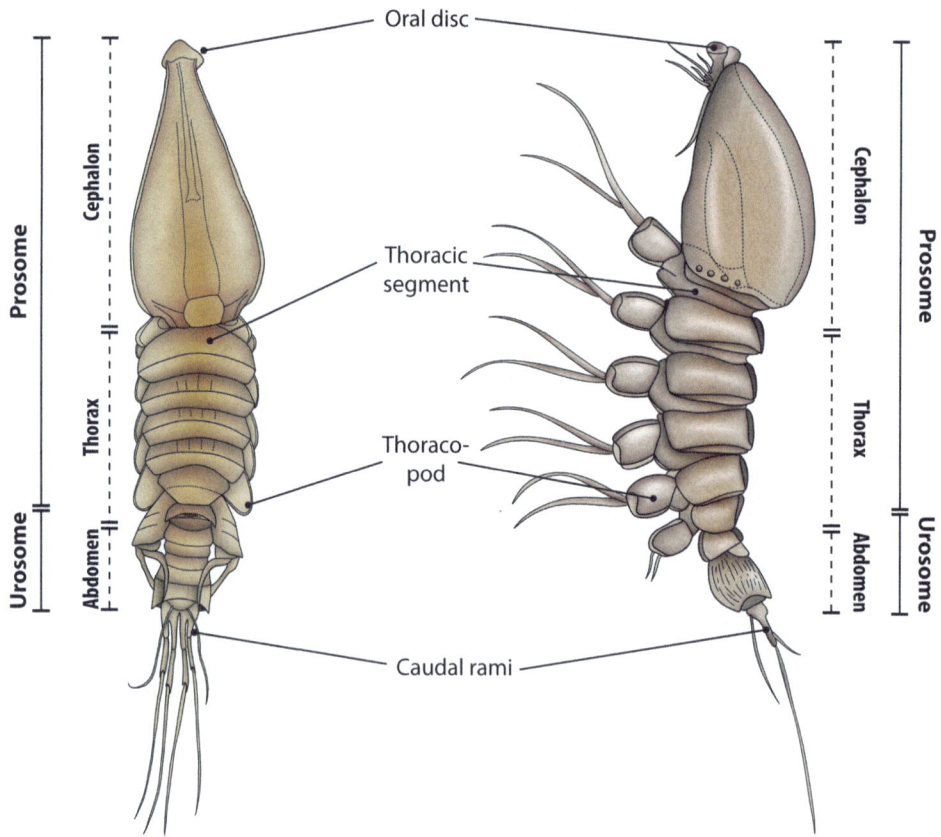

Fig. 6.18 Generalised morphology of the micro-crustacean tantulus larvae (Tantulocarida), dorsal (left) and lateral view (right)

distinct cycles: a sexual stage with free-swimming adults and an asexual (or parthenogenetic) stage that occurs on the host (Fig. 6.19). The tiny tantulus larva is present in both of these cycles. This permanent and obligatory larva undergoes significant enlargement and transforms into a cuticular sac where either sexual or asexual stages occur. A free tantulus exhibits active swimming capabilities, using six pairs of thoracopods, as it transitions through a benthic phase in search of its ideal host (Huys 1991). Using an array of sensory structures, it identifies a suitable epibenthic host in its aquatic environment. Once found, it secures itself to the host using a specialised oral disc and can be observed on various parts of the host body and appendages (Boxshall and Lincoln 1987).

In the sexual cycle (indicated in red in Fig. 6.19), the infective tantulus larva permanently attaches to a crustacean host, a bond that sustains it for the entirety of its existence. Nutrients are transported from the host into the larval cephalon via an intricate tubular rootlet system. Adults will begin to form inside the expanded trunk sac of the tantulus larva without conventional moulting. These adult males and females will develop separately, notably different in the placement of the trunk sac, and are supplied nutrients via an umbilical cord that appears to be continuous with the rootlet system. For the sexual females, the trunk sac forms immediately posterior to the cephalic shield, and the larval trunk alone is sloughed. For the males, the trunk sac forms at or near the back of the larval thorax, with the larval trunk retained, resulting in the ventral deflection of the urosome and thoracopods. Males are equipped with a range of sensors (of chemosensory aesthetascs and sensilla) and

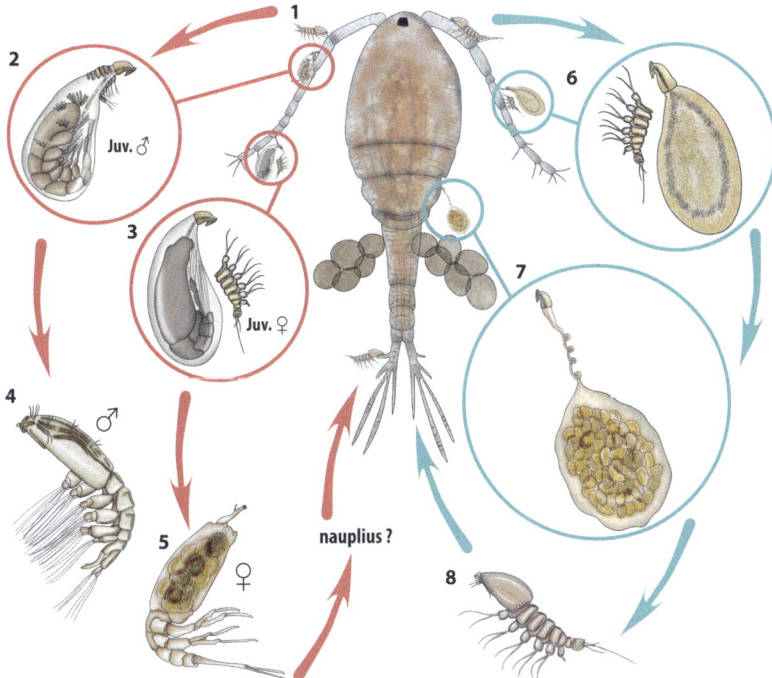

Fig. 6.19 Presumed sexual and asexual life cycle of tantulus larvae (Tantulocarida) infesting a copepod host. Sexual reproduction (indicated in red) occurs when the tantulocarid larvae (*1*) infests a copepod host and develops into a juvenile male (*2*) and female (*3*) through expansion of the trunk sac. The adult male (*4*) and female (*5*) emerge and potentially fertilise eggs to develop a nauplius stage (?). During the asexual life cycle (indicated in green), the tantulocarid larvae attach to the copepod host, form a large trunk sac and the larval trunk is shed (*6*). Development of the eggs (*7*) takes place internally, and the tantulus larvae (*8*) are subsequently released. Schematic, not to scale [Adapted from Huys et al. (2014)]

possess six well-developed thoracopods, presumably all beneficial for swimming and locating sexual females. Conversely, the female is poorly adapted for swimming, although she bears two biramous thoracopods possibly to grasp the male during copulation (Huys et al. 1993).

Mating occurs between the sexual adult stages. According to Huys et al. (1993), fertilisation of the eggs is internal, with the male depositing sperm directly into the female median copulatory pore using its well-developed intromittent organ (Huys et al. 1993). As the egg-filled sac expands, the trunk undergoes a noticeable swelling. Each female engages in a single reproductive event in her lifetime, a reproductive strategy known as semelparity, likely releasing her entire brood simultaneously. The sexual female requires adequate nourishment to support the development of fertilised eggs. In contrast, the sexual male, lacking mouthparts, does not feed and must store sufficient food reserves during its development to sustain itself during the subsequent free-swimming phase to find a suitable mate. It is believed that the eggs from the sexual female will hatch as free-living nauplii by rupturing the trunk wall. This benthic, non-feeding nauplius stage could possibly be the next stage, which metamorphosises into the infective tantulus larva, a hypothesis rooted in unpublished observations.

In the parthenogenetic female stage (which represents the asexual phase where females can produce embryos without fertilised eggs as with the sexual female; indicated in green in Fig. 6.19), the tantulus larva forms a large dorsal sac immediately behind the cephalon leading to the shedding of the larval trunk. The sac then differentiates into eggs that are later released as fully developed tantulus larvae.

Currently, mating between free-swimming adults, as well as the hatching and development of larvae from the sexual female, remains unobserved. There is also the possibility of yet undiscovered stages and the presence of a free-swimming and non-feeding female that mates with the sexual male. Additionally, the factors influencing the utilisation of these two distinct life cycles remain unknown. Notably, the different life stages, including sexual males, parthenogenetic females and tantulus larvae, can co-exist concurrently on a single host, effectively ruling out the likelihood of a seasonal cycle (Huys et al. 1993).

6.7 Thecostraca

6.7.1 Ascothoracida

The class Thecostraca has three subclasses: Ascothoracida, Cirripedia (Sect. 6.7.2) and Facetotecta (Sect. 6.7.3). Ascothoracidans are marine parasites inhabiting a wide range of marine environments, from intertidal regions to the depths of the deep sea. They have a diverse range of hosts, as well as morphology and biological characteristics, and parasitise echinoderms (excluding regular urchins and sea cucumbers), cnidarians (e.g. corals and gorgonians) and zoanthids (Grygier and Høeg 2005).

The typical ascothoracidan form is characterised by a bivalved carapace enclosing a body featuring an unsegmented head and segmented thorax and abdomen (Kolbasov and Petrunina 2019). The thorax has six pairs of biramous appendages, while the abdomen consists of four segments, culminating in a terminal telson and a caudal furca (constituting paired caudal or furcal rami) (Fig. 6.20). Adult females can be distinguished by their enlarged carapace, a consequence of their role in brooding eggs and larvae.

Ascothoracidans exhibit various parasitic behaviours, ranging from endo- to ecto- and mesoparasitic, contingent upon the species. Both juveniles and adults within this group are known to have parasitic lifestyles. In most cases, there are separate sexes with the exception of hermaphroditic species belonging to the family Petrarcidae. Typically, a larger petrarcid female co-exists with a significantly smaller, cypridiform-like male, often residing within her mantle cavity (Høeg et al. 2014b).

The life cycle of ascothoracidans is poorly known, but a typical life cycle is assumed to have up to six free-swimming naupliar instars, one or two a-cypris larvae (also known as ascothoracid

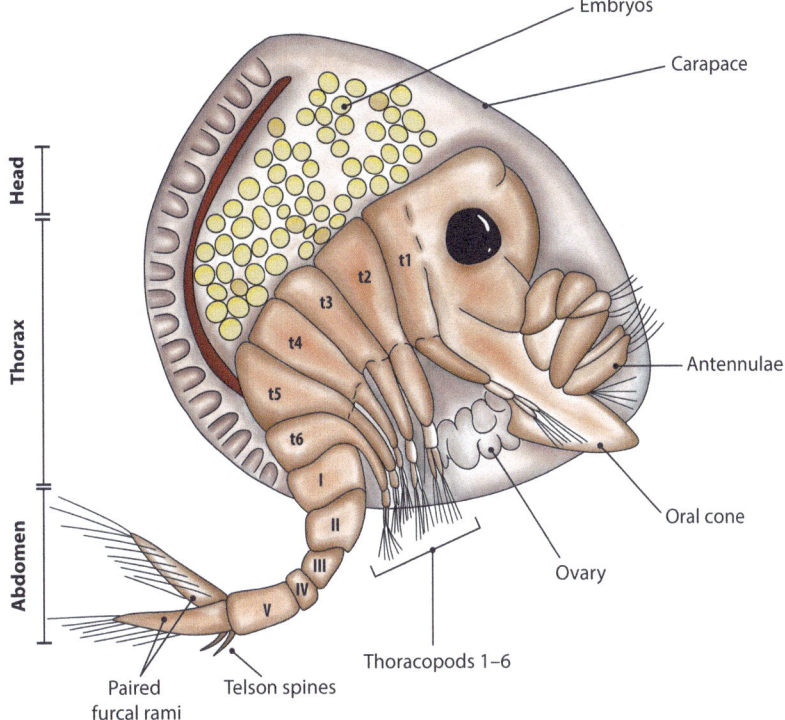

Fig. 6.20 Generalised morphology of Ascothoracida (Thecostraca), lateral view with one carapace valve removed. Abbreviations: *t1–t6* thoracic segments, *I–V* abdominal segments (telson) [Adapted from Kolbasov et al. (2019)]

larvae), a juvenile and adults. The naupliar instars are usually either lecithotrophic, refraining from feeding, or planktotrophic, actively foraging on phytoplankton and small zooplankton while suspended in the water column. Alternatively, some nauplii are brooded within the mantle cavity until they mature and are released as a-cyprids. In species featuring two a-cypris instars, the first typically remains within the mantle cavity, while only the second larva is released into the plankton, actively seeking a suitable host for attachment. The a-cyprid identifies and attaches to the host using chemosensory aesthetascs and grasping antennulae (first antennae) equipped with a retractable claw (Grygier and Høeg 2005). The manner in which these parasites acquire nutrients from their hosts involves either piercing-sucking mouthparts or the direct absorption of nutrients through the tegument (Høeg et al. 2014b). The settlement and metamorphosis of the a-cyprid into an adult has never been observed, and there seems to be minimal change from the a-cyprid to an adult. Additionally, copulation within this group remains unobserved.

A tentative life cycle was proposed by Grygier and Høeg (2005) based on the MSc thesis of Michael Hartmann. This cycle was of *Ulophysema oeresundense*, a parasite inhabiting the heart urchin *Echinocardium cordatum* (Fig. 6.21). This particular parasite species exhibits a life cycle featuring two brooded naupliar instars, two a-cyprid stages and an internal adult parasite. During development, the larvae are retained within the mantle cavity until the release of the second a-cyprid instar into the plankton, facilitated by a pore induced in the host test (outer skeleton or shell). Morphologically, male and female larvae display distinctive characteristics, with male a-cyprids possessing three antennular aesthetascs (females have only one) and mature sperm within the testes. Female entry into the host is presumed to occur through a gonopore, eventually leading to the penetration into the host body cavity. Within this cavity, the females exist freely but are enclosed by ciliated host cells. Ultimately, the female attaches itself to the host test by her apertural region and creates an opening to the exterior. This hole serves as a conduit

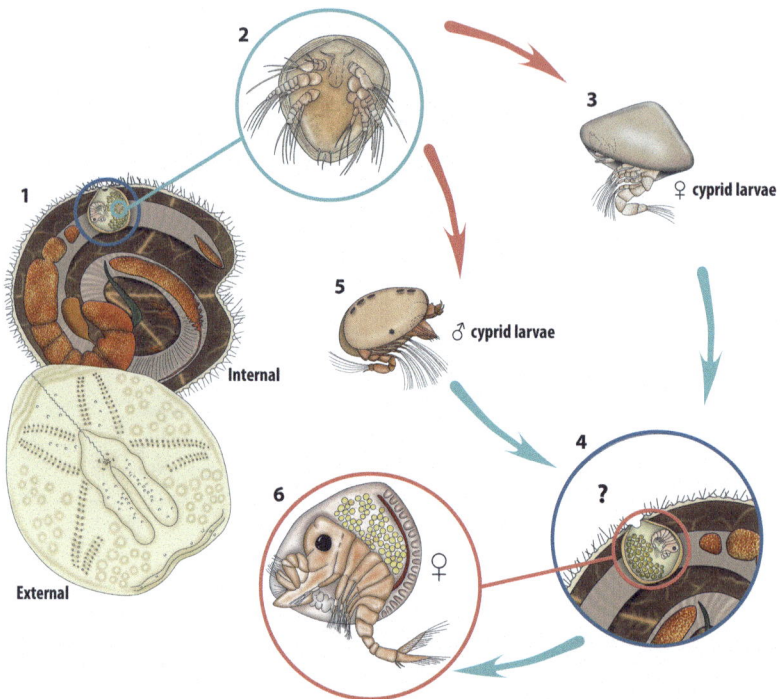

Fig. 6.21 Life cycle (partly hypothetical) of the ascothoracid (Thecostraca), *Ulophysema oeresundense*, infecting the heart urchin *Echinocardium cordatum* (*1*). Fertilised embryos develop in the mantle cavity of the host into a nauplius (*2*) (featuring two instars) until the second a-cyprid instar is released into the planktonic column through a hole created by the parasite in the host test (shell). Infective female a-cyprid larvae (*3*) possibly enter the host body through a gonopore (?) and eventually attach to the host shell (*4*). The female possibly creates a hole to the exterior (?) for the male a-cyprid larva (*5*) to fertilise the female. Adult females (*6*) then brood the embryos until they develop into nauplii. Schematic, not to scale. Red arrows indicate free-living stages [Adapted from Grygier and Høeg (2005)]

for male a-cyprids to fertilise the female. Notably, young females lacking such an opening for male entry, or females that are free in the host body cavity, never possess developing embryos or larvae.

Many parasites associated with echinoderms are believed to induce host castration. Furthermore, instances of hyperparasitism involving ascothoracidans and cryptoniscid isopods have been documented in a few families. These isopods occupy the brood chamber of the ascothoracid and feed on the host embryos (Kolbasov et al. 2021). A solitary cryptoniscid larva does not significantly impair the host reproductive capacity, whereas large female isopods effectively prevent the ascothoracid from depositing broods, rendering the isopods as parasitic castrators (Kolbasov and Petrunina 2019).

6.7.2 Cirripedia

The subclass Cirripedia, within the class Thecostraca, encompasses three distinct infraclass taxa: Acrothoracica, Rhizocephala and Thoracica. This group includes a variety of crustaceans, ranging from acorn and stalked (goose) barnacles to parasitic forms. They have mineralised shell plates and are characterised by sessile adult stages and free-swimming larvae. Typically, they possess six pairs of elongate, biramous feeding appendages known as cirri, which extend through the mantle opening. The body consists of a cephalic region and a trunk, exhibiting minimal external segmentation (Fig. 6.22). Barnacles exhibit a diverse lifestyle among different taxa and can be parasitic or commensal. These different taxa can be found within sponges, attached to corals or even associated with large hosts such as whales. Some species are known to act as castrating parasites in crabs, particularly within the family Sacculinidae. Generalised life cycle notes are provided below for the three infraclasses.

6.7.2.1 Biology and Life Cycle of Burrowing Barnacles (Acrothoracica)

This infraclass of barnacles lacks mineralised shell plates and exhibits only a partial parasitic nature, predominantly functioning as suspension feeders. These tiny barnacles, known as burrowing barnacles, possess the distinctive ability to burrow into calcareous substrates, which can range from marine molluscs and bryozoans to coralline algae and even the shells of thoracican barnacles. Acrothoracicans have separate sexes, characterised by the larger females sheltering dwarf males in the female mantle sac. The precise mechanism of their feeding is still unknown, but it has been observed that they can exert a significant negative influence on hermit crab reproduction. In a study by Murphy and Williams (2013), the authors suggested referring to these organisms as 'transient parasites' since they occasionally harm female hosts but have no harmful effects at other times. Development of the larval stages is similar to other Cirripedia groups and then proceeds directly into the adult. These barnacles have at least four free-swimming naupliar stages (when present), a cypris larva, a juvenile and a stationary adult in their life cycle.

6.7.2.2 Biology and Life Cycle of Parasitic Barnacles (Rhizocephala)

Rhizocephalans are highly specialised obligate parasites that primarily target crustaceans, especially those belonging to the order Decapoda.

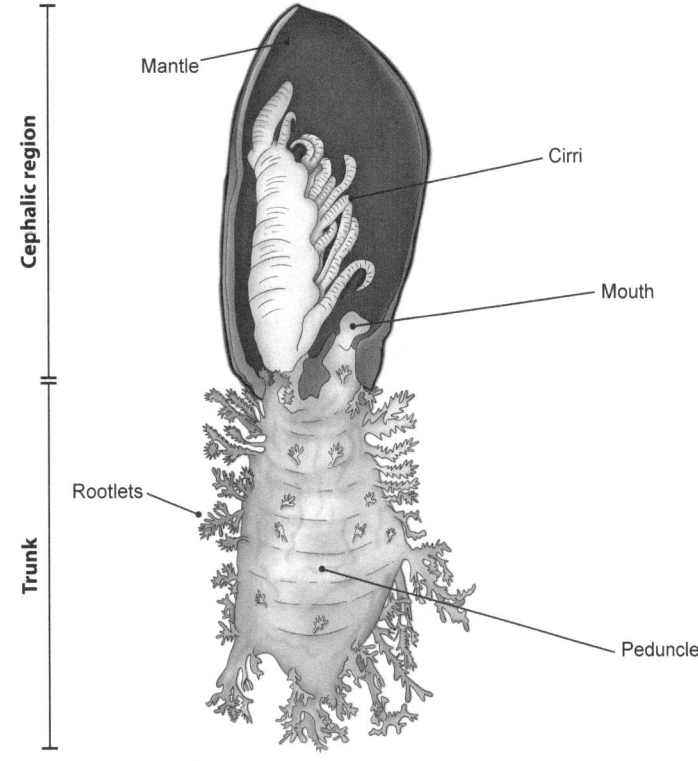

Fig. 6.22 Generalised morphology of a parasitic barnacle (Thecostraca: Cirripedia), front portion of the mantle removed showing internal structures

They inhabit a broad spectrum of marine environments, ranging from intertidal zones to the deep sea, and can occasionally be parasites of a few brackish and freshwater crustaceans. In some rare instances, they have been known to infest semiterrestrial crabs. These parasites exhibit a remarkable difference from typical crustacean anatomy, lacking most organs, segmentation and appendages, with only reproductive organs remaining intact.

In contrast to other barnacles, rhizocephalans have separate sexes. Adult females, being parasitic, have a remarkably simplified morphology. They have an external reproductive sac, commonly referred to as the 'externa', as well as an internal system of rootlets, known as the 'interna', which serves the purpose of nutrient absorption. The externa houses essential components such as the ovary, mantle cavity and specialised organs designed to host dwarf males. Larvae of rhizocephalans, resembling conventional barnacles, are free-living and can swim.

In most species, development begins with a nauplius stage, consisting of approximately four to six instars (Fig. 6.23). These nauplii are lecithotrophic, relying on maternal reserves supplied within the egg yolk, and they remain exceedingly small, allowing for larger brood sizes. The instars primarily differ in size, characterised by the gradual elongation of the post-cephalic trunk. The final instar has an enlarged head shield and a distinctive distal expansion of the antennula, which paves the way for the development of the cypris attachment organ and leads to the subsequent stage, the cyprid. The transformation from a nauplius into a cyprid typically spans two to seven days but can extend up to a month in colder waters. Several rhizocephalan species bypass the naupliar stages and hatch directly as cyprids. Cyprids must locate a suitable substrate for permanent attachment, initiating metamorphosis. This distinctive larval form does not engage in feeding, and male and female cyprids follow a different development path from this point onwards.

Female cyprids are thought to rely on their chemosensory ability and sensitivity to olfactory cues in the water column to locate a suitable host. Their preferred settling spots are characterised by a thin or soft cuticle for easy penetration. Most commonly, this occurs amidst the gill filaments within the branchial chamber or at the base of setae to avoid host grooming attempts. The host organisms may make efforts to dislodge cyprids during attachment, but once securely affixed, the female cyprid undergoes a transformation into a kentrogon.

The firmly anchored kentrogon is smaller in size compared to a cyprid and uses a hollow injection stylet to pierce the host integument, introducing a vermigon stage into the host haemocoel (Fig. 6.23). Once this migratory internal stage is inside the host (often 1–3 days after settlement), the parasite begins to take control of the host. The vermigon will migrate through the host circulatory system, ultimately reaching the abdomen where the interna will develop. The parasite's rootlet system impairs the host nervous system, assuming command over the hormonal regulation. Ultimately, the interna emerges underneath the host abdomen as a small, virgin female, serving as the target of male cyprids.

Male cyprids need to locate and inseminate the virgin females. To accomplish this, males have elongate aesthetasc setae on the antennulae for attachment as well as large, chemosensory aesthetascs to find a female parasite. Upon finding an infected host, males compete for the opportunity to inseminate the female (with a maximum of two males able to implant one female). Only the fastest and most robust males will successfully gain access. These male cyprids settle in proximity to the narrow mantle opening leading into the brood chamber of the virginal females (Fig. 6.23). Here, they undergo metamorphosis into a trichogon stage, which resembles the vermigon but is distinguished by prominent cuticular spines. The trichogon migrates through the mantle cavity and becomes inserted into one of the two receptacles of the virgin externa. Once a single male has been successfully implanted, the female externa immediately commences sexual maturation. The trichogon casts off its spiny cuticle, thereby preventing subsequent insemination by other males and is now recognised as a dwarf male. Once established

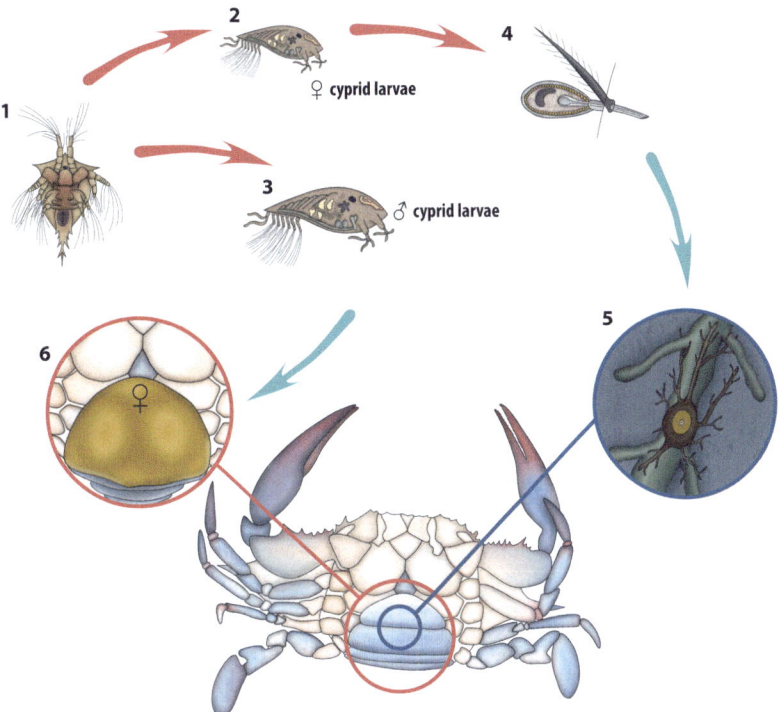

Fig. 6.23 Typical life cycle of parasitic barnacles *Sacculina* (Cirripedia: Rhizocephala), infecting a crab host. The free-swimming lecithotrophic nauplius stage (*1*), with approximately four to six instars, develops into a sexually dimorphic female (*2*) and male cyprid larvae (*3*) following different developmental paths onwards. The female settles at the base of host setae (or gill filaments), transforming into a kentrogon (*4*). The kentrogon injects a vermigon stage into the host haemocoel where it develops in the thoracic ganglion of the crab (*5*). The male cyprid larva then locates and inseminates the externalised female (*6*), undergoes metamorphosis, and degeneration of the gonads and deformities of the host occur. Red arrows indicate free-living stages. Schematic, not to scale [Adapted from Hickman et al. (2023)]

inside the female, the male undergoes spermatogenesis, and relies on the female parasite for nutrition for the duration of its life. It is worth mentioning that some rhizocephalan species employ the cypris antennula (antennule or antenna 1) to penetrate the host or the virginal female without the formation of kentrogons or trichogons (Høeg et al. 2014c).

A noteworthy characteristic of these rhizocephalan parasites is their ability to manipulate the morphology and biology of their hosts. Through the manipulation of the host nervous system and the release of specific hormones, these parasites can induce behavioural changes in the host, making it more susceptible to predation. Often referred to as 'parasitic castrators', these parasites have the power to diminish the host (both male and female) reproductive success, infiltrate and disrupt the functionality of the host reproductive organs and render it incapable of reproduction. Along with sterilisation, they can cause degeneration of the gonads, and induce the formation of abnormal growths and deformities within the host. Additionally, rhizocephalans can induce feminisation in male hosts, leading to the conversion of testes into ovaries, alterations in the host overall shape and size, and potential changes in its behavioural patterns (Høeg 1995).

6.7.2.3 Biology and Life Cycle of Gooseneck Barnacles (Thoracica)

Members of this infraclass are acorn or stalked (gooseneck) barnacles, characterised by their heavily calcified carapaces. The majority of thoracicans engage in symbiotic relationships with corals, crustaceans, echinoderms, molluscs, polychaetes, sea snakes, sponges, turtles and whales (Hadfield 2019). The typical life cycle of barnacles comprises two primary stages: a planktonic larval stage, encompassing six free-swimming nauplii and a cyprid that permanently attaches to a substrate, followed by benthic sessile stages involving juveniles and adults. Nonetheless, several interactions within this group exhibit mutualistic or parasitic attributes, possessing a slightly different life cycle.

One example of a parasitic interaction involves barnacles parasitising deep-water lantern sharks. *Anelasma squalicola* feeds on the shark host exclusively and has detrimental impacts on the health of its host. It can also impede the development of reproductive organs, potentially impacting fecundity. This relatively understudied parasite is presumed to progress through a free-living nauplius stage and a cypris stage, and then larvae somehow attach themselves to their shark hosts, ultimately maturing into adult forms. Upon attachment, these barnacles burrow into the host flesh, utilising their rootlets to extract nutrients, a behaviour reminiscent of rhizocephalans but distinctive within the Thoracica. Subsequently, these parasites settle, rapidly mature and commence reproduction. Parasitic thoracicans are hermaphroditic, equipped with a penis, which eliminates the presence of dwarf males within the reproductive system. Once they reach maturity as adults, fertilisation can take place through adjacent individuals.

6.7.3 Facetotecta

This most poorly understood subclass of Thecostraca is recognisable solely through its larvae, as the adult forms remain undiscovered. There are less than 20 known species of these distinctive and tiny planktonic crustaceans. The life cycle, although incomplete, involves a series of discernible stages, including at least five free-swimming naupliar instars (y-nauplii), a specialised attachment stage known as the y-cypris, and a stage called ypsigon that is probably a juvenile.

The y-nauplii (Fig. 6.24) are semi-transparent, featuring a cephalic anterior part covered by a

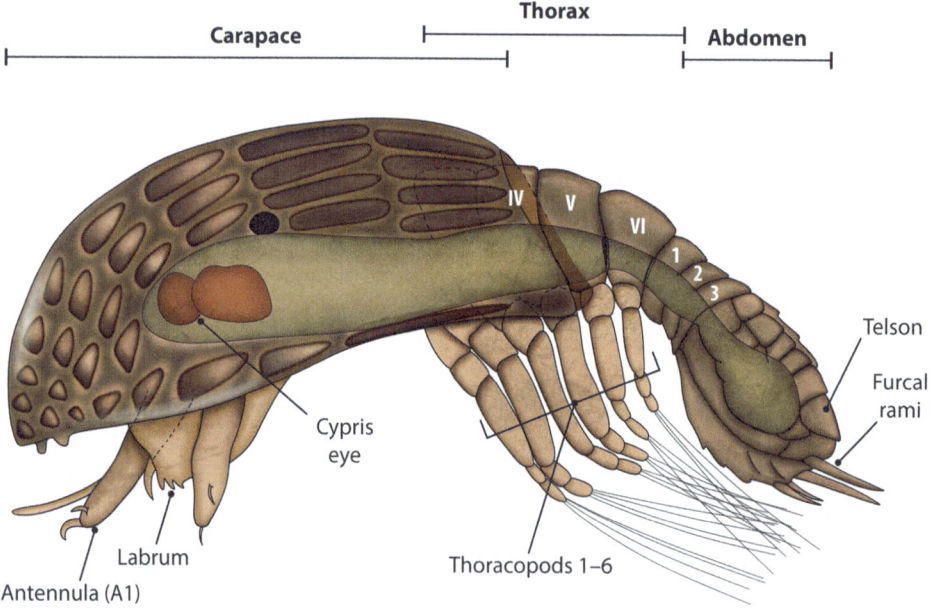

Fig. 6.24 Generalised morphology of y-cypris larvae of Facetotecta (Thecostraca), lateral view. Abbreviations: *A1* Antenna 1, *IV–VI* thoracic segments [Adapted from Kolbasov et al. (2007, 2022)]

dorsal faceted head shield, from which the group derives its name, and a posterior part or hind body. They can be either planktotrophic (feeding) or lecithotrophic (non-feeding), while the y-cyprid is consistently non-feeding. The y-cyprid possesses a univalved carapace partially covering the larval body, six pairs of natatory thoracopods, paired compound eyes, a segmented thorax and a limbless abdomen terminating with a telson bearing furcal rami. The ypsigon, resembling a limbless slug, is unsegmented and emerges from the y-cypris cuticle just behind the labrum. This stage is proposed as the initial parasitic phase of this group, as the earlier larval stages are free-living. The various life stages strongly suggest that the as-yet-unidentified adult life stages are likely advanced endoparasites (Høeg et al. 2014a).

6.8 Arthropods of Terrestrial Origin Parasitising Aquatic Hosts

The vast majority of parasitic arthropods undoubtedly belong to the crustaceans. However, a few terrestrial arthropod taxa within the Arachnida and the Hexapoda (insects) have also become important parasites of aquatic hosts. Mites (Arachnida), for example, have been identified as ectoparasites of marine mammals, including seals and sea lions. These parasites can live in hair follicles and skin glands, leading to symptoms such as alopecia and skin thickening. Moreover, lung mites can afflict these marine hosts, living within the nasal passages, lungs and respiratory systems of seals, sea lions, sea otters and walruses. Infestation by nasal mites can result in the production of copious amounts of mucus, nasal discharge, breathing difficulties and coughing. In addition to their impact on marine mammals, mites can also pose a threat to weakened or stressed fishes under certain environmental conditions. Various genera of mites have been isolated from the skin, fins, gills, intestines, mouth and oesophagus of fishes (and eels) worldwide. In severe cases, infestation by these mites can lead to high mortality rates among young hosts.

Galapagos marine iguanas (*Amblyrhynchus cristatus*) are susceptible to infestation by tick ectoparasites. Nymphs commonly inhabit the skin folds of the iguana neck, while adults tend to cluster on the tail, dorsal or ventral surfaces, particularly around the soft tissues near the cloaca, where they remain attached for extended periods. Additionally, certain ticks may also be found within the nasal fossae of these iguanas. These parasites demonstrate resilience to marine waters, specifically regarding the prolonged immersion, depths and temperatures of the sea. Furthermore, there is evidence to suggest that these ticks have the potential to transmit blood parasites, leading to a reduction in host haemoglobin levels and oxygen consumption. Consequently, this could diminish the stamina of the host and provoke alterations in its behaviour (Wikelski 1999).

Similar interactions have also been observed with parasites from the Hexapoda. While the majority of lice species (Phthiraptera) are obligate and permanent ectoparasites of vertebrate hosts, a few (e.g. those members of the family Echinophthiriidae) exhibit a remarkable ability to infest amphibious hosts such as otters, sea lions, seals and walruses. These parasites, commonly referred to as sucking lice, were initially believed to be terrestrial organisms incapable of withstanding aquatic environments, particularly the harsh conditions of the ocean. However, they have demonstrated a remarkable capacity to adapt and thrive on semi-aquatic hosts. Notably, some of these hosts are capable of diving to depths of up to 2,000 m and spending prolonged periods at sea, showcasing the parasite's resilience and adaptation to extreme environments characterised by high hydrostatic pressure, high salinity, hypoxia and low temperature (Leonardi et al. 2021).

Sucking lice have evolved various adaptations to thrive in underwater habitats. Specialised legs enable them to firmly cling to their hosts, while their spiracles can be sealed shut to prevent water from entering the tracheal system during submersion. Additionally, these lice possess scales and modified spines that likely serve a dual purpose: offering protection against desiccation and mechanical damage, and facilitating respiration while submerged (Leonardi et al. 2021). Sucking lice do not attach to cetaceans due to the absence of hair for egg attachment. For instance, in the case of the elephant seal louse, *Lepidophthirus macrorhini*, eggs are typically affixed to the base of the hair, hatching within a span of five to ten days. Nymphs undergo development, and within approximately two weeks they mature into adults. However, the eggs of these lice, along with early nymphal stages, cannot endure underwater conditions. Consequently, lice must reproduce while on land, thereby restricting the transmission and reproduction of these parasites to the biology and ecology of their hosts. In certain instances, the seal louse, *Echinophthirius horridus*, is capable of inducing anaemia in its host, and is suspected to serve as an intermediate host for the filarioid nematode, *Acanthocheilonema spirocauda*. Moreover, the presence of a common louse on sea lion pups can lead to alopecia, while the elephant seal louse possesses the potential to transmit viral infections between hosts.

Lastly, aquatic birds (e.g. penguins and ducks) have also been noted to be parasitised by chewing lice from the family Philopteridae. These lice have chewing or biting mouthparts and usually feed on skin secretions, dried blood, feathers and skin debris, causing discomfort to the hosts. Furthermore, seabird populations face challenges from the seabird tick, *Ixodes uriae* (Fig. 6.25). This ectoparasite commonly infects marine avian species, such as penguins, and serves as a vector for *Borrelia garinii* (see Petney and Pfäffle 2017), a spirochete responsible for Lyme disease in humans (see Chap. 2). This zoonotic pathogen is transmitted to avian and mammalian reservoir hosts, as well as humans, primarily by *Ixodes ricinus*. The impact of *Ixodes uriae* extends beyond disease transmission, influencing the population dynamics of seabirds through factors like nest desertion and chick mortality. This tick exhibits unique behaviour, alternating between feeding periods and off-host aggregation under rocks (Fig. 6.26). Furthermore, it demonstrates remarkable adaptability in synchronising its life cycle and parasitic phase precisely with the reproductive season. Additionally, *Ixodes uriae* has been linked to at least seven different viral pathogens, further highlighting its significance in marine ecosystems.

6.9 Concluding Remarks

This chapter serves as a preliminary overview of the extensive realm of parasitic arthropods, focusing particularly on the diverse array of different taxonomic groups of Crustacea that demonstrate extraordinary adaptations and highly complex life cycles. These parasites provide a distinctive vantage point for examining the intricate ecological interactions within diverse aquatic ecosystems.

Each parasite group within this chapter uses a distinct array of strategies that highlight the abundant biodiversity thriving in seemingly improbable niches. A wide spectrum of survival tactics and life histories are apparent, signifying the versatile and adaptable nature of these parasites. From specialised suctorial mouthparts and clawed appendages to the intricate art of mimicry and protelean strategies, each group of parasites reveals a wide array of evolutionary adaptations. Whether exploring the diverse ranges of direct or indirect life cycles, determining the complexities of monoxenous or heteroxenous development or considering the multiple roles of transmission, the adaptability of these parasites demonstrates the remarkable ingenuity seen within the aquatic Arthropoda.

Furthermore, the ecological ramifications of parasitism extend well beyond the individual organisms interlinked in these complex relationships. The intricate interplay between parasite and host gives rise to diverse ecological consequences that resonate throughout entire ecosystems. The transmission of pathogens, the modification of host behaviours and biological modifications all serve to verify the profound ecological implications of parasitic interactions.

In conclusion, this review of parasitic arthropods provides a glimpse into the vast diversity of life histories, adaptations and ecological roles that define these creatures.

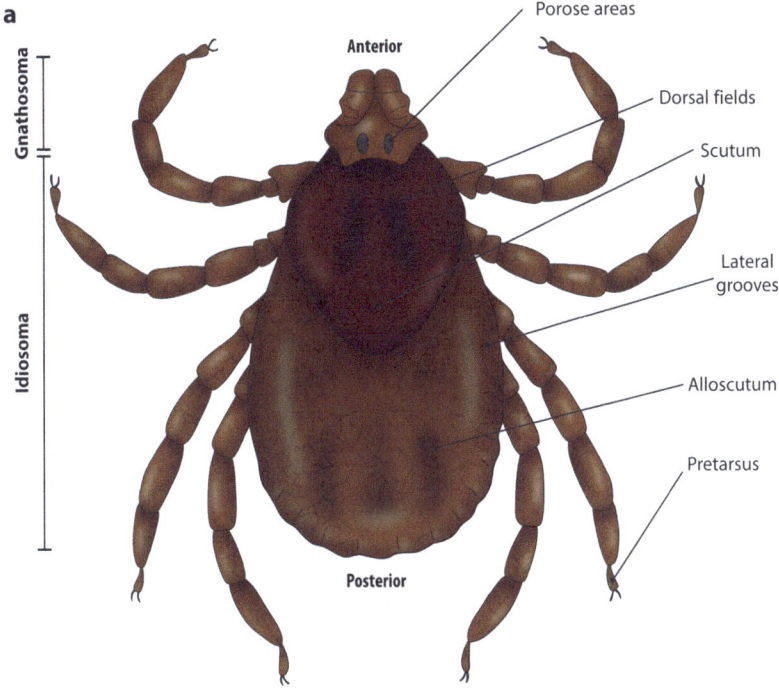

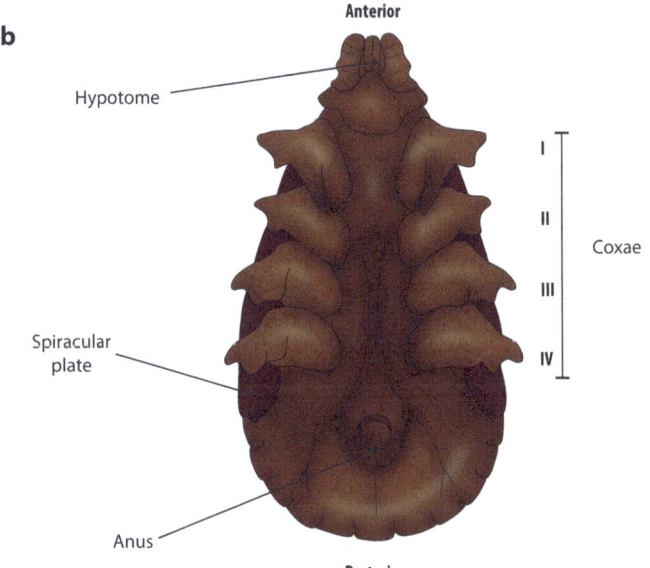

Fig. 6.25 General morphology of a female seabird tick from the family Ixodidae (Arachnida), dorsal (**a**) and ventral view (**b**). Abbreviations: *I–IV* coxal segments

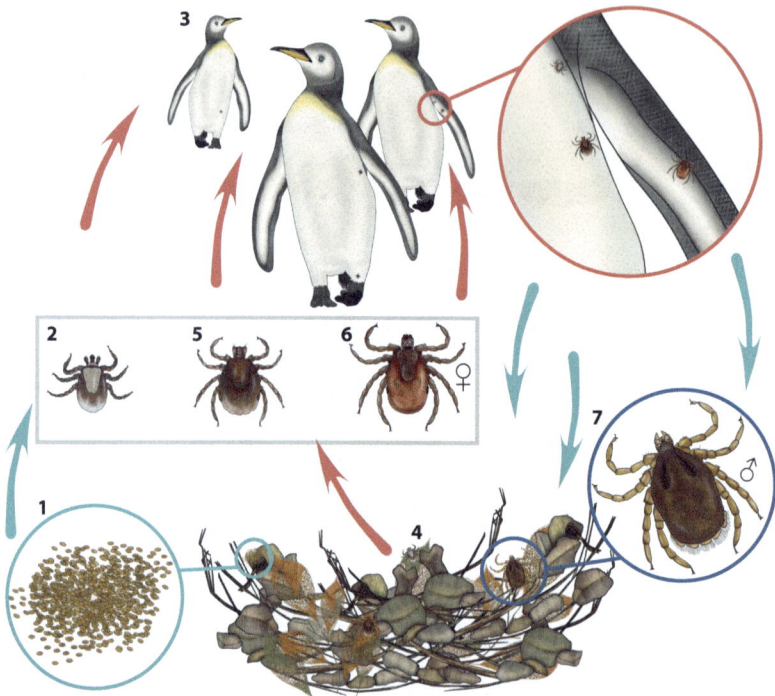

Fig. 6.26 Typical life cycle of seabird ticks (Ixodidae) infecting penguins. Eggs are laid on the ground, most likely in the penguin's nest (*1*), and develop into three active stages (red arrows indicate infective stages). The larvae (*2*) feed on the penguin host (*3*), then drop off (blue arrows indicate moulting stages) to develop into the next stage on the ground (*4*). The cycle repeats typically with a three-host life cycle (see Frenot et al. 2001), into the nymphal stage (*5*) and the adult stage [female (*6*) and male (*7*)]. The males do not gorge on blood. Schematic, not to scale

References

Avenant-Oldewage A, Everts L (2010) *Argulus japonicus*: sperm transfer by means of a spermatophore on *Carassius auratus* (L). Exp Parasitol 126(2):232–238

Baeza JA (2015) Crustaceans as symbionts: an overview of their diversity, host use and life styles. In: Watling L, Thiel M (eds) The life styles and feeding biology of the Crustacea. Oxford University Press, pp 163–189

Boxshall GA (2005) Copepoda (Copepods). In: Rhode K (ed) Marine parasitology. CSIRO Publishing, Collingwood, pp 123–138

Boxshall GA, Lincoln RJ (1987) The life cycle of the Tantulocarida (Crustacea). Philos Trans R Soc Lond B Biol Sci 315:267–303

Boxshall GA, Vader W (1993) A new genus of Tantulocarida (Crustacea) parasitic on an amphipod host from the North Sea. J Nat Hist 27:977–988

British Geological Survey (2024) Ostracods [online]. Discovering geology. https://www.bgs.ac.uk/discovering-geology/fossils-and-geological-time/ostracods/. Accessed 18 June 2024

Christoffersen ML, De Assis JE (2013) A systematic monograph of the recent pentastomida, with a compilation of their hosts. Zool Meded 87(1):1–206

Crossley SM, George AL, Keller CJ (2009) A method for eradicating amphipod parasites (Hyperiidae) from host jellyfish, *Chrysaora fuscescens* (Brandt, 1835), in a closed recirculating system. J Zoo Wildl Med 40(1):174–180

Dittrich B (1987) Post-embryonic development of the parasitic amphipod *Hyperia galba*. Helgoländer Meeresuntersuchungen [Helgoland Marine Research] 41:217–232

Frenot Y, De Oliveira E, Gauthier-Clerc M, Deunff J, Bellido A, Vernon P (2001) Life cycle of the tick *Ixodes uriae* in penguin colonies: relationships with host breeding activity. Int J Parasitol 31(10):1040–1047

Grygier MJ, Høeg JT (2005) Ascothoracida (ascothoracids). In: Rohde K (ed) Marine parasitology. CSIRO Publishing/CABI, Melbourne/Wallingford, pp 149–154

Hadfield KA (2019) History of discovery of parasitic Crustacea. In: Smit NJ, Bruce NL, Hadfield KA (eds) Parasitic Crustacea: state of knowledge and future trends, Zoological monographs, vol 30. Springer, Cham, pp 7–71

Hadfield KA, Smit NJ (2019) Parasitic Crustacea as vectors. In: Smit NJ, Bruce NL, Hadfield KA (eds) Parasitic Crustacea: state of knowledge and future trends, Zoological monographs, vol 30. Springer, Cham, pp 331–342

Hart DG, Hart CW (1974) The ostracod family Entocytheridae. The Academy of Natural Sciences of Philadelphia, Monograph, vol 18, pp 1–239

Hickman C, Keen S, Eisenhour D, Larson A, I'Anson H (2023) Integrated principles of zoology, vol 19. McGraw-Hill, New York, pp 421–440

Høeg JT (1995) The biology and life cycle of the Rhizocephala (Cirripedia). J Mar Biol Assoc UK 75:517–550

Høeg JT, Chan BK, Kolbasov GA, Grygier MJ (2014a) Facetotecta. In: Martin JW, Olesen J, Høeg JT (eds) Atlas of crustacean larvae. Johns Hopkins University Press, Baltimore, pp 100–103

Høeg JT, Chan BK, Kolbasov GA, Grygier MJ (2014b) Ascothoracida. In: Martin JW, Olesen J, Høeg JT (eds) Atlas of crustacean larvae. Johns Hopkins University Press, Baltimore, pp 104–106

Høeg JT, Chan BK, Rybakov AV (2014c) Rhizocephala. In: Martin JW, Olesen J, Høeg JT (eds) Atlas of crustacean larvae. Johns Hopkins University Press, Baltimore, pp 111–115

Huys R (1991) Tantulocarida (Crustacea: Maxillopoda): new taxon from the temporary meiobenthos. Mar Ecol 12:1–34. Pubblicazioni della Stazione Zoologica di Napoli I

Huys R (2014) Copepoda. In: Martin JW, Olesen J, Høeg JT (eds) Atlas of crustacean larvae. Johns Hopkins University Press, Baltimore, p 145

Huys R, Boxshall GA, Lincoln RJ (1993) The tantulocaridan life cycle: the circle closed? J Crustac Biol 13:432–442

Huys R, Olesen J, Petrunina AS, Martin JW (2014) Tantulocarida. In: Martin JW, Olesen J, Høeg JT (eds) Atlas of crustacean larvae. Johns Hopkins University Press, Baltimore, pp 122–127

Ismail N, Ohtsuka S, Maran BA, Tasumi S, Zaleha K, Yamashita H (2013) Complete life cycle of a pennellid *Peniculus minuticaudae* Shiino, 1956 (Copepoda: Siphonostomatoida) infecting cultured threadsail filefish, *Stephanolepis cirrhifer*. Parasite 20:42

Junker K, Boomker JDF, Booyse DG (1998) Experimental studies on the life-cycle of *Sebekia wedli* (Pentastomida: Sebekidae). Onderstepoort J Vet Res 65:233–237

Kearn GC (2004a) Cyclopoid copepods—the anchor worm. In: Kearn GC (ed) Leeches, lice and lampreys. Springer, Dordrecht, pp 208–213. https://doi.org/10.1007/978-1-4020-2926-4_11

Kearn GC (2004b) Siphonostomatoid copepods: (1) fish lice—caligids. In: Kearn GC (ed) Leeches, lice and lampreys. Springer, Dordrecht, pp 154–177. https://doi.org/10.1007/978-1-4020-2926-4_8

Kensley B, Schotte M (1989) Guide to the marine isopod crustaceans of the Caribbean. Smithsonian Institution Press, Washington D.C., p 308

Kolbasov GA, Petrunina AS (2019) The family Ascothoracidae Grygier, 1987, a review with descriptions of new abyssal taxa parasitizing ophiuroids and remarks on the invalidity of the genus *Parascothorax* Wagin, 1964 (Crustacea: Thecostraca: Ascothoracida). Mar Biodivers 49:1417–1447

Kolbasov GA, Grygier MJ, Ivanenko VN, Vagelli AA (2007) A new species of the y-larva genus *Hansenocaris* Itô, 1985 (Crustacea: Thecostraca: Facetotecta) from Indonesia, with a review of y-cyprids and a key to all their described species. Raffles Bull Zool 55:343–353

Kolbasov GA, Petrunina AS, Ho MJ, Chan BK (2019) A new species of Synagoga (Crustacea, Thecostraca, Ascothoracida) parasitic in an antipatharian from Green Island, Taiwan, with notes on its morphology. ZooKeys 876:55

Kolbasov GA, Savchenko AS, Newman WA, Chan BK (2021) A new species of *Waginella* (Crustacea: Thecostraca: Ascothoracida) parasitic on a stalked crinoid from Tasman Sea, with notes on morphology of related genera. Front Mar Sci 8:616001

Kolbasov GA, Savchenko AS, Dreyer N, Chan BK, Høeg JT (2022) A synthesis of the external morphology of cypridiform larvae of Facetotecta (crustacea: Thecostraca) and the limits of the genus *Hansenocaris*. Ecol Evol 12(11):e9488

Kretzler JE (1984) *Echinophilus xiphidion*, new species (Ostracoda: Paradoxostomatidae) parasitic on regular echinoids of the northeastern Pacific. J Crustac Biol 4:333–340

Legrand JJ (1952) Contribution B l'etude expérimentale et statistique de la biologie *d'Anilocra physodes* L. Archs Zool Exp Gén 89:1–55

Lehnert KI, Jsseldijk LL, Uy ML, Boyi JO, van Schalkwijk L, Tollenaar EA, Gröne A, Wohlsein P, Siebert U (2021) Whale lice (*Isocyamus deltobranchium* & *Isocyamus delphinii*; Cyamidae) prevalence in odontocetes off the German and Dutch coasts–morphological and molecular characterization and health implications. Int J Parasitol Parasites Wildl 15:22–30

Leonardi MS, Crespo JE, Soto F, Lazzari CR (2021) How did seal lice turn into the only truly marine insects? Insects 13(1):46

Leung YM (1976) Life cycle of Cyamus scammoni (Amphipoda: Cyamidae), ectoparasite of gray whale, with a remark on the associated species. Sci Rep Whales Res Inst 28:153–160

López-González PJ, Bresciani J, Huys R, Af G, Guerra A, Pascual S (2000) Description of *Genesis vulcanoctopusi* gen. et sp. nov. (Copepoda: Tisbidae) parasitic on a hydrothermal vent octopod and a reinterpretation of the life cycle of cholidyinid harpacticoids. Cah Biol Mar 41:241–253

Lützen J (2005) Amphipoda (amphipods). In: Rohde K (ed) Marine parasitology. CSIRO Publishing/CABI, Melbourne/Wallingford, pp 165–169

Maiti S (2022) Pentastomiasis. In: Parija SC, Chaudhury A (eds) Textbook of parasitic Zoonoses. Springer Nature Singapore, Singapore, pp 601–610

Mazda Y, Sasagawa T, Iinuma Y, Wakabayashi K (2019) Maternal care and juvenile feeding in a hyperiid amphipod (*Oxycephalus clausi* Bovallius, 1887) in association with gelatinous zooplankton. Mar Biol Res 15(10):541–547

Mikheev VN, Pasternak AF, Valtonen ET (2015) Behavioural adaptations of argulid parasites (Crustacea: Branchiura) to major challenges in their life cycle. Parasit Vectors 8:1–10

Møller OS, Olesen J (2014) Branchiura. In: Martin JW, Olesen J, Høeg JT (eds) Atlas of crustacean larvae. Johns Hopkins University Press, Baltimore, pp 128–134

Moravec F, Jirků M, Charo-Karisa H, Mašová Š (2009) *Mexiconema africanum* sp. n. (Nematoda: Daniconematidae) from the catfish *Auchenoglanis occidentalis* from Lake Turkana, Kenya. Parasitol Res 105:1047–1052

Murphy AE, Williams JD (2013) New records of two trypetesid burrowing barnacles (Crustacea: Cirripedia: Acrothoracica: Trypetesidae) and their predation on host hermit crab eggs. J Mar Biol Assoc UK 93:107–133

Neethling LAM, Avenant-Oldewage A (2016) Branchiura—a compendium of the geographical distribution and a summary of their biology. Crustaceana 89(11–12):1243–1446

Ota Y, Hoshino O, Hirose M, Tanaka K, Hirose E (2012) Third-stage larva shifts host fish from teleost to elasmobranch in the temporary parasitic isopod, *Gnathia trimaculata* (Crustacea; Gnathiidae). Mar Biol 159:2333–2347

Otake S, Wakabayashi K, Tanaka Y, Nagasawa K (2013) *Choniomyzon inflatus* n. sp. (Crustacea: Copepoda: Nicothoidae) associated with *Ibacus novemdentatus* (Crustacea: Decapoda: Scyllaridae) from Japanese waters. Syst Parasitol 84:157–165

Paré JA (2008) An overview of pentastomiasis in reptiles and other vertebrates. J Exot Pet Med 17(4):285–294

Petney TN, Pfäffle MP (2017) *Ixodes uriae* white, 1852 (figs. 38–40). In: Estrada-Peña A, Mihalca A, Petney T (eds) Ticks of Europe and North Africa. Springer, Cham. https://doi.org/10.1007/978-3-319-63760-0_23

Petrunina AS, Huys R (2020) A new species of Tantulocarida (Crustacea) parasitic on a deep-water cumacean host from the southwestern Atlantic, with a review of tantulocaridan host utilization, distribution, and diversity. J Crustac Biol 40(6):765–780

Pfeil-Putzien C (1978) Experimentelle übertragung der Frühjahrsvirämie (spring viraemia) der Karpfen durch Karpfenläuse (*Argulus foliaceus*). Zentralbl Veterinärmed B 25(4):319–323

Riley J (1986) The biology of pentastomids. Adv Parasitol 25:45–128

Rowntree VJ (1996) Feeding, distribution, and reproductive behavior of cyamids (Crustacea: Amphipoda) living on humpback and right whales. Can J Zool 74:103–109

Santhosh B, Anil MK, Muhammed Anzeer F, Aneesh KS, Abraham MV, Gopakumar G, George RM, Gopalakrishnan A, Unnikrishnan C (eds) (2018) Culture techniques of marine copepods. ICAR-Central Marine Fisheries Research Institute, Kochi, p 144

Schell DM, Rowntree VJ, Pfeiffer CJ (2000) Isotopic evidence that cyamids (Crustacea: Amphipoda) feed on whale skin. Can J Zool 78:721–727

Schwentner M, Combosch DJ, Nelson JP, Giribet G (2017) A phylogenomic solution to the origin of insects by resolving crustacean-hexapod relationships. Curr Biol 27:1818–1824

Smit NJ, Davies AJ (2004) The curious life-style of the parasitic stages of gnathiid isopods. Adv Parasitol 58:289–391

Smit NJ, Basson L, Van As JG (2003) Life cycle of the temporary fish parasite, *Gnathia africana* (Crustacea: Isopoda: Gnathiidae). Folia Parasitol 50:135–142

Smit NJ, Bruce NL, Hadfield KA (2014) Global diversity of fish parasitic isopod crustaceans of the family Cymothoidae. Int J Parasitol Parasites Wildl 3:188–197

Suárez-Morales E (2011) Diversity of the Monstrilloida (Crustacea: Copepoda). PLoS One 6(8):e22915

The Animal Files (2024) Amphipod body plan [online]. https://www.theanimalfiles.com/anatomy/amphipod_body_plan.html. Accessed 18 June 2024

The Robinson Library (2024) About Copepoda [online]. http://www.therobinsonlibrary.com/science/zoology/crustaceans/copepoda/about.htm. Accessed 18 June 2024

Wikelski M (1999) Influences of parasites and thermoregulation on grouping tendencies in marine iguanas. Behav Ecol 10(1):22–29

Williams JD, Boyko CB (2012) The global diversity of parasitic isopods associated with crustacean hosts (Isopoda: Bopyroidea and Cryptoniscoidea). PLoS One 7(4):e35350

Williams EH, Bunkley-Williams L (2019) Life cycle and life history strategies of parasitic Crustacea. In: Smit NJ, Bruce NL, Hadfield KA (eds) Parasitic Crustacea: state of knowledge and future trends, Zoological monographs, vol 30. Springer, Cham, pp 179–266

Open Access This chapter is licensed under the terms of the Creative Commons Attribution-NonCommercial-NoDerivatives 4.0 International License (http://creativecommons.org/licenses/by-nc-nd/4.0/), which permits any non-commercial use, sharing, distribution and reproduction in any medium or format, as long as you give appropriate credit to the original author(s) and the source, provide a link to the Creative Commons license and indicate if you modified the licensed material. You do not have permission under this license to share adapted material derived from this chapter or parts of it.

The images or other third party material in this chapter are included in the chapter's Creative Commons license, unless indicated otherwise in a credit line to the material. If material is not included in the chapter's Creative Commons license and your intended use is not permitted by statutory regulation or exceeds the permitted use, you will need to obtain permission directly from the copyright holder.

Part II

Ecological Principles and Latest Research Developments in Aquatic Parasitology

Unveiling the Hidden Players: Exploring the Intricate Dance of Aquatic Parasites, Host Biodiversity and Ecosystem Health

7

Clarisse Louvard, Kerry A. Hadfield, Maarten P. M. Vanhove, Bernd Sures, and Nico J. Smit

Abstract

This chapter delves into the multifaceted roles of aquatic parasites within natural ecosystems. It highlights both the negative and positive impacts these parasites can have on individual hosts, host populations, biodiversity and overall ecosystem health. The discussion covers how parasites influence various levels of biological organisation and ecosystem functions. It also explores how healthy ecosystems are defined and maintained, emphasising the roles of vigour, organisation and resilience. The complex interactions between parasites and their hosts are illustrated through numerous examples, spanning cases of behavioural modification, host–parasite coevolution, and broader ecological consequences stemming from those interactions. Understanding the interplay between parasites, hosts and ecosystems is presented as crucial for a comprehensive view of ecosystem dynamics.

C. Louvard · K. A. Hadfield · N. J. Smit (✉)
Water Research Group, Unit for Environmental Sciences and Management, North-West University, Potchefstroom, South Africa
e-mail: 55214770@mynwu.ac.za; clarisse.louvard@uqconnect.edu.au; kerry.malherbe@nwu.ac.za; nico.smit@nwu.ac.za

M. P. M. Vanhove
Research Group Zoology: Biodiversity and Toxicology, Centre for Environmental Sciences, Hasselt University - Campus Diepenbeek, Diepenbeek, Belgium
e-mail: maarten.vanhove@uhasselt.be

B. Sures
Water Research Group, Unit for Environmental Sciences and Management, North-West University, Potchefstroom, South Africa

Department of Aquatic Ecology and Centre for Water and Environmental Research (ZWU), University of Duisburg-Essen, Essen, Germany

Research Centre One Health Ruhr, Research Alliance Ruhr, University Duisburg-Essen, Essen, Germany
e-mail: bernd.sures@uni-due.de

7.1 Introduction

The impact of aquatic parasites on their hosts has been relatively well studied, especially for parasites of veterinary and medical importance (see Chap. 19) and for pathogens whose impact on aquaculture threatens food security (see Chap. 23). Within natural and modified aquatic ecosystems, parasites impact all levels of biological organisation (see Chap. 20). Although many of these effects are considered negative, research over the past three decades has also shed light on the positive impacts of aquatic parasites on individual host health and, ultimately, on the functioning and maintenance of ecosystems.

The aim of this chapter is to introduce the reader to the various ecological roles and impacts (positive and negative) of aquatic parasites in natural ecosystems. We focus on the effects of parasites on individual hosts, host populations, biodiversity and ecosystem health in general (see Chap. 8 for a detailed discussion on the impact of aquatic parasites on host community structures). To demonstrate the importance of aquatic parasites for ecological processes, we will first introduce the concept of a healthy ecosystem followed by how aquatic parasites both drive and bear the brunt of the state of ecosystems.

7.2 What Is a Healthy Ecosystem?

All systems, whether simple or complex, have a finite lifespan. They evolve as they age and as their smaller components are replaced (Costanza and Patten 1995; Costanza and Mageau 1999). A healthy system is thus a system *predicted* to be on track to achieving a full natural lifespan, with the outcome of that prediction being visible only retrospectively (Costanza and Patten 1995). Therefore, any process that prematurely reduces the predicted lifespan of a system beyond its normal evolution due to ageing can be considered detrimental to the health of that system (Costanza and Mageau 1999).

Historically, the health of an ecosystem was implicitly understood in light of the values promoted in human health, in a human-centred attempt to manage the environment (Science Advisory Board 1990). Since then, the definition of "ecosystem health" has in turn included notions of balance, complexity, stability and growth potential. Nowadays, a healthy ecosystem is defined as 'stable and sustainable' (Costanza et al. 1992), meaning it shows 'the ability to maintain its structure (organisation) and function (vigour) over time in the face of external stressors (resilience)' (Costanza and Mageau 1999) throughout its full predicted natural lifespan (Costanza and Patten 1995). These three attributes of ecosystems (i.e. organisation, vigour and resilience) have been characterised by various authors. The organisation of an ecosystem is a qualitative and quantitative measure of the interactions between species and with the surrounding habitat (Costanza and Mageau 1999). An ecosystem's organisation depends on its richness and diversity, the level of ecological specialisation of each of its species and the number of unique interactions between the ecosystem's components. Network analyses have been used to quantify this factor (see Leontief 1941; Ulanowicz 1986). The vigour of an ecosystem is defined by Costanza and Mageau (1999) as 'a measure of its activity, metabolism or primary productivity' that is reflected in a variety of quantifiable factors, for example, gross primary production (see also Odum 1971). Finally, an ecosystem's resilience is defined as its ability to keep its organisation intact when exposed to perturbations of biotic or abiotic origin (Costanza and Mageau 1999).

In twentieth-century ecology, these characteristics were mostly envisioned in relation to predator–prey relationships or interspecific competition. Parasites, whose biomass was deemed insignificant, were long excluded despite their extraordinary success (Horwitz and Wilcox 2005; Hudson et al. 2006b). Research has since demonstrated parasites can drive, as well as be a consequence of, the state of ecosystems (reviewed by Selbach et al. 2022) and contribute more to the biodiversity of an ecosystem than free-living organisms (see Fig. 7.1).

7.3 A Parasite's Effect Is Context Dependent

Contrary to popular belief, parasites are not always detrimental to their hosts (see Selbach et al. 2022; Chaps. 13 and 20) and their effects on any ecosystem vary over time. Their influence on both hosts and communities, positive or negative, depends on a wide range of environmental factors.

Parasitism usually incurs a cost to the hosts in terms of growth, survival and/or reproduction. The scale of damage from parasitic infections depends on factors related to the host–parasite relationship such as the life history of the para-

Fig. 7.1 An illustration of the diversity within an aquatic ecosystem highlighting the presence of parasites. All animal species living in this ecosystem serve as potential hosts for one or more parasite species at various developmental stages

site, site of infection, parasite load or efficiency of the host's immune response and on factors related to the ecosystem such as host population density, resource availability or overall parasite presence in the community (see Sects. 7.4 and 7.5). The interaction between these intrinsic and extrinsic factors determines the outcome of infections for the host, which can range from weight loss to heightened sensitivity to opportunistic diseases, reduced lifespan or parasitic castration (see Sect. 7.4.1).

In contrast, some studies suggest that phenotypic alterations provoked by parasite presence are not detrimental to hosts under specific environmental conditions. In these cases, the damage incurred by parasite presence is offset by the benefits of being infected, leading to increased net fitness and/or survival for the host and ensuring its reproductive success (see Chap. 21). For example, Richardson's ground squirrels (*Urocitellus richardsonii*) infected with *Trypanosoma otospermophili* show significantly higher mass gain when allowed to feed *ad libitum* under a vitamin B6-deficient diet compared to uninfected controls in laboratory conditions (Munger and Holmes 1988) because trypanosomes produce that vitamin for the host (Stoffel et al. 2006). Although trypanosomes are prevalent in aquatic habitats, no similar cases of function rescue are known from trypanosomes of aquatic hosts. The nature and extent of the effects a parasite exercises on its host also heavily depend on the characteristics of the surrounding biotope. Resource availability has been proposed to play a key role in how hosts manage their parasite loads. For instance, individuals of the limpet *Fissurella crassa* infected by metacercariae of *Proctoeces humboldti* in rich upwelling areas of the Chilean coast display a significantly higher gonadosomatic index (excluding parasite biomass in the gonads) than both infected and uninfected hosts in nutrient-poor areas. Moreover, infected hosts in upwelling areas consume significantly less oxygen than their uninfected counterparts, whereas those consumption rates are equivalent for both parasitised and unparasitised hosts in oligotrophic waters (high oxygen consumption indicates physiological stress) (Aldana et al. 2020). This example shows that access to resources (in this case: nutrient availability) mediates the physiological response of a host individual to parasite infection.

Historically, parasitism was described as a form of symbiosis (i.e. a long-term, intimate ecological relationship between individuals of distinct species; Combes 2001; Rózsa and Garay 2023) wherein one of the species involved incurs some form of fitness cost. However, the recognition of the existence of "conditionally helpful" parasites through the examples above has blurred the lines between mutualistic and parasitic relationships. This recognition forced changes to the definition of parasitism (e.g. Fellous and Salvaudon 2009; Parmentier and Michel 2013; Weinersmith and Earley 2016). If the host–parasite relationship between populations of two species is long-lasting enough, that relationship can evolve through evolutionary arms race and population dynamics to a more balanced interaction and, sometimes, a mutualistic one, thus reducing the cost of virulence for the parasite and that of resistance for the host (Antia et al. 1994; Combes 1997).

The benefits of parasitism can also extend to host populations. In ecosystems featuring species consuming potentially parasitised prey, selective feeding based on the probability of infection of a particular cohort of prey has been observed as a means to avoid parasites. Such behavioural adaptation tends to spare the whole prey cohort, including both parasitised and healthy animals. For example, in the UK, depending on the season, oystercatchers (*Haematopus ostralegus*) feeding on cockles infected by metacercariae of bird trematodes tend to maximise energy intake while minimising parasite loads by preying preferentially on the middle-sized cockle cohort, individuals of which are more nutritious than small cockles and statistically less infected than large (i.e. old) ones (Norris 1999). Consequently, both healthy and parasitised individuals from the two remaining cockle size cohorts may benefit from decreased predation rates and enhanced chances of reproduction (Thomas et al. 2000). In addition, the infection of cockles with echinostome trematodes can change the functional role of this dominant benthic organism, which in turn influences the surrounding benthic community (see Chap. 8 for a detailed discussion on a case study from the sand flats of Otago Harbour, New Zealand).

7.4 Parasites as Drivers of Ecosystem Processes

All ecosystem resources are finite. For example, *access* to sunlight is limited even if *sunlight* is not (Darlington Jr 1972). Thus, for a community of several species occupying distinct ecological niches and sharing common resources (e.g. space) to remain in equilibrium, each of these species must be limited in its growth by at least one biotic or abiotic factor to avoid complete resource depletion and subsequent biodiversity loss (MacArthur 1958; Levin 1970). Consequently, the population dynamics of each species in any ecosystem both affect and are affected by the dynamics of all other species within that ecosystem through mechanisms of direct or indirect species interactions: the presence of each species generates limiting factors affecting all the others (Paine 1966; Levin 1970). Parasites, like predators, prey or resources, provide limiting factors regulating species within ecosystems (Darlington Jr 1972). Limitation of species' fitness and population densities by parasitism operates either directly on their hosts or indirectly on non-host species, via density-mediated indirect effects (DMIE) and trait-mediated indirect effects (TMIE) (Figs. 7.2, 7.3) (see also Chaps. 8 and 18 for examples of DMIEs and TMIEs on host community structures).

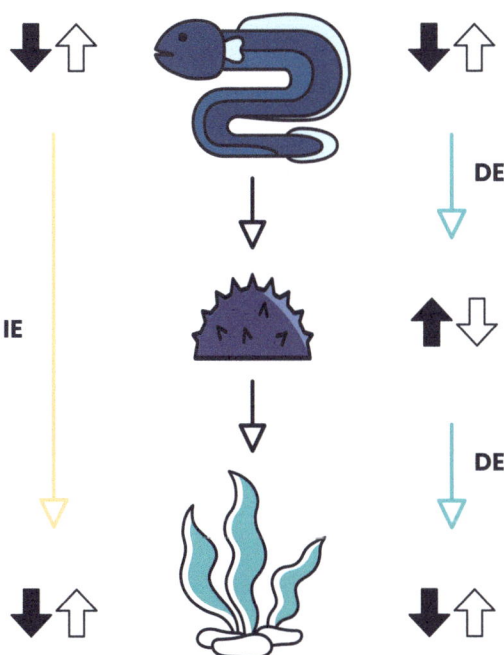

Fig. 7.2 The difference between direct and indirect effects. Wolf eels (top) feed on urchins (middle), which feed on macroalgae (bottom). Wolf eel populations thus have a direct effect (DE) on urchin populations, and urchins on algae populations. When wolf eel populations increase, urchin populations decrease, consequently favouring algal growth (white arrows). The opposite effect is observed in the case of a wolf eel population reduction (black arrows). Thus, wolf eel populations have an indirect effect (IE) on algae populations, mediated by urchins

7.4.1 Direct Effects on Hosts and Parasites

7.4.1.1 Parasitic Castration

An extreme example of parasitism-associated cost directly affecting fitness is parasitic castration (Fig. 7.3), a phenomenon recorded in a wide range of parasitic groups (reviewed by Lafferty and Kuris 2009b). Examples include digenean asexual stages castrating their first-intermediate mollusc (e.g. Sousa 1983) and polychaete hosts (Køie 1982; Cribb et al. 2011) (see Chap. 5); rhizocephalan cirripeds (e.g. species of *Sacculina*) castrating various decapod crustacean genera (e.g. Toyota et al. 2023; Chap. 6); some diphyllobothrium cestodes (e.g. *Schistocephalus solidus*) castrating fishes via nutrient deprivation (Heins 2017; Chap. 5); epicarid (e.g. *Hemioniscus balani*) and cymothoid isopods (e.g. *Riggia paranensis*) castrating cirripeds (Blower and Roughgarden 1988) and fishes (Azevedo et al. 2006), respectively (see Chap. 6); and pearlfishes of the genus *Encheliophis* (Ophidiiformes: Carapidae) permanently castrating their holothuroid hosts (Parmentier and Vandewalle 2005). At the individual level, the infected hosts usually become permanently unable to transmit their genetic material, although some host species have been shown to enhance their reproductive output before complete castration is attained (Minchella et al. 1985; Sorensen and Minchella 2001) or regain the ability to reproduce if they succeed in killing the parasite (Kuris et al. 1980). Following castration, the only genome from the host–parasite entity participating in natural selection (i.e. transmitted to the next generations) will be that of the parasite; from the point of view of natural selection, the castrated host is no more than a 'shell' (O'Brien and Van Wyk 1985).

At the population level, parasitic castration has been shown to significantly reduce host density, effectively removing part of the gene pool. However, the effects of castrators on host populations depend on the nature of the castrators themselves. In the case of trematodes, the main

Fig. 7.3 The four types of parasite-induced direct effects on hosts. (Top left) Direct parasite competition (spark) or synergy directly affects hosts; for example, the competition between digeneans *Ribeiroia ondatrae* ("R") and *Echinostoma trivolvis* ("E") decreases both their populations. (Top right) Parasitic castration, here of a crab by a rhizocephalan parasite (yellow crescent). (Bottom left) Parasite (acanthocephalan)-induced direct modification of a host's (amphipod) physiological, physical, or behavioural traits, here deformed exoskeleton and increased erratic activity in the movement from A to B. (Bottom right) Host morbidity or mortality. These four types of direct effects can have positive (green tick) and/or negative (red cross) outcomes for host fitness

castrators of molluscs, the effects of castration on intermediate-host population densities are mitigated by the complexity of trematode life cycles, the rates of parasite recruitment from definitive hosts (which depends on definitive-host mobility) and environmental conditions, so that rate of parasitic castration and intermediate host population density cannot be inferred from each other (Lafferty 1993). Host population densities might be more directly affected when confronted with castrators having direct life cycles. Castration impacts host populations on two levels. First, as unparasitised individuals compete with castrated hosts for resources and the energy consumed by castrated hosts is at least partly confiscated by parasites, competition for resources instead takes place between healthy snails and the parasites themselves, significantly affecting host populations (Lafferty 1993). Second, castration progressively reduces gamete production. This effect has been recorded in several host–parasite combinations, e.g. polpulations of the California horn snail *Cerithideopsis californica* parasitised by various trematode species (Lafferty 1993). The effect of castration on host populations is generally believed to be enhanced in the case of non-random selection of host individuals by parasites and depends on factors such as host size structure and sex ratios. For example, the isopod *Hemioniscus balani* preferentially targets the oldest and most fertile individuals of the barnacle *Chthamalus fissus* because more energy is allocated by older hosts to the reproduction effort, meaning greater gains for the castrator but significant reduction in barnacle reproductive output at the population level (Blower and Roughgarden 1988).

7.4.1.2 Host Morbidity and Mortality

Whether a host will die from being parasitised, or not, is hard to predict. Generally, parasitised individuals are significantly more likely to die prematurely than uninfected ones (Robar et al. 2010). Parasites can kill their hosts directly (e.g. through excessive infection burden) (Fig. 7.3) or indirectly (e.g. by making hosts more likely to be predated upon). Parasitic infections can also decrease host resistance to environmental stressors: in the study of Lafferty (1993), trematode-infected snails of *C. californica* subjected to an unidentified environmental stressor died significantly more than did unparasitised snails exposed to the same stressor. On the contrary, parasitism often does not affect host mortality rates (e.g. Friesen et al. 2017) and can even benefit the host. For example, in the laboratory, freshwater clams *Pisidium amnicum* parasitised by the trematode *Bunodera luciopercae* survive significantly longer to lethal pentachlorophenol (PCP) concentrations than unparasitised ones do, possibly because trematodes may sequester PCP in their fatty tissues to the benefit of their host (Heinonen et al. 2001).

A meta-analysis by Robar et al. (2010) showed that globally, infected-host mortality varies widely according to many factors, the most influential of which are parasite life-cycle characteristics (i.e. direct or non-predation-mediated trophic transmission vs. predation-mediated trophic transmission regardless of parasite lineage), host lineage and latitude.

Intermediate hosts of trophically transmitted parasites requiring predation at some stage of their life history (i.e. most acanthocephalan, trematode, nematode and cestode species) are 2.4 times more likely to die compared to hosts harbouring parasites not depending on predation-mediated transmission (Robar et al. 2010). In addition to the increased predation likelihood induced by physical damage to the hosts (e.g. Jonsson and Andé 1992), a high diversity of adaptive strategies of host morphological, sensorial and behavioural manipulation by parasites has been evolutionarily selected to ensure predation on infected hosts and successful trophic transmission (Fuller et al. 2003; Seppälä et al. 2004; Poulin et al. 2005) (see Sect. 7.4.2). Of the three parasite categories considered by Robar et al. (2010) (i.e. helminths, arthropods and microparasites), helminths were shown to have the strongest influence on host survival and microparasites the weakest, independently of transmission mode.

Globally and after correction for other significant factors, infected molluscs seem at increased risk of mortality directly or indirectly caused by parasites compared to other invertebrate groups (e.g. arthropods) as well as some vertebrate groups. This difference is explained in part by the commonly deleterious effect of digenean intermediate stages on mollusc first-intermediate hosts, as well as phenotypic modifications by parasites that render the host more vulnerable to predation (Robar et al. 2010) (see *Parasite-induced direct trait modification* below). An illustration of digenean-induced morbidity is given by Jonsson and Andé (1992), wherein a *Cerastoderma edule* cockle population parasitised by an unknown digenean in Sweden underwent a mass mortality event putatively caused by extensive tissue damage (also see Chap. 8). However, not all mollusc species display the same sensitivity to their digenean parasites, and not all digeneans have the same effects on their hosts. For example, Marchand et al. (2020) recorded a higher mortality (putatively caused by the parasite) in parasitised snails of *Ladislavella elodes* than in non-parasitised individuals, but equal rates of mortality in parasitised and non-parasitised snails of *Planorbella trivolvis*, when these two species were infected by *Echinostoma* spp. (Digenea: Echinostomatidae).

Infected fishes and amphibians are also more likely to die (directly or indirectly from their parasites) than parasitised invertebrates, birds and mammals. This discrepancy in mortality between vertebrate lineages might be linked, in part, to differences in host immune response to the presence of non-self across these lineages (Robar et al. 2010).

As vertebrates tend to feature sophisticated immune systems relying on highly efficient feedback loops, excessive triggering of their immune defences can induce deleterious immune cascades, ultimately resulting in organ damage and compromised host survival. Examples include chronic granulomatous inflammations in fishes following infection by metazoan endoparasites (Feist and Longshaw 2008) and cytokine storms in other vertebrate lineages (e.g. Tong et al. 2021). These processes may kill a host even if its parasite is not lethal on its own. Thus, a parasite tends to be the main cause of death only when the vertebrate host's immune response to that parasite is weak (Casadevall and Pirofski 1999).

Lastly, host mortality tends to vary on a latitudinal gradient. The meta-analysis of Robar et al. (2010) suggests that overall, and when accounting for all intrinsic host and parasite characteristics, host mortality tends to be significantly higher in equatorial areas compared to subpolar regions. Moreover, the influence of temperature on host survival is understood to add up to the effects of host taxon and predation-mediated transmission. In all ecosystems, parasites tend to be highly sensitive to abiotic factors, which partly depend on latitude (e.g. temperature, moisture and rainfall) (e.g. Thieltges and Rick 2006). Significant variations of these factors could be associated with parasite outbreaks, although the directions of trends appear different in aquatic and terrestrial systems (Harvell et al. 2002; Hudson et al. 2006a; Torchin et al. 2015). In migrating smolts of Atlantic salmon (*Salmo salar*), for example, sea temperature increase is predicted to accelerate the development of parasitic copepods *Lepeophtheirus salmonis* and *Caligus* spp., leading to higher levels of infestation and increased mortality (Vollset 2019). Outcomes of parasitism vary widely between host–parasite systems, rendering further generalisation difficult.

7.4.1.3 Parasite-Induced Direct Trait Modification

Many parasites can directly affect their host's physiological, physical or behavioural traits (Fig. 7.3). A "trait" is understood here as a specific phenotype, for example, fecundity level, body shape or size or swimming behaviour. In freshwater, an interesting example of parasite-induced trait modification is seen in the infection of the three-spined stickleback, *Gasterosteus aculeatus* by plerocercoid larvae of *Schistocephalus solidus*. Laboratory experiments demonstrated that fish infected by the plerocercoids spent significantly more time foraging near the surface than their uninfected counterparts (Quinn et al. 2012; Talarico et al. 2017), possibly because of an increased need for oxygen (Lester 1971). These trait modifications (i.e. higher oxygen consumption and higher occupation rates of surface layers) could potentially increase predation risk from piscivorous birds, helping the parasite complete its life cycle (Quinn et al. 2012). Procercoids of the same parasite also affect the behaviour of their first intermediate host, the copepod *Macrocyclops albidus*. Transmission of procercoid larvae from first to second (stickleback) intermediate hosts happening between 11 and 31 days post-copepod infection ensures the parasites have attained the critical mass necessary for optimal fitness inside the bird definitive host (Hammerschmidt et al. 2009). Procercoids were experimentally shown to manipulate their copepod hosts in order to minimise predation risk before Day 11 post-infection (by keeping the hosts less active and immobile for longer after a simulated "predator attack"), and maximise predation risk after Day 17 post-infection (by reducing immobility time after attacks and enhancing activity) (Hammerschmidt et al. 2009). *Schistocephalus solidus* is thus able to directly alter the behavioural traits of both of its intermediate hosts.

Significant differences seem to exist in the nature and degree of parasite-induced behavioural alterations across host types (i.e. vertebrate or invertebrate), parasite lineages and transmission strategies (i.e. trophic vs. non-trophic) (Lafferty and Shaw 2013). For instance, trematode and nematode parasites might induce analgesia in terrestrial vertebrates (Kavaliers et al. 1984; Pryor et al. 1998). In contrast, the trematode *Microphallus papillorobustus* supposedly alters serotonergic pathways and neuron morphology in the brain and optic neuropils of the amphipod *Gammarus insensibilis* (Helluy and Thomas 2003), provoking surface-seeking behaviours (Helluy 1983). As illustrated in the latter case, the infection site seems to play a critical role. Parasites invading a host's brain and nervous system (Klein 2003) or organs involved in hormone production, modulation of neuronal activity and immunity seem to have considerable effects on behaviour [see reviews by Beckage (1993), Adamo (2002), Thomas et al. (2005) and Helluy (2013)]. An instance of behavioural modification through alteration of a vertebrate's central nervous system is that of fathead minnows (*Pimephales promelas*) by the brain fluke *Posthodiplostomum ptychocheilus*. When cercariae of that species encyst at low intensities on the surface of the optic lobe of their hosts' brains, the infected fish follow moving objects significantly less and take significantly more time to respond to changes in the direction of those objects (Shirakashi and Goater 2001). Unlike many other cases [see Poulin (1994a) for review], such changes likely result from the impairment of visual organs rather than from any parasite-induced lethargy. Slow response to moving objects almost certainly has significant implications for fish survival, as it directly relates to predator avoidance (Shirakashi and Goater 2001).

The study of Acanthocephala-gammarid amphipod systems has led to particularly interesting observations on the evolution of host-behaviour manipulation strategies in parasite and anti-parasite counter reactions in hosts. Cystacanths of *Pomphorhynchus laevis* induce a strong attraction to light and the water surface in the amphipod *Gammarus pulex* by directly or indirectly manipulating serotonin levels in its brain (Tain et al. 2006, 2007), which supposedly increases the infected individuals' vulnerability to predation. However, infected individuals of the sympatric *Gammarus roeselii* are not significantly more photophilic than their uninfected counterparts and do not show any significant serotonin response to parasite presence (Bauer et al. 2000; Tain et al. 2007). In contrast, both *G. pulex* and *G. roeselii* infected by *Polymorphus minutus* swim significantly more often at the surface and cling significantly more to floating objects than do their uninfected conspecifics; however, in that case again, the magnitude of behaviour alteration is stronger in *G. pulex* than in *G. roeselii* (Bauer et al. 2005). Importantly, *G. roeselii* is an invasive species (Jażdżewski 1980) whereas *G. pulex* is native to the study areas investigated in the aforementioned studies. Thus, the lack of evolved ability of *P. laevis* to alter *G. roeselii*'s behaviour has been interpreted as a sign of maladaptation of *P. laevis* to that host (Bauer et al. 2005; Tain et al. 2007). Conversely, the strong effects of *P. laevis* on *G. pulex* in the above study areas [but not in water bodies where *P. laevis* was recently introduced; see Kennedy et al. (1989) and Kennedy (1996)] hint at the evolution of hyperspecialised manipulative abilities (Bauer et al. 2005; Tain et al. 2007) in a fish parasite with reduced dispersion potential (Kennedy 1996) [but see alternative explanation in Tain et al. (2007) and Poulin et al. (2005)]. Behaviour alteration by *P. minutus* in both native and invasive hosts may come from that parasite's adoption of birds as definitive hosts: high dispersal range of the birds might put the parasite in contact with a broader range of intermediate hosts and favour the evolution of host-manipulation abilities for a broader range of gammarids (Bauer et al. 2005).

The above example illustrates the evolution in host and parasite populations of specific phenotypes called "adaptive traits". These adaptations represent the visible outcome of the natural selection pressure applied by each antagonist against the other (see Chap. 11). An evolutionary adaptation refers to 'a genetically determined feature that has become or is becoming prevalent in a population because it confers a selective advantage to its bearer through an improvement in some function' (Poulin 1995). Importantly, the resulting physical, physiological or behavioural modification cannot be a by-product of the adaptation process, but must result directly from selection pressure to be called 'adaptive' (Ridley 1993; Poulin 1994b). As such, even beneficial changes to host or parasite phenotypes could be side effects of parasite infestation, not adaptations (Poulin 1995). First, changes in a parasite's trait (or, by extension, changes in its host) are more likely to be genuinely adaptive if they evolved independently more than once in distantly related parasite lineages (Poulin 1995) sharing closely related hosts (Cézilly and Perrot-

Minnot 2005). Second, true evolutionary adaptation is more likely if the trait under scrutiny predictably corresponds to the most efficient way of performing a task, for example, infecting the next host in line (for the parasite) or resisting a parasite (for a host) (Poulin 1995). One of the best examples of a host–parasite system fulfilling this requirement is that of the digenean *Dicrocoelium dendriticum* inducing ant intermediate hosts to remain out of the nest at night, climb up to the top of grass blades, clamp their mandibles shut around the stems and tetanise until morning, and to do so repeatedly, for the parasite to increase the likelihood of encounter with the definitive sheep hosts (Hohorst and Graefe 1961; Anokhin 1966; De Bekker et al. 2018). Third, an adaptive modification must confer fitness benefits to the species (Poulin 1995). Fourth, modifications imposed on the host are more likely to be truly adaptive if they target not only one but several host traits at once (Cézilly and Perrot-Minnot 2005). For example, *P. laevis* induces a wide range of behavioural and physical modifications in the intermediate host, *G. pulex*: infected individuals are significantly less fecund (females) (Bollache et al. 2002), significantly more active (Dezfuli et al. 2003), asymmetrical (Alibert et al. 2002) and phototactic (Cézilly et al. 2000). The presence of these four characteristics above in a host–parasite system is usually hard to prove (Poulin 1995).

In parasite populations, adaptive changes can happen to counter hosts' phenotypical or behavioural innovations; they are often expressed through the hosts as extended parasitic phenotypic traits, meaning natural selection pressure on a parasite is expressed in the effects induced on the hosts (Dawkins 1982). In host populations, adaptive changes can happen to counter the loss of fitness induced by parasite infection, or in reaction to parasite threat. Possible examples of behavioural adaptations by the host could be the premature egg production by young individuals of the snail *Biomphalaria glabrata*, the first-intermediate host of the blood fluke *Schistosoma mansoni*, following non-infective exposure to that parasite in the wild in order to compensate for the perceived imminent risk of castration (Minchella and Loverde 1981), or preference for middle-sized cockles by oystercatchers in order to maximise energy intake while minimising infection risk when the birds are not restricted by the nutritiousness of the prey (Norris 1999; see Sect. 7.3).

Adaptive changes in host behaviour in response to parasitism can go as far as influencing the outcomes of sexual selection. In vertebrates, sexual selection is based on behavioural and physical adaptations, sometimes taken to extravagant levels of sophistication. Tentative hypotheses from several angles have been provided to explain female choice (e.g. Ryan 1990; Nowicki and Searcy 2004). One of the most interesting hypotheses regarding reasons for male display is based on host–parasite coevolution and sexual selection of genetically resistant hosts: in natural ecosystems where parasites are assumed to be ubiquitous, males displaying the most attractive physical characters are energetically able to do so because their parasite burden is minimal, implying that they possess alleles conferring resistance to long-term parasite infections, acquired as part of an evolutionary arms race (Hamilton and Zuk 1982). Parasite-induced mate selection may thus, in some species, drive evolution within species and host populations, although mate selection has been proven to operate independently from parasite presence in some cases (Aguilar et al. 2008).

7.4.1.4 Direct Parasite Competition or Synergy Through Co-Infection

Just like many predator species can hunt the same prey, many parasite species or strains frequently share the same host individual (Pedersen and Fenton 2007). For hosts, the impacts of direct interactions between parasites depend on parasite life history, strain, virulence and population density, as well as on individual host immunity (Woolhouse et al. 2015). For parasites, sharing a host can result in benefits for one or more of the species involved or, on the contrary, in active hindrance of the establishment of a species by another via various forms of direct competition [see Mideo (2009) for details] (Fig. 7.3). An example of interaction with benefits for pathogens is found in aquaculture. The open wounds inflicted on the rainbow trout *Oncorhynchus mykiss* by the fish louse *Argulus coregoni* significantly facilitate colonisation by *Flavobacterium columnare*, leading to increased fish mortality (Bandilla et al. 2006).

Contrary to beneficial interactions, direct competition can lead to unpredictable outcomes for the host, ranging from survival due to strong interference competition between parasite species to death from excessive parasite burden, damage inflicted by each parasite species (Johnson and Buller 2011) or immunosuppression (Cox 2001). An example of direct parasite competition with positive outcomes for the host is that of co-infections by entomopathogenic enterobacteria like *Photorhabdus asymbiotica* and *Xenorhabdus nematophila* in the caterpillar *Galleria mellonella* (Massey et al. 2004). In vitro cultures of mixed infections on agar plates showed mutual growth inhibition of bacterial species on each other, confirming their ability for mutual allelopathic interference.[1] When inoculated together in *G. mellonella*, both strains together were less virulent than each species would have been if inoculated alone, resulting in significantly decreased host mortality.

The effects of co-infection are often more nuanced, especially when metazoan parasites able to modulate host immune defences are involved (see Maizels and Yazdanbakhsh 2003; Maizels et al. 2004). Importantly, a strong antagonistic interaction between infectious agents does not necessarily restore host fitness, as each parasite species can inflict its own damage in addition to that incurred by the host due to competing pathogens. For example, co-infections of *Echinostoma trivolvis* and *Ribeiroia ondatrae* in the Pacific chorus frog *Pseudacris regilla* in Californian wetlands result in direct negative effects on each parasite's survival (Fig. 7.3), but also in more damage to the host population than either species could inflict on its own (Johnson and Buller 2011).

7.4.2 Indirect Effects on (Other) Species and on Ecosystems

The presence of parasites in ecosystems affects not only their hosts but also the species linked to their hosts through predation or resource competition. For example, the local extirpation of a species (i.e. a 'resource') by parasitism will affect all the other species routinely interacting with that resource. These ripple effects, which can also be induced in predator–prey and competitive interactions, are classified as density- (DMIE) or trait-mediated indirect effects (TMIE). Both can coexist in the same trophic web.

The nature and intensity of indirect effects depend on the mode of parasite transmission (i.e. direct or indirect), parasite life stage, intrinsic and parasite-induced rates of host mortality and the efficiency of host immunity or behavioural adaptation against that parasite (see below). The level of complexity and the unpredictability of these interactions in any given ecosystem grow significantly with the number of parasite species and interactions between hosts, non-hosts and the environment (Hochberg et al. 1990; Hatcher et al. 2006; Keesing et al. 2006).

7.4.2.1 Parasite-Induced Density-Mediated Indirect Effects (DMIE)

The term "population density-mediated indirect effect" refers to cases where induced variations in the population density of a species in a system impact the population densities of other species that do not directly interact with it, for example, by propagation through the food web (trophic cascades) (Abrams et al. 1996). Population density variations can be mediated by parasites in various ways (see also Chaps. 8 and 18).

Parasite-Induced Apparent Competition When a parasite is shared by two or more host species,

---

[1] "Allelopathic interferenc" refers to a widely reported type of interference competition in which a species inhibits the development of another through the release of harmful chemical components. 'Interference competition' is a type of direct competition in which a parasite's establishment and/or life-cycle completion are directly impeached by mechanical or chemical attacks from another pathogen (Mideo 2009).

these species have the potential to up or down-regulate each other's population densities via parasite-induced apparent competition (Fig. 7.4). For apparent competition to take place, one of the species must have some competitive advantage over the other(s) regarding tolerance to, or population recovery from, parasitism. Such competitive advantages can range from more efficient immune systems to comparatively faster growth in uninfected individuals (Holt 1977; Holt and Pickering 1985). Apparent competition, often difficult to disentangle from interspecies direct resource competition (see below), is most obvious in the laboratory experiment of Bonsall and Hassell (1997): when two non-competing caterpillar populations of pyralid moths *Plodia interpunctella* and *Ephestia kuehniella* are parasitised by the parasitoid wasp *Venturia canescens*, *E. kuehniella* is always eradicated (Fig. 7.4). This three-species system systematically fails to reach equilibrium even while *P. interpunctella*-wasp and *E. kuehniella*-wasp systems perdure individually. Similarly, the success of clones of *Plasmodium chabaudi* of differing virulence inoculated together in mice depends on the hosts' immunity: the avirulent strain is immunosuppressed via heterologous reactivity[2] in immunocompetent mice and least affected in immunodeficient mice, implying that immunity against the weakest clone could be mediated by host immunity (Råberg et al. 2006). No such examples are known from aquatic host–parasite systems. However, as many cases exist for terrestrial hosts, similar rules are expected to apply in freshwater- and marine environments.

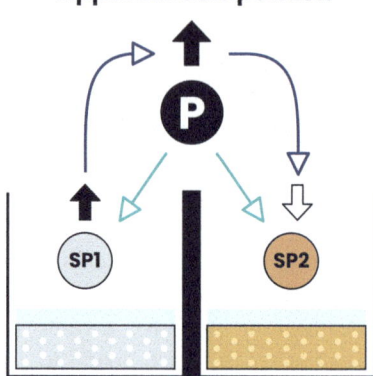

Fig. 7.4 The experiment of Bonsall and Hassell (1997) proves the existence of apparent competition. When two caterpillar populations of moths *Plodia interpunctella* ("SP1") and *Ephestia kuehniella* ("SP2"), feeding on independent unlimited resources (grey and orange substrates), are parasitised by the wasp *Venturia canescens* ("P" and green arrows), *E. kuehniella* (SP2) is always eradicated. Through its more efficient response to parasitism (black arrow), SP1 favours an increase in the population density of P (black arrow), which reduces the population density of SP2 (white arrow). The effect of SP1 on SP2 mediated by P is shown with long grey arrows

Parasite-Mediated Direct Competition Parasite-mediated direct competition has a profound influence on the respective population densities of both infected and non-infected species (Price et al. 1986). It can provoke the local extirpations of many native species populations, but also maintain balance between species that could not coexist otherwise (MacNeil and Dick 2011). This effect exists in contrast with apparent competition (see above) as host species or individuals actively compete for the same resource. In parasite-mediated direct competition, a parasite species can be shared by one (host-specific parasitism) or more host species. In the first case, depending on environmental conditions, the parasite can directly influence the fitness of its host and thus tip the scale in favour of or to the disadvantage of its non-infected competitors. For example, the native South African mussel *Perna perna* is easily outcompeted for space by the introduced mussel *Mytilus galloprovincialis*, potentially because the trematodes infecting the former significantly reduce its population density, hydration levels and growth while they do

---

[2] "Heterologous reactivity", also called heterologous- or cross-immunity, is a type of immune-cell cross-reactivity towards two or more antigens from different pathogen species or strains. In this phenomenon, a parasite species induces attacks against another species by triggering an immune response able to target both pathogens. Through this process, hosts can develop immunity towards many parasites sharing similar antigenic signatures if at least one of them triggers an effective immune response [see Agrawal (2019)]. On the contrary, a strong immune response against a single pathogen can be modulated by the presence of other parasites able to alter the host's immunity (Graham et al. 2005; Hardisty et al. 2022).

not infect the latter (Calvo-Ugarteburu and McQuaid 1998a, b) (Fig. 7.5). In the second case (i.e. that of a parasite shared by several host species), the parasite alters the resource exploitation efficiency of each host differently at the population level. The outcomes of parasite-mediated direct multi-host competitions depend on several factors intrinsic to both hosts and parasites, such as growth rate, magnitude of aggregation, or pathogenicity (Yan 1996). Parasite effects can be so strong as to suppress the competitive edge of more efficient, but more sensitive, species (see Chap. 8 for examples).

If the infected host species are predators of the same guild, parasite-mediated direct competition then becomes a case of parasite-mediated intraguild predation. For instance, in the British Isles, the crangonyctid amphipod *Crangonyx pseudogracilis* co-occurs significantly more often with two of its predators, the gammarid amphipods *Gammarus duebeni* and *G. pulex*, when they are parasitised by the microsporidian *Pleistophora mulleri* and the acanthocephalan *Polymorphus minutus* (see MacNeil and Dick 2011).

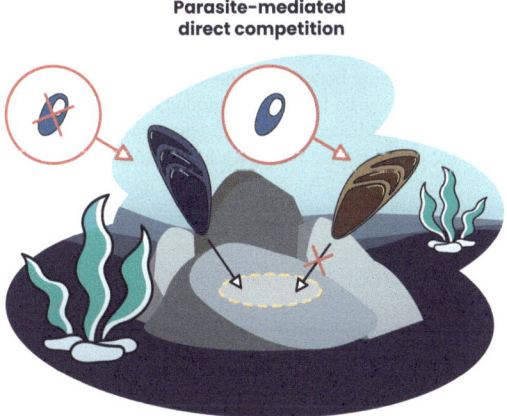

Fig. 7.5 An example of parasite-mediated direct resource competition. Here, competition for space (dashed circle) between the South African native mussel *Perna perna* (brown mussel), routinely affected by trematode larvae (white bubble), and the introduced mussel *Mytilus galloprovincialis* (black mussel), unaffected by trematodes (white bubble with crossed parasite), results in the competitive exclusion of *P. perna* by *M. galloprovincialis* (red cross on arrow)

Phenological and Population Synchrony The indirect effect of seasonal climatic fluctuations on parasite transmission can add up to the direct effects of climate on host populations, leading to synchrony in host abundance across those populations. An example is the spatial synchrony induced in populations of red grouse (*Lagopus lagopus*) by the terrestrial nematode *Trichostrongylus tenuis*, a one-host parasite with density-dependent transmission (Cattadori et al. 2005) and deleterious effects on brood production (Hudson 1986), when the effects of the parasite on grouse fertility and chick survival are enhanced by seasonal climatic conditions (Cattadori et al. 2005). A different phenomenon occurs in Californian aquatic systems (Fig. 7.6). Tadpoles of *P. regilla* at risk of developing limb malformations from infections by the frog flatworm *R. ondatrae* in the early stages of their development are significantly less likely to develop such malformations if they escape infection before or during the critical limb-development period (Johnson et al. 2011). Crucially, the risk of contracting infections starts rising in early spring in low-altitude ponds compared to mid-year in high-mountain ponds whereas tadpoles start developing at similar times in both habitats. At equal mean infection intensities, this difference results in 100 times more tadpoles malformed in earlier-warming ponds, where parasite- and host populations are most synchronised in their development and cercariae infect younger tadpoles, compared to later-warming ponds where most tadpoles have developed enough to avoid malformations by the time cercarial production takes off (McDevitt-Galles et al. 2020). This difference is particularly important in the context of climate change (see Chap. 22), for habitats might experience shifts in their temperature profiles that can synchronise parasite- and endangered-host populations, as with frogs (Yang and Rudolf 2010).

Parasite-Mediated Trophic Interactions Both predators and parasites consume other species and use the energy stolen from their 'prey' for their own survival and reproduction, leading to

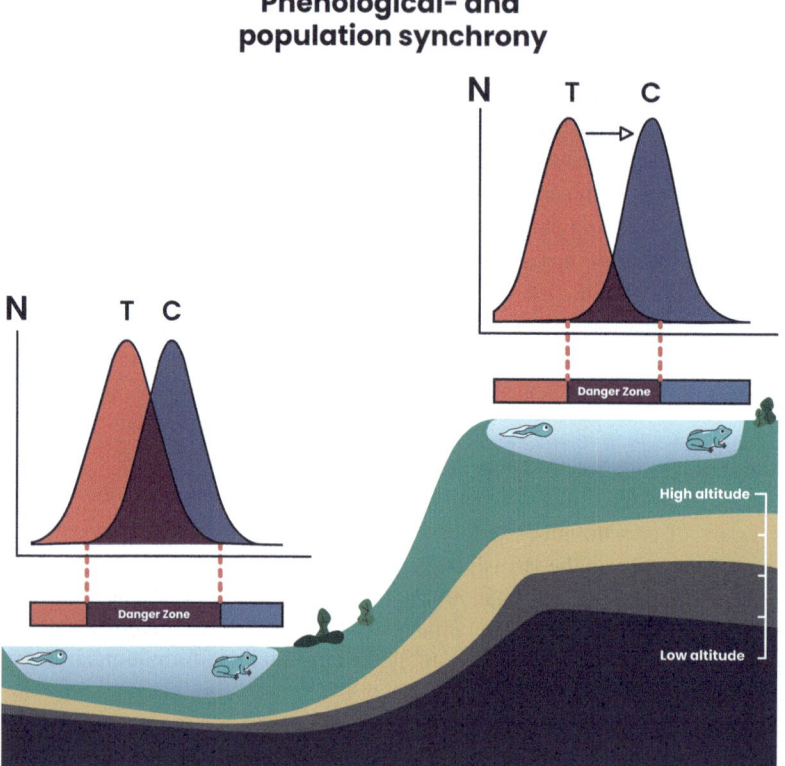

Fig. 7.6 An example of population synchrony adapted from McDevitt-Galles et al. (2020). Graphs represent population density curves of tadpoles of *Pseudacris regilla* and cercariae of the trematode *Ribeiroia ondatrae* in low- (bottom left) and high-altitude ponds (top right) (X-axis: time in a year; Y-axis: numbers of cercariae and tadpoles in each pond; C: cercariae; T: tadpoles). The purple area in each graph represents the area of synchrony between tadpole- and cercarial populations; the larger the area, the more tadpole and cercarial populations overlap and the higher the risk for tadpoles to encounter cercariae at any point during their development. Each purple area in the graphs is reported as a purple rectangle in the horizontal bar below; the longer that rectangle, the more likely tadpoles are to become infected before or during their critical window of limb development and to bear malformations as adults. Young tadpoles of *Pseudacris regilla* from high-altitude lakes (top graph and pond) avoid infection during their early lives and subsequent malformations due to a cold-induced delay in trematode emergence (arrow between F and C curves). In contrast, tadpoles of *P. regilla* from low-altitude lakes are more frequently infected because cercarial emergence occurs earlier in the year (population curves closer to each other). In low-altitude lakes, frog and trematode populations are more synchronised in their dynamics than in high-altitude lakes

an increase in consumer population (Hall et al. 2008) (but see Chap. 16). Just as in predation, consumption by parasites induces complex patterns of host population dynamics (Anderson and May 1979; May and Anderson 1979) with, in some cases, significant impacts on host population density (e.g. Hudson et al. 1998; Lafferty 2004) (Fig. 7.7). However, contrary to predators, each parasite individual can target only a single host individual per life stage. Moreover, whereas predation can only take three forms (micropredation, social predation and solitary predation) (Lafferty and Kuris 2002), host–parasite and host–parasitoid interactions are much more diverse (Hall et al. 2008). As every host–parasite system sits on a parasitism–predation gradient, each relationship will influence the surrounding ecosystem in a unique manner (Hall et al. 2008). Parasite-induced DMIEs on trophic interactions can take various forms. The hosts can be prey, predators (i.e. free-living carnivorous species) or both, in ecological assemblages incorporating many parasite species infecting animals at all trophic levels. The effects of infections sometimes cascade from predator level to resource level (Fig. 7.7).

Natural systems in which a parasite infects prey but not its predators result in highly diverse outcomes. Indeed, the effects of consumption of prey by both predators and parasites on prey, parasite and predator population dynamics differ depending on the system under study. The more species interact in that system, the more unpredictable the overall effects of parasitism will be (Hochberg et al. 1990; Banerji et al. 2015). In this context, the fact that a large part of the knowledge on parasite-induced DMIEs on prey populations has been obtained through mathematical modelling (e.g. Chattopadhyay and Arino 1999; Greenhalgh and Haque 2007) is problematic. The strength of the links between parasite and host population dynamics is hard to assess in real-world prey populations because of the presence of many confounding factors, like predation and co-occurring diseases. An example of such difficulty is seen in the *L. lagopus–T. tenuis* system. *Trichostrongylus tenuis* has a significant destabilising effect on red grouse populations in the UK and is at least partly responsible for marked cycles (Hudson et al. 1992, 1998, 2002). Additionally, red grouse are predated upon by, among others, the hen harrier (*Circus cyaneus*) (Thirgood et al. 2000), thus imposing a density-dependent limiting and stabilising effect on those populations (Redpath and Thirgood 1999; Thirgood et al. 2000). When culling of harriers was stopped in 1992 in Langholm Moor, Scotland, the red grouse population missed a cycle and kept declining. Although there is evi-

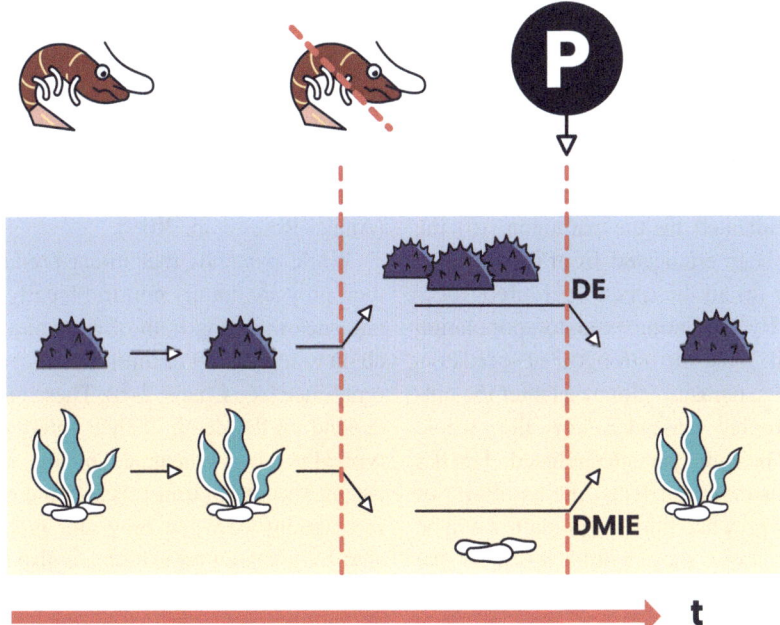

Fig. 7.7 An example of parasite-mediated trophic interactions on prey following the removal of natural predators, based on the study of Lafferty (2004). Lobsters (top) regulated urchin populations (middle) *via* predation, ensuring the stability of kelp populations (bottom) in the Channel Islands. The removal of lobsters through overfishing provoked urchin multiplication to the point that kelp forests became urchin barrens (three urchins, bare rocks). The emergence of disease (circled 'P' and downward arrow) in areas where urchin populations reached critical density levels became the main regulator of urchin populations in the absence of their natural predator, allowing kelp regrowth (one urchin, kelp). The direct effect of the parasite on urchin population density led to a density-dependent, parasite-induced density-mediated indirect effect (DMIE) on kelp forests

dence that combined effects of parasite-induced population cycles and increased hen harrier predation alone prevented the grouse population from recovering after its cyclic decline, other factors were identified as possibly being involved in the failed red grouse population recovery in the studied area (Hudson et al. 2002), those being louping ill virus (Reid et al. 1978; Hudson et al. 2002) and habitat loss (Thirgood et al. 2000). Unfortunately, no similar examples exist for aquatic ecosystems. The disappearance of predators and the subsequent rise in prey populations can itself be the source of increased density-dependent parasite-induced DMIEs on prey (see Chap. 8 for examples).

Parasite-mediated DMIEs on trophic webs tend to be strong but subtle, with complex outcomes. In contrast, predator-induced DMIEs are immediately visible, to the point that most known trophic cascades recorded in natural ecosystems are induced by predator pressure (Buck and Ripple 2017). When DMIEs induced by both parasites and predators are present in ecosystems, powerful prey- and resource-population responses can be observed (Fig. 7.8). Parasite–predator–prey systems in which a parasite infects a predator but not its prey are predicted to evolve into a state of equilibrium where (a) all predators have been eradicated by the pathogen, (b) the pathogen has been eradicated from the predator population, or (c) all the species perdure (Auger et al. 2009). By regulating predator population densities, and thus the strength of predator-induced DMIEs, parasites alone can alter the outcomes of trophic cascades for the whole ecosystem. Through parasite-induced DMIEs and direct consumptive effects, the equilibria of populations of prey and infected predators will be distinct from those they would reach in the absence of a parasite (Auger et al. 2009). A real-world illustration of this principle was observed in the Isle Royale National Park, USA. Following a canine parvovirus-induced crash of the wolf population in that area, the Western moose (*Alces alces*) population became primarily regulated, not by predation as before, but by resource availability. This change shifted the point of equilibrium of the moose population and rendered the latter vulnerable to changes in the Northern-Atlantic Oscillation (Wilmers et al. 2006, 2007).

When a parasite affects a predator, the strength of its top-down effects on the trophic web depends on two factors (Buck and Ripple 2017; Anaya-Rojas et al. 2019). First, the intensity of its top-down effects depends on how strongly this parasite regulates the predator's population density, which in turn depends on (1) the level of host resistance to parasitic invasion (Hudson et al. 1998); and (2) the extent of parasite-induced host trait alterations (see section below) (Anaya-Rojas et al. 2016). Second, a parasite's top-down effects depend on the trophic position of the predator it infects (Lafferty et al. 2006). Thus, parasite-induced DMIEs on the trophic chain are strongest when the parasite's direct impact is strongest (Anaya-Rojas et al. 2019). For example, mesocosm experiments on three-spined stickleback (*G. aculeatus*) infected by a natural combination of various parasites showed that the probability of stickleback-induced trophic cascades was significantly increased when stickleback parasite load was reduced. Moreover, fishes with natural parasite loads had much weaker direct and indirect effects on the zooplankton grazer population and the phytoplankton resource, respectively, than fishes with artificially reduced parasite loads (Anaya-Rojas et al. 2019).

While parasites that infect predators but not their prey are mainly non-trophically transmitted, parasites infecting both often exploit the trophic chain to reach their definitive hosts, in which they reproduce (see Chaps. 2–6). These parasites often depend on the death of their intermediate hosts, typically when consumed by the next host, to ensure successful transmission and reproduction. Interactions between prey and predators mediated by a shared parasite have also been mathematically modelled, allowing for the prediction of complex population dynamics with a wide variety of possible outcomes (e.g. Hsieh and Hsiao 2008). Importantly, the outcome of a trophically transmitted infection for host populations, predator or prey, is heavily dependent on the parasite's nature and life stage. Thus, although digenean intermediate-stage larvae are produced more numerously and longer-lived in healthy

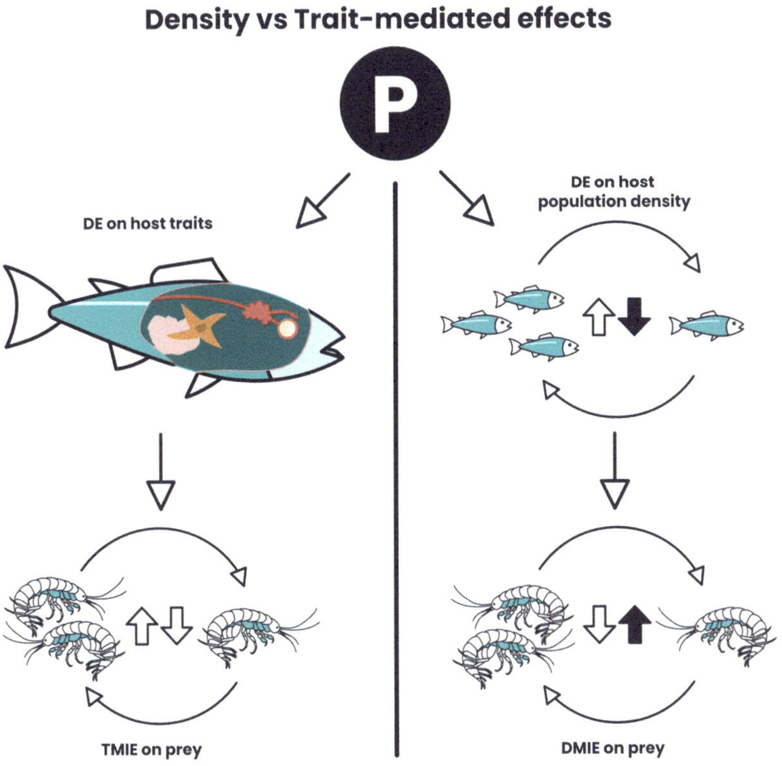

Fig. 7.8 Difference between trait- (TMIE) and density-mediated indirect effects (DMIE). (Left) Direct parasite (circled "P") effects on a species' (fish) behaviour (symbolised by brain, brainstem and eye), immunity (symbolised by a dendritic cell), other organs (symbolised by muscle tissues) or life history induces TMIEs impacting prey (amphipods) population density (thick white arrows). These TMIEs can in turn trigger more TMIEs in the ecosystem. (Right) Direct parasite effects on a species' (fish) population density induce population increases (white arrow) or decreases (black arrow) in that species, which decrease (white arrow) or increase (black arrow) prey populations, respectively, via DMIEs

first intermediate hosts (Seppälä et al. 2008), parasite-induced adaptive phenotypic changes (see section below) in digenean second-and-above intermediate hosts often tend to increase the likelihood of death by predation (e.g. Lafferty and Morris 1996; Seppälä et al. 2004; Seppälä et al. 2005; reviewed in Lafferty and Shaw 2013).

DMIEs can be manifested by all types of parasites but are strongest when parasites kill their hosts (Buck and Ripple 2017). Thus, the outcomes of both effects on the ecosystem depend on the parasite lineage and life stage (see Sect. 7.4.1). A widely known example of a parasite inducing strong consumptive and indirect effects is chytrid fungus (*Batrachochytrium dendrobatidis*) (see Chap. 4). Where chytridiomycosis outbreaks collapsed frog populations in Rio Guabal, Panama (i.e. direct consumptive effect), the absence of tadpoles resulted in a sharp increase in chlorophyll *a* and inorganic matter, and a drastic shift in the periphyton community composition from small diatom-dominated to larger diatom- and cyanobacteria-dominated (i.e. parasite-induced DMIE on the trophic web) (Connelly et al. 2008; Chap. 4).

7.4.2.2 Parasite-Induced Trait-Mediated Indirect Effects (TMIE) or Interactions (TMII)

The term "trait-mediated indirect effect" (TMIE) is used when alterations of a species' behaviour (see Sect. 7.4.1), phenotype or life history, because of the presence of another species, impact the population densities of still other species that do not directly interact with it (Abrams 1995; Werner and Peacor 2003, also see Chap. 8) (Fig. 7.8). The latter species become impacted by the propagation of TMIEs through the food web via predation or competition (Abrams 1995). TMIEs are understood to be at least as strong as DMIEs (Werner and Peacor 2003).

Predator-induced TMIEs stem from prey modifying their behaviour or physiology in order to remain alive. In contrast, most parasite-mediated TMIEs on the ecosystems happen when host populations have already been infected, that is, 'consumed' (Fig. 7.8). Thus, parasite-induced TMIEs can arise from both consumptive and non-consumptive effects depending on whether a host has been successfully infected (Buck 2019). An example of consumptive TMIE, the most frequent type of parasite-induced TMIE (Buck and Ripple 2017), is described by Toscano et al. (2014): crabs of *Eurypanopeus depressus* conspicuously infected by the castrating rhizocephalan barnacle *Loxothylacus panopaei* consumed significantly fewer mussels (*Brachidontes exustus*) than uninfected or inconspicuously-infected crabs, partly because they reacted significantly more slowly to the introduction of mussel prey and because conspicuously infected crabs seemed to become more inactive. The mussel population is thus indirectly affected by the rhizocephalan parasites. Although seemingly less common than in predator-prey dynamics (Buck and Ripple 2017), parasite-induced non-consumptive TMIEs can be observed when potential hosts modify their habits or phenotype to avoid infection, which in turn affects other species in their environment (Koprivnikar et al. 2021). Preventative self-defence, widely present in predator-prey interactions (e.g. Peacor et al. 2020), is termed risk-induced trait response (RITR). In host–parasite interactions, RITRs are expressed through five strategies that may incur associated fitness costs (Rigby et al. 2002; Daversa et al. 2021): (1) avoidance of diseased conspecifics (as seen in lobsters; Behringer et al. 2006); (2) active avoidance of infective parasites upon detection (e.g. tadpoles swimming explosively to avoid cercariae; Taylor et al. 2004); (3) avoidance of places or items where parasites tend to become concentrated (see Hutchings et al. 2000); (4) the use of specific counterattack behaviours when those items, or the parasites, cannot be avoided (e.g. fishes soliciting the services of cleaners against gnathiid isopods, parasitic copepods and monogeneans; Grutter 1999; Becker and Grutter 2004); and (5) the development of immune resistance (Råberg et al. 2009). Importantly, parasite- and predator-induced indirect effects are not mutually exclusive and can act on the same trophic chain (see Banerji et al. 2015).

Through prey, parasites can trigger bottom-up (from prey on predator) or top-down (from prey on resource) TMIEs at the same time in the same trophic chain. Whether more than one TMIE occurs in any predator-prey-resource system depends on the diversity of direct effects a parasite can induce in its prey host. For example, when moderately infected by the bacterium *Holospora undulata*, individuals of the ciliate *Paramecium caudatum* swim faster and feed more often than non-infected ones, thus impacting the primary resource more strongly; this impact reflects parasite-induced top-down TMIEs on the resource. In addition, a parasitised *P. caudatum* population induces lower peak population densities and higher mortality rates in the ciliate *Didinium nasutum*, the predator of *P. caudatum*. This population decline can reflect either a parasite-induced bottom-up DMIE on the predator mediated by higher prey mortality, or a parasite-induced TMIE on the predator mediated by the compromised nutritional value of the infected prey (Banerji et al. 2015).

Both TMIEs and DMIEs can also affect the traits or population densities of species outside of the direct trophic chain of the parasitised host. This type of interspecies link, first described in plant-insect systems and not usually included in trophic web models, is termed an "indirect interaction web" (Utsumi et al. 2010). An example in aquatic systems is the link between infestation of the cricket *Nemobius sylvestris* by parasitic larvae of the hairworm *Paragordius tricuspidatus* on land and benthic algal density in fresh water. Larval nematomorphs nearing maturity manipulate their cricket hosts' behaviour to jump into water, where they exit the host, mature and become free-living adults (Thomas et al. 2002). In season, the water-trapped crickets form most of the diet of Kirikuchi char (*Salvelinus leucomaenis*) (Sato et al. 2011). Here, "consumption" of *N. sylvestris* by *P. tricuspidatus* induces *S. leucomaenis* to temporarily shift its diet, thus amounting to a parasite-induced TMIE on the fish mediated by behavioural change in the cricket. In turn, preferential feeding on crickets by the fish reduces predation on invertebrate grazers, thus reducing benthic macroalgae density via predator-induced DMIE on the resource mediated by grazers (Sato et al. 2012).

When it acts on pathogens, host immunity amounts to predation in terms of trophic level (Pedersen and Fenton 2007). Unlike predation, however, both host immunity and host resource (i.e. energy) are contained within the ecosystem of the host's body (i.e. the 'inner' ecosystem), so that the impairment of one results in the impairment of the other (Johnson and Buller 2011). In turn, any failure of immunity or energy depletion can affect host behaviour and mortality, and thus the entire ecosystem. Therefore, host immunity is a major way through which parasite-induced TMIEs act on ecosystems at large. This is especially true in cases of co-infection. As parasite interactions with each other and their use of the host resource are mediated in part by host immunity, parasites induce TMIEs on each other and the host as well as direct effects [see Sect. 7.4.1 and review by Johnson and Buller (2011)]. Trait-mediated apparent competition between parasite species (top-down effect) takes place when one of the parasites stimulates host immune defences, hampering the growth and reproduction of the others through immune cross-reactivity (Christensen et al. 1987; Balmer et al. 2009) or cross-immunity between pathogens with similar antigenic signatures (see Curry et al. 1995). In contrast, host immune alteration induced by one of the parasites can also incidentally benefit co-infecting agents (Cattadori et al. 2008; Su et al. 2005). Thus, parasite assemblages in hosts tend to be non-random and not merely the consequence of parasite accumulation over time (Thomas et al. 1997; Dezfuli et al. 2000; Poulin and Valtonen 2001). In fact, parasites that are able to modify host traits can condition the successful invasion of a chain of other parasite species extending in both space (i.e. infection sites) and time (i.e. sequential infections throughout a host's life) via parasite-induced TMIEs on other parasites mediated by changes in host traits (Christensen et al. 1987; Poulin and Valtonen 2001; Jackson et al. 2006; Karvonen et al. 2019). Some parasites that are unable to manipulate those traits even depend on these associations to complete their life cycles. For example, the digenean *Maritrema subdolum* infecting the amphipod *Gammarus insensibilis*, its second intermediate host, benefits from coinfection with *Microphallus papillorobustus* as this parasite induces the amphipod to swim closer to the surface in order to increase transmission probability whereas *M. subdolum* cannot (Thomas et al. 1997). These TMIEs can in turn trigger more TMIEs in the other species in the ecosystem. The outcomes of TMIEs on both parasites and hosts depend on a wide variety of biotic (i.e. intrinsic to the pathogens, the hosts or the surrounding free-living species interacting with the host) and abiotic factors (reviewed by Herczeg et al. 2021). For that reason, synergistic relationships between parasites can be tricky to ascertain and difficult to disentangle from individual-host immunity and environmental variations (Karvonen et al. 2009).

7.4.2.3 Parasite-Induced Indirect Alteration of Habitat

In aquatic ecosystems, the influence of parasites extends not only to their hosts but also, indirectly, to the whole community via TMIEs. By altering their hosts' ability to perform their usual ecosystem functions through normal behaviour, parasites indirectly alter the characteristics of the surrounding physical habitat, influencing the living conditions of the other species and leading to profound changes in community composition and function (Mouritsen and Poulin 2010). Such influence on physical habitat characteristics is most visible in the case of the himasthlid trematode *Curtuteria australis* infecting the foot of the cockle *Austrovenus stutchburyi*, an ecosystem engineer of muddy substrates in New Zealand (see Chap. 8 for details). The influence of parasite presence on habitat and community structure is on par with those of more studied interspecific interactions such as competition and predation, especially when parasites affect keystone species. Parasites' effects are so important for ecosystems some authors consider them as ecosystem engineers in their own right (Hatcher et al. 2012).

7.5 Parasites Are Affected by Ecosystem Processes

7.5.1 A Parasite's Fate Is Linked to Those of Its Hosts

Parasites, like all other species in natural ecosystems, are affected by the processes arising from species and habitats interacting with each other. If we consider parasitism as a form of predation (see Hall et al. 2008 but see also Chap. 16), then the loss of a host species to a parasite amounts to the loss of a prey species to a predator. However, given parasitism is a long-term relationship, the "consumption" of the host by the parasite happens over a much longer time. Thus, and contrary to free-living species, parasites can be positively or negatively affected by what happens to them directly [e.g. predation on free-living stages or on ectoparasites; see Sect. 7.5.2 and also Rohr et al. (2008)], what happens to uninfected host populations [e.g. in case of local extirpation or extinction of a host species, thus preventing life-cycle completion; see below and Sures et al. (2023)], and what happens to the individual host while the parasite is consuming it [see Raffel et al. (2008) on cercarial encystment].

Broadly speaking, the more stressed an ecosystem is, the more affected parasite biodiversity is and *vice versa* (Huspeni and Lafferty 2004; Pérez-del Olmo et al. 2007). However, predicting how a parasite species will react to any specific environmental stressor applied to it or its hosts is challenging. The effects of stressors greatly depend on the life-cycle characteristics of each parasite species (e.g. heteroxenous or monoxenous, internal or external stages, trophic or non-trophic transmission, level of host-specificity, number of host species used); in particular, a parasite species' survival in a stressed environment is determined by the resilience of its most sensitive life stage (Sures et al. 2023). Parasites with indirect life cycles (i.e. heteroxenous) are thought to be more at risk than monoxenous ones (Wood and Lafferty 2015). Nevertheless, individual parasite groups exhibit varying sensitivities, even among monoxenous species (see Lafferty 1997), resulting in a complex scenario. For example, contrasting results have been obtained in studies on gyrodactylid monogeneans (Poléo et al. 2004; Pravdová et al. 2023) depending on the type of environmental damage. In addition, a parasite's resilience depends heavily on the tolerance level of each of its hosts towards environmental stressors (Lafferty and Kuris 2009a). As any species on Earth can have one or more symbionts (including parasites), parasites might be the group disappearing at the highest rate through mechanisms of co-extinction and extinction cascades (Koh et al. 2004; Dunn et al. 2009; Lafferty 2012) as well as the group for which extinctions might be the most underreported (Dunn et al. 2009). As parasites are seen as broadly undesirable by the wider community despite their vital importance (Stork and Lyal 1993; Dougherty et al. 2016), conservation efforts on these species are rare (see Chap. 13).

The threats to aquatic ecosystems are many, ranging from water pollution (Chap. 20) to the impacts of climate change (Chap. 22). When environmental damage leads to a significant population decrease of one link in a parasite's host chain, then that parasite may become extinct. Several cases have illustrated this point, such as that of the nematode *Cystidicola stigmatura*, pronounced locally extinct following a crash in the abundance of its definitive host, the lake trout *Salvelinus namaycush* in the North American Great Lakes (Black 1983). Furthermore, if host density decreases below a sustainable transmission threshold in an area where all of a parasite's life-stages are concentrated, then that parasite will also be at risk even if the host population is not itself threatened with extirpation (Lafferty and Kuris 1999). As such, a parasite is predicted to disappear before any of its hosts (De Castro and Bolker 2005). Parasite survival and host population fluctuations are so intricately linked that parasites can be used in wild commercial fish stock monitoring (Williams et al. 1992; Marcogliese 2002; Chap. 17).

7.5.2 Parasites as Prey for Incompatible Hosts

Parasites are commonly consumed as prey at both intermediate and adult stages by incompatible hosts that arrest trophic transmission (Johnson et al. 2010). Parasites can be detrimentally consumed in four different ways, namely, through concomitant predation, targeted predation on free adult or larval stages, grooming and hyperparasitism (Fig. 7.9) (reviewed by Johnson et al. 2010).

7.5.2.1 Concomitant Predation

Adaptive changes of parasites in response to an evolutionary arms race with their hosts can, in turn, induce profound changes in host phenotype (see Sect. 7.4.2). Such changes, however, are not always to the advantage of the parasite. Indeed, a behaviourally impaired, weakened or more conspicuous host is rendered vulnerable, not only to the next compatible host species in line, but also to the rest of the animal community, thus leading to concomitant predation and subsequent reduction in parasite transmission. A typical example of this kind of parasite consumption is that of trematode parasitism in the cockle *A. stutchburyi*. Although the definitive hosts of the trematode are birds, cockles stuck on the sediment surface are predated upon by fishes significantly more often. As the great majority of metacercariae are consumed by unsuitable hosts or lost to the environment (Mouritsen and Poulin 2003), the value of metacercarial accumulation in the cockle foot as an evolutionary adaptation facilitating transmission has been questioned (see Chap. 8).

7.5.2.2 Grooming

Whole groups of animals specialise in feeding on parasites already infecting their hosts. One of the most representative examples of this phenomenon is predation on gnathiid isopods on coral reefs by obligate cleaner species like cleaner wrasses (*Labroides* spp.) and shrimps [*Lysmata* spp. (Grutter 1999; Becker and Grutter 2004)] and facultative consumers like species of *Diproctacanthus* and *Thalassoma* (Labridae), *Saurogobio* (Cyprinidae) and *Gramma* (Grammatidae) (Randall 1967; Grutter and Feeney 2016). Interspecies grooming in reaction to parasitism has conditioned the adaptive behavioural complexification of a wide range of species towards investing time in queueing to get cleaned rather than in feeding or mating (Grutter 1995, 1996; Bshary and Grutter 2002). This phenomenon has important short- and long-term implications for parasite burdens both in the environment and on the host. As both host- and non-host species benefit from parasite removal (Bshary 2003; Waldie et al. 2011), the presence of cleaners regulating parasite populations is critical for maintaining free-living species biodiversity and abundance on coral reefs (Grutter et al. 2003).

7.5.2.3 Predation on Free-Living Life Stages

In some ecosystems, free-living stages of parasites (i.e. eggs, larvae, juveniles or adults) are purposely or incidentally consumed by predators (Thieltges et al. 2008a, b). For example, cercariae of the marine trematode *Himasthla elongata* targeting the keystone cockle *C. edule* as second-intermediate host are hunted by crabs, accidentally ingested by other compatible hosts not consumed by the definitive hosts (thus not allowing life-cycle completion), or incidentally consumed as planktonic food by incompatible filter-feeders. The dilution of cercarial output between all these non-target species significantly reduces metacercarial infection intensity in the whole cockle population (Thieltges et al. 2008a). In coral reef ecosystems, the free-living adults of gnathiid isopods are targeted by nocturnal predatory soldierfishes of the genera *Myripristis*, *Holocentrus* and *Sargocentron*. As these fishes are more numerous than specialist cleaner gobies (*Elacatinus evelynae*) in their

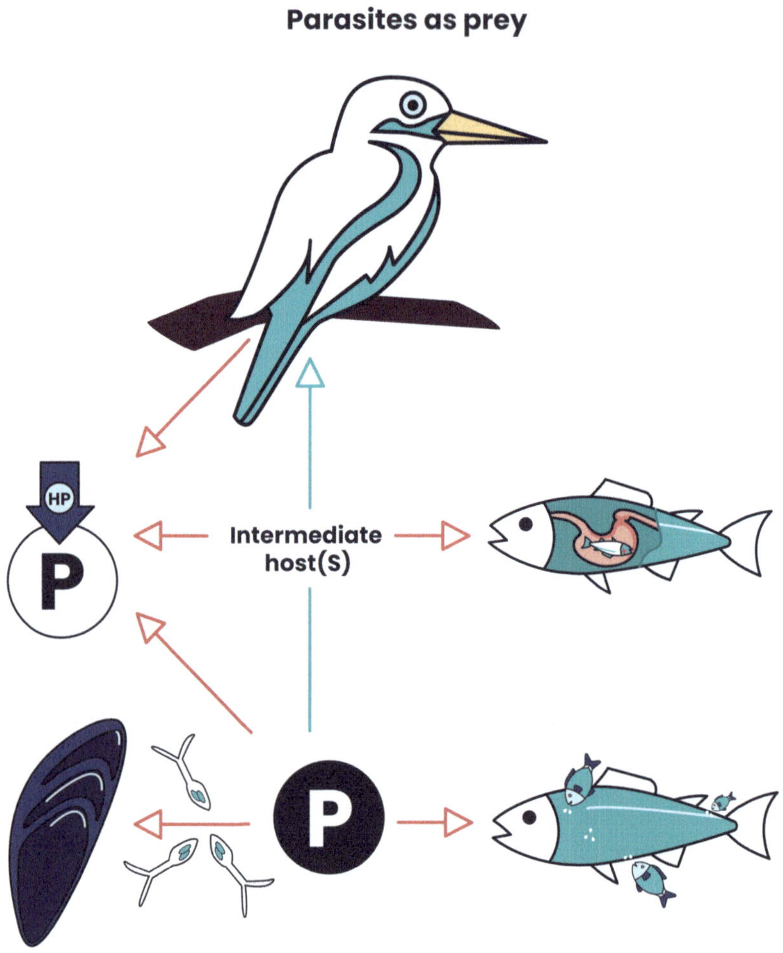

Parasites as prey

Fig. 7.9 An illustration of how parasites can become prey for incompatible hosts. A parasite individual (black-circled 'P') can fail to reach and reproduce in its final host (bird) in various ways (red arrows). (Bottom right) It can be directly eliminated from the intermediate (or definitive) host through consumption by cleaner species. (Bottom left) Its infective swimming stages (cercariae) can be consumed incidentally by filter feeders (mussel) or intentionally by other predators. (Top left) It can be killed by a hyperparasite at any stage of its development ('HP' in thick arrow). (Top right) It can be incidentally consumed by a predator along with its host. The green arrow symbolises the only pathway through which a parasite will successfully reach and reproduce in the definitive host

studied system, Artim et al. (2017) hypothesise that these facultative predators significantly affect larval gnathiid populations at ecosystem level. For incompatible predators targeting them, parasites are valuable prey (see Schultz and Koprivnikar 2019) with high reproductive output (Lafferty et al. 2006; Kuris et al. 2008). Thus, the consumption of parasitic free-living stages decreases the predation burden on other prey (Schultz and Koprivnikar 2019), triggering TMIEs and DMIEs on trophic chains (Lafferty et al. 2006).

7.5.2.4 Hyperparasitism

Hyperparasitism refers to the consumption of parasites by other parasites. One of the most studied cases of hyperparasitism is that of bacteriophages (Węgrzyn 2022) infesting pathogenic bacteria (reviewed by Ye et al. 2019). However, hyperparasitism encompasses many more interaction types often involving metazoan organisms. In aquatic systems, ciliates of the subclass Peritrichia (phylum Ciliophora) infect parasitic crustaceans (reviewed by van As 2019), while parasitic nematomorph larvae parasitise trematode rediae in their snail hosts (Hanelt 2009). In the marine environment, the cryptoniscid isopod *Liriopsis pygmaea* attaches to the rhizocephalan barnacle *Briarosaccus callosus*, which in turn infects the false king crab *Paralomis granulosa* (Peresan and Roccatagliata 2005); many more cryptoniscid genera are known to parasitise rhizocephalans (van As 2019). Aquatic microsporidians (Fungi) and myxosporeans (Cnidaria: Myxozoa) also include several hyperparasitic species. For instance, microsporidians of the genus *Unikaryon* parasitise species of *Microphallus* Ward, 1901 (Digenea: Microphallidae), the trematodes whose metacercariae encyst in the shrimp *Panopeus herbstii* (Sokolova et al. 2021); and the myxozoan *Myxidium giardi* parasitises the monogenean *Pseudodactylogyrus bini* which in turn infects the European eel *Anguilla anguilla* (Aguilar et al. 2004). Hyperparasitism contributes significant direct and indirect effects on parasite- and host populations, respectively, regulating the proliferation of the former and favouring the growth of the latter (Gleason et al. 2014). This phenomenon is also predicted to drive the evolution of both hyperparasite and parasite virulence (Parratt and Laine 2016; Northrup et al. 2024), with potentially significant consequences for host–parasite relationships (Springer et al. 2013).

7.5.3 High Biodiversity Favours Parasite Dilution Effects

A host–parasite relationship never exists in a void. It is surrounded by a multitude of other species that enact a wide range of direct and indirect effects on both the host and the parasite. In many ecosystems, those species that are not part of the parasite's life cycle can hinder transmission to definitive hosts and the spread of a disease in a host population. This phenomenon is called a "dilution effect" (Ostfeld and Keesing 2012; see also Chap. 18).

Dilution effects are enabled by dead-end predation on free-living stages, competition dynamics between compatible and incompatible hosts, and downregulation of compatible host populations via predation (Hall et al. 2009). In addition, the more the obligate hosts are regulated by other incompatible species, the more likely the dilution effects are (Ostfeld and Keesing 2012). For example, the targeted removal of sick and weakened hosts by predator populations (e.g. Genovart et al. 2010) contributes to the health of the host population through concomitant parasite predation (see Sect. 7.5.2) (Packer et al. 2003). For example, freshwater crayfish (*Faxonius limosus*) and water scorpions (*Nepa cinerea*) catch significantly more individuals of *G. pulex* infected by the acanthocephalan *P. laevis* than they do uninfected ones, destroying those parasites and preventing their transmission to the fish definitive hosts (Kaldonski et al. 2008). Thus, the higher the biodiversity of an ecosystem is, the less likely it is that parasites will be able to reach their definitive hosts. The inverse is also true: in prey populations mainly regulated by disease outbreaks, factors negatively affecting predator populations could also favour parasite transmission (Packer et al. 2003). The lack of dilution effects has been most apparent in recent decades as human disease emergence has intensified together with habitat- and biodiversity loss (Civitello et al. 2015). Indeed, reservoir-species of pathogens shared by both humans and wild animals, notably rats, bats and birds, are more present and abundant in habitats heavily damaged by humans (Gibb et al. 2020).

Natural ecosystems are delicate, intricate webs. Dilution effects can appear or disappear, not only because of the removal of predators but also *via* the introduction of invasive competitors. An example of such is the sharp decline in infections of native St Lawrence River fishes by trematodes of the genus *Diplostomum* following the introduction of the Eurasian round goby, *Neogobius melanostomus*. This invasive species is thought to act as a dead-end host for the trematodes, thereby reducing the numbers of cercariae entering the intermediate hosts that would allow infection of the definitive host, the ring-billed gull *Larus delawarensis*. In addition, a shift in the gull's diet towards invasive gobies may have reduced predation on native fishes and thus trematode transmission (Gendron and Marcogliese 2017).

7.6 Conclusion and Future Direction

The intricate interactions between aquatic parasites, their hosts and the ecosystems they inhabit reveal a complex web of ecological dynamics. Aquatic parasites play multifaceted roles, influencing individual host health, population dynamics, biodiversity and overall ecosystem health. These roles often manifest in both direct and indirect ways, shaping ecosystems in subtle yet profound manners.

The exploration of terrestrial examples within this chapter highlights the paucity of comparable data in aquatic environments. While terrestrial ecosystems provide numerous case studies of parasite-induced ecological interactions, aquatic systems remain underexplored. This disparity underscores the critical need for targeted research in aquatic environments to uncover the processes and interactions analogous to terrestrial systems.

Future research should prioritise filling this knowledge gap by investigating the specific ways in which aquatic parasites influence their ecosystems. This includes studying parasite-induced TMIEs and DMIEs within aquatic food webs, as well as the broader ecological consequences of parasite-host dynamics in aquatic settings. Understanding these interactions is essential for developing comprehensive models of aquatic ecosystem health and resilience. Additionally, future studies should focus on the role of parasites as ecosystem engineers in aquatic environments. Given their potential to alter ecosystem characteristics and community structures, understanding how parasites interact with key species in aquatic ecosystems will provide deeper insights into their ecological significance. By advancing research in these areas, it will be possible to develop a more holistic understanding of aquatic parasitology, contributing to the conservation and management of aquatic ecosystems at large. The insights gained will not only enhance scientific knowledge but also inform practical approaches to maintaining ecosystem health in the face of environmental changes and challenges.

References

Abrams PA (1995) Implications of dynamically variable traits for identifying, classifying, and measuring direct and indirect effects in ecological communities. Am Nat 146(1):112–134

Abrams PA, Menge BA, Mittelbach GG et al (1996) The role of indirect effects in food webs. In: Polis GA, Winemiller KO (eds) Food webs: integration of patterns & dynamics. Springer, Boston, pp 371–395

Adamo SA (2002) Modulating the modulators: parasites, neuromodulators and host behavioral change. Brain Behav Evol 60(6):370–377

Agrawal B (2019) Heterologous immunity: role in natural and vaccine-induced resistance to infections. Front Immunol 10:2631

Aguilar A, Aragort W, Álvarez M et al (2004) Hyperparasitism by *Myxidium giardi* Cépède 1906 (Myxozoa: Myxosporea) in *Pseudodactylogyrus bini* (Kikuchi, 1929) Gussev, 1965 (Monogenea: Dactylogyridae), a parasite of the European eel *Anguilla anguilla* L. Bull Eur Assoc Fish Pathol 24:287–292

Aguilar TM, Maia R, Santos ESA et al (2008) Parasite levels in blue-black grassquits correlate with male displays but not female mate preference. Behav Ecol 19(2):292–301. https://doi.org/10.1093/beheco/arm130

Aldana M, Pulgar J, Hernández B et al (2020) Context-dependence in parasite effects on keyhole limpets. Mar Environ Res 157:104923

Alibert P, Bollache L, Corberant D et al (2002) Parasitic infection and developmental stability: fluctuating asymmetry in *Gammarus pulex* infected with two acanthocephalan species. J Parasitol 88(1):47–54

Anaya-Rojas JM, Brunner FS, Sommer N et al (2016) The association of feeding behaviour with the resistance and tolerance to parasites in recently diverged sticklebacks. J Evol Biol 29(11):2157–2167

Anaya-Rojas JM, Best RJ, Brunner FS et al (2019) An experimental test of how parasites of predators can influence trophic cascades and ecosystem functioning. Ecology 100(8):e02744

Anderson RM, May RM (1979) Population biology of infectious diseases: part I. Nature 280(5721):361–367

Anokhin IA (1966) Daily rhythm in ants infected with metacercariae of *Dicrocoelium lanceatum*. Dokl Akad Nauk SSSR 166:757–759

Antia R, Levin BR, May RM (1994) Within-host population dynamics and the evolution and maintenance of microparasite virulence. Am Nat 144(3):457–472

Artim JM, Hook A, Grippo RS et al (2017) Predation on parasitic gnathiid isopods on coral reefs: a comparison of Caribbean cleaning gobies with non-cleaning microcarnivores. Coral Reefs 36:1213–1223

Auger P, Mchich R, Chowdhury T et al (2009) Effects of a disease affecting a predator on the dynamics of a predator–prey system. J Theor Biol 258(3):344–351

Azevedo JS, Silva LG, Bizerri CRSF et al (2006) Infestation pattern and parasitic castration of the crustacean *Riggia paranensis* (Crustacea: Cymothoidea) on the fresh water fish *Cyphocharax gilbert* (Teleostei: Curimatidae). Neotrop Ichthyol 4:363–369

Balmer O, Stearns SC, Schötzau A et al (2009) Intraspecific competition between co-infecting parasite strains enhances host survival in African trypanosomes. Ecology 90(12):3367–3378

Bandilla M, Valtonen E, Suomalainen L-R et al (2006) A link between ectoparasite infection and susceptibility to bacterial disease in rainbow trout. Int J Parasitol 36(9):987–991

Banerji A, Duncan AB, Griffin JS et al (2015) Density- and trait-mediated effects of a parasite and a predator in a tri-trophic food web. J Anim Ecol 84(3):723–733

Bauer A, Trouvé S, Grégoire A et al (2000) Differential influence of *Pomphorhynchus laevis* (Acanthocephala) on the behaviour of native and invader gammarid species. Int J Parasitol 30(14):1453–1457

Bauer A, Haine ER, Perrot-Minnot M-J et al (2005) The acanthocephalan parasite *Polymorphus minutus* alters the geotactic and clinging behaviours of two sympatric amphipod hosts: the native *Gammarus pulex* and the invasive *Gammarus roeseli*. J Zool 267(1):39–43

Beckage N (1993) Endocrine and neuroendocrine host-parasite relationships. Receptor 3(3):233–245

Becker JH, Grutter AS (2004) Cleaner shrimp do clean. Coral Reefs 23(4):515–520. https://doi.org/10.1007/s00338-004-0429-3

Behringer DC, Butler MJ, Shields JD (2006) Avoidance of disease by social lobsters. Nature 441(7092):421–421

Black G (1983) Taxonomy of a swimbladder nematode, *Cystidicola stigmatura* (Leidy), and evidence of its decline in the Great Lakes. Can J Fish Aquat Sci 40(5):643–647. https://doi.org/10.1139/f83-085

Blower S, Roughgarden J (1988) Parasitic castration: host species preferences, size-selectivity and spatial heterogeneity. Oecologia 75:512–515

Bollache L, Rigaud T, Cézilly F (2002) Effects of two acanthocephalan parasites on the fecundity and pairing status of female *Gammarus pulex* (Crustacea: Amphipoda). J Invertebr Pathol 79(2):102–110

Bonsall MB, Hassell MP (1997) Apparent competition structures ecological assemblages. Nature 388(6640):371–373. https://doi.org/10.1038/41084

Bshary R (2003) The cleaner wrasse, *Labroides dimidiatus* is a key organism for reef fish diversity at Ras Mohammed National Park, Egypt. J Anim Ecol 72(1):169–176

Bshary R, Grutter AS (2002) Experimental evidence that partner choice is a driving force in the payoff distribution among cooperators or mutualists: the cleaner fish case. Ecol Lett 5(1):130–136

Buck JC (2019) Indirect effects explain the role of parasites in ecosystems. Trends Parasitol 35(10):835–847

Buck JC, Ripple WJ (2017) Infectious agents trigger trophic cascades. Trends Ecol Evol 32(9):681–694. https://doi.org/10.1016/j.tree.2017.06.009

Calvo-Ugarteburu G, McQuaid C (1998a) Parasitism and introduced species: epidemiology of trematodes in the intertidal mussels *Perna perna* and *Mytilus galloprovincialis*. J Exp Mar Biol Ecol 220:47–65

Calvo-Ugarteburu G, McQuaid C (1998b) Parasitism and invasive species: effects of digenetic trematodes on mussels. Mar Ecol Prog Ser 169:149–163

Casadevall A, Pirofski L-a (1999) Host-pathogen interactions: redefining the basic concepts of virulence and pathogenicity. Infect Immun 67(8):3703–3713

Cattadori IM, Haydon DT, Hudson PJ (2005) Parasites and climate synchronize red grouse populations. Nature 433(7027):737–741

Cattadori IM, Boag B, Hudson PJ (2008) Parasite co-infection and interaction as drivers of host heterogeneity. Int J Parasitol 38(3):371–380. https://doi.org/10.1016/j.ijpara.2007.08.004

Cézilly F, Perrot-Minnot M-J (2005) Studying adaptive changes in the behaviour of infected hosts: a long and winding road. Behav Process 68(3):223–228. https://doi.org/10.1016/j.beproc.2004.08.013

Cézilly F, Gregoire A, Bertin A (2000) Conflict between co-occurring manipulative parasites? An experimental study of the joint influence of two acanthocephalan parasites on the behaviour of *Gammarus pulex*. Parasitology 120:625–630

Chattopadhyay J, Arino O (1999) A predator-prey model with disease in the prey. Nonlinear Anal 36:747–766

Christensen NØ, Nansen P, Fagbemi BO et al (1987) Heterologous antagonistic and synergistic interactions between helminths and between helminths and protozoans in concurrent experimental infection of mammalian hosts. Parasitol Res 73(5):387–410. https://doi.org/10.1007/BF00538196

Civitello DJ, Cohen J, Fatima H et al (2015) Biodiversity inhibits parasites: broad evidence for the dilution effect. PNAS 112(28):8667–8671

Combes C (1997) Fitness of parasites: pathology and selection. Int J Parasitol 27(1):1–10. https://doi.org/10.1016/S0020-7519(96)00168-3

Combes C (2001) Parasitism: the ecology and evolution of intimate interactions. University of Chicago Press, Chicago

Connelly S, Pringle CM, Bixby RJ et al (2008) Changes in stream primary producer communities resulting from large-scale catastrophic amphibian declines:

can small-scale experiments predict effects of tadpole loss? Ecosystems 11:1262–1276

Costanza R, Mageau M (1999) What is a healthy ecosystem? Aquat Ecol 33:105–115

Costanza R, Patten BC (1995) Defining and predicting sustainability. Ecol Econ 15(3):193–196

Costanza R, Norton BG, Haskell BD (1992) Ecosystem health: new goals for environmental management. Island Press

Cox FEG (2001) Concomitant infections, parasites and immune responses. Parasitology 122(S1):S23–S38. https://doi.org/10.1017/S003118200001698X

Cribb TH, Adlard RD, Hayward CJ et al (2011) The life cycle of *Cardicola forsteri* (Trematoda Aporocotylidae), a pathogen of ranched southern bluefin tuna. Int J Parasitol 41:861–870

Curry A, Else K, Jones F et al (1995) Evidence that cytokine-mediated immune interactions induced by *Schistosoma mansoni* alter disease outcome in mice concurrently infected with *Trichuris muris*. J Exp Med 181(2):769–774

Darlington P Jr (1972) Competition, competitive repulsion, and coexistence. PNAS 69(11):3151–3155

Daversa D, Hechinger RF, Madin E et al (2021) Broadening the ecology of fear: non-lethal effects arise from diverse responses to predation and parasitism. Proc Biol Sci 288(1945):20202966

Dawkins R (1982) The extended phenotype: the gene as the unit of selection. W. H. Freeman, San Francisco

De Bekker C, Will I, Das B et al (2018) The ants (Hymenoptera: Formicidae) and their parasites: effects of parasitic manipulations and host responses on ant behavioral ecology. Myrmecol News 28:1–24

Dezfuli BS, Giari L, Poulin R (2000) Species associations among larval helminths in an amphipod intermediate host. Int J Parasitol 30(11):1143–1146. https://doi.org/10.1016/S0020-7519(00)00093-X

Dezfuli BS, Maynard BJ, Wellnitz TA (2003) Activity levels and predator detection by amphipods infected with an acanthocephalan parasite, *Pomphorhynchus laevis*. Folia Parasitol (Praha) 50(2):129–134

Dougherty ER, Carlson CJ, Bueno VM et al (2016) Paradigms for parasite conservation. Conserv Biol 30(4):724–733

Dunn RR, Harris NC, Colwell RK et al (2009) The sixth mass coextinction: are most endangered species parasites and mutualists? Proc Biol Sci 276(1670):3037–3045

Feist S, Longshaw M (2008) Histopathology of fish parasite infections–importance for populations. J Fish Biol 73(9):2143–2160

Fellous S, Salvaudon L (2009) How can your parasites become your allies? Trends Parasitol 25(2):62–66

Friesen OC, Poulin R, Lagrue C (2017) Differential impacts of shared parasites on fitness components among competing hosts. Ecol Evol 7(13):4682–4693

Fuller CA, Rock P, Philips T (2003) Behavior, color changes, and predation risk induced by acanthocephalan parasitism in the Caribbean termite *Nasutitermes acujutlae*. Caribb J Sci 39(1):128–135

Gendron AD, Marcogliese DJ (2017) Enigmatic decline of a common fish parasite (*Diplostomum* spp.) in the St. Lawrence River: evidence for a dilution effect induced by the invasive round goby. Int J Parasitol Parasites Wildl 6(3):402–411. https://doi.org/10.1016/j.ijppaw.2017.04.002

Genovart M, Negre N, Tavecchia G et al (2010) The young, the weak and the sick: evidence of natural selection by predation. PLoS One 5(3):e9774

Gibb R, Redding DW, Chin KQ et al (2020) Zoonotic host diversity increases in human-dominated ecosystems. Nature 584(7821):398–402. https://doi.org/10.1038/s41586-020-2562-8

Gleason FH, Lilje O, Marano AV et al (2014) Ecological functions of zoosporic hyperparasites. Front Microbiol 5:244

Graham AL, Lamb TJ, Read AF et al (2005) Malaria-filaria coinfection in mice makes malarial disease more severe unless filarial infection achieves patency. J Infect Dis 191(3):410–421. https://doi.org/10.1086/426871

Greenhalgh D, Haque M (2007) A predator–prey model with disease in the prey species only. Math Methods Appl Sci 30(8):911–929

Grutter AS (1995) Relationship between cleaning rates and ectoparasite loads in coral reef fishes. Mar Ecol Prog Ser 118:51–58

Grutter A (1996) Parasite removal rates by the cleaner wrasse *Labroides dimidiatus*. Mar Ecol Prog Ser 130:61–70

Grutter AS (1999) Cleaner fish really do clean. Nature 398:672–673

Grutter AS, Feeney WE (2016) Equivalent cleaning in a juvenile facultative and obligate cleaning wrasse: an insight into the evolution of cleaning in labrids? Coral Reefs 35:991–997

Grutter AS, Murphy JM, Choat JH (2003) Cleaner fish drives local fish diversity on coral reefs. Curr Biol 13(1):64–67

Hall SR, Lafferty KD, Brown JH et al (2008) Is infectious disease just another type of predator-prey interaction. In: Ostfeld R, Keesing F, Eviner VT (eds) Infectious disease ecology: the effects of ecosystems on disease and of disease on ecosystems. Princeton University Press, Princeton, pp 223–241

Hall SR, Becker CR, Simonis JL et al (2009) Friendly competition: evidence for a dilution effect among competitors in a planktonic host–parasite system. Ecology 90(3):791–801

Hamilton WD, Zuk M (1982) Heritable true fitness and bright birds: a role for parasites? Science 218(4570):384–387

Hammerschmidt K, Koch K, Milinski M et al (2009) When to go: optimization of host switching in parasites with complex life cycles. Evolution 63:1976–1986. https://doi.org/10.1111/j.1558-5646.2009.00687.x

Hanelt B (2009) Hyperparasitism by *Paragordius varius* (Nematomorpha: Gordiida) larva of monostome redia (Trematoda: Digenea). J Parasitol 95(1):242–243

Hardisty GR, Knipper JA, Fulton A et al (2022) Concurrent infection with the filarial helminth *Litomosoides sigmodontis* attenuates or worsens influenza A virus pathogenesis in a stage-dependent manner. Front Immunol 12:819560

Harvell CD, Mitchell CE, Ward JR et al (2002) Climate warming and disease risks for terrestrial and marine biota. Science 296(5576):2158–2162

Hatcher MJ, Dick JT, Dunn AM (2006) How parasites affect interactions between competitors and predators. Ecol Lett 9(11):1253–1271

Hatcher MJ, Dick JT, Dunn AM (2012) Diverse effects of parasites in ecosystems: linking interdependent processes. Front Ecol Environ 10(4):186–194

Heinonen J, Kukkonen JV, Holopainen IJ (2001) Temperature-and parasite-induced changes in toxicity and lethal body burdens of pentachlorophenol in the freshwater clam *Pisidium amnicum*. Environ Toxicol Chem 20:2778–2784. https://doi.org/10.1002/etc.5620201217

Heins DC (2017) The cestode parasite *Schistocephalus pungitii*: castrator or nutrient thief of ninespine stickleback fish? Parasitology 144(6):834–840

Helluy S (1983) Relations hôtes-parasite du trematode *Microphallus papillorobustus* (Rankin, 1940)-II—modifications du comportement des *Gammarus* hôtes intermédiaires et localisation des métacercaires. Ann Parasitol Hum Comp 58(1):1–17

Helluy S (2013) Parasite-induced alterations of sensorimotor pathways in gammarids: collateral damage of neuroinflammation? J Exp Biol 216(1):67–77

Helluy S, Thomas F (2003) Effects of *Microphallus papillorobustus* (Platyhelminthes: Trematoda) on serotonergic immunoreactivity and neuronal architecture in the brain of *Gammarus insensibilis* (Crustacea: Amphipoda). Proc Biol Sci 270(1515):563–568

Herczeg D, Ujszegi J, Kásler A et al (2021) Host–multiparasite interactions in amphibians: a review. Parasites Vectors 14(1):296. https://doi.org/10.1186/s13071-021-04796-1

Hochberg M, Hassell M, May R (1990) The dynamics of host-parasitoid-pathogen interactions. Am Nat 135(1):74–94

Hohorst W, Graefe G (1961) Ameisen—Obligatorische Zwischenwirte des Lanzettegels (*Dicrocoelium dendriticum*). Sci Nat 48:229–230

Holt RD (1977) Predation, apparent competition, and the structure of prey communities. Theor Popul Biol 12(2):197–229

Holt RD, Pickering J (1985) Infectious disease and species coexistence: a model of Lotka-Volterra form. Am Nat 126(2):196–211

Horwitz P, Wilcox BA (2005) Parasites, ecosystems and sustainability: an ecological and complex systems perspective. Int J Parasitol 35(7):725–732

Hsieh Y-H, Hsiao C-K (2008) Predator-prey model with disease infection in both populations. Math Med Biol 25(3):247–266. https://doi.org/10.1093/imammb/dqn017

Hudson PJ (1986) The effect of a parasitic nematode on the breeding production of red grouse. J Anim Ecol 55(1):85–92. https://doi.org/10.2307/4694

Hudson PJ, Newborn D, Dobson AP (1992) Regulation and stability of a free-living host-parasite system: *Trichostrongylus tenuis* in red grouse. I. Monitoring and parasite reduction experiments. J Anim Ecol 61(2):477–486

Hudson PJ, Dobson AP, Newborn D (1998) Prevention of population cycles by parasite removal. Science 282(5397):2256–2258

Hudson PJ, Dobson AP, Newborn D (2002) Parasitic worms and population cycles of red grouse. In: Berriman A (ed) Population cycles: the case for trophic interactions. Oxford Academic, New York, pp 109–129

Hudson P, Cattadori I, Boag B et al (2006a) Climate disruption and parasite–host dynamics: patterns and processes associated with warming and the frequency of extreme climatic events. J Helminthol 80(2):175–182

Hudson PJ, Dobson AP, Lafferty KD (2006b) Is a healthy ecosystem one that is rich in parasites? Trends Ecol Evol 21(7):381–385

Huspeni TC, Lafferty KD (2004) Using larval trematodes that parasitize snails to evaluate a saltmarsh restoration project. Ecol Appl 14(3):795–804

Hutchings MR, Kyriazakis I, Papachristou TG et al (2000) The herbivores' dilemma: trade-offs between nutrition and parasitism in foraging decisions. Oecologia 124:242–251

Jackson JA, Pleass RJ, Cable J et al (2006) Heterogenous interspecific interactions in a host–parasite system. Int J Parasitol 36(13):1341–1349. https://doi.org/10.1016/j.ijpara.2006.07.003

Jażdżewski K (1980) Range extensions of some gammaridean species in European inland waters caused by human activity. In: Meijering MPD (ed) Studies on Gammaridea, vol 2. 4th International Colloquium on *Gammarus* and *Niphargus*, Blacksburg, VA, 10–16 September 1978. Crustaceana (Supplements), vol 6. Brill, Leiden, pp 84–107

Johnson PT, Buller ID (2011) Parasite competition hidden by correlated coinfection: using surveys and experiments to understand parasite interactions. Ecology 92(3):535–541

Johnson PTJ, Dobson A, Lafferty KD et al (2010) When parasites become prey: ecological and epidemiological significance of eating parasites. Trends Ecol Evol 25(6):362–371. https://doi.org/10.1016/j.tree.2010.01.005

Johnson PT, Kellermanns E, Bowerman J (2011) Critical windows of disease risk: amphibian pathology driven by developmental changes in host resistance and tolerance. Funct Ecol 25(3):726–734

Jonsson PR, Andé C (1992) Mass mortality of the bivalve Cerastoderma edule on the Swedish west coast caused by infestation with the digenean trematode *Cercaria cerastodermae* I. Ophelia 36(2):151–157

Kaldonski N, Perrot-Minnot M-J, Motreuil S et al (2008) Infection with acanthocephalans increases the vulnerability of *Gammarus pulex* (Crustacea, Amphipoda) to non-host invertebrate predators. Parasitology 135(5):627–632

Karvonen A, Seppälä O, Tellervo Valtonen E (2009) Host immunization shapes interspecific associations in trematode parasites. J Anim Ecol 78(5):945–952

Karvonen A, Jokela J, Laine A-L (2019) Importance of sequence and timing in parasite coinfections. Trends Parasitol 35(2):109–118

Kavaliers M, Podesta RB, Hirst M et al (1984) Evidence for the activation of the endogenous opiate system in hamsters infected with human blood flukes, *Schistosoma mansoni*. Life Sci 35(23):2365–2373

Keesing F, Holt RD, Ostfeld RS (2006) Effects of species diversity on disease risk. Ecol Lett 9(4):485–498

Kennedy C (1996) Colonization and establishment of *Pomphorhynchus laevis* (Acanthocephala) in an isolated English river. J Helminthol 70(1):27–31

Kennedy C, Bates R, Brown A (1989) Discontinuous distributions of the fish acanthocephalans *Pomphorhynchm laevis* and *Acanthocephalus anguillae* in Britain and Ireland: an hypothesis. J Fish Biol 34(4):607–619

Klein SL (2003) Parasite manipulation of the proximate mechanisms that mediate social behavior in vertebrates. Physiol Behav 79(3):441–449. https://doi.org/10.1016/S0031-9384(03)00163-X

Koh LP, Dunn RR, Sodhi NS et al (2004) Species coextinctions and the biodiversity crisis. Science 305(5690):1632–1634

Køie M (1982) The redia, cercaria and early stages of *Aporocotyle simplex* Odhner, 1900 (Sanguinicolidae) – a digenetic trematode which has a polychaete annelid as the only intermediate host. Ophelia 21:115–145

Koprivnikar J, Rochette A, Forbes MR (2021) Risk-induced trait responses and non-consumptive effects in plants and animals in response to their invertebrate herbivore and parasite natural enemies. Front Ecol Evol 9:667030

Kuris AM, Poinar GO, Hess RT (1980) Post-larval mortality of the endoparasitic isopod castrator *Portunion conformis* (Epicaridea: Entoniscidae) in the shore crab, *Hemigrapsus oregonensis*, with a description of the host response. Parasitology 80(2):211–232

Kuris AM, Hechinger RF, Shaw JC et al (2008) Ecosystem energetic implications of parasite and free-living biomass in three estuaries. Nature 454(7203):515–518. https://doi.org/10.1038/nature06970

Lafferty KD (1993) Effects of parasitic castration on growth, reproduction and population dynamics of the marine snail *Cerithidea californica*. Mar Ecol Prog Ser 96:229–237

Lafferty KD (1997) Environmental parasitology: what can parasites tell us about human impacts on the environment? Trends Parasitol 13(7):251–255

Lafferty KD (2004) Fishing for lobsters indirectly increases epidemics in sea urchins. Ecol Appl 14(5):1566–1573

Lafferty KD (2012) Biodiversity loss decreases parasite diversity: theory and patterns. Philos Trans R Soc Lond Ser B Biol Sci 367(1604):2814–2827

Lafferty KD, Kuris AM (1999) How environmental stress affects the impacts of parasites. Limnol Oceanogr 44(3):925–931

Lafferty KD, Kuris AM (2002) Trophic strategies, animal diversity and body size. Trends Ecol Evol 17(11):507–513

Lafferty KD, Kuris AM (2009a) Parasites reduce food web robustness because they are sensitive to secondary extinction as illustrated by an invasive estuarine snail. Philos Trans R Soc Lond Ser B Biol Sci 364(1524):1659–1663

Lafferty KD, Kuris AM (2009b) Parasitic castration: the evolution and ecology of body snatchers. Trends Parasitol 25(12):564–572

Lafferty KD, Morris AK (1996) Altered behavior of parasitized killifish increases susceptibility to predation by bird final hosts. Ecology 77(5):1390–1397

Lafferty KD, Shaw JC (2013) Comparing mechanisms of host manipulation across host and parasite taxa. J Exp Biol 216(1):56–66

Lafferty KD, Dobson AP, Kuris AM (2006) Parasites dominate food web links. PNAS 103(30):11211–11216

Leontief W (1941) The structure of the American economy, 1919–1929. Harvard University Press, Cambridge

Lester RJG (1971) The influence of *Schistocephalus* plerocercoids on the respiration of *Gasterosteus* and a possible resulting effect on the behavior of the fish. Can J Zool 49(3):361–366. https://doi.org/10.1139/z71-052

Levin SA (1970) Community equilibria and stability, and an extension of the competitive exclusion principle. Am Nat 104(939):413–423

MacArthur RH (1958) Population ecology of some warblers of northeastern coniferous forests. Ecology 39(4):599–619

MacNeil C, Dick JTA (2011) Parasite-mediated intraguild predation as one of the drivers of co-existence and exclusion among invasive and native amphipods (Crustacea). Hydrobiologia 665(1):247–256. https://doi.org/10.1007/s10750-011-0627-2

Maizels RM, Yazdanbakhsh M (2003) Immune regulation by helminth parasites: cellular and molecular mechanisms. Nat Rev Immunol 3(9):733–744

Maizels RM, Balic A, Gomez-Escobar N et al (2004) Helminth parasites—masters of regulation. Immunol Rev 201(1):89–116

Marchand J, Robinson SA, Forbes MR (2020) Size and survival of two freshwater snail species in relation to shedding of cercariae of castrating *Echinostoma* spp. Parasitol Res 119(9):2917–2925

Marcogliese DJ (2002) Food webs and the transmission of parasites to marine fish. Parasitology 124:S83–S99

Massey RC, Buckling A, ffrench-Constant R (2004) Interference competition and parasite virulence. Proc Biol Sci 271(1541):785–788

May RM, Anderson RM (1979) Population biology of infectious diseases: part II. Nature 280(5722):455–461

McDevitt-Galles T, Moss WE, Calhoun DM et al (2020) Phenological synchrony shapes pathology in host–parasite systems. Proc Biol Sci 287(1919):20192597

Mideo N (2009) Parasite adaptations to within-host competition. Trends Parasitol 25(6):261–268

Minchella DJ, Loverde PT (1981) A cost of increased early reproductive effort in the snail *Biomphalaria glabrata*. Am Nat 118(6):876–881

Minchella DJ, Leathers BK, Brown KM et al (1985) Host and parasite counteradaptations: an example from a freshwater snail. Am Nat 126:843–854

Mouritsen KN, Poulin R (2003) Parasite-induced trophic facilitation exploited by a non-host predator: a manipulator's nightmare. Int J Parasitol 33(10):1043–1050

Mouritsen KN, Poulin R (2010) Parasitism as a determinant of community structure on intertidal flats. Mar Biol 157(1):201–213. https://doi.org/10.1007/s00227-009-1310-2

Munger JC, Holmes JC (1988) Benefits of parasitic infection: a test using a ground squirrel-trypanosome system. Can J Zool 66(1):222–227. https://doi.org/10.1139/z88-032

Norris K (1999) A trade-off between energy intake and exposure to parasites in oystercatchers feeding on a bivalve mollusc. Proc Biol Sci 266(1429):1703–1709

Northrup GR, White A, Parratt SR et al (2024) The evolutionary dynamics of hyperparasites. J Theor Biol 582:111741. https://doi.org/10.1016/j.jtbi.2024.111741

Nowicki S, Searcy WA (2004) Song function and the evolution of female preferences: why birds sing, why brains matter. Ann N Y Acad Sci 1016(1):704–723

O'Brien J, Van Wyk P (1985) Effects of crustacean parasitic castrators (epicaridean isopods and rhizocephalan barnacles) on growth of crustacean hosts. In: Wenner AM (ed) Factors in adult growth. A. A. Balkema, Boston, pp 191–218

Odum H (1971) Environment, power, and society. Wiley, New York

Ostfeld RS, Keesing F (2012) Effects of host diversity on infectious disease. Annu Rev Ecol Evol Syst 43:157–182

Packer C, Holt RD, Hudson PJ et al (2003) Keeping the herds healthy and alert: implications of predator control for infectious disease. Ecol Lett 6(9):797–802

Paine RT (1966) Food web complexity and species diversity. Am Nat 100(910):65–75

Parmentier E, Michel L (2013) Boundary lines in symbiosis forms. Symbiosis 60:1–5

Parmentier E, Vandewalle P (2005) Further insight on carapid-holothuroid relationships. Mar Biol 146:455–465

Parratt SR, Laine A-L (2016) The role of hyperparasitism in microbial pathogen ecology and evolution. ISME J 10(8):1815–1822. https://doi.org/10.1038/ismej.2015.247

Peacor SD, Barton BT, Kimbro DL et al (2020) A framework and standardized terminology to facilitate the study of predation-risk effects. Ecology 101(12):e03152

Pedersen AB, Fenton A (2007) Emphasizing the ecology in parasite community ecology. Trends Ecol Evol 22(3):133–139

Peresan L, Roccatagliata D (2005) First record of the hyperparasite *Liriopsis pygmaea* (Cryptoniscidae, Isopoda) from a rhizocephalan parasite of the false king crab *Paralomis granulosa* from the Beagle Channel (Argentina), with a redescription. J Nat Hist 39(4):311–324. https://doi.org/10.1080/0022293042000200103

Pérez-del Olmo A, Raga JA, Kostadinova A et al (2007) Parasite communities in *Boops boops* (L.) (Sparidae) after the prestige oil-spill: detectable alterations. Mar Pollut Bull 54(3):266–276

Poléo ABS, Schjolden J, Hansen H et al (2004) The effect of various metals on *Gyrodactylus salaris* (Platyhelminthes, Monogenea) infections in Atlantic salmon (*Salmo salar*). Parasitology 128(2):169–177. https://doi.org/10.1017/S0031182003004396

Poulin R (1994a) Meta-analysis of parasite-induced behavioural changes. Anim Behav 48(1):137–146. https://doi.org/10.1006/anbe.1994.1220

Poulin R (1994b) The evolution of parasite manipulation of host behaviour: a theoretical analysis. Parasitology 109(S1):S109–S118

Poulin R (1995) "Adaptive" changes in the behaviour of parasitized animals: a critical review. Int J Parasitol 25(12):1371–1383

Poulin R, Valtonen ET (2001) Interspecific associations among larval helminths in fish. Int J Parasitol 31(14):1589–1596. https://doi.org/10.1016/S0020-7519(01)00276-4

Poulin R, Fredensborg BL, Hansen E et al (2005) The true cost of host manipulation by parasites. Behav Process 68(3):241–244

Pravdová M, Kolářová J, Grabicová K et al (2023) Response of parasite community composition to aquatic pollution in common carp (*Cyprinus carpio* L.): a semi-experimental study. Animals 13(9):1464

Price PW, Westoby M, Rice B et al (1986) Parasite mediation in ecological interactions. Annu Rev Ecol Evol Syst 17(1):487–505

Pryor SC, Carter C, Mendes M et al (1998) Opioid involvement in behavior modifications of mice infected with the parasitic nematode, *Nippostrongylus brasiliensis*. Life Sci 63(18):1619–1628

Quinn TP, Kendall NW, Rich HB et al (2012) Diel vertical movements, and effects of infection by the cestode *Schistocephalus solidus* on daytime proximity of three-spined sticklebacks *Gasterosteus aculeatus* to the surface of a large Alaskan lake. Oecologia 168(1):43–51. https://doi.org/10.1007/s00442-011-2071-4

Råberg L, De Roode JC, Bell AS et al (2006) The role of immune-mediated apparent competition in genetically diverse malaria infections. Am Nat 168(1):41–53

Råberg L, Graham AL, Read AF (2009) Decomposing health: tolerance and resistance to parasites in animals. Philos Trans R Soc Lond Ser B Biol Sci 364(1513):37–49

Raffel TR, Martin LB, Rohr JR (2008) Parasites as predators: unifying natural enemy ecology. Trends Ecol Evol 23(11):610–618

Randall JE (1967) Food habits of reef fishes of the West Indies. Hawaii Institute of Marine Biology – University of Miami, Coral Gables

Redpath SM, Thirgood SJ (1999) Numerical and functional responses in generalist predators: hen harriers and peregrines on Scottish grouse moors. J Anim Ecol 68(5):879–892

Reid H, Duncan J, Phillips J et al (1978) Studies on louping-ill virus (Flavivirus group) in wild red grouse (*Lagopus lagopus scoticus*). Epidemiol Infect 81(2):321–329

Ridley M (1993) Evolution. Blackwell Scientific Publications, Oxford

Rigby MC, Hechinger RF, Stevens L (2002) Why should parasite resistance be costly? Trends Parasitol 18(3):116–120. https://doi.org/10.1016/S1471-4922(01)02203-6

Robar N, Burness G, Murray DL (2010) Tropics, trophics and taxonomy: the determinants of parasite-associated host mortality. Oikos 119(8):1273–1280

Rohr JR, Raffel TR, Sessions SK et al (2008) Understanding the net effects of pesticides on amphibian trematode infections. Ecol Appl 18(7):1743–1753

Rózsa L, Garay J (2023) Definitions of parasitism, considering its potentially opposing effects at different levels of hierarchical organization. Parasitology 150(9):761–768

Ryan MJ (1990) Sexual selection, sensory systems and sensory exploitation. In: Futuyma D, Antonovics J (eds) Oxford surveys in evolutionary biology. Oxford University Press, Oxford, pp 157–195

Sato T, Watanabe K, Kanaiwa M et al (2011) Nematomorph parasites drive energy flow through a riparian ecosystem. Ecology 92(1):201–207

Sato T, Egusa T, Fukushima K et al (2012) Nematomorph parasites indirectly alter the food web and ecosystem function of streams through behavioural manipulation of their cricket hosts. Ecol Lett 15(8):786–793

Schultz B, Koprivnikar J (2019) Free-living parasite infectious stages promote zooplankton abundance under the risk of predation. Oecologia 191(2):411–420. https://doi.org/10.1007/s00442-019-04503-z

Science Advisory Board (1990) Reducing risk: setting priorities and strategies for environmental protection. US Environmental Protection Agency, Washington DC

Selbach C, Mouritsen KN, Poulin R et al (2022) Bridging the gap: aquatic parasites in the one health concept. Trends Parasitol 38(2):109–111

Seppälä O, Karvonen A, Valtonen ET (2004) Parasite-induced change in host behaviour and susceptibility to predation in an eye fluke-fish interaction. Anim Behav 68(2):257–263

Seppälä O, Karvonen A, Valtonen ET (2005) Impaired crypsis of fish infected with a trophically transmitted parasite. Anim Behav 70(4):895–900. https://doi.org/10.1016/j.anbehav.2005.01.021

Seppälä O, Liljeroos K, Karvonen A et al (2008) Host condition as a constraint for parasite reproduction. Oikos 117:749–753. https://doi.org/10.1111/j.0030-1299.2008.16396.x

Shirakashi S, Goater CP (2001) Brain-encysting parasites affect visually-mediated behaviours of fathead minnows. Ecoscience 8(3):289–293

Sokolova YY, Overstreet RM, Heard RW et al (2021) Two new species of *Unikaryon* (Microsporidia) hyperparasitic in microphallid metacercariae (Digenea) from Florida intertidal crabs. J Invertebr Pathol 182:107582. https://doi.org/10.1016/j.jip.2021.107582

Sorensen R, Minchella D (2001) Snail-trematode life history interactions: past trends and future directions. Parasitology 123(7):S3–S18

Sousa WP (1983) Host life history and the effect of parasitic castration on growth: a field study of *Cerithidea californica* Haldeman (Gastropoda: Prosobranchia) and its trematode parasites. J Exp Mar Biol Ecol 73(3):273–296

Springer JC, Baines ALD, Fulbright DW et al (2013) Hyperparasites influence population structure of the chestnut blight pathogen, *Cryphonectria parasitica*. Phytopathology 103(12):1280–1286

Stoffel SA, Rodenko B, Schweingruber A-M et al (2006) Biosynthesis and uptake of thiamine (vitamin B1) in bloodstream form *Trypanosoma brucei* brucei and interference of the vitamin with melarsen oxide activity. Int J Parasitol 36(2):229–236

Stork N, Lyal C (1993) Extinction or "co-extinction" rates? Nature 366:307

Su Z, Segura M, Morgan K et al (2005) Impairment of protective immunity to blood-stage malaria by concurrent nematode infection. Infect Immun 73(6):3531–3539. https://doi.org/10.1128/iai.73.6.3531-3539.2005

Sures B, Nachev M, Schwelm J et al (2023) Environmental parasitology: stressor effects on aquatic parasites. Trends Parasitol 39(6):461–474

Tain L, Perrot-Minnot M-J, Cézilly F (2006) Altered host behaviour and brain serotonergic activity caused by acanthocephalans: evidence for specificity. Proc Biol Sci 273(1605):3039–3045

Tain L, Perrot-Minnot M-J, Cézilly F (2007) Differential influence of *Pomphorhynchus laevis* (Acanthocephala) on brain serotonergic activity in two congeneric host species. Biol Lett 3(1):69–72

Talarico M, Seifert F, Lange J et al (2017) Specific manipulation or systemic impairment? Behavioural changes of three-spined sticklebacks (*Gasterosteus aculeatus*) infected with the tapeworm *Schistocephalus solidus*. Behav Ecol Sociobiol 71:1–10

Taylor CN, Oseen KL, Wassersug RJ (2004) On the behavioural response of *Rana* and *Bufo* tadpoles to echinostomatoid cercariae: implications to synergistic factors influencing trematode infections in anurans. Can J Zool 82(5):701–706. https://doi.org/10.1139/z04-037

Thieltges DW, Rick J (2006) Effect of temperature on emergence, survival and infectivity of cercariae of the marine trematode *Renicola roscovita* (Digenea: Renicolidae). Dis Aquat Org 73(1):63–68

Thieltges DW, Bordalo M, Hernández AC et al (2008a) Ambient fauna impairs parasite transmission in a marine parasite-host system. Parasitology 135(9):1111–1116

Thieltges DW, Jensen KT, Poulin R (2008b) The role of biotic factors in the transmission of free-living endohelminth stages. Parasitology 135(4):407–426. https://doi.org/10.1017/S0031182007000248

Thirgood SJ, Redpath SM, Haydon DT et al (2000) Habitat loss and raptor predation: disentangling long- and short-term causes of red grouse declines. Proc Biol Sci 267(1444):651–656

Thomas F, Mete K, Helluy S et al (1997) Hitch-hiker parasites or how to benefit from the strategy of another parasite. Evolution 51(4):1316–1318

Thomas F, Poulin R, Guégan JF et al (2000) Are there pros as well as cons to being parasitized? Trends Parasitol 16(12):533–536. https://doi.org/10.1016/S0169-4758(00)01790-7

Thomas F, Schmidt-Rhaesa A, Martin G et al (2002) Do hairworms (Nematomorpha) manipulate the water seeking behaviour of their terrestrial hosts? J Evol Biol 15(3):356–361

Thomas F, Adamo S, Moore J (2005) Parasitic manipulation: where are we and where should we go? Behav Process 68(3):185–199

Tong ZWM, Karawita AC, Kern C et al (2021) Primary chicken and duck endothelial cells display a differential response to infection with highly pathogenic avian influenza virus. Genes (Basel) 12(6):901

Torchin ME, Miura O, Hechinger RF (2015) Parasite species richness and intensity of interspecific interactions increase with latitude in two wide-ranging hosts. Ecology 96(11):3033–3042

Toscano BJ, Newsome B, Griffen BD (2014) Parasite modification of predator functional response. Oecologia 175(1):345–352. https://doi.org/10.1007/s00442-014-2905-y

Toyota K, Ito T, Morishima K et al (2023) *Sacculina*-induced morphological feminization in the grapsid crab *Pachygrapsus crassipes*. Zool Sci 40(5):367–374

Ulanowicz RE (1986) Growth and development: ecosystems phenomenology. Springer, New York

Utsumi S, Ando Y, Miki T (2010) Linkages among trait-mediated indirect effects: a new framework for the indirect interaction web. Popul Ecol 52(4):485–497. https://doi.org/10.1007/s10144-010-0237-2

van As LL (2019) Hypersymbionts and hyperparasites of parasitic Crustacea. In: Smit N, Bruce N, Hadfield KA (eds) Parasitic Crustacea: state of knowledge and future trends. Springer Nature, pp 343–385

Vollset KW (2019) Parasite induced mortality is context dependent in Atlantic salmon: insights from an individual-based model. Sci Rep 9(1):17377. https://doi.org/10.1038/s41598-019-53871-2

Waldie PA, Blomberg SP, Cheney KL et al (2011) Long-term effects of the cleaner fish *Labroides dimidiatus* on coral reef fish communities. PLoS One 6(6):e21201

Węgrzyn G (2022) Should bacteriophages be classified as parasites or predators? Pol J Microbiol 71(1):3–9

Weinersmith KL, Earley RL (2016) Better with your parasites? Lessons for behavioural ecology from evolved dependence and conditionally helpful parasites. Anim Behav 118:123–133

Werner EE, Peacor SD (2003) A review of trait-mediated indirect interactions in ecological communities. Ecology 84(5):1083–1100

Williams HH, MacKenzie K, McCarthy AM (1992) Parasites as biological indicators of the population biology, migrations, diet, and phylogenetics of fish. Rev Fish Biol Fish 2(2):144–176. https://doi.org/10.1007/BF00042882

Wilmers CC, Post E, Peterson RO et al (2006) Predator disease out-break modulates top-down, bottom-up and climatic effects on herbivore population dynamics. Ecol Lett 9(4):383–389

Wilmers CC, Post E, Hastings A (2007) A perfect storm: the combined effects on population fluctuations of autocorrelated environmental noise, age structure, and density dependence. Am Nat 169(5):673–683

Wood CL, Lafferty KD (2015) How have fisheries affected parasite communities? Parasitology 142(1):134–144

Woolhouse ME, Thumbi SM, Jennings A et al (2015) Co-infections determine patterns of mortality in a population exposed to parasite infection. Sci Adv 1(2):e1400026

Yan G (1996) Parasite-mediated competition: a model of directly transmitted macroparasites. Am Nat 148(6):1089–1112

Yang LH, Rudolf V (2010) Phenology, ontogeny and the effects of climate change on the timing of species interactions. Ecol Lett 13(1):1–10

Ye M, Sun M, Huang D et al (2019) A review of bacteriophage therapy for pathogenic bacteria inactivation in the soil environment. Environ Int 129:488–496. https://doi.org/10.1016/j.envint.2019.05.062

De Castro F, Bolker B (2005) Mechanisms of disease-induced extinction. Ecol Lett 8(1):117–126

Open Access This chapter is licensed under the terms of the Creative Commons Attribution-NonCommercial-NoDerivatives 4.0 International License (http://creativecommons.org/licenses/by-nc-nd/4.0/), which permits any non-commercial use, sharing, distribution and reproduction in any medium or format, as long as you give appropriate credit to the original author(s) and the source, provide a link to the Creative Commons license and indicate if you modified the licensed material. You do not have permission under this license to share adapted material derived from this chapter or parts of it.

The images or other third party material in this chapter are included in the chapter's Creative Commons license, unless indicated otherwise in a credit line to the material. If material is not included in the chapter's Creative Commons license and your intended use is not permitted by statutory regulation or exceeds the permitted use, you will need to obtain permission directly from the copyright holder.

Impact of Aquatic Parasites on Host Community Structure

Kim N. Mouritsen

Abstract

Parasitism is increasingly recognised as a significant determinant of composition and structure of natural communities across ecosystems. In this chapter, a large amount of evidence from the aquatic realm is brought together to highlight the multitude of processes through which parasites impact aquatic community structure. There are three main ways by which parasites affecting community structure are emphasised: (1) direct effects through differential regulation of host populations; (2) density-mediated indirect effects through modification of biotic processes such as predation, herbivory, facilitation and inhibition; and (3) trait-mediated indirect effects through altered phenotype of infected host species that often act as keystone species or ecosystem engineers in the community. Selected studies are treated in-depth to illustrate the ways parasites modify aquatic communities, involving a variety of host organisms such as monocotyledons, gastropods, bivalves, amphipods, echinoderms and fishes, as well as parasites ranging from protists to trematodes and nematodes, among others. This synthesis concludes by stressing that the available evidence does not equivocally support the contemporary theory that parasitism generally acts to maintain biodiversity. Rather, the qualitative effect is context-dependent, determined by the ecological roles of involved host species.

8.1 Introduction

The evidence that parasites are able to reduce host survival and fecundity is overwhelming, though it does not necessarily invoke additive regulation of host populations below the carrying capacity of the habitat (e.g. Morand and Deter 2009). In aquatic ecosystems, the inference of such a regulating role of parasitism is of an indirect nature and experimental tests, carried out under natural settings incorporating the multitude of interactions with the surrounding environment, are generally lacking. Upon the available evidence, should we thus accept that parasites are potent regulators of host populations? This does not necessarily entail measurable impact on community structure in terms of species composition, richness or diversity. Evidence from terrestrial systems does, however, suggest this to be the case.

In a seminal paper, Park (1948) demonstrated experimentally that the dominance by one of two co-occurring flour beetle species (*Tribolium* spp.)

K. N. Mouritsen (✉)
Department of Biology, Aquatic Biology, University of Aarhus, Aarhus C, Denmark
e-mail: kim.mouritsen@bio.au.dk

was reversed in the presence of a protozoan parasite, *Adelina tribolii*. In the absence of the parasite, the competitively dominant *T. castaneum* generally out-competed the less-dominant species *T. confusum*, whereas in its presence, the two flour beetles coexisted, or the less dominant species prevailed, in its presence. Because the parasite inflicts asymmetrical pathology on the two host species, affecting the superior competitor more severely than the inferior, it facilitates coexistence or determines species composition.

Outside the laboratory we can find similar evidence. For instance, an effective vaccination programme on domestic East African cattle in the 1950s managed to eradicate the virulent rinderpest virus that was introduced to the region in the nineteenth century and spread to both the local cattle and the wild populations of ungulates. The establishment of the virus had a detrimental impact on the native artiodactylid wildlife, disrupting the entire ecosystem, and the subsequent removal of the disease changed not only the assemblage of ungulates but also the community of primary producers on which these herbivores were grazing as well as the community of predators relying on them as prey (Plowright 1982; Dobson and Hudson 1986; Dobson 1995a, b; Thomas et al. 2005a; Lebarbenchon et al. 2009).

Whereas the *Tribolium* example represents the addition of a single shared parasite species to a simple but controlled two-host laboratory community, the rinderpest example represents uncontrolled removal of a single parasite species from an entire natural grassland ecosystem. Together, however, these two very different lines of evidence suggest a general and potent community structuring role of parasitism that have been increasingly recognised as supportive evidence has accumulated over the years (Freeland 1983; Price et al. 1986; Dobson and Hudson 1986; Minchella and Scott 1991; Sousa 1991; Hudson and Greenman 1998; Poulin 1999; Combes 2001; Mouritsen and Poulin 2002; Thomas et al. 2005b; Hatcher and Dunn 2011; Hatcher et al. 2012; Frainer et al. 2018).

Before we turn to specific lines of evidence from aquatic communities, it will be useful to address three basic ways in which parasites appear to influence the organisation of natural communities: (1) through direct effects, (2) through density-mediated indirect effects, and (3) through trait-mediated indirect effects (Fig. 8.1; also see Chaps. 7 and 18).

Direct Effects Firstly, community structure can be affected by parasites through their direct effects on the size of the various host populations that together constitute the full community. Most, if not all, living species of plants and animals are host to one or more species of parasites. As mentioned, parasites are usually harmful to their hosts and hence likely to regulate their host population. Virulence differs between different types of parasites and their infection pattern and parasite-induced effects are often asymmetrical as susceptibility and tolerance to parasites differs among host species. Host A may be less affected or less infected by parasite A than host B is affected or infected by parasite B. Such differential impact of parasitism on hosts may even manifest itself between closely-related host species sharing the same species of parasite, a phenome-

---

Fig. 8.1 (continued) hosts sharing a parasite may ultimately become locally extinct (apparent competition). (**b**) *The density-mediated indirect effects*. A specialist parasite (circle) infecting an abundant keystone species (K). Apart from reduced host abundance due to parasite pathology, the community structure will also be affected by changes in the abundance of the non-host organisms following from a cascade of interspecific interactions (predation, competition, facilitation and inhibition). If the host (species A) is a predator or superior competitor, the affected species B will increase in relative frequency. If species C likewise is affected negatively by species A, but also inhibited by species B, that species' abundance may remain unchanged in presence of parasites. Finally, if species C is facilitated by species A, it will decrease in abundance when the population of species A decreases. (**c**) *The trait-mediated indirect effects*. A specialist parasite (circle) infecting an abundant keystone species (K). Albeit unaltered in abundance, the parasite-induced changes in the phenotype of the key stone species will, for instance, keep one of its prey species or inferior competitor unaffected (species B), whereas another may be more strongly affected (species C). Finally, the way in which the keystone species facilitates species D may improve, resulting in increased abundance

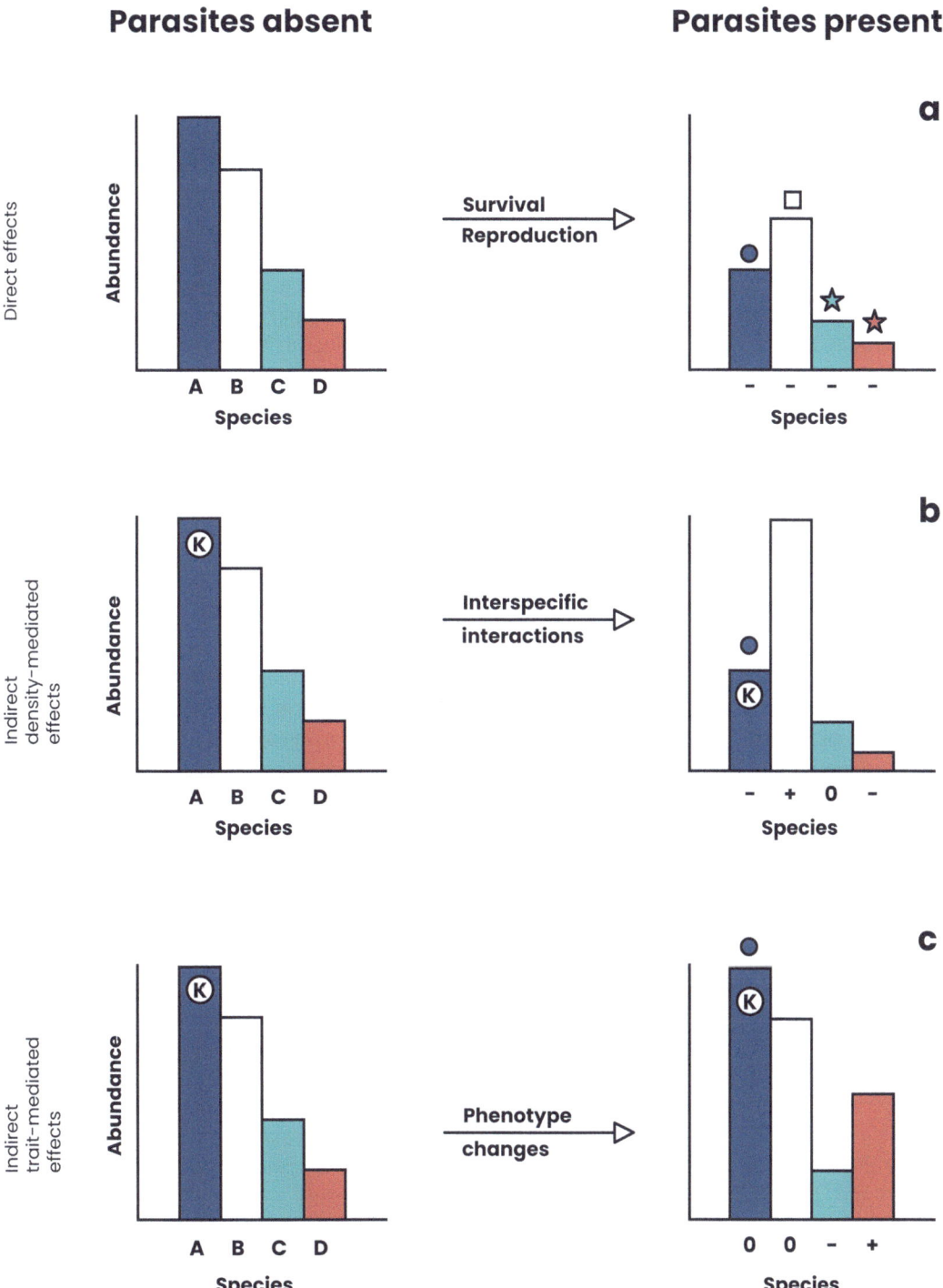

Fig. 8.1 The three basic ways parasitism impacts the structure of a hypothetical community of four free-living species. Letters (A–D) indicate different species in the community. Increase, no change and decrease in abundance of given species is indicated by +, 0 and –, respectively. (**a**) *The direct effects*. Two host species infected by two specialist parasite species (circle, square) and two host species sharing a generalist parasite (asterisk). As a consequence of asymmetrical parasite-induced negative effects on survival and reproduction, the host community will in the presence of parasites not only decrease in abundance as a whole, but also occur in different relative proportions. The less abundant of the two

non that has fuelled the field of parasite-induced apparent competition, parasite-mediated competition and, in turn, the whole concept of parasitism as a community structuring process (see e.g. Hatcher and Dunn 2011). A community of hosts infected by a parallel community of parasites (some virulent, some benign and some shared among groups of hosts) will, partly as a consequence of these parasite infections, exist in specific relative proportions of abundance below the carrying capacity of the habitat. Released from their parasites and the negative impact of virulent parasites on survival and reproduction, the host community will not only increase in abundance, their relative frequencies are also bound to change, all else being equal (Fig. 8.1a).

Density-Mediated Indirect Effects Secondly, community members never exist as isolated identities but generally interact through processes of, for example, predation, competition and facilitation, or inhibition. Apart from the direct effect of parasites on the abundance of a given host population, a parasite may also indirectly affect the occurrence of a range of non-host organisms in the community through interaction cascades. This may be particularly evident if the host species acts as a key-stone species or is a bioengineer in the community. For instance, if the population size of a keystone predator is diminished by a species-specific parasite, the predator's prey species are likely to increase in abundance. This, in turn, may also affect non-prey community members through interference competition between prey and non-prey organisms (Fig. 8.1b). This is just one example out of many possible interaction cascades mediated by parasitism, and a full account of such envisaged parasite-mediated interactions is reviewed in Hatcher and Dunn (2011).

Trait-Mediated Indirect Effects Thirdly, even if a parasite has little or no measurable effect on the population size of its host, it may still have a profound impact on the structure of the natural community in which the host–parasite system is embedded. Parasites often change the phenotype of their hosts, and this may alter the hosts' ecological role in the community. For instance, if the consumption rate of a keystone predator, or the activity level of a superior interference competitor, is changed due to parasitism, it will likely cascade through the community in ways similar to those expected to result from decreased host abundance (Fig. 8.1c; Hatcher and Dunn 2011; Hatcher et al. 2014).

The above three basic processes of parasite-mediated community structure are clearly not mutually exclusive. They will under in situ conditions act in concert and are thus difficult to separate. Furthermore, their relative importance will intrinsically depend on the actual functional composition of hosts and parasites in the community, that is, the ways and the strength by which the free-living species interact and the pathology of the parasites infecting them, and therefore also differ between ecosystems. Although there is increasing evidence suggesting that all three processes are active in shaping natural communities, the specific way in which they influence—either in isolation or together—standard community metrics such as species diversity and richness in intertidal habitats is still poorly understood.

8.2 Direct Effects, Apparent and Parasite-Mediated Competition

No one has yet managed to remove all parasites from a full host community—terrestrial, freshwater or marine—and monitored what such a manipulation might bring about in terms of structural changes to the species assemblages. Nor has any study, for the same purpose, introduced a full guild of parasites to a previously parasite-free community of potential hosts. Although such approaches would undoubtedly be enlightening and could definitively demonstrate the full-scale community impact of parasitism if replicated, they are either prohibitively impractical or simply unethical. What little is known comes from natural or accidental introductions of usually a single species of parasite to a natural community of host and non-host organisms (e.g. the rinderpest example) or the addition of a common parasite to a simple two-host community under controlled laboratory settings (e.g. the *Tribolium* example). The latter represents apparent or parasite-mediated competition, in that host species sharing the same parasite may occur in different relative proportions in presence rather than in absence of the parasite due to its often differential impact on the different host species survival, fecundity or competitive ability (see Holt 1977; Holt and Lawton 1994; Hudson and Greenman 1998; Tompkins et al. 2002; Hatcher and Dunn 2011 for discussions and definitions).

There exist several terrestrial and freshwater examples that are comparable to the *Tribolium* study. In terrestrial habitats, dominance of Ring-necked pheasant (*Phasianus colchicus*) over Grey partridge (*Perdix perdix*) in UK is due to a shared nematode parasite, *Heterakis gallinarum* (Tompkins et al. 2000a, b, 2001); replacement of native red squirrels (*Sciurus vulgaris*) by the introduced grey squirrel (*S. carolinensis*) is due to Squirrelpox Virus (SQPV), a viral disease detrimental to the native but not the invasive species (Tompkins et al. 2003); and elimination of one of two sympatric species of *Drosophila* fruit flies is due to the parasitoid wasp *Leptopilina boulardi* (Boulétreau et al. 1991). From freshwater ecosystems, we find evidence of reversal of the competitive hierarchy between two frog species, the Cascades frog (*Rana cascadae*) and the Pacific treefrog (*Pseudacris regilla*) due to a pathogenic water mould, *Saprolegnia ferax* (Kiesecker and Blaustein 1999), differential parasite-induced mortality between two competing species of crayfish (the native *Austropotamobius pallipes* and the invasive *Pacifastacus leniusculus*) infested by a fungus (*Aphanomyces astaci*) as well as a microsporidian parasite (*Thelohania contejeani*) in the United Kingdom (Dunn 2009; Dunn et al. 2009; Hatcher et al. 2012), and spatial segregation of two species of boatmen (Hemiptera: Corixidae) due to parasitic water mites (Acari) (Scudder 1983; Bennett and Scudder 1998). In the latter case, two closely-related boatmen, *Cenocorixa bifida* and *C. expleta*, coexist in saline lakes in British Columbia (Canada) that have higher salinities. In the lower-salinity lakes, on the other hand, *C. expleta* is generally absent. Physiologically and ecologically, both species have similar ability to live in low-salinity lakes and there is no indication of competitive exclusion. However, the two species of boatmen show marked differences in susceptibility to water mite parasitism (genera *Eylais* and *Hydrachna*) that is particularly abundant in low-salinity lakes and appears lethal to *C. expleta* but not *C. bifida*.

There is similar evidence from coastal ecosystems, albeit the number of reports is limited. The mud shrimp (amphipod) *Corophium volutator* occurs abundantly in soft-bottom intertidal habitats on both sides of the north Atlantic. In its eastern distributional range, *C. volutator* co-occurs with the competitively inferior and much less abundant *C. arenarium*. The latter species is consequently often limited to marginal microhabitats such as the more sandy fringe of the upper tidal zone or other areas of the mudflat generally avoided by *C. volutator*. Both amphipod species act as secondary hosts to a range of microphallid trematodes that use the sympatric mud snail *Peringia ulvae* as first intermediate hosts. Because *C. arenarium* is less affected than *C. volutator* by trematode infections (Jensen et al. 1998), *C. arenarium* may experience a competitive release and invade areas previously occupied by *C. volutator* if the population of the latter is decimated by parasitism. This was unequivocally demonstrated by an incidence of rapid microphallid-induced decline in a population of *C. volutator* in the Danish Wadden Sea, in the spring of 1990 (Jensen and Mouritsen 1992; see Sect. 8.3.2 for details). *Corophium volutator* went locally extinct and did not return to the area until the following year. In its absence, however, the competitively inferior but parasitically robust *C. arenarium* increased in abundance, only to disappear again when the superior competitor eventually regained its former population strength two years later (Larsen et al. 2011; Fig. 8.2).

At a larger spatial scale, the bucephalid trematode *Prosorhynchus squamatus* has been held

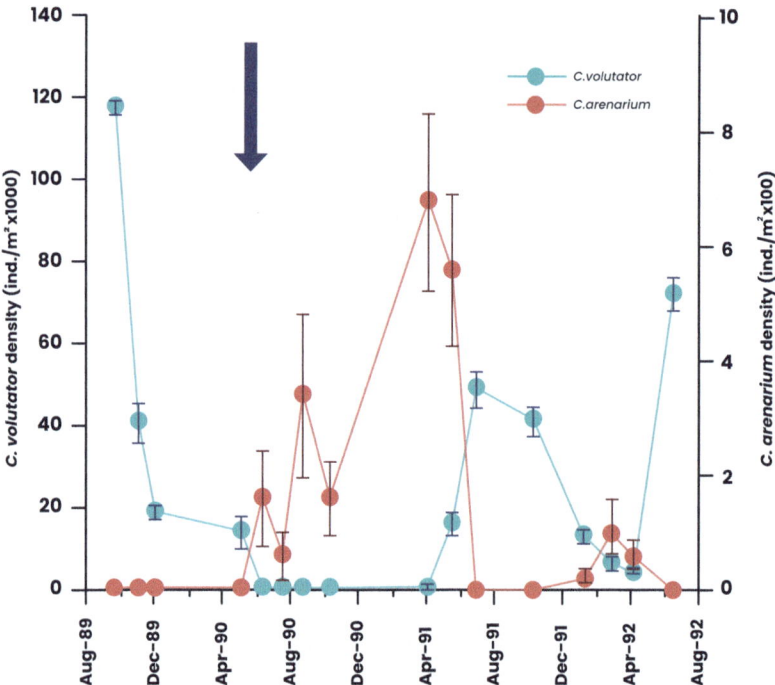

Fig. 8.2 Parasite-mediated competitive release in sympatric amphipod populations. The population dynamics (density m^{-2} ± SE) of the competitive superior *Corophium volutator* and the less abundant inferior competitor, *C. arenarium*, in the southern Danish Wadden Sea, 1989–1992. Arrow denotes the onset of the microphallid-induced die-off in the *C. volutator* population in spring 1990 [After Larsen et al. (2011)]

partly responsible for the latitudinal position of the hybrid zone at the European Atlantic coast between the northern mussel *Mytilus edulis* and the more southerly distributed *M. galloprovincialis* (Coustau et al. 1990, 1991). The parasite uses the mussels as first intermediate hosts, which results in total castration of infected hosts and possible death. The *Mytilus edulis* genotype seems particularly susceptible to these bucephalid infections, and the parasite hence configures *M. galloprovincialis*'s distribution in a northward direction by keeping the relative frequency of *M. edulis* lower than it would have been in its absence.

Looking at a slightly more complex host community sharing a common parasite, Poulin and FitzGerald (1987) proposed that the fish louse *Argulus stizostethii* (as *Argulus canadensis*) plays a role in structuring a Canadian saltmarsh community of three sympatric species of sticklebacks (Gasterosteidae). Laboratory experiments showed that fish exposed to the parasite suffered higher mortality than unexposed control fish, and that one of the sticklebacks, *Pungitius pungitius*, was significantly more susceptible to infection than the two other species, *Gasterosteus aculeatus* and *G. wheatlandi*. However, the relative frequency of *P. pungitius* can be expected to be particularly low at sites where fish lice are abundant.

Mouritsen et al. (2018) also showed, through an outdoor mesocosm experiment, the differential impact of the microphallid trematode *Maritrema novaezealandense* on six sympatric species of amphipods from New Zealand soft-bottom intertidal habitats. Although all potential host species clearly suffered from exposure to these trematodes, the abundant epifaunal *Paramoera chevreuxi* was particularly sensitive, whereas the benthic species *Proharpinia stephenseni* and *Paracorophium lucasi* were least affected (Fig. 8.3a). As a consequence of this species-specific effect of parasitism, both mean species richness and diversity of the experimental amphipod host community decreased significantly. The implication of these findings is underlined by the existence of a negative relationship between amphipod species richness and the abundance of *M. novaezealandense* infections in field samples collected from a range of sand flats neighbouring the site of experimentation (Fig. 8.3b). From New Zealand lake ecosystems

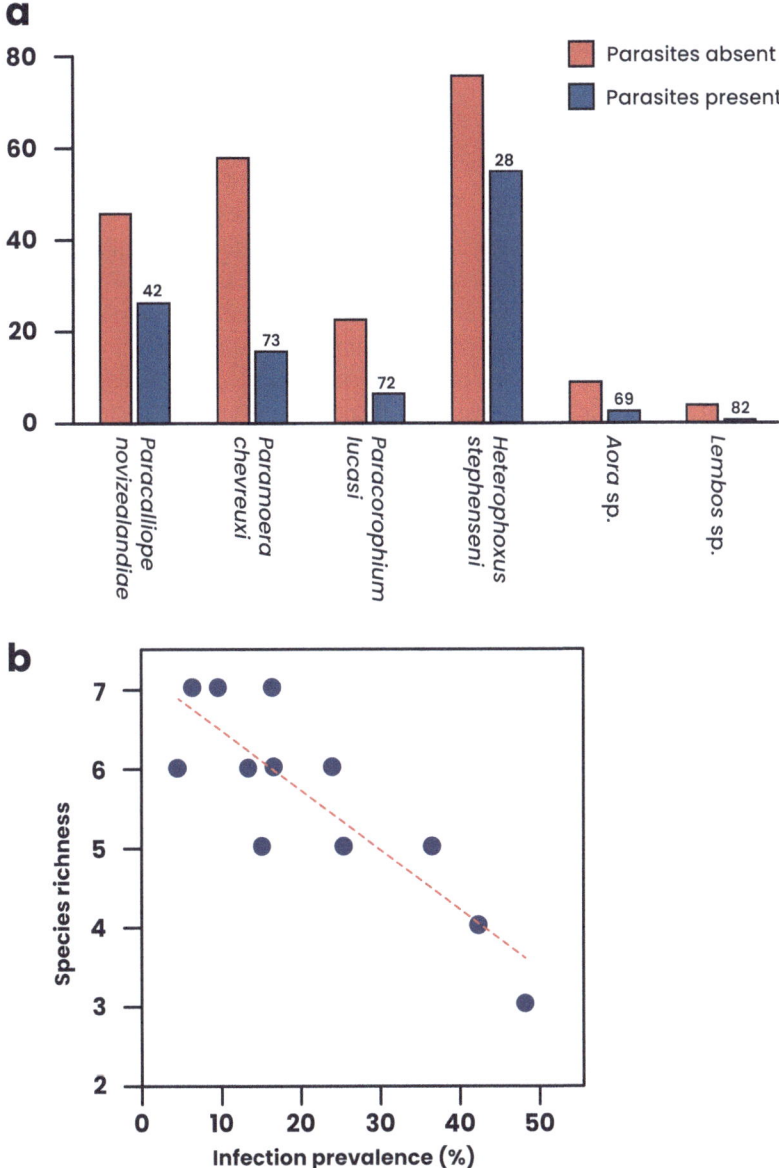

Fig. 8.3 (**a**) Post-experimental abundance (mean no. individuals per experimental container ± SE) of six different benthic species of amphipods respectively unexposed and exposed to *Maritrema novaezealandense* cercariae during a three-week outdoor mesocosm experiment at Portobello Marine Laboratory, Otago Harbour, New Zealand. Experimental abundance of amphipods and parasites (in terms of added cercariae producing first intermediate snail hosts, *Zeacumantus subcarinatus*) corresponded to densities found in situ. Percent parasite-induced decline is indicated. (**b**) The relationship between mean amphipod species richness and arcsine-transformed infection prevalence of *M. novaezealandense* (%) in the first intermediate host population of *Zeacumantus subcarinatus* from 12 different intertidal sand flats in Otago Harbour ($r^2 = 0.742$, $P < 0.0005$) [After Mouritsen et al. (2018) and unpublished data]

we find similar evidence. In a mesocosm experiment, Friesen et al. (2020) followed the population dynamic of four sympatric freshwater crustacean species (two amphipods and two isopods) sharing two to four parasite species (two trematodes, one acanthocephalan and one nematode), and found that the resulting relative abundances of host species were influenced by the parasites.

8.2.1 Lessons from Invasions

The community structuring potential of parasitism may also become evident in association with invasion by hosts, parasites or both to areas previously not occupied by these organisms (see Chap. 18). For instance, the regulating parasites of an invading host species are often lost, and possibly for that reason, the hosts attain far greater population densities in the new distributional range (Torchin et al. 2002, 2003; Torchin and Mitchell 2004; Dunn 2009). Aside from demonstrating a population-regulating role for these parasites, it also suggests a different community structure in the indigenous region had the parasites been absent, as well as in the invaded area had the parasites here been present. If the invading species is simultaneously forced into competitive interactions with members of the new community and happens to be less susceptible to parasites residing there (e.g. Cornet et al. 2010), it will provide the invader with an advantage over the parasite-regulated competitors in the new environment. This may ultimately result in replacement of resident species by the invader, further emphasising parasites' community structuring role. These topics are dealt with in detail in Chap. 18.

8.2.2 Behavioural Aspects and Parasites as a Resource

So far, we have discussed the ability of parasitism to directly modify the relative abundance of species—and consequently community structure—solely on the basis of the pathology that infective agents inflict on their hosts in terms of reduced survival and fecundity. Parasites may also directly affect community organisation in total absence of parasite-induced casualties in the host population. Parasites commonly change the behaviour of their hosts either facilitating their own transmission or as a side effect of infection. Regardless of the reason, parasitism divides a host population into infected and uninfected cohorts, each expressing different behavioural patterns. When such phenotypic change involves dispersal, it is bound to affect the spatial distribution of host individuals and in turn community structure. The impact of these processes on invertebrate hosts of generally limited dispersal ability may occur on a small spatial scale, for instance, along environmental gradients or among neighbouring sites supporting different levels of parasitism. Trematode infected populations of intertidal gastropods provide good examples of this phenomenon, as the infected individuals' home range and direction of dispersal are often considerably altered. For instance, the mud snails *Batillaria cumingi* (Japan), *Ilyanassa obsoleta* (USA) and *Diloma subrostratum* (New Zealand) as well as periwinkles *Littorina littorea* (Europe, USA) move—depending on host species, trematode species and locality in question—to either lower or higher tidal levels when infected (Lambert and Farley 1968; Williams and Ellis 1975; Curtis 1987, 1993; McCarthy et al. 2000; McCurdy et al. 2000; Miller and Poulin 2001; Miura et al. 2006). Hence, the density of snail hosts will increase at lower or higher tidal levels when there is an increase in parasitism. Because these species of host snails are often abundant as well as heavily parasitised, the impact of modified distributional patterns on community metrics along a depth gradient should indeed be measurable. Studies addressing this potential have not yet been carried out in aquatic systems.

The same holds true with respect to direct community effects of parasites as a food source for free-living predatory or omnivorous organisms. Helminths with free-living larval stages in their life cycle may experience a considerable loss of larvae to a wide range of non-host organisms that deliberately or accidentally ingest them (Thieltges et al. 2008b; Johnson and Thieltges 2010; Johnson et al. 2010; Koprivnikar et al. 2023). Trematodes may be especially relevant in this context. Considering the often high abundance of molluscs in aquatic habitats, the often high prevalence of trematode infections in these first intermediate host populations, and the continuous asexual production and release of larvae (cercariae) from every infected host individual, the abundance of these free-living larvae can be enormous, potentially representing a valuable as well as predictable source of energy. Accordingly, Kuris et al. (2008) estimated the annual cercarial production from gastropod intermediate host populations in a Californian estuary to surpass the biomass of the combined shorebird community also present in the estuary. Similarly, Thieltges et al. (2008c) found that the mean annual production of cercariae across several species of coastal trematodes and snail hosts was within the same order of magnitude as that of free-living benthic taxa such as molluscs, annelids, arthropods and echinoderms. Preston et al. (2013, 2021), Soldánová et al. (2016) and Rosenkranz et al. (2018) posit that the substantial biomass of trematode cercariae in pond and stream habitats may exceed that of the coexisting insect fauna. The link to community structure here is straight forward, as non-host organisms managing to exploit this significant food source may receive a fitness benefit that potentially affects their population density positively. In a parasite-free environment, this opportunity would be absent. Although there exists several reports of cercarial consumption by aquatic predators (Koprivnikar et al. 2023), investigations specifically targeting how this path of energy flow affects consumer abundance and community are largely lacking.

Indications that the impact is far from negligible comes from the New Zealand mud flat anemone *Anthopleura hermaphroditica*, that has been shown to heavily prey on cercariae released from intertidal mud snails and whelks (Mouritsen and Poulin 2003a; Hopper et al. 2008). In a correlative study across several intertidal mudflats, Mouritsen and Poulin (2010) showed that the density of mudflat anemones was positively related to the abundance of larval echinostome trematodes, even when other population regulating factors were corrected for. Mouritsen and Haun (2008) also argued that a positive relationship established between the abundance of capitellid polychaetes and cercariae releasing mud snails (*Peringia ulvae*) in a field experiment in the Danish Wadden Sea was possibly linked to the worms' ingesting of larval trematodes. Schultz and Koprivnikar (2019) showed experimentally that the presence of larval trematodes (*Plagiorchis* sp.) increased the population of freshwater zooplankton (*Daphnia* sp.) by ~50% when exposed to larval dragonflies (*Leucorrhinia intacta*) known to prey on both cercariae and *Daphnia*. Presumably, the parasite larvae acted as alternative prey for the larval odonates, which in turn mitigated the predation pressure on *Daphnia*. Not only invertebrates but also intertidal vertebrates prey on free-swimming cercariae. Juvenile killifish *Fundulus parvipinnis* in Carpinteria Salt Marsh, California, have been found to ingest a range of different species of cercariae emitted from the California horn snail, *Cerithideopsis californica* (Kaplan et al. 2009), and it was estimated that this food source potentially could account for 2–3% of the energetic needs for the local fish community. In what way, if at all, this additional source of food influences the structure of intertidal fish communities remains to be established.

8.3 Density-Mediated Indirect Effects: Lessons from Epidemics

Whereas subtle regulation of host populations by parasites and the resulting indirect ramifications for the surrounding community may be widespread in aquatic ecosystems, such interaction cascades will, in most cases, pass unnoticed to the scientific world. A visit to any given tidal habitat will, however, quickly reveal the presence of a potential keystone predator or herbivore that can readily be envisaged to play an important structuring role in the community; from there, investigating any hypothesised indirect impact on community organisation should be straightforward. The situation is somewhat different when dealing with parasites that are often well hidden in the bodies of their hosts, and whose impact on the surrounding community therefore is far from obvious. We see predators kill, but rarely parasites (Price 1980). Hence, our understanding of how parasites indirectly influence the community through regulation of keystone host abundance originates to a large extent from their full-scaled manifestations: epidemics.

8.3.1 Eelgrass Meadows and Protists

One of the most spectacular changes to the intertidal habitat in historic time involved the eradication of the widespread seagrass meadows in the early 1930s, during which approximately 90% of the temperate eelgrass *Zostera marina* disappeared in the North Atlantic region (Tutin 1938; Rasmussen 1977). The die-off has been partly attributed to 'wasting disease' caused by infection by a virulent species or strain of slime mould of the genus *Labyrinthula* (Protista) (Rasmussen 1977; den Hartog 1987; Short et al. 1987; Muehlstein 1989).

Eelgrass is a key bioengineer of shallow subtidal and lower intertidal soft-bottom habitats, where its presence facilitates the occurrence of a long array of other plant and animal species. It plays an important role in providing structural complexity both below and above the sediment surface, stabilising the substrate and stimulating precipitative sedimentation of finer particles, enhancing larval settlement and thereby also presence of infaunal species, providing refuge for fishes and epibenthic organisms against predators, serving as substrate for a rich community of epiphytes and sessile invertebrates, forming attractive and vital feeding grounds for various herbivorous and predatory shorebirds such as swans, geese, ducks, gulls and herons (Milne and Milne 1951; Cottam and Munro 1954; Orth 1975, 1977; Peterson 1979; Bertness 2007; Boström and Bonsdorff 2000; Valdez et al. 2020). Sea grass beds therefore play a critical role in enhancing the soft-bottom community, and both the diversity and abundance of many functional and taxonomical different groups of organisms are generally greater inside than outside seagrass beds (Connolly 1994; Bertness 2007; Mattila et al. 1999; Boström and Bonsdorff 2000; Carvalho et al. 2006; Kelly et al. 2008; Valdez et al. 2020).

By almost wiping out a keystone species of the coastal soft bottom community, the 1930s *Labyrinthula* sp. epidemic prompted a regime shift in the benthic assemblages. Fishes and birds depending on eelgrass beds rapidly decreased in numbers (Milne and Milne 1951; Cottam and Munro 1954) and the muddy soft bottom indirectly created by the eelgrass canopies through increased sedimentation eroded into a coarser and stony seafloor (Rasmussen 1977). This in turn changed the benthic community of invertebrates. Generally, deposit-feeders were replaced by filter-feeders, and the overall abundance and species richness declined (Stauffer 1937; Rasmussen 1977). Indeed, a reanalysis of the data from a west Atlantic locality, collected prior to and after the onset of the wasting disease (Stauffer 1937), shows a statistically significant 40% decline in species richness and that 55% of all benthic species recorded prior to the incident had decreased in abundance (Fig. 8.4). Although the wasting disease generally caused a functional shift in the benthic community towards suspension-feeders, some of these were also strongly negatively affected. For instance, along western Atlantic shorelines, the recruitment of the bay scallop *Argopecten irradians* is directly dependent on eelgrass, and this suspension-feeding bivalve experienced dramatic population declines after the eelgrass die-off (Thayer and Stuart 1974). Similarly, and perhaps even more spectacularly, it appears that the close-to-full eradication of *Zostera marina* beds along north Atlantic coastlines caused the complete extinction of the eelgrass limpet *Lottia alveus* (Carlton et al. 1991). This gastropod limpet was entirely dependent on eelgrass, found only on leaves of *Zostera marina*, and was last recorded immediately prior to the *Labyrinthula* sp. epidemic. The parasite-induced eelgrass die-off therefore not only reduced the local alpha diversity in the coastal zone but also negatively affected global biodiversity.

This example clearly demonstrates the potential of parasitism to determine aquatic commu-

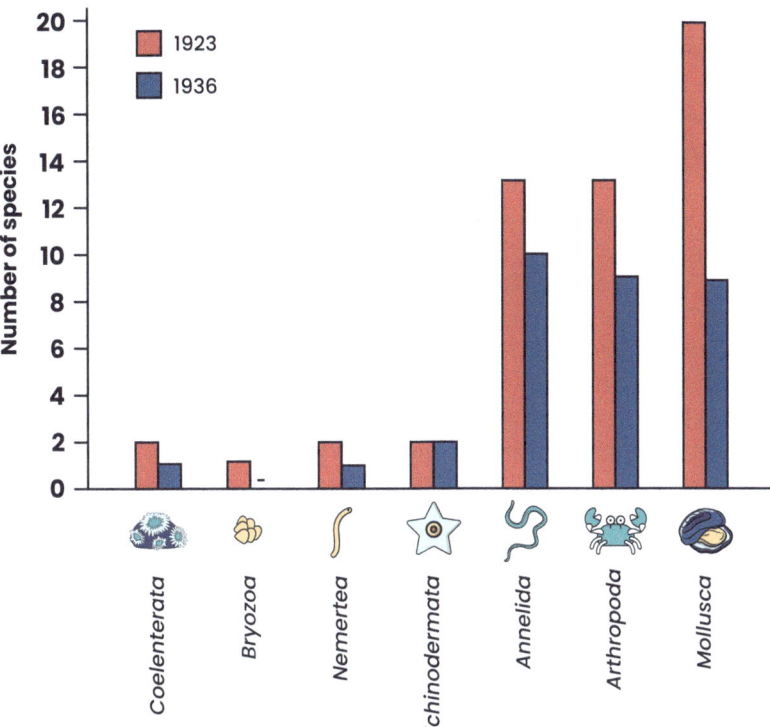

Fig. 8.4 The number of benthic species of invertebrates (distributed on taxonomical classes) recorded in an east Atlantic locality (Woods Hole) prior to (1923) and after (1936) the *Labyrinthula*-induced epidemic in eelgrass *Zostera marina* during the early 1930s. The species distribution among classes differs significantly between years (Two-way contingency test, $\chi^2_1 = 5.19$, $P = 0.023$) [Reanalysed after Stauffer (1937)]

nity structure and function through host density-dependent indirect effects. Not surprisingly, when the host species is an important facilitator in the ecosystem, the role of the parasite will be to decrease diversity. The next example shows that the community impact of parasitism can be the exact opposite, that is, increasing diversity, when the host acts mainly as an inhibitor.

8.3.2 Amphipod Beds and Microphallid Trematodes

In the Wadden Sea, southeastern North Sea, the widespread infaunal tube-building mud shrimp (amphipod) *Corophium volutator* may, following a summers burst of reproduction, achieve densities exceeding 100,000 individuals per square metre. The tube-building activity of this surface deposit-feeding amphipod often results in a soft, silty but also stable substrate, which manifests itself as elevated emerged areas or plateaux on the intertidal mudflats during low tide (see Mouritsen et al. 1998). Such amphipod-beds are very attractive to smaller shorebirds, particularly sandpipers (*Calidris* spp.) that feed on the beds in densities several fold higher than on the surrounding flats (Mouritsen 1994). Apart from *C. volutator*, the small herbivorous mud snail *Peringia ulvae* also occurs abundantly on these tidal flats, and together, snails, amphipods and birds constitute the first intermediate, second intermediate and definitive hosts, respectively, in the life cycles of a range of microphallid trematode species (Deblock 1980; Jensen and Mouritsen 1992; Mouritsen et al. 1997). *Maritrema subdolum* and *Microphallus claviformis* are particularly abundant and have been shown to substantially elevate the mortality rate of *Corophium* both under laboratory conditions and in situ (Bick 1994; Mouritsen and Jensen 1997; Meissner and Bick 1997, 1999; Jensen et al. 1998; Meissner 2001).

One of the most significant microphallid-induced die-offs recorded in an amphipod population took place in the spring of 1990 in the Danish Wadden Sea. Here, the density of microphallid-infected mud snails exceeded 10,000 per square metre, and the vast amount of parasite larvae (cercariae) continuously released from these snails inflicted, upon their transmission, a mass mortality in the sympatric population of *C. volutator*. In effect, about 15,000 amphipod individuals per square metre vanished, together with a majority of the infected population of snails in a mere five weeks from May to June; *C. volutator* went locally extinct for approximately one year (Jensen and Mouritsen 1992; Fig. 8.2). Aside from the dramatic decline in the populations of first and particularly the second intermediate hosts, the surrounding benthic community also changed character because of this massive parasite attack (Fig. 8.5). Both *P. ulvae* and *Corophium* spp. are important grazers on benthic diatoms, exerting a significant top-down control at high densities (e.g. Gerdol and Hughes 1994; Hagerthey et al. 2002), and now partly released from this grazing pressure, the microphytobenthos increased in abundance (Fig. 8.5b). The predatory nemertean *Lineus ruber* disappeared together with its main prey, *C. volutator*, and so did in part the ragworm *Hediste (Nereis) diversicolor* (Fig. 8.5c). This omnivorous polychaete occasionally preys on *C. volutator* but may also have been indirectly favoured by the amphipods' substrate stabilising effect. In contrast to *C. volutator*, the relatively rare amphipod *Corophium arenarium* increased in density (Fig. 8.5c). As an inferior competitor to *C. volutator* (Jensen and Kristensen 1990), this parasite-resistant species must have experienced a competitive release and thus invaded, albeit in much smaller numbers, the areas previously occupied by the dominant sister species (see Sect. 8.2). Because of its surface deposit-feeding activity, *C. volutator* is also known to inhibit the recruitment of other benthic invertebrates (Ambrose 1984; Ólafsson and Persson 1986; Limia and Raffaelli 1997) and this may also, in part, justify the increases in abundance of the polychaetes *Pygospio elegans* and the bivalve *Macoma balthica* following the extermination of the amphipod population (Fig. 8.5). Together with the unaffected species, these immediate responses to the amphipod die-off resulted in twofold increases in the species diversity of the macrozoobenthic community (Fig. 8.5d).

On a longer time scale, the benthic community may have been subject to even more radical changes: as the amphipods vanished, so did their stabilising effect on the substrate, and the plateaux of the former sea bed area were subject to significant erosion that eventually changed the entire topography of the seafloor (see Mouritsen et al. 1998). It is likely that the disappearance of *C. volutator* also affected higher trophic levels of the ecosystem, although this was not documented. For instance, shorebirds like Dunlin *Calidris alpina* often prey on benthic amphipods and is therefore strongly attracted to *Corophium* amphipod-beds while feeding during diurnal and nocturnal low tides (Mouritsen 1994; MacDonald et al. 2014). Dunlin often occur in several-fold higher densities on these beds than on the ordinary sand flats and, being one of the most abundant shorebird species in the Wadden Sea, it seems inevitable that a prolonged absence of benthic amphipods would affect the overall structure of feeding shorebirds at the site.

8.3.3 Sea Urchins, Amoeba and Nematodes

Sea urchins (*Strongylocentrotus* spp.) also enforce a top-down control of benthic primary producers in soft-bottom intertidal habitats and lower fringes of the intertidal rocky shore. Here, the sea urchins graze heavily on kelp forests (i.e. *Laminaria* spp.) and the associated macroalgal community and, if not held in check by predators such as sea otters and lobsters (Estes et al. 1998; Lafferty 2004; Kriegisch et al. 2019; Bernal-Ibáñez et al. 2021), the devastating grazing activity of a rapidly increasing population of sea urchins will eventually transform the species-rich kelp ecosystem into a species-poor barren ground. Because sea urchins can subsist solely on detritus as a food source, these echinoderms may persist in high densities following the disappearance of the macroalgal community, and thus maintain this barren, low-diversity steady state for years (Hagen 1995a). In the

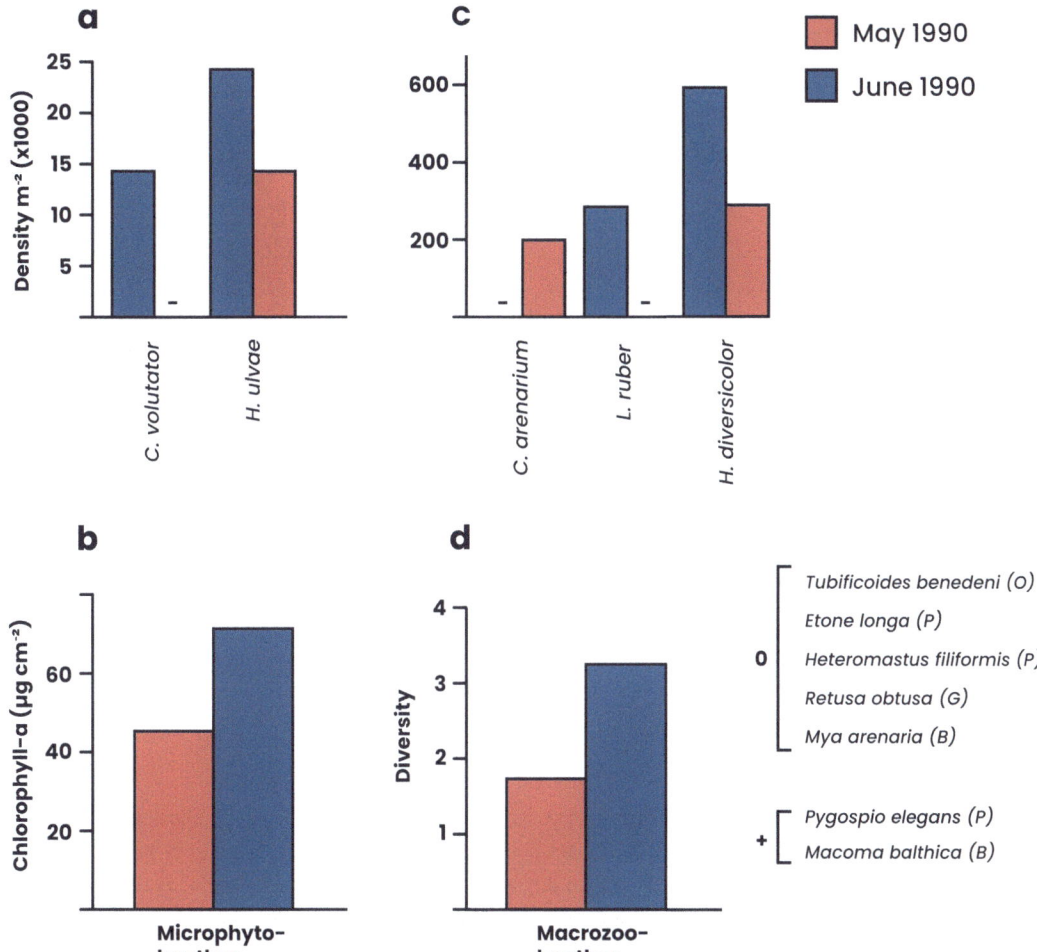

Fig. 8.5 Significant immediate changes to the benthic community during the microphallid-induced die-off in the population of the amphipod *Corophium volutator* in the Danish Wadden Sea, Højer, Maj-June 1990. (**a**) Changes in mean abundance (no. m^{-2}) of first (*Hydrobia (Peringia) ulvae*) and second (*Corophium volutator*) intermediate hosts. (**b**) Change in mean abundance of microphytobenthos (mainly diatoms) in terms of chlorophyll-a (µg cm^{-2}). (**c**) Changes in mean abundance (no. m^{-2}) of the inferior amphipod competitor *Corophium arenarium*, the nemertean predator *Lineus ruber* and the omnivorous polychaete *Hediste (Nereis) diversicolor*. (**d**) Change in the species diversity (Simpsons diversity index) of macrozoobenthic organisms. The increase from May to June is statistically significant (Student's *t*-test, $t_{18} = 7.34$, $P < 0.0005$). 0 denotes benthic species of macroinvertebrates unaffected by the die-off; + denotes species that increased in abundance. O oligochaete, P polychaete, G gastropod, B bivalve. Redrawn and reanalysed after Jensen and Mouritsen (1992) and Mouritsen et al. (1998)

absence of predators, parasites appear to be the key agents that eventually terminate the sea urchin outbreaks on these hard-bottom habitats (Scheibling 1984, 1986; Scheibling and Stephenson 1984; Hagen 1995a, b; Feehan and Scheibling 2014). Epizootics in dense populations of sea urchins have been documented in warmer waters of the north-west Atlantic due to infections of the amoeba *Paramoeba invadens*, and in colder waters due to the endoparasitic nematode *Echinomermella matsi*. These directly-transmitted and virulent parasites manage to almost eradicate the sea urchin populations and, released from the urchin grazers, the kelp forest will re-establish itself on the former barren ground and once again develop into the high-diversity steady state. Hence, in these ecosystems there seems to exist a cyclic develop-

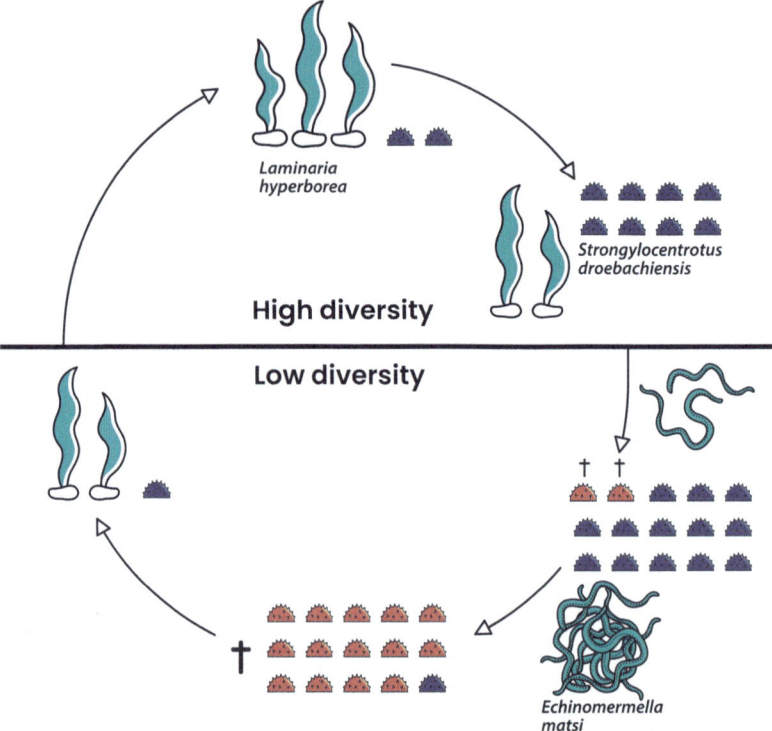

Fig. 8.6 The cyclic development between the high-diversity kelp forest community (i.e. *Laminaria hyperborea*) and the sea urchin *Strongylocentrotus droebachiensis* dominated low-diversity barren ground, in this case driven by recurrent outbreaks of the endoparasitic nematode *Echinomermella matsi* in dense sea urchin populations

ment between two principal community configurations, kelp forests and urchin barren ground, in which parasitism plays a decisive role in elevating species diversity by decimating the population of a keystone grazer (Fig. 8.6).

8.3.4 Lessons Learned

Whereas the above three examples of in situ epidemics clearly emphasises the general community structuring potential of parasitism, they do also represent 'natural experiments' where causes and effects are not always well-established. Overlooked processes and factors unrelated to parasitism may, in principle, have participated significantly in producing the observed community impact. Furthermore, such lessons from epidemics naturally represent the extreme cases, which raises the question whether a more subtle 'everyday' down regulation of a keystone species by its parasites also has a measurable effect on community structure. To answer this question, experiments, particularly in the field, are surely needed. In this respect, we are better off when it comes to assessing the community impact of parasitic modification of host traits, as the following section will demonstrate.

8.4 Trait-Mediated Indirect Effects: The Cryptic Interactions

Aside from pathological effects directly reducing a host's prospects of survival, parasite infections quite often also modify their phenotypic expression. This includes behavioural changes of infected hosts due to the presence of a parasite, wherein the change in the behaviour will increase the potential of the parasite to complete its complex life cycle (i.e. intermediate hosts being eaten by the definitive host), although this is not always the case. Whatever the adaptive value of these parasite-induced changes to host traits might be, they are likely to ramify to higher levels of biological organisation if a host acts as a keystone species or bioengineer in the ecosystem in question (Fig. 8.1). Three examples of such trait-mediated indirect impacts on aquatic community structure display such changes: two from the coastal soft-bottom habitats involving both frequency- and intensity-dependent effects of parasites, and one from coastal hard-bottom habitats involving frequency-dependent effects.

8.4.1 Surfacing of the New Zealand Cockle

The suspension-feeding New Zealand cockle, the venerid *Austrovenus stutchburyi* (Veneridae), is probably the most abundant mollusc on New Zealand sand flats. It often attains densities of several hundred individuals per square metre (Stewart and Creese 2002; Mouritsen and Poulin 2010), and due to its relatively large size, the sand flats just below the sediment surface may literally be paved by this species. By occupying space, serving as substrate for attachment by other benthic organisms, and inducing significant bioturbation due to frequent relocation, *A. stutchburyi* qualifies as an important bioengineer (Whitlatch et al. 1997; Thomas et al. 1998, 2021; Mouritsen 2004; Mouritsen and Poulin 2006, 2010; Adkins et al. 2014). This bioengineering role is, however, altered radically when cockles are infected by echinostome trematodes that preferentially attack the functionally important foot of the host. The bivalve then becomes notably more sessile and is unable to rebury if dislodged to the sediment surface. This, in turn, reduces its bioturbation activity and affects its suitability as substrate for attachment, aside from creating new surface structures that change near-seabed hydrodynamics and sedimentation patterns (see Fig. 8.7; Thomas et al. 1998; Mouritsen 2004; Mouritsen and Poulin 2005a, b). Clearly, such significant changes of the functional role of a dominant benthic organism must influence the surrounding benthic community, which indeed is the case on the sand flats of Otago Harbour. Here, Mouritsen and Poulin (2005a, 2006) performed two separate long-term field experiments, one addressing the impact of reduced mobility and bioturbation of infected but still buried cockles,

Fig. 8.7 New Zealand cockle *Austrovenus stutchburyi* stranded on the sediment due to infection by echinostome trematodes targeting especially the foot tissue of the cockle

and one addressing the impact of increasing density of cockles upon the sediment surface (none, 30 and 100 per square metre). Prevalence of infection often reaches 100% and at sites where cockles are heavily infested by echinostomes; the natural density of individuals on the surface may exceed 50 per square metre (Mouritsen and Poulin 2005a, 2010).

In the first experiment, the bioturbation activity of cockles from heavily infected plots was reduced to one-fifth of the activity in lightly infected control plots, which positively affected the abundance of several other zoobenthic species amounting to a 30% increase in overall animal density. The species richness also increased significantly by more than 10% in comparison to control plots. In the second experiment, increasing numbers of cockles on the sediment surface likewise resulted in a general increase in the abundance of benthic animals by 18%. Focusing on major taxonomic groups, the density of nemertines, polychaetes and gastropods increased, crustaceans peaked, and bivalves remained unaffected with increasing numbers of surfaced cockles (Fig. 8.8a–d). Thus, increasing intensity of echinostome infections in the cockle population, as reflected by the increasing numbers of surfaced cockles, appeared to change overall taxonomic composition, and hence, function of the zooben-

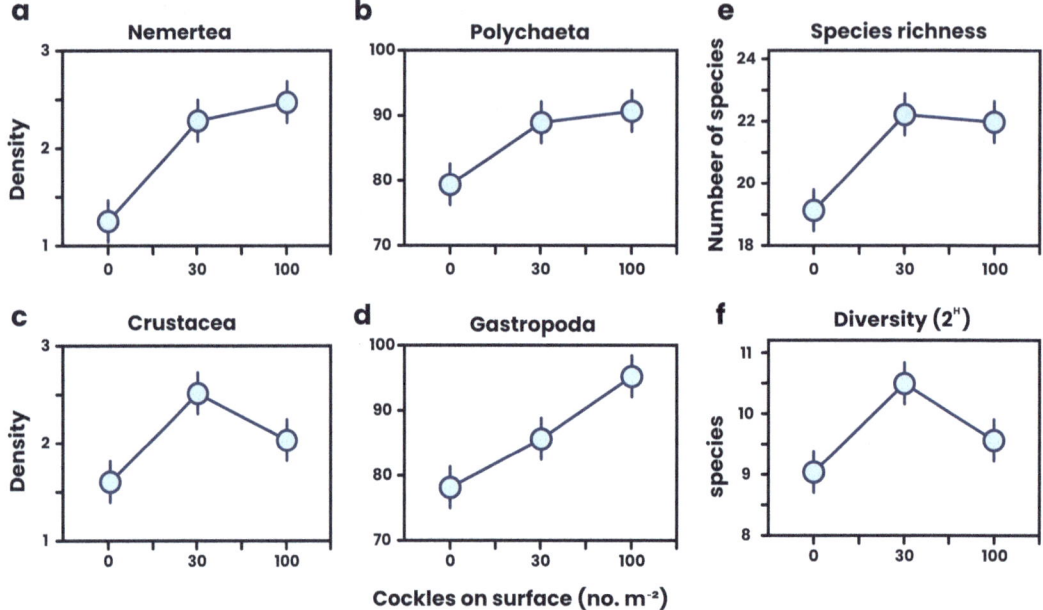

Fig. 8.8 The impact of surfaced cockles *Austrovenus stutchburyi* on the macrozoobenthic community in Otago Harbour, South Island, New Zealand, following a six months experimental period. (**a–d**) Abundance of major animal classes (no. individuals) as a function of cockles on the sediment surface (**a**: Nemertea, **b**: Polychata, **c**: Crustacea, **d**: Gastopoda). (**e–f**) Faunal diversity (all recorded species combined) as a function of cockles on the sediment surface (**e**: species richness, **f**: species diversity, Shannon-Wiener index, 2^H). Values are mean sample^{-1} (0.012 m^{-2}) ± SE, $n = 35$) [Modified after Mouritsen and Poulin (2005a)]

thic community. Changes in the abundance of primary producers (microphytobenthos) and the biomass of secondary producers were evident as well. Apart from these numerical responses, species richness and diversity of the benthic community were positively influenced by surfacing cockles. Species richness was found significantly higher (16%) in presence of surfaced cockles, whereas species diversity peaked at the intermediate level of surfacing by just less than 30% in comparison to the control (no cockles on surface) (Fig. 8.8e–f).

It is evident from these experiments that echinostome infections, indirectly through altered host behaviour, may substantially boost the intertidal community in terms of animal abundance and diversity, at least where infection intensities are relatively high. However, the experiments were carried out at a single site under a specific set of environmental conditions, and in light of the also great spatial variation seen in cockle parasitism (Mouritsen and Poulin 2010), it raises the question whether echinostomes are an important determinant of macrozoobenthic community structure also on a larger spatial scale. The indirect impact of these parasites could well be overridden by environmental conditions and more potent biotic processes fluctuating in strength and character from site to site. Interestingly, this appears not to be the case in the Otago Harbour ecosystem. Based on data on cockle parasitism and benthic animal assemblages collected from the intertidal zone of 17 different bays, multiple regression analyses—corrected for the influence of several other structuring forces such as abundance of dominant fauna, primary producers and sediment characteristics—echinostome parasitism in cockles is a significant determinant of overall benthic community structure (Mouritsen and Poulin 2010). Cockle parasitism was one of the best predictors of animal abundance, significantly affecting 16% of the 49 most widespread species of benthic invertebrates (including anthozoans, polychaetes and crustaceans) and the biomass of the animal classes Anthozoa, Nemertea and Bivalvia. Regarding overall community metrics, species diversity and total species richness across the 17 investigated sites was positively related to parasitism.

The combined evidence from both correlative and experimental field investigations emphasises echinostome trematodes as important indirect determinants of benthic community structure in Otago Harbour, maintaining local diversity, but the specific interaction cascades resulting in this role remains largely unknown. Two studies do reveal a glimpse of one of the processes involved. Thomas et al. (1998) showed that the small limpet *Notoacmea elongata* (as *Notoacmea helmsi*) and the similarly small mud flat anemone *Anthopleura hermaphroditica* (as *Anthopleura aureoradiata*) both use the New Zealand cockle as a substrate for grazing and attachment, occur in different relative proportions on heavily and weakly infected cockles collected at different sites in Otago Harbour. Limpets are more frequent on the most heavily-infected and hence surfaced cockles, whereas the anemone is most frequent on the less-infected buried ones. Mirroring this finding, Mouritsen and Poulin (2005b) showed that limpets increase and anemones decrease in abundance, shoreward along the intertidal gradient, a pattern related in part to the corresponding shoreward increase in cockle parasitism and therefore also density of surfaced cockles. The mechanism governing the complementary distributions of limpets and anemones is most probably higher anemone mortality on surfaced cockles due to desiccation during low tide and the development of microalgal films on the shell of surfaced cockles that attract the herbivorous limpets. Echinostome trematodes hence indirectly modify the intertidal zonation of non-host organisms. This phenomenon is not restricted to limpets and anemones. The intertidal distribution of the predatory whelk *Cominella glandiformis* is also shifted shoreward, towards higher densities of surfaced cockles that represent easy prey, and away from peak densities of buried cockles at lower tidal levels (Mouritsen and Poulin 2005b). Interestingly, not only the macrozoobenthic invertebrates appear affected by the intertidal gradient in cockle parasitism. Underwater video recordings have revealed that the rocky-shore fish, the spotty *Notolabrus celidotus*, has specialised in partial predation of surfaced cockles on the intertidal sand flats, the fish cropping off the feet of these heavily infected cockles when they try in vain to rebury during high tide (Mouritsen and Poulin 2003b, 2005b). Because the density of heavily-infected and surfaced cockles increases shoreward, so does the frequency of foot cropping by spotties, despite a much shorter immersion time at these upper shore levels.

The bivalve *Austrovenus stutchburyi* is endemic to New Zealand and the intriguing interaction cascades stemming from this host's interactions with its trematode fauna detailed above might then be viewed as a special case limited to the New Zealand soft-bottom intertidal ecosystem. This is, however, not the case. Along sedimentary shores of western Europe, the edible (or European) cockle *Cerastoderma edule* appears in several aspects to occupy a similar ecological position. Although taxonomically distinct from the venerid New Zealand cockle, the European cockle (Cardiidae) is ecologically identical to the New Zealand cockle and acts as second intermediate host to at least three different species of echinostomes (Lauckner 1983; Thieltges et al. 2006; Richard et al. 2022). As was the case in the New Zealand cockle, these trematodes preferentially attack the foot of the European cockle with potentially the same consequences: reduced mobility when still buried in the substrate and surfacing of heavily infected individuals (Lauckner 1983, 1987; Desclaux et al. 2002; Richard et al. 2021, 2022). Because the European cockle generally exerts a strong negative impact on the abundance of the sympatric community of macrozoobenthos due to its bioturbation activity (e.g. Flach 1996), it is straightforward to assume that echinostome infections will mitigate this disturbance and thus indirectly favour the surrounding benthos. However, this link between the host–parasite interaction and overall community structure has not yet been fully experimentally demonstrated on European shores.

8.4.2 Bioturbation by Mud Snails

Cockles thrive mainly in sandy substrates, whereas there exists another mollusc in the more muddy realms of the northeast Atlantic intertidal that similarly affects the surrounding benthic community through its surface activity. The about half a centimetre large mud snail *Peringia ulvae* is very abundant in western European soft-bottom habitats where it usually reaches adult densities of 10–25,000 individuals per square metre (Fig. 8.9). The grazing activity of these mud snails on benthic diatoms reworks the upper sediment strata several times a day, and that significantly affects the community of both microphytobenthos and smaller faunal organisms living in these sediments (Mouritsen and Haun 2008 and references therein). *Peringia ulvae* also acts as the first intermediate host to a range of trematode species including echinostomes, microphallids and heterophyids that can reduce the overall activity level of their snail host by about 50% (Mouritsen and Jensen 1994). As emphasised in the preceding sections, parasitism is very rarely evenly distributed across a host population. This applies also to prevalence of trematode infection in mud snail populations that show great spatial as well as seasonal variation ranging between a few to more than 80% (Huxham et al. 1995; Field and Irwin 1999; Mouritsen et al. 1997; de Montaudouin et al. 2003; Poulin and Mouritsen 2003; Thieltges et al. 2006; Levakin et al. 2013). Therefore, there will be foci in time and space where parasitism reaches levels that seriously decrease the bioturbation activity of the snail population, with possible implications for the benthic community of plants and animals. Such an indirect relationship between trematodes and community structure was established in field experiments carried out in the Danish part of the Wadden Sea (Mouritsen and Haun 2008). Both the abundance and diversity of microphytobenthos (mainly diatoms) decreased at increasing trematode prevalence in experimental snails, whereas the density and species richness of benthic animals peaked at intermediate levels of parasitism (Fig. 8.10). The community of primary and secondary producers hence responded differently to the trematode-induced change in mud snail activity. The exact mechanisms governing these responses are, as in the New Zealand cockle example above, poorly understood. Differential grazing of primary producers (algae) by infected and uninfected snails cannot account for the observed pattern, because the two groups of snails are similar in this respect (Mouritsen and Jensen 1994). Rather, the impoverishment of the microalgal assemblage at high levels of mud snail parasitism may follow from reduced release of nutrients regenerated by bacterial activity in the uppermost sediment strata and made available for the diatoms growing on the sediment surface by snail bioturbation (Mouritsen and Haun 2008). In contrast, the preferred animal community at intermediate levels of snail parasitism is believed to follow directly from relaxed disturbance, together with the availability of an additional food source to the predatory fauna in the form of trematode larvae continuously released from infected snails. However, other more subtle interaction cascades are probably also in action.

The mechanisms behind the indirect effects of trematodes infections in *P. ulvae* on the Wadden Sea soft-bottom community need further unravelling (see Dairain et al. 2019 for additional examples of parasite-mediated bioturbation), but the processes involved in a comparable rocky-shore system described in the following section appear much more straightforward.

Fig. 8.9 Mud snails *Peringia ulvae* grazing on the sediment surface in the Danish part of the Wadden Sea

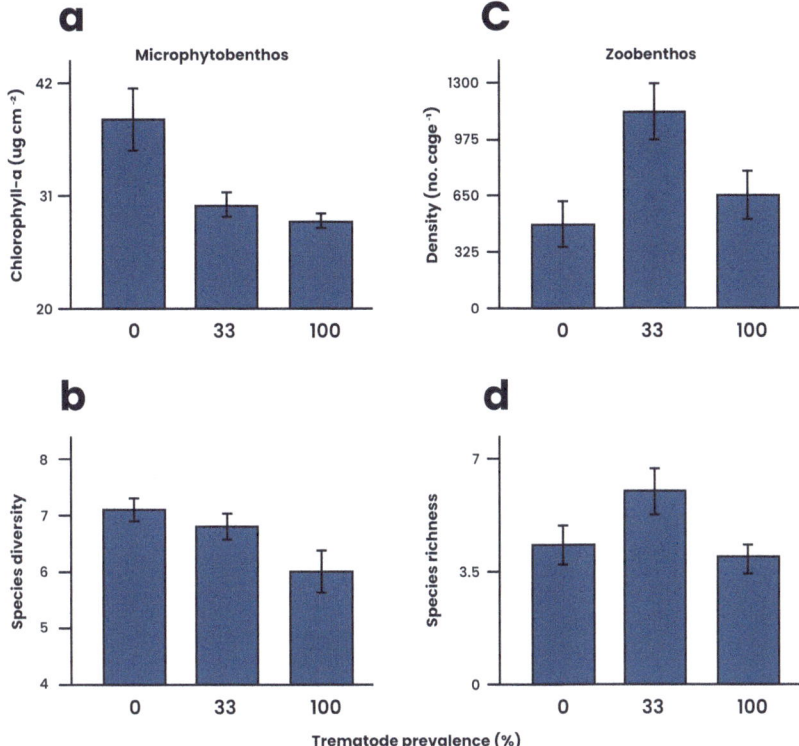

Fig. 8.10 Changes to the benthic community of primary and secondary producers in the Danish Wadden Sea as a function of trematode prevalence (% infected) in caged experimental populations of mud snails *Hydrobia (Peringia) ulvae*. (**a**) the abundance of microalgae (mainly diatoms) in terms of chlorophyll-*a* concentration ($\mu g\ cm^{-2}$) in the substrate. (**b**) Simpson's diversity index (1/D) on epipelic diatoms (no. species). (**c**) Density of benthic invertebrates (macro- and meiofauna combined, individuals per cage). (**d**) Zoobenthic species richness (number of meiofaunal species per cage). Values are means ± SE ($n = 7$–8) and the mud snail density in experimental cages (50 cm^2 large) corresponds to 15,000 individuals per square metre [Reanalysed and modified after Mouritsen and Haun (2008)]

8.4.3 Grazing by Periwinkles

The periwinkle *Littorina littorea* is one of the most abundant herbivorous gastropods on North Atlantic rocky shores, where its grazing activity exerts strong regulation of the intertidal community of macroalgae (Lubchenco 1978; Lein 1980; Bertness 2007; Wood et al. 2007). The periwinkle prefers rapidly growing ephemeral algal species (mostly green and filamentous red algae) over the less palatable perennials (brown and other red algae), and therefore affects the relative abundance of these two ecologically rather different groups of macroalgae. This functional role is modified by trematode infections. Similar to *Peringia ulvae*, periwinkles serve as first intermediate hosts to a range of trematode species of which echinostomes, heterophyids and renicolids are most frequent (Werding 1969; Lauckner 1980; Mouritsen et al. 1999; Thieltges et al. 2006; Blakeslee and Byers 2008; Lambert et al. 2012). When infected by these parasites, the hosts' consumption of ephemeral algae is substantially reduced (Wood et al. 2007; Clausen et al. 2008; but see Díaz-Morales et al. 2023), which has ramifications for the overall structure of the intertidal macroalgal community. In a field experiment manipulating the prevalence of trematode infections in caged periwinkles, Wood et al. (2007) showed that the cover of ephemeral algae increased more rapidly in cages harbouring infected snails than uninfected snails (Fig. 8.11a). Not only did ephemeral algal species increase in abundance at the expense of perennial species such as Irish moss *Chondrus crispus*, coral weed

Corallina officinalis and false Irish moss *Mastocarpus stellatus*, the different ephemeral species were also differentially affected by the parasite treatments. Whereas the cover of laver (*Porphyra* sp.) was unaffected, sea lettuce *Ulva lactuca* and the filamentous red algae *Spermnothamnion* sp. and *Melanothamnus harveyi* were increasingly favoured by trematode infections in experimental herbivores (Fig. 8.11b). Not only did trematode infections affect the relative abundance of ephemerals and perennials, these parasites also indirectly modified, through trait-mediated interactions, the structure of the ephemeral algal community. Interestingly, because species of macroalgae and a range of sessile species of invertebrates often compete for space on rocky shores, the observed indirect impact of trematodes on the algal community might also affect the zoobenthic community (see Janke 1990; McQuaid 1996; Bertness 2007; Bommarito et al. 2023). However, such an intriguing interaction cascade remains yet to be experimentally demonstrated.

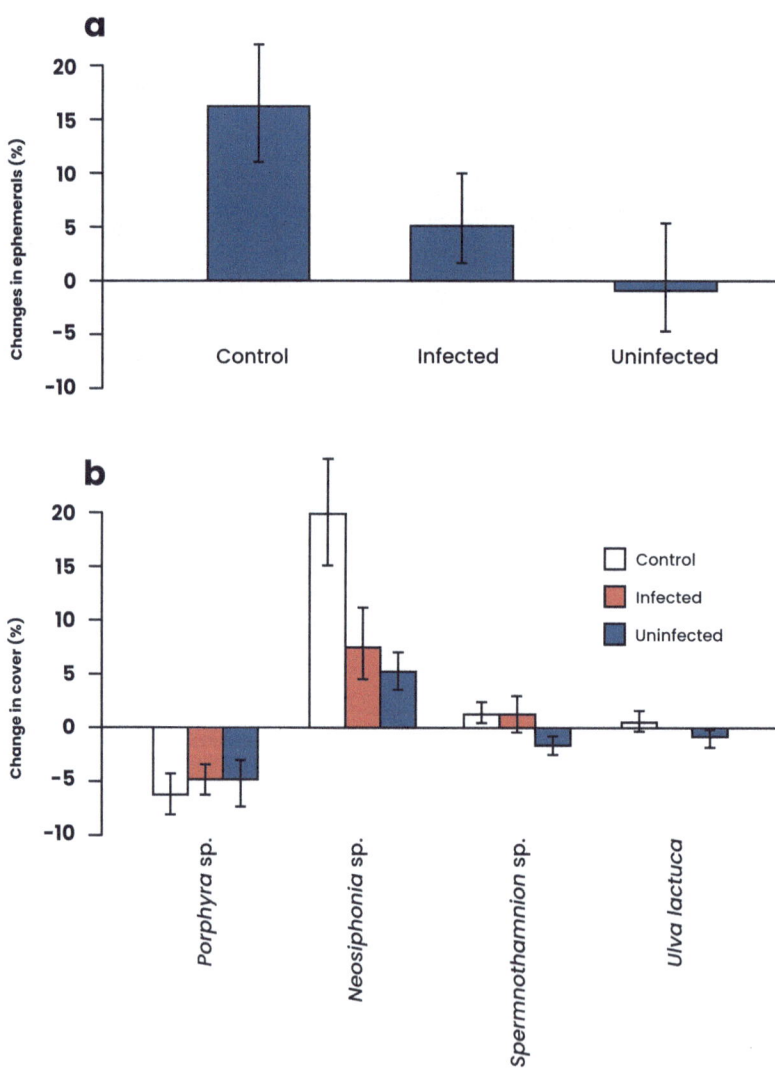

Fig. 8.11 Change in percent cover of (**a**) ephemeral macroalgae combined, and (**b**) ephemeral species contrasted, during an approximately three weeks snail enclosure experiment carried out on a rocky shore in Maine, USA. Control: no snails present; Infected and Uninfected: enclosed snails (*Littorina littorea*), established in a density corresponding to 161 individuals per square metre, were mainly infected and mainly uninfected by trematodes, respectively. Columns are means ± SE ($n = 14$) [Modified after Wood et al. (2007)]. Previously unpublished data at species level (**b**) were kindly provided by CL Wood and JE Byers

8.4.4 Lessons Learned

The above three studies unequivocally demonstrate the community structuring potential of parasite-induced trait-mediated indirect interactions in intertidal habitats. This recently recognised process appears to result in a community impact comparable to if not overriding that of other parasite-mediated processes (direct and host density-mediated indirect). These studies presently are among the most thorough on the topic, although similar evidence exist for systems outside the marine realm (e.g. MacNeil et al. 2003; Bernot and Lamberti 2008; Hernandez and Sukhdeo 2008; Hatcher et al. 2012; Giari et al. 2020), suggesting the importance of parasitic interactions to community organisation across ecosystems.

The magnitude of the community impact resulting from trait-mediated interaction is, as opposed to direct effects but in line with host density-mediated indirect effects, intimately linked to the abundance and functional role of the host species for which phenotypic expression has been altered. If it involves a keystone or bioengineering species, we may expect a great impact cascading far into the community, whereas the behavioural change of an ecologically more cryptic or rare organism, on the other hand, should pass relatively unnoticed.

Low species diversity in a given community is expected to increase the probability that a single species acts as a keystone species (Power et al. 1996). This may readily apply, for example, to intertidal habitats that support relatively dense populations of the few species managing to adapt to the severe environmental constrains that prevail there (Bertness 2007). Significant community ramifications of parasite-induced trait-mediated interactions may be particularly prone to manifest themselves in the intertidal zone. As there are few studies that have been carried out on this subject, there seems to be an almost unlimited potential for investigating the indirect community impact of parasitism.

8.5 Conclusions and Future Directions

That parasitism can play a significant role in structuring aquatic communities is incontestable based on the available evidence. Parasitism is generally believed to *maintain* biodiversity (also see Chap. 7) as long as the involved parasites are specialists or generalists with asymmetrical impact on coexisting host species particularly targeting the dominant competitors (Holt and Pickering 1985; Dobson and Hudson 1986; Begon et al. 2004; Chesson 2000; Hudson et al. 2006; Fenton and Brockhurst 2008). This is not unequivocally corroborated. The ways species diversity and richness are affected qualitatively in the examples given here appear to be context-dependent, varying according to the specific functional role of the host species in the community. If the affected host species is functioning as an inhibitor, parasitism will tend to boost diversity; if the host serves as a facilitator, parasitism tends to affect biodiversity negatively. This pattern seems to emerge regardless of the type of effect in action: direct, density-mediated indirect or trait-mediated indirect. The apparent discrepancy between contemporary ideas and existing empirical evidence for aquatic habitats may follow on from the fact that the most significant, and therefore most easily-observed, community impacts of parasitism arise from indirect ramifications in terms of interaction cascades that lead to the parasitic decimation of a keystone or bioengineering species. However, these processes do not rule out the simultaneous existence of an underlying subtle and immediate boosting of diversity due to competitive release of the keystone species' closest competitors. In other words, parasitism may generally serve to maintain diversity, but this effect can be overridden by the ecological roles of the target hosts, which ultimately determine the character of the host–parasite system's full community impact. Such context-dependence may indeed be the real picture, but it may also be that we are currently lacking the data for a clear, integrated directional pattern to emerge. To resolve this uncertainty, more investigations specifically addressing the community impact of parasitism on a broader range of host–parasite systems are clearly needed.

It would be advantageous to conduct field-experiments targeting the consequences of also the more subtle 'every-day' regulation of host populations, instead of merely assessing the more severe community ramifications following from occasional epidemic-like mass mortalities. It will also be useful to integrate different host–parasite systems in one experimental setup, allowing for an analysis of their combined community impact. Although different host–parasite systems coexist in natural settings, they are typically studied in isolation. Thus, it is far from obvious that their individual impacts can be cumulative. Such an integrative approach will be the first necessary step towards understanding how communities manifest in absence or presence of parasitism. Investigations of trait-mediated indirect effects appear to be a particularly fruitful and non-exhausted avenue to follow in this respect. Whereas most documented direct and host density-mediated indirect effects of parasitism, such as impacts via dramatic changes in host abundance, could just as well have been induced by predation, herbivory or competition, trait-mediated effects represent the true fingerprint of parasitism on community structure that cannot easily be established through other processes.

These considerations aside, it will also be valuable to better scrutinise the community impact following from direct and indirect numerical and functional responses to parasitism. For instance, larval parasites are often preyed upon by non-host organisms, and the surprisingly high productivity of parasites, particularly trematodes, suggests that this overlooked but significant source of energy could be important to a range of intertidal predators and omnivores (see Sect. 8.2.2). Tidal communities especially beset by parasitism may support higher densities of such consumers. Intertidal predators may also benefit more indirectly from parasitism. As in the case of surfacing New Zealand cockles (Sect. 8.4.1), predatory whelks and fishes targeting these stranded cockles were attracted to sites of heavy cockle parasitism. Such numerical (or aggregational) responses may be more widespread than presently recognised. The guild of waterbirds acting as definitive hosts for a wide range of parasites could here be a relevant focus. The evolution of parasitic alternation of intermediate host behaviour has often enabled the acceleration of trophic transmission of parasites to their (usually avian) definitive host. The strategy is clearly successful, and disregarding the possibility that parasitic manipulation merely determines which food items are preyed upon, the generally elevated accessibility of specific prey organisms that follows from such manipulations should attract more birds specialised in feeding upon these particular prey species. This might well result in measurable effects on the structure of the avian community as well.

Future studies that integrate multiple host–parasite systems within experimental frameworks are essential to elucidate the important roles of parasites in shaping community dynamics. Such research will not only clarify the existing uncertainties but also pave the way for a deeper understanding of how parasitism fundamentally influences biodiversity and ecosystem stability.

Acknowledgements The author is in depth to Robert Poulin, University of Otago, for discussions, corrections and editing of an earlier version of this chapter.

References

Adkins SC, Marsden ID, Pirker JG (2014) Variation in population structure and density of *Austrovenus stutchburyi* (Veneridae) from Canterbury, New Zealand. J Shellfish Res 33:343–354. https://doi.org/10.2983/035.033.0204

Ambrose WG (1984) Influence of predatory polychaetes and epibenthic predators on the structure of a soft–bottom community in a Maine estuary. J Exp Mar Biol Ecol 81:115–145. https://doi.org/10.1016/0022-0981(84)90002-9

Begon M, Townsend CR, Harper JL (2004) Ecology: from individuals to communities. Wiley–Blackwell, Oxford

Bennett AMR, Scudder GGE (1998) Differences in attachment of water mites on water boatmen: further evidence of differential parasitism and possible exclusion of host from part of its potential range. Can J Zool 76:824–834. https://doi.org/10.1139/z98-014

Bernal-Ibáñez A, Cacabelos E, Melo R et al (2021) The role of sea-urchins in marine forests from Azores, Webbnesia, and Cabo Verde: human pressures, climate-change effects and restoration opportunities. Front Mar Sci 8:649873. https://doi.org/10.3389/fmars.2021.649873

Bernot RJ, Lamberti GA (2008) Indirect effects of a parasite on a benthic community: an experiment with trematodes, snails and periphyton. Freshw Biol 53:322–329. https://doi.org/10.1111/j.1365-2427.2007.01896.x

Bertness MB (2007) Atlantic shorelines: natural history and ecology. Princeton University Press, Princeton

Bick A (1994) *Corophium volutator* (Corophiidae: Amphipoda) as an intermediate host of larval Digenea: an ecological analysis in a coastal region of the southern Baltic. Ophelia 40:27–36. https://doi.org/10.1080/00785326.1994.10429548

Blakeslee AMH, Byers JE (2008) Using parasites to inform ecological history: comparisons among three congeneric marine snails. Ecology 89:1068–1078. https://doi.org/10.1890/07-0832.1

Bommarito C, Díaz-Morales DM, Guy-Haim T et al (2023) Warming and parasitism impair the performance of Baltic native and invasive macroalgae and their associated fauna. Limnol Oceanogr 68:1852–1864. https://doi.org/10.1002/lno.12390

Boström C, Bonsdorff E (2000) Zoobenthic community establishment and habitat complexity—the importance of seagrass shoot-density, morphology and physical disturbance for faunal recruitment. Mar Ecol Prog Ser 205:123–138. https://doi.org/10.3354/meps205123

Boulétreau M, Fouillet P, Allemand R (1991) Parasitoids affect competitive interactions between the sibling species, *Drosophila melanogaster* and *D. simulans*. Redia 84:171–177

Carlton JT, Vermeij GJ, Lindberg DR et al (1991) The first historical extinction of a marine invertebrate in an ocean basin: the demise of the eelgrass limpet *Lottia alveus*. Biol Bull 180:72–80. https://doi.org/10.2307/1542430

Carvalho S, Moura A, Sprung M (2006) Ecological implications of removing seagrass beds (*Zostera noltii*) for bivalve aquaculture in southern Portugal. Cah Biol Mar 47:321–329

Chesson P (2000) Mechanisms of maintenance of species diversity. Annu Rev Ecol Syst 31:343–366. https://doi.org/10.1146/annurev.ecolsys.31.1.343

Clausen KT, Larsen MH, Iversen NK et al (2008) The influence of trematodes on the macroalgae consumption by the common periwinkle *Littorina littorea*. J Mar Biol Assoc UK 88:1481–1485. https://doi.org/10.1017/S0025315408001744

Combes C (2001) Parasitism: the ecology and evolution of intimate interactions. University of Chicago Press, Chicago

Connolly RM (1994) Removal of seagrass canopy: effects on small fish and their prey. J Exp Mar Biol Ecol 184:99–110. https://doi.org/10.1016/0022-0981(94)90168-6

Cornet S, Sorci G, Moret Y (2010) Biological invasion and parasitism: invaders do not suffer from physiological alterations of the acanthocephalan *Pomphorhynchus laevis*. Parasitology 137:137–147. https://doi.org/10.1017/S0031182009991077

Cottam C, Munro DA (1954) Eelgrass status and environmental relations. J Wildlife Man 18:449–460. https://doi.org/10.2307/3797080

Coustau C, Combes C, Maillard C et al (1990) *Prosorhynchus squamatus* (Trematoda) parasitosis in the *Mytilus edulis-Mytilus galloprovincialis* complex: specificity and host–parasite relationships. In: Perkins FO, Cheng TC (eds) Pathology in marine sciences. Academic Press, New York, pp 291–298. https://doi.org/10.1016/C2009-0-02735-4

Coustau C, Renaud F, Maillard C et al (1991) Differential susceptibility to a trematode parasite among genotypes of the *Mytilus edulis/Mytilus galloprovincialis* complex. Genet Res 57:207–212. https://doi.org/10.1017/s0016672300029359

Curtis LA (1987) Vertical distribution of an estuarine snail altered by a parasite. Science 235:1509–1511. https://doi.org/10.1126/science.3823901

Curtis LA (1993) Parasite transmission in the intertidal zone—vertical migrations, infective stages, and snail trails. J Exp Mar Biol Ecol 173:197–209. https://doi.org/10.1016/0022-0981(93)90053-Q

Dairain A, Legeay A, de Montaudouin X (2019) Influence of parasitism on bioturbation: from host to ecosystem functioning. Mar Ecol Prog Ser 619:201–214. https://doi.org/10.3354/meps12967

de Montaudouin X, Blanchet H, Kisielewski I et al (2003) Digenean trematodes moderately alter *Hydrobia ulvae* population size structure. J Mar Biol Assoc UK 83:297–305. https://doi.org/10.1017/S0025315403007112h

Deblock S (1980) Inventaire des trematodes larvaires parasites des mollusques *Hydrobia* (Prosobranchia) des côtes de France. Parassitologia 22:1–105

den Hartog C (1987) 'Wasting disease' and other dynamic phenomena in *Zostera* beds. Aquat Bot 27:3–14. https://doi.org/10.1016/0304-3770(87)90082-9

Desclaux C, de Montaudouin X, Bachelet G (2002) Cockle emergence at the sediment surface: 'favourization' mechanism by digenean parasites? Dis Aquat Org 52:137–149. https://doi.org/10.3354/dao052137

Díaz-Morales DM, Bommarito C, Knol J et al (2023) Parasitism enhances gastropod feeding on invasive and native algae while altering essential energy reserves for organismal homeostasis upon warming. Sci Tot Environ 863:160727. https://doi.org/10.1016/j.scitotenv.2022.160727

Dobson AP (1995a) The ecology and epidemiology of rinderpest virus in Serengeti and Ngorongoro crater conservation area. In: Sinclair ARE, Arcese P (eds) Serengeti II: research, management and conservation of an ecosystem. University of Chicago Press, Chicago, pp 485–505

Dobson AP (1995b) Rinderpest in the Serengeti ecosystem: the ecology and control of a keystone virus. In: Junge RE (ed) Proceedings of a joint conference of the American Association of Zoo Veterinarians, Wildlife Disease Association, and American Association of Wildlife Veterinarians. AAZV, Michigan, pp 518–519

Dobson AP, Hudson PJ (1986) Parasites, disease and the structure of ecological communities. Trends Ecol Evol 1:11–15. https://doi.org/10.1016/0169-5347(86)90060-1

Dunn AM (2009) Parasites and biological invasions. Adv Parasitol 68:161–184. https://doi.org/10.1016/S0065-308X(08)00607-6

Dunn JC, McClymont HE, Christmas M et al (2009) Competition and parasitism in the native white clawed crayfish *Austropotamobius pallipes* and the invasive signal crayfish *Pacifastacus leniusculus* in the UK. Biol Invasions 11:315–324. https://doi.org/10.1007/s10530-008-9249-7

Estes JA, Tinker MT, Williams TM et al (1998) Killer whale predation on sea otters linking oceanic and nearshore ecosystems. Science 282:473–476. https://doi.org/10.1126/science.282.5388.473

Feehan CJ, Scheibling RE (2014) Effects of sea urchin disease on coastal marine ecosystems. Mar Biol 161:1467–1485. https://doi.org/10.1007/s00227-014-2452-4

Fenton A, Brockhurst MA (2008) The role of specialist parasites in structuring host communities. Ecol Res 23:795–804. https://doi.org/10.1007/s11284-007-0440-6

Field LC, Irwin SWB (1999) Digenean larvae in *Hydrobia ulvae* from Belfast Lough (Northern Ireland) and the Ythan Estuary (north-east Scotland). J Mar Biol Assoc UK 79:431–435. https://doi.org/10.1017/S0025315498000551

Flach EC (1996) The influence of the cockle, *Cerastoderma edule*, on the macrozoobenthic community of tidal flats in the Wadden Sea. Mar Ecol 17:87–98. https://doi.org/10.1111/j.1439-0485.1996.tb00492.x

Frainer A, McKie BG, Amundsen P-A et al (2018) Parasitism and the biodiversity–functioning relationship. Trends Ecol Evol 33:260–268. https://doi.org/10.1016/j.tree.2018.01.011

Freeland WJ (1983) Parasites and the coexistence of animal host species. Am Nat 121:223–236. https://doi.org/10.1086/284052

Friesen OC, Goellner S, Poulin R et al (2020) Parasites shape community structure and dynamics in freshwater crustaceans. Parasitology 147:182–193. https://doi.org/10.1017/S0031182019001483

Gerdol V, Hughes RG (1994) Effect of *Corophium volutator* on the abundance of benthic diatoms, bacteria and sediment stability in two estuaries in southeastern England. Mar Ecol Prog Ser 114:109–115. https://doi.org/10.3354/meps114109

Giari L, Fano EA, Castaldelli G et al (2020) The ecological importance of amphipod-parasite associations for aquatic ecosystems. Water 12:2429. https://doi.org/10.3390/w12092429

Hagen NT (1995a) Sea urchin outbreaks and epizootic disease as regulating mechanism in coastal ecosystems. In: Eleftheriou AD, Ansell AD, Smith CJ (eds) Biology and ecology of shallow water. Olsen & Olsen, Fredensborg, pp 303–308

Hagen NT (1995b) Recurrent destructive grazing of successionally immature kelp forests by green sea urchin in Vestfjorden, Northern Norway. Mar Ecol Prog Ser 123:95–106. https://doi.org/10.3354/MEPS123095

Hagerthey SE, Defew EC, Paterson DM (2002) Influence of *Corophium volutator* and *Hydrobia ulvae* on intertidal benthic diatom assemblages under different nutrient and temperature regimes. Mar Ecol Prog Ser 254:47–59. https://doi.org/10.3354/meps245047

Hatcher MJ, Dunn AM (2011) Parasites in ecological communities: from interactions to ecosystems. Cambridge University Press, Cambridge

Hatcher MJ, Dick JTA, Dunn AM (2012) Diverse effects of parasites in ecosystems: linking interdependent processes. Front Ecol Environ 10:186–194. https://doi.org/10.1890/110016

Hatcher MJ, Dick JTA, Dunn AM (2014) Parasites that change predator or prey behaviour can have keystone effects on community composition. Biol Lett 10:20130879. https://doi.org/10.1098/rsbl.2013.0879

Hernandez AD, Sukhdeo MVK (2008) Parasite effects on isopod feeding rates can alter the host's functional role in a natural stream ecosystem. Int J Parasitol 38:683–690. https://doi.org/10.1016/j.ijpara.2007.09.008

Holt RD (1977) Predation, apparent competition, and the structure of prey communities. Theor Popul Biol 12:197–229. https://doi.org/10.1016/0040-5809(77)90042-9

Holt RD, Lawton JH (1994) The ecological consequences of shared natural enemies. Ann Rev Ecol Syst 25:495–520. https://doi.org/10.1146/annurev.es.25.110194.002431

Holt RD, Pickering J (1985) Infectious disease and species coexistence: a model of Lotka-Volterra form. Am Nat 126:196–211. https://doi.org/10.1086/284409

Hopper JV, Poulin R, Thieltges DW (2008) Buffering role of the intertidal anemone *Anthopleura aureoradiata* in cercarial transmission from snails to crabs. J Exp Mar Biol Ecol 367:153–156. https://doi.org/10.1016/j.jembe.2008.09.013

Hudson PJ, Greenman J (1998) Competition mediated by parasites: biological and theoretical progress. Trends Ecol Evol 13:387–390. https://doi.org/10.1016/S0169-5347(98)01475-X

Hudson PJ, Dobson AP, Lafferty KD (2006) Is a healthy ecosystem one that is rich in parasites? Trends Ecol Evol 21:381–385. https://doi.org/10.1016/j.tree.2006.04.007

Huxham M, Raffaelli D, Pike A (1995) The effect of larval trematodes on the growth and burrowing behaviour of *Hydrobia ulvae* (Gastropoda: Prosobranchia) in the Ythan estuary, north-east Scotland. J Exp Mar Biol Ecol 185:1–17. https://doi.org/10.1016/0022-0981(94)00119-X

Janke K (1990) Biological interactions and their role in community structure in the rocky intertidal of Helgoland (German Bight, North Sea). Helgoländer Meeresunters 44:219–263. https://doi.org/10.1007/BF02365466

Jensen KT, Kristensen LD (1990) A field experiment on competition between *Corophium volutator* (Pallas) and *Corophium arenarium* Crawford (Crustacea: Amphipoda): effects on survival, reproduction and recruitment. J Exp Mar Biol Ecol 137:1–24. https://doi.org/10.1016/0022-0981(90)90057-J

Jensen KT, Mouritsen KN (1992) Mass mortality in two common soft-bottom invertebrates, *Hydrobia ulvae* and *Corophium volutator*—the possible role of trematodes. Helgoländer Meeresunters 46:329–339. https://doi.org/10.1007/BF02367103

Jensen T, Jensen KT, Mouritsen KN (1998) The influence of the trematode *Microphallus claviformis* on two congeneric intermediate host species (*Corophium*): infection characteristics and host survival. J Exp Mar Biol Ecol 227:35–48. https://doi.org/10.1016/S0022-0981(97)00260-8

Johnson PTJ, Thieltges DW (2010) Diversity, decoys and the dilution effect: how ecological communities affect disease risk. J Exp Biol 213:961–970. https://doi.org/10.1242/jeb.037721

Johnson PTJ, Dobson A, Lafferty KD et al (2010) When parasites become prey: ecological and epidemiological significance of eating parasites. Trends Ecol Evol 25:362–371. https://doi.org/10.1016/j.tree.2010.01.005

Kaplan AT, Rebhal S, Lafferty KD et al (2009) Small estuarine fishes feed on large trematode cercariae: lab and field investigations. J Parasitol 95:477–480. https://doi.org/10.1645/GE-1737.1

Kelly JR, Proctor H, Volpe JP (2008) Intertidal community structure differs significantly between substrates dominated by native eelgrass (*Zostera marina* L.) and adjacent to the introduced oyster *Crassostrea gigas* (Thunberg) in British Columbia, Canada. Hydrobiologia 596:57–66. https://doi.org/10.1007/s10750-007-9057-6

Kiesecker JM, Blaustein AR (1999) Pathogen reverses competition between larval amphibians. Ecology 80:2442–2448. https://doi.org/10.1890/0012-9658(1999)080

Koprivnikar J, Thieltges DW, Johnson PTJ (2023) Consumption of trematode parasite infectious stages: from conceptual synthesis to future research agenda. J Helminthol 97:e33. https://doi.org/10.1017/S0022149X23000111

Kriegisch N, Reeves SE, Johnson CR et al (2019) Top-Down Sea urchin overgrazing overwhelms bottom-up stimulation of kelp beds despite sediment enhancement. J Exp Mar Biol Ecol 514–515:48–58. https://doi.org/10.1016/j.jembe.2019.03.012

Kuris AM, Hechinger RF, Shaw JC et al (2008) Ecosystem energetic implications of parasite and free-living biomass in three estuaries. Nature 454:515–518. https://doi.org/10.1038/nature06970

Lafferty KD (2004) Fishing for lobsters indirectly increases epidemics in sea urchins. Ecol Appl 14:1566–1573. https://doi.org/10.1890/03-5088

Lambert TC, Farley J (1968) The effect of parasitism by the trematode *Cryptocotyle lingua* (Creplin) on zonation and winter migration of the common periwinkle, *Littorina littorea* (L.). Can J Zool 46:1139–1147. https://doi.org/10.1017/S003118200004453X

Lambert WJ, Corliss E, Sha J et al (2012) Trematode infections in *Littorina littorea* on the New Hampshire Coast. Northeast Nat 19:461–474. https://doi.org/10.1656/045.019.0308

Larsen M, Jensen KT, Mouritsen KN (2011) Climate influences parasite–mediated competitive release. Parasitology 138:1436–1441. https://doi.org/10.1017/S0031182011001193

Lauckner G (1980) Diseases of mollusca: gastropoda. In: Kinne O (ed) Diseases of marine animals, vol 1. Biologische Anstalt Helgoland, Hamburg, pp 311–424

Lauckner G (1983) Diseases of mollusca: bivalvia. In: Kinne O (ed) Diseases of marine animals, vol 2. Biologische Anstalt Helgoland, Hamburg, pp 477–961

Lauckner G (1987) Ecological effects of larval trematode infestations on littoral marine invertebrate populations. Int J Parasitol 17:391–398. https://doi.org/10.1016/0020-7519(87)90114-7

Lebarbenchon C, Poulin R, Thomas F (2009) Parasitism, biodiversity, and conservation biology. In: Thomas F, Guégan J-F, Renaud F (eds) Ecology and evolution of parasitism. Oxford University Press, Oxford, pp 149–160. https://doi.org/10.1093/oso/9780199535323.003.0010

Lein TE (1980) The effects of *Littorina littorea* L. (Gastropoda) grazing on littoral green algae in the inner Oslofjord, Norway. Sarsia 65:87–92. https://doi.org/10.1080/00364827.1980.10431477

Levakin IA, Nikolaev KE, Galaktionov KV (2013) Long-term variation in trematode (Trematoda, Digenea) component communities associated with intertidal gastropods is linked to abundance of final hosts. Hydrobiologia 706:103–118. https://doi.org/10.1007/s10750-012-1267-x

Limia J, Raffaelli D (1997) The effects of burrowing by the amphipod *Corophium volutator* on the ecology of intertidal sediments. J Mar Biol Assoc UK 77:409–423. https://doi.org/10.1017/S0025315400071769

Lubchenco J (1978) Plant species diversity in a marine intertidal community: importance of herbivore food preference and algal competitive abilities. Am Nat 112:23–29. https://doi.org/10.1086/283250

MacDonald EC, Frost EH, MacNeil SM et al (2014) Behavioral response of *Corophium volutator* to shorebird predation in the upper Bay of Fundy, Canada. PLoS One 9(10):e110633. https://doi.org/10.1371/journal.pone.0110633

MacNeil C, Fielding NJ, Dick JT et al (2003) An acanthocephalan parasite mediates intraguild predation between invasive and native freshwater amphipods (Crustacea). Freshw Biol 48:2085–2093. https://doi.org/10.1046/j.1365-427.2003.01145.x

Mattila J, Chaplin G, Eilers MR et al (1999) Spatial and diurnal distribution of invertebrate and fish fauna of a *Zostera marina* bed and nearby unvegetated sediments in Damariscotta River, Maine (USA). J Sea Res 41:321–332. https://doi.org/10.1016/S1385-1101(99)00006-4

McCarthy HO, Fitzpatrick SM, Irwin SWB (2000) A transmissible trematode affects the direction and rhythm of movements in a marine gastropod. Anim Behav 59:1161–1166. https://doi.org/10.1006/anbe.2000.1414

McCurdy DG, Boates JS, Forbes MR (2000) Spatial distribution of the intertidal snail *Ilyanassa obsoleta* in relation to parasitism by two species of trematodes. Can J Zool 78:1137–1143. https://doi.org/10.1139/z00-038

McQuaid CD (1996) Biology of the gastropod family Littorinidae. II. Role in the ecology of intertidal and shallow marine ecosystems. Oceanogr Mar Biol Annu Rev 34:263–302

Meissner K (2001) Infestation patterns of microphallid trematodes in *Corophium volutator* (Amphipoda). J Sea Res 45:141–151. https://doi.org/10.1016/S1385-1101(01)00055-7

Meissner K, Bick A (1997) Population dynamics and ecoparasitological surveys of *Corophium volutator* in coastal waters in the Bay of Mecklenburg (southern Baltic Sea). Dis Aquat Org 29:169–179. https://doi.org/10.3354/dao029169

Meissner K, Bick A (1999) Mortality of *Corophium volutator* (Amphipoda) caused by infestation with *Maritrema subdolum* (Digenea, Microphallidae)—laboratory studies. Dis Aquat Org 35:47–52. https://doi.org/10.3354/dao035047

Miller AA, Poulin R (2001) Parasitism, movements and distribution of the snail *Diloma subrostrata* (Trochidae) in a soft-sediment intertidal zone. Can J Zool 79:2029–2035. https://doi.org/10.1139/cjz-79-11-2029

Milne LJ, Milne MJ (1951) The eelgrass catastrophe. Sci Am 184:52–55

Minchella DJ, Scott ME (1991) Parasitism: a cryptic determinant of animal community structure. Trends Ecol Evol 8:250–254. https://doi.org/10.1016/0169-5347(91)90071-5

Miura O, Kuris AM, Torchin ME et al (2006) Parasites alter host phenotype and may create a new ecological niche for snail hosts. Proc Biol Sci 273:1323–1328. https://doi.org/10.1098/rspb.2005.3451

Morand S, Deter J (2009) Parasitism and regulation of the host population. In: Thomas F, Guégan J-F, Renaud F (eds) Ecology and evolution of parasitism. Oxford University Press, Oxford, pp 83–105. https://doi.org/10.1093/oso/9780199535323.003.0007

Mouritsen KN (1994) Day and night feeding in Dunlins *Calidris alpina*: choice of habitat, foraging technique and prey. J Avian Biol 25:55–62

Mouritsen KN (2004) Facilitation and indirect effects: causes and consequences of crawling in the New Zealand cockle. Mar Ecol Prog Ser 271:207–220. https://doi.org/10.3354/meps271207

Mouritsen KN, Haun SCB (2008) Community regulation by herbivore parasitism and density: trait-mediated indirect interactions in the intertidal. J Exp Mar Biol Ecol 367:236–246. https://doi.org/10.1016/j.jembe.2008.10.009

Mouritsen KN, Jensen KT (1994) The enigma of gigantism: effect of larval trematodes on growth, fecundity, egestion and locomotion in *Hydrobia ulvae* (pennant) (Gastropoda: Prosobranchia). J Exp Mar Biol Ecol 181:53–66. https://doi.org/10.1016/0022-0981(94)90103-1

Mouritsen KN, Jensen KT (1997) Parasite transmission between soft-bottom invertebrates: temperature mediated infection rates and mortality in *Corophium volutator*. Mar Ecol Prog Ser 151:123–134. https://doi.org/10.3354/meps151123

Mouritsen KN, Poulin R (2002) Parasitism, community structure and biodiversity in intertidal ecosystems. Parasitology 124:S101–S117. https://doi.org/10.1017/s0031182002001476

Mouritsen KN, Poulin R (2003a) The mud flay anemone-cockle association: mutualism in the intertidal zone? Oecologia 135:131–137. https://doi.org/10.1007/s00442-003-1183-x

Mouritsen KN, Poulin R (2003b) Parasite-induced trophic facilitation exploited by a non-host predator: a manipulator's nightmare. Int J Parasitol 33:1043–1050. https://doi.org/10.1016/S0020-7519(03)00178-4

Mouritsen KN, Poulin R (2005a) Parasites boost biodiversity and change animal community structure by trait-mediated indirect effects. Oikos 108:344–350. https://doi.org/10.1111/j.0030-1299.2005.13507.x

Mouritsen KN, Poulin R (2005b) Parasitism can influence the intertidal zonation of non-host organisms. Mar Biol 148:1–11. https://doi.org/10.1007/s00227-005-0060-z

Mouritsen KN, Poulin R (2006) A parasite indirectly impacts both abundance of primary producers and biomass of secondary producers in an intertidal benthic community. J Mar Biol Assoc UK 86:221–226. https://doi.org/10.1017/S0025315406013063

Mouritsen KN, Poulin R (2010) Parasitism as a determinant of community structure on intertidal flats. Mar Biol 157:201–213. https://doi.org/10.1007/s00227-009-1310-2

Mouritsen KN, Jensen T, Jensen KT (1997) Parasites on an intertidal *Corophium*-bed: factors determining the phenology of microphallid trematodes in the intermediate host populations of the mud-snail *Hydrobia ulvae* and the amphipod *Corophium volutator*. Hydrobiologia 355:61–70. https://doi.org/10.1023/A:1003067104516

Mouritsen KN, Mouritsen LT, Jensen KT (1998) Change of topography and sediment characteristics on an intertidal mud-flat following mass-mortality of the amphipod *Corophium volutator*. J Mar Biol Assoc UK 78:1167–1180. https://doi.org/10.1017/S0025315400044404

Mouritsen KN, Gorbushin A, Jensen KT (1999) Influence of trematode infections on in situ growth rates of *Littorina littorea*. J Mar Biol Assoc UK 79:425–430. https://doi.org/10.1017/S002531549800054X

Mouritsen KN, Sørensen MM, Poulin R et al (2018) Coastal ecosystems on a tipping point: global warming and parasitism combine to alter community structure and function. Glob Change Biol 24:4340–4356. https://doi.org/10.1111/gcb.14312

Muehlstein LK (1989) Perspectives on the wasting disease of eelgrass *Zostera marina*. Dis Aquat Org 7:211–221. https://doi.org/10.3354/dao007211

Ólafsson EB, Persson L-E (1986) The interaction between *Nereis diversicolor* (O. F. Müller) and *Corophium volutator* Pallas as a structuring force in a shallow brackish sediment. J Exp Mar Biol Ecol 103:103–117. https://doi.org/10.1016/0022-0981(86)90135-8

Orth R (1975) Destruction of eelgrass, *Zostera marina*, by the cownose ray, *Rhinoptera bonasus*, in the Chesapeake Bay. Chesap Sci 16:205–208. https://doi.org/10.2307/1350896

Orth R (1977) The importance of sediment stability in seagrass communities. In: Coull BC (ed) Ecology of marine benthos. South Carolina Press, Columbia, pp 281–300

Park T (1948) Interspecific competition in populations of *Trilobium confusum* Duval and *Trilobium castaneum* Herbst. Ecol Monogr 18:265–307. https://doi.org/10.2307/1948641

Peterson CH (1979) Predation, competitive exclusion, and diversity in the soft-sediment benthic communities of estuaries and lagoons. In: Livingston RJ (ed) Ecological processes in coastal and marine systems. Plenum Press, New York, pp 233–263. https://doi.org/10.1007/978-1-4615-9146-7_12

Plowright W (1982) The effect of rinderpest and rinderpest control on wildlife in Africa. Symp Zool Soc Lond 50:1–28

Poulin R (1999) The functional importance of parasites in animal communities: many roles at many levels? Int J Parasitol 29:903–914. https://doi.org/10.1016/S0020-7519(99)00045-4

Poulin R, FitzGerald GJ (1987) The potential of parasitism in the structuring of a salt marsh stickleback community. Can J Zool 65:2793–2798. https://doi.org/10.1139/z87-421

Poulin R, Mouritsen KN (2003) Large-scale determinants of trematode infections in intertidal gastropods. Mar Ecol Prog Ser 254:187–198. https://doi.org/10.3354/meps254187

Power ME, Tilman D, Estes JA et al (1996) Challenges in the quest for keystones. Bioscience 46:609–620. https://doi.org/10.2307/1312990

Preston DL, Orlofske SA, Lambden JP et al (2013) Biomass and productivity of trematode parasites in pond ecosystems. J Anim Ecol 82:509–517. https://doi.org/10.1111/1365-2656.12030

Preston DL, Layden TJ, Segui LM et al (2021) Trematode parasites exceed aquatic insect biomass in Oregon stream food webs. J Anim Ecol 90:766–775. https://doi.org/10.1111/1365-2656.13409

Price PW (1980) Evolutionary biology of parasites. Princeton University Press, Princeton

Price PW, Westoby M, Rice B et al (1986) Parasite mediation in ecological interactions. Annu Rev Ecol Sys 17:487–505. https://doi.org/10.1146/annurev.es.17.110186.002415

Rasmussen E (1977) The wasting disease of eelgrass (*Zostera marina*) and its effects on environmental factors and fauna. In: McRoy CP, Helfferich C (eds) Seagrass ecosystems: a scientific perspective. Marcel Dekker Inc, New York, pp 1–51

Richard A, de Montaudouin X, Rubiello A et al (2021) Cockle as second intermediate host of trematode parasites: consequences for sediment bioturbation and nutrient fluxes across the benthic interface. J Mar Sci Eng 9:749. https://doi.org/10.3390/jmse90707

Richard A, Maire O, Daffe G et al (2022) *Himasthla* spp. (Trematoda) in the edible cockle *Cerastoderma edule*: review, long-term monitoring and new molecular insights. Parasitology 149:878–892. https://doi.org/10.1017/S0031182022000373

Rosenkranz M, Lagrue C, Poulin R et al (2018) Small snails, high productivity? Larval output of parasites from an abundant host. Freshw Biol 63:1602–1609. https://doi.org/10.1111/fwb.13189

Scheibling RE (1984) Echinoids, epizootics and ecological stability in the rocky subtidal off Nova Scotia, Canada. Helgoländer Meeresunters 37:233–242. https://doi.org/10.1007/BF01989308

Scheibling RE (1986) Increased macroalgal abundance following mass mortalities of sea urchins (*Strongylocentrotus droebachiensis*) along the Atlantic coast off Nova Scotia. Oecologia 68:153–164. https://doi.org/10.1007/BF00384786

Scheibling RE, Stephenson RL (1984) Mass mortality of *Strongylocentrotus droebachiensis* (Echinodermata: Echinoidea) off Nova Scotia, Canada. Mar Biol 78:153–164. https://doi.org/10.1007/BF00394695

Schultz B, Koprivnikar J (2019) Free-living parasite infectious stages promote zooplankton abundance under the risk of predation. Oecologia 191:411–420. https://doi.org/10.1007/s00442-019-04503-z

Scudder GGE (1983) A review of factors governing the distribution of two closely related corixids in the saline lakes of British Columbia. Hydrobiologia 105:143–154. https://doi.org/10.1007/BF00025184

Short FT, Muehlstein LK, Porter D (1987) Eelgrass wasting disease: causes and recurrence of a marine epidemic. Biol Bull 173:557–562. https://doi.org/10.2307/1541701

Soldánová M, Selbach C, Sures B (2016) The early worm catches the bird? Productivity and patterns of *Trichobilharzia szidati* cercarial emission from

Lymnaea stagnalis. PLoS One 11:e0149678. https://doi.org/10.1371/journal.pone.0149678

Sousa WP (1991) Can models of soft-sediment community structure be complete without parasites? Am Zool 31:821–830. https://doi.org/10.1093/icb/31.6.821

Stauffer RC (1937) Changes in the invertebrate community of a lagoon after disappearance of the eel grass. Ecology 18:427–431

Stewart MJ, Creese RG (2002) Transplants of intertidal shellfish for enhancement of depleted populations: preliminary trails with the New Zealand little neck clam. J Shellfish Res 21:21–27

Thayer GW, Stuart HH (1974) The bay scallop makes its bed of seagrass. Mar Fish Rev 36:27–30

Thieltges DW, Krakau M, Andresen H et al (2006) Macroparasite community in molluscs of a tidal basin in the Wadden Sea. Helgoland Mar Res 60:307–316. https://doi.org/10.1007/s10152-006-0046-3

Thieltges DW, Jensen KT, Poulin R (2008b) The role of biotic factors in the transmission of free-living endohelminth stages. Parasitology 135:407–426. https://doi.org/10.1017/S0031182007000248

Thieltges DW, de Montaudouin X, Fredensborg B et al (2008c) Production of marine trematode cercariae: a potentially overlooked path of energy flow in benthic systems. Mar Ecol Prog Ser 372:147–155. https://doi.org/10.3354/meps07703

Thomas F, Renaud F, de Meeüs T et al (1998) Manipulation of host behaviour by parasites: ecosystem engineering in the intertidal zone? Proc Biol Soc 265:1091–1096. https://doi.org/10.1098/rspb.1998.0403

Thomas F, Bonsall MB, Dobson AP (2005a) Parasitism, biodiversity, and conservation. In: Thomas F, Renaud F, Guégan J-F (eds) Parasitism and ecosystems. Oxford University Press, Oxford, pp 124–139. https://doi.org/10.1093/acprof:oso/9780198529873.003.0009

Thomas F, Renaud F, Guégan J-F (2005b) Parasitism and ecosystems. Oxford University Press, Oxford

Thomas S, Pilditch CA, Thrush SF et al (2021) Does the size structure of venerid clam populations affect ecosystem functions on intertidal sandflats? Estuar Coast 44:242–252. https://doi.org/10.1007/s12237-020-00774-5

Tompkins DM, Draycott RAH, Hudson PJ (2000a) Field evidence for apparent competition mediated via the shared parasites of two gamebird species. Ecol Lett 3:10–14. https://doi.org/10.1046/j.1461-0248.2000.00117.x

Tompkins DM, Greenman JV, Robertson PA et al (2000b) The role of shared parasites in the exclusion of wildlife hosts: *Heterakis gallinarum* in the ring-necked pheasant and the grey patridge. J Anim Ecol 69:829–840. https://doi.org/10.1046/j.1365-2656.2000.00439.x

Tompkins DM, Greenman JV, Hudson PJ (2001) Differential impact of a shared nematode parasite on two gamebird hosts: implications for apparent competition. Parasitology 122:187–193. https://doi.org/10.1017/S0031182001007247

Tompkins DM, Dobson AP, Arneberg P et al (2002) Parasites and host population dynamics. In: Hudson PJ, Rizzoli A, Grenfell BT et al (eds) The ecology of wildlife diseases. Oxford University Press, Oxford, pp 45–62. https://doi.org/10.1093/oso/9780198506201.003.0003

Tompkins DM, White AR, Boots M (2003) Ecological replacement of native red squirrels by invasive greys driven by disease. Ecol Lett 6:189–196. https://doi.org/10.1046/j.1461-0248.2003.00417.x

Torchin ME, Mitchell CE (2004) Parasites, pathogens, and invasions by plants and animals. Front Ecol Environ 2:183–190. https://doi.org/10.1890/1540-9295(2004)002[0183:PPAIBP]2.0.CO;2

Torchin ME, Lafferty KD, Kuris AM (2002) Parasites and marine invasions. Parasitology 124:S137–S151. https://doi.org/10.1017/S0031182002001506

Torchin ME, Lafferty KD, Dobson AP et al (2003) Introduced species and their missing parasites. Nature 421:628–630. https://doi.org/10.1038/nature01346

Tutin TG (1938) The autecology of *Zostera marina* in relation to its wasting disease. New Phytol 37:50–71. https://doi.org/10.1111/j.1469-8137.1938.tb06926.x

Valdez SR, Zhang YS, van der Heide T et al (2020) Positive ecological interactions and the success of seagrass restoration. Front Mar Sci 7:91. https://doi.org/10.3389/fmars.2020.00091

Werding B (1969) Morphologie, entwicklung und ökologie digener trematoden-larven der strandschnecke *Littorina littorea*. Mar Biol 3:306–333. https://doi.org/10.1007/BF00698861

Whitlatch RB, Hines AH, Trush SF et al (1997) Benthic faunal responses to variation in patch density and patch size of a suspension-feeding bivalve. J Exp Mar Biol Ecol 216:171–189. https://doi.org/10.1016/S0022-0981(97)00095-6

Williams IC, Ellis C (1975) Movements of the common periwinkle, *Littorina littorea* (L.), on the Yorkshire coast in winter and the influence of infection with larval Digenea. J Exp Mar Biol Ecol 17:47–58. https://doi.org/10.1016/0022-0981(75)90079-9

Wood CL, Byers JE, Cottingham KL et al (2007) Parasites alter community structure. Proc Natl Acad Sci USA 104:9335–9339. https://doi.org/10.1073/pnas.0700062104

Open Access This chapter is licensed under the terms of the Creative Commons Attribution-NonCommercial-NoDerivatives 4.0 International License (http://creativecommons.org/licenses/by-nc-nd/4.0/), which permits any non-commercial use, sharing, distribution and reproduction in any medium or format, as long as you give appropriate credit to the original author(s) and the source, provide a link to the Creative Commons license and indicate if you modified the licensed material. You do not have permission under this license to share adapted material derived from this chapter or parts of it.

The images or other third party material in this chapter are included in the chapter's Creative Commons license, unless indicated otherwise in a credit line to the material. If material is not included in the chapter's Creative Commons license and your intended use is not permitted by statutory regulation or exceeds the permitted use, you will need to obtain permission directly from the copyright holder.

Molecular Approaches for Investigating the Population Genetic Structure of Aquatic Parasites

Simonetta Mattiucci, Paolo Cipriani, Daniele Canestrelli, Giuseppe Nascetti, and Marialetizia Palomba

Abstract

Studying the population genetics of a parasite species provides insights into the ecological and evolutionary implications for the host–parasite interactions. In this context, this chapter provides an overview of the genetic and molecular approaches applied to study the genetic variation of aquatic helminth parasites and their advantages and disadvantages in resolving their genetic structure over spatial scales. It then explores the investigations into parasite and host traits that influence the shape of the observed genetic population structure of certain parasite species. The spatial structure of a parasite population may also reflect the population genetic structure of its hosts; in this regard, examples of congruence and incongruence between the genetic population structure of the two interacting organisms are presented. Additionally, future trends, such as the genome-wide approach in estimating and understanding genome-scale variability in parasite populations, that is, population genomics, are discussed. Population genomics will enable addressing of the microevolutionary aspects of parasite–host associations and gaining of insights into adaptive coevolutionary processes of parasite species at the ecological level.

9.1 Introduction

Investigating the genetic architecture of parasites in aquatic environments can reveal their microevolutionary history in terms of adaptation to both biotic and abiotic factors. The population genetic structure of an aquatic parasite is shaped by its reproductive mode, the number of host species involved in the life cycle and the aquatic ecosystem. The amount of genetic variation within a specific parasite population could also result from the speciation mode, demographic changes occurring in time and space, bottleneck events and other factors that have affected the population density, particularly during the species' colonisation process into host populations. Furthermore, genetic variation can be the result of migration rates and dispersal capability of the hosts, as well as local adaptation to different microhabitat conditions. These phenomena, in

S. Mattiucci (✉) · P. Cipriani
Department of Public Health and Infectious Diseases - Section of Parasitology, Sapienza University of Roma, Rome, Italy
e-mail: simonetta.mattiucci@uniroma1.it; paolo.cipriani@uniroma1.it

D. Canestrelli · G. Nascetti · M. Palomba
Department of Ecological and Biological Sciences, Tuscia University, Viterbo, Italy
e-mail: canestrelli@unitus.it; nascetti@unitus.it; marialetizia.palomba@unitus.it

turn, mediate the effects of genetic drift, natural selection and inbreeding (Nadler 1995), resulting in local genetic variation of the parasite species and consequently structuring populations throughout a parasite's distribution range (Louhi et al. 2010). In other words, all these factors significantly influence the genetic patterns within populations as well as microevolutionary processes of parasites (Nadler 1995; Huyse et al. 2005; Mattiucci and Nascetti 2008; Blasco-Costa and Poulin 2013). Studying the population genetics of a parasite species provides insights into its infection dynamics and population density across its geographic and host range. Such studies can also provide information on changes in the aquatic ecosystem where the parasite undergoes its life cycle and how it might be influenced by anthropogenic disturbances (Palomba et al. 2023). Finally, parasite species of aquatic organisms often reflect their host phylogeny and biogeography (see Chap. 17), given that they are ecologically driven by their hosts in terms of demography, distribution and dispersal history. Genetic patterns observed in parasite phylogeographic structures are expected to partially overlap with those of their hosts, especially in highly host-specific species. Due to this intimate association between parasites and hosts, especially for highly host-specific parasites, the genetic data of the parasite could shed light on host evolutionary history. This can help us better understand the evolutionary traits that have accompanied the host-parasite co-speciation processes in host–parasite associations of aquatic ecosystems.

This chapter aims to introduce the genetic and molecular approaches and genetic markers applied so far in studying the genetic variation and genetic structure of aquatic helminth parasite populations, primarily in marine environments, across spatial and temporal scales. The advantages and disadvantages of each genetic marker are discussed, with examples given of their proper usage. The drivers responsible for shaping the genetic structure of a parasite population are also reviewed. Finally, perspectives on the use of genome-wide approaches in studying host–parasite co-phylogeographical aspects are discussed.

9.2 What Types of Markers Are Best for Unravelling the Genetic Variation of Marine Parasites?

The first step in studying the genetic variation of a certain parasite species is choosing appropriate molecular tools and genetic interpretation to verify the 'species identity'. Because speciation in parasites is not always accompanied by corresponding morphological changes, the actual number of biological species is inevitably greater than the current tally of nominal species, most of which are delineated on morphological grounds. Convergent evolution can often result in near identical phenotypes adapted to a similar host environment or habitat, leading to a strong degree of morphological similarity between closely related parasite species. Therefore, initially, an appropriate genetic approach and interpretation is needed to avoid the misidentification of cryptic (sibling) species within populations, whether sympatric or allopatric. The correct identification of an aquatic parasitic species is the absolute 'baseline' of any further investigation of population genetics of a biological species at its intraspecific level. In sexually reproducing diecious organisms (i.e. having separate sexes), the biological species concept (BSC) (Mayr 1963) can be tested using the following definition: '…species are systems of populations: the gene exchange between these systems is limited or prevented by a reproductive isolating mechanism (RIM), i.e., by a combination of prezygotic and postzygotic barriers'. In other words, species are 'groups of actually or potentially interbreeding natural populations which are reproductively isolated from other such groups' (Mayr 1963).

The possible presence of sibling species in anisakid nematodes, that is, reproductively isolated taxa but very similar from a morphological point of view, can confound the intraspecific population genetic structure at the host and geographic level of a species.

In the last few decades, parasitologists have challenged the discovery of several sibling species through studies of the genetic variation of parasites over large geographical ranges and several host species. The demonstration of reproductive isolation and the absence of gene flow between sympatric and allopatric sibling or cryptic species within nominal species of gonochoristic parasites can be inferred from nuclear loci allowing the testing of Hardy-Weinberg (H-W) equilibrium expectations and estimates of population genetic structure. Indeed, the application of H-W equilibrium can be used to test the hypothesis of panmixia for individuals sampled from a sympatric distribution, potentially revealing reproductively isolated biological species. A statistically significant deviation from the H-W equilibrium at a certain number of nuclear loci may indicate the existence of genetic heterogeneity within a population, which could be explained by the existence of different gene pools without gene flow between them, likely corresponding to different species within a nominal species. As a consequence, those loci are considered as 'diagnostic loci', allowing differentiation between the discovered species. The corollary of testing the biological species concept (BSC) is the high number of loci and individuals to be tested over a large number of host and geographic areas to support the existence of biological species.

Another supporting method for determining the existence of 'biological species' in parasites, particularly in the case of hermaphroditic organisms, involves demonstrating molecular evidence of their historical lineage independence. Thus, under the phylogenetic species concept (PSC), 'species are retained monophyletic, and speciation results from cladogenesis' (Avise 1994). Species delimitation based on PSC cannot be confused with intraspecific genetic differentiation within a species, as distinct lineages could exist from different geographical areas or hosts. This could confound the result in the interpretation of that genetic variation as corresponding to a distinct biological species, leading to a misinterpretation in considering those as historically independent species lineages. To overcome this challenge, the concordance principle (CP) (Avise 1994) states that results should be acquired from sequence analysis of several nuclear and mitochondrial gene loci. The evidence of a concordant phylogenetic partition among multiple genes (considering both conservative and polymorphic ones) should be evaluated (Avise 1994). This is why the BSC and PSP concepts in biological species delimitation should be applied, with the support of a multigene molecular approach, particularly when using genetic variation to delimit parasite species of aquatic organisms from different host species and geographical areas.

It often happens that closely related biological species have adjacent (parapatric) or even overlapping (sympatric) geographical distribution. Evolutionary boundaries between pairs of closely related parasite species are important for studying microevolutionary processes, speciation mechanisms, genetic architecture and gene flow between interacting species. These boundaries are often permeable, leading to natural hybridisation and introgression in contact zones of populations. Disentangling and characterising patterns of ongoing hybridisation and introgression across a species genome requires the application of a multi-locus approach. Indeed, to identify F1 and Fn hybrids (i.e. backcrossed individuals), a high number of nuclear diagnostic markers, as well as the sampling of many specimens, is necessary. For instance, a wide genotyping approach incorporating multiple nuclear and mitochondrial loci and analysis of Bayesian population structure have been validated and used in studies of two sibling nematode species, *Anisakis pegreffii* and *A. simplex* (sensu stricto) to define the occurrence and pattern of on-going hybridisation and bi-directional introgressive hybridisation in the two interacting species in sympatry, as well as to detect mitochondrial introgression in both allopatric and sympatric populations (Palomba et al 2025).

9.2.1 Nuclear DNA Loci Used to Study Microevolutionary Aspects of Aquatic Parasites

9.2.1.1 Internal Transcribed Spacer (ITS rDNA)

In the last two decades, the internal transcribed spacer (ITS) region of ribosomal DNA (rDNA) has become a standard nuclear marker in studies focusing on the recognition of various species of marine parasites. Sequences analysis of this gene region, combined with sequences from mitochondrial DNA gene loci, is commonly used in combined phylogenetic analyses. This combined approach supports the identification of operational taxonomic units (OTUs) of marine parasites as reciprocal monophyletic units (i.e. to support PSC).

The ITS region of rDNA situated between the SSU rDNA and the LSU rDNA coding regions is divided into ITS1 and ITS2 subregions, separated by the 5.8S rDNA subregion. The ITS subregions are less conserved than the slower-evolving 5.8S rDNA, with ITS1 being generally more variable than ITS2 due to the presence of variable repeat units (Nolan and Cribb 2005). This characteristic has made the ITS subregions suitable for detecting diagnostic nucleotide sites, which are crucial for recognising various species of marine parasites. For example, the Restriction Fragment Length Polymorphism (RFLP) of ITS rDNA has been developed for species recognition of anisakid nematodes of the genus *Anisakis* (D'Amelio et al. 2000). Similarly, the ITS region of rDNA was chosen to disentangle cryptic species of salmon trematodes (Criscione and Blouin 2004), as well as in the cases of monogeneans of the genera *Lamellodiscus* (Desdevises et al. 2000) and *Gyrodactylus* (Meinilä et al. 2004), and trematodes of the genus *Cryptocotyle* (Duflot et al. 2021).

However, several limitations have been identified in population genetics studies based on ITS markers. For one, it has been suggested that the ITS region of rDNA (and the rDNA complex in general) is a multicopy gene, undergoing concerted evolution (Ganley and Kobayashi 2007) which violates the assumption of Hardy-Weinberg. Therefore, genotypes inferred from PCR-RFLP analysis of ITS rDNA cannot be used to test H-W assumptions when describing 'biological species'. It has also been demonstrated that a single molecular marker based on the ITS region of rDNA, when used exclusively, is not always able to recognise hybrid categories (i.e. F1 hybrids and backcrosses) between cryptic species, as in the case of *A. pegreffii* and *A. simplex* (sensu stricto) (Mattiucci et al. 2016, 2019). Vilas et al. (2005) discussed the uncertainty when using sequence analysis of ITS region of rDNA in cases of incomplete lineage sorting and introgressive hybridisation.

However, the low occurrence of single nucleotide polymorphisms (SNPs) in the ITS rDNA means that, at the intraspecific level, this gene locus appears not useful to study the population genetic structure of certain parasite species. Indeed, generally, parasitic helminths accumulate much higher levels of nucleotide substitutions in certain other nuclear (i.e. microsatellites, also known as SSRs DNA) and mitochondrial (*cox*1 mtDNA and *cox*2 mtDNA) loci (see following paragraphs), than in the ITS region.

9.2.1.2 Small (SSU rDNA) and Large Subunit (LSU rDNA)

Molecular analyses of the SSU (18S) and LSU (28S) rDNA have been performed in intraspecific-level studies of several marine parasite taxa. For instance, SSU rDNA and LSU rDNA have been used to investigate potential intraspecific variations in selected species of trypanorhynch cestodes from diverse oceanographic regions (Palm et al. 2009); however, the study employed a small sample size.

Low intraspecific genetic variation has been found in the SSU gene locus between distinct populations of sibling species of the genus *Anisakis* (Mattiucci et al. 2018a). Indeed, due to its low level of genetic polymorphism, this gene locus has not been regarded as suitable for studying population genetic structure at the intraspecific level in species of *Anisakis* throughout the range of geographic and host distribution (Mattiucci et al. 2018a).

In contrast, LSU rDNA has been used to elucidate population genetic structure of several parasitic species, including didymozoid trematodes in the bluefin tuna *Thunnus thynnus* and Pacific tuna *Thunnus orientalis* (Mladineo et al. 2010), as well as in investigations with respect to host and geographic ranges for species of the cryptogonimid trematode of the genus *Retrovarium* (Miller and Cribb 2007). In 2023, the LSU rDNA gene was employed to examine the population genetic structure of larvae of the trypanorhynch cestode *Grillotia adenoplusia* parasitising the velvet belly, *Etmopterus spinax* across various oceanographic regions. The analysis identified three distinct populations in the northeastern Atlantic, western Mediterranean and eastern Mediterranean regions (Isbert et al. 2023).

9.2.1.3 Microsatellites (SSR DNA)

Microsatellites, also known as SSR DNA, are codominant markers with repetitive DNA sequences, widely dispersed along and among chromosomes. These markers have been considered to be under natural selection (Laine et al. 2012) and exhibit a consistent mutation rate (10^2–10^6 mutations per generation). Microsatellites are mostly employed for fine-scale genetic population studies, including the assessment of genetic structure at the intraspecific level. They can also be used to detect genetic erosion in natural populations of parasites (Louhi et al. 2010). However, so far, these nuclear markers have not been extensively used in studying the population genetic structure of aquatic parasites. Indeed, despite their significant advantages in population genetics studies, investigating SSR DNA loci still incurs a high cost, as it involves constructing a DNA library of the target parasite species and detecting polymorphic loci. In addition, a high number of specimens need to be tested to have a reliable population genetic analysis gathered from these loci, representing a time and cost limitation in their application.

A notable success in applying SSR DNA as molecular tools to discern microevolutionary aspects of marine parasites involves the digenean trematode *Plagioporus shawi*, investigated by Criscione and Blouin (2007). Their study revealed genetic sub-structuring, even showing some congruence with distinct host population structure. Similarly, significant genetic differentiation was detected using multiple SSR loci in the copepod *Lepeophtheirus salmonis*, sampled from salmon in various geographical areas of the NE Atlantic waters (Glover et al. 2011).

Several microsatellite DNA loci have been developed by Mladineo et al. (2017) for the anisakid species *A. simplex* (sensu stricto) and *A. pegreffii*; however, the analysis was limited to a low number of specimens. A panel of a further seven DNA microsatellite loci has been developed and investigated for hundreds of specimens of various populations of *A. pegreffii*, *A. simplex* (sensu stricto) and *A. berlandi* (Mattiucci et al. 2019; Bello et al. 2020), in both sympatry and allopatry. The existence of fixed alternative alleles at certain SSR loci (i.e. diagnostic loci) among the three sibling species of the *A. simplex* (sensu lato) complex enabled the use of these markers at the interspecific level. In addition, the extensive representative sample of *Anisakis* species demonstrated that these nuclear loci are robust and polymorphic enough to serve as valuable population markers for these parasites at the intraspecific level. In addition, allele frequencies gathered at these loci at the intraspecific level allow the estimation of gene flow between geographically distant populations of these species, and to calculate the genetic differentiation and variability at intrapopulation level (Mattiucci et al. 2019; Bello et al. 2020). A significant genetic differentiation was inferred from the analysis of polymorphic SSR loci at the intraspecific level, particularly between the Mediterranean populations of *A. pegreffii* and its population from southern Pacific waters (off the New Zealand coast) (Mattiucci et al. 2019) and southwestern Atlantic Ocean (off the Argentine coast) (Palomba et al. 2023; Mattiucci et al. unpublished data) (Fig. 9.1). Similarly, a consistent genetic sub-structuring was revealed through the analysis of 11 SSR loci, between the austral populations of the sibling species *A. berlandi* collected from both New Zealand and Argentine waters ($Fst \approx 0.05$) (Mattiucci et al. unpublished data) (Fig. 9.2). Finally, the combined use of these seven DNA microsatellite loci with other nuclear markers (i.e. nas 10 nDNA and *EF-1a* nDNA) allowed the genotyping of parental specimens of *A. simplex* (sensu stricto), *A. pegreffii* and *A. berlandi*. The multilocus approach also allowed the discernment of signals indicating admixture genotypes, enabling the discrimination between F1 hybrids and introgression (Mattiucci et al. 2019; Bello et al. 2020, 2021).

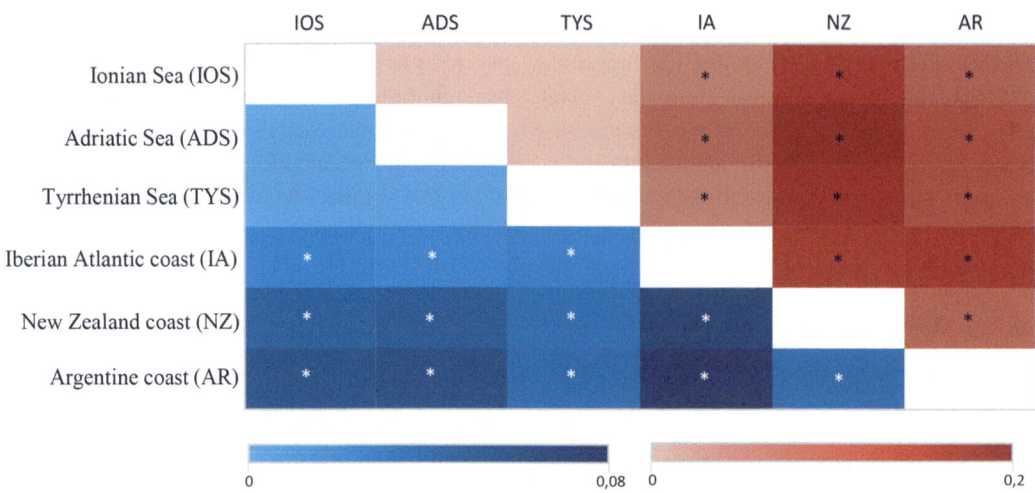

Fig. 9.1 Heat map of pairwise *Fst* values matrix among six *Anisakis pegreffii* populations. Microsatellite loci *Fst* values are depicted in blue, while *cox*2 mtDNA sequence FST values are given in red. Asterisks denote significant values

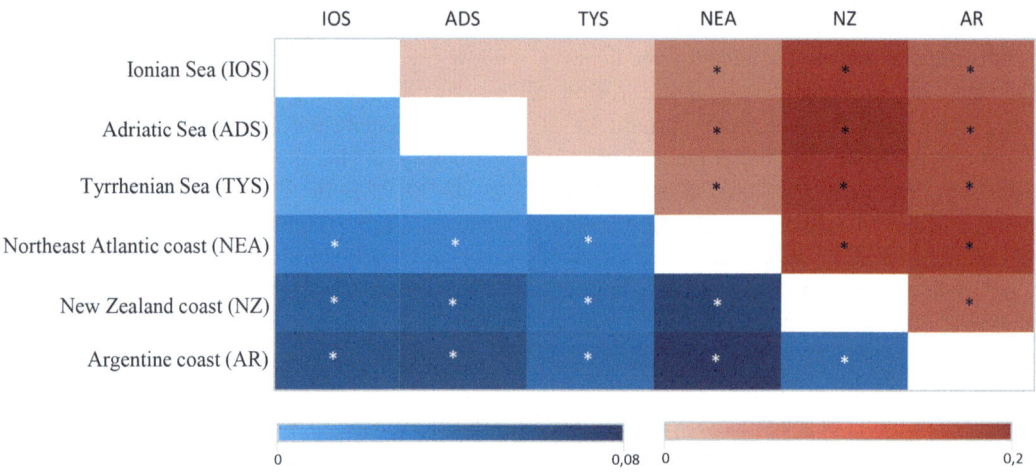

Fig. 9.2 Principal coordinate analysis (PCoA) plot of nematode individuals, belonging to six populations of *Anisakis pegreffii*, genotyped at 11 nuclear DNA microsatellite loci. Individuals are clustering according to distinct populations of the parasite species by geographic region

The application of seven newly developed SSRs DNA loci for studying the genetic variation of anisakid populations of the five sibling species of the *Contracaecum osculatum* (sensu lato) complex from both Arctic and Antarctic waters was performed to monitor genetic variability over a temporal scale in relation to anthropogenic habitat disturbance. Tracking of anthropogenic impacts on the population size of trophically transmitted helminth (TTH) parasites such as anisakid nematodes has been recently attempted using genetic variability estimates of several polymorphic nuclear loci as a 'metric' (Palomba et al. 2023). Furthermore, allele variation of these nematodes, inferred by microsatellites, can offer a way to answer questions related to the detection of certain genotypes associated with anisakiasis disease in humans, as well as the existence of differential genotype adaptation to distinct hosts and/or geographical clines.

9.2.1.4 Double-Digest Restriction Site-Associated DNA Sequencing (dd-RADseq)

An increasingly important aspect of population genetics is represented by the shift towards a much wider genetic dataset analysis comprising multiple loci, that is 'population genomics'. This approach allows the accumulation of informative genetic variation from an extensive set of markers. Advances in DNA sequencing technology have led to the development of methods that enable the analysis of hundreds to thousands of genomic loci in a rapid and cost-effective manner. In addition to providing multi-locus analysis in population genetics of a parasite species, high-throughput genomic methods may offer sufficient coverage of the genome to detect gene regions involved in phenotypic traits and speciation. The possibility of obtaining multi-locus datasets has, in turn, led to the development of new phylogenetic analysis protocols to assess population genetic structure based on a wide genotyping approach for estimating the phylogeny of closely-related taxa. However, the application of this approach in macro- and micro-evolutionary studies of marine parasites is currently limited. This limitation is likely attributed to the requirement for methods that consistently recover a set of homologous loci across multiple samples. This challenge is particularly pronounced in parasite species, due to difficulties in obtaining enough uncontaminated DNA for genome sequencing. Among high-throughput genomic methods, the double-digest restriction site-associated DNA sequencing (ddRADseq) method provides an alternative approach, centred on restriction-site associated DNA sequencing (RAD-seq). This method overcomes constraints associated with total genome analysis (Baird et al. 2008). It employs one or more restriction enzymes to target homologous loci among a set of samples, eliminating the need for a priori genomic resources. The method has an advantage in that it provides cost-effective means of broadly sampling the genome. Therefore, it has emerged as a powerful tool for various population genomics applications in molecular ecology (Davey and Blaxter 2010). However, while this method has gained widespread use in aquatic organisms, few studies have explored the utility of RAD-seq data for population genetics analysis in parasites. One of the few examples of the application of this method in aquatic parasites is the ddRAD sequencing of 341 tapeworms from 12 *Schistocephalus solidus* populations infecting three-spined sticklebacks (*Gasterosteus aculeatus*) on Vancouver Island (Shim et al. 2022). The study revealed a significant genetic sub-structuring of the tapeworm populations collected from different sites (Shim et al. 2022). Notably, the tapeworm genetic differences were significantly smaller than those of the fish hosts, suggesting that the parasite disperses more readily than its fish hosts (Shim et al. 2022). Another example involves the assessment of population structure and genetic diversity of two trypanorhynch cestode species, namely *Rhinoptericola megacantha* and *Callitetrarhynchus gracilis*, using a RAD sequencing approach. Both species exhibit relaxed host specificity. For *R. megacantha*, the population structure aligned with geographical distribution rather than the host species, while limited population structure was identified for *C. gracilis*, implying a potential association between the degree of host specificity and population structure (Herzog et al. 2023).

9.2.2 Mitochondrial DNA Loci to Study Microevolutionary Aspects of Aquatic Parasites

Mitochondrial functional gene loci serve as suitable markers for studying the systematics and phylogeny of various marine parasite taxa. Mitochondrial DNA is commonly used in combination with nuclear datasets in phylogenetic analyses to support the existence of operational taxonomic units as reciprocal monophyletic units (i.e. to support PSC). Additionally, mitochondrial DNA is used for examining population genetics structure of a parasite species over a geographical and host scale, that is, for phylogeographical and host-parasite co-phylogeographic inference studies.

9.2.2.1 Cytochrome Oxidase c Subunit II Gene Locus (cox2 mtDNA)

Several studies on marine parasites have used sequences analysis of the *cox*2 mtDNA region. This locus has been developed and widely applied as a diagnostic molecular tool for the recognition of all anisakid species (Mattiucci and Nascetti 2008) and it has been further used in studies investigating population genetic structure of certain target species. For instance, Ding et al. (2022) performed a population genetics analysis and examined the demographic history of 206 mitochondrial *cox*2 mtDNA sequences from 12 metapopulations of *A. pegreffii* collected from cutlassfish (*Trichiurus japonicus*) captured along the coasts of mainland China and Taiwan. The results revealed the presence of genetic sub-structuring of the parasite species from the two geographical areas. Notably, a genetic differentiation (on average, $F_{st} \approx 0.08$) was observed between the Taiwan population and those from off mainland China. Despite the detection of two haplogroups in the network analysis, a high gene flow was observed between the different geographical areas, suggesting the presence of a panmictic population of *A. pegreffii*, with a recent past population expansion in the Pacific during the late Pleistocene (100,000–120,000 years ago) (Ding et al. 2022). Similar genetic sub-structuring patterns were found in *A. pegreffii* populations in European waters between the Mediterranean Sea and the northeast Atlantic (Palomba et al. 2023). Specifically, while intraspecific genetic variation between metapopulations from the Adriatic and Tyrrhenian Seas was not significant ($F_{st} \approx 0.015$, $p = 0.10$), significant genetic sub-structuring was observed when comparing Mediterranean (Adriatic and Tyrrhenian) populations of *A. pegreffii* with those from the Iberian Atlantic coast ($F_{st} \approx 0.07$, $p = 0.0001$ and $F_{st} \approx 0.04$, $p = 0.001$, respectively). The analysis of *cox*2 mtDNA haplotypes of *A. pegreffii* showed Hap3 being the most frequent haplotype, shared by all metapopulations of the parasite (Fig. 9.3). A high frequency of certain unique haplotypes in parasite specimens from definitive hosts (cetaceans) stranded on the Spanish Atlantic coast was observed in *A. pegreffii* (Cipriani et al. 2022).

Furthermore, higher significant values of genetic differentiation were observed between the metapopulation of *A. pegreffii* from European waters and those from austral regions such as from off New Zealand (on average $F_{st} \approx 0.10$) and Argentina (on average, $F_{st} \approx 0.09$) (Mattiucci et al. unpublished) (Fig. 9.1). Several private haplotypes exist in the distinct geographical regions, in spite of the similar number of parasite individuals tested (Fig. 9.3).

A population genetics study of 873 specimens of *A. simplex* (sensu stricto) infecting European hake *Merluccius merluccius* from fishing grounds in Iberian waters of the NE Atlantic was recently carried out by Ramilo et al. (2023) employing the *cox*2 mtDNA gene. The study revealed a high haplotype diversity, likely related to the high rate of polymorphism observed at the same locus in all anisakid species (Mattiucci and Nascetti 2008). However, no genetic structure was detected among the parasite species in the sampled populations. Dominant haplotypes were shared among different hake fishing grounds, suggesting a high level of gene flow between parasite populations and the existence of a panmictic population of *A. simplex* (sensu stricto) in this fish host and across the sampling area. However, a certain level of genetic sub-structuring was observed between the southern populations of *A. simplex* (sensu stricto) from the Spanish and Portuguese coasts relative to those from the northwest coast of Scotland (on average, $F_{st} \approx 0.04$) (Ramilo et al. 2023). This finding seems to be congruent with the genetic population sub-structuring observed at the same mitochondrial marker between the population of *A. simplex* (sensu stricto), collected from stranded cetaceans along the Atlantic Iberian coast, and those from Scotland and the Norwegian coast (Cipriani et al. 2022). An average level of differentiation of $F_{st} \approx 0.07$ between samples from Scotland versus those from Iberian Atlantic waters, and $F_{st} \approx 0.12$ between the Iberian population versus those collected from the Norwegian coast, suggests that genetic structure of populations of this heteroxenous parasite in these areas is also shaped by the population genetic structure of its definitive and intermediate/paratenic hosts

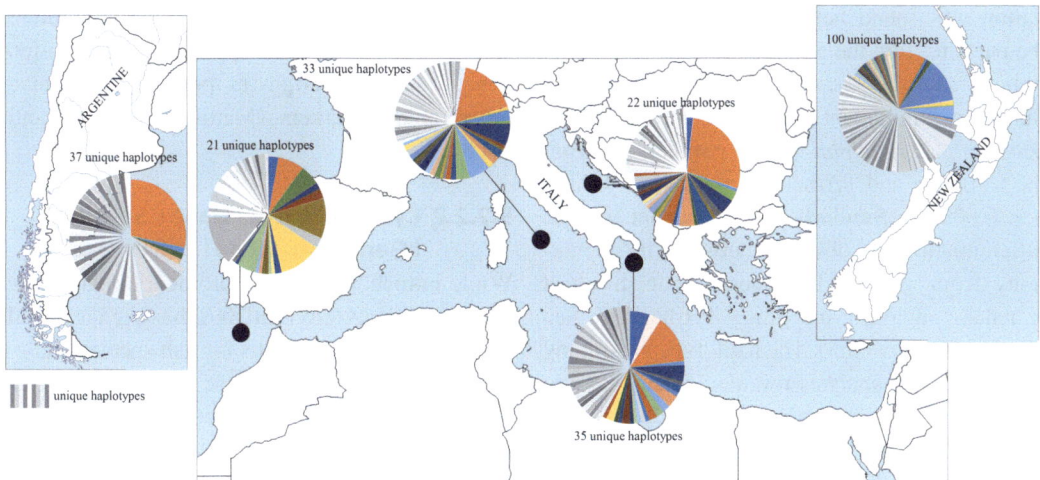

Fig. 9.3 Distribution of *cox*2 mtDNA haplotypes observed in populations of *Anisakis pegreffii* collected from different sampling localities. Distinct unique haplotypes found in the different populations of the parasite species in the geographic regions, are represented through a spectrum of grey tones

across its range in the northeast Atlantic. In support of this hypothesis, there is also evidence of significant genetic sub-structuring in *A. simplex* (sensu stricto) larval populations in herrings (*Clupea harengus*) from the northeast Atlantic (Mattiucci et al. 2018b). A comprehensive sequence network analysis of L3 stage *A. simplex* (sensu stricto) larvae revealed haplotype diversity values that fell within the same range as those of Cipriani et al. (2022). Notably, the population showing the highest differentiation was the most northern area, specifically the Norwegian Sea (Mattiucci et al. 2018b). The haplotype parsimony network (TCS) inferred from mtDNA *cox*2 haplotypes of *A. simplex* (sensu stricto) showed the presence of two major haplotypes, that is, Hap25 and Hap48. Hap25 was shared by several *A. simplex* (sensu stricto) individuals collected from cetaceans from the Norwegian Sea and Scottish coast in the northeast Atlantic, whereas Hap48 was shared mostly by the metapopulations collected from off the Iberian and Scottish Atlantic coasts. The haplotype TCS analysis also revealed the existence of several private haplotypes in the Norwegian and Iberian Atlantic metapopulations of *A. simplex* (sensu stricto), as well as concordance between haplotype clustering and geographic origin of the endoparasite samples (Cipriani et al. 2022). Given that herring are an important prey source for cetaceans in those waters, it is reasonable to expect that the Norwegian *A. simplex* (sensu stricto) population detected in herrings overlaps with the identified adult population found in those cetacean predators.

The *cox*2 mitochondrial gene locus has been demonstrated to be valuable in population genetic analysis of another anisakid species, 'species A' of the '*Contracaecum rudolphii* complex' which, as an adult, is a parasite of avian hosts, specifically cormorants such as Great cormorant, *Phalacrocorax carbo* and the European shag, *Gulosus aristotelis*. It has been hypothesised that the life cycle of this parasite is adapted to brackish and marine coastal ecosystems (Mattiucci et al. 2020). Comparison of the sequence analysis of '*C. rudolphii* sp. A' collected from the western Mediterranean coast of Spain with those of other western Mediterranean populations from west Sardinia and the Tyrrhenian Sea provided insights into the migration routes of fish-eating bird populations across European brackish and marine waters. Significant genetic differentiation ($Fst = 0.08$, $p < 0.00001$) was observed between the metapopulations of '*C. rudolphii* sp. A' recovered from *Ph. aristotelis* from the Spanish Catalan coast and those from *Ph. carbo* from Sardinian brackish waters. However, no differentiation was found between metapopulations of the parasite from the Spanish and the Tyrrhenian Italian coasts (Roca-Geronès et al. 2023). Similarly, significant Fst values were found between '*C. rudolphii* sp. A' populations from the Sardinian and Tyrrhenian coasts ($Fst = 0.06$, $p < 0.00001$). Similarly, no significant genetic differentiation was observed between the populations from the Mediterranean Spanish coast (Catalan region) and the Tyrrhenian Sea ($Fst = 0.005$, $p > 0.05$). The haplotype parsimony network (TCS) analysis revealed a star-like tree (Fig. 9.4) (Roca-Geronès et al. 2023). In the Spanish coast population, 40 haplotypes were detected, while 80 and 28 haplotypes were identified from the Tyrrhenian and Sardinian populations, respectively (Fig. 9.4). The observed genetic differentiation between the Spanish Catalan population of '*C. rudolphii* sp. A' and the other populations could be attributed to the relatively high number of private haplotypes ($n = 26$) present in this area. It was also hypothesised that the low mobility of *Ph. aristotelis* from Spanish Catalan coast, the most common definitive host of the parasite in that part of the western Mediterranean, could contribute to maintaining the genetic differentiation of the '*C. rudolphii* sp. A' population (Fig. 9.4).

The genetic structure of the pinniped-infecting anisakid worm *Phocanema bulbosum*, a species within the *P. decipiens* (sensu lato) complex (see Mattiucci and Nascetti 2008), that matures in the bearded seal *Erignathus barbatus* in the Arctic Sea waters, was recently investigated using the *cox*2 mtDNA gene. The pairwise genetic differences and the median-joining haplotype network of sequences obtained from Atlantic cod (*Gadus morhua*) and American plaice (*Hippoglossoides platessoides*) from the Barents Sea have not indicated genetic sub-structuring between the anisakid sampled from the different fish hosts. A low number of haplotypes (10) were obtained, likely due to the limited number of specimens analysed ($n = 68$) (Gordeev et al. 2023). However, these genetic data, when compared with populations of the same parasite species from other geographical areas such as the Arctic Boreal region of Svalbard and the Canadian waters where the parasite is also distributed, could reveal in future studies using multigene approaches, population genetic sub-structuring, as previously revealed by allozyme markers (Mattiucci and Nascetti 2008).

9.2.2.2 Cytochrome Oxidase c Subunit I Gene Locus (cox1 mtDNA)

While mitochondrial cytochrome c oxidase subunit I gene locus (*cox*1 mtDNA) has been revealed to be a barcode gene locus for fish species recognition, less known was its use in parasite species recognition and population genetic analysis.

One prime example is the study of intraspecific genetic diversity of the trypanorhynch cestode *Tentacularia coryphaenae* (Palm et al. 2007). Despite specimens being sourced from diverse fish inhabiting disparate environments and originating from locations over 1000 km apart, a notable genetic homogeneity at cox1 mtDNA locus was observed among populations of this parasite. This genetic homogeneity can be attributed to a high gene flow facilitated by extensive migrations of both the second intermediate and paratenic hosts and the final hosts (respectively teleost and elasmobranchs) of *T. coryphaenae* (Palm et al. 2007).

Among other recent investigations, the *cox*1 mtDNA gene locus was employed to explore the genetic structure of two acanthocephalan species, the specialist *Hexaglandula corynosoma* and the generalist *Southwellina hispida*, coexisting in Mexico. Phylogeographic analyses revealed that populations of both species lacked a discernible phylogeographic structure, displaying elevated haplotype diversity, low nucleotide diversity and very low Fst values across biogeographic provinces. Furthermore, negative values on neutrality tests suggest ongoing population expansions in these acanthocephalan populations (García-Varela et al. 2023).

In other cases, such as in the molecular study of the fish tapeworms *Caryophyllaeus laticeps* and *C. chondrostomi*, mitochondrial gene analysis did not provide molecular support for distinguishing between the two species. In contrast, the analysis of six DNA microsatellites proved to be better discriminative markers than rDNA and

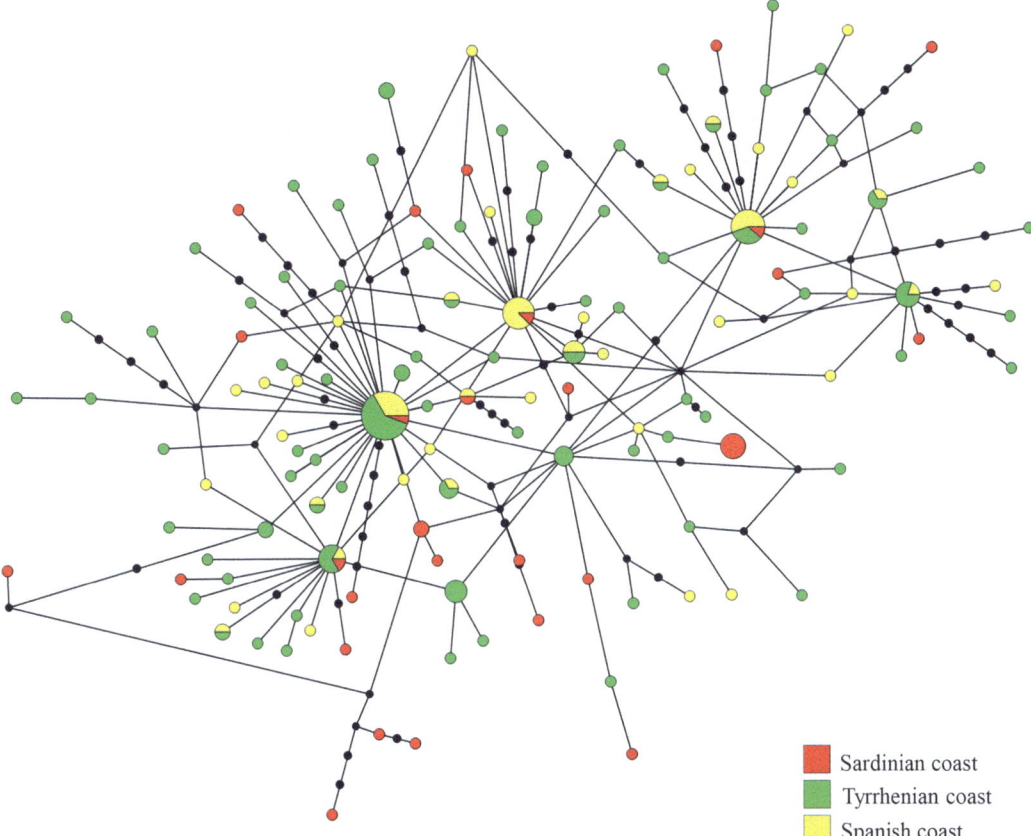

Fig. 9.4 TCS network of 'Contracaecum rudolphii sp. A' sequences of cox2 mtDNA dataset. Hatch marks indicate mutations. Circle sizes are proportional to the number of individuals showing the specific haplotype

mtDNA for genetic differentiation at both interspecific and intraspecific levels (Bazsalovicsová et al. 2020).

The genetic architecture of the pinniped hookworm *Uncinaria lucasi*, investigated at the intraspecific level from two different host species, that is, the Steller sea lion *Eumetopias jubatus* and the northern fur seal *Callorhinus ursinus*, revealed no genetic differentiation between those populations. Indeed, despite high cox1 haplotype (h = 0.96–0.98) and nucleotide (π = 0.014) diversity observed, no significant *Fst* value was found, suggesting high gene flow between the two populations of this parasite species (Davies et al. 2020).

Recently, a global mtDNA *cox1* genetic variation of the zoonotic parasite *Dibothriocephalus nihonkaiensis* (syn. *Diphyllobothrium nihonkaiense*) was carried out by a network analysis of *cox1* sequences from 14 isolates of *D. nihonkaiensis* from 12 patients and 79 previously published sequences from definitive and second intermediate hosts. A total of 48 haplotypes in three haplotype groups (Types A, B and C) were identified, highlighting co-infections with genetically different *D. nihonkaiensis* in humans and Pacific salmon (Abe et al. 2021). Further comparative haplotype analysis of this mitochondrial gene locus on isolates from definitive and second intermediate hosts collected in the Pacific and Atlantic areas may clarify ecological aspects of *D. nihonkaiensis*.

In recent years, the mitochondrial *cox1* gene locus has proven useful for barcoding recognition of Monogenea of the Family Mazocraeidae represented by *Kuhnia scombri*, *K. scombercolias* and *Pseudokuhnia minor* and the Family Gastrocotylidae represented by *Gastrocotyle kurra* and *Allogastrocotyle bivaginalis*. There was no comparable sequence available in GenBank for *K. scombercolias* and limited sequences were available in GenBank for the gastrocotylid Monogenea identified in this study. Phylogenetic analyses of mtDNA *cox*1 sequences of the identified Monogenea clustered according to their familial groups (Hossen et al. 2022).

Finally, in the case of anisakid nematodes, less attention has been paid to the mtDNA cox1 gene locus, as the mtDNA cox2 has been more commonly used for genetic comparison due to the limited number of anisakid species sequenced at the mtDNA *cox*1 locus.

A population genetics analysis was performed on third stage larvae of *A. simplex* (sensu stricto) from Atlantic herring, *Clupea harengus* caught off the north-west coast of Scotland. A total of 161 *A. simplex* (sensu stricto) *cox*1sequences were analysed from 37 herring representing three spawning periods over two years. Overall, very high haplotype and low nucleotide diversities were observed ($h = 0.997$ and $\pi = 0.008$, respectively). The genetic analysis indicated little variation related to spawning events or years, which may suggest localised temporal stability. Temporal homogeneity in the *cox*1 gene, coupled with the ubiquitous and widespread nature of the parasite, indicates both the potential and limitations for its use in stock-separation studies of its hosts (Cross et al. 2007).

9.3 What Are the Drivers Shaping the Genetic Structure of Marine Parasites Over Space and Time?

Understanding the drivers that shape genetic diversity in aquatic parasites has significant ecological and evolutionary implications for host–parasite interactions. This interaction is fascinating because both hosts and parasites contribute co-evolutionary traits over space and time. This aspect offers challenging investigations into the traits that may influence the observed population genetic structure of a certain parasite species.

Host dispersal capacity, effective population size (host and parasite density), the parasite reproductive strategy and the presence of free-living stages in the parasite's life cycle can play fundamental roles in generating and maintaining genetic variation in both interacting organisms (i.e. parasites and hosts) on which natural selection acts, allowing coadaptation and coevolution. It is well-established, for example, that the reciprocal selective pressure exerted by parasites and hosts can result in the selection of more adaptive genotypes in parasite species. This occurs within the framework of evolutionary dynamics, such as the 'arms races' proposed by the Red Queen hypothesis (Van Valen 1973).

9.3.1 Parasite Life Cycle

Parasite host-range and life cycle features could be significant in sub-structuring the population genetics of a parasite species. Parasite host range refers to the capacity of parasites to use host species with different dispersal abilities, representing different microenvironments for the parasite species. According to their life cycle, aquatic parasites can be categorised as: (1) parasites with a direct life cycle infecting a single host species (i.e. specialist), or able to infect several host species belonging to the same genus or family (i.e. generalist); (2) parasites with an indirect or complex life cycle (i.e. several successive intermediate and definitive host species that generally include organisms belonging to different families or phyla), often linked by a trophic web (see Chap. 5). The ability of parasites to use not only different host species but also hosts with different dispersal capacities can play a direct role in their distribution and range expansion, generating chances of geographical shaping of their genetic structure. It has been generally assumed that the type of life cycle (i.e. allogenic versus autogenic, sensu Esch et al. 1988) can predict the parasite population's genetic structure (Blasco-Costa and Poulin 2013). Autogenic parasites, being more confined to aquatic ecosystems than allogenic parasites whose life cycles involve both aquatic and terrestrial hosts, would exhibit a higher level of genetic differentiation between populations from different localities. Aquatic parasites using host organisms with limited dispersal or mobility, or that are confined by hydrological barriers, have less chance to continuously spread, limiting gene flow between parasite metapopulations. Host ecology contributes to the limitation of the dispersal capacity of the parasite on a large spatial scale. This phenomenon is more pronounced in freshwater parasites than in marine ones, and trematodes with autogenic life cycles generally exhibit a greater degree of population genetic structure (Criscione and Blouin 2004). However, among trematodes, some species use only rather sedentary aquatic hosts to complete their life cycle, while others infect more vagile definitive hosts such as birds or mammals, allowing dispersal between separate aquatic habitats. Thus, in the case of allogenic aquatic parasites having an avian species as the definitive host could play a role in reducing the population genetic complexity of its endoparasite species, allowing the maintenance of gene flow between geographically distant populations.

Blasco-Costa and Poulin (2013) performed a meta-analysis to investigate how host traits influenced the genetic structure of 16 trematode species, mainly from freshwater ecosystems. They found that the type of parasite life cycle (simple or complex; specialist or generalist; allogenic or autogenic) to be the best predictor of a population's genetic structure. Autogenic trematode species completing their life cycles in organisms living in a specific water body showed significantly higher genetic structuring, in terms of F_{st} value, than the allogenic parasites leaving the water bodies through a migratory bird or mammal host (Blasco-Costa and Poulin 2013).

Parasite species that have multi-host life cycles, but with a definitive host having a limited range of distribution, generally show a higher degree of intraspecific genetic differentiation between populations. For instance, among the anisakids maturing in pinnipeds such as those belonging to the *Contracaecum osculatum* complex or *Phocanema decipiens* complex (Mattiucci et al. 1998; Bao et al. 2023), the genetic differentiation values at the intraspecific level, analysed across numerous populations and a substantial number of individuals using a wide array of polymorphic nuclear loci (allozymes), were higher than those of other anisakids whose adult stages infect cetacean species. Gene flow value was estimated to be, on average, $Nm = 4.90$ and $Nm = 3.90$ in the pinniped-infecting species *C. osculatum* sp. B and *C. osculatum* sp. A, respectively (Nascetti et al. 1993), as well as $Nm = 4.38$ and $Nm = 3.66$ respectively in *Phocanema krabbei* and *P. decipiens* sensu stricto (Mattiucci et al. 1998). Conversely, values of $Nm = 15.0$ and 9.00 were reported for the cetacean-infecting species *A. pegreffii* and *A. berlandi* respectively (Mattiucci et al. 1997). Thus, despite the generally high level of gene flow found across anisakid populations, differential gene exchange linked to host ecology, host range of distribution and host mobility among different populations has also been demonstrated. Even if these parasite taxa share some common and migratory fish species (in this case, gadids) as paratenic and transport hosts, their definitive hosts contribute differentially to shaping their population genetics.

9.3.2 Parasite and Host Dispersal

The specific mode of host dispersal has complex and variable consequences in structuring metapopulations of parasites. Dispersal limitation of the hosts involved in the parasite life cycle remains a main driving force behind the observed genetic structure of parasite species. A certain degree of genetic differentiation and spatial structure of parasite populations can be detected when host populations are restricted to isolated water bodies or when host mobility is constrained. Indeed, it is generally assumed, that a high dispersal capacity of a parasite species is maintained by the high vagility/and dispersal ability of its interacting hosts; this phenomenon, in turn, is responsible for a high level of gene flow and low genetic differentiation between its populations (Criscione 2008). This is exemplified by marine anisakid nematodes belonging to the genus *Anisakis*, whose distribution range is enabled by the dispersal capacity of their definitive hosts, mainly cetaceans (Mattiucci and Nascetti 2008; Kuhn et al. 2011), or in the case of the sibling species of the *Contracaecum rudolphii* complex, parasites as adults of migratory birds, for example, the Great cormorant, *Phalacrocorax carbo* (Mattiucci et al. 2020). Conversely, the low dispersal capacity of a definitive host and dependence by a specific sessile and less mobile intermediate host involved in the life cycle of a parasite species may result in a higher level of genetic differentiation with a restricted gene flow. An example of this is represented by the trematode *Maritrema novaezealandense*, which infects the snail *Zeacumantus subcarinatus* and crab *Macrophthalmus hirtipes* in New Zealand, displaying high genetic structure (Keeney et al. 2007).

The definitive host's mobility can also have an effect on the population genetic structure of hermaphroditic parasites (see also Sect. 9.3.3). When the dispersal rate in self-fertilising parasites is high, a rapid genetic homogenisation of populations at the meta-population scale would result as a consequence of the host's migration capacity. For example, in trematode species that infect fish or aquatic avian species that undertake frequent long-distance dispersal events, a quick homogenisation of gene pools can occur (Feis et al. 2015). For example, for the bird-infecting opisthorchiid *Cryptocotyle lingua*, no significant level of differentiation was observed between five geographic areas in which populations have been studied along the French coast, and only a slightly significant genetic differentiation between *C. lingua* from the western North Sea and the eastern English Channel (Fst = 0.01199; p = 0.033) (Duflot et al. 2023).

9.3.3 Parasite Reproductive Strategies

The parasite's mode of sexual reproduction can affect genetic variation at the intraspecific level in parasite species. Parasite species can reproduce sexually, with unisexual (i.e. gonochoristic or dioecious species) or bisexual individuals (i.e. hermaphroditic or monoecious species), or asexually (e.g. parthenogenetic species), or both. We can generally assume that the ability of a species to generate and maintain genetic differentiation among its populations is higher in parasites that are gonochoristic or parthenogenetic species, rather than in hermaphroditic ones. Dioecious species of anisakid nematodes are a good example of the first category. They generally exhibit high levels of genetic polymorphism on a large number of alleles at nuclear loci (allozymes, haplotype diversity and SSRs DNA) (Mattiucci and Nascetti 2008; Mattiucci et al. 2015a, 2017, 2019). This diversity is likely maintained by a high recombination rate at the intraspecific level, a result of sexual reproduction. In contrast, hermaphroditic cestode species typically show comparatively lower levels of genetic polymorphism, even at genes considered to have a moderate level of polymorphism, such as the $cox1$ mtDNA region. As a consequence, in hermaphroditic parasites, while a low level of genetic differentiation has been observed within the same population, a higher level between different populations can be observed (Huyse et al. 2005). Indeed, self-fertilisation is expected to generate more genetic variation (Ingvarsson 2002; Huyse et al. 2005). Cross-fertilisation has been documented in some hermaphroditic cestodes, one example being the cestode *Schistocephalus solidus*, which has been shown to both self- and cross-fertilise. Interestingly, in experimental infections wherein second intermediate hosts (three-spined sticklebacks, *Gasterosteus aculeatus*) were exposed to cestode larvae, produced either by self-fertilisation or cross-fertilisation, the genetic quality of the cestodes resulting from self-fertilisation was found to be inferior to those produced by cross-fertilisation (Christen and Milinski 2003).

9.3.4 Parasite Demography

Beyond host dispersal capacity, other host biological factors may influence the population structure of parasites. Among the major drivers, the parasite species' density in a specific geographical area, a proxy for its effective population size, can also affect pairwise intraspecific genetic differentiation between populations from different geographical areas. It has been suggested that parasite density in a certain geographical area may be closely linked to the demography of the host species involved in its life cycle (Palomba et al. 2023). For instance, it has been demonstrated that the demographic dynamics of host populations and environmental abiotic factors play a role in shaping the genetic structure and diversity of trophically-transmitted nematodes and cestodes (Mattiucci and Nascetti 2008; Isbert et al. 2023; Palomba et al. 2023). A significant positive correlation between the genetic variability values of a parasite population and its abundance (as a proxy for the population size of the parasite species) has been documented (Mattiucci et al. 2015a; Palomba et al. 2023). Taking this assumption into consideration, genetic differentiation can be higher between populations of the same species with distinct densities in different geographical areas within its distribution range. For example, the significant genetic differentiation observed between the boreal and austral populations of *A. pegreffii* within its geographical range can also be attributed to different densities (i.e. infection intensities) of this parasite in the austral region (i.e. New Zealand and Argentine waters) compared to the Mediterranean Sea (Palomba et al. 2023). The higher level of genetic polymorphism detected in the population of *A. pegreffii* from New Zealand waters, as well as the presence of unique haplotypes in that population, could explain the genetic differentiation found in this metapopulation (Fig. 9.3) (Palomba et al. 2023). According to an analysis of 11 polymorphic DNA-SSR loci, estimates of the genetic variability of the Mediterranean population of *A. pegreffii* showed remarkably lower mean heterozygosity values (He) in the Mediterranean populations (average $He \approx 0.67$) compared to those populations from an extra-Mediterranean area, such as New Zealand (average $He \approx 0.75$). Similarly, assessment of intraspecific genetic diversity at the $cox2$ mtDNA locus of *A. pegreffii* from the Adriatic Sea showed lower nucleotide diversity (π) and haplotype number (Nh) ($\pi \approx 0.0059$; $Nh = 52$) compared to populations from New Zealand waters ($\pi \approx 0.0102$; $Nh = 113$) (Palomba et al. 2023). These findings suggest that the Mediterranean populations of *A. pegreffii* are facing genetic erosion, with the loss of genetic polymorphism linked to lower parasite demography (i.e. density) in that region compared to the austral populations, which show higher levels of demography and consequently lower probability of genetic drift in the population's gene pool.

A recent molecular epidemiological study from different geographical areas of the Mediterranean Sea and Atlantic waters revealed substantial genetic variability in the cestode *Grillotia adenoplusia*. Specimens of *G. adenoplusia* were obtained from the benthic shark *Etmopterus spinax* in the eastern and western Mediterranean, as well as the Atlantic Ocean (Isbert et al. 2023). The study revealed noteworthy intraspecific variation with the occurrence of numerous haplotypes ($Nh = 17$) with 18 polymorphic sites in the large subunit ribosomal DNA (28S rDNA) gene (Isbert et al. 2023). Interestingly, in the Mediterranean, the parasite also showed high abundance in intermediate hosts sampled from the Balearic Sea and off Cyprus, also corresponding with the highest number of polymorphic sites and haplotypes (the latter from the Balearic Sea) (Isbert et al. 2023). This observation supports the hypothesis that, even in the case of this heteroxenous cestode, genetic variability is positively correlated with the parasite population size at the local level (Palomba et al. 2023).

Additional genetic markers and further sampling will be necessary to fully elucidate the genetic structure and intraspecific genetic variation of this taxon.

Mitochondrial DNA loci better reflect the effects of demographic processes than nuclear DNA due to the lower mutation rate than nuclear loci and the fact that they are maternally inherited. On the other hand, it is likely that dispersal has more of an effect on nuclear loci, because of the possibly lower or possibly higher recombination rate. In other words, each set of markers gives more or less weight to demographic and dispersal processes. We therefore cannot recommend exclusive use of one or the other when conducting genetic assessments of traits affecting populations, but rather suggest using a combination of polymorphic mtDNA and nuclear loci and take care when interpreting results obtained by the respective markers.

9.3.5 Host Ecology

Another driver influencing the genetic structure of a parasite population is the ecology and population genetic structure of its hosts, both definitive and intermediate. Aspects of definitive host ecology such as migration and vagility can have a homogenising effect on the genetic structure of parasite populations at the intraspecific level, allowing the maintenance of high gene flow between geographically distant populations. For example, genetic studies of the anisakid nematode *Sulcascaris sulcata* at three gene loci (the ITS region of rDNA, and *cox*1 and *cox*2 mtDNA) revealed an overall genetic similarity between populations from the Adriatic and Tyrrhenian Seas on either side of the Italian Peninsula, without geographical clustering of parasites from the different basins (Marcer et al. 2020). This is likely the consequence of the long-span migrations recorded for loggerhead turtles (*Caretta caretta*, the definitive hosts of *S. sulcata*) throughout the Mediterranean Sea (Marcer et al. 2020). However, the existence of a certain number of haplotypes (Nh = 38) observed at the *cox*1 mtDNA gene locus in the Mediterranean population of *S. sulcata* are positively correlated with the parasite population size maintained by both juveniles and adult loggerheads. Although a spatial and temporal comparison of the genetic diversity of *S. sulcata* from extra-Mediterranean waters has yet to be performed, these findings represent the basis for future genetic comparisons, possibly including nuclear polymorphic markers. Such a comparison may be feasible with the possible existence of a sibling species of *S. sulcata* from the Eastern China Sea off Japan, showing a high level of genetic differentiation from Mediterranean isolates (Sata and Nakano 2023).

Conversely, in other endoparasites, host ecology may contribute to maintaining a high level of differentiation between populations/species of parasites. For instance, the ecology of the great cormorant *Phalacrocorax carbo* can affect the spatio-temporal heterogeneity observed in the dispersal and density of infection of the two sibling nematode species '*Contracaecum rudolphii* sp. A' and '*C. rudolphii* sp. B'. These sibling species were detected and studied in the same definitive host species from inland and coastal waters of Italy (Mattiucci et al. 2020). It has been hypothesised that the different feeding ecology and behaviour of two wintering populations of *P. carbo* in the Mediterranean brackish coastal lagoons and freshwater ecosystems shape the life cycles of the two sibling species. Individuals of *P. carbo* commence migration from their breeding areas in northern Europe, where they had certain fixed foraging behaviour. Upon arrival in wintering areas on the Mediterranean coast of southern Europe, they settle and feed in their preferred ecosystems where they probably maintain the same foraging behaviour, but effectively divide into two populations: the 'brackish cormorants' and 'freshwater cormorants'. In this way, they contribute to maintain the life cycles of the two parasite species '*C. rudolphii*' in brackish and freshwater basins in the southern wintering areas of distribution, with '*C. rudolphii* sp. A' adapted to brackish waters and '*C. rudolphii* sp. B' to freshwater. Due to this behaviour of the cormorants and further ecological adaptation to different coastal habitats, the two parasite species, which exist in sympatry in the cormorants' breeding areas and share the same individual definitive hosts, maintain isolated gene pools, with no gene flow between them. Indeed, a higher level of genetic differentiation at interspecific level, inferred from the same $cox2$ mitochondrial locus ($Fst = 0.08$) was recorded, compared to the pairwise genetic differentiation found at the intraspecific level (on average, $Fst = 0.8$) (Mattiucci et al. 2020).

9.3.6 Host Evolutionary History and Phylogeography

Natural history factors can influence the genetic structure of parasite populations. The genetic structure of definitive hosts, which are shaped by their natural and evolutionary history, is also important in shaping parasite genetic structure. This has been demonstrated for anisakid species, for which it has been shown that the evolutionary history of their definitive cetacean and pinniped hosts is mirrored by the parasites (see Mattiucci and Nascetti 2008).

The spatial structure of a parasite population may reflect the population genetic structure of its hosts. At intraspecific level, it has been suggested that the genetic sub-structuring (average $Fst \approx 0.08$) of *Anisakis pegreffii* from two geographical areas of Taiwan Sea and others from off the mainland China coast reflect the biogeography of the area (Ding et al. 2022). The authors suggested that the Taiwan Strait represented a barrier between the populations in Taiwan and those in the China Sea during the last glacial period, as well as allowing for secondary contact between them during the glacial retreat. Despite the detection of two haplogroups in the network analysis of the $cox2$ mtDNA sequences, a high gene flow was observed between samples of *A. pegreffii* from different geographical areas (Ding et al. 2022).

Analogously, the absence of genetic differentiation in analyses of mtDNA $cox1$ data at the intraspecific level, in the pinniped hookworm *Uncinaria lucasi* of two host species (the northern fur seal, *Callorhinus ursinus* and the Steller's sea lion, *Eumetopias jubatus*) from three widely separated geographic regions (the Sea of Okhotsk, eastern Bering Sea and eastern Pacific Ocean) (Davies et al. 2020), was consistent with the absence of genetic structure of the parasite species. The authors suggested that the finding was likely due to the high gene flow between the parasite populations maintained by *C. ursinus* (but not *E. jubatus*) and also to the recent (postglacial) population expansion (during the last 10,000 years) of *C. ursinus,* as revealed by the studies on its genetic structure (Dickerson et al. 2010).

Parasites can reveal cryptic phylogeographic information of their specific hosts (Nieberding et al. 2004). It has been suggested that if a parasite has a similar or more complex genetic structure compared to its host, then the genotypes/haplotype of the parasite species can be used to assign hosts to their population of origin, with higher accuracy than by using the host's genotype alone (Criscione et al. 2006; Froeschke and Von der Heyden 2014; see also Chap. 17). A common approach in investigating the congruence in host–parasite genetic structure has been to describe and compare the patterns of genetic differentiation for each interacting species, that is, the parasite and its hosts, mainly the definitive hosts. This approach is nowadays possible by co-phylogeographical analysis. Generally, congruence in the shape of co-phylogeographic structure of the host–parasite association is expected when parasites and their hosts have the same rate of dispersal capacity. The parasite reproductive mode may explain why the genetic differentiation of hermaphroditic parasites over a geographical scale can be higher than that of the hosts in certain host–parasite interactions. This is because parasites usually have shorter generation times relative to their hosts. As a consequence, demographic changes and mutation rates would lead to higher levels of genetic differentiation in parasite gene pools than those of their hosts. The evolutionary history of parasite populations and species may reveal the hosts evolutionary history before the host DNA has coalesced (Rannala and Michalakis 2003). As a result, distinct genotypes and haplotypes of parasites can be used to infer the assignment of parasite populations to host populations; equal or higher genetic structure in parasites compared to that of their hosts would occur. This has been detected in the association formed by the trematode *Plagioporus shawi* and the rainbow trout *Onchorhynchus mykiss,* using SSR DNA genotypes of parasites that proved to be more accurate than those from the fish to assign individuals to their population of origin (Criscione et al. 2006). In other cases, the occurrence of co-evolutionary host–parasite population associations, in terms of co-phylogeographic events, was demonstrated between fish of the genus *Pagellus* and the monogenean *Lamellodiscus* sp. (Desdevises et al. 2000). A population structure of parasites congruent with the geographically different host populations was also found between salmonid host fishes and the myxozoan *Parvicapsula minibicornis* (Atkinson et al. 2011).

In other host–parasite associations, the high dispersal ability of the host can lead to the genetic homogenisation of parasite gene pools. In the case of the conger eel *Conger conger* and the trematode *Leicithochirium grandiporum*, low genetic differentiation between different geographic populations was detected (Vilas et al. 2003). Autogenic parasite species such as *Deropegus aspina* were found to have more highly structured populations compared to allogenic parasites such as *Nanophyetus salmincola* and *Plagioporus shawi* infecting the same species of *Oncorhynchus* (Criscione and Blouin 2004).

In gonochoristic parasites, the dispersal rates of the host may allow genetic sub-structuring more similar to that of their definitive hosts. The two sexes of the parasite must successfully disperse and meet in their suitable definitive host, whose mobility allows gene flow to occur. As a consequence, a similar population genetic structure in both organism of the host–parasite associations would occur. For instance, the absence of genetic sub-structuring of *Anisakis* spp. larvae from the pelagic fish species *Sardinops sagax* from the Pacific Ocean, as detected by haplotype analysis of *cox*2 mtDNA (Baldwin et al. 2011), was likely congruent with the existence of a single panmictic population of the fish host in that fishing ground, as well as the high dispersal capacity of the parasite definitive hosts in the Pacific Ocean.

In other gonochoristic parasites, the differential migration routes of definitive host populations likely maintain genetic sub-structuring in the parasites' gene pools. The migration pattern of the great cormorant, *P. carbo*, the definitive host of the anisakid '*C. rudolphii* sp. A', might explain the genetic differentiation found between the metapopulation of '*C. rudolphii* sp. A' from cormorants of Sardinia Island with respect to those from the Spanish Mediterranean Sea coast (i.e. Catalan coast) and Tyrrhenian Sea (Latium region coast) (Roca-Geronès et al. 2023). The great cormorant exhibits highly seasonal migration, inhabiting brackish and freshwater coastal environments along the Mediterranean coast during its non-breeding season (Frederiksen et al. 2018). The observed genetic differentiation on the *cox*2 mtDNA locus between populations of '*C. rudolphii* sp. A' from Mediterranean Spanish and Tyrrhenian coast of Latium Region was explained by the authors as possibly due to the existence of a population of *P. carbo* living in Sardinia during the winter season. This population could have a different breeding area to those populations of *P. carbo* wintering in other areas of the western Mediterranean. Marion and Le Gentil (2006) detected, using *cox*2 mtDNA data, the presence of two genetically distinct wintering cormorant populations of *P. carbo* in the western Mediterranean Sea, one along the Sardinian coast and the other along the Italian Peninsula. In addition, '*C. rudolphii* sp. A' collected from cormorants on the Sardinian coast showed a high number of private haplotypes ($n = 23$) not shared with other populations (Roca-Geronès et al. 2023). These results support the notion that '*C. rudolphii* sp. A' migration routes of wintering populations of cormorants in the Mediterranean Sea influence the genetic structure of '*C. rudolphii*sp. A'.

The long-range dispersal of certain cetacean and fish species may maintain high levels of gene flow between *Anisakis* populations throughout their distribution range (Mattiucci and Nascetti

2008). However, cetaceans often exhibit cryptic population genetic structure over small geographical ranges, more than expected for such highly migratory marine mammals. Interestingly, a similar degree of genetic differentiation and genetic sub-structuring has been found in both *A. simplex* (sensu stricto) and *A. pegreffii*, despite the generally high gene flow values observed between populations of these species (Cipriani et al. 2022). This observation is supported by haplotype network analysis showing geographic separation in *A. simplex* (sensu stricto), with significant differentiation between individuals collected from cetaceans of the Norwegian Sea and from the northeast Atlantic Ocean (on average, $Fst \approx 0.08$) (Cipriani et al. 2022). The existence of microevolutionary events occurring in cetacean population units, along with differences in their ecology, may be responsible for maintaining the genetic differentiation observed in *A. simplex* (sensu stricto). Indeed, a significant genetic sub-structure of the parasite species appears to exist among parasite metapopulations sampled in different geographic areas of the host species (Cipriani et al. 2022). In turn, this seems to suggest that the geographic origin of the host species could be a major factor in explaining the sub-structuring of the parasite populations. Interestingly, in a previous population genetic study of *A. simplex* (sensu stricto) larvae in herring (*Clupea harengus*) from several northeast Atlantic fishing grounds based on *cox*2 mtDNA, the haplotype diversity values were, on average, in the same range as those observed in that of Cipriani et al. (2022) (see Mattiucci et al. 2018b).

Similarly, in the northeast Atlantic, significant genetic differentiation on the *cox*2 mtDNA locus was observed (average $Fst \approx 0.05$) between parasite populations of *A. pegreffii* from dolphins in the Mediterranean Sea and Iberian Atlantic waters (Cipriani et al. 2022). Interestingly, congruent significant genetic differentiation was reported between populations of the short-beaked common dolphin (*Delphinus delphis*) from the Mediterranean Sea and adjacent Atlantic waters (Galicia and Portugal) (Natoli et al. 2008). This suggests that the population sub-structuring of *A. pegreffii* across their distribution overlaps with different cetacean populations from different European waters.

Finally, the congruence between subpopulations of fish species and the differential distribution of larval stages of *Anisakis* was investigated by comparing Mediterranean and Atlantic fish stocks in a multidisciplinary framework, using generalised Procrustes Rotation (PR). This analysis assessed the association between host genetics and different species of *Anisakis* on demersal (hake) and pelagic (horse mackerel, swordfish) species (Mattiucci et al. 2015b). When discordant results emerged from the two data sets (i.e. host populations genetics and *Anisakis* spp. parasite assemblage at geographical), it was likely due to the different features of the data. While fish population genetics reflecting changes over a microevolutionary timescale do provide indications of the high gene flow between the different subpopulations of the fish host, the distribution of the different species of *Anisakis* in the host subpopulations served as suitable biomarkers for the detection of fish stocks over smaller temporal and spatial scales. The parasite species assemblages are thus able to provide information on the movements of fish subpopulations over their lifespan (Mattiucci et al. 2015b; also see Chap. 17).

Instead, when comparing the fish genetic structure and the population genetics of its endoparasites, some congruence was detected. This is the case of the population genetic differentiation of metapopulations of *A. simplex* (sensu stricto) sampled in the herring *Clupea harengus* between the different fishing areas of the northeastern Atlantic Ocean. The differentiation was estimated at the intraspecific level, based on *cox*2 mtDNA sequences analysis. Spatial comparison based on molecular variance analysis, *Fst* values and haplotype network construction showed relevant differences in haplotype frequencies between samples of *A. simplex* (sensu stricto) from the different geographical areas. Results indicate a genetic sub-structuring of *A. simplex* (sensu stricto) obtained from herring in different areas, with the population from the Norwegian Sea being the most differentiated one from the others, whereas the North Sea and Baltic Sea populations of *A. simplex* (sensu stricto) from herrings were phylogenetically most similar.

The population genetic structure of *A. simplex* (sensu stricto) was in accordance with the herring population genetic structure throughout the host geographical range in the northeastern Atlantic, which was inferred based on SSR DNA loci (Mariani et al. 2005). It has been supposed that the slight, but significant, level of differentiation found between the *A. simplex* (sensu stricto) metapopulations of the Norwegian Sea and Baltic Sea could be related to the different distribution the Norwegian and Baltic herring stocks. On the contrary, the *Fst* values found among the populations of the parasite species from herring fished in the North Sea, the English Channel and the Baltic Sea revealed a lower degree of differentiation between these populations. This could be explained by the herring population in that area being characterised by a long temporal persistence of the three different herring stocks in the same feeding area. Finally, the weak genetic differentiation between the North Sea and English Channel populations of *A. simplex* (sensu stricto) was in agreement with data obtained by SSR DNA analysis carried out on the herring which suggested the existence of low levels of genetic differentiation between the two corresponding herring stocks, that is, the North Sea 'Banks' and 'Buchan' stocks (NSAS) and the English Channel 'Downs' stock (Mariani et al. 2005).

9.4 Concluding Remarks and Future Trends

We have seen that both parasite and host factors can explain the genetic composition and structure of parasite populations over geographical scales. We have expounded upon the paradigm that genetic structure of parasite populations depends on host dispersal and have also highlighted several alternative factors that are also important in driving the co-distribution of host and parasite genetic variation.

It has been suggested that different life cycle strategies have a major role in shaping the genetic variation of parasites species. For instance, it has been postulated that multi-host parasites with life cycles completed in both poikilothermic and homeothermic hosts (i.e. anisakid nematodes) exhibit higher levels of genetic polymorphism compared to those which infect single homeothermic host species (Bullini et al. 1986). Such intraspecific genetic variation is fundamental for the species adaptive capacity to changing environments (May 1994). In the case of multi-host parasites, it was hypothesised that, within the gene pool of a parasite species, a given allele may confer greater fitness upon one stage of the life cycle, while another allele may function optimally in another group of hosts of its life cycle (Bullini et al. 1986). Despite the crucial role of genotypes and haplotypes (i.e. genetic variation) in adaptation to different hosts, geographical areas and infection sites, patterns of genetic structure in parasite populations of aquatic organisms are yet to be extensively investigated regarding epidemiological processes, dynamics and evolutionary adaptation throughout their range of distribution.

In this context, it has been observed that distinct genetic clades in the phylogenetic analysis of a parasite species are correlated with different sites of infection in hosts, as preliminarily shown by the association between some teleostean fish species and species of *Philometra* (de Buron et al. 2011). Furthermore, the usefulness of studying functional phenotypic variation in different metapopulations of parasite species, due to differential gene expression of functional genes under selection pressure, should be investigated in future studies using tools such as transcriptomics in order to provide data on both local and host molecular adaptations of a parasite species.

It has also been emphasised that, to infer the population genetic structure of aquatic parasites (i.e. potential genetic differentiation based on different hosts and geography), a combined dataset of mitochondrial and nuclear markers should be applied. Besides this, a wide sampling effort is required to strengthen results reliability. So far, numerous genetic markers (i.e. mitochondrial and ribosomal DNA gene loci, and genotyping of a panel of codominant genetic markers such as allozymes and DNA-SSRs), have been used to investigate the population genetic structure of parasites. With the advent and rapid improvement of next-generation sequencing (NGS) technologies, thousands of single nucleotide polymorphisms (SNPs) can now be obtained in a short time frame and at reduced cost. Additionally, obtaining the complete genome of a single parasite species would be more feasible. This will allow for expansion of 'population genomics', defined by Charlesworth (2010) as the study of a large amount of gene variation and the causes of genome-wide variability found in natural populations of parasites. Topics to be investigated in aquatic parasites also include hybridisation, loci under selection and linkage mapping (Thorn et al. 2023). Thus, following Luikart et al. (2018), the future trend will also be the widespread use of 'population genomics' to investigate host–parasite associations in the context of co-phylogeography. This novel approach will enable addressing evolutionary and ecological aspects of the parasite–host association over space and time.

References

Abe N, Baba T, Nakamura Y et al (2021) Global analysis of cytochrome c oxidase subunit 1 (cox1) gene variation in *Dibothriocephalus nihonkaiensis* (Cestoda: Diphyllobothriidae). Curr Res Parasitol Vector Borne Dis 1:100042. https://doi.org/10.1016/j.crpvbd.2021.100042

Atkinson SD, Jones SRM, Adlard R et al (2011) Geographical and host distribution patterns of *Parvicapsula minibicornis* (Myxozoa) small subunit ribosomal RNA genetic types. Parasitology 138(8):960–968. https://doi.org/10.1017/S003118201100073

Avise JC (1994) Speciation and Hybridization. In: Molecular Markers, Natural History and Evolution. Springer, Boston, MA. https://doi.org/10.1007/978-1-4615-2381-9_7

Baird NA, Etter PD, Atwood TS et al (2008) Rapid SNP discovery and genetic mapping using sequenced RAD markers. PLoS One 3(10):e3376. https://doi.org/10.1371/journal.pone.0003376

Baldwin RE, Rew MB, Johansson ML et al (2011) Population structure of three species of *Anisakis* nematodes recovered from Pacific sardines (*Sardinops sagax*) distributed throughout the California Current System. J Parasitol 97(4):545–554. https://doi.org/10.1645/GE-2690.1

Bao M, Giulietti L, Levsen A et al (2023) Resurrection of genus *Phocanema* Myers, 1959, as a genus independent from *Pseudoterranova* Mozgovoĭ, 1953, for nematode species (Anisakidae) parasitic in pinnipeds and cetaceans, respectively. Parasitol Int 97:102794

Bazsalovicsová E, Králová-Hromadová I, Juhásová L et al (2020) Comparative analysis of monozoic fish tapeworms *Caryophyllaeus laticeps* (Pallas, 1781) and recently described *Caryophyllaeus chondrostomi* Barčák, Oros, Hanzelová, Scholz, 2017, using microsatellite markers. Parasitol Res 119:3995–4004. https://doi.org/10.1007/s00436-020-06898-8

Bello E, Paoletti M, Webb SC (2020) Cross-species utility of microsatellite loci for the genetic characterisation of *Anisakis berlandi* (Nematoda: Anisakidae). Parasite 27:9. https://doi.org/10.1051/parasite/2020004

Bello E, Palomba M, Webb S (2021) Investigating the genetic structure of the parasites *Anisakis pegreffii* and *A. berlandi* (Nematoda: Anisakidae) in a sympatric area of the southern Pacific Ocean waters using a multilocus genotyping approach: first evidence of their interspecific hybridisation. Infect Genet Evol 92:104887. https://doi.org/10.1016/j.meegid.2021.104887

Blasco-Costa I, Poulin R (2013) Host traits explain the genetic structure of parasites: a meta-analysis. Parasitology 140(10):1316–1322. https://doi.org/10.1017/S0031182013000784

Bullini L, Nascetti G, Paggi L et al (1986) Genetic variation of ascaridoid worms with different life cycles. Evolution 40:437–440. https://doi.org/10.1111/j.1558-5646.1986.tb00488.x

Charlesworth B (2010) Molecular population genomics: a short history. Genet Res 92(5–6):397–411. https://doi.org/10.1017/S0016672310000522

Christen M, Milinski M (2003) The consequences of self-fertilisation and outcrossing of the cestode *Schistocephalus solidus* in its second intermediate host. Parasitology 126(4):369–378. https://doi.org/10.1017/s0031182003002956

Cipriani P, Palomba M, Giulietti L et al (2022) Distribution and genetic diversity of *Anisakis* spp. in cetaceans from the Northeast Atlantic Ocean and the Mediterranean Sea. Sci Rep 12:13664. https://doi.org/10.1038/s41598-022-17710-1

Criscione CD (2008) Parasite co-structure: broad and local scale approaches. Parasite 15:439–443. https://doi.org/10.1051/parasite/2008153439

Criscione CD, Blouin MS (2004) Life cycle shape parasite evolution: comparative population genetics of salmon trematodes. Evolution 58:198–202. https://doi.org/10.1111/j.0014-3820.2004.tb01587.x

Criscione CD, Blouin MS (2007) Parasite phylogeographical congruence with salmon host evolutionarily significant units: implications for salmon conservation. Mol Ecol 16(5):993–1005. https://doi.org/10.1111/j.1365-294X.2006.03220.x

Criscione CD, Cooper B, Blouin MS (2006) Parasite genotypes identify source populations of migratory fish more accurately than fish genotypes. Ecology 87:823–827. https://doi.org/10.1890/0012-9658(2006) 87[823:pgispo]2.0.co;2

Cross MA, Collins C, Campbell N et al (2007) Levels of intra-host and temporal sequence variation in a large CO1 sub-units from *Anisakis simplex* sensu stricto (Rudolphi 1809) (Nematoda: Anisakidae): implications for fisheries management. Mar Biol 151:695–702. https://doi.org/10.1007/s00227-006-0516-9

D'Amelio S, Mathiopoulos KD, Santos CP et al (2000) Genetic markers in ribosomal DNA for the identification of members of the genus *Anisakis* (Nematoda: Ascaridoidea) defined by polymerase-chain-reaction-based restriction fragment length polymorphism. Int J Parasitol 30(2):223–226. https://doi.org/10.1016/s0020-7519(99)00178-2

Davey JW, Blaxter ML (2010) RADSeq: next-generation population genetics. Brief Funct Genomics 9(56):416–423. https://doi.org/10.1093/bfgp/elq031

Davies K, Pagan C, Nadler SA (2020) Host population expansion and the genetic architecture of the pinniped hookworm *Uncinaria lucasi*. J Parasitol 106:383–391. https://doi.org/10.1645/19-172

De Buron I, France SG, Connors VA et al (2011) Philometrids of the southern flounder *Paralichthys lethostigma*: a multidimensional approach to determine their diversity. J Parasitol 97:466–475. https://doi.org/10.2307/23019109

Desdevises Y, Jovelin R, Jousson O et al (2000) Comparison of ribosomal DNA sequences of *Lamellodiscus* spp. (Monogenea, Diplectanidae) parasitising *Pagellus* (Sparidae, Teleostei) in the North Mediterranean Sea: species divergence and coevolutionary interactions. Int J Parasitol 30(6):741–746. https://doi.org/10.1016/s0020-7519(00)00051-5

Dickerson BR, Ream RR, Vignieri SN, Bentzen P (2010) Population structure as revealed by mtDNA and microsatellites in Northern Fur Seals, *Callorhinus ursinus*, throughout their range. PLoS One 5(5):e10671. https://doi.org/10.1371/journal.pone.0010671

Ding F, Gu S, Yi MR et al (2022) Demographic history and population genetic structure of *Anisakis pegreffii* in the cutlassfish *Trichiurus japonicus* along the coast of mainland China and Taiwan. Parasitol Res 121(10):2803–2816. https://doi.org/10.1007/s00436-022-07611-7

Duflot M, Gay M, Midelet G et al (2021) Morphological and molecular identification of *Cryptocotyle lingua* metacercariae isolated from Atlantic cod (*Gadus morhua*) from Danish seas and whiting (*Merlangius merlangus*) from the English Channel. Parasitol Res 120(10):3417–3427. https://doi.org/10.1007/s00436-021-07278-6

Duflot M, Cresson P, Julien M et al (2023) Black spot diseases in seven commercial fish species from the English Channel and the North Sea: infestation levels, identification and population genetics of *Cryptocotyle* spp. Parasite 30:28. https://doi.org/10.1051/parasite/2023028

Esch GW, Kennedy CR, Bush AO et al (1988) Patterns in helminth communities in freshwater fish in Great Britain: alternative strategies for colonization. Parasitology 96(3):519–532. https://doi.org/10.1017/S003118200008305X

Feis ME, Thieltges DW, Olsen JL et al (2015) The most vagile host as the main determinant of population connectivity in marine macroparasites. Mar Ecol Prog Ser 520:85–99. https://doi.org/10.3354/meps11096

Frederiksen M, Korner-Nievergelt F, Marion L et al (2018) Where do wintering Cormorants come from? Long-term changes in the geographical origin of a migratory bird on a continental scale. J Appl Ecol 55:2019–2032. https://doi.org/10.1111/1365-2664.13106

Froeschke G, von der Heyden S (2014) A review of molecular approaches for investigating patterns of coevolution in marine host–parasite relationships. Adv Parasitol 84:209–252. https://doi.org/10.1016/B978-0-12-800099-1.00004-1

Ganley AR, Kobayashi T (2007) Highly efficient concerted evolution in the ribosomal DNA repeats: total rDNA repeat variation revealed by whole-genome shotgun sequence data. Genome Res 17(2):184–191. https://doi.org/10.1101/gr.5457707

García-Varela M, López-Jiménez A, González-García MT (2023) Contrasting the population genetic structure of a specialist (*Hexaglandula corynosoma*: Acanthocephala: Polymorphidae) and a generalist parasite (*Southwellina hispida*) distributed sympatrically in Mexico. Parasitology 150(4):348–358. https://doi.org/10.1017/S0031182023000033

Glover KA, Stolen AB, Messmer A et al (2011) Population genetic structure of the parasitic copepod *Lepeophtheirus salmonis* throughout the Atlantic. Mar Ecol Prog Ser 427:161–172. https://doi.org/10.3354/meps09045

Gordeev II, Bakay YI, Kalashnikova MY et al (2023) Genetic structure of juvenile stages of *Phocanema bulbosum* (Nematoda, Chromadorea: Anisakidae) parasitising commercial fish, Atlantic cod *Gadus morhua*, and American plaice *Hippoglossoides platessoides* in the Barents Sea. Diversity 15(10):1036. https://doi.org/10.3390/d15101036

Herzog KS, Hackett JL, Hime PM et al (2023) First insights into population structure and genetic diversity versus host specificity in trypanorhynch tapeworms using multiplexed shotgun genotyping. Genome Biol Evol 15(10):evad190. https://doi.org/10.1093/gbe/evad190

Hossen MS, Barton DP, Wassens S et al (2022) Molecular (cox1), geographical, and host record investigation of monogeneans *Mazocraes australis* (Mazocraeidae), *Polylabris sillaginae*, and *P. australiensis* (Microcotylidae). Parasitol Res 121(12):3427–3442. https://doi.org/10.1007/s00436-022-07590-7

Huyse T, Poulin R, Théron A (2005) Speciation in parasites: a population genetics approach. Trends Parasitol 21:469–475. https://doi.org/10.1016/j.pt.2005.08.009

Ingvarsson PK (2002) A metapopulation perspective on genetic diversity and differentiation in partially self-fertilising plants. Evolution 56:2368–2373. https://doi.org/10.1111/j.0014-3820.2002.tb00162.x

Isbert W, Dallarés S, Grau A et al (2023) A molecular and epidemiological study of *Grillotia* (Cestoda: Trypanorhyncha) larval infection in *Etmopterus spinax* (Elasmobranchii: Squaliformes) in the Mediterranean Sea and Northeast Atlantic Ocean. Deep-Sea Res I Oceanogr Res Pap 199:104102. https://doi.org/10.1016/j.dsr.2023.104102

Keeney DB, Waters JM, Poulin R (2007) Clonal diversity of the marine trematode *Maritrema novaezealandensis* within intermediate hosts: the molecular ecology of parasite life cycles. Mol Ecol 16(2):431–439. https://doi.org/10.1111/j.1365-294X.2006.03143.x

Kuhn T, García-Màrquez J, Klimpel S (2011) Adaptive radiation within marine anisakid nematodes: a zoogeographical modelling of cosmopolitan, zoonotic parasites. PLoS One 6(12):e28642. https://doi.org/10.1371/journal.pone.0028642

Laine VN, Herczeg G, Shikano T et al (2012) Heterozygosity-behaviour correlations in nine-spined stickleback (*Pungitius pungitius*) populations: contrasting effects at random and functional loci. Mol Ecol 21(19):4872–4884. https://doi.org/10.1111/j.1365-294X.2012.05741.x

Louhi KR, Karvonen A, Rellstab C et al (2010) Is the population genetic structure of complex life cycle parasites determined by the geographic range of the most motile host? Infect Genet Evol 10(8):1271–1277. https://doi.org/10.1016/j.meegid.2010.08.013

Luikart G, Kardos M, Hand BK et al (2018) Population genomics: advancing understanding of nature. In: Rajora O (ed) Population genomics. Springer, Cham

Marcer F, Tosi F, Franzo G et al (2020) Updates on ecology and life cycle of *Sulcascaris sulcata* (Nematoda: Anisakidae) in Mediterranean grounds: molecular identification of larvae infecting edible scallops. Front Vet Sci 14(7):64. https://doi.org/10.3389/fvets.2020.00064

Mariani S, Hutchinson WF, Hatfield EM et al (2005) North Sea herring population structure revealed by microsatellite analysis. Mar Ecol Prog Ser 303:245–257. https://doi.org/10.3354/meps303245

Marion L, Le Gentil J (2006) Ecological segregation and population structuring of the cormorant *Phalacrocorax carbo* in Europe, in relation to the recent introgression of continental and marine subspecies. Evol Ecol 20:193–216. https://doi.org/10.1007/s10682-005-5828-6

Mattiucci S, Nascetti G (2008) Advances and trends in the molecular systematics of anisakid nematodes, with implications for their evolutionary ecology and host-parasite co-evolutionary processes. Adv Parasitol 66:47–168. https://doi.org/10.1016/S0065-308X(08)00202-9

Mattiucci S, Nascetti G, Cianchi R et al (1997) Genetic and ecological data on the *Anisakis simplex* complex, with evidence for a new species (Nematoda, Ascaridoidea, Anisakidae). J Parasitol 83(3):401–416. https://doi.org/10.2307/3284402

Mattiucci S, Paggi L, Nascetti G et al (1998) Allozyme and morphological identification of *Anisakis*, *Contracaecum* and *Pseudoterranova* from Japanese waters (Nematoda: Ascaridoidea). Syst Parasitol 40:81–92. https://doi.org/10.1023/a:1005914926720

Mattiucci S, Cipriani P, Paoletti M et al (2015a) Temporal stability of parasite distribution and genetic variability values of *Contracaecum osculatum* sp. D and *C. osculatum* sp. E (Nematoda: Anisakidae) from fish of the Ross Sea (Antarctica). Int J Parasitol Parasites Wildl 4(3):356–367. https://doi.org/10.1016/j.ijppaw.2015.10.004

Mattiucci S, Cimmaruta R, Cipriani P et al (2015b) Integrating *Anisakis* spp. parasites data and host genetic structure in the frame of a holistic approach for stock identification of selected Mediterranean Sea fish species. Parasitology 142(1):90–108. https://doi.org/10.1017/S0031182014001103

Mattiucci S, Acerra V, Paoletti M et al (2016) No more time to stay 'single' in the detection of *Anisakis pegreffii*, *A. simplex* (s. s.) and hybridisation events between them: a multi-marker nuclear genotyping approach. Parasitology 143:998–1011. https://doi.org/10.1017/S0031182016000330

Mattiucci S, Paoletti M, Cipriani P et al (2017) Inventorying biodiversity of anisakid nematodes from the austral region: a hotspot of genetic diversity. In: Klimpel S, Kuhn T, Mehlhorn H (eds) Biodiversity and evolution of parasitic life in the Southern Ocean, Parasitology research monographs. Springer Nature, pp 109–140. https://doi.org/10.1007/978-3-319-46343-8

Mattiucci S, Cipriani P, Levsen A et al (2018a) Molecular epidemiology of *Anisakis* and Anisakiasis: an ecological and evolutionary road map. Adv Parasitol 99:93–263. https://doi.org/10.1016/bs.apar.2017.12.001

Mattiucci S, Giulietti L, Paoletti M et al (2018b) Population genetic structure of the parasite *Anisakis simplex* (s. s.) collected in *Clupea harengus* L. from North East Atlantic fishing grounds. Fish Res 202:103–111. https://doi.org/10.1016/j.fishres.2017.08.002

Mattiucci S, Bello E, Paoletti M et al (2019) Novel polymorphic microsatellite loci in *Anisakis pegreffii* and *A. simplex* (s. s.) (Nematoda: Anisakidae): implications for species recognition and population genetic analysis. Parasitology 146(11):1387–1403. https://doi.org/10.1017/S003118201900074X

Mattiucci S, Sbaraglia GL, Palomba M et al (2020) Genetic identification and insights into the ecology of *Contracaecum rudolphii* A and *C. rudolphii* B (Nematoda: Anisakidae) from cormorants and fish of aquatic ecosystems of Central Italy. Parasitol Res 119(4):1243–1257. https://doi.org/10.1007/s00436-020-06658-8

May RM (1994) Biological diversity: differences between land and sea. Philos Trans R Soc Lond Ser B Biol Sci 343:105–111. https://doi.org/10.1098/rstb.1994.0014

Mayr E (1963) Animal species and evolution. Harvard University Press, Cambridge. https://doi.org/10.4159/harvard.9780674865327

Meinilä M, Kuusela J, Ziętara MS et al (2004) Initial steps of speciation by geographic isolation and host switch in salmonid pathogen *Gyrodactylus salaris* (Monogenea: Gyrodactylidae). Int J Parasitol 34(4):515–526. https://doi.org/10.1016/j.ijpara.2003.12.002

Miller TL, Cribb TH (2007) Coevolution of *Retrovarium* n. gen. (Digenea: Cryptogonimidae) in Lutjanidae and Haemulidae (Perciformes) in the Indo-West Pacific. Int J Parasitol 37(8–9):1023–1045. https://doi.org/10.1016/j.ijpara.2007.01.006

Mladineo I, Bott NJ, Nowak BF et al (2010) Multilocus phylogenetic analyses reveal that habitat selection drives the speciation of Didymozoidae (Digenea) parasitising Pacific and Atlantic bluefin tunas. Parasitology 137(6):1013–1025. https://doi.org/10.1017/S0031182009991703

Mladineo I, Trumbić Ž, Radonić I et al (2017) Anisakis simplex complex: ecological significance of recombinant genotypes in an allopatric area of the Adriatic Sea inferred by genome-derived simple sequence repeats. Int J Parasitol 47:215–223. https://doi.org/10.1016/j.ijpara.2016.11.003

Nadler SA (1995) Microevolution and the genetic structure of parasite populations. J Parasitol 81:395–403. https://doi.org/10.2307/3283821

Nascetti G, Cianchi R, Mattiucci S et al (1993) Three sibling species within *Contracaecum osculatum* (Nematoda, Ascaridida, Ascaridoidea) from the Atlantic Arctic-boreal region: reproductive isolation and host preferences. Int J Parasitol 23:105–120. https://doi.org/10.1016/0020-7519(93)90103-6

Natoli A, Canadas A, Vaquero C et al (2008) Conservation genetics of the short-beaked common dolphin (*Delphinus delphis*) in the Mediterranean Sea and in the eastern North Atlantic Ocean. Conserv Genet 9:1479–1487. https://doi.org/10.1007/s10592-007-9481-1

Nieberding C, Morand S, Libois R et al (2004) A parasite reveals cryptic phylogeographic history of its host. Proc Biol Sci 271:2559–2568. https://doi.org/10.1098/rspb.2004.2930

Nolan MJ, Cribb T (2005) The use and implications of ribosomal DNA sequencing for the discrimination of digenean species. Adv Parasitol 60:101–163. https://doi.org/10.1016/S0065-308X(05)60002-4

Palm HW, Waeschenbach A, Littlewood DT (2007) Genetic diversity in the trypanorhynch cestode *Tentacularia coryphaenae* Bosc, 1797: evidence for a cosmopolitan distribution and low host specificity in the teleost intermediate host. Parasitol Res 101(1):153–159. https://doi.org/10.1007/s00436-006-0435-1

Palm HW, Waeschenbach A, Olson PD et al (2009) Molecular phylogeny and evolution of the Trypanorhyncha Diesing, 1863 (Platyhelminthes: Cestoda). Mol Phylogenet Evol 52(2):351–367. https://doi.org/10.1016/j.ympev.2009.01.019

Palomba M, Marchiori E, Tedesco P et al (2023) An update and ecological perspective on certain sentinel helminth endoparasites within the Mediterranean Sea. Parasitology 13:1–19. https://doi.org/10.1017/S0031182023000951

Palomba M, Mattiucci S, Belli B (2025) Hybridization and introgression of mitochondrial genome between the two species Anisakis pegreffii and A. simplex (s. s.) using a wide genotyping approach: evolutionary and ecological implications. Parasitology, in press https://doi.org/10.1017/S0031182025000228

Ramilo A, Rodríguez H, Pascual S et al (2023) Population genetic structure of *Anisakis simplex* infecting the European hake from Northeast Atlantic fishing grounds. Animals 13(2):197. https://doi.org/10.3390/ani13020197

Rannala B, Michalakis Y (2003) Population genetics and cospeciation: from process to pattern. In: Page RDM (ed) Tangled trees: phylogeny, cospeciation and coevolution. University of Chicago Press, Chicago, pp 120–143

Roca-Geronès X, Fisa R, Montoliu I et al (2023) Genetic diversity of *Contracaecum rudolphii* sp. A (Nematoda: Anisakidae) parasitising the European Shag *Phalacrocorax aristotelis desmarestii* from the Spanish Mediterranean coast. Front Vet Sci 10:1122291. https://doi.org/10.3389/fvets.2023.1122291

Sata N, Nakano T (2023) Molecular analysis of larvae suggests the existence of a second species of *Sulcascaris* (Nematoda: Anisakidae: Anisakinae) in the Japanese moon scallop (*Ylistrum japonicum*) from Japanese waters. Parasitol Int 92:102674. https://doi.org/10.1016/j.parint.2022.102674

Shim KC, Weber JN, Hernandez CA et al (2022) Population genomics of a three-spine stickleback tapeworm in Vancouver Island. bioRxiv. https://doi.org/10.1101/2022.05.15.491937

Thorn CS, Maness RW, Hulke JM, Delmore KE, Criscione CD (2023) Population genomics of helminth parasites. J Helminthol 97:e29. https://doi.org/10.1017/S0022149X23000123

Van Valen L (1973) A new evolutionary law. Evol Theory 1:1–30. https://doi.org/10.7208/9780226115504-022

Vilas R, Paniagua E, Sanmartín ML (2003) Genetic variation within and among infrapopulations of the marine digenetic trematode *Lecithochirium fusiforme*. Parasitology 126(5):465–472. https://doi.org/10.1017/s0031182003003081

Vilas R, Criscione CD, Blouin MS (2005) A comparison between mitochondrial DNA and the ribosomal internal transcribed regions in prospecting for cryptic species of platyhelminth parasites. Parasitology 131(6):839–846. https://doi.org/10.1017/S0031182005008437

Open Access This chapter is licensed under the terms of the Creative Commons Attribution-NonCommercial-NoDerivatives 4.0 International License (http://creativecommons.org/licenses/by-nc-nd/4.0/), which permits any non-commercial use, sharing, distribution and reproduction in any medium or format, as long as you give appropriate credit to the original author(s) and the source, provide a link to the Creative Commons license and indicate if you modified the licensed material. You do not have permission under this license to share adapted material derived from this chapter or parts of it.

The images or other third party material in this chapter are included in the chapter's Creative Commons license, unless indicated otherwise in a credit line to the material. If material is not included in the chapter's Creative Commons license and your intended use is not permitted by statutory regulation or exceeds the permitted use, you will need to obtain permission directly from the copyright holder.

Chemosensory Behaviour of Free-Living Stages of Aquatic Parasites

Clayton Vondriska and Paul C. Sikkel

Abstract

Chemical signals are essential for organisms living in aquatic environments and are likely used by every aquatic organism at some point in their life cycle. Although microfauna remains understudied in this regard, they comprise a significant portion of global biodiversity and can be extremely influential in ecosystem processes. This group of influential organisms includes parasites that attach to the inside (endoparasites) or outside (ectoparasites) of their host's body and can have a variety of effects on individual hosts and ecosystems. Ectoparasites typically have free-living stages where they must rely heavily on chemical cues to navigate complex environments in order to locate hosts for infestation, conspecifics for mating and suitable habitat, while avoiding predators and other hazards. Additionally, some endoparasites experience free-swimming stages where they similarly rely on chemical signals. However, chemosensory behaviour has only been investigated for a small subset of aquatic parasites. This chapter summarises the research that has been conducted on the chemosensory behaviour of aquatic ectoparasites and free-swimming stages of endoparasites, including their ability to detect chemical cues and how this chemical detection proficiency influences their behaviours at key points in their life cycle. A brief background of commonly used methodology used in chemosensory research is also provided. Finally, this chapter identifies the gaps in knowledge that exist and directions that future studies should prioritise.

C. Vondriska
Rosenstiel School of Marine, Atmospheric, and Earth Science, University of Miami, Miami, FL, USA

Department of Biology, College of Science and Mathematics, University of the Virgin Islands, St Thomas, VI, USA
e-mail: cmv148@earth.miami.edu

P. C. Sikkel (✉)
Rosenstiel School of Marine, Atmospheric, and Earth Science, University of Miami, Miami, FL, USA

Water Research Group, Unit of Environmental Sciences, North-West University, Potchefstroom, South Africa

10.1 Introduction

Chemical cues play critical roles in the dynamics of aquatic ecosystems, with all aquatic organisms relying on chemical signals at some point during their life cycles (Hay 2009; Brönmark and Hansson 2012a). These signals can be particularly important in aquatic ecosystems compared to terrestrial ones. Visual signals are more prone to disruption through environmental factors such as turbidity and increasing absorption of light waves with depth, resulting in longer periods of lower light in deeper waters (Brönmark and Hansson 2012b). Acoustic cues, while able to travel equally far distances as chemical cues, are much shorter lived than chemical cues and therefore much less effective for locating-behaviours (Nummela and Thewissen 2008). Additionally, acoustic cues can get lost in the complex aquatic soundscape as well as be affected by anthropogenic noise (Cox et al. 2018).

Chemical signals can influence food- (Derby et al. 2001; Dove 2015; Kamio and Derby 2017), habitat- (Pawlik 1992; Paul et al. 2006; Bilodeau and Hay 2022), and mate-finding decisions (Saunders et al. 2010; Paul et al. 2011; Schwartz et al. 2016), as well as predator avoidance strategies (Jacobson and Stabell 2004; Hay 2009; Crane et al. 2022). Additionally, chemical cues can allow individuals to establish dominance hierarchies (Breithaupt and Thiel 2010), initiate metamorphosis (Williamson et al. 2004), assess the sex and reproductive stage of potential mates (Sato and Goshima 2007) and guide long-distance migrations and navigation (Brönmark and Hansson 2012b). However, most studies on the ecological significance of chemical signals have been rudimentary, only observing the organism's behaviour in the presence of a cue and assuming the behaviour is caused by the cue. Very few studies identify the chemical structures involved and validate their significance (von Elert 2012).

Although chemical cues are important to all aquatic organisms, research into chemosensory ecology and behaviour on smaller organisms (<5 cm) has been limited relative to larger, more charismatic ones. This discrepancy persists even though small organisms dominate biodiversity and biomass metrics in aquatic ecosystems and are critical to ecosystem function (Vidondo et al. 1997; Hudson et al. 2006; Knowlton et al. 2010). Multiple factors have contributed to the fact that this group of organisms remains understudied. Small organisms are often difficult to collect, rear and observe. Additionally, their small size has given them an assumed lack of influence or importance in aquatic ecosystems.

Many small organisms are parasitic in nature (Hudson et al. 2002, 2006; Sures et al. 2017). These parasitic organisms can be broadly divided into two categories based on site of host attachment: endo- and ectoparasites. Endoparasites are those that parasitise the inside of their host's body and are typically transmitted via consumption of the parasite while ectoparasites attach to the surface of the body (Hopla et al. 1994; Goater et al. 2014). All ectoparasites and some internal parasites include at least one temporary free-swimming stage of their life cycle, where they are detached from the body of their host until they either locate and attach to their host, are ingested or otherwise manage to enter their host's body. Ectoparasites and endoparasites that spend some time outside their host must often actively locate at least one host, locate mates away from their host and occasionally find suitable habitat to transition to another life stage, reproduce or wait for their next host to be nearby. Many of these behaviours are chemically mediated.

Most descriptions of studies into the chemosensory behaviour of aquatic parasites provided in reviews are brief and limited to specific taxonomic groups (e.g. reviews of copepod behaviour). In this chapter, we synthesise the known literature on the chemosensory behaviour of the free-living stages of aquatic parasites, focusing on key stages in their life cycles including host- and mate-finding and cues associated with life cycle progression, and how these behaviours are commonly studied. We begin with an overview of the methodology used in chemosensory research, followed by reviews of the chemosensory behaviour of different parasitic groups, arranged by taxonomic class. Finally, we expose the significant gaps in knowledge that exist for aquatic ectoparasites and highlight areas for future research.

10.2 Common Methodology for Behavioural Responses to Chemical Stimuli

When testing for behavioural responses to chemical stimuli in aquatic organisms, a small number of similar devices have become standards in the field. These include two- and multi-channel choice flumes and Y- or T-tubes (Fig. 10.1). Choice flumes (Fig. 10.1a) are designed to allow two or more sources of water to flow through a central choice arena without mixing (Jutfelt et al. 2017). To achieve this laminar flow, the water sources must be at similar temperatures and densities; otherwise, they will mix within the choice arena. An organism can move freely throughout the choice arena and often can move up one of the arms of the flume. A preference or avoidance of a stimulus can then be inferred by the organism spending more or less time, respectively, in a water source.

Y- and T-tubes (Fig. 10.1b) work similarly to choice flumes with slight but important differences. Y- and T-tubes typically have two arms from which two separate water sources flow into a choice arena (Davenport 1950). However, unlike two- and multi-channel choice flumes, water will mix throughout the choice arena. For these experimental set-ups, preference can be inferred by an organism swimming up one of the arms and spending more time there than in the other arm or initial choice arena. Avoidance behaviours are typically difficult to test using these methods, as an individual would have to first move towards an unfavourable cue to reach the arm with no stimuli.

In all instruments, the position of the organism can be recorded by a researcher, often by noting which water source the organism is in at predetermined intervals or, in more recent studies, by camera recordings. Using cameras has the advantages of reducing recorder bias, decreasing the potential influence of recorder presence on behaviour and allowing the organism's position to be recorded continuously (Johnson et al. 2020; Vondriska et al. 2020; Vondriska and Sikkel 2023). This also allows for later verification of data that has been recorded and the use of video analysis software when recording data.

10.3 Monogenea

Monogeneans are a group of parasitic flatworms primarily infecting freshwater and marine fish, although they have also been reported to infect some amphibians and aquatic reptiles (Whittington and Kearn 2011; Goater et al. 2014; also see Chap. 5). Most species are extremely host- and site-specific, infecting only one location (e.g. part of the gills) on a single species or genus of organism (Kearn 1986; Goater et al.

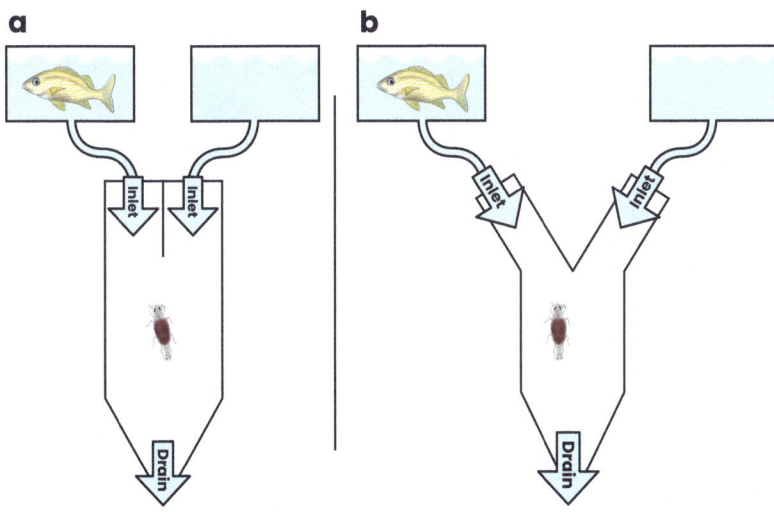

Fig. 10.1 Diagrams of historically used instruments to test chemoreceptive behaviour in marine organisms. (**a**) Two-channel choice flume and (**b**) Y-Tube

2014). However, some species of *Neobenedenia* that attach to the body of their host appear to have broader host ranges (Whittington 2012).

Chemosensory research has focused exclusively on the hatching behaviour of these parasites. In the absence of chemical stimuli, most monogeneans hatch rhythmically, with eggs hatching in pulses in the first few hours after dawn (reviewed in Whittington and Kearn 2011). However, when experimentally introduced to the mucus of their host, over ten species spanning eight genera show rapid hatching regardless of the time of day (Whittington and Kearn 2011 and references within).

10.4 Trematoda

Trematodes are a large group of parasitic flukes that are typically endoparasitic (Goater et al. 2014). All trematodes have complex life cycles, and the group contains a wide variety of life strategies (see Chap. 5). While typically endoparasitic, digenean trematodes have two to three free-living stages where they are outside of their host's body. These free-living stages are called the miracidium and cercaria and can be either active or passive in their manner of infection (Niewiadomska and Pojmańska 2011). The miracidial stage of the trematode will parasitise their first intermediate host (typically a gastropod mollusc) and the cercarial stage will infect a second intermediate or final host (Haas 2003).

During the miracidial and cercarial stages, digeneans have only a short amount of time to locate a new host. Typically, when not exposed to any chemical stimuli, active species in this stage will swim erratically or remain relatively stationary (Niewiadomska and Pojmańska 2011; Haas 2003). However, exposure to numerous types of chemical stimuli can initiate various responses in both miracidial and cercarial stages. In general, during miracidial stages individuals are attracted to a wide variety of small organic compounds, many of which associated with snail tissue and mucus (Haas 2003). When exposed to chemical attractants of an intermediate mollusc host, many

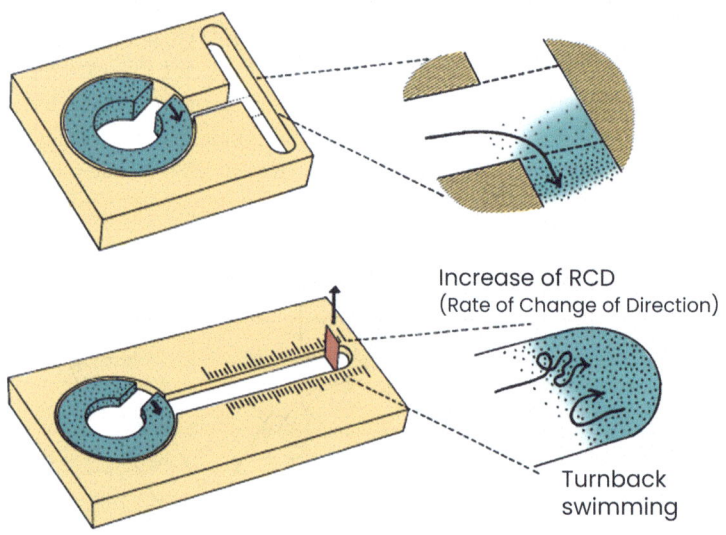

Fig. 10.2 Mechanisms of chemosensory orientation used by miracidium and cercaria stages of digeneans. Chemotaxis involves a directional response towards a higher concentration of chemical signal. Chemokinesis involves the initiation or increase of 'searching behaviours' when exposed to chemical signals [Image from Haas (2003)]

miracidia exhibit a chemokinetic response, identified as an increase in random turning and turn-back swimming (Fig. 10.2). This includes the miracidia of *Schistosoma mansoni* (Haberl et al. 1995), *S. haematobium* (Haberl et al. 1995), *Trichobilharzia ocellata* (Haas et al. 1995a; Hertel et al. 2006), *Fasciola hepatica* (Haas et al. 1995a) and *Echinostoma caproni* (Haberl et al. 2000). Chemical fractionation has determined that *S. mansoni, S. haematobium, T. ocellata, F. hepatica* and *E. caproni* respond exclusively to soluble macromolecules termed 'miracidia-attracting glycoconjugates' or MAGs (Kalbe et al. 1997; Haberl et al. 2000; Haas 2003). Additionally, host-specific responses to chemical stimuli appear to vary by species and even in some cases strains of species. *Trichobilharzia ocellata, T. franki, Fasciola hepatica* and an Egyptian strain of *Schistosoma mansoni* exhibit chemokinetic responses when exposed to snail-conditioned water from their specific host snails. In contrast, *Echinostoma caproni* and a Brazilian strain of *S. mansoni* show chemokinetic responses to water conditioned by both host and non-host snail species (Kalbe et al. 1997; Haberl et al. 2000; Kock 2001; Hassan et al. 2003). While rarer, some species such as *S. japonicum* and *Hypoderaeum conoideum* were capable of a chemotaxis response, where they are able to identify the direction of a chemical attractant and move towards the source of the cue (Haas et al. 1995a, b) (Fig. 10.2).

While the cercariae of several species rely on random chance when locating their second intermediate or definitive host, numerous other species appear to respond to chemical stimuli to achieve this. For cercariae infecting a second intermediate host (i.e. another snail), chemoreception has been observed. The cercariae of *Echinostoma revolutum* and *Pseudechinoparyphium echinatum* exhibit a chemokinetic response to amino acids, urea and ammonia originating from their hosts (Haas et al. 1995b; Körner and Haas 1998), while *S. mansoni* exhibits a similar response when exposed to human-skin-surface extracts (Brachs and Haas 2008). *Hypoderaeum conoideum* exhibits a chemotaxis response to peptides (Haas et al. 1995b). However, chemo-orientation seems to be less common in cercariae that infect faster moving hosts, such as mammals, birds and fish. For these species, chemical stimuli are more often used in inducing and encouraging attachment and penetration. *Schistosoma mansoni* and *haematobium* show attachment responses when exposed to L-arginine and penetration responses when exposed to fatty acids. Similar fatty acid-related penetration responses are seen in several other species including *S. spindale, S. japonicum, Orientobilharzia turkestanica, Trichobilharzia ocellata, Austrobilharzia variglandis, Diplostomum spathaceum* and *Acanthostomum brauni*. *O. turkestanica* shows an attachment response when exposed to skin surface extracts, with *T. ocellata* being responsive to ceramides and cholesterol, *A. brauni* being responsive to glycoproteins with sialic acids and *Opisthorchis viverrini* being responsive to glycosaminoglycans (Haas 2003 and references therein).

10.5 Crustacea

Crustacea is one of the most diverse groups in nature, consisting of over 66,000 species (Ahyong et al. 2011). It is estimated that over 7000 of those species are parasitic (Boxshall and Hayes 2019). These crustacean parasites exploit an even more diverse group of both invertebrate and vertebrate hosts (Boxshall and Hayes 2019, see Chap. 6). Their presence on or in a host can have numerous effects, at both individual and community levels (reviewed in Sikkel and Welicky 2019). For individuals, the presence of parasites has been shown to affect growth and physiological functions (Johnson et al. 2004; Jones and Grutter 2005; Östlund-Nilsson et al. 2005; Roche et al. 2013; Triki et al. 2016; Welicky et al. 2019) as well as host behaviour (Grutter 2001; Meadows and Meadows 2003; Fogelman et al. 2009; Binning et al. 2018). Additionally, crustacean parasites can affect host and cleaner populations (Cheney and Côté 2003; Hayes et al. 2011; Sellers et al. 2019) and act as vectors for bloodborne parasites and diseases (Dunlap et al. 2013; Curtis et al. 2013; Sikkel et al. 2020).

10.5.1 Thecostraca

The crustacean class Thecostraca comprises over 2100 species from over 65 families (Chan et al. 2021). The infraclass Rhizocephala is a highly specialised group of over 250 species of barnacle that parasitise decapod crustaceans (Goater et al. 2014; Chan et al. 2021; see Chap. 6). Unlike most of the previously discussed parasites, female larvae of rhizocephalans must locate a previously uninfected host before metamorphosing to later life stages, often resembling the egg sacs of the female host's egg mass (Goater et al. 2014). Many times, if a male host is parasitised, the presence of the parasite will suppress gonad maturation (Kamio et al. 2022). While infection by rhizocephalans likely costs crab fisheries large amounts of money annually, primarily from reduced growth rates causing crabs to not meet legal collection sizes (Kamio et al. 2022), only *Loxothylacus texanus* has had any aspects of its host-finding behaviour studied.

Loxothylacus texanus parasitises *Callinectes sapidus* throughout the Gulf of Mexico. These barnacles have only an estimated three to four days to locate and settle on an available host before dying (Boone et al. 2003). In laboratory settings, *L. texanus* can locate and settle on pieces of exoskeleton from *C. sapidus* within three days (Boone et al. 2003). Carbohydrate- and glycoprotein-related cues originating from the exoskeleton appear to be particularly important for host location and settlement. Removal of these compounds from the exoskeleton causes a significant decrease in settlement rates (Boone et al. 2003). Additionally, lipid-related cues appear to be unimportant settlement cues as their removal does not reduce settlement rates (Boone et al. 2003).

10.5.2 Maxillopoda

10.5.2.1 Branchiura

Branchiurans are a group of parasites, also often referred to as fish lice, which occur in fresh, brackish and marine waters (Møller 2009; Suárez-Morales 2015; see Chap. 6). Branchiurans include at least 125 species comprising four genera, all within the family Argulidae (Møller 2009). Chemosensory research has focused on only a few species in the genus *Argulus*, which are obligate ectoparasites of fish. Parasites in this genus tend to be host-generalist and have been reported to infect a variety of fishes and even some amphibians and aquatic reptiles (Ringuelet 1943; Møller 2009).

Argulus coregoni primarily parasitises salmonid fishes and is typically more successful in lighted environments compared to other arguilid parasites (Mikheev et al. 2004). When exposed to the chemical stimuli of a host fish (*Oncorhynchus mykiss*) within a Y-maze, *A. coregoni* shows a positive directional response to the fish cues (Bandilla et al. 2007). In similar experiments, males are attracted to the chemical cues of females but not those of other males. Additionally, females do not show significant attraction to the cues of males or females (Bandilla et al. 2007). When comparing potential mates and hosts, male *A. coregoni* shows a significant preference for the chemical cues of the host fish over cues originating from females, indicating a potential preference for food over mates (Bandilla et al. 2007). However, this conclusion is likely biased as a higher concentration of fish cues was introduced in these experiments than for mates (five *O. mykiss* in source water vs 10 adult female parasites).

Argulus japonicus similarly appears to use chemical stimuli when locating a fish host. When exposed to water conditioned with chemical cues of a common host (*Cyprinus carpio*) *A. japonicus* shows increased swimming activity associated with host-finding behaviours compared to when exposed to control water (Galarowicz and Cochran 1991).

10.5.2.2 Copepoda

The Copepoda (copepods) is an extremely diverse group of invertebrates with over 14,000 identified species (Suárez-Morales 2015, see Chap. 6). Of these copepods, at least 20% are considered parasitic (Williams and Bunkley-Williams 1996). These parasitic copepods infect a wide variety of hosts spanning at least 13 phyla (Boxshall and Halsey 2004; Boxshall and Hayes 2019). Copepods also exhibit a wide variety of life histories that can be complex and the complete life cycle of many species has yet to be described.

The best-studied aquatic ectoparasites are the parasitic copepods in the genera *Lepeophtheirus* and *Caligus*. These copepods, often called sea lice, are external parasites of farmed and wild salmonid fishes (Tully and Nolan 2002; Wagner et al. 2007). *Lepeophtheirus salmonis* is the most economically significant sea lice species. Infection by this species of copepod alone costs the Norwegian salmon farming industry hundreds of millions of dollars annually (Abolofia et al. 2017). Because of this, *L. salmonis* has been the focus of many studies. *Lepeophtheirus salmonis* starts their lives as free-swimming, non-parasitic larvae that survive off energy reserves for one to two weeks (Boxaspen 2006). Near the end of this period, individuals will begin actively searching for a host (Pike and Wadsworth 1999).

In their host-finding behaviour, chemical cues appear to be important. In the absence of any chemical stimuli, *L. salmonis* swims in a manner that conserves energy, simply maintaining depth (Heuch et al. 1995). Once host cues are detected, individuals exhibit host-searching behaviours which include swimming in circles or helices and swimming towards the source of the cue (Genna 2002; Bailey et al. 2006). Chemical cues are likely important for directing *L. salmonis* towards

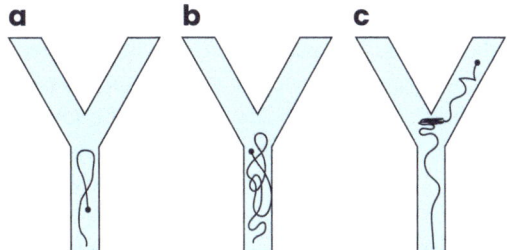

Fig. 10.3 Diagram of representative behaviours of *Lepeophtheirus salmonis* within a Y-tube. (**a**) Representation of low activity associated with no exposure to chemical signals. (**b**) and (**c**) Representation of high activity when exposed to chemical stimuli. Additionally, (**c**) represents a strong directional response to chemical stimuli [Adapted from Ingvarsdóttir et al. (2002a)]

groups of salmon at a larger scale, and orienting and activating searching behaviours at a much smaller scale (Mordue Luntz and Birkett 2009). Additionally, when given the choice between the chemical cues of preadult virgin females and control seawater in a Y-tube, adult males spend significantly more time in the water containing female signals (Ingvarsdóttir et al. 2002a) (Fig. 10.3).

Electron microscopy and electrophysiological recordings have helped indicate that the olfactory receptors of *L. salmonis* are, as with many other arthropods, located on their antennae, specifically the first antennae (Hull 1997; Ingvarsdóttir et al. 2002b). Electrophysiological recordings and behavioural trials have also indicated that, in addition to being attracted to water conditioned with the chemical cues of a host, *L. salmonis* is responsive to non-polar, low-molecular weight compounds (isophorone; 6-methyl-5-hepten-2-one), low-molecular weight water-soluble compounds (Ingvarsdóttir et al. 2002b; Fields et al. 2007) and antimicrobial peptides (cathelicidin-2) originating from *Salmo salar* (Núñez-Acuña et al. 2018). Furthermore, *L. salmonis* is not attracted to chemical compounds (4-methylquinazoline; 2-aminoacetophenone) originating from the non-host *Scophthalmus maximus* when exposed to them in a Y-tube (Bailey et al. 2006).

Caligus spp. are parasitic copepods that can be found worldwide parasitising multiple species of fish, including commercially important farmed fishes. While the genus *Caligus* contains over 250 recognised species, *Caligus rogercresseyi* has been the primary focus of chemosensory research. This is because, like *L. salmonis* in the northeast Atlantic, *C. rogercresseyi* costs the Chilean salmon farming industry millions of dollars each year (Costello 2009). The life cycle of *C. rogercresseyi* includes eight stages, five of which are parasitic. Upon reaching their parasitic stages, individuals must locate and attach to a host (González and Carvajal 2003).

In laboratory experiments, *C. rogercresseyi* is attracted to water conditioned with chemical cues of *Salmo salar*, a common host (Pino-Marambio et al. 2007). Additionally, *C. rogercresseyi* avoids water conditioned by the non-host *Hypsoblennius sordidus* (Pino-Marambio et al. 2007). This behaviour appears to be influenced by the feeding behaviours of previous generations. The offsprings of individuals reared on a less common host (*Oncorhynchus mykiss*) are attracted to water conditioned with that host but show avoidance behaviours when exposed to their traditionally preferred host, *S. salar* (Pino-Marambio et al. 2007).

Another parasitic copepod, *Lernaeocera branchialis*, is unique when compared to the previous examples in that it has a complex life cycle, infecting an intermediate host (typically the flatfishes *Platichthys flesus* or *Microstomus kitt*), where it matures and mates before infecting a definitive host to metamorphose and releasing its fertilised eggs (Whitfield et al. 1988; Brooker et al. 2007). Possible definitive hosts include a wide variety of gadoid fishes such as *Gadus morhua*, *Merlangius merlangus* and *Pollachius pollachius* (reviewed in Brooker et al. 2007). When exposed to definitive host (*G. morhua* and *M. merlangus*) cues in a Y-tube, *L. branchialis* does not appear to be directly attracted to them, as they spend similar time in each of the three 'zones' of the Y-tube (each arm and central mixed-cue zone) as control trials without chemical stimuli (Brooker et al. 2013). *Lernaeocera branchialis* does show increased movement and swim speed, often associated with host-searching behaviours, only when exposed to *M. merlangus* (Brooker et al. 2013). Brooker et al. suggest this may indicate a cryptic subspecies of *L. branchialis* genetically predisposed to prefer *M. merlangus* or a phenomenon similar to *C. rogercresseyi*, with offspring preferring the host earlier generations had parasitised. However, this may also be a result of the previously stated limitations of the Y-tube method with smaller organisms.

10.5.3 Malacostraca

Malacostraca is one of the most diverse groups of organisms and the most numerous and diverse group of crustaceans (Poore 2002). Although difficult to estimate, at least 850 species within Malacostraca are parasitic, with the true number likely being much higher (Poulin and Morand 2000, see Chap. 6). However, information on their chemosensory behaviour is still severely limited. While the number of parasites is large, chemosensory behaviours have only been examined in three main groups: cymothoid isopods, gnathiid isopods and bivalve pea crabs (family Pinnotheridae), and studies have focused nearly exclusively on host-finding behaviours.

10.5.3.1 Cymothoidae

Cymothoid isopods are a large group of parasitic isopods that infect a wide variety of benthic-associated fishes in warm temperate and tropical waters, typically attaching near or in the mouth or gills of their host, but occasionally may attach to other parts of the body (Brusca 1981; Smit et al. 2014; Williams and Bunkley-Williams 2019). They are sequential hermaphrodites, with their sex being determined by the presence or absence of a conspecific on a host (Cook and Munguia 2013; also see Chap. 6). Because of this, host-finding is tightly associated with mate-finding, as reproduction occurs on the host (Smit et al. 2014).

Cymothoa excisa is, to date, the only cymothoid species for which studies on sensory behaviour have been conducted. *Cymothoa excisa* commonly infects the Atlantic croaker, *Micropogonias undulatus*, in the Gulf of Mexico. If this parasite is the first of its species on a host, it will settle on the tongue, with subsequent individuals settling behind the tongue and on the gills (Cook and Munguia 2015). *Cymothoa excisa* shows an increased time spent swimming rather than resting when placed in water that has chemical cues of both familiar (*M. undulatus*) and unfamiliar (*Lagodon rhomboides*) hosts compared to control water but does not show a difference in swim activity when comparing the two species (Cook and Munguia 2013). Similarly, *C. excisa* spends an increased proportion of time orientating itself towards water conditioned with *M. undulatus* over control water but does not show a preference for one host over the other (Cook and Munguia 2013). At least some cymothoids have been known to select intermediate hosts on their way to their definitive host (Fogelman and Grutter 2008), which may explain the lack of a preference for one species over another. These results suggest that, although chemical stimuli are likely important in host-finding for *C. excisa*, other mechanisms also influence their definitive host-finding and settlement behaviours.

10.5.3.2 Gnathiidae

Gnathiid isopods are found globally from polar to tropical oceans and from intertidal to abyssal depths where they are temporary ectoparasites of a wide variety of fishes (Smit and Davies 2004; Tanaka 2007; Coile and Sikkel 2013; Hendrick et al. 2023; Quattrini and Demopoulos 2016, see Chap. 6). Gnathiids are unusual among aquatic parasites in that they are parasitic only during their three larval stages and must locate a separate host to feed on during each of these stages before metamorphosing to adults, where they are no longer parasitic (Smit and Davies 2004; Tanaka 2007; Williams and Bunkley-Williams 2019). In the Indo-Pacific, an unidentified nocturnal gnathiid species is heavily attracted to fish mucus rather than a control paste made of flour and water when in dark conditions, suggesting these gnathiids can use only chemical cues to locate hosts in the absence of any other cues (i.e. visual) (Nagel et al. 2008).

In the north-eastern Caribbean, *Gnathia marleyi* (Farquharson et al. 2012), a primarily nocturnal species of gnathiid, can locate hosts in fish-baited traps using only chemical cues (Sikkel et al. 2011; Vondriska et al. 2020), and injured fish that are more chemically 'leaky' are more attractive than uninjured fish (Jenkins et al. 2018). Gnathiid traps that release only chemical cues can attract similar numbers of gnathiids as traps releasing both visual and chemical cues (Sikkel et al. 2011). However, traps that release only visual cues but not chemical attract fewer gnathiids than those releasing visual and chemical cues (Sikkel et al. 2011), further supporting the notion that chemical cues are more significant in host-finding than visual. Similar results have been seen in lab experiments, where *G. marleyi* spends significantly more time in water conditioned with chemical cues from the common host *Haemulon flavolineatum* than control seawater in a two-channel choice flume (Vondriska et al. 2020). When comparing multiple familiar host species (*H. flavolineatum*, *Lutjanus synagris*, *Holocentrus rufus* and *Stegastes diencaeus*), *G. marleyi* also spends similar amounts of time in each water source within the flume regardless of which host species' cues are present, suggesting the use of (a) chemical compound(s) common to different hosts (Vondriska et al. 2020) (Fig. 10.4).

Caecognathia sp. is an undescribed species of gnathiid common to waters near the Ryukyu Archipelago in Japan. These gnathiids appear to rely at least in part on chemical cues when locating conspecifics with which to mate. Female *Caecognathia* sp. take significantly longer to metamorphose to their adult, reproductive stage than males (Hayashi et al. 2020). In laboratory trials, fed, final-stage juvenile *Caecognathia* sp. are attracted to filter paper conditioned with the chemical cues of fully developed males (Hayashi et al. 2020). Additionally, adult females do not show a similar attraction to the cues of other females, suggesting that an aggregation of females is due to attraction to a male rather than each other (Hayashi et al. 2020).

10.5.3.3 Pinnotheridae

The family Pinnotheridae is a group of small, soft-bodied crabs that parasitise numerous invertebrates, primarily bivalve molluscs

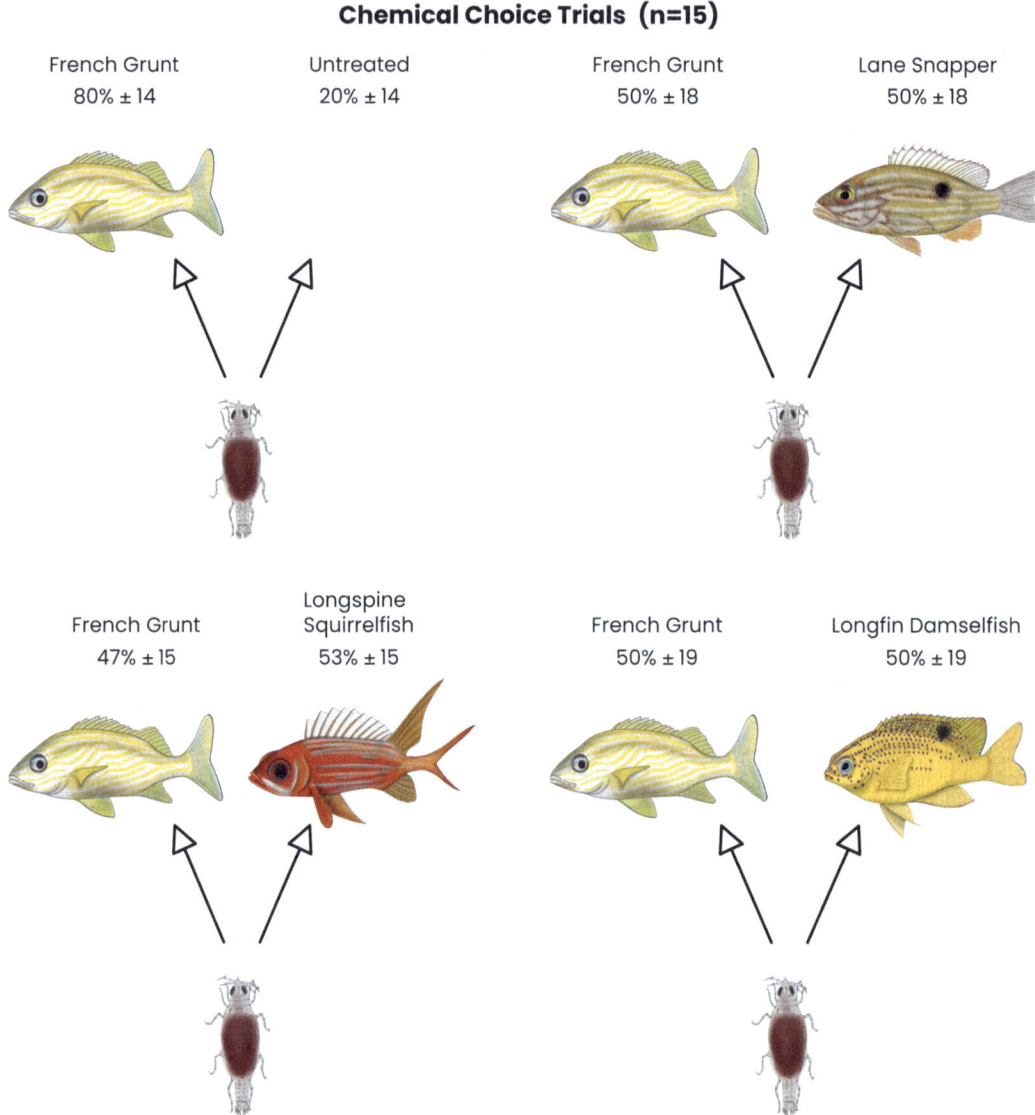

Fig. 10.4 Diagram of the behavioural responses of *Gnathia marleyi* when exposed to the chemical cues of host fish in a two-choice channel flume. *Gnathia marleyi* experiences a strong attraction to the chemical cues of the common host, *Haemulon flavolineatum* (French grunt). They did not show a preference for one host over another in similar channel flume trial [Adapted from Vondriska et al. (2020)]

(Castro 2015). Their infection can slow the growth of molluscs and costs the mollusc farming industry millions of US dollars every year (Trottier and Jeffs 2015). *Nepinnotheres novaezelandiae* is a parasitic pea crab found in New Zealand commonly parasitising *Perna canaliculus* (see Jones 1977). When exposed to the chemical cues of conspecific females sheltered in *P. canaliculus* within a modified flume, male *N. novaezelandiae* will actively exit their host, a behaviour associated with mate searching (Trottier and Jeffs 2015). While females are present upstream in the flume, this behaviour will occur in the absence of any light, supporting the notion that chemical cues are important in initiating conspecific-finding behaviours in *N. novaezelandiae* (Trottier and Jeffs 2015).

10.6 Hyperoartia

While most aquatic parasites are small, invertebrate species, parasitic lampreys (e.g. Petromyzontidae) are notable exceptions. Lampreys include at least 38 species that are all jawless vertebrates found in temperate waters, with most species temporarily parasitising the bodies of fishes and occasionally marine mammals (Hardisty and Potter 1971; Miočić-Stošić et al. 2020). Using their sucking mouth and tongue-like structures they feed on the blood and tissue of their host (Hardisty and Potter 1971).

Nearly all lampreys reproduce in streams and parasitic lampreys can often be displaced large distances from these streams by their host fishes (Johnson et al. 2014). To relocate these streams, chemical signals originating from larvae within the streams appear to be crucial (Buchinger et al. 2015; Fissette et al. 2021) (Fig. 10.5). Exposure to chemicals from larvae can both orientate and encourage searching-associated behaviours, such as sweeping swimming movements, in reproductive-age lampreys (Vrieze et al. 2011; Fissette et al. 2021). Several potentially important chemical cue compounds have been identified through both laboratory and field experiments, including petromyzonol sulphate (PZS) (Bjerselius et al. 2000; Vrieze and Sorensen 2001), allocholic acid (ACA) (Bjerselius et al.

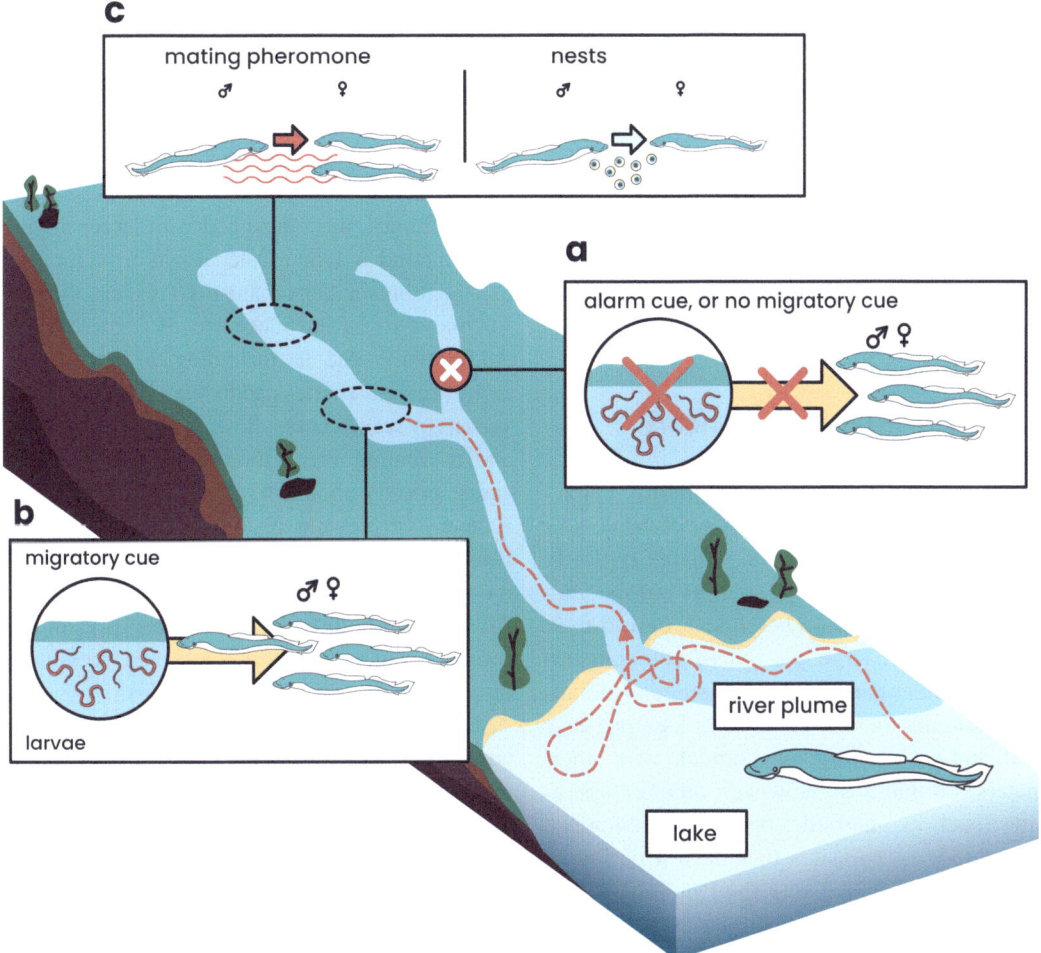

Fig. 10.5 Diagram of the various hypothesised functions of migratory cues, alarm cues, and mating pheromones of reproductive sea lamprey [Adapted from Buchinger et al. (2015)]

2000; Vrieze and Sorensen 2001), petromyzonamine disulphate (PADS) (Fine and Sorensen 2008) and petromyzosterol disulphate (PSDS) (Fine and Sorensen 2008). While these chemical cue compounds have been shown to attract lampreys to streams, they were unsuccessful in attracting lampreys upstream (Meckley et al. 2014). For upstream swimming behaviours, an entirely different set of chemical compounds appears to be important, including (−)- and (+)-petromyric acid A (PMA) (Li et al. 2018).

10.7 Conclusions and Future Directions

Chemically mediated behaviours are likely present at some point in the life cycles of all aquatic ectoparasites, with many relying heavily on them for multiple key fitness-related processes. However, these behaviours have been studied for very few species. The focus has primarily centred around commercially important species, such as *L. salmonis*. Host-finding-associated behaviours have, for obvious reasons, been the focus. This has left many gaps in our current knowledge of the chemosensory behaviours.

10.7.1 Expansion of Behaviours

For most ectoparasites, host-finding behaviours are typically the only type of behaviour investigated by researchers. This is typically because other behaviours such as conspecific-finding are assumed to be closely tied to their host-finding. A large portion of ectoparasites are permanent parasites with a singular definitive host and, therefore, focus on locating their host before coming in contact with their mate(s) on that host. However, other ectoparasites, such as gnathiid isopods and lampreys, are temporary parasites that have a period of time between parasitising their definitive host and mating. In these cases, they must navigate complex ecosystems to locate each other, likely using a variety of sensory cues to do so. While studies examining specifically the use of chemical signals in conspecific-location exist, they are rare.

Additionally, there are settlement and habitat-finding behaviours. While only common in specific groups of parasites, some must find suitable habitats either to shelter until a host is near or to progress to a future life history stage. This could be closely tied to conspecific-locating cues or be a separate cue entirely. To date, no studies have been conducted on the influence of chemical signals on settlement outside of those associated with host-finding. Future studies should expand to investigate the chemical ecology of both conspecific and habitat location in applicable species.

10.7.2 Improved Methodology

Many of the past studies conducted on chemosensory behaviour have used equipment or techniques that are now outdated. Many instruments used to test these behavioural responses, such as choice flumes, were designed with larger organisms in mind and then scaled down in attempts to use them on smaller organisms This has allowed for biases to be introduced to the experiments. Additionally, due to technological limitations, many previous studies had to rely on direct observations by researchers, allowing additional biases and preventing verification of results. In these cases, priority should be given to verifying the accuracy of previous results by retesting organisms using updated instruments as well as video recordings and behavioural tracking software. Advances in both have now made it possible to record smaller organisms easily and accurately on proper temporal and spatial scales.

Recently, a novel aquatic olfactometer has been introduced which improves on the classic methodology (Alkhafaji et al. 2021; Vondriska and Sikkel 2023). Choice flumes and tubes have many limitations. Both have the risk of allowing the organism to become 'behaviourally trapped' by swimming too far up one arm and not being able to move from one cue source to another. Choice flumes are limited in the fact that they provide unnatural chemical landscapes lacking

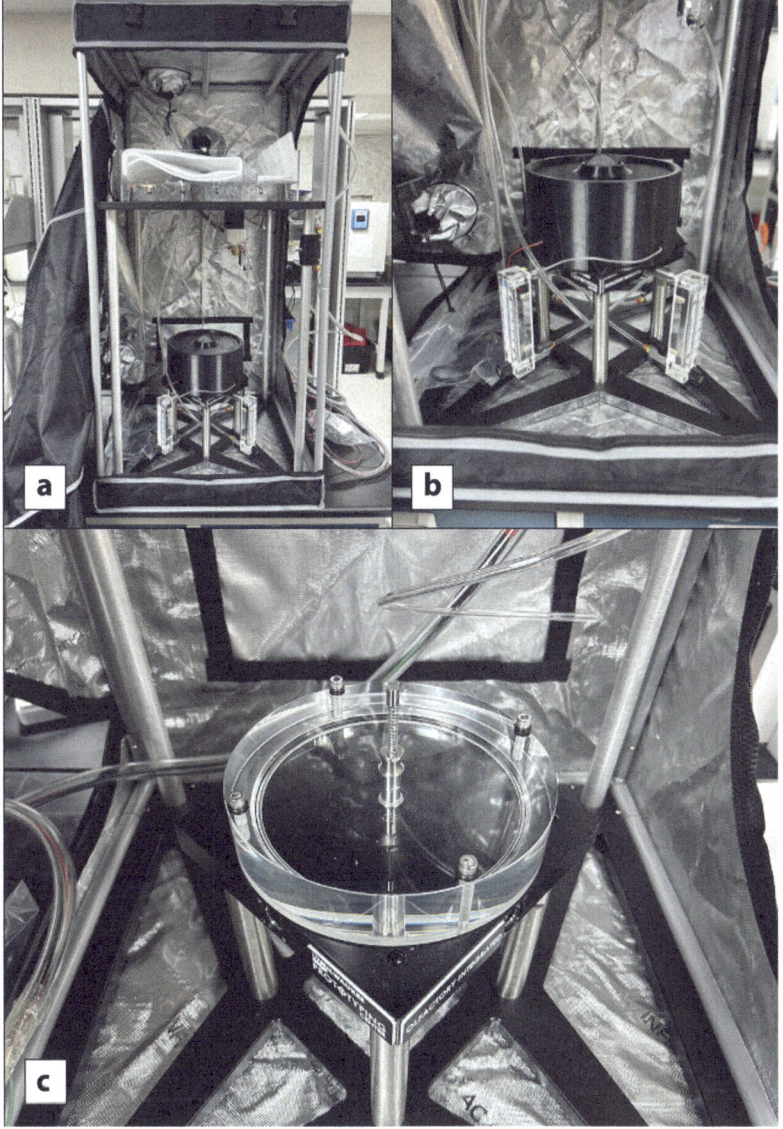

Fig. 10.6 Example of a newly developed aquatic olfactometer. (**a**) Overview of the entire olfactometer; (**b**) main testing instrument of the olfactometer. Choice arena is covered with a light blocking cover, and (**c**) choice arena covered with glass cylinder to create watertight seal within arena. Photo Credit: Clayton Vondriska

any chemical gradient which is likely important in most organisms' locating behaviours. Y- and T-tubes cannot effectively test avoidance behaviours.

This aquatic olfactometer (Fig. 10.6), inspired by those used in terrestrial systems, was designed specifically to test behaviours in small aquatic organisms. It allows for increased freedom of movement while still completely dividing water sources (Fig. 10.7). Additionally, it provides a more natural chemical environment for the organism (Alkhafaji et al. 2021; Vondriska and Sikkel 2023). The aquatic olfactometer shows promise to replace choice flumes and tubes as the new standard when conducting research on the chemosensory behaviour of small organisms.

New techniques to study the sensory systems of organisms have emerged and should be used to examine aquatic parasites. One example of these are 'omics' technologies, which can allow for the identification of chemosensory-related genes. These technologies have already been used in a host of insects and other terrestrial invertebrates (see Tram et al. 2020; Hu et al. 2022; Tang and

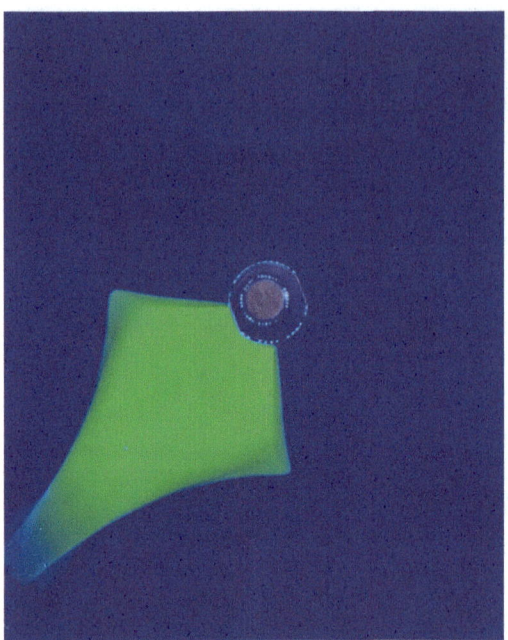

Fig. 10.7 Fluorescence test showing the divided sections of the aquatic olfactometer. This allows the test organism to move unimpeded between zones and provides a chemical gradient within zones (Image with permission from Vondriska and Sikkel 2023)

Zhao 2022 for examples). Advanced microscopy techniques and improved microscopy technologies have made examining chemosensory structures easier and inexpensive as ever. Instruments such as scanning electron microscopes have been used to examine the chemosensory structures of a host of terrestrial and aquatic invertebrates (see Aruna et al. 2019; Gellert et al. 2022; Ma et al. 2022). Additionally, other recording techniques such as calcium imaging can allow researchers to study the neural activity of organisms (Chartier et al. 2018; Buiu et al. 2019).

10.7.3 Chemical Compounds Associated with Chemosensory Behaviours

For those organisms that have shown to be responsive to chemical cues, the next priority should be to determine the chemical compound(s) associated with those behaviours. The majority of studies have only investigated whether an ectoparasite has a positive or negative directional response to a general chemical cue or shows an increase in movement when exposed to that cue. Very few have attempted to determine what chemical compound or group of compounds is important in eliciting those responses. While these initial experiments have been an important first step in determining behavioural responses, they only begin to give us information about the parasite's chemical ecology and behaviour. Future studies should follow the few more comprehensive studies such as those reviewed in Mordue Luntz and Birkett (2009) on *L. salmonis* and those by Haas and colleagues (many reviewed in Haas 2003) on miracidia and cercariae of trematodes, the most intensively studied parasites in this regard, examining which chemical compounds are important for these different behavioural responses.

10.7.4 Anthropogenic Changes

Anthropogenic changes have been shown to affect the chemosensory behaviour of many aquatic organisms (Roggatz et al. 2016; Wilke 2021; Roggatz et al. 2021). Ocean pH is projected to decline by 0.08 to 0.37 by 2100 (Cooley et al. 2022). Ocean water temperature is projected to rise between 0.85 and 2.89 °C and the frequency of marine heatwaves is projected to increase 2- to 15-fold in the same period (Cooley et al. 2022). More severe changes are expected to occur in freshwater ecosystems, with some areas projected to see similar acidification (Leduc et al. 2013) and an increase in temperature by up to 10 °C (Heino et al. 2009; Liu et al. 2020). A changing climate is particularly concerning for invertebrates, of which most ectoparasites are (Wang and Wang 2020).

Ocean acidification has been shown to negatively affect marine invertebrates' ability to detect and respond to chemical signals (Kroeker et al. 2014; Zupo et al. 2015; Roggatz et al. 2016). Similar effects, while rarer, have been seen in association with increasing water temperatures (Ross and Behringer 2019) To date, no study has been conducted on any ectoparasite's response to chemical stimuli under acidified water or increasing water temperature. In order to properly understand how our changing climate will affect these organisms this area of research needs to begin and become a priority.

10.7.5 Conclusions

Research into the chemosensory behaviours in aquatic parasites is trending in a positive direction. However, there remain gaps in knowledge that need to be addressed in future studies. Most aquatic parasites have yet to have a single study conducted to attempt to understand any aspect of their sensory biology.

There is still significant work required to bring the understanding of the chemosensory-related behaviours and mechanisms of these parasites to the level seen in their terrestrial counterparts or larger aquatic organisms. Advancements in technology and methodology have begun to allow for the easier and more accurate study of aquatic parasites. In particular, the previously described aquatic olfactometer promises to significantly improve the quality of research conducted on small aquatic organisms. Additionally, improvements in recording technology and behavioural analysis software have allowed researchers to more accurately record and interpret the behaviours of even the smallest of organisms.

Aquatic parasites have shown to be important and influential in ecosystem processes. For example, modern ecological models have revealed that parasites have a disproportionate influence in food web processes relative to their individual size. They can have both direct and indirect effects on not just their hosts, but the community as a whole. For reasons such as these, it is essential to gain a better understanding of the behaviour and sensory biology of aquatic parasites.

Acknowledgement We thank A. Packard, Nico J. Smit and Bernd Sures for helpful comments on this review. We are grateful to the National Science Foundation for their generous support of our research programme over the past 15 years. Additional support was provided by the International Coral Reef Society (ICRS Graduate Fellowship), and the American Museum of Natural History (Lerner-Gray Grant for Marine Research).

References

Abolofia J, Asche F, Wilen JE (2017) The cost of lice: quantifying the impacts of parasitic sea lice on farmed salmon. Mar Resour Econ 32(3):329–349. https://doi.org/10.1086/691981

Ahyong ST, Lowry JK, Alonso M, Bamber RN, Boxshall GA, Castro P, Gerken S, Karaman GS, Goy JW, Jones DS, Meland K, Rogers DC, Svavarsson J (2011) Subphylum Crustacea Brünnich, 1772. In: Zhang ZQ (ed) Animal biodiversity: an outline of higher-level classification and survey of taxonomic richness. Mongolia Press, Auckland, pp 165–191

Alkhafaji AA, Selim OM, Amano RS, Strickler JR, Hinow P, Jiang H, Sikkel PC, Kohls N (2021) Mass transfer performance of a marine zooplankton olfactometer. J Energy Resour Technol 143(11):e112102. https://doi.org/10.1115/1.4049602

Aruna R, Jeryarani S, Mohankumar S, Durairaj C (2019) Documentation and validation of chemosensory structures in antennae of *Spodoptera litura* (Fabricius) through scanning electron microscope (SEM). Indian J Agric Res 53(5):554–559. https://doi.org/10.18805/IJARe.A-5084

Bailey RJE, Birkett MA, Ingvarsdóttir A, Mordue Luntz AJ, Mordue W, O'Shea B, Pickett JA, Wadhams LJ (2006) The role of semiochemicals in host location and non-host avoidance by salmon louse (*Lepeophtheirus salmonis*) copepodids. Can J Fish Aquat Sci 63(2):448–456. https://doi.org/10.1139/f05-231

Bandilla M, Hakalahti-Sirén T, Valtonen ET (2007) Experimental evidence for a hierarchy of mate and host-induced cues in a fish ectoparasite, *Argulus coregoni* (Crustacea: Branchiura). Int J Parasitol 37(12):1343–1349. https://doi.org/10.1016/j.ijpara.2007.04.004

Bilodeau SM, Hay ME (2022) Chemical cues affecting recruitment and juvenile habitat selection in marine versus freshwater systems. Aquat Ecol 56(2):339–360. https://doi.org/10.1007/s10452-021-09905-x

Binning SA, Roche DG, Grutter AS, Colosio S, Sun D, Miest J, Bshary R (2018) Cleaner wrasse indirectly affect the cognitive performance of a damselfish through ectoparasite removal. Proc R Soc B Biol Sci 285(1874):e20172447. https://doi.org/10.1098/rspb.2017.2447

Bjerselius R, Li W, Teeter JH, Seelye JG, Johnsen PB, Maniak PJ, Grant GC, Polkinghorne CN, Sorensen PW (2000) Direct behavioural evidence that unique bile acids released by larval sea lamprey (*Petromyzon marinus*) function as a migratory pheromone.

Can J Fish Aquat Sci 57(3):557–569. https://doi.org/10.1139/f99-290

Boone EJ, Boettcher AA, Sherman TD, O'Brien JJ (2003) Characterization of settlement cues used by the rhizocephalan barnacle *Loxothylacus texanus*. Mar Ecol Prog Ser 252:187–197. https://doi.org/10.3354/meps252187

Boxaspen K (2006) A review of the biology and genetics of sea lice. ICES J Mar Sci 63(7):1304–1316. https://doi.org/10.1016/j.icesjms.2006.04.017

Boxshall GA, Halsey SH (2004) An introduction to copepod diversity. Ray Society, Andover

Boxshall G, Hayes P (2019) Biodiversity and taxonomy of the parasitic Crustacea. In: Smit NJ, Bruce NL, Hadfield KA (eds) Parasitic Crustacea: state of knowledge and future trends. Springer, Cham, pp 73–134

Brachs S, Haas W (2008) Swimming behaviour of *Schistosoma mansoni* cercariae: responses to irradiance changes and skin attractants. Parasitol Res 102(4):685–690. https://doi.org/10.1007/s00436-007-0812-4

Breithaupt T, Thiel M (2010) Chemical communication in crustaceans. Springer, New York

Brönmark C, Hansson LA (2012a) Chemical ecology in aquatic systems. Oxford University Press, Oxford

Brönmark C, Hansson LA (2012b) Chemical ecology in aquatic systems-an introduction. In: Brönmark C, Hansson LA (eds) Chemical ecology in aquatic systems. Oxford University Press, Oxford, pp 14–19

Brooker AJ, Shinn AP, Bron JE (2007) A review of the biology of the parasitic copepod *Lernaeocera branchialis* (L., 1767) (Copepoda: Pennellidae). Adv Parasitol 65:297–341. https://doi.org/10.1016/S0065-308X(07)65005-2

Brooker AJ, Shinn AP, Souissi S, Bron JE (2013) Role of kairomones in host location of the pennelid copepod parasite, *Lernaeocera branchialis* (L. 1767). Parasitology 140(6):756–770. https://doi.org/10.1017/S0031182012002119

Brusca RC (1981) A monograph on the Isopoda Cymothoidae (Crustacea) of the eastern Pacific. Zool J Linnean Soc 73(2):114–199. https://doi.org/10.1111/j.1096-3642.1981.tb01592.x

Buchinger TJ, Siefkes MJ, Zielinski BS, Brant CO, Li W (2015) Chemical cues and pheromones in the sea lamprey (*Petromyzon marinus*). Front Zool 12:e32. https://doi.org/10.1186/s12983-015-0126-9

Buiu M, Shyong J, Lutz EK, Yang T, Li M, Truong K, Arvidson R, Buchman A, Riffell JA, Akbari OS (2019) Live calcium imaging of *Aedes aegypti* neuronal tissues reveals differential importance of chemosensory systems for life-history-specific foraging strategies. BMC Neurosci 20:e27. https://doi.org/10.1186/s12868-019-0511-y

Castro P (2015) Symbiotic brachyura. In: Castro P, Davie P, Guinot D, Schram F, von Vaupel Klein C (eds) Treatise on zoology - anatomy, taxonomy, biology. The Crustacea, vol 9, Part C. Brill Publishers, Leiden, pp 543–581

Chan BKK, Dreyer N, Gale AS, Glenner H, Ewers-Saucedo C, Pérez-Losada M, Kolbasov GA, Crandall KA, Høeg JT (2021) The evolutionary diversity of barnacles, with an updated classification of fossil and living forms. Zool J Linnean Soc 193(3):789–846. https://doi.org/10.1093/zoolinnean/zlaa160

Chartier TF, Deschamps J, Dürichen W, Jékely G, Arendt D (2018) Whole-head recording of chemosensory activity in the marine annelid *Platynereis dumerilii*. Open Biol 8(10):e180139. https://doi.org/10.1098/rsob.180139

Cheney KL, Côté IM (2003) Do ectoparasites determine cleaner fish abundance? Evidence on two spatial scales. Mar Ecol Prog Ser 263:189–196. https://doi.org/10.3354/meps263189

Coile AM, Sikkel PC (2013) An experimental field test of susceptibility to ectoparasitic gnathiid isopods among Caribbean reef fishes. Parasitology 140(7):888–896. https://doi.org/10.1017/S0031182013000097

Cook C, Munguia P (2013) Sensory cues associated with host detection in a marine parasitic isopod. Mar Biol 160:867–875. https://doi.org/10.1007/s00227-012-2140-1

Cook C, Munguia P (2015) Sex change and morphological transitions in a marine ectoparasite. Mar Ecol 36(3):337–346. https://doi.org/10.1111/maec.12144

Cooley S, Schoeman D, Bopp L, Boyd P, Donner S, Ghebrehiwet DY, Ito SI, Kiessling W, Martinetto P, Ojea E, Racault MF, Rost B, Skern-Mauritzen M (2022) Oceans and coastal ecosystems and their services. In: Pörtner HO, Roberts DC, Tignor M, Poloczanska ES, Mintenbeck K, Alegría A, Craig M, Langsdorrf S, Löschke S, Möller V, Okem A, Rama B (eds) Climate change 2022: impacts, adaptations and vulnerability. Cambridge University Press, Cambridge, pp 379–550

Costello MJ (2009) The global economic cost of sea lice to the salmonid farming industry. J Fish Dis 32(3):115–118. https://doi.org/10.1086/691981

Cox K, Brennan LP, Gerwing TG, Dudas SE, Juanes F (2018) Sound the alarm: a meta-analysis on the effect of aquatic noise on fish behaviour and physiology. Glob Change Biol 24(7):3105–3116. https://doi.org/10.1111/gcb.14106

Crane AL, Bairros-Novak KR, Goldman JA, Brown GE (2022) Chemical disturbance cues in aquatic systems: a review and prospectus. Ecol Monogr 92(1):e01487. https://doi.org/10.1002/ecm.1487

Curtis LM, Grutter AS, Smit NJ, Davies AJ (2013) *Gnathia aureamaculosa* a likely definitive host of *Haemogregarina balistapi* and potential vector for *Haemogregarina bigemina* between fishes of the great barrier reef, Australia. Int J Parasitol 43(5):361–370. https://doi.org/10.1016/j.ijpara.2012.11.012

Davenport D (1950) Studies in the physiology of commensalism. I. The polynoid genus *Arctonë*. Biol Bull 98(2):81–93. https://doi.org/10.2307/1538570

Derby CD, Steullet P, Harner AJ, Cate HS (2001) The sensory basis of feeding behaviour in the Caribbean spiny lobster, *Panulirus argus*. Mar Freshwater Res 52(8):1339–1350. https://doi.org/10.1071/MF01099

Dove AMD (2015) Foraging and ingestive behaviours of whale sharks, *Rhincodon typus*, in response to chemical stimulus cues. Biol Bull 228(1):65–74. https://doi.org/10.1086/BBLv228n1p65

Dunlap DS, Ng TF, Rosario K, Barbosa JG, Greco AM, Breitbar M, Hewson I (2013) Molecular and microscopic evidence of viruses in marine copepods. Proc Natl Acad Sci 110(4):1375–1380. https://doi.org/10.1073/pnas.1216595110

Farquharson C, Smit NJ, Sikkel PC (2012) *Gnathia marleyi* sp. nov. (Crustacea, Isopoda, Gnathiidae) from the eastern Caribbean. Zootaxa 3381:47–61. https://doi.org/10.11646/zootaxa.3381.1.3

Fields DM, Weissburg MJ, Browman HI (2007) Chemoreception in the salmon louse *Lepeophtheirus salmonis*: an electrophysiology approach. Dis Aquat Org 78(2):161–168. https://doi.org/10.3354/dao01870

Fine JM, Sorensen PW (2008) Isolation and biological activity of the multi-component sea lamprey migratory pheromone. J Chem Ecol 34:1259–1267. https://doi.org/10.1007/s10886-008-9535-y

Fissette SD, Buchinger TJ, Wagner CM, Johnson NS, Scott AM, Li W (2021) Progress towards integrating an understanding of chemical ecology into sea lamprey control. J Great Lakes Res 47:S660–S672. https://doi.org/10.1016/j.jglr.2021.02.008

Fogelman RM, Grutter AS (2008) Mancae of the parasitic cymothoid isopod, *Anilocra apogonae*: early life history, host-specificity, and effect on growth and survival of preferred young cardinal fishes. Coral Reefs 27:685–693. https://doi.org/10.1007/s00338-008-0379-2

Fogelman RM, Kuris AM, Grutter AS (2009) Parasitic castration of a vertebrate: effect of the cymothoid isopod, *Anilocra apogonae*, on the five-lined cardinalfish, *Cheilodipterus quinquelineatus*. Int J Parasitol 39(5):577–583. https://doi.org/10.1016/j.ijpara.2008.10.013

Galarowicz T, Cochran PA (1991) Response by the parasitic crustacean *Argulus japonicus* to host chemical cues. J Freshw Ecol 6:455–456. https://doi.org/10.1080/02705060.1991.9665326

Gellert HR, Halley DC, Sieb ZJ, Smith JC, Pask GM (2022) Microstructures at the distal tip of ant chemosensory sensilla. Sci Rep 12:e19328. https://doi.org/10.1038/s41598-022-21507-7

Genna RL (2002) The chemical ecology, physiology and infection dynamics of the sea louse copepodids *Lepeophtheirus salmonis* Krøyer. Dissertation. University of Aberdeen

Goater TM, Goater CP, Esch GW (2014) Parasitism: the diversity and ecology of animal parasites, 2nd edn. Cambridge University Press, Cambridge

González L, Carvajal J (2003) Life cycle of *Caligus rogercresseyi*, (Copepoda: Caligidae) parasite of Chilean reared salmonids. Aquaculture 220(1–4):101–117. https://doi.org/10.1016/S0044-8486(02)00512-4

Grutter AS (2001) Parasite infection rather than tactile stimulation is the proximate cause of cleaning behaviour in reef fish. Proc Biol Soc 268(1474):1361–1365. https://doi.org/10.1098/rspb.2001.1658

Haas W (2003) Parasitic worms: strategies of host-finding, recognition and invasion. Zoology 106(4):349–364. https://doi.org/10.1078/0944-2006-00125

Haas W, Haberl B, Kalbe M, Krömer M (1995a) Snail-host-finding by miracidia and cercariae: chemical host cues. Parasitol Today 11(12):468–472. https://doi.org/10.1016/0169-4758(95)80066-2

Haas W, Körner M, Hutterer E, Wegner M, Haberl B (1995b) Finding and recognition of the snail intermediate hosts by 3 species of echinostome cercariae. Parasitology 110:133–142. https://doi.org/10.1017/s0031182000063897

Haberl B, Kalbe M, Fuchs H, Ströbel M, Schmalfuss G, Haas W (1995) *Schistosoma mansoni* and *S. haematobium*: miracidial host-finding behaviour is stimulated by macromolecules. Int J Parasitol 25(5):551–560. https://doi.org/10.1016/0020-7519(94)00158-k

Haberl B, Körner M, Spengler Y, Hertel J, Kalbe M, Haas W (2000) Host-finding in *Echinostoma caproni*: miracidia and cercariae use different signals to identify the same snail species. Parasitology 120:479–486. https://doi.org/10.1017/s0031182099005697

Hardisty MW, Potter IC (1971) The biology of lampreys, vol 1. Academic, London

Hassan AHM, Haberl B, Hertel J, Haas W (2003) Miracidia of an Egyptian strain of *Schistosoma mansoni* differentiate between sympatric snail species. J Parasitol 89(6):1248–1250. https://doi.org/10.1645/GE-85R

Hay ME (2009) Marine chemical ecology: chemical signals and cues structure marine populations, communities, and ecosystems. Annu Rev Mar Sci 1:193–212. https://doi.org/10.1146/annurev.marine.010908.163708

Hayashi C, Tanaka K, Hirose E (2020) Larvae of female *Caecognathia* sp. (Isopoda: Gnathiidae) are attracted to male adults and prolong their larval phase in the absence of males. J Crust Biol 40(2):156–161. https://doi.org/10.1093/jcbiol/ruz094

Hayes PM, Smit NJ, Grutter AS, Davies AJ (2011) Unexpected response of a captive blackeye thicklip, *Hemigymnus melapterus* (Bloch), from Lizard Island, Australia, exposed to juvenile *Gnathia aureamaculosa* Ferreira & Smit, isopods. J Fish Dis 34(7):563–566. https://doi.org/10.1111/j.1365-2761.2011.01261.x

Heino J, Erkinaro J, Huusko A, Luoto M (2009) Climate change effects on freshwater fishes, conservation and management. In: Closs G, Krkosek M, Olden JD (eds) Conservation of freshwater fishes. Cambridge University Press, Cambridge, pp 76–106

Hendrick GC, Nicholson MD, Pagan JA, Artmi JM, Dolan MC, Sikkel PC (2023) Blood meal identification reveals extremely broad host range and host-bias in a temporary ectoparasite of coral reef fishes. Oecologia 203:349–360. https://doi.org/10.1007/s00442-023-05468-w

Hertel J, Holweg A, Haberl B, Kalbe M, Haas W (2006) Snail odour-clouds: spreading and contribution to the transmission success of *Trichobilharzia ocellata* (Trematoda, Digenea) miracidia. Behav Ecol 147:173–180. https://doi.org/10.1007/s00442-005-0239-5

Heuch PA, Parson A, Boxaspen K (1995) Diel vertical migration: a possible host finding mechanism in salmon louse (*Lepeophtheirus salmonis*) copepodids? Can J Fish Aquat Sci 52(4):681–689. https://doi.org/10.1139/f95-069

Hopla CE, Durden LA, Keirans JE (1994) Ectoparasites and classification. Rev Sci Tech Off Int Epizoot 13(4):985–1034. https://doi.org/10.20506/rst.13.4.815

Hu J, Wang XY, Tan LS, Lu W, Zheng XL (2022) Identification of chemosensory genes, including candidate pheromone receptors, in *Phauda flammans* (Walker) (Lepidoptera: Phaudidae) through transcriptomic analyses. Front Physiol 12(13):e907694. https://doi.org/10.3389/fphys.2022.907694

Hudson PJ, Rizzoli AP, Grenfell BT, Heesterbeek JAP, Dodson AP (2002) Ecology of wildlife diseases. In: Hudson PJ, Rizzoli AP, Grenfell BT, Heesterbeek H, Dodson AP (eds) The ecology of wildlife diseases, 1st edn. Oxford University Press, Oxford, pp 1–5

Hudson PJ, Dobson AP, Lafferty KD (2006) Is a healthy ecosystem one that is rich in parasites? Trends Ecol Evol 21(7):381–385. https://doi.org/10.1016/j.tree.2006.04.007

Hull MQ (1997) The role of semiochemicals in the behaviour and biology of *Lepeophtheirus salmonis* (Krøyer, 1837): potential for control? Dissertation. University of Aberdeen

Ingvarsdóttir A, Birkett MA, Duce I, Mordue W, Pickett JA, Wadhams LJ, Mordue Luntz AJ (2002a) Role of semiochemicals in mate location by parasitic sea louse, *Lepeophtheirus salmonis*. J Chem Ecol 28:2107–2117. https://doi.org/10.1023/A:1020762314603

Ingvarsdóttir A, Birkett MA, Duce I, Genna RL, Mordue W, Pickett JA, Wadhams LJ, Mordue Luntz AJ (2002b) Semiochemical strategies for sea louse control: host location cues. Pest Manag Sci 58(6):537–545. https://doi.org/10.1002/ps.510

Jacobson HP, Stabell OB (2004) Antipredator behaviour mediated by chemical cues: the role of conspecific alarm signalling and predator labelling in the avoidance response of a marine gastropod. Oikos 104(1):43–50. https://doi.org/10.1111/j.0030-1299.2004.12369.x

Jenkins WG, Demopoulos AWJ, Sikkel PC (2018) Effects of host injury on susceptibility of marine reef fishes to ectoparasitic gnathiid isopods. Symbiosis 75:112–121. https://doi.org/10.1007/s13199-017-0518-z

Johnson SC, Treasurer JW, Bravo S, Nagasawa K, Kabata Z (2004) A review of the impact of parasitic copepods on marine aquaculture. Zool Stud 43(2):229–243

Johnson NS, Buchinger TJ, Li W (2014) Reproductive ecology of lampreys. In: Docker MF (ed) Lampreys: biology, conservation and control. Springer, Dordrecht, pp 265–303

Johnson NS, Miehls SM, Haro AJ, Wagner CM (2020) Push and pull of downstream moving juvenile sea lamprey (*Petromyzon marinus*) exposed to chemosensory and light cues. Conserv Physiol 7(1):coz080. https://doi.org/10.1093/conphys/coz080

Jones JB (1977) Natural history of the pea crab in Wellington Harbour, New Zealand. N Z J Mar Freshw Res 11(4):667–676. https://doi.org/10.1080/00288330.1977.9515704

Jones CM, Grutter AS (2005) Parasitic isopods (*Gnathia* sp.) reduce haematocrit in captive *Hemigymnus melapterus* (Bloch) (Pisces: Labridae) on the great barrier reef. J Fish Biol 66(3):860–864. https://doi.org/10.1111/j.0022-1112.2005.00640.x

Jutfelt F, Sundin J, Raby GD, Krång AS, Clark TD (2017) Two-current choice flumes for testing avoidance and preference in aquatic animals. Methods Ecol Evol 8(3):379–390. https://doi.org/10.1111/2041-210x.12668

Kalbe M, Haberl B, Haas W (1997) Miracidial host-finding in *Fasciola hepatica* and *Trichobilharzia ocellata* is stimulated by species-specific glycoconjugates from the host-snail. Parasitol Res 83:806–812. https://doi.org/10.1007/s004360050344

Kamio M, Derby CD (2017) Finding food: how marine invertebrates use chemical cues to track and select food. Nat Prod Rep 34(5):514–528. https://doi.org/10.1039/C6NP00121A

Kamio M, Yambe H, Fusetani N (2022) Chemical cues for intraspecific chemical communication and interspecific interactions in aquatic environments: applications for fisheries and aquaculture. Fish Sci 88:203–239. https://doi.org/10.1007/s12562-021-01563-0

Kearn GC (1986) Role of chemical substances from fish hosts in hatching and host-finding in monogeneans. J Chem Ecol 12(8):1651–1658. https://doi.org/10.1007/BF01022371

Knowlton N, Brainard RE, Fisher R, Moews M, Plaisance L, Caley MJ (2010) Coral reef biodiversity. In: McIntyre AD (ed) Life in the world's oceans. Blackwell Publishing Ltd., Hoboken

Kock S (2001) Investigations on intermediate host specificity help to elucidate the taxonomic status of *Trichobilharzia ocellata* (Digenea: Schistosomatidae). Parasitology 123(1):67–70. https://doi.org/10.1017/S0031182001008101

Körner M, Haas W (1998) Chemo-orientation of echinostome cercariae towards their snail hosts: amino acids signal a low host-specificity. Int J Parasitol 28(3):511–516. https://doi.org/10.1016/s0020-7519(97)00196-3

Kroeker KJ, Sanford E, Jellison BM, Gaylord B (2014) Predicting the effects of ocean acidification on predator-prey interactions: a conceptual framework based on coastal molluscs. Biol Bull 226(3):211–222. https://doi.org/10.1086/BBLv226n3p211

Leduc AO, Munday PL, Brown GE, Ferrari MC (2013) Effects of acidification on olfactory-mediated behaviour in freshwater and marine ecosystems: a synthesis. Philo Trans R Soc Lond B: Biol Sci

368(1627):e20120447. https://doi.org/10.1098/rstb.2012.0447

Li K, Brant CO, Huertas M, Hessler EJ, Mezei G, Scott AM, Hoye TR, Li W (2018) Fatty-acid derivative acts as a sea lamprey migratory pheromone. Proc Natl Acad Sci 115(34):8603–8608. https://doi.org/10.1073/pnas.1803169115

Liu S, Xie Z, Liu B, Wange Y, Gao J, Zeng Y, Xie J, Xie Z, Jia B, Qin P, Li R, Wang L, Chen S (2020) Global river water warming due to climate change and anthropogenic heat emission. Glob Planet Change 193:e103289. https://doi.org/10.1016/j.gloplacha.2020.103289

Ma C, Yue Y, Zhang Y, Tian ZY, Chen HS, Guo JY, Zhou ZS (2022) Scanning electron microscopic analysis of antennal sensilla and tissue-expression profiles of chemosensory protein genes in *Ophraella communa* (Coleoptera: Chrysomelidae). Insects 13(2):e183. https://doi.org/10.3390/insects13020183

Meadows DW, Meadows CM (2003) Behavioural and ecological correlates of foureye butterflyfish, *Chaetodon capistratus*, (Perciformes: Chaetodontidae) infected with *Anilocra chaetodontis* (Isopoda: Cymothoidae). Rev Biol Trop 51:77–81

Meckley TD, Wagner CM, Gurarie E (2014) Coastal movements of migrating sea lamprey (*Petromyzon marinus*) in response to a partial pheromone added to river water: implications for management of invasive populations. Can J Fish Aquat Sci 71(4):533–544. https://doi.org/10.1139/cjfas-2013-0487

Mikheev VN, Pasternak AF, Valtonen ET (2004) Tuning host specificity during the ontogeny of a fish ectoparasite: behavioural responses to host-induced cues. Parasitol Res 92(3):220–224. https://doi.org/10.1007/s00436-003-1044-x

Miočić-Stošić J, Pleslić G, Holcer D (2020) Sea lamprey (*Petromyzon marinus*) attachment to the common bottlenose dolphin (*Tursiops truncatus*). Aquat Mamm 46:152–166. https://doi.org/10.1578/AM.46.2.2020.152

Møller OS (2009) Branchiura (Crustacea) - survey of historical literature and taxonomy. Arthropod Syst Phylogeny 67(1):41–55. https://doi.org/10.3897/asp.67.e31687

Mordue Luntz AJ, Birkett MA (2009) A review of host finding behaviour in the parasitic sea louse, *Lepeophtheirus salmonis* (Caligidae: Copepoda). J Fish Dis 32(1):3–13. https://doi.org/10.1111/j.1365-2761.2008.01004.x

Nagel L, Montgomerie R, Lougheed SC (2008) Evolutionary divergence in common marine ectoparasites *Gnathia* spp. (Isopoda: Gnathiidae) on the great barrier reef: phylogeography, morphology, and behaviour. Biol J Linn Soc 94(3):569–587. https://doi.org/10.1111/j.1095-8312.2008.00997.x

Niewiadomska K, Pojmańska T (2011) Multiple strategies of digenean trematodes to complete their life cycles. Wiad Parazytol 57(4):233–241

Nummela S, Thewissen JGM (2008) The physics of sound in air and water. In: Thewissen JGM, Nummela S (eds) Sensory evolution on the threshold: adaptations in secondarily aquatic vertebrates. University of California Press, Berkley, pp 175–182

Núñez-Acuña G, Gallardo-Escárate C, Fields DM, Shema S, Skiftesvik AB, Ormazábal I, Browman HI (2018) The Atlantic salmon (*Salmo salar*) antimicrobial peptide cathelicidin-2 is a molecular host-associated cue for the salmon louse (*Lepeophtheirus salmonis*). Sci Rep 8:e13738. https://doi.org/10.1038/s41598-018-31885-6

Östlund-Nilsson S, Curtis L, Nilsson GE, Grutter AS (2005) Parasitic isopod *Anilocra apogonae*, a drag for the cardinal fish *Cheilodipterus quinquelineatus*. Mar Ecol Prog Ser 287:209–216. https://doi.org/10.3354/meps287209

Paul VJ, Puglisi MP, Ritson-Williams R (2006) Marine chemical ecology. Nat Prod Rep 28(2):153–180. https://doi.org/10.1039/B404735B

Paul VJ, Ritson-Williams R, Sharp K (2011) Marine chemical ecology in benthic environments. Nat Prod Rep 28(2):345–387. https://doi.org/10.1039/C0NP00040J

Pawlik JR (1992) Chemical ecology of the settlement of benthic marine invertebrates. In: Barnes M, Ansell AD, Gibson RN (eds) Oceanography and marine biology: an annual review, vol 30. UCL Press, London, pp 273–335

Pike AW, Wadsworth SL (1999) Sealice on salmonids: their biology and control. Adv Parasitol 44:233–337. https://doi.org/10.1016/S0065-308X(08)60233-X

Pino-Marambio J, Mordue AJ, Birkett M, Carvajal J, Asencio G, Mellado A, Quiroz A (2007) Behavioural studies of host, non-host and mate location by the sea louse *Caligus rogercresseyi* Boxshall and Bravo, 2000 (Copepoda: Caligidae). Aquaculture 271:70–76. https://doi.org/10.1016/j.aquaculture.2007.05.025

Poore GCB (2002) Zoological catalogue of Australia, Crustacea: malacostraca, vol 19.2A. CSIRO Publishing, Clayton

Poulin R, Morand S (2000) The diversity of parasites. Q Rev Biol 75(3):277–293. https://doi.org/10.1086/393500

Quattrini AM, Demopoulos AWJ (2016) Ectoparasitism on deep-sea fishes in the western North Atlantic: in situ observation from ROV surveys. Int J Parasitol Parasite Wildl 5(3):217–228. https://doi.org/10.1016/j.ijppaw.2016.07.004

Ringuelet RA (1943) Revisión de los argúlidos (Crustácea. Branchiura) con el catálogo de las especies neotropicales. Rev Mus Plata (Nueva Serie) 3(19):43–99

Roche DG, Binning SA, Strong LE, Davies JN, Jennions MD (2013) Increased behavioural lateralization in parasitized coral reef fish. Behav Ecol Sociobiol 67:1339–1344. https://doi.org/10.1007/s00265-013-1562-1

Roggatz CC, Lorch M, Hardege JD, Benoit DM (2016) Ocean acidification affects marine chemical communication by changing structure and function of peptide signalling molecules. Glob Change Biol 22(12):3914–3926. https://doi.org/10.1111/gcb.13354

Roggatz CC, Saha M, Blanchard S, Schirrmacher P, Fink P, Verheggen F, Hardege JD (2021) Becoming nose-blind - climate change impacts on chemical communication. Glob Change Biol 28(15):4495–4505. https://doi.org/10.1111/gcb.16209

Ross E, Behringer D (2019) Changes in temperature, pH, and salinity affect the sheltering responses of Caribbean spiny lobsters to chemosensory cues. Sci Rep 9:e4375. https://doi.org/10.1038/s41598-019-40832-y

Sato T, Goshima S (2007) Female choice in response to risk of sperm limitation by the stone crab, *Hapalogaster dentata*. Anim Behav 73:331–338. https://doi.org/10.1016/j.anbehav.2006.05.016

Saunders KM, Brockmann HJ, Watson WH III, Jury SH (2010) Male horseshoe crabs *Limulus polyphemus* use multiple sensory cues to locate mates. Curr Zool 56(1):485–498. https://doi.org/10.1093/czoolo/56.5.485

Schwartz ER, Poulin RX, Mojib N, Kubanek J (2016) Chemical ecology of marine plankton. Nat Prod Rep 33(7):843–860. https://doi.org/10.1039/C6NP00015K

Sellers JC, Holstein DJ, Botha T, Sikkel PC (2019) Lethal and sublethal impacts of a micropredator on post-settlement Caribbean reef fishes. Oecologia 189:293–305. https://doi.org/10.1007/s00442-018-4262-8

Sikkel PC, Welicky RL (2019) The ecological significance of parasitic crustaceans. In: Smit NJ, Bruce NL, Hadfield KA (eds) Parasitic Crustacea: state of knowledge and future trends. Springer, Cham, pp 421–477

Sikkel PC, Sears WT, Weldon B, Tuttle BC (2011) An experimental field test of host-finding mechanisms in a Caribbean gnathiid isopod. Mar Biol 158:1075–1083. https://doi.org/10.1007/s00227-011-1631-9

Sikkel PC, Pagan JA, Santos JL, Hendrick GC, Nicholson MD, Xavier R (2020) Molecular detection of apicomplexan blood parasites of coral reef fishes from free-living stages of ectoparasitic gnathiid isopods. Parasitol Res 119:1975–1980. https://doi.org/10.1007/s00436-020-06676-6

Smit NJ, Davies AJ (2004) The curious life-style of the parasitic stages of gnathiid isopods. Adv Parasitol 58:289–391. https://doi.org/10.1016/S0065-308X(04)58005-3

Smit NJ, Bruce NL, Hadfield KA (2014) Global diversity of fish parasitic isopod crustaceans of the family Cymothoidae. Int J Parasitol Parasites Wildl 3(2):188–197. https://doi.org/10.1016/j.ijppaw.2014.03.004

Suárez-Morales E (2015) Class Maxillopoda. In: Thorp JH, Rogers DC (eds) Thorp and Covich's freshwater invertebrates: ecology and general biology, 4th edn. Academic, Cambridge, pp 709–755

Sures B, Nachev M, Pahl M, Grabner D, Selbach C (2017) Parasites as drivers of key processes in aquatic ecosystems: facts and future directions. Exp Parasitol 180:141–147. https://doi.org/10.1016/j.exppara.2017.03.011

Tanaka K (2007) Life history of gnathiid isopods - current knowledge and future directions. Plankton Benthos Res 2(1):1–11. https://doi.org/10.3800/pbr.2.1

Tang R, Zhao XC (2022) Editorial: the morphology and physiology of insect chemosensory systems—its origin and evolution. Front Neuroanat 16:e1024927. https://doi.org/10.3389/fnana.2022.1024927

Tram G, Klare WP, Cain JA, Mourad B, Cordwell SJ, Day CJ, Korolik V (2020) Assigning a role for chemosensory signal transduction in *Campylobacter jejuni* biofilms using a combined omics approach. Sci Rep 10(1):e6829. https://doi.org/10.1038/s41598-020-63569-5

Triki Z, Grutter AS, Bshary R, Ros AF (2016) Effects of short-term exposure to ectoparasites on fish cortisol and hematocrit levels. Mar Biol 163:e187. https://doi.org/10.1007/s00227-016-2959-y

Trottier O, Jeffs AG (2015) Mate locating and access behaviour of the parasitic pea crab, *Nepinnotheres novaezelandiae*, an important parasite of the mussel *Perna canaliculus*. Parasite 22:e13. https://doi.org/10.1051/parasite/2015013

Tully O, Nolan DT (2002) A review of the population biology and host-parasite interactions of the sea louse *Lepeophtheirus salmonis* (Copepoda: Caligidae). Parasitology 124:S165–S182. https://doi.org/10.1017/s0021182002001889

Vidondo B, Prairie YT, Blanco JM, Duarte CM (1997) Some aspects of the analysis of size spectra in aquatic ecology. Limnol Oceanogr 42(1):184–192. https://doi.org/10.4319/lo.1997.42.1.0184

von Elert E (2012) Information conveyed by chemical cues. In: Brönmark C, Hansson LA (eds) Chemical ecology in aquatic systems. Oxford University Press, Oxford, pp 19–38

Vondriska C, Sikkel PC (2023) Assessment of a new olfactometer for study of sensory ecology in small aquatic organisms. Limnol Oceanogr Methods 21(2):98–105. https://doi.org/10.1002/lom3.10531

Vondriska C, Dixson DL, Packard AJ, Sikkel PC (2020) Differentially susceptible host fishes exhibit similar chemo-attractiveness to a common coral reef ectoparasite. Symbiosis 81:247–253. https://doi.org/10.1007/s13199-020-00700-0

Vrieze LA, Sorensen PW (2001) Laboratory assessment of the role of a larval pheromone and natural stream odor in spawning stream localization by migratory sea lamprey (*Petromyzon marinus*). Can J Fish Aquat Sci 58(12):2374–2385. https://doi.org/10.1139/f01-179

Vrieze LA, Bergstedt RA, Sorensen PW (2011) Olfactory-mediated stream-finding behavior of migratory adult sea lamprey (*Petromyzon marinus*). Can J Fish Aquat Sci 68(3):523–533. https://doi.org/10.1139/F10-169

Wagner GN, Fast MD, Johnson SC (2007) Physiology and immunology of *Lepeophtheirus salmonis* infections of salmonids. Trends Parasitol 24(4):176–183. https://doi.org/10.1016/j.pt.2007.12.010

Wang T, Wang Y (2020) Behavioural responses to ocean acidification in marine invertebrates: new insights and future directions. J Oceanol Limnol 38:759–772. https://doi.org/10.1007/s00343-019-9118-5

Welicky RL, Malherbe W, Hadfield KA, Smit NJ (2019) Understanding growth relationships of African cymothoid fish parasitic isopods using specimens from museum and field collections. Int J Parasitol Parasites Wildl 8:182–187. https://doi.org/10.1016/j.ijppaw.2019.02.002

Whitfield PJ, Plicher MW, Grant HJ, Riley J (1988) Experimental studies on the development of *Lernaeocera branchialis* (Copepoda: Pennellidae): population processes from egg production to maturation on the flatfish host. In: Boxshall GA, Schminke HK (eds) Biology of copepods. Springer, Dordrecht, pp 579–586

Whittington ID (2012) *Benedenia seriolae* and *Neobenedenia* species. In: Woo PTK, Buchmann K (eds) Fish parasites: pathobiology and protection. CAB International, Wallingford

Whittington ID, Kearn GC (2011) Hatching strategies in monogenean (Platyhelminth) parasites that facilitate host infection. Integr Comp Biol 51(1):91–99. https://doi.org/10.1093/iicb/icr003

Williams EH, Bunkley-Williams L (1996) Parasites of off shore, big game sport fishes of Puerto Rico and the Western North Atlantic. Department of Biology, University of Puerto Rico, Mayaguez

Williams EH, Bunkley-Williams L (2019) Life cycle and life history strategies of parasitic Crustacea. In: Smit NJ, Bruce NL, Hadfield KA (eds) Parasitic Crustacea: state of knowledge and future trends. Springer, Cham, pp 179–266

Williamson JE, Carson DG, de Nys R, Steinberg PD (2004) Demographic consequences of an ontogenetic shift by a sea urchin in response to host plant chemistry. Ecology 85(5):1355–1371. https://doi.org/10.1890/02-4083

Wilke C (2021) Climate change could alter undersea chemical communication. ACS Cen Sci 7(7):1091–1094. https://doi.org/10.1021/acscentsci.1c00819

Zupo V, Maibam C, Buia MC, Gambi MC, Patti FP, Scipione MB, Lorenti M, Fink P (2015) Chemoreception of the seagrass *Posidonia oceanica* by benthic invertebrates is altered by seawater acidification. J Chem Ecol 41(8):766–779. https://doi.org/10.1007/s10886-015-0610-x

Open Access This chapter is licensed under the terms of the Creative Commons Attribution-NonCommercial-NoDerivatives 4.0 International License (http://creativecommons.org/licenses/by-nc-nd/4.0/), which permits any non-commercial use, sharing, distribution and reproduction in any medium or format, as long as you give appropriate credit to the original author(s) and the source, provide a link to the Creative Commons license and indicate if you modified the licensed material. You do not have permission under this license to share adapted material derived from this chapter or parts of it.

The images or other third party material in this chapter are included in the chapter's Creative Commons license, unless indicated otherwise in a credit line to the material. If material is not included in the chapter's Creative Commons license and your intended use is not permitted by statutory regulation or exceeds the permitted use, you will need to obtain permission directly from the copyright holder.

Evolutionary Biology of Aquatic Parasites

11

Robert Poulin and Jerusha Bennett

Abstract

Aquatic habitats exert some unique selective pressures on parasites, from the properties of water as a medium to the fragmented nature of freshwater bodies dispersed across a terrestrial landscape. We first discuss some of the key features of aquatic habitats that may impact parasite evolution. We then summarise what is known of microevolutionary processes affecting aquatic parasites, from the forces maintaining the genetic structuring of parasite populations to the emergence of local adaptation and cryptic species. Next, we discuss host–parasite macroevolution and how different historical events have shaped patterns of host specificity. Finally, we address the impact of anthropogenic changes to aquatic habitats on current and future parasite evolution, with a focus on global climate change, fisheries and aquaculture.

R. Poulin (✉) · J. Bennett
Department of Zoology, University of Otago, Dunedin, New Zealand
e-mail: robert.poulin@otago.ac.nz; jerusha.bennett@otago.ac.nz

11.1 Introduction

Life had its origins in water and, not surprisingly, so did parasitism. Of the more than 200 separate transitions from a free-living existence to a parasitic one during the evolutionary history of animals (Weinstein and Kuris 2016), practically all occurred in aquatic habitats, even among many lineages presently inhabiting terrestrial environments. A convergent feature in the evolution of many parasite lineages has been the increasing complexity of life cycles, that is, the addition of life stages and host species needed for the completion of one parasite generation. Driven by selection for greater lifetime fecundity or improved transmission success (Parker et al. 2003), complex life cycles have also evolved primarily in aquatic environments prior to the invasion of land in most parasite taxa (e.g. Cribb et al. 2003). As a consequence, the initial adaptation of parasites for host finding and exploitation, and much of their subsequent evolution, took place in water.

Aquatic parasites show some general differences from those that colonised land (McCallum et al. 2004). For example, parasitic castration is a common host exploitation strategy in marine systems (see Chap. 7), but it is rare among terrestrial parasites. In contrast, vector-borne dispersal and the faecal–oral transmission route are used by many terrestrial parasites but are uncommon in marine environments. It is therefore appropriate to specifically consider the evolutionary biology of aquatic parasites, in terms of the unique selective pressures they face.

In this chapter, we first discuss some of the features of aquatic habitats relevant to parasite evolution and host–parasite coevolution. We then summarise what is known regarding both the micro- and macro-evolutionary patterns and processes characterising host–parasite interactions in aquatic habitats. Finally, we discuss the evolution of aquatic parasites under the new selective pressures exerted by the recent environmental changes associated with the late Anthropocene, and what this might mean for the future of aquatic parasite biology.

11.2 Aquatic Habitats: Theatre for Evolution

Aquatic environments differ from terrestrial ones in many ways. Over both short (hours, days) and long (months, years) timescales, temperature and other abiotic factors can be orders of magnitude less variable in many aquatic environments than on land (Steele et al. 2019). Nevertheless, aquatic habitats are far from homogeneous: they range from freshwater to marine, from small temporary ponds to vast oceans, from still lakes to fast-flowing rivers. The living conditions they present and the selective pressures they impose on organisms vary accordingly. For instance, organisms living in estuaries and intertidal mudflats can be exposed to huge thermal shifts over short periods of time, as well as substantial variation in salinity and high levels of ultraviolet radiation. In contrast, those in the deep-sea experience relatively constant conditions of cold temperatures, stable salinity and absence of light.

Some properties of water apply to all aquatic habitats and have no doubt played a key role in parasite evolution. For example, the viscosity and density of water and the buoyancy it provides (slightly higher in seawater than freshwater) have no doubt shaped the dispersal and transmission strategies of infective stages, such as tail propulsion in trematode cercariae (Koehler et al. 2012), as well as the attachment structures and drag-reducing body shapes of fish ectoparasites (Kearn 2004).

There are also differences between aquatic and terrestrial biota. Although many eukaryotic phyla are exclusively or almost exclusively found in freshwater or marine habitats, both currently known species richness and estimated total species richness are much higher in terrestrial ecosystems (Mora et al. 2011; Grosberg et al. 2012), mostly due to insects. Certain types of animals are also more common, if not almost entirely restricted, to either aquatic or terrestrial environments. For instance, large ruminant herbivores are an almost exclusive component of terrestrial habitats. However, it is the way in which the structure and complexity of animal communities and food webs differ between aquatic and terrestrial environments that can drive the evolutionary biology of parasites along different paths. In particular, food chains tend to be longer in aquatic habitats (Briand and Cohen 1987). As a consequence, many trophically transmitted aquatic parasites, such as acanthocephalans (Kennedy 2006) and nematodes (Anderson 2000), have added paratenic (or transport) hosts to their life cycle, to facilitate their transmission from small intermediate hosts to apex predators serving as definitive hosts. Such longer life cycles involving paratenic hosts are much less common in terrestrial habitats.

The main features of aquatic habitats that have received attention in the context of parasite evolution are those that can affect dispersal and gene flow. First, a clear contrast exists between the fragmented nature of freshwater habitats (lakes, ponds, rivers), all separated by inhospitable terrestrial environments, versus the open and interconnected nature of marine habitats (Fig. 11.1). Oceans are not fully open habitats, of course, as biogeographic barriers restrict the long-range dispersal of most marine organisms (e.g. Cowman and Bellwood 2013). Nevertheless, on any spatial scale, movement and exchanges of individual parasites among freshwater habitats should be much more challenging, leading to greater genetic structuring of freshwater parasite populations than marine ones, and more opportunities

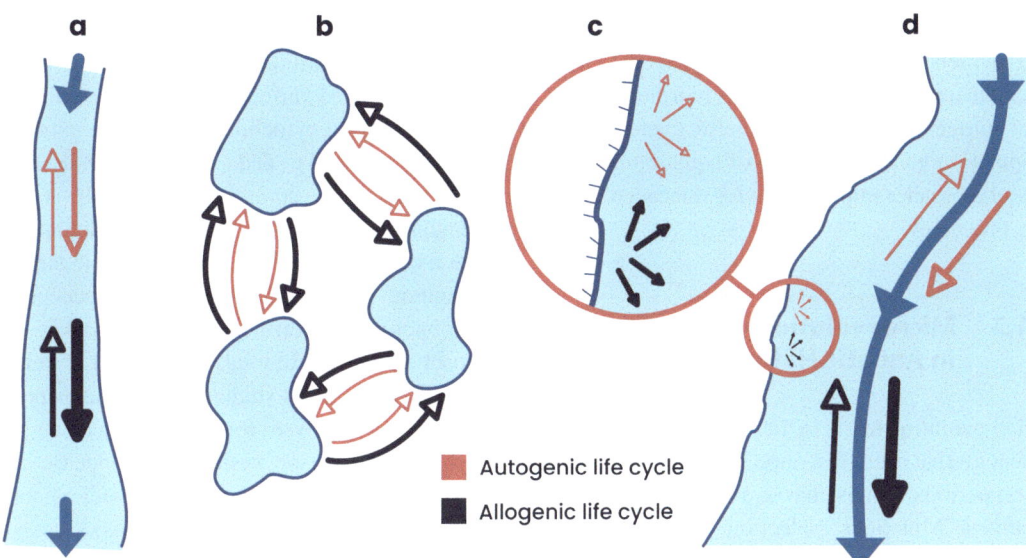

Fig. 11.1 Predicted direction (arrow) and strength (proportional to line thickness) of gene flow for aquatic parasites (**a**) along a river continuum, (**b**) among lakes on a continental landscape, (**c**) in coastal marine environments, and (**d**) on a larger oceanic continuum influenced by currents. The expected gene flow patterns are shown separately for autogenic parasites with aquatic definitive hosts, and allogenic parasites with bird definitive hosts. Blue arrows indicate water flow

for ecological differentiation and local adaptation. The physical isolation of freshwater habitats from each other can be overcome by the co-dispersal of parasites along with their hosts, if their life cycle includes a mobile host. Some freshwater parasites complete their entire life cycle within the same aquatic habitat (autogenic life cycle), whereas others have a definitive host that lives outside of water, such as waterfowl or highly vagile terrestrial mammals (allogenic life cycle), and that is in principle capable of releasing the parasite's eggs in a different habitat (Esch et al. 1988). The latter parasites can presumably achieve greater gene flow among isolated populations and should therefore have weaker genetic structuring across their geographical range than the former parasites (Criscione and Blouin 2004; Blasco-Costa and Poulin 2013).

Second, the unidirectional water flow of streams and rivers should favour, if not actually enforce, the predominantly one-way upstream-to-downstream dispersal of parasites, a situation that contrasts with the potential omnidirectional dispersal in lakes or still-water habitats (Fig. 11.1). The importance of water flow as a structuring force shaping population genetics of river organisms has been confirmed for free-living species with limited mobility (Davis et al. 2018). It has also been demonstrated for ectoparasitic copepods, which complete their entire life cycle in water and lack a free-swimming stage capable of fighting river currents (Prunier et al. 2021). In a comparative study of two trematode species sampled along the same river, the autogenic species with a fish definitive host exhibited moderate genetic structure and a decrease in genetic diversity in an upstream direction, whereas the allogenic species with avian definitive hosts showed neither of these patterns (Blasco-Costa et al. 2012). These findings are consistent with the relative dispersal potential of the parasites' definitive hosts, with the greater mobility of bird hosts counteracting the unidirectional passive dispersal of parasite eggs and larvae via the river current. Oceanic currents may similarly impose a dominant direction on parasite gene flow, but on a larger spatial scale.

In the next section, we synthesise the empirical evidence regarding the influence of fragmented freshwater versus open marine habitats, and autogenic versus allogenic life cycles, on the population genetic structure of aquatic parasites, and the broader implications for microevolutionary processes.

11.3 Microevolution of Parasites in Aquatic Habitats

Microevolution refers to the changes in allele frequencies that occur in a population over relatively short period of times, that is, a few to several generations. Mutations, selection, gene flow and genetic drift are the key drivers of microevolution. In turn, the main outcomes of microevolution on short ecological timescales are (1) genetic structuring or differentiation of populations, (2) local adaptation and (3) incipient or cryptic speciation. Here, we discuss these three phenomena in aquatic parasites.

To determine whether genetic structuring of parasite populations is more common in fragmented freshwater systems than in open marine ones, and whether the use of a mobile host (allogenic life cycle) makes genetic structuring less likely, we conducted a literature review focusing on helminth endoparasites with complex life cycles. We searched the Web of Science for relevant publications published between 2000 and February 2023, using the topic search string: ('population structure' OR 'genetic structure' OR phylogeograph* OR 'gene flow') AND (aquatic OR freshwater OR marine) AND parasit* AND (helminth* OR trematode* OR digene* OR cestod* OR tapeworm* OR acanthocephala* OR nematod*). The search returned 106 articles. These were examined individually to exclude studies that did not meet our criteria. We focused exclusively on studies of natural patterns, excluding studies on parasites of humans, or studies of wildlife parasites that have been introduced to new geographical areas by recent human activities. Also, in the case of studies on freshwater parasites, we considered only those comparing populations sampled from different lakes or rivers, and not those sampled at different locations within the same continuous water body. The studies recovered used various gene markers (mostly the mitochondrial cytochrome c oxidase subunit 1 gene, or CO1) and different analytical approaches (most commonly by analysis of molecular variance, or AMOVA, and calculation of the fixation index F_{st}) to detect genetic structure among parasite populations; we accepted their conclusion regardless of what marker or approach they used. In total, we retained 45 relevant analyses from 34 studies (some studies presented separate analyses for different species): 18 analyses from marine systems (13 species, as some were investigated in separate studies) and 27 from freshwater systems (22 species).

Overall, there is no obvious difference in the likelihood that parasite populations show significant genetic structuring between those living in fragmented freshwater systems and those living in open marine ones (Table 11.1). However, parasites with allogenic life cycles are much less likely to exhibit genetic structuring among populations than those with autogenic cycles, in both freshwater and marine ecosystems (Fig. 11.2). This confirms that concomitant dispersal of parasites by bird definitive hosts can overcome distance and maintain gene flow and genetic homogeneity among distinct parasite populations. We considered parasites using marine mammals as definitive hosts as autogenic parasites, since many cetaceans and pinnipeds remain within particular coastal habitats or near breeding colonies. Interestingly, among allogenic parasites only, studies in which significant genetic structuring among populations was observed were conducted on larger spatial scales than those where no structuring was observed (Fig. 11.3). This suggests that the dispersal potential of birds has limits: the most distant parasite populations are somewhat genetically isolated from each other in spite of the high mobility of their avian hosts. Nevertheless, the signature of allogenic life cycles can even be detected following trans-continental species introductions. In comparisons of trematode species with either fish or bird definitive hosts, all co-introduced from the east coast of North

Table 11.1 Summary of studies investigating genetic structure in aquatic helminth parasites, classified according to their environment (freshwater or marine) and the definitive host used. The spatial scale represents the maximum linear distance between sampling localities in the original study, estimated from maps. The existence of genetic structure (yes or no) among parasite populations is based on the conclusions of the authors of the original studies

Environment	Parasite species	Higher taxon	Definitive host	Spatial scale (km)	Genetic structure?	Reference
Freshwater	*Neoechinorhynchus emyditoides*	Acanthocephalan	Fish	940	Yes	Sereno-Uribe et al. (2022)
Freshwater	*Pomphorhynchus laevis*	Acanthocephalan	Fish	3100	Yes	Perrot-Minnot et al. (2018)
Freshwater	*Pomphorhynchus tereticollis*	Acanthocephalan	Fish	2700	No	Perrot-Minnot et al. (2018)
Freshwater	*Wenyonia virilis*	Cestode	Fish	1300	Yes	Jirsová et al. (2017)
Freshwater	*Rhabdochona lichtenfelsi*	Nematode	Fish	500	Yes	Mejía-Madrid et al. (2007)
Freshwater	*Procamallanus neocaballeroi*	Nematode	Fish	1230	Yes	Santacruz et al. (2020)
Freshwater	*Deropegus aspina* A	Trematode	Fish	750	Yes	Criscione and Blouin (2004)
Freshwater	*Deropegus aspina* B	Trematode	Fish	750	Yes	Criscione and Blouin (2004)
Freshwater	*Plagioporus shawi*	Trematode	Fish	750	Yes	Criscione and Blouin (2004)
Freshwater	*Plagioporus shawi*	Trematode	Fish	750	Yes	Criscione and Blouin (2007)
Freshwater	*Paralechriorchis syntomentera*	Trematode	Fish	900	Yes	Johnson et al. (2020)
Freshwater	*Stegodexamene anguillae*	Trematode	Fish	200	No	Herrmann et al. (2014)
Freshwater	*Coitocaecum parvum*	Trematode	Fish	180	Yes	Lagrue et al. (2016)
Freshwater	*Ligula intestinalis*	Cestode	Bird	240	No	Nazarizadeh et al. (2022)
Freshwater	*Ligula intestinalis*	Cestode	Bird	2500	No	Bouzid et al. (2008)
Freshwater	*Ligula intestinalis*	Cestode	Bird	10,200	Yes	Bouzid et al. (2008)
Freshwater	*Schistocephalus solidus*	Cestode	Bird	750	Yes	Sprehn et al. (2015)
Freshwater	*Schistocephalus solidus*	Cestode	Bird	700	Yes	Strobel et al. (2019)
Freshwater	*Cotylurus* sp.	Trematode	Bird	90	Yes	Keeney et al. (2023)
Freshwater	*Posthodiplostomum* sp.3	Trematode	Bird	260	No	Boone et al. (2018)
Freshwater	*Posthodiplostomum* sp.8	Trematode	Bird	260	No	Boone et al. (2018)
Freshwater	*Diplostomum pseudospathaceum*	Trematode	Bird	1660	No	Enabulele and Awharitoma (2018)

(continued)

Table 11.1 (continued)

Environment	Parasite species	Higher taxon	Definitive host	Spatial scale (km)	Genetic structure?	Reference
Freshwater	*Diplostomum pseudospathaceum*	Trematode	Bird	300	No	Louhi et al. (2010)
Freshwater	*Diplostomum sp.6*	Trematode	Bird	45	No	Rahn et al. (2016)
Freshwater	*Nanophyetus salmincola*	Trematode	Bird	750	No	Criscione and Blouin (2004)
Freshwater	*Atriophallophorus winterbourni*	Trematode	Bird	450	Yes	Feijen et al. (2022)
Freshwater	*Ribeiroia ondatrae*	Trematode	Bird	900	No	Johnson et al. (2020)
Marine	*Bucephalus minimus*	Trematode	Fish	2350	Yes	Feis et al. (2015)
Marine	*Profilicollis altmani*	Acanthocephalan	Bird	8700	No	Goulding and Cohen (2014)
Marine	*Profilicollis novaezelandensis*	Acanthocephalan	Bird	460	No	Hay et al. (2018)
Marine	*Renicola parvicaudatus*	Trematode	Bird	8400	No	Galaktionov et al. (2023)
Marine	*Gymnophallus choledochus*	Trematode	Bird	2850	No	Feis et al. (2015)
Marine	*Microphallus pygmaeus*	Trematode	Bird	6800	No	Galaktionov et al. (2012)
Marine	*Himasthla leptosoma*	Trematode	Bird	2700	No	Galaktionov et al. (2021)
Marine	*Tristriara anatis*	Trematode	Bird	6870	Yes	Gonchar and Galaktionov (2017a)
Marine	*Tristriara anatis*	Trematode	Bird	8200	Yes	Gonchar and Galaktionov (2017b)
Marine	*Maritrema novaezealandensis*	Trematode	Bird	8	No	Keeney et al. (2008)
Marine	*Maritrema novaezealandensis*	Trematode	Bird	470	No	Keeney et al. (2009)
Marine	*Philophthalmus attenuatus*	Trematode	Bird	470	No	Keeney et al. (2009)
Marine	*Anisakis simplex*	Nematode	Mammal	3340	Yes	Cipriani et al. (2022)
Marine	*Anisakis simplex*	Nematode	Mammal	2320	No	Ramilo et al. (2023)
Marine	*Anisakis simplex*	Nematode	Mammal	1500	Yes	Mattiucci et al. (2018)
Marine	*Anisakis pegreffii*	Nematode	Mammal	2200	Yes	Cipriani et al. (2022)
Marine	*Anisakis pegreffii*	Nematode	Mammal	1900	Yes	Ding et al. (2022)
Marine	*Anisakis physeteris*	Nematode	Mammal	1200	Yes	Cipriani et al. (2022)

America to the west coast along with their common snail intermediate host, founder effects were stronger and genetic diversity lower in species using fish than those using birds as definitive hosts even a full century after introduction (Blakeslee et al. 2020). The main conclusion from existing studies (Table 11.1, Figs. 11.2 and 11.3) is that on scales of a few thousand kilometres, bird definitive hosts can usually maintain sufficient gene flow between helminth popula-

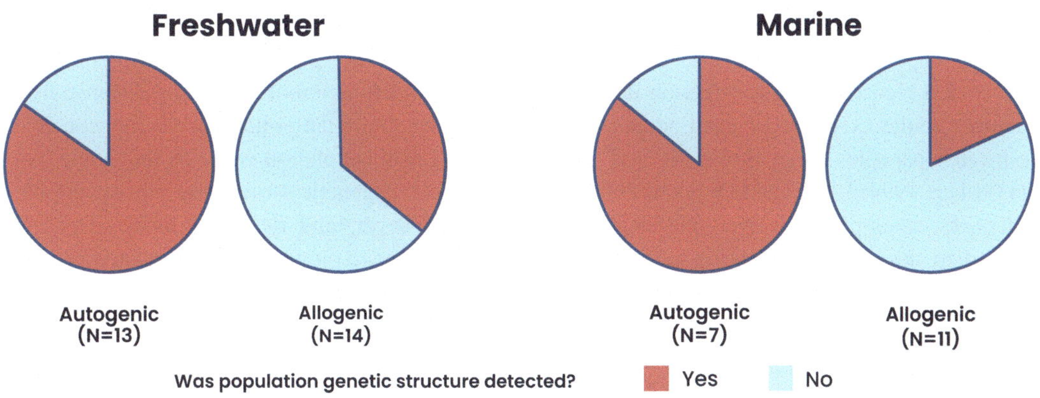

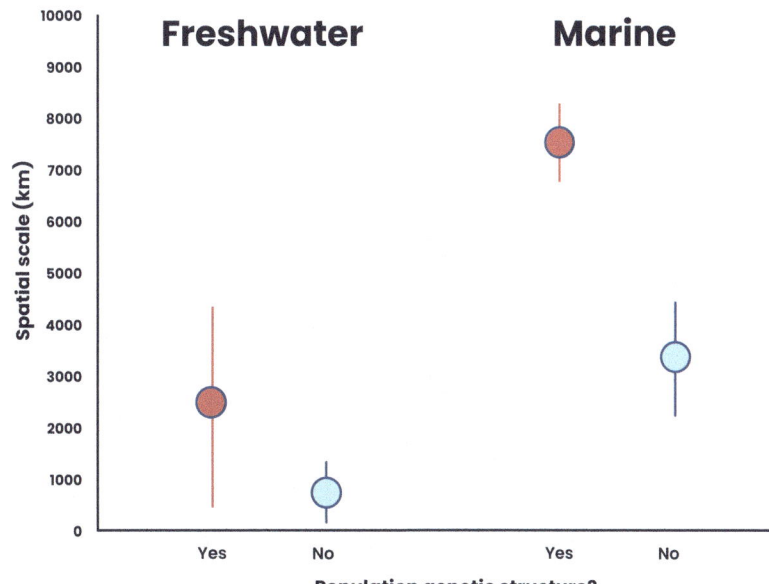

Fig. 11.2 Frequency at which genetic structuring was observed among conspecific populations of helminth parasites with either autogenic life cycles (water-bound definitive host) or allogenic life cycles (bird definitive host), in both freshwater and marine ecosystems. See Table 11.1 for list of studies included

Fig. 11.3 Mean (± standard error) spatial scale of studies on helminth parasites in freshwater and marine ecosystems, shown separately based on whether or not genetic structuring was observed among populations. Only studies on parasites with allogenic life cycles (bird definitive host) are included. Spatial scale is defined as the maximum linear distance between two sampled populations. See Table 11.1 for list of studies included

tions and thereby prevent significant genetic structuring.

This has important microevolutionary implications. In principle, the absence of genetic structuring among populations of the same parasite species indicates high levels of gene flow leading to a homogeneous genetic composition, whereas strong genetic structuring suggests the existence of dispersal barriers and limited gene flow. Although not necessarily the case (see Tigano and Friesen 2016), genetic structuring with little gene flow can favour local adaptation, defined as different populations of the same species possessing traits that confer higher fitness in their local environment than elsewhere in the species range (Blanquart et al. 2013). Local adaptation results from local selective pressures matching genetic variation with environmental variation. For parasites, individuals of the same species may experience different suites of potential host species, or different genetic variants of the same host species, as well as different environmental conditions in each locality they occupy. If these local selective pressures outweigh gene flow and genetic drift, each local population should evolve distinct adaptations. Several classical studies have used a reciprocal cross-infection design to demonstrate the existence of local adaptation in freshwater parasites (e.g. Ballabeni and Ward 1993; Ebert 1994; Lively and Dybdahl 2000). In such cases, when parasites from one locality are exposed to hosts from either the same locality (sympatric hosts) or from a different locality (allopatric hosts), the parasites often perform better on local hosts in terms of infection success, within-host replication, or other fitness components (see Fig. 11.4 for examples).

The roles of autogenic versus allogenic life cycles, limited gene flow and genetic structuring in driving local adaptation are beautifully illustrated in a comparison between two trematode species infecting the same frog populations along the Pacific coast of the United States (Johnson et al. 2020). For *Paralechriorchis syntomentera*, which uses snakes as definitive hosts, cross-infections of tadpoles in sympatric-allopatric combinations revealed that the more geographically distant the allopatric host population used, the lower the infection success of the parasite. In contrast, for *Ribeiroia ondatrae*, which uses bird definitive hosts with greater dispersal potential, there was no such pattern of decreasing infection success as a function of increasing distance from the source of allopatric hosts used (Johnson et al. 2020). Local adaptation is not an inevitable consequence of genetic structuring among populations, however, and detecting local adaptation can also depend on which fitness-related parasite trait is examined (Lagrue et al. 2016). In addition, hosts themselves may also be locally adapted, showing inter-population differences in their ability to resist infection that reflect variation in parasite pressure among localities (Bryan-Walker et al. 2007). Still, genetic differentiation among parasite populations determined by host-mediated dispersal is probably a key factor shaping microevolution of aquatic parasites.

Strong genetic structuring combined with local adaptation may also promote incipient speciation, that is, the early steps towards divergence of different populations into distinct species. Indeed, limited gene flow and specialisation on the most locally common host genotypes or host species are bound to create increasing genetic variation among parasite populations, eventually resulting in reproductive isolation. There are

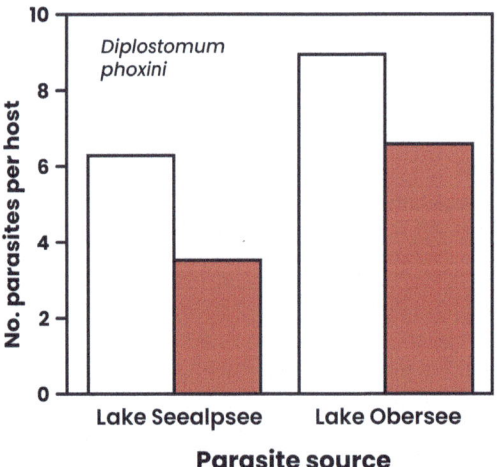

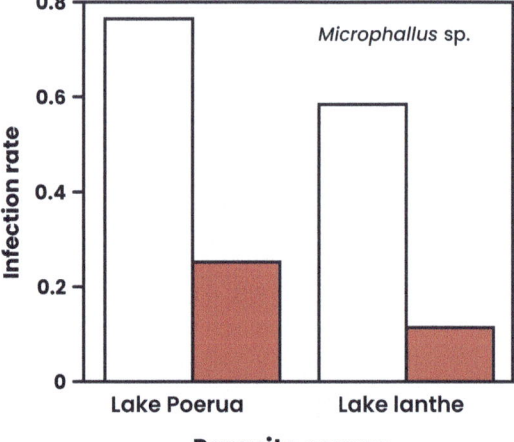

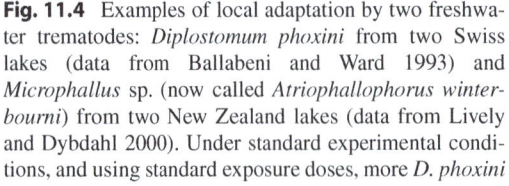

Fig. 11.4 Examples of local adaptation by two freshwater trematodes: *Diplostomum phoxini* from two Swiss lakes (data from Ballabeni and Ward 1993) and *Microphallus* sp. (now called *Atriophallophorus winterbourni*) from two New Zealand lakes (data from Lively and Dybdahl 2000). Under standard experimental conditions, and using standard exposure doses, more *D. phoxini* cercariae successfully establish in fish second intermediate hosts from their lake of origin (sympatric hosts; open bars) than in hosts from another lake (allopatric hosts; black bars). Similarly, a greater proportion of sympatric snail first intermediate hosts (open bars) become infected by miracidia of *Microphallus* than allopatric snails (black bars)

increasing numbers of studies reporting cryptic parasite species, in other words, parasite species that are genetically distinct but morphologically indistinguishable from each other (Nadler and Pérez-Ponce de León 2011). Cryptic species appear to be particularly common among aquatic trematodes, possibly because of the lack of hard body structures and substantial host-induced morphological plasticity in these helminths making it challenging to distinguish species without genetic data (Pérez-Ponce de León and Poulin 2018). Cryptic speciation is a direct consequence of either (or both) genetic structuring among geographically distinct populations of the same parasite species, or local adaptation of parasites to the most locally common host species. It provides the starting point for macroevolutionary processes, which we discuss in the following section.

11.4 Host–Parasite Macroevolution in Aquatic Habitats

As argued above, the fragmented nature of freshwater ecosystems should in principle lead to greater physical isolation of conspecific populations, as well as to greater environmental heterogeneity among localities, than in open marine ecosystems. As a consequence, limited gene flow and variable local selection pressures may result in higher speciation and net diversification rates in freshwater than in marine systems. Indeed, despite marine environments covering a much higher proportion of the Earth's surface than does freshwater, the two environments have very similar overall species richness (Wiens 2015), suggesting that high speciation rates might compensate for the small overall dimensions of freshwater habitats. This may also apply to parasites: the average number of species per genus is generally higher for freshwater taxa of metazoan parasites of fish than for marine ones (Poulin 2016), hinting at higher diversification rates in freshwater habitats. The processes shaping parasite diversification are tightly linked to fidelity versus switches in host use over evolutionary time, and the associated patterns of host specificity. Let's look at these in turn.

The dependence of parasites on their hosts constrains their evolution in many ways. Across generations, host–parasite coevolution can proceed through a gene-for-gene process, where changes in host allele frequencies are matched by changes in parasite alleles, as the parasites gradually adapt to evolving host physiology and immune defences (Sasaki 2000). This can strengthen the dependence of the parasite on a particular host species at a given stage of its life cycle and prevent its use of other host species. Therefore, over longer evolutionary timescales, the diversification of a parasite lineage might occur mostly as a result of host diversification: each time a host species undergoes a speciation event, so do its closely associated parasites. In this case, repeated host–parasite cospeciation would result in a number of parasite species associated by descent to a matching number of host species, with the phylogenetic tree of a parasite clade being a mirror image of that of their hosts (Page 2003; Clayton et al. 2016). This idealised scenario has been known to parasitologists as Fahrenholz's rule, and has long provided a null hypothesis against which to test alternative ideas regarding host–parasite coevolution (Klassen 1992). Empirical evidence from a growing number of host–parasite co-phylogenetic studies has revealed that perfect congruence between host and parasite phylogenetic trees is never observed (Hayward et al. 2021). Instead, host–parasite co-phylogenetic analyses typically uncover a mixture of past cospeciation events leading to similarities between the trees, and other processes causing differences in the branching patterns of the two trees. For example, a co-phylogenetic analysis between *Lamellodiscus* monogeneans and their sparid fish hosts shows multiple mismatches between the two phylogenies against a general background of parallel evolution through cospeciation events (Fig. 11.5). Several evolutionary events can disconnect the evolution of parasites from the phylogenetic history of their ancestral host (Page 2003; Clayton et al. 2016). Gene flow can be maintained between parasite populations exploiting recently

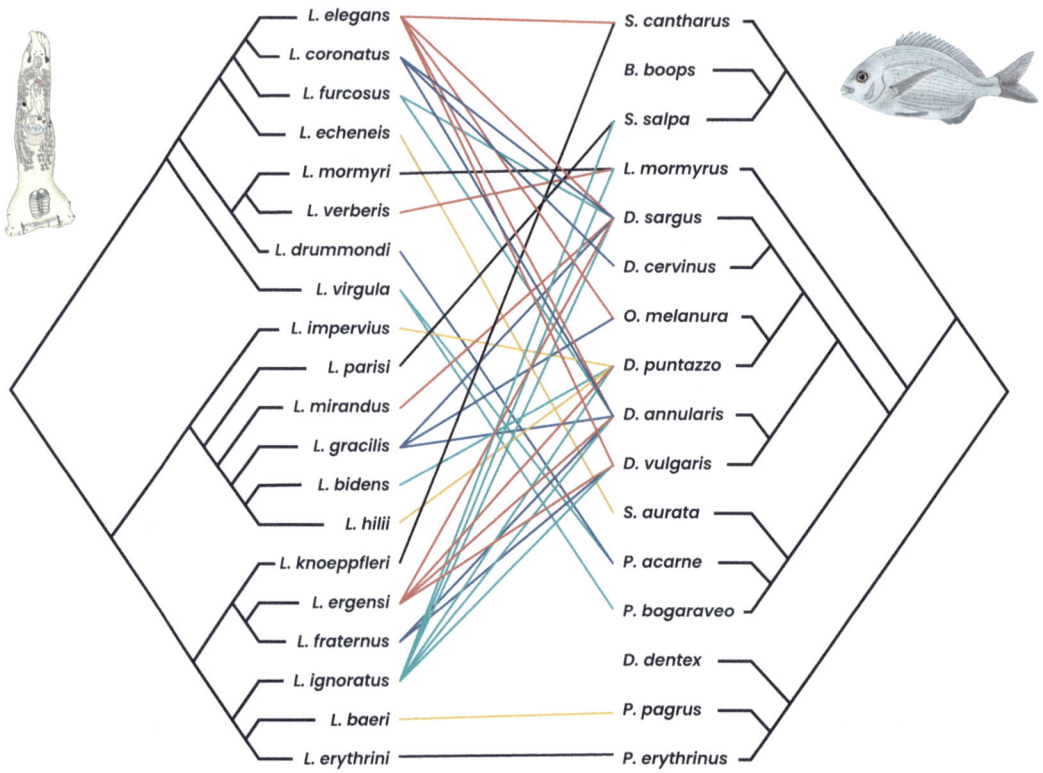

Fig. 11.5 Tanglegram depicting the phylogenetic trees of *Lamellodiscus* monogeneans and their sparid fish hosts, and their host–parasite associations (indicated by lines connecting host and parasite species), based on Desdevises et al. (2002). The maximum-likelihood trees were derived from genetic data

diverged host species, resulting in a parasite occurring on more than one host species. Alternatively, parasites can speciate independently of their hosts, or they can go extinct, also leading to mismatches between host–parasite phylogenetic patterns. Parasites can also host-switch, that is, they can colonise new host species, adding them to the repertoire of host species they can exploit with or without ceasing to use their original host. This includes the process of replacing one host by another one that may not be phylogenetically related but provides equivalent and compatible resources to the parasite, a process referred to as ecological fitting (Araujo et al. 2015). Despite these evolutionary processes that may disconnect the coevolutionary trajectory of host–parasite interactions, a recent meta-analysis (Hayward et al. 2021) of 212 co-phylogenetic studies, including many that involve aquatic parasites, found that some degree of congruence between host and parasite phylogenies is almost always detectable, regardless of the type of parasite or their mode of transmission.

One of the main outcomes of these evolutionary processes is the substantial variation in host specificity that is observed among parasite species today. On the one hand, the signature of tight cospeciation is visible in the tendency for most parasite species to be quite host-specific (Fig. 11.6). On the other hand, the fact that several species are generalists supports a role for host-switching or other processes that broaden the range of hosts a parasite can use. Some small but consistent differences have been observed in the extent of host specificity shown by parasites belonging to different higher taxa. For instance, monogeneans often appear a little more host-specific than other helminths, with more of them being restricted to only one host species (Poulin 1992). This pattern is not always apparent (see

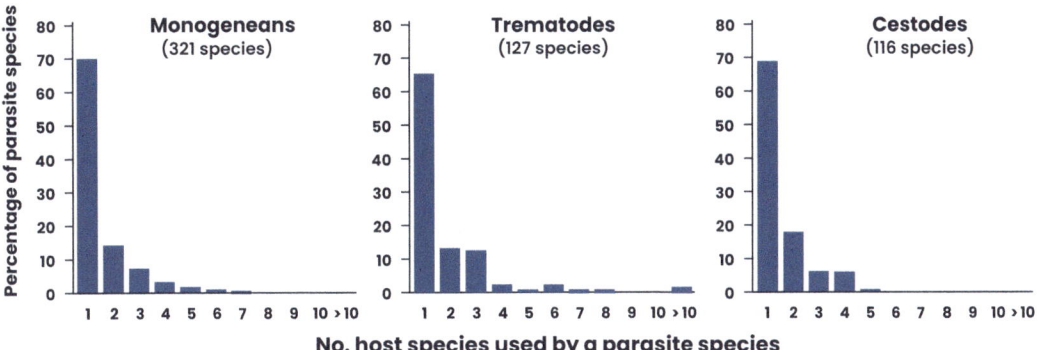

Fig. 11.6 Frequency distribution of host specificity among helminth species from three higher taxa parasitic in freshwater fishes in Brazil (data from Eiras et al. 2010). Most parasite species are found in only one host species. Note: data for trematodes include only species using fish as definitive hosts

Fig. 11.6) and may be related to differences among helminth taxa in the complexity of the life cycle, the site of attachment or other taxon-specific properties. However, there are no obvious differences in host specificity between parasites in freshwater versus marine environments (Sasal et al. 1998). Within given parasite species, there may be variation in host specificity throughout the life cycle. For instance, trematodes tend to be highly specific for their first intermediate host, typically being capable of infecting only one species of snail, but are much less choosy for their second intermediate hosts and definitive hosts (Cribb 2005). All large-scale patterns of host specificity must be interpreted with caution, however, because assessments of host specificity have too often been based entirely on the morphological identification of parasites found in different host species. When molecular markers are applied, in many cases what was thought to be a single species of generalist parasite exploiting multiple host species turns out to consist of a few similar but genetically distinct specialist species each using a single host species (Poulin and Keeney 2008). Additionally, rarely are all potential host taxa investigated in any given system to provide a representative overview of parasite host specificity. Therefore, host specificity may have been underestimated in several cases.

Host specificity should also be considered as more than merely the number of host species used by a parasite. If a parasite uses several host species but achieves high abundance on one host species and very low abundance on the others, it could still be seen as specialised for its main host species. For example, several species of helminths and copepods infecting sympatric flatfish species along the Portuguese coast achieve hugely different abundances on their various host species (Marques et al. 2011). From an ecological point of view, the host species supporting the highest total abundance of parasites is the main driver of parasite population dynamics. Coevolutionary pressures in such a case would probably favour more strongly the parasite adaptations that improve a parasite's infectivity and compatibility with the most frequently used host species. Similarly, from a phylogenetic perspective, a parasite that exploits four congeneric host species can be said to be more specialised than a parasite that exploits four host species belonging to different genera, even if both parasites use the same number of host species (Poulin et al. 2011). Host-switching may be more likely to occur between phylogenetically closely related host species, since the adaptations that a parasite possesses to exploit its original host should predispose it to also successfully exploit related host species that present similar physiology and immune defences (Poulin 2005). The signature of past coevolutionary events, from ancient cospeciation to more recent host-switching, is therefore visible in how the current range of host species available in any aquatic ecosystem is exploited by extant species of parasites.

11.5 Aquatic Parasite Evolution Under Anthropogenic Changes

The micro- and macroevolutionary processes discussed in previous sections have occurred independently of human activities, and the same patterns would have arisen even if people had never been around. However, human activities have indirectly exerted a different set of evolutionary pressures on aquatic parasites, which are currently driving the evolution of several facets of parasite biology. The Anthropocene is presently characterised by extensive habitat change, species translocated to new areas where they become invasive, accelerated climate change, pollution and overexploitation of natural resources, among other human impacts. The net results for wildlife are that many species are going extinct, others are in decline, and the geographical range of many species is changing (see Poloczanska et al. 2013; Díaz et al. 2019; Albert et al. 2021, see also Chap. 18). Human activities are also imposing new selective pressures on wild and domesticated species, which are likely to evolve in response to changing conditions, most likely very rapidly given the strength of some of these new pressures (Hendry et al. 2017). In turn, parasites will experience strong selection to track changes in their hosts and coevolve in parallel. They will also be directly exposed to human impacts such as pollution and rising temperatures. Given the short generation times and high fecundity of many parasites, evolutionary responses of parasites to human-driven selective pressures can be rapid. Little is currently known regarding the direct and indirect impacts of anthropogenic changes on the evolution of aquatic parasites, either at the genomic or phenotypic level. Here, we provide three hypothetical scenarios for the possible evolution of parasites in response to different types of human impacts, as examples of what is possibly happening right now.

First, climate change is widely seen as a major factor affecting the abundance, distribution and severity of aquatic parasites and diseases (Marcogliese 2001; Rohr and Cohen 2020; Byers 2021). However, climate change in general, and global warming in particular, can also select for different optimal trait values, especially those relating to phenology or physiology (Hendry et al. 2017), in both free-living and parasitic species. As a general example, consider trematodes. These parasites are notoriously sensitive to their thermal environment. Small changes in temperature can have profound effects on the infectivity of miracidia to the snail first intermediate host (Morley and Lewis 2015), the rate at which cercariae are produced asexually in the snail host (Poulin 2006), as well as the survival and infectivity of cercariae to the second intermediate host (Morley 2011; Morley and Lewis 2015). The sensitivity of trematodes to temperature shows substantial intraspecific variation, however, with some genotypes performing much better at warmer temperatures and others performing better at cooler temperatures (Berkhout et al. 2014). This standing genetic variation in trematode populations provides the raw material for natural selection. We might therefore expect genotypes that do better in slightly warmer waters to gradually increase in frequency and replace those that perform better in cooler conditions as global warming progresses. The net effects of such microevolutionary changes may be that the trematodes' thermal performance curves for a range of fitness-related traits will shift towards a higher temperature range (Fig. 11.7). Thus, trematodes in a warmer future might not achieve higher transmission and infection rates than today's trematodes, but instead simply adjust to the new conditions. Of course, this prediction ignores host adaptation to global warming, as well as the potential adaptation of trematodes to other environmental changes occurring in parallel, such as ocean acidification (see Harland et al. 2016). However, it highlights the possibility of rapid evolution and adjustment to changing conditions by parasites rather than major climate-driven changes in epidemiological dynamics and impacts on hosts.

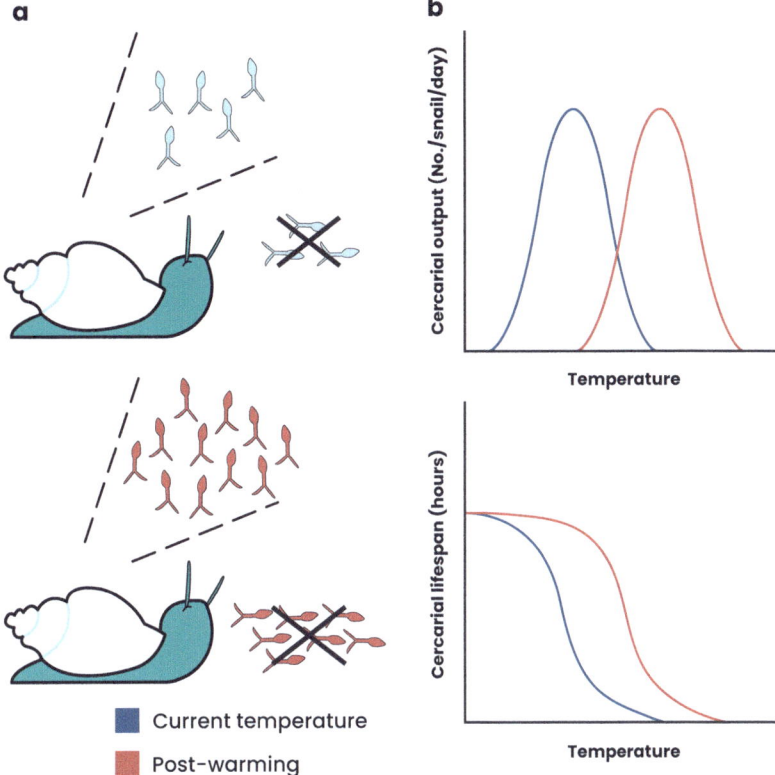

Fig. 11.7 Hypothetical evolutionary trends in a trematode parasite under a global warming scenario. (**a**) At higher temperature, the typical trematode genotype in today's world produces more cercariae which exhaust their stored energy faster and have shorter lifespan than at normal temperatures. (**b**) The average population-level performance versus temperature curves for two trematode traits related to fitness, that is, cercarial production rate within snail hosts and cercarial lifespan, in today's populations of aquatic trematodes and in a warmer future

Second, overexploitation of fish populations and ineffective fisheries management are anthropogenic drivers likely influencing parasite evolution. Combined, these have caused large declines in marine fish populations over the past century (Murawski 2010; Hilborn et al. 2020). Fisheries represents an important source of fish mortality that can exert strong selective pressures and drive the evolution of fish populations, for example through the biased harvesting of particular size classes (Allendorf and Hard 2009). From the perspective of parasites, reduced host densities and changes in host size distributions have epidemiological consequences (Wood et al. 2010), but they also pose new challenges that can lead to selection for new optimal trait values. Consider a population of ectoparasitic copepods infecting an intensively harvested fish population whose density is rapidly dwindling. A central tenet of life history theory (Stearns 1992) is that resources allocated to one function cannot simultaneously be used for a different one. Thus, for a given investment into reproduction, life history theory predicts the existence of a trade-off between the number of offspring produced and the size of individual offspring. Across different species of parasitic copepods, there is indeed a negative correlation between relative egg size and the number of eggs per clutch (Poulin 1995), with some evidence that the trade-off also applies within species (Timi et al. 2005). When transmission success is high, that is, at high host density when infective stages have a good chance of encountering a suitable host, genotypes that produce large eggs may be favoured: producing offspring well-provisioned with energy can give them a head start once they attach to a host. However, at low host density when the probability of successfully finding a host is greatly reduced, we might expect that investing in the production of many offspring becomes the strategy favoured by selection, even if it comes at the expense of offspring size (Fig. 11.8). In those conditions, the latter strategy should improve the fitness of female copepods: a greater number of offspring increases the probability that at least a few will find a host and

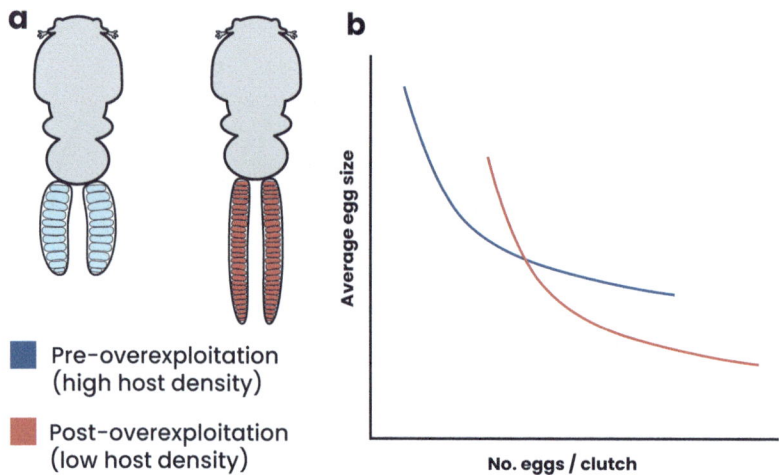

Fig. 11.8 Hypothetical evolutionary trends in a copepod ectoparasitic on fish as overexploitation causes a large decline in host population density. (**a**) For a given body size, female copepods (here, with uniseriate eggs sacs) may be expected to produce more and smaller eggs after the overexploitation of host stocks. (**b**) The trade-off curve between egg numbers and egg sizes among individual females may be expected to shift towards the many-small-eggs end of the spectrum

survive. Therefore, we may expect the size-versus-number of eggs in copepods parasitic on overexploited fish stocks to gradually shift towards the many-small-eggs end of the continuum. This evolutionary response to overfishing may lead to longer developmental periods for copepods following attachment on their host, but it would maintain transmission success.

Third, aquaculture conditions may drive the evolution of parasites. We know that rapid parasite evolution can take place in aquaculture facilities, as evidenced by the rapid evolution of antibiotic and anthelmintic resistance in parasites of fish and shellfish exposed to chemotherapy (Treasurer et al. 2000; Miranda et al. 2013). The typical aquaculture situation involves hosts of a single species, often genetically much more homogeneous than those in natural populations, maintained at high stocking densities and with regular introductions of immunologically naïve hosts. These conditions can have huge impacts on the probability of transmission for any parasite invading an aquaculture, with serious epidemiological and evolutionary consequences (Nowak 2007). For instance, high virulence is generally associated with high rates of pathogen replication and transmission, but also with high host mortality; because of this trade-off, optimal virulence levels are rarely very high in nature. Only an abundance of hosts within easy reach of infective stages can lead to highly virulent genotypes achieving high overall fitness. Aquaculture facilities provide such conditions: the facilitated host-to-host transmission experienced within aquaculture relaxes the constraints normally preventing the evolution of increased virulence (Pulkkinen et al. 2010; Kennedy et al. 2016). We might therefore expect more virulent strains of parasites to spread within aquaculture populations and spill over to adjacent natural populations (Kennedy et al. 2016).

Other parasite life history traits may also experience selection under aquaculture conditions. For example, fast growth and early transmission may also be favoured under conditions of high host density (Mennerat et al. 2010). There is empirical evidence that this has indeed happened to sea lice, *Lepeophtheirus salmonis*, an ectoparasitic copepod of Atlantic salmon, *Salmo salar*. When lice from either salmon farms or from wild fish caught in locations at least 200 km from the nearest salmon farm were allowed to infect hosts from the same cohort, lice originating from salmon farms produced more eggs in their first clutch, fewer eggs in later clutches, and died earlier than lice from the wild (Mennerat et al. 2017). These findings suggest that aquaculture conditions have selected for greater investment into early reproduction at the expense of future fecundity and survival. Sea lice in salmon farms have also evolved towards greater virulence, as measured by the extent of skin lesions and reduced growth incurred by fish hosts (Ugelvik et al. 2017). Therefore, it seems that a whole suite of traits are rapidly evolving in new directions, in this parasite and possibly many others, under aquaculture conditions. The accelerated evolution of parasites exposed to ideal transmission conditions will need to be monitored and mitigated (see Kennedy et al. 2016) in order for aquaculture to sustain the high yield needed to feed a growing human population.

11.6 Concluding Remarks

Aquatic parasites span multiple phyla, each with its own evolutionary history. They display a wide variety of life cycles and transmission pathways, and huge differences in morphology and general biology. In spite of these apparent disparities, they have been exposed to similar evolutionary forces. On a microevolutionary scale, the role of host dispersal is crucial in maintaining gene flow among distant parasite populations, especially those from freshwater bodies. In freshwater parasites without a mobile definitive host, confinement to a single water body for completion of the life cycle results in significant genetic structure among populations, favours local adaptation, and possibly as well the emergence of cryptic species. At a macroevolutionary scale, a tendency towards host fidelity and close coevolution with their hosts, in spite of frequent host-switching events, has resulted in very similar patterns of host specificity among different higher taxa of parasites. Undoubtedly, the most pressing questions and wide knowledge gaps regarding the evolution of aquatic parasites concern their responses to anthropogenic pressures. The scenarios presented above are only meant as examples of what might happen; they may never ensue and instead parasites under human-mediated pressures may follow other evolutionary trajectories that we cannot foresee. The increasing adoption of -omics tools to study parasites (Selbach et al. 2019) will no doubt provide powerful approaches to track rapid evolutionary changes in aquatic parasite populations of economic or medical importance, and better prepare us to deal with emerging disease problems resulting from our own environmental footprint.

Acknowledgements Jerusha Bennett was supported through the Cawthron Institute's Ministry of Business, Innovation and Employment, *Emerging Aquatic Diseases* Endeavour grant, award number CAWX2207.

References

Albert JS, Destouni G, Duke-Sylvester SM, Magurran AE, Oberdorff T, Reis RE, Winemiller KO, Ripple WJ (2021) Scientists' warning to humanity on the freshwater biodiversity crisis. Ambio 50:85–94

Allendorf FW, Hard JJ (2009) Human-induced evolution caused by unnatural selection through harvest of wild animals. Proc Natl Acad Sci USA 106:9987–9994

Anderson RC (2000) Nematode parasites of vertebrates: their development and transmission, 2nd edn. CABI Publishing, Wallingford

Araujo SBL, Braga MP, Brooks DR, Agosta SJ, Hoberg EP, von Hartenthal FW, Boeger WA (2015) Understanding host-switching by ecological fitting. PLoS One 10:e0139225

Ballabeni P, Ward PI (1993) Local adaptation of the trematode *Diplostomum phoxini* to the European minnow *Phoxinus phoxinus*, its second intermediate host. Funct Ecol 7:84–90

Berkhout BW, Lloyd MM, Poulin R, Studer A (2014) Variation among genotypes in responses to increasing temperature in a marine parasite: evolutionary potential in the face of global warming? Int J Parasitol 44:1019–1027

Blakeslee AMH, Haram LE, Altman I, Kennedy K, Ruiz GM, Miller AW (2020) Founder effects and species introductions: a host versus parasite perspective. Evol Appl 13:559–574

Blanquart F, Kaltz O, Nuismer SL, Gandon S (2013) A practical guide to measuring local adaptation. Ecol Lett 16:1195–1205

Blasco-Costa I, Poulin R (2013) Host traits explain the genetic structure of parasites: a meta-analysis. Parasitology 140:1316–1322

Blasco-Costa I, Waters JM, Poulin R (2012) Swimming against the current: genetic structure, host mobility and the drift paradox in trematode parasites. Mol Ecol 21:207–217

Boone EC, Laursen JR, Colombo RE, Meiners SJ, Romani MF, Keeney DB (2018) Infection patterns and molecular data reveal host and tissue specificity of *Posthodiplostomum* species in centrarchid hosts. Parasitology 145:1458–1468

Bouzid W, Štefka J, Hypša V, Lek S, Scholz T, Legal L, Ben Hassine OK, Loot G (2008) Geography and host specificity: two forces behind the genetic structure of the freshwater fish parasite *Ligula intestinalis* (Cestoda: Diphyllobothriidae). Int J Parasitol 38:1465–1479

Briand F, Cohen JE (1987) Environmental correlates of food chain length. Science 238:956–960

Bryan-Walker K, Leung TLF, Poulin R (2007) Local adaptation of immunity against a trematode parasite in marine amphipod populations. Mar Biol 152:687–695

Byers JE (2021) Marine parasites and disease in the era of global climate change. Annu Rev Mar Sci 13:397–420

Cipriani P, Palomba M, Giulietti L, Marcer F, Mazzariol S, Santoro M, Alburqueque RA, Covelo P, López A, Santos MB, Pierce GJ, Brownlow A, Davison NJ, McGovern B, Frantzis A, Alexiadou P, Højgaard DP, Mikkelsen B, Paoletti M, Nascetti G, Levsen A, Mattiucci S (2022) Distribution and genetic diversity of *Anisakis* spp. in cetaceans from the Northeast Atlantic Ocean and the Mediterranean Sea. Sci Rep 12:13664

Clayton DH, Bush SE, Johnson KP (2016) Coevolution of life on hosts: integrating ecology and history. University of Chicago Press, Chicago

Cowman PF, Bellwood DR (2013) Vicariance across major marine biogeographic barriers: temporal concordance and the relative intensity of hard versus soft barriers. Proc R Soc B 280:20131541

Cribb TH (2005) Digenea (endoparasitic flukes). In: Rohde K (ed) Marine parasitology. CABI Publishing, Wallingford, pp 76–87

Cribb TH, Bray RA, Olson PD, Littlewood DTJ (2003) Life cycle evolution in the Digenea: a new perspective from phylogeny. Adv Parasitol 54:197–254

Criscione CD, Blouin MS (2004) Life cycles shape parasite evolution: comparative population genetics of salmon trematodes. Evolution 58:198–202

Criscione CD, Blouin MS (2007) Parasite phylogeographical congruence with salmon host evolutionarily significant units: implications for salmon conservation. Mol Ecol 16:993–1005

Davis CD, Epps CW, Flitcroft RL, Banks MA (2018) Refining and defining riverscape genetics: how rivers influence population genetic structure. WIREs Water 5:e1269

Desdevises Y, Morand S, Jousson O, Legendre P (2002) Coevolution between *Lamellodiscus* (Monogenea: Diplectanidae) and Sparidae (Teleostei): the study of a complex host-parasite system. Evolution 56:2459–2471

Díaz S, Settele J, Brondízio ES, Hien Ngo T, Agard J, Arneth A, Balvanera P, Brauman KA, Butchart SHM, Chan KMA, Garibaldi LA, Ichii K, Liu J, Subramanian SM, Midgley GF, Miloslavich P, Molnár Z, Obura D, Pfaff A, Polasky S, Purvis A, Razzaque J, Reyers B, Chowdhury RR, Shin Y-J, Visseren-Hamakers I, Willis KJ, Zayas CN (2019) Pervasive human-driven decline of life on earth points to the need for transformative change. Science 366:eaax3100

Ding F, Gu S, Yi M-R, Yan Y-R, Wang W-K, Tung K-C (2022) Demographic history and population genetic structure of *Anisakis pegreffii* in the cutlassfish *Trichiurus japonicus* along the coast of mainland China and Taiwan. Parasitol Res 121:2803–2816

Ebert D (1994) Virulence and local adaptation of a horizontally transmitted parasite. Science 265:1084–1086

Eiras JC, Takemoto RM, Pavanelli GC (2010) Diversidade dos parasitas de peixes de água doce do Brasil. Clichetec, Maringá

Enabulele EE, Awharitoma AO (2018) First molecular identification of an agent of diplostomiasis, *Diplostomum pseudospathaceum* (Niewiadomska 1984) in the United Kingdom and its genetic relationship with populations in Europe. Acta Parasitol 63:444–453

Esch GW, Kennedy CR, Bush AO, Aho JM (1988) Patterns in helminth communities in freshwater fish in Great Britain: alternative strategies for colonization. Parasitology 96:519–532

Feijen F, Zajac N, Vorburger C, Blasco-Costa I, Jokela J (2022) Phylogeography and cryptic species structure of a locally adapted parasite in New Zealand. Mol Ecol 31:4112–4126

Feis ME, Thieltges DW, Olsen JL, de Montaudouin X, Jensen KT, Bazaïri H, Culloty SC, Luttikhuizen PC (2015) The most vagile host as the main determinant of population connectivity in marine macroparasites. Mar Ecol Progr Ser 520:85–99

Galaktionov KV, Blasco-Costa I, Olson PD (2012) Life cycles, molecular phylogeny and historical biogeography of the '*pygmaeus*' microphallids (Digenea: Microphallidae): widespread parasites of marine and coastal birds in the Holarctic. Parasitology 139:1346–1360

Galaktionov KV, Solovyeva AI, Miroliubov A (2021) Elucidation of *Himasthla leptosoma* (Creplin, 1829) Dietz, 1909 (Digenea, Himasthlidae) life cycle with insights into species composition of the North Atlantic *Himasthla* associated with periwinkles *Littorina* spp. Parasitol Res 120:1649–1668

Galaktionov KV, Solovyeva AI, Blakeslee AMH, Skirnisson K (2023) Overview of renicolid digeneans (Digenea, Renicolidae) from marine gulls of northern Holarctic with remarks on their species statuses, phylogeny and phylogeography. Parasitology 150:55–77

Gonchar A, Galaktionov KV (2017a) Life cycle and biology of *Tristriata anatis* (Digenea: Notocotylidae): morphological and molecular approaches. Parasitol Res 116:45–59

Gonchar A, Galaktionov KV (2017b) New data support phylogeographic patterns in a marine parasite *Tristriata anatis* (Digenea: Notocotylidae). J Helminthol 94:e79

Goulding TC, Cohen CS (2014) Phylogeography of a marine acanthocephalan: lack of cryptic diversity in a cosmopolitan parasite of mole crabs. J Biogeogr 41:965–976

Grosberg RK, Vermeij GJ, Wainwright PC (2012) Biodiversity in water and on land. Curr Biol 22:R900–R903

Harland H, MacLeod CD, Poulin R (2016) Lack of genetic variation in the response of a trematode parasite to ocean acidification. Mar Biol 163:1

Hay E, Jorge F, Poulin R (2018) The comparative phylogeography of shore crabs and their acanthocephalan parasites. Mar Biol 165:69

Hayward A, Poulin R, Nakagawa S (2021) A broad-scale analysis of host-symbiont cophylogeny reveals the drivers of phylogenetic congruence. Ecol Lett 24:1681–1696

Hendry AP, Gotanda KM, Svensson EI (2017) Human influences on evolution, and the ecological and societal consequences. Philos Trans R Soc B 372:20160028

Herrmann KK, Poulin R, Keeney DB, Blasco-Costa I (2014) Genetic structure in a progenetic trematode: signs of cryptic species with contrasting reproductive strategies. Int J Parasitol 44:811–818

Hilborn R, Amoroso RO, Anderson CM, Ye Y (2020) Effective fisheries management instrumental in improving fish stock status. Proc Natl Acad Sci USA 117:2218–2224

Jirsová D, Štefka J, Jirků M (2017) Discordant population histories of host and its parasite: a role for ecological permeability of extreme environment? PLoS One 12:e0175286

Johnson P, Calhoun DM, Moss WE, McDevitt-Galles T, Riepe TB, Hallas JM, Parchman TL, Feldman CR, Achatz TJ, Tkach VV, Cropanzano J, Bowerman J, Koprivnikar J (2020) The cost of travel: how dispersal ability limits local adaptation in host-parasite interactions. J Evol Biol 34:512–524

Kearn GC (2004) Leeches, lice and lampreys: a natural history of skin and gill parasites of fishes. Springer, Cham

Keeney DB, Bryan-Walker K, King TM, Poulin R (2008) Local variation of within-host clonal diversity coupled with genetic homogeneity in a marine trematode. Mar Biol 154:183–190

Keeney DB, King TM, Rowe DL, Poulin R (2009) Contrasting mtDNA diversity and population structure in a direct-developing marine gastropod and its trematode parasites. Mol Ecol 18:4591–4603

Keeney DB, Cobb SA, Jadin RC, Orlofske SA (2023) Atypical life cycle does not lead to inbreeding or selfing in parasites despite clonemate accumulation in intermediate hosts. Mol Ecol 32:1777–1790

Kennedy CR (2006) Ecology of the Acanthocephala. Cambridge University Press, Cambridge

Kennedy DA, Kurath G, Brito IL, Purcell MK, Read AF, Winton JR, Wargo AR (2016) Potential drivers of virulence evolution in aquaculture. Evol Appl 9:344–354

Klassen GJ (1992) Coevolution: a history of the macroevolutionary approach to studying host-parasite associations. J Parasitol 78:573–587

Koehler AV, Brown B, Poulin R, Thieltges DW, Fredensborg BL (2012) Disentangling phylogenetic constraints from selective forces in the evolution of trematode transmission stages. Evol Ecol 26:1497–1512

Lagrue C, Joannes A, Poulin R, Blasco-Costa I (2016) Genetic structure and host-parasite co-divergence: evidence for trait-specific local adaptation. Biol J Linn Soc 118:344–358

Lively CM, Dybdahl MF (2000) Parasite adaptation to locally common host genotypes. Nature 405:679–681

Louhi K-R, Karvonen A, Rellstab C, Jokela J (2010) Is the population genetic structure of complex life cycle parasites determined by the geographic range of the most motile host? Infect Gen Evol 10:1271–1277

Marcogliese DJ (2001) Implications of climate change for parasitism of animals in the aquatic environment. Can J Zool 79:1331–1352

Marques JF, Santos MJ, Teixeira CM, Batista MI, Cabral HN (2011) Host-parasite relationships in flatfish (Pleuronectiformes): the relative importance of host biology, ecology and phylogeny. Parasitology 138:107–121

Mattiucci S, Giulietti L, Paoletti M, Cipriani P, Gay M, Levsen A, Klapper R, Karl H, Bao M, Pierce GJ, Nascetti G (2018) Population genetic structure of the parasite *Anisakis simplex* (s. s.) collected in *Clupea harengus* L. from North East Atlantic fishing grounds. Fish Res 202:103–111

McCallum HI, Kuris A, Harvell CD, Lafferty KD, Smith GW, Porter J (2004) Does terrestrial epidemiology apply to marine systems? Trends Ecol Evol 19:585–591

Mejía-Madrid HH, Vázquez-Domínguez E, Pérez-Ponce de León G (2007) Phylogeography and freshwater basins in Central Mexico: recent history as revealed by the fish parasite *Rhabdochona lichtenfelsi* (Nematoda). J Biogeogr 34:787–801

Mennerat A, Nilsen F, Ebert D, Skorping A (2010) Intensive farming: evolutionary implications for parasites and pathogens. Evol Biol 37:59–67

Mennerat A, Ugelvik MS, Jensen CH, Skorping A (2017) Invest more and die faster: the life history of a parasite on intensive farms. Evol Appl 10:890–896

Miranda CD, Tello A, Keen PL (2013) Mechanisms of antimicrobial resistance in finfish aquaculture environments. Front Microbiol 4:233

Mora C, Tittensor DP, Adl S, Simpson AGB, Worm B (2011) How many species are there on earth and in the ocean? PLoS One 9:e1001127

Morley NJ (2011) Thermodynamics of cercarial survival and metabolism in a changing climate. Parasitology 138:1442–1452

Morley NJ, Lewis JW (2015) Thermodynamics of trematode infectivity. Parasitology 142:585–597

Murawski SA (2010) Rebuilding depleted fish stocks: the good, the bad, and, mostly, the ugly. ICES J Mar Sci 67:1830–1840

Nadler SA, Pérez-Ponce de León G (2011) Integrating molecular and morphological approaches for characterizing parasite cryptic species: implications for parasitology. Parasitology 138:1688–1709

Nazarizadeh M, Peterka J, Kubečka J, Vašek M, Juza T, Ribeiro de Moraes K, Čech M, Holubová M, Souza AT, Blabolil P, Muška M, Tsering L, Barton D, Říha M, Šmejkal M, Tušer M, Vejřík L, Frouzová J, Jarić I, Prchalová M, Vejříková I, Štefka J (2022) Different hosts in different lakes: prevalence and population genetic structure of plerocercoids of *Ligula intestinalis* (Cestoda) in Czech water bodies. Folia Parasitol 69:018

Nowak BF (2007) Parasitic diseases in marine cage culture: an example of experimental evolution of parasites? Int J Parasitol 37:581–588

Page RDM (2003) Tangled trees: phylogeny, cospeciation, and coevolution. University of Chicago Press, Chicago

Parker GA, Chubb GC, Ball MA, Roberts GN (2003) Evolution of complex life cycles in helminth parasites. Nature 425:480–484

Pérez-Ponce de León G, Poulin R (2018) An updated look at the uneven distribution of cryptic diversity among parasitic helminths. J Helminthol 92:197–202

Perrot-Minnot MJ, Špakulová M, Wattier R, Kotlík P, Düsen S, Aydogdu A, Tougard C (2018) Contrasting phylogeography of two Western Palaearctic fish parasites despite similar life cycles. J Biogeogr 45:101–115

Poloczanska ES, Brown CJ, Sydeman WJ, Kiessling W, Schoeman DS, Moore PJ, Brander K, Bruno JF, Buckley LB, Burrows MT, Duarte CM, Halpern BS, Holding J, Kappel CV, O'Connor MI, Pandolfi JM, Parmesan C, Schwing F, Thompson SA, Richardson AJ (2013) Global imprint of climate change on marine life. Nat Clim Chang 3:919–925

Poulin R (1992) Determinants of host-specificity in parasites of freshwater fishes. Int J Parasitol 22:753–758

Poulin R (1995) Clutch size and egg size in free-living and parasitic copepods: a comparative analysis. Evolution 49:325–336

Poulin R (2005) Relative infection levels and taxonomic distances among the host species used by a parasite: insights into parasite specialization. Parasitology 130:109–115

Poulin R (2006) Global warming and temperature-mediated increases in cercarial emergence in trematode parasites. Parasitology 132:143–151

Poulin R (2016) Greater diversification of freshwater than marine parasites of fish. Int J Parasitol 46:275–279

Poulin R, Keeney DB (2008) Host specificity under molecular and experimental scrutiny. Trends Parasitol 24:24–28

Poulin R, Krasnov BR, Mouillot D (2011) Host specificity in phylogenetic and geographic space. Trends Parasitol 27:355–361

Prunier J, Saint-Pé K, Blanchet S, Loot G, Rey O (2021) Molecular approaches reveal weak sibship aggregation and a high dispersal propensity in a non-native fish parasite. Ecol Evol 11:6080–6090

Pulkkinen K, Suomalainen L-R, Read AF, Ebert D, Rintamäki P, Valtonen ET (2010) Intensive fish farming and the evolution of pathogen virulence: the case of columnaris disease in Finland. Proc R Soc B 277:593–600

Rahn AK, Krassmann J, Tsobanidis K, MacColl ADC, Bakker TCM (2016) Strong neutral genetic differentiation in a host, but not in its parasite. Infect Gen Evol 44:261–271

Ramilo A, Rodríguez H, Pascual S, González AF, Abollo E (2023) Population genetic structure of *Anisakis simplex* infecting the European hake from North East Atlantic fishing grounds. Animals 13:197

Rohr JR, Cohen JM (2020) Understanding how temperature shifts could impact infectious disease. PLoS Biol 18:e3000938

Santacruz A, Ornelas-García CP, Pérez-Ponce de León G (2020) Incipient genetic divergence or cryptic speciation? *Procamallanus* (Nematoda) in freshwater fishes (*Astyanax*). Zool Script 49:768–778

Sasaki A (2000) Host-parasite coevolution in a multilocus gene-for-gene system. Proc R Soc B 267:2183–2188

Sasal P, Desdevises Y, Morand S (1998) Host-specialization and species diversity in fish parasites: phylogenetic conservatism? Ecography 21:639–643

Selbach C, Jorge F, Dowle E, Bennett J, Chai X, Doherty J-F, Eriksson A, Filion A, Hay E, Herbison R, Lindner J, Park E, Presswell B, Ruehle B, Sobrinho PM, Wainwright E, Poulin R (2019) Parasitological research in the molecular age. Parasitology 146:1361–1370

Sereno-Uribe AL, López-Jiménez A, González-García MT, Pinacho-Pinacho CD, Ríos RM, García-Varela M (2022) Phenotypic plasticity, genetic structure and systematic position of *Neoechinorhynchus emyditoides* Fisher, 1960 (Acanthocephala: Neoechinorhynchidae): a parasite of emydid turtles from the Nearctic and Neotropical regions. Parasitology 149:991–1002

Sprehn CG, Blum MJ, Quinn TP, Heins DC (2015) Landscape genetics of *Schistocephalus solidus* parasites in threespine stickleback (*Gasterosteus aculeatus*) from Alaska. PLoS One 10:e0122307

Stearns SC (1992) The evolution of life histories. Oxford University Press, Oxford

Steele JH, Brink KH, Scott BE (2019) Comparison of marine and terrestrial ecosystems: suggestions of an evolutionary perspective influenced by environmental variation. ICES J Mar Sci 76:50–59

Strobel HM, Hays SJ, Moody KN, Blum MJ, Heins DC (2019) Estimating effective population size for a cestode parasite infecting three-spined sticklebacks. Parasitology 146:883–896

Tigano A, Friesen VL (2016) Genomics of local adaptation with gene flow. Mol Ecol 25:2144–2164

Timi JT, Lanfranchi AL, Poulin R (2005) Is there a trade-off between fecundity and egg volume in the parasitic copepod *Lernanthropus cynoscicola*? Parasitol Res 95:1–4

Treasurer JW, Wadsworth S, Grant A (2000) Resistance of sea lice, *Lepeophtheirus salmonis* (Kroyer), to hydrogen peroxide on farmed Atlantic salmon, *Salmo salar* L. Aquac Res 31:855–860

Ugelvik MS, Skorping A, Moberg O, Mennerat A (2017) Evolution of virulence under intensive farming: salmon lice increase skin lesions and reduce host growth in salmon farms. J Evol Biol 30:1136–1142

Weinstein SB, Kuris AM (2016) Independent origins of parasitism in Animalia. Biol Lett 12:20160324

Wiens JJ (2015) Faster diversification on land than sea helps explain global biodiversity patterns among habitats and animal phyla. Ecol Lett 18:1234–1241

Wood CL, Lafferty KD, Micheli F (2010) Fishing out marine parasites? Impacts of fishing on rates of parasitism in the ocean. Ecol Lett 13:761–775

Open Access This chapter is licensed under the terms of the Creative Commons Attribution-NonCommercial-NoDerivatives 4.0 International License (http://creativecommons.org/licenses/by-nc-nd/4.0/), which permits any non-commercial use, sharing, distribution and reproduction in any medium or format, as long as you give appropriate credit to the original author(s) and the source, provide a link to the Creative Commons license and indicate if you modified the licensed material. You do not have permission under this license to share adapted material derived from this chapter or parts of it.

The images or other third party material in this chapter are included in the chapter's Creative Commons license, unless indicated otherwise in a credit line to the material. If material is not included in the chapter's Creative Commons license and your intended use is not permitted by statutory regulation or exceeds the permitted use, you will need to obtain permission directly from the copyright holder.

Parasites Under Extreme Conditions

Christian Selbach and Rachel A. Paterson

Abstract

Habitats with extreme conditions, such as the deep sea or polar regions, were long considered 'retreats' where organisms could escape and evade their parasites. However, parasitic organisms occur in literally every habitat on Earth, and many aquatic parasites have successfully adapted their complex and intricate lifestyles to extreme conditions, from ice-covered arctic regions to the scathing temperatures of deep-sea vents. In this chapter, we illustrate the distribution and specialised adaptations of metazoan parasites to different naturally occurring extreme conditions in aquatic environments. Although few of the metazoan hosts and parasites exist at the extreme chemical and physical boundaries of life under which only truly extremophile microbial organisms can survive, parasitic organisms have evolved to co-exist with their hosts in aquatic habitats that pose unique challenges and conditions for their lifestyles. Parasites in extreme environments constitute diverse, highly specialised, and ecologically important components of these unique ecosystems. Moreover, studying parasites under naturally occurring extreme conditions can help us better understand how host-parasite systems might respond to environmental changes and newly developing anthropogenic extremes.

12.1 Introduction

Despite Aristotle's view that excellence lies in the middle or 'golden mean' between two extremes, excess and deficiency, humans have long been fascinated with and drawn to extreme environments, from traversing the polar ice caps to summiting the highest mountains, diving into the deepest marine trenches, or dreaming about colonising Mars and other planets. When we think of 'extreme environments', associations such as inhospitable, hostile, barren, bleak, or lifeless might come to mind. Yet, as we know today, this absence of eukaryotic life in many extreme environments could not be further from the truth. Nature contains many extreme aquatic environments, each with specific and often diverse ecological communities, ranging from

C. Selbach (✉)
Department of Arctic and Marine Biology, UiT The Arctic University of Norway, Tromsø, Norway

Water Research Group, Unit for Environmental Sciences and Management, North-West University, Potchefstroom, South Africa
e-mail: christian.selbach@uit.no

R. A. Paterson
Norwegian Institute for Nature Research, Trondheim, Norway
e-mail: rachel.paterson@nina.no

deep seas that cover more than half of the surface of our planet, underground caves and rivers, to the vast amounts of frozen water of the Arctic regions and polar caps.

The first life forms developed on our planet some 4–5 billion years ago under extreme conditions, likely around deep-sea hydrothermal vents that harboured rich chemical complexity that might have given rise to primitive metabolic life, or in an anoxic prebiotic soup in which the first self-replicating entities emerged (Bada 2004; Martin et al. 2008). Although the exact mechanisms and environmental conditions under which life on Earth developed are not known, these conditions can be considered as highly extreme by our anthropocentric definitions (see Rothschild and Mancinelli 2001). In fact, by our standards, we would classify these first prehistoric life forms as extremophiles ('loving' and thriving under such extreme conditions), or polyextremophile (preferring multiple extremes). However, for most of our planet's history, these organisms might have been the dominant life forms (Knoll 2015). The current atmospheric and climate conditions on Earth are a rather recent (and possibly short-lived) development that allowed the Cambrian explosion to occur some 500–600 million years ago, giving rise to the vast diversity of eukaryotic organisms and the majority of animal phyla we know today, most of which thrive in 'normal' or 'moderate' conditions in benign ambient habitats (Seckbach 2013).

Most of the known extremophile or extremotolerant organisms today belong to prokaryotes, i.e., archaea and bacteria. Discoveries of those life forms under the most extreme conditions at the boundaries of life have repeatedly forced us to extend and widen the boundary parameters under which life can succeed (Rothschild and Mancinelli 2001; Merino et al. 2019). For example, the archaean *Pyrolobus fumarii* can grow at temperatures of up to 113 °C (and possibly even greater than 120 °C), and some enzymes remain functional at temperatures beyond 140 °C, while *Sulfolobus acidocaldarius* can flourish under the simultaneous stress of high acidity and temperatures (pH 3 and 80 °C) in geysers of Yellowstone National Park (Stetter 2006; Seckbach 2013; Rothschild and Mancinelli 2001). Eukaryotes and metazoan organisms can be found under extreme conditions, such as the nematode *Panagrolaimus davidi*, which can endure freezing of all bodily water (Rothschild and Mancinelli 2001). Another example, the Pompeii worm *Alvinella pompejana*, lives on the sea floor near hydrothermal vents and can withstand temperatures up to 105 °C thanks to symbiotic bacteria that live on the worm's body surface, making it one of the most heat-tolerant animals on the planet (Seckbach 2013). Perhaps most impressive, tardigrades ('water bears') can survive extreme temperature ranges (from −273 to 180 °C), almost a decade without water, high atmospheric pressure, and radiation, allowing these unassuming animals to live in nearly all regions on Earth (Seckbach 2013).

During the first explorations of life in extreme environments, these habitats were considered 'retreats' where organisms could escape and evade their parasites or predators, at the cost of the harsh environmental conditions (see Bray et al. 1999). However, as we know today, this is not universally true, and we find parasites in literally every habitat on Earth, and parasitic organisms may comprise almost half of animal life on the planet and have independently evolved hundreds of times (Weinstein and Kuris 2016). This raises the question of which (protist and metazoan) parasites we can expect in extreme habitats and under such conditions, and how extreme environments can shape and modify the complex interactions between parasites and their hosts.

By definition, parasites are masters of the extreme, having evolved to live in or on highly hostile environments, their hosts. For instance, cestodes living inside a mammalian digestive system will experience a relatively stable food supply and abiotic conditions; however, this environment remains extreme in terms of the acidic, anaerobic, and relatively high temperature characteristics of the digestive system. At the same time, parasites live in environments that actively try to kill, remove, or suppress them, e.g., via immune responses or mechanical defence mechanisms, and habitats that continuously co-evolve with the parasite in an arms race to keep the upper hand. Since parasites are a major force that 'drives' the host's immune response, it has been argued that parasites not only invade and survive extreme environments within their hosts, but rather actively create them (Combes and Morand 1999). Tinsley (1999a, 2005) provides an in-depth review of the hostile environments created by parasites within their hosts (and by parasites competing with each other inside their hosts) as well as an overview of abiotic factors shaping extreme environments for parasitic lifestyles.

In this chapter, we explore and illustrate how and to what extent macroparasites have been able to establish themselves in some of the most extreme aquatic environments on our planet, and which parasitic groups can maintain their life cycles or even thrive under the various extreme conditions these environments provide. Furthermore, we ask if there are limitations in these environments that could reduce the likelihood that metazoan parasites will occur in such regions and if there are any 'final frontiers' that parasitic organisms have not yet been able to conquer. Naturally, not all extreme environments and conditions will be covered in this chapter, nor will all parasites that have been described from these various ecosystems. Rather, we aim to highlight and illustrate the distribution range and wide adaptation of metazoan parasites to different naturally occurring extreme environmental conditions in aquatic habitats. Moreover, we include some habitats that may not represent the most extreme chemical and physical parameters per se, but pose uniquely 'extreme' and stressful challenges to host and parasite communities living there (e.g., intertidal zones or temporary aquatic systems; see also Tinsley 2005). In the following, we will explore host-parasite systems in the deep sea, including hydrothermal vents, hypersaline and soda lakes, lakes in areas of volcanic activity, arctic and polar regions, as well as subsurface waterbodies, temporary aquatic systems (e.g., in arid environments), and systems with extreme fluctuations and variations, such as intertidal zones (Fig. 12.1). In a synthesis, we attempt to provide a unifying framework for the different challenges these various environments pose for parasitic organisms and explore how knowledge about host–parasite interactions under extreme pressure can help us better understand how parasites might respond to environmental changes and 'new' anthropogenic extremes.

12.2 Deep Sea

Nearly 11,000 m below the surface of the Pacific Ocean, the Mariana Trench contains the world's deepest point in the ocean. However, the deep sea is by no means a unique or rare habitat. The abyssal (3000–6000 m) and hadal (>6000 m) seafloor regions[1] comprise the majority of the global seafloor area and cover more than half of the Earth's surface. More than 90% of oceans are considered deep sea, making them the most common environment and the largest biome on this planet (Scheckenbach et al. 2010; Ramirez-Llodra et al. 2010; Woolley et al. 2016). Whilst the deep seas have long been considered to be devoid of any life (apart from mythological sea monsters), over the last two centuries, hundreds of expeditions have shed light on the fascinating and vast diversity of life that exists in the depths of our planet's oceans, which can even surpass that of species-rich shallow-water habitats (Klimpel et al. 2006). Yet, despite the technological advancements and stimulating discoveries in deep-sea research since the nineteenth century that have changed and reshaped our understanding of life on this planet, vast portions of the deep sea remain unknown and discovery rates of new species are high (Ramirez-Llodra et al. 2010). To date, the largest ecosystem on our planet might be our least explored, in particular at the greatest depths below 6000 m.

Two hundred metres below the surface of the ocean, the deep sea is largely characterised by a total absence of sunlight, low but stable temperatures (1–2 °C), extremely high pressure, and a physical homogeneity of the environment

---

[1] Referring to 'a deep or seemingly bottomless chasm' (abyssal), and named after Hades, the Greek mythological god of the underworld, where souls go after death, which highlights both our fascination and unease with the dark and vast depths of our oceans (see Jamieson et al. 2021).

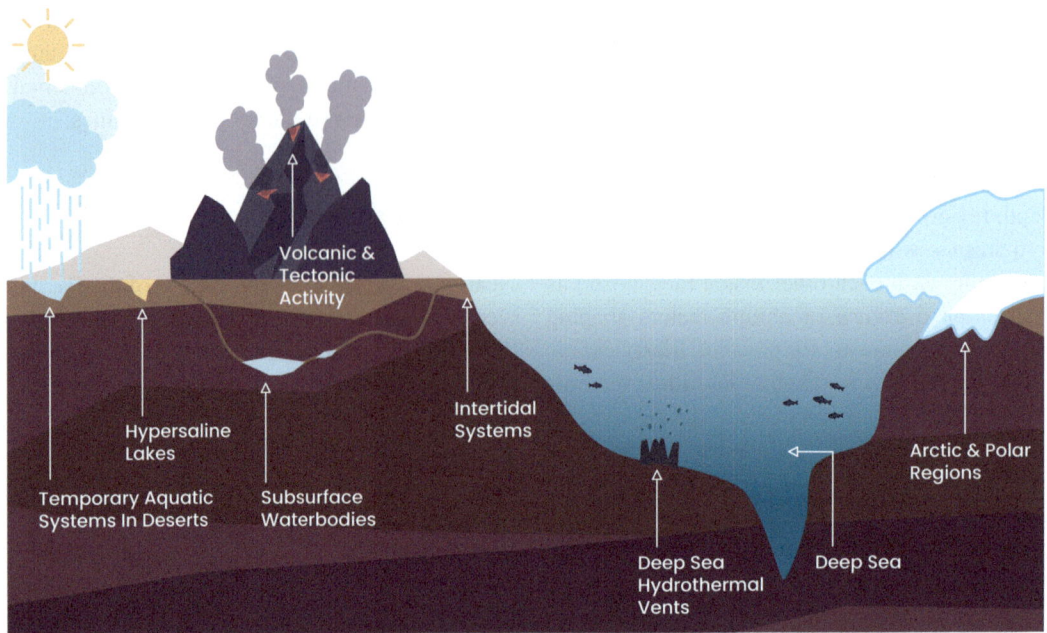

Fig. 12.1 Extreme aquatic habitats that pose unique conditions and challenges for hosts and their parasites

(Rothschild and Mancinelli 2001; Klimpel et al. 2006; Fig. 12.2). Apart from hydrothermal deep-sea vents (see below), the deep sea is energy-poor with almost all energy slowly descending as 'marine snow' from the productive upper layers, where sunlight allows primary production, sometimes taking weeks to reach the sea floor (Turner 2015). Yet, it is estimated that less than 2% of the primary production eventually reaches depths below 2000 m (Buesseler et al. 2007). Although the vast abyssal plains covered with soft, muddy sediment are among the planet's most homogenous and flattest regions (Scheckenbach et al. 2010), the seabed topography is much more varied in places and contains complex and sometimes unique and extreme habitats, including steep ocean ridges, frozen methane hydrates, and hydrothermal vents (Rothschild and Mancinelli 2001; Bray 2020). Likewise, the benthic deep sea is also a highly dynamic environment with currents that move huge volumes of water in the deep, seasonal variation in marine snow, or irregular disturbances, such as benthic storms (Ramirez-Llodra et al. 2010).

Even though the deep pelagic is a vast environment with generally low abundances of organisms, the deep sea supports one of the highest levels of biodiversity and the greatest biomass on Earth. Distinct populations can contain millions of individuals in the three-dimensional space without solid boundaries (Herring 2002; Snelgrove and Smith 2002; Robison 2009; Ramirez-Llodra et al. 2010). Although there is no phylum that lives exclusively in deep sea, these habitats harbour a highly specialised and distinctive fauna that is adapted to the low availability of energy and the unique and extreme abiotic features of the environment. For instance, many deep-sea organisms display dwarfism or gigantism as an adaptation to the limited food availability at greater depths, have evolved highly sophisticated sensory organs to detect prey, and many pelagic animals show a reduced tissue density to improve buoyancy in the water (Ramirez-Llodra et al. 2010). Gelatinous animals (e.g., cnidarians) comprise a major portion of the metazoan deep pelagic fauna, while annelids, arthropods, nematodes, and echinoderms dominate the deep-sea floors (Rose et al. 2005; Robison 2009; Rabone et al. 2023).

Recent surveys and reviews of digenean trematodes in deep-sea teleost fish have reported 25

Fig. 12.2 More than 90% of the oceans' volume is two hundred metres or more below the surface, making the deep sea the most common environment and the largest biome on Earth. Organisms like the animals in this crinoid field (*Bathycrinus carpenterii*) on the Mohn's Treasure mound at 2800 m depth need to adapt to extremely high pressure, low but stable temperatures (1–2 °C), the absence of sunlight, and organic nutrients that slowly descend from the surface. Photo: ©MarMine/NTNU/Eva Ramirez-Llodra

families of these parasites thus far, with depth ranges described for 246 species (Bray et al. 1999; Bray 2004, 2020; Klimpel et al. 2006). The greatest confirmed depths were for *Gonocerca phycidis* at 8450 m (with prevalence exceeding 30%), *Steringophorus thulini* at 4865 m, and *Lepidapedon zubchenkoi* at 4877 m. Yet, the majority of recorded species occurred well above the abyssal zone, i.e., above 3000 m (Bray et al. 1999; Bray 2020). Overall, the known parasite species represent only a fraction of the ca. 150 known families of digenean trematodes, indicating that the deep-sea trematode fauna uncovered so far may be considerably less diverse than the parasite fauna in shallow marine systems, with trematodes largely absent from the oceans' greatest depths (Bray 2020). In this, trematodes and other common fish parasites appear to mirror the decline in overall species richness and diversity with increasing ocean depth, and the limited number of species restricted to deeper than 3500 m. However, higher parasite diversity close to the sea floor due to nutrient availability has been documented (Marcogliese 2002; Klimpel et al. 2006; Costello and Chaudhary 2017). Although little is known about the life cycles of deep-sea trematodes, suitable invertebrate hosts are present in deep waters of the North Atlantic Ocean. Gastropod, bivalve, and polychaete diversity in the deep-sea peaks between 2000 and 3000 m, though suitable populations can exist down to 5000 m, potentially allowing trematodes to complete their complex multi-host life cycle in these habitats (Rex and Etter 2010; Bray 2020). It therefore remains to be explored whether digeneans can maintain their life cycles in the deep sea and which transmission modes and strategies these parasites might employ.

Some species of deep-sea lanternfish (Myctophidae) are known to be infected with juvenile digeneans of the family Didymozoidae at depths of 400 m, and it has been suggested that transmission of the parasites to piscivorous final fish hosts likely occurs in shallower waters where the lanternfishes migrate at night (Mateu et al. 2015; Bray 2020). Vertical diel migration of fish, usually upward at night to feed on the greater biomass closer to the surface and back down during the day to avoid predation, is typical for many deep-sea fish and other organisms and is considered the largest migration on the planet (Ramirez-Llodra et al. 2010). These vertical migration dynamics lead to biologically generated turbulence that plays a central role in nutrient transport and cycling in the ocean (Kunze et al. 2006) and likely also plays a crucial role in the transmission dynamics, life cycles, and distribution of parasite stages across the various depth zones and layers of the ocean.

Not surprisingly, parasitic nematodes, whose free-living conspecifics are extremely abundant and diverse in deep-sea sediment (10 cm^3 of seafloor can harbour up to 100 different morphologically distinct species; Ramirez-Llodra et al. 2010), are widely distributed parasites in deep-sea fishes, even though prevalence and abundance can be low, indicating that deep-sea fish might not be instrumental in the completion of their life cycles. Many of these nematode parasites might be generalist taxa also using fish from shallower water zones (Klimpel et al. 2006; Moravec and Justine 2011). On the other hand, Klimpel et al. (2011) demonstrated a high prevalence of *Anisakis paggiae* in deep-sea fish that suggest an oceanic deep-water life cycle for this parasite in the north Atlantic.

In contrast to endoparasites, ectoparasites in deep-sea fishes have received less research attention, with most existing records originating from opportunistic trawling and dredging efforts during which external parasites can easily be dislodged or removed (Quattrini and Demopoulos 2016). A survey of ectoparasites on deep-sea fishes using remotely operated submersibles uncovered that ectoparasite communities were dominated by isopods and copepods, and diversity was highest at depths between 500 and 1400 m, thereafter declining with increasing depth with the exception of gnathiid isopods that were observed at depths down to 3260 m (Quattrini and Demopoulos 2016). However, the survey based on the analyses of underwater cameras might underestimate the presence of smaller ectoparasites, such as monogeneans, that are less visible and will need to be assessed via other methods. The authors found ectoparasite diversity to be highest in undersea canyons, probably due to the increased habitat heterogeneity at these sites. This is not surprising, as deep-sea canyons can concentrate biological material descending from the surface, and thereby support greater diversity and biomass than surrounding areas (Ramirez-Llodra et al. 2010), with fishes from these regions also found to harbour higher endoparasite loads (Campbell et al. 1980). Parasitic copepods have been recorded from fish hosts at depths down to 5440 m (*Chondracanthus deflexus*), the deepest confirmed record of a parasitic copepod from a fish host (Boxshall 1998).

Copepods parasitising invertebrates have been recorded at similar depths, e.g. *Ophelicola kurambia* at 4987–4991 m (Conradi et al. 2015), or the aptly named *Abyssotaurus vermiambatus* at 5395 m in the Angola Basin of the Atlantic Ocean where they live as commensals or parasites of polychaetes (Brenke et al. 2018). Since free-living benthic copepods make up the second most common and abundant metazoan group in deep-sea sediments (after nematodes, Ramirez-Llodra et al. 2010), a large diversity of parasitic taxa in this group is likely still unknown and undescribed. Moreover, recent molecular analyses of abyssal and hadal deep-sea sediments in the Atlantic and Pacific revealed a great and unknown diversity of protist communities, including many parasitic or potentially parasitic groups at depths of 5000 m or more (Scheckenbach et al. 2010; Schoenle et al. 2021). The deep-sea floor appears to offer a rich habitat for metazoan and protist parasites that are deeply embedded in marine food webs and ecological functions of the largest ecosystem on our planet (Schoenle et al. 2021).

12.3 Deep-Sea Hydrothermal Vents, Methane Seeps, and Brine Lakes

In the vast cold and energy-poor space of deep sea, hydrothermal vents are island-like, habitats rich in energy, and with distinct chemical compositions (Dykman et al. 2023; Fig. 12.3). These vents are underwater geysers or hot springs that

Fig. 12.3 In contrast to the energy-poor and cold deep sea, hydrothermal vents are island-like energy-rich habitats with distinct chemical compositions that drive chemotroph lifestyles around volcanic activities on the sea floor, like the Ganymede black smoker on the Aurora Vent Field, Gakkel Ridge. Free-living and parasitic organisms living in these environments must be able to withstand the extreme water temperature gradients around vents that can release water reaching up to 350 °C. Photo: HACON21/REV Ocean, Eva Ramirez-Llodra

typically form around active volcanic regions and release hot water of up to 400 °C that remains fluid under the intense pressure of the deep sea (Rothschild and Mancinelli 2001). In contrast to the diverse but sparsely populated communities of the deep sea, hydrothermal vent communities are characterised by low species richness, simple food web structures, and high abundance and biomass, often dominated by few invertebrate species that can account for up to 90% of the total free-living abundance (Van Dover 2000; Voight 2000). The high productivity of deep-sea vent ecosystems is largely driven by chemoautotroph microbes that serve the same role as green plants and other photosynthetic primary producers in areas with sunlight (Nakagawa and Takai 2008). Organisms at deep-sea vents can grow large in size, often driven by chemoautotroph symbiotic bacteria, with mussels and clams reaching 20–30 cm, and giant tube worms even growing up to 3 m in length (Ramirez-Llodra et al. 2010).

A first in-depth literature review of parasitic diversity from deep-sea hydrothermal vents was conducted by De Buron and Morand (2004) who reported the presence of fewer than 20 parasitic taxa including a leech species, a nematode, as well as several taxa of copepods, acanthocephalans, and digeneans at depths down to 3600 m. The authors contrasted this with the significantly higher diversity of 126 parasitic species that represent almost every group of macroparasites that had been identified in the open ocean at depths of 1000 m or below. However, rather than indicating a generally low suitability at hydrothermal vents for a parasitic lifestyle, the low parasite species richness probably reflects the low study effort for parasites in these habitats that are even less sampled than many other deep-sea habitats (De Buron and Morand 2004).

Discovery rates of parasitic taxa at hydrothermal vents have remained low, with just five trematodes and copepod species, and one isopod and nematode described in the last 20 years (in contrast to the more than 40 new free-living organisms described per year since life was discovered at deep-sea vents in 1977, see Dykman et al.

2023). Even though the total number of parasite species at vent systems may be relatively low (in particular due to the low number of top predators that typically accumulate a high number of parasites), comparative analyses showed that parasite species richness within hosts did not differ significantly from other marine environments (atolls and kelp forests), and, surprisingly, the presence of some parasites with complex multi-host life cycles, such as digenean trematodes, was not lower at vents (Dykman et al. 2023). The authors conclude that trematodes might be specifically successful in energy-rich systems that maintain a high abundance of gastropod intermediate hosts that the parasites require for their asexual multiplication.

Due to their volcanic nature, vent systems can be rather short-lived, and eruptions can wipe out whole communities in relatively short timespans (Rubin et al. 2012). Consequently, parasitic and free-living organisms that successfully invade and thrive in these habitats are likely good colonisers that can quickly establish their life cycles in these isolated and extreme habitats. Yet, despite the ultimate risk of catastrophic disruptions of the habitat, hydrothermal vents can be expected to provide temporally stable and predictable environments with a continuously high energy supply and host abundance, and free of any seasonal or daily fluctuations that characterise most other habitats on this planet. The thriving host-parasite communities at deep-sea vents are diverse and unique components of these ecosystems and offer a window into the conditions that might have shaped the earliest biological communities and interactions on our planet (Rothschild and Mancinelli 2001; Nakagawa and Takai 2008).

The deep oceans contain a range of other insular habitat types with highly specialised and extreme conditions such as hypersaline undersea brine lakes or methane seeps where highly specialised free-living and parasitic organisms can occur, yet knowledge of host–parasite interactions in these environments remains even more scarce (e.g., Ward et al. 2004, Sapir et al. 2014). Hypersaline aquatic conditions also occur in many inland waterbodies that pose challenging environments for unique and highly adapted free-living and parasitic communities, which are discussed below.

12.4 Arctic and Polar Regions

The biosphere of our planet is cold and 90% of the oceans' volume consists of cold water, and almost three quarters of all freshwater occurs as ice, mostly at the polar caps (Seckbach et al. 2013). These high latitude systems are characterised by typically low temperatures, strong seasonality with long and cold winters and short productive summers, and a dynamic interaction between frozen and unfrozen fresh- and saltwater (Hoberg et al. 2013, Fig 12.4). For a long time, these vast polar and arctic regions were considered barren landscapes devoid of life and fuelled the romantic imaginations of adventurers, armchair travellers, and artists, from the paintings of Casper David Friedrich to the gothic novels of Mary Shelley. Scientific explorations of the nineteenth century, such as Fridtjof Nansen's *Fram* expedition, uncovered an unexpected diversity of organisms in and on the ice and 'established' these extreme habitats as viable ecosystems for life (Seckbach et al. 2013), an understanding that had long existed in indigenous communities that have lived in these regions for thousands of years and accumulated extensive knowledge of cold environments and their biota (Eerkes-Medrano and Huntington 2021).

Today, high latitude/polar habitats represent perhaps one of the best-studied extreme habitats for aquatic parasites, partially driven by our fascination for exploring such relatively accessible, yet highly extreme, environments (MacKenzie 2017), with several historical Arctic and Antarctic expeditions including parasitological surveys, not least Captain Robert F. Scott's ill-fated expedition to the South Pole between 1910 and 1912 (Campbell and Overstreet 1994). Whilst access to parasitological specimens from commercially obtained fish has supported much of our current understanding of high latitude marine parasites in later years, there is increasing interest in the

Fig. 12.4 Arctic and polar regions, such as the Barents Sea and the Arctic Ocean, offer a wide variety of freshwater and marine habitats on our planet and are characterised by low temperatures, a pronounced seasonality (long cold winters and short productive summers), and a dynamic interaction between frozen and unfrozen fresh- and saltwater. Photo: Olaf Schneider, Norwegian Polar Institute

response of aquatic parasites in both high latitude marine and freshwater environments to the effects of climate change (e.g., Kutz et al. 2005, 2009).

Freshwater ecosystems of the Arctic and Antarctica are not created equal in terms of habitats that allow metazoan parasites to exist, let alone, thrive. In freshwaters of Antarctica, not only are insects and fish absent (i.e., the common second intermediate and final hosts of metazoan parasites), but food webs are also generally truncated, with biota dominated by crustaceans and microscopic organisms (e.g., rotifers, tardigrades; Vincent and James 1996; Janiec 1996). Although nematode species are present in Antarctic freshwater habitats, they are only represented by free-living as opposed to parasitic forms (Janiec 1996).

In contrast, Arctic (and sub-Arctic) freshwater habitats support relatively diverse assemblages of fish (127 species; Nelson 2006), invertebrates (e.g., Coulson 2000), and their metazoan parasite communities (e.g., 24 trematode species [Takvatn, Norway], Soldánová et al. 2017; 6 *Diplostomum* spp. lineages [Iceland], Blasco-Costa et al. 2014). For example, Arctic charr *Salvelinus alpinus*, the northernmost freshwater fish species, is regarded as a hotspot of metazoan parasite diversity in Arctic and sub-Arctic freshwater habitats (e.g., Paterson et al. 2019; Siwertsson et al. 2016). However, parasite diversity may become constrained near the northern limits of this fish species. In a study of endoparasites of Arctic charr from Svalbard, Sobecka and Piasecki (1993) noted that parasite communities were comprised of cestodes (copepod-transmitted), nematodes (copepod-transmitted), and the parasitic crustacean *Salmincola* sp., with a distinct absence of trematode species commonly found in northern Norwegian Arctic charr populations (e.g., *Diplostomum* spp., *Crepidostomum* spp., *Phyllodistomum umblae*). This pattern fits with the known freshwater macroinvertebrate communities of Svalbard being dominated by zooplankton species, and the absence of molluscs that act as the first intermediate hosts of trematodes. However, trematodes are not completely absent from the parasite fauna of Arctic fish species, with one trematode species (*Crepidostomum farionis*) noted among the seven metazoan parasites reported from nine-spined stickleback *Pungitius pungitius* from Baffin Island (Canada; Gallagher and Dick 2011).

Whilst the diversity of freshwater parasites that use hosts adapted to Arctic environments may be reduced, those parasite species that have successfully navigated this challenging environment may occur at higher intensities in the Arctic than elsewhere in their southern range. For instance, *Dibothriocephalus* spp. (formerly *Diphyllobothrium* spp.) plerocercoids (cestode second intermediate larval stage, see Chap. 5) occur in very high intensities in Arctic charr populations in the high Arctic as a result of ontogenetic niche shifts whereby the fish switch from zooplanktivorous/insectivorous diets to cannibalism, with infections acquired throughout the lifetime of the hosts (Borgstrøm et al. 2015; Hammar 2000). Other cestodes trophically transmitted via copepods, though representing yearly/seasonal infections, namely *Eubothrium salvelini* (Schrank, 1790) Nybelin, 1922 and *Proteocephalus* sp., which also appear to occur at much higher intensities than in more southerly populations, are thought to benefit from the restricted diet range of high Arctic charr populations (Hammar 2000).

Although intermediate and/or final host-derived limitations may restrict the metazoan parasite communities of high latitude freshwater environments to varying degrees, the low diversity of parasites supported in Arctic and Antarctic marine waters appears largely related to the research attention that they have received thus far, with each new study uncovering new parasite taxa (e.g., Zdzitowiecki 1990, 2001; Palm et al. 2007). In addition, what Antarctica lacks in terms of metazoan parasite diversity in freshwater habitats, it makes up for in the marine environment where diverse fish parasite communities are known (e.g., Rocka 2006; Palm et al. 2007; Gordeev and Sokolov 2016, 2017). A recent review suggests that more than 320 marine metazoan parasite species are currently recognised from Antarctic and sub-Antarctic waters (Polyakova and Gordeev 2021). This should not be considered as a complete list of parasites that have successfully adapted to Antarctic environments given that parasitological studies have only begun to explore ~36% of the 374 fish species known from this region, and most often from relatively common fish species captured in surface waters (Polyakova and Gordeev 2021).

As the number of parasitological studies continues to increase, some broad-scale patterns in high latitude fish parasite diversity begin to appear. For example, a recent study by Muñoz and Cartes (2020) showed that the diversity of fish parasites in Antarctic marine waters may even exceed that of sub-Antarctic waters. Meanwhile, in Arctic marine waters, the high parasite diversity observed in demersal fish did not extend to bathy- and meso-pelagic fish species, where low productivity is thought to limit the density of intermediate hosts and thus restrict the parasite community (Klimpel et al. 2006). Studies also demonstrated the tendency for aquatic parasites of the Arctic and Antarctica to be generalist species (e.g., Galaktionov 2017, Galaktionov et al. 2019; Rocka 2006; Muñoz and Cartes 2020), capable of parasitising a wide range of intermediate, paratenic, and/or final hosts. Furthermore, there is increasing evidence of endemism in metazoan fish parasite from Arctic and Antarctic marine waters, with very few species noted to be globally widespread or shared with fish from sub-Arctic/Antarctic waters (Polyakova and Gordeev 2021; Hemmingsen and Mackenzie 2001; Rocka 2006).

Parasite transmission dynamics are strongly influenced by temperature. For instance, in temperate regions, trematode cercariae emergence may cease as water temperatures drop below 4 °C (e.g., Lyholt and Buchmann 1996). However, in high latitude environments, some marine trematode species remain active at water temperatures close to or below zero degrees Celsius. For example, Graefe (1971) experimentally demonstrated that cercariae emergence from marine snails infected with an unidentified opecoelid trematode may occur at temperatures as low as 0–1 °C. Similarly, Nikolaev et al. (2020) observed that periwinkles (*Littorina saxatilis*, *L. obtusata*) of the White Sea contained motile cercariae of *Himasthla littorinae* and *Renicola parvicaudatus* despite winter water temperatures of −1 °C. This remarkable ability to remain active, though at a lower rate than at higher water temperatures, demonstrates that some larval parasite stages can successfully persist in marine environments that freeze below −1.85 °C (Haumann et al. 2020).

However, perhaps more remarkable are the endoparasites with complex life cycles that must withstand not only the low temperature environments in which their ectothermic first/second intermediate/paratenic (invertebrates, fish) hosts occur, but also the internal temperatures of their endothermic final hosts (birds, mammals). For instance, larval stages of the notocotylid trematode *Paramonostomum antarcticum* must tolerate winter temperatures of −2 °C in the first intermediate snail host *Laevilitorina caliginosa*, in addition to successfully transitioning to the 40.2 °C internal temperature of its final host, the snowy sheathbill (*Chionis albus*). Similar abilities of endoparasites to adapt to large temperature shifts can also be expected across endoparasites of all birds (González-Acuña et al. 2013; Diaz et al. 2016), pinnipeds (Johansen et al. 2010), and cetaceans (Dailey and Vogelbein 1991) occurring in Antarctic and Arctic marine waters.

In contrast, ectoparasites are constantly exposed to the abiotic conditions that make any environment so-called 'extreme'. For instance, the semi-aquatic lice parasitising marine mammals in Arctic and Antarctic waters not only have to contend with the consequences of low water temperature, but also high water pressures as a consequence of being attached to the surface of a host that may dive to >500 m (Mehlhorn and Mehlhorn 2017; Garde et al. 2018; Nachtsheim et al. 2017). Suggested adaptations to protect against low water temperatures include the presence of spines on the body surface of the louse, which function to hold a protective layer of their host's sebum and dorsally thickened cuticles to reduce heat loss (as opposed to generally thin cuticles of terrestrial lice species; Mehlhorn and Mehlhorn 2017).

Life cycle truncation may also be an effective strategy to increase transmission success when parasites and/or their hosts approach their lower thermal limits in high latitude waters. For instance, two host (dixenous) trematode life cycles, in which free-living stages (i.e., miracidia, cercariae) are absent, are frequently observed in Arctic intertidal environments (Galaktionov et al. 2015, 2019). Such truncated life cycles may enhance transmission success by removing vulnerable free-living life stages from the lifecycle and removing the limitations of scarce or absent second intermediate host populations that may not tolerate high latitude environments (Galaktionov 2017). However, there is also evidence to suggest that high latitude marine waters pose little challenge for transmission for some metazoan parasites, with anisakid nematodes, for instance maintaining the same life cycle as their relatives in lower latitude marine waters (Klimpel et al. 2010). This is perhaps not surprising given that nematodes are widely recognised to be highly adaptable to extreme environmental conditions (Aleuy and Kutz 2020).

12.5 Subsurface Waterbodies

Although surface freshwater lakes and rivers contain less than one percent of our planet's freshwater, approximately 97% of the unfrozen freshwater occurs below the surface, either as groundwater or in subsurface streams, rivers or ponds (Gibert and Deharveng 2002). This makes underground waterbodies the most widely distributed environment on Earth after marine systems (Mammola et al. 2019). While subsurface aquatic ecosystems are typically characterised by stable temperatures, organisms occupying these habitats must also contend with the absence of sunlight, a high degree of physical fragmentation, and often the low availability of food and organic resources (Gibert and Deharveng 2002; Culver et al. 2009; Fig. 12.5). In the absence of primary producers that rely on light for photosynthesis (photoautotrophs), all energy is limited to chemoautotroph organisms that can obtain

Fig. 12.5 Most of our planet's unfrozen freshwater occurs below the surface as groundwater, or in underground rivers, ponds, or lakes, such as the Gruta do Lago Azul in Brazil. These environments typically offer stable temperatures, but organisms need to adjust to a total absence of sunlight and low availability of food and organic resources as well as a high degree of physical habitat fragmentation. Photo: Maria Elina Bichuette

energy from chemical reactions (Northup and Lavoie 2001) or to detritus, food and energy resources that are transported or get washed in from the productive surface (Simon et al. 2003). In this regard, subsurface aquatic ecosystems are similar to deep-sea environments where organisms typically experience constant and absolute darkness and rely on energy that sinks down from the productive and biodiverse surface (Gibert and Deharveng 2002). Despite the absence of primary producers and the oligotrophic nature of aquatic underground habitats, these ecosystems can support distinct and surprisingly species-rich communities of metazoan animals that have adapted to their subsurface surroundings and often lack eyes or pigmentation (Culver et al. 2009; Hlebec et al. 2023). While terrestrial subsurface ecosystems are typically dominated by arachnids and insects, the aquatic underground fauna is dominated by crustaceans, in particular copepods, isopods, and amphipods, as well as oligochaetes, nematodes, acari, and molluscs (Culver and Sket 2000; Stein et al. 2012). Subterranean crustacean communities can even exceed the diversity of crustaceans in surface waterbodies in the same region (Stoch 1995). In total, 7000 groundwater invertebrates have been described to date, but the true number is expected to be far higher due to our poor knowledge of these ecosystems (Gibert et al. 2009; Stein et al. 2012). Subterranean aquatic vertebrate taxa, on the other hand, seem to be mainly limited to various species of fish and amphibians that have independently adapted to an underground lifestyle (Botosaneanu 1986; Culver and Sket 2000; Sánchez-Fernández et al. 2021).

In contrast to free-living diversity in subsurface ecosystems, the diversity and distribution of parasites in these habitats remain much less studied and understood, and our knowledge often relies on individual or anecdotal findings. For instance, Moravec and Huffman (1988) described a new nematode species, *Rhabdochona longleyi*, from blind catfishes in an underground water system in Texas. The parasite appears to have adapted its life cycle to subterranean conditions by using crustaceans instead of insects as intermediate hosts. North American cavefish of the family Amblyopsidae have been reported to harbour two species of cestodes (*Proteocephalus* spp.); however, these parasites have never been re-encountered since their original description 50 years ago and remain insufficiently described and unconfirmed to date (Whittaker and Hill 1968; Whittaker and Zober 1978; Scholz et al. 2021). Likewise, the one-off finding of a juvenile

Acuaria sp. nematode in Australian cave fish *Milyeringa veritas* may only apply to a single fish, apparently due to an accidental infection after the parasite was washed into the underground system (Whittaker and Kritsky 1973).

This overall lower parasite diversity and abundance in cave ecosystems has important implications for host–parasite interactions and co-evolutionary adaptation over time. Mexican cavefish, *Astyanax mexicanus*, have developed a drastically simplified innate immune system and shifted their immune investment to the adaptive immune system as an adaptation to lower parasitic pressure in contrast to closely related surface fishes (Peuß et al. 2020). Such isolated vertebrate populations that have lost parasite diversity for thousands of generations offer unique opportunities to study and assess the role of parasites as evolutionary drivers of host defence and immune system development (Peuß et al. 2020). However, recent investigations of *A. mexicanus* populations revealed surprisingly diverse and species-rich macroparasite communities in cave environments that included trematodes, monogeneans, nematodes, copepods, and mites (Santacruz et al. 2023). Monogeneans dominated the parasitic communities, most likely due to their direct one-host life cycle that allows these parasites to colonise isolated habitats and their ability to complete their direct life cycle in the confined space of underground systems (see Chap. 5 for more on monogenean life cycles). Moreover, the presence of the invasive copepod *Lernaea cyprinacea* Linnaeus, 1758, also known as anchor worms, in the cave system implies connectivity to surface waterbodies. This indicates a higher vulnerability of cave fish to invasive parasites if their underground habitats are connected to surface waterbodies, and the possibility of differences in the co-evolutionary adaptation of host immune response within the same species in different cave systems. Likewise, contrasting findings are available from the cave fish *Poecilia mexicana* and their ability to evade parasites in underground systems with high toxic hydrogen sulphide concentrations. While some studies show that fish can find a refuge from parasite infections under these extreme conditions (Tobler et al. 2007), studies from similar systems fail to support this Pathogen Refuge Hypothesis (Riesch et al. 2020).

Other taxa that contain many parasitic species, such as nematodes and leeches, have been found in cave systems around the world (e.g., Phillips et al. 2020; Du Preez et al. 2023). However, many of these species appear to be free-living and predatory rather than parasitic in underground systems such as the deep cave-dwelling leech *Croatobranchus mestrovi* that depends on chemo- and mechanoreceptors to locate its prey (Sket et al. 2001; Phillips et al. 2020). However, since we are far from a full understanding of the diversity of subsurface ecosystems, with an estimated 50,000–100,000 obligate subterranean species (Sánchez-Fernández et al. 2021), a much larger number of parasite taxa with life cycles and transmission strategies adapted to their perpetually dark environments are yet to be described. Metabarcoding samplings from cave systems in the North American Appalachian Mountains revealed a high diversity of microbial eukaryotes, in particular Rhizaria and Alveolata (dinoflagellates, ciliates, apicomplexans) that live as bacterial grazers and parasites (Cahoon and VanGundy 2022). Together with recent macroparasite findings from cave fish populations (Santacruz et al. 2023), this indicates that underground systems might support a much higher diversity of parasitic taxa than we currently know. It therefore remains to be investigated whether the low number of described macroparasites in these systems is accurate, or simply reflective of insufficient sampling efforts in these inaccessible environments.

12.6 Extreme Variation in Intertidal Systems

There are more than 600,000 km of coastline along the world's oceans (NASA 2023), making coastal and intertidal zones a major biome on our planet. Although intertidal systems are not per se extreme environments with harsh chemical or physical conditions, organisms living in this transitory zone between aquatic and terrestrial habitats will experience drastic changes in environmental conditions on a daily basis during the shifting tidal phases. Sessile organisms, such as mussels that occur in high abundances in the intertidal, will experience extreme shifts in their aquatic habitat from wave action and water movement and rather stable ambient temperatures to much colder or hotter aerial temperatures (in winter or summer, respectively) within a matter of minutes and must be able to withstand the physical conditions of both (Helmuth 1998; Fig. 12.6). These sudden changes make the intertidal zone one of the thermally most variable and stressful habitats on earth (Harley 2008).

This highly dynamic and diverse coastline comprises a large variety of different habitats, from rocky shores with tidal pools to mudflat ecosystems or tidal wetlands and estuaries. Many of these habitats, such as mudflats, not only harbour large numbers of polychaetes, bivalves, crustaceans, fish, and birds, but also abundant and diverse communities of parasites (Thieltges et al. 2018). The diverse metazoan parasites commonly found in intertidal systems include a wide range of helminth parasites that use intertidal invertebrates as intermediate hosts and infect fish and birds as definitive hosts, as well as rhizocephalans infecting crabs. Trematodes represent the most common and abundant parasites of intertidal animals and can make up large amounts of the biomass in these systems (Mouritsen and Poulin 2002; Kuris et al. 2008). Overall, some of these parasite groups from intertidal habitats are among the best-studied host-parasite systems and have served as important model systems to investigate the many integral ecological roles of parasitic organisms in ecological communities, including the potential role of global climate change (see Chap. 22) on host–parasite interactions (e.g., Lafferty et al. 2006; Kuris et al. 2008; Prokofiev et al. 2016; Thieltges et al. 2018; Mouritsen et al. 2022). This detailed knowledge of parasites in intertidal ecosystems is probably due to the easy access to these habitats for

Fig. 12.6 Parasites and their hosts in intertidal zones, for example along the rocky shoreline of northern Norway, regularly experience rapid transitions from (**a**) aquatic to (**b**) terrestrial conditions during high and low tide, respectively. Organisms in this habitat must be able to withstand exposure to a wide variety of stressors including temperature fluctuations and wave action. Photos: Erling Svensen (**a**), and Femke Emma de Ruiter (**b**)

surveys or experimental approaches, compared to other more inaccessible 'extreme' environments, such as deep-sea or polar systems.

Many of these studies illustrate how diverse, abundant, and ecologically relevant parasitic organisms can be in stressful and dynamic environments, and how successful parasites with complex life cycles that require multiple hosts (e.g., trematodes) can be under such extremely variable conditions. However, organisms in the intertidal zone that are exposed to sudden temperature fluctuations and thermal stress are particularly vulnerable to heat wave and climate change impacts that have led to reported mass mortality events for mussels, limpets, or barnacles on rocky shores (Seuront et al. 2019 and references therein). Accordingly, the rich and diverse parasite communities in these extremely variable ecosystems are equally susceptible to such environmental changes (see Sures et al. 2023). Moreover, the synergistic effects of parasitism and heat waves have been shown to drastically alter community structure and function in intertidal sand flats in New Zealand, due to the regulating role of the microphallid trematode *Maritrema novaezealandense* on amphipod communities in coastal environments (Mouritsen et al. 2018). These examples highlight that parasites in extreme environments are not only subject to environmental changes but can actively mediate interspecific interactions in changing ecosystems.

12.7 Temporary Aquatic Systems in Deserts and Arid Regions

Deserts and other arid environments seem unlikely places to look for aquatic parasites. Yet, these regions are home to parasitic organisms that rely on aquatic environments to complete their life cycles. Many of these taxa have evolved highly specialised adaptations to overcome the harsh and extreme conditions of their surroundings, namely the extreme limitations of water during most of the year as well as the often extreme temperatures. Therefore, this extreme environment only allows short transmission windows when water becomes available, often during sudden and unpredictable heavy rain events, which requires the parasites to remain in a constant state of readiness until favourable conditions arise (Fig. 12.7).

Fig. 12.7 Temporary waterbodies in arid regions or deserts, like this pond in the Pongola floodplain in northern KwaZulu-Natal, South Africa, often only offer short transmission windows for aquatic parasites after seasonal rainfall. Photo: Christian Selbach

One highly fascinating example of this is the monogenean *Pseudodiplorchis americanus* that parasitises the desert toad *Scaphiopus couchii* (see Tinsley 1999b for a full overview of the adaptations of both host and parasite to their habitat). Both the amphibian host and the monogenean parasites require aquatic habitats for their reproduction and transmission. Desert toads hibernate in the cool ground for most of the year (more than 10 months) until they become active during a brief but intense period of summer rains that create temporary pools and ponds where the amphibians breed. It is during this short window (typically limited to seven hours per night due to the host's nocturnal behaviour, and often for only a few days) that parasites need to release all of their transmission stages into the water to find and infect new hosts (Tinsley 1999b). Despite this short transmission window and annual fluctuations in the onset and duration of rain periods, this parasite has been able to maintain stable populations from year to year in their extreme habitat (Tinsley 1995).

Amphibians inhabiting ephemeral waterbodies in dry regions can also serve as hosts to species-rich and diverse parasite assemblages. The South American frog *Lepidobatrachus llanensis* was shown to harbour 17 helminth species, mostly trematodes and cestodes with an aquatic life cycle, and to a lesser degree nematodes with a terrestrial life cycle (Hamann et al. 2022). Since *L. llanensis* spends most of the dry season buried in a cocoon underground and only emerges during the rainy season, the majority of infections must occur during this short transmission period, highlighting these parasites' high degree of specialisation to complete their life cycle under such conditions. Moreover, the high number of parasite stages using these frogs as intermediate or paratenic hosts (with mammals, reptiles, and birds serving as final hosts, Hamann et al. 2022) show the high connectivity and trophic transmission of parasites between terrestrial and ephemeral aquatic systems in extremely dry environments.

The generally low host densities and the harsh abiotic conditions pose unique challenges for the transmission of parasites in arid environments, especially for parasites with short-lived and fragile dispersal stages, such as digenean trematodes (see Chap. 5). Aquatic parasites in these regions have evolved two solutions to overcome this obstacle, a close synchronisation between the parasite life cycle and host behaviour, and the use of intermediate hosts and mobile vectors that can protect against the unforgiving environment and increase the chances of encountering a suitable host (Warburton 2020). In an increasingly warmer and dryer world (see Chap. 22), extreme conditions are likely to occur more frequently and will increasingly shape parasite communities in other aquatic systems (e.g., in intermittent rivers, Lymbery et al. 2020). The patterns, processes, and transmission strategies of parasites that have adapted to the extreme conditions in deserts and arid regions might therefore help us better anticipate which parasites might be better adapted to survive or even thrive under such circumstances. Yet, both the free-living and parasitic fauna of these ecosystems remain largely understudied and we will require further specific research efforts to better understand the mechanisms of host–parasite interactions in dry regions (Warburton 2020).

12.8 Hypersaline and Soda Lakes

The Dead Sea in Israel is one of the most saline waterbodies on Earth, and it is gradually increasing in salinity as water continues to evaporate from the lake (Gavrieli and Oren 2004). Anyone who has ever experienced the buoyancy or burning sensation of the Dead Sea's hypersaline water will be aware of the extreme conditions of this lake. At salt concentrations of 350 g/l, the lake is almost devoid of life and can only support microbial algal blooms and archaea after heavy rainfall that dilutes the surface waters of the lake (Saccò et al. 2021), highlighting the extreme challenges of dehydration and osmotic stress that high salt concentrations pose to organisms (Rothschild and Mancinelli 2001). However, at lower salt concentrations, many hypersaline waterbodies (those exceeding 35 g/l of dissolved salt) can serve as ecosystems for some unique and highly adapted biological communities (Saccò et al. 2021). Although many hypersaline waterbodies are characterised by a near-neutral pH, they can also range from slightly acidic to high alkaline, as in the case of soda lakes, creating polyextreme conditions for the biota (Seckbach et al. 2013; Saccò et al. 2021).

Metazoan communities in hypersaline waterbodies show a low complexity and simple food web structures, as well as a lower species richness than freshwater communities that typically decrease with increasing levels of salinity or alkalinity. However, these species-poor communities can achieve extremely high abundance due to high biotic productivity in some systems (Seckbach et al. 2013; Saccò et al. 2021). Hypersaline communities are dominated by invertebrates, in particular crustaceans, such as species of the brine shrimp genus *Artemia* that are globally distributed (Seckbach et al. 2013; Saccò et al. 2021). These crustacean communities play a central role in the food webs of these ecosystems, both as filter feeders and as prey items for diverse bird communities (Sánchez et al. 2016; Redón et al. 2020).

A recent review of the parasite fauna in hypersaline systems summarised the presence of at least 85 taxa (not all of which are known to species level) in these extreme environments, including Myxozoa, Nematoda, Acanthocephala, Arthropoda, and Platyhelminthes, with cestodes dominating the parasite communities (Kornyychuk et al. 2023). Overall, parasite diversity was shown to decrease exponentially with increases in salinity levels (Kornyychuk et al. 2023), which appears to mirror the diversity distribution of free-living species in these habitats. Not surprisingly, crustaceans and in particular the widespread and abundant *Artemia* spp. harboured the most species-rich parasite assemblages (22 taxa), making brine shrimp a key host in hypersaline habitats and an important intermediate host for the trophic transmission of parasites to the abundant bird populations at these waterbodies (e.g., flamingos; Redón et al. 2020; Kornyychuk et al. 2023). Extreme saline environments can support both ectoparasites with direct life cycles, such as the monogenean *Gyrodactylus salinae* on fish hosts (Paladini et al. 2011), as well as digenean trematodes with complex indirect life cycles that require the presence of multiple hosts and whose free-swimming transmission stages need to reach the next host under these adverse conditions (Louizi et al. 2022). Moreover, a species of the genus *Gyrodactylus* was described from tilapian fish (*Alcolapia grahami* synonym *Oreochromis grahami*) in Lake Magadi (Kenya), a hypersaline soda lake with pH levels reaching up to 11, and water temperatures reaching up to 42 °C (Dos Santos et al. 2019), showing the polyextreme conditions these ectoparasites can withstand.

Free-living organisms that can tolerate a wide range of salinity levels and live in habitats with a salinity gradient might be able to escape some of their parasites at the cost of more extreme abiotic conditions. For instance, the pupfish *Cyprinodon tularosa* can evade trematode infections in increasingly saline desert streams (Rogowski and Stockwell 2006). However, contrasting evidence has shown that some parasite taxa (e.g., the trematode *Maritrema novaezealandense*) might have a wider tolerance to increasing salinity conditions than their amphipod host in evaporating tide pools (Studer and Poulin 2012; but see Studer and Poulin 2013 for contrasting results). Altogether, this highlights the complexity of host–parasite interactions in habitats with extreme and potentially limiting abiotic conditions.

Similar to freshwater and coastal systems, parasites play central structuring and regulating roles in extreme hypersaline habitats, from affecting survival or feeding rates of their hosts, to regulating free-living populations and energy flow in these ecosystems (Kornyychuk et al. 2023 and references therein). In fact, crucial knowledge on the ecological importance of parasites in aquatic habitats was gained from Californian estuarine salt marsh systems that reach hypersaline conditions in various zones and seasons (Callaway et al. 1990). In this estuarine ecosystem, parasites, in particular digenean trematodes, have been shown to dominate the biotic productivity and standing stock biomass and play crucial roles in the structure and complexity of food webs and energy flow in these systems with high salt concentrations (Kuris et al. 2008; Lafferty et al. 2008). Hypersaline habitats therefore offer interesting model systems to study and investigate host–parasite dynamics along a gradient of abiotic stressors, from the lower range in saltmarsh ecosystems to the upper ranges of salt concentrations in the Dead Sea.

12.9 Other Extreme Habitats: Volcanic Activity and Asphalt Lakes

Waterbodies associated with volcanic activity are typically subject to a combination of high temperatures, extreme pH levels, and high ion concentrations, which presents highly challenging conditions for metazoan parasites and their hosts. However, there is evidence to suggest that some parasites may tolerate or even benefit from being associated with aquatic environments influenced by volcanic activity. For example, three-spined stickleback (*Gasterosteus aculeatus*) inhabiting the warm waters of two geothermal Icelandic lakes (Myvatn and Thingvallavatn) were found to support greater trematode and cestode abundance than sticklebacks from cool water habitats (Karvonen et al. 2013). As the observed pattern occurred despite contrasting differences in the physical habitat (mud or lava rock) associated with the cool and warm water habitats between each lake, the authors suggested that water temperature rather than substrate type had a greater influence on parasite abundance. In contrast, in a study investigating the effect of geothermal influences on the occurrence of *Tubifex* worms infected with *Myxobolus cerebralis*, which causes whirling disease in salmonids, the abundance of *M. cerebralis* was observed to be high in stream reaches with intermediate geothermal influences, in terms of substrate type, temperature, conductivity, and pH, compared to reaches with high or no geothermal influences (Alexander et al. 2011). The authors suggested that the observed patterns are likely due to the combined effects of these geothermal influences on the distribution of the hosts and potentially the ability of parasitised individuals to tolerate the abiotic conditions associated with strong geothermal influences.

Besides volcanic activities that move hot molten rock and gases to the Earth's surface, tectonic activity can also push oil to the surface where it evaporates into sticky asphalt lakes. Schelkle et al. (2011, 2012) investigated ectoparasites of freshwater fish species from the hostile aquatic habitat of the world's largest asphalt lake, Pitch Lake (Trinidad). Such lakes are largely considered toxic to aquatic life due to their characteristic natural asphalt upwellings consisting of hydrocarbons, sulphur, metals, and volcanic ash oils (Schelkle et al. 2012). A single gyrodactylid monogenean, *Ieredactylus rivulus*, was recovered from two fish species known to occur in this lake, Hart's Rivulus *Rivulus hartii* (synonym: *Anablepsoides hartii*) and guppy (*Poecilia reticulata*), indicating that some parasites may be able to survive in these environments (Schelkle et al. 2012). However, such infections may be temporary in nature and reflect the dispersal characteristics of Hart's Rivulus, which is able to migrate overland between waterways, and thus introduce the monogenean to new host species and locations (Schelkle et al. 2011).

12.10 Discussion

The examples discussed here show how aquatic parasites can successfully complete their often complex and intricate life cycles under various extreme conditions, from cold polar regions to the scathing temperatures of deep-sea vents. Although none of the metazoan hosts and parasites in these systems reach the extreme chemical and physical boundaries of life under which only truly extremophile microbial organisms can survive (see Rothschild and Mancinelli 2001), they illustrate how parasitic organisms have evolved to co-exist with their hosts in aquatic environments that pose unique challenges and conditions for their lifestyles. Even though individual hosts can evade less-tolerant parasites under some conditions, e.g., in waters with salinity gradients

(Rogowski and Stockwell 2006), the overall notion that extreme environments can serve as refuges for hosts to escape parasites is not supported (Pathogen Refuge Hypothesis; Riesch et al. 2020). Wherever free-living animals have successfully colonised extreme aquatic habitats, it can be expected that parasites will eventually reach and exploit these resources, showing that parasite richness is tightly coupled to host richness (Hechinger and Lafferty 2005).

However, our ability to evaluate whether hosts in some habitats are easier for parasites to exploit than in other regions remains limited by the patchy and uneven research effort on host–parasite communities, let alone host–parasite interactions, in many extreme environments. For example, are isolated cave systems with their high degree of physical fragmentation and truncated food webs inherently more difficult for a parasitic lifestyle than the highly diverse but not very densely populated vast deep-sea regions, or are the different species discovery rates rather due to unequal research efforts in these systems? Likewise, do parasites fulfil equally important structuring functions in the highly specialised food webs around hot deep-sea vents as they do in the much more accessible and extremely well-studied hypersaline saltmarshes (e.g., Kuris et al. 2008), or along coastal and intertidal zones (e.g., Mouritsen et al. 2018)?

Moreover, the way in which extreme abiotic conditions shape life strategies that are successful under these circumstances can be surprising. For instance, we would expect parasites with simple life cycles and direct transmission to dominate in extreme environments with high host densities but a low species richness (e.g., around hot vents), whereas parasites with trophic transmission and paratenic hosts should be expected to be most successful where host encounters are sparse. On the other hand, ectoparasites with direct transmission should be at a disadvantage in habitats with low host densities (e.g., the vast deep sea). While there are good examples to support these hypotheses, for instance ectoparasite diversity appears to decline at greater depths of the ocean and increases in regions where more hosts are concentrated, such as undersea canyons (Quattrini and Demopoulos 2016), other successful parasite life strategies in extreme environments are more surprising. For instance, trematodes with complex multi-host life cycles and free-swimming transmission stages appear to be highly successful parasites in the superheated waters around geothermal vents (Dykman et al. 2023), and monogeneans have successfully adapted to survive and complete their life cycles on amphibians that spend most of their lifetime burrowed underground in the desert and only access waterbodies for a few hours (Tinsley 1999b).

The aquatic ecosystems discussed in this chapter all present unique environmental conditions for the parasitic and free-living organisms inhabiting them. Not only are the physical and chemical properties (e.g., temperature, pressure, salinity, pH) of some of these habitats radically different; the ecosystems also present varying

Table 12.1 Overview of extreme aquatic habitats and the unique abiotic conditions these environments pose to host-parasite systems

Habitat	Unique environmental conditions
Deep sea	High pressure, low temperature, no sunlight, low energy availability
Deep-sea hydrothermal vents, methane seeps, and brine lakes	High pressure, high temperature, extreme chemical conditions, isolated 'islands'
Volcanic and tectonic activity	High temperature, extreme chemical conditions (pH, ions)
Arctic/polar regions	Low temperature, pronounced seasonality (especially in freshwater systems)
Subsurface waterbodies	No/limited sunlight, low energy availability, highly fragmented habitat
Temporary aquatic systems in deserts and arid environments	Dry environment with short-term waterbodies, often appearing spontaneously
Intertidal systems	Regular, extreme, transition from aquatic to terrestrial conditions, wave exposure, high temperature fluctuations
Hypersaline lakes and soda lakes	High salinity (>35 g/l of dissolved salt), high pH in soda lakes

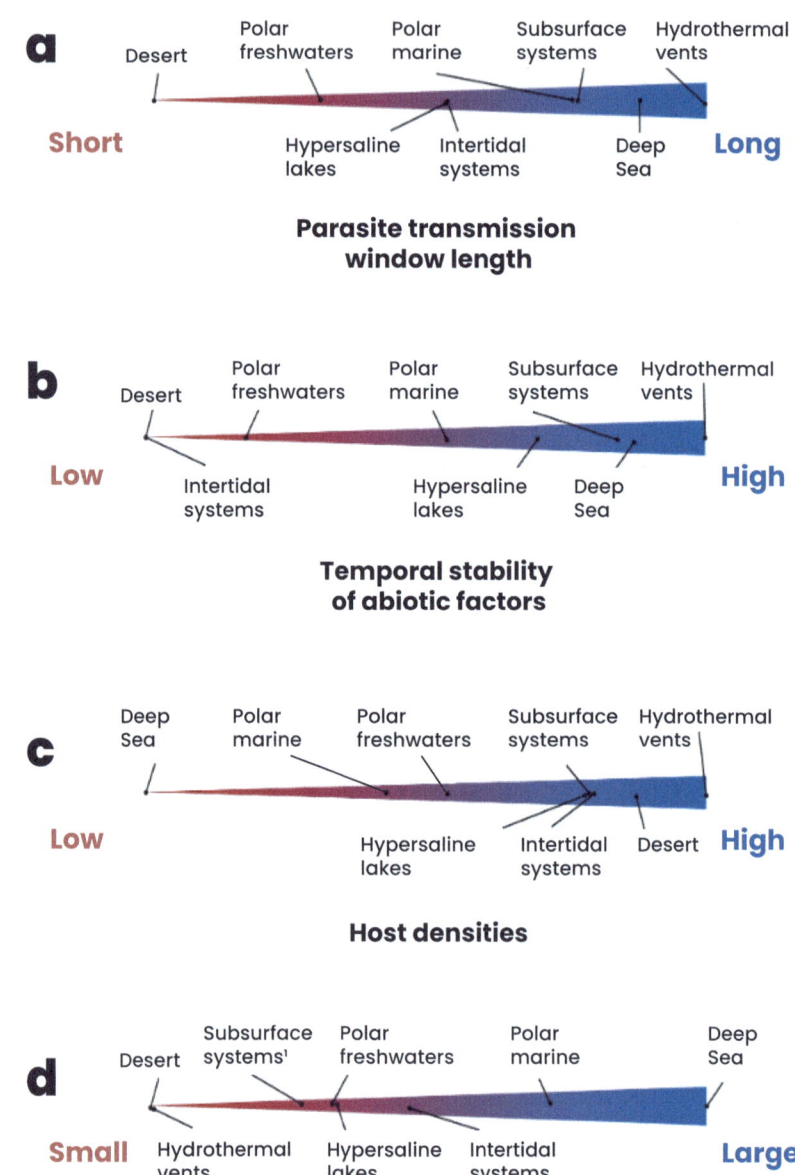

Fig. 12.8 Relative comparison of (**a**) the parasite transmission window length, (**b**) temporal (diurnal and/or seasonal) stability of the abiotic factors, (**c**) host densities, and (**d**) available habitat size for host-parasite communities in extreme environments. All points are based on the habitat descriptions in the available literature (see main text). [a]Even though subsurface waterbodies form vast networks, the systems are characterised by a high degree of fragmentation, hence the individual habitats are considered to be rather small in size

spatio-temporal conditions for the parasites' life cycles and transmission strategies (Table 12.1, Fig. 12.8). For instance, both intertidal and temporary waterbodies in deserts are characterised by a high degree of fluctuation of the abiotic factors due to strong tidal and seasonal effects, respectively. This can either result in extremely limited and suddenly occurring transmission windows of just a few hours per year in ephemeral desert ponds (Tinsley 1999b), or in very predictable and cyclical transmission phases during tidal shifts twice per day (e.g., Studer and Poulin 2013). On the other end of the continuum, the highly stable abiotic conditions around hydrothermal deep-sea vents, free of any diurnal or seasonal fluctuations, should offer near-constant transmission opportunities in these very concentrated habitats with high densities of host organisms. Altogether, the various aquatic habitats offer highly varied physio-chemical and spatio-temporal settings under which metazoan parasites have successfully been able to co-evolve with the, often highly specialised, hosts in these environments.

However, under current and predicted global climate change scenarios, many ecosystems around the world are changing at an increasingly rapid rate, with wide reaching implications for the free-living and parasitic communities (Altizer et al. 2013; Marcogliese 2023; Sures et al. 2023). These changes also drastically impact some extreme ecosystems, such as coastal and intertidal communities, where organisms already experience highly stressful fluctuations and might more easily reach a tipping point (Mouritsen et al. 2018), or in polar and Arctic regions, where the impacts of global warming are already felt and expected to increase much further (Overland et al. 2019). Climate change-related impacts in other extreme environments, such as the deep sea or in isolated caves might be less obvious or will possibly only manifest themselves with some delay. At the same time, anthropogenic actions will not only amplify the conditions in some naturally existing extreme environments, e.g., more regular drying of temporary waterbodies (Lymbery et al. 2020) or increasing salt concentrations in hypersaline systems due to evaporation (Saccò et al. 2021) but can also create new 'extremes' and stressful conditions for free-living and parasitic taxa. For instance, mining run-off can create acid and metal pollution in extreme concentrations or chemical combinations that do not occur naturally, nuclear waste pollution can put extreme radiation stress on organisms, or synthetic materials or nanoparticles can expose hosts and parasites to new substances or particles that create unprecedented extremes in aquatic habitats (see also Chap. 20). For this reason, understanding host–parasite assemblages under the various challenging conditions in naturally occurring extreme environments discussed in this chapter can provide insights into how parasites might respond to newly emerging and upcoming 'anthropogenic' extremes, from pollution to ocean acidification, or more frequent heat waves. Arctic and polar regions may be particularly useful models or systems for understanding the impacts of climate change on parasite ecology, transmission, or host–parasite interactions (Kutz et al. 2005, 2009).

Despite their ecological importance and usefulness to predict future changes, we lack baseline parasite biodiversity data in aquatic systems of the Arctic (Hoberg et al. 2013). Likewise, Poulin et al. (2016) called for a stronger integration of parasitology and marine ecology to better understand the functioning of these natural ecosystems that harbour some of the largest extreme habitats on the planet. In fact, we would argue that such an integrated understanding of parasites and their interactions with hosts and abiotic environments is needed for all the extreme, and often underexplored, aquatic habitats. Focussing on a small and narrow range of host–parasite systems from relatively few (and often easily accessible) habitats and environments can constrain our understanding of the roles of parasites in other habitats and the functioning of these complex ecosystems (Poulin et al. 2016; Quattrini and Demopoulos 2016).

12.11 Conclusion

The examples of parasites in various extreme environments in this chapter highlight the vast diversity of parasitic lifestyles that have evolved to successfully overcome the challenging abiotic conditions in these habitats. In fact, for most of these species, living outside the specific ranges that we consider 'extreme' and challenging might be impossible. For the parasites of deep-sea fishes, the high-pressure environment with low temperatures offers a benign and stable surrounding to complete their life cycle (Bray et al. 1999), just like the short-term waterbodies in deserts offer the perfect transmission conditions for the highly adapted monogeneans of amphibians in these regions. As with other aquatic systems, parasites in more 'extreme' environments constitute diverse, highly specialised, ecologically important components of these ecosystems, and last but not least, offer highly fascinating insights into the life on our planet.

Acknowledgements We thank Robert Poulin for his very helpful and constructive feedback and comments on an earlier version of the chapter, and Ingrid Wiedmann, Markus Molis, Rodrigo Salvador, and Zoe Koenig for their help in finding photographs of the various extreme habitats.

References

Aleuy OA, Kutz S (2020) Adaptations, life-history traits and ecological mechanisms of parasites to survive extremes and environmental unpredictability in the face of climate change. Int J Parasitol Parasites Wildl 12:308–317. https://doi.org/10.1016/j.ijppaw.2020.07.006

Alexander JD, Kerans BL, Koel TM, Rasmussen C (2011) Context-specific parasitism in *Tubifex tubifex* in geothermally influenced stream reaches in Yellowstone National Park. J North Am Benthol Soc 30(3):853–867. https://doi.org/10.1899/10-043.1

Altizer S, Ostfeld RS, Johnson PT, Kutz S, Harvell CD (2013) Climate change and infectious diseases: From evidence to a predictive framework. Science 341(6145):514–519. https://doi.org/10.1126/science.1239401

Bada JL (2004) How life began on Earth: a status report. Earth Planet Sci Lett 226:1–15. https://doi.org/10.1016/j.epsl.2004.07.036

Blasco-Costa I, Faltýnková A, Georgieva S, Skírnisson K, Scholz T, Kostadinova A (2014) Fish pathogens near the Arctic Circle: molecular, morphological and ecological evidence for unexpected diversity of *Diplostomum* (Digenea: diplostomidae) in Iceland. Int J Parasitol 44:703–715. https://doi.org/10.1016/j.ijpara.2014.04.009

Borgstrøm R, Isdahl T, Svenning MA (2015) Population structure, biomass, and diet of landlocked Arctic charr (*Salvelinus alpinus*) in a small, shallow High Arctic lake. Polar Biol 38:309–317. https://doi.org/10.1007/s00300-014-1587-6

Botosaneanu L (ed) (1986) Stygofauna Mundi. EJ Brill and Dr W Backhuys, Leiden

Boxshall GA (1998) Host specificity in copepod parasites of deep-sea fishes. J Mar Syst 15:215–223. https://doi.org/10.1016/S0924-7963(97)00058-4

Bray RA (2004) The bathymetric distribution of the digenean parasites of deep-sea fishes. Folia Parasitol 51(2/3):268–274. https://doi.org/10.14411/fp.2004.032

Bray RA (2020) Digenean parasites of deep-sea teleosts: A progress report. Int J Parasitol Parasites Wildl 12:251–264. https://doi.org/10.1016/j.ijppaw.2020.01.007

Bray RA, Littlewood DTJ, Herniou EA, Williams B, Henderson RE (1999) Digenean parasites of deep-sea teleosts: A review and case studies of intrageneric phylogenies. Parasitology 119(1):125–144. https://doi.org/10.1017/s0031182000084687

Brenke N, Fanenbruch M, George KH (2018) A new parasitic deep-sea copepod from the Angola Basin (southeast Atlantic Ocean): *Abyssotaurus vermiambatus* gen. et sp. nov. (Copepoda: Cyclopoida: Serpulidicolidae Stock, 1979), with remarks on serpulidicolid systematics and a key to the species. Mar Biodivers 48:517–530. https://doi.org/10.1007/s12526-017-0724-1

Buesseler KO, Lamborg CH, Boyd PW, Lam PJ, Trull TW, Bidigare RR, Bishop JK, Casciotti KL, Dehairs F, Elskens M, Honda M (2007) Revisiting carbon flux through the ocean's twilight zone. Science 316(5824):567–571. https://doi.org/10.1126/science.1137959

Cahoon AB, VanGundy RD (2022) Alveolates (dinoflagellates, ciliates and apicomplexans) and Rhizarians are the most common microbial eukaryotes in temperate Appalachian karst caves. Environ Microbiol Rep 14(4):538–548. https://doi.org/10.1111/1758-2229.13060

Callaway RM, Jones S, Ferren WR Jr, Parikh A (1990) Ecology of a Mediterranean climate estuarine wetland at Carpinteria. California: plant distribution and soil salinity in the upper marsh. Can J Bot 68(5):1139–1145. https://doi.org/10.1139/b90-144

Campbell WC, Overstreet RM (1994) Historical basis of binomials assigned to helminths collected on Scott's last Antarctic expedition. J Helminthol Soc Wash 61:1–11

Campbell RA, Haedrich RL, Munroe TA (1980) Parasitism and ecological relationships among deep-sea benthic fishes. Mar Biol 57:301–313

Combes C, Morand S (1999) Do parasites live in extreme environments? Constructing hostile niches and living in them. Parasitology 119(S1):S107–S110. https://doi.org/10.1017/s0031182000084663

Conradi M, Bandera ME, Marin I, Martin D (2015) Polychaete-parasitizing copepods from the deep-sea Kuril-Kamchatka Trench (Pacific Ocean), with the description of a new Ophelicola species and comments on the currently known annelidicolous copepods. Deep Res Part II Top Stud Oceanogr 111:147–165. https://doi.org/10.1016/j.dsr2.2014.08.018

Costello MJ, Chaudhary C (2017) Marine biodiversity, biogeography, deep-sea gradients, and conservation. Curr Biol 27(11):R511–R527. https://doi.org/10.1016/j.cub.2017.04.060

Coulson SJ (2000) A review of the terrestrial and freshwater invertebrate fauna of the High Arctic archipelago of Svalbard. Nor J Entomol 47:41–63

Culver DC, Sket B (2000) Hotspots of subterranean biodiversity in caves and wells. J Cave Karst Stud 62(1):11–17

Culver D, Pipan T, Schneider K (2009) Vicariance, dispersal and scale in the aquatic subterranean fauna of karst regions. Freshw Biol 54(4):918–929. https://doi.org/10.1111/j.1365-2427.2007.01856.x

Dailey MD, Vogelbein WK (1991) Parasite fauna of three species of Antarctic whales with reference to their use as potential stock indicators. Fish Bull 89(3):355–365. https://scholarworks.wm.edu/vimsarticles/607

De Buron I, Morand S (2004) Deep-sea hydrothermal vent parasites: Why do we not find more? Parasitology 128(1):1–6. https://doi.org/10.1017/S0031182003004347

Diaz JI, Fusaro B, Longarzo L, Coria NR, Vidal V, D'Amico V, Barbosa A (2016) Gastrointestinal helminths of Adélie penguins (*Pygoscelis adeliae*) from Antarctica. Polar Res 35(1):28516. https://doi.org/10.3402/polar.v35.28516

Dos Santos QM, Maina JN, Avenant-Oldewage A (2019) *Gyrodactylus magadiensis* n. sp. (Monogenea, Gyrodactylidae) parasitising the gills of *Alcolapia grahami* (Perciformes, Cichlidae), a fish inhabiting the extreme environment of Lake Magadi, Kenya TT—*Gyrodactylus magadiensis* n. sp. (Monogenea, Gyrodactyli). Parasite 26:76. https://doi.org/10.1051/parasite/2019077

Du Preez GC, Silva MS, Fourie H, Girgan C, Netherlands EC, Swart A, Ferreira RL (2023) Nematode dynamics in an African dolomite cave: What is the role of environmental filtering in spatial and temporal distribution? Basic Appl Ecol 71:18–32. https://doi.org/10.1016/j.baae.2023.05.002

Dykman LN, Tepolt CK, Kuris AM, Solow AR, Mullineaux LS (2023) Parasite diversity at isolated, disturbed hydrothermal vents. Proc R Soc B Biol Sci 290: 20230877. https://doi.org/10.1098/rspb.2023.0877

Eerkes-Medrano L, Huntington HP (2021) Untold stories: indigenous knowledge beyond the changing Arctic cryosphere. Front Clim 3:1–16. https://doi.org/10.3389/fclim.2021.675805

Galaktionov KV (2017) Patterns and processes influencing helminth parasites of Arctic coastal communities during climate change. J Helminthol 91(4):387–408. https://doi.org/10.1017/S0022149X17000232

Galaktionov KV, Bustnes JO, Bårdsen BJ, Wilson JG, Nikolaev KE, Sukhotin AA, Skírnisson K, Saville DH, Ivanov MV, Regel KV (2015) Factors influencing the distribution of trematode larvae in the blue mussels *Mytilus edulis* in the North Atlantic and Arctic Oceans. Mar Biol 162:193–206. https://doi.org/10.1007/s00227-014-2586-4

Galaktionov KV, Nikolaev KE, Aristov DA, Levakin IA, Kozminsky EV (2019) Parasites on the edge: patterns of trematode transmission in the Arctic intertidal at the Pechora Sea (South-Eastern Barents Sea). Polar Biol 42:1719–1737. https://doi.org/10.1007/s00300-018-2413-3

Gallagher CP, Dick TA (2011) Ecological characteristics of ninespine stickleback *Pungitius pungitius* from southern Baffin Island, Canada. Ecol Freshw Fish 20(4):646–655. https://doi.org/10.1111/j.1600-0633.2011.00516.x

Garde E, Jung-Madsen S, Ditlevsen S, Hansen RG, Zinglersen KB, Heide-Jørgensen MP (2018) Diving behavior of the Atlantic walrus in high Arctic Greenland and Canada. J Exp Mar Bio Ecol 500:89–99. https://doi.org/10.1016/j.jembe.2017.12.009

Gavrieli I, Oren A (2004) The dead sea as a dying lake. In: Nihoul JCJ, Zavialov PO, Micklin PP (eds) Dying and dead seas climatic versus anthropic causes, NATO science series: IV: Earth and environmental sciences, vol 36. Springer, Dordrecht, pp 287–305. https://doi.org/10.1007/978-94-007-0967-6_11

Gibert J, Deharveng L (2002) Subterranean ecosystems: A truncated functional biodiversity. BioScience 52(6):473–481. https://doi.org/10.1641/0006-3568(2002)052[0473:SEATFB]2.0.CO;2

Gibert J, Culver DC, Dole-Olivier MJ, et al (2009) Assessing and conserving groundwater biodiversity: Synthesis and perspectives. Freshw Biol 54:930–941. https://doi.org/10.1111/j.1365-2427.2009.02201.x

González-Acuña D, Hernández J, Moreno L, Herrmann B, Palma R, Latorre A, Medina-Vogel G, Kinsella MJ, Martín N, Araya K, Torres I (2013) Health evaluation of wild gentoo penguins (Pygoscelis papua) in the Antarctic Peninsula. Polar Biol 36:1749–1760. https://doi.org/10.1007/s00300-013-1394-5

Gordeev II, Sokolov SG (2016) Parasites of the Antarctic toothfish (*Dissostichus mawsoni* Norman, 1937) (Perciformes, Nototheniidae) in the Pacific sector of the Antarctic. Polar Res 35(1):29364. https://doi.org/10.3402/polar.v35.29364

Gordeev II, Sokolov SG (2017) Helminths and the feeding habits of the marbled moray cod *Muraenolepis marmorata* Günther, 1880 (Gadiformes, Muraenolepididae) in the Ross Sea (Southern Ocean). Polar Biol 40(6):1311–1318. https://doi.org/10.1007/s00300-016-2055-2

Graefe G (1971) Die Temperatur des Lebensraumes und ihre Wirkung auf Cercarien. Überlegungen und Versuche im Anschluß an Beobachtungen in der Antarktis. Parasitologische Schriftenreihe 21:151–156

Hamann MI, González CE, Duré MI, Palomas YS (2022) Helminth community in the Llanos frog, *Lepidobatrachus llanensis* (Ceratophryidae), from the Dry Chaco. South Am J Herpetol 25(1):12–17. https://doi.org/10.2994/SAJH-D-20-00054.1

Hammar J (2000) Cannibals and parasites: conflicting regulators of bimodality in high latitude Arctic char, *Salvelinus alpinus*. Oikos 88(1):33–47. https://doi.org/10.1034/j.1600-0706.2000.880105.x

Harley CDG (2008) Tidal dynamics, topographic orientation, and temperature-mediated mass mortalities on rocky shores. Mar Ecol Prog Ser 371:37–46. https://doi.org/10.3354/meps07711

Haumann FA, Moorman R, Riser SC, Smedsrud LH, Maksym T, Wong AP, Wilson EA, Drucker R, Talley LD, Johnson KS, Key RM (2020) Supercooled southern ocean waters. Geophys Res Lett 47(20). https://doi.org/10.1029/2020GL090242

Hechinger RF, Lafferty KD (2005) Host diversity begets parasite diversity: bird final hosts and trematodes in snail intermediate hosts. Proc R Soc B 272(1567):1059–1066. https://doi.org/10.1098/rspb.2005.3070

Helmuth BST (1998) Intertidal mussel microclimates: predicting the body temperature of a sessile invertebrate. Ecol Monogr 68(1):51–74. https://doi.org/10.2307/2657143

Hemmingsen W, MacKenzie K (2001) The parasite fauna of the Atlantic cod, *Gadus morhua* L. Adv Mar Biol 40:1–80. https://doi.org/10.1016/S0065-2881(01)40002-2

Herring P (2002) The biology of the deep ocean. Oxford University Press, Oxford

Hlebec D, Podnar M, Kučinić M, Harms D (2023) Molecular analyses of pseudoscorpions in a subterranean biodiversity hotspot reveal cryptic diversity and microendemism. Sci Rep 13(1):430. https://doi.org/10.1038/s41598-022-26298-5

Hoberg EP, Kutz SJ, Cook JA, Galaktionov K, Haukisalmi V, Henttonen H, Laaksonen S, Makarikov A, Marcogliese DJ (2013) Parasites in terrestrial, freshwater and marine systems. In: Meltofte H (ed) Arctic biodiversity assessment status and trends in Arctic biodiversity. Conservation of Arctic Flora and Fauna, Iceland, pp 476–505

Jamieson AJ, Singleman G, Linley TD, Casey S (2021) Fear and loathing of the deep ocean: why don't people care about the deep sea? ICES J Mar Sci 78(3):797–809. https://doi.org/10.1093/icesjms/fsaa234

Janiec K (1996) The comparison of freshwater invertebrates of Spitsbergen (Arctic) and King George Island (Antarctic). Pol Polar Res 17:173–202

Johansen CE, Lydersen C, Aspholm PE, Haug T, Kovacs KM (2010) Helminth parasites in ringed seals (*Pusa hispida*) from Svalbard, Norway with special emphasis on nematodes: variation with age, sex, diet, and location of host. J Parasitol 96(5):946–953. https://doi.org/10.1645/GE-1685.1

Karvonen A, Kristjánsson BK, Skúlason S, Lanki M, Rellstab C, Jokela J (2013) Water temperature, not fish morph, determines parasite infections of sympatric Icelandic threespine sticklebacks (*Gasterosteus aculeatus*). Ecol Evol 3(6):1507–1517. https://doi.org/10.1002/ece3.568

Klimpel S, Palm HW, Busch MW, Kellermanns E, Rückert S (2006) Fish parasites in the Arctic deep-sea: Poor diversity in pelagic fish species vs. heavy parasite load in a demersal fish. Deep Res Part I Oceanogr Res Pap 53(7):1167–1181. https://doi.org/10.1016/j.dsr.2006.05.009

Klimpel S, Busch MW, Kuhn T, Rohde A, Palm HW (2010) The *Anisakis simplex* complex off the South Shetland Islands (Antarctica): endemic populations versus introduction through migratory hosts. Mar Ecol Prog Ser 403:1–11. https://doi.org/10.3354/meps08501

Klimpel S, Kuhn T, Busch MW, Karl H, Palm HW (2011) Deep-water life cycle of *Anisakis paggiae* (Nematoda: Anisakidae) in the Irminger Sea indicates kogiid whale distribution in north Atlantic waters. Polar Biol 34:899–906. https://doi.org/10.1007/s00300-010-0946-1

Knoll AH (2015) Life on a young planet: the first three billion years of evolution on earth. Princeton University Press, Princeton, NJ

Kornyychuk Y, Anufriieva E, Shadrin N (2023) Diversity of parasitic animals in hypersaline waters: A review. Diversity 15(3):409. https://doi.org/10.3390/d15030409

Kunze E, Dower JF, Beveridge I, Dewey R, Bartlett KP (2006) Observations of biologically generated turbulence in a coastal inlet. Science 313(5794):1768–1770. https://doi.org/10.1126/science.1129378

Kuris AM, Hechinger RF, Shaw JC, Whitney KL, Aguirre-Macedo L, Boch CA, Dobson AP, Dunham EJ, Fredensborg BL, Huspeni TC, Lorda J (2008) Ecosystem energetic implications of parasite and free-living biomass in three estuaries. Nature 454(7203):515–518. https://doi.org/10.1038/nature06970

Kutz S, Hoberg E, Polley L, Jenkins E (2005) Global warming is changing the dynamics of Arctic host–parasite systems. Proc R Soc B Biol Sci 272(1581):2571–2576. https://doi.org/10.1098/rspb.2005.3285

Kutz SJ, Jenkins EJ, Veitch AM, Ducrocq J, Polley L, Elkin B, Lair S (2009) The Arctic as a model for anticipating, preventing, and mitigating climate change impacts on host–parasite interactions. Vet Parasitol 163(3):217–228. https://doi.org/10.1016/j.vetpar.2009.06.008

Lafferty KD, Hechinger RF, Shaw JC, Whitney KL, Kuris AM (2006) Food webs and parasites in a salt marsh ecosystem. In: Collinge S, Ray C (eds) Disease ecology: Community structure and pathogen dynamics. Oxford University Press, Oxford, pp 119–134

Lafferty KD, Allesina S, Arim M, Briggs CJ, De Leo G, Dobson AP, Dunne JA, Johnson PT, Kuris AM, Marcogliese DJ, Martinez ND (2008) Parasites in food webs: the ultimate missing links. Ecol Lett 11(6):533–546. https://doi.org/10.1111/j.1461-0248.2008.01174.x

Louizi H, Hill-Spanik KM, Qninba A, Connors VA, Belafhaili A, Agnèse JF, Pariselle A, de Buron I (2022) Parasites of Moroccan desert *Coptodon guineensis* (Pisces, Cichlidae): Transition and resilience in a simplified hypersaline ecosystem. Parasite 29:64. https://doi.org/10.1051/parasite/2022064

Lyholt HCK, Buchmann K (1996) *Diplostomum spathaceum*: effects of temperature and light on cercarial shedding and infection of rainbow trout. Dis Aquat Org 25(3):169–173. https://doi.org/10.3354/dao025169

Lymbery AJ, Lymbery SJ, Beatty SJ (2020) Fish out of water: Aquatic parasites in a drying world. Int J

Parasitol Parasites Wildl 12:300–307. https://doi.org/10.1016/j.ijppaw.2020.05.003

MacKenzie K (2017) The history of Antarctic parasitological research. In: Klimpel S, Kuhn T, Mehlhorn H (eds) Biodiversity and evolution of parasitic life in the Southern Ocean. Springer, Heidelberg, pp 13–31

Mammola S, Cardoso P, Culver DC, Deharveng L, Ferreira RL, Fišer C, Galassi DM, Griebler C, Halse S, Humphreys WF, Isaia M (2019) Scientists' warning on the conservation of subterranean ecosystems. BioScience 69(8):641–650. https://doi.org/10.1093/biosci/biz064

Marcogliese DJ (2002) Food webs and the transmission of parasites to marine fish. Parasitology 124(7):83–99. https://doi.org/10.1017/S003118200200149X

Marcogliese DJ (2023) Major drivers of biodiversity loss and their impacts on helminth parasite populations and communities. J Helminthol 97:e34. https://doi.org/10.1017/S0022149X2300010X

Martin W, Baross J, Kelley D, Russell MJ (2008) Hydrothermal vents and the origin of life. Nat Rev Microbiol 6:805–814. https://doi.org/10.1038/nrmicro1991

Mateu P, Nardi V, Fraija-Fernández N, Mattiucci S, de Sola LG, Raga JA, Fernández M, Aznar FJ (2015) The role of lantern fish (Myctophidae) in the life-cycle of cetacean parasites from western Mediterranean waters. Deep Res Part I Oceanogr Res Pap 95:115–121. https://doi.org/10.1016/j.dsr.2014.10.012

Mehlhorn B, Mehlhorn H (2017) Lice on seals in the Antarctic waters and lice in temperate climates. In: Klimpel S, Kuhn T, Mehlhorn H (eds) Biodiversity and evolution of parasitic life in the Southern Ocean. Springer, Heidelberg, pp 205–215

Merino N, Aronson HS, Bojanova DP, et al (2019) Living at the extremes: extremophiles and the limits of life in a planetary context. Front Microbiol 10. https://doi.org/10.3389/fmicb.2019.00780

Moravec F, Huffman DG (1988) *Rhabdochona longleyi* sp. n. (Nematoda: Rhabdochonidae) from blind catfishes, *Trogloglanis pattersoni* and *Satan eurystomus* (Ictaluridae) from the subterranean waters of Texas. Folia Parasitol 35:235–243

Moravec F, Justine J-L (2011) Cucullanid nematodes (Nematoda: Cucullanidae) from deep-sea marine fishes off New Caledonia, including *Dichelyne etelidis* n. sp. Syst Parasitol 78:95–108. https://doi.org/10.1007/s11230-010-9281-8

Mouritsen KN, Poulin R (2002) Parasitism, community structure and biodiversity in intertidal ecosystems. Parasitology 124(7):101–S117. https://doi.org/10.1017/S0031182002001476

Mouritsen KN, Sørensen MM, Poulin R, Fredensborg BL (2018) Coastal ecosystems on a tipping point: Global warming and parasitism combine to alter community structure and function. Glob Chang Biol 24(9):4340–4356. https://doi.org/10.1111/gcb.14312

Mouritsen KN, Dalsgaard NP, Flensburg SB, Madsen JC, Selbach C (2022) Fear of parasitism affects the functional role of ecosystem engineers. Oikos 2022(5):e08965. https://doi.org/10.1111/oik.08965

Muñoz G, Cartes FD (2020) Endoparasitic diversity from the Southern Ocean: is it really low in Antarctic fish? J Helminthol 94:e180. https://doi.org/10.1017/S0022149X20000590

Nachtsheim DA, Jerosch K, Hagen W, Plötz J, Bornemann H (2017) Habitat modelling of crabeater seals (*Lobodon carcinophaga*) in the Weddell Sea using the multivariate approach Maxent. Polar Biol 40:961–976. https://doi.org/10.1007/s00300-016-2020-0

Nakagawa S, Takai K (2008) Deep-sea vent chemoautotrophs: Diversity, biochemistry and ecological significance. FEMS Microbiol Ecol 65(1):1–14. https://doi.org/10.1111/j.1574-6941.2008.00502.x

NASA (2023) Living ocean. https://web.archive.org/web/20230806020252/https://science.nasa.gov/earth-science/oceanography/living-ocean. Accessed 14 Aug 2023

Nelson JS (2006) Fishes of the world, 4th edn. Wiley, Hoboken, NJ

Nikolaev KE, Levakin IA, Galaktionov KV (2020) Seasonal dynamics of trematode infection in the first and the second intermediate hosts: A long-term study at the subarctic marine intertidal. J Sea Res 164:101931. https://doi.org/10.1016/j.seares.2020.101931

Northup DE, Lavoie KH (2001) Geomicrobiology of caves: A review. Geomicrobiol J 18(3):199–222. https://doi.org/10.1080/01490450152467750

Overland J, Dunlea E, Box JE, Corell R, Forsius M, Kattsov V, Olsen MS, Pawlak J, Reiersen LO, Wang M (2019) The urgency of Arctic change. Polar Sci 21:6–13. https://doi.org/10.1016/j.polar.2018.11.008

Paladini G, Huyse T, Shinn AP (2011) *Gyrodactylus salinae* n. sp. (Platyhelminthes: Monogenea) infecting the south European toothcarp *Aphanius fasciatus* (Valenciennes) (Teleostei, Cyprinodontidae) from a hypersaline environment in Italy. Parasit Vectors 4:1–12. https://doi.org/10.1186/1756-3305-4-100

Palm HW, Klimpel S, Walter T (2007) Demersal fish parasite fauna around the South Shetland Islands: High species richness and low host specificity in deep Antarctic waters. Polar Biol 30:1513–1522. https://doi.org/10.1007/s00300-007-0312-0

Paterson RA, Knudsen R, Blasco-Costa I, et al (2019) Determinants of parasite distribution in arctic charr populations: catchment structure versus dispersal potential. J Helminthol 93:559–566. https://doi.org/10.1017/S0022149X18000482

Peuß R, Box AC, Chen S, Wang Y, Tsuchiya D, Persons JL, Kenzior A, Maldonado E, Krishnan J, Scharsack JP, Slaughter BD (2020) Adaptation to low parasite abundance affects immune investment and immunopathological responses of cavefish. Nat Ecol Evol 4(10):1416–1430. https://doi.org/10.1038/s41559-020-1234-2

Phillips AJ, Govedich FR, Moser WE (2020) Leeches in the extreme: Morphological, physiological, and behavioral adaptations to inhospitable habitats. Int J Parasitol Parasites Wildl 12:318–325. https://doi.org/10.1016/j.ijppaw.2020.09.003

Polyakova TA, Gordeev II (2021) Parasites as an inseparable part of Antarctic and Subantarctic marine biodiversity. In: Morozov EG, Flint MV, Spiridonov VA (eds) Antarctic Peninsula region of the Southern Ocean: Oceanography and ecology. Springer, Heidelberg, pp 321–354

Poulin R, Blasco-Costa I, Randhawa HS (2016) Integrating parasitology and marine ecology: seven challenges towards greater synergy. J Sea Res 113:3–10. https://doi.org/10.1016/j.seares.2014.10.019

Prokofiev VV, Galaktionov KV, Levakin IA (2016) Patterns of parasite transmission in polar seas: Daily rhythms of cercarial emergence from intertidal snails. J Sea Res 113:85–98. https://doi.org/10.1016/j.seares.2015.07.007

Quattrini AM, Demopoulos AWJ (2016) Ectoparasitism on deep-sea fishes in the western North Atlantic: In situ observations from ROV surveys. Int J Parasitol Parasites Wildl 5(3):217–228. https://doi.org/10.1016/j.ijppaw.2016.07.004

Rabone M, Wiethase JH, Simon-Lledó E, Emery AM, Jones DO, Dahlgren TG, Bribiesca-Contreras G, Wiklund H, Horton T, Glover AG (2023) How many metazoan species live in the world's largest mineral exploration region? Curr Biol 33(12):2383–2396. https://doi.org/10.1016/j.cub.2023.04.052

Ramirez-Llodra E, Brandt A, Danovaro R, De Mol B, Escobar E, German CR, Levin LA, Martinez Arbizu P, Menot L, Buhl-Mortensen P, Narayanaswamy BE (2010) Deep, diverse and definitely different: Unique attributes of the world's largest ecosystem. Biogeosciences 7(9):2851–2899. https://doi.org/10.5194/bg-7-2851-2010

Redón S, Vasileva GP, Georgiev BB, Gajardo G (2020) Exploring parasites in extreme environments of high conservational importance: *Artemia franciscana* (Crustacea: Branchiopoda) as intermediate host of avian cestodes in Andean hypersaline lagoons from Salar de Atacama, Chile. Parasitol Res 119:3377–3390. https://doi.org/10.1007/s00436-020-06768-3

Rex MA, Etter RJ (2010) Deep-sea biodiversity. Harvard University Press, Cambridge

Riesch R, Morley NJ, Jourdan J, Arias-Rodriguez L, Plath M (2020) Sulphide-toxic habitats are not refuges from parasite infections in an extremophile fish. Acta Oecol 106:103602. https://doi.org/10.1016/j.actao.2020.103602

Robison BH (2009) Conservation of deep pelagic biodiversity. Conserv Biol 23(4):847–858. https://doi.org/10.1111/j.1523-1739.2009.01219.x

Rocka A (2006) Helminths of Antarctic fishes: Life cycle biology, specificity and geographical distribution. Acta Parasitol 51:26–35. https://doi.org/10.2478/s11686-006-0003-y

Rogowski DL, Stockwell CA (2006) Parasites and salinity: Costly tradeoffs in a threatened species. Oecologia 146:615–622. https://doi.org/10.1007/s00442-005-0218-x

Rose A, Seifried S, Willen E, George KH, Veit-Köhler G, Bröhldick K, Drewes J, Moura G, Arbizu PM, Schminke HK (2005) A method for comparing within-core alpha diversity values from repeated multicorer samplings, shown for abyssal Harpacticoida (Crustacea: Copepoda) from the Angola Basin. Org Divers Evol 5:3–17. https://doi.org/10.1016/j.ode.2004.10.001

Rothschild LJ, Mancinelli RL (2001) Life in extreme environments. Nature 409(6823):1092–1101. https://doi.org/10.1038/35059215

Rubin KH, Soule SA, Chadwick WW Jr, Fornari DJ, Clague DA, Embley RW, Baker ET, Perfit MR, Caress DW, Dziak RP (2012) Volcanic eruptions in the deep sea. Oceanography 25(1):142–157. https://doi.org/10.5670/oceanog.2012.12

Saccò M, White NE, Harrod C, Salazar G, Aguilar P, Cubillos CF, Meredith K, Baxter BK, Oren A, Anufriieva E, Shadrin N (2021) Salt to conserve: A review on the ecology and preservation of hypersaline ecosystems. Biol Rev 96(6):2828–2850. https://doi.org/10.1111/brv.12780

Sánchez MI, Paredes I, Lebouvier M, Green AJ (2016) Functional role of native and invasive filter-feeders, and the effect of parasites: Learning from hypersaline ecosystems. PLoS One 11(8):1–19. https://doi.org/10.1371/journal.pone.0161478

Sánchez-Fernández D, Galassi DM, Wynne JJ, Cardoso P, Mammola S (2021) Don't forget subterranean ecosystems in climate change agendas. Nat Clim Chang 11(6):458–459. https://doi.org/10.1038/s41558-021-01057-y

Santacruz A, Hernández-Mena D, Miranda-Gamboa R, Gerardo PPDL, Claudia POG (2023) Host-parasite interactions in perpetual darkness: Macroparasite diversity in the cavefish *Astyanax mexicanus*. Zool Res 44(4):782–792. https://doi.org/10.24272/j.issn.2095-8137.2022.376

Sapir A, Dillman AR, Connon SA, Grupe BM, Ingels J, Mundo-Ocampo M, Levin LA, Baldwin JG, Orphan VJ, Sternberg PW (2014) Microsporidia-nematode associations in methane seeps reveal basal fungal parasitism in the deep sea. Front Microbiol 5:1–12. https://doi.org/10.3389/fmicb.2014.00043

Scheckenbach F, Hausmann K, Wylezich C, Weitere M, Arndt H (2010) Large-scale patterns in biodiversity of microbial eukaryotes from the abyssal sea floor. Proc Natl Acad Sci USA 107(1):115–120. https://doi.org/10.1073/pnas.0908816106

Schelkle B, Paladini G, Shinn AP, King S, Johnson M, van Oosterhout C, Mohammed RS, Cable J (2011) *Ieredactylus rivuli* gen. et sp. nov. (Monogenea, Gyrodactylidae) from *Rivulus hartii* (Cyprinodontiformes, Rivulidae) in Trinidad. Acta Parasitol 56:360–370. https://doi.org/10.2478/s11686-011-0081-3

Schelkle B, Mohammed RS, Coogan MP, McMullan M, Gillingham EL, Van Oosterhout C, Cable J (2012) Parasites pitched against nature: Pitch Lake water protects guppies (*Poecilia reticulata*) from microbial and gyrodactylid infections. Parasitology 139(13):1772–1779. https://doi.org/10.1017/S0031182012001059

Schoenle A, Hohlfeld M, Hermanns K, Mahé F, de Vargas C, Nitsche F, Arndt H (2021) High and specific diversity of protists in the deep-sea basins dominated by diplonemids, kinetoplastids, ciliates and foraminiferans. Commun Biol 4(1):501. https://doi.org/10.1038/s42003-021-02012-5

Scholz T, Choudhury A, Reyda F (2021) The *Proteocephalus* species-aggregate (Cestoda) in cyprinoids, pike, eel, smelt and cavefish of the Nearctic region (North America): diversity, host associations and distribution. Syst Parasitol 98:255–275. https://doi.org/10.1007/s11230-021-09975-3

Seckbach J (2013) Life on the Edge and Astrobiology: Who Is who in the polyextremophiles world? In: Seckbach J, Oren A, Stan-Lotter H (eds) Polyextremophiles. Cellular origin, life in extreme habitats and astrobiology, vol 27. Springer, Dordrecht. https://doi.org/10.1007/978-94-007-6488-0_2

Seckbach J, Oren A, Stan-Lotter H (2013) Polyextremophiles—life under multiple forms of stress, vol 27. Springer, Dordrecht

Seuront L, Nicastro KR, Zardi GI, Goberville E (2019) Decreased thermal tolerance under recurrent heat stress conditions explains summer mass mortality of the blue mussel *Mytilus edulis*. Sci Rep 9(1):17498. https://doi.org/10.1038/s41598-019-53580-w

Simon KS, Benfield EF, Macko SA (2003) Food web structure and the role of epilithic biofilms in cave streams. Ecology 84(9):2395–2406. https://doi.org/10.1890/02-334

Siwertsson A, Refsnes B, Frainer A, Amundsen PA, Knudsen R (2016) Divergence and parallelism of parasite infections in Arctic charr morphs from deep and shallow lake habitats. Hydrobiologia 783:131–143. https://doi.org/10.1007/s10750-015-2563-z

Sket B, Dovč P, Jalžić B, Kerovec M, Kučinić M, Trontelj P (2001) A cave leech (Hirudinea, Erpobdellidae) from Croatia with unique morphological features. Zool Scr 30(3):223–229. https://doi.org/10.1046/j.1463-6409.2001.00065.x

Snelgrove PVR, Smith CR (2002) A riot of species in an environmental calm: The paradox of the species-rich deep sea. In Oceanography and Marine Biology vol. 40, eds Gibson RN, Barnes M, Atkinson RJA (Taylor & Francis, London and New york), pp 311–342

Sobecka E, Piasecki W (1993) Parastic fauna of Arctic charr, *Salvelinus alpinus* (L., 1758) from the Hornsund region (Spitsbergen). Acta Ichthyol Piscat 23(S):99–106

Soldánová M, Georgieva S, Roháčová J, Knudsen R, Kuhn JA, Henriksen EH, Siwertsson A, Shaw JC, Kuris AM, Amundsen PA, Scholz T (2017) Molecular analyses reveal high species diversity of trematodes in a sub-Arctic lake. Int J Parasitol 47(6):327–345. https://doi.org/10.1016/j.ijpara.2016.12.008

Stein H, Griebler C, Berkhoff S, Matzke D, Fuchs A, Hahn HJ (2012) Stygoregions—a promising approach to a bioregional classification of groundwater systems. Sci Rep 2(1):673. https://doi.org/10.1038/srep00673

Stetter KO (2006) Hyperthermophiles in the history of life. Philos Trans R Soc Lond B Biol Sci 361(1474): 1837–1843. https://doi.org/10.1098/rstb.2006.1907

Stoch F (1995) The ecological and historical determinants of crustacean diversity in groundwaters, or: Why are there so many species? Mémoires de Biospéologie 22:139–160

Studer A, Poulin R (2012) Effects of salinity on an intertidal host-parasite system: Is the parasite more sensitive than its host? J Exp Mar Biol Ecol 412:110–116. https://doi.org/10.1016/j.jembe.2011.11.008

Studer A, Poulin R (2013) Cercarial survival in an intertidal trematode: A multifactorial experiment with temperature, salinity and ultraviolet radiation. Parasitol Res 112:243–249. https://doi.org/10.1007/s00436-012-3131-3

Sures B, Nachev M, Schwelm J, Grabner D, Selbach C (2023) Environmental parasitology: stressor effects on aquatic parasites. Trends Parasitol 39:461–474. https://doi.org/10.1016/j.pt.2023.03.005

Thieltges DW, Mouritsen KN, Poulin R (2018) Ecology of parasites in mudflat ecosystems. Mudflat Ecol 213–242. https://doi.org/10.1007/978-3-319-99194-8_9

Tinsley RC (1995) Parasitic disease in amphibians: control by the regulation of worm burdens. Parasitology 111(S1):SI53–SI78

Tinsley RC (1999a) Overview: Extreme environments. Parasitology 119(S1):S1–S6. https://doi.org/10.1017/s0031182000084602

Tinsley RC (1999b) Parasite adaptation to extreme conditions in a desert environment. Parasitology 119(S1):S31–S56. https://doi.org/10.1017/s0031182000084638

Tinsley RC (2005) Parasitism and hostile environments. In: Thomas F, Renaud F, Guegan J-F (eds) Parasitism and ecosystems. Oxford University Press, Oxford, pp 85–112

Tobler M, Schlupp I, García de León FJ, et al (2007) Extreme habitats as refuge from parasite infections? Evidence from an extremophile fish. Acta Oecologica 31:270–275. https://doi.org/10.1016/j.actao.2006.12.002

Turner JT (2015) Zooplankton fecal pellets, marine snow, phytodetritus and the ocean's biological pump. Prog Oceanogr 130:205–248. https://doi.org/10.1016/j.pocean.2014.08.005

Van Dover C (2000) The ecology of deep-sea hydrothermal vents. Princeton University Press, Princeton, NJ

Vincent WF, James MR (1996) Biodiversity in extreme aquatic environments: lakes, ponds and streams of the Ross Sea sector, Antarctica. Biodivers Conserv 5:1451–1471

Voight JR (2000) A review of predators and predation at deep-sea hydrothermal vents. Cahiers de Biologie Mar 41(2):155–166

Warburton EM (2020) Untapped potential: The utility of drylands for testing eco-evolutionary relationships between hosts and parasites. Int J Parasitol Parasites Wildl 12:291–299. https://doi.org/10.1016/j.ijppaw.2020.04.003

Ward ME, Shields JD, Van Dover CL (2004) Parasitism in species of *Bathymodiolus* (Bivalvia: Mytilidae) mussels from deep-sea seep and hydrothermal vents. Dis Aquat Organ 62(1–2):1–16. https://doi.org/10.3354/dao062001

Weinstein SB, Kuris AM (2016) Independent origins of parasitism in Animalia. Biol Lett 12(7):20160324. https://doi.org/10.1098/rsbl.2016.0324

Whittaker FH, Hill LG (1968) *Proteocephalus chologasteri* sp. n. (Cestoda: Proteocephalidae) from the spring cavefish *Chologaster agassizi* Putman, 1782 (Pisces: Amblyopsidae) of Kentucky. Proc Helminthol Soc Wash 35(1):15–18

Whittaker FH, Kritsky DC (1973) An examination of the Australian blind cavefish, *Milyeringa veritas* Whitley, 1945, for helminth parasites. Proc Helminthol Soc Wash 40:297

Whittaker FH, Zober SJ (1978) *Proteocephalus poulsoni* n. sp. (Cestoda: Proteocephalidae) from the northern cavefish *Amblyopsis spelaea* DeKay (Pisces: Amblyopsidae) of Kentucky. Folia Parasitol 25:277–280

Woolley SNC, Tittensor DP, Dunstan PK, et al (2016) Deep-sea diversity patterns are shaped by energy availability. Nature 533:393–396. https://doi.org/10.1038/nature17937

Zdzitowiecki K (1990) Occurrence of acanthocephalans in fishes of the open sea off the South Shetlands and South Georgia (Antarctic). Acta Parasitol Pol 35(2):131–141

Zdzitowiecki K (2001) Occurrence of endoparasitic worms in a fish, *Parachaenichthys charcoti* (Bathydraconidae), off the South Shetland Islands (Antarctica). Acta Parasitol 46(1):18–23

Open Access This chapter is licensed under the terms of the Creative Commons Attribution-NonCommercial-NoDerivatives 4.0 International License (http://creativecommons.org/licenses/by-nc-nd/4.0/), which permits any non-commercial use, sharing, distribution and reproduction in any medium or format, as long as you give appropriate credit to the original author(s) and the source, provide a link to the Creative Commons license and indicate if you modified the licensed material. You do not have permission under this license to share adapted material derived from this chapter or parts of it.

The images or other third party material in this chapter are included in the chapter's Creative Commons license, unless indicated otherwise in a credit line to the material. If material is not included in the chapter's Creative Commons license and your intended use is not permitted by statutory regulation or exceeds the permitted use, you will need to obtain permission directly from the copyright holder.

Aquatic Parasite Conservation

Marliese Truter, Bjoern C. Schaeffner, and Nico J. Smit

Abstract

Global biodiversity has been in decline for several decades, marked by species extinctions and population losses that have cascading negative impacts on ecosystems. Despite the growing list of described species, much of the actual threatened biodiversity remains unknown, particularly among invertebrates, including the ones with a parasitic mode of life. Parasites, which may outnumber free-living species and constitute a significant proportion of ecosystem biomass, have largely been overlooked in conservation efforts. The extinction of free-living host species often leads to co-extinction events, where dependent parasite species also face extinction. This chapter highlights the need to integrate parasite conservation into broader conservation science and management practices. It provides an overview of the evolution of species conservation from a focus solely on free-living species to one that includes associated parasitic species. A case study on chondrichthyan parasites illustrates the critical role of parasites in ecosystem health and the importance of their conservation. The chapter also outlines future research directions to better understand and conserve aquatic parasites, advocating for their inclusion in biodiversity monitoring programmes and conservation agendas. Addressing these gaps is essential for maintaining the integrity of natural systems and the biodiversity they support.

M. Truter · N. J. Smit (✉)
Water Research Group, Unit for Environmental Sciences and Management, North-West University, Potchefstroom, South Africa

NRF-South African Institute for Aquatic Biodiversity, Makhanda, South Africa
e-mail: 23378123@mynwu.ac.za; nico.smit@nwu.ac.za

B. C. Schaeffner
Water Research Group, Unit for Environmental Sciences and Management, North-West University, Potchefstroom, South Africa

Institute for Experimental Pathology at Keldur, University of Iceland, Reykjavík, Iceland

Department of Anatomy, Physiology, and Pharmacology, St. George's University, St. George's, Grenada

13.1 Introduction

The Anthropocene (Crutzen 2002; Steffen et al. 2011; Hamilton 2016; Waters et al. 2016), which depicts the human-influenced time period, has forced the World into the sixth mass extinction event (Wake and Vredenburg 2008; Barnosky et al. 2011; Ceballos et al. 2017, 2020; Díaz et al. 2019). Although the number of described species continues to increase, global biodiversity has been in decline for several decades with species extinctions and population extirpations that result in cascading detrimental consequences to natural environments. However, in this time of unprecedented threats to global biodiversity and rapid incline in the number of threatened species, the reality of species losses/extinctions is masked by a lack of information and strong bias towards charismatic target groups, predominantly terrestrial vertebrates (Régnier et al. 2015).

The International Union for Conservation of Nature (IUCN) Red List of Threatened Species currently lists about 7% of described species as threatened (IUCN 2023b). This estimate is undoubtedly low, as it does not include the immense number of invertebrate species that would almost certainly qualify under their criteria (Régnier et al. 2015), many of which have a parasitic mode of life (Poulin and Morand 2000). Although the exact number of parasite species is unknown (Dobson et al. 2008), they might outnumber free-living species (Windsor 1998). Estimates on the proportion of parasites range from approximately 40% to up to 70% of total species richness (Price 1980; Rohde 1982; Poulin and Morand 2004). Parasites also comprise a large proportion of the biomass in many ecosystems (Kuris et al. 2008). However, they have been mostly ignored as eligible conservation targets (Carlson et al. 2020b; Gómez and Nichols 2013). A threatened, free-living host species that faces extinction, mainly through anthropogenic processes, will almost certainly be host to a number of parasite species (Koh et al. 2004; Dunn 2009; Dunn et al. 2009). Every parasite that becomes extinct represents a loss for the composition of natural systems (Dobson et al. 2008; Gómez and Nichols 2013). These co-extinction events, i.e. a parasite species facing extinction due to the extinction or local extirpation of its host species (Colwell et al. 2012), are considered the predominant cause of biodiversity loss to date (Brodie et al. 2014; Colwell et al. 2012; Dunn et al. 2009; Strona and Bradshaw 2018). Therefore, it becomes essential to incorporate parasite and affiliate species in academic conservation science and updated conservation agendas, large-scale scientific studies of biodiversity, monitoring programmes, and in the management of wildlife and natural resources. However, integrating these poorly known and often undescribed taxa into conservation initiatives presents a significant challenge for scientists, managers and conservationists (reviewed in Lymbery and Smit 2023).

This chapter provides an overview of the development of species conservation from a focus on free-living species to a conservation approach where the species associated with the free-living species are also included. This will be done through the critical analysis of the available literature, by presenting a case study of the parasites of chondrichthyans and providing a way forward for future research on the conservation of aquatic parasites.

13.2 Global Species Conservation to Parasite Conservation

The earliest conservation movement emerged in the 1660s when John Evelyn published his work concerning the conservation of forests due to near exhaustion of timber resources in England (Evelyn 1664). During the following two centuries, the development and practice of conservation globally experienced what could be seen as a political rollercoaster ride, dependent on government agendas and military expansion.

It was only towards the mid-nineteenth century that the International Union for the Protection of Nature was founded (now the International Union for the Conservation of Nature) (IUCN 2023a), with its sole purpose to assess the impact of human activities on nature. These assessments led to the investigation of the damaging effects of chemical pollutants (such as pesticides) and encouraged the use of impact assessments. The focus on the conservation of biodiversity and species-specific assessment came into play in 1956 with the establishment of the IUCN Species Survival Commission (IUCN SSC) and the joining of forces with the World Wide Fund for Nature (WWF) (then World Wildlife Fund, est. 1961) that aimed at developing management policies and guidelines that facilitate conservation planning based on knowledge on the status and threats of species of the world. Within the first two decades of its existence, the IUCN was at the forefront of developing the framework for conservation and the driving force behind endorsements for environmental law, mobilisation of science, implementation of policy and governance of protected species and their habitats. The commissioning of the IUCN Red List of Threatened Species in 1964 was probably the most powerful contribution of this organisation to the modern species conservation movement, and policy makers, conservation managers and scientists are guided by these policies for the assessment, monitoring and protection of species and their habitats (IUCN 2023a).

Today, direct action towards species-specific conservation is based on data from the IUCN Red List of Threatened Species and is implemented through the Save Our Species initiative (SOS, est. 2010) together with experts from the various SSCs. Notwithstanding more than seven decades of dedication to global biodiversity conservation, associated species (i.e. parasites, commensals or mutualists) of the diverse groups of free-living species already assessed are sparse or even non-existent. This is evident by the only fully parasitic species listed on the IUCN Red List, the Pygmy hog sucking louse, *Haematopinus oliveri* due to the endangered status of its only known host, the Pygmy hog *Porcula salvania* (Wells et al. 1983; Gerlach 2014).

In the World Conservation Strategy of 1980, the IUCN defined conservation as 'the management of human use of the biosphere so that it may yield greatest sustainable benefit to present generations while maintaining its potential to meet the needs and aspirations of future generations' (IUCN 1980). In the same report 'parasites' were only mentioned twice in chapters on the 'maintenance of essential ecological processes and life-support systems', where parasites of pest species are regarded as important components in preserving the productivity of agricultural ecosystems, and on the 'preservation of genetic diversity' (see IUCN 1980), where parasites are seen as pests introduced with alien and invasive species. Little attention and few resources have been directed at assessing and understanding parasites, commensals or mutualists of free-living species.

The consideration of parasite conservation became prominent in the 1990s and action towards their conservation manifested when Donald Windsor expressed concern on the exclusion of parasites in the conservation of biodiversity, claiming 'Equal rights for parasites!' (Windsor 1995). He later described hosts and their associated parasites as 'biocartels' and ultimately, in dismay at the neglect of recognition for parasites by ecologists and conservationists, he robustly declared that parasites are 'a cohesive force that holds ecosystems together' and that it seems as if 'biodiversity' is exclusively used for, and defined by, free-living animals and plants (Windsor 1990, 1995, 1997a, b, 1998). Windsor's advocacy for parasitic species seems to have thrown some life into parasite conservation, and the few studies that focus on this ever-present but neglected group of organisms have increased substantially since the 1990s. Brian (2023) discusses the misleading nature (or rather 'false moral dichotomy') of the question on whether parasites are our friends or enemies in biodiversity conservation through an analysis of publications on parasites. This indicates that the focus in the literature is still on the negative effects of parasites and management strategies of this important group in ecosystems for purposes of controlling them, with little consideration for their diversity, function and role in communities or ecosystems. Stronger advocacy and a more explicit inclusion and recognition of parasitic organisms in ecology and conservation are needed. It must be noted that, different to the considerations needed for assessing free-living species, the considerations required for assessing parasitic species for conservation is much more complex. Parasites do not merely exist; these remarkable organisms provide important linkages of several components within ecosystems (Dunne et al. 2013), may be responsible for host population control within its distribution range, free-living stages (i.e. cercariae) can act as food source while other parasites manipulate host behaviour such as favouring predation to ensure a natural equilibrium within any given ecosystem (see Chaps. 7 and 8). At its core, advocating for parasite conservation means departing from the one-dimensional, species-by-species approach to conservation, and arguing in favour of conservation models that prioritise preserving ecosystem complexity.

13.3 Critical Roles of Parasites

An often-immediate reaction towards parasites is that of disgust or fear, due to the indoctrine of their complete harmful nature. It is undeniable that there are some aquatic parasites of serious concern to human-, wildlife- or livestock health, such as *Dracunculus medinensis*, *Fasciola* spp. and *Schistosoma* spp. that cause dracunculiasis, fascioliasis and schistosomiasis, respectively. However, these are exceptions to the rule since most aquatic parasites are harmless or non-lethal to their hosts. In many instances, disease-causing parasites are managed through medicinal or ecological treatments (Jones et al. 2018; Arostegui et al. 2019). In an era of focused efforts towards disease eradication, one should consider the potential of unknown outcomes in the instance of parasite eradication for ecosystems where the role and functionality of these parasites remain largely undetermined. Instead, focus must shift towards managing rather than eradicating and it should be recognised that parasites constitute instrumental and unique components that warrant complete exploration of their contribution to functionality and complexity of the very ecosystems we depend on, such as the antagonistic interaction between cercarial stages of *Schistosoma mansoni* and *Calicophoron sukari* infecting *Biomphalaria pfeifferi* that aid in limiting transmission of *S. mansoni* to humans (see Laidemitt et al. 2019). In other instances, the benefits and functional role of parasites should not be ignored or overlooked. It has been noted that parasites contribute to the survival success of their host in providing selective pressures that ensure the hosts' genetic diversity based on knowledge of the evolutionary ecology and productivity of the host (Rózsa 1992; Ebert and Hamilton 1996; Sheldon and Verhulst 1996). Inclusion of parasites in biodiversity conservation, in particular those in aquatic ecosystems, proves advantageous at genetic-, individual-, component- and ecosystem- level and as such parasitic species have been defined as prominent ecosystem engineers. Aquatic parasites can drive functional ecosystems through creation of food-web links between intermediate and definitive host species, and free-living stages (i.e. digeneans, larval nematodes, and metacestodes) can be a food source (see Artim et al. 2017; Gopko et al. 2017), while some monoxenous parasites, and the intermediate stages of heteroxenous parasites, also form part of the free-living biomass (Dobson et al. 2008; Hatcher et al. 2012; Lambden and Johnson 2013; Dunne et al. 2013; Sures et al. 2017). Some species even assist in host population regulation through reproductive castration (such as in freshwater and marine snails, mussels or fishes; also see Chaps. 5 and 7) and feminisation (May and Anderson 1978; Lafferty 1993; Hatcher et al. 1999; Brian and Aldridge 2020), while others promote more diverse community assemblages and co-existence of species or can serve as bioindicators (Thompson 1996; Brian and Aldridge 2022 and references therein; also see Chap. 20). The significance in the contribution of parasitic species to biodiversity, trophic organisation and overall ecosystem complexity is thus of great importance during assessment and conservation planning of a host–parasite system.

13.4 Ecological Aspects of Parasite Extinction

The evaluation of the susceptibility of parasites to extinction is not as straightforward as it may be for hosts that lead independent lifestyles. Co-extinction is a highly complex process, and many different factors influence and accelerate the risks and rate of species losses, respectively. Extinction or local extirpation of a key host species can trigger extinction cascades within complex trophic networks (Colwell et al. 2012). Parasites will be impacted differently to declining populations, local species extirpations and the extinction of host species. While a high parasite species diversity and susceptibility to secondary and tertiary extinctions overall increase extinction rates (Colwell et al. 2012), some parasite species face extinction even prior to their hosts due to their uneven distribution (Moir et al. 2010, Poulin 2011), narrow geographical range (Dunn et al. 2009) or transmission impediments due to host population declines and decreased density and abundance of available hosts (Altizer and Pedersen 2008; Cizauskas et al. 2017; Moir et al. 2010, 2012; Powell 2011). Conversely to the extinction of a principal host species, it is speculated that a loss of parasites may also negatively impede and potentially even accelerate the extinction of a threatened host species (Dunn et al. 2009). Host specificity and complexity of parasite life cycles have been considered dominant drivers for parasite co-extinctions (Cizauskas et al. 2017; Moir et al. 2010; Poulin et al. 2011). Host specificity is a direct result of historical associations between parasites and their hosts, mediated by host-switching and co-speciation events (Page 1993; De Vienne et al. 2013; Wells and Clark 2019) and is generally well-conserved among parasite groups (Mouillot et al. 2006). The host specificity among individual groups of chondrichthyan parasites varies but is generally high (Benz and Bullard 2004). In theory, highly host-specific parasites (i.e. specialists) that infect a single host or closely related host species are inflexible and strongly depend on the fate of their host. Specialists may hence be more endangered and face higher extinction risks than parasites with a lower specificity (i.e. generalists). These in turn utilise a broader host spectrum, acquire better survival and reproductive traits, and might therefore be less susceptible to co-extinction (Bush and Kennedy 1994; Dunn et al. 2009; Koh et al. 2004; Lafferty 2012; Poulin et al. 2006; Vázquez et al. 2005). Upon closer observation, the complexity of species co-extinctions might be mirrored by host specificity as a dominant driver and specialist parasites might have a lower vulnerability to co-extinction than previously thought (Cizauskas et al. 2017; Farrell et al. 2015, 2021; Strona 2015; Strona et al. 2013). Strona et al. (2013) analysed the co-extinction risk of helminths infecting fishes and found that specialist parasites have a reduced co-extinction risk due to an evolutionary strategy of infecting fewer but less vulnerable host species. The tendency to specialise in host species with a broader geographic distribution, high abundance in ecosystems and increased persistence to environmental perturbations might further explain the higher parasite species richness in these common hosts (Strona 2015; Strona et al. 2013). In contrast, generalist parasites infecting various host species with a lower abundance and more restricted distribution occur in much lower intensities in the environment (Poulin 2011). Consequently, specialists have a higher risk of co-extinction if the host species becomes threatened, which ultimately impacts parasite richness (Dunn et al. 2009). Generalist parasites, on the other hand, might be more strongly affected by host population declines and reduced host availability, which would make them more vulnerable to co-extinction, particularly in threatened host species and with increased extinction levels (Cizauskas et al. 2017; Farrell et al. 2015, 2021). Although host specificity greatly influences the likelihood of parasite co-extinctions, the outcomes and potential escape trajectories of parasites following host decline and/or disappearance are hard to predict. One rarely documented mechanism of parasites to reduce their co-extinction risk is host shifting. Parasites which can colonise new host species and increase their geographical range are more likely to avoid extinction with their host (Cizauskas et al. 2017; Dunn et al. 2009; Hoberg and Brooks 2008; Moir et al. 2010; Strona 2015). However, this might not be the case for parasites with complex life cycles, which implement multiple host species for each developmental stage and with varying degrees of specificity (Adamson and Caira 1994). In contrast to parasite species with direct (i.e. single host) life cycles, species with complex life cycles have an increased risk of co-extinction (Colwell et al. 2012; Koh et al. 2004; Lafferty 2012; Poulin and Morand 2004; Strona 2015). Extinction risk of multi-host parasites is intricately tied to the vulnerability of each host involved in their life cycle. Population declines or extinction of a single intermediate host can lead to the disruption of parasite transmission and impede its ability to complete the life cycle. In light of this, life cycle complexity emerges as a compelling and potentially more significant indicator for assessing parasite extinction risk compared to host specificity (Strona 2015). However, despite the critical importance of understanding parasite life cycles, information on intermediate hosts is sparse, which poses considerable challenges in accurately predicting parasite extinction risks.

13.5 Quantitative Analysis of Aquatic Parasite Conservation Literature

To determine the progress on the inclusion and consideration of parasites in conservation, we conducted an intensive literature search on the Web of Science and Google Scholar. Publications addressing the topics of parasite conservation (or extinction) and aquatic species were included. A final total of 255 articles (including four dissertations and eight theses) spanning from 1990 to 2023 were inspected and were grouped into categories: 'Aquatic', 'Aquatic/Terrestrial', 'Terrestrial' and 'Theory' to determine which proportion of the total literature focused on parasites and their conservation or extinction in the aquatic environment. Closer inspection was only done for 130 of the total number of initial publications as these exclusively discussed aquatic hosts and their parasites. Publications addressing extinction of parasites were included since factors driving extinction directly relate to the conservation effort and feasibility for a group of organisms for which survival potential and ecology are not fully understood.

Half of the initial publications ($n = 125$) focus on parasites (or parasitoids) of terrestrial hosts (i.e. mammals and plants), the extinction and conservation of parasitoids in pest management (i.e. pesticide application or plant and host eradication), or proteins and genes related to parasites concerned with human health. Of the remaining literature addressing conservation or extinction, or a combination thereof, of parasites and their hosts in aquatic environments, 22% ($n = 56$) focus on parasite conservation, extinction and co-extinction using mathematical prediction models and theoretical concepts, including frameworks on how to approach the conservation of invertebrates (including parasites) or parasites of vertebrate hosts (Fig. 13.1). Another 25% of the publications ($n = 66$) focus exclusively on aquatic ecosystems and include explicit concern for parasites and their persistence or co-extinction with their host, and 3% ($n = 8$) included parasites in both aquatic and terrestrial environments. The year 2023 saw the most publications on parasite conservation ($n = 9$), followed by five published in 2018, four in 2020 and 2022 and three in 2021. Earlier publications (1990–2012) advocate for the inclusion of parasites as components of biodiversity in conservation schemes and explore potential factors threatening parasite persistence (or extinction risk) in general and in aquatic ecosystems, advocate for parasite conservation and highlight bias in conservation efforts and the lack of recognition for parasites as important components of ecosystem biodiversity. Later publications (2012–2023) address and discuss the same concerns; however, these utilise all available datasets to more directly appeal for better understanding of parasite ecology and empirical conservation implementation, and to make predictions on the fate of parasites in the looming sixth Mass Extinction event.

It is clear from the above analysis that we are slowly moving away from pure conservation activism and focusing on intentional conservation development with directional, empirical and quantitative processes in action for parasite conservation through multidisciplinary approaches to (1) identify parasitic species associated with threatened hosts; (2) understand the ecology of these parasitic species and (3) assessing and identifying factors that can contribute to conservation of specific host–parasite systems or developing conservation work flows for taxon-specific conservation within existing conservation management plans.

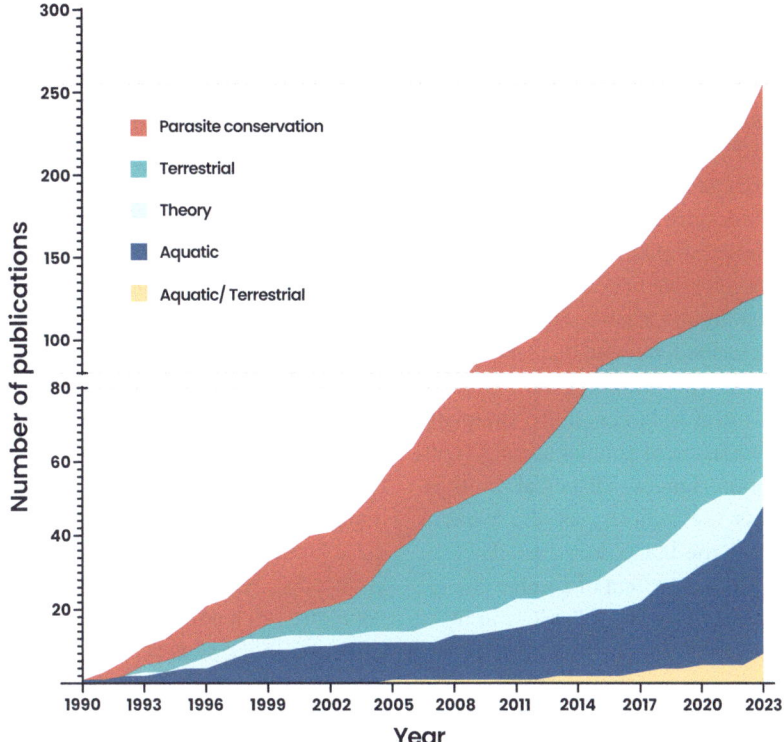

Fig 13.1 Cumulative number of publications concerning parasite conservation included in the analysis. Of the 255 publication that focus on 'Parasite Conservation', half (50%, n = 128) of them address parasites or parasitoids of mammals or plants, including humans ('Terrestrial'). The remaining publications discuss parasite extinction using prediction models and theoretical concepts applied to non-specific parasite–host systems ('Theory', 22%, n = 56), while the remainder focus on parasites in 'Aquatic' (25%, n = 66) or 'aquatic and terrestrial' ('Aquatic/Terrestrial', 3%, n = 8) ecosystems

13.6 Advocating for Parasite Conservation

The recognition of the importance of parasites and their consideration in conservation strategies is belatedly appealing to the greater conservation community, and biodiversity and conservation managers. However, parasitic species are still under-represented in international threatened species lists. The parasitic species included in current international conservation registries (i.e. IUCN) consist of one louse, a flea (both terrestrial) and a number of freshwater bivalves with obligate parasitic larval stages on fish hosts (Jørgensen 2015; Kwak et al. 2019, 2020). Several obstacles continue to hinder the inclusion of any parasitic species in conservation (see below) and the reality is that a large proportion of parasites, potentially the group of organisms on the planet most at risk of facing extinction, may become extinct before their discovery, much less before their conservation status can be assessed (Dunn et al. 2009; Lees and Pimm 2015; Carlson et al. 2020a). In the early 2000s, it was estimated that parasites comprise approximately 30–50% of all species on the planet (Poulin and Morand 2004). Dobson et al. (2008) suggested that between 3 and 5% of parasitic helminths on earth are threatened with extinction, while Carlson et al. (2017a) indicated that 5–10% of 457 analysed parasite species across eight major clades (Acanthocephala, Astigmata, Cestoda, Ixodida, Nematoda, Phthiraptera, Siphonaptera and Trematoda) are threatened with extinction by 2070 (also see Okamura et al. 2018). It is predicted that up to 30% of parasitic worms may become extinct, and that ectoparasites will fare worse than endoparasites (Carlson et al. 2017a; Bellay et al. 2020). Although approximately 28% of all known species have been listed as threatened by the IUCN, Cardoso et al. (2011) indicated that approximately only 0.5% of the more than one million described arthropods and only 4% of the almost 80,000 molluscan species described worldwide have been evaluated by the IUCN, compared to the near-complete assessments for all described mammalian, avian and amphibian species. This supports the notion of taxonomic bias towards larger and charismatic species and the omission of smaller species (with narrow distribution ranges, dispersal abilities which encompass the majority of the planet's fauna and flora) on the IUCN Red List. There is empirical evidence of the effect of climate change on parasite biodiversity through the loss in hosts (diversity and density) due to habitat degradation, fragmentation or complete loss thereof, and presence of invasive host species (see Chap. 18). The persistence of parasite species being dependent on multiple factors (competitive interaction between hosts, parasite host-specificity and host-switching capability, transmission strategies and success) is a key driver of their vulnerability to extinction (Carlson et al. 2017a; Kwak et al. 2020; Brian 2023).

Predictions strictly based on aquatic datasets are sparse and are faced with a number of challenges associated with (1) prioritisation of parasitic species (in both aquatic and terrestrial environments) being biased towards certain species or those infecting charismatic host species; (2) restrictions in expertise, time and effort to describe and study parasites and (3) the instinctive prioritisation of species threatening economies and human health (Brian and Aldridge 2022). A central hurdle to studying and understanding the extent of parasite diversity in aquatic ecosystems (not exclusively) is the Linnaean shortfall, which delineates the small proportion of parasitic fauna described to date creating lack of knowledge regarding actual richness (Fig. 13.2) (see Lomolino 2004; Hortal et al. 2015). Little is known about most parasite species' geographic range (Wallacean shortfall, see Lomolino 2004) or the population trends, sizes and abundances of already described parasite species (Prestonian shortfall, see Cardoso et al. 2011), further hindering sufficient and efficient monitoring or detection of species. From an ecological point of view, there is limited knowledge or understanding of the ecological requirements (Hutchinsonian shortfall, see Cardoso et al. 2011) and interactions (Eltonian shortfall, see Hortal et al. 2015) of most known parasite species in aquatic systems and, finally, a diverse and complete visual archive does not exist for many known species and their unique morphological features, behaviour and habitat (Keartonian impediment, Marshall et al. 2024). Addressing these shortfalls is confounded by the absence of knowledge regarding intermediate hosts, host population dynamics, parasite specialisation and ability to host switch, and potential for local extinction and re-emergence due to environmental changes or host availability (Kennedy 1993; Bush and Kennedy 1994; Cardoso et al. 2011; Brian and Aldridge 2022). Identifying and scrutinising the traits and functional ecology of the parasitic species also poses great challenges (Raunkiæran shortfall, Hortal et al. 2015). The dearth of knowledge on all the above shortfalls can bring the viability of parasites in both parasite and host conservation programmes into question, in particular where the links between parasites and hosts remain unexplored or limited in information, as is true for the majority of parasite–host systems (Darwinian shortfall, see Diniz-Filho et al. 2013). It has been suggested that extinction of many parasite species will precede that of their hosts due to the requirement of minimum host population sizes to sustain a parasite population. It is impossible to know how many species have already become extinct due to a deficiency in data for any given geographic region (e.g. isolated populations that are difficult to sample), challenges in studying threatened hosts, and, most importantly, simple neglect. Furthermore, many species and historic communities will remain uncatalogued and unknown due to the lack of studies over timeframes that

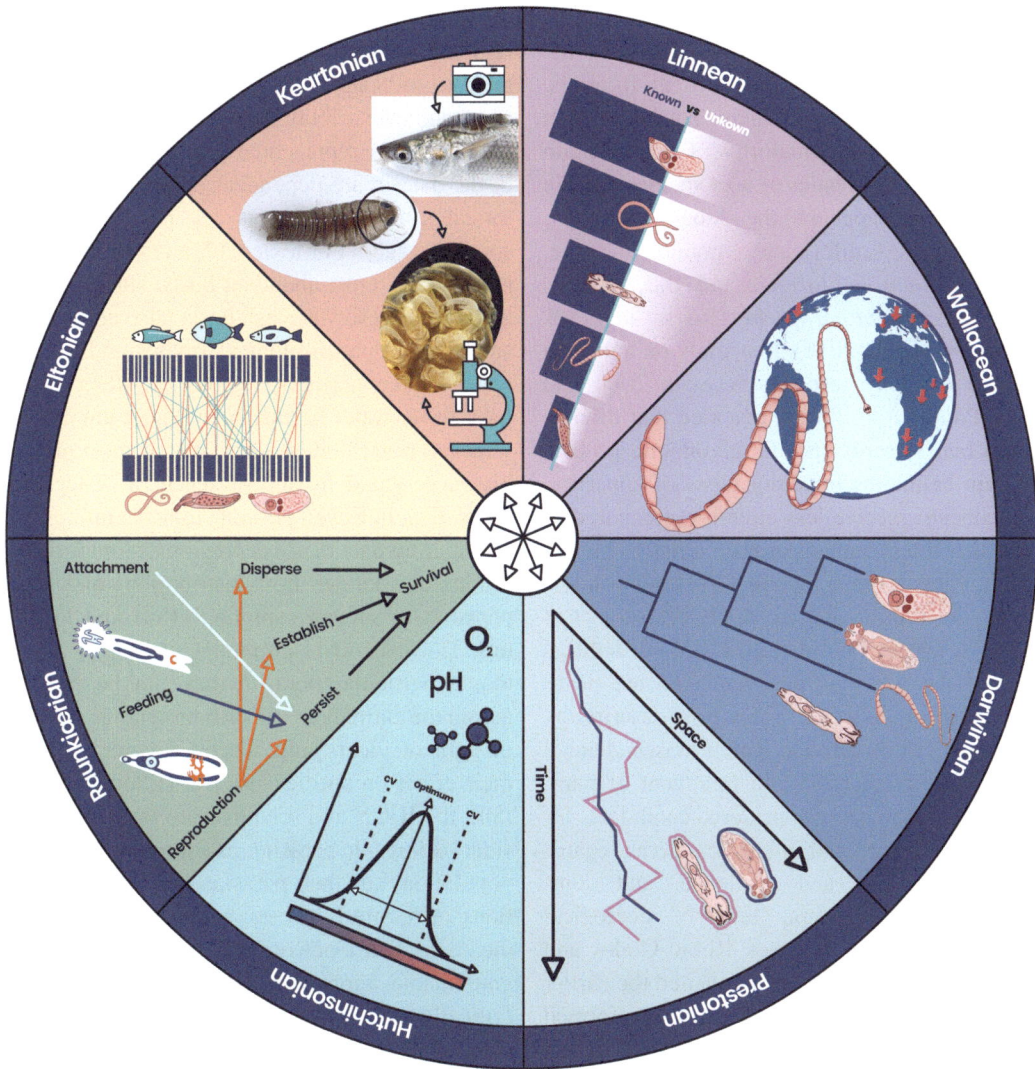

Fig 13.2 Diagram illustrating the inherent complexity and persistent knowledge shortfalls that inform species conservation efforts

capture biodiversity before and after the extinction of hosts following anthropogenic or natural extirpation or dispersal events (Rubio-Godoy and Pérez-Ponce de León 2023). The reality of parasite–host co-extinctions and the lack of proactive inclusion of parasites as components of biodiversity and their inclusion in conservation (or co-conservation) processes lead to Stork and Lyal's (1993) deep condemnation to the scientific community by stating that we must deal with this ignorance '…in the full knowledge of what is being lost'.

13.7 Progress of Global Aquatic Parasite Species Conservation

Attention towards conserving parasites is growing. However, a substantial effort is still needed to match knowledge and information on aquatic parasites to that of terrestrial counterparts such as ticks and lice of birds and mammals, which have a substantial head start (Rózsa and Vas 2015; Spencer and Zuk 2016). Unfortunately, parasites do not enter an even playing field. Of the 28% of species globally classified as threatened with extinction, parasites are among the 72% that are almost entirely absent from the IUCN Red List and assessment. Furthermore, it is also clear that the majority of conservation efforts focus on parasites of host species that are charismatic and appeal to the global community, e.g., the Californian condor louse, *Colpocephalum californici*, and the platypus tick, *Ixodes ornithorhynchi* (Rózsa and Vas 2015; Kwak et al. 2018). In a study on helminth parasite diversity research between 2000 and 2018, Poulin et al. (2023) revealed that there is also a bias towards specific taxonomic groups, wherein helminths infecting hosts of conservation concern receive less attention compared to those that infect humans or pose a risk to human health. Parasites also receive little attention following description and are rarely the subject of further research. Poulin et al. (2023) found that 60% of the parasites they analysed are never mentioned or collected again post-description. The negativity in attitude which biases channel towards the eradication and treatment of parasites, rather than being considered candidates for conservation, is rooted in base concerns regarding pathogenicity and virulence, and compounded by a general lack of ecological information for most species. Rubio-Godoy and Pérez-Ponce de León (2023) revisited the earlier literature on parasite conservation and performed an analysis on the Eltonian-, Darwinian-, Linnean-, and Wallacean shortfalls that directly concern the knowledge and conservation of parasites, and indirectly to the conservation of their hosts and ecosystems using a dataset from García-Prieto et al. (2022) on fishes and their helminths from Mexico. A great leap towards breaching the Linnean shortfall in known freshwater fish parasite diversity of Mexican fish was achieved, with the inclusion of 70% of the known Mexican freshwater fish and their helminths in the analysed dataset. The authors noted that no indication is provided on the number of potential new discoveries for Mexican freshwater fish parasites. Further, it was highlighted that there were more parasite–host interactions reported from southern Mexico compared to the northern regions (Wallacean shortfall), but it was not possible to indicate with certainty if the complete geographic distribution of the fishes and their parasites were represented. The records included in the dataset are dependent on (1) a dense historical sampling of hosts and parasites in the southern region of Mexico; (2) an extensive sampling of the most abundant hosts, since 30% of the known freshwater fish hosts have never been subjected to parasitological studies and (3) the water scarce northern regions of Mexico that have fewer water basins and a higher number of seasonal water bodies where data is most probably represented by the presence of water and hosts. It is, however, possible to make inferences considering data dense ecoregions, climatic conditions and known host distributions and their parasites to inform sampling effort and localities. However, it is important to keep in mind that a particular parasite may not be present across the entire host's distribution range. Lastly, in light of identifying species interactions and their effect on species survival (Eltonian shortfall), the IUCN and scientists are invited to consider pathogenic parasites as a threat to wild host populations and their parasites, since the interactions of harmful species, such as that between the Asian fish tapeworm, *Schyzocotyle acheilognathi*, and approximately 28 endangered and critically endangered fish species of Mexico, have not been considered.

Rubio-Godoy and Pérez-Ponce de León (2023) further concluded that there was a need to 'make parasites as charismatic as whales' and proposed the implementation of an IUCN Red list for parasites of zoonotic and host conservation concern, and an IUCN 'Green List' for parasites that pose low to no threat to their host organisms and substantially contribute to ecosystem function and services. Brian (2023) provides a summary on earlier considerations or shortfalls in knowledge of parasites, validates their worthiness for conservation in addressing threats to parasites, risks and benefits to hosts, complexity in food webs, and pointing out that the 'positive' or 'negative' ecosystem value of a parasite is in the eye of the beholder, or in this case, the researcher (i.e. biologist, conservationist and parasitologist).

Aquatic biodiversity encompasses a wide range of animal life, including species across several taxonomic groups such as amphibians, birds, fishes, insects, mammals, molluscs and reptiles, all constituting species that are aquatic or semi-aquatic and utilising or residing in freshwater and marine ecosystems intermittently or permanently. For some of these groups, such as the amphibians and fishes, 93 and 70% of all the estimated described species (respectively) have been evaluated under the IUCN Red List version 2023-1 (IUCN 2023b; Re:wild 2023; Fricke et al. 2024), however almost nothing is known about the parasitic communities that most species of these groups host.

13.7.1 Parasites of Molluscs

It is estimated that approximately 67% of the parasite richness of freshwater mussels in Europe and North America is yet to be described. However, more than 80% of these are host-specific taxa belonging to digeneans and ciliates. Furthermore, a total of 21% of the total parasite fauna of freshwater molluscs from these regions are at risk of extinction and will, in all likelihood, never be formally described (Brian and Aldridge 2022). Despite these morbid estimations, not much has been done to advance the conservation of parasites of molluscs, and the primary focus is on the risk and management of parasitic infection of mollusc hosts in freshwater, marine and aquaculture systems (Reichard et al. 2005; Coen and Bishop 2016; Carnegie et al. 2016; FMCS 2016; McElwain 2019; Brian and Aldridge 2019; Taskinen et al. 2021). There is not a single example advocating for the conservation of any known molluscan parasite along with its host. In fact, not a single known endosymbiont of unionid mussels from Europe or North America has been considered in the IUCN assessment to date (Brian and Aldridge 2019). This may be due to the overwhelming negative associations of these symbionts, such as trematodes, which are known to cause castration and lowering fecundity in their hosts, and mites responsible for the destruction of the mantle and gill tissue (Thomas et al. 2010). Notwithstanding, the effects of a relatively small number of endosymbionts of molluscs have to date been quantified (McElwain 2019). For example, the association between nematodes in the mantle of molluscs remains unknown and the potential of a positive defensive relationship between oligochaetes that devour trematode cercariae in mussels needs to be clarified for molluscan species of conservation concern (Hopkins et al. 2016; Brian and Aldridge 2019).

13.7.2 Parasites of Aquatic Arthropods

Many insects have aquatic larval stages that serve as intermediate hosts for larval parasitic stages in aquatic ecosystems. Some examples include the larval stages of dragonflies (Odonata), midges (Chironomidae) and caddisflies (Trichoptera) that serve as second intermediate hosts for metacercariae of plagiorchiid trematodes (Ponomareva et al. 2022). While metacestodes are known from dragonfly larvae (Anisoptera), and mermithid nematode larvae often utilise aquatic larvae of caddisflies (Trichoptera), flies (Diptera), beetles (Coleoptera) and mayflies (Ephemeroptera) as paratenic hosts (Molloy et al. 1999 and references therein; Regel et al. 2013). Another interesting group are the hairworms (Nematomorpha), and in particular members of the Gordiida that are parasites of terrestrial insects, which are sexually dimorphic, and mate and oviposit outside of their host. Their life cycle involves the pairing and mating, laying and development of eggs and hatching of larvae in an aquatic environment. In aquatic environments, the larvae penetrate and encyst in a wide range of paratenic or dead-end hosts (i.e. fishes, snails and crustaceans), however, only aquatic insect larvae successfully transport the cysts to terrestrial environments after metamorphosis followed by ingestion by definitive insect hosts (i.e. crickets, grasshoppers and locusts, cockroaches, mantids and beetles). In the definitive insect host the larvae excyst, penetrate the hemocoel and manipulate the host to jump in the water where the final transition of the gordiid worm from terrestrial to aquatic environment occurs (Hanelt et al. 2012; Bolek et al. 2015). This unique parasitic mode of life provides an opportunity to study the efficacy of a selection of genetic and analytical tools for the application in parasite conservation studies. These tools can be used to estimate the demographic history and genetic structure of a species that could inform declines and dispersal traits of parasitic taxa across geographic scales and time, as has been done for *Chordodes formosanus* that utilises the praying mantis as a definitive host, and *Acutogordius taiwanensis* and *Gordius chiashanus* that both infect millipede definitive hosts (De Vivo et al. 2023). Such approaches along with the understanding of a species' ecology could assist in establishing monitoring or reintroduction programmes for parasitic species that are currently geographically restricted but possess the ability to disperse through host-switching or in the absence of range restricting stressors.

13.7.3 Parasites of Teleost Fishes

Teleosts represent the most diverse group of fishes and approximately 30% of all known species are considered at risk of extinction (WWF 2021a, b). Very few studies have investigated the parasitic communities associated with these threatened species, primarily due to ethical constraints. Nonetheless, in rare cases, opportunities arose to document the parasitic fauna of threatened fishes and advocate for their conservation (see Baruš et al. 1997; Truter et al. 2023). To date, two studies are known to consider freshwater fish parasites for conservation. The first study encompasses the proposal of conservation categories for the adult forms of a number of host-specific helminths, inferred from the threat degree of their respective cypriniform, siluriform and gadiform fish hosts in the Czech Republic and Slovakia (formerly Czechoslovakia) (Baruš et al. 1997). More recently, an assessment was completed for parasites of endemic freshwater fishes in South Africa. The branchiuran *Chonopeltis minutus*, known from two endemic and threatened cyprinids in the Cape Fold ecoregion in South Africa has been proposed for inclusion in the IUCN Red List under the Critically Endangered (CR) category. Additionally, ten parasitic species from five endemic cyprinids in the Olifants-Doorn River system in South Africa have been assessed using the conservation assessment methodology for animal parasites (CAMAP) (Kwak et al. 2020; Truter et al. 2023). The CR status for *C. minutus* is based on its apparent absence from its two known hosts for more than 50 years, while four monoxenic species (three monogeneans of the genera *Dactylogyrus*, *Gyrodactylus* and *Paradiplozoon*, and a pre-metamorphic lernaeid copepod) (Fig. 13.3a–d) and three polyxenic species (an indeterminate myxozoan species *Myxobolus* sp., a nematode of the genus *Rhabdochona*, and Caryophyllidea gen. sp.) (Fig. 13.4a–e), all undescribed, were assessed as Near Threatened, and larval forms of *Acanthogyrus* sp. (Acanthocephala), *Contracaecum* sp. (Nematoda) and a diplostomid (Trematoda) were all assessed as Data Deficient (Fryer 1977; Van As and Van As 1999; Truter et al. 2023; Přikrylová et al. 2024). Although no other fish parasitic species has been assessed to date, concerns for host–parasite co-extinction have been raised. Leis et al. (2023) report on the increased mortality of endangered Pallid sturgeon, *Scaphirhynchus albus*, parasitised by *Gyrodactylus conei* from a hatchery in South Dakota (USA). These authors discuss the complex implications of conserving *G. conei* considering the increased risk it poses to the survival of *S. albus*. Similarly, Mathews et al. (2020) raise concerns about the conservation of the myxozoan *Myxobolus iquitoensis* known only from the endangered loricariid catfish *Otocinclus cocama* in Peru. Another example is the mesocercaria of a strigeid trematode infecting wild silver perch, *Bidyanus bidyanus*, from Australia (Barton et al. 2023). *Bidyanus bidyanus* is listed as Critically Endangered according to the Australian Environmental Protection and Biodiversity Conservation (EPBC) Act 1999, and is listed as Near Threatened in the IUCN Red List of 2019.

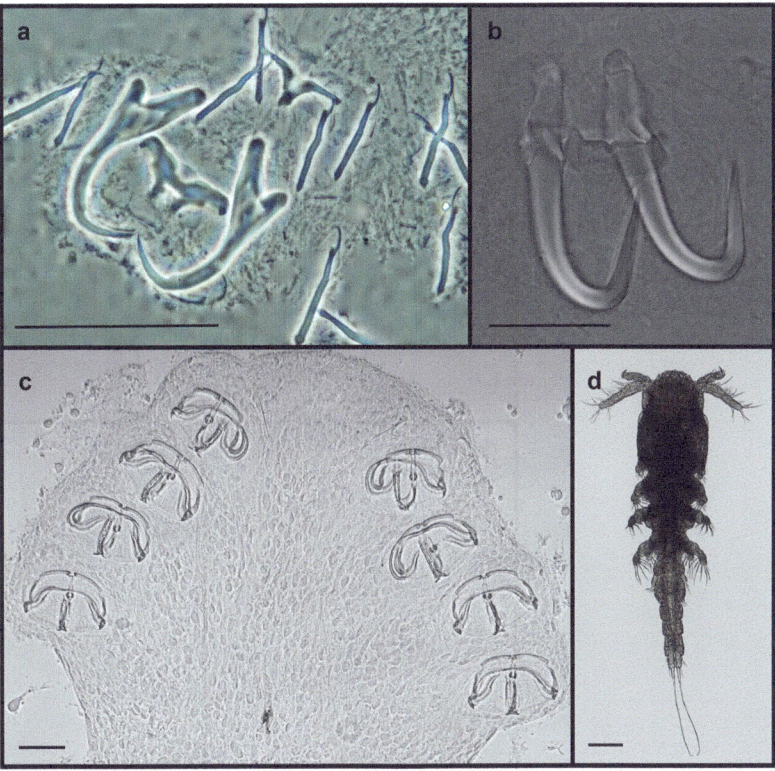

Fig. 13.3 Near-threatened monoxenic parasitic species from endemic cyprinids in the Cape Fold ecoregion, South Africa. (**a**) Hamulus complex of *Dactylogyrus* sp.; (**b**) hamulus complex of *Gyrodactylus serrai*; (**c**) haptoral section with attachment clamps of *Paradiplozoon* sp.; (**d**) pre-matamorphic lerneaid copepod. Scale bars: 10 µm (**d**), 25 µm (**c**), 50 µm (**a**)

Strigeids are known to negatively impact fish host health and may limit recovery and recruitment of adult host populations thus impeding conservation management efforts. The impact of this particular strigeid, *B. bidyanus*, on its fish host, however, is at present unknown (Barton et al. 2023).

Paterson et al. (2021) conducted a review on the global parasite diversity of galaxiid fish, a group of fishes threatened by numerous anthropogenic stressors reducing their geographic range, and for which many species are now considered vulnerable to critically endangered. They found that only 50% of galaxiid species have been screened for parasites. Crucially, these studies predominantly focus on single parasitic taxa and are biased towards a single, relatively abundant and widely distributed species, *Galaxias maculatus* (Paterson et al. 2021). No definitive consensus has been reached by the global aquatic parasitological community on whether hosts or parasites should take precedence in conservation planning. However, there is increasing support for the view that hosts and parasites should be equally considered in conservation and eradication actions. The consequences of irreversible biodiversity loss should be given weight if it is decided that a parasitic species be eradicated. This ultimately emphasises that, in cases where the survival of a parasite is dependent on the host but poses a risk to the survival of the host, a carefully constructed and managed treatment plan for host and parasite as one ecological unit is given precedence over total eradication of the parasite (see Paterson et al. 2021; Barton et al. 2023; Leis et al. 2023).

The larval stages (glochidia) of unionid molluscs are considered temporary parasites that attach to the gill filaments of fish until they develop into juveniles and are ready for dispersal (Chowdhury et al. 2019; Rock et al. 2022). A study searching for a suitable fish host for the critically endangered Spengler's freshwater mussel *Pseudunio auricularius* in Spain identified two fish species that can aid in the recovery of native populations of *P. auricularius*. They first

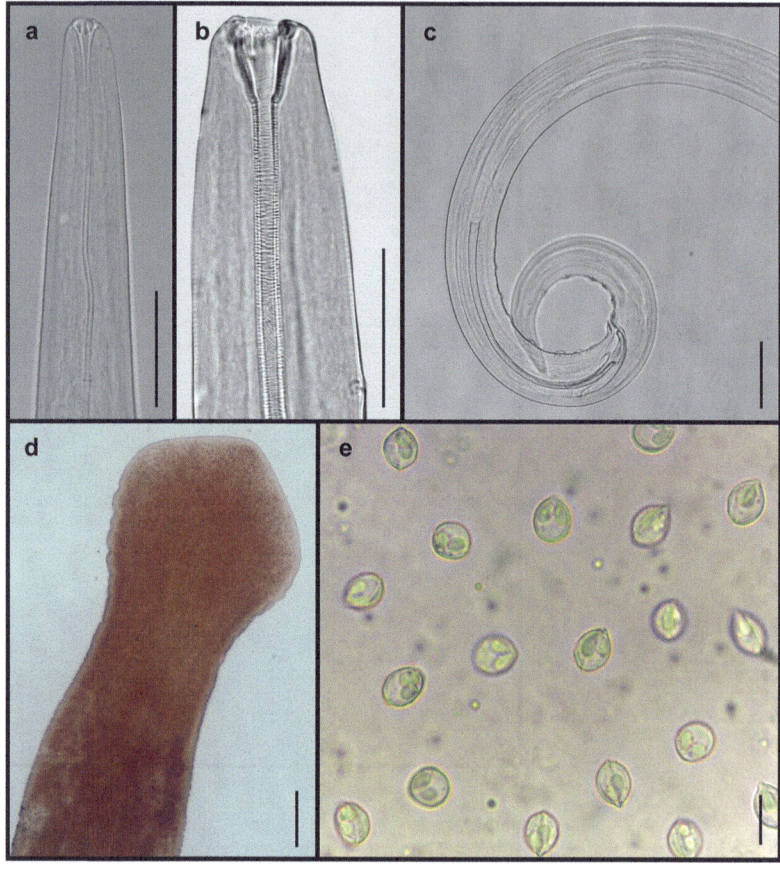

Fig. 13.4 Near-threatened polyxenic parasitic species from cyprininds in the Cape Fold ecoregion, South Africa. Anterior (**a**, **b**) and posterior (**c**) ends of *Rhabdochona* spp.; (**d**) Caryophyllidea gen. sp. and (**e**) *Myxobolus* sp. Scale bars: 10 μm (**a–c**), 100 μm (**e**), 500 μm (**d**)

concluded that the declines in the critically endangered European sturgeon, *Acipenser sturio*, from Spain was most probably the cause of population declines in *P. auricularius*, and identified the critically endangered Siberian sturgeon, *Acipenser baerii*, as an alternative host. However, the latter species of Asian origin is an alien in European waters and would prove difficult to implement in a recovery plan for *P. auricularius* in natural waters (Araujo and Ramos 2000). Alternatively, native fish species that historically co-inhabited freshwaters with *P. auricularius* were experimentally infected with mature glochidia of the bivalve and the regionally threatened freshwater blenny, *Salariopsis fluviatilis*, proved a suitable host (Araujo et al. 2001). Although the experimental infection of *S. fluviatilis* shows the way forward to conserve the only fertile population of *P. auricularius* in Europe, careful consideration on infection and release in natural systems should be considered, as the tolerance of *S. fluviatilis* to glochidia infection has not been determined and could lead to unprecedented morbidity and mortality of an equally threatened host species.

In other instances, the absence of novel knowledge on parasitic communities of threatened and endangered species should be reconciled in a manner that does not come at the expense of the host or parasites, or the extinction of either one. There is opportunity in using specimens in natural history museum collections to augment or replace current sampling. The field of historical ecology of parasites can thus be utilised for threatened species, should material be available in collections (Wood and Vanhove 2022; Wood et al. 2023a). The use of fluid-preserved material has proven to be trustworthy in reconstructing historic parasite communities and changes in parasite abundances of particular host species, as was done for the English sole, *Parophrys vetulus*, in Puget Sound, Washington, USA (Welicky et al. 2021). Wood et al. (2023b) utilised the same approach to determine the trajectory of metazoan parasite abundances of eight fish species over a century and identified potential factors that could have contributed to declines or increases in parasitic species with different life cycle strategies (i.e. one host, two obligatory hosts or three or more obligatory hosts). Although the conservation status of none of these hosts were assessed in these studies, the value of tracking the population trajectories of such parasites is apparent, indicating that this approach is invaluable for the future direction of parasite conservation. Immediate knowledge on the parasitic communities of all known species is not available, thus collecting and depositing sufficient and exemplary representatives of any host species, even those not currently of conservation concern, could aid in future mining of parasitological data to assess trajectories of host–parasite systems that may become threatened. This approach can also help resolve ethical constraints when using threatened species for research and reduce the burden of live sampling from these populations when they are under severe pressure.

13.7.4 Parasites of Amphibians

Amphibians (caecilians, frogs and salamanders) are collectively one of the most threatened aquatic organismal groups globally (Mendelson et al. 2006; Martins et al. 2021). There is no apparent single cause for amphibian declines; however, the most common threats are lethal diseases caused by pathogens (chytridiomycosis, see also Chap. 4; co-infections of several parasites), the combination of parasitic infection with a suite of stressors (population fragmentation, invasive species, human impacts and pollution) and climate change (Wake and Vredenburg 2008; Blaustein et al. 2012; Bower et al. 2019; Herczeg et al. 2021; Re:wild 2023; also see Chap. 22).

The primary focus has been on directly conserving this host group, but some advances in knowledge of a select few amphibian parasites and pathogens have been made that can aid in host conservation. For countering pathogens like *Batrachochytrium dendrobatidis*, eradication of infected populations, rigorous quarantine, *ex situ* breeding before pathogen emergence and intensive monitoring of uninfected populations against the introduction of reservoir species with subclinical infections are essential for protecting threatened amphibians (Weldon et al. 2004; Mendelson et al. 2006; Pessier 2008; Voyles et al. 2009; Yap et al. 2017; also see Chap. 4). In that light, understanding the evolution of host immunopathology and enhanced detection methods play a critical role in aiding the aforementioned conservation measures. Cryopreservation of known pathogens would advance development of sensitive detection kits and treatments for pathogens that currently have no known treatment, such as iridoviruses. The isolation and archiving of strains of *B. dendrobatidis*, the agent of chytridiomycosis, have also been suggested (Voyles et al. 2009).

The use of museum collections to detect parasites and their range and host expansions has also proven valuable. Du Preez et al. (2008) described two new species, *Nanopolystoma brayi* and *N. lynchi* from the urinary bladder and phallodeum of preserved *Caecilia gracilis* from Demerara, Guyana and *Caecilia* cf. *pachynema* from an unknown locality in South America, respectively. Alongside their collection of live Madagascar jumping frog *Aglyptodactylus madagascariensis* in Madagascar, Landman et al. (2023) investigated preserved specimens of *A. madagascariensis* that were collected from several localities in Madagascar in 1971 and 2005. From these, the authors described a novel polystomatid, *Metapolystoma ohlerianum*, finding different life history stages parasitising the urinary bladder and Müllerian ducts. Hartigan et al. (2016) detected the myxozoan *Cystodiscus anoxis* in preserved material of captive rubber eel, *Typhlonectes natans* from Columbia and wild-caught individuals of the Cayenne caecilian, *Typhlonectes compressicaudata*, from French

Guiana. Interestingly, *C. anoxis* was previously only known from a number of Australian frogs, indicating that, despite 300 million years of divergence between frogs and caecilians, *C. anoxis* is either capable of exceptional host-switching, or the host environment remained relatively unchanged since their divergence. Once more, the value of mining natural history collections highlights potential for species discovery and documenting historical presence and distributions of parasitic species, including species that are now threatened (see Landman et al. 2023).

13.7.5 Parasites of Aquatic Mammals

Aquatic mammals are among the more charismatic groups of hosts that are often prioritised in conservation actions, including beavers, otters and lemmings in freshwater ecosystems, and cetaceans, pinnipeds, sirenians and marine fissipeds in the marine realm (also see Chap. 15). As mentioned earlier, the ideal approach to parasite conservation is an integrated process wherein hosts and parasites are considered in conservation action and management. Concern is usually only raised, however, where parasitic infection is seen as a threat for aquatic mammals such as the critically endangered harbour porpoise, *Phocoena phocoena*, the endangered Mediterranean monk seal, *Monachus monachus*, and the sea otter, *Enhydra lutris*.

Despite being known from porpoises globally, the high intensity and parasite load for eight parasitic taxa, including four previously undetected parasites in Baltic subpopulations, is thought to be a health concern for the subpopulation of *P. phocoena* from the Baltic Proper (Dzido et al. 2021). The detection of nematodes of the genus *Pseudoterranova* and appearance of a previously locally extinct mite, *Halarachne halichoeri*, in and on *M. monachus* in the Mediterranean recently raised concerns of increased parasite transmission rates and susceptibility for the survival of this host that is thought to be one of the most genetically depauperate marine mammals on the planet (Reckendorf et al. 2019, Galatius et al. 2020, Koitsanou et al. 2022a, b). The locally extinct *H. halichoeri* in the southern Baltic Sea, Danish Straits and Kattegat is naturally absent from the Mediterranean and its recent emergence and expansion in host range, along with its potential to facilitate lethal pathogen transmission (previously reported in stranded threatened Southern sea otters, *Enhydra lutris nereis*, along the coast of California, USA) (see Pesapane et al. 2018; Seguel et al. 2018 and references therein; Koitsanou et al. 2022b) underscore the importance of (1) understanding the pathology and ecology of respiratory mites; (2) determining the effect it might have on small and genetically homozygous populations and (3) the complexity of parasite conservation.

Another interesting case for parasite conservation is that of the re-extension in distribution of the rare specialist chewing louse, *Lutridia exilis* (Phthiraptera), in northern Germany. This chewing louse, along with its Near-Threatened host, the Eurasian otter, *Lutra lutra*, has rarely been reported in the past century, but was recently detected during a population health monitoring programme (Rózsa 1993; Rhoner et al. 2023). The louse is highly host-specific, being restricted to Eurasian otters and is among a group of lice that have specialised morphological adaptations that allow them to successfully attach and survive on semi-aquatic hosts (see Leidenberger et al. 2007; Leonardi et al. 2022; Chap. 15). The fact that the comeback (or 'reverse effect') of a host species aided in the return of specialist parasite species is both crucial for biodiversity and function within local ecosystems (also see Raga et al. 1997). Apart from the above examples, until recently, no genuine example existed for the implementation of an integrated conservation event for any aquatic mammal and its parasites. Whinfield et al. (2024) executed one of the first integrated disease risk analyses prior to a translocation of the Near-Threatened platypus *Ornithorhynchus anatinus* in Australia, in the process demonstrating the feasibility of including parasites in the conservation process.

13.8 Case Study: Chondrichthyans and Their Parasites Under Threat

In comparison to biodiversity losses on land, where population declines and extinction of species have been recorded and actions taken to prevent and counteract species losses, it is harder to assess and act in oceanic systems. Multiple human-induced threats, such as climate change, overharvesting and habitat degradation drive marine defaunation and increase the extinction risks for marine wildlife (McCauley et al. 2015). However, marine extinctions have rarely been reported (Dulvy et al. 2003, 2009). There is, however, growing concern on accelerated extinction rates and an imminent global extinction crisis. Cartilaginous fishes of the class Chondrichthyes (i.e. sharks, rays and chimaeras) are among the world's most threatened marine vertebrate groups, with roughly one-third of species regarded as threatened with extinction (Dulvy et al. 2021; IUCN 2023b). Of the 1,479 described species of chondrichthyans (Fricke et al. 2024), the IUCN Red List of Threatened Species currently lists 1,234 species. Collectively, 397 species (32.2%) are placed in the three highest categories: Critically Endangered (CR; 91 species, 7.4%), Endangered (EN; 124 species, 10.0%) and Vulnerable (VU; 182 species, 14.7%). An additional 125 species (10.1%) are categorised as Near Threatened (NT), while less than half (i.e. 539 species, 43.7%) have the status of Least Concern (LC), signifying a lack of immediate risk of extinction (IUCN 2023b). In the past 10 years, members of the IUCN SSC Shark Specialist Group and international experts have reduced the number of species for which insufficient data are available (data deficient, DD) to 173 species (i.e. 14.0%). Chondrichthyans are currently the only marine fish lineage for which a risk determination can be formulated for the entire clade. Dulvy et al. (2021) estimated that around 37.5% (range 32.6–45.5%) of cartilaginous species are threatened given that DD species are proportionally distributed along the five categories of the IUCN Red List assessment system. Only sea turtles (Cheloniidae and Dermochelyidae), which directly engage with terrestrial environments (and consequently humans) during specific life stages, exhibit a comparatively greater proportion of threatened species (IUCN 2023b).

Cartilaginous fishes face several human-induced threats (i.e. over-exploitation, habitat loss and degradation, climate change, pollution, etc.) (Dulvy et al. 2014, 2021; Pacoureau et al. 2021; Sherman et al. 2023). Among these intertwined threats, overfishing—either targeted or unintentional—has been identified as the most significant cause of decreasing chondrichthyan populations (Dulvy et al. 2021; Sherman et al. 2023). The level of risk for chondrichthyans is intricately connected to their life history sensitivity and ecological specialisation. Certain features of their life history, such as their large size, low fecundity, slow growth and late maturity, have guaranteed their perseverance in aquatic food webs for millions of years. However, these life characteristics, allied with slow population growth and low resilience to external pressures, make them extremely vulnerable to the ever-growing anthropogenic impacts, resulting in drastic population declines and local extinctions. Depth range constitutes an additional factor which greatly influences the threat level of chondrichthyans (Dulvy et al. 2021). While shallow-water (depth <200 m) coastal species are more threatened by their accessibility to fisheries and increased fishing pressure, deep-water chondrichthyans avoid these threats but are more susceptible to overfishing due to a lower productivity and slow growth (Dulvy et al. 2014, 2021).

Euryhaline species such as sawfishes (family Pristidae) are among the most threatened groups of chondrichthyans because of their large sizes, restricted ranges, habitat specialisation in coastal, estuarine and riverine habitats, narrow depth distributions and high exposures to fisheries (Dulvy et al. 2014, 2016). Recent studies (Pacoureau et al. 2021; Sherman et al. 2023) provide additional insights on the increased extinction risk of sharks and rays based on their habitat preference, which is higher than the average threat level for the entire clade identified by Dulvy et al. (2021). Coral reef-associated sharks and rays are facing higher threat levels through human interference (i.e. fishing pressure and habitat destruction, amongst others), driving nearly two-thirds of species (59%; 79 of 134) to the brink of extinction (Sherman et al. 2023). The scenario is further exacerbated for widely distributed oceanic pelagic sharks. Due to a high exposure to fisheries, more than three-quarters of oceanic sharks (77.4%; 24 of 30 species) experience an increased risk of extinction (Pacoureau et al. 2021). Absence of high-trophic level species through severe population declines and local extinctions has major implications on aquatic food webs leading to ecosystem perturbations and the loss of critical function and services. Nonetheless, the occurrence of chondrichthyan extinction events remains speculative and currently lacks confirmatory evidence. Certain species have undergone localised extinctions, while others have been documented solely on the basis of a single specimen and might be highly threatened or even possibly extinct although they have been categorised as being of least concern (see Dulvy et al. 2021 and references therein). Dulvy et al. (2021) list four species that were classified as critically endangered (possibly extinct) [CR(PE)]: the Java stingaree *Urolophus javanicus*, the lost shark *Carcharhinus obsoletus*, the Pondicherry shark *Carcharhinus hemiodon* and the Red sea torpedo *Torpedo suessii*. No parasites have been recorded for any of these CR(PE)-listed species.

13.8.1 Parasites of Chondrichthyans

Chondrichthyans and their specific parasite fauna display an ancient host–parasite system with a strongly interconnected co-evolution. Given the current threat levels and future projections for the host group, this unique host–parasite system represents one of the most threatened on this planet. It is highly likely that parasite species, particularly those exhibiting high host specificity and complex life cycles, are limited in their connectivity and may face even higher threat levels than their hosts. Persistent global threats and rising extinction rates, combined with the high specificity and uniqueness of their parasite fauna, make chondrichthyans an ideal target group to assess the threats to their parasites and affiliate species with the goal to not only preserve predators along with symbiont assemblages in updated conservation agendas but also to maintain ecological and evolutionary processes along with species.

Each chondrichthyan host serves as a macro-habitat for a variety of parasitic organisms (Caira and Healy 2004). The number of known metazoan parasites infecting chondrichthyans is constantly growing and currently larger than the number of host species. Metazoan parasites of several phyla infect essentially every organ and organ system of chondrichthyans (Caira et al. 2012). However, many chondrichthyan genera have never been observed for parasites (Caira and Jensen 2017) or (if observed) only selected organs (such as the spiral intestine, gills and/or skin) were assessed. Full necropsies of the entire chondrichthyan host are sparse and many parasites might not have been detected. Therefore, our current knowledge on chondrichthyan parasites is vastly incomplete and the total parasite fauna remains underestimated with a majority of species awaiting scientific discovery. The most dominant groups of chondrichthyan parasites are cestodes followed by parasitic copepods and monogeneans which, in combination, comprise more than 85% of reported species (Caira et al. 2012). The remaining parasite groups infecting chondrichthyans have a lower species richness (i.e. annelids, isopods, myxozoans, nematodes and trematodes) or exhibit infrequent occurrences or near-complete absence (i.e. acanthocephalans, acari, amphipods, branchiurans, cirripeds, molluscs, ostracods and triclads). An overview is presented below on the most dominant chondrichthyan parasite groups, delineating their current state of knowledge and proposing plausible conservation measures.

Cestodes
The most diversified group of metazoan parasites infecting chondrichthyans are cestodes (tapeworms). With well over 1,000 described species (Caira and Jensen 2017), cestodes represent the largest and most species-rich group of chondrich-

thyan parasites, accounting for more than 60% of the known parasite fauna. The number of described species surpasses the collective count of metazoan chondrichthyan parasites of other taxonomic groups. Although not all chondrichthyans have been assessed for cestode infections, Caira and Jensen (2017) estimated the species diversity to exceed 5,000 species in chondrichthyans. This estimate is based on the extrapolation of known cestode species from investigated hosts over the chondrichthyan diversity (Caira and Jensen 2017). Cartilaginous fishes and cestodes share an extensive time of co-evolution. Cestode eggs were recovered in a shark coprolite dating back 270 million years, hence establishing it as the earliest recorded instance of cestode infections in vertebrates (Dentzien-Dias et al. 2013). For millions of years, cestodes diversified with and within their marine hosts invaded freshwater environments and survived five mass extinction events with their high trophic counterparts. At present, ten out of the 19 currently recognised cestode orders infect cartilaginous fishes. With the exception of the 57 holocephalan taxa (Fricke et al. 2024) which harbour a relatively depauperate cestode fauna (the enigmatic cestode order Gyrocotylidea and a single species of the order Phyllobothriidea), most cestode assemblages of sharks and batoids exhibit higher species richness and consist of representatives from multiple orders that co-infect the same host species (Caira and Healy 2004). Randhawa and Poulin (2010) investigated the cestode assemblages of sharks and rays and found that the diversity is positively correlated with host size (and most likely the size of the spiral intestine), host diversity as well as environmental temperature. The diet of chondrichthyans might be another influencing factor for the high diversity (Rasmussen and Randhawa 2018). The life cycles of true tapeworms (Eucestoda) in marine systems are virtually unknown. Transmission is facilitated over the food chain involving one or several intermediate and/or paratenic hosts (i.e. invertebrates, teleosts, marine mammals or other chondrichthyans) rendering life cycles elaborate and complex (Caira and Littlewood 2013). In contrast, the life cycle of the most basal cestodes, the Gyrocotylidea, may be facilitated through direct transmission of decacanth larvae to holocephalan hosts. However, information on this enigmatic group infecting exclusively chimaeras is sparse (Kuchta et al. 2017). Barčák et al. (2021) estimated that gyrocotylideans have only been reported from 25% of extant chimaera species known to date. Chondrichthyan cestodes tend to infect a single or closely related host species, with the exception of the order Trypanorhyncha (Caira and Jensen 2001; Palm and Caira 2008). This high degree of host specificity indicates a strongly linked association between chondrichthyans and their cestode parasites. Due to these unique life characteristics (i.e. complex life cycles and high degree of host specificity), cestodes might represent the single-most threatened group of chondrichthyan parasites. To date, conservation efforts targeting this immensely diverse and dependent lineage of chondrichthyan parasites have received limited attention within scientific circles. To our knowledge, only a single study identified threatened cestode species, indicating a significant gap in addressing the conservation needs of this group. Van der Spuy et al. (2022) described three cestode species new to science from the IUCN Red Listed (Endangered) white skate, *Rostroraja alba* in South Africa (Fig. 13.5). This skate host occurs in the Mediterranean Sea and eastern Atlantic Ocean from the British Isles, along the African coast to South Africa and in the south-western Indian Ocean towards Mozambique (Froese and Pauly 2023) (Fig. 13.5a). Due to severe population declines across its entire geographical range, this species is currently listed on the IUCN Red List as Endangered (Dulvy et al. 2006). However, regional assessments in the Mediterranean and north-eastern Atlantic (i.e. Celtic Seas) provide an even more dire picture of *R. alba*, with severely depleted populations and even local extirpations in its northern-most range, making it (at least regionally) a critically endangered species (Ellis et al. 2015) (Fig. 13.5a). Overall, parasite records from this EN(CR)-listed host are sparse, with 15 parasites (six cestodes, four monogeneans, two copepods and isopods, respectively, and one trematode) (Euzet 1959; Sproston 1946; Williams

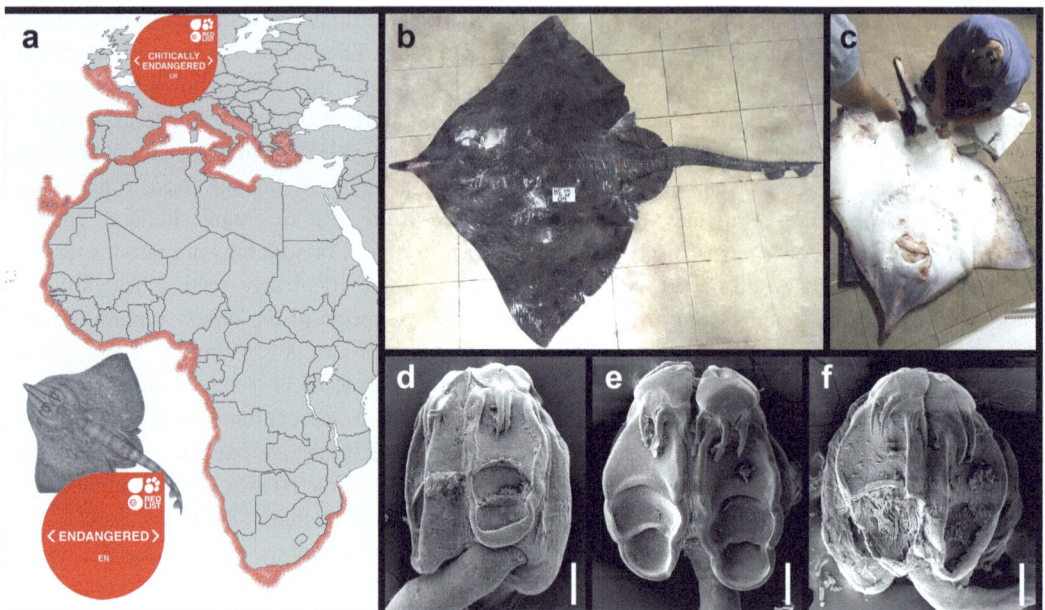

Fig. 13.5 Biogeography, appearance and parasite records of the white skate, *Rostroraja alba*. (**a**) Distribution records of *R. alba* (in red) and the indication of the IUCN Red List assessment ('endangered') and regional assessment ('critically endangered'); (**b**) female specimen of *R. alba* in dorsal view; (**c**) female specimen of *R. alba* in ventral view during dissection; (**d**) scanning electron micrograph of scolex of *Acanthobothrium umbungus*; (**e**) scolex micrograph of *Acanthobothrium usengozinius*; (**f**) scolex micrograph of *Acanthobothrium ulondolozus*. Scale bars: 100 μm (**d**–**f**). Photos: (**b**, **c**) © Ruan Gerber; (**d**–**f**) © Linda van der Spuy

1969; Dollfus and Trilles 1976; Moreira and Sadowsky 1978; Dippenaar 2004, 2016; Tyler 2006; Schaeffner and Smit 2019; Zaragoza-Tapia et al. 2020; Derbel et al. 2022; Vaughan et al. 2023). The detailed descriptions of Van der Spuy et al. (2022) mark the first cestode records of this host species in the Southern Hemisphere. The three species (Fig. 13.5d–f) belong to the species-rich, synhospitalic tapeworm genus *Acanthobothrium* (Cestoda: Onchoproteocephalidea II, sensu Caira and Jensen 2017). Given the strict host specificity ('oioxenous' sensu Euzet and Combes 1980) and the fact that several congeneric species of *Acanthobothrium* can co-infect the same host species (Fyler 2009), Van der Spuy et al. (2022) pointed out that, with declining host populations, these cestodes face higher extinction risks than their hosts. The authors recommended that future conservation actions should incorporate parasite species along with the threatened host to minimise the risks of parasite (co-)extinctions and protection of ancient and potentially beneficial host–parasite interrelationships.

Monogeneans

The second-largest group of platyhelminths infecting chondrichthyans are monogeneans, with over 200 described species (Yamaguti 1963; Caira et al. 2012). Whittington (1998) hypothesised that worldwide more than 80% of the monogenean fauna remains unknown and most species await discovery, including the ones infecting chondrichthyans. Most monogeneans are ectoparasites with a low fecundity and direct life cycles (Whittington et al. 2000; Whittington and Kearn 2011). As ectoparasites, they attach to outer surfaces, primarily the skin and the gills and to a lesser extent the nasal fossae and buccal cavity; a smaller number have adapted to an endoparasitic life, infecting organs like the oviduct, rectum/ rectal gland, circulatory system and inner body cavity wall (Chisholm and Whittington 1998; Whittington et al. 2000). Monogeneans are considered the parasite group with the highest degree of host specificity, with the majority of species infecting just a single host species (Whittington 1998; Whittington et al. 2000). Several members of the family Monocotylidae

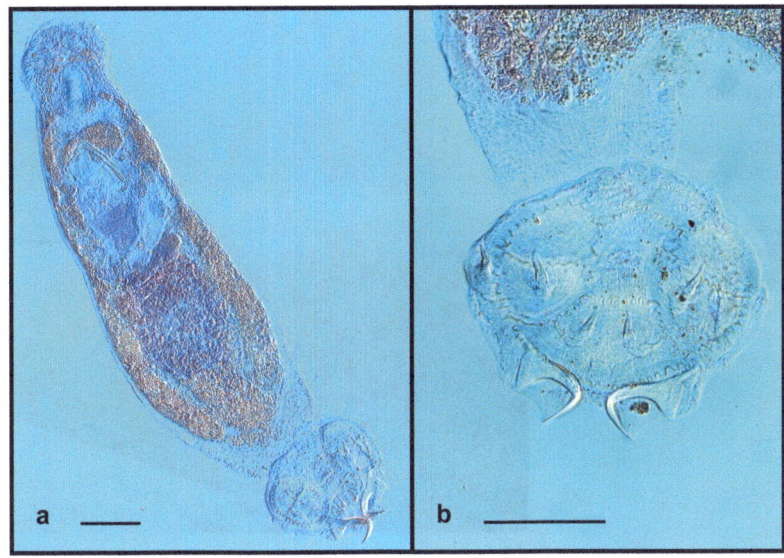

Fig. 13.6 *Neoheterocotyle darwinensis*. (**a**) Whole specimen; (**b**) haptor with hamuli and dorsal haptor accessory sclerites. Scale bars: 100 μm. Photos © Russell Q-Y. Yong

(*Calicotyle*, *Heterocotyle*, *Merizocotyle* and *Neoheterocotyle*) (Fig. 13.6a, b), however, infect different chondrichthyan hosts, which are either closely related or have similar life history traits (Chisholm and Whittington 1996, 1997, 1999, 2012; Chisholm et al. 1997, 2001a, b; Whittington 1998; Whittington et al. 2000). Monogeneans with a high or strict host specificity generally infect large-bodied hosts (Sasal et al. 1999). Larger hosts occupy higher trophic positions, have a higher longevity and offer a multitude of niches for parasites in a more stable environment which is reflected in a greater parasite diversity and higher levels of specificity (Guégan et al. 1992; Sasal et al. 1999; Ezenwa et al. 2006; Lindenfors et al. 2007; Harris and Dunn 2010; Colwell et al. 2012). However, these large-sized hosts are more vulnerable to extinction (Cardillo and Bromham 2001; Sodhi et al. 2008; Stork et al. 2009) that ultimately increases the risk of co-extinction for parasites (Purvis et al. 2000; Estes et al. 2011; Cizauskas et al. 2017). The life cycles of monogeneans would theoretically lower their overall vulnerability to extinction due to the direct transmission and absence of intermediate hosts. This apparent advantage, however, is hampered by their strict host specificity and tendency to infect larger hosts, which elevates their extinction risk due to practices such as 'fishing down the food web' (Pauly et al. 1998). In addition, adult monogeneans are incapable of swimming which greatly limits their transmission to other, non-threatened hosts if primary host populations become depleted. To our knowledge, only a couple of studies identified monogeneans that might face extinction with their host species (Bakenhaster et al. 2018; Ingelbrecht et al. 2022). Ingelbrecht et al. (2022) described a new microbothriid species of the genus *Dermopristis* from the critically endangered longcomb sawfish, *Pristis zijsron*, in Australia. The authors demonstrated that, given the high host specificity and thus dependency upon a single host species that is on the brink of extinction, their new species, *Dermopristis pterophila* (Fig. 13.7a, b), as well as other host-specific parasites of threatened species, should be regarded as having the same risk level as their host species (Ingelbrecht et al. 2022). The monogenean family Microbothriidae, to which *D. pterophila* belongs, is currently composed of 12 genera and 24 species (WoRMS 2023). This group of ectoparasites infects the skin of sharks and rays, where they cement themselves to the denticles using adhesive secretions (Kearn 1965; Whittington and Chisholm 2008; Whittington and Kearn 2011). Most microbothriids display strict host specificity (17 species, 71% of total family richness), with fewer species utilising two (five species), three or four host species (a single species each). The host spectrum

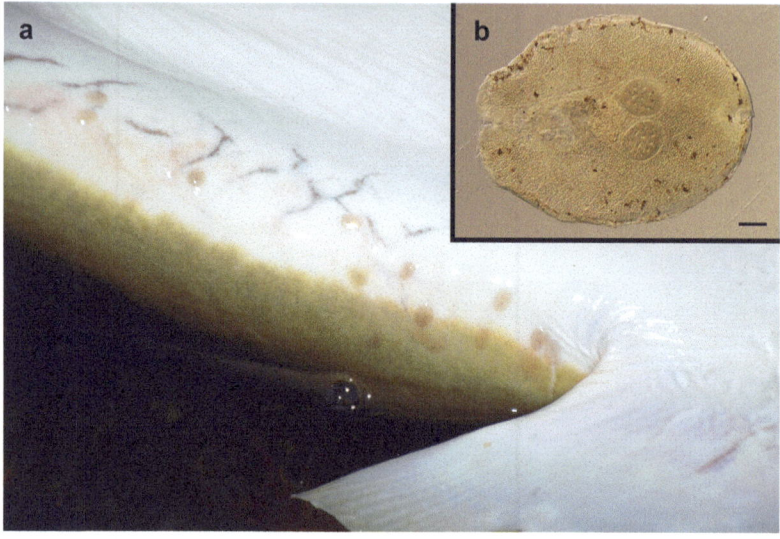

Fig. 13.7 *Dermopristis pterophila*. (**a**) *In situ* on the endangered longcomb sawfish *Pristis zijsron*; (**b**) ventral view of lactophenol cleared and mounted specimen. Scale bar: 500 μm. Photos: (**a**) © Dave Morgan; (**b**) © Jack Ingelbrecht

(29 species, of which four are infected by two to three microbothriid species) contains numerous threatened species, amounting to roughly 59% of the microbothriid family (six CR, three EN, eight VU, assuming equal level of threat status), while only 38% of hosts are considered non- or near-threatened (Froese and Pauly 2023). One species, the largespine velvet dogfish *Scymnodon macracanthus*, is currently listed as Data Deficient. Applying Ingelbrecht et al.'s (2022) proposition, more than half of all known microbothriid species (i.e. 13 of 24 species) face an increased risk of extinction, including six species (25% of the fauna) whose hosts are listed as critically endangered. Two additional microbothriids infect two threatened host species, resulting in a marginal reduction in their respective threat levels without substantial improvement. Among the microbothriids, only nine species (38%) exhibit a reduced vulnerability to extinction, attributed to their association with one or several non- or near-threatened host species. This dire picture only focuses on a single family and does not account for other groups of monogeneans infecting chondrichthyans. In accordance with Whittington's (1998) hypothesis and under the assumption that the current threat levels of microbothriids are representative for all monogeneans infecting chondrichthyans, the number of threatened species could potentially surpass the number of known monogeneans from this host group.

Parasitic Copepods

After cestodes, parasitic copepods represent the second-most diverse group of metazoan parasites infecting chondrichthyans (Caira et al. 2012). With roughly 290 species and spanning over three cyclopoid and nine siphonostomatid families (Boxshall and Hayes 2019), these parasites infect a wide variety of chondrichthyans in marine and estuarine environments worldwide (Yamaguti 1963). The overwhelming majority of taxa are ectoparasites inhabiting external body surfaces and orifices (i.e. skin, gills, branchial chamber, buccal cavity, eyes, nasal fossae and cloaca), either superficially attaching to or deeply penetrating the host (Benz 1993; Schaeffner and Smit 2019). Most species have direct life cycles that involve several larval stages (Kabata 1979, 1981; Benz 1993). Host specificity seems variable within genera but appears generally high. Pollerspöck and Straube (2023) list 243 copepod species infecting chondrichthyans, belonging to 58 genera and 12 families. Almost half of the species (49%) exhibit strict specificity to a single host species. Each additional host leads to a gradual reduction in the number of copepod species present. A total of 19% of copepods utilise two host species, while the proportion of species with a larger host range is 7% (three hosts), 4% (four hosts), 2% (six to eight hosts) and less than 1% for those infecting more than ten hosts. Most families exhibit high host specificity, with more than 50% of species infecting a single host. Sphyriids appear particularly host specific, with 79% of species (i.e. 11 of 14 species) presenting a strict specificity. Conversely, the family Pandaridae exhibits the lowest host specificity with 77% of species (i.e. 36 of 47 species) infecting two or more host species. One species, *Perissopus dentatus*, infects more than 30 different hosts. While the dataset may not be exhaustive and does not account for sampling bias or potential species misidentifications, it illustrates the general tendency of parasitic copepods to limit the number of hosts, frequently confined to a single species. A small number of parasitic copepods may become problematic for their chondrichthyan hosts. In high intensities, they may cause tissue lesions at the sites of attachment or even destructive disabilities that severely impair hosts' respiratory or feeding capabilities (Benz and Bullard 2004). However, the majority of species do not appear to cause health problems in their chondrichthyan hosts, with generally benign host–parasite associations. The high species diversity and generally high level of specificity of parasitic copepods raise concerns regarding the increased levels of vulnerability and elevated extinction risks of their hosts. Up to this point, the issue of copepods being threatened together with their chondrichthyan hosts has been addressed in only two studies (Morgan et al. 2010; Norman et al. 2021). The whale shark copepod *Pandarus rhincodonicus* is a benign and highly host-specific species, which occurs exclusively on the endangered whale shark, *Rhincodon typus* (Norman et al. 2000; Pierce and Norman 2016) (Fig. 13.8a, b). Norman et al. (2021) proposed that, given the inextricable nature of and inherent threats to this host–parasite system, the whale shark copepod merits classification within the same risk category as its host and warrants recognition as an endangered species. In another study, Morgan et al. (2010) reported the first species of *Caligus*, *C. furcisetifer*, from the critically endangered common sawfish, *Pristis pristis* (as *Pristis microdon*), in Australia (see Faria et al. 2013). Low copepod prevalence at one of their sampling localities made the authors speculate that this parasite faces a high vulnerability to extinction thus necessitating conservation measures.

13.8.2 Threatened Pristids and Their Parasites

Pristis pristis belongs to the most threatened group of marine fishes, sawfishes of the family Pristidae (Dulvy et al. 2016). The family consists of two genera, *Anoxypristis* and *Pristis*, and includes five species: *A. cuspidata*, *P. clavata*, *P. pectinata*, *P. pristis* and *P. zijsron*. With the exception of the pointed sawfish, *A. cuspidata*, which is classified as Endangered, the remaining four species of *Pristis* are categorised as Critically Endangered (Froese and Pauly 2023). Sawfishes occur in shallow coastal areas, both in marine and

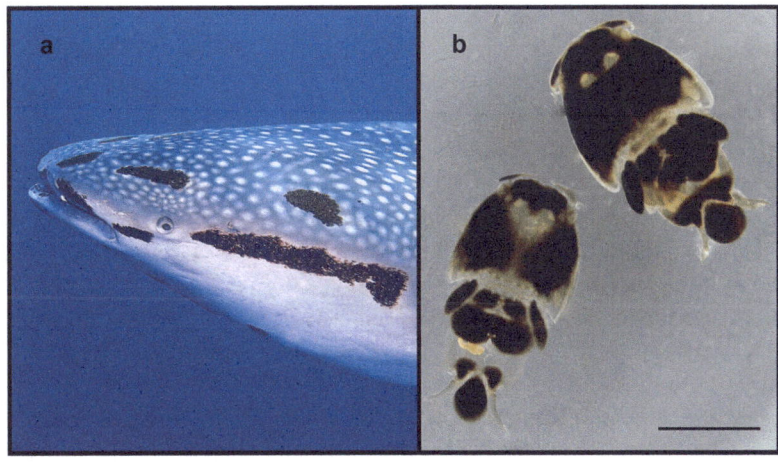

Fig. 13.8 *Pandarus rhincodonicus*. (**a**) *In situ* on the endangered whale shark *Rhincodon typus*; (**b**) Whole specimens, dorsal view. Scale bar: 2 mm. Photos: (**a**) © Albert Kang; (**b**) © Andrew Hosie, Western Australian Museum

riverine habitats, with relatively shallow depth distributions (<100 m for most species; frequently <10 m) (Carlson et al. 2014). Although protected in 16 countries, high susceptibility to fishing activities, habitat degradation, low population increase and limited recovery potential have led to severe population declines, range contractions and possible regional extinctions in several countries (Dulvy et al. 2016). Decreasing population trends and elevated risks of extinction of this imperilled chondrichthyan group will unequivocally have dire consequences for a multitude of parasites and affiliate species. Parasites of the most dominant groups (cestodes, monogeneans and copepods) and more depauperate groups (trematodes, nematodes, isopods and hirudineans) have been recorded from sawfishes. However, it is expected that these reports only represent a small proportion of the actual parasite diversity of pristid hosts, given the obvious constraints of collecting endoparasites (i.e. lethal sampling of protected species), sampling bias of researchers for a particular parasite group as well as the historically broad distribution of sawfishes (see Dulvy et al. 2016 for distribution areas of species) and knowledge of parasites from only few localities. In line with expectations, records of cestodes (17 species) and monogeneans (ten species) from five pristid hosts vastly outnumber members of the remaining parasite groups (i.e. ten species combined). Most pristid-infecting cestodes (14 species or 82% of all recorded species) exhibit a high host specificity, with all but three taxa infecting either a single pristid host or several members of Pristidae ('metastenoxenous' sensu Caira et al. 2003). In total, 11 species (65%) exhibit oioxenous host specificity, including two species of *Floriparicapitus* that await formal description (Cielocha et al. 2014). Three metastenoxenous taxa (18%) infect either two (*Fossobothrium perplexum* and *Pristiorhynchus palmi*) or three (*Pterobothrium australiense*) pristid host species (Beveridge and Campbell 2005; Schaeffner and Beveridge 2012, 2013). The remaining three species (18%) are euryxenous (sensu Euzet and Combes 1980) and infect hosts of different chondrichthyan families, including two trypanorhynch taxa, *Proemotobothrium linstowi* and *Prochristianella clarkeae*, and a phyllobothriid species that is considered *incertae sedis* ('*Anthobothrium/Phyllobothrium*' *pristis*; see Caira et al. 2023). '*Anthobothrium/Phyllobothrium*' *pristis* is regarded as one of the least host-specific cestode species, having been reported from 41 different chondrichthyan hosts across 14 genera and seven families (Schaeffner and Beveridge 2014). Among the monogeneans, all ten species exhibit a pronounced degree of host specificity, being confined strictly to a single pristid host (Watson and Thorson 1976; Cheung and Nigrelli 1983; Ogawa 1991; Chisholm and Whittington 2000; Kearn et al. 2010; Kritsky et al. 2017; Ingelbrecht et al. 2022, 2024). Remarkably, pristids are associated with only a single copepod species (Morgan et al. 2010; Ingelbrecht et al. 2024), a sharp contrast to the typically extensive species diversity exhibited by this parasite group in chondrichthyans, as discussed earlier. However, the absence of copepods in pristids might not accurately reflect the true picture. Bakenhaster et al. (2018) reported the parasite component community of the smalltooth sawfish, *P. pectinata*, in Florida. Long-term, extensive sampling efforts of 277 living (for ectoparasite screening) and 13 deceased animals (for gill- and endoparasites) revealed 22 parasite species, including eight known species and 14 taxa whose identification to species level has been pending (Bakenhaster et al. 2018). The findings also comprised five species of copepods, which, among other species, await taxonomic evaluations. This demonstrates that our current knowledge on copepods of pristid hosts is limited and will almost certainly be augmented through additional sampling efforts of copepods from catch and release hosts. The remaining parasite groups (i.e. trematodes, nematodes, isopods and hirudineans) that exhibit a lower diversity in chondrichthyans are less numerous with only ten species reported from pristids (Moreira and Sadowsky 1978; Bruce et al. 1994; Bakenhaster et al. 2018; Burreson 2020; Ingelbrecht et al. 2024). Apart from the unidentified gnathiid isopods and the blood fluke, *Achorovermis testisinuosus*, which exhibit a high host specificity [infecting only the critically-endangered green sawfish, *P. zijsron* (Ingelbracht et al. 2024) and the smalltooth sawfish *P. pectinata* (Warren et al. 2020), respectively], the seven members of the remaining groups (i.e. nematodes, isopods and hirudineans) are euryxenous. The current knowledge on parasitic organisms of pristids provides an unambiguous pattern, effectively portraying the broader parasite fauna of the imperilled chondrichthyan host group. Merely eleven pristid parasite species, equivalent to 29.7% of the known diversity, are capable of infecting two or more hosts, which might counteract imminent risks faced by a single host and might play a pivotal role amidst the ongoing extinction crisis. In contrast, 70.3% of parasites (i.e. 26 species) exhibit a high host specificity either to a single host species (i.e. 62%) or closely related hosts (i.e. 8.1%). Coupled with highly complex life cycles, 14 of these highly host-specific taxa, accounting for 38% of the known parasite diversity of pristids, are currently confronted with an increased risk of extinction. However, many of the most host-specific parasites, some of which might have eluded detection, might have already been lost. Extending this scenario to all cartilaginous fishes currently confronting an elevated risk to extinction underscores the precarious state of their associated and host-dependent parasites.

13.8.3 Chondrichthyan Parasite Conservation: The Way Forward

The estimation of numbers and scale of threatened parasitic organisms associated with aquatic hosts remains speculative due to the lack of knowledge on their diversity. However, given the rising vulnerability and elevated extinction risk of the chondrichthyan host group, it is conceivable that the number of endangered parasite species could potentially exceed 1,000. The evident knowledge deficit becomes apparent in light of Carlson et al.'s (2020a) estimation, which suggests that only 23% of helminth species (i.e. acanthocephalans, cestodes, nematodes and trematodes) associated with chondrichthyans have been formally documented. The same situation applies to other taxonomic groups, particularly monogeneans and parasitic copepods. Yet, how do we protect aquatic parasite diversity and proceed with conservation plans when only a small proportion of their actual diversity is known? This is the obvious question, albeit it remains somewhat insufficient in addressing the real challenge at hand. Parasites have not been in the spotlight of conservation biologists and have rarely been considered in assessments of threatened species or conservation agendas. Fortunately, this situation has been changing in recent years. Numerous research papers and review articles have embraced the concept of imperilled parasites, advocating for their conservation (Windsor 1995; Koh et al. 2004; Dunn et al. 2009; Gómez and Nichols 2013; Dougherty et al. 2015; Strona 2015; Spencer and Zuk 2016; Carlson et al. 2017a; Kwak et al. 2020; Lymbery and Smit 2023). Therefore, more than 30 years after Windsor's (1990) plea ('Equal rights for parasites') mentioned earlier, parasite conservation takes a step further, with elaborate mechanisms for the assessment of threatened species and Red List classifications (Carlson et al. 2020a, b). To convey the importance of parasite conservation and augment the presence of parasitic organisms in Red List assessments, the IUCN SSC Parasite Specialist Group has been recently established with the main purpose of evaluating the extinction risk of metazoan parasites infecting vertebrate hosts. Previous IUCN Red List assessments were contingent upon criteria encompassing the geographical distribution and population size and trends to classify species; yet this data has mostly been unavailable or unachievable for invertebrates, including metazoan parasites and co-dependent species rendering them data deficient (Cardoso et al. 2011; Carlson et al. 2017b). Recent modifications to these criteria have been suggested to encompass parasites and other affiliate species by considering their threat status in relation to that of their hosts (Kwak et al. 2020; Moir and Brennan 2020; Lymbery and Smit 2023). Considering parasites as conservation targets relies on effective management and implementation of conservation action plans for species recovery. One rapid approach is to link parasites, particularly those with higher host specificity, to the conservation status of their endangered host species and consider them as threatened ecological communities for subsequent recovery plans (Moir et al. 2012; Lymbery and Smit 2023). This host-centred conservation approach for parasites based on co-dependency deviates from previous conservation plans for single, free-living species to accommodate symbiont assemblages. However, threatened parasites may exhibit a higher risk of extinction than their hosts, which is attributed to their unique life history traits as detailed above. A different approach is focused on the conservation of entire ecosystems (Keith 2009; Wilson et al. 2009; Gómez and Nichols 2013). This conservation approach facilitates the management of protected areas to sustain biodiversity, including threatened host species and their parasites, along with ecosystem services and processes. However, this regionally restricted approach may not be suitable for wide-ranging host species and their parasitofauna, particularly in the marine realm. A new system for the assessment of threatened parasite species has been proposed by Kwak et al. (2020). The conservation assessment methodology for animal parasites (CAMAP) utilises modified IUCN criteria on the ecological data of infected hosts (criteria 1–5) as well as newly implemented criteria for the determination of parasite vulnerability based on the IUCN status of definitive (criterion 6) and intermediate host species (criterion 7). Given the paucity of ecological data (e.g. data on the geographical distribution, infection levels and host decline) for most host–parasite systems, the new criteria facilitate the rapid and accurate determination of the minimum conservation status of any parasite species (Kwak et al. 2020). A prospective approach for the implementation of this novel framework entails prioritising parasites of highly threatened and rare hosts, thereby using them as a paradigmatic model group.

For example, as shown above, cartilaginous fishes and their parasites are among the most threatened host–parasite systems on Earth. Several parasite groups exhibit an enormous diversity and abundance in cartilaginous fishes which reflects long-lasting and strongly interconnected associations and further highlights the parasites' critical importance as integral components for host individuals and populations and the perpetuation of functional and healthy aquatic systems. Although precise predictions on co-extinction trajectories remain challenging, the likelihood of species losses and elevated co-extinction risks is accentuated by the prevailing high host specificity observed across several groups, alongside life cycle complexity, as exemplified by cestodes. Additionally, chondrichthyan parasites are subject to direct influences from anthropogenic factors that amplify their levels of vulnerability.

The preservation of symbiotic relationships and host–specific interactions necessitates a comprehensive approach to conservation as detailed above, including all symbiotic partners in conservation actions and thus deviating from the concept of focusing solely on the threatened host species. These parasites (or rather their ancestral lineages) have endured and survived five mass extinction events together with their cartilaginous hosts. The responsibility now lies with us to contribute towards the protection and ultimate survival of these organisms during the ongoing sixth mass extinction event and mitigate their catastrophic decline and resulting repercussions for the aquatic biota and ecosystems.

13.9 Conclusion and Future Perspectives

This chapter summarises the importance of aquatic parasitic species in ecosystems, the intricate links they facilitate and niche they occupy in ecosystems as well as the advances made from advocacy to implementation of their conservation as an ecological unit with its host. Further, several challenges to an inclusive conservation approach for parasites and hosts as an ecological unit was highlighted and it is evident that although parasite conservation is still in its infancy, momentum, drive, expertise and tool development exist across a global multidisciplinary scientific and management community. The way forward is clear. Enrichment of ecological knowledge and the heritage of local, regional and global aquatic systems that serve biodiversity as well as humans can be achieved. This can be done through focused efforts on ecological units (the parasite, its host and the contribution to an ecosystem) by characterising and documenting intrinsic and ecologically important components and interactions within vulnerable or even non-threatened ecosystems. In addition, comprehensive and integrative approaches can shift away from the often one-dimensional (single species) focus of past conservation action and can contribute to informed decision-making during the management and mitigation of often neglected species and the protection of the biosphere of the future.

Acknowledgements The authors would like to thank Jessica Schwelm (University of Duisburg-Essen, Essen Germany) and Anja Erasmus [North-West University NWU, Potchefstroom, South Africa] for creating components representing parasitic taxa in Fig. 13.2 using BioRender.com, agreement numbers: DX26MPSWER, IQ26MPVR00, DF26MPWBWM and HP26MPX15L. Further thanks to Linda van der Spuy (NWU) and Ruan Gerber (NWU) for their photos used in Fig. 13.5, Russell Q-Y. Yong (NWU) for the photos used in Fig. 13.6, and Albert Kang, Andrew Hosie and the Western Australian Museum for photos used in Fig. 13.8.

References

Adamson ML, Caira JN (1994) Evolutionary factors influencing the nature of parasite specificity. Parasitology 109:85–95

Altizer S, Pedersen AB (2008) Host-pathogen evolution, biodiversity and disease risks for natural populations. In: Carroll SP, Fox CW (eds) Conservation biology: evolution in action. Oxford University Press, Oxford

Araujo R, Ramos MA (2000) Status and conservation of the giant European freshwater pearl mussel (*Margaritifera auricularia*) (Spengler, 1793) (Bivalvia: Unionidea). Biol Conserv 96:233–239

Araujo R, Bragado D, Ramos MA (2001) Identification of the river blenny, *Salaria fluviatilis*, as a host to the glochidia of *Margaritifera auricularia*. J Molluscan Stud 67:128–129

Arostegui MC, Wood CL, Jones IJ, Chamberlin AJ, Jouanard N, Faya DS, Kuris AM, Riveau G, De Leo GA, Sokolow SH (2019) Potential biological control of schistosimiasis by fishes in the Lower Senegal River basin. Am J Trop Med Hyg 100(1):117–126

Artim JM, Hook A, Grippo RS, Sikkel PC (2017) Predation on parasitic gnathiid isopods on coral reefs: a comparison of Caribbean cleaning gobies with non-cleaning microcarnivores. Coral Reefs 36:1213–1223

Bakenhaster MD, Bullard SA, Curran SS, Kritsky DC, Leone EH, Partridge LK, Ruiz CF, Scharer RM, Poulakis GR (2018) Parasite component community of smalltooth sawfish off Florida: diversity, conservation concerns, and research applications. Endanger Species Res 35:47–58

Barčák D, Fan CK, Sonko P, Kuchta R, Scholz T, Orosová M, Chen H-W, Oros M (2021) Hidden diversity of the most basal tapeworms (Cestoda, Gyrocotylidea), the enigmatic parasites of holocephalans (Chimaeriformes). Sci Rep 11:5492

Barnosky AD, Matzke N, Tomiya S, Wogan GO, Swartz B, Quental TB, Marshall C, McGuire JL, Lindsey EL, Maguire KC, Mersey B, Ferrer EA (2011) Has the earth's sixth mass extinction already arrived? Nature 471:51–57

Barton DP, Kopf RK, Zhu X, Shamsi S (2023) The presence of a parasite in the head tissues of a threatened fish (*Bidyanus bidyanus*, Terapontidae) from southeastern Australia. Pathogens 12:1296

Baruš V, Moravec F, Špakulová M (1997) The red data list of helminths parasitizing fishes of the orders cypriniformes, siluriformes and gadiformes in the Czech Republic and Slovak Republic. Helminthologia 34(1):35–44

Bellay SB, de Oliveira EF, Almeida-Neto M, Takemoto RM (2020) Ectoparasites are moe vulnerable to host extinction than co-occurring endoparasites: evidence from metazoan parasites of freshwater and marine fishes. Hydrobiologia 847:2873–2882

Benz GW (1993) Evolutionary biology of Siphonostomatoida (Copepoda) parasitic on vertebrates. Dissertation, The University of British Columbia

Benz GW, Bullard SA (2004) Metazoan parasites and associates of chondrichthyans with emphasis on taxa harmful to captive hosts. In: Smith M, Warmolts D, Thoney DA, Hueter R (eds) Elasmobranch husbandry manual: captive care of sharks, rays, and their relatives. Ohio Biological Survey, Columbus, pp 325–416

Beveridge I, Campbell RA (2005) Three new genera of trypanorhynch cestodes from Australian elasmobranch fishes. Syst Parasitol 60(3):211–224

Blaustein AR, Gervasi SS, Johnson PTJ, Hoverman JT, Belden LK, Bradley PW, Xie GY (2012) Ecophysiology meets conservation: understanding the role of disease in amphibian population declines. Philos Trans R Soc B 367:1688–1707

Bolek MG, Schmidt-Rhaesa A, De Villalobos LC, Hanelt B (2015) Phylum Nematophora. In: Thorp JH, Roger DC (eds) Thorp and Covich's freshwater invertebrates, 4th edn. Academic, London, pp 303–326

Bower DS, Brannelly LA, McDonald CA, Webb RI, Greenspan SE, Vickers M, Gardner MG, Greenlees MJ (2019) A review of the role of parasites in the ecology of reptiles and amphibians. Anim Ecol 44:433–448

Boxshall G, Hayes P (2019) Chapter 3 Biodiversity and taxonomy of the parasitic Crustacea. In: Smit NJ, Bruce NL, Hadfield KA (eds) Parasitic Crustacea. State of knowledge and future trends, Zoological monographs, vol 3. Springer Nature, Cham, pp 73–134

Brian JI (2023) Parasites in biodiversity conservation: friend or foe? Trends Parasitol 39:618–621

Brian JI, Aldridge DC (2020) An efficient photograph-based quantitative method for assessing castrating trematode parasites in bivalve molluscs. Parasitology 147:1375–1380

Brian JI, Aldridge DC (2019) Endosymbionts: An overlooked threat in the conservation of freshwater mussels? Biol Conserv 237:155–165

Brian JI, Aldridge DC (2022) Mussel parasite richness and risk of extinction. Conserv Biol 36:e13979

Brodie JF, Aslan CE, Rogers HS, Redford KH, Maron JL, Bronstein JL, Groves CR (2014) Secondary extinctions of biodiversity. Trends Ecol Evol 29(12):664–672

Bruce NL, Cannon LRG, Adlard R (1994) Synoptic checklist of ascaridoid parasites (Nematoda) from fish hosts. Invertebr Taxon 8:583–674

Burreson EM (2020) Marine and estuarine leeches (Hirudinida : Ozobranchidae and Piscicolidae) of Australia and New Zealand with a key to the species. Invertebr Syst 34(3):235–259

Bush AO, Kennedy CR (1994) Host fragmentation and helminth parasites: hedging your bets against extinction. Int J Parasitol 24:1333–1343

Caira JN, Healy CJ (2004) Elasmobranchs as hosts of metazoan parasites. In: Carrier JC, Musick JA, Heithaus MR (eds) Biology of sharks and their relatives, 2nd edn. CRC, Boca Raton, pp 523–551

Caira JN, Jensen K (2001) An investigation of the coevolutionary relationships between onchobothriid tapeworms and their elasmobranch hosts. Int J Parasitol 31:960–975

Caira JN, Jensen K (2017) Planetary Biodiversity Inventory (2008-2017): Tapeworms from the vertebrate bowels of the earth. University of Kansas, Natural History Museum, Special Publication No. 25, Lawrence

Caira JN, Littlewood DTJ (2013) Worms, Platyhelminthes. In: Levin SA (ed) Encyclopedia of biodiversity, 2nd edn. Academic, Waltham, pp 437–469

Caira JN, Jensen K, Holsinger KE (2003) On a new index of host specificity. In: Combes C, Jourdane J (eds) Taxonomy, ecology and evolution of metazoan parasites. Presses Universitaires de Perpignan, Perpignan, pp 161–201

Caira JN, Healy CJ, Jensen K (2012) An updated look at elasmobranchs as hosts of metazoan parasites. In: Carrier JC, Musick JA, Heithaus MR (eds) Biology of sharks and their relatives. CRC, Boca Raton, pp 547–578

Caira JN, Jensen K, Barbeau E (2023) Global Cestode Database. https://tapewormdb.uconn.edu/. Accessed 15 Aug 2023

Cardillo M, Bromham L (2001) Body size and risk of extinction in Australian mammals. Conserv Biol 15:1435–1440

Cardoso P, Borges PAV, Triantis KA, Ferrández MA, Martín JL (2011) Adapting the IUCN red list criteria for invertebrates. Biol Conserv 144:2432–2440

Carlson JK, Gulak SJB, Simpfendorfer CA, Grubbs RD, Romine JG, Burgess GH (2014) Movement patterns and habitat use of Smalltooth sawfish, *Pristis pectinata*, determined using pop-up satellite archival tags. Aquat Conserv 24(1):104–117

Carlson CJ, Burgio KR, Dougherty ER, Phillips AJ, Bueno VM, Clements CF, Castaldo G, Dallas TA, Cizauskas CA, Cumming GS, Doña J, Harris NC, Jovani R, Mironov S, Muellerklein OC, Proctor HC, Getz WM (2017a) Parasite biodiversity faces extinction and redistribution in a changing climate. Sci Adv 3(9):e1602422

Carlson CJ, Muellerklein OC, Phillips AJ, Burgio KR, Castaldo G, Cizauskas CA, Cumming GS, Dallas TA, Doña J, Harris N, Jovani R (2017b) The Parasite Extinction Assessment & Red List: an open-source, online biodiversity database for neglected symbionts. bioRxiv. https://doi.org/10.1101/192351

Carlson CJ, Dallas TA, Alexander LW, Phelan AL, Phillips AJ (2020a) What would it take to describe the global diversity of parasites? Proc R Soc B 287:20201841

Carlson CJ, Hopkins S, Bell KC, Doña J, Godfrey SS, Kwak ML, Lafferty KD, Moir ML, Speer KA, Strona G, Torchin M, Wood CL (2020b) A global parasite conservation plan. Biol Conserv 250:108596

Carnegie RB, Arzul I, Bushek D (2016) Managing marine mollusc disease in the context of regional and international commerce: policy issues and emerging concerns. Philos Trans R Soc Lond B Biol Sci 371:20150215

Ceballos G, Ehrlich PR, Dirzo R (2017) Biological annihilation via the ongoing sixth mass extinction signaled by vertebrate population losses and declines. Proc Natl Acad Sci USA 114:E6089–E6096

Ceballos G, Ehrlich PR, Raven PH (2020) Vertebrates on the brink as indicators of biological annihilation and the sixth mass extinction. Proc Natl Acad Sci USA 117(24):13596–13602

Cheung PJ, Nigrelli RF (1983) *Dermophthirioides pristidis* n. gen., n.sp. (Microbothriidae) from the skin and *Neoheterocotyle ruggierii* n. sp. (Monocotylidae) from the gills of the small tooth sawfish, *Pristis pectinata*. Trans Am Microsc Soc 102(4):366–370

Chisholm LA, Whittington ID (1996) A revision of *Heterocotyle* (Monogenea: Monocotylidae) with a description of *Heterocotyle capricornensis* n. sp. from *Himantura fai* (Dasyatididae) from Heron Island, Great Barrier Reef, Australia. Int J Parasitol 26(11):1169–1190

Chisholm LA, Whittington ID (1997) A revision of *Neoheterocotyle* (Monogenea: Monocotylidae) with descriptions of the larvae of *N. rhinobatis* and *N. rhynchobatis* from Heron Island, Great Barrier Reef, Australia. Int J Parasitol 27(9):1041–1060

Chisholm LA, Whittington ID (1998) Morphology and development of the haptors among the Monocotylidae (Monogenea). Hydrobiologia 383:251–261

Chisholm LA, Whittington ID (1999) A revision of the *Merizocotylinae* Johnston and Tiegs, 1922

(Monogenea: Monocotylidae) with descriptions of the new species of *Empruthotrema* Johnston and Tiegs, 1922 and *Merizocotyle* Cerfontaine, 1894. J Nat Hist 33(1):1–28

Chisholm LA, Whittington ID (2000) A new species of *Neoheterocotyle* Hargis, 1955 (Monogenea: Monocotylidae) from the gills of *Pristis clavata* Garman (Pristidae) from Darwin, Australia. Syst Parasitol 46(2):93–98

Chisholm LA, Whittington ID (2012) Three new species of *Merizocotyle* Cerfontaine, 1894 (Monogenea: Monocotylidae) from the nasal tissues of dasyatid rays collected off Malaysian and Indonesian Borneo. Syst Parasitol 82(2):167–176

Chisholm LA, Hanskneckt T, Whittington ID, Overstreet RM (1997) A revision of the *Calicotylinae* Monticelli, 1903 (Monogenea: Monocotylidae). Syst Parasitol 38(3):159–183

Chisholm LA, Morgan JAT, Adlard RD, Whittington ID (2001a) Phylogenetic analysis of the Monocotylidae (Monogenea) inferred from 28S rDNA sequences. Int J Parasitol 31(11):1253–1263

Chisholm LA, Whittington ID, Morgan JAT, Adlard RD (2001b) The *Calicotyle* conundrum: do molecules reveal more than morphology? Syst Parasitol 49(2):81–87

Chowdhury MMR, Marjomäki TJ, Taskinen J (2019) Effect of glochidia infection on growth of fish: freshwater pearl mussel *Margaritifera margaritifera* and brown trout *Salmo trutta*. Hydrobiologia 848:3179–3189

Cielocha JJ, Jensen K, Caira JN (2014) *Floriparicapitus*, a new genus of lecanicephalidean tapeworm (Cestoda) from sawfishes (Pristidae) and guitarfishes (Rhinobatidae) in the Indo-West Pacific. J Parasitol 100(4):485–499

Cizauskas CA, Carlson CJ, Burgio KR, Clements CF, Dougherty ER, Harris NC, Phillips AJ (2017) Parasite vulnerability to climate change: an evidence-based functional trait approach. R Soc Open Sci 4(1):160535

Coen LD, Bishop MJ (2016) The ecology, evolution, impacts and management of host-parasite interactions of marine molluscs. J Invertebr Pathol 131:177–211

Colwell RK, Dunn RR, Harris NC (2012) Coextinction and persistence of dependent species in a changing world. Annu Rev Ecol Evol Syst 43:183–203

Crutzen P (2002) Geology of mankind. Nature 415:23

Dentzien-Dias PC, Poinar G Jr, de Figueiredo AEQ, Pacheco ACL, Horn BLD, Schultz CL (2013) Tapeworm eggs in a 270 million-year-old shark coprolite. PLoS One 8(1):e55007

Derbel H, Chaari M, Neifar L (2022) Checklist of the Monogenea (Platyhelminthes) parasitic in Tunisian aquatic vertebrates. Helminthologia 59(2):179–199

De Vienne DM, Refrégier G, López-Villavicencio M, Tellier A, Hood ME, Giraud T (2013) Cospeciation vs host-shift speciation: methods for testing, evidence from natural associations and relation to coevolution. New Phytol 198:347–385

De Vivo M, Chen W-Y, Huang J-P (2023) Testing the efficacy of different molecular tools for parasite conservation genetics: a case study using horsehair worms (Phylum: Nematomorpha). Parasitology 150:842–851

Díaz S, Settele J, Brondízio ES, Ngo HT, Agard J, Arneth A, Balvanera P, Brauman KA, Butchart SHM, Chan KMA, Garibaldi LA, Ichii K, Liu J, Subramanian SM, Midgley GF, Miloslavich P, Molnár Z, Obura D, Pfaff A, Polasky S, Purvis A, Razzaque J, Reyers B, Chowdhury RR, Shin Y-J, Visseren-Hamakers I, Willis KJ, Zayas CN (2019) Pervasive human-driven decline of life on Earth points to the need for transformative change. Science 366:eaax3100

Diniz-Filho JAF, Loyola RD, Raia P, Mooers AO, Bini LM (2013) Darwinian shortfalls in biodiversity conservation. Trends Ecol Evol 28(12):689–695

Dippenaar SM (2004) Reported siphonostomatoid copepods parasitic on marine fishes of southern Africa. Crustaceana 77(11):1281–1328

Dippenaar SM (2016) *Schistobrachia kabata* sp. nov. (Siphonostomatoida: Lernaeopodidae) from rajiform hosts off South Africa. Zootaxa 4174(1):104–113

Dobson A, Lafferty KD, Kuris AM, Hechinger RF, Jetz W (2008) Homage to Linnaeus: How many parasites? How many hosts? Proc Natl Acad Sci USA 105:11482–11489

Dollfus RP, Trilles JP (1976) A propos de la collection RP Dollfus, mise au point sur les Cymothoidiens jusqu'à présent récoltés sur des Téléostéens du Maroc et de l'Algérie. Bull Mus Natl Hist Nat Paris, 3e sér, 390, Zool 272, 821pp

Dougherty ER, Carlson CJ, Bueno VM, Burgio KR, Cizauskas CA, Clements CF, Seidel DP, Harris NC (2015) Paradigms for parasite conservation. Conserv Biol 30(4):724–733

Dulvy NK, Sadovy Y, Reynolds JD (2003) Extinction vulnerability in marine populations. Fish Fish 4:25–64

Dulvy NK, Pasolini P, Notarbartolo di Sciara G, Serena F, Tinti F, Ungaro N, Mancusi C, Ellis JE (2006) *Rostroraja alba*. IUCN Red List Threatened Species: e.T61408A12473706

Dulvy NK, Pinnegar JK, Reynolds JD (2009) Holocene extinctions in the sea. In: Turvey ST (ed) Holocene extinctions. Oxford University Press, Oxford, pp 129–150

Dulvy NK, Fowler SL, Musick JA, Cavanagh RD, Kyne PM, Harrison LR, Carlson JK, Davidson LNK, Fordham SV, Francis MP, Pollock CM, Simpfendorfer CA, Burgess GH, Carpenter KE, Compagno LJ, Ebert DA, Gibson C, Heupel MR, Livingstone SR, Sanciangco JC, Stevens JD, Valenti S, White WT (2014) Extinction risk and conservation of the world's sharks and rays. Elife 3:e00590

Dulvy NK, Davidson LNK, Kyne PM, Simpfendorfer CA, Harrison LR, Carlson JK, Fordham SV (2016) Ghosts of the coast: global extinction risk and conservation of sawfishes. Aquat Conserv 26:134–153

Dulvy NK, Pacoureau N, Rigby CL, Pollom RA, Jabado RW, Ebert DA, Finucci B, Pollock CM, Cheok J, Derrick DH, Herman KB, Sherman CS, VanderWright WJ, Lawson JM, Walls RHL, Carlson JK, Charvet P, Bineesh KK, Fernando D, Ralph GM, Matsushiba JH, Hilton-Taylor C, Fordham SV, Simpfendorfer CA

(2021) Overfishing drives over one-third of all sharks and rays toward a global extinction crisis. Curr Biol 31(21):4773–4787

Dunn RR (2009) Coextinction: anecdotes, models, and speculation. In: Turvey ST (ed) Holocene extinctions. Oxford University Press, Oxford, pp 167–180

Dunn RR, Harris NC, Colwell RK, Koh LP, Sodhi NS (2009) The sixth masscoextinction: are most endangered species parasites and mutualists? Proc R Soc B Biol Sci 276:3037–3045

Dunne JA, Lafferty KD, Dobson AP, Hechinger RF, Kuris AM, Martinez ND, McLaughlin JP, Mouritsen KN, Poulin R, Reise K, Stouffer DB, Thieltges DW, Williams RJ, Zander CD (2013) Parasites affect food web structure primarily through increased diversity and complexity. PLoS Biol 11(6):e1001579

Du Preez LH, Wilkinson M, Huyse T (2008) The first record of polystomes (Monogenea: Polystomatidae) from caecilian hosts (Amphibia: Gymnophiona), with the description of a new genus and two new species. Syst Parasitol 69:201–209

Dzido J, Rolbiecki L, Izdebska JN, Rokicki J, Kuczkowski T, Pawliczka I (2021) A global checklist of the parasites of the harbor porpoise *Phocoena phocoena*, a critically-endangered species, including new findings from the Baltic Sea. Int J Wildl Parasit Wildl 15:290–302

Ebert D, Hamilton WD (1996) Sex against virulence: the coevolution of parasitic disease. TREE 11(2):79–82

Ellis J, Morey G, Walls R (2015) *Rostroraja alba* (Europe assessment). IUCN Red List Threatened Species: e.T61408A48954174

Estes JA, Terborgh J, Brashares JS, Power ME, Berger J, Bond WJ, Carpenter SR, Essington TE, Holt RD, Jackson JB, Marquis RJ, Oksanen L, Oksanen T, Paine RT, Pikitch EK, Ripple WJ, Sandin SA, Scheffer M, Schoener TW, Shurin JB, Sinclair AR, Soulé ME, Virtanen R, Wardle DA (2011) Trophic downgrading of planet Earth. Science 333:301–306

Euzet L (1959) Recherches sur les cestodes tétraphyllides des sélaciens des cotes de France. Doctoral Dissertation, Université de Montpellier

Euzet L, Combes C (1980) Les problèmes de l'espèce chez les animaux parasites. In: Bocquet C, Genermont J, Lamotte M (eds) Les problèmes de l'espèce dans le règne animal. Société Zoologique de France, Paris, pp 239–285

Evelyn J (1664) Sylva, or a discourse on forestry-trees and the propagation of timber in His Majesty's Dominions. London, pp 335

Ezenwa VO, Price SA, Altizer S, Vitone ND, Cook KC (2006) Host traits and parasite species richness in even and odd-toed hoofed mammals, *Artiodactyla* and *Perissodactyla*. Oikos 115:526–536

Faria VV, McDavitt MT, Charvet P, Wiley TR, Simpfendorfer CA (2013) Species delineation and global population structure of Critically Endangered sawfishes (Pristidae). Zool J Linn Soc 167(1):136–164

Farrell MJ, Stephens PR, Berrang-Ford L, Gittleman JL, Davies TJ (2015) The path to host extinction can lead to loss of generalist parasites. J Anim Ecol 84:978–984

Farrell MJ, Park AW, Cressler CE, Dallas T, Huang S, Mideo N, Morales-Castilla I, Davies TJ, Stephens P (2021) The ghost of hosts past: impacts of host extinction on parasite specificity. Philos Trans R Soc B 376:20200351

FMCS—Freshwater Mollusk Conservation Society (2016) A national strategy for the conservation of native freshwater mollusks. Freshw Mollusk Biol Conserv 19:1–21

Fricke R, Eschmeyer WN, Van der Laan R (2024) Eschmeyer's catalog of fishes: genera, species, references. https://researcharchive.calacademy.org/research/ichthyology/catalog/fishcatmain.asp. Accessed 19 June 2024

Froese R, Pauly D (2023) FishBase, version 02/2024. https://www.fishbase.se/search.php. Accessed 19 June 2024

Fryer G (1977) On some parasites of *Chonopeltis* (Crustacea: Branchiura) from the rivers of the extreme South West Cape region of Africa. J Zool Lond 182:411–455

Fyler CA (2009) Systematics, biogeography and character evolution in the tapeworm genus *Acanthobothrium* Van Beneden, 1850. Dissertation, University of Conneticut

Galatius A, Teilmann J, Dähne M, Ahola M, Westphal L, Kyhn LA, Pawliczka I, Olsen MT (2020) Diets R (2020) Grey seal *Halichoerus grypus* recolonisation of the southern Baltic Sea, Danish Straits and Kattegat. Wildl Biol 4:1–8

García-Prieto L, Dáttilo W, Rubio-Godoy M, Pérez-Ponce de León G (2022) Fish-parasite interactions: A dataset of continental waters in Mexico involving fishes and their helminth fauna. Ecology 103(12):E3815

Gerlach J (2014) *Haematopinus oliveri*. The IUCN Red List of Threatened Species 2014:e.T9621A21423551. https://doi.org/10.2305/IUCN.UK.2014-1.RLTS.T9621A21423551.en. Accessed 5 June 2024

Gómez A, Nichols E (2013) Neglected wildlife: parasitic biodiversity as a conservation target. Int J Parasitol Parasites Wildl 2:222–227

Gopko M, Mironova E, Pasternak A, Mikheev V, Taskinen J (2017) Freshwater mussels (*Anodonta anatine*) reduce transmission of a common fish trematode (eye fluke, *Diplostomum pseudospathaceum*). Parasitology 144:1971–1979

Guégan JF, Lambert A, Lévêque C, Combes C, Euzet L (1992) Can host body size explain the parasite species richness in tropical freshwater fishes? Oecologia 90:197–204

Hamilton C (2016) The Anthropocene as rupture. Anthropocene Rev 3(2):93–106

Hanelt B, Bolek MG, Schmidt-Rhaesa A (2012) Going solo: discovery of the first parthenogenetic gordiid (Nematomorpha: Gordiida). PLoS One 7(4):e34472

Harris NC, Dunn RR (2010) Using host associations to predict spatial patterns in the species richness of the parasites of North American carnivores. Ecol Lett 13:1411–1418

Hartigan A, Wilkinson M, Gower DJ, Streicher JW, Holzer AS, Okamura B (2016) Myxozoan infections of caecilians demonstrate broad host specificity and indicate a link with human activity. Int J Parasitol 46:375–381

Hatcher MJ, Taneyhill DE, Dunn AM (1999) Population dynamics under parasitic sex ratio distortion. Theor Popul Biol 56:11–28

Hatcher MJ, Dick JTA, Dunn AM (2012) Diverse effects of parasites in ecosystems: linking interdependent processes. Front Ecol Environ 10(4):186–194

Herczeg D, Ujszegi J, Kásler A, Holly D, Hettyey A (2021) Host-multiparasite interactions in amphibians: a review. Parasit Vectors 14:296

Hoberg EP, Brooks DR (2008) A macroevolutionary mosaic: episodic host-switching, geographical colonization and diversification in complex host–parasite systems. J Biogeogr 35:1533–1550

Hopkins SR, Ocampo JM, Wojdak JM, Belden LK (2016) Host community composition and defensive symbionts determine trematode parasite abundance in host communities. Ecosphere 7(3):e01278

Hortal J, de Bello F, Diniz-Filho JAF, Lewinsohn TM, Lobo JM, Ladle RJ (2015) Seven shortfalls that beset large-scale knowledge of biodiversity. Annu Rev Ecol Evol Syst 46:523–549

Ingelbrecht J, Morgan DL, Lear KO, Fazeldean T, Lymbery AJ, Norman BM, Martin SB (2022) A new microbothriid monogenean *Dermopristis pterophilus* n. sp. from the skin of the Critically Endangered green sawfish *Pristis zijsron* Bleeker, 1851 (Batoidea: Pristidae) in Western Australia. Int J Parasitol Parasites Wildl 17:185–193

Ingelbrecht J, Lear KO, Martin SB, Lymbery AJ, Norman BM, Boxshall GA, Morgan DL (2024) Ectoparasites of the Critically Endangered green sawfish *Pristis zijsron* and sympatric elasmobranchs in Western Australia. Parasitol Int 101:102900

IUCN (1980) World Conservation Strategy: Living resource conservation for sustainable development. Gland, Switzerland, pp 50

IUCN (2023a) Seventy five years of experience. https://www.iucn.org/about-iucn/history. Accessed 19 June 2024

IUCN (2023b) Barometer of life. https://www.iucnredlist.org/about/barometer-of-life. Accessed 19 June 2024

Jones I, Lund A, Riveua G, Jouanard N, Ndione RA, Sokolow SH, De Leo GA (2018) Ecological control of schistosomiasis in Sub-Saharan Africa: restoration of predator-prey dynamics to reduce transmission. In: Roche B, Broutin H, Simard F (eds) Ecology and evolution of infection diseases: pathogen control and public health management in low-income countries. Oxford Academic, Oxford, p 336

Jørgensen D (2015) Conservation implications of parasites co-reintroduction. Conserv Biol 29(2):602–604

Kabata Z (1979) Parasitic Copepoda of British fishes. The Ray Society, London

Kabata Z (1981) Copepoda (Crustacea) parasitic on fishes: Problems and perspectives. Adv Parasitol 19:1–71

Kearn GC (1965) The biology of *Leptocotyle minor*, a skin parasite of the dogfish, *Scyliorhinus canicula*. Parasitology 55:473–480

Kearn GC, Whittington ID, Evans-Gowing R (2010) A new genus and new species of microbothriid monogenean (Platyhelminthes) with a functionally enigmatic reproductive system, parasitic on the skin and mouth lining of the largetooth sawfish, *Pristis microdon*, in Australia. Acta Parasitol 55(2):115–122

Keith DA (2009) The interpretation, assessment and conservation of ecological communities. Ecol Manag Restor 10:S3–S15

Kennedy CR (1993) The dynamics of intestinal helminth communities in eels *Anguilla anguilla* in a small stream: long-term changes in richness and structure. Parasitology 107:71–78

Koh LP, Dunn RR, Sodhi NS, Colwell RK, Proctor HC, Smith VS (2004) Species coextinctions and the biodiversity crisis. Science 305:1632–1634

Koitsanou E, Akritopoulou E, Athinaiou N, Sarantopoulou J, Komnenou A, Dendrinos P, Exadactylos A, Gkafa GA (2022a) Molecular identification of a parasitic mite found in the respiratory system of a stranded Mediterranean monk seal in the area of Pagasitikos Gulf. In: Marine and inland waters research symposium

Koitsanou E, Sarantopoulou J, Komnenou A, Exadactylos A, Dendrinos P, Papadopoulos E, Gkafas GA (2022b) First report of the parasitic nematode *Pseudoterranova* spp. found in Mediterranean monk seal (*Monachus monachus*) in Greece: conservation implications. Conservation 2:122–133

Kritsky DC, Bullard SA, Bakenhaster MD, Scharer RM, Poulakis GR (2017) Resurrection of *Mycteronastes* (Monogenoidea: Monocotylidae), with description of *Mycteronastes caalusi* n. sp. from olfactory sacs of the Smalltooth sawfish, *Pristis pectinata* (Pristiformes: Pristidae), in the Gulf of Mexico off Florida. J Parasitol 103(5):477–485

Kuchta R, Choudhury A, Scholz T (2018) Asian fish tapeworm: the most successful invasive parasite in freshwaters. Trends Parasitol 34(6):511–523

Kuris AM, Hechinger RF, Shaw JC, Whitney KL, Aguirre-Macedo L, Boch CA, Dobson AP, Dunham EJ, Fredensborg BL, Huspeni TC, Lorda J, Mababa L, Mancini FT, Mora AB, Pickering M, Talhouk NL, Torchin ME, Lafferty KD (2008) Ecosystem energetic implications of parasite and free-living biomass in three estuaries. Nature 454:515–518

Kwak ML, Griffiths J, Barry D, Begent M, Hoang T, Taafua L, Chiovitti A (2018) The first DNA barcodes for the Australian platypus tick Ixodes ornithorhynchi Lucas, 1846 (Acari: Ixodidae) to facilitate conservation efforts for a declining parasite and its host. Acarologia 58(4):845–849

Kwak ML, Heath ACG, Palma RL (2019) Saving the Manx shearwater flea *Ceratophyllus* (*Emmareus*)

fionnus (Insecta: Siphonaptera): the road to developing a recovery plan for a threatened ectoparasite. Acta Parasitol 64:903–910

Kwak ML, Heath AC, Cardoso P (2020) Methods for the assessment and conservation of threatened animal parasites. Biol Conserv 248:108696

Lafferty KD (1993) Effects of parasitic castration on growth, reproduction and population dynamics of the marine snail *Cerithidea californica*. Mar Ecol Prog Ser 96:229–237

Lafferty KD (2012) Biodiversity loss decreases parasite diversity: theory and patterns. Philos Trans R Soc B Biol Sci 367:2814–2827

Laidemitt MR, Anderson LC, Wearing HJ, Mutuku MW, Mkoji GM, Loker ES (2019) Antagonism between parasites within snail hosts impacts the transmission of human schistosomiasis. eLife 8:e50095

Lambden J, Johnson PTJ (2013) Quantifying the biomass of parasites to understand their role in aquatic communities. Ecol Evol 3(7):2310–2321

Landman W, Verneau O, Vences M, du Preez L (2023) *Metapolystoma ohlerianum* n. sp. (Monogenea: Polystomatidae) from *Aglyptodactylus madagascariensis* (Anura: Mantellidae). Acta Parasitol 68:344–358

Lees AC, Pimm SL (2015) Species, extinct before we know them? Curr Biol 25(5):177–180

Leidenberger S, Harding K, Härkönen T (2007) Pochid seals, seal lice and heartworms: a terrestrial host-parasite system conveyed to the marine environment. Dis Aquat Org 77:235–253

Leis E, Bailey J, Katona R, Standish I, Dziki S, McCann R, Perkins J, Eckert N, Baumgartner W (2023) A mortality event involving endangered Pallid sturgeon (*Scaphirhynchus albus*) associated with *Gyrodactylus conei* n. sp. (Monogenea: Gyrodactylidae) effectively treated with Parasite-S (Formalin). Parasitologia 3:205–214

Leonardi MS, Crespo JE, Soto F, Lazzari CR (2022) How did seal lice turn into the only truly marine insects? Insects 13:46

Lindenfors P, Nunn CL, Jones KE, Cunningham AA, Sechrest W, Gittleman JL (2007) Parasite species richness in carnivores: effects of host body mass, latitude, geographical range and population density. Glob Ecol Biogeogr 16:496–509

Lomolino MV (2004) Conservation biogeography. In: Lomolino MV, Heaney LR (eds) Frontiers of biogeography: New directions in the geography of nature. Sunderland, MA, Sinauer, pp 293–296

Lymbery AJ, Smit NJ (2023) Conservation of parasites: A primer. Int J Parasitol Parasites Wildl 21:255–263

Marshall L, Leclerq N, Calvalheiro LG, Dathe HH, Jacobi B, Kuhlmann M, Potts SG, Rasmont P, Roberts SPM, Vereecken NJ (2024) Understanding and addressing shortfalls in European wild bee data. Biol Conserv 290:110455

Martins PM, Pouin R, Gonçalves-Souza T (2021) Integrating climate and host richness as drivers of global parasite diversity. Global Ecol Biogeogr 30:196–204

Mathews PD, Mertins O, Milanin T, Espinoza LL, Flores-Gonzales AP, Audebert F, Moranidini AC (2020) Molecular phylogeny and taxonomy of a new *Myxobolus* species from the endangered ornamental fish, *Otocinclus cocana* endemic to Peru: a host-parasite coextinction approach. Acta Tropica 210:105545

May RM, Anderson RM (1978) Regulation and stability of host-parasite population interactions: II. Destabilizing processes. J Anim Ecol 47:249–267

McCauley DJ, Pinsky ML, Palumbi SR, Estes JA, Joyce FH, Warner RR (2015) Marine defaunation: Animal loss in the global ocean. Science 347:1255641

McElwain A (2019) Are parasites and diseases contributing to the decline of freshwater mussels (Bivalvia, Unionida)? Fresh Mollusk Biol Conserv 22:85–89

Mendelson JR, Lips KR, Gagliardo RW, Rabb GB, Collins JP, Diffendorfer JE, Daszak P, Ibáñez RD, Zippel KC, Lawson DP, Wright KM, Stuart SN, Gascon C, da Silva JR, Burrowes PA, Joglar RL, La Marca E, Lötters S, du Preez LH, Weldon C, Hyatt A, Rodriguez-Mahecha JV, Hunt S, Robertson H, Lock B, Raxworthy CJ, Frost DR, Lacy RC, Alford RA, Campbell JA, Parra-Olea G, Bolaños F, Domingo JJC, Halliday T, Murphy JB, Wake MH, Coloma LA, Kuzmin SL, Price MS, Howell KM, Lau M, Pethiyagoda R, Boone M, Lannoo MJ, Blaustein AR, Dobson A, Griffiths RA, Crump ML, Wake DB, Brodie ED (2006) Confronting amphibian declines and extinctions. Science 313:48

Moir ML, Brennan KE (2020) Incorporating coextinction in threat assessments and policy will rapidly improve the accuracy of threatened species lists. Biol Conserv 249:108715

Moir ML, Vesk PA, Brennan KE, Keith DA, Hughes L, McCarthy MA (2010) Current constraints and future directions in estimating coextinction. Conserv Biol 24:682–690

Moir ML, Vesk PA, Brennan KEC, Poulin R, McCarthy MA, Coates DJ (2012) Considering extinction of dependent species during translocation, ex situ conservation and assisted migration of threatened hosts. Conserv Biol 26:199–207

Molloy DP, Vinikour WS, Anderson RV (1999) New North American records of aquatic insects as paratenic hosts of *Pheromermis* (Nematoda: Mermithidae). J Invertebr Pathol 74:89–95

Moreira PS, Sadowsky V (1978) An annotated bibliography of parasitic Isopoda (Crustacea) of Chondrichthyes. Braz J Oceanogr 27(2):95–152

Morgan DL, Tang D, Peverell SC (2010) Critically endangered *Pristis microdon* (Elasmobranchii), as a host for the Indian parasitic copepod, *Caligus furcisetifer* Redkar, Rangnekar et Murti, 1949 (Siphonostomatoida): new records from northern Australia. Acta Parasitol 55:419–423

Mouillot D, Krasnov BR, Shenbrot GI, Gaston KJ, Poulin R (2006) Conservatism of host specificity in parasites. Ecography 29:596–602

Norman BM, Newbound DR, Knott B (2000) A new species of Pandaridae (Copepoda), from the whale shark *Rhincodon typus* (Smith). J Nat Hist 34:355–366

Norman BM, Reynolds S, Morgan DL (2021) Three-way symbiotic relationships in whale sharks. Pac Conserv Biol 28:80–83

Ogawa K (1991) Ectoparasites of sawfish, *Pristis microdon*, caught in freshwaters of Australia and Papua New Guinea. In: Shimizu M, Taniuchi T (eds) Studies on elasmobranchs collected from seven river systems in Northern Australia and Papua New Guinea. Nature Culture, Tokyo, pp 91–102

Okamura B, Hartigan A, Naldoni J (2018) Extensive uncharted biodiversity: the parasite dimension. Intergr Compar Biol 58(6):1132–1145

Pacoureau N, Rigby CL, Kyne PM, Sherley RB, Winker H, Carlson JK, Fordham SV, Barreto R, Fernando D, Francis MP, Jabado RW, Herman KB, Liu KM, Marshall AD, Pollom RA, Romanov EV, Simpfendorfer CA, Yin JS, Kindsvater HK, Dulvy NK (2021) Half a century of global decline in oceanic sharks and rays. Nature 589:567–571

Page RDM (1993) Parasites, phylogeny and cospeciation. Int J Parasitol 23:499–506

Palm HW, Caira JN (2008) Host specificity of adult versus larval cestodes of the elasmobranch tapeworm order Trypanorhyncha. Int J Parasitol 38:381–388

Paterson RA, Viozzi GP, Rauque CA, Flores VR, Poulin R (2021) A global assessment of parasite diversity in galaxiid fishes. Diversity 13:27

Pauly D, Christensen V, Dalsgaard J, Froese R, Torres F (1998) Fishing down marine food webs. Science 279:860–863

Pesapane R, Dodd E, Javeed N, Miller M, Foley J (2018) Molecular characterisation and prevalence of *Halarachne halichoeri* in threatened southern sea otters (*Enhydra lutris nereis*). Int J Parasitol Parasit Wildl 7:386–390

Pessier AP (2008) Management of disease as a threat to amphibian conservation. Int Zoo Yb 42:30–39

Pierce SJ, Norman B (2016) *Rhincodon typus*. IUCN Red List Threatened Species: e.T19488A2365291

Pollerspöck J, Straube N (2023) Shark references, version 2023. https://www.shark-references.com/. Accessed 15 Aug 2023

Ponomareva NM, Popova ON, Yurlova NI (2022) Odonata (Insecta) larvae as second intermediate host of the trematodes of genus *Plagiorchis* in the basin of Chany Lake, Western Siberia. Contemp Probl Ecol 15(6):631–641

Poulin R (2011) Evolutionary ecology of parasites. Princeton University Press, Princeton

Poulin R, Morand S (2000) The diversity of parasites. Q Rev Biol 75:277–293

Poulin R, Morand S (2004) Parasite biodiversity. Smithsonian Institution Press, Washington, DC

Poulin R, Krasnov BR, Shenbrot GI, Mouillot D, Khokhlova IS (2006) Evolution of host specificity in fleas: Is it directional and irreversible? Int J Parasitol 36:185–191

Poulin R, Krasnov BR, Mouillot D (2011) Host specificity in phylogenetic and geographic space. Trends Parasitol 27:355–361

Poulin R, Presswell B, Benner J, de Angeli DD, Salloum PM (2023) Biases in parasite biodiversity research: why some helminth species attract more research than others. Int J Parasitol Parasites Wildl 21:89–98

Powell FA (2011) Can early loss of affiliates explain the coextinction paradox? An example from *Acacia* inhabiting psyllids (Hemiptera: Psylloidea). Biodivers Conserv 20:1533–1544

Price PW (1980) Evolutionary biology of parasites, vol 15. Princeton University Press, Princeton

Přikrylová I, Truter M, Luus-Powell WJ, Chakona A, Smit NJ (2024) Gyrodactylus serrai n. sp. (Gyrodactylidae), from the Near-Threatened Clanwilliam Sawfin, *Cheilobarbus serra* (Peters) (Cyprinidae, Smilogastrinae), in the Cape Fold Ecoregion, South Africa. Syst Parasitol 101:67

Purvis A, Jones KE, Mace GM (2000) Extinction. BioEssays 22:1123–1133

Raga JA, Balbuena JA, Aznar J, Fernández M (1997) The impact of parasites on marine mammals: a review. Parasitologia 39:293–296

Randhawa HS, Poulin R (2010) Determinants of tapeworm species richness in elasmobranch fishes: untangling environmental and phylogenetic influences. Ecography 33(5):866–877

Rasmussen TK, Randhawa HS (2018) Host diet influences parasite diversity: a case study looking at tapeworm diversity among sharks. Mar Ecol Prog Ser 605:1–16

Reckendorf A, Wohlsein P, Lakemeyer J, Stokhol I, von Vietinghoff V, Lehnert K (2019) There and back again—the return of the nasal mite *Halarachne halichoeri* to seals in German waters. Int J Parasitol Parasit Wildl 9:112–118

Regel KV, Guliaev VD, Pospekhova NA (2013) On the life cycle and morphology of metacestodes *Dioecocestus asper* (Cyclophyllidea: Dioecocestidae). Parazitologiia 47(1):3–22

Régnier C, Achaz G, Lambert A, Cowie RH, Bouchet P, Fontaine B (2015) Mass extinction in poorly known taxa. Proc Natl Acad Sci USA 112:7761–7766

Reichard M, Ondračková M, Przyblyski M, Lius H, Smith C (2005) The cost and benefits in an unusual symbiosis: experimental evidence that bitterling fish (*Rhodus sericeus*) are parasites of unionid mussels in Europe. J Evol Biol 19(3):788–796

Re:wild, Synchronicity Earth, IUCN SSC Amphibian Specialist Group (2023) State of the world's amphibians: The second global amphibian assessment. RE:wild, Texas, pp 91

Rhoner S, Boyi JO, Artemeva V, Zinke O, Kiendl A, Siebert U, Lehnert K (2023) Back from exile? First records of chewing lice (*Lutridia exilis*; Ischnocera; Mallophaga) in growing Eurasian otter (*Lutra lutra*) populations from northern Germany. Pathogens 12:587

Rock SL, Watz J, Nilsson A, Österling M (2022) Effects of parasitic freshwater mussels on their host fishes: a review. Parasitology 149:1958–1975

Rohde K (1982) Ecology of marine parasites. University of Queensland Press, St Lucia

Rózsa L (1992) Points in question endangered parasite species. Int J Parasitol 22:265–266

Rózsa L (1993) Speciation patterns of ectoparasites and "straggling" lice. Int J Parasitol 23(7):859–864

Rózsa L, Vas Z (2015) Co-extinction and critically co-endangered species of parasitic lice, and conservation-induced extinction: should lice be reintroduced to their hosts? Oryx 49:107–110

Rubio-Godoy M, Pérez-Ponce de León G (2023) Equal rights for parasites: Windsor 1995, revisited after ecological parasitology has come of age. Biol Conserv 284:110174

Sasal P, Trouvé S, Müller-Graf C, Morand S (1999) Specificity and host predictability: a comparative analysis among monogenean parasites of fish. J Anim Ecol 68(3):437–444

Schaeffner BC, Beveridge I (2012) *Cavearhynchus*, a new genus of tapeworm (Cestoda: Trypanorhyncha: Pterobothriidae) from *Himantura lobistoma* Manjaji-Matsumoto & Last, 2006 (Rajiformes) off Borneo, including redescriptions and new records of species of *Pterobothrium* Diesing, 1850. Syst Parasitol 82(2):147–165

Schaeffner BC, Beveridge I (2013) *Pristiorhynchus palmi* n. g., n. sp. (Cestoda: Trypanorhyncha) from sawfishes (Pristidae) off Australia, with redescriptions and new records of six species of the Otobothrioidea Dollfus, 1942. Syst Parasitol 84(2):97–121

Schaeffner BC, Beveridge I (2014) The trypanorhynch cestode fauna of Borneo. Zootaxa 3900(1):021–049

Schaeffner BC, Smit NJ (2019) Parasites of cartilaginous fishes (Chondrichthyes) in South Africa—a neglected field of marine science. Folia Parasitol 66(2019):002

Seguel M, Gutiérrez J, Hernández C, Montalva F, Verdugo C (2018) Respiratory mites (*Orthohalarachne dimunata*) in β-hemolytic *Streptococci*-associated bronchopneumonia outbreak in South American Fur seal pups (*Arctocephalus australis*). J Wildl Dis 54(2):380–385

Sheldon BC, Verhulst S (1996) Ecological immunology: costly parasite defences and trade-offs in evolutionary ecology. TREE 11:317–321

Sherman CS, Simpfendorfer CA, Pacoureau N, Matsushiba JH, Yan HF, Walls RHL, Rigby CL, VanderWright WJ, Jabado RW, Pollom RA, Carlson JK, Charvet P, Bin Ali A, Fahmi CJ, Derrick DH, Herman KB, Finucci B, Eddy TD, Palomares MLD, Avalos-Castillo CG, Kinattumkara B, Blanco-Parra MD, Dharmadi EM, Fernando D, Haque AB, Mejía-Falla PA, Navia AF, Pérez-Jiménez JC, Utzurrum J, Yuneni RR, Dulvy NK (2023) Half a century of rising extinction risk of coral reef sharks and rays. Nat Commun 14:15

Sodhi NS, Bickford D, Diesmos AC, Lee TM, Koh LP, Brook BW, Sekercioglu CH, Bradshaw CJ (2008) Measuring the meltdown: drivers of global amphibian extinction and decline. PLoS One 3:e1636

Spencer HG, Zuk M (2016) For host's sake: the pluses of parasite preservation. Trends Ecol Evol 31(5):341–343

Sproston NG (1946) A synopsis of the monogenetic trematodes. Trans Zool Soc Lond 25:185–600

Steffen W, Grinevald J, Crutzen PJ, McNeill JR (2011) The Anthropocene: conceptual and historical perspectives. Philos Trans R Soc Ser A 369:842–867

Stork NE, Lyal CHC (1993) Extinction or 'co-extinction' rates? Nature 366:307

Stork NE, Coddington JA, Colwell RK, Chazdon RL, Dick CW, Peres CA, Sloan S, Willis K (2009) Vulnerability and resilience of tropical forest species to land-use change. Conserv Biol 23:1438–1447

Strona G (2015) Past, present and future of host-parasite co-extinctions. Int J Parasitol Parasites Wildl 4:431–441

Strona G, Bradshaw CJA (2018) Co-extinctions annihilate planetary life during extreme environmental change. Sci Rep 8:16724

Strona G, Galli P, Fattorini S (2013) Fish parasites resolve the paradox of missing coextinctions. Nat Commun 4:1718

Sures B, Nachev M, Pahl M, Grabner D, Selbach C (2017) Parasites as drivers of key processes in aquatic ecosystems: facts and future directions. Exp Parasitol 180:141–147

Taskinen J, Urbańska M, Ercoli F, Andrzejewski W, Ozgo M, Deng B, Choo JM, Riccardi N (2021) Parasites in sympatric populations of native and invasive freshwater bivalves. Hydrobiologia 848:3167–3178

Thomas GR, Taylor J, de Leaniz CG (2010) Captive breeding of the endangered freshwater pearl mussel *Margaritifera margaritifera*. Endanger Species Res 12:1–9

Thompson JN (1996) Evolutionary ecology and the conservation of biodiversity. TREE 11:300–303

Truter M, Přikrylová I, Hadfield KA, Smit NJ (2023) Working towards a conservation plan for fish parasites: Cyprinid parasites from the south African cape fold freshwater ecoregion as a case study. Int J Parasitol Parasites Wildl 21:277–286

Tyler GA (2006) Tapeworms of elasmobranchs (Part II). A monograph on the Diphyllidea (Platyhelminthes, Cestoda). Bull Univ Nebr State Mus 20:1–142

Van As LL, Van As JG (1999) Aspects of the morphology and a review of the taxonomic status of three species of the genus *Chonopeltis* (Crustacea: Branchiura) from the orange-vaal and South west Cape river systems, South Africa. Folia Parasitol 46:221–228

Van Der Spuy L, Smit NJ, Schaeffner BC (2022) Threatened, host-specific affiliates of a red-listed host: Three new species of *Acanthobothrium* van Beneden, 1849 (Cestoda: Onchoproteocephalidea) from the endangered white skate, *Rostroraja alba* (Lacépède). Int J Parasitol Parasites Wildl 17:114–126

Vaughan DB, Christison KW, Hansen H (2023) *Rajonchocotyle* Cerfontaine, 1899 (Monogenea: Hexabothriidae) species from South Africa, with discussion of the literary accounts of *R. emarginata* (Olsson, 1876). J Parasitol 109(3):148–168

Vázquez DP, Poulin R, Krasnov BR, Shenbrot GI (2005) Species abundance and the distribution of specialization in host–parasite interaction networks. J Anim Ecol 74:946–955

Voyles J, Cashins SD, Rosenblum EB, Puschendorf R (2009) Preserving pathogens for wildlife conservation: a case for action on amphibian declines. Oryx 43(4):527–529

Wake DB, Vredenburg VT (2008) Are we in the midst of the sixth mass extinction? A view from the world of amphibians. Proc Natl Acad Sci USA 105:11466–11473

Warren MB, Bakenhaster MD, Scharer RM, Poulakis GR, Bullard SA (2020) A new genus and species of fish blood fluke, *Achorovermis testisinuosus* gen. et sp. n. (Digenea: Aporocotylidae), infecting critically endangered smalltooth sawfish, *Pristis pectinata* (Rhinopristiformes: Pristidae) in the Gulf of Mexico. Folia Parasitol 67:2020–2009

Waters CN, Zalasiewicz J, Summerhayes C, Barnosky AD, Poirier C, Gałuszka A, Cearreta A, Edgeworth M, Ellis EC, Ellis M, Jeandel C, Leinfelder R, McNeill JR, Richter DD, Steffen W, Syvitski J, Vidas D, Wagreich M, Williams M, Zhisheng A, Grinevald J, Odada E, Oreskes N, Wolfe AP (2016) The Anthropocene is functionally and stratigraphically distinct from the Holocene. Science 351(6269):aad2622

Watson DE, Thorson TB (1976) Helminths from elasmobranchs in Central American fresh waters. In: Thorson TB (ed) Investigation of the ichthyofauna of Nicaraguan lakes. University of Nebraska Press, Lincoln, pp 629–640

Weldon C, du Preez LH, Hyatt AD, Muller R, Speare R (2004) Origin of the amphibian chytrid fungus. Emerg Infect Dis 10(12):2100–2105

Welicky RL, Preisser WC, Leslie KL, Mastick N, Fiorenza E, Maslenikov KP, Tornabene L, Kinsella JM, Wood CL (2021) Parasites of the past: 90 years of change in parasitism for English sole. Front Ecol Environ 19(8):470–477

Wells K, Clark NJ (2019) Host specificity in variable environments. Trends Parasitol 35:452–465

Wells SM, Pyle RM, Collins NM (1983) The IUCN invertebrate Red Data book. IUCN, Gland, Switzerland, pp 632

Whinfield J, Warren K, Vogelnest L, Vaughan-Higgins R (2024) Applying a modified streamlined disease risk analysis framework to a platypus conservation translocation, with special consideration for the conservation of ecto- and endoparasites. Int J Wildl Parasit Wildl 24:100948

Whittington ID (1998) Diversity "down under:" monogeneans in the Antipodes (Australia) with a prediction of monogenean biodiversity worldwide. Int J Parasitol 28:1481–1493

Whittington I, Chisholm L (2008) Diseases caused by Monogenea. In: Eiras J, Segner H, Wahli T, Kapoor G (eds) Fish diseases, vol 2. Science, New Hampshire, pp 683–816

Whittington ID, Kearn GC (2011) A new species of *Dermopristis* Kearn, Whittington & Evans-Gowing, 2010 (Monogenea: Microbothriidae), with observations on associations between the gut diverticula and reproductive system and on the presence of denticles in the nasal fossae of the host *Glaucostegus typus* (Bennett) (Elasmobranchii: Rhinobatidae). Syst Parasitol 80:41–51

Whittington ID, Cribb BW, Hamwood TE, Halliday JA (2000) Host-specificity of monogenean (platyhelminth) parasites: a role for anterior adhesive areas? Int J Parasitol 30:305–320

Williams HH (1969) The genus *Acanthobothrium* Beneden 1849 (Cestoda: Tetraphyllidea). Nytt Magasin Zoologi 17:1–56

Wilson KA, Carwardine J, Possingham HP (2009) Setting conservation priorities. Ann NY Acad Sci 1162:237–264

Windsor DA (1990) Heavenly hosts. Nature 348:104

Windsor DA (1995) Equal rights for parasites. Conserv Biol 9:1–2

Windsor DA (1997a) Stand up for parasites. TREE 12:32

Windsor DA (1997b) The basic unit of evolution is the host-symbiont "Biocartel". Evol Theory 11:275

Windsor DA (1998) Most of the species on Earth are parasites. Int J Parasitol 28(12):1939–1941

Wood CL, Vanhove MPM (2022) Is the world wormier than it used to be? We'll never know without natural history collections. J Anim Ecol 92:250–262

Wood CL, Leslie KL, Claar D, Mastick N, Preisser W, MPM V, Welicky R (2023a) How to use natural history collections to resurrect information on historical parasite abundances. J Helminthol 97(e6):1–13

Wood CL, Welicky RL, Preisser WC, Leslie KL, Mastick N, Greene C, Maslenikov KP, Tornabene L, Kinsella JM, Essington TE (2023b) A reconstruction of parasite burden reveals one century of climate-associated parasite decline. Proc Natl Acad Sci USA 120:e2211903120

WoRMS Editorial Board (2023) World Register of Marine Species. https://www.marinespecies.org/. Accessed 15 Aug 2024

WWF, World Wide Fund for Nature (2021a) The world's forgotten fishes, pp 1–47

WWF, World Wide Fund for Nature (2021b) One-third of freshwater fish face extinction and other freshwater fish facts. https://www.worldwildlife.org/stories/one--third-of-freshwater-fish-face-extinction-and-other-freshwater-fish-facts. Accessed 3 Nov 2023

Yamaguti S (1963) Systema helminthum, Volume IV. Monogenea and Aspidocotylea. Interscience, New York

Yap TA, Nguyen NT, Serr M, Shepack A, Vredenburg VT (2017) *Batrachochytrium salamandrivorans* and the risk of a second amphibian pandemic. EcoHealth 14:851–864

Zaragoza-Tapia F, Pulido-Flores G, Gardner SL, Monks S (2020) Host relationships and geographic distribution of species of *Acanthobothrium* Blanchard, 1848 (Onchoproteocephalidea, Onchobothriidae) in elasmobranchs: a metadata analysis. ZooKeys 940:1–49

Open Access This chapter is licensed under the terms of the Creative Commons Attribution-NonCommercial-NoDerivatives 4.0 International License (http://creativecommons.org/licenses/by-nc-nd/4.0/), which permits any non-commercial use, sharing, distribution and reproduction in any medium or format, as long as you give appropriate credit to the original author(s) and the source, provide a link to the Creative Commons license and indicate if you modified the licensed material. You do not have permission under this license to share adapted material derived from this chapter or parts of it.

The images or other third party material in this chapter are included in the chapter's Creative Commons license, unless indicated otherwise in a credit line to the material. If material is not included in the chapter's Creative Commons license and your intended use is not permitted by statutory regulation or exceeds the permitted use, you will need to obtain permission directly from the copyright holder.

Detecting and Assessing Aquatic Parasite Diversity Using Environmental DNA

14

Kamil Hupało, Isabel Blasco-Costa, Alejandro Trujillo-González, and Florian Leese

Abstract

Environmental DNA (eDNA) methodology has emerged as a groundbreaking tool for non-invasive biodiversity assessment, particularly in aquatic ecosystems. Although slightly lagging behind the progress made on other organismal groups, recently researchers started to recognise the great potential of using eDNA for parasite detection. Although eDNA-based methods show great potential for studying aquatic parasites, they are still neither routinely used nor implemented in a regular biomonitoring. In this chapter we review current applications of eDNA-based methods for research approaches in aquatic parasitology. Then we critically discuss the challenges and limitations of eDNA methodology for studying aquatic parasites and propose solutions for overcoming them. Finally, we provide an outlook and guidelines for future implementation of the eDNA methodology for bioassessment of aquatic parasites. With that, we show that eDNA is a very capable tracer for studying aquatic parasites and can be a promising solution for both biodiversity surveys of aquatic parasites and more effective risk assessment strategies.

14.1 Introduction

The emergence of methods based on DNA extracted from environmental samples has revolutionised biodiversity assessment and monitoring (Thomsen and Willerslev 2015; Deiner et al.

K. Hupało (✉)
Aquatic Ecosystem Research, University of Duisburg-Essen, Essen, Germany

Centre for Water and Environmental Research (ZWU), Essen, Germany

Aquatic Ecology, University of Duisburg-Essen, Essen, Germany
e-mail: kamil.hupalo@uni-due.de

I. Blasco-Costa
Department of Invertebrates, Natural History Museum of Geneva, Geneva, Switzerland
e-mail: isabel.blasco-costa@geneve.ch

A. Trujillo-González
Centre for Conservation Ecology and Genomics, Institute for Applied Ecology, University of Canberra, Bruce, ACT, Australia
e-mail: Alejandro.Trujillogonzalez@canberra.edu.au

F. Leese
Aquatic Ecosystem Research, University of Duisburg-Essen, Essen, Germany

Centre for Water and Environmental Research (ZWU), Essen, Germany
e-mail: florian.leese@uni-due.de

© The Author(s) 2025
N. J. Smit, B. Sures (eds.), *Aquatic Parasitology: Ecological and Environmental Concepts and Implications of Marine and Freshwater Parasites*, https://doi.org/10.1007/978-3-031-83903-0_14

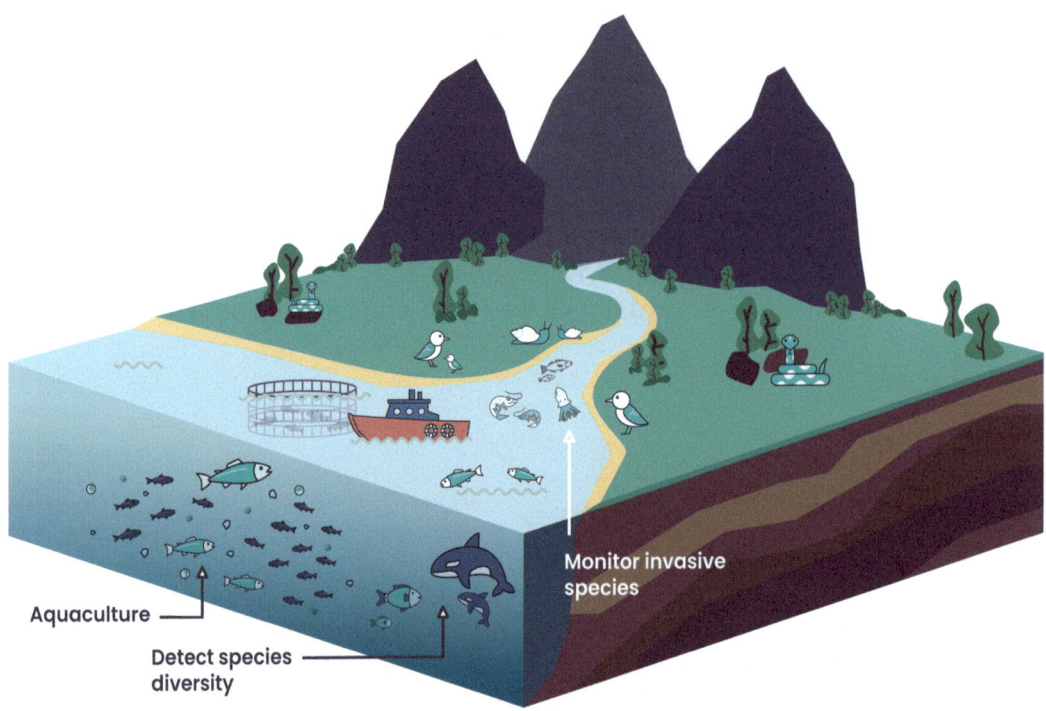

Fig. 14.1 Applications of environmental DNA (eDNA) in the aquatic ecosystems

2017; Taberlet et al. 2018). Detection of species by targeting environmental DNA (= eDNA) that originates from cells, tissue, excretions to whole organisms, created multiple opportunities for less- to non-invasive bioassessment of a wide range of environments including soil, air and water (Ficetola et al. 2008; Minamoto et al. 2012; Kirse et al. 2021; Jo et al. 2022; Bohmann and Lynggaard 2023). Environmental DNA-based methods embrace a variety of applications used mostly to study aquatic ecosystems (Fig. 14.1) (Deiner et al. 2016; Stewart 2019; Macher et al. 2021). These approaches range from targeted applications that focus on a single species, to broad non-targeted approaches that aim at detecting a broad range of taxa simultaneously, e.g. all fish species or all insect species (Fig. 14.2). Targeted approaches include highly specific single-species assays, e.g. digital PCR (Vogelstein and Kinzler 1999; Pohl and Shih 2004), real-time quantitative PCR (Heid et al. 1996; Arya et al. 2005) or loop-mediated isothermal amplification methods (Notomi et al. 2000, 2015). Targeted approaches involve amplification of a specific DNA fragment using single-species-specific primers and thus do not require sequencing. Using these targeted approaches require thorough validation and testing of these assays prior to implementation (Thalinger et al. 2021). However, when successfully established, targeted approaches can quickly deliver information on species presence and absence while also conveying quantitative information (Thomsen et al. 2012). The most popular non-targeted approaches are DNA barcoding (Hebert et al. 2003) and DNA metabarcoding (Taberlet et al. 2012), which rely on PCR amplification of marker genes and amplicon sequencing of a single or multiple species of interest at once (Coble et al. 2019; Tsuji et al. 2019). Independent of whether a targeted or non-target approach is selected, the key innovation of these eDNA-based methods is that information on species occurrence, distribution and ecology can be inferred without direct observation of the studied organisms. This aspect of "sight unseen" detection (Jerde et al. 2011) is of particular importance if species are difficult to observe or rare and in applications where early

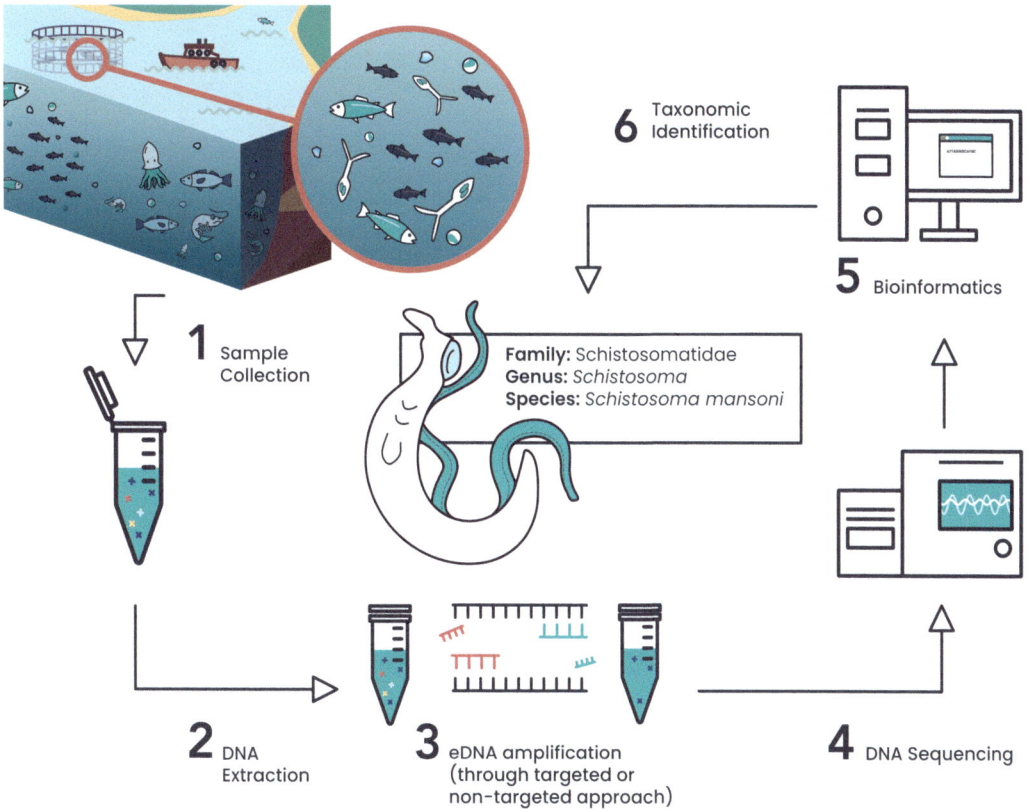

Fig. 14.2 Environmental samples from aquatic ecosystems are processed through a standardised workflow starting from sample collection to taxonomic identification

detection is important to consider management options. All these aspects often apply to a so-far understudied, ecologically and economically highly relevant yet highly heterogeneous target group: parasites.

Investigation of parasites began in ancient times when metazoan parasites were reported from direct observations in fishes, stools and domesticated animals (Cox 2004). The study of parasites in aquatic ecosystems started about two centuries ago with modern parasitology and has since documented a rich and diverse parasite fauna. The study effort has been uneven across host species and highly scattered geographically (Poulin and Presswell 2016). A substantial part of the work focused on basic natural history, providing taxonomic descriptions, surveys and elucidation of life cycles. Parasite ecological studies started to appear in the 1960s causing a "tsunami" of publications on fish parasites (Holmes 1990; Kennedy 1990). Those findings, which form the basis of our current knowledge for most common groups of aquatic parasites, are covered in Chaps. 2–6 of this book. Traditional diversity assessments were mostly based on morphological studies, which are often time-consuming and require trained personnel and meticulous handling of the organisms. The popularisation of molecular techniques in the last 50 years turned DNA sequences into a fundamental resource, together with morphological analysis, for the study of parasite diversity and species discovery, but also a valuable tool to resolve their phylogenetic relationships, distribution range and life cycles (Fig. 14.3) (Caira 2011; Perkins et al. 2011; Blasco-Costa et al. 2016; Blasco-Costa and Poulin 2017). Sampling effort for most reference genes, like the mitochondrial COI (cytochrome c oxidase subunit 1), nuclear 18S ribosomal (r)RNA or ITS (internal transcribed

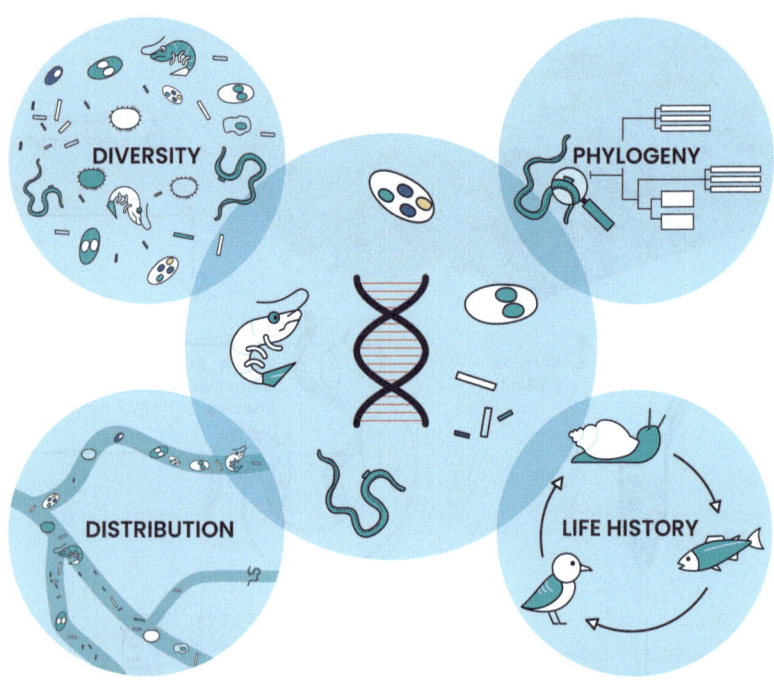

Fig. 14.3 Multiple applications of DNA-based methods for studying parasites ranging from diversity and distribution to phylogeny and life history

spacers), is uneven across parasite taxonomic groups, with no universal genetic markers applicable to all parasitic groups in place (Blasco-Costa et al. 2016; Poulin et al. 2019). This is problematic because DNA-based approaches rely on the availability of molecular sequence data in public repositories for comparative analysis. The problem is also magnified because the number of new parasite species discovered every year steadily increases (Poulin and Presswell 2016). At the same time, recent estimates predict that 85–95% of all parasite species still await scientific discovery and formal description, e.g. endoparasitic helminths (Carlson et al. 2020). Thus, many undescribed parasite species as well as species not yet represented with sequence data limit our ability to trace parasite diversity through space and time (Poulin et al. 2019). With the shortage of taxonomic experts (Brooks and Hoberg 2001), new diagnostic methods for large-scale multi-taxa surveys such as eDNA-based approaches, may not only reveal novel parasitic taxa, but could also provide additional information about their distribution, function and ecology (Bass et al. 2023). Well-curated molecular sequence data repositories are key to the success of such eDNA-based approaches and should be developed in parallel.

In recent years, eDNA-based methodology has been widely applied in biomonitoring, conservation biology, community ecology and risk assessment (Rees et al. 2014; Valentini et al. 2016; Ruppert et al. 2019; Takahashi et al. 2023). Although the relevance and benefits of using eDNA are clear for parasitological research, eDNA-based methodology has been rarely used for studying parasites (Bass et al. 2015). The method offers a sensitive and non-invasive solution to detect cryptic and microscopic parasite species that would otherwise go unnoticed (Trujillo-González et al. 2019), detect asymptomatic parasitic infestations (Sieber et al. 2022) and monitor parasite populations both in wild (Rusch et al. 2018; Alzaylaee et al. 2020a, b) and agricultural environments (Gomes et al. 2019) (Fig. 14.4). Complementing conventional detection methods with eDNA-based monitoring has greatly improved early detection of parasitic diseases and allowed for the detection of novel aquatic parasite species in the environment (Bass et al. 2023). Within the context of aquatic environmental surveillance, given that parasites

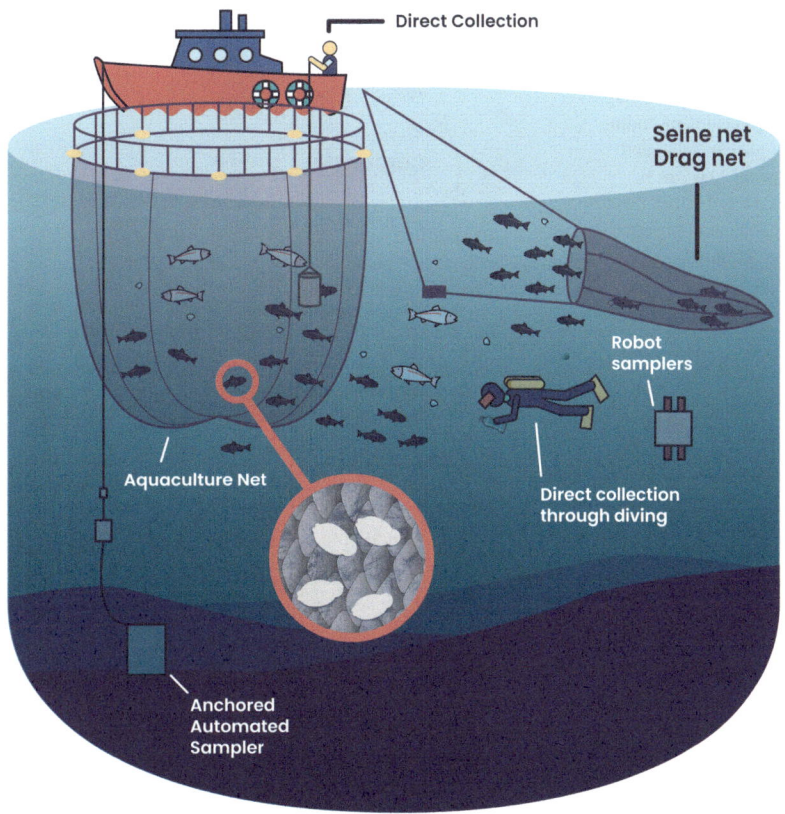

Fig. 14.4 Environmental samples for studying aquatic parasites can be collected through a variety of methods

commonly function at a microscopic scale with an incredible versatility to increase their numbers exponentially with little warning, research beyond the scope of assay design for species detection has made great advances in the understanding how eDNA-based detection can inform on parasitic disease and surveillance. The capacity to detect aquatic parasites using eDNA-based methods has so far depended on capturing parasite life stages from the water column rather than parasite trace DNA (Bass et al. 2023) (Fig. 14.5). Eggs, immature and mature life stages have been captured using filtration and precipitation methods with ranging outcomes and success, both in marine and freshwater ecosystems across a diverse array of parasite species (Amarasiri et al. 2021; Farrell et al. 2021). This is in contrast to the fields of human and veterinary parasitology, where eDNA application is marginal and primarily focuses on a few selected species of interest (Sengupta et al. 2022). Lastly, inferring parasite prevalence and infestation load using eDNA remains a challenging gap to breach in order to inform on disease risk and provide diagnostic evidence. Research has investigated the potential of targeted approaches (see above) to assess parasite abundance in wild (Fossøy et al. 2020) and farmed environments (Tsang et al. 2021) as well as explored the correlation of eDNA concentration with parasite abundance, prevalence or load, with varying success (Berger and Aubin-Horth 2018; Trujillo-González et al. 2020; Duval et al. 2021). Despite methodological limitations, eDNA research has greatly advanced applications for monitoring of parasites in aquatic environments. However, the application of eDNA monitoring to inform management decisions on the risk of parasite disease still remains underexplored. Therefore, the aim of this chapter is to (1) review the current applications of eDNA-based methods in aquatic parasitology, (2) critically discuss the challenges and limitations of eDNA methodology in studying aquatic parasites and (3) provide outlook and guidelines for future

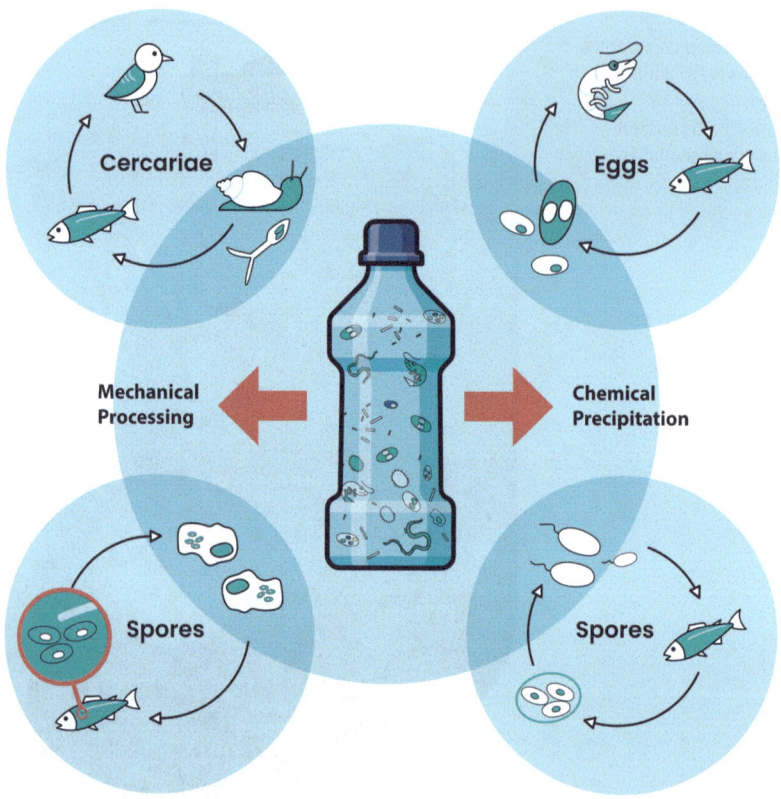

Fig. 14.5 eDNA samples mostly capture DNA signal from free-flowing life stages of parasites including eggs, cercariae and spores that can be extracted either through mechanical processing or chemical precipitation

implementation of eDNA methodology for bioassessment of aquatic parasites.

14.2 Applications of eDNA in Aquatic Parasitology

To identify the applications of eDNA-based methods in aquatic parasitology, we set up a query in Google Scholar and the Web of Science search engines (Query: ALL=(("eDNA" OR "environmental DNA") AND "parasit*" AND "water")) searching for publication entries until May 2023 (17.05.2023). To date, we retrieved 78 case studies, three PhD and master theses and nine reviews using eDNA-based methods to study aquatic parasites. The first study dates from 2006 and focused on *Vibrio penaeicida* (Bacteria: Pseudomonadota), the etiological agent of Syndrome 93 in New Caledonian shrimp, detecting presence of the parasite through a targeted approach from the shrimp haemolymph, as well as water and sediment samples from a shrimp farm (Goarant and Merien 2006). It was not until 2013 when the first study on a metazoan parasite, *Ceratomyxa puntazzi* (Cnidaria: Myxozoa) was published. Its aim was to assess the dynamics of transmission of the studied parasite in the marine environment (Alama-Bermejo et al. 2013). Since then, the number of aquatic eDNA-based parasitological studies continuously increased (Fig. 14.6). As the number of studies increased, so did the number of reviews. Those have covered the diverse applications of eDNA-based methodology in parasitology in a broader context (Bass et al. 2015, 2023), but in most cases, the reviews focused on target parasite groups of medical or veterinary importance (Archer et al. 2020; Bohara et al. 2022). Furthermore, research efforts have proliferated in Europe and North America, with approximately only a third of studies conducted elsewhere, confirming geographical bias and underrepresentation of Global South already observed in other biodiversity studies (Tydecks et al. 2018; Maas et al. 2021).

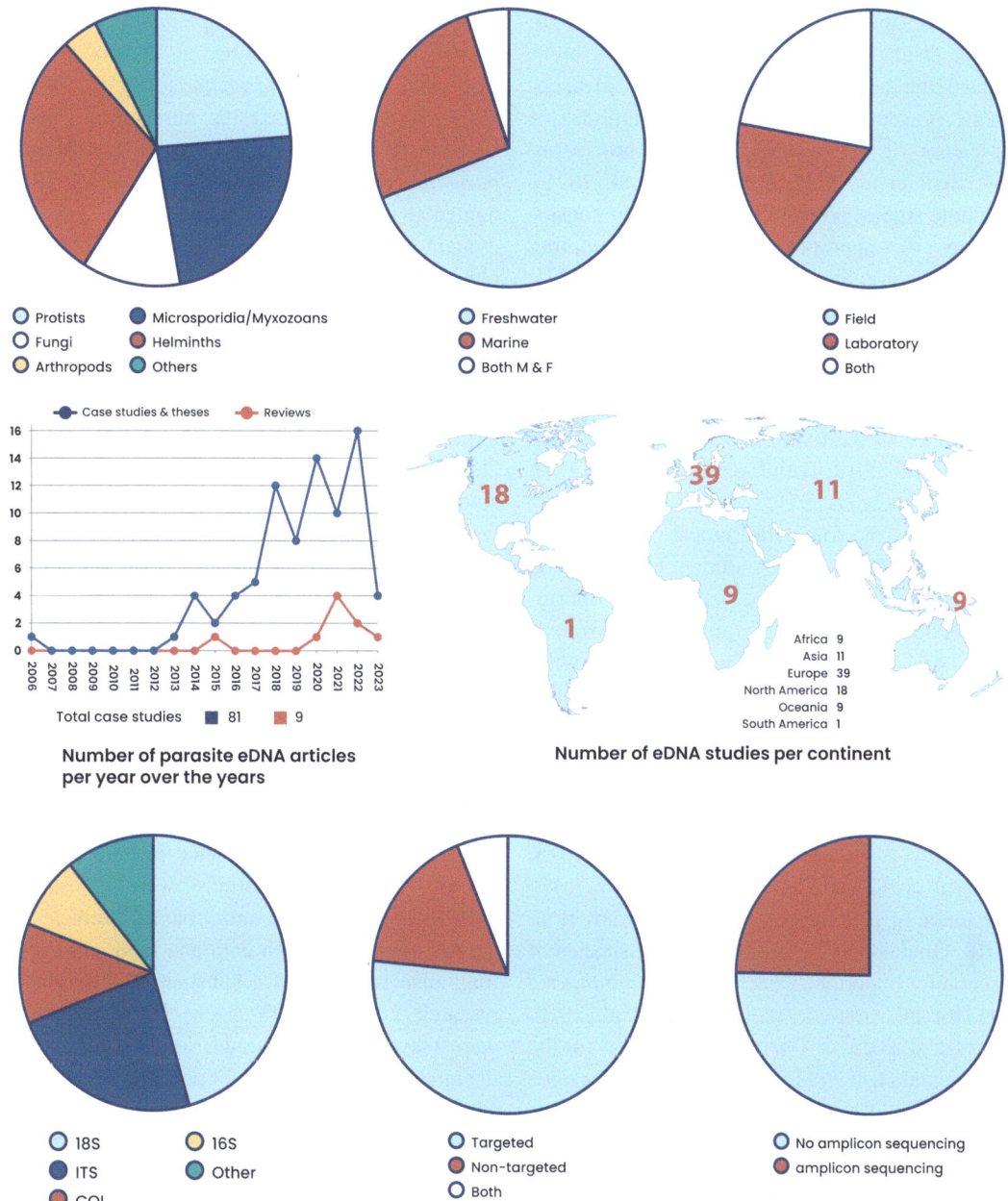

Fig. 14.6 Results of the literature review performed in this study including the number of studies on using eDNA for studying parasites, its geographical distribution, aquatic habitat studied, type of study, parasite group considered, molecular marker used, eDNA scope and eDNA methodology applied

The showcase of eDNA applications in parasitology is marked by numerous studies aiming at improving our understanding of the methodological parameters including eDNA dynamics under different ecological or methodological settings (Peters et al. 2018; Krolicka et al. 2022; Rusch et al. 2022), eDNA detectability and detection limits (Strepparava et al. 2014; Huver et al. 2015; Douchet et al. 2022) and its correspondence with classical detection methods (Hartikainen et al.

2014a, b; Richey et al. 2018; Norris et al. 2020; Sieber et al. 2022). Some of the studies highlighted the advantages of equipment advancements presenting the automated solutions for upscaling of eDNA sampling and processing (Marshall et al. 2022; Sepulveda et al. 2021). Multiple studies pointed out the need for consideration of appropriate time of sampling (Hershberger et al. 2019; Matsuoka et al. 2021), its frequency (Sepulveda et al. 2021), importance of replication and sampled water volume (Fontes et al. 2017; Rathinasamy et al. 2018; Hutchins et al. 2021; Oredalen et al. 2022) as well as the importance of the biology of the targeted group (Rudko et al. 2019; Trujillo-González et al. 2020; Rathinasamy et al. 2021).

Amongst the five main parasite groups covered by this book, eDNA studies have targeted mostly helminths and myxozoans/microsporidians, followed by protists, fungi and arthropods (Fig. 14.6). The diversity of helminth and myxozoan/microsporidian taxa investigated is dominated by a few pathogen species of each group, e.g. *Opisthorchis viverrini* (e.g. Hashizume et al. 2017), *Schistosoma* spp. (e.g. Sato et al. 2018; Fornillos et al. 2019; Alzaylaee et al. 2020a, b), *Gyrodactylus salaris* (e.g. Rusch et al. 2018; Hansen et al. 2022), *Tetracapsuloides bryosalmonae* (e.g. Fontes et al. 2017; Carraro et al. 2018; Oredalen et al. 2022) including species of economic importance such as *Trichobilharzia* spp. for lake tourism (e.g. Rudko et al. 2018; Soper et al. 2023) or *Dactylogyrus* species for the ornamental fish trade (e.g. Trujillo-González et al. 2019). Their goals were oriented towards surveillance and biosecurity as well as identifying transmission foci and dynamics (e.g. Eyre et al. 2020; Trujillo-González et al. 2019). The studies frequently relied on capturing life cycle stages (e.g. spores, cercariae, eggs) more than their eDNA traces (Bass et al. 2023). In contrast, studies on protists have mainly focused on assessing the species diversity in multiple environments, often implementing broad, non-targeted approaches for multispecies surveys (e.g. Hartikainen et al. 2014a, b; Chen et al. 2020). For fungi, studies aimed at detecting the presence of important pathogens like *Aphanomyces astaci* (Rusch et al. 2022; Sieber et al. 2022) and *Batrachochytrium dendrobatidis* (Sieber et al. 2020), or even quantify the number of spores based on DNA concentration, e.g., in *Saprolegnia parasitica* (Korkea-Aho et al. 2022), but also fundamental questions such as assessing seasonal dynamics (Matsuoka et al. 2021). Arthropods represent the least studied group, with only three studies, two of which focused on the salmon lice, *Lepeophtheirus salmonis* (Peters et al. 2018; Krolicka et al. 2022) and a third study that reported *Demodex folliculorum* in a marine, non-targeted multispecies survey (Ríos-Castro et al. 2021). Our literature review also included seven studies in which other aquatic parasite groups were investigated, including viruses (Johnson and Brunner 2014; Miaud et al. 2019, Shea et al. 2020), bacteria (Goarant and Merien 2006; Strepparava et al. 2014; Mahon et al. 2018; Shea et al. 2020) and multiple potentially pathogenic eukaryotic species across the tree of life (Ríos-Castro et al. 2021).

Overall, nearly 70% of the reviewed studies were conducted in freshwater environments, with the remainder focusing on marine ecosystems and very few studies conducted in both aquatic realms. Field investigations were dominant (~60%) whereas laboratory-based studies or studies combining both field and laboratory settings also represented a substantial part of the research efforts (~20% each). Notably, there were a few studies that used field data to calibrate the modelling approaches for predictions of parasite species distribution and abundance (Carraro et al. 2017, 2018; Sepulveda et al. 2021). Targeted approaches using single-species-specific assays aiming at species detection or quantification accounted for nearly 95% of the studies with very few studies having a broader, non-targeted focus. For a great majority of targeted studies, no amplicon sequencing was involved (Fig. 14.6). The amplicon sequencing was employed mostly within the scope of non-targeted studies which aimed at evaluating alpha-diversity and/or its seasonal dynamics, where sequence data were generated either through Sanger sequencing (single species) or high-throughput sequencing (HTS; multiple species) (e.g. Matsuoka et al. 2021; Ríos-Castro et al. 2021; Thomas et al. 2022; Lisnerová et al. 2023). Genetic markers used depended greatly on the parasite group and mostly involved a fragment of the 18S ribosomal RNA gene followed by the ITS (internal transcribed spacers) and the mitochondrial COI (cytochrome c oxidase subunit 1) and 16S ribosomal RNA gene, respectively, with a handful of other genetic markers also employed.

The vast majority of case studies, reviews and theses have focused on a specific group of interest with very few studies collectively assessing parasite diversity across different taxonomic groups (but see e.g. Ríos-Castro et al. 2021; Douchet et al. 2022). To the best of our knowledge, to date there is no eDNA-based survey that studies the entire parasitological biome in an aquatic environment.

14.3 Methodological Limitations and How to Overcome Them

As the application of eDNA-based approaches for studying aquatic parasites has become increasingly common, researchers have identified certain limitations of this methodology. Several major challenges associated with the collection, PCR amplification and interpretation of eDNA data were identified for studying aquatic organisms from environmental samples (Barnes et al. 2014; Strickler et al. 2015; Deiner et al. 2017). These technical limitations also apply for studying parasites using eDNA. Key limitations include the scarcity and degradation of parasite DNA in the water column, the limited detectability of parasites due to seasonality and the availability of detectable life stages, PCR inhibition, lack of universal primers for amplifying parasites across the tree of life, limited quantitative information and incomplete DNA sequence reference databases (Fig. 14.7). Studies focusing on eDNA applications for aquatic parasitology have highlighted and discussed each of these limitations. Below, we focus on the limitations that currently hinder the efficient implementation of eDNA methodology as a routine biomonitoring tool for studying parasites in aquatic ecosystems. We also discuss possible solutions that could be implemented now or in the future.

14.3.1 DNA Availability and Detectability

The detection of organismal DNA in environmental samples depends on the amount of available template DNA. This, in turn, relies on the origin and nature of the parasite DNA present in the water column, which can range from extracellular DNA shed from a parasite to organelles, cells and the entire parasitic organism (Bass et al. 2023). Even though some of the parasite life stages can reach up to several metres, given the often rather small size of many aquatic parasites, the amount of DNA shed into the environment is expected to be very low and, in many cases, virtually undetectable. Even if the amount of shed DNA was sufficient, its persistence in the water column depends on various environmental factors, including UV radiation, salinity, chemicals and microbial activity, which can lead to the rapid degradation of the DNA signal (Barnes et al. 2014, Strickler et al. 2015). Therefore, in many cases, the most reliable source of input DNA is the entire viable life cycle stage (Bass et al. 2023). Thus, the detectability of a particular parasite largely depends on the nature of its life cycle and the availability of free-living stages shed into the water column. Detailed studies on the protist parasite *Marteilia refringens* have shown that the parasite can only be detected in the water column if its sporangium life stages are released (Mérou et al. 2022). Similarly, the detectability of trematode species based on DNA showed diurnal fluctuations and correlated with the number of cercariae present in the water column at the time of sampling (Rudko et al. 2018). Moreover, the abundance and prevalence of certain parasite groups are known to vary seasonally, which also affects the detectability of parasite DNA. For example, the detection of the myxozoan species *Ceratomyxa puntazzi* in seawater varied strongly throughout the year

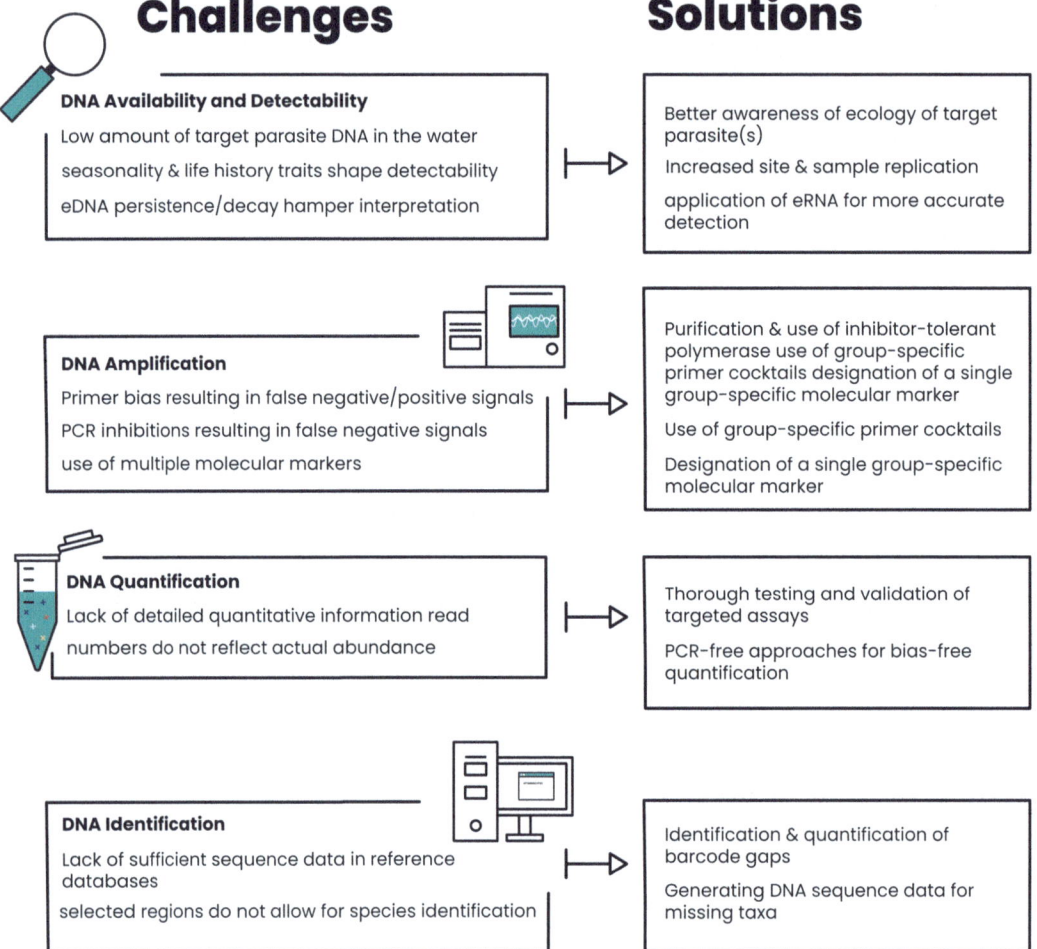

Fig. 14.7 Methodological challenges and proposed solutions to overcome them

(Alama-Bermejo et al. 2013). Similar seasonal fluctuations in detectability were observed for helminths (Rathinasamy et al. 2021), parasitic fungi (Matsuoka et al. 2021) and dinoflagellates (Davies et al. 2019).

Therefore, detailed knowledge about the ecology of parasites proves beneficial for selecting the appropriate time for sampling and obtaining reliable estimates of parasite presence (Rudko et al. 2019). Additionally, increased site and sample replication is highly recommended to decrease variation in detection (Sieber et al. 2020). It is also crucial to be aware of the ecology of the eDNA signal itself for proper interpretation of the observed detection (Mauvisseau et al. 2022). An eDNA signal can persist in the water column for weeks, depending on abiotic factors such as temperature, pH and UV radiation (Barnes et al. 2014; Strickler et al. 2015). In the case of parasites, the detection of the presence of certain parasites could indicate a signal accumulated over time, which could derive from dead specimens shedding their DNA into the water column (Kamel et al. 2021). For a more accurate indication of viable parasite presence, eRNA-based approaches offer a promising solution (Amarasiri et al. 2021; Farrell et al. 2021; Bass et al. 2023).

14.3.2 DNA Amplification

After eDNA was successfully isolated, detection of parasite presence largely depends on the successful amplification of the target DNA. Established DNA polymerase-based techniques such as conventional PCR, real-time quantitative PCR (qPCR) and digital PCR (dPCR) have become key methods for studying aquatic biodiversity, including parasites. One of the main analytical challenges, particularly when working with samples that contain a low amount of target DNA, is the presence of other molecules in the sample matrix that might affect the amplification success. These compounds, collectively referred to as PCR inhibitors, have been frequently observed in eDNA samples, with detrimental effects on the successful application of species-specific qPCR assays (McKee et al. 2015; Hunter et al. 2019). In eDNA-based studies focusing on parasite detection, PCR inhibition is rarely considered, yet it can affect the success rate of established qPCR assays for selected fungal and myxozoan species (Sieber et al. 2020). Another significant challenge lies in designing primers or probes that specifically target parasite DNA while avoiding the amplification of non-target DNA. As parasites do not form a monophyletic group (see Chaps. 2–6), their genomes can be highly diverse due to their complex phylogenetic relationships. Thus, to date, there are no universal primers that allow for the simultaneous detection of all parasite species in the environment for a non-targeted bioassessment. Several universal or group-specific primers have been widely used in DNA-based parasitological studies. However, even these primers may fail to amplify widely distributed species (Ward et al. 2016) or create a bias in the amplification of specific taxa (Matsuoka et al. 2021; Ríos-Castro et al. 2021). Similarly, using seemingly species-specific probes can result in co-amplification of closely related species (Fossøy et al. 2020). Both PCR inhibition and issues with primer specificity can lead to false positive or false negative results, which have been reported in nearly 20% of the evaluated studies. Additionally, selecting appropriate target regions within parasite genomes is crucial to ensure accurate diversity estimates. We found that up to 11 different genomic target regions have been used as molecular markers in eDNA-based studies of aquatic parasites, which hampers the comparability of observed genetic patterns across and within parasite groups. The choice of the markers used for a particular group did not always correspond to the most recommended and commonly used marker for this group (e.g. for trematodes see Blasco-Costa et al. 2016). Failing to use a broadly established genetic marker could mislead assay specificity and hamper species identification.

The most effective way to optimise the performance of already validated targeted eDNA-based assays is to consider PCR inhibitors. While purification of the obtained DNA extracts is a common approach to deal with PCR inhibition, its effectiveness largely depends on the specific inhibition mechanisms. Therefore, alternative solutions have been proposed, including the use of internal amplification controls, inhibitor-tolerant DNA polymerases and blend of different fluorescent dyes (Sidstedt et al. 2020). As single-species detection approaches based on eDNA are already well established, the next step would involve the detection of the entire parasite group present in the ecosystem, preferably the whole parasite biome. However, due to the heterogeneity, polyphyletic nature and phylogenetic divergence of aquatic parasites, designing a single universal primer that targets the entire aquatic parasite biome is not feasible. A potential solution to facilitate multispecies surveys would be to apply a group-specific primer cocktail that targets multiple species within a given parasite group, rather than using a single primer pair. However, this approach would require the availability of substantial genetic reference material for proper primer design and a collective agreement on the choice of a single, most appropriate marker for the target parasite group.

14.3.3 DNA Quantification

Monitoring parasite dynamics and assessing infection risk heavily relies on the quantification of parasite DNA input. eDNA-based methods have been used to estimate species abundance and/or biomass, with eDNA concentrations in the water column providing reliable estimates for species abundance particularly in fish communities (Dougherty et al. 2016; Hänfling et al. 2016; Lacoursière-Roussel et al. 2016; Doi et al. 2017; Ushio et al. 2018). However, reliable abundance estimates derived from eDNA data have not always been fully comparable to traditional bioassessments, as species-specific DNA persistence/decay rates and primer- and PCR-based bias can result in unequal amplification rates for different species (Elbrecht and Leese 2015; Nichols et al. 2018; Harrison et al. 2019). Currently, eDNA-based species assessment is mostly limited to presence/absence data, which also has its own limitations (see the section above on DNA detectability). So far, most studies using eDNA for parasite detection focus on evaluating the presence of parasites in the studied aquatic system. Some targeted studies have attempted to quantify the amount of obtained DNA, mostly to provide detection limits for developed qPCR assays (e.g. Jørgensen et al. 2020; Barry et al. 2021; Sieber et al. 2022). A few studies have gone further by providing relative abundance estimates using detected DNA copy numbers and demonstrating a positive correlation between DNA concentration and observed parasite abundance (Gomes et al. 2017; Sieber et al. 2020; Korkea-Aho et al. 2022). However, only a limited number of studies have been able to estimate the actual number of parasites present in the system (Goarant and Merien 2006; Alama-Bermejo et al. 2013; Huver et al. 2015; Mérou et al. 2022).

For now, reliable estimation of parasite abundance can only be achieved with targeted approaches and requires meticulous testing and validation of the assay (Broeders et al. 2014; Hays et al. 2022). The ongoing development of PCR-free approaches involving metagenomics through shotgun sequencing provides hope for bias-free quantification of parasite taxa (Bass et al. 2023).

14.3.4 DNA Identification

Accurate estimation of parasite diversity in a studied aquatic system relies on the correct taxonomic identification of detected DNA traces. The accuracy of DNA-based taxonomic assignment, however, heavily depends on the completeness and quality of the sequence reference databases as well as the availability of species-specific DNA barcodes. Recent reviews on European aquatic biota have revealed considerable gaps in public repositories, significantly hampering the taxonomic assignment of DNA sequence data from eDNA-based studies (Weigand et al. 2019). Although no similar barcode gap assessment has been done for aquatic parasites to date, it is likely that the gaps are even larger than for freshwater fishes or macroinvertebrates, as parasites are generally understudied. Several parasitological studies have highlighted the lack of sufficient sequence data in public repositories, resulting in major gaps for various parasite groups e.g. protists (Chambouvet et al. 2015; Hartikainen et al. 2014a, b; Ward et al. 2016, 2018), myxozoans (Hartikainen et al. 2016; Lisnerová et al. 2023) and helminths (Thomas et al. 2022).

To address this issue, collective efforts should be made to identify and estimate the aquatic parasite-specific barcode gaps in reference databases. Subsequently, future efforts can focus on gradually filling up the reference databases by providing DNA sequences for the study parasites from direct observations. As best practice for increasing the quality of sequence data in public repositories, it is essential that submitted DNA sequences are linked to voucher specimens deposited in museum collections, which remain publicly available and can be re-examined and their identification re-assessed at any time. This can also be done through image series that allow assessing diagnostic characters. Simultaneously, all studies describing novel parasite species should include a corresponding DNA barcode sequence, as already recommended for new species descriptions. Adding missing DNA sequence information is of crucial importance. More complete DNA reference databases would provide an important step towards implementation of eDNA-based screening of the parasites and pathogens in an entire aquatic ecosystem.

14.4 Towards Implementation of eDNA-Based Aquatic Parasite Detection

A decade of research shows the value and utility of eDNA to detect and study parasites in aquatic environments. However, implementing eDNA-based tools, regardless of the target, is ultimately a question of how reliably these tools can inform risk management. Within this context, implementation lies on how much rigour is given to developing, testing and optimising eDNA-based monitoring methods and how sovereign nations recognise and regulate eDNA-based methods. So far, implementation has been rarely considered in eDNA-based studies on parasites barring only few exceptions (Trujillo-González 2018; Eyre et al. 2020; Jørgensen et al. 2020; Schuster et al. 2022) with only a single case where parasite detection using environmental sample has been officially recognised (WOAH 2023); *Gyrodactylus salaris*, https://www.woah.org/en/disease/gyrodactylosis-g-salaris/). To enable implementation of eDNA-based methods in parasitology, there are three important aspects that should be considered: (1) assay development and validation, (2) proficiency testing and (3) governing policy and regulation.

14.4.1 Assay Development and Validation

Molecular assays and protocols must be designed, optimised and tested for eDNA applications to ensure reproducibility. The range of applications and assays designed to determine the occurrence of species by targeting eDNA has without a doubt increased over the years (Takahashi et al. 2023; Fig. 14.6). However, research has shown that a vast majority of assays used for eDNA-based lack validation, resulting in little reproducibility and reliability of results (Thalinger et al. 2021). As such, assays must undergo extensive validation using high-quality standards to evaluate how well they suit a desired application. Validation frameworks for molecular assays are readily available, with many drawing certain validation steps used in diagnostic frameworks recognised internationally [(De Brauwer et al. 2023), see chapter 1.1.2 of the Manual of Diagnostic Tests for Aquatic Animals, (WOAH 2023)]. Of importance, assay limits must be assessed to fully understand what boundaries define how reliable assays are and when they would fall short of producing accurate results. To do so, assays must not only be optimised in laboratory conditions for specificity and sensitivity, but they must also be field tested within the context of their intended purpose to examine how environmental matrices, inhibiting factors and representative DNA concentrations of the target would affect assay limits of detection and their suitability within their defined purpose (Rusch et al. 2018). Validation of eDNA assays must go beyond testing scenarios wherein it is clear assays would work as intended (e.g. clean water matrices spiked with ranging concentrations of genomic DNA) and truly test where assays falter and fail to consider reproducibility. Developing a high-quality framework to validate molecular assays will provide greater trust in eDNA results, regardless of their application and intended purpose.

14.4.2 Proficiency Testing

Proficiency testing provides an important layer of quality and competency in delivering reliable results to the end users. It must be considered that all molecular workflows and assays, when used routinely, will encounter a range of limitations on how outcomes are delivered (e.g. conforming or non-conforming) and inaccurate results that can occur in analyses (i.e. type I and II errors, human error). International standards on how to conduct proficiency tests exist (especially ISO/IEC 17043:2010) and are broadly applied in classical and molecular diagnostics. Within this context, proficiency testing provides service providers with a mechanism to demonstrate proficiency in their workflows, as well as a suitable framework to identify potential caveats in their methods. There are multiple ways in which proficiency testing can be undertaken. Service providers can purchase proficiency tests from companies worldwide and be assessed independently in their competency. For example, Fapas® from Fera Science Ltd (https://fapas.com/) offers an extensive portfolio of proficiency tests designed to assess species detection, which in the context of environmental testing includes a proficiency testing scheme designed for the detection of great crested newt, *Triturus cristatus* (test WC1067; DEFRA 2019) through environmental DNA, which offers a framework for future eDNA-based proficiency testing schemes. Similarly, service providers can take part in independent ring tests designed to test competency in testing by comparing results against a predefined standard or amongst facilities, in which case any significant deviation from a mean standard would indicate a lack of proficiency (Zaiko et al. 2022). Lastly, service providers can undertake internal audits of their workflows and assays to assess performance and proficiency. Auditing internal processes, assays and workflows is a common practice in human health and veterinary applications, wherein competency in molecular testing is essential to uphold regulatory requirements (Mejia et al. 2020).

14.4.3 Governing Policy and Regulation

Regulation and policies are required to better manage how eDNA-based results can be used for the purpose of management. For this purpose, end users require trust in eDNA results as well as mechanisms to understand and manage the proliferation of mistakes and errors. A significant body of work has been dedicated to outline how risks can be managed in eDNA analyses in the event of false positive detections (Darling et al. 2021), sample cross-contamination (Furlan and Gleeson 2016; Sepulveda et al. 2020b) and low abundances of target species (Lugg et al. 2018; Rusch et al. 2018). These considerations, although common to eDNA experts, must be translated into implementable actions suitable for regulators and policy makers (Sepulveda et al. 2020a, b). Therefore, a strong collaboration is needed between parasitologists using eDNA to inform management on suitable protocols and assays, for establishing mechanisms to manage the risk of type I & II errors and defining frameworks for end users and managers to highlight when eDNA results are actionable (Fig. 14.8). Actionable results in turn should entail a well-understood decision tree describing all possible actions following detection of a target species or lack thereof (Sepulveda et al. 2023). To do so, however, it is essential that experts that develop eDNA-based methods for parasitology understand what governing policies and regulations are present for their purposes and what requirements are needed to inform non-regulatory and regulatory applications for management (Fig. 14.8).

Regulatory frameworks may require accreditation for eDNA service providers to provide actionable results. Laboratories can have their workflows assessed and certified through accreditation following guidelines of the globally recognised International Organisation for Accreditation (ISO; www.iso.org) and the International Electrotechnical Commission (IEC; www.iec.ch). Similarly, there are recognised standards for laboratory best practices that outline minimum quality requirements for test and calibration purposes (e.g. OECD Principles of Good Laboratory Practice). Service providers interested in obtaining accreditation should consider which bodies govern accreditation and its certification for their purposes, as well as considering if accreditation is truly required for their functions and services (Trujillo-González et al. 2021). As such, it is recommended that eDNA service providers consider what regulatory framework, if any, would govern the application of eDNA methods at early stages of development of assays and protocols. In instances where no regulatory framework is present, it is recom-

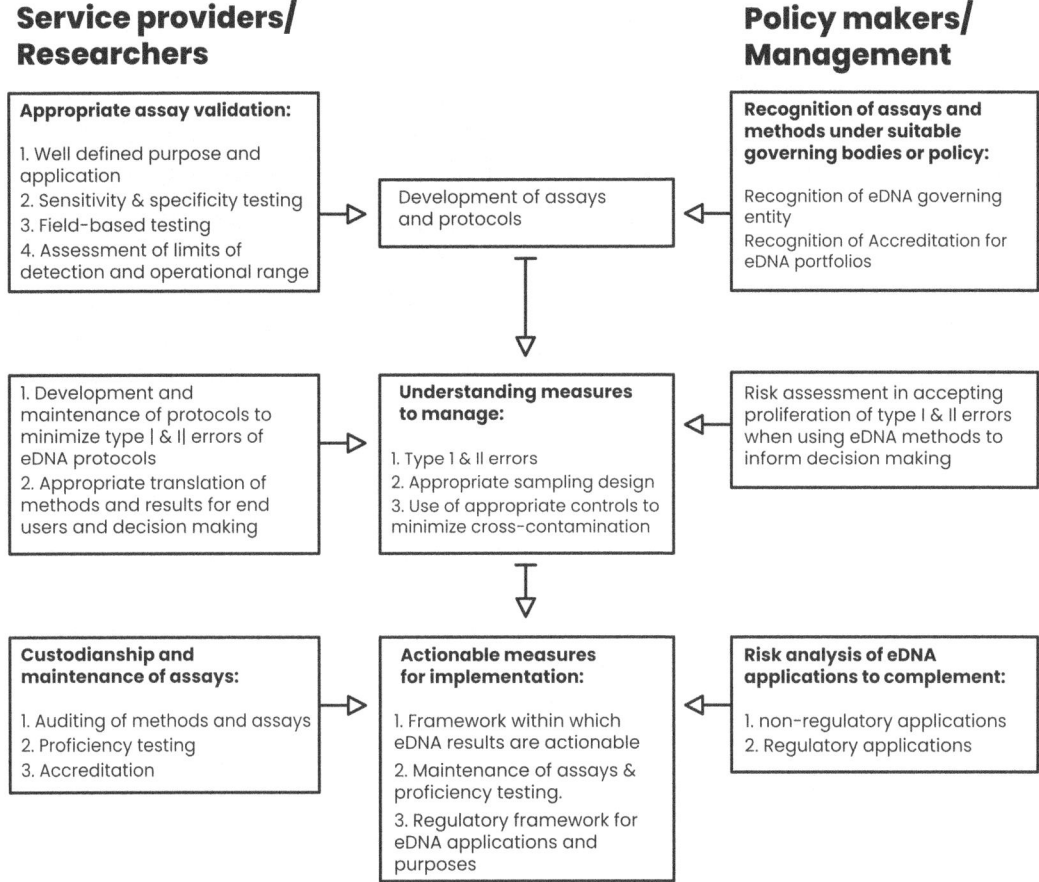

Fig. 14.8 Pathway towards implementation of eDNA-based parasite detection in routine monitoring

mended that policy makers first consider how and when eDNA assays are actionable for risk management (Fig. 14.8).

14.5 Conclusions and Outlook

Given the remaining knowledge gaps surrounding aquatic parasites and the mounting evidence of their importance in ecosystem functioning, the assessment and monitoring of aquatic parasite biodiversity emerge as critical endeavours. In this context, the application of environmental DNA (eDNA) proves highly promising as a non-invasive tracer, including simple sample collection and cost-efficient data analyses. While there is a rapidly expanding body of work showcasing its use in studying aquatic parasites directly from water samples, the widespread implementation of eDNA-based methodologies for routine parasite monitoring is still some distance away. Hence, the immediate focus should be on validating the assays, implementing quality control measures and establishing quality assurance routines to facilitate the integration of eDNA techniques into global biodiversity monitoring efforts. Additionally, it is imperative to make the collected data readily accessible not only to decision-makers and researchers but also to the interested public. Therefore, the creation of common platforms for data access becomes a priority. Once these essential steps are in place, a range of novel solutions can be implemented. These may include routine monitoring of parasite diversity, screening for alien parasites and early warning systems for detecting harmful pathogens in recreational waters, drinking water sources and globally relevant food items (Fig. 14.9). By harnessing the

Fig. 14.9 Applications of environmental DNA (eDNA) for studying aquatic parasites

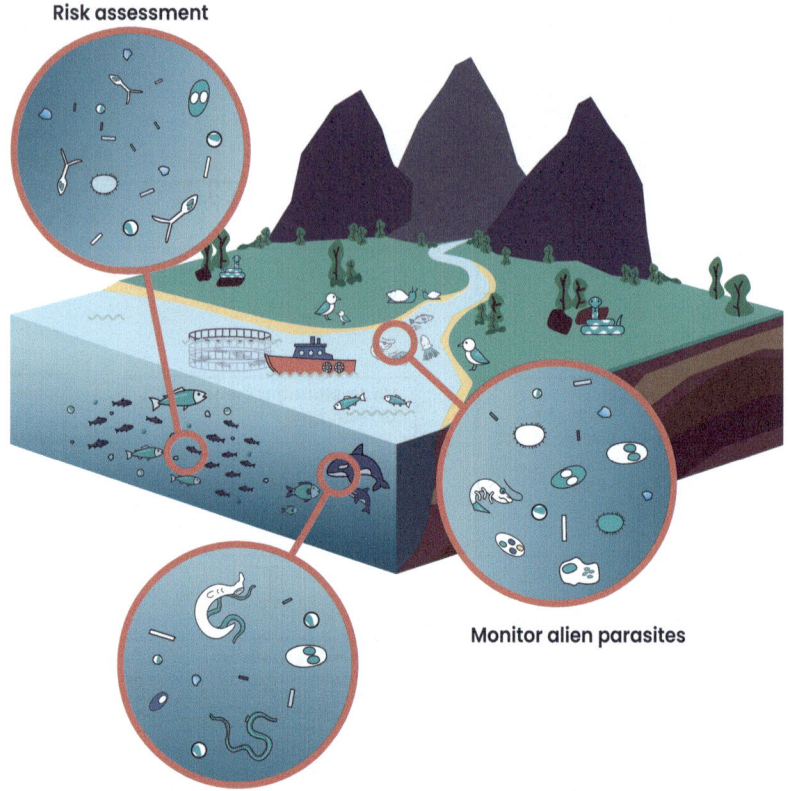

power of environmental DNA, these solutions pave the way for mitigating biodiversity loss and enabling more effective risk assessment strategies.

Acknowledgments & data availability The authors would like to thank Dakeishla M. Díaz Morales for providing valuable input for the literature review. The database containing all the records analyzed for this study alongside all the relevant information used for the inferences throughout the chapter is publicly available on the ZENODO repository under the following DOI 10.5281/zenodo.14749485.

References

Alama-Bermejo G, Šíma R, Raga JA, Holzer AS (2013) Understanding myxozoan infection dynamics in the sea: seasonality and transmission of *Ceratomyxa puntazzi*. Int J Parasitol 43(9):771–780

Alzaylaee H, Collins RA, Rinaldi G, Shechonge A, Ngatunga B, Morgan ER, Genner MJ (2020a) *Schistosoma* species detection by environmental DNA assays in African freshwaters. PLoS Negl Trop Dis 14(3):e0008129

Alzaylaee H, Collins RA, Shechonge A, Ngatunga BP, Morgan ER, Genner MJ (2020b) Environmental DNA-based xenomonitoring for determining *Schistosoma* presence in tropical freshwaters. Parasit Vectors 13(1):1–11

Amarasiri M, Furukawa T, Nakajima F, Sei K (2021) Pathogens and disease vectors/hosts monitoring in aquatic environments: Potential of using eDNA/eRNA based approach. Sci Total Environ 796:148810

Archer J, O'Halloran L, Al-Shehri H, Summers S, Bhattacharyya T, Kabaterine NB, Atuhaire A, Adriko M, Arianaitwe M, Stewart M (2020) Intestinal schistosomiasis and giardiasis co-infection in sub-Saharan Africa: can a one health approach improve control of each waterborne parasite simultaneously? Trop Med Infect Dis 5(3):137

Arya M, Shergill IS, Williamson M, Gommersall L, Arya N, Patel HR (2005) Basic principles of real-time quantitative PCR. Expert Rev Mol Diagn 5(2):209–219

Barnes MA, Turner CR, Jerde CL, Renshaw MA, Chadderton WL, Lodge DM (2014) Environmental conditions influence eDNA persistence in aquatic systems. Environ Sci Technol 48(3):1819–1827

Barry DE, Veillard M, James CT, Brummelhuis L, Pila EA, Turnbull A, Oddy-van Oploo A, Han X, Hanington PC (2021) qPCR-based environmental monitoring of

Myxobolus cerebralis and phylogenetic analysis of its tubificid hosts in Alberta, Canada. Dis Aquat Org 145:119–137

Bass D, Stentiford GD, Littlewood D, Hartikainen H (2015) Diverse applications of environmental DNA methods in parasitology. Trends Parasitol 31(10):499–513

Bass D, Christison KW, Stentiford GD, Cook LS, Hartikainen H (2023) Environmental DNA/RNA for pathogen and parasite detection, surveillance, and ecology. Trends Parasitol 39(4):285–304

Berger CS, Aubin-Horth N (2018) An eDNA-qPCR assay to detect the presence of the parasite *Schistocephalus solidus* inside its threespine stickleback host. J Exp Biol 221(9):jeb178137

Blasco-Costa I, Poulin R (2017) Parasite life-cycle studies: a plea to resurrect an old parasitological tradition. J Helminthol 91(6):647–656

Blasco-Costa I, Cutmore SC, Miller TL, Nolan MJ (2016) Molecular approaches to trematode systematics: 'best practice' and implications for future study. Syst Parasitol 93:295–306

Bohara K, Yadav AK, Joshi P (2022) Detection of fish pathogens in freshwater aquaculture using eDNA methods. Diversity 14(12):1015

Bohmann K, Lynggaard C (2023) Transforming terrestrial biodiversity surveys using airborne eDNA. Trends Ecol Evol 38:119–121

Broeders S, Huber I, Grohmann L, Berben G, Taverniers I, Mazzara M, Roosens N, Morisset D (2014) Guidelines for validation of qualitative real-time PCR methods. Trends Food Sci Technol 37(2):115–126

Brooks DR, Hoberg EP (2001) Parasite systematics in the 21st century: opportunities and obstacles. Trends Parasitol 17(6):273–275

Caira JN (2011) Synergy advances parasite taxonomy and systematics: an example from elasmobranch tapeworms. Parasitology 138(13):1675–1687

Carlson CJ, Dallas TA, Alexander LW, Phelan AL, Phillips AJ (2020) What would it take to describe the global diversity of parasites? Proc Roy Soc B 287(1939):20201841

Carraro L, Bertuzzo E, Mari L, Fontes I, Hartikainen H, Strepparava N, Schmidt-Posthaus H, Wahli T, Jokela J, Gatto M (2017) Integrated field, laboratory, and theoretical study of PKD spread in a Swiss prealpine river. Proc Natl Acad Sci USA 114(45):11992–11997

Carraro L, Hartikainen H, Jokela J, Bertuzzo E, Rinaldo A (2018) Estimating species distribution and abundance in river networks using environmental DNA. Proc Natl Acad Sci USA 115(46):11724–11729

Chambouvet A, Gower DJ, Jirků M, Yabsley MJ, Davis AK, Leonard G, Maguire F, Doherty-Bone TM, Bittencourt-Silva GB, Wilkinson M (2015) Cryptic infection of a broad taxonomic and geographic diversity of tadpoles by Perkinsea protists. Proc Natl Acad Sci USA 112(34):E4743–E4751

Chen T, Liu Y, Xu S, Song S, Li C (2020) Variation of *Amoebophrya* community during bloom of *Prorocentrum donghaiense* Lu in coastal waters of the East China Sea. Estuar Coast Shelf Sci 243:106887

Coble AA, Flinders CA, Homyack JA, Penaluna BE, Cronn RC, Weitemier K (2019) eDNA as a tool for identifying freshwater species in sustainable forestry: A critical review and potential future applications. Sci Total Environ 649:1157–1170

Cox FE (2004) History of human parasitic diseases. Infect Dis Clin 18(2):171–188

Darling JA, Jerde CL, Sepulveda AJ (2021) What do you mean by false positive? eDNA 3(5):879–883

Davies CE, Batista FM, Malkin SH, Thomas JE, Bryan CC, Crocombe P, Coates CJ, Rowley AF (2019) Spatial and temporal disease dynamics of the parasite *Hematodinium* sp. in shore crabs, *Carcinus maenas*. Parasit Vectors 12:1–15

De Brauwer M, Clarke LJ, Chariton A, Cooper MK, De Bruyn M, Furlan E, MacDonald AJ, Rourke ML, Sherman CD, Suter L (2023) Best practice guidelines for environmental DNA biomonitoring in Australia and New Zealand. eDNA 5(3):417–423

DEFRA (2019) Analytical and methodological development for improved surveillance of the Great Crested Newt, and other pond vertebrates - WC1067. United Kingdom of Great Britain and Northern Ireland. Accessed 16 April 2019. Retrieved from http://sciencesearch.defra.gov.uk/Default.aspx?Module=More&Location=None&ProjectID=18650

Deiner K, Fronhofer EA, Mächler E, Walser J-C, Altermatt F (2016) Environmental DNA reveals that rivers are conveyer belts of biodiversity information. Nat Commun 7(1):12544

Deiner K, Bik HM, Mächler E, Seymour M, Lacoursière-Roussel A, Altermatt F, Creer S, Bista I, Lodge DM, De Vere N (2017) Environmental DNA metabarcoding: Transforming how we survey animal and plant communities. Mol Ecol 26(21):5872–5895

Doi H, Inui R, Akamatsu Y, Kanno K, Yamanaka H, Takahara T, Minamoto T (2017) Environmental DNA analysis for estimating the abundance and biomass of stream fish. Freshw Biol 62(1):30–39

Douchet P, Boissier J, Mulero S, Ferté H, Doberva M, Allienne JF, Toulza E, Bethune K, Rey O (2022) Make visible the invisible: Optimized development of an environmental DNA metabarcoding tool for the characterization of trematode parasitic communities. eDNA 4(3):627–641

Dougherty MM, Larson ER, Renshaw MA, Gantz CA, Egan SP, Erickson DM, Lodge DM (2016) Environmental DNA (eDNA) detects the invasive rusty crayfish *Orconectes rusticus* at low abundances. J Appl Ecol 53(3):722–732

Duval E, Blanchet S, Quéméré E, Jacquin L, Veyssière C, Lautraite A, Garmendia L, Yotte A, Parthuisot N, Côte J (2021) Urine DNA (uDNA) as a non-lethal method for endoparasite biomonitoring: development and validation. eDNA 3(5):1035–1045

Elbrecht V, Leese F (2015) Can DNA-based ecosystem assessments quantify species abundance? Testing primer bias and biomass—sequence relationships with an innovative metabarcoding protocol. PLoS One 10(7):e0130324

Eyre M, Stanton M, Macklin G, Bartoníček Z, O'Halloran L, Ombede DE, Chuinteu G, Stewart M, LaCourse E, Tchuenté LT (2020) Piloting an integrated approach for estimation of environmental risk of *Schistosoma haematobium* infections in pre-school-aged children and their mothers at Barombi Kotto, Cameroon. Acta Trop 212:105646

Farrell JA, Whitmore L, Duffy DJ (2021) The promise and pitfalls of environmental DNA and RNA approaches for the monitoring of human and animal pathogens from aquatic sources. BioScience 71(6):609–625

Ficetola GF, Miaud C, Pompanon F, Taberlet P (2008) Species detection using environmental DNA from water samples. Biol Lett 4(4):423–425

Fontes I, Hartikainen H, Holland JW, Secombes CJ, Okamura B (2017) *Tetracapsuloides bryosalmonae* abundance in river water. Dis Aquat Org 124(2):145–157

Fornillos RJC, Sato MO, Tabios IKB, Sato M, Leonardo LR, Chigusa Y, Minamoto T, Kikuchi M, Legaspi ER, Fontanilla IKC (2019) Detection of *Schistosoma japonicum* and *Oncomelania hupensis quadrasi* environmental DNA and its potential utility to *Schistosomiasis japonica* surveillance in the Philippines. PLoS One 14(11):e0224617

Fossøy F, Brandsegg H, Sivertsgård R, Pettersen O, Sandercock BK, Solem Ø, Hindar K, Mo TA (2020) Monitoring presence and abundance of two gyrodactylid ectoparasites and their salmonid hosts using environmental DNA. eDNA 2(1):53–62

Furlan EM, Gleeson D (2016) Improving reliability in environmental DNA detection surveys through enhanced quality control. Mar Freshw Res 68(2):388–395

Goarant C, Merien F (2006) Quantification of *Vibrio penaeicida*, the etiological agent of Syndrome 93 in New Caledonian shrimp, by real-time PCR using SYBR Green I chemistry. J Microbiol Methods 67(1):27–35

Gomes GB, Hutson KS, Domingos JA, Chung C, Hayward S, Miller TL, Jerry DR (2017) Use of environmental DNA (eDNA) and water quality data to predict protozoan parasites outbreaks in fish farms. Aquaculture 479:467–473

Gomes GB, Hutson KS, Domingos JA, Villamil SI, Huerlimann R, Miller TL, Jerry DR (2019) Parasitic protozoan interactions with bacterial microbiome in a tropical fish farm. Aquaculture 502:196–206

Hänfling B, Lawson Handley L, Read DS, Hahn C, Li J, Nichols P, Blackman RC, Oliver A, Winfield IJ (2016) Environmental DNA metabarcoding of lake fish communities reflects long-term data from established survey methods. Mol Ecol 25(13):3101–3119

Hansen H, Ieshko E, Rusch J, Samokhvalov I, Melnik V, Mugue N, Sokolov S, Parshukov A (2022) *Gyrodactylus salaris* Malmberg, 1957 (Monogenea, Gyrodactylidae) spreads further—a consequence of rainbow trout farming in Northern Russia. Aquat Invasions 17(2):224–237

Harrison JB, Sunday JM, Rogers SM (2019) Predicting the fate of eDNA in the environment and implications for studying biodiversity. Proc Roy Soc B 286(1915):20191409

Hartikainen H, Ashford OS, Berney C, Okamura B, Feist SW, Baker-Austin C, Stentiford GD, Bass D (2014a) Lineage-specific molecular probing reveals novel diversity and ecological partitioning of haplosporidians. ISME J 8(1):177–186

Hartikainen H, Stentiford GD, Bateman KS, Berney C, Feist SW, Longshaw M, Okamura B, Stone D, Ward G, Wood C (2014b) Mikrocytids are a broadly distributed and divergent radiation of parasites in aquatic invertebrates. Curr Biol 24(7):807–812

Hartikainen H, Bass D, Briscoe AG, Knipe H, Green AJ, Okamura B (2016) Assessing myxozoan presence and diversity using environmental DNA. Int J Parasitol 46(12):781–792

Hashizume H, Sato M, Sato MO, Ikeda S, Yoonuan T, Sanguankiat S, Pongvongsa T, Moji K, Minamoto T (2017) Application of environmental DNA analysis for the detection of *Opisthorchis viverrini* DNA in water samples. Acta Trop 169:1–7

Hays A, Islam R, Matys K, Williams D (2022) Best practices in qPCR and dPCR validation in regulated bioanalytical laboratories. AAPS J 24(2):36

Hebert PD, Cywinska A, Ball SL, DeWaard JR (2003) Biological identifications through DNA barcodes. Proc Roy Soc Lond B Biol Sci 270(1512):313–321

Heid CA, Stevens J, Livak KJ, Williams PM (1996) Real time quantitative PCR. Genome Res 6(10):986–994

Hershberger P, Powers R, Besijn BL, Rankin J, Wilson M, Antipa B, Bjelland J, MacKenzie A, Gregg J, Purcell M (2019) Intra-annual changes in waterborne *Nanophyetus salmincola*. J Aquat Anim Health 31(3):259–265

Holmes JC (1990) Helminth communities in marine fishes. In: Esch GW, Bush AO, Aho JM (eds) Parasite communities: patterns and processes. Springer, Dordrecht, pp 101–130

Hunter ME, Ferrante JA, Meigs-Friend G, Ulmer A (2019) Improving eDNA yield and inhibitor reduction through increased water volumes and multi-filter isolation techniques. Sci Rep 9(1):5259

Hutchins PR, Sepulveda AJ, Hartikainen H, Staigmiller KD, Opitz ST, Yamamoto RM, Huttinger A, Cordes RJ, Weiss T, Hopper LR (2021) Exploration of the 2016 Yellowstone River fish kill and proliferative kidney disease in wild fish populations. Ecosphere 12(3):e03436

Huver J, Koprivnikar J, Johnson P, Whyard S (2015) Development and application of an eDNA method to detect and quantify a pathogenic parasite in aquatic ecosystems. Ecol Appl 25(4):991–1002

Jerde CL, Mahon AR, Chadderton WL, Lodge DM (2011) "Sight-unseen" detection of rare aquatic species using environmental DNA. Conserv Lett 4(2):150–157

Jo T, Takao K, Minamoto T (2022) Linking the state of environmental DNA to its application for biomonitoring and stock assessment: Targeting mitochondrial/nuclear genes, and different DNA fragment lengths and particle sizes. eDNA 4(2):271–283

Johnson A, Brunner J (2014) Persistence of an amphibian ranavirus in aquatic communities. Dis Aquat Org 111(2):129–138

Jørgensen LVG, Nielsen JW, Villadsen MK, Vismann B, Dalvin S, Mathiessen H, Madsen L, Kania PW, Buchmann K (2020) A non-lethal method for detection of *Bonamia ostreae* in flat oyster (*Ostrea edulis*) using environmental DNA. Sci Rep 10(1):16143

Kamel B, Laidemitt MR, Lu L, Babbitt C, Weinbaum OL, Mkoji GM, Loker ES (2021) Detecting and identifying *Schistosoma* infections in snails and aquatic habitats: A systematic review. PLoS Negl Trop Dis 15(3):e0009175

Kennedy C (1990) Helminth communities in freshwater fish: structured communities or stochastic assemblages? In: Esch GW, Bush AO, Aho JM (eds) Parasite communities: patterns and processes. Springer, Dordrecht, pp 131–156

Kirse A, Bourlat SJ, Langen K, Fonseca VG (2021) Unearthing the potential of soil eDNA metabarcoding—towards best practice advice for invertebrate biodiversity assessment. Front Ecol Evol 9:337

Korkea-Aho T, Wiklund T, Engblom C, Vainikka A, Viljamaa-Dirks S (2022) Detection and quantification of the oomycete *Saprolegnia parasitica* in aquaculture environments. Microorganisms 10(11):2186

Krolicka A, Mæland Nilsen M, Klitgaard Hansen B, Wulf Jacobsen M, Provan F, Baussant T (2022) Sea lice (*Lepeophtherius salmonis*) detection and quantification around aquaculture installations using environmental DNA. PLoS One 17(9):e0274736

Lacoursière-Roussel A, Côté G, Leclerc V, Bernatchez L (2016) Quantifying relative fish abundance with eDNA: a promising tool for fisheries management. J Appl Ecol 53(4):1148–1157

Lisnerová M, Holzer A, Blabolil P, Fiala I (2023) Evaluation and optimization of an eDNA metabarcoding assay for detection of freshwater myxozoan communities. eDNA 5(2):312–325

Lugg WH, Griffiths J, van Rooyen AR, Weeks AR, Tingley R (2018) Optimal survey designs for environmental DNA sampling. Methods Ecol Evol 9(4):1049–1059

Maas B, Pakeman RJ, Godet L, Smith L, Devictor V, Primack R (2021) Women and Global South strikingly underrepresented among top-publishing ecologists. Conserv Lett 14(4):e12797

Macher T-H, Schütz R, Arle J, Beermann AJ, Koschorreck J, Leese F (2021) Beyond fish eDNA metabarcoding: Field replicates disproportionately improve the detection of stream associated vertebrate species. MBMG 5

Mahon AR, Horton DJ, Learman DR, Nathan LR, Jerde CL (2018) Investigating diversity of pathogenic microbes in commercial bait trade water. PeerJ 6:e5468

Marshall WL, MacWilliam T, Williams K, Reinholt H, VanVliet H, New D, Mills M, Morrison D (2022) Detection of *Kudoa thyrsites* (Myxozoa) eDNA by real-time and digital PCR from high seawater volumes. J Fish Dis 45(9):1403–1407

Matsuoka S, Sugiyama Y, Shimono Y, Ushio M, Doi H (2021) Evaluation of seasonal dynamics of fungal DNA assemblages in a flow-regulated stream in a restored forest using eDNA metabarcoding. Environ Microbiol 23(8):4797–4806

Mauvisseau Q, Harper LR, Sander M, Hanner RH, Kleyer H, Deiner K (2022) The multiple states of environmental DNA and what is known about their persistence in aquatic environments. Environ Sci Technol 56(9):5322–5333

McKee AM, Spear SF, Pierson TW (2015) The effect of dilution and the use of a post-extraction nucleic acid purification column on the accuracy, precision, and inhibition of environmental DNA samples. Biol Conserv 183:70–76

Mejia R, Cuellar M, Salyards J (2020) Implementing blind proficiency testing in forensic laboratories: Motivation, obstacles, and recommendations. Forensic Sci Int Synergy 2:293–298

Mérou N, Lecadet C, Billon T, Chollet B, Pouvreau S, Arzul I (2022) Investigating the environmental survival of *Marteilia refringens*, a marine protozoan parasite of the flat oyster Ostrea edulis, through an environmental DNA and microscopy-based approach. Front Mar Sci 9:811284

Miaud C, Arnal V, Poulain M, Valentini A, Dejean T (2019) eDNA increases the detectability of ranavirus infection in an alpine amphibian population. Viruses 11(6):526

Minamoto T, Yamanaka H, Takahara T, Honjo MN, Kawabata Z (2012) Surveillance of fish species composition using environmental DNA. Limnology 13:193–197

Nichols RV, Vollmers C, Newsom LA, Wang Y, Heintzman PD, Leighton M, Green RE, Shapiro B (2018) Minimizing polymerase biases in metabarcoding. Mol Ecol Resour 18(5):927–939

Norris L, Lawler N, Hunkapiller A, Mulrooney DM, Kent ML, Sanders JL (2020) Detection of the parasitic nematode, *Pseudocapillaria tomentosa*, in zebrafish tissues and environmental DNA in research aquaria. J Fish Dis 43(9):1087–1095

Notomi T, Okayama H, Masubuchi H, Yonekawa T, Watanabe K, Amino N, Hase T (2000) Loop-mediated isothermal amplification of DNA. Nucleic Acids Res 28(12):e63–e63

Notomi T, Mori Y, Tomita N, Kanda H (2015) Loop-mediated isothermal amplification (LAMP): principle, features, and future prospects. J Microbiol 53(1):1–5

Oredalen TJ, Mo TA, Jenkins A, Haugan N, Sæbø M (2022) Use of environmental DNA to detect the myxozoan endoparasite *Tetracapsuloides bryosalmonae* in large Norwegian lakes. eDNA 4(6):1294–1322

Perkins S, Martinsen E, Falk B (2011) Do molecules matter more than morphology? Promises and pitfalls in parasites. Parasitology 138(13):1664–1674

Peters L, Spatharis S, Dario MA, Dwyer T, Roca IJ, Kintner A, Kanstad-Hanssen Ø, Llewellyn MS, Praebel K (2018) Environmental DNA: A new low-cost monitoring tool for pathogens in salmonid aquaculture. Front Microbiol 9:3009

Pohl G, Shih I-M (2004) Principle and applications of digital PCR. Expert Rev Mol Diagn 4(1):41–47

Poulin R, Presswell B (2016) Taxonomic quality of species descriptions varies over time and with the number of authors, but unevenly among parasitic taxa. Syst Biol 65(6):1107–1116

Poulin R, Hay E, Jorge F (2019) Taxonomic and geographic bias in the genetic study of helminth parasites. Int J Parasitol 49(6):429–435

Rathinasamy V, Hosking C, Tran L, Kelley J, Williamson G, Swan J, Elliott T, Rawlin G, Beddoe T, Spithill TW (2018) Development of a multiplex quantitative PCR assay for detection and quantification of DNA from *Fasciola hepatica* and the intermediate snail host, *Austropeplea tomentosa*, in water samples. Vet Parasitol 259:17–24

Rathinasamy V, Tran L, Swan J, Kelley J, Hosking C, Williamson G, Knowles M, Elliott T, Rawlin G, Spithill TW (2021) Towards understanding the liver fluke transmission dynamics on farms: Detection of liver fluke transmitting snail and liver fluke-specific environmental DNA in water samples from an irrigated dairy farm in Southeast Australia. Vet Parasitol 291:109373

Rees HC, Maddison BC, Middleditch DJ, Patmore JR, Gough KC (2014) The detection of aquatic animal species using environmental DNA—a review of eDNA as a survey tool in ecology. J Appl Ecol 51(5):1450–1459

Richey CA, Kenelty KV, Van Stone HK, Stevens BN, Martínez-López B, Barnum SM, Hallett SL, Atkinson SD, Bartholomew JL, Soto E (2018) Distribution and prevalence of *Myxobolus cerebralis* in postfire areas of Plumas National Forest: utility of environmental DNA sampling. J Aquat Anim Health 30(2):130–143

Ríos-Castro R, Romero A, Aranguren R, Pallavicini A, Banchi E, Novoa B, Figueras A (2021) High-throughput sequencing of environmental DNA as a tool for monitoring eukaryotic communities and potential pathogens in a coastal upwelling ecosystem. Front Vet Sci 8:765606

Rudko SP, Reimink RL, Froelich K, Gordy MA, Blankespoor CL, Hanington PC (2018) Use of qPCR-based cercariometry to assess swimmer's itch in recreational lakes. EcoHealth 15:827–839

Rudko SP, Turnbull A, Reimink RL, Froelich K, Hanington PC (2019) Species-specific qPCR assays allow for high-resolution population assessment of four species avian schistosome that cause swimmer's itch in recreational lakes. Int J Parasitol Parasites Wildl 9:122–129

Ruppert KM, Kline RJ, Rahman MS (2019) Past, present, and future perspectives of environmental DNA (eDNA) metabarcoding: A systematic review in methods, monitoring, and applications of global eDNA. Glob Ecol Conserv 17:e00547

Rusch JC, Hansen H, Strand DA, Markussen T, Hytterød S, Vrålstad T (2018) Catching the fish with the worm: a case study on eDNA detection of the monogenean parasite *Gyrodactylus salaris* and two of its hosts, Atlantic salmon (*Salmo salar*) and rainbow trout (*Oncorhynchus mykiss*). Parasit Vectors 11(1):1–12

Rusch J, Strand D, Laurendz C, Andersen T, Johnsen SI, Edsman L, Vrålstad T (2022) Exploring the eDNA dynamics of the host-pathogen pair *Pacifastacus leniusculus* (Decapoda) and *Aphanomyces astaci* (Saprolegniales) under experimental conditions. NeoBiota 79:1–29

Sato MO, Rafalimanantsoa A, Ramarokoto C, Rahetilahy AM, Ravoniarimbinina P, Kawai S, Minamoto T, Sato M, Kirinoki M, Rasolofo V (2018) Usefulness of environmental DNA for detecting *Schistosoma mansoni* occurrence sites in Madagascar. Int J Infect Dis 76:130–136

Schuster CJ, Kent ML, Peterson JT, Sanders JL (2022) Multi-state occupancy model estimates probability of detection of an aquatic parasite using environmental DNA: *Pseudoloma neurophilia* in zebrafish aquaria. J Parasitol 108(6):527–538

Sengupta M, Lynggaard C, Mukaratirwa S, Vennervald B, Stensgaard A (2022) Environmental DNA in human and veterinary parasitology-Current applications and future prospects for monitoring and control. Food Waterborne Parasitol 29:e00183

Sepulveda AJ, Birch JM, Barnhart EP, Merkes CM, Yamahara KM, Marin R III, Kinsey SM, Wright PR, Schmidt C (2020a) Robotic environmental DNA biosurveillance of freshwater health. Sci Rep 10(1):14389

Sepulveda AJ, Hutchins PR, Forstchen M, Mckeefry MN, Swigris AM (2020b) The elephant in the lab (and field): Contamination in aquatic environmental DNA studies. Front Ecol Evol 8:609973

Sepulveda AJ, Hoegh A, Gage JA, Caldwell Eldridge SL, Birch JM, Stratton C, Hutchins PR, Barnhart EP (2021) Integrating environmental DNA results with diverse data sets to improve biosurveillance of river health. Front Ecol Evol 9:620715

Sepulveda AJ, Dumoulin CE, Blanchette DL, McPhedran J, Holme C, Whalen N, Hunter ME, Merkes CM, Richter CA, Neilson ME (2023) When are environmental DNA early detections of invasive species actionable? J Environ Manag 343:118216

Shea D, Bateman A, Li S, Tabata A, Schulze A, Mordecai G, Ogston L, Volpe JP, Neil Frazer L, Connors B (2020) Environmental DNA from multiple pathogens is elevated near active Atlantic salmon farms. Proc Roy Soc B 287(1937):20202010

Sidstedt M, Rådström P, Hedman J (2020) PCR inhibition in qPCR, dPCR and MPS—mechanisms and solutions. Anal Bioanal Chem 412(9):2009–2023

Sieber N, Hartikainen H, Vorburger C (2020) Validation of an eDNA-based method for the detection of wildlife pathogens in water. Dis Aquat Org 141:171–184

Sieber N, Hartikainen H, Krieg R, Zenker A, Vorburger C (2022) Parasite DNA detection in water samples enhances crayfish plague monitoring in asymptomatic invasive populations. Biol Invasions 24(1):281–297

Soper D, Raffel T, Sckrabulis J, Froelich K, McPhail B, Ostrowski M, Reimink R, Romano D, Rudko S, Hanington P (2023) A novel schistosome species hosted by *Planorbella (Helisoma) trivolvis* is the most widespread swimmer's itch-causing parasite in Michigan inland lakes. Parasitology 150(1):88–97

Stewart KA (2019) Understanding the effects of biotic and abiotic factors on sources of aquatic environmental DNA. Biodivers Conserv 28(5):983–1001

Strepparava N, Wahli T, Segner H, Petrini O (2014) Detection and quantification of *Flavobacterium psychrophilum* in water and fish tissue samples by quantitative real time PCR. BMC Microbiol 14:1–10

Strickler KM, Fremier AK, Goldberg CS (2015) Quantifying effects of UV-B, temperature, and pH on eDNA degradation in aquatic microcosms. Biol Conserv 183:85–92

Taberlet P, Coissac E, Pompanon F, Brochmann C, Willerslev E (2012) Towards next-generation biodiversity assessment using DNA metabarcoding. Mol Ecol 21(8):2045–2050

Taberlet P, Bonin A, Zinger L, Coissac E (2018) Environmental DNA: For biodiversity research and monitoring. Oxford University Press, Oxford

Takahashi M, Saccò M, Kestel JH, Nester G, Campbell MA, Van Der Heyde M, Heydenrych MJ, Juszkiewicz DJ, Nevill P, Dawkins KL (2023) Aquatic environmental DNA: A review of the macro-organismal biomonitoring revolution. Sci Total Environ 873:162322

Thalinger B, Deiner K, Harper LR, Rees HC, Blackman RC, Sint D, Traugott M, Goldberg CS, Bruce K (2021) A validation scale to determine the readiness of environmental DNA assays for routine species monitoring. eDNA 3(4):823–836

Thomas LJ, Milotic M, Vaux F, Poulin R (2022) Lurking in the water: testing eDNA metabarcoding as a tool for ecosystem-wide parasite detection. Parasitology 149(2):261–269

Thomsen PF, Willerslev E (2015) Environmental DNA—An emerging tool in conservation for monitoring past and present biodiversity. Biol Conserv 183:4–18

Thomsen PF, Kielgast J, Iversen LL, Wiuf C, Rasmussen M, Gilbert MTP, Orlando L, Willerslev E (2012) Monitoring endangered freshwater biodiversity using environmental DNA. Mol Ecol 21(11):2565–2573

Trujillo-González A (2018) Parasite threats from the ornamental fish trade. James Cook University

Trujillo-González A, Edmunds R, Becker J, Hutson K (2019) Parasite detection in the ornamental fish trade using environmental DNA. Sci Rep 9(1):5173

Trujillo-González A, Becker J, Huerlimann R, Saunders R, Hutson K (2020) Can environmental DNA be used for aquatic biosecurity in the aquarium fish trade? Biol Invasions 22:1011–1025

Trujillo-González A, Villacorta-Rath C, White NE, Furlan EM, Sykes M, Grossel G, Divi UK, Gleeson D (2021) Considerations for future environmental DNA accreditation and proficiency testing schemes. eDNA 3(6):1049–1058

Tsang HH, Domingos JA, Westaway JA, Kam MH, Huerlimann R, Bastos Gomes G (2021) Digital droplet PCR-based environmental DNA tool for monitoring *Cryptocaryon irritans* in a marine fish farm from Hong Kong. Diversity 13(8):350

Tsuji S, Takahara T, Doi H, Shibata N, Yamanaka H (2019) The detection of aquatic macroorganisms using environmental DNA analysis—A review of methods for collection, extraction, and detection. eDNA 1(2):99–108

Tydecks L, Jeschke JM, Wolf M, Singer G, Tockner K (2018) Spatial and topical imbalances in biodiversity research. PLoS One 13(7):e0199327

Ushio M, Murakami H, Masuda R, Sado T, Miya M, Sakurai S, Yamanaka H, Minamoto T, Kondoh M (2018) Quantitative monitoring of multispecies fish environmental DNA using high-throughput sequencing. MBMG 2:e23297

Valentini A, Taberlet P, Miaud C, Civade R, Herder J, Thomsen PF, Bellemain E, Besnard A, Coissac E, Boyer F (2016) Next-generation monitoring of aquatic biodiversity using environmental DNA metabarcoding. Mol Ecol 25(4):929–942

Vogelstein B, Kinzler KW (1999) Digital PCR. Proc Natl Acad Sci USA 96(16):9236–9241

Ward GM, Bennett M, Bateman K, Stentiford GD, Kerr R, Feist SW, Williams ST, Berney C, Bass D (2016) A new phylogeny and environmental DNA insight into paramyxids: an increasingly important but enigmatic clade of protistan parasites of marine invertebrates. Int J Parasitol 46(10):605–619

Ward GM, Neuhauser S, Groben R, Ciaghi S, Berney C, Romac S, Bass D (2018) Environmental sequencing fills the gap between parasitic haplosporidians and free-living giant amoebae. J Eukaryot Microbiol 65(5):574–586

Weigand H, Beermann AJ, Čiampor F, Costa FO, Csabai Z, Duarte S, Geiger MF, Grabowski M, Rimet F, Rulik B (2019) DNA barcode reference libraries for the monitoring of aquatic biota in Europe: Gap-analysis and recommendations for future work. Sci Total Environ 678:499–524

WOAH (2023) Manual of diagnostic tests for aquatic animals

Zaiko A, Greenfield P, Abbott C, von Ammon U, Bilewitch J, Bunce M, Cristescu ME, Chariton A, Dowle E, Geller J (2022) Towards reproducible metabarcoding data: Lessons from an international cross-laboratory experiment. Mol Ecol Resour 22(2):519–538

Open Access This chapter is licensed under the terms of the Creative Commons Attribution-NonCommercial-NoDerivatives 4.0 International License (http://creativecommons.org/licenses/by-nc-nd/4.0/), which permits any non-commercial use, sharing, distribution and reproduction in any medium or format, as long as you give appropriate credit to the original author(s) and the source, provide a link to the Creative Commons license and indicate if you modified the licensed material. You do not have permission under this license to share adapted material derived from this chapter or parts of it.

The images or other third party material in this chapter are included in the chapter's Creative Commons license, unless indicated otherwise in a credit line to the material. If material is not included in the chapter's Creative Commons license and your intended use is not permitted by statutory regulation or exceeds the permitted use, you will need to obtain permission directly from the copyright holder.

Ecology of Marine Mammal Parasites

15

Kristina Lehnert and Jesús S. Hernández-Orts

Abstract

Marine mammals exhibit unique adaptations for life in aquatic environments worldwide. They play crucial roles in ecosystem dynamics but face threats from human activities, leading to population declines and endangerment. Despite their significance, marine mammal parasites are often overlooked in research, though they play vital roles in host health and ecosystem function. Both protists and metazoan parasites can impact host populations, leading to decreased fitness and mortality of marine mammals. However, research on marine mammal parasites is hindered by limited opportunities for sampling free-ranging and vulnerable wildlife. While traditional morphological analyses and long-term datasets aid parasite study, molecular methods are increasingly utilised. Understanding marine mammal parasites is crucial from a One Health perspective, not only for ecosystem health but also for public health, given the potential for zoonotic disease, yet knowledge gaps persist, particularly in certain geographic areas and for certain marine mammal host and parasite species. The changing marine environment can accelerate parasitism dynamics, necessitating ongoing research. This chapter reviews marine mammal parasite diversity, life cycles and pathogenicity and highlights ecological differences between hemispheres, underscoring the importance of continued investigation in this field.

15.1 Introduction

Marine mammals comprise a diverse group of species, including cetaceans, pinnipeds, sea otters, sirenians and polar bears. They have evolved unique morphological, behavioural and ecological adaptations to live in a variety of aquatic environments worldwide (Berta et al. 2015). Marine mammals play key roles in ecosystem functioning, as they consume large amounts of biomass, modify habitats, transport nutrients and connect ocean ecosystems (Pimiento et al. 2020). Commercial hunting, fisheries interactions, pollution and human alteration of the environment have reduced their

K. Lehnert (✉)
Institute for Terrestrial and Aquatic Wildlife Research, University of Veterinary Medicine Hannover Foundation, Büsum, Germany
e-mail: Kristina.Lehnert@tiho-hannover.de

J. S. Hernández-Orts
Natural History Museum, London, UK

Institute of Parasitology, Biology Centre, Czech Academy of Sciences,
České Budějovice, Czech Republic

populations, with numerous marine mammal species listed as threatened (Avila et al. 2018).

Marine mammals host a broad variety of endo- and ectoparasites (Aznar et al. 2001a). Several groups of parasites, including protists, helminths and arthropods have been reported in marine mammals (Delyamure 1955; Aznar et al. 2001a, b; Miller et al. 2018). Some of these parasites can negatively impact host individuals and populations (Geraci and St Aubin 1987; van Bressem et al. 2009; Seguel and Gottdenker 2017) and may be involved in increased mortality (Measures 2001; Osinga et al. 2012), which is a concern for threatened marine mammal species. However, marine mammal parasites are often neglected in research, even though they are important components of biodiversity and are increasingly threatened by environmental change (Reckendorf et al. 2019; Rohner et al. 2023). Specialist parasites and those with complex heteroxenous life cycles are particularly at risk from environmental changes in their ecosystems and the decline of their hosts (Fiorenza et al. 2020; Wood et al. 2023). Larval forms of some marine mammal parasites can influence the behaviour and health of their intermediate hosts, play a key role in ecosystem functioning and are important components of animal biomass (Hudson et al. 2006; Klimpel and Palm 2011; Kent et al. 2020).

Marine mammal parasites can constitute a substantial amount of biomass, e.g. the nematode *Placentonema gigantissima*, which infects the uterus of female sperm whales (*Physeter macrocephalus*), can grow up to 9 m (Gubanov 1951; Lucius et al. 2018), and the cestode *Tetragonoporus calyptocephalus*, which infects the bile duct of sperm whales, can reach a length of up to 30 m, making it one of the longest invertebrates on earth (Delyamure and Skryabin 1968). The adaptation of marine mammal ancestors to the aquatic environment enabled the evolution of new relationships with marine invertebrates, such as crustaceans, which colonised cetaceans when insects lost their niche on the hairless and pelagic whales (Sedlak-Weinstein 1992; Pfeiffer 2009). On the other hand, some parasites of the terrestrial ancestors underwent co-evolution with their host in the new environment. Insect seal lice (order Phthiraptera), which share a long co-evolutionary history with their hosts, were able to co-evolve with the amphibious lifestyle of their pinniped hosts (Leidenberger et al. 2007; Leonardi et al. 2019; see also Chap. 6). While the numerous endoparasitic species commonly rely on trophic transmission via the food web, the relatively few ectoparasite species are directly transmitted via body contact between host individuals. Many endoparasites have complex, multi-stage life cycles including several invertebrate intermediate and vertebrate fish paratenic hosts (see Chap. 5) which can influence food web structure and ecosystem integrity. Consequently, they are important bioindicators for diet and population dynamics of their hosts (see Chap. 17) and, especially in the case of direct transmission by bodily contact, can reflect on distribution, migration, and social behaviour of their marine mammal hosts (Balbuena et al. 1995; Fraija-Fernández et al. 2017b).

Ecological studies on marine mammal parasites face major obstacles, mainly because free-ranging wildlife are difficult to access for sampling and are subject to strict legal and ethical regulations. Most studies on marine mammal parasite ecology are limited to a few mammal species that are found stranded, as by-catch in fisheries, or harvested for subsistence purposes, resulting in small sample sizes and sampling bias (Raga et al. 1997; Lehnert et al. 2014; Siebert et al. 2020). Parasite surveys of marine mammals often have to rely on opportunistic sampling due to restricted access as a result of ethical and juridical constraints (Sikes et al. 2016), and the challenging oceanic environments where they reside. Parasites, however, can be collected from marine mammals with minimal invasiveness during dedicated surveys of free-ranging wildlife (Kleinertz et al. 2014; Leonardi 2014; Hermosilla et al. 2015) and routine medical examinations during rehabilitation of animals in human care (Martins et al. 2021). Most endoparasite and tissue samples originate from stranded animals collected by dedicated stranding networks or were obtained during health monitoring or postmortem investigations (Osinga et al. 2012; IJsseldijk et al. 2019; Lakemeyer et al. 2020b; Pool et al. 2020a;

Hernández-Orts et al. 2021a; Wund et al. 2023). Long-term datasets generated from stranding data at research institutions and museum collections are valuable treasure troves for analyses on time trends in prevalence and epidemiology of parasites in marine mammals (Siebert et al. 2007; Reckendorf et al. 2019; Wood et al. 2023). Traditional morphological analyses for identification of the species are increasingly complemented by molecular methods (Hernández-Orts et al. 2015; Lehnert et al. 2015; Pool et al. 2020b) and experiments with parasites in in vitro culture are performed to better understand aspects of their biology such as their attachment biomechanics (Mladineo et al. 2023; Preuss et al. 2024).

Despite the role of marine mammals as sentinels of ecosystem health (Bossart 2011; see Chap. 20), as carriers of zoonotic diseases (see Chap. 19) and their importance from a "One Health" perspective (Tryland 2022; Waltzek et al. 2012; Godfroid 2017), knowledge on the parasite fauna of many marine mammal host species and geographic areas, as well as regarding many of the parasite groups, is still scarce, with many of the few existing studies being highly dated or published in languages that are not readily available for a broader scientific community (Reckendorf et al. 2019; Lehnert et al. 2023a). The marine environment undergoes constant and increasingly rapid changes which influence the parasite fauna, diversity and pathological impact in marine mammals (van Bressem et al. 2009; Shanebeck and Lagrue 2020; Selbach et al. 2022). In mammals from landlocked freshwater ecosystems or the Baltic Sea with brackish waters, environmental characteristics such as salinity and anoxic conditions influence prey fish abundance and trophically transmitted helminth parasite fauna (Neimanis et al. 2016; Giari et al. 2022). While there are host-specific and specialist parasites which have developed relationships with their hosts over long evolutionary timescales, there are also emerging parasites that spill over from terrestrial ecosystems or benefit from environmental change by gaining access to new host species. Despite logistical and methodological limitations, interest in ecological studies of marine mammal parasites is increasing, especially in the last two decades. In this chapter, we provide an updated review of the diversity of the major group of parasites of marine mammals. We then explore selected specific groups to highlight ecological traits of, and differences between, the Northern and Southern hemispheres.

15.2 Protists

Protists infect marine mammals worldwide; some species move between terrestrial and marine ecosystems, while others seem to permanently reside in the marine environment. Unidentified novel species have been recorded using molecular tools, but their taxonomy, ecology and life cycles remain unknown, and new marine mammal protist parasites are continually being discovered (Miller et al. 2018). *Cryptosporidium* spp. and *Giardia* spp. are ubiquitous enteric protists occurring globally in humans, domestic animals and wildlife (see Chap. 2). *Giardia* cysts and *Cryptosporidium* oocysts are transmitted by the faecal-oral route from humans and animals and dispersal in the environment can infect pinnipeds and cetaceans acting as carriers of human and domestic animal derived species (Reboredo-Fernández et al. 2014). The genotypes and assemblages of these protistan parasites are often of human origin, but evidence suggests that marine-exclusive strains of these pathogens also exist (Gaydos et al. 2008; Rengifo-Herrera et al. 2013). Infections by both *Cryptosporidium* spp. and *Giardia* spp. in multiple aquatic mammal species have been reported from different geographic regions (Reboredo-Fernández et al. 2015; Borges et al. 2017; Grilo et al. 2018; also see Chap. 2). Although little is known about their effect on marine mammal health, they may cause direct disease or reduction of host fitness (Grilo et al. 2018). There are public health concerns due to zoonotic transmissions related to recreational water use and food safety (Iqbal et al. 2015; Moratal et al. 2020).

Toxoplasma gondii oocysts are excreted from felids, can survive in moist environments for months and are environmentally robust to a range of temperatures, salinities and disinfectants

(Shapiro et al. 2019; Dubey et al. 2020; also see Chap. 2). Rainfall and wastewater, as well as agricultural and urban runoff, have caused dispersal of *T. gondii* to marine mammals in coastal waters worldwide for more than 50 years (Fayer et al. 2004; van Bressem et al. 2009), emphasising the role of marine mammals as indicators of environmental contamination (Bossart 2011). *Toxoplasma gondii* has caused pathogenicity and mortality in free-ranging marine mammals (Mazzariol et al. 2021; Díaz-Delgado et al. 2020), with higher pathogenicity in immune-suppressed and debilitated animals indicated (Mazzariol et al. 2012). Dual infections of *T. gondii* and *Sarcocystis neurona* were more frequently associated with mortality and protist encephalitis than single infections, indicating an effect of polyparasitism on disease severity (Gibson et al. 2011), though diagnosis and interpretation are still limited (van de Velde et al. 2016). Prevention of faecal contamination recurrently threatening human and wildlife health along densely populated coasts is important in the face of emerging virulent *T. gondii* strains in wildlife, with further implications for food safety (Moratal et al. 2020; Miller et al. 2023). Aside from easing egress for marine mammals by melting frozen waterways, increasing global temperatures also allow pathogens to recruit new hosts in previously unexposed Arctic and Antarctic waters (Dubey et al. 2020): *Toxoplasma gondii* has been found in western Arctic beluga (*Delphinapterus leucas*) whale meat, a common food of Inuit people. Antibodies against *T. gondii* were also found in several pinniped species around the Antarctic Peninsula (Rengifo-Herrera et al. 2012). Higher prevalence of antibodies against *T. gondii* in polar bears (*Ursus maritimus*), ringed seals (*Pusa hispida*) and bearded seals (*Erignathus barbatus*) around Svalbard in the Norwegian high Arctic were linked to increasing warm-water influx and point to a marine transmission pathway or increased anthropogenic traffic in the area (Jensen et al. 2010). *Toxoplasma gondii* and potentially *Sarcocystis* sp. and *Neospora* sp. infecting marine mammals of the Arctic can be a threat to humans in the region (Reiling et al. 2019), where seal and whale meat are crucial for subsistence. Humans and their animals (Tryland et al. 2014) can become infected by consuming raw or undercooked infected meat (Dubey et al. 2020; see Chap. 19).

15.3 Digenea

Marine mammals are final hosts of about 110 digenetic trematode species, belonging to 49 genera in 18 families. Most digeneans of marine mammals parasitise the digestive tracts of cetaceans, pinnipeds and, to a lesser extent, sirenians and sea otters (Fig. 15.1). To date, there are no records of digeneans in polar bears. Nearly half of all digenean species reported in marine mammals belong to members of the family Brachycladiidae, including species of the genera *Brachycladium* (9 spp.), *Campula* (2 spp.), *Cetitrema* (2 spp.), *Odhneriella* (3 spp.), *Oschmarinella* (5 spp.) (Figs. 15.1a and 15.2), *Orthosplanchnus* (7 spp.), *Synthesium* (8 spp.) and *Zalophotrema* (2 spp.). These flukes occur in the intestines, gallbladder, hepatic and pancreatic ducts of cetaceans, and to a lesser degree in pinnipeds (Dailey 1975; Raga et al. 2009; Andersen-Ranberg et al. 2018; Fraija-Fernández et al. 2016). Brachycladiid species of the genera *Nasitrema* (10 spp.) and *Hunterotrema* (2 spp.) occur in, respectively, the air sinuses and lungs of cetaceans (e.g. Dailey 1971; Ebert and Valentere 2013). *Orthosplanchnus fraterculus* is the only brachycladiid reported from the gallbladder of sea otters (*Enhydra lutris*) (Rausch 1953). The first life cycle of a brachycladiid species was recently elucidated by Kremnev et al. (2020) for *Orthosplanchnus arcticus* and includes a caenogastropoda as the first intermediate host, benthic bivalves as the second intermediate hosts and seals as the final hosts. However, information derived from the feeding habits of several cetaceans and pinnipeds suggests that cephalopods and fishes might serve as second intermediate or transport hosts for other *Orthosplanchnus* species. Brachycladiids might cause significant pathological alterations and lesions of the liver and bile ducts of marine mammals (Siebert et al. 2001; Andersen-Ranberg et al. 2018; Nakagun et al. 2018).

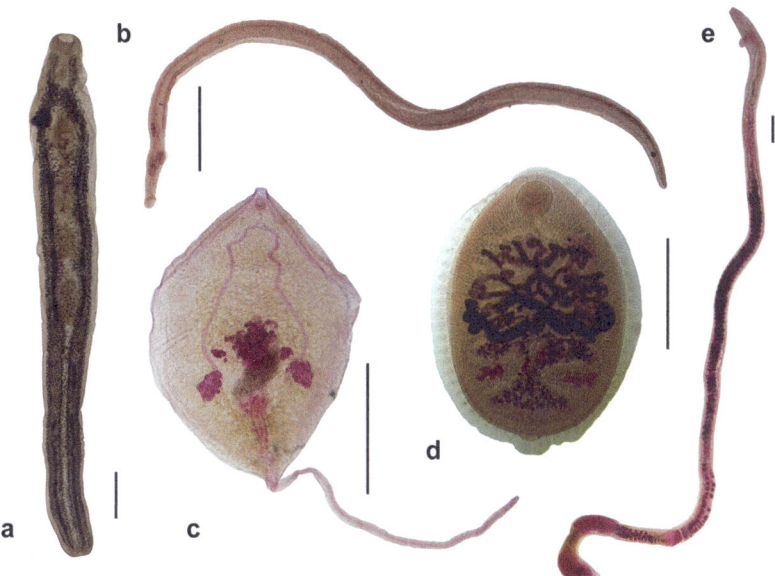

Fig. 15.1 Photomicrographs of digenetic trematodes from marine mammals (**a**) *Oschmarinella rochebruni* (Brachycladiidae) from the hepatic duct of common dolphin (*Delphinus delphis*), Argentina (Río Negro), ventral view; (**b**) *Schistosoma mansoni* (Schistosomatidae) from the mesenteric veins of California sea lion (*Zalophus californianus*) (Natural History Museum, London (NHMUK): 1965.1.20.11–100), Egypt (Giza), lateral view; (**c**) *Opisthotrema australe* (Opisthotrematidae) from the ear and Eustachian tube of a dugong (*Dugong dugon*) (Paratype, NHMUK 1979.5.24.23–25), Australia (North Queensland), ventral view; (**d**) *Pulmonicola pulmonalis* (syn. *Cochleotrema indicum*) (Opisthotrematidae) from the lungs of a dugong (*Dugong dugon*) (NHMUK 1979.5.24.26–27), Australia (North Queensland), ventral view; (**e**) *Labicola elongata* (Labicolidae) from the upper lip of a dugong (*Dugong dugon*) (paratype, NHMUK 1979.1.31.1–2), Australia (North Queensland), lateral view. Scale bars: (**a–c**) 1 mm; (**d**) 500 μm; (**e**) 2 mm

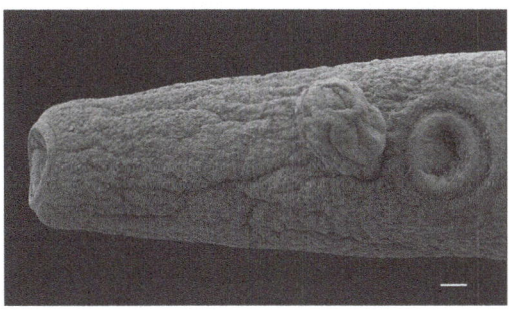

Fig. 15.2 Scanning electron micrographs of the anterior end of *Oschmarinella rochebruni* (Brachycladiidae) from the hepatic duct of common dolphin (*Delphinus delphis*), Argentina (Río Negro), ventral view. Scale bar: 100 μm

Pinnipeds harbour a diverse fauna of digeneans belonging to the family Opisthorchiidae, including species of six genera: *Apophallus* (2 spp.), *Cryptocotyle* (4 spp.), *Opisthorchis* (2 spp.), *Metorchis* (1 sp.), *Pricetrema* (1 sp.), and *Pseudamphistomum* (1 sp.). They occur in the gallbladder, bile ducts and intestines of seals from the Northern Hemisphere. Two opisthorchiid species belonging to *Amphimerus* and *Delphinicola* infect the bile ducts of toothed whales (Odontoceti). Currently, the only completely-elucidated life cycle for a marine mammal-infecting opisthorchiid is that of *Metorchis bilis* (syn. *Metorchis albidus*), which involves bithyniid snails as the first intermediate hosts (Serbina 2016), roach (*Rutilus rutilus*), a freshwater fish, as the second intermediate hosts (Näreaho et al. 2017), and a wide range of terrestrial and aquatic carnivores (including seals) as the final hosts (Delyamure 1955; Sherrard-Smith et al. 2016). In the Baltic Sea, metacercariae of *Pseudamphistomum truncatum* have been reported in cyprinid fishes (Näreaho et al. 2017), while adults occur in the hepatobiliary system of seals (Neimanis et al. 2016). Opisthorchiids are recognised pathogens of seals and are responsible for damage and obstruction of hepatic ducts and vessels, necrosis, liver failure, haemorrhages and extensive degeneration of hepatic lobules, pancreas and pancreatic ducts (Kuiken et al. 2006; Heckmann et al. 2014; Neimanis et al. 2016).

Pseudamphistomum truncatum is an emerging parasite with zoonotic potential in the Baltic Sea ecosystem (Neimanis et al. 2016). This trematode parasitises the liver of grey seals (*Halichoerus grypus*) along the Swedish Baltic Sea coast, with increasing prevalences recorded over time. Hepatobiliary infections with *P. truncatum* can cause chronic, severe cholangiohepatitis in the seal host. If consumed in undercooked fish, the larvae can also infect and cause pathology in humans (see Fig. 5.6 in Chap. 5). The emergence of this parasite in grey seals along the Swedish coast is assumed to be related to recent freshwater influx and subsequent distribution and abundance of freshwater organisms serving as intermediate or paratenic hosts in the area, thereby facilitating the dispersion among seal final hosts.

Six species of *Ogmogaster* (Notocotylidae) have been found in the intestines of marine mammals. Baleen whales are the typical hosts for these flukes (Dailey et al. 2000), although *Ogmogaster heptalineata* and *O. antarctica* have been reported in pinnipeds from the southeast Pacific and Antarctic, respectively (Rausch and Fay 1966; Ebmer et al. 2020). The life cycle of these worms is not known, but some studies suggest that prosobranch snails serve as intermediate hosts and zooplanktonic crustaceans (krill) as transport hosts for notocotylids associated with baleen whales. Small toothed whales harbour *Braunina cordiformis* (Brauninidae) in the stomach and the duodenal ampulla (Fraija-Fernández et al. 2015), where it causes focal lesions and inflammation of the mucosa and gastric wall, extensive fibrosis and gastritis (Hrabar et al. 2017; Lombardini et al. 2019). Intermediate hosts in the life cycles of these worms are still unknown.

Trematodes of the family Heterophyidae from marine mammals consist of species from six genera: *Ascocotyle* (4 spp.), *Galactosomum* (1 sp.), *Heterophyes* (2 spp.), *Heterophyopsis* (1 sp.), *Pholeter* (1 sp.) and *Phocitrema* (1 sp.). *Pholeter gastrophilus* lives encysted within submucosal fibrotic nodules in the stomach wall of small odontocetes (Lehnert et al. 2005), while other heterophyids inhabit the intestines of pinnipeds and (rarely) dolphins and sea otters (Dailey et al. 2004; Fraija-Fernández et al. 2015; Kuzmina et al. 2018). The complete life cycle of most marine mammal-infecting heterophyids is unknown. For *Ascocotyle longa*, a cochliopid snail (*Heleobia australis*) serves as the first intermediate host, mullets (*Mugil liza*) as the second intermediate host, and dolphins and sea lions as final hosts (Simões et al. 2010; Pereira et al. 2013; Ebert et al. 2021). Adult forms of *Ascocotyle patagoniensis* have been reported in South American sea lions (*Otaria flavescens*) and metacercariae of this species in silversides (*Odontesthes* spp.) from Patagonia, Argentina (Hernández-Orts et al. 2019a). The life cycle of *Ascocotyle patagoniensis* appears to be restricted to coastal waters, as metacercariae of this species were not observed in marine fishes from the Patagonian shelf in Argentina (Hernández-Orts et al. 2012). *Pholeter gastrophilus* infections cause moderate to extensive lesions of the submucosa, fibrosis and granulomatous gastritis of the stomach (Siebert et al. 2001; Lehnert et al. 2005; Jaber et al. 2006).

Philophthalmus zalophi (Philophthalmidae) infects the ocular cavity of juvenile Galapagos sea lions (*Zalophus wollebaeki*). The life cycle is unknown but marine snails are thought to act as first intermediate hosts. Young sea lions are likely to become infected by ingesting the cercariae or by direct contact of the eye with water bearing infective intermediate stages. Conjunctival hyperaemia, conjunctivitis, corneal ulcers, exophthalmia, globe rupture and phthisis bulbi are associated with *Philophthalmus zalophi* infections of the eyes (Phillips et al. 2018), which may have long-term consequences for the survival of juvenile sea lions (Meise and Garcia-Parra 2015). Pinnipeds and sea otters are infected with members of the Microphallidae belonging to the genera *Maritrema* (2 spp.), *Microphallus* (2 spp.) and *Plenosoma* (1 sp.). These marine mammals likely become infected when they feed on benthic crustaceans that are known intermediate hosts in the life cycles of other microphallid trematodes. The trematode fauna of pinnipeds also includes species belonging to the families Echinochasmidae (*Stephanoprora*; 3 spp.), Echinostomatidae (*Echinostoma*; 1 sp.), Schistosomatidae (*Schistosoma*; 2 spp.) (Fig. 15.1b) and Troglotrematidae (*Nanophyetus*; 1 sp.). Little is known about the life cycles and pathogenicity of most of these trematodes, and they are considered accidental parasites of pinnipeds. Salmonids are the second intermediate host in the life cycle of *Nanophyetus salmincola*. This trematode species is the vector of the endosymbiotic bacteria *Neorickettsia* spp. that causes salmon poisoning disease in bears and dogs (Greiman et al. 2016). Californian sea lions (*Zalophus californianus*) and Northern fur seals (*Callorhinus ursinus*) are final hosts for *Nanophyetus salmincola* in the northern Pacific, but it is not known whether these seals are susceptible to salmon poisoning disease.

Sirenians harbour a diverse trematode fauna represented by seven families and 15 genera found almost exclusively in these aquatic mammals. The family Opisthotrematidae is widely distributed in sirenians, with species of *Folitrema* (1 sp.), *Lankatrema* (4 spp.), *Lankatrematoides* (1 sp.) and *Moniligerum* (1 sp.) occurring in the gastrointestinal tract, liver and pancreas. *Lankatrematoides gardneri* is associated with ulcerative and fibrinosuppurative inflammation of the pancreatic duct (Woolford et al. 2015). Other opisthotrematids belonging to *Opisthotrema* (2 spp.) (Fig. 15.1c) and *Pulmonicola* (2 spp.) (Fig. 15.1d) infect the auditory system and the lungs, but the pathogenicity of these trematodes is unknown. Trematodes belonging to the families Cladorchiidae (*Chiorchis*; 1 sp.), Nudacotylidae (*Nudacotyle*; 1 sp.) and Rhabdiopoeidae (monotypic genera *Faredifex*, *Haerator*, *Rhabdiopoeus* and *Taprobanella*) occur in different sites of the small and large intestine. *Labicola elongata* (Labicolidae) is known from the upper lip (Fig. 15.1e) and *Cardicola dhangali* (Aporocotylidae) the circulatory system of dugong (*Dugong dugon*) from Australia. The life cycles of any sirenian trematodes have not been elucidated and further research is needed to clarify their intermediate hosts and transmission strategies.

15.4 Cestoda

Nearly 64 species from 15 genera and two families of tapeworms (Cestoda) parasitise the intestines of marine mammals as adults. Most cestode species associated with pinnipeds and cetaceans belong to the Diphyllobothriidae. Recent molecular phylogenetic studies have shown that the genus *Diphyllobothrium sensu stricto* (s.s.) includes eight species that occur worldwide in toothed whales and more rarely in baleen whales, while about 20 *Diphyllobothrium* species that parasitise pinnipeds, mainly in polar regions, are considered *incertae sedis* (Waeschenbach et al. 2017; Hernández-Orts et al. 2021b). Otariids serve as hosts for the Pacific broad tapeworm (*Adenocephalus pacificus*) (Fig. 15.3a, b), which appears to be widely distributed in both hemispheres except in the North Atlantic, where sea lions and fur seals are absent. Recently, plerocercoids and immature forms of *A. pacificus* have been found in the intestines of common dolphins (*Delphinus delphis*) from the southwest Atlantic, suggesting that cetaceans are "dead-end" hosts in the life cycle of this tapeworm (Hernández-Orts et al. 2021c). *Diphyllobothrium stemmacephalum* has been reported from harbour porpoises (*Phocoena phocoena*) and *Diphyllobothrium* spp. from grey and harbour seals in the North and Baltic Seas, as well as from Norwegian and Icelandic waters (Borgsteede et al. 1991; Strauss et al. 1991; Siebert et al. 2001, 2006) (Fig. 15.4). Diphyllobothriids of the genera *Baylisia* (2 spp.), *Baylisiella* (1 sp.), *Flexobothrium* (1 sp.) and *Glandicephalus* (2 spp.) (Fig. 15.3c, d) occur in Antarctic seals, and the sole species of *Pyramicocephalus*, *P. phocarum*, is only known from seals in the Arctic and subarctic (Kuchta and Scholz 2017). *Ligula intestinalis* and *Schistocephalus solidus* have also been reported from seals in northeastern Europe (Nyman et al. 2021). These tapeworms typically occur in fish-eating birds (Scholz et al. 2019), but adult forms

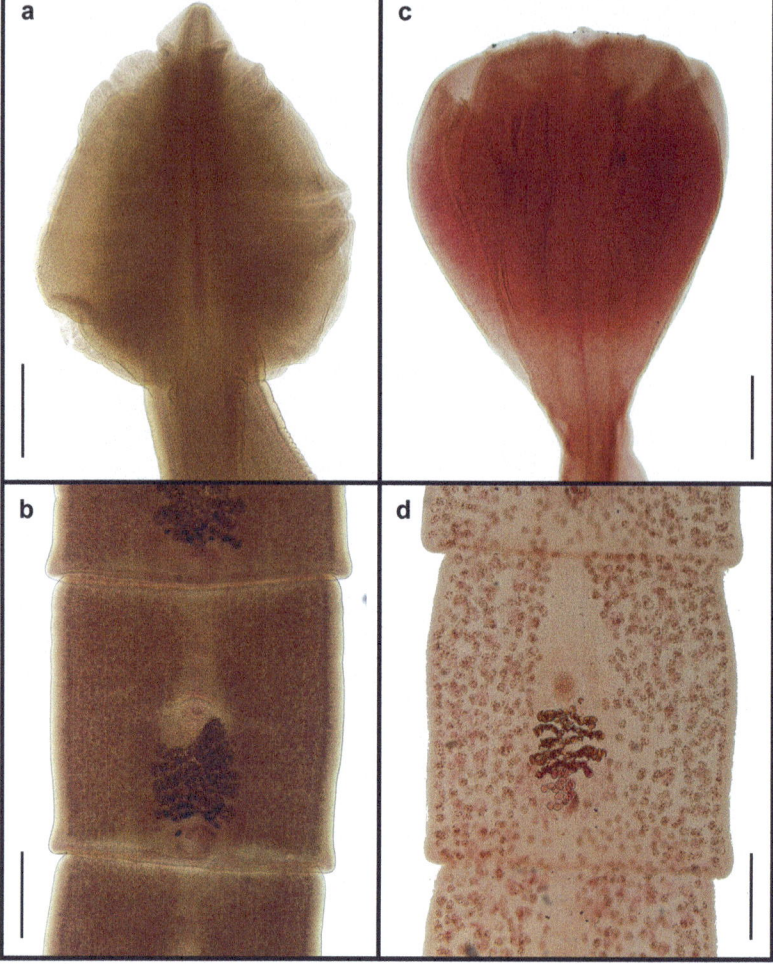

Fig. 15.3 Photomicrographs of diphyllobothriidean cestodes from marine mammals. (**a, b**) *Adenocephalus pacificus* from the intestine of northern fur seal (*Callorhinus ursinus*) (NHMUK 2017.5.11.1–6), USA (Alaska); (**a**) Scolex, lateral view; (**b**) gravid proglottid, ventral view; (**c, d**) *Glandicephalus perfoliatus* from the intestine Weddell seal (*Leptonychotes weddellii*) (NHMUK 1995.8.21.19–23), Antarctica (Weddell Sea); scolex (**c**) and gravid proglottid (**d**). Scale bars: 500 µm

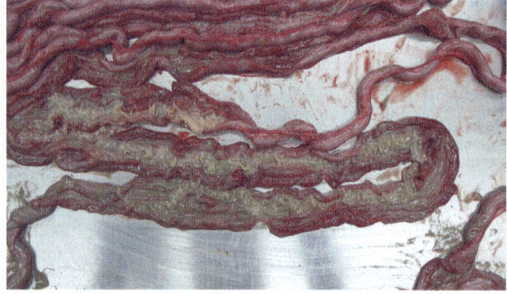

Fig. 15.4 *Diphyllobothrium* sp. (Diphyllobothriidea) infecting the intestine of a harbour porpoise (*Phocoena phocoena*), Germany. ©ITAW/TiHo

of *S. solidus* are known to occasionally occur in ringed seals in the Baltic Sea (Chubb et al. 1995), whereas *L. intestinalis* has recently been detected in landlocked ringed seal populations (Nyman et al. 2021; Fraija-Fernández et al. 2021). Gravid forms of *Dibothriocephalus latus* (syn. *Diphyllobothrium latum*) were collected from captive polar bears, although these are most likely accidental infections. *Plicobothrium globicephalae* is known from pilot whales (*Globicephala* spp.) and rarely from other dolphins. Sperm whales host *Tetragonoporus calyptocephalus*, the only polygonoporal diphyllobothriid recognised today. Diphyllobothriids are often considered non-pathogenic and are rarely associated with clinically severe illness. Massive infections may result in enteritis or intestinal obstruction. Large nodules have been reported at the site of attachment of the scolex of *Baylisiella tecta* in the rectal wall of southern elephant seals (*Mirounga leonina*).

No complete life cycle is known for these marine mammal tapeworms, but available data suggest that marine copepods serve as the first intermediate host and a wide range of fishes as the second intermediate or paratenic host. Plerocercoids of the Pacific broad tapeworm (*A. pacificus*) have been reported encysted in the body cavity and on the surface of the internal organs of 25 marine fish species off the Pacific and Atlantic coasts of South America (Kuchta et al. 2015; Cantatore et al. 2023; Mondragón-Martínez et al. 2024). Identification of plerocercoids in marine fishes is problematic due to their morphological similarity, and reliable species identification is only possible using molecular methods, particularly sequences for nuclear RNA (28S) and mitochondrial (*cox*1) genes (Hernández-Orts et al. 2015). At least 11 diphyllobothriid species described from marine mammals have been reported infecting humans (Scholz and Kuchta 2016); however, species identification of several records needs to be verified. *Adenocephalus pacificus* is the most common agent of diphyllobothriosis in humans in South America, and some human infections by this tapeworm have also been reported in Australia and Spain (Hernández-Orts et al. 2021c). Humans become infected with these tapeworms by eating raw or poorly cooked marine fish infected with plerocercoids (see Chap. 19). In recent years, this zoonotic infection has become an emerging foodborne zoonosis of global significance.

Species of five genera of cestodes belonging to the family Tetrabothriidae have been reported mainly from cetaceans and occasionally from pinnipeds. Tetrabothriid species from cetaceans occur worldwide and belong to the genera *Priapocephalus* (3 spp.), *Strobilocephalus* (1 sp.), *Tetrabothrius* (10 sp.) and *Trigonocotyle* (5 spp.) (Fig. 15.5a). Only *Anophryocephalus* (7 spp.) (Fig. 15.5b) have speciated in Arctic and subarctic pinnipeds, although an unidentified species of this genus was reported in the Southern Hemisphere from South American fur seals (*Arctocephalus australis*) off Argentina and Uruguay. The tetrabothriids remains one of the most poorly known groups of marine mammal parasites. Very little is known about their host associations, pathogenicity and phylogenetic relationships. No complete life cycle is known for any species, although invertebrates (crusta-

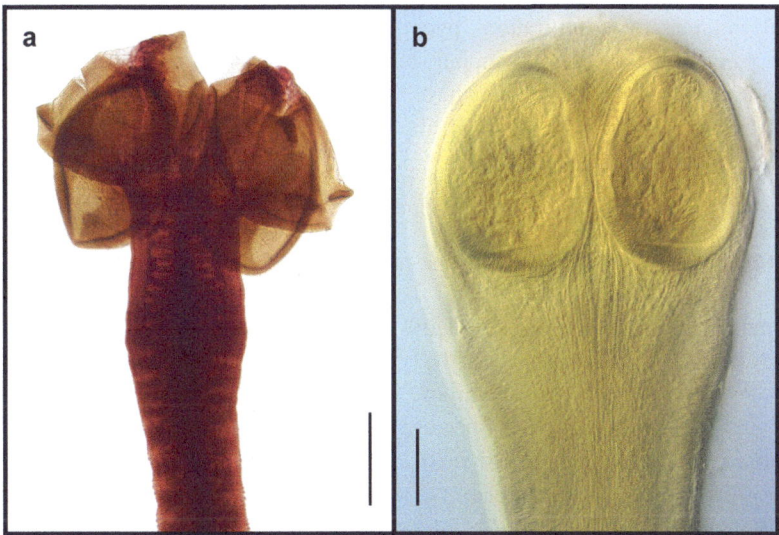

Fig. 15.5 Photomicrographs of scoleces of tetrabothriid cestodes. (**a**) *Trigonocotyle prudhoei* from the intestine of rough-toothed dolphin (*Steno bredanensis*) (Cotype, NHMUK 1956.5.16.65–71), Falkland Islands, ventral view; (**b**) *Anophryocephalus anophrys* from the small intestine of ringed seal (*Pusa hispida*) (Type, NHMUK 1922.5.3.1–6), Norway (Svalbard), ventral view. Scale bars: (**a**) 500 μm; (**b**) 100 μm

ceans or cephalopods) and fish have been suggested as first and second intermediate/paratenic hosts, respectively (Mariaux et al. 2017).

Cestode larvae of the family Phyllobothriidae have been routinely reported parasitising cetaceans and pinnipeds worldwide. Recent molecular studies have shown that *Phyllobothrium delphini* and *Monorygma grimaldii* are actually species of *Clistobothrium* (see Caira et al. 2020). Merocercoids of *C. delphini* and *C. grimaldii* occur encapsulated in the blubber of cetaceans or in the liver, mesenteries and anal canal of cetaceans and pinnipeds, respectively (Agustí et al. 2005; Lehnert et al. 2014, 2017; IJsseldijk et al. 2018; Klotz et al. 2018; Hernández-Orts et al. 2021a). Plerocercoids of an unidentified *Clistobothrium* species occur free in the lumen of the intestine, bile ducts, anal crypts and, rarely, in the lumen of the stomach (Aznar et al. 2007; Caira et al. 2020). Other unidentified phyllobothriid plerocercoids have been found in the digestive tract of cetaceans (Aznar et al. 2007; Hernández-Orts et al. 2021a), for which sequence data are needed for their identification. Adults of *Clistobothrium* spp. occur in large pelagic sharks, and marine mammals serve as intermediate hosts. Merocercoid infections in marine mammals are not considered a serious disease problem, although focal inflammatory infiltration, pyogranulomatous panniculitis and severe suppurative response associated with *C. grimaldii* have been reported (Lehnert et al. 2014; Klotz et al. 2018).

15.5 Nematoda

Marine mammals harbour a diverse fauna of roundworms (Nematoda), including about 100 species from 19 genera across 13 families. Cetaceans are infected with adults of species of the families Anisakidae (whaleworms), Pseudaliidae (30 spp.) and Tetrameridae (14 spp.). Common nematodes of pinnipeds belong to the families Ancylostomatidae (hookworms), Anisakidae (sealworms), Crenosomatidae (lungworms; 1 sp.), Filaroididae (lungworms; 9 spp.) and Onchocercidae (heartworms; 2 pp.). Roundworms of the families Ascarididae (1 sp.) and Heterocheilidae (1 sp.) occur in the digestive tract of sirenians, while members of the Anisakidae (2 spp.) and Capillariidae (1 sp.) have been reported in sea otters. Additionally, larval forms of *Trichinella* sp. (Trichinellidae) have been reported in cetaceans, pinnipeds and polar bears.

15.5.1 Hookworms

Four named and three indeterminate taxa of pinniped hookworms, all belonging to the genus *Uncinaria* (Ancylostomatidae), have been reported in otariids and (rarely) phocids. Seal hookworms are soil-transmitted nematodes with a land-dependent life cycle similar to that of hookworms of terrestrial carnivores. Adult worms mature and reproduce in the intestines of otariid pups. Eggs are shed onto the rockery surface within pup faeces. Free-living third-stage larvae hatch from the eggs and develop on the soil where they can survive for several months. These larvae infect seals of all ages by boring into the skin of the flippers, but also orally, and migrate through the circulatory system into the ventral blubber and mammary tissue. Pups are infected by transmammary transmission of the larvae in the immediate postnatal period.

Hookworm infections cause chronic anaemia, haemorrhagic enteritis, intestinal perforation, bacteraemia, peritonitis, diarrhoea and retarded growth in newborns of sea lions and fur seals. These nematodes are a major cause of pup mortality in several otariid species. For instance, up to 70% of the natural mortality rate of California sea lion (*Zalophus californianus*) pups in some rookeries in southern California is accounted for by infection with *Uncinaria lyonsi* (Spraker et al. 2007). Hookworms therefore potentially represent an important factor regulating the population of some otariid species. Evidence of possible density-dependent effects between parasite-induced mortality and population size exists for hookworms infecting Northern fur seals and California sea lions in the north Pacific (Fowler 1990; Spraker et al. 2007). However, it is not regarded as conclusive and additional surveys are needed to establish whether parasite-mediated population regulation actually occurs.

The geographic distribution of seal hookworms is still largely unknown for most species. Available data suggest a high degree of host-specificity and a wide geographical range for some hookworm species (Seguel and Gottdenker 2017). *Uncinaria lucasi* infects Steller sea lions (*Eumetopias jubatus*) and Northern fur seals from the North Pacific, as well as California sea lions from California (also see Chap. 9). *Uncinaria hamiltoni* has been recorded from the South American sea lion (*Otaria flavescens*) and the South American fur seal (*Arctocephalus australis*) from South America. The latter hookworm species has recently been reported in pups of the Mediterranean monk seal (*Monachus monachus*) from Greece (Komnenou et al. 2021), although this record needs to be confirmed. Finally, *Uncinaria sanguinis* occurs in Australian sea lions (*Neophoca cinerea*) from southern Australia. One species-level genetic lineage of *Uncinaria* has been reported from the South American fur seal, the Australian fur seal (*Arctocephalus pusillus*) and the New Zealand fur seal (*Arctocephalus forsteri*) along the Pacific coast of Peru and Oceania (González et al. 2018). Molecular data also revealed two genetically distinct hookworm lineages in phocids, one infecting southern elephant seals from Macquarie Island in the southwestern Pacific and other from Mediterranean monk seals from Greece. The diversity and geographical distribution of hookworms infecting otariids and phocids in Antarctic and subantarctic regions are still unknown.

15.5.2 Anisakids

Anisakid nematodes are common gastric parasites of marine mammals (also see Chap. 9). Those belonging to the genera *Anisakis* (8 spp.), *Pseudoterranova* (2 spp.) and *Skrjabinisakis* (4 spp.) infect cetaceans as final hosts, and those of the genera *Contracaecum* (10 spp.) (Fig. 15.6a, b), *Phocanema* (5 spp.) (Fig. 15.6c, d) and *Phocascaris* (3 spp.) mainly infect pinnipeds as

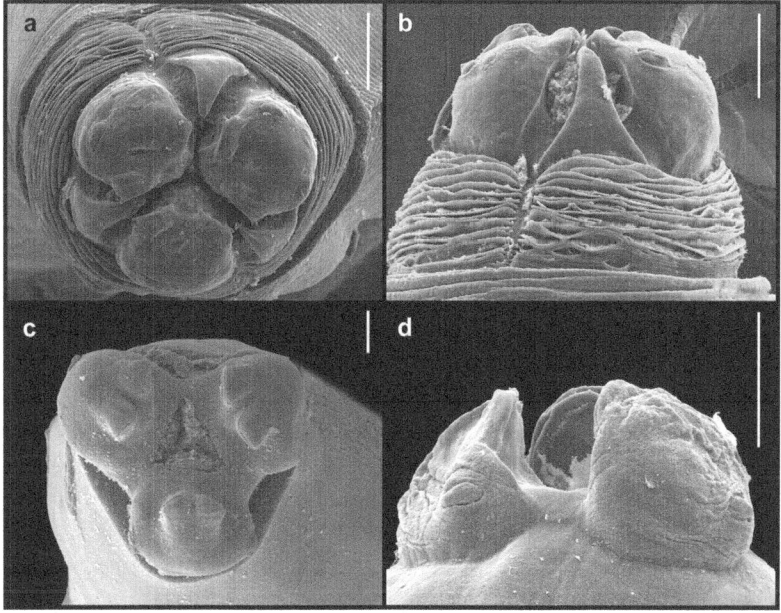

Fig. 15.6 Scanning electron micrographs of anisakid nematodes from pinnipeds. (**a, b**) Adult male of *Contracaecum ogmorhini* from the stomach of South American sea lion (*Otaria flavescens*), Argentina (Patagonia). (**a**) Anterior end, apical view; (**b**) Anterior end, lateral view. (**c, d**) Adult male of *Phocanema cattani* from the stomach of South American sea lion (*Otaria flavescens*), Argentina (Patagonia); (**c**) anterior end, apical view; (**d**) anterior end, lateral view. Scale bars: (**a, b**) 50 μm; (**c**) 100 μm; (**d**) 30 μm

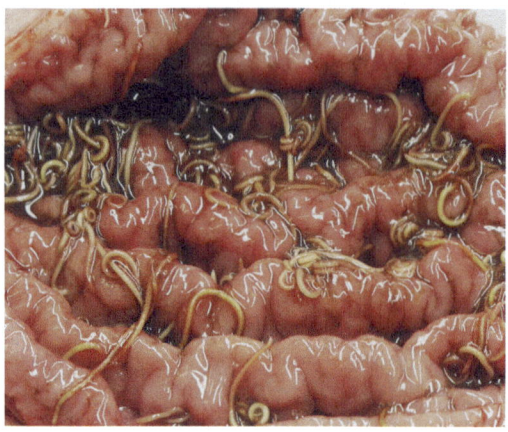

Fig. 15.7 Anisakid nematodes from the gastric mucosa of a harbour seal (*Phoca vitulina*). ©ITAW/TiHo

final hosts, (Mattiucci and Nascetti 2008; Takano and Sata 2022) (Fig. 15.7). In seals, dolphins and porpoises, gastric infections by anisakids can cause ulcerations and various degrees of gastritis in the mucosa of the stomach (Siebert et al. 2001, Lehnert et al. 2007; Hrabar et al. 2021; Pons-Bordas et al. 2020; Katahira et al. 2021). In contrast to many other marine mammal parasites, the life cycle of anisakid nematodes is relatively well understood (Smith and Wootten 1978; Nagasawa 1990; Klimpel et al. 2004; also see Chap. 5). Third-stage larvae hatch from eggs excreted with faeces by their marine mammal hosts. The larvae are ingested by invertebrates, such as euphausiids (krill) (Smith 1983) and copepods (Klimpel et al. 2004) as intermediate hosts. Infected crustaceans can be ingested by fish or cephalopods, which serve as paratenic (transport) hosts. Larvae of *Anisakis simplex* sensu lato (s.l.) infect the body cavity, musculature and various organs, *Phocanema* spp. (syn. *Pseudoterranova* spp.) occur in the muscle and *Contracaecum osculatum* (s.l.) predominantly infect the liver, body cavity, mesenteries and pyloric caeca of fish (McClelland 2002; Levsen and Berland 2012; Buchmann and Mehrdana 2016). Anisakid larvae accumulate in piscivorous fish (Hochberg and Hamer 2010) or cephalopods (Gutiérrez et al. 2023) and are transmitted via the food chain to marine mammal final hosts where they mature and reproduce in the stomach (McClelland 2002; Mattiucci and Nascetti 2008; Aibinu et al. 2019).

The taxonomy of *Anisakis* is intricate and has undergone several changes in recent years (discussed in detail in Chap. 9). The genus comprises a complex of morphologically and genetically highly similar sibling species (Mattiucci and Nascetti 2008; Takano and Sata 2022) and additional species, which were often only recognised as independent species by genetic studies (Mattiucci et al. 2004, 2009, 2014; Nadler et al. 2005; Mattiucci and Nascetti, 2006). The taxonomic situation of the genera *Phocanema* and *Contracaecum* is equally complex, due to conserved morphologies between sibling species. These sibling species frequently inhabit distinct geographical ranges and infect a wide range of host species (Paggi et al. 1991; Nascetti et al. 1993; Lymbery and Cheah 2007; Mattiucci and Nascetti 2008). Anisakid nematodes can be difficult, even impossible to differentiate morphologically, especially when sibling species or larval stages are targeted (Mattiucci and Nascetti 2007; Zhu et al. 2007; Mattiucci et al. 2017). Anisakid nematodes in harbour porpoises from the North and Baltic Seas as well as the North Atlantic have previously been identified morphologically as *A. simplex* sensu stricto (s.s.) or *A. simplex* sensu lato (s.l.) (Siebert et al. 2001; Lehnert et al. 2014), and subsequently using a restriction fragment length polymorphism (RFLP) molecular sequencing method (Lakemeyer et al. 2020a). Analyses of partial or complete sequences of the internal transcribed spacer cluster (ITS1-5.8S-ITS2) region of ribosomal DNA (Zhu et al. 1998; Abollo et al. 2003) as well as the mitochondrial cytochrome *c* oxidase II (*cox*2) gene (Cipriani et al. 2022) by PCR amplification have proven useful to complement morphology for an unambiguous discrimination between anisakid nematode species (Mattiucci et al. 2018). *Phocanema decipiens* s.l. may infect cetaceans like harbour porpoises because they are sympatric hosts with seals in the eastern North Sea (Herreras et al. 1997), but rarely becomes sexually mature in cetaceans (Lick 1991).

The increased presence of anisakid larvae in fish and a deterioration in fish condition, particularly in commercially important cod (*Gadus morhua*), has been linked to the recovery and increase of seal populations in the Baltic Sea (Jensen and Idås 1992; Hauksson 2011; Buchmann and Mehrdana 2016). Comprehensive health assessments of fish are required to assess the viability of larval stages and their dispersal in the aquatic environment, and environmental and anthropogenic factors influencing fish immunity such as low salinity, increasing temperatures, prey availability in anoxic areas and pollutant exposure (such as from ammunition dump sites) need to be considered (Baršienė et al. 2014; Lang et al. 2017, 2018). Fluctuations in definite pinniped host densities did not influence the intestinal helminth abundance in a study on three acanthocephalan species in Baltic Sea ringed seals (Valtonen et al. 2004). Larval fate in multi-stage life cycles needs to be investigated, as developmental stages may be lost by predation (Johnson et al. 2010; Orlofske et al. 2015).

Anisakidosis is a frequent foodborne zoonosis worldwide (Thompson 2023). Previously more prevalent in Asia and North and South America, where the consumption of raw fish dishes is an important component of traditional cuisine, e.g. sushi, sashimi, ceviche, etc., it has also become an emerging infectious disease in Europe with the rise in popularity of such dishes (Bao et al. 2017, 2019; Rahmati et al. 2020; also see Chap. 19). Anisakid species pose a zoonotic threat as infective third-stage larvae can infect humans when ingested with infected raw or undercooked fish and cause severe pathology (Mattiucci et al. 2013; Baptista-Fernandes et al. 2017; Muñoz-Caro et al. 2022). Varying population size and abundance of marine mammals in the study areas, as well as complexity and functionality of life cycles, have to be considered to evaluate the role of porpoises and seals in the dispersal of anisakid nematodes (Valtonen et al. 2004), especially in concert with environmental change (Mastick et al. 2024).

15.5.3 Lungworms

In marine mammals, nematodes infecting the respiratory tract are among the most pathogenic, causing severe lesions and considerable mortality, e.g. in harbour seals and harbour porpoises in the North and Baltic Sea (Siebert et al. 2001, 2007). Marine mammals rely on unimpeded respiration for efficient oxygen transport during diving and foraging, as well as for travelling and social interactions. Lungworm infections and subsequent secondary bacterial infections (Neimanis et al. 2022; Siebert et al. 2020) pose a threat to animals already debilitated by habitat degradation.

The superfamily Metastrongyloidea includes multiple species of lungworms that are specific to the respiratory tract, cranial and auditory sinuses, as well as the circulatory system of toothed whales (members of the family Pseudaliidae; Figs. 15.8 and 15.9), while other nematodes of the families Crenosomatidae and Parafilaroididae infect the airways of pinnipeds and terrestrial carnivores (Anderson 1982, 2000; Lehnert et al. 2010; Rojano-Doñate 2018). Pseudaliids are specific to odontocetes and have become almost extinct in the terrestrial realm (Durette-Desset et al. 1994), with the exception of *Stenuroides herpestis* in the mongoose (*Herpestes ichneumon*), while the Parafilaroididae, which includes the genus *Parafilaroides,* were recently shown to be paraphyletic (Lehnert et al. 2023a; Pool et al. 2023). Metastrongyloid lungworm infections can have negative impacts on odontocete and phocid health, often causing bronchopneumonia and secondary bacterial infections (Measures 2001;

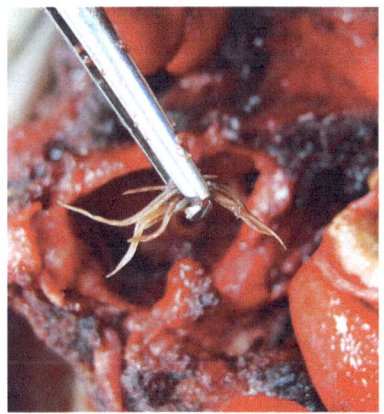

Fig. 15.8 Pseudaliid nematodes (*Stenurus minor*) from the tympanic cavity of a harbour porpoise (*Phocoena phocoena*), Germany. ©ITAW/TiHo

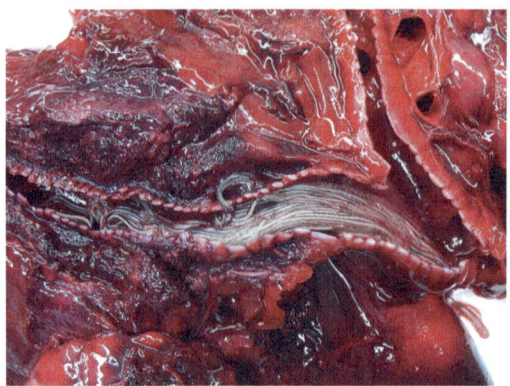

Fig. 15.9 Lung nematodes (Pseudaliidae, Metastrongyloidea) from the bronchial tree of. ©ITAW/TiHo

Houde et al. 2003; Lehnert et al. 2005) resulting in mortality. Severe lung nematode burdens result in respiratory distress, obstruction of airways and inhibit the capability to dive and forage successfully (Siebert et al. 2001; Rojano-Doñate et al. 2018). Transmission pathways of metastrongyloids in marine mammals are not completely understood. There is evidence of benthic fish intermediate hosts (Dailey 1970; Houde et al. 2003; Lehnert et al. 2010) for lungworms of pinnipeds and cetaceans, but other studies have indicated that direct infections by *Halocercus* species are possible in bottlenose dolphins (*Tursiops truncatus*) (Dailey et al. 1991; Fauquier et al. 2009) and common dolphins (Tomo et al. 2010). In two stranded neonatal orcas (*Orcinus orca*) that were days to weeks old and had been feeding on milk, mature and gravid female lungworms with eggs and larvae in utero were detected in histology (Reckendorf et al. 2018) indicating direct transmission in utero or during lactation. In the case of *Halocercus delphini* (as *Skrjabinalius guevarai*) infecting striped dolphins (*Stenella coeruleoalba*) in the Mediterranean, evidence points towards vertical as well as horizontal transmission (Pool et al. 2020a). Because of frequent high-intensity infections in odontocetes like harbour porpoises and pilot whales, they have been speculated to contribute to mass strandings by impeding echolocation. So far, this hypothesis remains doubtful, as few associated lesions have been recorded (Morell et al. 2017)

and severely infected whales seem to still be able to effectively hunt and navigate (Faulkner et al. 1998). Lungworm infections accumulate with age in porpoises, while there are strong age-related infection patterns with high prevalence and intensities in harbour seals, especially in young of the year and yearlings. Adults seem to develop protective immunity after first infections (Ulrich et al. 2015). Grey seals, in contrast, are seldom and only mildly infected, highlighting species-specific immune and genetic traits that need to be further investigated.

15.5.4 Trichinellosis

Trichinella nematodes have a wide host range, causing severe zoonotic disease worldwide. Trichinellosis is common among carnivorous and omnivorous Arctic marine mammals like polar bears and walruses (*Odobenus rosmarus*), which sometimes prey on ringed and bearded seals (Rausch et al. 1956; Madsen 1961). Furthermore, *Trichinella spiralis* was reported in a South American sea lion from Patagonia, Argentina, comprising the first record of these parasitic nematodes from a marine mammal in the Southern Hemisphere (Pasqualetti et al. 2018). However, *Trichinella* species most often cause outbreaks in the circumpolar Arctic with the occurrence of freeze-tolerant *Trichinella nativa* (Skírnisson et al. 2010; Seymour et al. 2014). Trichinellosis has been found to be prevalent among indigenous people in the Arctic, sometimes causing mortality (Rausch 1970; Møller et al. 2005), due to traditional food preparation and consumption of infected polar bear, walrus or beluga whale meat (Tryland 2000; Tryland et al. 2014).

15.6 Acanthocephala

Polymorphid acanthocephalans belonging to the genera *Bolbosoma* and *Corynosoma* are common parasites of marine mammals in coastal and pelagic ecosystems. Species of *Bolbosoma* (12 spp.) (Fig. 15.10) infect whales and dolphins,

Fig. 15.10 Polymorphid acanthocephalans from cetaceans. (**a**) *Bolbosoma capitatum* attached to the small intestine of false killer whales (*Pseudorca crassidens*), Argentina (Chubut); (**b–e**) Scanning electron micrographs of an adult female of *B. capitatum* from the intestine of false killer whales, Argentina (Chubut); (**b**) proboscis, lateral view; (**c**) proboscis, apical view; (**d**) anterior end, ventral view; (**e**) anterior end, subapical view. Scale bars: (**b**, **c**) 100 μm; (**d**, **e**) 1 mm

whereas adults of *Corynosoma* are typically found in pinnipeds (22 spp.), but some species are specific to cetaceans (4 spp.) and sea otters (1 sp.). These parasites exhibit clear patterns of host-specificity in their final hosts, with members of *Corynosoma* associated with pinnipeds unable to mature in cetaceans, whereas immature specimens of *Bolbosoma* spp. have been reported in seals (Ionita et al. 2008; Aznar et al. 2012). Marine mammal acanthocephalans have a global distribution, but most *Corynosoma* species are found in circumpolar regions. *Bolbosoma capitatum* has been recorded in the intestines of toothed whales from all oceans except Antarctica (Fig. 15.10), and also in sperm whales in the Northern Hemisphere (IJsseldijk et al. 2018). *Corynosoma australe* is widely distributed mainly in sea lions and fur seals throughout the Southern Hemisphere, including Antarctica, South Africa, South America, and South Australia. Recent studies, based on morphological and molecular data, have shown that this acanthocephalan also occurs in California sea lions on the Pacific coast of the United States and Mexico.

Most acanthocephalan species inhabit the intestines of marine mammals, but *Corynosoma hamanni* and *Corynosoma cetaceum* are found in the stomach of Antarctic seals and dolphins from the Southern Hemisphere, respectively (Aznar et al. 2001a, b; Laskowski and Zdzitowiecki 2017). Other polymorphid acanthocephalans of the genera *Andracantha* and *Profilicollis*, which primarily infect seabirds, occur as immature forms in pinnipeds and sea otters. Pathogenic effects of *Corynosoma* spp. are often limited to local irritation at the mucosal anchoring location, but can cause inflammation, ulcerations and haemorrhage in large numbers, leading to intestinal perforations (Geraci and St Aubin 1987; Lakemeyer et al. 2020b). Severe infections with *Profilicollis* are a major cause of mortality for young and old sea otters off California due to acanthocephalan peritonitis. *Profilicollis major* and *P. altmani* are known to cause enteritis and peritonitis with subsequent mortality in otters along the Alaskan and Californian coast in the USA. In contrast, *Corynosoma enhydri* infections are perceived as minor health concern with mild lesions in sea otters, although prevalence of *C. enhydri* in Northern and Southern sea otters can be high in all age classes, with some co-infections of *Profilicollis* spp. that were observed at similarly high intensities (Shanebeck et al. 2020).

The life cycle is well documented for some polymorphid acanthocephalan species. Pelagic euphausiids (krill) and probably copepods are intermediate hosts for *Bolbosoma*, while offshore amphipods are known as intermediate hosts for some *Corynosoma* species. Cystacanth larvae have been reported in several fish species that serve as paratenic or transport hosts. In the fish host, cystacanths are found as encysted resting stages that may have a long life span of several years (Comiskey and MacKenzie 2000). Humans are accidental hosts for some species of *Corynosoma* and *Bolbosoma*, with few human cases reported in Japan (Mathison et al. 2021).

Three acanthocephalan species predominate in pinnipeds of the North and Baltic seas. *Corynosoma strumosum* and *C. magdaleni*, which are morphologically very similar, inhabit the seals' small intestine, whereas *C. semerme* infects their caecum, colon and rectum (Nickol et al. 2002; Siebert et al. 2007). However, a recent study using genomic methods on *Corynosoma* spp. from northern European seals revealed the presence of putative new species (Waindok et al. 2018; Sromek et al. 2023). Comprehensive taxonomic studies combining morphologically precisely identified adult acanthocephalans with high-quality molecular sequences are still needed to elucidate the diversity of *Corynosoma* in the Baltic Sea. Amphipods are known as intermediate hosts, and the fourhorn sculpin (*Myoxocephalus quadricornis*) is assumed to be crucial in the life cycle of *C. semerme* in the Baltic (Sinisalo and Valtonen 2003) (see Fig. 5.12 in Chap. 5). The Raneya (*Raneya brasiliensis*) apparently play a key role in the transmission of *C. australe* to sea lions on the Patagonian shelf of Argentina (Hernández-Orts et al. 2019a, b).

Since the 1970s, grey seals in the Baltic Sea have displayed characteristic lesions in their reproductive organs, intestines and endocrine system that became known as the Baltic seal disease complex (BSDC) and that were linked to persistent organic pollutant (POPs) exposure, which was assumed to affect the immune and endocrine system (Olsson 1994; Bergman 2007). Lesions consistent with BSDC were not found in harbour seals from the German North Sea examined between 1996 and 2005 (Siebert et al. 2007) or in North Sea and Atlantic grey seals investigated in 1997 (O'Neill and Whelan 2002), making this complex of pathological signs unique to the Baltic Sea. While the occurrence of reproductive tract lesions decreased since the 1990s, colonic ulcers and thickened intestinal walls, seen as BSDC clinical signs, are still found and often associated with acanthocephalan infection (Bergman 1999, 2007; Britt-Marie et al. 2021), with lesions already appearing in younger grey seals (Bäcklin et al. 2003). Grey seals with colonic infections of acanthocephalans display chronic colitis with tunica muscularis hypertrophy (Lakemeyer et al. 2020b). The level of acanthocephalan infection and intestinal inflammation in grey and harbour seals was suggested as effective indicators of seal population health (Lakemeyer et al. 2020b). The link between infection by *Corynosoma* spp., exposure to pollutants and subsequent immunosuppression, as well as *Corynosoma* spp. prevalence and intestinal lesions, need to be investigated further.

15.7 Arthropoda: Insecta, Acari and Crustacea

When ancestors of recent marine mammals colonised the sea in the Eocene and Miocene epochs, their parasites had to adapt to a new environment

(Anderson 1982; Raga et al. 2009). Skin and fur as well as mucosal linings and body cavities are difficult to attach to on fast-moving and deep-diving hosts in the ocean. While maintaining their grip on a host specimen, parasites must also ensure transmission and reproduction. In some cases, e.g. seal lice, which are in fact hematophagous insects of terrestrial origin (order Psocodea), they co-evolved and managed to adapt to the marine habitat (Leidenberger et al. 2007; Leonardi et al. 2019; also see Chap. 6). On the other hand, marine arthropods, such as the whale lice (family Cyamidae), which are in fact amphipod crustaceans of marine origin, established new parasitic relationships with marine mammals (Iwasa-Arai et al. 2018; Lehnert et al. 2021; also see Chap. 6). Whale lice have adapted to their hosts' lifestyles by developing dorsoventrally compressed bodies and modified back three pairs of legs with large claws to cling to their cetacean host (Fig. 15.11a) (Rowntree 1996). All developmental stages are spent on their hosts (Roussel de Vauzème 1834; St. Leger et al. 2018) and they rely on physical contact between host individuals during reproduction, nursing or aggressive behaviour in males for transmission (Balbuena and Raga 1991; St. Leger et al. 2018). Whale lice tend to aggregate in the genital slit, corners of the mouth or the blowhole (Rowntree, 1996; Fraija-Fernández et al. 2017a) and in lesions with thickened edges in faster swimming hydrodynamic odontocetes (Lehnert et al. 2007) (Fig. 15.11b). Whale lice ingest small pieces of skin and scar tissue (Schell et al. 2000), but may also feed on plankton (Rowntree 1996). Cyamids of the species *Isocyamus deltobranchium* are found on harbour porpoises, True's beaked whale (*Mesoplodon mirus*), common dolphin (*Delphinus delphis*) (Martinez et al. 2008) and pilot whales. They occurr with different prevalences due to geographic regions and prevailing health status or lesions (Lehnert et al. 2021). Stalked barnacles (primarily of the families Coronulidae and Lepadidae) and ectoparasitic copepods, such as *Pennella balaenopterae*, are also common ectoparasites of cetaceans (Hanninger et al. 2023). As directly transmitted parasites, they have potential as indicators for population structure and social behaviour as well as many traits of host ecology (Balbuena et al. 1995; Kaliszewska et al. 2005; Ten et al. 2022a, b).

The seal louse *Echinophthirius horridus* (Insecta: Psocodea) co-evolved with its hosts and remained throughout its transition to the aquatic environment (Leidenberger et al. 2007; Herzog et al. 2024) (Fig. 15.12). These lice developed claws for attachment, use host sebum for insulation, are believed to retain air within their spines or scales (Mehlhorn et al. 2002; Leonardi et al. 2012), and may be able to store air in their tracheal system during host dives. Seal lice disperse either vertically or horizontally during haul-out (Leonardi et al. 2013). Their nits are glued to seal

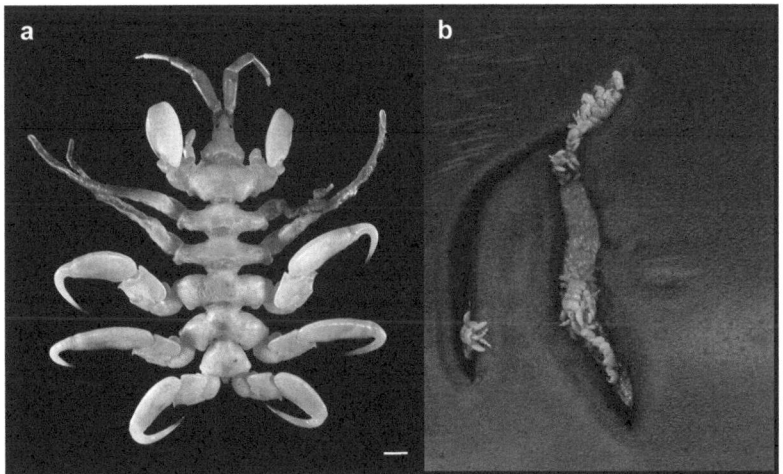

Fig. 15.11 Ectoparasitic crustaceans/whale lice (Cyamidae) (**a**) dorsal view of a male whale lice, unknown host, South Africa, ©Willie Landman; (**b**) located in skin lesions of a harbour porpoise (*Phocoena phocoena*), Germany, ©ITAW/TiHo. Scale bar: (**a**) 1 mm

Fig. 15.12 Seal louse *Echinophthirius horridus* (Echinophthiriidae), a bloodsucking obligate ectoparasite of amphibious phocid seals. ©Insa Herzog/TiHo

hair, from which the first mobile nymphal stage hatches and subsequently moults successively into two more nymph stages characterised by increasing size and sclerotised structures before finally moulting into adults. Seal lice may cause anaemia and alopecia in their hosts during severe infections and can act as vectors for filarial nematodes and viral disease (La Linn et al. 2001; Lehnert et al. 2015; Leonardi et al. 2021). During attachment to the skin and hair infundibula they cause purulent folliculitis with intralesional bacteria, as well as hyperkeratitis and dermatitis in grey and harbour seals (Herzog et al. 2024). Seal lice are believed to be vectors for the transmission of heartworm filariae (*Acanthocheilonema spirocauda*) in seals (Geraci et al. 1981; Ebmer et al. 2022); filarial stages have been detected in the pharynx, intestine and hemacoel of seal lice (Herzog et al. 2024), thus transmitting a typically terrestrial parasite to marine hosts (Hirzmann et al. 2021; Lehnert et al. 2023b).

Parasitic mites (Arachnida: Acari) are usually ectoparasitic on terrestrial vertebrates. However, the naso-pharyngeal-infecting mite *Halarachne halichoeri* (Halarachnidae) infects the respiratory tract of harbour and grey seals, and has developed specific morphological adaptations to life in a semi-aquatic host (Furman 1979; Fain 1994; Chap. 6). Motile hexapod *H. halichoeri* larvae with spider-like appearance are concentrated in the anterior nasal passages and transmitted through convulsive coughing and social interactions (Furman and Smith 1973). The sessile adults with claws are true endoparasites, anchored to mucosal surfaces deeper inside the nares and cannot survive for long outside their host (Furman and Smith 1973; Geraci and St Aubin 1987). *Halarachne halichoeri* can cause different levels of respiratory inflammatory disease by damaging mucous membranes and impairing respiration (Alonso-Farré et al. 2012; Reckendorf et al. 2019). Mild to severe sinusitis, as well as hyperaemia and mild sinus inflammation in the upper respiratory tract, along with respiratory symptoms, observed in conjunction with *H. halichoeri* infection of grey seals from the Spanish and German coasts. These mites can be vectors for bacterial pathogens (Pesapane et al. 2021) in Californian pinnipeds (Pesapane et al. 2021). Recently, individuals were reported from Mediterranean monk seal (Athinaiou et al. 2023). Nasal mites belonging to the genus *Orthohalarachne* are reported from Pacific walrus (Fravel et al. 2016) and South American fur seals (Gastal et al. 2016). Due to their reliance on direct transmission through bodily contact during host social interactions, these endoparasitic arthropods can highlight host population dynamics and inter-species interactions, as well as showcasing how marine mammal parasites are important indicators for host ecology and behaviour.

15.8 Marine Mammal Parasites as Biological Markers

Few investigations into the use of parasites as biological markers in marine mammals have been published since Balbuena et al. (1995) and

Aznar et al. (2001a, b) summarised and reviewed the potential utility of parasites as biological markers for studying populations, social behaviour, movements and phylogeny of marine mammals (also see Chap. 17). Marigo et al. (2002) examined the helminth fauna of franciscanas (*Pontoporia blainvillei*) from three geographic areas along the coast of southern Brazil. They selected the intestinal trematode *Synthesium pontoporiae* (Brachycladiidae) as a potentially useful tag, because it showed a trend of variation with latitude, with a progressive increase, especially in mean intensity, from north to south. Recently, Kuzmina et al. (2021) carried out an extensive survey of the gastrointestinal helminth fauna of Northern fur seal males from five rookeries on St. Paul Island, Alaska. The authors found no significant differences in parasite faunal composition and the absence of separate stocks. The study of Iwasa-Arai et al. (2018) provided interesting conclusions on social behaviour of baleen whales. They analysed the genetic population structure of the whale louse *Cyamus boopis* (Cyamidae) in humpback whales (*Megaptera novaeangliae*) from different breeding stocks in the Southern Hemisphere. Using sequences of the cytochrome *c* oxidase subunit I (*cox*1) gene, they found that the genetic diversity of *C. boopis* did not differ between breeding stocks of humpback whales from the southwestern Atlantic and southwestern Pacific but differed significantly from stocks in the southeastern Atlantic and North Pacific. The homogeneity of the genetic population structure in *C. boopis* observed between the southwestern Atlantic and southwestern Pacific was interpreted as a possible result of contact between humpback whales of these two breeding stocks during their migratory cycle or in foraging areas in Antarctic waters. In a recent study, Ten et al. (2022a) used seven species of epibiotic crustaceans, i.e. the mesoparasitic pennellid copepod *Pennella balaenoptera*, the ectoparasitic amphipod *Balaenocyamus balaenopterae*, and three obligate coronulid and two facultative lepadid barnacle species to track the seasonal movements of Antarctic minke whales (*Balaenoptera bonaerensis*). They found that recruitment of coronulid barnacles, particularly of *Xenobalanus globicipitis*, appeared to be largely restricted to low latitudes, as all barnacles were found dead and exhibited varying degrees of tissue degradation in whales from Antarctic waters. The authors suggest that *X. globicipitis* may be used to chronologically track the migrations of these whales from their breeding grounds in subtropical waters to their feeding grounds in polar regions, as this barnacle apparently does not withstand the low temperatures in polar regions.

15.9 Parasites of Marine Mammals Pose Zoonotic Disease Risks

Because of their role as indicators of ecosystem viability (Bossart 2011; Reddy et al. 2001) and some physiological similarities to humans, marine mammals reflect ocean ecosystem health and can inform human health risks with regard to zoonotic disease. Infectious diseases, including many marine mammal parasites, have significant health implications for humans. Multiple marine mammal parasites can cause zoonoses or be reservoirs and vectors of zoonotic agents along densely populated coastal regions. Marine wildlife and their parasite fauna can be relevant sentinels for evaluation of human health risks, relating to chemical or microplastics pollution but also zoonotic infectious disease (Stewart et al. 2008; Reif 2011) like the highly pathogenic influenza strains H5N1 and H3N8 (Karlsson et al. 2014; Puryear et al. 2023). Because host–parasite interactions are accelerated by environmental change, emerging helminth foodborne zoonosis anisakidosis and diphyllobothriasis are of increasing relevance globally (Thompson 2023; also see Chap. 19). Although some intestinal helminths of marine mammals, like tapeworms of the genus *Diphyllobothrium*, are known to be emerging foodborne zoonoses (Scholz and Kuchta 2016), cestodes infecting grey and harbour seals of the North and Baltic Seas are still lacking identification to the species level and there is no information on their epidemiology and temporal trends in prevalence.

15.10 Metabarcoding Methods for the Detection of Marine Mammal Parasites in Faecal Samples

The introduction of high-throughput sequencing technology has imparted greater detection sensitivity for exploring the parasite diversity in threatened, difficult-to-sample animals (Gogarten et al. 2020). These techniques allow for practical, rapid and standardised methods such as metabarcoding for characterising parasite diversity of specific hosts (see Chap. 14). To date, few studies have applied metabarcoding methods to detect marine mammal parasite operational taxonomic units (OTUs). McCosker et al. (2020) applied a DNA metabarcoding approach to faecal samples from grey seals and detected parasites ambiguously identified as Nematoda, Cestoda and Trematoda. Günther et al. (2022) detected OTUs assigned to *Anisakis simplex* and Platyhelminthes (probably cestodes or trematodes) in faecal samples from free-ranging beluga on the Norwegian coast. These authors also reported an OTU identified as *Echinorhynchus gadi*, a globally distributed acanthocephalan in marine fishes. Couch et al. (2022) detected OTUs of platyhelminths, including an unidentified species of *Diphyllobothrium*, nematodes and acanthocephalans in metabarcoding analyses of faecal samples of Pacific walrus from Alaska. These studies demonstrate that the metabarcoding approach is an excellent tool for non-invasive and sensitive detection of parasite diversity in marine mammals using environmental samples. However, the development of a comprehensive database of reference sequences of marine mammal parasites derived from accurate identifications of vouchers based on morphological characteristics remains necessary for proper delineation of OTUs in metabarcoding research.

15.11 Conclusions and Outlook

This comprehensive review provides insights into the diversity and ecological traits of marine mammal parasites, highlighting the need for continued research and conservation efforts to safeguard these iconic species and the ecosystems they inhabit. In conclusion, the intricate relationship between marine mammals and their parasites underscores the importance of understanding the ecological dynamics within ocean ecosystems. While marine mammals contribute significantly to ecosystem function, their populations face numerous threats, including habitat destruction, pollution and climate change, which can also impact their parasite fauna. Despite challenges in studying marine mammal parasites, such as limited access to specimens and ethical considerations, research in this field is essential for monitoring ecosystem health and understanding the implications for both marine mammal populations and human health. With advancements in molecular techniques and increasing interest in ecological studies including parasites, there is a growing opportunity to enhance our understanding of marine mammal–parasite interactions and their implications for global conservation efforts.

Acknowledgements The authors thank countless colleagues who contributed to the body of knowledge with their research. The authors would like to highlight the importance of stranding networks to collect data on marine mammals, and especially thank those who help to collect marine mammals on the German and Argentinean coasts and help to assess pathological findings. We also thank Nico Smit and Kerry Malherbe (Hadfield) for their assistance in editing and designing the figures. We are grateful to Rocío Loizaga and to Maitá A. Barrena for providing photographs and parasites of dolphins from Río Negro and Chubut, Argentina. The authors would like to stress the value of museum collections and archives at research institutions for marine mammal parasitology research.

References

Abollo E, Paggi L, Pascual S et al (2003) Occurrence of recombinant genotypes of *Anisakis simplex* s.s. and *Anisakis pegreffii* (Nematoda: Anisakidae) in an area of sympatry. Infect Genet Evol 3(3):175–181. https://doi.org/10.1016/s1567-1348(03)00073-x

Agustí C, Aznar FJ, Raga JA (2005) Tetraphyllidean plerocercoids from Western Mediterranean cetaceans and other marine mammals around the world: a comprehensive morphological analysis. J Parasitol 91(1):83–92. https://doi.org/10.1645/GE-372R

Aibinu IE, Smooker PM, Lopata AL (2019) Anisakis nematodes in fish and shellfish- from infection to allergies. Int J Parasitol Parasites Wildl 9:384–393. https://doi.org/10.1016/j.ijppaw.2019.04.007

Alonso-Farré JM, D'Silva JI, Gestal C (2012) Nasopharyngeal mites Halarachne halichoeri (Allman, 1847) in Grey seals stranded on the NW Spanish Atlantic Coast. Vet Parasitol 183(3–4):317–322. https://doi.org/10.1016/j.vetpar.2011.08.002

Andersen-Ranberg E, Lehnert K, Leifsson PS et al (2018) Morphometric, molecular and histopathologic description of hepatic infection by Orthosplanchnus arcticus (Trematoda: Digenea: Brachycladiidae) in ringed seals (Pusa hispida) from Northwest Greenland. Polar Biol 41:1019–1025. https://doi.org/10.1007/s00300-017-2245-6

Anderson RC (1982) Host-parasite relations and evolution of the Metastrongyloidea (Nematoda)[mammals]. Mém Mus Natl Hist Nat Sér A Zool 123:129–133

Anderson RC (2000) Nematode parasites of vertebrates: Their development and transmission. CABI, Egham

Athinaiou N, Sarantopoulou J, Komnenou A et al (2023) From Atlantic to Greece. The case of Nasal Mite in Mediterranean Monk Seal.

Avila IC, Kaschner K, Dormann CF (2018) Current global risks to marine mammals: Taking stock of the threats. Biol Conserv 221:44–58. https://doi.org/10.1016/j.biocon.2018.02.021

Aznar FJ, Balbuena JA, Fernández M et al (2001a) Living together: the parasites of marine mammals. In: Evans PGH, Raga JA (eds) Marine mammals: biology and conservation. Springer, New York, pp 385–423

Aznar FJ, Bush AO, Balbuena JA et al (2001b) Corynosoma cetaceum in the stomach of franciscanas, Pontoporia blainvillei (Cetacea): an exceptional case of habitat selection by an acanthocephalan. J Parasitol 87(3):536–541. https://doi.org/10.1645/0022-3395(2001)087[0536:CCITSO]2.0.CO;2

Aznar FJ, Agustí C, Littlewood DT et al (2007) Insight into the role of cetaceans in the life cycle of the tetraphyllideans (Platyhelminthes: Cestoda). Int J Parasitol 37(2):243–255. https://doi.org/10.1016/j.ijpara.2006.10.010

Aznar FJ, Hernández-Orts J, Suárez AA et al (2012) Assessing host-parasite specificity through coprological analysis: a case study with species of Corynosoma (Acanthocephala: Polymorphidae) from marine mammals. J Helminthol 86(2):156–164. https://doi.org/10.1017/S0022149X11000149

Bäcklin BM, Eriksson L, Olovsson M (2003) Histology of uterine leiomyoma and occurrence in relation to reproductive activity in the Baltic gray seal (Halichoerus grypus). Vet Pathol 40(2):175–180. https://doi.org/10.1354/vp.40-2-175

Balbuena JA, Raga JA (1991) Ecology and host relationships of the whale-louse Isocyamus delphini (Amphipoda: Cyamidae) parasitizing long-finned pilot whales (Globicephala melas) off the Faroe Islands (Northeast Atlantic). Can J Zool 69(1):141–145. https://doi.org/10.1139/z91-021

Balbuena JA, Aznar FJ, Fernandez M et al (1995) Parasites as indicators of social structure and stock identity of marine mammals. In: Blix AS, Walloe L, Ulltang O (eds) Whales, seals, fish and man. Elsevier, Amsterdam, pp 133–140. https://doi.org/10.1016/S0163-6995(06)80017-X

Bao M, Pierce G, Pascual S et al (2017) Assessing the risk of an emerging zoonosis of worldwide concern: anisakiasis. Sci Rep 7:43699. https://doi.org/10.1038/srep43699

Bao M, Pierce GJ, Strachan NJC et al (2019) Human health, legislative and socioeconomic issues caused by the fish-borne zoonotic parasite Anisakis: Challenges in risk assessment. Trends Food Sci Technol 86:298–310. https://doi.org/10.1016/j.tifs.2019.02.013

Baptista-Fernandes T, Rodrigues M, Castro I et al (2017) Human gastric hyperinfection by Anisakis simplex: A severe and unusual presentation and a brief review. Int J Infect Dis 64:38–41. https://doi.org/10.1016/j.ijid.2017.08.012

Baršienė J, Butrimavičienė L, Grygiel W et al (2014) Environmental genotoxicity and cytotoxicity in flounder (Platichthys flesus), herring (Clupea harengus) and Atlantic cod (Gadus morhua) from chemical munitions dumping zones in the southern Baltic Sea. Mar Environ Res 96:56–67. https://doi.org/10.1016/j.marenvres.2013.08.012

Bergman A (1999) Health condition of the Baltic grey seal (Halichoerus grypus) during two decades. Gynaecological health improvement but increased prevalence of colonic ulcers. Acta Pathol Microbiol Scand 107(3):270–282. https://doi.org/10.1111/j.1699-0463.1999.tb01554.x

Bergman A (2007) Pathological changes in seals in Swedish waters: the relation to environmental pollution (No. 2007: 131)

Berta A, Sumich JL, Kovacs KM (2015) Marine mammals evolutionary biology, 3rd edn. Academic, Cambridge

Borges JCG, Lima DS, da Silva EM et al (2017) Cryptosporidium spp. and Giardia sp. in aquatic mammals in northern and northeastern Brazil. Dis Aquat Org 126:25–31. https://doi.org/10.3354/dao03156

Borgsteede FHM, Bus HGJ, Verplanke JAW, Van Burg WPJ (1991) Endoparasitic helminths of the harbour seal, Phoca vitulina, in the Netherlands. Neth J Sea Res 28(3):247–250. https://doi.org/10.1016/0077-7579(91)90022-S

Bossart GD (2011) Marine mammals as sentinel species for oceans and human health. Vet Pathol 48(3):676–690. https://doi.org/10.1177/0300985810388525

Britt-Marie B, Sara P, Suzanne F et al (2021) Temporal and geographical variation of intestinal ulcers in grey seals (Halichoerus grypus) and environmental contaminants in Baltic biota during four decades. Animals 11(10):2968

Buchmann K, Mehrdana F (2016) Effects of anisakid nematodes *Anisakis simplex* (s.l.), *Pseudoterranova decipiens* (s.l.) and *Contracaecum osculatum* (s.l.) on fish and consumer health. Food Waterborne Parasitol 4:13–22. https://doi.org/10.1016/j.fawpar.2016.07.003

Caira JN, Jensen K, Pickering M et al (2020) Intrigue surrounding the life-cycles of species of *Clistobothrium* (Cestoda: Phyllobothriidea) parasitising large pelagic sharks. Int J Parasitol 50(13):1043–1055. https://doi.org/10.1016/j.ijpara.2020.08.002

Cantatore DMP, Lanfranchi AL, Canel D et al (2023) Plerocercoids of *Adenocephalus pacificus* in Argentine hakes: Broad distribution, low zoonotic risk. Int J Food Microbiol 391–393:110142. https://doi.org/10.1016/j.ijfoodmicro.2023.110142

Chubb JC, Valtonen ET, McGeorge J et al (1995) Characterisation of the external features of *Schistocephalus solidus* (Müller, 1776) (Cestoda) from different geographical regions and an assessment of the status of the Baltic ringed seal *Phoca hispida botnica* (Gmelin) as a definitive host. Syst Parasitol 32:113–123. https://doi.org/10.1007/BF00009510

Cipriani P, Palomba M, Giulietti L et al (2022) Distribution and genetic diversity of *Anisakis* spp. in cetaceans from the Northeast Atlantic Ocean and the Mediterranean Sea. Sci Rep 12:13664. https://doi.org/10.1038/s41598-022-17710-1

Comiskey P, MacKenzie K (2000) *Corynosoma* spp. may be useful biological tags for saithe in the northern North Sea. J Fish Biol 57:525–528. https://doi.org/10.1111/j.1095-8649.2000.tb02190.x

Couch C, Sanders J, Sweitzer D et al (2022) The relationship between dietary trophic level, parasites and the microbiome of Pacific walrus (*Odobenus rosmarus divergens*). Proc Biol Sci 289(1972):20220079. https://doi.org/10.1098/rspb.2022.0079

Dailey MD (1970) The Transmission of *Parafilaroides decorus* (Nematoda: Metastrongyloidea) in the California Sea Lion (*Zalophus californianus*). Proc Helminthol Soc Wash 37(2):215–222

Dailey MD (1971) A new species of *Hunterotrema* (Digenea: Campulidae) from the Amazon River dolphin (*Inia geoffrensis*). Bull S Calif Acad Sci 70(2):79–80

Dailey MD (1975) The distribution and intraspecific variation of helminth parasites in pinnipeds. Rapp P-V Réun Cons Int Explor Mer 169:338–352

Dailey M, Walsh M, Odell D et al (1991) Evidence of prenatal infection in the bottlenose dolphin (*Tursiops truncatus*) with the lungworm *Halocercus lagenorhynchi* (Nematoda: Pseudaliidae). J Wildl Dis 27(1):164–165. https://doi.org/10.7589/0090-3558-27.1.164

Dailey MD, Gulland FM, Lowenstine LJ et al (2000) Prey, parasites and pathology associated with the mortality of a juvenile gray whale (*Eschrichtius robustus*) stranded along the northern California coast. Dis Aquat Organ 42(2):111–117. https://doi.org/10.3354/dao042111

Dailey MD, Kliks MM, Demaree RS (2004) *Heterophyopsis hawaiiensis* n. sp. (Trematoda: Heterophyidae) from the Hawaiian Monk Seal, *Monachus schauinslandi* Matschie, 1905 (Carnivora: Phocidae). Comp Parasitol 71(1):9–12. https://doi.org/10.1654/4071

Delyamure SL (1955) Helminthofauna of marine mammals (ecology and phylogeny). Academy of Science of the U.S.S.R, Moscow

Delyamure SL, Skryabin AS (1968) Origin and systematic position of diplogonal and polygonal diphyllobothriids. In: Tabarin VG (ed) Gelminty cheloveka, zhivotnych i rastenii i mery borby s nimi. Izdatel'stvo Nauka, Moscow, pp 159–166. (in Russian)

Díaz-Delgado J, Groch KR, Ramos HG et al (2020) Fatal systemic toxoplasmosis by a novel non-archetypal *Toxoplasma gondii* in a Bryde's whale (*Balaenoptera edeni*). Front Mar Sci 7:336. https://doi.org/10.3389/fmars.2020.00336

Dubey JP, Murata FHA, Cerqueira-Cézar CK et al (2020) Recent epidemiologic and clinical importance of *Toxoplasma gondii* infections in marine mammals: 2009-2020. Vet Parasitol 288:109296. https://doi.org/10.1016/j.vetpar.2020.109296

Durette-Desset MC, Beveridge I, Spratt DM (1994) The origins and evolutionary expansion of the Strongylida (Nematoda). Int J Parasitol 24(8):1139–1165. https://doi.org/10.1016/0020-7519(94)90188-0

Ebert M, Valentere A (2013) New records of *Nasitrema atenuatta* and *Nasitrema globicephalae* (Trematoda: Brachycladiidae) Neiland, Rice and Holden, 1970 in delphinids from South Atlantic. Check List 9(6):1538–1540. https://doi.org/10.15560/9.6.1538

Ebert MB, Fernández M, Valente ALS et al (2021) *Ascocotyle longa* (Digenea: Heterophyidae) infecting dolphins from the Atlantic Ocean. Parasitol Res 120(1):347–353. https://doi.org/10.1007/s00436-020-06956-1

Ebmer D, Navarrete MJ, Muñoz P et al (2020) Anthropozoonotic parasites circulating in synanthropic and Pacific colonies of South American sea lions (*Otaria flavescens*): Non-invasive techniques data and a review of the literature. Front Mar Sci 7:543829. https://doi.org/10.3389/fmars.2020.543829

Ebmer D, Handschuh S, Schwaha T et al (2022) Novel 3D in situ visualization of seal heartworm (*Acanthocheilonema spirocauda*) larvae in the seal louse (*Echinophthirius horridus*) by X-ray microCT. Sci Rep 12(1):14078. https://doi.org/10.1038/s41598-022-18418-y

Fain A (1994) Adaptation, specificity and host-parasite coevolution in mites (ACARI). Int J Parasitol 24(8):1273–1283. https://doi.org/10.1016/0020-7519(94)90194-5

Faulkner J, Measures LN, Whoriskey FG (1998) *Stenurus minor* (Metastrongyloidea: Pseudaliidae) infections of the cranial sinuses of the harbour porpoise, *Phocoena*

phocoena. Can J Zool 76(7):1209–1216. https://doi.org/10.1139/z98-057

Fauquier DA, Kinsel MJ, Dailey MD et al (2009) Prevalence and pathology of lungworm infection in bottlenose dolphins *Tursiops truncatus* from southwest Florida. Dis Aquat Organ 88(1):85–90. https://doi.org/10.3354/dao02095

Fayer R, Dubey JP, Lindsay DS (2004) Zoonotic protozoa: from land to sea. Trends Parasitol 20(11):531–536. https://doi.org/10.1016/j.pt.2004.08.008

Fiorenza EA, Wendt CA, Dobkowski KA et al (2020) It's a wormy world: Meta-analysis reveals several decades of change in the global abundance of the parasitic nematodes *Anisakis* spp. and *Pseudoterranova* spp. in marine fishes and invertebrates. Glob Chang Biol 26(5):2854–2866. https://doi.org/10.1111/gcb.15048

Fowler CW (1990) Density dependence in northern fur seals (*Callorhinus ursinus*). Mar Mamm Sci 6(3):171–195. https://doi.org/10.1111/j.1748-7692.1990.tb00242.x

Fraija-Fernández N, Olson PD, Crespo EA et al (2015) Independent host switching events by digenean parasites of cetaceans inferred from ribosomal DNA. Int J Parasitol 45(2-3):167–173. https://doi.org/10.1016/j.ijpara.2014.10.004

Fraija-Fernández N, Aznar FJ, Fernández A et al (2016) Evolutionary relationships between digeneans of the family Brachycladiidae Odhner, 1905 and their marine mammal hosts: A cophylogenetic study. Parasitol Int 65(3):209–217. https://doi.org/10.1016/j.parint.2015.12.009

Fraija-Fernández N, Fernández M, Gozalbes P et al (2017a) Living in a harsh habitat: epidemiology of the whale louse, *Syncyamus aequus* (Cyamidae), infecting striped dolphins in the Western Mediterranean. J Zool 303(3):199–206. https://doi.org/10.1111/jzo.12482

Fraija-Fernández N, Fernández M, Lehnert K et al (2017b) Long-distance travellers: phylogeography of a generalist parasite, *Pholeter gastrophilus*, from cetaceans. PLoS One 12(1):e0170184. https://doi.org/10.1371/journal.pone.0170184

Fraija-Fernández N, Waeschenbach A, Briscoe AG et al (2021) Evolutionary transitions in broad tapeworms (Cestoda: Diphyllobothriidea) revealed by mitogenome and nuclear ribosomal operon phylogenetics. Mol Phylogenet Evol 163:107262. https://doi.org/10.1016/j.ympev.2021.107262

Furman DP (1979) Specificity, adaptation, and parallel evolution in the endoparasitic Mesostigmata of mammals. In: Rodriguez JG (ed) Recent advances in acarology, vol 2. Academic, New York, pp 329–337

Furman DP, Smith AW (1973) In vitro development of two species of *Orthohalarachne* (Acarina: Halarachnidae) and adaptations of the life cycle for endoparasitism in mammals. J Med Entomol 10(4):415–416. https://doi.org/10.1093/jmedent/10.4.415

Gastal SB, Mascarenhas CS, Ruas JL (2016) Infection rates of Orthohalarachne attenuata and Orthohalarachne diminuata (Acari: Halarachnidae) in Arctocephalus australis (Zimmermann, 1783)(pinipedia: Otariidae). Comp. Parasitol 83(2):245–249

Gaydos JK, Miller WA, Johnson C et al (2008) Novel and canine genotypes of *Giardia duodenalis* in harbor seals (*Phoca vitulina richardsi*). J Parasitol 94(6):1264–1268. https://doi.org/10.1645/GE-1321.1

Geraci JR, St Aubin DJ (1987) Effects of parasites on marine mammals. Int J Parasitol 17(2):407–414. https://doi.org/10.1016/0020-7519(87)90116-0

Geraci JR, Fortin JF, St. Aubin DJ et al (1981) The seal louse, *Echinophthirius horridus*: an intermediate host of the seal heartworm, *Dipetalonema spirocauda* (Nematoda). Can J Zool 59(7):1457–1459. https://doi.org/10.1139/z81-197

Giari L, Castaldell G, Timi JT (2022) Ecology and effects of metazoan parasites of fish in transitional waters. Parasitology 149(14):1829–1841. https://doi.org/10.1017/S0031182022001068

Gibson AK, Raverty S, Lambourn DM et al (2011) Polyparasitism is associated with increased disease severity in *Toxoplasma gondii*-infected marine sentinel species. PLoS Negl Trop Dis 5(5):e1142. https://doi.org/10.1371/journal.pntd.0001142

Godfroid J (2017) Brucellosis in livestock and wildlife: zoonotic diseases without pandemic potential in need of innovative one health approaches. Arch Public Health 75:34. https://doi.org/10.1186/s13690-017-0207-7

Gogarten JF, Calvignac-Spencer S, Nunn CL et al (2020) Metabarcoding of eukaryotic parasite communities describes diverse parasite assemblages spanning the primate phylogeny. Mol Ecol Resour 20(1):204–215. https://doi.org/10.1111/1755-0998.13101

González MT, López Z, Nuñez JJ et al (2018) Morphometrical and molecular evidence suggests cryptic diversity among hookworms (Nematoda: *Uncinaria*) that parasitize pinnipeds from the southeastern Pacific coasts. J Helminthol 94:e8. https://doi.org/10.1017/S0022149X18000950

Greiman SE, Kent ML, Betts J et al (2016) *Nanophyetus salmincola*, vector of the salmon poisoning disease agent *Neorickettsia helminthoeca*, harbors a second pathogenic *Neorickettsia* species. Vet Parasitol 229:107–109. https://doi.org/10.1016/j.vetpar.2016.10.003

Grilo ML, Gomes L, Wohlsein P et al (2018) *Cryptosporidium* species and *Giardia* species prevalence in marine mammal species present in the German North and Baltic Seas. J Zoo Wildl Med 49(4):1002–1006. https://doi.org/10.1638/2017-0255.1

Gubanov NM (1951) Giant nematoda from the placenta of Cetacea; *Placentonema gigantissima* nov. gen., nov. sp. Doklady Akademii Nauk SSSR 77(6):1123–1125. (in Russian)

Günther B, Jourdain E, Rubincam L et al (2022) Feces DNA analyses track the rehabilitation of a free-ranging beluga whale. Sci Rep 12:6412. https://doi.org/10.1038/s41598-022-09285-8

Gutiérrez MP, Canel D, Braicovich PE et al (2023) Parasite assemblages in volatile host stocks: Inter- and intra-cohort variability restrict their value as biological tags for squid stock assessment. Parasitology 150(13):1254–1262. https://doi.org/10.1017/S0031182023001051

Hanninger EM, Selling J, Heyer K et al (2023) Skin conditions, epizoa, ectoparasites and emaciation in cetaceans in the Strait of Gibraltar: An update for the period 2016-2020. J Cetacean Res Manag 24(1):121–142

Hauksson E (2011) The prevalence, abundance, and density of *Pseudoterranova* sp. (p) larvae in the flesh of cod (*Gadus morhua*) relative to proximity of grey seal (*Halichoerus grypus*) colonies on the coast off Drangar, Northwest Iceland. J Mar Sci 2011:235832. https://doi.org/10.1155/2011/235832

Heckmann R, Halajian A, El-Naggar A et al (2014) A histopathology study of Caspian Seal (*Pusa caspica*) (Phocidae, Mammalia) liver infected with trematode, *Pseudamphistomum truncatum* (Rudolphi, 1819) (Opisthorchidae, Trematoda). Iran J Parasitol 9(2):266–275

Hermosilla C, Silva LMR, Prieto R et al (2015) Endo- and ectoparasites of large whales (Cetartiodactyla: Balaenopteridae, Physeteridae): Overcoming difficulties in obtaining appropriate samples by non- and minimally-invasive methods. Int J Parasitol Parasites Wildl 4(3):414–420. https://doi.org/10.1016/j.ijppaw.2015.11.002

Hernández-Orts JS, Montero FE, Crespo EA et al (2012) A new species of *Ascocotyle* (Trematoda: Heterophyidae) from the South American sea lion, *Otaria flavescens*, off Patagonia, Argentina. J Parasitol 98(4):810–816. https://doi.org/10.1645/GE-2959.1

Hernández-Orts JS, Scholz T, Brabec J et al (2015) High morphological plasticity and global geographical distribution of the Pacific broad tapeworm *Adenocephalus pacificus* (syn. *Diphyllobothrium pacificum*): Molecular and morphological survey. Acta Tropica 149:168–178. https://doi.org/10.1016/j.actatropica.2015.05.017

Hernández-Orts JS, Georgieva S, Landete DN et al (2019a) Heterophyid trematodes (Digenea) from penguins: A new species of *Ascocotyle* Looss, 1899, first description of metacercaria of *Ascocotyle* (*A.*) *patagoniensis* Hernández-Orts, Montero, Crespo, García, Raga and Aznar, 2012, and first molecular data. Int J Parasitol Parasites Wildl 8:94–105. https://doi.org/10.1016/j.ijppaw.2018.12.008

Hernández-Orts JS, Montero FE, García NA et al (2019b) Transmission of *Corynosoma australe* (Acanthocephala: Polymorphidae) from fishes to South American sea lions *Otaria flavescens* in Patagonia, Argentina. Parasitol Res 118(2):433–440. https://doi.org/10.1007/s00436-018-6177-z

Hernández-Orts JS, Hernández-Mena DI, Pantoja C et al (2021a) A visitor of tropical waters: First record of a Clymene dolphin (*Stenella clymene*) off the Patagonian coast of Argentina, with comments on diet and metazoan Parasites. Front Mar Sci 8:658975. https://doi.org/10.3389/fmars.2021.658975

Hernández-Orts JS, Kuzmina TA, Gomez-Puerta LA et al (2021b) *Diphyllobothrium sprakeri* n. sp. (Cestoda: Diphyllobothriidae): a hidden broad tapeworm from sea lions off North and South America. Parasit Vectors 14(1):219. https://doi.org/10.1186/s13071-021-04661-1

Hernández-Orts JS, Scholz T, Loizaga R et al (2021c) Marine fish imported from Argentina as source of human diphyllobothriosis in Europe? Ecological evidence from dolphins. Zoonoses Public Health 68(6):691–695. https://doi.org/10.1111/zph.12838

Herreras MV, Kaarstad SE, Balbuena JA et al (1997) Helminth parasites of the digestive tract of the harbour porpoise *Phocoena phocoena* in Danish waters: a comparative geographical analysis. Dis Aquat Organ 28:163–167. https://doi.org/10.3354/dao028163

Herzog I, Wohlsein P, Preuss A et al (2024) Heartworm and seal louse: Trends in prevalence, characterisation of impact and transmission pathways in a unique parasite assembly on seals in the North and Baltic Sea. Int J Parasitol Parasites Wildl 23:100898. https://doi.org/10.1016/j.ijppaw.2023.100898

Hirzmann J, Ebmer D, Sánchez-Contreras GJ et al (2021) The seal louse (*Echinophthirius horridus*) in the Dutch Wadden Sea: investigation of vector-borne pathogens. Vectors 14:1–10. https://doi.org/10.1186/s13071-021-04586-9

Hochberg NS, Hamer DH (2010) Anisakidosis: Perils of the deep. Clin Infect Dis 51(7):806–812. https://doi.org/10.1086/656238

Houde M, Measures LN, Hout J (2003) Lungworm (*Pharurus pallasii*: Metastrongyloidea: Pseudaliidae) infection in the endangered St. Lawrence beluga whale (*Delphinapterus leucas*). Can J Zool 81(3):543–551. https://doi.org/10.1139/z03-033

Hrabar J, Bočina I, Gudan Kurilj A et al (2017) Gastric lesions in dolphins stranded along the Eastern Adriatic coast. Dis Aquat Organ 125(2):125–139. https://doi.org/10.3354/dao03137

Hrabar J, Smodlaka H, Rasouli-Dogaheh S et al (2021). Phylogeny and pathology of anisakids parasitizing stranded California sea lions (*Zalophus californianus*) in Southern California. Front Mar Sci 8:636626

Hudson PJ, Dobson AP, Lafferty KD (2006) Is a healthy ecosystem one that is rich in parasites? Trends Ecol Evol 21(7):381–385. https://doi.org/10.1016/j.tree.2006.04.007

IJsseldijk LL, Van Neer A, Deaville R et al (2018) Beached bachelors: An extensive study on the largest recorded sperm whale *Physeter macrocephalus* mortality event in the North Sea. PLoS One 13(8):e0201221. https://doi.org/10.1371/journal.pone.0201221

IJsseldijk LL, Brownlow AC, Mazzariol S (2019) European best practice on cetacean post-mortem investigation and tissue sampling. ACCOBAMS-MOP7/2019/Doc 33. https://doi.org/10.31219/osf.io/zh4ra

Ionita M, García-Varela M, Lyons ET et al (2008) Hookworms (*Uncinaria lucasi*) and acanthocephalans (*Corynosoma* spp. and *Bolbosoma* spp.) found in dead northern fur seals (*Callorhinus ursinus*) on St. Paul Island, Alaska in 2007. Parasitol Res 103(5):1025–1029. https://doi.org/10.1007/s00436-008-1087-0

Iqbal A, Goldfarb DM, Slinger R et al (2015) Prevalence and molecular characterization of *Cryptosporidium* spp. and *Giardia duodenalis* in diarrhoeic patients in the Qikiqtani Region, Nunavut, Canada. Int J Circumpolar Health 74:27713. https://doi.org/10.3402/ijch.v74.27713

Iwasa-Arai T, Serejo CS, Siciliano S et al (2018) The host-specific whale louse (*Cyamus boopis*) as a potential tool for interpreting humpback whale (*Megaptera novaeangliae*) migratory routes. J Exp Mar Bio Ecol 505:45–51. https://doi.org/10.1016/j.jembe.2018.05.001

Jaber JR, Pérez J, Arbelo M et al (2006) Pathological and immunohistochemical study of gastrointestinal lesions in dolphins stranded in the Canary Islands. Vet Rec 159(13):410–414. https://doi.org/10.1136/vr.159.13.410

Jensen T, Idås K (1992) Infection with *Pseudoterranova decipiens* (Krabbe, 1878) larvae in cod (*Gadus morhua*) relative to proximity of seal colonies. Sarsia 76:227–223. https://doi.org/10.1080/00364827.1992.10413478

Jensen SK, Aars J, Lydersen C et al (2010) The prevalence of *Toxoplasma gondii* in polar bears and their marine mammal prey: evidence for a marine transmission pathway? Polar Biol 33:599–606. https://doi.org/10.1007/s00300-009-0735-x

Johnson PT, Dobson A, Lafferty KD et al (2010) When parasites become prey: ecological and epidemiological significance of eating parasites. Trends Ecol Evol 25(6):362–371. https://doi.org/10.1016/j.tree.2010.01.005

Kaliszewska ZA, Seger JON, Rowntree VJ et al (2005) Population histories of right whales (Cetacea: Eubalaena) inferred from mitochondrial sequence diversities and divergences of their whale lice (Amphipoda: *Cyamus*). Mol Ecol 14(11):3439–3456. https://doi.org/10.1111/j.1365-294X.2005.02664.x

Karlsson EA, Ip HS, Hall JS et al (2014) Respiratory transmission of an avian H3N8 influenza virus isolated from a harbour seal. Nat Commun 5:4791. https://doi.org/10.1038/ncomms5791

Katahira H, Matsuda A, Banzai A et al (2021) Gastric ulceration caused by genetically identified Anisakis simplex sensu stricto in a harbor porpoise from the Western Pacific stock. Parasitol Int 83:102327

Kent AJ, Pert CC, Briers RA et al (2020) Increasing intensities of *Anisakis simplex* third-stage larvae (L3) in Atlantic salmon of coastal waters of Scotland. Parasit Vectors 13:62. https://doi.org/10.1186/s13071-020-3942-5

Kleinertz S, Hermosilla C, Ziltener A et al (2014) Gastrointestinal parasites of free-living Indo-Pacific bottlenose dolphins (*Tursiops aduncus*) in the Northern Red Sea, Egypt. Parasitol Res 113(4):1405–1415. https://doi.org/10.1007/s00436-014-3781-4

Klimpel S, Palm HW (2011) Anisakid nematode (Ascaridoidea) life cycles and distribution: increasing zoonotic potential in the time of climate change? In: Mehlhorn H (ed) Progress in parasitology, Parasitology research monographs 2. Springer, Berlin, pp 201–222

Klimpel S, Palm HW, Rückert S et al (2004) The life cycle of *Anisakis simplex* in the Norwegian Deep (northern North Sea). Parasitol Res 94(1):1–9. https://doi.org/10.1007/s00436-004-1154-0

Klotz D, Hirzmann J, Bauer C et al (2018) Subcutaneous merocercoids of *Clistobothrium* sp. in two Cape fur seals (*Arctocephalus pusillus pusillus*). Int J Parasitol Parasites Wildl 7(1):99–105. https://doi.org/10.1016/j.ijppaw.2018.02.003

Komnenou AT, Gkafas GA, Kofidou E et al (2021) First report of *Uncinaria hamiltoni* in orphan eastern Mediterranean monk seal pups in Greece and its clinical significance. Pathogens 10(12):1581. https://doi.org/10.3390/pathogens10121581

Kremnev G, Gonchar A, Krapivin V et al (2020) First elucidation of the life cycle in the family Brachycladiidae (Digenea), parasites of marine mammals. Int J Parasitol 50(12):997–1009. https://doi.org/10.1016/j.ijpara.2020.05.011

Kuchta R, Scholz T (2017) Diphyllobothriidea. In: Caira JN, Jensen K (eds) Planetary biodiversity inventory (2008–2017): tapeworms from vertebrate bowels of the earth, Natural History Museum, Special publication no. 25. University of Kansas, Lawrence, pp 167–189

Kuchta R, Serrano-Martínez ME, Scholz T (2015) Pacific broad tapeworm *Adenocephalus pacificus* as a causative agent of globally reemerging Diphyllobothriosis. Emerg Infect Dis 21(10):1697–1703. https://doi.org/10.3201/eid2110.150516

Kuiken T, Kennedy S, Barrett T et al (2006) The 2000 canine distemper epidemic in Caspian seals (*Phoca caspica*): pathology and analysis of contributory factors. Vet Pathol 43(3):321–338. https://doi.org/10.1354/vp.43-3-321

Kuzmina TA, Tkach VV, Spraker TR et al (2018) Digeneans of northern fur seals *Callorhinus ursinus* (Pinnipedia: Otariidae) from five subpopulations on St. Paul Island, Alaska. Parasitol Res 117(4):1079–1086. https://doi.org/10.1007/s00436-018-5784-z

Kuzmina TA, Kuzmin Y, Dzeverin I et al (2021) Review of metazoan parasites of the northern fur seal (*Callorhinus ursinus*) and the analysis of the gastrointestinal helminth community of the population on St. Paul Island, Alaska. Parasitol Res 120(1):117–132. https://doi.org/10.1007/s00436-020-06935

Lakemeyer J, Siebert U, Abdulmawjood A et al (2020a) Anisakid nematode species identification in harbour porpoises (*Phocoena phocoena*) from the North Sea, Baltic Sea and North Atlantic using RFLP analysis. Int J Parasitol Parasites Wildl 12:93–98. https://doi.org/10.1016/j.ijppaw.2020.05.004

Lakemeyer J, Lehnert K, Woelfing B et al (2020b) Pathological findings in North Sea and Baltic grey seal and harbour seal intestines associated with acanthocephalan infections. Dis Aquat Organ 138:97–110. https://doi.org/10.3354/dao03440

La Linn M, Gardner J, Warrilow D et al (2001) Arbovirus of marine mammals: a new alphavirus isolated from the elephant seal louse, *Lepidophthirus macrorhini*. J Virol 75(9):4103–4109. https://doi.org/10.1128/JVI.75.9.4103-4109.2001

Lang T, Kruse R, Haarich M et al (2017) Mercury species in dab (*Limanda limanda*) from the North Sea, Baltic Sea and Icelandic waters in relation to host-specific variables. Mar Environ Res 124:32–40. https://doi.org/10.1016/j.marenvres.2016.03.001

Lang T, Kotwicki L, Czub M et al (2018) The health status of fish and benthos communities in chemical munitions dumpsites in the Baltic Sea. In: Bełdowski J, Been R, Turmus EK (eds) Towards the monitoring of dumped munitions threat (MODUM): a study of chemical munitions dumpsites in the Baltic Sea. Springer, Dordrecht, pp 129–152

Laskowski Z, Zdzitowiecki K (2017) Acanthocephalans in Sub-Antarctic and Antarctic. In: Klimpel S, Kuhn T, Mehlhorn H (eds) Biodiversity and evolution of parasitic life in the Southern Ocean, Parasitology research monographs, vol 9. Springer, Cham, pp 141–182. https://doi.org/10.1007/98-3-319-46343-8_8

Lehnert K, Raga JA, Siebert U (2005) Macroparasites in stranded and bycaught harbour porpoises from German and Norwegian waters. Dis Aquat Organ 64(3):265–269. https://doi.org/10.3354/dao064265

Lehnert K, Fonfara S, Wohlsein P, Siebert U (2007) Whale lice (*Isocyamus delphinii*) on a harbour porpoise (*Phocoena phocoena*) from German waters. Vet Rec 161(15):526–528. https://doi.org/10.1136/vr.161.15.526

Lehnert K, von Samson-Himmelstjerna G, Schaudien D et al (2010) Transmission of lungworms of harbour porpoises and harbour seals: Molecular tools determine potential vertebrate intermediate hosts. Int J Parasitol 40(7):845–853. https://doi.org/10.1016/j.ijpara.2009.12.008

Lehnert K, Seibel H, Hasselmeier I et al (2014) Increase in parasite burden and associated pathology in harbour porpoises (*Phocoena phocoena*) in West Greenland. Polar Biol 37:321–331. https://doi.org/10.1007/s00300-013-1433-2

Lehnert K, Schwanke E, Hahn K et al (2015) Heartworm (*Acanthocheilonema spirocauda*) and seal louse (*Echinophthirius horridus*) infections in harbour seals (*Phoca vitulina*) from the North and Baltic Seas. J Sea Res 113:65–72. https://doi.org/10.1016/j.seares.2015.06.014

Lehnert K, Randhawa H, Poulin R (2017) Metazoan parasites from odontocetes off NewZealand: new records. Parasitol Res 116(10):2861–2868. https://doi.org/10.1007/s00436-017-5573-0

Lehnert K, IJsseldijk LL, Uy ML et al (2021) Whale lice (*Isocyamus deltobranchium* & *Isocyamus delphinii*; Cyamidae) prevalence in odontocetes off the German and Dutch coasts—morphological and molecular characterization and health implications. Int J Parasitol Parasites Wildl 15:22–30. https://doi.org/10.1016/j.ijppaw.2021.02.015

Lehnert K, Boyi JO, Siebert U (2023a) Potential new species of pseudaliid lung nematode (Metastrongyloidea) from two stranded neonatal orcas (*Orcinus orca*) characterized by ITS-2 and COI sequences. Ecol Evol 13(5):e10036. https://doi.org/10.1002/ece3.10036

Lehnert K, Herzog I, Boyi JO et al (2023b) Heartworms in *Halichoerus grypus*: first records of *Acanthocheilonema spirocauda* (Onchocercidae; Filarioidea) in 2 grey seals from the North Sea. Parasitology 150(9):781–785. https://doi.org/10.1017/S0031182023000501

Leidenberger S, Harding K, Härkönen T (2007) Phocid seals, seal lice and heartworms: a terrestrial host–parasite system conveyed to the marine environment. Dis Aquat Organ 77(3):235–253. https://doi.org/10.3354/dao01823

Leonardi MS (2014) Faster the better: a reliable technique to sample anopluran lice in large hosts. Parasitol Res 113(6):2015–2018. https://doi.org/10.1007/s00436-014-3890-0

Leonardi MS, Crespo EA, Raga JA et al (2012) Scanning electron microscopy of *Antarctophthirus microchir* (Phthiraptera: Anoplura: Echinophthiriidae): studying morphological adaptations to aquatic life. Micron 43(9):929–936. https://doi.org/10.1016/j.micron.2012.03.009

Leonardi MS, Crespo EA, Raga JA et al (2013) Lousy mums: patterns of vertical transmission of an amphibious louse. Parasitol Res 112(9):3315–3123. https://doi.org/10.1007/s00436-013-3511-3

Leonardi MS, Virrueta Herrera S, Sweet A et al (2019) Phylogenomic analysis of seal lice reveals codivergence with their hosts. Syst Entomol 44(4):699–708. https://doi.org/10.1111/syen.12350

Leonardi MS, Krmpotic C, Barbeito C et al (2021) I've got you under my skin: inflammatory response to elephant seal's lice. Med Vet Entomol 35(4):658–662. https://doi.org/10.1111/mve.12538

Levsen A, Berland B (2012) Anisakis species. In: Woo PTK, Buchmann K (eds) Fish parasites, pathobiology and protection 18. CAB, Wallingford, pp 298–309

Lick RR (1991) Untersuchungen zu Lebenszyklus (Krebse-Fische-marine Säuger) und Gefrierresistenz anisakider Nematoden in Nord-und Ostsee. Doctoral dissertation. Christian-Albrechts-Universität, Kiel, Germany

Lombardini E, Haetrakul T, Kuit SH et al (2019) Gastric *Braunina cordiformis* and a review of helminth parasitism in the finless porpoise (*Neophocaena phocaenoides*). Braz J Vet Pathol 12(1):24–26. https://doi.org/10.24070/bjvp.1983-0246.v12i1p24-26

Lucius R, Loos-Frank B, Lane RP (2018) Biologie von Parasiten. Springer, Berlin. https://doi.org/10.1007/978-3-662-54862-2

Lymbery A, Cheah FY (2007) Anisakid nematodes and Anisakiasis. In: Murrell KD, Fried B (eds) Foodborne parasitic zoonoses, World class parasites, vol 11. Springer, New York, pp 185–207

Madsen H (1961) The distribution of *Trichinella spiralis* in sledge dogs and wild mammals in Greenland under a global aspect. Medd Grønl 159(7):1–124

Mariaux J, Kuchta R, Hoberg EP (2017) Tetrabothriidea. In: Caira JN, Jensen K (eds) Planetary biodiversity inventory (2008–2017): tapeworms from vertebrate bowels of the earth, Natural History Museum, Special publication no. 25. University of Kansas, Lawrence, pp 357–370

Marigo J, Rosas FCW, Andrade ALV et al (2002) Parasites of franciscana *Pontoporia blainvillei* from São Paulo and Paraná States, Brazil. Lat Am J Aquat Mamm 1(1):115–122. https://doi.org/10.5597/lajam00015

Martínez R, Segade P, Martínez-Cedeira JA et al (2008) Occurrence of the ectoparasite *Isocyamus deltobranchium* (Amphipoda: Cyamidae) on cetaceans from Atlantic waters. J Parasitol 94(6):1239–1242. https://www.jstor.org/stable/40059190

Martins M, Urbani N, Flanagan C et al (2021) Seroprevalence of *Toxoplasma gondii* in pinnipeds under human care and in wild pinnipeds. Pathogens 10(11):1415. https://doi.org/10.3390/pathogens10111415

Mastick NC, Fiorenza E, Wood CL (2024) Meta-analysis suggests that, for marine mammals, the risk of parasitism by anisakids changed between 1978 and 2015. Ecosphere 15(3):e4781. https://doi.org/10.1002/ecs2.4781

Mathison BA, Mehta N, Couturier MR (2021) Human acanthocephaliasis: a thorn in the side of parasite diagnostics. J Clin Microbiol 59(11):e0269120. https://doi.org/10.1128/JCM.02691-20

Mattiucci S, Nascetti G (2006) Molecular systematics, phylogeny and ecology of anisakid nematodes of the genus *Anisakis* Dujardin, 1845: an update. Parasite 13(2):99–113. https://doi.org/10.1051/parasite/2006132099

Mattiucci S, Nascetti G (2007) Genetic diversity and infection levels of anisakid nematodes parasitic in fish and marine mammals from Boreal and Austral hemispheres. Vet Parasitol 148(1):43–57. https://doi.org/10.1016/j.vetpar.2007.05.009

Mattiucci S, Nascetti G (2008) Advances and trends in the molecular systematics of anisakid nematodes, with implications for their evolutionary ecology and host-parasite co-evolutionary processes. Adv Parasitol 66:47–148. https://doi.org/10.1016/S0065-308X(08)00202-9

Mattiucci S, Abaunza P, Ramadori L et al (2004) Genetic identification of *Anisakis* larvae in European hake from Atlantic and Mediterranean waters for stock recognition. J Fish Biol 65:495–510. https://doi.org/10.1111/j.0022-1112.2004.00465.x

Mattiucci S, Paoletti M, Webb SC (2009) *Anisakis nascettii* n. sp. (Nematoda: Anisakidae) from beaked whales of the southern hemisphere: morphological description, genetic relationships between congeners and ecological data. Syst Parasitol 74(3):199–217. https://doi.org/10.1007/s11230-009-9212-8

Mattiucci S, Fazii P, De Rosa A et al (2013) Anisakiasis and gastroallergic reactions associated with *Anisakis pegreffii* infection, Italy. Emerg Infect Dis 19(3):496–499. https://doi.org/10.1007/s00436-020-06892-0

Mattiucci S, Cipriani P, Webb SC et al (2014) Genetic and morphological approaches distinguish the three sibling species of the *Anisakis simplex* species complex, with a species designation as *Anisakis berlandi* n. sp. for *A. simplex* sp. C (Nematoda: Anisakidae). J Parasitol 100(2):199–214. https://doi.org/10.1645/12-120.1

Mattiucci S, Paoletti M, Colantoni A et al (2017) Invasive anisakiasis by the parasite *Anisakis pegreffii* (Nematoda: Anisakidae): diagnosis by real-time PCR hydrolysis probe system and immunoblotting assay. BMC Infect Dis 17(1):530. https://doi.org/10.1186/s12879-017-2633-0

Mattiucci S, Cipriani P, Levsen A et al (2018) Molecular epidemiology of *Anisakis* and anisakiasis: An ecological and evolutionary road map. Adv Parasitol 99:93–263. https://doi.org/10.1016/bs.apar.2017.12.001

Mazzariol S, Marcer F, Mignone W et al (2012) Dolphin Morbillivirus and *Toxoplasma gondii* coinfection in a Mediterranean fin whale (*Balaenoptera physalus*). BMC Vet Res 8:20. https://doi.org/10.1186/1746-6148-8-20

Mazzariol S, Centelleghe C, Petrella A et al (2021) Atypical *Toxoplasmosis* in a Mediterranean Monk Seal (*Monachus monachus*) Pup. J Comp Pathol 184:65–71. https://doi.org/10.1016/j.jcpa.2021.02.005

McClelland G (2002) The trouble with sealworms (*Pseudoterranova decipiens* species complex, Nematoda): a review. Parasitology 124(Suppl):S183–S203. https://doi.org/10.1017/s0031182002001658

McCosker C, Flanders K, Ono K et al (2020) Metabarcoding fecal DNA reveals extent of *Halichoerus grypus* (Gray Seal) foraging on invertebrates and incidence of parasite exposure. Northeast Nat 27(4):681–700. https://doi.org/10.1656/045.027.0409

Measures LN (2001) Lungworms of Marine Mammals. In: Samuel WM, Pybus MJ, Kocan AA (eds) Parasitic diseases of wild mammals, 2nd edn. Iowa State University Press, Ames, pp 279–300

Mehlhorn B, Mehlhorn H, Plötz J (2002) Light and scanning electron microscopical study on *Antarctophthirus ogmorhini* lice from the Antarctic seal *Leptonychotes weddellii*. Parasitol Res 88(7):651–660. https://doi.org/10.1007/s00436-002-0630-7

Meise K, García-Parra C (2015) Behavioural and environmental correlates of *Philophthalmus zalophi* infections and their impact on survival in juvenile

Galapagos sea lions. Mar Biol 162:2107–2117. https://doi.org/10.1007/s00227-015-2740-7

Miller M, Shapiro K, Murray MJ et al (2018) Protozoan parasites of marine mammals. In: Gulland FMD, Dierauf LA, Whitman KL (eds) CRC handbook of marine mammal medicine, 3rd edn. CRC, Boca Raton, pp 425–470

Miller MA, Newberry CA, Sinnott DM et al (2023) Newly detected, virulent *Toxoplasma gondii* COUG strain causing fatal steatitis and toxoplasmosis in southern sea otters (*Enhydra lutris nereis*). Front Mar Sci 10:1116899. https://doi.org/10.3389/fmars.2023.1116899

Mladineo I, Charouli A, Jelić F et al (2023) In vitro culture of the zoonotic nematode *Anisakis pegreffii* (Nematoda, Anisakidae). Parasit Vectors 16(1):51. https://doi.org/10.1186/s13071-022-05629-5

Møller LN, Petersen E, Kapel CM et al (2005) Outbreak of trichinellosis associated with consumption of game meat in West Greenland. Vet Parasitol 132(1–2):131–136. https://doi.org/10.1016/j.vetpar.2005.05.041

Mondragón-Martínez A, Marroquin-Vilchez D, Martínez-Rojas R et al (2024) Molecular identification and prevalence of plerocercoid larvae (Cestoda: Diphyllobothriidae) in some commercial fish species from Peru. Parasitol Res 123:243. https://doi.org/10.1007/s00436-024-08267-1

Moratal S, Dea-Ayuela MA, Cardells J et al (2020) Potential risk of three zoonotic Protozoa (*Cryptosporidium* spp., *Giardia duodenalis*, and *Toxoplasma gondii*) transmission from fish consumption. Foods 9(12):1913. https://doi.org/10.3390/foods9121913

Morell M, Lehnert K, IJsseldijk LL et al (2017) Parasites in the inner ear of harbour porpoise: cases from the North and Baltic Seas. Dis Aquat Organ 127(1):57–63. https://doi.org/10.3354/dao03178

Muñoz-Caro T, Machuca A, Morales P et al (2022) Prevalence and molecular identification of zoonotic *Anisakis* and *Pseudoterranova* species in fish destined to human consumption in Chile. Parasitol Res 121(5):1295–1304. https://doi.org/10.1007/s00436-022-07459-x

Nadler SA, D'Amelio S, Dailey MD et al (2005) Molecular phylogenetics and diagnosis of *Anisakis*, *Pseudoterranova*, and *Contracaecum* from northern Pacific marine mammals. J Parasitol 91(6):1413–1429. https://doi.org/10.1645/GE-522R.1

Nagasawa K (1990) The life cycle of *Anisakis simplex*: A review. In: Ishikura H, Kikuchi K (eds) Intestinal anisakiasis in Japan. Infected fish, sero-immunological diagnosis, and prevention. Springer, Tokyo, pp 31–40

Nakagun S, Shiozaki A, Ochiai M et al (2018) Prominent hepatic ductular reaction induced by *Oschmarinella macrorchis* in a Hubbs' beaked whale *Mesoplodon carlhubbsi*, with biological notes. Dis Aquat Organ 127(3):177–192. https://doi.org/10.3354/dao03201

Näreaho A, Eriksson-Kallio AM, Heikkinen P et al (2017) High prevalence of zoonotic trematodes in roach (*Rutilus rutilus*) in the Gulf of Finland. Acta Vet Scand 59(1):75. https://doi.org/10.1186/s13028-017-0343-7

Nascetti G, Cianchi R, Mattiucci S et al (1993) Three sibling species within *Contracaecum osculatum* (Nematoda, Ascaridida, Ascaridoidea) from the Atlantic Arctic-Boreal region: reproductive isolation and host preferences. Int J Parasitol 23(1):105–120. https://doi.org/10.1016/0020-7519(93)90103-6

Neimanis AS, Moraeus C, Bergman A et al (2016) Emergence of the zoonotic biliary trematode *Pseudamphistomum truncatum* in grey seals (*Halichoerus grypus*) in the Baltic Sea. PLoS One 11(10):e0164782. https://doi.org/10.1371/journal.pone.0164782

Neimanis A, Stavenow J, Ågren EO et al (2022) Causes of death and pathological findings in stranded harbour porpoises (*Phocoena phocoena*) from Swedish waters. Animals 12(3):369. https://doi.org/10.3390/ani12030369

Nickol BB, Helle E, Valtonen ET (2002) *Corynosoma magdaleni* in Gray seals from the Gulf of Bothnia, with emended descriptions of *Corynosoma strumosum* and *Corynosoma magdaleni*. J Parasitol 88(6):222–229. https://doi.org/10.2307/3285497

Nyman T, Papadopoulou E, Ylinen E et al (2021) DNA barcoding reveals different cestode helminth species in northern European marine and freshwater ringed seals. Int J Parasitol Parasites Wildl 15:255–261. https://doi.org/10.1016/j.ijppaw.2021.06.004

Olsson M (1994) Effects of persistent organic pollutants on biota in the Baltic Sea. In: Bolt HM, Hellman B, Dencker L (eds) Use of mechanistic information in risk assessment. Archives of toxicology. Supplement 16, vol 16. Springer, Berlin, pp 43–52. https://doi.org/10.1007/978-3-642-78640-2_5

O'Neill G, Whelan J (2002) The occurrence of *Corynosoma strumosum* in the grey seal, *Halichoerus grypus*, caught off the Atlantic coast of Ireland. J Helminthol 76(3):231–234. https://doi.org/10.1079/JOH2002117

Orlofske SA, Jadin RC, Johnson PT (2015) It's a predator-eat-parasite world: how characteristics of predator, parasite and environment affect consumption. Oecologia 178(2):537–547. https://doi.org/10.1007/s00442-015-3243-4

Osinga N, Shahi Ferdous MM, Morick D et al (2012) Patterns of stranding and mortality in common seals (*Phoca vitulina*) and grey seals (*Halichoerus grypus*) in The Netherlands between 1979 and 2008. J Comp Pathol 147(4):550–565. https://doi.org/10.1016/j.jcpa.2012.04.001

Paggi L, Nascetti G, Cianchi R et al (1991) Genetic evidence for three species within *Pseudoterranova decipiens* (Nematoda, Ascaridida, Ascaridoidea) in the North Atlantic and Norwegian and Barents Seas. Int J Parasitol 21(2):195–212. https://doi.org/10.1016/0020-7519(91)90010-5

Pasqualetti MI, Fariña FA, Krivokapich SJ et al (2018) *Trichinella spiralis* in a South American sea lion (*Otaria flavescens*) from Patagonia, Argentina. Parasitol Res 117(12):4033–4036. https://doi.org/10.1007/s00436-018-6116-z

Pereira EM, Müller G, Secchi E et al (2013) Digenetic trematodes in South American sea lions from southern Brazilian waters. J Parasitol 99(5):910–913. https://doi.org/10.1645/GE-3216.1

Pesapane R, Archibald W, Norris T et al (2021) Nasopulmonary mites (Halarachnidae) of coastal Californian pinnipeds: Identity, prevalence, and molecular characterization. International Journal for Parasitology: Parasites and Wildlife 16:113–119.

Pfeiffer CJ (2009) Whale lice. In: Perrin WF, Würsig B, Thewissen JGM (eds) Encyclopedia of marine mammals, 2nd edn. Academic, Cambridge, pp 1220–1223. https://doi.org/10.1016/B978-0-12-373553-9.00279-0

Phillips BE, Páez-Rosas D, Flowers JR et al (2018) Evaluation of the ophthalmic disease and histopathologic effects due to the ocular trematode *Philophthalmus zalophi* on juvenile Galapagos sea lions (*Zalophus wollebaeki*). J Zoo Wildl Med 49(3):581–590. https://doi.org/10.1638/2017-0096.1

Pimiento C, Leprieur F, Silvestro D et al (2020) Functional diversity of marine megafauna in the Anthropocene. Sci Adv 6(16):eaay7650. https://doi.org/10.1126/sciadv.aay7650

Pons-Bordas C, Hazenberg A, Hernandez-Gonzalez A et al (2020) Recent increase of ulcerative lesions caused by Anisakis spp. in cetaceans from the northeast Atlantic. J Helminthol 94:e127

Pool R, Chandradeva N, Gkafas G et al (2020a) Transmission and predictors of burden of lungworms of the striped dolphin (*Stenella coeruleoalba*) in the Western Mediterranean. J Wildl Dis 56(1):186–191. https://doi.org/10.7589/2018-10-242

Pool R, Fernández M, Chandradeva N et al (2020b) The taxonomic status of *Skrjabinalius guevarai* Gallego & Selva, 1979 (Nematoda: Pseudaliidae) and the synonymy of *Skrjabinalius* Delyamure, 1942 and *Halocercus* Baylis & Daubney, 1925. Syst Parasitol 97:389–401. https://doi.org/10.1007/s11230-020-09921-9

Pool R, Shiozaki A, Raga JA et al (2023) Molecular phylogeny of the Pseudaliidae (Nematoda) and the origin of associations between lungworms and marine mammals. Int J Parasitol Parasites Wildl 20:192–202. https://doi.org/10.1016/j.ijppaw.2023.03.002

Preuss A, Büscher TH, Herzog I et al (2024) Attachment performance of the ectoparasitic seal louse Echinophthirius horridus. Communications biology, 7(1):36.

Puryear W, Sawatzki K, Hill N et al (2023) Highly pathogenic avian influenza A(H5N1) virus outbreak in New England seals, United States. Emerg Infect Dis 29(4):786–791. https://doi.org/10.3201/eid2904.221538

Raga JA, Balbuena JA, Aznar J et al (1997) The impact of parasites on marine mammals: a review. Parassitologia 39(4):293–296

Raga JA, Fernández M, Balbuena JA, Aznar FJ (2009) Parasites. In: Perrin WF, Würsig B, Thewissen JGM (eds) Encyclopedia of marine mammals, 2nd edn. Academic, Cambridge, pp 821–830. https://doi.org/10.1016/B978-0-12-373553-9.00193-0

Rahmati AR, Kiani B, Afshari A et al (2020) World-wide prevalence of *Anisakis* larvae in fish and its relationship to human allergic anisakiasis: a systematic review. Parasitol Res 119(11):3585–3594. https://doi.org/10.1007/s00436-020-06892-0

Rausch RL (1953) Studies on the helminth fauna of Alaska. XIII. Disease in the sea otter, with special reference to helminth parasites. Ecology 34(3):584–604. https://doi.org/10.2307/1929729

Rausch RL (1970) Trichinosis in the arctic. In: Gould SE (ed) Trichinosis in man and animals. Charles C. Thomas, Springfield, pp 348–373

Rausch RL, Fay FH (1966) Studies of the helminth fauna of Alaska. XLIV. Revision of *Ogmogaster* Jägerskiöld, 1891, with a description of *O. pentalineatus* sp. n. (Trematoda: Notocotylidae). J Parasitol 52(1):26–38

Rausch R, Babero BB, Rausch VR et al (1956) Studies on the helminth fauna of Alaska. XXVII. The occurrence of larvae of *Trichinella spiralis* in Alaskan mammals. J Parasitol 42(3):259–271. https://doi.org/10.2307/3274850

Reboredo-Fernández A, Prado-Merini Ó, García-Bernadal T et al (2014) Benthic macroinvertebrate communities as aquatic bioindicators of contamination by *Giardia* and *Cryptosporidium*. Parasitol Res 113:1625–1628. https://doi.org/10.1007/s00436-014-3807-y

Reboredo-Fernández A, Ares-Mazás E, Martínez-Cedeira JA et al (2015) *Giardia* and *Cryptosporidium* in cetaceans on the European Atlantic coast. Parasitol Res 114(2):693–698. https://doi.org/10.1007/s00436-014-4235-8

Reckendorf A, Ludes-Wehrmeister E, Wohlsein P et al (2018) First record of *Halocercus* sp. (Pseudaliidae) lungworm infections in two stranded neonatal orcas (*Orcinus orca*). Parasitology 145(12):1553–1557. https://doi.org/10.1017/S0031182018000586

Reckendorf A, Wohlsein P, Lakemeyer J et al (2019) There and back again—the return of the nasal mite *Halarachne halichoeri* to seals in German waters. Int J Parasitol Parasites Wildl 9:112–118. https://doi.org/10.1016/j.ijppaw.2019.04.003

Reddy ML, Dierauf LA, Gulland FMD (2001) Marine mammals as sentinels of ocean health. In: Dierauf L, Gulland F (eds) Marine mammal medicine. CRC, Boca Raton, pp 3–13

Reif JS (2011) Animal sentinels for environmental and public health. Public Health Rep 126(Suppl 1):50–57. https://doi.org/10.1177/00333549111260S108

Reiling SJ, Measures L, Feng S et al (2019) *Toxoplasma gondii*, *Sarcocystis* sp. and *Neospora caninum*-like parasites in seals from northern and eastern Canada: potential risk to consumers. Food Waterborne Parasitol 17:e00067. https://doi.org/10.1016/j.fawpar.2019.e00067

Rengifo-Herrera C, Ortega-Mora LM, Álvarez-García G et al (2012) Detection of *Toxoplasma gondii* antibodies in Antarctic pinnipeds. Vet Parasitol 190(1–2):259–262. https://doi.org/10.1016/j.vetpar.2012.05.020

Rengifo-Herrera C, Ortega-Mora LM, Gómez-Bautista M et al (2013) Detection of a novel genotype of *Cryptosporidium* in Antarctic pinnipeds. Vet Parasitol 191(1–2):112–118. https://doi.org/10.1016/j.vetpar.2012.08.021

Rohner S, Boyi JO, Artemeva V et al (2023) Back from exile? First records of chewing lice (*Lutridia exilis*; Ischnocera; Mallophaga) in growing Eurasian otter (*Lutra lutra*) populations from Northern Germany. Pathogens 12(4):587. https://doi.org/10.3390/pathogens12040587

Rojano-Doñate L, McDonald BI, Wisniewska DM et al (2018) High field metabolic rates of wild harbour porpoises. J Exp Biol 221(23):jeb185827. https://doi.org/10.1242/jeb.185827

Roussel de Vauzème M (1834) Mémoire sur le Cyamus ceti (Latr.) de la classe des Crustacés. Ann Sci Nat Zoo 2e Série 1:239–265

Rowntree VJ (1996) Feeding, distribution, and reproductive behavior of cyamids (Crustacea: Amphipoda) living on humpback and right whales. Can J Zool 74(1):103–109. https://doi.org/10.1139/z96-014

Schell DM, Rowntree VJ, Pfeiffer CJ (2000) Stable-isotope and electron-microscopic evidence that cyamids (Crustacea: Amphipoda) feed on whale skin. Can J Zool 78(5):721–727. https://doi.org/10.1139/z99-249

Scholz T, Kuchta R (2016) Fish-borne, zoonotic cestodes (*Diphyllobothrium* and relatives) in cold climates: A never-ending story of neglected and (re)-emergent parasites. Food Waterborne Parasitol 4:23–38. https://doi.org/10.1016/j.fawpar.2016.07.002

Scholz T, Kuchta R, Brabec J (2019) Broad tapeworms (Diphyllobothriidae), parasites of wildlife and humans: Recent progress and future challenges. Int J Parasitol Parasites Wildl 9:359–369. https://doi.org/10.1016/j.ijppaw.2019.02.001

Sedlak-Weinstein E (1992) The occurrence of a new species of *Isocyamus* (Crustacea, Amphipoda) from Australian and Japanese pilot whales, with a key to species of *Isocyamus*. J Nat Hist 26(5):937–946. https://doi.org/10.1080/00222939200770561

Seguel M, Gottdenker N (2017) The diversity and impact of hookworm infections in wildlife. Int J Parasitol Parasites Wildl 6(3):177–194. https://doi.org/10.1016/j.ijppaw.2017.03.007

Selbach C, Mouritsen KN, Poulin R et al (2022) Bridging the gap: aquatic parasites in the One Health concept. Trends Parasitol 38(2):109–111. https://doi.org/10.1016/j.pt.2021.10.007

Serbina EA (2016) Cercariae *Opisthorchis felineus* and *Metorchis bilis* from first intermediate hosts for the first time in basin of Chany lake (Novosibirsk region, Russia) is found. Russ J Parasitol 37(3):421–429. (in Russian)

Seymour J, Horstmann-Dehn L, Rosa C et al (2014) Occurrence and genotypic analysis of *Trichinella* species in Alaska marine-associated mammals of the Bering and Chukchi seas. Vet Parasitol 200(1–2):153–164. https://doi.org/10.1016/j.vetpar.2013.11.015

Shanebeck KM, Lagrue C (2020) Acanthocephalan parasites in sea otters: Why we need to look beyond associated mortality…. Mar Mamm Sci 36(2):676–689. https://doi.org/10.1111/mms.12659

Shanebeck K, Lakemeyer J, Siebert U et al (2020) Habitat selection and populations of *Corynosoma* (Acanthocephala) in the intestines of sea otters (*Enhydra lutris*) and seals. J Helminthol 94:e211. https://doi.org/10.1017/S0022149X20000747

Shapiro K, Bahia-Oliveira L, Dixon B et al (2019) Environmental transmission of *Toxoplasma gondii*: Oocysts in water, soil and food. Food Waterborne Parasitol 15:e00049. https://doi.org/10.1016/j.fawpar.2019.e00049

Sherrard-Smith E, Stanton DW, Cable J et al (2016) Distribution and molecular phylogeny of biliary trematodes (Opisthorchiidae) infecting native *Lutra lutra* and alien *Neovison vison* across Europe. Parasitol Int 65(2):163–170. https://doi.org/10.1016/j.parint.2015.11.007

Siebert U, Wünschmann A, Weiss R et al (2001) Post-mortem findings in harbour porpoises (*Phocoena phocoena*) from the German North and Baltic seas. J Comp Pathol 124(2–3):102–114. https://doi.org/10.1053/jcpa.2000.0436

Siebert U, Tolley K, Vikingsson GA, Olafsdottir D, Lehnert K, Weiss R, Baumgärtner W (2006) Pathological findings in harbour porpoises (*Phocoena phocoena*) from Norwegian and Icelandic waters. J. Comp. Pathol. 134(2–3):134–142

Siebert U, Wohlsein P, Lehnert K et al (2007) Pathological findings in harbour seals (*Phoca vitulina*): 1996-2005. J Comp Pathol 137(1):47–58. https://doi.org/10.1016/j.jcpa.2007.04.018

Siebert U, Pawliczka I, Benke H et al (2020) Health assessment of harbour porpoises (*Phocoena phocoena*) from Baltic area of Denmark, Germany, Poland and Latvia. Environ Int 143:105904. https://doi.org/10.1016/j.envint.2020.105904

Sikes RS, The Animal Care and Use Committee of the American Society of Mammalogists (2016) Guidelines of the American Society of Mammalogists for the use of wild mammals in research and education. J Mammal 97(3):663–688. https://doi.org/10.1093/jmammal/gyw098

Simões SBE, Barbosa HS, Santos CP (2010) The life cycle of *Ascocotyle* (*Phagicola*) *longa* (Digenea: Heterophyidae), a causative agent of fish-borne trematodosis. Acta Tropica 113(3):226–233. https://doi.org/10.1016/j.actatropica.2009.10.020

Sinisalo T, Valtonen ET (2003) *Corynosoma* acanthocephalans in their paratenic fish hosts in the northern Baltic Sea. Parasite 10(3):227–233. https://doi.org/10.1051/parasite/2003103227

Skírnisson K, Marucci G, Pozio E (2010) *Trichinella nativa* in Iceland: an example of *Trichinella* dispersion in a frigid zone. J Helminthol 84(2):182–185. https://doi.org/10.1017/S0022149X09990514

Smith JW (1983) *Anisakis simplex* (Rudolphi, 1809, det. Krabbe, 1878) (Nematoda: Ascaridoidea): morphology and morphometry of larvae from euphausiids and fish, and a review of the life-history and ecology. J Helminthol 57(3):205–224. https://doi.org/10.1017/s0022149x00009512

Smith JW, Wootten R (1978) Anisakis and anisakiasis. Adv Parasitol 16:93–163. https://doi.org/10.1016/s0065-308x(08)60573-4

Spraker TR, DeLong RL, Lyons ET et al (2007) Hookworm enteritis with bacteremia in California sea lion pups on San Miguel Island. J Wildl Dis 43(2):179–188. https://doi.org/10.7589/0090-3558-43.2.179

Sromek L, Ylinen E, Kunnasranta M et al (2023) Loss of species and genetic diversity during colonization: Insights from acanthocephalan parasites in northern European seals. Ecol Evol 13(10):e10608. https://doi.org/10.1002/ece3.10608

Stewart JR, Gast RJ, Fujioka RS et al (2008) The coastal environment and human health: microbial indicators, pathogens, sentinels and reservoirs. Environ Health 7(Suppl 2):S3. https://doi.org/10.1186/1476-069X-7-S2-S3

St. Leger J, Raverty S, Mena A (2018) Cetacea. In: Terio KA, McAloose D, St. Leger J (eds) Pathology of wildlife and zoo animals. Academic, Cambridge, pp 565–568. https://doi.org/10.1016/B978-0-12-805306-5.00022-5

Strauss V, Claussen D, Jäger M, Ising S, Schnieder T, Stoye M (1991) The helminth fauna of the common seal (*Phoca vitulina vitulina*, Linné, 1758) from the Wadden sea in lower saxony Part 1: Trematodes, cestodes and acantocephala. J Vet Med B 38(1-10):641–648

Takano T, Sata N (2022) Multigene phylogenetic analysis reveals non-monophyly of *Anisakis* s.l. and *Pseudoterranova* (Nematoda: Anisakidae). Parasitol Int 91:102631. https://doi.org/10.1016/j.parint.2022.102631

Ten S, Konishi K, Raga JA et al (2022a) Epibiotic fauna of the Antarctic minke whale as a reliable indicator of seasonal movements. Sci Rep 12:22214. https://doi.org/10.1038/s41598-022-25929-1

Ten S, Raga JA, Aznar FJ (2022b) Epibiotic fauna on cetaceans worldwide: A systematic review of records and indicator potential. Front Mar Sci 9:846558

Thompson RCA (2023) Zoonotic helminths—why the challenge remains. J Helminthol 97:e21. https://doi.org/10.1017/S0022149X23000020

Tomo I, Kemper CM, Lavery TJ (2010) Eighteen-year study of South Australian dolphins shows variation in lung nematodes by season, year, age class, and location. J Wildl Dis 46(2):488–498. https://doi.org/10.7589/0090-3558-46.2.488

Tryland M (2000) Zoonoses of arctic marine mammals. Infect Dis Rev 2(2):55–64

Tryland M (Ed.) (2022) Arctic one health: Challenges for northern animals and people. Springer Nature

Tryland M, Nesbakken T, Robertson L et al (2014) Human pathogens in marine mammal meat–a northern perspective. Zoonoses Public Health 61(6):377–394. https://doi.org/10.1111/zph.12080

Ulrich SA, Lehnert K, Siebert U et al (2015) A recombinant antigen-based enzyme-linked immunosorbent assay (ELISA) for lungworm detection in seals. Parasit Vectors 8:443. https://doi.org/10.1186/s13071-015-1054-4

Valtonen ET, Helle E, Poulin R (2004) Stability of *Corynosoma* populations with fluctuating population densities of the seal definitive host. Parasitology 129:635–642. https://doi.org/10.1017/s0031182004005839

Van Bressem MF, Raga JA, Di Guardo G et al (2009) Emerging infectious diseases in cetaceans worldwide and the possible role of environmental stressors. Dis Aquat Organ 86(2):143–157. https://doi.org/10.3354/dao02101

van de Velde N, Devleesschauwer B, Leopold M et al (2016) *Toxoplasma gondii* in stranded marine mammals from the North Sea and Eastern Atlantic Ocean: Findings and diagnostic difficulties. Vet Parasitol 230:25–32. https://doi.org/10.1016/j.vetpar.2016.10.021

Waeschenbach A, Brabec J, Scholz T et al (2017) The catholic taste of broad tapeworms—multiple routes to human infection. Int J Parasitol 47(13):831–843. https://doi.org/10.1016/j.ijpara.2017.06.004

Waindok P, Lehnert K, Siebert U et al (2018) Prevalence and molecular characterisation of Acanthocephala in pinnipedia of the North and Baltic Seas. Int J Parasitol Parasites Wildl 7(1):34–43. https://doi.org/10.1016/j.ijppaw.2018.01.002

Waltzek TB, Cortés-Hinojosa G, Wellehan JFX Jr et al (2012) Marine mammal zoonoses: a review of disease manifestations. Zoonoses Public Health 59(8):521–535. https://doi.org/10.1111/j.1863-2378.2012.01492.x

Wood CL, Welicky RL, Preisser WC et al (2023) A reconstruction of parasite burden reveals one century of climate-associated parasite decline. Proc Natl Acad Sci USA 120(3):e2211903120. https://doi.org/10.1073/pnas.2211903120

Woolford L, Franklin C, Whap T et al (2015) Pathological findings in wild harvested dugongs *Dugong dugon* of central Torres Strait, Australia. Dis Aquat Organ 113(2):89–102. https://doi.org/10.3354/dao02825

Wund S, Meheust E, Dars C et al (2023) Strengthening the health surveillance of marine mammals in the waters of metropolitan France by monitoring strandings. Front Mar Sci 10:1116819. https://doi.org/10.3389/fmars.2023.1116819

Zhu X, Gasser RB, Podolska M et al (1998) Characterisation of anisakid nematodes with zoonotic potential by nuclear ribosomal DNA sequences. Int J Parasitol 28(12):1911–1921. https://doi.org/10.1016/s0020-7519(98)00150-7

Zhu XQ, Podolska M, Liu JS et al (2007) Identification of anisakid nematodes with zoonotic potential from Europe and China by single-strand conformation polymorphism analysis of nuclear ribosomal DNA. Parasitol Res 101(6):1703–1707. https://doi.org/10.1007/s00436-007-0699-0

Open Access This chapter is licensed under the terms of the Creative Commons Attribution-NonCommercial-NoDerivatives 4.0 International License (http://creativecommons.org/licenses/by-nc-nd/4.0/), which permits any non-commercial use, sharing, distribution and reproduction in any medium or format, as long as you give appropriate credit to the original author(s) and the source, provide a link to the Creative Commons license and indicate if you modified the licensed material. You do not have permission under this license to share adapted material derived from this chapter or parts of it.

The images or other third party material in this chapter are included in the chapter's Creative Commons license, unless indicated otherwise in a credit line to the material. If material is not included in the chapter's Creative Commons license and your intended use is not permitted by statutory regulation or exceeds the permitted use, you will need to obtain permission directly from the copyright holder.

16. Host–Parasite Trophic Interactions as Revealed by Stable Isotope Analyses: Determinants for Trophic and Isotopic Niches of Hosts and Their Associated Parasites

Milen Nachev, Philip M. Riekenberg, Maik A. Jochmann, Ana Born-Torrijos, Marcel T. J. van der Meer, Nico J. Smit, Torsten C. Schmidt, David W. Thieltges, and Bernd Sures

Abstract

The analysis of stable isotopes (SIA) in organisms, especially those of carbon ($^{13}C/^{12}C$) and nitrogen ($^{15}N/^{14}N$), is a well-established and powerful tool for analysing trophic interactions between free-living organisms and energy and nutrient fluxes in ecosystems. However, SIA has only been increasingly used for parasitological studies in the last two decades. In this chapter, the basics of stable isotope chemistry and isotope ecology are first introduced in order to reach scientists who are not familiar with this field of research. It then reviews the main studies on host–parasite interactions and summarises the fractionation patterns observed between different parasitic taxa and their associated hosts, also highlighting the potential impact of parasitism on the isotopic composition of the hosts. In addition, different approaches to study nutrient uptake in parasites using mixing models are presented. Another focus is on nutrient fluxes within host-parasite systems using compound-specific stable isotope analysis, which has been used in very few studies to date. At the end, we present a short conclusion and outlook addressing urgent research needs.

16.1 Introduction

Parasites are important components of aquatic ecosystems, as they account for a considerable proportion of biomass and biodiversity (Kuris et al. 2008; Soldánová et al. 2016; see also Chaps. 2–6) and are intricately involved in food webs (Dobson et al. 2008; Kuris et al. 2008). To characterise the involvement of parasites in food webs as well as to unravel trophic relationships with their hosts, the use of stable isotope analysis (SIA) is increasing (Sabadel et al. 2019; Thieltges et al. 2019; Born-Torrijos et al. 2023). For approximately 40 years, stable isotope analyses have played a crucial role in the ecology of free-living species by providing valuable insights into various ecological processes (e.g. Peterson and Fry 1987; Fry 2006; Michener and Lajtha 2008; Boecklen et al. 2011). These analyses involve measuring the ratios of stable isotopes (different forms of an element with the same number of protons and electrons, but a different number of neutrons, mainly carbon ($^{13}C/^{12}C = \delta^{13}C$) and nitrogen ($^{15}N/^{14}N = \delta^{15}N$) in ecological samples involving plants, animals, or environmental materials. Stepwise differences (called discrimination factor, Δ) arise in the isotopic ratios of naturally occurring stable isotopes of carbon ($\delta^{13}C$) and nitrogen ($\delta^{15}N$) between consumers at different trophic levels, resulting from a process called isotope fractionation (Fry 2006). By analysing stable isotope ratios, information about an organism's diet, trophic position, migration patterns, and habitat use can be gained. For example, carbon isotopes can reveal the primary source of an organism's food, while nitrogen isotopes can indicate an organism's position in the food chain. Accordingly, SIA has been successfully applied to explore the underlying resources supporting food webs (Demopoulos et al. 2007; Woodland and Secor 2013; Christianen et al. 2017) and to identify the diet resources that consumers rely upon (Fry and Arnold 1982; Then et al. 2021; Riekenberg et al. 2022). Despite this long history, the application of SIA in ecological studies is still growing exponentially because the information it provides is independent of other, more traditional, methods to study food webs such as stomach content analysis or direct observation of trophic interactions. This independent information comes from the elements incorporated into consumer tissue from their diet, providing an indication of diet integrated across the period during which that tissue was formed (Vander Zanden et al. 2015). This technique has become particularly useful given new analytical advancements that allow for the isolation and measurement of carbon and nitrogen from individual

---

D. W. Thieltges
Department of Coastal Systems, NIOZ Royal Netherlands Institute for Sea Research,
Den Burg, Texel, The Netherlands

Groningen Institute for Evolutionary Life Sciences, GELIFES, University of Groningen,
Groningen, The Netherlands
e-mail: david.thieltges@nioz.nl

B. Sures
Department of Aquatic Ecology, University of Duisburg-Essen, Essen, Germany

Centre for Water and Environmental Research, University of Duisburg-Essen, Essen, Germany

Water Research Group, Unit for Environmental Sciences and Management, North-West University, Potchefstroom, South Africa

Research Center One Health Ruhr, Research Alliance Ruhr, University Duisburg-Essen, Essen, Germany
e-mail: bernd.sures@uni-due.de

compounds such as amino acids and fatty acids that give further insight to sources of organic carbon (Kürten et al. 2013; Larsen et al. 2013; see Sect. 16.4) and exploring trophic relationships within food webs (McClelland and Montoya 2002; Chikaraishi et al. 2009). Despite these advances, ecological studies examining food webs using SIA techniques still rarely consider or include parasites, partly due to the complexity and multi-host nature of host–parasite interaction networks (Gómez-Díaz and González-Solís 2010).

In this chapter, we give a basic introduction to the principles of stable isotope techniques and provide a starting point for parasitologists to make the most of their efforts in applying these tools to elucidate parasite–host interactions and the impact of these relationships on food webs. In addition, we summarise the main results of the studies published so far on stable isotope fractionation in parasites and their hosts and show how the patterns of stable isotope fractionation differ in various parasite groups and to what extent the parasites, for their part, can also influence the patterns of stable isotope distribution in the tissues of their hosts.

16.2 Stable Isotope Analyses: Understanding the Basics

There are multiple forms of elemental C, N, H, O, and S that have the same number of electrons and protons, but differing numbers of neutrons (see Fig. 16.1). This results in masses that are different between the different forms of the same element. Some of the forms, such as hydrogen and carbon with two additional neutrons and a mass of 3 (^{3}H) and 14 (^{14}C), respectively, are subject to radioactive decay; others do not decay and are therefore called stable isotopes (e.g. ^{2}H, ^{13}C, ^{15}N, ^{18}O, ^{34}S). There are two other, very rare, stable forms of sulphur containing additional neutrons, ^{33}S and ^{36}S; these are normally not included in stable isotope measurements. On Earth, the most abundant forms are the lighter isotopes (e.g. > 95% for ^{1}H, ^{12}C, ^{14}N, ^{16}O, ^{32}S). Although the chemical properties of both the lighter and heavier isotopes are the same, they slightly differ in the rate of formation and breaking of bonds in biochemical reactions, resulting in isotope effects. Thus, molecules with different numbers of heavy and light isotopes (isotopologues) will go through reactions at different rates.

The mass-dependent processes affecting isotope values result in isotope fractionation that can be divided into equilibrium isotope effects and kinetic isotope effects. Equilibrium isotope effects are a result of thermodynamic processes, usually between two substances in chemical equilibrium exchange reactions that simultaneously proceed both forward and backward and will eventually come to an equilibrium, with the heavier isotopes concentrating where the bond strengths are strongest. Kinetic isotope effects are typically larger than equilibrium effects and have their origin in the energy differences between isotopologues during the formation of an active transitional complex in unidirectional, rate-dependent reactions such as diffusion and chemical reactions. A thorough introduction to isotope fractionation through various isotope effects is beyond the scope of this chapter, but further information can be found in Wolfsberg et al. (2009) and Meier-Augenstein (2004). As a result of isotope effects, measurable isotopic fractionation occurs as the lighter and heavier forms proceed through reactions at different rates due to the relative difference in their masses, giving rise to differences that are large enough to result in unique and measurably different stable isotopic compositions.

Biological processes are generally unidirectional and dominated by kinetic isotope effects. Due to the distinctly different energy costs associated with breaking chemical bonds in these molecules, bond cleavage of the lighter isotopic species (^{12}C) occurs more quickly than bond cleavage involving the heavier isotopic form (^{13}C), resulting in accumulation of differences between the two forms (fractionation) as the lighter form proceeds more quickly through the reaction. This fractionation leaves the substrate heavier in comparison with the biologically conveyed product (the isotopically lighter one) in situations where the substrate is not completely

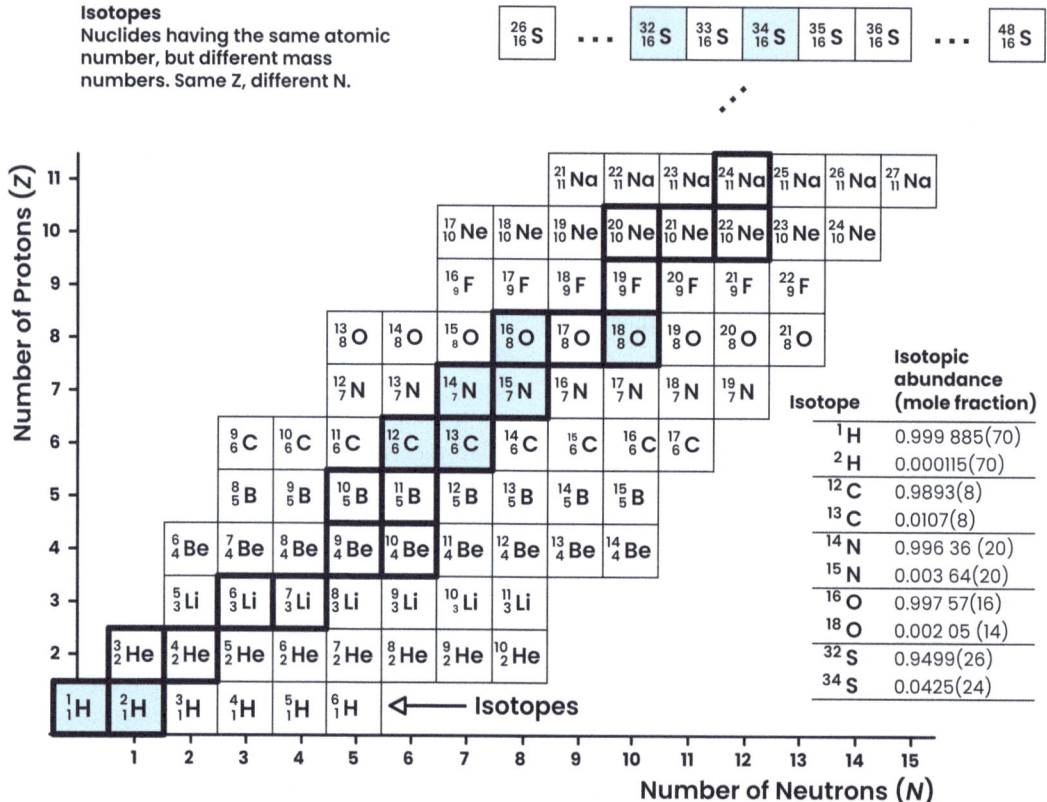

Fig. 16.1 Definition of isotopes and schematic presentation of the chart of nuclides for the light stable isotopes that are mainly used in ecological applications (highlighted in blue). The stable isotopes are represented by black squares. The table on the right contains the abundance of the light isotopes relevant in ecological and parasitological applications

converted into a single product. If the substrate pool is completely converted, then no fractionation occurs as the product pool now has the same composition as the substrate.

In addition to fractionation, mixing of material pools with distinct ratios of isotopic forms also influences the distribution of isotopes. The combination of fractionation and mixing results in characteristic patterns for each element when organic material of different origin is processed. Stable isotope measurements for individual elements are generally reported as the ratios of the heavy and the light isotope $R(^hE/^lE)$ (e.g. $^2H/^1H$, $^{13}C/^{12}C$, $^{15}N/^{14}N$, $^{18}O/^{16}O$, etc.) in a sample or individual compound and not as absolute contents. In practice, the most common form for reporting stable isotope values is the δ-notation or δ-scale (Fig. 16.2, e.g. $\delta^{13}C$ would measure the ratio of $^{13}C/^{12}C$). This notation indicates measurements made against a standard material using a ratio of ratios; that is, the relative difference between the unknown ratio of light to heavy isotopes contained in an unknown sample versus the well-characterised ratio of the isotopic standard that is appropriate for the element being measured. To standardise the measurement of these ratios across periods of time and between laboratories, standards are scaled against an international reference material. In Table 16.1 the primary reference materials as well as their accepted isotope ratios are presented. R_{stad} represents the ratio of a known international reference standard for each element (Vienna PeeDee Belemnite for C and O, atmospheric N_2 for N, Vienna Standard Mean Ocean Water for H and O, and Vienna Canyon Diablo Troilite for S). Isotope ratios are very

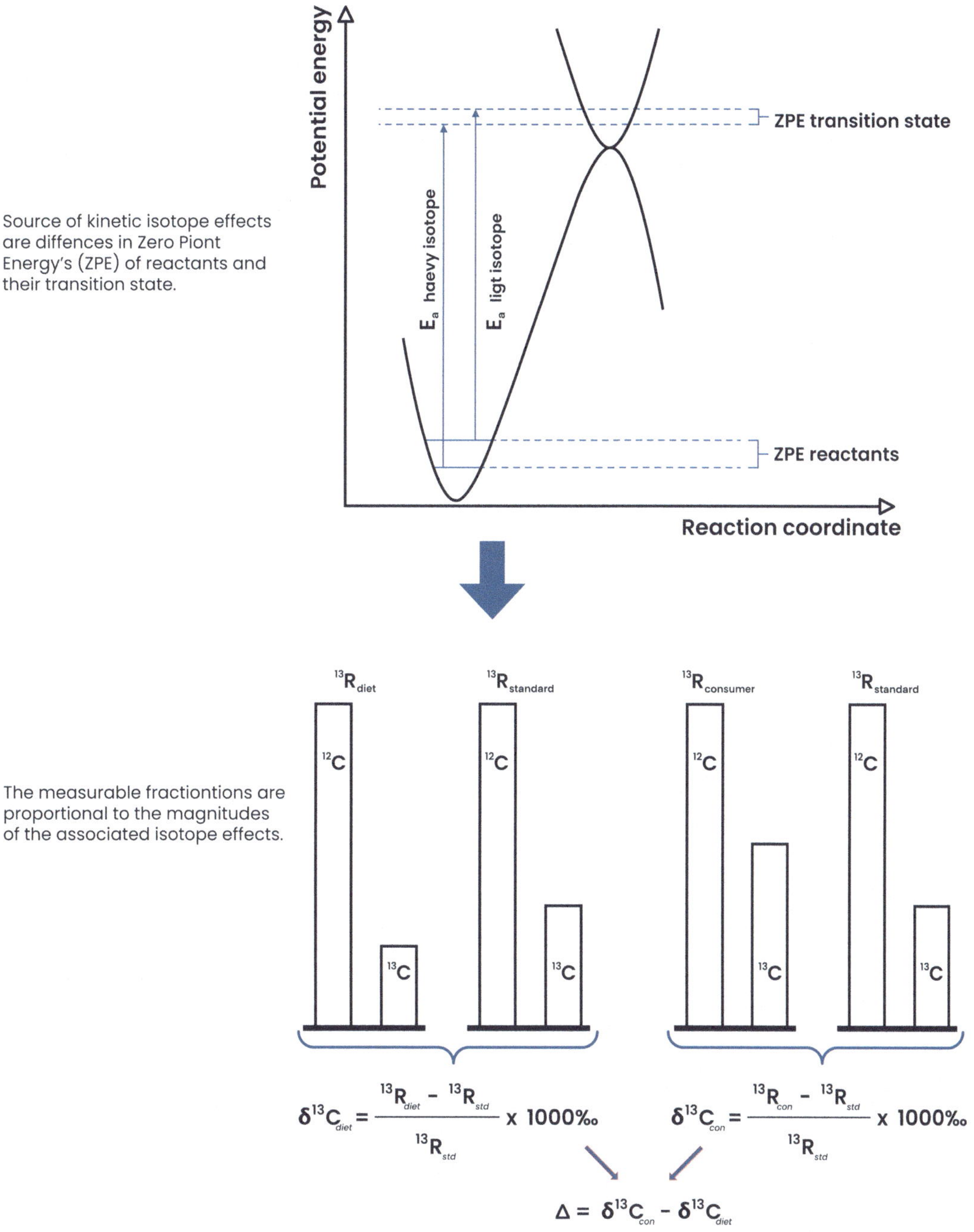
Source of kinetic isotope effects are diffences in Zero Piont Energy's (ZPE) of reactants and their transition state.
Potential energy
ZPE transition state
Ea haevy isotope
Ea ligt isotope
ZPE reactants
Reaction coordinate
The measurable fractiontions are proportional to the magnitudes of the associated isotope effects.
13R diet
13R standard
13R consumer
13R standard
12C
13C
δ13C diet = (13R diet - 13R std) / 13R std x 1000‰
δ13C con = (13R con - 13R std) / 13R std x 1000‰
Δ = δ13C con - δ13C diet

Fig. 16.2 Origin of kinetic isotope effects and isotopic fractionation

small and difficult to handle. Due to this, the values from the ratio of differences between samples and standard are amplified by a factor of 1000 to bring the differences into a range that is practical for comparison in ‰ or per mil (see equations in Fig. 16.2), instead of dealing with differences in the 0.0001 range.

Trophic fractionation, the difference between $\delta^{13}C$ or $\delta^{15}N$ of a consumer's tissues and its diet, is generally referred to as enrichment. This is

Table 16.1 Accepted isotope ratios of the δ-notation for selected elements

Element	Ratio	International scale	Kind of international standard material	Accepted isotope ratio (×10^6) of the standard material
Hydrogen	^{2}H/^{1}H	Vienna Standard Mean Ocean Water (VSMOW)	Water	155.75 ± 0.08
Carbon	^{13}C/^{12}C	Vienna Pee Dee Belemnite (VPDB)	Carbonate	11,180.2 ± 2.8
Nitrogen	^{15}N/^{14}N	AIR-N$_2$	Nitrogen gas	3678.2 ± 1.5
Oxygen	^{18}O/^{16}O	Vienna Standard Mean Ocean Water (VSMOW)	Water	2005.2 ± 0.45
		Vienna Pee Dee Belemnite (VPDB)	Carbonate	2067.2
Sulphur	^{34}S/^{32}S	Vienna Canyon Diablo Troilite (V-CDT)	Troilite	44,159.9 ± 11.7

denoted by the symbol Δ (see equation in Fig. 16.2), where $δ^{13}C_{con}$ and $δ^{13}C_{diet}$ are the carbon isotope ratios of the consumer and its diet, respectively (Fry 1988; Vanderklift and Ponsard 2003). The magnitude of the kinetic isotope fractionation of biological processes can vary with the concentrations of reactants and products, reaction rate, environmental conditions, and between taxa. In contrast to plants, animals fully rely on external, organic carbon sources from plant or animal matter, thus their stable carbon isotopic composition is governed by their diet. For this reason, one of the fundamentals in food chemistry has been coined "you are (isotopically) what you eat, plus a few per mil". Organic matter processing during digestion in consumers causes isotopic fractionation, as enzymes involved in modifications of proteins and amino acids, for instance, more quickly process molecules containing ^{14}N, thereby increasing their biomass in ^{15}N relative to their food source as the lighter isotope is preferentially transferred. The initial production of organic matter by fixation of inorganic carbon in ecosystems takes place in primary producers, which comprise photosynthetic (bacteria, algae and plants) and chemosynthetic (micro) organisms. They also provide the isotopic baseline for the entire food web (primary consumers up to top predators).

16.3 Unravelling Ecological Context with Stable Isotope Analysis

Food webs are complex structures that usually require a variety of simplifications and assumptions to meaningfully convey the underlying ecological relationships between species. One common strategy is to sort organisms into trophic levels and then rank them into a food chain based on feeding relationships. This allows for the exploration of underlying resource use as well as the fractionations occurring between consumers and their diets. However, this endeavour often quickly becomes complicated by factors such as fractionally assigned trophic levels (also called trophic positions), or omnivory. Estimating food chain lengths from simple groupings of species sorted into discrete trophic levels often underestimates trophic complexity and the frequency of omnivores in food webs (Lau et al. 2009). There have been a variety of ways to explore the status of an organism within a food web: (1) estimating the mean number of trophic links between a consumer and a basal species, such as a primary producer; (2) accounting for energy flows across multiple food chains describing feeding relationships; or (3) accounting and comparing network complexity between ecosystems involving foundation species and those without. However, complete characterisation of food webs and energy flows are time-consuming and remain difficult, as these efforts often require huge sampling efforts with repeated campaigns and have to deal with temporal effects as species are acquired across a single or multiple field seasons.

Stable isotope analysis of bulk carbon and nitrogen (BSIA) combusts the entire sample (e.g. tissue, leaves, sediment) and measures all the resulting carbon and nitrogen it contained. With some additional calculations and modelling, these values can be applied to characterise food web structures and nutrient flow within food webs. Since the values provided from BSIA describe a signal of carbon and nitrogen that was incorporated into the tissues across a known period of time, this method provides dietary information that was integrated across time (2 weeks to 2 years, depending on tissue type) rather than more traditional techniques such as gut content analysis. Gut contents tend to provide a snapshot in time that is limited to the last meal and often skew towards overrepresentation of hard parts that are preferentially preserved throughout digestion while easily digested organisms can be missed. Differing rates of turnover for carbon and nitrogen in a consumer's different tissues (e.g. whole blood, 1 month vs muscle, 2–3 months) opens the possibility for temporally integrated studies through examination of multiple tissue types from the same organism. However, the turnover rates in a particular tissue might be affected by various factors such as parasitism (e.g. Yohannes et al. 2017).

An organism's trophic position can be calculated from $δ^{15}N$ values taken from tissues that have assimilated the consumers diet (Beaudoin et al. 1999; Post 2002). The trophic position (TP_x) is defined as a non-integer value, which represents an energy weighted number of trophic energy transfers:

$$TP_x = \left(\frac{\delta^{15}N_x - \delta^{15}N_{base}}{\Delta \delta^{15}N} \right) + \lambda_{base}$$

where $\delta^{15}N_x$ is the isotope ratio of the measured individual and $\delta^{15}N_{base}$ the baseline value of $\delta^{15}N$ of the primary producers. $\Delta\delta^{15}N$ is the trophic discrimination factor (TDF) which represents the average difference between consumer and their diet throughout the food web and λ is the trophic position of the organism that was used to estimate $\delta^{15}N_{base}$ ($\lambda = 1$ for primary producers, Vander Zanden and Rasmussen 2001; Nilsen et al. 2008). Typical TDFs account for increases in consumer nitrogen of ~2.3 ± 0.18‰ and carbon of ~0.5 ± 0.13‰ (McCutchan et al. 2003). These values are slightly lower than those reported by DeNiro and Epstein (1978) and O'Leary (1988) who found $\Delta\delta^{15}N$~ +3‰ and $\Delta\delta^{13}C$~+1‰ between trophic levels. Values widely used in ecology estimated an average increase in $\delta^{15}N$ of 3.4‰ for every trophic level (Post 2002). Multiple meta-analyses of TDFs across several taxa and tissue types have recommended use of tissue-specific offsets that account for biosynthetic effects as well as TDFs specific to the taxa being examined, with the most recent examining relative effects of dietary mismatches with consumer TDFs (Caut et al. 2009; Stephens et al. 2023).

16.4 Gaining an In-Depth Understanding of Trophic Interactions by Using Compound-Specific Isotope Analysis (CSIA)

The analytical developments and refinements in gas and liquid chromatography peripherals have allowed for the purification and isolation of compound classes or single compounds from the sample matrix (e.g. tissues, leaves, sediment) prior to analysis on the isotope ratio mass spectrometer (IRMS). This has been quite powerful, as the isotope values of individual compounds can provide additional information about the ecosystem interactions and metabolic pathways (Boecklen et al. 2011; Ohkouchi et al. 2017; Kruger et al. 2016). Examples for such individual substances are the isotopic patterns of amino acids (CSIA-AA) and fatty acids (CSIA-FA) (Boecklen et al. 2011; Chikaraishi et al. 2011; Sabadel et al. 2016). Amino acids and fatty acids play important roles as essential compounds that are transferred through food webs and can give insights into organism interactions as isotope patterns in various tissues change as metabolism occurs and material is transferred across primary producers to higher trophic levels (Barreto-Curiel et al. 2017; Hanz et al. 2022). McMahon et al. (2010) showed that the average AA carbon isotope pattern mirrors that of bulk tissue but differences among AAs are more pronounced (see Fig. 16.3). This is especially useful for the examination of primary producers at the base of the food web, as tracing amino acids can identify unique patterns of carbon isotope values and distinguish unique sources from plants, fungi, and bacteria in communities (Larsen et al. 2009, 2013). Among the twenty biologically relevant amino acids, glutamic acid plays a key role in biochemical synthesis. Carbon isotope signatures of essential amino acids (EAAs) show only slight fractionation between diet and consumer and thus can be used as conservative tracers for the origin of resources (Boecklen et al. 2011). Plants, bacteria, and fungi EAAs show distinctive patterns of $\delta^{13}C$ values that can be applied as naturally occurring fingerprints of biosynthetic origin of EAAs in food webs. Larsen et al. (2013) showed that the investigation of carbon isotope AA patterns in food webs for freshwater, pelagic, and estuarine consumers was carried out. Carbon isotope patterns of essential amino acids from consumers largely matched those of the dominant primary producers in each system. Since animals cannot synthesise EAAs and must obtain them from food, their tissues reflect carbon isotope patterns found in diet, but it is not known how microbes responsible for hindgut fermentation in some herbivores influence the $\delta^{13}C$ values of EAAs in their hosts' tissues (Arthur et al.

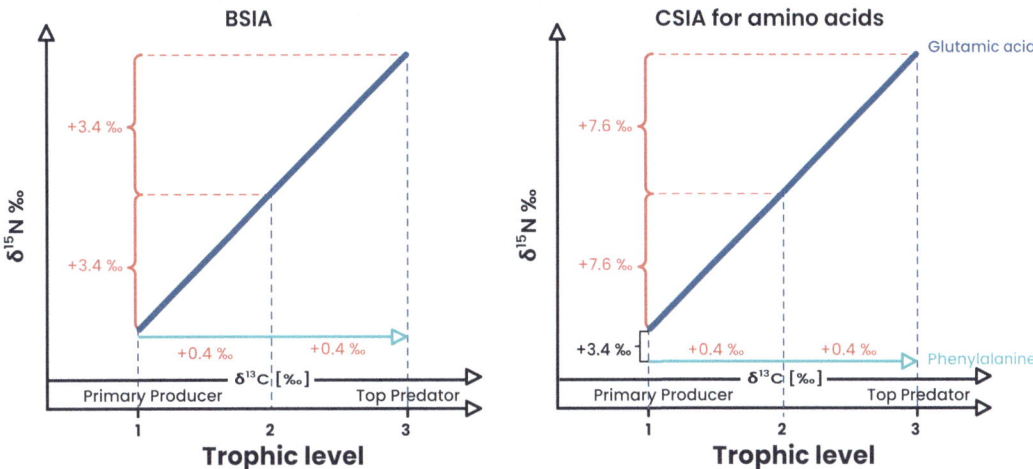

Fig. 16.3 Schematic illustration of the magnification effect of BSIA and CSIA according to their stable isotope composition to estimate trophic levels of organisms from primary producers to top predators. In case of CSIA-AA the relationship between source AA (phenylalanine) and trophic AA (glutamic acid) is considered. Figure adapted from Sabadel et al. (2019)

2014). In contrast to EAAs, the isotope values of nonessential AAs show a large fractionation between diet and consumer tissue, which is associated with metabolic activities such as catabolism or synthesis (McMahon et al. 2010).

Chikaraishi et al. (2009) and Ishikawa et al. (2014) found that CSIA of N from AAs offers substantial advantages over traditional bulk methods for food web analysis, because it defines the food web structure based on the metabolic pathway of amino acid groups and can be used to estimate food web structures under conditions where the bulk stable isotope methods are not reliable. The main reason for this is that trophic position calculations by using isotopic analysis data of bulk tissue often require the knowledge of baseline isotopic values to correct for spatial and temporal variations in the values of compounds such as nitrate, nitrite, ammonium, and total dissolved inorganic carbon (DIC), which are supporting the food webs. As an example, McClelland and Montoya (2002) found that amino acids like phenylalanine showed no significant change in nitrogen isotopic signatures with trophic position and therefore preserve information about nitrogen sources at the baseline of the food webs; this was further confirmed by subsequent authors (Chikaraishi et al. 2007; Décima et al. 2017; Won et al. 2018; see also Fig. 16.3). Thus, the nitrogen isotope values of individual AAs can help to better differentiate trophic levels since they provide additional information. From each examined individual, AA-N simultaneously captures the large isotope fractionation between diet and consumer in some AAs (trophic AAs) but also an indication of isotopic baseline supporting the ecosystem, as isotope values are relatively well preserved in another subset of AAs due to a lower associated fractionation (source AAs; Popp et al. 2007; Boecklen et al. 2011). Gutiérrez-Rodríguez et al. (2014) showed that nitrogen AA isotope values indicate changes in the isotopic baseline that propagate rapidly through the protist food chain (see Fig. 16.3). They also highlighted the need to account for this variability at an ecologically relevant time scale (days to years).

Due to its ability to simultaneously evaluate trophic change from metabolism and underlying baseline nutrient values from single organisms, AA-N analysis has received increasing attention in a variety of ecological studies due to the novel output of a baseline integrated trophic position. Thus, CSIA of AA allows for further investigation of ecosystem interactions, gathered from discrete samples of targeted species with a reduced need for exhaustive surveys to collect every possible dietary resource that are typically required to characterise ecosystem baselines when applying bulk isotope analyses (Bradley et al. 2014). Additional information on diets can be obtained from fatty acid profiles and isotope signatures obtained by CSIA-FA (Boecklen et al. 2011). Individual carbon isotope ratios for FAs have been investigated in feeding experiments of, for example, birds and invertebrates. However, it is more difficult to predict isotope patterns of individual FAs than for AAs because of the greater diversity of FA structure, the variety of potential post-assimilation FA modifications, and the prominent dual roles of FAs as structural molecules and as metabolic energy source (Whiteman et al. 2019).

16.5 Dynamics of Isotopic Incorporation and Turnover Rates in Biological Matrices

The isotopic composition in the tissues of a consumer reflects those of its diet. However, different tissue types in the same organism can have a large variation in composition of stable isotopes, as a result of the differences in time that is needed for a tissue to completely incorporate and turn over that tissue from the nutritional pool. This is expressed as a turnover rate which addresses the dynamic range of this process between tissue types. Thus, high turnover rates are to be expected in metabolically active tissues (e.g. liver, blood, plasma), as their isotopic composition changes rapidly in the course of biosynthesis under normal feeding conditions (Hobson and Clark 1992; Caut et al. 2009). In contrast, tissues such as bones, collagen, or baleen have relatively long turnover times, or are metabolically inert and can provide a record of feeding at the moment of formation up to several years prior to sampling (or death) for bone and baleen, for instance. For example, studying birds and their diet, Hobson and Clark (1992) reported approximately 3 days of half-life of carbon in tissues with high turnover rate (liver and plasma proteins), in contrast to collagen, where the half-life turnover time was estimated to be 173 days. Turnover rates have been found to scale with the body mass of selected organisms as well as tissue type. Thus, consideration of comparable sizes between the same tissue type between different bird species is necessary, as smaller species have higher turnover rates. High growth rates during ontogeny and periods with higher activity (seasonal activity) also impact isotope incorporation and result in higher turnover rates in tissues formed during those periods. Dalerum and Angerbjörn (2005) investigated the influence of elevated temperature on diet and tissue turnover rates as well as discrimination of carbon and nitrogen isotopes in omnivorous fish. It was shown that temperature and diet affected bulk tissue $\delta^{15}N$ turnover and discrimination factors, with increased turnover and smaller discrimination factors at higher temperatures. The AAs, glutamic acid, aspartic acid, and leucine changed most over the course of the experiment and results mirrored those of treatment effects in bulk $\delta^{15}N$ tissue values. Accordingly, when selecting a tissue for comparison between species or within a single species, it is best to choose a tissue type that fits the period of the research question under investigation. Transferring this background knowledge to parasites, it can be challenging to take into account their taxonomic affiliation and their different ontogeny (developmental strategies, size, microhabitat preferences and nutrient/energy use in the host, etc.). Accordingly, when conducting SIA studies on parasites and their associated hosts, it is important to check whether the tissues and the uptake periods are comparable and whether there are differences in metabolism between the parasites and the host tissues. This highlights the need to carefully consider all aspects of parasite biology, as they could be driving factors affecting the isotopic composition of parasites during their life cycle and of host-parasite systems in general.

16.6 Trophic Relationship Between Parasites and Their Hosts

Ecological studies on food webs have rarely included parasites, partly due to the complexity of host–parasite interaction networks (Gómez-Díaz and González-Solís 2010) and studies using bulk SIA techniques to examine parasite-host associations are still rather limited (Born-Torrijos et al. 2023). Sabadel et al. (2019) have discussed the potential of SIA in placing parasites in the food web and emphasised the additional information gained from CSIA-AA, since it allows for calculation of trophic positions for species without having values for primary producers at the food web baseline. Additionally, the application of SIA might also be helpful to elucidate the trophic relations between different parasite taxa and their hosts. The general assumption is that parasites behave similarly to predators, which feed on their prey (Raffel et al. 2008). However, this is not necessarily true from an ecological point of view, as parasites have a much lower biomass compared to their prey (hosts) and usually assimilate energy from a single host organism instead of feeding on several prey individuals, as would be expected for predators in a classical sense (Lafferty and Kuris 2002). Moreover, different parasite taxa, as well as different developmental stages of the same species, have divergent feeding strategies, usually related to their phylogeny and ontogenetic development, so that a uniform classification of the trophic position of (all) parasites in relation to their hosts is not possible or practical, as is often the case for interactions between free-living consumers and their food. In general, parasites pursue some combination of three main strategies for nutrient assimilation: (a) active feeding directly on host tissues, which could be considered a typical predator–prey relationship; (b) passive assimilation of products that derive from the host's metabolism; and (c) sharing the same food source with the host, which represents a form of commensalism (see Nachev et al. 2017). Comparison of published data on different host-parasite systems shows that a large number of associations do not fit a general consumer-resource diet discrimination pattern (e.g. Boag et al. 1998; Iken et al. 2001; Pinnegar et al. 2001; Deudero et al. 2002; Power and Klein 2004; Neilson et al. 2005; Persson et al. 2007; Doi et al. 2008; Dubois et al. 2009; Navarro et al. 2014; Behrmann-Godel and Yohannes 2015; Nachev et al. 2017; Gilbert et al. 2020a, b; Fig. 16.4). Several parasite groups are not enriched in ^{15}N with respect to their hosts (see Thieltges et al. 2019; Riekenberg et al. 2021b; Fig. 16.4), which should be expected for a consumer-predator feeding on its resource-prey (average trophic level increase 3.4‰, see Minagawa and Wada (1984); Post (2002). More specifically, all records for acanthocephalans and cestodes showed that they are depleted in ^{15}N in comparison with the host (e.g. Nachev et al. 2017; Gilbert et al. 2020a; Riekenberg et al. 2021b) with approximately –2.0‰ for acanthocephalans and values ranging between –0.3 and –4.4‰ (mean –2.6‰) for cestodes, in comparison with host muscle (Fig. 16.4). The currently available data for trematodes include larval stages, with sporocysts and cercariae (see Chap. 5) being mostly depleted in ^{15}N in a range between –0.6 and –3.3‰ (Dubois et al. 2009; Doi et al. 2010), but also in rare cases enriched by up to 1.2‰ (see Yurlova et al. 2014) with respect to snail host tissue. Adults trematodes were reported to be δ^{15}N-depleted between –0.3 and –2.23‰ when compared to the fish host tissues (Iken et al. 2001; Fig. 16.4). Considering other parasitic taxa, only larval nematodes were consistently depleted in ^{15}N with a δ^{15}N value of –3.0‰ on average (–1.1 to –6.3‰) (Fig. 16.4). Concerning copepods, many host-parasite records showed lower isotopic discrimination values for the parasites, whereas some showed clear consumer-resource patterns by being enriched with a δ^{15}N value ranging between 0.8 and 2.8‰ (Fig. 16.4). Data regarding other parasitic crustaceans such as amphipods and cirripeds are limited and the data available show no clear trends in differences with respect to host tissues (Fig. 16.4). In contrast, parasitic insects (micropredators, parasitoids), arachnids, pearlfish (Carapidae: tribe Carapini) and helminths such as monogeneans and adult nematodes showed typical consumer-resource discrimination patterns (Fig. 16.4). All available data pointed to a clear isotopic shift of more than 1‰, with the highest differences observed for parasitic arachnids (7.6‰) and adult nematodes (6.7‰) followed by pearlfish (6.3‰) and insects (4.7‰).

The different fractionation patterns observed among different parasitic taxa/groups are likely

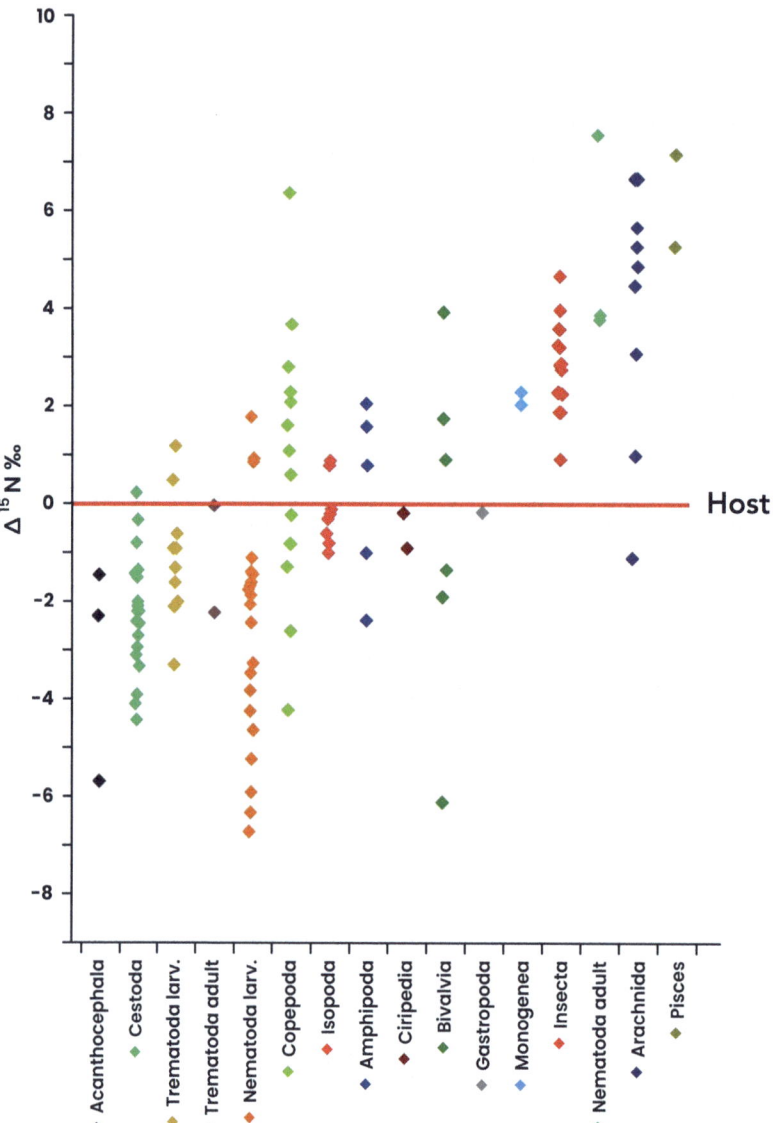

Fig. 16.4 Differences in isotope discrimination values ($\Delta^{15}N_{parasite\text{-}host}$) of different host-parasite systems. Data were extracted from published studies and comprise 134 host-parasite associations (Boag et al. 1998; Doucett et al. 1999; Iken et al. 2001; Pinnegar et al. 2001; Deudero et al. 2002; Baud et al. 2004; Butterworth et al. 2004; Parmentier and Das 2004; Power and Klein 2004; Langellotto et al. 2005; Neilson et al. 2005; Yarnes et al. 2005; O'Grady and Dearing 2006; Towanda and Thuesen 2006; Voigt and Kelm 2006; Persson et al. 2007; Xu et al. 2007; Catenazzi and Donnelly 2008; Dubois et al. 2009; Stapp and Salkeld 2009; Doi et al. 2010; Gómez-Díaz and González-Solís 2010; Dean et al. 2011; Schmidt et al. 2011; Clark and Hawke 2012; Fritts et al. 2013; Sánchez et al. 2013; Baillon et al. 2014; Fleming et al. 2014; Navarro et al. 2014; Yurlova et al. 2014; Behrmann-Godel and Yohannes 2015; Denic et al. 2015; Demopoulos and Sikkel 2015; Eloranta et al. 2015; McGrew et al. 2015; Riascos et al. 2015; Nachev et al. 2017; Sures et al. 2019; Gilbert et al. 2020a, b; Kamiya et al. 2020; Sabadel et al. 2022)

to be related to the diversity of feeding strategies, i.e. how parasites obtain nutrients and energy from their host. Some parasitic taxa independently evolved mechanisms for direct assimilation of nutrients from the host. The development of a syncytium (tegument), which is characteristic of trematodes, cestodes, and acanthocephalans (see Chap. 5), enables the organisms to directly take up pre-digested and metabolised substances from the host. As a separate gastrovascular system is also missing in cestodes and acanthocephalans, their tegument shows morphological adaptations in order to enlarge the absorptive surface (microtriches and canal zone). Morphological adaptations for direct uptake of nutrients are also known for some ectoparasites, such as parasitic copepods and barnacles. Some species develop a system of rootlets, which penetrate the host tissues and are capable of extracting and absorbing nutrients from the host (see Riekenberg et al. 2021c; Sabadel et al. 2022). Additionally, the structure of the body integument of some mesoparasitic copepods (i.e. partly embedded in the host tissues) seems to facilitate the assimilation of nutrients via the body surface (Bresciani 1986; Boxshall and Halsey 2004). The cuticle of juvenile nematodes has a similar function in allowing the uptake of nutrients (Bird and Bird 1991) and is not as complex as that of the adults. If organisms absorb nutrients which are processed or provided by the host, these molecules are commonly depleted in the heavier nitrogen isotope (^{15}N) due to the associated kinetic isotope effect (see Sect. 16.2). Adsorptive feeding is facilitated by different mechanisms such as passive transport of nonpolar molecules, pino(endo)cytosis and predominantly by active selective membrane transport that incorporates a high variety of carrier proteins with different levels of specificity (Rothman 1967; Pappas and Read 1975; Barrett 1981). The selective transport through the tegument or assimilative surfaces is assumed to favour isotopically lighter molecules and additionally contributes to the irregular fractionation patterns observed for some parasitic taxa. Enzymatic complexes bound to tegumental membranes, the release of histolytic secretions and adsorbed enzymes from the host are supposed to be involved in extracorporeal (pre-) degradation of macromolecules (Rothman 1967; Taylor and Thomas 1968; Read 1973; summarised also in Barrett 1981; Smyth and McManus 1989). These mechanisms are essential for digestion of host tissues during, for example, host invasion or predigestion of nutrients which are subsequently assimilated. It has been demonstrated that cestodes and acanthocephalans can adsorb α-amylase in vitro leading to higher amylolytic activity. Such contact digestion is associated with enzymatic reactions that results in isotopically depleted reaction products (oligomers and monomers), which are subsequently selectively assimilated. Carbohydrates represent the most important source of energy during catabolism in endohelminths, whereas intermediates of glycolysis serve as precursor for biosynthesis of amino acids. As various helminths can re-use ^{15}N-depleted ammonia to synthesise amino acids (Barrett 1981) the use of this pool of ammonia may additionally contribute to the observed patterns for parasitic taxa with assimilative feeding strategies.

As previously mentioned, absorptive feeding (extracorporeal digestion) on selected compounds favours isotopically depleted compounds that need lower activation energy in further biochemical reactions. It also takes place in the host environment without the need for excretion of molecules (energy dependent if the digestion takes place in the parasite environment) that are a result of the breakdown of complex compounds and are not required for a parasite's metabolism or biosynthesis. These aspects might increase the general metabolic efficiency of parasites with assimilative feeding strategies and have evolved several times in various parasitic lineages. Assimilative feeding is usually less virulent for the definitive host, resulting in a better general host condition of the host, which then provides a reliable nutritional base for the parasite. Although virulence is usually infection intensity-dependent, parasites such as acanthocephalans and cestodes appear less virulent for the definitive host than parasitoids, parasitic castrators (virulence on intermediate host level) and micropredators (Ewald 1995; Poulin 2011).

In contrast to absorbing specific nutrients and compounds, other parasites feed on tissue or blood of their hosts and behave more like micropredators than adsorptive feeders. This is usually the case for both ecto- and endoparasitic taxa that have a digestive tract and actively feed on host tissues, as well as phytophagous organisms (nematodes, insects) and parasitoids. After active feeding, digestion leads to positive enrichment of the heavier N-isotope, similar to the ones typical of non-parasitic consumer–resource interactions. Accordingly, these parasites (e.g. parasitic insects, arachnids, gastropods, and fishes) were clearly enriched in ^{15}N in comparison with host tissues, with fractionation ranges found comparable to free-living predator–prey interactions. The example of digenean trematodes shows clear effects of feeding strategy for trophic fractionation within the same taxon. While adult trematodes as well as rediae feed mainly actively and are enriched in ^{15}N compared to host tissues, sporocysts, being assimilative feeders, are depleted in ^{15}N. However, adult digeneans have a mixed way of feeding as they also assimilate part of their nutrients via the body surface and therefore the fractionation patterns of ^{15}N in comparison with their host are not as pronounced as for taxa with an active feeding strategy.

16.7 Identification of Resources Used by Parasites

Nutritional transfer within host-parasite associations is still poorly studied, with most research effort focused on economically relevant parasitic taxa or those that are important for human health (Chappell 1979). For the latter, research has predominately concentrated on nutrient requirements in the form of macromolecules, uptake routes, and metabolic pathways with an eye towards the development of antiparasitic drugs, and not taking into consideration any potential relevance of these findings for other closely related taxa lacking human relevance (Chappell 1979). As previously mentioned (see Sect. 16.2), the processes shaping the composition of SI in consumers are (1) fractionation that occurs between a consumer and their diet and (2) mixing if a consumer feeds upon a variety of resources (Fry 2006). Thus, SIA can provide valuable information about the transfer and flow of food within the food webs, whereas applying isotopic mixing models allows translation of isotopic data into estimates for food source contribution to consumers. Mixing models and, in particular, those that are based on a Bayesian statistical framework have gained increasing attention among researchers (see e.g. Phillips et al. 2014), with a strong focus on trophic interactions of free-living organisms (Fig. 16.5). This approach seems to be very promising for accessing the diet composition of parasites but, at the same time, very challenging due to the variety of feeding strategies (see Sect. 16.6).

As was noted for free-living organisms, when applying mixing models to studying trophic interactions (reviewed by Phillips et al. 2014), a proper sampling strategy plays a decisive role for achieving reliable results when studying the diets of parasites. An appropriate sampling strategy should address all potential food sources of the consumer/parasite as well as the consumer/parasite tissues. Recent models using $\delta^{13}C$ and $\delta^{14}N$ can be applied to assess the contributions of many dietary resources to a consumer, whereas for host–parasite interactions, the potential food sources are mainly limited to the host tissues or compounds derived from the microhabitat where they occur. These locally acquired resources might include body fluids of hosts and gut contents (for interstitial parasites), as parasites exhibit various nutrient uptake strategies that should be considered for each relationship (see Sect. 16.6). In general, a mixing model implies that both C and N of the food sources contribute to some extent to the isotope mixing in the consumer and integrates concentration dependency (e.g. MixSIAR, SIMMR), as the contribution of a dietary source to the isotopic signal of a consumer depends on its C and N concentrations and on the relative efficiency of assimilation (routing; Phillips and Koch 2002). Thus, concentration-dependent mixing models should also be applied where possible for host–parasite trophic interactions, especially in cases where concentrations of C and N in different sources (host tissues) vary widely.

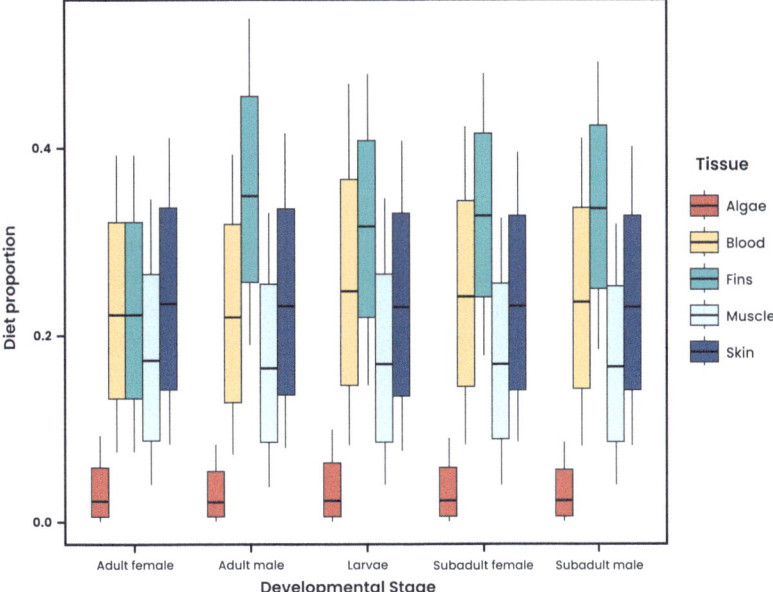

Fig. 16.5 Example output of mixing based on data of a fish–branchiura interaction model run using package MixSIAR for R (Stock et al. 2018). Due to the partially free-living lifestyle of the larvae algae as a potential food source were also considered (Gilbert et al. 2025)

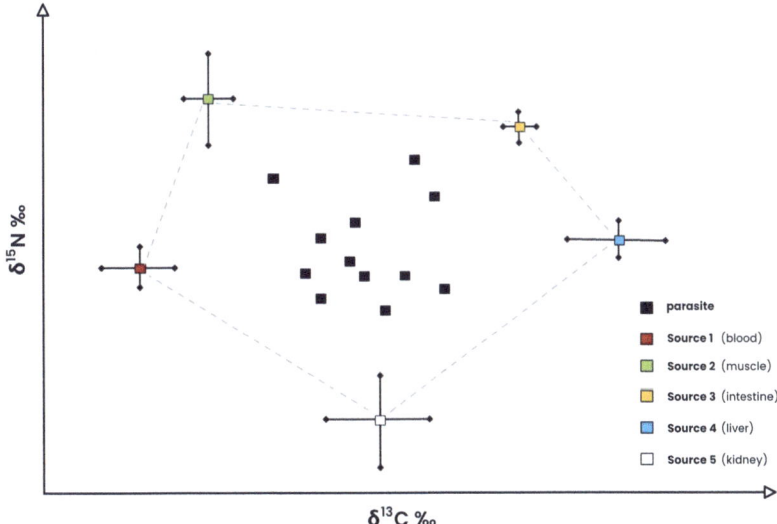

Fig. 16.6 Isoplot representing optimal placement of consumer/parasite signatures within the isospace of its potential dietary sources after applying TDF correction

Mixing models are also sensitive to missing food sources (if a potential source is not included) or default to the null hypothesis of equal contribution if sources with similar isotope composition are insufficiently resolved (e.g. muscle and blood), as summarised by Phillips et al. (2014). In both scenarios, a carefully designed sampling strategy after considering details about an organism's life strategy (all available nutritional sources), and which elements are likely to be most useful for separation between sources within the organism's environment (C, N, or S), will help improve resolution of dietary resources.

Similar to trophic interactions of free-living organisms, it is essential that parasite isotopic values as consumers fall within the isospace of the mixing isotopic envelope created by their food sources (Fig. 16.6). Excluding isotopically distinct food sources, which are far outside the range of the majority of other sources, would improve the certainty of estimates and will avoid additionally "overfeeding" the model. Therefore, plotting and exploring the isospace inhabited by both consumer and dietary sources in a C-N plot should be conducted before parameterising the model, while considering TDF of particular taxa is also important. Based on empirical data (see Sect. 16.6; Fig. 16.4), different parasitic taxa showed a wide range of fractionations with respect to their host (dietary sources) (Thieltges et al. 2019).

As with free-living consumers, tissue selection in parasites is important as tissue types may have very different turnover times representing dietary resource across different time periods and should be selected according to the research aims of the project (Kamiya et al. 2020; Riekenberg et al. 2021a). Using different tissues or body parts of a parasite, it is possible to provide estimates for its long- or short-term dietary patterns, as different tissues may vary in their metabolic activity and therefore exhibit different isotopic turnover rates (see Sect. 16.5). Thus, dietary patterns provided after correcting for turnover times may represent different periods of life history for the individual (Hobson and Clark 1992). In the case of parasites, considering body parts with lower turnover rates (e.g. extremities of parasitic arthropods) should cover long-term diet, and including larval and subadult stages and those with higher turnover rates (thorax that includes metabolically active organs) should provide the estimates for most recent diet of parasites. Estimation of the diet of larger parasitic taxa with a differentiated digestive tract can be biased by gut content, which should be removed prior to analysis (parasitic arthropods, nematodes). However, this is often impossible or very challenging due to their small size and propensity for death after removal from their hosts.

Apart from mixing models based on bulk isotope data, CSIA can also provide valuable information about nutrient and energy transfer from the host to the parasite. For example, Hesse et al. (2023) provided evidence that in the stickleback-*Schistocephalus solidus* host-parasite system, the plerocercoids obtain essential amino acids from the host and retain similar essential amino acid values to those of the host liver.

16.8 Trophic Positions of Parasites in Food Webs Depend on Their Ontogenetic Development

The isotopic niche occupied by a parasite may vary in accordance with various factors such as isotope turnover rates in the host tissues/microhabitat (Deudero et al. 2002; Butterworth et al. 2004; Yohannes et al. 2017), feeding duration of parasites (Dean et al. 2011; Fritts et al. 2013), variance in metabolism across different life stages (Deudero et al. 2002), and the variety of host species from which nutrients are derived through its life cycle (Taccardi et al. 2020). Generally, a parasite with a multiple-host life cycle (heteroxenous) occupies different trophic positions covering at least two or more trophic levels during its development. This is especially true for trophically transmitted helminths such as acanthocephalans and cestodes (and partly also trematodes and nematodes), where their developmental stages infect hosts from different trophic levels and therefore are expected to differ in their isotopic composition. So far, isotopic studies addressing the entire life cycle of a parasite species are rare with just one, a study on the (monoxenous) crustacean *Argulus japonicus* infecting goldfish, *Carassius auratus* (Gilbert et al. 2025).

Parasitic taxa that develop in only one host are also expected to undergo ontogenetic isotopic shifts as they switch their food sources during host-switching events (as occurs in ectoparasites with low host-specificity) (e.g. Gómez-Díaz and González-Solís 2010; Riascos et al. 2015), a microhabitat shift within the host or a switch from free-living to parasitic life-phases. A good example is *A. japonicus*, where differences in stable isotope enrichment occur during different developmental stages (see Fig. 16.7), reflecting changes in the parasite's diet during the transition from the metanauplius stage to adulthood, as it ceases deriving nutrition exclusively from its yolk and becomes completely dependent on the host fish for its diet. The observed shifts in nitrogen and carbon isotope ratios in *A. japonicus*

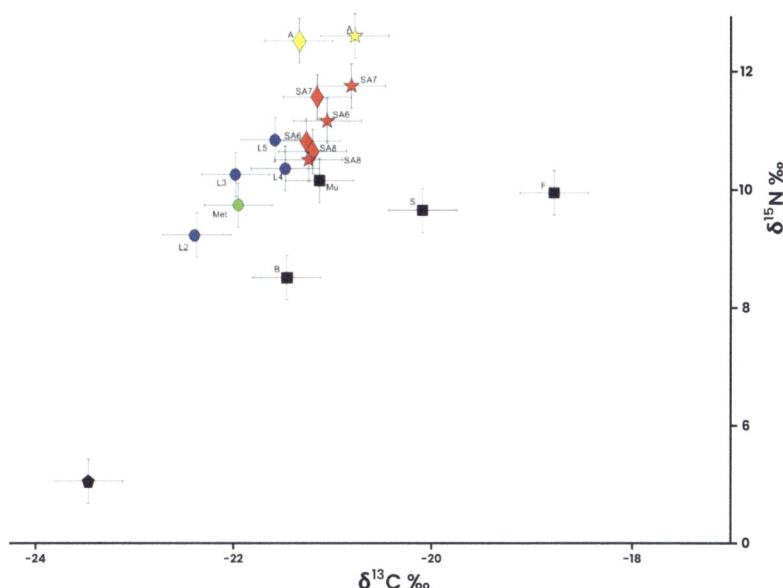

Fig. 16.7 $\delta^{15}N$ and $\delta^{13}C$ mean and standard deviation values for algae, tissues of *Carassius auratus* and developmental and adult stages of *Argulus japonicus*. Black shaded objects represent potential dietary sources of *A. japonicus*—algae (black polygon), blood (B), fins (F), muscle (Mu), and skin (S) of *C. auratus*. For *A. japonicus*, the green circle represents the metanauplii (Met). Blue circles represent larval stages (L2–L5), red diamonds and stars represent subadult stages (SA6–SA8), and yellow stars and diamonds the adults. For subadult and adult stages, males are represented by stars and females by diamond shapes

align with the development of their mouthparts and digestive system. Absence of a mouth and gut in early life suggests that the parasite's feeding on host blood is restricted to later developmental stages. By comparing the differences in $\delta^{15}N$ between parasites and host tissues, it was found that adult male and female parasites are enriched in $\delta^{15}N$ by 2–4‰, which aligns with parasites feeding on host tissues (Sect. 16.6; Fig. 16.4).

16.9 Parasite Effects on Hosts

Besides tracing the resources used by parasites and identifying their trophic position relative to their hosts, stable isotopes can also be used to identify certain effects of parasites on their hosts. Many parasites can cause trait changes in infected hosts, either as an unintended side-effect of infections or as adaptive host manipulations that increase transmission to the next host (Poulin 2007). Some of these infection-mediated trait changes will lead to differences in resource intake between infected and uninfected hosts (see Fig. 16.8), and the resulting shifts in trophic niches can be reflected by changes in the isotopic niche of infected hosts (Britton and Andreou 2016; Born-Torrijos et al. 2023).

In general, wild-caught individuals can host a variety of parasites and should therefore arguably be regarded as akin to whole ecosystems rather than just single organisms. Parasites are not usually considered in ecological and food web studies, despite them often altering their hosts' resource intake and internal metabolic use. This can potentially lead to changes in the trophic and/or isotopic niche of their hosts, resulting in niche separation between infected and uninfected individuals. The trophic niche of an organism is determined by the composition of the different resources in its diet, which also determines its trophic position within a food chain. The isotopic niche, defined as the position of an organism based on its isotopic composition (often $\delta^{15}N$ and $\delta^{13}C$), can be a proxy for the trophic niche of an organism as long as resource use or metabolic processing is not affected.

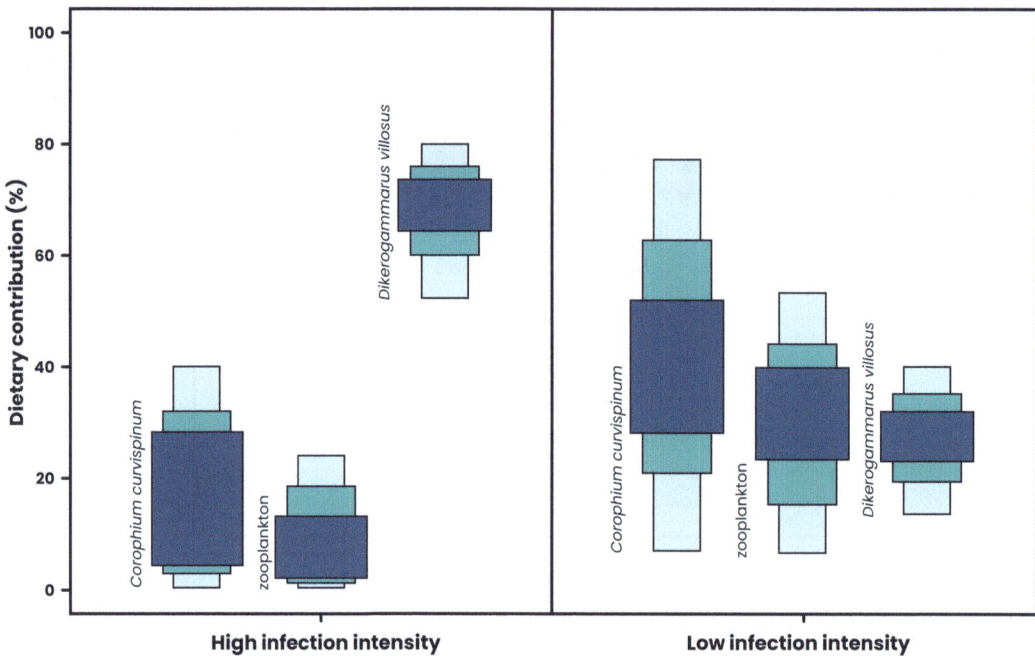

Fig. 16.8 Output of Bayesian mixing model demonstrating differences in prey choice with respect to the level of infection of young European perch (*Perca fluviatilis*) with eye flukes (Vivas Muñoz et al. 2021). High numbers of eye flukes impair the vision of fish, shifting their preferences to larger prey items (*Dikerogammarus villosus*). Dietary items of infected fish: *Chelicorophium curvispinum*, zooplankton and *Dikerogammarus villosus*

Within the scenarios of impact on the isotopic niche of infected hosts, three potential outcomes of parasite infections may occur (Born-Torrijos et al. 2023). In the first scenario, parasite infections might not alter resource intake or use, which might result in overlapping isotopic niches of infected and uninfected hosts (Fig. 16.9a). In the second scenario, infections might alter the resource intake of infected hosts but not affect resource use, leading to isotopic niche shifts of infected hosts and a niche similar to that of other consumers feeding on an alternative resource, in this case a primary producer (Fig. 16.9b). Infection-induced behavioural changes can have an impact on resource intake, for example, from the infected host starting to feed on a different resource community (e.g. primary producer versus predatory zooplankton). In the third scenario, where the internal use changes but the intake remains the same, the trophic position of the infected host may remain the same (Fig. 16.9c). However, infection-induced changes in internal resource use in infected individuals can alter their isotopic niche compared to uninfected individuals (Fig. 16.9a, c). For example, in Fig. 16.9c, the isotopic niche of the infected host suggests an intermediate trophic level while infected and uninfected individuals actually feed at the same top trophic level.

A diverse range of factors can impact an organism's resource intake and internal resource use, i.e. how the energy pool is filled and utilised, respectively, thereby affecting its metabolism. Although resource intake depends on availability and composition of external food, internal resource use is influenced by energy-demanding processes such as reproduction, growth, or immunity, amongst others. Parasite infections can have an impact at both levels, through single or multiple pathways ranging from behavioural to physiological changes in infected hosts.

The behavioural, morphological, or physiological changes that can lead to alterations in the quantity or quality of the host's intake are reflected in the isotopic composition or niche, as has been shown for parasites that affect habitat

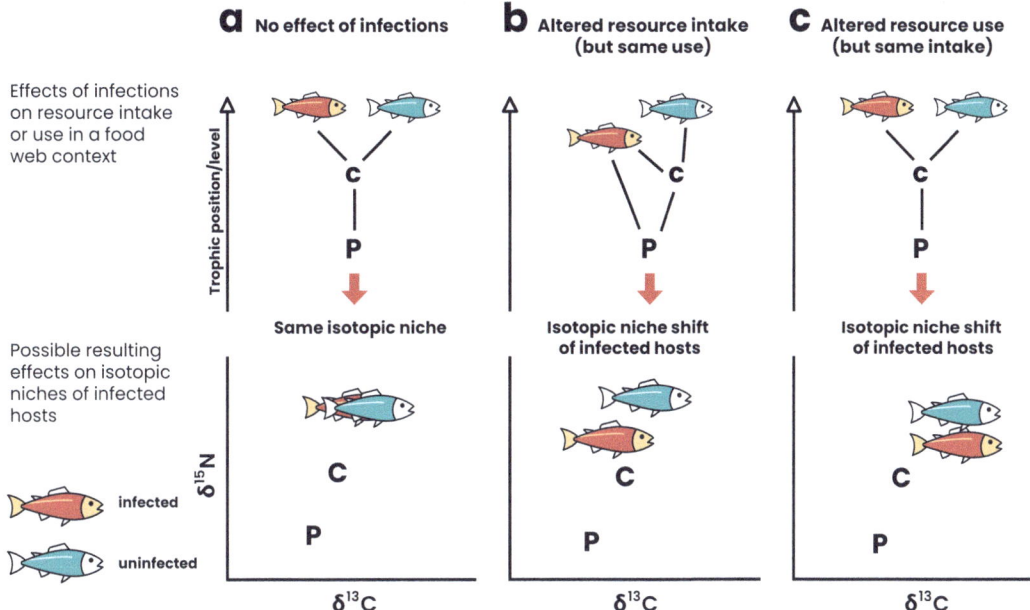

Fig. 16.9 Potential scenarios for impacts on the isotopic niches of infected hosts, with (**a**) no parasite-induced effects resulting in overlapping niches, (**b**) effects only on the resource intake or (**c**) only on the resource use, resulting in the isotopic niche shift of infected hosts. *P* primary producer, *C* consumer. Figure modified from Born-Torrijos et al. (2023)

use or prey choice of their hosts (Britton and Andreou 2016; Welicky et al. 2017; Vivas Muñoz et al. 2021) (Fig. 16.9b). Other parasite-induced effects might alter the isotopic niche of infected hosts (Fig. 16.9c) compared to uninfected ones (Fig. 16.9a), even if they do not cause a change in their trophic position. There are several internal mechanisms that might cause this shift in host resource use, ranging from infection-induced host-starvation (Eloranta et al. 2015; Karlson et al. 2018) to nutrient-demanding processes, such as immune response, growth, pathological damage, or reproduction (Doi et al. 2008; Parker and Booth 2013; Brace et al. 2015; Pegg et al. 2015; Taylor et al. 2019).

Stable carbon and nitrogen isotope analyses have already been used to illustrate the impacts of parasite infections on resource use and the resulting trophic niches of their hosts (Britton and Andreou 2016; Vivas Muñoz et al. 2021; Sabadel and MacLeod 2022); however, they fail to account for the multiple mechanisms by which host–parasite interactions can alter host isotopic values. To help identify the presence of parasite-induced effects on host resource intake or use, stable isotopes of infected and uninfected individuals should be compared, something that is currently very limited in the stable isotope literature (reviewed by Born-Torrijos et al. 2023). Furthermore, to identify which underlying mechanisms induce these changes and whether they are related to altered resource intake or use, experimentally-infected organisms should be used to avoid the multiple parasite infections that wild-caught organisms usually harbour and that might potentially have diverging effects on the stable isotope values of their hosts. Feeding studies using controlled diets would further help to disentangle changes in stable isotope values due to diet, due to parasite infections. Accounting for parasite-induced changes in resource intake and use will therefore grant a more realistic scenario of how parasites alter the flow of biomass and energy through food webs.

16.10 Tracing Parasite-Induced Alterations of Resource Flows in Food Webs

Parasites not only affect the resource use of individual hosts, they can also affect the resource flow through food webs which SIA can help to reveal (also see Chap. 7). Such resource flow alterations can be knock-on effects from infection-induced sublethal changes of resource intake or use of individual hosts (see Sect. 16.9). For example, if body mass or nutritional value of hosts is reduced by parasite infections, this will also lower the contribution of infected hosts to energy flows through food chains. Alternatively, parasite effects on resource flows in food webs can arise from parasites being a resource themselves, e.g. when infective stages are consumed by non-hosts. As the parasites themselves derive their energetic demands from their hosts, consumption of infective stages can indirectly create energy flows between consumers on different trophic levels. Isotope labelling studies have been used to quantify these indirect effects on resource flows. In isotope labelling studies, ^{15}N or ^{13}C-labelled resources are experimentally added and then traced through the food chain, often referred to as a pulse-chase study (Galvan et al. 2011). This approach has been used to investigate the effect of chytrid fungal infections on the energy transfer through planktonic food webs (Sánchez Barranco et al. 2020). Chytrids are aquatic fungal parasites of phytoplankton (see also Chap. 4) that have been suggested to make

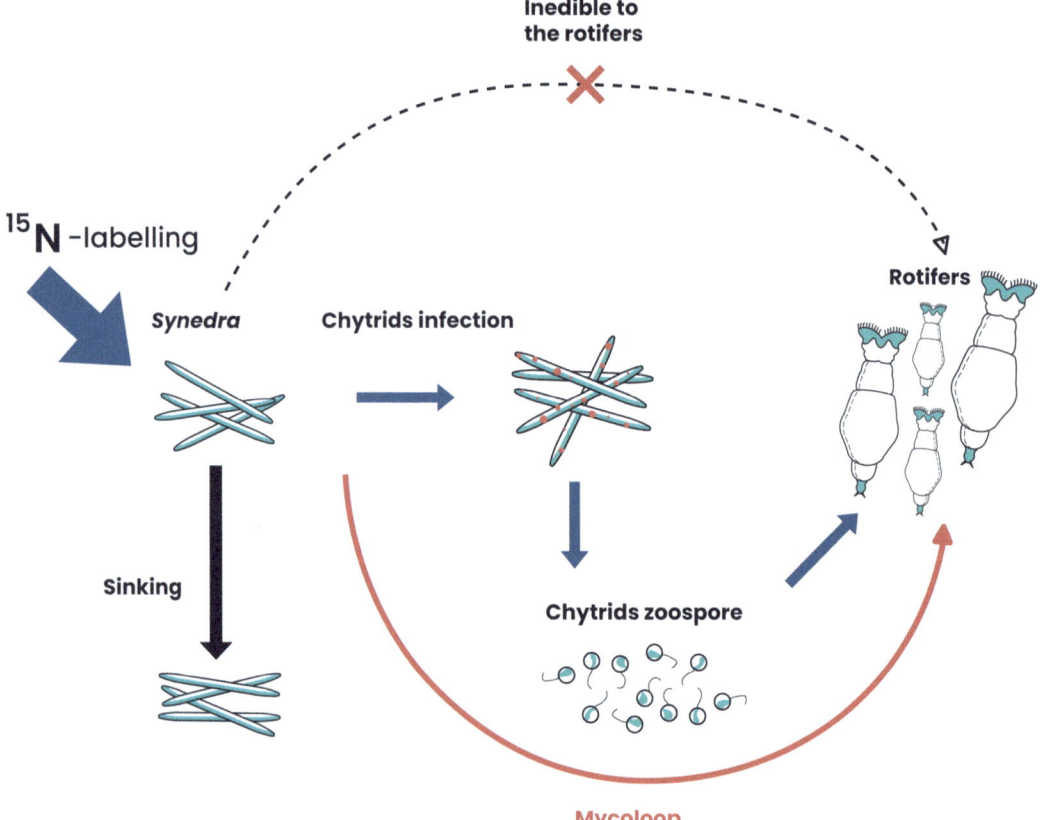

Fig. 16.10 Isotope labelling studies can trace the indirect resource flow from diatoms to rotifers induced by chytrid infections of phytoplankton (mycoloop): chytrids that infect inedible diatoms can take up nutrients from their hosts which are released into the water through their zoospores that can be consumed by rotifers. Figure adapted from Sánchez Barranco et al. (2020)

nutrients from inedible phytoplankton available to zooplankton via a phenomenon called the *mycoloop* (Fig. 16.10; Kagami et al. 2014). They do so through the production of edible infective fungal spores that are released from infected phytoplankton. For example, phytoplankton species such as large *Synedra* diatoms (25–400 μm) are inedible for most zooplankton species such as the rotifer *Keratella* sp. Upon the death of the diatoms, they sink to the bottom and the nutrients they include are thus considered to be lost from the pelagic food web (Ibelings et al. 2004). Chytrids that infect the diatoms can take up nutrients from their hosts which then end up in infective zoospores that are released into the water and can be consumed by zooplankton such as *Keratella* sp. (Kagami et al. 2007; Frenken et al. 2018, 2020). In this way, chytrid infections can introduce an indirect resource link between inedible phytoplankton and zooplankton (Kagami et al. 2014). By isotopically labelling their diatom cultures with fertiliser (10 atom% ^{15}N nitrate; ^{15}N NaNO$_3$), Sánchez Barranco et al. (2020) were able to trace and verify the indirect resource flow from diatoms to rotifers via chytrid zoospores. Moreover, they could quantify the first step from diatoms to zoospores and found that chytrids took up about 14% of the host's standing stock of N per day (Sánchez Barranco et al. 2020).

In another study, Schultz and Koprivnikar (2021) used stable isotope labelling to trace the flow of trematode cercarial biomass within an aquatic food web. They produced cercariae (*Plagiorchis* sp.) labelled with ^{13}C by offering their gastropod hosts food to which sodium bicarbonate (NaH^{13}CO$_3$) containing 99% ^{13}C was added. The labelled cercariae were then added to mesocosms that included a simplified freshwater food web consisting of dragonfly larvae and diving beetles, oligochaete worms, and zooplankton. Their experiment revealed that the oligochaete worms showed the highest uptake of ^{13}C while, for the other taxa, no significant uptake could be found. This result indicates that benthic detritivores such as oligochaetes can be significant consumers of cercarial biomass, something which had previously not been recognised. Oligochaetes are typically unselective detritus feeders, and they probably consume cercarial biomass in the form of detritus after dead cercariae have sunk to the bottom. As the consumption of cercariae and other infective stages of parasites by various organisms is very common (Thieltges et al. 2008; Johnson et al. 2010; Koprivnikar et al. 2023), labelling studies using stable isotopes are a promising approach to quantify the resulting resource flows through food webs.

16.11 Conclusion and Outlook

The use of SIA in parasitological studies helps to gain a fundamental understanding of parasite biology, particularly in relation to the interaction of parasites with their hosts. Furthermore, the application of CSIA allows us to obtain information on the nutrient uptake of parasites. Comparison of the fractionation patterns of different taxonomic parasite groups reveals large differences in the nutrient uptake of the respective parasites in their hosts. In addition to more predatory species actively feeding on host tissues, the fractionation patterns also show a commensal mode of feeding through passive assimilation of nutrients or sharing the same food with the host. The application of SIA and CSIA will provide further important insights into the uptake of specific nutrients in parasites in the future. In addition, parasite infections can affect both the resource uptake and internal resource use of their hosts, leading to changes in the trophic and isotopic niches of infected compared to uninfected individuals. This provides an opportunity to quantify the impact of parasite infections on resource flow in food webs, a topic that remains largely underexplored (see Chap. 7). However, it also shows that there are still major gaps in our understanding of the extent of infection-related effects on host resource uptake and use. This chapter also highlights the need for more studies using SIA to compare the stable isotope fractionation of infected and uninfected individuals from the same ecosystem. Comparing the isotopic niches of infected and uninfected individuals helps to determine the presence of infection-related effects on host resource uptake or utilisation. Further experimental studies are needed to disentangle the two pathways and to identify possible complex interactive effects. Finally, we hope that the information provided in this chapter will motivate scientists to use stable isotope analysis in their future research and to conduct more stable isotope studies that take into account the entire parasite life cycle, including all associated hosts and developmental stages, as these types of studies are largely lacking.

References

Arthur KE, Kelez S, Larsen T, Choy CA, Popp BN (2014) Tracing the biosynthetic source of essential amino acids in marine turtles using $\delta^{13}C$ fingerprints. Ecology 95(5):1285–1293. https://doi.org/10.1890/13-0263.1

Baillon S, Hamel JF, Mercier A (2014) Diversity, distribution and nature of faunal associations with deep-sea pennatulacean corals in the northwest Atlantic. PLoS One 9(11). https://doi.org/10.1371/journal.pone.0111519

Barreto-Curiel F, Focken U, D'Abramo LR, Viana MT (2017) Metabolism of *Seriola lalandi* during starvation as revealed by fatty acid analysis and compound-specific analysis of stable isotopes within amino acids. PLoS One 12(1). https://doi.org/10.1371/journal.pone.0170124

Barrett J (1981) Biochemistry of parasitic helminths. Red Globe, London. https://doi.org/10.1007/978-1-349-86119-4

Baud A, Cuoc C, Grey J, Chappaz R, Alekseev V (2004) Seasonal variability in the gut ultrastructure of the parasitic copepod *Neoergasilus japonicus* (Copepoda, Poecilostomatoida). Can J Zool 82(10):1655–1666. https://doi.org/10.1139/z04-149

Beaudoin CP, Tonn WM, Prepas EE, Wassenaar LI (1999) Individual specialization and trophic adaptability of northern pike (*Esox lucius*): An isotope and dietary analysis. Oecologia 120(3):386–396. https://doi.org/10.1007/s004420050871

Behrmann-Godel J, Yohannes E (2015) Multiple isotope analyses of the pike tapeworm *Triaenophorus nodulosus* reveal peculiarities in consumer-diet discrimination patterns. J Helminthol 89(2):238–243. https://doi.org/10.1017/S0022149X13000849

Bird AF, Bird J (1991) The structure of nematodes, 2nd edn. Academic, London. https://doi.org/10.1016/C2009-0-02659-2

Boag B, Neilson R, Robinson D, Scrimgeour CM, Handley LL (1998) Wild rabbit host and some parasites show trophic-level relationships for $\delta^{13}C$ and $\delta^{15}N$: A first report. Isot Environ Health Stud 34(1–2):81–85. https://doi.org/10.1080/10256019708036335

Boecklen WJ, Yarnes CT, Cook BA, James AC (2011) On the use of stable isotopes in trophic ecology. In: Futuyma DJ, Shaffer HB, Simberloff D (eds) Annual review of ecology, evolution, and systematics, vol 42. Annual Reviews, Palo Alto, pp 411–440. https://doi.org/10.1146/annurev-ecolsys-102209-144726

Born-Torrijos A, Riekenberg P, van der Meer MT, Nachev M, Sures B, Thieltges DW (2023) Parasite effects on host's trophic and isotopic niches. Trends Parasitol 39(9):749–759. https://doi.org/10.1016/j.pt.2023.06.003

Boxshall GA, Halsey SH (2004) An introduction to copepod diversity. Ray Society, London

Brace AJ, Sheikali S, Martin LB (2015) Highway to the danger zone: Exposure-dependent costs of immunity in a vertebrate ectotherm. Funct Ecol 29(7):924–930. https://doi.org/10.1111/1365-2435.12402

Bradley CJ, Madigan DJ, Block BA, Popp BN (2014) Amino acid isotope incorporation and enrichment factors in pacific bluefin tuna, *Thunnus orientalis*. PLoS One 9(1). https://doi.org/10.1371/journal.pone.0085818

Bresciani J (1986) The fine structure of the integument of free-living and parasitic copepods. A review. Acta Zool 67(3):125–145. https://doi.org/10.1111/j.1463-6395.1986.tb00857.x

Britton JR, Andreou D (2016) Parasitism as a driver of trophic niche specialisation. Trends Parasitol 32(6):437–445. https://doi.org/10.1016/j.pt.2016.02.007

Butterworth KG, Li W, McKinley RS (2004) Carbon and nitrogen stable isotopes: A tool to differentiate between *Lepeophtheirus salmonis* and different salmonid host species? Aquaculture 241(1–4):529–538. https://doi.org/10.1016/j.aquaculture.2004.07.021

Catenazzi A, Donnelly MA (2008) Sea lion *Otaria flavescens* as host of the common vampire bat *Desmodus rotundus*. Mar Ecol Prog Ser 360:285–289. https://doi.org/10.3354/meps07393

Caut S, Angulo E, Courchamp F (2009) Variation in discrimination factors ($\Delta^{15}N$ and $\Delta^{13}C$): The effect of diet isotopic values and applications for diet reconstruction. J Appl Ecol 46(2):443–453. https://doi.org/10.1111/j.1365-2664.2009.01620.x

Chappell LH (1979) Physiology of parasites. Springer, New York. https://doi.org/10.1007/978-1-4684-7808-2

Chikaraishi Y, Kashiyama Y, Ogawa NO, Kitazato H, Ohkouchi N (2007) Metabolic control of nitrogen isotope composition of amino acids in macroalgae and gastropods: implications for aquatic food web studies. Mar Ecol Prog Ser 342:85–90

Chikaraishi Y, Ogawa NO, Kashiyama Y, Takano Y, Suga H, Tomitani A, Miyashita H, Kitazato H, Ohkouchi N (2009) Determination of aquatic food-web structure based on compound-specific nitrogen isotopic composition of amino acids. Limnol Oceanogr Meth 7(11):740–750

Chikaraishi Y, Ogawa NO, Doi H, Ohkouchi N (2011) $^{15}N/^{14}N$ ratios of amino acids as a tool for studying terrestrial food webs: A case study of terrestrial insects (bees, wasps, and hornets). Ecol Res 26(4):835–844. https://doi.org/10.1007/s11284-011-0844-1

Christianen MJA, Middelburg JJ, Holthuijsen SJ, Jouta J, Compton TJ, van der Heide T, Piersma T, Sinninghe Damsté JS, van der Veer HW, Schouten S, Olff H (2017) Benthic primary producers are key to sustain the Wadden Sea food web: stable carbon isotope analysis at landscape scale. Ecology 98(6):1498–1512. https://doi.org/10.1002/ecy.1837

Clark JM, Hawke DJ (2012) A new epizoic laelapid mite from the New Zealand sand scarab *Pericoptus truncatus* larvae and its isotopic ecology. N Z J Zool 39(3):187–199. https://doi.org/10.1080/03014223.2011.628997

Dalerum F, Angerbjörn A (2005) Resolving temporal variation in vertebrate diets using naturally occurring stable isotopes. Oecologia 144(4):647–658. https://doi.org/10.1007/s00442-005-0118-0

Dean S, DiBacco C, McKinley RS (2011) Assessment of stable isotopic signatures as a means to track the exchange of sea lice (*Lepeophtheirus salmonis*) between host fish populations. Can J Fish Aquat Sci 68(7):1243–1251. https://doi.org/10.1139/f2011-039

Décima M, Landry MR, Bradley CJ, Fogel ML (2017) Alanine $\delta^{15}N$ trophic fractionation in heterotrophic protists. Limnol Oceanogr 62(5):2308–2322. https://doi.org/10.1002/lno.10567

Demopoulos AWJ, Sikkel PC (2015) Enhanced understanding of ectoparasite-host trophic linkages on coral reefs through stable isotope analysis. Int J Parasitol Parasites Wildl 4(1):125–134. https://doi.org/10.1016/j.ijppaw.2015.01.002

Demopoulos AWJ, Fry B, Smith CR (2007) Food web structure in exotic and native mangroves: A Hawaii-Puerto Rico comparison. Oecologia 153(3):675–686. https://doi.org/10.1007/s00442-007-0751-x

Denic M, Taeubert JE, Geist J (2015) Trophic relationships between the larvae of two freshwater mussels and their fish hosts. Invertebr Biol 134(2):129–135. https://doi.org/10.1111/ivb.12080

DeNiro MJ, Epstein S (1978) Influence of diet on the distribution of carbon isotopes in animals. Geochim Cosmochim Acta 42(5):495–506

Deudero S, Pinnegar JK, Polunin NVC (2002) Insights into fish host-parasite trophic relationships revealed by stable isotope analysis. Dis Aquat Org 52(1):77–86. https://doi.org/10.3354/dao052077

Dobson A, Lafferty KD, Kuris AM, Hechinger RF, Jetz W (2008) Homage to Linnaeus: How many parasites? How many hosts? Proc Natl Acad Sci USA 105(Suppl 1):11482–11489. https://doi.org/10.1073/pnas.0803232105

Doi H, Yurlova NI, Vodyanitskaya SN, Kikuchi E, Shikano S, Yadrenkina EN, Zuykova EI (2008) Parasite-induced changes in nitrogen isotope signatures of host tissues. J Parasitol 94(1):292–295. https://doi.org/10.1645/GE-1228.1

Doi H, Yurlova NI, Vodyanitskaya SN, Kanaya G, Shikano S, Kikuchi E (2010) Estimating isotope fractionation between cercariae and host snail with the use of isotope measurement designed for very small organisms. J Parasitol 96(2):314–317. https://doi.org/10.1645/GE-2245.1

Doucett RR, Giberson DJ, Power G (1999) Parasitic association of *Nanocladius* (Diptera:Chironomidae) and *Pteronarcys biloba* (Plecoptera:Pteronarcyidae): Insights from stable-isotope analysis. J N Am Benthol Soc 18(4):514–523. https://doi.org/10.2307/1468383

Dubois SY, Savoye N, Sauriau PG, Billy I, Martinez P, De Montaudouin X (2009) Digenean trematodes-marine mollusc relationships: A stable isotope study. Dis Aquat Org 84(1):65–77. https://doi.org/10.3354/dao2022

Eloranta AP, Knudsen R, Amundsen PA, Merilä J (2015) Consistent isotopic differences between *Schistocephalus* spp. parasites and their stickleback hosts. Dis Aquat Org 115(2):121–128. https://doi.org/10.3354/dao02893

Ewald PW (1995) The evolution of virulence: A unifying link between parasitology and ecology. J Parasitol 81(5):659–669. https://doi.org/10.2307/3283951

Fleming NEC, Harrod C, Griffin DC, Newton J, Houghton JDR (2014) Scyphozoan jellyfish provide short-term reproductive habitat for hyperiid amphipods in a temperate near-shore environment. Mar Ecol Prog Ser 510:229–240. https://doi.org/10.3354/meps10896

Frenken T, Wierenga J, van Donk E, Declerck SAJ, de Senerpont Domis LN, Rohrlack T, Van de Waal DB (2018) Fungal parasites of a toxic inedible cyanobacterium provide food to zooplankton. Limnol Oceanogr 63(6):2384–2393. https://doi.org/10.1002/lno.10945

Frenken T, Miki T, Kagami M, Van de Waal DB, Van Donk E, Rohrlack T, Gsell AS (2020) The potential of zooplankton in constraining chytrid epidemics in phytoplankton hosts. Ecology 101(1). https://doi.org/10.1002/ecy.2900

Fritts MW, Fritts AK, Carleton SA, Bringolf RB (2013) Shifts in stable-isotope signatures confirm parasitic relationship of freshwater mussel glochidia attached to host fish. J Molluscan Stud 79(2):163–167. https://doi.org/10.1093/mollus/eyt008

Fry B (1988) Food web structure on Georges Bank from stable C, N, and S isotopic compositions. Limnol Oceanogr 33(5):1182–1190. https://doi.org/10.4319/lo.1988.33.5.1182

Fry B (2006) Stable isotope ecology, vol 521. Springer, New York

Fry B, Arnold C (1982) Rapid $^{13}C/^{12}C$ turnover during growth of brown shrimp (*Penaeus aztecus*). Oecologia 54(2):200–204. https://doi.org/10.1007/BF00378393

Galvan K, Fleeger JW, Peterson B, Drake D, Deegan LA, Johnson DS (2011) Natural abundance stable isotopes and dual isotope tracer additions help to resolve resources supporting a saltmarsh food web. J Exp Mar Biol Ecol 410:1–11. https://doi.org/10.1016/j.jembe.2011.08.007

Gilbert BM, Nachev M, Jochmann MA, Schmidt TC, Köster D, Sures B, Avenant-Oldewage A (2020a) Stable isotope analysis spills the beans about spatial variance in trophic structure in a fish host–parasite system from the Vaal River System, South Africa. Int J Parasitol Parasites Wildl 12:134141. https://doi.org/10.1016/j.ijppaw.2020.05.011

Gilbert BM, Nachev M, Jochmann MA, Schmidt TC, Köster D, Sures B, Avenant-Oldewage A (2020b) You are how you eat: differences in trophic position of two parasite species infecting a single host according to stable isotopes. Parasitol Res 119(4):1393–1400. https://doi.org/10.1007/s00436-020-06619-1

Gilbert B, Nachev M, Sures B, Avenant-Oldewage A (2025) Dietary shifts among the developmental stages of the ectoparasite, *Argulus japonicus* (Crustacea; Branchiura), mirror ontogeny as shown through differences in stable isotope ratios of carbon ($\delta^{13}C$) and nitrogen ($\delta^{15}N$). Ecol Evol 5(1):e70652. https://doi.org/10.1002/ece3.70652

Gómez-Díaz E, González-Solís J (2010) Trophic structure in a seabird host-parasite food web: Insights from stable isotope analyses. PLoS One 5(5). https://doi.org/10.1371/journal.pone.0010454

Gutiérrez-Rodríguez A, Décima M, Popp BN, Landry MR (2014) Isotopic invisibility of protozoan trophic steps in marine food webs. Limnol Oceanogr 59(5):1590–1598. https://doi.org/10.4319/lo.2014.59.5.1590

Hanz U, Riekenberg P, de Kluijver A, van der Meer M, Middelburg JJ, de Goeij JM, Bart MC, Wurz E, Colaço A, Duineveld GCA, Reichart GJ, Rapp HT, Mienis F (2022) The important role of sponges in carbon and nitrogen cycling in a deep-sea biological hotspot. Funct Ecol 36(9):2188–2199. https://doi.org/10.1111/1365-2435.14117

Hesse T, Nachev M, Khaliq S, Jochmann MA, Franke F, Scharsack JP, Kurtz J, Sures B, Schmidt TC (2023) A new technique to study nutrient flow in host-parasite systems by carbon stable isotope analysis of amino acids and glucose. Sci Rep 13(1):1054

Hobson KA, Clark RG (1992) Assessing avian diets using stable isotopes II: Factors influencing diet-tissue fractionation. Condor 94(1):189–197. https://doi.org/10.2307/1368808

Hobson KA, Welch HE (1992) Determination of trophic relationships within a high Arctic marine food web using $\delta^{13}C$ and $\delta^{15}N$ analysis. Mar Ecol Prog Ser 84:9–18

Ibelings BW, De Bruin A, Kagami M, Rijkeboer M, Brehm M, Van Donk E (2004) Host parasite interactions between freshwater phytoplankton and chytrid fungi (Chytridiomycota). J Phycol 40(3):437–453. https://doi.org/10.1111/j.1529-8817.2004.03117.x

Iken K, Brey T, Wand U, Voigt J, Junghans P (2001) Food web structure of the benthic community at the Porcupine Abyssal Plain (NE Atlantic): A stable isotope analysis. Prog Oceanogr 50(1–4):383–405. https://doi.org/10.1016/S0079-6611(01)00062-3

Ishikawa NF, Kato Y, Togashi H, Yoshimura M, Yoshimizu C, Okuda N, Tayasu I (2014) Stable nitrogen isotopic composition of amino acids reveals food web structure in stream ecosystems. Oecologia 175(3):911–922. https://doi.org/10.1007/s00442-014-2936-4

Johnson PTJ, Dobson A, Lafferty KD, Marcogliese DJ, Memmott J, Orlofske SA, Poulin R, Thieltges DW (2010) When parasites become prey: Ecological and epidemiological significance of eating parasites. Trends Ecol Evol 25(6):362–371. https://doi.org/10.1016/j.tree.2010.01.005

Kagami M, Von Elert E, Ibelings BW, De Bruin A, Van Donk E (2007) The parasitic chytrid, *Zygorhizidium*, facilitates the growth of the cladoceran zooplankter, *Daphnia*, in cultures of the inedible alga. Asterionella. P Roy Soc B-Biol Sci 274(1617):1561–1566. https://doi.org/10.1098/rspb.2007.0425

Kagami M, Miki T, Takimoto G (2014) Mycoloop: Chytrids in aquatic food webs. Front Microbiol. https://doi.org/10.3389/fmicb.2014.00166

Kamiya E, Urabe M, Okuda N (2020) Does atypical 15N and 13C enrichment in parasites result from isotope ratio variation of host tissues they are infected? Limnology 21(1):139–149. https://doi.org/10.1007/s10201-019-00596-w

Karlson AML, Reutgard M, Garbaras A, Gorokhova E (2018) Isotopic niche reflects stress-induced variability in physiological status. Roy Soc Open Sci 5(2). https://doi.org/10.1098/rsos.171398

Koprivnikar J, Thieltges DW, Johnson PTJ (2023) Consumption of trematode parasite infectious stages: from conceptual synthesis to future research agenda. J Helminthol 97. https://doi.org/10.1017/S0022149X23000111

Kruger BR, Werne JP, Branstrator DK, Hrabik TR, Chikaraishi Y, Ohkouchi N, Minor EC (2016) Organic matter transfer in Lake Superior's food web: Insights from bulk and molecular stable isotope and radiocarbon analyses. Limnol Oceanogr 61(1):149–164. https://doi.org/10.1002/lno.10205

Kuris AM, Hechinger RF, Shaw JC, Whitney KL, Aguirre-Macedo L, Boch CA, Dobson AP, Dunham EJ, Fredensborg BL, Huspeni TC, Lorda J, Mababa L, Mancini FT, Mora AB, Pickering M, Talhouk NL, Torchin ME, Lafferty KD (2008) Ecosystem energetic implications of parasite and free-living biomass in three estuaries. Nature 454(7203):515–518. https://doi.org/10.1038/nature06970

Kürten B, Painting SJ, Struck U, Polunin NVC, Middelburg JJ (2013) Tracking seasonal changes in North Sea zooplankton trophic dynamics using stable isotopes. Biogeochemistry 113(1–3):167–187. https://doi.org/10.1007/s10533-011-9630-y

Lafferty KD, Kuris AM (2002) Trophic strategies, animal diversity and body size. Trends Ecol Evol 17(11):507–513. https://doi.org/10.1016/S0169-5347(02)02615-0

Langellotto GA, Rosenheim JA, Williams MR (2005) Enhanced carbon enrichment in parasitoids (Hymenoptera): A stable isotope study. Ann Entomol Soc Am 98(2):205–213. https://doi.org/10.1603/0013-8746(2005)098[0205:ECEIPH]2.0.CO;2

Larsen T, Lee Taylor D, Leigh MB, O'Brien DM (2009) Stable isotope fingerprinting: A novel method for identifying plant, fungal, or bacterial origins of amino acids. Ecology 90(12):3526–3535. https://doi.org/10.1890/08-1695.1

Larsen T, Ventura M, Andersen N, O'Brien DM, Piatkowski U, McCarthy MD (2013) Tracing carbon sources through aquatic and terrestrial food webs using amino acid stable isotope fingerprinting. PLoS One 8(9):e73441

Lau DCP, Leung KMY, Dudgeon D (2009) What does stable isotope analysis reveal about trophic relationships and the relative importance of allochthonous and autochthonous resources in tropical streams? A synthetic study from Hong Kong. Freshwater Biol 54(1):127–141. https://doi.org/10.1111/j.1365-2427.2008.02099.x

McClelland JW, Montoya JP (2002) Trophic relationships and the nitrogen isotopic composition of amino acids in plankton. Ecology 83(8):2173–2180

McCutchan JH, Lewis WM Jr, Kendall C, McGrath CC (2003) Variation in trophic shift for stable isotope ratios of carbon, nitrogen, and sulfur. Oikos 102(2):378–390

McGrew AK, O'Hara TM, Stricker CA, Margaret Castellini J, Beckmen KB, Salman MD, Ballweber LR (2015) Ecotoxicoparasitology: Understanding mercury concentrations in gut contents, intestinal helminths and host tissues of Alaskan gray wolves (*Canis lupus*). Sci Total Environ 536:866–871. https://doi.org/10.1016/j.scitotenv.2015.07.106

McMahon KW, Fogel ML, Elsdon TS, Thorrold SR (2010) Carbon isotope fractionation of amino acids in fish muscle reflects biosynthesis and isotopic routing from dietary protein. J Anim Ecol 79(5):1132–1141

Meier-Augenstein W (2004) GC and IRMS technology for ^{13}C and ^{15}N analysis on organic compounds and related gases. In: de Groot PA (ed) Handbook of stable isotope analytical techniques. Elsevier, Amsterdam, pp 153–176

Michener R, Lajtha K (2008) Stable isotopes in ecology and environmental science, 2nd edn. Blackwell, Oxford. https://doi.org/10.1002/9780470691854

Minagawa M, Wada E (1984) Stepwise enrichment of ^{15}N along food chains: further evidence and the relation between δ^{15}N and animal age. Geochim Cosmochim Acta 48(5):1135–1140

Nachev M, Jochmann MA, Walter F, Wolbert JB, Schulte SM, Schmidt TC, Sures B (2017) Understanding trophic interactions in host-parasite associations using stable isotopes of carbon and nitrogen. Parasit Vectors 10(1). https://doi.org/10.1186/s13071-017-2030-y

Navarro J, Albo-Puigserver M, Coll M, Saez R, Forero MG, Kutcha R (2014) Isotopic discrimination of stable isotopes of nitrogen (δ^{15}N) and carbon (δ^{13}C) in a host-specific holocephalan tapeworm. J Helminthol 88(3):371–375. https://doi.org/10.1017/S0022149X13000126

Neilson R, Boag B, Hartley G (2005) Temporal host-parasite relationships of the wild rabbit, *Oryctolagus cuniculus* (L.) as revealed by stable isotope analyses. Parasitology 131(2):279–285. https://doi.org/10.1017/S0031182005007717

Nilsen M, Pedersen T, Nilssen EM, Fredriksen S (2008) Trophic studies in a high-latitude fjord ecosystem—a comparison of stable isotope analyses (δ^{13}C and δ^{15}N) and trophic-level estimates from a mass-balance model. Can J Fish Aquat Sci 65(12):2791–2806. https://doi.org/10.1139/F08-180

O'Grady SP, Dearing MD (2006) Isotopic insight into host-endosymbiont relationships in Liolaemid lizards. Oecologia 150(3):355–361. https://doi.org/10.1007/s00442-006-0487-z

Ohkouchi N, Chikaraishi Y, Close HG, Fry B, Larsen T, Madigan DJ, McCarthy MD, McMahon KW, Nagata T, Naito YI, Ogawa NO, Popp BN, Steffan S, Takano Y, Tayasu I, Wyatt ASJ, Yamaguchi YT, Yokoyama Y (2017) Advances in the application of amino acid nitrogen isotopic analysis in ecological and biogeochemical studies. Org Geochem 113:150–174. https://doi.org/10.1016/j.orggeochem.2017.07.009

O'Leary MH (1988) Carbon isotopes in photosynthesis. Bioscience 38:328–336. https://doi.org/10.2307/1310735

Pappas PW, Read CP (1975) Membrane transport in helminth parasites: A review. Exp Parasitol 37(3):469–530. https://doi.org/10.1016/0014-4894(75)90016-8

Parker D, Booth AJ (2013) The tongue-replacing isopod *Cymothoa borbonica* reduces the growth of largespot pompano *Trachinotus botla*. Mar Biol 160(11):2943–2950. https://doi.org/10.1007/s00227-013-2284-7

Parmentier E, Das K (2004) Commensal vs. parasitic relationship between Carapini fish and their hosts: Some further insight through δ^{13}C and δ^{15}N measurements. J Exp Mar Biol Ecol 310(1):47–58. https://doi.org/10.1016/j.jembe.2004.03.019

Pegg J, Andreou D, Williams CF, Britton JR (2015) Temporal changes in growth, condition and trophic niche in juvenile *Cyprinus carpio* infected with a non-native parasite. Parasitology 142(13):1579–1587. https://doi.org/10.1017/S0031182015001237

Persson ME, Larsson P, Stenroth P (2007) Fractionation of δ^{15}N and δ^{13}C for Atlantic salmon and its intestinal cestode *Eubothrium crassum*. J Fish Biol 71(2):441–452. https://doi.org/10.1111/j.1095-8649.2007.01500.x

Peterson BJ, Fry B (1987) Stable isotopes in ecosystem studies. Annu Rev Ecol Syst 18:293–320. https://doi.org/10.1146/annurev.es.18.110187.001453

Phillips DL, Koch PL (2002) Incorporating concentration dependence in stable isotope mixing models. Oecologia 130(1):114–125. https://doi.org/10.1007/s004420100786

Phillips DL, Inger R, Bearhop S, Jackson AL, Moore JW, Parnell AC, Semmens BX, Ward EJ (2014) Best practices for use of stable isotope mixing models in food-web studies. Can J Zool 92(10):823–835. https://doi.org/10.1139/cjz-2014-0127

Pinnegar JK, Campbell N, Polunin NVC (2001) Unusual stable isotope fractionation patterns observed for fish host-parasite trophic relationships. J Fish Biol 59(3):494–503. https://doi.org/10.1006/jfbi.2001.1660

Popp BN, Graham BS, Olson RJ, Hannides CC, Lott MJ, López-Ibarra GA, Galván-Magaña F, Fry B (2007) Insight into the trophic ecology of yellowfin tuna, *Thunnus albacares*, from compound-specific nitrogen isotope analysis of proteinaceous amino acids. Terrestrial Ecol 1:173–190

Post DM (2002) Using stable isotopes to estimate trophic position: Models, methods, and assumptions. Ecology 83(3):703–718. https://doi.org/10.1890/0012-9658(2002)083[0703:Usitet]2.0.Co;2

Poulin R (2007) Evolutionary ecology of parasites, 2nd edn. Princeton University Press, Princeton

Poulin R (2011) The many roads to parasitism. A tale of convergence. Adv Parasit 74:1–40. https://doi.org/10.1016/B978-0-12-385897-9.00001-X

Power M, Klein GM (2004) Fish host-cestode parasite stable isotope enrichment patterns in marine, estuarine and freshwater fishes from northern Canada. Isot Environ Health Stud 40(4):257–266. https://doi.org/10.1080/10256010410001678062

Raffel TR, Martin LB, Rohr JR (2008) Parasites as predators: unifying natural enemy ecology. Trends Ecol Evol 23(11):610–618. https://doi.org/10.1016/j.tree.2008.06.015

Read CP (1973) Contact digestion in tapeworms. J Parasitol 59(4):672–677. https://doi.org/10.2307/3278861

Riascos JM, Docmac F, Reddin C, Harrod C (2015) Trophic relationships between the large scyphomedusa *Chrysaora plocamia* and the parasitic amphipod *Hyperia curticephala*. Mar Biol 162(9):1841–1848. https://doi.org/10.1007/s00227-015-2716-7

Riekenberg P, Joling T, Ijsseldijk L, Waser A, Van der Meer M, Thieltges D (2021a) Stable nitrogen isotope analysis of amino acids as a new tool to clarify complex parasite-host interactions within marine food webs. Oikos 130:1650–1664. https://doi.org/10.1101/2021.03.04.433913

Riekenberg PM, Briand MJ, Moléana T, Sasal P, van der Meer MTJ, Thieltges DW, Letourneur Y (2021b) Isotopic discrimination in helminths infecting coral reef fishes depends on parasite group, habitat within host, and host stable isotope value. Sci Rep 11(1). https://doi.org/10.1038/s41598-021-84255-0

Riekenberg PM, Joling T, Ijsseldijk LL, Waser AM, van der Meer MTJ, Thieltges DW (2021c) Stable nitrogen isotope analysis of amino acids as a new tool to clarify complex parasite–host interactions within food webs. Oikos 130(10):1650–1664. https://doi.org/10.1111/oik.08450

Riekenberg PM, van der Heide T, Holthuijsen SJ, van der Veer HW, van der Meer MTJ (2022) Compound-specific stable isotope analysis of amino acid nitrogen reveals detrital support of microphytobenthos in the Dutch Wadden Sea benthic food web. Front Ecol Evol 10. https://doi.org/10.3389/fevo.2022.951047

Rothman AH (1967) Ultrastructural enzyme localization in the surface of *Moniliformis dubius* (Acanthocephala). Exp Parasitol 21(1):42–46. https://doi.org/10.1016/0014-4894(67)90065-3

Sabadel AJM, MacLeod CD (2022) Stable isotopes unravel the feeding mode–trophic position relationship in trematode parasites. J Anim Ecol 91(2):484–495. https://doi.org/10.1111/1365-2656.13644

Sabadel AJM, Woodward EMS, Van Hale R, Frew RD (2016) Compound-specific isotope analysis of amino acids: A tool to unravel complex symbiotic trophic relationships. Food Webs 6:9–18. https://doi.org/10.1016/j.fooweb.2015.12.003

Sabadel AJM, Stumbo AD, Macleod CD (2019) Stable-isotope analysis: A neglected tool for placing parasites in food webs. J Helminthol 93(1):1–7. https://doi.org/10.1017/S0022149X17001201

Sabadel AJM, Cresson P, Finucci B, Bennett J (2022) Unravelling the trophic interaction between a parasitic barnacle (*Anelasma squalicola*) and its host Southern lanternshark (*Etmopterus granulosus*) using stable isotopes. Parasitology 149(14):1976–1984. https://doi.org/10.1017/S0031182022001299

Sánchez MI, Varo N, Matesanz C, Ramo C, Amat JA, Green AJ (2013) Cestodes change the isotopic signature of brine shrimp, *Artemia*, hosts: Implications for aquatic food webs. Int J Parasitol 43(1):73–80. https://doi.org/10.1016/j.ijpara.2012.11.003

Sánchez Barranco V, Van der Meer MTJ, Kagami M, Van den Wyngaert S, Van de Waal DB, Van Donk E, Gsell AS (2020) Trophic position, elemental ratios and nitrogen transfer in a planktonic host–parasite–consumer food chain including a fungal parasite. Oecologia 194(4):541–554. https://doi.org/10.1007/s00442-020-04721-w

Schmidt O, Dautel H, Newton J, Gray JS (2011) Natural isotope signatures of host blood are replicated in moulted ticks. Ticks Tick Borne Dis 2(4):225–227. https://doi.org/10.1016/j.ttbdis.2011.09.006

Schultz B, Koprivnikar J (2021) The contributions of a trematode parasite infectious stage to carbon cycling in a model freshwater system. Parasitol Res 120(5):1743–1754. https://doi.org/10.1007/s00436-021-07142-7

Smyth JD, McManus DP (1989) The physiology and biochemistry of cestodes. Cambridge University Press, Cambridge. https://doi.org/10.1017/CBO9780511525841

Soldánová M, Selbach C, Sures B (2016) The early worm catches the bird? Productivity and patterns of *Trichobilharzia szidati* cercarial Emission from *Lymnaea stagnalis*. PLoS One 11(2):e0149678. https://doi.org/10.1371/journal.pone.0149678

Stapp P, Salkeld DJ (2009) Inferring host-parasite relationships using stable isotopes: Implications for disease transmission and host specificity. Ecology 90(11):3268–3273. https://doi.org/10.1890/08-1226.1

Stephens RB, Shipley ON, Moll RJ (2023) Meta-analysis and critical review of trophic discrimination factors (Δ^{13}C and Δ^{15}N): Importance of tissue, trophic level and diet source. Funct Ecol 37(9):2535–2548. https://doi.org/10.1111/1365-2435.14403

Stock BC, Jackson AL, Ward EJ, Parnell AC, Phillips DL, Semmens BX (2018) Analyzing mixing systems using a new generation of bayesian tracer mixing models. Peer J 6:e5096. https://doi.org/10.7717/peerj.5096

Sures B, Nachev M, Gilbert BM, Dos Santos QM, Jochmann MA, Köster D, Schmidt TC, Avenant-Oldewage A (2019) The monogenean *Paradiplozoon ichthyoxanthon* behaves like a micropredator on two of its hosts, as indicated by stable isotopes. J Helminthol 93(1):71–75. https://doi.org/10.1017/S0022149X17001195

Taccardi EY, Bricknell IR, Byron CJ (2020) Stable isotopes reveal contrasting trophic dynamics between host–parasite relationships: a case study of Atlantic salmon (Salmo salar) and parasitic lice (Lepeophtheirus salmonis and Argulus foliaceus). J Fish Biol 97(6):1821-1832. https://doi.org/10.1111/jfb.14546

Taylor EW, Thomas JN (1968) Membrane (contact) digestion in the three species of tapeworm *Hymenolepis diminuta*, *Hymenolepis microstoma* and *Moniezia expansa*. Parasitology 58(3):535–546. https://doi.org/10.1017/S0031182000028845

Taylor CH, Young S, Fenn J, Lamb AL, Lowe AE, Poulin B, MacColl ADC, Bradley JE (2019) Immune state is associated with natural dietary variation in wild mice *Mus musculus* domesticus. Funct Ecol 33(8): 1425–1435. https://doi.org/10.1111/1365-2435.13354

Then AYH, Adame MF, Fry B, Chong VC, Riekenberg PM, Mohammad Zakaria R, Lee SY (2021) Stable isotopes clearly track mangrove inputs and food web changes along a reforestation gradient. Ecosystems 24(4):939–954. https://doi.org/10.1007/s10021-020-00561-0

Thieltges DW, Jensen KT, Poulin R (2008) The role of biotic factors in the transmission of free-living endohelminth stages. Parasitology 135(4):407–426. https://doi.org/10.1017/S0031182007000248

Thieltges DW, Goedknegt MA, O'Dwyer K, Senior AM, Kamiya T (2019) Parasites and stable isotopes: a comparative analysis of isotopic discrimination in parasitic trophic interactions. Oikos 128(9):1329–1339. https://doi.org/10.1111/oik.06086

Towanda T, Thuesen EV (2006) Ectosymbiotic behavior of *Cancer gracilis* and its trophic relationships with its host *Phacellophora camtschatica* and the parasitoid *Hyperia medusarum*. Mar Ecol Prog Ser 315:221–236. https://doi.org/10.3354/meps315221

Vander Zanden MJ, Rasmussen JB (2001) Variation in δ^{15}N and δ^{13}C trophic fractionation: implications for aquatic food web studies. Limnol Oceanogr 46(8):2061–2066

Vander Zanden MJ, Clayton MK, Moody EK, Solomon CT, Weidel BC (2015) Stable isotope turnover and half-life in animal tissues: A literature synthesis. PLoS One 10(1). https://doi.org/10.1371/journal.pone.0116182

Vanderklift MA, Ponsard S (2003) Sources of variation in consumer-diet δ^{15}N enrichment: a meta-analysis. Oecologia 136(2):169–182. https://doi.org/10.1007/s00442-003-1270-z

Vivas Muñoz JC, Feld CK, Hilt S, Manfrin A, Nachev M, Köster D, Jochmann MA, Schmidt TC, Sures B, Ziková A, Knopf K (2021) Eye fluke infection changes diet composition in juvenile European perch (*Perca fluviatilis*). Sci Rep 11(1). https://doi.org/10.1038/s41598-021-81568-y

Voigt CC, Kelm DH (2006) Host preferences of bat flies: Following the bloody path of stable isotopes in a host-parasite food chain. Can J Zool 84(3):397–403. https://doi.org/10.1139/z06-007

Welicky RL, Demopoulos AWJ, Sikkel PC (2017) Host-dependent differences in resource use associated with *Anilocra* spp. parasitism in two coral reef fishes, as revealed by stable carbon and nitrogen isotope analyses. Mar Ecol. https://doi.org/10.1111/maec.12413

Whiteman JP, Smith EAE, Besser AC, Newsome SD (2019) A guide to using compound-specific stable isotope analysis to study the fates of molecules in organisms and ecosystems. Diversity. https://doi.org/10.3390/d11010008

Wolfsberg M, Alexander Van Hook W, Paneth P, Paulo L, Rebelo N (2009) Isotope effects in the chemical, geological, and bio sciences. Springer, Heidelberg

Won EJ, Choi B, Hong S, Khim JS, Shin KH (2018) Importance of accurate trophic level determination by nitrogen isotope of amino acids for trophic magnification studies: A review. Environ Pollut 238:677–690. https://doi.org/10.1016/j.envpol.2018.03.045

Woodland RJ, Secor DH (2013) Benthic-pelagic coupling in a temperate inner continental shelf fish assemblage. Limnol Oceanogr 58(3):966–976. https://doi.org/10.4319/lo.2013.58.3.0966

Xu J, Zhang M, Xie P (2007) Trophic relationship between the parasitic isopod *Ichthyoxenus japonensis* and the fish *Carassius auratus auratus* as revealed by stable isotopes. J Freshw Ecol 22(2):333–338. https://doi.org/10.1080/02705060.2007.9665055

Yarnes CT, Rockwell JN, Boecklen WJ (2005) Patterns of trophic shift in δ^{15}N and δ^{13}C through a cynipid gall wasp community (*Neuroterus* sp.) in *Quercus turbinella*. Environ Entomol 34(6):1471–1476. https://doi.org/10.1603/0046-225X-34.6.1471

Yohannes E, Grimm C, Rothhaupt KO, Behrmann-Godel J (2017) The effect of parasite infection on stable isotope turnover rates of δ^{15}N, δ^{13}C and δ^{34}S in multiple tissues of Eurasian perch *Perca fluviatilis*. PLoS One. https://doi.org/10.1371/journal.pone.0169058

Yurlova NI, Shikano S, Kanaya G, Rastyazhenko NM, Vodyanitskaya SN (2014) The evaluation of snail host-trematode parasite trophic relationships using stable isotope analysis. Parazitologiya 48(3):193–205

Open Access This chapter is licensed under the terms of the Creative Commons Attribution-NonCommercial-NoDerivatives 4.0 International License (http://creativecommons.org/licenses/by-nc-nd/4.0/), which permits any non-commercial use, sharing, distribution and reproduction in any medium or format, as long as you give appropriate credit to the original author(s) and the source, provide a link to the Creative Commons license and indicate if you modified the licensed material. You do not have permission under this license to share adapted material derived from this chapter or parts of it.

The images or other third party material in this chapter are included in the chapter's Creative Commons license, unless indicated otherwise in a credit line to the material. If material is not included in the chapter's Creative Commons license and your intended use is not permitted by statutory regulation or exceeds the permitted use, you will need to obtain permission directly from the copyright holder.

Part III

Environmental Aquatic Parasitology and Ecological Applications: Impacts on Animal and Human Life

Parasites as Biological Tags in Aquatic Hosts

17

Juan T. Timi

Abstract

Extant host-parasite associations are the result of tight reciprocal adaptations that allow parasites to exploit specific biological features of their hosts. Due to the intimacy of their relationships, parasites may provide important ecological information on their host's individuals, populations, communities, and even ecosystems, with their variability in any dimension of the host niche being interpretable as an indicator of such changes. This chapter compiles and updates the published research on parasites as indicators of ecological and distributional patterns and processes of their hosts and environments, mostly from the perspective of fisheries, which has promoted this methodology the most. These patterns are discussed at spatial and temporal scales. The rationale for using parasites as biological tags and the routine statistical techniques used are analysed, along with the criteria for the selection of suitable parasite tags, and their advantages and limitations in light of recent research, including that on non-teleost hosts. Those host traits affecting the use of parasite tags are also considered together with the increasingly recognised necessity for integrative multidisciplinary studies on fish parasites, combining both classical methods and modern techniques. The analysis is extended to the use of parasite genetics as a source of data on spatial and temporal distribution of hosts. Finally, the scope of the methodology is extended from the level of host populations to host communities and biogeography, with application for ecosystem approaches to fisheries management and conservation of resources, respectively.

17.1 Introduction

Ecological bioindicators are taxa or biological assemblages that, by their presence, number, or condition, are indicative of patterns and processes of biological systems and their environments (see also Chap. 20). Parasites meet many of the conditions to be considered as ecological indicators, as they are ubiquitous components of biological systems, comprising a significant proportion of world biodiversity (Dobson et al. 2008) and achieving substantial biomass, abundance, and productivity in some ecosystems (Kuris et al. 2008; Hechinger et al. 2011). Host-parasite associations are the result of antag-

J. T. Timi
Departamento de Biología, Instituto de Investigaciones Marinas y Costeras (IIMyC), Universidad Nacional de Mar del Plata – CONICET, Mar del Plata, Argentina
e-mail: jtimi@mdp.edu.ar

onistic interactions that have led to tight reciprocal adaptations (Timi and Poulin 2020). As a result of the often strict interdependence with their host, parasites may provide important ecological information on host individuals, populations, communities, and even ecosystems. Therefore, parasites can be valuable as indicators for aspects of host biology at these levels. Indeed, the composition and structure of parasite communities are the results of multiple biotic and abiotic factors interacting in an ecosystem (Poulin 2007). These parasite assemblages are hence highly dynamic, displaying heterogeneities in the properties of their components at various scales, both spatially and temporally. Consequently, parasites can provide a broad range of suitable tools for studying ecological patterns and processes of their hosts. In that sense, the use of parasite tags has been largely applied to fisheries science, probably fuelled by the role of fisheries as paramount contributors to global food security and nutrition (FAO 2022) and the increasing necessity of assessment of biological status and population structure (Costello et al. 2012; Kleisner et al. 2012). This chapter is mostly presented in that context.

In fisheries, variability of parasite populations and parasite assemblages has been used as an indicator of differences among host stocks, populations, and communities, but also to indicate migrations, spawning areas and other distribution-related processes (Williams et al. 1992; MacKenzie and Abaunza 2014; Timi and MacKenzie 2015; Timi and Buchmann 2023). On the other hand, a prerequisite for fishery management is the identification of past or ongoing changes in resources, in order to facilitate intensive management fisheries so that stocks can reach above target levels, rebuild and restored to their original conditions as much as possible (Hilborn et al., 2020). Therefore, the application of area-appropriate recommendations and management tools are needed for sustaining fisheries in many regions throughout the world. Consequently, changes in parasite abundance or diversity at temporal scales may indicate the effect of fisheries and other anthropogenic stressors on natural resources that require management and monitoring (Marcogliese 2001, 2016; Wood et al. 2010; Wood and Lafferty 2015).

Since the pioneering works of Dogiel and Bychovsky (1939) and Herrington et al. (1939), parasite tags have been increasingly applied to fishery management (Timi and MacKenzie 2015; Pita et al. 2016; Pascual et al. 2016; Timi and Buchmann 2023). Over the subsequent 80 years or so, consensus on a series of methodologies for the use of parasite tags has been achieved, but with a variable degree of application at a global scale. These methods are periodically reviewed and adjusted (Timi and MacKenzie 2015). A number of reviews have been published on the overall criteria for the selection of suitable indicators, for different groups of parasites, hosts, and geographical regions (Timi and MacKenzie 2015; Pascual et al. 2016; Timi and Buchmann 2023 and references therein). This compilation of literature provides an 'evolutionary view' of the development undergone in the use of parasites as biological markers, resulting in an increasingly efficient and reliable methodology (Timi and Buchmann 2023). A meta-analysis by Poulin and Kamiya (2015) corroborated that parasites have a real usefulness as tags for marine fish stocks, showing that the probability of correctly assigning a fish to its original stock based on parasite data was twice that which can be achieved by chance alone. Conversely, however, aside from anadromous fishes, very few studies have used parasite tags to discriminate stocks of freshwater fish (Marcogliese and Jacobson 2015). For example, in some freshwater systems, parasites are commonly used to determine the river of origin for anadromous salmonids and to distinguish anadromous from resident individuals (Marcogliese and Jacobson 2015 and references therein).

In this chapter, concepts, procedures, and methods of the use of parasite tags in aquatic, mostly marine hosts, are analysed in the light of recent findings and novel perspectives, developed in recent times to improve, reinforce, or complement this methodology, with an emphasis in stock identification of marine fish. These topics include: (1) a critical review of the proposed criteria for the selection of suitable parasite tags, with an analysis of the effect of host traits on their value as tags, and the advantages and limitation of the method; (2) the use of parasites and their genetic information to provide insights into the spatial distribution of their hosts and about temporal changes in their populations and communities; (3) the extension of the use of parasite indicators to host groups other than bony fishes; (4) and the increasingly recognised necessity for integrative, multidisciplinary studies; and (5) a proposal to expand the use of parasite tags beyond the host population level to host community and biogeographical scales.

17.2 Rationale and Methodology for Using Parasite Tags

The use of parasite tags for spatial comparisons is underlined by the rationale that hosts harbouring a given parasite inhabit or once inhabited the endemic areas of the parasite. The conditions in such areas, both biotic and abiotic, are those necessary and suitable for an effective transmission of the parasite and to viably sustain their populations (MacKenzie and Abaunza 1998).

Changes in such conditions, for example, in the density of definitive or intermediate hosts or in physical characteristics, such as temperature and salinity, are expected to produce concomitant changes in parasite abundance, especially when the distance between hosts increases. Such changes in parasite populations have consequent repercussions on the assemblages they compose, a process underpinned by the almost universal decay in the similarity in species composition as a function of the increasing distance separating the assemblages (Nekola and White 1999; Poulin and Kamiya 2015). Whereas geographical distances between host populations are key determinants of how many parasite species they share, decay in similarity should also occur with increasing distance along any other dimension that characterises any form of separation between communities (Timi et al. 2010a). Thus, beyond spatial patterns, the same rationale can be applied to analyse temporal, ontogenetic, evolutionary, genetic, or any other distance of interest between hosts by means of measuring changes in the parasitism.

Combining the analysis of two or more dimensions of such variability can also provide useful information on host biology. For example, spatial and temporal variability combined can be used to trace trophic or reproductive migrations, whereas by adding the variation in age or size of hosts to the analyses, ontogenetic and asynchronous migrations can explain the observed patterns (Canel et al. 2021).

Different statistical methods, applied to different parasitological parameters, are commonly used for comparing samples of interest (Fig. 17.1). The simplest one is comparing the abundance (mean/median abundance or intensity) of a given parasite species (Fig. 17.1a) by means of appropriate parametric or non-parametric tests between two groups of infrapopulations (Bush et al. 1997). This can also be applied to additional parasite species, as well as to community descriptors, such as diversity, species richness, or evenness (Magurran 2004), whereas analyses of multivariate response data can be applied to groups of infracommunities (Bush et al. 1997). The latter comparisons can be made in terms of composition or multivariate abundance using adequate similarity indices, for example Jaccard and Bray-Curtis, respectively (Magurran 2004). At the level of host sample or component community (Bush et al. 1997), population parameters as prevalence, or community descriptors as diversity, can be also compared, for example by Chi-squared test or Hutcheson t-test for the Shannon index, respectively (Magurran 2004). Of course, the analytical capabilities increase with the number of samples analysed, and spatial patterns such as gradients can be detected (Fig. 17.1b). When parasites with low specificity are studied (Fig. 17.1c), samples from different host species can be analysed to reveal higher scale distribution patterns.

The abundance and distribution of parasites depend on factors related to the parasites themselves, their hosts, and ultimately to their environment; consequently, analytical procedures should include as much information as possible. Long-lived parasites are recommended as bio-

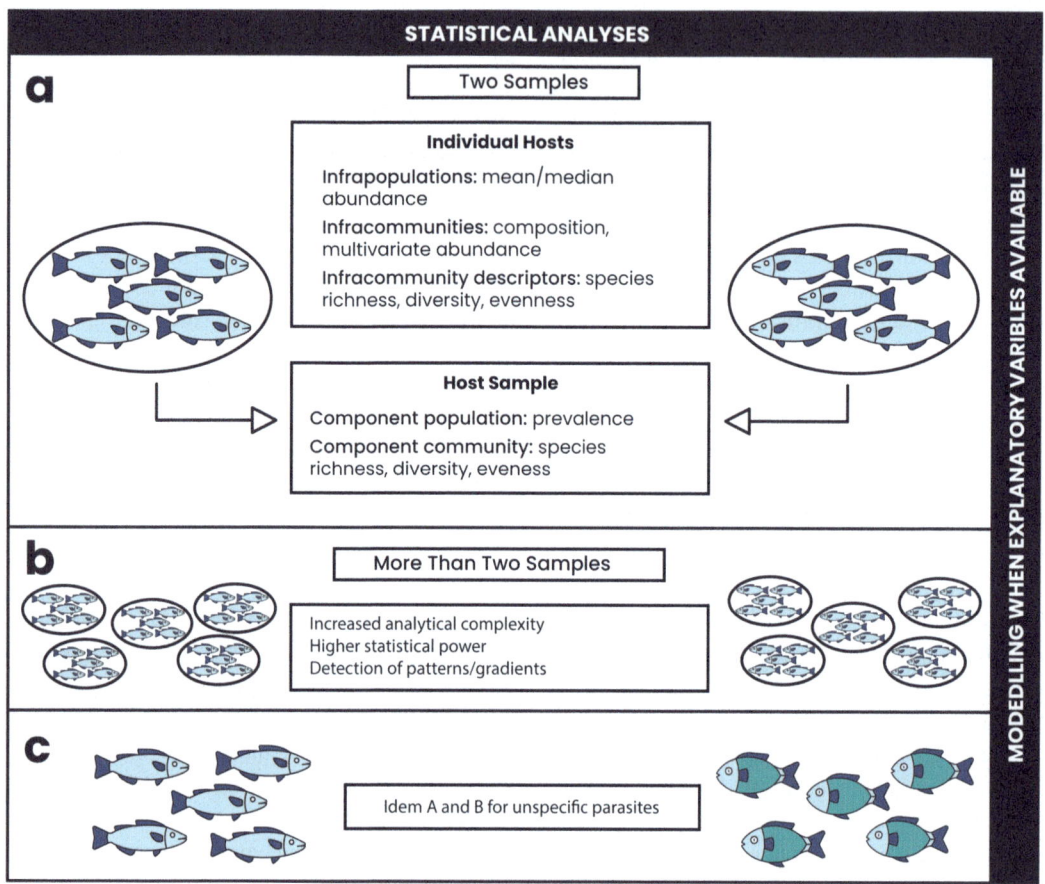

Fig. 17.1 Summary of statistics commonly used for comparisons of parasite burdens at different levels of organisation and scales of study

logical tags for most studies (Lester and MacKenzie 2009). In this guild, parasites are mainly represented by the larval stages of several groups of helminths, which constitute major components of parasite assemblages in many fish-parasite systems (Poulin and Valtonen 2001; Luque and Poulin 2004; Cantatore and Timi 2015). Often, these helminths are transmitted trophically (Poulin and Valtonen 2001) and consequently their loads in the host tend to increase with the age or size of fishes (Poulin 2000). Cumulative patterns of parasite abundance occur because larger hosts can accommodate more parasite species and sustain a greater absolute number of parasites than smaller hosts. Larger hosts also display larger surface areas for parasite attachment and can ingest larger quantities of food, resulting in a higher exposure to infective stages (Poulin 2000; Valtonen et al. 2010). Large fishes can feed on large prey items, increasing the number and broadening the set of potential parasites still further (Timi et al. 2010a, 2011). Fish length must therefore be considered in comparisons of samples from different localities to avoid attributing ontogenetic variability in parasite loads to a locality effect (Cantatore and Timi 2015). This is especially relevant when fishes of different length or age are included in the comparison (Fig. 17.2). Cumulative processes of long-lived parasites depend, not only on fish size, but also on age and longevity (Cantatore and Timi 2015). In most comparative studies on parasite tags controlling for fish growth, host age is not contemplated. However, for conspecific hosts of the same size, but different age, a higher parasite load should be expected for the older

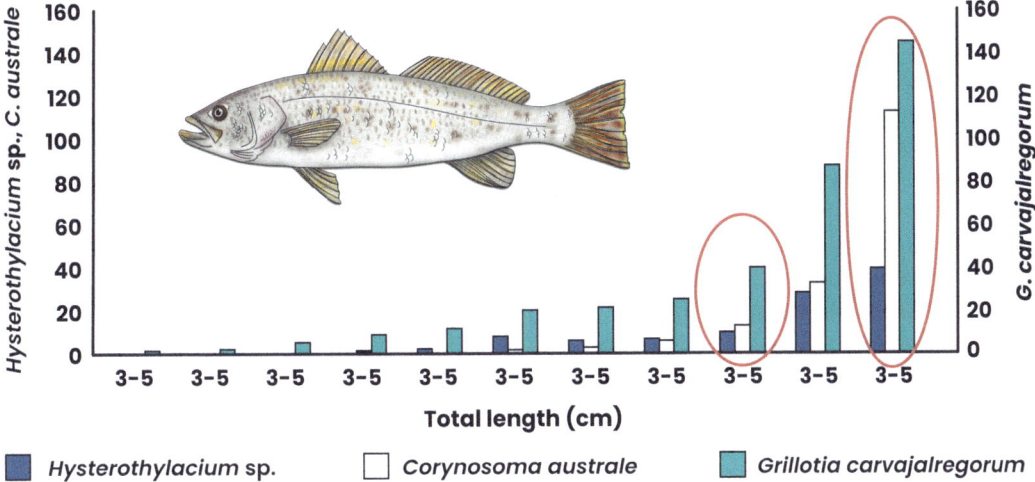

Fig. 17.2 Influence of host size on parasite burdens. Columns represent the mean abundance of three long-lived larval parasites of *Cynoscion guatucupa* (Sciaenidae) grouped by size classes. For *G. carvajalregorum* mean abundance was transformed dividing by 10. Comparisons between two size classes (red circles) of parasite population or community data probably will render significant statistical differences, even when all fishes came from the same population, with the risk of attributing a geographic cause to an ontogenetic factor

specimens. Braicovich et al. (2016) analysed the effect of these host variables on parasite abundance and species richness of *Percophis brasiliensis* caught in the Argentine Sea. Length and sex consistently appeared in the most parsimonious models suggesting that fish length seems to be a slightly better predictor than age or mass. Thus, fish size was confirmed as a suitable measure of growth, and it is recommended to restrict comparisons to fish of similar length or to incorporate length as covariate when comparing parasite burdens. Host sex should also be considered for fish that are sexually dimorphic in terms of morphology, metabolism, behaviour, or growth rates (Duneau and Ebert 2012), especially when sex ratios are not balanced across samples.

Beyond the effect of characteristics of individual fish on their parasite burdens, ecological and evolutionary traits may have a considerable influence on comparative analyses. For example, Levy et al. (2019) demonstrated that parasites can be excellent models for comparative research to assess fine-scale population structure, when site fidelity and strong adaptations to local conditions prevail and/or where physical heterogeneity needs to be revealed. For fish with these features, incipient speciation processes were revealed from their parasites for fish living in contiguous, but contrasting environments (Levy et al. 2021). On the other hand, such information could not be obtained from host-parasite systems where fish are highly vagile or migratory.

The interaction between host traits, such as size or age, and ecology must be also considered in certain studies. For example, for migratory species that alternate their foraging and spawning habitats seasonally during their life cycles, ontogenetic changes in the structure of parasite assemblages must be considered, since differences in size between hosts can lead to a misinterpretation of the patterns, especially when migrations are asynchronous among cohorts. This was observed for fish migrating between coastal waters of southern Brazil and northern Argentina (Canel et al. 2021). Parasite assemblages differed more noticeably for young fish, indicating possible variations in migratory routes, distance travelled or latitude reached, depending on environmental conditions and age. Consequently, differences among fish length classes need to be considered when using parasite tags for resources with temporally and spatially variable migratory behaviour, especially when different cohorts are compared. In case environmental or host-related variables are available, their inclusion as explanatory variables for modelling parasite data can provide the best results for interpreting the observed patterns and estimating the relative contributions of their determinants.

17.3 Criteria for the Selection of Suitable Parasite Tags: Advantages and Limitations

The selection criteria for ideal parasite tags have been coined, discussed, and refined over the years (Kabata 1963; Sindermann 1983; MacKenzie 1983, 1987a; Lester 1990; Moser 1991; Williams et al. 1992; Catalano et al. 2014, MacKenzie and Abaunza 2014). However, parasites fulfilling all of these criteria are not frequently found and compromises often have to be made, resulting in the criteria being used just as guidelines (MacKenzie and Abaunza 2014).

As indicated by Catalano et al. (2014) and MacKenzie and Abaunza (2014), a candidate parasite should meet the properties of a general tag:

(i) *Remain in the host for an appropriate length of time*
(ii) *Do not affect the marked organism (not increasing exposure to predators, non-toxic, no effect on mobility, behaviour, growth, or survival)*
(iii) *Ability to mark large numbers of individuals in a cost-effective way, and*
(iv) *Be relatively quick and inexpensive to detect*

These properties confer advantages in the use of parasites over artificial markers. Parasites are useful markers for fragile organisms such as small and delicate fish or invertebrates; they cannot be lost, for example, during moulting of crustacean hosts, and they may provide information on the period previous to capture, whereas artificial tags only inform what occurred between capture and recapture (Williams et al. 1992; MacKenzie and Abaunza 2014). Compared to genetic markers, parasite tags are inexpensive to use and can often be applied for the identification of host population structure or substructure where environmental or behavioural differences occur, but between which there is still a considerable amount of gene flow (Williams et al. 1992; MacKenzie and Abaunza 2014).

Beyond the advantages of using parasite indicators, the main limitations of biological tagging include the need to count many parasite individuals, especially when multiple taxa are analysed and infections are massive. Taxonomic resolution is crucial for this kind of study, even if the species name identity cannot be confirmed. Identification can not only be time-consuming but, in many cases, difficult to achieve (Poulin and Leung 2010), especially for larval parasites, which are among the most suitable parasite tags (Timi 2007). The existence of cryptic species can be an additional source of misinterpretation of comparative results, as in any other field of ecology.

For parasites, the following specific requirements to be a suitable biological tag candidate are compiled from those available in the literature (Kabata 1963; Sindermann 1983; MacKenzie 1983, 1987a; Lester 1990; Moser 1991; Williams et al. 1992; Catalano et al. 2014, MacKenzie and Abaunza 2005, 2014):

1. *Selected parasites should have significant differences in their levels of infection in the subject host in different parts of the study area.*

Such differences are indicative of stock or population structure from a spatial perspective. Conversely, the lack of them does not necessarily indicate the existence of a single stock or population. Indeed, similarity in parasite load can be a 'convergence' due to equivalent environmental and biological characteristics at both sites, between which there is no host gene flow. Beyond such spatial perspective, the lack of significant differences can be useful for tracing migration routes or determining geographic origin in a space-time context.

2. *Remaining in the host over an appropriate length of time means that the lifespan of the selected parasite needs to be as long as that of*

the host, or else their past presence should be detectable throughout the timescale of the study. For stock identification and recruitment studies, only parasites with lifespans of more than 1 year should be considered, whereas for studies of seasonal migrations, species with lifespans of less than 1 year are acceptable.

The strict dependence of parasitism on some host characteristics (size, age, or sex; see previous sections) requires special attention since it can cloud the actual spatial-temporal patterns of parasitism. For example, finding significant differences in parasitism between hosts of different size or age coming from two localities can lead to the erroneous attribution of a geographic origin to an ontogenetic cause in case these parasites increase or decrease in number as the hosts grow (Braicovich et al. 2016). The residence time (infection duration) of a parasite in the subject host has been considered as the most important criterion for judging whether it is a suitable marker (Lester 1990; Lester and MacKenzie 2009). Some studies have evaluated these recommendations by comparing long-lived vs. short-lived parasite guilds, with results confirming the validity of these criteria. For example, Lester et al. (1985), recorded 26 types of parasites in skipjack tuna, *Katsuwonus pelamis*, from the southwestern Pacific. Transitory species (ectoparasites and gastrointestinal species) comprising several different communities across the study area reflected the different habitats recently occupied by the fish. Permanent parasites (larval anisakid nematodes and trypanorhynch metacestodes), in contrast, provided little evidence for the existence of separate stocks, as corroborated by artificial tagging studies. In the same sense, long-lived larval endoparasites were used as tags to discriminate between different stocks of red porgy, *Pagrus pagrus* along the southwestern Atlantic coasts, whereas short-lived guilds exhibited significant differences between samples from a single stock (Soares et al. 2018). Similar results were obtained for *Sympterygia bonapartii*, a seasonally migratory skate from the same region; however, in this case, variations of transient parasites reflected the migratory behaviour of the fish, thereby constituting suitable markers for such purposes (Irigoitia et al. 2017). In addition, temporal and short-scale spatial variations in parasitism have also been analysed in order to avoid possible pseudoreplication (Ferrer-Castelló et al. 2007; Espínola-Novelo and Oliva 2021). As previously mentioned, cumulative patterns are common for permanent or long-lived parasites, leading to the additional requirement of including host growth parameters in the analyses for proper comparisons.

3. *The life cycle of the parasite species should preferably be monoxenous, reducing sources of variability that could affect its transmission between subsequent hosts, as well as population parameters in these other hosts. For the opposite reasons, parasites with complex life cycles, with two or more developmental stages in different hosts, are more difficult to use.*

In aquatic parasites, direct life cycles are common, for example, in monogeneans and most copepods, both represented mainly by ectoparasitic forms which are generally short-lived. Therefore, their value as markers depends on the nature of the study (see point 2). On the other hand, larvae of parasites with complex life cycles, such as larval anisakid nematodes, trypanorhynch cestodes, polymorphic acanthocephalans, or trematode metacercariae, which normally are long-lived (see Chap. 5 for detailed life cycles), are among the recommended markers for discriminating stocks and assessing population structure (Williams et al. 1992; Lester and MacKenzie 2009; Catalano et al. 2014, Cantatore and Timi 2015). As suggested by Lester (1990), this criterion should be omitted.

4. *The level of infection of the parasite species should remain relatively stable between seasons and years.*

The abundance and distribution of parasites depend on a combination of factors, related to the parasites themselves, to their hosts and to the environment where they live, plus the interactions between them. Such factors act differentially, depending on the complexity of parasite life cycle (Wood et al. 2014), on the different parasite life stages studied (Warburton et al. 2022), and at multiple scales (Brian and Aldridge 2023); consequently, temporal changes are not only permanent but often contrasting. This criterion refers to a desirable, although often-unrealistic, situation, especially when samples from long-term periods are analysed (see following sections). On the other hand, using samples taken at different seasons and years, which may vary considerably, can help avoid or reduce the occurrence of pseudoreplication (Ferrer-Castelló et al. 2007). It has been suggested that, to be representative, samples should be taken at different seasons and over a period of at least 2 years (MacKenzie and Abaunza 2014).

5. *The parasite should be easily detected and identified, preferably by morphological examination and should require a minimum of host dissection. Site-specificity in the subject host is desirable.*
6. *Parasites should not cause large-scale selective mortality or behavioural changes to their host.*

Despite the rather obvious requirements of criteria 5 and 6, parasites that are difficult to detect can be as valuable as more obvious parasites, whereas pathogenic parasites sometimes can be excellent markers, especially for studies on host migrations (Lester 1990). Their selection as markers should depend on the availability of other species having the desired features or, alternatively, on the possibility of their joint analyses by means of multivariate statistics.

17.4 Parasite Genetics as Source of Information on Spatial Distribution of Hosts

Genetic markers are increasingly used to resolve parasitological taxonomic issues (see also Chaps. 9 and 16), allowing for better taxonomic resolution in ecological studies and more accurately estimating actual species richness, which in turn improves the value of parasites as indicators. For example, the genetic identification of several species of larval *Anisakis* spp. contributed to the identification of stocks of the European hake, *Merluccius merluccius*, of the Atlantic horse mackerel, *Trachurus trachurus* and of the skipjack tuna, *Katsuwonus pelamis* (Mattiucci et al. 2004, 2008; Takano et al. 2021). It also confirmed the relative contribution of oceanographically contrasting masses of water to the parasite fauna of the silvery John dory *Zenopsis conchifer*, living in an environmentally transitional zone (Lanfranchi et al. 2016, 2018).

Genetic markers comprise innate tags whose advantages are that all members of a population are inherently marked, do not affect the behaviour or survival of the organisms, and may provide information for conservation by identifying the origin of catches and elucidating migratory routes (Antoniou and Magoulas 2014). The use of genetics to ascertain population membership of individuals is a well-established methodology that includes a variety of approaches, known as assignment methods, also used to identify the number of populations coexisting in a given area, mixed-stock analysis or the origin of migrant individuals (Manel et al. 2005; Criscione et al. 2006). Both mitochondrial and nuclear genetic markers are used in such studies (Antoniou and Magoulas 2014), although, unfortunately, these assignments can often be inaccurate when there is little or no neutral genetic differentiation among source populations (Criscione et al. 2006).

In the case of parasitic organisms, their phylogeographic patterns and those of their hosts may be congruent in time and space for specific and obligate species (Nieberding et al. 2004). Beverley-Burton (1978) was the first to propose the use of parasite genetic data to assign fish to their source populations, by comparing the frequencies of allozymes in *Anisakis simplex* parasitising Atlantic salmon (*Salmo salar*). Nevertheless, and despite the potential of parasite genetics to reveal many aspects of host phylogeography and population structure, molecular methodologies have not been widely applied to fish parasites (Criscione et al. 2005; Pascual et al. 2016).

The rationale is that if a parasite is more finely genetically differentiated than its host, then the genotypes of that parasite could potentially be used to assign hosts to their population of origin with higher probabilities than by using the host genotypes (Criscione et al. 2005). This concept is supported by the fact that the rate of molecular evolution is faster in parasite DNA and RNA relative to that within the homologous loci of their hosts, yielding genetic sequences that are more informative sources of data (Whiteman and Parker 2005). In addition to differences in mutation rates, most parasites display shorter generation times than their hosts, resulting in faster evolutionary processes, allowing favourable genotypes of the population obtained by mutation, recombination, or migration to rapidly increase in frequency, rendering them powerful inferential tools (Gandon and Michalakis 2002; Whiteman and Parker 2005). Furthermore, the genetic structure of parasites may be higher than in their dispersing hosts. This is due to different rates of gene flow of effective population sizes between them, because parasites may be locally adapted in different ways than their hosts and because not all hosts are infected (Criscione et al. 2006). This concept was proved by Criscione et al. (2006), who used microsatellite markers to compare the accuracy of assignment back to known source populations between the strictly freshwater parasitic trematode *Plagioporus shawi* and the steelhead trout *Oncorhynchus mykiss*. Their research showed that the genotypes of the parasite rendered four times greater odds of correct assignment than host genotypes. This kind of analysis is also useful for identifying individuals originating from protected areas and for tracing dispersal patterns or feeding grounds for migratory species.

In recent years, several studies have used mitochondrial genetic markers to assess the population genetic structure of aquatic parasites in order to infer that of their fish hosts (Cross et al. 2007, Baldwin et al. 2011, Sepúlveda and González 2015, Klapper et al. 2016, Mattiucci et al. 2018, Marigo et al. 2013). Most studies were carried out on larval *Anisakis* spp. on fish, with one study performed on cetaceans (Marigo et al. 2013). Their success varied with the identity of both host and parasite species, the host ecology, the region of origin and the geographical scale, thus indicating the need for selecting an appropriate scale of sampling to account for the different life histories and geographic distributions of the species under study (Cross et al. 2007). In addition, Baldwin et al. (2012) proposed that the features and approaches necessary for suitable parasite molecular markers include:

(i) *Ease of recovery and identification.*
(ii) *Use of more than one genetic marker to verify if a parasite species is cryptic.*
(iii) *Assessment of mutation rates for specific genetic marker types.*
(iv) *Temporal and geographic stability in a parasite population to enable long-term monitoring.*

17.5 Parasites as Indicators of Temporal Changes in Host Biology and Environment

As previously stated, the composition, structure, and distribution of parasite communities are the results of a combination of multiple biotic and abiotic factors interacting in an ecosystem (Poulin 2007). Consequently, these assemblages are highly dynamic systems, showing heterogeneities in their properties at different scales, both spatially and temporally. Whereas variability in parasitism at spatial scales has been increasingly used as an indicator of population structure or migrations for different kinds of hosts (see previous sections), temporal changes have been comparatively almost ignored, with some exceptions where parasites have been used as indicators of the success of protection measures (Timi and Buchmann 2023) or of the recovery of sites after natural and anthropogenic catastrophic events (Aguirre-Macedo et al. 2011; Overstreet 2007; Perez-del-Olmo et al. 2022).

Temporal changes in parasitism are expected to result from climate change and overfishing, among other anthropogenic stressors affecting aquatic biota (Marcogliese 2001, 2016; Vidal-

Martínez et al. 2010; Wood et al. 2010; Morley and Lewis 2014; Wood and Lafferty 2015). Many fishing practices, in combination with other anthropogenic factors, have led to a radical and rapid degradation of marine ecosystems, driving complex changes in the physical structure, chemistry, biology, and ecological functioning of oceans (Agardy 2000; Lubchenco et al. 2003). Fisheries not only affect marine biota, but also compound the effects of climate change in marine ecosystems, sometimes dramatically so (Bundy et al. 2021; Gissi et al. 2021). Ocean warming, deoxygenation, acidification, reductions in near-surface nutrients, and changes to primary production are predicted processes (Kwiatkowski et al. 2020), with concomitant expected declines in fish stocks (Lotze et al. 2019) and in the parasite assemblages they harbour. Despite long-term research being fundamental to understanding complex ecological and evolutionary dynamics (Kuebbing et al. 2018), it has been rarely performed for parasites, particularly in marine environments (but see Chap. 22). Therefore, long-term data on parasite abundance and host use are necessary to answer many questions on parasite ecology (Wood et al. 2023a), as well as to prove their value as indicators of changes caused by anthropogenic stressors. For example, competing hypotheses about levels of parasitism rising (Harvell et al. 2004; Lafferty et al. 2004; Harvell 2019) or declining (Carlson et al. 2017; Wood et al. 2023b) with anthropogenic climate change are currently under discussion (Tracy et al. 2019).

Unfortunately, historical parasitological studies are extremely scarce for wild hosts (Welicky et al. 2021; Wood et al. 2023b), being restricted to a small number of parasite species (Wood and Vanhove 2023). Meta-analyses have been used as an alternative to elucidate temporal trends in parasitism (Ward and Lafferty 2004; Tracy et al. 2019; Fiorenza et al. 2020) and, more recently, the parasitological examination of preserved hosts held in natural history collections is increasingly seen as excellent source of information (Howard et al. 2019; Welicky et al. 2021; Preisser et al. 2022; Wood and Vanhove 2023; Wood et al. 2023a, b) and the only way to understand the environmental correlates of temporal change in parasite abundance (Wood et al. 2023a).

As a consequence of the commonly high but variable number of factors affecting the abundance of each parasite species in an assemblage, as well as their possible interactions, temporal trends are often contrasting. For example, using historical data and comparing parasite population and community attributes at very different periods and frequencies, significant changes (mostly declines) have been observed in different systems and attributed to the effect of oceanographic anomalies (MacKenzie 1987b), anthropogenic changes in the habitat (Kennedy 1993; Keas and Blankespoor 1997), or climate change (Rokicki 2009), while other studies detected no significant changes over time (e.g. Bentley and Burgner 2011; Kuhn et al. 2016). Similarly, meta-analyses of trends over time of diseases in different marine taxa showed increases, decreases, or no changes in recent times (Tracy et al. 2019; Fiorenza et al. 2020). Equally, the examination of hosts preserved in natural history collections yielded contrasting results depending on the host and period (Howard et al. 2019; Welicky et al. 2021; Preisser et al. 2022; Wood et al. 2023b). Thus, independently of the methodology, there is a need to intensify studies of temporal parasite variability in aquatic organisms. Equally important is to identify the drivers of such changes, which at present seems to be a challenge for parasite ecologists.

Wood et al. (2023b) analysed the parasite communities of eight fluid-preserved fish species collected between 1880 and 2019 at Puget Sound, United States. Additionally, they used long-term environmental datasets (sea surface temperature, heavy metal and organic pollutants, and fish host density) to assess the correlation between environmental variables and parasite burdens. The authors found that the abundance of parasite taxa using three or more obligately required host species declined at a rate of 10.9% per decade, whereas no changes were recorded for those species using one or two obligately required host species. Among the possible mechanisms driving such declines is sea surface temperature, although other factors may also have contributed to the observed pattern.

To relate fishing activity to changes in parasitism, several lines of evidence are necessary, including continuous monitoring, evaluation along gradients of fishing pressure, and large-scale exclusion experiments to reveal possible relationships between parasitism and fishing-induced changes to host populations and communities (Wood et al. 2010). Unfortunately, in most cases these conditions are difficult to meet, especially those requiring previous data as part of time series of parasitological descriptors and their related environmental data, which generally are not available. The available research shows that our understanding of the long-term variability of parasitism in aquatic systems and its forcing factors is in its infancy and requires new hypotheses, techniques, tools, and more importantly, datasets in order to improve the use of parasites as indicators and our understanding of the ongoing and future changes and their causes.

Overfishing, as a cause of ecological extinctions, has preceded all other pervasive human disturbances such as pollution, habitat degradation, and climate change (Jackson et al. 2001) and is consequently predicted to affect the levels of parasitism (Wood et al. 2010). Nevertheless, in only a few cases has overfishing been examined as a possible cause of long-term changes in parasitism (Koch et al. 2014; Klapper et al. 2019). Evidence has been provided that intensive commercial fishing of the definitive host, the thornback ray *Raja clavata*, caused the decline and local extinction of the parasite *Stichocotyle nephropis* in the North Sea (MacKenzie and Pert 2018). This is a clear example of a parasite that would be an excellent indicator of the state of conservation of host populations. Because fishing can drive host populations below the density threshold required for parasite transmission (Dobson and May 1987), changes in parasitism can be used to monitor their correlates in host density. Some studies have supported these predictions in freshwater habitats (e.g. Amundsen et al. 2002). Furthermore, beyond the direct effect of decreased host density, which reduces transmission efficiency of parasites, indirect consequences of overfishing, such as changing population structure or size at maturation or changes at the host community, level can have synergistic or antagonistic effects on parasite populations (Wood et al. 2010), especially if changes in the topology of food webs take place (Marcogliese 2005; Lafferty et al. 2008).

Among the signs of alteration to the world's oceans, those related to fisheries include abrupt changes in species composition, habitat degradation, epidemics, mass mortalities, and fisheries collapses (Lubchenco et al. 2003). In order to prevent or reverse these widespread declines in natural populations, the implementation of Marine-Protected Areas (MPAs) is an effective mechanism to protect, maintain, and restore threatened ecosystems (Pelletier et al. 2008; Watson et al. 2014). Indeed, MPAs allow the recovery of depleted stocks of exploited resources and constitute a source of individuals to fished areas (Lubchenco et al. 2003). An efficient management of MPAs requires continuous information on objectives achieved and suitable indicators for assessing their effectiveness (Pomeroy et al. 2005).

Recent studies have demonstrated that in an MPA the whole fish assemblage achieves a biomass 670% greater than in adjacent unprotected areas and 343% greater than in partially-protected MPAs; furthermore, once the abundance of large animals recovers enough, the reserves help restore the ecosystem complexity through trophic cascades (Sala and Giakoumi 2018; Sonnenholzner et al. 2009). Due to the strict dependence of parasites on the population density of every host species involved in their life cycles, as well as on the complexity of food webs for trophically transmitted species (Wood et al. 2010), protective measures yielding differences in free living communities are expected to have similar effects on parasite assemblages.

As an example, Huspeni and Lafferty (2004) evaluated the success of an ecological restoration project at a degraded site, in a 'Before-After-Control-Impact' study in the USA using intermediate-stage digeneans infecting the California horn snail, *Cerithideopsis californica*. Over a period of 6 years, trematode prevalence nearly quadrupled and species richness doubled at restored sites, whereas both remained unchanged at control sites. These variations were attributed to changes in bird use of the restored habitats (birds being the definitive hosts of many of these digenean species; see Chap. 5).

The success of MPAs using parasites as indicators has been evaluated several times in different regions of the world (Sasal et al. 1996, 2004; Bartoli et al. 2005; Loot et al. 2005; Lafferty et al. 2008; Ternengo et al. 2009; Marzoug et al. 2012; Wood et al. 2013, 2014; Isbert et al. 2018; Navarro-Barranco et al. 2019; Braicovich et al. 2021). As a result of the implementation of different protection measures, most of these studies recorded higher abundance, species richness, and/or diversity of parasites belonging to different taxa. Only a few studies were unable to detect changes in parasite loads (Loot et al. 2005; Ternengo et al. 2009). Most studies evaluated the effect of MPAs on parasites of fish hosts and by comparing between protected and unprotected areas; only Braicovich et al. (2021) evaluated temporal changes in the composition and structure of parasite communities of the flathead *Percophis brasiliensis*, caught at the beginning of the implementation of protection measures (temporal and spatial closures to fishery) and after a period of 13 years, in a coastal region of northern Argentina.

Biotic and abiotic characteristics in other, even adjacent environments may affect the structure of parasite populations and assemblages, even at small spatial scales (Levy et al. 2019). It should therefore be necessary to complement the evaluation of temporal changes in parasitism as a consequence of MPAs with that of spatial ones, thus avoiding erroneously attributing the observed changes to the protection measures. The most reliable results would be expected with a sampling scheme comparing 'before and after', coupled with simultaneous sampling 'inside and outside' the MPAs, which will allow identifying and differentiating any possible regional change from those local ones resulting from the protective measures.

17.6 Parasite Genetics as Source of Information on Temporal Changes in Host Populations and Environment

Many factors drive the population structure of parasites, including their evolutionary and ecological histories, modes of transmission, host dispersal patterns, life-cycle complexity, as well as diverse anthropogenic factors (Zarlenga et al. 2014). Overfishing in particular can reduce the host population size (Wood et al. 2010), with a concomitant decline in parasite populations and consequently in their genetic diversity. Therefore, a minimal genetic structure among spatially isolated populations may be interpreted as a hallmark of human activity (Zarlenga et al. 2014). However, a possible correlation between genetic diversity of parasite populations and habitat degradation has been little studied. Nevertheless, clear-cut examples about the effects of habitat disturbance on the genetic diversity of parasites are available for anisakid nematodes (Mattiucci and Nascetti 2007, 2008). Different anisakid genera (*Anisakis*, *Contracaecum*, *Pseudoterranova*) showed values of genetic variability significantly higher in Austral populations than in Boreal regions, correlating with a lower level of habitat disturbance in the Southern Hemisphere (Mattiucci et al. 2017). Indeed, increased overfishing, by-catch of cetaceans, increased mortality of seals by hunting and diseases, pollution and acidification in Boreal regions, all reduce host population sizes and, consequently, anisakid population sizes, with an increased probability of genetic drift in the parasite gene pools (Mattiucci et al. 2015). Therefore, comparisons of genetic variability among parasite populations inhabiting ecosystems with different degrees of disruption is a promising tool to evaluate the effect of overfishing or, alternatively, of protection measures in any kind of host and environment.

17.7 Parasite Tags in Integrative Multidisciplinary Studies

Integrative collaborative approaches have successfully met the challenging objectives of interdisciplinary stock identification for several fisheries and are routinely being applied in several fishery management organisations (Cadrin 2020). The complexity of spatial and temporal patterns in nature poses different challenges to stock identification, including contradictory information from different methodologies. Therefore, complementary perspectives offered by parasite tags can help to reconcile these apparent contradictions.

Integrating parasite tags within multidisciplinary studies for stock identification has been a recurrent topic for improving tools for fishery management (McClelland et al. 2005; Abaunza et al. 2008; Niklitschek et al. 2010; Baldwin et al. 2012; Mattiucci et al. 2015; Van der Lingen et al. 2015; Welch et al. 2015; de Moor et al. 2017; Brickle et al. 2021; Zhang et al. 2021). However, only a few studies appear to have actually contributed to management decisions, such as the redefinition of stock boundaries or the setting of catch limits (Mosquera et al. 2003). An increase of integrated collaborative efforts, coupled with sophisticated statistical approaches, will facilitate resolving the pressing issues of resource management and conservation by providing more robust information on stock structure to policymakers (Cadrin et al. 2014). Despite the fact that parasites have been included in multidisciplinary research with different alternative methodologies for decades, integrative studies for the purposes of tagging and monitoring have remained proportionally underrepresented in the literature (Fig. 17.3).

A combination of different sources of information, including host genetics, morphometry and life-history traits, otolith microchemistry, and parasite data can result in higher discriminatory power and increased accuracy of stock assignment (Baldwin et al. 2012; Brickle et al. 2021; Zhang et al. 2021). Even over smaller geographic areas, simultaneously increasing the number of spatial and temporal scales studied can lead to better methods for successful management of marine fish species (Baldwin et al. 2012).

The advantages of integrating information include the elucidation of connectivity patterns of host populations over different spatial and temporal scales (Taillebois et al. 2017). Additionally, phylogeographic analysis of parasites, as well as of variations in genetic diversity resulting from the effect of natural factors or anthropogenic stressors, represents valuable information to be integrated in multidisciplinary studies dealing with fisheries management and stock identification from a holistic approach (Begg and Waldman 1999), aiming for long-term maintenance and sustainability of fishery resources.

Simultaneous phylogeographic analysis of parasites and their fish hosts, performed on the same genes, is a promising tool to be included in multidisciplinary studies on stock structure (Mattiucci et al. 2015). Such co-phylogeographic studies, under the 'magnifying glass hypothesis' (the increased chance to track the genealogical history of the host with genetic data of the parasite) (Huyse et al. 2017; Geraerts et al. 2022), along with recent advances in next-generation DNA sequencing technologies and genome-wide genotyping applications to fish parasites, will undoubtedly be of paramount relevance in the near-future for fisheries and aquaculture.

17.8 Parasite Tags in Studies of Non-Teleost Hosts

Parasites occur naturally in practically all living species, and they can therefore be used as indicators of diverse ecological features even for those species that are very fragile or difficult to mark or study by other methodologies, such as small pelagic fishes or many invertebrates (Williams et al. 1992). Aquatic invertebrates are mostly exploited through extractive activities on many natural populations, requiring strict management measures to ensure sustainability. These fisheries, e.g. squid, scallops, prawns, lobsters, or oysters, are of great economic importance, and their

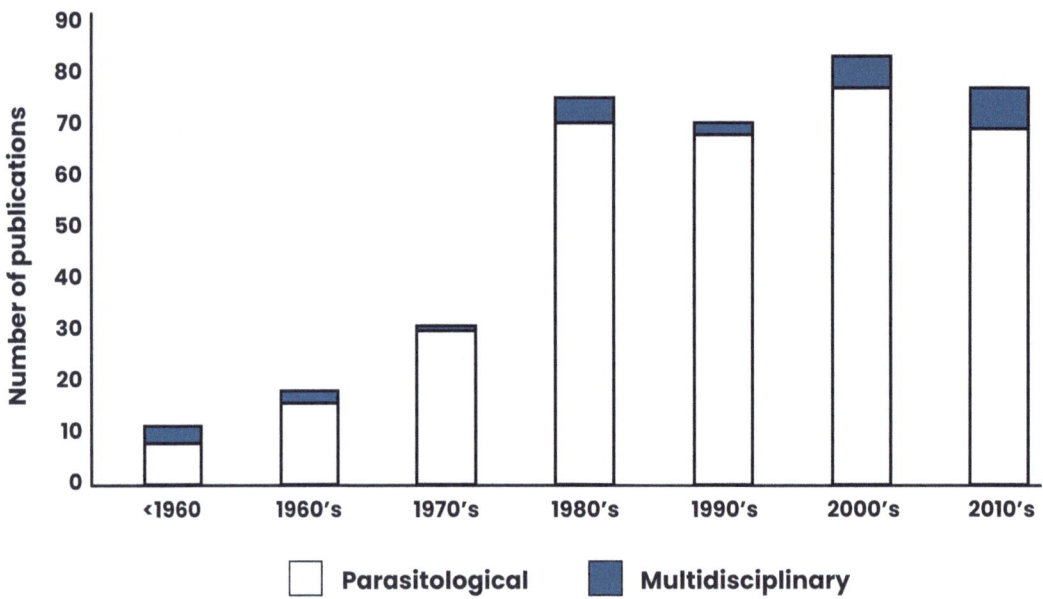

Fig. 17.3 Number of publications on parasite tags by decade, and proportions of exclusively parasitological vs multidisciplinary research

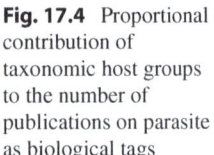

Fig. 17.4 Proportional contribution of taxonomic host groups to the number of publications on parasite as biological tags

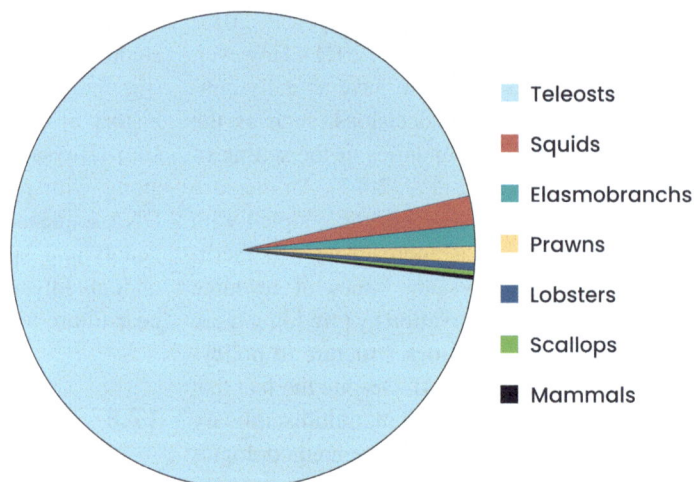

parasites are commonly known. However, few studies have used parasitological evidence for stock or population assessment. Similarly, the overwhelming majority of papers dealing with parasite tags focus on teleost fishes as hosts, and only a minor proportion of studies have been carried out on elasmobranchs and invertebrates of commercial interest (Timi and MacKenzie 2015) (Fig. 17.4). This situation has remained unchanged in recent years, with only four publications on elasmobranchs (Isbert et al. 2015; Irigoitia et al. 2017, 2022; Gérard et al. 2022) and none on invertebrates added to the list. Little research has been conducted on exploited invertebrates, despite the value of natural tags being recognised for many years (Williams et al. 1992; Pascual and Hochberg 1996).

The consumption of cephalopods has historical and cultural importance throughout the world, and cephalopod fisheries have rapidly expanded worldwide over the last 50 years (Ainsworth et al. 2023). Squid have the second-highest number of studies to do with parasite tags, although still far fewer than for teleosts and similar to that of elasmobranchs (Fig. 17.4). Characteristics of squid biology, such as short (annual) life cycles and variable growth rates, make them highly susceptible to overfishing, and their stock dynamics are naturally stochastic and difficult to assess. Research on parasites as tags for squid species has been misleading as an ad hoc tool for elucidating the discreteness of stock units (Pascual and Hochberg 1996). Indeed, Dawe et al. (1984) found that the larval parasites of the short-finned squid, *Illex illecebrosus* were of little value as tags due to their broad geographic distribution, generalist life-history, and difficulty of taxonomic identification. Other researchers used parasites of squids as indicators of stock assessment, with the caveat that differences in squid life-history characteristics, including age at maturity, growth rate, and feeding ecology also vary between regions, casting doubt on whether the differences in parasitism are real or artefacts of the variability in host features (Pascual et al. 1995). On the other hand, persistent larval parasites were acknowledged as suitable markers for discriminating stocks of the neon flying squid, *Ommastrephes bartramii*, in North Pacific waters (Bower and Margolis 1991).

Formal stock assessment has not been performed for several commercially targeted cephalopod species but, on a whole, increasing attention is being paid to the value of parasite tags of ecological stocks of other species, providing information on trophic relationships, zoogeographic patterns, ecological distributions, intraspecific groupings, and seasonal migration patterns (Pascual and Hochberg 1996). As an example, González et al. (2003) analysed the risk of parasitic infection related to abiotic and biotic factors for 2000 individuals of 10 cephalopod species collected at Galician waters, Spain. Their results revealed the existence of three ecological groupings (coastal, intermediate, and nerito-oceanic) of cephalopods, suggesting that the ecological niche of a given cephalopod species is more important in determining its risk of parasitic infection than its phylogeny.

Similar difficulties in the parasitological research of other invertebrates, such as scallops, prawns, and lobsters, and that of elasmobranchs (i.e. sharks, rays and skates, and chimaeras) may be the cause of the low number of studies carried out. In the case of elasmobranchs, there are no obvious candidates for using natural tags for stock assessment due to their position of top predators, which make them unsuitable hosts for long-lived larval parasites (MacKenzie and Hemmingsen 2015). In most systems, short-lived parasites have little value for stock assessment because they are acquired and lost as hosts move geographically (MacKenzie and Abaunza 2005) and their temporal variability could be wrongly interpreted as having a geographical cause. The numerical dominance of short-lived parasites that characterises the parasitism in elasmobranchs, however, allows, on the other hand, to trace feeding or reproductive movements in migratory species (Irigoitia et al. 2017). The few papers that have used parasite tags for population assessment of elasmobranchs have mainly focused on selachian hosts (Moore 2001; Yamaguchi et al. 2003; Isbert et al. 2015, Gérard et al. 2022) with only two studies dealing with batoid hosts (Irigoitia et al. 2017, 2022). Therefore, the value of parasites as indicators of population structure in elasmobranchs remains largely unexplored, and the value of many short-lived parasites as indicators needs to be assessed.

17.9 Expanding the Scope for Parasite Tags: Host Community and Zoogeography

This chapter is mainly focused on the use of parasite tags for assessing variability in the spatial and temporal distribution of their hosts; however, the variability of parasitism in any dimension of the host niche can be used to indicate such changes. For example, Sue et al. (2022) analysed data on infections by the monogenean *Diclidophora merlangi*, parasite of whiting *Merlangius merlangus* in relation to the locations of active oil installations. The mean abundance of these parasites increased significantly and at an accelerated rate with increasing proximity to the nearest oil field. The authors concluded that *D. merlangi* is an indicator of hydrocarbon pollution. The concept of parasite indicators, therefore, can be expanded to any field of ecology where changes in parasitism might be a response to specific environmental stimulus. The corresponding response patterns can then be used to estimate the amplitude of the induced change. The effects of environmental impacts such as pollution and climate change on parasites are covered in Chaps. 20, 21, and 22 in this book; therefore, the potential value of parasite markers for informing understanding of such processes is not further discussed here.

The concept of biological tags is here expanded from a local to a regional scale, and from the level of fish and parasite populations to the level of communities. By shifting the focus in this way, parasites are proposed as markers of the regions or water masses they inhabit and, consequently, as ecosystem indicators. A direct consequence is the interpretation of these patterns from a zoogeographic perspective, with assemblages of parasitic organisms as indicators of biogeographic units.

Some of the parasite taxa considered among the best tag candidates, such as anisakid nematodes, trypanorhynch metacestodes, and some polymorphic acanthocephalans (Timi 2007; Catalano et al. 2014), display very low specificity in intermediate or paratenic fish hosts, allowing comparative analyses across host species that share them. In the southwestern Atlantic, the occurrence of recurrent latitudinal patterns of a set of these parasites and their capability to discriminate the stocks of fish species with different ecological habits across the same regions (Timi 2003; Sardella and Timi 2004; Timi et al. 2005) led to the proposition that these species were good for identifying populations of other fish species in further studies (Timi 2007). Such patterns were later confirmed with the inclusion of other parasite species and increasing the number and ecological diversity of fish hosts (Timi et al. 2008, 2010b; Braicovich and Timi 2008; Timi and Lanfranchi 2009a; Vales et al. 2011; Braicovich et al. 2012; Alarcos and Timi 2013; Cantatore and Timi 2015; Pereira et al. 2014; Alarcos et al. 2016). A direct consequence of these results is the consideration of that assemblage of parasites as an indicator of the assemblage of hosts in the ecosystem (Fig. 17.5).

Since most of these parasites with low host specificity are obtained by passive ingestion of infective stages in fish food, they lead to predictable regional assemblages with a non-random composition and structure, determined mainly by ecological filters rather than by host phylogeny (Timi and Lanfranchi 2009b). Unrelated fish species of similar trophic level, diet, and size display equivalent and, therefore, predictable assemblages of long-lived larval parasites in the region (Timi et al. 2011). Such recurrence in composition of assemblages composed by long-lived larval endoparasites is identifiable in the southwestern Atlantic because, there, they are the dominant guild, as documented for most fish species so far examined (Timi and Lanfranchi 2009b; Rossin and Timi 2010). Similar patterns should be expected in other regions where the same groups of parasites are common components of parasite faunas, and a considerable proportion thereof shared by several host species. The study on ten cephalopod species in Galician waters, Spain, that revealed the existence of three ecological host groupings (coastal, intermediate, and nerito-oceanic; mentioned in Sect. 17.8 above) suggests that the ecological niche of a cephalopod species is more important in determining its risk of parasitic infection than its phylogeny (González et al. 2003).

Parasite assemblages are suitable indicators not only for fish stock assessment, but also for identifying the assemblages their hosts compose and therefore the ecosystems they inhabit (Fig. 17.5). Therefore, their use can be expanded to delineating the ecosystem boundaries for host communities whose components differ in their temporal and spatial dynamics. This is relevant since fisheries have shifted their focus towards Ecosystem-Based Fisheries Management and an Ecosystem Approach to Fisheries Management (García et al. 2003; Griffith and Fulton 2014). Parasites can also help to establish the limits for

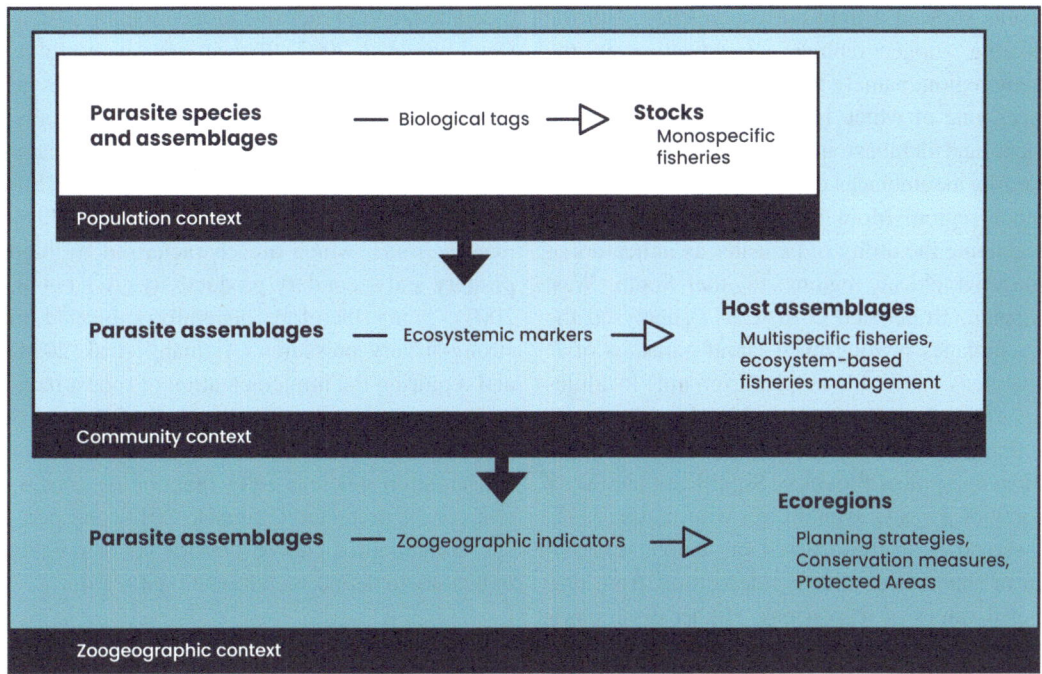

Fig. 17.5 Hierarchical levels of analysis using parasite tags, their targets, and applications

multi-specific fisheries, to determine the origin of captures, even illegal ones, and to trace host migration through ecosystems (Carballo et al. 2012).

Another application of parasite markers lies in the field of biogeography. At present, relatively few studies have used parasites as indicators of zoogeographical regions in the marine realm (Rohde 2002), with some studies failing to correlate parasite distributions with known zoogeographical regions (Byrnes and Rohde 1992; González and Moreno 2005; Marques et al. 2009), whereas others succeeded in defining zoogeographic areas congruent with those previously established on the basis of the distribution of free living organisms (Blaylock et al. 1998, 2003; González et al. 2006). The success of finding such patterns probably relies on the geographical scale and the parasite group studied.

Ecoregions are areas with a relatively homogeneous species composition, which is likely to be determined by the predominance of a small number of ecosystems and a distinct suite of oceanographic or topographic features (Spalding et al. 2007). Therefore, when a set of parasite species characterises a given host community or ecosystem, the next step is to expand their use as markers to zoogeographic studies (Fig. 17.5). This is the case of parasite communities of fish in the southwestern Atlantic. In this region, when parasite communities of long-lived taxa with low host-specificity are compared between several localities along the coasts, those belonging to the same ecoregion, as defined by Spalding et al., (2007), show no significant differences. This was observed for the Brazilian codling *Urophycis brasiliensis* (Pereira et al. 2014) and for the white-mouthed croaker *Micropogonias furnieri* (Luque et al. 2010). Lanfranchi et al. (2016) studied four fish species, the Argentine hake, *Merluccius hubbsi*, the Brazilian flathead, *Percophis brasiliensis*, the flounder, *Paralichthys isosceles* and the rough scad, *Trachurus lathami*, which harboured parasite assemblages representative of three ecoregions in the southwestern Atlantic. The parasite fauna of a fifth species, the silvery John dory, *Zenopsis conchifer* showed characteristics of occupying an ecotone zone, indicating that the study population represented the convergence of three masses of water. The results showed a tight correspondence with the existing zoogeographical classification in the study region, namely two zoogeographical provinces, one of which is subdivided into two districts, and demonstrated the ecotonal nature of parasite assemblages of fish living in the convergence region. More recently, with the aim of evaluating the utility of parasites as indicators of zoogeographical regions in the South West Atlantic, Braicovich et al. (2017) analysed the assemblages of long-lived larval parasites of *P. brasiliensis* from 11 samples from nine localities covering a distance of nearly 3300 km, almost the entire distribution of the species in the Argentine Biogeographical Province. Significant similarity decay of parasite assemblages over distance was observed, with those based on abundances and mean abundances showing departures from predicted values of regressions. Higher dissimilarities were observed between samples coming from different zoogeographical regions than between those caught within the same region, statistically independently of the distance separating them, thus identifying zoogeographical units in a distance-decay context which were then corroborated by separate multivariate analyses.

Many zoogeographical schemes use higher-level taxa, such as phyla or classes, to represent individual lineages. In the case of parasite assemblages, the potential derived from the combined evidence of the evolutionary history and biogeography of multiple lineages results in an efficient tool for capturing recurrent patterns of biodiversity. As parasite distributions greatly depend on that of their hosts, distributional variability information of the latter can be used synergically to enhance understanding of distributional patterns.

17.10 Conclusion

The conservation and sustainable use of marine resources is a primary goal for a growing number of national and international policy agendas. Among strategic plans and priorities for marine conservation measures, the implementation of protected areas has been hampered by the lack of a detailed, comprehensive biogeographic system (Spalding et al. 2007). Parasites can be useful as complementary tools for integrative studies to classify the oceans biogeographically. Furthermore, the identification of ecotonal parasite assemblages could be relevant for the delineation of those interface regions defined by marine fronts, which are characterised by high primary and secondary productivity (Acha et al. 2004), being therefore generally subjected to strong fishery pressures (Alemany et al. 2014) and requiring the implementation of robust management programmes. The borders of biological communities are suggested as priority areas for conservation where a fully functioning ecosystem can be protected (Primack, 2014) and parasite communities can be considered as reliable indicators to define such transitional regions.

References

Abaunza P, Murta AG, Campbell N et al (2008) Stock identity of horse mackerel (*Trachurus trachurus*) in the Northeast Atlantic and Mediterranean Sea: Integrating the results from different stock identification approaches. Fish Res 89:196–209. https://doi.org/10.1016/j.fishres.2007.09.022

Acha EM, Mianzán HW, Guerrero RA et al (2004) Marine fronts at the continental shelves of austral South America physical and ecological process. J Mar Syst 44:83–105. https://doi.org/10.1016/j.jmarsys.2003.09.005

Agardy T (2000) Effects of fisheries on marine ecosystems: a conservationist's perspective. ICES J Mar Sci 57:761–765. https://doi.org/10.1006/jmsc.2000.0721

Aguirre-Macedo ML, Vidal-Martínez VM, Lafferty KD (2011) Trematode communities in snails can indicate impact and recovery from hurricanes in a tropical coastal lagoon. Int J Parasitol 41:1403–1408. https://doi.org/10.1016/j.ijpara.2011.10.002

Ainsworth GB, Pita P, Garcia Rodrigues J et al (2023) Disentangling global market drivers for cephalopods to foster transformations towards sustainable seafood systems. People Nat 5:508–528. https://doi.org/10.1002/pan3.10442

Alarcos AJ, Timi JT (2013) Stocks and seasonal migrations of the flounder *Xystreurys rasile* as indicated by its parasites. J Fish Biol 83:531–534. https://doi.org/10.1111/jfb.12190

Alarcos AJ, Pereira NA, Taboarda NL et al (2016) Parasitological evidence of stocks of *Paralichthys isosceles* (Pleuronectiformes: Paralichthyidae) at small and large geographical scales in South American

Atlantic coasts. Fish Res 173:221–228. https://doi.org/10.1016/j.fishres.2015.07.018

Alemany D, Acha ME, Iribarne OO (2014) Marine fronts are important fishing areas for demersal species at the Argentine Sea (Southwest Atlantic Ocean). J Sea Res 87:56–67. https://doi.org/10.1016/j.seares.2013.12.006

Amundsen PA, Kristoffersen R, Knudsen R et al (2002) Long-term effects of a stock depletion programme: the rise and fall of a rehabilitated whitefish population. Ergeb Limnol 57:577–588. https://doi.org/10.1111/j.1365-3148.2006.00696.x

Antoniou A, Magoulas A (2014) Application of mitochondrial DNA in stock identification. In: Cadrin SX, Kerr LA, Mariani S (eds) Stock identification methods: applications in fishery science, 2nd edn. Academic, London, pp 257–295

Baldwin RE, Rew MB, Johansson ML et al (2011) Population structure of three species of *Anisakis* nematodes recovered from Pacific sardines (*Sardinops sagax*) distributed throughout the California Current System. J Parasitol 97:545–554. https://doi.org/10.1645/GE-2690.1

Baldwin RE, Banks MA, Jacobson KC (2012) Integrating fish and parasite data as a holistic solution for identifying the elusive stock structure of Pacific sardines (*Sardinops sagax*). Rev Fish Biol Fish 22:137–156. https://doi.org/10.1007/s11160-011-9227-5

Bartoli P, Gibson DI, Bray RA (2005) Digenean species diversity in teleost fish from a nature reserve off Corsica, France (Western Mediterranean), and a comparison with other Mediterranean regions. J Nat Hist 39:47–70. https://doi.org/10.1080/00222930310001613557

Begg GA, Waldman JR (1999) An holistic approach to fish stock identification. Fish Res 43:35–44. https://doi.org/10.1016/S0165-7836(99)00065-X

Bentley KT, Burgner RL (2011) An assessment of parasite infestation rates of juvenile sockeye salmon after 50 years of climate warming in southwest Alaska. Environ Biol Fish 92:267–273. https://doi.org/10.1007/s10641-011-9830-2

Beverley-Burton M (1978) Population genetics of *Anisakis simplex* (Nematoda: Ascaridoidea) in Atlantic salmon (*Salmo salar*) and their use as biological indicators of host stocks. Environ Biol Fish 3:369–377. https://doi.org/10.1007/BF00000529

Blaylock RB, Margolis L, Holmes JC (1998) Zoogeography of the parasites of Pacific halibut (*Hippoglossus stenolepis*) in the northeast Pacific. Can J Zool 76:2262–2273. https://doi.org/10.1139/z98-172

Blaylock RB, Margolis L, Holmes JC (2003) The use of parasites in discriminating stocks of Pacific halibut (*Hippoglossus stenolepis*) in the northeast Pacific. Fish Bull 1:1–9. https://doi.org/10.1139/cjz-76-12-2262

Bower SM, Margolis L (1991) Potential use of helminth parasites in stock identification of flying squid, *Ommastrephes bartrami*, in North Pacific waters. Can J Zool 69:1124–1126. https://doi.org/10.1139/z91-158

Braicovich PE, Timi JT (2008) Parasites as biological tags for stock discrimination of the Brazilian flathead, *Percophis brasiliensis* in the south-west Atlantic. J Fish Biol 73:557–571. https://doi.org/10.1111/j.1095-8649.2008.01948.x

Braicovich PE, Luque JL, Timi JT (2012) Geographical patterns of parasite infracommunities in the rough scad, *Trachurus lathami* Nichols off southwestern Atlantic Ocean. J Parasitol 98:768–771. https://doi.org/10.1645/GE-2950.1

Braicovich PE, Ieno EN, Sáez M et al (2016) Assessing the role of host traits as drivers of the abundance of long-lived parasites in fish stock assessment studies. J Fish Biol 89:2419–2433. https://doi.org/10.1111/jfb.13127

Braicovich PE, Pantoja C, Pereira AN et al (2017) Parasites of the Brazilian flathead *Percophis brasiliensis* reflect West Atlantic biogeographic regions. Parasitology 144:169–178. https://doi.org/10.1017/S0031182016001050

Braicovich PE, Irigoitia MM, Bovcon ND et al (2021) Parasites of *Percophis brasiliensis* (Percophidae) benefited from fishery regulations: Indicators of success for marine protected areas? Aquat Conserv 31:139–152. https://doi.org/10.1002/aqc.3436

Brian JI, Aldridge DC (2023) Factors at multiple scales drive parasite community structure. J Anim Ecol 92:377–390. https://doi.org/10.1111/1365-2656.13853

Brickle P, Randhawa HS, Reid MR et al (2021) Otolith trace elemental analyses and parasites provide useful tools for the stock discrimination of *Patagonotothen ramsayi* (Regan, 1913) (Nototheniidae) on the southern Patagonian Shelf. Fish Res 244:106129. https://doi.org/10.1016/j.fishres.2021.106129

Bundy A, Renaud PE, Coll M et al (2021) Editorial: Managing for the future: challenges and approaches for disentangling the relative roles of environmental change and fishing in marine ecosystems. Front Mar Sci 8:753459. https://doi.org/10.3389/fmars.2021.753459

Bush AO, Lafferty KD, Lotz JM et al (1997) Parasitology meets ecology on its own terms: Margolis et al. Revisited. J Parasitol 83:575–583. https://doi.org/10.2307/3284227

Byrnes T, Rohde K (1992) Geographical distribution and host specificity of ectoparasites of Australian bream, *Acanthopagrus* spp. (Sparidae). Folia Parasitol 39:249–264

Cadrin SX (2020) Defining spatial structure for fishery stock assessment. Fish Res 221:105397. https://doi.org/10.1016/j.fishres.2019.105397

Cadrin SX, Kerr LA, Mariani S (2014) Interdisciplinary evaluation of spatial population structure for definition of fishery management units. In: Cadrin SX, Kerr LA, Mariani S (eds) Stock identification methods: applications in fishery science, 2nd edn. Academic, London, pp 535–552

Canel D, Levy E, Braicovich PE et al (2021) Ontogenetic asynchrony in fish migrations may lead to disparate parasite assemblages: Implications for its use as biological tags. Fish Res 239:105941. https://doi.org/10.1016/j.fishres.2021.105941

Cantatore DMP, Timi JT (2015) Marine parasites as biological tags in South American Atlantic waters, current status and perspectives. Parasitology 142:5–24. https://doi.org/10.1017/S0031182013002138

Carballo MC, Cremonte F, Navone GT et al (2012) Similarity in parasite community structure may be used to trace latitudinal migrations of *Odontesthes smitti* along Argentinean coasts. J Fish Biol 80:15–28. https://doi.org/10.1111/j.1095-8649.2011.03125.x

Carlson CJ, Burgio KR, Dougherty ER et al (2017) Parasite biodiversity faces extinction and redistribution in a changing climate. Sci Adv 3:e1602422. https://doi.org/10.1126/sciadv.1602422

Catalano SR, Whittington ID, Donnellan SC et al (2014) Parasites as biological tags to assess host population structure: Guidelines, recent genetic advances and comments on a holistic approach. Int J Parasitol Parasites Wildl 3:220–226. https://doi.org/10.1016/j.ijppaw.2013.11.001

Costello C, Ovando D, Hilborn R et al (2012) Status and solutions for the world's unassessed fisheries. Science 338:517–520. https://doi.org/10.1126/science.1223389

Criscione CD, Poulin R, Blouin MS (2005) Molecular ecology of parasites: elucidating ecological and microevolutionary processes. Mol Ecol 14:2247–2257. https://doi.org/10.1111/j.1365-294X.2005.02587.x

Criscione CD, Cooper B, Blouin MS (2006) Parasite genotypes identify source populations of migratory fish more accurately than fish genotypes. Ecology 87:823–828. https://doi.org/10.1890/0012-9658(2006)87[823:PGISPO]2.0.CO;2

Cross MA, Collins C, Campbell N et al (2007) Levels of intra-host and temporal sequence variation in a large CO1 sub-units from *Anisakis simplex sensu stricto* (Rudolphi 1809) (Nematoda: Anisakisdae): implications for fisheries management. Mar Biol 151:695–702. https://doi.org/10.1007/s00227-006-0509-8

Dawe EG, Mercer MC, Threlfall W (1984) On the stock identity of short-finned squid (*Illex illecebrosus*) in the Northwest Atlantic. NAFO Sci Coun Stud 7:77–86

de Moor CL, Butterworth DS, van der Lingen CD (2017) The quantitative use of parasite data in multistock modelling of South African sardine *(Sardinops sagax)*. Can J Fish Aquat Sci 74:1895–1903. https://doi.org/10.1139/cjfas-2016-0280

Dobson AP, May RM (1987) The effects of parasites on fish populations-theoretical aspects. Int J Parasitol 17:363–370. https://doi.org/10.1016/0020-7519(87)90111-1

Dobson A, Lafferty KD, Kuris AM et al (2008) Homage to Linnaeus: How many parasites? How many hosts? Proc Natl Acad Sci USA 105:11482–11489. https://doi.org/10.1073/pnas.0803232105

Dogiel VA, Bychovsky BE (1939) Parasites of fishes of the Caspian Sea (In Russian). Trud Kompl Izuch Kasp Morya 7:1–150

Duneau D, Ebert D (2012) Host sexual dimorphism and parasite adaptation. PLoS Biol 10:e1001271. https://doi.org/10.1371/journal.pbio.1001271

Espínola-Novelo JF, Oliva ME (2021) Spatial and temporal variability of parasite communities: implications for fish stock identification. Fishes 6:71. https://doi.org/10.3390/fishes6040071

FAO (2022) The state of world fisheries and aquaculture 2022. Towards blue transformation. FAO, Rome. https://doi.org/10.4060/cc0461en

Ferrer-Castelló E, Raga JA, Aznar FJ (2007) Parasites as fish population tags and pseudoreplication problems: the case of striped red mullet *Mullus surmuletus* in the Spanish Mediterranean. J Helminthol 81:169–178. https://doi.org/10.1017/S0022149X07729553

Fiorenza EA, Wendt CA, Dobkowski KA et al (2020) It's a wormy world: Meta-analysis reveals several decades of change in the global abundance of the parasitic nematodes *Anisakis* spp. and *Pseudoterranova* spp. in marine fishes and invertebrates. Glob Change Biol 26:2854–2866. https://doi.org/10.1111/gcb.15048

Gandon S, Michalakis Y (2002) Local adaptation, evolutionary potential and host–parasite coevolution: interactions between migration, mutation, population size and generation time. J Evol Biol 15:451–462. https://doi.org/10.1046/j.1420-9101.2002.00402.x

García SM, Zerbi A, Aliaume C et al (2003) The ecosystem approach to fisheries. Issues, terminology, principles, institutional foundations, implementation and outlook. FAO fisheries technical paper no. 443. FAO, Rome

Geraerts M, Huyse T, Barson M et al (2022) Mosaic or melting pot: the use of monogeneans as a biological tag and magnifying glass to discriminate introduced populations of Nile tilapia in sub-Saharan Africa. Genomics 114:110328. https://doi.org/10.1016/j.ygeno.2022.110328

Gérard C, Hervé MR, Hamel H et al (2022) Metazoan parasite community as a potential biological indicator in juveniles of the starry smooth-hound *Mustelus asterias* Cloquet, 1819 (Carcharhiniformes Triakidae). Aquat Living Resour 35:3. https://doi.org/10.1051/alr/2022002

Gissi E, Manea E, Mazaris AD et al (2021) A review of the combined effects of climate change and other local human stressors on the marine environment. Sci Total Environ 755:142564. https://doi.org/10.1016/j.scitotenv.2020.142564

González MT, Moreno CA (2005) The distribution of the ectoparasite fauna of *Sebastes capensis* from the southern hemisphere does not correspond with zoogeographical provinces of free-living marine animals. J Biogeogr 32:1539–1547. https://doi.org/10.1111/j.1365-2699.2005.01323.x

González AF, Pascual S, Gestal C et al (2003) What makes a cephalopod a suitable host for parasite? The case of Galician waters. Fish Res 60:177–183. https://doi.org/10.1016/s0165-7836(02)00059-0

González MT, Barrientos C, Moreno CA (2006) Biogeographical patterns in endoparasite communities of a marine fish (*Sebastes capensis* Gmelin) with extended range in the Southern Hemisphere. J Biogeogr 33:1086–1095. https://doi.org/10.1111/j.1365-2699.2006.01488.x

Griffith GP, Fulton EA (2014) New approaches to simulating the complex interaction effects of multiple human impacts on the marine environment. ICES J Mar Sci 71:764–774. https://doi.org/10.1093/icesjms/fst196

Harvell CD (2019) Ocean outbreak: Confronting the rising tide of marine disease. University of California Press, Oakland, CA

Harvell D, Aronson R, Baron N et al (2004) The rising tide of ocean diseases: unsolved problems and research priorities. Front Ecol Environ 2:375–382. https://doi.org/10.1890/1540-9295(2004)002[0375:TRTOOD]2.0.CO;2

Hechinger RF, Lafferty KD, Dobson AP et al (2011) A common scaling rule for abundance, energetics, and production of parasitic and free-living species. Science 333:445–448. https://doi.org/10.1126/science.1204337

Herrington WC, Bearse HM, Firth FE (1939) Observations on 748 the life history, occurrence and distribution of the redfish parasite *Sphyrion lumpi*. US Bur Fish Spec Rep 5:1–18

Hilborn R, Amoroso RO, Anderson CM et al (2020) Effective fisheries management instrumental in improving fish stock status. Proc Nat Acad Sci USA 117:2218–2224. https://doi.org/10.1073/pnas.1909726116

Howard I, Davis E, Lippert G et al (2019) Abundance of an economically important nematode parasite increased in Puget Sound between 1930 and 2016: Evidence from museum specimens confirms historical data. J Appl Ecol 56:190–200. https://doi.org/10.1111/1365-2664.13264

Huspeni TC, Lafferty K (2004) Using larval trematodes that parasitize snails to evaluate a saltmarsh restoration project. Ecol Appl 14:795–804. https://doi.org/10.1890/01-5346

Huyse T, Oeyen M, Larmuseau MH et al (2017) Co-phylogeographic study of the flatworm *Gyrodactylus gondae* and its goby host *Pomatoschistus minutus*. Parasitol Int 66:119–125. https://doi.org/10.1016/j.parint.2016.12.008

Irigoitia MM, Incorvaia IS, Timi JT (2017) Evaluating the usefulness of natural tags for host population structure in chondrichthyans: parasite assemblages of *Sympterygia bonapartii* (Rajiformes: Arhynchobatidae) in the Southwestern Atlantic. Fish Res 195:80–90. https://doi.org/10.1016/j.fishres.2017.07.006

Irigoitia MM, Levy E, Canel D et al (2022) Parasites as tags for stock identification of a highly exploited vulnerable skate *Dipturus brevicaudatus* (Chondrichthyes: Rajidae) in the south-western Atlantic Ocean, a complementary tool for its conservation. Aquat Conserv 32:1634–1646. https://doi.org/10.1002/aqc.3869

Isbert W, Rodríguez-Cabello C, Frutos I et al (2015) Metazoan parasite communities and diet of the velvet belly lantern shark *Etmopterus spinax* (Squaliformes: Etmopteridae): a comparison of two deep-sea ecosystems. J Fish Biol 86:687–706. https://doi.org/10.1111/jfb.12591

Isbert W, Montero FE, Pérez-del-Olmo A et al (2018) Parasite communities of the white seabream *Diplodus sargus sargus* in the marine protected area of Medes Islands, north-west Mediterranean Sea. J Fish Biol 93:586–596. https://doi.org/10.1111/jfb.13729

Jackson JBC, Kirby MX, Berger WH et al (2001) Historical overfishing and the recent collapse of coastal ecosystems. Science 293:629–638. https://doi.org/10.1126/science.1059199

Kabata Z (1963) Parasites as biological tags. International Commission for the Northwest Atlantic Fisheries, Special Publication 4, pp 31–37

Keas BC, Blankespoor HD (1997) The Prevalence of cercariae from *Stagnicola emarginata* (Lymnaeidae) over 50 Years in Northern Michigan. J Parasitol 83:536–540. https://doi.org/10.2307/3284427

Kennedy CR (1993) The dynamics of intestinal helminth communities in eels *Anguilla anguilla* in a small stream: long-term changes in richness and structure. Parasitology 107:71–78. https://doi.org/10.1017/s0031182000079427

Klapper R, Kochmann J, O'Hara RB et al (2016) Parasites as biological tags for stock discrimination of beaked redfish (*Sebastes mentella*): parasite infra-communities vs. limited resolution of cytochrome markers. PLoS One 11:e0153964. https://doi.org/10.1371/journal.pone.0153964

Klapper R, Bernreuther M, Wischnewski J et al (2019) Long-term stability of *Sphyrion lumpi* abundance in beaked redfish *Sebastes mentella* of the Irminger Sea and its use as biological marker. Parasitol Res 116:1561–1572. https://doi.org/10.1007/s00436-017-5433-y

Kleisner K, Froese R, Zeller D et al (2012) Using global catch data for inferences on the world's marine fisheries. Fish Fish 14:293–311. https://doi.org/10.1111/j.1467-2979.2012.00469

Koch W, Boer P, Witte JI et al (2014) Inventory and comparison of abundance of parasitic copepods on fish hosts in the western Wadden Sea (North Sea) between 1968 and 2010. J Mar Biol Assoc UK 94:547–555. https://doi.org/10.1017/s0025315413001677

Kuebbing SE, Reimer AP, Rosenthal SA et al (2018) Long-term research in ecology and evolution: a survey of challenges and opportunities. Ecol Monogr 88:245–258. https://doi.org/10.1002/ecm.1289

Kuhn JA, Knudsen R, Kristoffersen R et al (2016) Temporal changes and between-host variation in the intestinal parasite community of Arctic charr in a subarctic lake. Hydrobiologia 783:79–91. https://doi.org/10.1007/s10750-016-2731-9

Kuris AM, Hechinger RF, Shaw JC et al (2008) Ecosystem energetic implications of parasite and free-living biomass in three estuaries. Nature 454:515–518. https://doi.org/10.1038/nature06970

Kwiatkowski L, Torres O, Bopp L et al (2020) Twenty-first century ocean warming, acidification, deoxygenation, and upper-ocean nutrient and primary production decline from CMIP6 model projections. Biogeosciences 17:3439–3470. https://doi.org/10.5194/bg-17-3439-2020

Lafferty KD, Porter JW, Ford SE (2004) Are diseases increasing in the ocean? Annu Rev Ecol Evol Syst 35:31–54. https://doi.org/10.1146/annurev.ecolsys.35.021103.105704

Lafferty KD, Shaw JC, Kuris AM (2008) Reef fishes have higher parasite richness at unfished Palmyra Atoll compared to fished Kiritimati Island. EcoHealth 5:338–345. https://doi.org/10.1007/s10393-008-0196-7

Lanfranchi AL, Braicovich PE, Cantatore DMP et al (2016) Ecotonal marine regions—ecotonal parasite communities: helminth assemblages in the convergence of masses of water in the southwestern Atlantic Ocean. Int J Parasitol 46:809–818. https://doi.org/10.1016/j.ijpara.2016.07.004

Lanfranchi AL, Braicovich PE, Cantatore DM et al (2018) Influence of confluent marine currents in an ecotonal region of the South-West Atlantic on the distribution of larval anisakids (Nematoda: Anisakidae). Parasit Vectors 11:1–13. https://doi.org/10.1186/s13071-018-3119-7

Lester RJG (1990) Reappraisal of the use of parasites for fish stock identification. Aust J Mar Freshw Res 41:855–864. https://doi.org/10.1071/MF9900855

Lester RJG, MacKenzie K (2009) The use and abuse of parasites as stock markers for fish. Fish Res 97:1–2. https://doi.org/10.1016/j.fishres.2008.12.016

Lester RJG, Barnes A, Habib G (1985) Parasites of skipjack tuna: fishery implications. Fish Bull 83:343–356

Levy E, Canel D, Rossin MA et al (2019) Parasites as indicators of fish population structure at two different geographical scales in contrasting coastal environments of the south-western Atlantic. Estuar Coast Shelf Sci 229:106400. https://doi.org/10.1016/j.ecss.2019.106400

Levy E, Canel D, Rossin MA et al (2021) Parasite assemblages as indicators of an incipient speciation process of *Odontesthes argentinensis* in an estuarine environment. Estuar Coast Shelf Sci 250:107168. https://doi.org/10.1016/j.ecss.2021.107168

Loot G, Aldana M, Navarrete SA (2005) Effects of human exclusion on parasitism in intertidal food webs of central Chile. Conserv Biol 19:203–212. https://doi.org/10.1111/j.1523-1739.2005.00396.x

Lotze HK, Tittensor DP, Bryndum-Buchholz A et al (2019) Global ensemble projections reveal trophic amplification of ocean biomass declines with climate change. Proc Nat Acad Sci USA 116:2907–12912. https://doi.org/10.1073/pnas.1900194116

Lubchenco J, Palumbi SR, Gaines SD et al (2003) Plugging a hole in the ocean: the emerging science of marine reserves. Ecol Appl 13:S3–S7. https://doi.org/10.1890/1051-0761(2003)013[0003:PAHITO]2.0.CO;2

Luque JL, Poulin R (2004) Use of fish as intermediate hosts by helminth parasites: a comparative analysis. Acta Parasitol 49:353–361

Luque JL, Cordeiro AS, Oliva ME (2010) Metazoan parasites as biological tags for stock discrimination of whitemouth croaker *Micropogonias furnieri*. J Fish Biol 76:591–600. https://doi.org/10.1111/j.1095-8649.2009.02515.x

MacKenzie K (1983) Parasites as biological tags in fish population studies. Adv Appl Biol 7:251–331. https://doi.org/10.1016/S1383-5769(98)80066-4

MacKenzie K (1987a) Parasites as indicators of host populations. Int J Parasitol 17:345–352. https://doi.org/10.1016/0020-7519(87)90109-3

MacKenzie K (1987b) Long-term changes in the prevalence of two helminth parasites (Cestoda: Trypanorhyncha) Infecting marine fish. J Fish Biol 31:83–87. https://doi.org/10.1111/j.1095-8649.1987.tb05216.x

MacKenzie K, Abaunza P (1998) Parasites as biological tags for stock discrimination of marine fish: a guide to procedures and methods. Fish Res 38:45–56. https://doi.org/10.1016/S0165-7836(98)00116-7

MacKenzie K, Abaunza P (2005) Parasites as biological tags. In: Cadrin SX, Friedland KD, Waldman JR (eds) Stock identification methods. Applications in fisheries science. Elsevier Academic, San Diego, pp 211–226

MacKenzie K, Abaunza P (2014) Parasites as biological tags. In: Cadrin SX, Kerr LA, Mariani S (eds) Stock identification methods: applications in fishery science, 2nd edn. Academic, London, pp 185–204

MacKenzie K, Hemmingsen W (2015) Parasites as biological tags in marine fisheries research: European Atlantic waters. Parasitology 142:54–67. https://doi.org/10.1017/S0031182014000341

MacKenzie K, Pert C (2018) Evidence for the decline and possible extinction of a marine parasite species caused by intensive fishing. Fish Res 198:63–65. https://doi.org/10.1016/j.fishres.2017.10.014

Magurran AE (2004) Measuring biological diversity. Blackwell, Oxford

Manel S, Gaggiotti OE, Waples RS (2005) Assignment methods: matching biological questions with appropriate techniques. Trends Ecol Evol 20:136–142. https://doi.org/10.1016/j.tree.2004.12.004

Marcogliese DJ (2001) Implications of climate change for parasitism of animals in the aquatic environment. Can J Zool 79:1331–1352. https://doi.org/10.1139/z01-067

Marcogliese DJ (2005) Parasites of the superorganism: Are they indicators of ecosystem health? Int J Parasitol 35:705–716. https://doi.org/10.1016/j.ijpara.2005.01.015

Marcogliese DJ (2016) The distribution and abundance of parasites in aquatic ecosystems in a changing climate: more than just temperature. Integr Comp Biol 56:611–619. https://doi.org/10.1093/icb/icw036

Marcogliese DJ, Jacobson KC (2015) Parasites as biological tags of marine, freshwater and anadromous fishes in North America from the tropics to the Arctic. Parasitology 142:68–89. https://doi.org/10.1017/S0031182014000110

Marigo J, Cunha HA, Bertozzi CP et al (2013) Genetic diversity and population structure of *Synthesium pontoporiae* (Digenea, Brachycladiidae) linked to its definitive host stocks, the endangered Franciscana dolphin, *Pontoporia blainvillei* (Pontoporiidae) off the coast of Brazil and Argentina. J Helminthol 89:19–27. https://doi.org/10.1017/S0022149X13000540

Marques JF, Santos MJ, Cabral HN (2009) Zoogeographical patterns of flatfish (Pleuronectiformes) parasites in the Northeast Atlantic and the importance of the Portuguese coast as a transitional area. Sci Mar 73:461–471. https://doi.org/10.3989/scimar.2009.73n3461

Marzoug D, Boutiba Z, Kostadinova A et al (2012) Effects of fishing on parasitism in a sparid fish: contrasts between two areas of the Western Mediterranean. Parasitol Int 61:414–420. https://doi.org/10.1016/j.parint.2012.02.002

Mattiucci S, Nascetti G (2007) Genetic diversity and infection levels of anisakid nematodes parasitic in fish and marine mammals from Boreal and Austral hemispheres. Vet Parasitol 148:43–57. https://doi.org/10.1016/j.vetpar.2007.05.009

Mattiucci S, Nascetti G (2008) Advances and trends in the molecular systematics of anisakid nematodes, with implications for their evolutionary ecology and host-parasite co-evolutionary processes. Adv Parasitol 66:47–148. https://doi.org/10.1016/S0065-308X(08)00202-9

Mattiucci S, Abaunza P, Ramadori L et al (2004) Genetic identification of *Anisakis* larvae in European hake from Atlantic and Mediterranean waters for stock recognition. J Fish Biol 65:495–510. https://doi.org/10.1111/j.0022-1112.2004.00465.x

Mattiucci S, Farina V, Campbell N et al (2008) *Anisakis* spp. larvae (Nematoda: Anisakidae) from Atlantic horse mackerel: their genetic identification and use as biological tags for host stock characterization. Fish Res 89:146–151. https://doi.org/10.1016/j.fishres.2007.09.032

Mattiucci S, Cimmaruta R, Cipriani P et al (2015) Integrating *Anisakis* spp. parasites data and host genetic structure in the frame of a holistic approach for stock identification of selected Mediterranean Sea fish species. Parasitology 142:90–108. https://doi.org/10.1017/S0031182014001103

Mattiucci S, Paoletti M, Cipriani P et al (2017) Inventorying biodiversity of anisakid nematodes from the Austral Region: a hotspot of genetic diversity? In: Klimpel S, Kuhn T, Melhorn H (eds) Biodiversity and evolution of parasitic life in the Southern Ocean. Springer, Cham, pp 109–140

Mattiucci S, Giulietti L, Paoletti M et al (2018) Population genetic structure of the parasite *Anisakis simplex* (s. s.) collected in *Clupea harengus* L. from North East Atlantic fishing grounds. Fish Res 202:103–111. https://doi.org/10.1016/j.fishres.2017.08.002

McClelland G, Melendy J, Osborne J (2005) Use of parasite and genetic markers in delineating populations of winter flounder from the central and south-west Scotian Shelf and north-east Gulf of Maine. J Fish Biol 66:1082–1100. https://doi.org/10.1111/j.0022-1112.2005.00659.x

Moore ABM (2001) Metazoan parasites of the lesser-spotted dogfish *Scyliorhinus canicula* and their potential as stock discrimination tools. J Mar Biol Assoc UK 81:1009–1013. https://doi.org/10.1017/s0025315401004982

Morley NJ, Lewis JW (2014) Temperature stress and parasitism of endothermic hosts under climate change. Trends Parasitol 30:221–227. https://doi.org/10.1016/j.pt.2014.01.007

Moser M (1991) Parasites as biological tags. Parasitol Today 7:8–185. https://doi.org/10.1016/0169-4758(91)90128-B

Mosquera J, De Castro M, Gómez-Gesteira M (2003) Parasites as biological tags of fish populations: advantages and limitations. Comments Theor Biol 8:69–91. https://doi.org/10.1080/08948550390181612

Navarro-Barranco C, Tierno de Figueroa JM, Ros M et al (2019) Influence of marine protected areas on parasitic prevalence: the case of the isopod *Anilocra physodes* as a parasite of the fish *Lithognathus mormyrus*. J Zool 308:280–292. https://doi.org/10.1111/jzo.12674

Nekola JC, White PS (1999) The distance decay of similarity in biogeography and ecology. J Biogeogr 26:867–878. https://doi.org/10.1046/j.1365-2699.1999.00305.x

Nieberding C, Morand S, Libois R et al (2004) A parasite reveals cryptic phylogeographic history of its host. Proc Roy Soc Lond B Biol Sci 271:2559–2568. https://doi.org/10.1098/rspb.2004.2930

Niklitschek EJ, Secor DH, Toledo P et al (2010) Segregation of SE Pacific and SW Atlantic southern blue whiting stocks: integrating evidence from complementary otolith microchemistry and parasite assemblage approaches. Environ Biol Fish 89:399–413. https://doi.org/10.1007/s10641-010-9695-9

Overstreet RM (2007) Effects of hurricanes on fish parasites. Parassitologia 49:161–168

Pascual S, Hochberg FG (1996) Marine parasites as biological tags of cephalopod hosts. Parasitol Today 12:324–327. https://doi.org/10.1016/0169-4758(96)40004-7

Pascual S, González A, Arias C et al (1995) Helminth infection in the short-finned squid *Illex coindetii* (Cephalopoda, Ommastrephidae) off NW Spain. Dis Aquat Org 23:71–75. https://doi.org/10.3354/dao023071

Pascual S, Abollo E, González AF (2016) Biobanking and genetic markers for parasites in fish stock studies. Fish Res 173:214–220. https://doi.org/10.1016/j.fishres.2015.10.001

Pelletier D, Claudet J, Ferraris J et al (2008) Models and indicators for assessing conservation and fisheries-related effects of marine protected areas. Can J Fish Aquat Sci 65:765–779. https://doi.org/10.1139/F08-026

Pereira AN, Pantoja C, Luque JL (2014) Parasites of *Urophycis brasiliensis* (Gadiformes: Phycidae) as indicators of marine ecoregions in coastal areas of the South American Atlantic. Parasitol Res 113:4281–4292. https://doi.org/10.1007/s00436-014-4106-3

Perez-del-Olmo A, Raga JA, Kostadinova A (2022) Parasite communities in a marine fish indicate ecological recovery from the impacts of the Prestige oil-spill 12–13 years after the disaster. Sci Total Environ 847:157354. https://doi.org/10.1016/j.scitotenv.2022.157354

Pita A, Casey J, Hawkins SJ et al (2016) Conceptual and practical advances in fish stock delineation. Fish Res 173:185–193. https://doi.org/10.1016/j.fishres.2015.10.029

Pomeroy RS, Watson LM, Parks JE et al (2005) How is your MPA doing? A methodology for evaluating the management effectiveness of marine protected areas. Ocean Coast Manag 48:485–502. https://doi.org/10.1016/j.ocecoaman.2005.05.004

Poulin R (2000) Variation in the intraspecific relationship between fish length and intensity of parasitic infection: biological and statistical causes. J Fish Biol 56:123–137. https://doi.org/10.1111/j.1095-8649.2000.tb02090.x

Poulin R (2007) Are there general laws in parasite ecology? Parasitology 134:763–776. https://doi.org/10.1017/S0031182006002150

Poulin R, Kamiya T (2015) Parasites as biological tags of fish stocks: a meta-analysis of their discriminatory power. Parasitology 142:145155. https://doi.org/10.1017/s0031182013001534

Poulin R, Leung TLF (2010) Taxonomic resolution in parasite community studies: are things getting worse? Parasitology 137:1967–1973. https://doi.org/10.1017/s0031182010000910

Poulin R, Valtonen ET (2001) Nested assemblages resulting from host size variation: the case of endoparasite communities in fish hosts. Int J Parasitol 31:1194–1204. https://doi.org/10.1016/S0020-7519(01)00262-4

Preisser WC, Welicky RL, Leslie KL (2022) Parasite communities in English Sole (*Parophrys vetulus*) have changed in composition but not richness in the Salish Sea, Washington, USA since 1930. Parasitology 149:786–798. https://doi.org/10.1017/S0031182022000233

Primack RB (2014) Essentials of conservation biology, 6th edn. Sinauer, Sunderland, MA

Rohde K (2002) Ecology and biogeography of marine parasites. Adv Mar Biol 43:1–83. https://doi.org/10.1016/s0065-2881(02)43002-7

Rokicki J (2009) Effects of climatic changes on anisakid nematodes in polar regions. Polar Sci 3:197–201. https://doi.org/10.1016/j.polar.2009.06.002

Rossin MA, Timi JT (2010) Parasite assemblages of *Nemadactylus bergi* (Pisces: Latridae): the role of larval stages in the short-scale predictability. Parasitol Res 107:1373–1379. https://doi.org/10.1007/s00436-010-2011-y

Sala E, Giakoumi S (2018) No-take marine reserves are the most effective protected areas in the ocean. ICES J Mar Sci 75:1166–1168. https://doi.org/10.1093/icesjms/fsx059

Sardella NH, Timi JT (2004) Parasites of Argentine hake in the Argentine Sea: population and infracommunity structure as evidences for host stock discrimination. J Fish Biol 65:1472–1488. https://doi.org/10.1111/j.1095-8649.2004.00572.x

Sasal P, Faliex E, Morand S (1996) Parasitism of *Gobius bucchichii* Steindachner, 1870 (Teleostei, Gobiidae) in protected and unprotected marine environments. J Wild Dis 32:607–613. https://doi.org/10.7589/0090-3558-32.4.607

Sasal P, Desdevises Y, Durieux E et al (2004) Parasites in marine protected areas: success and specificity of monogeneans. J Fish Biol 64:370–379. https://doi.org/10.1111/j.0022-1112.2004.00297.x

Sepúlveda FA, González MT (2015) Patterns of genetic variation and life history traits of *Zeuxapta seriolae* infesting *Seriola lalandi* across the coastal and oceanic areas in the southeastern Pacific Ocean: potential implications for aquaculture. Parasit Vectors 8:282. https://doi.org/10.1186/s13071-015-0892-4

Sindermann CJ (1983) Parasites as natural tags for marine fish: a review. NAFO Sci Counc Stud 6:63–71

Soares IA, Lanfranchi AL, Luque JL, Haimovici M, Timi JT (2018) Are different parasite guilds of Pagrus pagrus equally suitable sources of information on host zoogeography?. Parasitol Res 117:1865–1875. https://doi.org/10.1007/s00436-018-5878-7

Sonnenholzner JI, Ladah LB, Lafferty KD (2009) Cascading effects of fishing on Galapagos rocky reef communities: reanalysis using corrected data. Mar Ecol Prog Ser 375:209–218. https://doi.org/10.3354/meps07890

Spalding MD, Fox HE, Allen GR et al (2007) Marine ecoregions of the world: a bioregionalization of coastal and shelf areas. Bioscience 57:773–583. https://doi.org/10.1641/B570707

Sue H, MacKenzie K, Ives C et al (2022) *Diclidophora merlangi* (Kuhn, 1829) Krøyer, 1838 (Monogenea: Diclidophoridae) as an indicator of hydrocarbon pollution in the North Sea. Mar Poll Bull 185:114268. https://doi.org/10.1016/j.marpolbul.2022.114268

Taillebois L, Barton DP, Crook DA et al (2017) Strong population structure deduced from genetics, otolith chemistry and parasite abundances explains vulnerability to localized fishery collapse in a large Sciaenid fish, *Protonibea diacanthus*. Evol Appl 10:978–993. https://doi.org/10.1111/eva.12499

Takano T, Iwaki T, Waki T (2021) Species composition and infection levels of Anisakis (Nematoda: Anisakidae) in the skipjack tuna *Katsuwonus pelamis* (Linnaeus) in the Northwest Pacific. Parasitol Res 120:1605–1615. https://doi.org/10.1007/s00436-021-07144-5

Ternengo S, Levron C, Mouillot D et al (2009) Site influence in parasite distribution from fishes of the Bonifacio Strait Marine Reserve (Corsica Island, Mediterranean Sea). Parasitol Res 104:1279–1287. https://doi.org/10.1007/s00436-008-1323-7

Timi JT (2003) Parasites of Argentine anchovy in the Southwest Atlantic: latitudinal patterns and their use for discrimination of host populations. J Fish Biol 63:90–107. https://doi.org/10.1046/j.1095-8649.2003.00131.x

Timi JT (2007) Parasites as biological tags for stock discrimination in marine fish from South American Atlantic waters. J Helminthol 81:107–111. https://doi.org/10.1017/S0022149X07726561

Timi JT, Buchmann K (2023) A century of parasites in fisheries and aquaculture. J Helminhol 97(e4):1–18. https://doi.org/10.1017/S0022149X22000797

Timi JT, Lanfranchi AL (2009a) The metazoan parasite communities of the Argentinean sandperch *Pseudopercis semifasciata* (Pisces: Perciformes) and their use to elucidate the stock structure of the host. Parasitology 136:1209–1219. https://doi.org/10.1017/S0031182009990503

Timi JT, Lanfranchi AL (2009b) The importance of the compound community on the parasite infracommunity structure in a small benthic fish. Parasitol Res 104:295–302. https://doi.org/10.1007/s00436-008-1191-1

Timi JT, Mackenzie K (2015) Parasites in fisheries and mariculture. Parasitology 142:1–4. https://doi.org/10.1017/S0031182014001188

Timi JT, Poulin R (2020) Why ignoring parasites in fish ecology is a mistake. Int J Parasitol 50:755–761. https://doi.org/10.1016/j.ijpara.2020.04.007

Timi JT, Luque JL, Sardella NH (2005) Parasites of *Cynoscion guatucupa* along South American Atlantic coasts: evidence for stock discrimination. J Fish Biol 67:1603–1618. https://doi.org/10.1111/j.1095-8649.2005.00867.x

Timi JT, Lanfranchi AL, Etchegoin JA et al (2008) Parasites of the Brazilian sandperch, *Pinguipes brasilianus*: a tool for stock discrimination in the Argentine Sea. J Fish Biol 72:1332–1342. https://doi.org/10.1111/j.1095-8649.2008.01800.x

Timi JT, Luque JL, Poulin R (2010a) Host ontogeny and the temporal decay of similarity in parasite communities of marine fish. Int J Parasitol 40:963–968. https://doi.org/10.1016/j.ijpara.2010.02.005

Timi JT, Lanfranchi AL, Luque JL (2010b) Similarity in parasite communities of the teleost fish *Pinguipes brasilianus* in the southwestern Atlantic: infracommunities as a tool to detect geographical patterns. Int J Parasitol 40:243–254. https://doi.org/10.1016/j.ijpara.2009.07.006

Timi JT, Rossin MA, Alarcos AJ et al (2011) Fish trophic level and the similarity of larval parasite assemblages. Int J Parasitol 41:309–316. https://doi.org/10.1016/j.ijpara.2010.10.002

Tracy AM, Pielmeier ML, Yoshioka RM et al (2019) Increases and decreases in marine disease reports in an era of global change. Proc R Soc B 286:20191718. https://doi.org/10.1098/rspb.2019.1718

Vales DG, García NA, Crespo EA et al (2011) Parasites of a marine benthic fish in the Southwestern Atlantic: searching for geographical recurrent patterns of community structure. Parasitol Res 108:261–272. https://doi.org/10.1007/s00436-010-2052-2

Valtonen ET, Marcogliese DJ, Julkunen M (2010) Vertebrate diets derived from trophically transmitted fish parasites in the Bothnian Bay. Oecologia 162:139–152. https://doi.org/10.1007/s00442-009-1451-5

Van der Lingen CD, Weston LF, Ssempa NN et al (2015) Incorporating parasite data in population structure studies of South African sardine *Sardinops sagax*. Parasitology 142:156–167. https://doi.org/10.1017/S0031182014000018

Vidal-Martínez VM, Pech D, Sures B et al (2010) Can parasites really reveal environmental impact? Trends Parasitol 26:44–51. https://doi.org/10.1016/j.pt.2009.11.001

Warburton EM, Budischak SA, Jolles AE et al (2022) Within-host and external environments differentially shape β-diversity across parasite life stages. J Anim Ecol 92:665–676. https://doi.org/10.1111/1365-2656.13877

Ward JR, Lafferty KD (2004) The elusive baseline of marine disease: Are diseases in ocean ecosystems increasing? PLoS Biol 2:E120. https://doi.org/10.1371/journal.pbio.0020120

Watson JE, Dudley N, Segan DB (2014) The performance and potential of protected areas. Nature 515:67–73. https://doi.org/10.1038/nature13947

Welch DJ, Newman SJ, Buckworth RC et al (2015) Integrating different approaches in the definition of biological stocks: A northern Australian multijurisdictional fisheries example using grey mackerel, *Scomberomorus semifasciatus*. Mar Policy 55:73–80. https://doi.org/10.1016/j.marpol.2015.01.010

Welicky RL, Preisser WC, Leslie KL et al (2021) Parasites of the past: 90 years of change in parasitism for English sole. Front Ecol Environ 19:470–477. https://doi.org/10.1002/fee.2379

Whiteman NK, Parker PG (2005) Using parasites to infer host population history: a new rationale for parasite conservation. Anim Conserv 8:175–181. https://doi.org/10.1017/S1367943005001915

Williams HH, MacKenzie K, McCarthy AM (1992) Parasites as biological indicators of the population biology, migrations, diet, and phylogenetics of fish. Rev Fish Biol Fish 2:144–176. https://doi.org/10.1007/BF00042882

Wood CL, Lafferty KD (2015) How have fisheries affected parasite communities? Parasitology 142:134–144. https://doi.org/10.1017/S003118201400002X

Wood CL, Lafferty KD, Micheli F (2010) Fishing out marine parasites? Impacts of fishing on rates of parasitism in the ocean. Ecol Lett 13:761–775. https://doi.org/10.1111/j.1461-0248.2010.01467.x

Wood CL, Micheli F, Fernandez M et al (2013) Marine protected areas facilitate parasite populations among four fished host species of central Chile. J Anim Ecol 82:1276–1287. https://doi.org/10.1111/1365-2656.12104

Wood CL, Sandin SA, Zgliczynski B et al (2014) Fishing drives declines in fish parasite diversity and has variable effects on parasite abundance. Ecology 95:1929–1946. https://doi.org/10.1890/13-1270.1

Wood CL, Leslie KL, Claar D et al (2023a) How to use natural history collections to resurrect information on historical parasite abundances. J Helminthol 97:1–13. https://doi.org/10.1017/S0022149X2200075X

Wood CL, Vanhove MP (2023) Is the world wormier than it used to be? We'll never know without natural historycollections. J Animal Ecol 92:250–262. https://doi.org/10.1111/1365-2656.13794

Wood CL, Welicky RL, Preisser WC et al (2023b) A reconstruction of parasite burden reveals one century of climate-associated parasite decline. PNAS 120:e2211903120. https://doi.org/10.1073/pnas.2211903120

Yamaguchi A, Yokoyama H, Ogawa K et al (2003) Use of parasites as biological tags for separating stocks of the starspotted dogfish *Mustelus manazo* in Japan and Taiwan. Fish Sci 69:337–342. https://doi.org/10.1046/j.1444-2906.2003.00626.x

Zarlenga DS, Hoberg E, Rosenthal B et al (2014) Anthropogenics: human influence on global and genetic homogenization of parasite populations. J Parasitol 100:756–772. https://doi.org/10.1645/14-622.1

Zhang S, Zhu J, Xu S et al (2021) An integrated approach to determine the stock structure of spinyhead croaker *Collichthys lucidus* (Sciaenidae) in Chinese coastal waters. Front Mar Sci 8:693954. https://doi.org/10.3389/fmars.2021.693954

Open Access This chapter is licensed under the terms of the Creative Commons Attribution-NonCommercial-NoDerivatives 4.0 International License (http://creativecommons.org/licenses/by-nc-nd/4.0/), which permits any non-commercial use, sharing, distribution and reproduction in any medium or format, as long as you give appropriate credit to the original author(s) and the source, provide a link to the Creative Commons license and indicate if you modified the licensed material. You do not have permission under this license to share adapted material derived from this chapter or parts of it.

The images or other third party material in this chapter are included in the chapter's Creative Commons license, unless indicated otherwise in a credit line to the material. If material is not included in the chapter's Creative Commons license and your intended use is not permitted by statutory regulation or exceeds the permitted use, you will need to obtain permission directly from the copyright holder.

Invasion Biology in the Context of Aquatic Host–Parasite Interactions

18

Dakeishla M. Díaz-Morales, Bernd Sures, E. Rosa Jolma, and David W. Thieltges

Abstract

Biological invasions have become a topic of concern due to their increasing occurrence and ecological and economic consequences. This chapter examines the stages, mechanisms and effects of biological invasions from a parasitological point of view, focusing on parasitic protists, microsporidians, fungi, oomycetes, helminths and arthropods. We present hypotheses and mechanisms underlying biological invasions of parasites and hosts, such as the enemy release hypothesis, spillover, spillback, disease acquisition and invasional meltdown. We illustrate these mechanisms and the effects of biological invasions on parasite–host interactions and native ecosystems with examples from aquatic environments. In terms of effects, we discuss direct competitive and consumer–resource interactions and indirect density- and trait-mediated effects by invasive parasites, ranging from the individual to the community and ecosystem level. We conclude the chapter with some remarks on management and mitigation aspects.

D. M. Díaz-Morales (✉)
Aquatic Ecology and Centre for Water and Environmental Research, University of Duisburg-Essen, Essen, Germany

Research Centre One Health Ruhr, Research Alliance Ruhr, University of Duisburg-Essen, Essen, Germany

School of Aquatic and Fishery Sciences, University of Washington, Seattle, WA, USA

B. Sures
Aquatic Ecology and Centre for Water and Environmental Research, University of Duisburg-Essen, Essen, Germany

Research Centre One Health Ruhr, Research Alliance Ruhr, University of Duisburg-Essen, Essen, Germany

Water Research Group, Unit for Environmental Sciences and Management, North-West University, Potchefstroom, South Africa
e-mail: bernd.sures@uni-due.de

E. R. Jolma
Department of Coastal Systems, NIOZ Royal Netherlands Institute for Sea Research, Den Burg, Texel, The Netherlands

Department of Population Health Sciences, Veterinary Medicine, Utrecht University, Utrecht, The Netherlands
e-mail: rosa.jolma@nioz.nl

D. W. Thieltges
Department of Coastal Systems, NIOZ Royal Netherlands Institute for Sea Research, Den Burg, Texel, The Netherlands

Groningen Institute for Evolutionary Life-Sciences, GELIFES, University of Groningen, Groningen, The Netherlands
e-mail: david.thieltges@nioz.nl

© The Author(s) 2025
N. J. Smit, B. Sures (eds.), *Aquatic Parasitology: Ecological and Environmental Concepts and Implications of Marine and Freshwater Parasites*, https://doi.org/10.1007/978-3-031-83903-0_18

18.1 Introduction

In recent decades, biological invasions have increased worldwide, sometimes with devastating ecological and economic consequences. Globally, invasive species are estimated to have caused cumulative costs of up to $345 billion since 1971, primarily due to *Aedes* spp. mosquitoes, followed by fish (mainly *Gymnocephalus* spp.) and bivalves (*Dreissena* spp.) (Cuthbert et al. 2021). All aforementioned species are associated with water, emphasising how important and vulnerable aquatic ecosystems are to invasions due to the interconnectedness of macro- and microhabitats, and human activities associated with water bodies, such as shipping, aquaculture and aquarium trade, among others. In other words, aquatic environments are under constant pressure from biological invasions facilitated by anthropogenic activities (Padilla and Williams 2004; Havel et al. 2015). Given all the ecosystem services aquatic systems provide and the expectation of biological invasion rates to continue increasing, understanding the mechanisms and effects of invasions becomes essential.

In addition to parasitic interactions, our understanding of the mechanisms and effects of biological invasions on aquatic systems is rather limited. All free-living organisms, including invasive species, can serve as hosts for parasite species (Windsor 1998; Goater et al. 2014). Parasites likely account for roughly half of all (metazoan) biodiversity, making them ubiquitous (Windsor 1998; Poulin and Morand 2000; Lafferty and Kuris 2002; Dobson et al. 2008), and they are well-known to be capable of modulating ecological interactions (Mouritsen and Poulin 2005; Dunne et al. 2013; Sures et al. 2017; Giari et al. 2020). Parasites frequently appear in invasion events as both invaders and parasitic agents of invasive and native hosts. Some of the roles that parasites play in host invasions include influencing the invasion success of their hosts or modulating the magnitude of the impact of invasive hosts on the ecosystem. However, the introduction of new hosts can affect the transmission of native parasites in the invaded habitat. Given the ubiquity of parasites in aquatic systems and the vulnerability of these habitats to invasion events, we will dedicate this chapter to collating the most up-to-date knowledge on invasion biology and aquatic parasites. As mentioned earlier, we focus on parasitic taxa that are traditionally covered by parasitology as a discipline, including protists, microsporidians, fungi, oomycetes, helminths and arthropods (see Chaps. 2–6). We will first unify and clarify relevant invasion biology terminology. Then, we will present some of the most prominent hypotheses and mechanisms explaining biological invasions and discuss the effects of such invasions on aquatic systems, all from a parasitological perspective. Although we focus on aquatic systems, the general concepts presented herein are valid for all ecosystems. We will conclude the chapter with a view on relevant evolutionary implications and an outlook on management implications and open venues for future research.

18.2 Invasion Biology and Parasites: Terminology and Basic Concepts

The road towards successful invasions is a bumpy one, with many obstacles to overcome and many different definitions for non-native species. Although several researchers have tried to map this road by identifying potential bottlenecks, stages and categorisations (Williamson and Fitter 1996, Richardson et al. 2000), the terminology used has not been unified due to the different perspectives (individual- versus population-based) and expertise (botanical versus zoological). To combat this confusion and lack of consistent terminology, Blackburn et al. (2011) revised previous approaches and proposed an overarching framework that considers all potential stages and barriers an alien species faces on its way to invasion and suggested appropriate terminology and management recommendations for these different stages (Fig. 18.1). Although this framework was designed with free-living species in mind, it can also be applied to parasitic organisms. However, some additional barriers are relevant to

18 Invasion Biology in the Context of Aquatic Host–Parasite Interactions

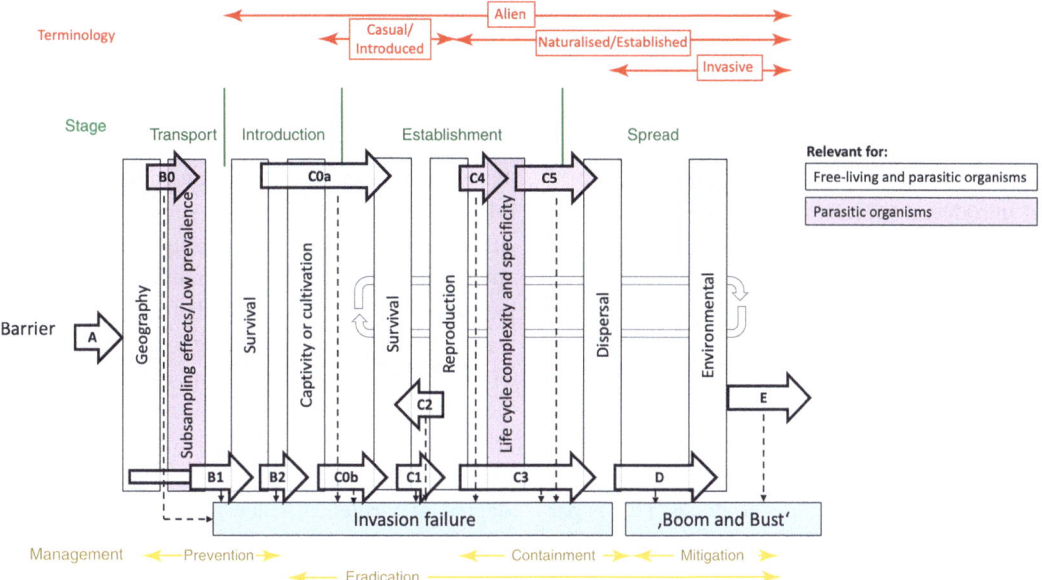

Fig. 18.1 Unified framework for biological invasions developed originally by Blackburn et al. (2011) and adapted to invasion processes for parasitic organisms. The framework includes categories (represented as arrows) for populations, their stages of invasion, the barriers relevant to each stage, and potential management solutions. Alien populations might not be transported at all to novel environments or might die on the way (A1). For parasites, they might fail to be transported due to subsampling effects, especially for parasites with naturally low infection prevalence (A2). If transport is successful, alien species might experience captivity, which hinders their introduction (B). Nevertheless, species can also be introduced directly (C0a) or released from captivity (C0b), but die soon after. An introduced species might survive in the wild but do not reproduce (C1) or reproduce without the capacity to be a self-sustaining population (C2), thus failing to establish itself. As such, a successfully established alien species is one that can reproduce in the wild in a self-sustaining way (C3). Establishment might be more challenging for alien parasites, especially those with complex life cycles, since some species might reproduce (asexually or sexually) for an extended period but fail to encounter suitable downstream hosts (C4). Therefore, a parasite can be considered established when it can survive, reproduce, and complete the life cycle in the wild (C5). The last barrier that alien species face is their capacity to spread. Populations that spread successfully are defined by those species that can establish at a significant distance from the point of introduction (D). Finally, "fully invasive species" are those that have spread successfully at several sites across different habitats (E). Adapted from Blackburn et al. (2011)

parasite invasion success (Fig. 18.1), which we added to the original concept.

Under natural conditions, the dispersal of species to new areas and subsequent spread can occur through ecological and environmental processes such as migration, storms or flooding events. However, if introductions of new species into a system, intentionally or unintentionally, are assisted by human intervention, we consider these new species to be alien species (also called non-indigenous or non-native species). The year 1492 is often used as a baseline because it marks the beginning of intercontinental trade, with Christopher Columbus' arrival on the American continent and Caribbean islands, introducing alien species from Europe, such as rats, cats and wild pigs to the Caribbean and the naval shipworm (*Teredo navalis*) to Europe (Sures and Boenigk 2021).

The framework described by Blackburn et al. (2011) classifies alien species as 'casual/introduced', 'naturalised/established' or 'invasive' depending on their stage of invasion and the barriers they have successfully overcome (Fig. 18.1). An established species is one that has persisted over several generations without human intervention, while a casual/introduced species occurs only occasionally and sparsely but is unable to build self-reproducing populations. An invasive species (IS) is an alien organism (free-living or parasitic) that has been able to reproduce and spread at a significant distance from the original point of introduction (Blackburn et al. 2011). Success rates during an invasion process are roughly summarised by the 'Tens rule' hypothesis, which suggests that upon transport, only 10% of introduced species are able to establish, from which again 10% will be able to build permanent, self-reproducing populations (without human intervention) and of which, in turn, only 10% become invasive (Jeschke 2014). It is important to note that a successful or failed establishment and progress towards the invasion stage of a host species does not imply the same for its parasites. A parasite can become a successful invader, even if the host does not manage to do so. This is particularly true for parasites that can be transported and transmitted *via* resistant stages such as spores or cysts in the environment (e.g. several protist species, fungi, oomycetes and microsporidians; see Chaps. 2–4) and that are not very host-specific, that is, parasites that can successfully use other hosts instead. In addition, a host might become a successful invader while losing its parasites upon introduction. These processes are described in greater detail in Sect. 18.3.

The progression throughout the stages of invasion depends on overcoming barriers such as geography, captivity, survival, reproduction, dispersal and environmental conditions (Fig. 18.1). The general stages of biological invasions described by Blackburn et al. (2011) include transport, introduction, establishment and spread.

Stage I: Transport

During the first invasion stage (the transport stage), the first obstacle that a putative alien species (*i.e.* a free-living or parasitic species) must overcome is the geographical barrier. Geographical barriers such as mountains or oceans are often crossed by transport through human activities such as legal and illegal trade and shipping. For parasites, there are additional barriers, such as a low prevalence of infection in the indigenous host species in the original habitat. Even if the possibility of crossing geographical barriers is available, the parasite must be lucky enough to be present in the subsample of transported hosts (Blackburn and Ewen 2017). This is not a significant issue for highly prevalent parasites. However, for parasites with a very low prevalence of infection, this barrier might be difficult to overcome, causing them to 'miss the boat' during transport (Paterson and Gray 1997). Moreover, during transport, some hosts might eliminate their parasites through an artificial quarantine process. This is mainly relevant for ocean vessel transport, which can take several weeks or months, depending on the distance. However, as intercontinental trade becomes faster (i.e. transportation by air and faster vessel transport *via* new and shorter routes), the latter becomes less relevant. Another factor is the degree of virulence, since highly virulent parasites might kill the host during transport and, therefore, fail to be co-transported (Lymbery et al. 2014; Blackburn and Ewen 2017). Furthermore, even organisms usually considered to be commensals can cause host mortality when transport stress decreases host immunocompetence, and overcrowding can facilitate transmission, resulting in abnormally high infection loads (Sainsbury and Vaughan-Higgins 2012).

Stage II: Introduction

Since the first indication of an invasion is usually the detection of an alien species in the new environment, and because the transport stage is rarely detected, the stages of transport and introduction are often jointly referred to as translocation (Blackburn and Ewen 2017). However, for clarity in terms of the barriers to take on in each stage, we kept them separate. Free-living species can be directly introduced into the new region (e.g. through trade, ship ballast, or intentional introductions). Parasites can also be directly introduced even without their host, particularly those with environmentally resistant stages as part of their life cycle (e.g. several protist species, fungi, microsporidia and oomycetes). However, parasite introductions of this type are usually accidental. Some alien hosts (infected or uninfected) may be transported over a geographical barrier but then go through a containment process, such as when imported for aquaculture or pet trade. This barrier may be passed during the introduction stage by owners who intentionally or accidentally release exotic pets, their parasites or both into the environment. An example of a parasite being introduced in this way is the release of aquatic chytrid fungus *Batrachochytrium salamandrivorans*, from captive Asian salamanders and newts to wild urodele amphibians in Europe (Martel et al. 2013, 2014; see also Chap. 4).

Stage III: Establishment
Following an introduction, several cycles of survival and successful reproduction are required to establish a self-sustaining population (Blackburn et al. 2011). Therefore, the establishment of alien free-living and parasitic species will depend on factors such as reproduction rate, presence of competitors and enemies, propagule pressure and environmental conditions. For alien parasites, their establishment success also depends on factors such as their life cycle complexity, host-specificity and the environmental conditions in the novel habitat, which could be inhospitable for infective free-living larval stages. Parasites with complex life cycles (heteroxenous parasites; that is, having more than one host and, in some instances, even four hosts, Fig. 18.2) can experience a bottleneck since they need all these hosts to complete their life cycles. In contrast, parasites with a simple life cycle (monoxenous parasites, Fig. 18.2) only require a single host, and the chances of completing the life cycle are higher. However, even for parasites with simple life cycles, their establishment can be challenging if the parasite has a high degree of host-specificity (Fig. 18.2). Specialist parasites need a particular host species to survive; in that case, the alien parasite would completely depend on the establishment success of the alien host that co-introduced the parasite. Meanwhile, generalist parasites are more versatile and have more opportunities (*i.e.* more suitable host species) for establishment in the new environment. All in all, the establishment process for both free-living and parasitic alien species is complex, often including interactions among the various factors and barriers. Blackburn et al. (2011) stress that, due to the complexity of the barriers and their dynamic state (some of these factors may change intermittently or in the future), a failure to establish does not imply the establishment failure of subsequently introduced species.

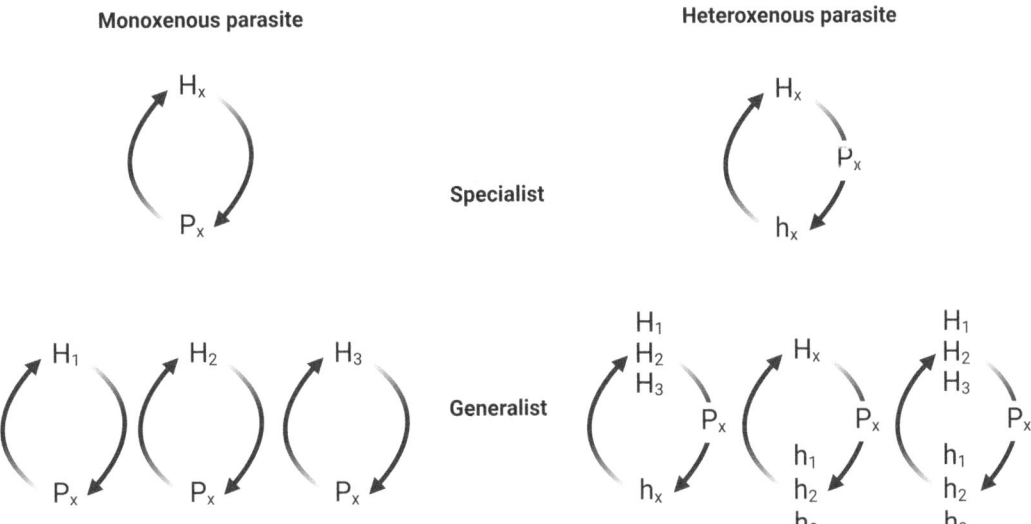

Fig. 18.2 Parasites (P_x) with simple life cycles (monoxenous parasites) require only one host, while those with complex life cycles (heteroxenous parasites) require at least two hosts (a definitive host "H_x" and at least one intermediate host "h_x") to complete the life cycle. Specialist parasites are restricted to a single host species, while generalist parasites can infect several host species. Adapted from Sures (2011)

Stage IV: Spread

An established alien species is not necessarily invasive. Before being considered an invasive species, an established alien species must find ways of significant dispersal beyond the point of introduction. Only after a species has managed to disperse and sustain its population in multiple habitats and regions can it be considered a fully invasive species (Blackburn et al. 2011). The distinction is important because an established alien species will encounter new obstacles (e.g. enemies, environmental conditions) impeding or challenging its spread upon dispersal. The invasion process can therefore be considered a series of follow-up establishment events, since distance from the origin of introduction usually correlates with environmental dissimilarity and expanding environmental gradients (Blackburn et al. 2011). In other words, the obstacles and barriers observed during the establishment stage are still valid during the spreading process. Also, environmental barriers can be seen as the wall that an invasion process might hit once the invasive species cannot spread further because it cannot adapt to new environmental conditions. Finally, an additional consideration of the unified framework by Blackburn et al. (2011) is the 'boom and bust' process in which the population size of a successful invader collapses after an initial spread, sometimes even to extinction. All of these concepts apply to both free-living and parasitic alien species. However, in the case of established alien parasites, host presence/absence and host density, which are positively correlated with parasite transmission (Lafferty et al. 2004; Bommarito et al. 2021), act as another barrier that is likely to be more limiting in the invasion phase than in the establishment phase.

18.3 Parasite–Host Interactions in Invaded Ecosystems

Several mechanisms and hypotheses on invasion biology and parasitism have been presented in the literature that contribute to a better understanding of the processes involved in each invasion stage. These processes include potential interactions between invasive hosts and parasites with native hosts and parasites (Fig. 18.3). In this section, we will summarise some of these processes and provide some examples illustrating each concept.

18.3.1 Enemy Release Hypothesis

The Enemy Release Hypothesis has been used to explain the success of some free-living invasive species. This hypothesis maintains that an invasive species can succeed due to the lack or release of enemies in the recipient environment (Torchin et al. 2003) (Fig. 18.3). These enemies include predators and parasites that could potentially limit the proliferation and subsistence of an introduced species. With respect to parasites as enemies, they can be lost during transport or when entering the transport stage. The latter can happen when only uninfected individuals of free-living species, from the natural range where they are usually infected by parasites, make it to the transport stage (subsampling effects; see also Sect. 20.2). Furthermore, in new environments, invasive hosts may not become infected with native parasites because they lack successful host–parasite coevolution, resulting in highly host-specific parasite species that cannot switch to new host species. With the invasive host lacking parasites from their original environment and evading parasites in their new environment, their population is not under biological control from parasites and can therefore spread and become invasive.

In the context of parasitology, there are several examples of enemy release of invaders. For instance, in aquatic systems, enemy release can explain the interactions between the invasive shrimp *Artemia franciscana* and native '*Artemia parthenogenetica*' and *Artemia salina* (Georgiev et al. 2007) in the Mediterranean. These shrimps serve as intermediate hosts for cyclophyllidean cestodes. This invasive crustacean species has consistently been found to have a lower prevalence and abundance of cestodes compared to the native species (Georgiev et al. 2007). These cestodes can have several negative effects on their

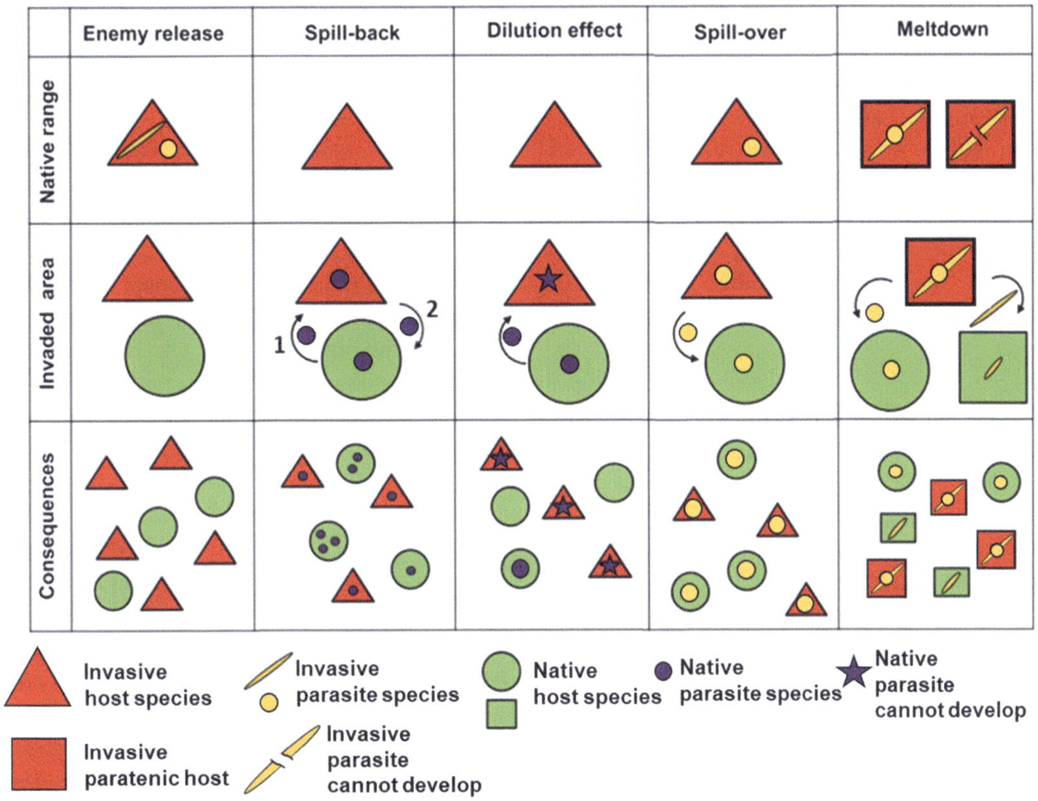

Fig. 18.3 Selected general concepts and processes in invasion biology under the umbrella of Parasitology. Adapted from Sures et al. (2019)

native shrimp hosts, including a decrease in fecundity (Amat et al. 1991) and an increase in positive phototaxis, making them more susceptible to predation (Sánchez et al. 2007). Georgiev et al. (2007) argued that this parasite-induced decrease in fecundity and positive phototaxis could contribute to the replacement rate of native brine shrimp with invasive *A. franciscana*. Nevertheless, the extent to which introduced hosts have a lower parasite burden than their conspecifics in the native range is sometimes questionable, particularly when comparisons with the source population in the native range are biased by sampling effects. To that end, Colautti et al. (2005) suggest the concepts of realised versus apparent enemy reduction, in which a correction for subsampling effects is applied. The authors contend that, because the source population is typically sampled from a specific point rather than randomly across an entire region, apparent enemy reduction may be inflated if caution is not exercised regarding the reference population for comparison (Colautti et al. 2005).

18.3.2 Parasite Spillback

An invasive host may acquire parasites in the novel environment without transmitting them back to native species, or it can become infected with a native parasite and transmit it forward to a native host, resulting in disease amplification, a process called parasite spillback (Kelly et al. 2009; Fig. 18.3). Spillback is exemplified by salmonid introductions in Lake Moreno in Argentina (Poulin et al. 2011). Invasive salmonids (*Oncorhynchus mykiss* and *Salvelinus fontinalis*) introduced from North America to Argentina serve as suitable hosts for the native acanthocephalan *Acanthocephalus tumescens* (Rauque et al. 2003). The reproductive capacity of this acanthocephalan has been found to be higher in the invasive host than in some native fish species (Rauque et al. 2003), suggesting that the presence of these invasive fish may increase transmission rates (Poulin et al. 2011). Disease amplification can also take place by invasive hosts acting as vectors or reservoirs through their indirect involvement in other parasites' life cycles (Chalkowski et al. 2018).

18.3.3 Dilution Effect

On the flip side, host–parasite interactions can go the other way and result in parasite dilution. The dilution effect occurs when an invasive host acts as a dead end for a parasite, preventing further transmission and resulting in a net lower infection prevalence in native hosts (Dunn 2009; Fig. 18.3). This process is perfectly illustrated by the invasion of freshwater gastropods in New Zealand (Dunn 2009). The native gastropod *Potamopyrgus antipodarum* presents lower prevalence of infection with *Microphallus* sp. in the presence of the invasive gastropod *Lymnaea stagnalis* (Kopp and Jokela 2007). *Lymnaea stagnalis* serves as a decoy to the trematode *Microphallus* sp., thus attracting larval stages that could potentially infect *P. antipodarum*. However, due to the resistance of the invasive host to the parasite, it represents a sink and a dead end for the parasite. Similarly, some invasive species can act as parasite diluters by ingesting free-living infective larval stages or physically hindering them from reaching their host (Welsh et al. 2014). In laboratory settings, a reduction of up to 89% in infection loads of the trematode *Himasthla elongata* in the northeast Atlantic native blue mussel, *Mytilus edulis*, has been observed in the simultaneous presence of invasive Pacific oysters *Magallana gigas* (previously *Crassostrea gigas*; Fig. 18.4a) and American slipper limpets, *Crepidula fornicata* (Thieltges et al. 2009). These examples show how invasive species alter not only parasite transmission dynamics but also how an invader can provide certain benefits to indigenous populations.

18.3.4 Parasite Spillover

When a host is translocated, its parasites are not always lost; sometimes, they are co-introduced into the new environment and become established themselves (Lymbery et al. 2014; Foster et al. 2021). This can have different consequences. For instance, if an alien parasite is co-introduced with the host, it can either keep using the introduced host as its only host (particularly parasites with high specificity towards a particular host species; Fig. 18.2) or infect native species, a process known as parasite spillover (Daszak et al. 2000; Fig. 18.3). A well-known example is the spillover of the crayfish plague (caused by the oomycete *Aphanomyces astaci*) into European native white-clawed crayfish, *Austropotamobius pallipes* (Dunn and Hatcher 2015). This parasite was co-introduced to Europe with its host, the invasive signal crayfish, *Pacifastacus leniusculus*, and has caused a severe decline in the population of native *A. pallipes*. The spread of this parasite in European waters has been so severe that it has caused the local extinction of native crayfish populations, and it is considered one of the 'World's Worst Invasive Species' (Lowe et al. 2000).

18.3.5 Invasional Meltdown

An invasive host could bring multiple parasites, resulting in a potential invasional meltdown in certain instances. According to the invasional meltdown hypothesis, a group of alien species can aid one another through the process of invasion by increasing their chances of survival and overall invasion success, which can result in synergistic negative effects on native species (Simberloff and Von Holle 1999; Fig. 18.3). Although this hypothesis has rarely been applied to parasites and requires further testing, Emde et al. (2014) identified a system that could be representative of this process. In this case, the invasive Ponto-Caspian Round Goby, *Neogobius melanostomus*, serves as a paratenic host to the invasive Ponto-Caspian acanthocephalan *Pomphorhynchus laevis* (Kvach and Skóra 2007),

Fig. 18.4 Examples of invasive host and parasite species, with invasive Pacific oysters *Magallana gigas* (previously *Crassostrea gigas*) next to the native blue mussel *Mytilus edulis* (**a**). The first invasive copepod infecting mussels (*Mytilicola intestinalis*, red arrow) arrived without an invasive host (**b**, **c**), whereas *Mytilicola orientalis* (blue asterisk) spilled-over from oysters (**c**). The invasive swim bladder nematode of eels, *Anguillicola crassus* (**d**), from a naturally infected swim bladder of a European eel from the river Rhine

which later turned out to be most likely *Pomphorhynchus bosniacus* (Reier et al. 2019; Sures et al. 2019). The preadult larvae of this invasive acanthocephalan encyst in the goby, which often contain another species of invasive parasite, the swim bladder nematode *Anguillicola crassus* (Emde et al. 2014; Hohenadler et al. 2018; Fig. 18.4d). The nematode can benefit from the cyst by avoiding and protecting itself from the immune response of the paratenic host, allowing it to be still infective to its final host, the European eel, *Anguilla anguilla*. Moreover, *N. melanostomus* lacks a swim bladder, which would be the preferred host organ of third-stage larvae of *A. crassus* in all known paratenic hosts of this nematode (Sures et al. 1999). Experimental infections of chub (*Squalius cephalus*) with *Pomphorhynchus* sp. cystacanths harbouring *A. crassus* larvae resulted in successful *Pomphorhynchus* sp. infection, whereas no *A. crassus* individuals could be detected in chub (Nachev et al. 2024). On the other hand, after infection of eels with *Pomphorhynchus* sp. cystacanths harbouring *A. crassus* larvae, adult *A. crassus* were present in the swim bladder of eels, but none of the acanthocephalan individuals was able to establish in the intestine (Hohenadler et al. 2018; Honka and Sures 2021). Therefore, by shielding *A. crassus* from an otherwise unsuitable host (*i.e. Neogobius* sp.), *Pomphorhynchus* sp. enables the invasive nematode to exploit a transmission strategy (*i.e.* the use of paratenic hosts) that promotes its spread in the invaded region. This 'Trojan-horse' strategy (Emde et al. 2014) is an excellent illustration of how interspecific interactions among invasive parasite species can result in potential and fascinating mechanisms for parasite transmission in an invaded region.

18.4 Effects of Parasites in Aquatic Bioinvasions

After discussing the processes and mechanisms underlying biological invasions within the context of parasitology, the next logical question is: what happens when alien species become invasive, and what are the implications for native populations, communities and ecosystems? In this section, we will discuss the effects of invasions by hosts and parasites at the individual and population levels and the cascading effects of these on native communities and the ecosystem.

18.4.1 Direct Effects of Parasite and Host Invasions

From a parasitological standpoint, the direct effects that parasite invasions have on the native fauna result from a parasite's inherent metabolic dependence on its host (Thieltges and Goedknegt 2023). This dependence can manifest as physiological effects (*e.g.* changes in growth rate, fertility, immunocompetence), changes in phenology, and, in extreme scenarios, death of the host (Lymbery et al. 2014; Poulin et al. 2011). At the individual host level, the loss of energy to the parasite and to immunological reactions can jeopardise the energy available to hosts for fitness-related processes (Thieltges and Goedknegt 2023). Overall, the outcome and direction of these effects will be dictated by traits such as host resistance and tolerance and parasite virulence and infectivity (Råberg et al. 2009). This 'give and take' between host and parasite traits is a very complex process, and in established host-parasite systems, it is usually the result of a long coevolutionary history. Invasion events frequently lack this coevolutionary complex dynamic, and thus novel interactions between hosts and parasites (native or invasive) can often result in severe pathogenicity and emerging diseases (Blakeslee et al. 2021).

An example of a highly pathogenic invasive parasite is the case of the myxozoan *Myxobolus cerebralis,* the causative agent of whirling disease (Hoffman 1990; see Chap. 3). This parasite is native to Eurasia and originally was a parasite of the brown trout, *Salmo trutta.* This parasite does not have significant negative impacts on its native host *S. trutta* due to a long co-evolutionary history. However, after spreading worldwide, it has come to infect naïve fish (e.g. *Oncorhynchus* spp.), which do not have immunological resistance to it (Sarker et al. 2015). Since the novel host does not effectively regulate the parasite, *M. cerebralis* can have drastic direct impacts, such as erratic swimming behaviour (*i.e.* whirling), and can ultimately result in the deaths of more than 85% of fish stocks (Thompson et al. 1999). Since infections can occur in both captive and wild fish, this invasive parasite represents a great economic and ecological burden (see also Chap. 23).

Although the fatal effects of invasive parasites on native hosts are the most striking, invasive parasites can also have subtler sublethal effects on native hosts; influencing traits such as behaviour, tissue integrity, fertility, growth rates and immunocompetence (summarised in Thieltges and Goedknegt 2023). For instance, the parasitic copepod *Mytilicola orientalis* was co-introduced with the Pacific oyster, *M. gigas,* to European coastal waters (Fig. 18.4b, c). After introduction, the parasite spilled over to several native bivalves, including the blue mussel *M. edulis* (Goedknegt et al. 2017). This parasite feeds on the host's gut tissue and, to a lesser extent, on the host's gut content (Goedknegt et al. 2018b). Goedknegt et al. (2018a) determined that in its juvenile stages, this parasite can reduce the condition index (between 11–13%) of mussels under laboratory conditions. Because infection prevalences in blue mussels are increasing and sometimes exceeding those in Pacific oysters (Pogoda et al. 2012; Goedknegt et al. 2017), this parasite is an important invader to further research and monitor.

The effects of co-introduced parasites on native hosts are not always negative. In some situations where there is no spillover to native hosts, invasive parasites can be a burden only for the invasive host. For instance, the cane toad, *Rhinella marina,* serves as host to the lungworm *Rhabdias pseudosphaerocephala*. Both were co-introduced in Australia from Hawaii in 1935 and successfully established as invaders (Shine 2010). The invasive toad is poisonous and potentially lethal to native predators (Shine 2010). However, the parasite *R. pseudosphaerocephala* can negatively affect the survival and locomotor performance (swimming speed and endurance) of juvenile toads (Kelehear et al. 2009) and the growth rates of both juvenile and adult toads (Kelehear et al. 2009, 2011). After more than 75 years of invasion, this lungworm has not been found to spillover into native anurans (Pizzatto et al. 2012). As a result, even though these poisonous cane toads still harm Australia's native wildlife (Shine 2010), their effects might have been worse if *R. pseudosphaerocephala* did not have such an impact on this invasive host.

18.4.2 Indirect Effects of Parasite and Host Invasions

Indirect effects of invasions by parasites and hosts can be observed in the way that species interact with each other, either in competitive interactions or interactions between consumers and their resources (Dunn et al. 2012; Fig. 18.5). For instance, parasites can impose a burden on the invasive host and make it less competitive to the native ones, thereby reducing the host's invasion success. In this case, parasites can act as equaliser when superior competitors are most affected and thus enable coexistence (Fig. 18.5). However, if the parasite effects are stronger on the inferior competitor (*e.g.* a native host), the stronger competitor (*e.g.* an invasive host) can possibly replace the inferior one. The mechanisms by which these effects occur can be categorised as density-mediated indirect effects (DMIEs) and trait-mediated indirect effects (TMIEs) (also see Chaps. 7 and 8). DMIEs are mainly driven by effects on the survival and reproduction of hosts, while TMIEs are driven by impacts on behavioural, morphological, and physiological aspects, including changes in the life-history of the hosts (Fig. 18.6) (Dunn et al. 2012).

An excellent system to illustrate the indirect effects of invasive parasites on competitive dynamics is that of acanthocephalans infecting amphipods (Giari et al. 2020). Invasive *Pomphorhynchus laevis* is known to use invasive and native gammarids in eastern France as intermediate hosts. *P. laevis* has been observed to have an immunological cost on its native host *Gammarus pulex*, while no physiological impact has been detected in its invasive host *Gammarus roeseli* (Cornet et al. 2010). This immunosuppression makes *G. pulex* more susceptible to secondary opportunistic infections (*i.e.* bacteria), while *G. roeseli* was able to fight such infections (Cornet et al. 2010). This immunological cost could imply a decrease in the competitive ability of native amphipods towards invasive ones through TMIEs. Another example of a TMIE of invasive parasites is the displacement of the native California horn snail, *Cerithideopsis californica*, by the invasive Asian mud snail *Batillaria attramentaria* (Byers 2000; Dunn et al. 2012). The displacement of the native gastropod is driven by physiological effects induced by the co-introduced trematode *Cercaria batillariae*. This trematode induces behavioural and phenotypical changes such as gigantism and castration in the invasive gastropod and with it a change in the resources used by the Asian mud snail in the invasion range (Miura et al. 2006). This change in the use of resources induces exploitation of a particular niche in the ecosystem by the invasive snails, displacing and forcing smaller native snails to migrate to other microhabitats (Miura et al. 2006).

Consumer-resource effects are particularly interesting from a parasitological perspective since parasites are widespread across food webs (Lafferty et al. 2008; Morton et al. 2021) and are known to often modulate the behaviour and phenotypical traits of their hosts to facilitate transmission (Poulin 1994; Cezilly et al. 2000). Parasites with complex life cycles often infect hosts at several trophic levels (i.e. infecting both resources and their consumers). Dunn et al. (2012) conducted a thorough review on this topic, decomposing these various levels according to whether the resource, the consumer, or both are being parasitised. Parasites (native or invasive) can alter consumer–resource (native or invasive) interactions in various ways (Fig. 18.5). Indirect effects of native parasites on invaders *via* infected resources can be illustrated with the native shell-boring parasitic polychaete *Polydora* spp. in New England infecting native whelks. Invasive crabs *Carcinus maenas* usually feed on whelks (*Nucella lapillus*) of a certain size (Fisher 2010). However, *Polydora* sp. infections significantly increase the size range of whelks that *C. maenas* can predate (Fisher 2010), as the holes and burrows excavated

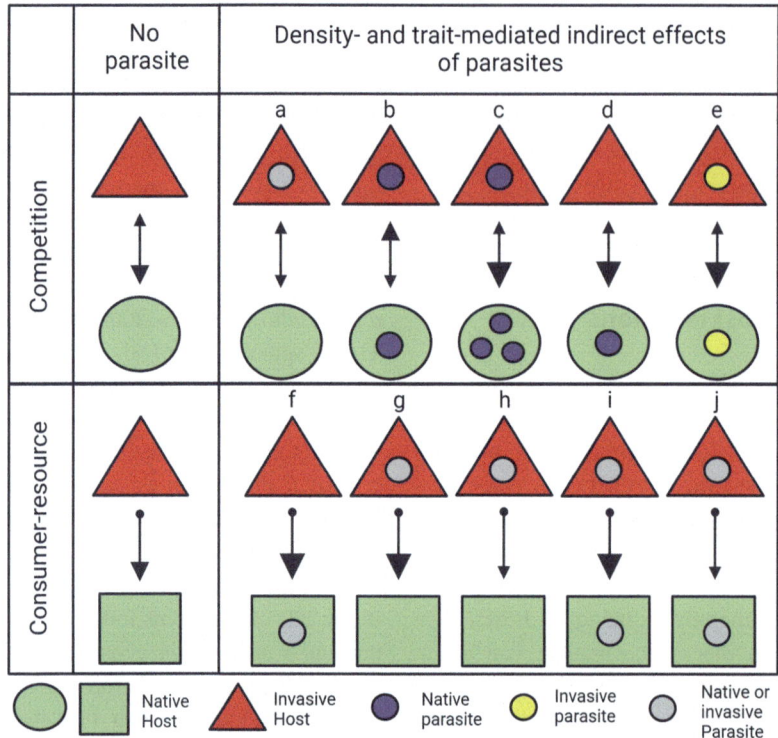

Fig. 18.5 Density- and trait-mediated indirect effects of parasites on competitive and consumer-resource interactions. For simplicity, we use a hypothetical scenario in which the native host represents the inferior competitor and the resource. However, we make the caveat that these interactions can occur in both directions, where the roles between the native and invasive hosts and the outcomes can be exchanged depending on the ecological context. Competition between invasive and native hosts is represented by a double-headed arrow, with the largest arrow pointing at the inferior competitor. Parasites can lower the competition exerted by the stronger competitor and allow co-existence (a). Also, through parasite acquisition (b), spillback (c), enemy-release (d), or spillover (e) events, the parasite can either enhance or reduce host competitive ability, thus shifting the outcome of competitive interactions. In the case of consumer-resource interactions, parasites can make resources more prone to consumption by changing specific traits of infected resources (f). Alternatively, when consumers are infected, parasites can lead to increased (g) or reduced consumption rates of consumers (h). These last two scenarios can also take place with a parasite that infects both the consumer and the resource, but has different effects on each host (i, j). The trophic interaction is represented by a one-headed arrow, with the arrow pointing towards the resource. The size of the arrowheads represents changes in the intensity or direction of the interactions

by the worms destabilise the gastropod shells, making it possible for crabs to crack larger gastropods. The presence of this polychaete hence modulates the prey–predator interaction and removes a size-related predation refuge. An example of indirect effects of invasive parasites on native species *via* infected resources is the case of the invasive parasitic barnacle *Loxothylacus panopaei* infecting the native North American flatback mud crab, *Eurypanopeus depressus*. Using a mesocosm setting, Brothers and Blakeslee (2021) evaluated the influence of parasite infection and habitat heterogeneity on the survival and escape behaviour of *E. depressus* from predatory crabs. After 24 h in homogeneous gravel-dominated habitats and under predatory pressure, nearly all infected *E. depressus* showed reduced activity inducing them to succumb to predation, while 50% of uninfected crabs managed to survive (Brothers and Blakeslee 2021). From these two examples, we can see how parasites can act as third-players in highly complex interactions between invasive and native species, by infecting and modulating the behaviour and susceptibility of resources.

Changes in the interaction between consumers and resources can also occur when the consumer is infected (Fig. 18.5). This type of mediation can be observed in the same host–parasite system just described, the parasitic invasive barnacle *L. panopaei* and the native mud crab *E. depressus* in North America. However, in this case, the functional response of the crab is affected. Infected crabs showed an eightfold decrease in maximum consumption of prey by affecting the time that it took crabs to handle prey and by delaying the predatory reaction time (i.e. the amount of time it took crabs to react to prey; Toscano et al. 2014). This resonates with the study by Brothers and Blakeslee (2021), in which infected crabs showed decreased activity. Another example is the infection of the invasive amphipod *G. pulex* in Ireland by the native acanthocephalan *Echinorhynchus truttae* (Dunn et al. 2012). This acanthocephalan induced an increase of 30% in the feeding rate of this amphipod to compensate for the parasite-induced increase in the host's metabolic demand and activity (Dick et al. 2010).

Invasive parasites and hosts can also alter the transmission dynamics of parasites and the competition between parasites in the invasion range (Fig. 18.6). This can happen through (1) changes in host densities mediated by invasive hosts or parasites, (2) invasive hosts acting as sinks and therefore diluting free-living larval stages of parasites, (3) competition between parasites within the host for space and nutrients or (4) parasite-induced changes in host immunity (Thieltges and Goedknegt 2023). To explain this, we will go back to some examples mentioned earlier in this chapter. First, let us focus on the impact of invasive hosts on the transmission of parasites in the invasion range using one of the host–parasite systems already mentioned. Invasive cane toads in Australia are associated with severe impacts on their native predators (Shine 2010). However, this invasive species might actually provide certain benefits to native species. Although cane toads serve as hosts for the invasive lungworm *R. pseudosphaerocephala*, they are resistant to the native lungworm *Rhabdias hylae*, which exclusively infects native anurans (Nelson et al. 2015a). Although native lungworms have been found in invasive toads, these nematodes are killed by the host's immune system before reaching the target site (Nelson et al. 2015a). As a result, the invasive toad acts as a sink for the native lungworms, resulting in reduced parasite

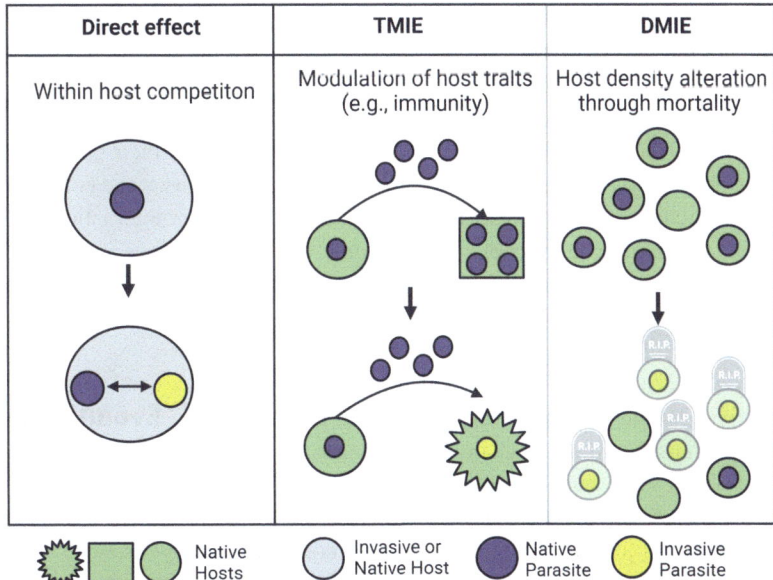

Fig. 18.6 Invasive parasites can affect the biology and transmission dynamics of native parasites. They can affect native parasites through direct competition for resources and space in the host, or through trait-mediated indirect effects (TMIE) such as immunocompetence or by reducing native host densities (Density-Mediated Indirect Effects; DMIE) *via* highly virulent host-parasite interactions (Thieltges and Goedknegt 2023)

burdens in sympatric native anurans (Lettoof et al. 2013; Nelson et al. 2015b). A similar dynamic can be observed with the invasive Pacific oyster, *M. gigas*, which exerts a dilution effect on parasites with free-swimming larval stages, such as trematodes (Thieltges et al. 2008). However, in the case of *M. gigas*, this example can be further expanded. As mentioned above, the Pacific oyster co-introduced the invasive intestinal copepod *M. orientalis* (Fig. 18.4b, c). This invasive parasite is a generalist parasite with a wide host range and has spilled over to native bivalves (Goedknegt et al. 2017). In contrast, the specialist and invasive copepod *M. intestinalis* is known to infect mostly blue mussels, thus having a narrower host range (Elsner et al. 2011). Although there is no evidence of these two species directly competing inside the bivalve hosts, it has been observed that, since the introduction of *M. orientalis,* the infection prevalences of *M. intestinalis* have decreased. Feis et al. (2022) suggest that this dominance by *M. orientalis* can be explained by recurrent spillover events of *M. orientalis* propagules that have higher fitness compared to *M. intestinalis*. Moreover, the invasive Pacific oyster may act as a sink host for *M. intestinalis*, thereby diluting infective propagules of this species (Goedknegt et al. 2019).

18.4.3 Effects of Biological Invasions on Communities and Ecosystems

Direct and indirect effects of parasite and host invasions can go beyond impacts on individuals and populations and can also manifest in community- and ecosystem-level repercussions. On the one hand, these repercussions can arise from threats to the survival of functionally important organisms (Thieltges and Goedknegt 2023). Particularly, threats to the survival of a species can have implications for biodiversity, and ecosystem energetics through changes in biomass. In less severe cases, changes in the density of organisms or in some of their ecologically relevant traits can lead to alterations in community composition. These alterations can translate into changes in the stability and robustness of ecological dynamics, as well as the functionality and services provided by ecosystems (Thieltges and Goedknegt 2023).

DMIEs, through parasite-induced mortality of native hosts after spillover events, can have ecological impacts. For instance, the spillover of crayfish plague into European native white-clawed crayfish has caused local extinctions of this native crayfish (Edgerton et al. 2002, 2004). Crayfish are detritivorous, herbivorous and adaptable predators (Abrahamsson 1966; Matthews and Reynolds 1992; Huner and Lindqvist 1995) that shape benthic communities from different angles. Therefore, their extirpation from ecosystems can have drastic ecological consequences. For instance, in Ireland, local extinctions of native crayfish have resulted in an increase in macroinvertebrates, particularly planorbid snails, and the growth of macrophytes (Matthews and Reynolds 1992). An example of TMIEs is the displacement of the native California horn snail by the invasive Asian mud snail. The parasite-induced increase in the size and competitive ability of the invasive host reduces not only native gastropod populations but also the diversity of native trematode species (Torchin et al. 2005). This is due to the high specificity of native trematodes, as the more than ten species of trematodes found in native snails do not spillover to the invasive gastropod (Torchin et al. 2005). Given their complex life cycles that often involve birds, molluscs, crustaceans and fish as hosts, trematodes are present throughout several trophic levels and, as such, their extirpation can have further repercussions at the community level, such as shifts in trophic interactions and changes in standing stock biomass of the ecosystem (Torchin et al. 2005; Rosenkranz et al. 2018; Morton et al. 2021).

18.5 Evolutionary Aspects of Biological Invasions

Although an entire chapter is devoted to the evolutionary aspects of aquatic parasitology (see Chap. 11), we would like to touch briefly on some evolutionary concepts relevant to biological invasions. Biological invasions create strong selective pressures since introduced species must adapt to new challenges, such as novel environmental conditions and species interactions (Hänfling and Kollmann 2002) and native species must adapt to the changes brought upon by the invasive species (Sakai et al. 2001). After a species (parasitic and free-living) becomes invasive, it will undergo repeated reintroductions as it spreads further because, with increasing distance from the introduction point, new biotic and abiotic challenges will arise to which it must adapt. Particularly at the edge of an invasion range, invader evolution is at its fastest due to assortative mating and selection for phenotypes with the highest offspring production at low population densities (Perkins et al. 2013). These evolutionary dynamics will affect not only invasive species but also those native species that directly or indirectly interact with invaders (Sakai et al. 2001). In other words, newly arrived species will have to adapt to local conditions, and they will introduce new selective pressures on native species.

The capacity to adapt to novel environments depends on several features, such as population size and density (host and parasite), phenotypic plasticity, genetic variation and generation times (De Meester et al. 2018; Merilä and Hendry 2013). One feature that decreases the evolutionary capacity of invaders is their typically limited initial variation due to a small number of founders in the invaded range (Sakai et al. 2001). With differences in the life histories of hosts and parasites, we should also anticipate differences in their adaptation strategies. Introduced parasite populations (especially those with a low prevalence of infection) are frequently smaller than the populations of the hosts with which they were co-introduced, thus resulting in even more reduced genetic variation in the invasion range (Goedknegt et al. 2016). An example is the invasion by the nematode *Anguilllicola crassus* originating from Japan (Taraschewski et al. 1987; Fig. 18.4d). After spilling over to European eels (*Anguilla anguilla*), a significant decrease in their genetic variation compared to those infecting Japanese eels (*Anguilla japonica*) was observed (Wielgoss et al. 2008). However, reduced genetic variation has not impeded this parasite from becoming a successful invader. Furthermore, parasites can adapt through changes in their phenotypes to compensate for the low host density at the invasion front. For instance, in the example of invasive cane toads and invasive lungworms in Australia discussed above, it has been observed that lungworms developed larger eggs, larger free-living adults and infective larvae, and shorter ages to maturity at the edge of their invasion range, in order to improve transmissibility in a low-host density environment (Kelehear et al. 2012). Although phenotypic changes do not necessarily originate from genetic changes but can represent plastic responses of an organism to change (*i.e.* acclimation instead of adaptation; Merilä and Hendry 2013), this change in fitness traits of the parasite has allowed its persistence throughout its spread. In cases of parasites with complex life cycles, they may rapidly adapt by skipping a host in its lifecycle in the absence of cues for the presence of a downstream host (Poulin 2003). Parasites typically have faster generation times than the host; therefore, such drastic adaptations are plausible.

Hosts are not exempt from the necessity to adapt. Invasive mammals, birds and marine fish, for instance, generally have more efficient life characteristics than native species (Liu et al. 2017). In contrast, invasive freshwater fish are an exception, with high fecundity but long generation times, which could jeopardise their capacity for adaptation (Liu et al. 2017). Nevertheless, in mammals, a fast pace of life is also associated with carrying more pathogens than species with slower paces, which can increase the risk of microparasite co-introduction and spillover (Gibb et al. 2020). Furthermore, in amphibians, species with faster paces of life are more likely to be infected with the trematode *Ribeiroia ondatrae* (Johnson et al. 2012) and the fungus *B. dendrobatidis* (Greenberg et al. 2017), making them also likely hosts for invasive parasites (also see Chap. 4). Therefore, although having a faster pace of life and shorter generation times might be beneficial from an adaptation point of view, this could render the hosts more susceptible to parasitic infections and more likely for them to invade with their parasites.

18.6 Conclusion and Outlook

The primary goal of studying biological invasions, both of free-living and parasitic species, is to understand their effects on native ecosystems and biodiversity. This knowledge can enable us to successfully prevent and mitigate potential harm caused by them. To understand the wider impacts of parasite and host invasions on native populations, we need to collect data on population densities of native and invasive species and share it openly across organisations and national borders (Lawson et al. 2021). Baseline information about pre-existing parasite communities, both in native and invasive hosts, is also necessary for correctly connecting possible observed impacts with the arrival of a new parasite. Modern methods such as environmental DNA, metagenomics and quantitative real-time PCR can help in collecting these baseline and invasion data of both parasites and their hosts in the aquatic environment where it is otherwise challenging to do so (also see Chap. 14; Bass et al. 2015; Burge et al. 2016).

It is often impossible to eradicate invasive species once they have established in aquatic ecosystems (Green and Grosholz 2021). Prevention is thus more realistic and cost-effective than managing already established species (Dunn and Hatcher 2015). One preventative step is to perform a rigorous disease risk analysis before any animal translocation (Sainsbury and Vaughan-Higgins 2012). If prevention in the first place fails, the best approach is to catch the presence of a new invader before it establishes. To this end, a systematic wildlife health surveillance programme that includes actionable threat mitigation and invasive species management plans is recommended to minimise the response time when a threat is detected (Lawson et al. 2021). In aquatic systems, the spread and establishment of invasive species occur rapidly, and human access to monitor all habitats is limited, so detecting species before their establishment is rarely possible (Green and Grosholz 2021). For these reasons, in aquatic systems, functional eradication, where an invasive species is brought to a level at which its harmful impact is limited, is often a more realistic goal than full eradication (Green and Grosholz 2021).

Because invasive species often have more efficient life characteristics than indigenous species (Liu et al. 2017), they may be more adaptable to changing environments. Climate change (see Chap. 22) and other anthropogenic disturbances in ecosystems can favour invaders. Chronic stress from disturbance and non-optimal temperature can harm host homeostasis, including the immune system (Hing et al. 2016), and thus destabilise the balance between a host and a parasite, making hosts more susceptible to both invading and native parasites.

Aquaculture activities are an important source of invasive species (see Chap. 23). Aquaculture is expanding rapidly to meet some of the increase in the global demand for animal protein (Ray et al. 2019). Bivalve aquaculture, in particular, causes much less greenhouse gas emissions than terrestrial livestock production, and thus a transition towards it could alleviate some effects of climate change (Ray et al. 2019). Most species used in aquaculture and many of their parasites are potentially invasive, and they can also be susceptible to indigenous parasites, thus both causing an invasion risk and being at risk of infection themselves (Bouwmeester et al. 2021). Therefore, aquatic bioinvasions are a threat to food security and can hinder the transition towards more sustainable protein production.

It is our hope that this chapter has demonstrated the importance of invasive parasites in aquatic ecosystems and the urgent need to better understand and control them to minimise ecosystem impacts and food production risks.

Acknowledgements The authors would like to thank Antoine Grenier-Journé for the use of his photo in Fig. 18.4c. The following figures were created with BioRender.com: Figure 18.2 agreement number: WB27JRM978 (Díaz Morales, D. (2025) https://BioRender.com/c12t327), Fig. 18.5, agreement number: FS27JRK630 (Díaz Morales, D. (2025) https://BioRender.com/s97e785); Fig. 18.6, agreement number: RR27JRLQ2R (Díaz Morales, D. (2025) https://BioRender.com/c19f272).

References

Abrahamsson S (1966) Dynamics of an isolated population of the crayfish *Astacus astacus* Linné. Oikos 17:96–107. https://doi.org/10.2307/3564784

Amat F, Gozalbo A, Navarro JC et al (1991) Some aspects of *Artemia* biology affected by cestode parasitism. Hydrobiologia 212:39–44. https://doi.org/10.1007/BF00025985

Bass D, Stentiford GD, Littlewood DTJ, Hartikainen H (2015) Diverse applications of environmental DNA methods in parasitology. Trends Parasitol 31:499–513. https://doi.org/10.1016/j.pt.2015.06.013

Blackburn TM, Ewen JG (2017) Parasites as drivers and passengers of human-mediated biological invasions. EcoHealth 14:61–73. https://doi.org/10.1007/s10393-015-1092-6

Blackburn TM, Pyšek P, Bacher S et al (2011) A proposed unified framework for biological invasions. Trends Ecol Evol 26:333–339. https://doi.org/10.1016/j.tree.2011.03.023

Blakeslee AMH, Pochtar DL, Fowler AE et al (2021) Invasion of the body snatchers: the role of parasite introduction in host distribution and response to salinity in invaded estuaries. Proc R Soc B Biol Sci 288. https://doi.org/10.1098/rspb.2021.0703

Bommarito C, Thieltges DW, Pansch C et al (2021) Effects of first intermediate host density, host size and salinity on trematode infections in mussels of the south-western Baltic Sea. Parasitology 148:486–494. https://doi.org/10.1017/S0031182020002188

Bouwmeester MM, Goedknegt MA, Poulin R, Thieltges DW (2021) Collateral diseases: aquaculture impacts on wildlife infections. J Appl Ecol 58:453–464. https://doi.org/10.1111/1365-2664.13775

Brothers CA, Blakeslee AMH (2021) Alien vs predator play hide and seek: how habitat complexity alters parasite mediated host survival. J Exp Mar Bio Ecol 535:151488. https://doi.org/10.1016/j.jembe.2020.151488

Burge CA, Friedman CS, Getchell R et al (2016) Complementary approaches to diagnosing marine diseases: a union of the modern and the classic. Philos Trans R Soc B Biol Sci 371:1–11. https://doi.org/10.1098/rstb.2015.0207

Byers J (2000) Competition between two estuarine snails: implications for invasions of exotic species. Ecology 81:1225–1239. https://doi.org/10.2307/177203

Cezilly F, Gregoire A, Bertin A (2000) Conflict between co-occurring manipulative parasites? An experimental study of the joint influence of two acanthocephalan parasites on the behaviour of *Gammarus pulex*. Parasitology 120:625–630. https://doi.org/10.1017/s0031182099005910

Chalkowski K, Lepczyk CA, Zohdy S (2018) Parasite ecology of invasive species: conceptual framework and new hypotheses. Trends Parasitol 34:655–663. https://doi.org/10.1016/j.pt.2018.05.008

Colautti RI, Muirhead JR, Biswas RN, Macisaac HJ (2005) Realized vs apparent reduction in enemies of the European starling. Biol Invasions 7:723–732. https://doi.org/10.1007/s10530-004-0998-7

Cornet S, Sorci G, Moret Y (2010) Biological invasion and parasitism: Invaders do not suffer from physiological alterations of the acanthocephalan *Pomphorhynchus laevis*. Parasitology 137:137–147. https://doi.org/10.1017/S0031182009991077

Cuthbert RN, Pattison Z, Taylor NG et al (2021) Global economic costs of aquatic invasive alien species. Sci Total Environ 775:145238. https://doi.org/10.1016/j.scitotenv.2021.145238

Daszak P, Cunningham AA, Hyatt AD (2000) Emerging infectious diseases of wildlife - threats to biodiversity and human health. Science(80-) 287:443–449. https://doi.org/10.1126/science.287.5452.443

De Meester L, Stoks R, Brans KI (2018) Genetic adaptation as a biological buffer against climate change: potential and limitations. Integr Zool 13:372–391. https://doi.org/10.1111/1749-4877.12298

Dick JTA, Armstrong M, Clarke HC et al (2010) Parasitism may enhance rather than reduce the predatory impact of an invader. Biol Lett 6:636–638. https://doi.org/10.1098/rsbl.2010.0171

Dobson A, Lafferty KD, Kuris AM et al (2008) Homage to Linnaeus: How many parasites? How many hosts? Proc Natl Acad Sci USA 105:11482–11489. https://doi.org/10.1073/pnas.0803232105

Dunn AM (2009) Parasites and biological invasions. In: Webster JP (ed) Advances in parasitology, vol 68, 1st edn. Academic Press, Elsevier Inc., Burlington, pp 161–184. https://doi.org/10.1016/S0065-308X(08)00607-6

Dunn AM, Hatcher MJ (2015) Parasites and biological invasions: parallels, interactions, and control. Trends Parasitol 31:189–199. https://doi.org/10.1016/j.pt.2014.12.003

Dunn AM, Torchin ME, Hatcher MJ et al (2012) Indirect effects of parasites in invasions. Funct Ecol 26:1262–1274. https://doi.org/10.1111/j.1365-2435.2012.02041.x

Dunne JA, Lafferty KD, Dobson AP et al (2013) Parasites affect food web structure primarily through increased diversity and complexity. PLoS Biol 11. https://doi.org/10.1371/journal.pbio.1001579

Edgerton BF, Watt H, Becheras JM, Bonami JR (2002) An intranuclear bacilliform virus associated with near extirpation of *Austropotamobius pallipes* Lereboullet from the Nant watershed in Ardéche, France. J Fish Dis 25:523–531. https://doi.org/10.1046/j.1365-2761.2002.00395.x

Edgerton BF, Henttonen P, Jussila J et al (2004) Understanding the causes of disease in European freshwater crayfish. Conserv Biol 18:1466–1474. https://doi.org/10.1111/j.1523-1739.2004.00436.x

Elsner NO, Jacobsen S, Thieltges DW, Reise K (2011) Alien parasitic copepods in mussels and oysters of the Wadden Sea. Helgol Mar Res 65:299–307. https://doi.org/10.1007/s10152-010-0223-2

Emde S, Rueckert S, Kochmann J et al (2014) Nematode eel parasite found inside acanthocephalan cysts – a "Trojan horse" strategy? Parasit Vectors 7:1–5. https://doi.org/10.1186/s13071-014-0504-8

Feis ME, Gottschalck L, Ruf LC et al (2022) Invading the occupied niche: how a parasitic copepod of introduced oysters can expel a congener from native mussels. Front Mar Sci 9:915841. https://doi.org/10.3389/fmars.2022.915841

Fisher JAD (2010) Parasite-like associations in rocky intertidal assemblages: Implications for escalated gastropod defenses. Mar Ecol Prog Ser 399:199–209. https://doi.org/10.3354/meps08352

Foster R, Peeler E, Bojko J et al (2021) Pathogens co-transported with invasive non-native aquatic species: implications for risk analysis and legislation. NeoBiota 69:79–102. https://doi.org/10.3897/neobiota.69.71358

Georgiev B, Sánchez M, Vasileva G et al (2007) Cestode parasitism in invasive and native brine shrimps (*Artemia* spp.) as a possible factor promoting the rapid invasion of *A. franciscana* in the Mediterranean region. Parasitol Res 101:1647–1655. https://doi.org/10.1007/s00436-007-0708-3

Giari L, Fano EA, Castaldelli G et al (2020) The ecological importance of amphipod-parasite associations for aquatic ecosystems. Water 12:2429. https://doi.org/10.3390/w12092429

Gibb R, Redding DW, Chin KQ et al (2020) Zoonotic host diversity increases in human-dominated ecosystems. Nature 584:398–402. https://doi.org/10.1038/s41586-020-2562-8

Goater TM, Goater CP, Esch GW (eds) (2014) Parasitism: the diversity and ecology of animal parasites, 2nd edn. Cambridge University Press. https://doi.org/10.1017/s0031182015000645

Goedknegt MA, Feis ME, Wegner KM et al (2016) Parasites and marine invasions: ecological and evolutionary perspectives. J Sea Res 113:11–27. https://doi.org/10.1016/j.seares.2015.12.003

Goedknegt MA, Schuster AK, Buschbaum C et al (2017) Spillover but no spillback of two invasive parasitic copepods from invasive Pacific oysters (*Crassostrea gigas*) to native bivalve hosts. Biol Invasions 19:365–379. https://doi.org/10.1007/s10530-016-1285-0

Goedknegt MA, Bedolfe S, Drent J et al (2018a) Impact of the invasive parasitic copepod *Mytilicola orientalis* on native blue mussels *Mytilus edulis* in the western European Wadden Sea. Mar Biol Res 14:497–507. https://doi.org/10.1080/17451000.2018.1442579

Goedknegt MA, Shoesmith D, Jung AS et al (2018b) Trophic relationship between the invasive parasitic copepod *Mytilicola orientalis* and its native blue mussel (*Mytilus edulis*) host. Parasitology 145:814–821. https://doi.org/10.1017/S0031182017001779

Goedknegt MA, Nauta R, Markovic M et al (2019) How invasive oysters can affect parasite infection patterns in native mussels on a large spatial scale. Oecologia 190:99–112. https://doi.org/10.1007/s00442-019-04408-x

Green SJ, Grosholz ED (2021) Functional eradication as a framework for invasive species control. Front Ecol Environ 19:98–107. https://doi.org/10.1002/fee.2277

Greenberg DA, Palen WJ, Mooers A (2017) Amphibian species traits, evolutionary history and environment predict *Batrachochytrium dendrobatidis* infection patterns, but not extinction risk. Evol Appl 10:1130–1145. https://doi.org/10.1111/eva.12520

Hänfling B, Kollmann J (2002) An evolutionary perspective of biological invasions. Trends Ecol Evol 17:545–546. https://doi.org/10.1016/S0169-5347(02)02644-7

Havel JE, Kovalenko KE, Thomaz SM et al (2015) Aquatic invasive species: challenges for the future. Hydrobiologia 750:147–170. https://doi.org/10.1007/s10750-014-2166-0

Hing S, Narayan EJ, Thompson RCA, Godfrey SS (2016) The relationship between physiological stress and wildlife disease: consequences for health and conservation. Wildl Res 43:51–60. https://doi.org/10.1071/WR15183

Hoffman GL (1990) *Myxobolus cerebralis*, a worldwide cause of salmonid whirling disease. J Aquat Anim Health 2:30–37. https://doi.org/10.1577/1548-8667(1990)002<0030:MCAWCO>2.3.CO;2

Hohenadler MAA, Honka KI, Emde S et al (2018) First evidence for a possible invasional meltdown among invasive fish parasites. Sci Rep 8:15085. https://doi.org/10.1038/s41598-018-33445-4

Honka KI, Sures B (2021) Mutual adaptations between hosts and parasites determine stress levels in eels. Int J Parasitol Parasites Wildl 14:179–184. https://doi.org/10.1016/j.ijppaw.2021.02.001

Huner J, Lindqvist O (1995) Physiological adaptations of freshwater crayfishes that permit successful aquacultural enterprises. Am Zool 35:12–19. https://doi.org/10.1093/icb/35.1.12

Jeschke JM (2014) General hypotheses in invasion ecology. Divers Distrib 20:1229–1234. https://doi.org/10.1111/ddi.12258

Johnson PTJ, Rohr JR, Hoverman JT et al (2012) Living fast and dying of infection: host life history drives interspecific variation in infection and disease risk. Ecol Lett 15:235–242. https://doi.org/10.1111/j.1461-0248.2011.01730.x

Kelehear C, Webb JK, Shine R (2009) *Rhabdias pseudosphaerocephala* infection in *Bufo marinus*: lung nematodes reduce viability of metamorph cane toads. Parasitology 136(8):919–927. https://doi.org/10.1017/S0031182009006325

Kelehear C, Brown GP, Shine R (2011) Influence of lung parasites on the growth rates of free-ranging and captive adult cane toads. Oecologia 165(3):585–592. https://doi.org/10.1007/s00442-010-1836-5

Kelehear C, Brown GP, Shine R (2012) Rapid evolution of parasite life history traits on an expanding range-edge. Ecol l 15:329–337. https://doi.org/10.1111/j.1461-0248.2012.01742.x

Kelly D, Paterson R, Townsend C et al (2009) Parasite spillback: a neglected concept in invasion ecology?

Concepts Synth 90:2047–2056. https://doi.org/10.1890/08-1085.1

Kopp K, Jokela J (2007) Resistant invaders can convey benefits to native species. Oikos 116:295–301. https://doi.org/10.1111/j.2006.0030-1299.15290.x

Kvach Y, Skóra KE (2007) Metazoa parasites of the invasive round goby *Apollonia melanostoma* (*Neogobius melanostomus*) (Pallas) (Gobiidae: Osteichthyes) in the Gulf of Gdańsk, Baltic Sea, Poland: a comparison with the Black Sea. Parasitol Res 100:767–774. https://doi.org/10.1007/s00436-006-0311-z

Lafferty K, Kuris A (2002) Trophic strategies, animal diversity and body size. Trends Ecol Evol 17:507–513. https://doi.org/10.1109/SMC.2016.7844557

Lafferty KD, Porter JW, Ford SE (2004) Are diseases increasing in the ocean? Annu Rev Ecol Evol Syst 35:31–54. https://doi.org/10.1146/annurev.ecolsys.35.021103.105704

Lafferty KD, Allesina S, Arim M et al (2008) Parasites in food webs: the ultimate missing links. Ecol Lett 11:533–546. https://doi.org/10.1111/j.1461-0248.2008.01174.x

Lawson B, Neimanis A, Lavazza A et al (2021) How to start up a national wildlife health surveillance programme. Animals 11:1–12. https://doi.org/10.3390/ani11092543

Lettoof DC, Greenlees MJ, Stockwell M, Shine R (2013) Do invasive cane toads affect the parasite burdens of native Australian frogs? Int J Parasitol Parasites Wildl 2:155–164. https://doi.org/10.1016/j.ijppaw.2013.04.002

Liu C, Comte L, Olden JD (2017) Heads you win, tails you lose: Life-history traits predict invasion and extinction risk of the world's freshwater fishes. Aquat Conserv Mar Freshw Ecosyst 27:773–779. https://doi.org/10.1002/aqc.2740

Lowe S, Browne M, Boudjelas S, De Poorter M (2000) 100 of the world's worst invasive alien species: a selection from the global invasive species database. https://doi.org/10.1525/9780520948433-159

Lymbery AJ, Morine M, Kanani HG et al (2014) Co-invaders: the effects of alien parasites on native hosts. Int J Parasitol Parasites Wildl 3:171–177. https://doi.org/10.1016/j.ijppaw.2014.04.002

Martel A, Spitzen-Van Der Sluijs A, Blooi M et al (2013) *Batrachochytrium salamandrivorans* sp. nov. causes lethal chytridiomycosis in amphibians. Proc Natl Acad Sci USA 110:15325–15329. https://doi.org/10.1073/pnas.1307356110

Martel A, Blooi M, Adriaensen C et al (2014) Recent introduction of a chytrid fungus endangers Western Palearctic salamanders. Science 346(6209):630–631. https://doi.org/10.1126/science.1258268

Matthews M, Reynolds JD (1992) Ecological impact of crayfish plague in Ireland. Hydrobiologia 234:1–6. https://doi.org/10.1007/BF00010773

Merilä J, Hendry AP (2013) Climate change, adaptation, and phenotypic plasticity: the problem and the evidence. Evol Appl 7:1–14. https://doi.org/10.1111/eva.12137

Miura O, Kuris AM, Torchin ME et al (2006) Parasites alter host phenotype and may create a new ecological niche for snail hosts. Proc R Soc B 273:1323–1328. https://doi.org/10.1098/rspb.2005.3451

Morton DN, Antonino CY, Broughton FJ et al (2021) A food web including parasites for kelp forests of the Santa Barbara Channel, California. Sci Data 8:1–14. https://doi.org/10.1038/s41597-021-00880-4

Mouritsen KN, Poulin R (2005) Parasites boost biodiversity and change animal community structure by trait-mediated indirect effects. Oikos 108:344–350. https://doi.org/10.1111/j.0030-1299.2005.13507.x

Nachev M, Hohenadler M, Bröckers N et al (2024) Suitability of invasive gobies as paratenic hosts for *Pomphorhynchus* sp. Parasitol 1–8. https://doi.org/10.1017/S0031182024001197

Nelson F, Brown G, Shilton C, Shine R (2015a) Helpful invaders: Can cane toads reduce the parasite burdens of native frogs? Int J Parasitol Parasites Wildl 4:295–300. https://doi.org/10.1016/j.ijppaw.2015.05.004

Nelson FBL, Brown GP, Dubey S, Shine R (2015b) The effects of a nematode lungworm (*Rhabdias hylae*) on its natural and invasive anuran hosts. J Parasitol 101:290–296. https://doi.org/10.1645/14-657.1

Padilla DK, Williams SL (2004) Beyond ballast water: aquarium and ornamental trades as sources of invasive species in aquatic ecosystems. Front Ecol Environ 2:131–138. https://doi.org/10.1890/1540-9295(2004)002[0131:BBWAAO]2.0.CO;2

Paterson A, Gray R (1997) Host-parasite co-speciation, host switching, and missing the boat. In: Clayton DH, Moore J (eds) Host - parasite evolution: general principles & avian models. Oxford University Press, Oxford, pp 236–250. https://doi.org/10.1093/oso/9780198548935.003.0012

Perkins TA, Phillips BL, Baskett ML, Hastings A (2013) Evolution of dispersal and life history interact to drive accelerating spread of an invasive species. Ecol Lett 16:1079–1087. https://doi.org/10.1111/ele.12136

Pizzatto L, Kelehear C, Dubey S et al (2012) Host-parasite relationships during a biologic invasion: 75 years postinvasion, cane toads and sympatric Australian frogs retain separate lungworm faunas. J Wildl Dis 48:951–961. https://doi.org/10.7589/2012-02-050

Pogoda B, Jungblut S, Buck BH, Hagen W (2012) Infestation of oysters and mussels by mytilicolid copepods: differences between natural coastal habitats and two offshore cultivation sites in the German Bight. J Appl Ichthyol 28:756–765. https://doi.org/10.1111/jai.12025

Poulin R (1994) Meta-analysis of parasite-induced behavioural changes. Anim Behav 48:137–146. https://doi.org/10.1006/anbe.1994.1220

Poulin R (2003) Information about transmission opportunities triggers a life-history switch in a parasite. Evolution 57(12):2899–2903. https://doi.org/10.1554/03-378

Poulin R, Morand S (2000) The diversity of parasites. Q Rev Biol 75:277–293. https://doi.org/10.1086/393500

Poulin R, Paterson RA, Townsend CR et al (2011) Biological invasions and the dynamics of endemic diseases in freshwater ecosystems. Freshw Biol 56:676–688. https://doi.org/10.1111/j.1365-2427.2010.02425.x

Råberg L, Graham AL, Read AF (2009) Decomposing health: tolerance and resistance to parasites in animals. Philos Trans R Soc B Biol Sci 364:37–49. https://doi.org/10.1098/rstb.2008.0184

Rauque CA, Viozzi GP, Semenas LG (2003) Component population study of *Acanthocephalus tumescens* (Acanthocephala) in fishes from Lake Moreno, Argentina. Folia Parasitol (Praha) 50:72–78. https://doi.org/10.14411/fp.2003.013

Ray NE, Maguire TJ, Al-Haj AN et al (2019) Low greenhouse gas emissions from oyster aquaculture. Environ Sci Technol 53:9118–9127. https://doi.org/10.1021/acs.est.9b02965

Reier S, Sattmann H, Schwaha T et al (2019) An integrative taxonomic approach to reveal the status of the genus *Pomphorhynchus* Monticelli, 1905 (Acanthocephala: Pomphorhynchidae) in Austria. Int J Parasitol Parasites Wildl 8:145–155. https://doi.org/10.1016/j.ijppaw.2019.01.009

Richardson DM, Pyšek P, Rejmánek M et al (2000) Naturalization and invasion of alien plants: concepts and definitions. Divers Distrib 6:93–107. https://doi.org/10.1046/j.1472-4642.2000.00083.x

Rosenkranz M, Lagrue C, Poulin R, Selbach C (2018) Small snails, high productivity? Larval output of parasites from an abundant host. Freshw Biol 63:1602–1609. https://doi.org/10.1111/fwb.13189

Sainsbury AW, Vaughan-Higgins RJ (2012) Analyzing disease risks associated with translocations. Conserv Biol 26:442–452. https://doi.org/10.1111/j.1523-1739.2012.01839.x

Sakai AK, Allendorf FW, Holt JS et al (2001) The population biology of invasive species. Annu Rev Ecol Evol Syst 32:305–332. https://doi.org/10.1146/annurev.ecolsys.32.081501.114037

Sánchez MI, Georgiev BB, Green AJ (2007) Avian cestodes affect the behaviour of their intermediate host *Artemia parthenogenetica*: an experimental study. Behav Process 74:293–299. https://doi.org/10.1016/j.beproc.2006.11.002

Sarker S, Kallert DM, Hedrick RP, El-matbouli M (2015) Whirling disease revisited: pathogenesis, parasite biology and disease intervention. Dis Aquat Org 114:155–175. https://doi.org/10.3354/dao02856

Shine R (2010) The ecological impact of invasive cane toads (*Bufo marinus*) in Australia. Q Rev Biol 85:253–291. https://doi.org/10.1086/655116

Simberloff D, Von Holle B (1999) Positive interactions of nonindigenous species: invasional meltdown? Biol Invasions 1:21–32. https://doi.org/10.1023/A:1010086329619

Sures B (2011) Parasites of animals. In: Simberloff D, Rejmánek M (eds) Encyclopedia of biological invasions. University of California Press, Berkeley and Los Angeles, pp 500–503. https://doi.org/10.1525/9780520948433-112

Sures B, Boenigk J (2021) Angewandte Ökologie. In: Boenigk J (ed) Biologie: Der Begleiter in und durch das Studium. Springer, Heidelberg, pp 989–1014. https://doi.org/10.1007/978-3-662-61270-5_38

Sures B, Knopf K, Taraschewski H (1999) Development of *Anguillicola crassus* (Dracunculoidea, Anguillicolidae) in experimentally infected Balearic congers *Ariosoma balearicum* (Anguilloidea, Congridae). Dis Aquat Org 39:75–78. https://doi.org/10.3354/dao039075

Sures B, Nachev M, Pahl M et al (2017) Parasites as drivers of key processes in aquatic ecosystems: Facts and future directions. Exp Parasitol 180:141–147. https://doi.org/10.1016/j.exppara.2017.03.011

Sures B, Nachev M, Grabner D (2019) The rhine as hotspot of parasite invasions. In: Mehlhorn H, Klimpel S (eds) Parasite and disease spread by major rivers on earth. Springer Nature Switzerland AG, pp 409–429. https://doi.org/10.1007/978-3-030-29061-0_19

Taraschewski H, Moravec F, Lamah T, Anders K (1987) Distribution and morphology of two helminths recently introduced into European eel populations: *Anguillicola crassus* (Nematoda, Dracunculoidea) and *Paratenuisentis ambiguus* (Acanthocephala, Tenuisentidae). Dis Aquat Org 3:167–176. https://doi.org/10.3354/dao003167

Thieltges DW, Goedknegt MA (2023) Ecological consequences of parasite invasions. In: Bojko J, Dunn A, Blakeslee A (eds) Parasites in Biological Invasions. CAB International, Wallingford, pp 100–114. https://doi.org/10.1079/9781789248135.0007

Thieltges DW, Bordalo MD, Caballero Hernández A et al (2008) Ambient fauna impairs parasite transmission in a marine parasite-host system. Parasitology 135:1111–1116. https://doi.org/10.1017/S0031182008004526

Thieltges DW, Reise K, Prinz K, Jensen KT (2009) Invaders interfere with native parasite-host interactions. Biol Invasions 11:1421–1429. https://doi.org/10.1007/s10530-008-9350-y

Thompson KG, Barry Nehring R, Bowden DC, Wygant T (1999) Field exposure of seven species or subspecies of salmonids to *Myxobolus cerebralis* in the Colorado river, Middle park, Colorado. J Aquat Anim Health 11:312–329. https://doi.org/10.1577/1548-8667(1999)011<0312:FEOSSO>2.0.CO;2

Torchin ME, Lafferty KD, Dobson AP et al (2003) Introduced species and their missing parasites. Nature 421:628–630. https://doi.org/10.1038/nature01346

Torchin ME, Byers JE, Huspeni TC (2005) Differential parasitism of native and introduced snails: replacement of a parasite fauna. Biol Invasions 7:885–894. https://doi.org/10.1007/s10530-004-2967-6

Toscano BJ, Newsome B, Griffen BD (2014) Parasite modification of predator functional response.

Oecologia 175:345–352. https://doi.org/10.1007/s00442-014-2905-y

Welsh JE, Van Der Meer J, Brussaard CPD, Thieltges DW (2014) Inventory of organisms interfering with transmission of a marine trematode. J Mar Biol Assoc United Kingdom 94:697–702. https://doi.org/10.1017/S0025315414000034

Wielgoss S, Taraschewski H, Meyer A, Wirth T (2008) Population structure of the parasitic nematode *Anguillicola crassus*, an invader of declining North Atlantic eel stocks. Mol Ecol 17:3478–3495. https://doi.org/10.1111/j.1365-294X.2008.03855.x

Williamson M, Fitter A (1996) The varying success of invaders. Ecology 77:1661–1666. https://doi.org/10.2307/2265769

Windsor DA (1998) Controversies in parasitology: most of the species on Earth are parasites. Int J Parasitol 28:1939–1941. https://doi.org/10.1016/s0020-7519(98)00153-2

Open Access This chapter is licensed under the terms of the Creative Commons Attribution-NonCommercial-NoDerivatives 4.0 International License (http://creativecommons.org/licenses/by-nc-nd/4.0/), which permits any non-commercial use, sharing, distribution and reproduction in any medium or format, as long as you give appropriate credit to the original author(s) and the source, provide a link to the Creative Commons license and indicate if you modified the licensed material. You do not have permission under this license to share adapted material derived from this chapter or parts of it.

The images or other third party material in this chapter are included in the chapter's Creative Commons license, unless indicated otherwise in a credit line to the material. If material is not included in the chapter's Creative Commons license and your intended use is not permitted by statutory regulation or exceeds the permitted use, you will need to obtain permission directly from the copyright holder.

Aquatic Foodborne Zoonoses

19

Shokoofeh Shamsi

Abstract

For reasons such as significantly increasing human populations and medical research into the health benefits of seafood, global seafood consumption is increasing rapidly; however, seafood safety guidelines are lagging in many countries. The primary objective of this chapter is to provide information about some of the most prevalent seafood-borne parasites, including both well-recognised and frequently neglected species found in both freshwater and marine seafood products. When addressing foodborne parasitic diseases arising from aquatic animals, it is crucial to acknowledge that, unlike some other parasitic diseases, their prevalence is not necessarily limited to individuals with poor hygiene or from lower socio-economic backgrounds. They are common in developed countries as much as elsewhere, if not more. Given the impacts of global change, adopting a 'One Health' approach is paramount in ensuring safe seafood for all.

S. Shamsi (✉)
School of Agricultural, Environmental and Veterinary Sciences, Gulbali Institute, Charles Sturt University, Wagga Wagga, NSW, Australia
e-mail: sshamsi@csu.edu.au

19.1 Introduction

Aquatic parasitic zoonoses are parasites that have the potential to be transmitted to humans, often through the consumption of seafood that contain their infectious stages. This route of transmission is commonly associated with ingesting raw or undercooked seafood, which can harbour parasitic larvae or cysts.

Parasitic zoonoses are often discussed in the context of aquatic foodborne (seafood, freshwater, fish and shellfish) parasitic diseases in the literature. However, it is important to recognise that other routes of transmission can also occur for aquatic parasitic zoonoses. These include dermal transmission, where parasites can penetrate the skin of individuals who come into contact with water resources containing infective parasite stages (e.g. cercariae; see also Chap. 5). This can happen during activities such as swimming, wading or bathing in waters inhabited by zoonotic parasites. In some cases, inhalation can also serve as a route of transmission for certain aquatic parasites. For example, *Cryptosporidium* sp. present in freshwater environments can become airborne through activities such as water sports, irrigation or water-related occupational tasks (Sponseller et al. 2014).

It is important to understand and consider these alternative routes of transmission for aquatic parasitic zoonoses, as they highlight the importance of practising proper hygiene, safe food handling and maintaining appropriate water sanitation measures. By being aware of the potential transmission pathways, individuals can take necessary precautions to minimise the risk of infection and protect their health when engaging in activities involving aquatic environments.

It is also important to note that the term 'seafood' does not only refer to marine organisms used for human consumption but instead, according to the USA Food and Drug Administration (FDA; https://www.fda.gov/food/resources-you-food/seafood), includes all commercially obtained freshwater and saltwater fish as well as edible molluscs and crustaceans. Seafood products have gained significantly more popularity in more regions and continue to grow in demand (FAO 2022). They are a sought-after choice for consumers seeking delicious, nutritious and diverse food options for several reasons, such as high nutritional value, culinary versatility offering a diverse range of flavours, textures and cooking possibilities. As people become more health-conscious and seek out nutritious food options, seafood stands out as a healthier alternative to other protein sources. Unsurprisingly, seafood has gained prominence in culinary trends and has been embraced by renowned chefs, culinary influencers and food enthusiasts worldwide. The incorporation of seafood in innovative recipes, fusion cuisines and fine dining experiences has contributed to its popularity and exposure. The association of seafood with a balanced diet and potential health benefits (Rimm et al. 2018; ANON 2020) contributes to its increasing popularity among individuals who prioritise their well-being. Seafood also holds cultural significance in many regions and communities around the world (Palomino 2022; Shamsi et al. 2020) and plays a central role in traditional cuisines, rituals and celebrations, creating a deep connection to cultural heritage and identity. Advancements in transportation, aquaculture practices, logistics and trade have facilitated the availability of a wider variety of seafood products in different regions (Chai et al. 2005). Improved distribution networks and global supply chains ensure that seafood is more accessible, even in landlocked areas, allowing consumers to enjoy diverse options.

With increased consumption of seafood among humans, there is a higher risk of aquatic foodborne diseases, including those caused by parasites. Seafood can harbour various parasites, such as protists (see Chap. 2) and helminths (see Chap. 5), which can pose health risks to humans, particularly if consumed raw or undercooked.

To date, over 40 species of parasites associated with seafood, including protists, myxozoans, cestodes, trematodes, nematodes and acanthocephalans, have been reported in humans (Shamsi 2019). The aim of this chapter is to provide information about some of the most important aquatic parasitic zoonoses. This chapter focuses on parasitic infections that can be transmitted to humans through aquatic environments, including both marine and freshwater ecosystems, which have the potential to cause diseases in humans. It will highlight the major parasitic zoonoses associated with seafood consumption.

19.2 Protista

Among the protistan zoonotic parasites associated with water, *Cryptosporidium* species are the most important. The genus *Cryptosporidium* belongs to the phylum Apicomplexa and species infect the gastrointestinal tract epithelium of various vertebrate hosts, including humans, domestic animals and wildlife (see Chap. 2). Currently, there are 44 recognised species and more than 70 genotypes of *Cryptosporidium* that can infect fish, amphibians, reptiles, birds and mammals (Golomazou et al. 2021; Couso-Pérez et al. 2022). *Cryptosporidium* spp. has been recognised as the second most prevalent contributor to diarrhoea and fatalities in children globally and continues to be the sole member among diarrheal diseases lacking a reliable and successful treatment (Tamomh et al. 2021).

Waterborne transmission is a major route in the epidemiology of *Cryptosporidium* spp. Transmission of *Cryptosporidium* occurs through the faecal–oral route. Outbreaks associated with swimming pools are common. Livestock (mainly cattle) and wildlife (deer and migrating geese) contribute to the presence of zoonotic *Cryptosporidium* oocysts in natural waters such as streams and rivers (Widmer et al. 2020). Typically, humans become infected after accidentally swallowing oocysts of the parasite. This might happen when swallowing recreational water contaminated with sewage or faeces from humans or animals shedding oocysts. Another dominant source is eating uncooked food such as fruits and vegetables that had been watered with water contaminated with *Cryptosporidium* spp. oocysts. *Cryptosporidium* infection can cause symptoms such as watery diarrhoea, abdominal pain and vomiting. Although some individuals with a healthy immune system may have asymptomatic infections, immunocompromised patients can experience severe or even life-threatening cryptosporidiosis.

Understanding the epidemiology, transmission, clinical manifestations and diagnostic methods of *Cryptosporidium* infection is crucial for effective management and prevention. Ongoing research and surveillance efforts continue to contribute to our knowledge of the diverse *Cryptosporidium* species, their host range and the development of improved diagnostic tools and treatment strategies.

Another significant parasites within the Protista kingdom are *Giardia spp.* (Fig. 19.1), commonly known as travellers' disease. This parasitic infection has a global distribution and is responsible for causing chronic diarrhoea with malabsorption in humans (Ryan and Zahedi 2019). While many infections are asymptomatic, especially in children, these asymptomatic carriers can serve as major reservoirs for the parasite within the population. *Giardia duodenalis* is also prevalent among various mammals, including companion animals like dogs. An effective strategy involves treating both human and animal hosts to curb the spread of infective stages through faeces. Waterborne outbreaks, such as those resulting from the consumption of raw oysters, are common but not the exclusive mode of transmission. Chlorination of water, a common preventive measure for waterborne diseases, is not consistently effective against *Giardia* sp. due to the parasite's cysts' high resistance to chlorination. As this chapter primarily focuses on seafood-borne parasitic diseases, the discussion of *G. duodenalis*, which is a waterborne parasite, will be brief and not detailed.

19.3 Metazoa

19.3.1 Myxozoa

Myxozoans have an indirect life cycle involving lower vertebrate hosts, primarily teleostean fish and occasionally amphibians, as well as invertebrates such as annelids (see Chap. 3). Detection of infections in humans with these parasites usually occurs as an incidental finding during examination of human faecal samples collected for investigating other parasitic infections. There are several myxozoans that can infect humans, including species of *Kudoa, Unicapsula, Henneguya* and *Myxobolus* (Ohnishi et al. 2018; dos Reis et al. 2019).

Until now, only two *Kudoa* species have been documented in humans, specifically *K. septempunctata* in Korea and *K. hexapunctata* in Japan (Yoshihito et al. 2021; Sung et al. 2023). The typical presentation of food poisoning attributed to these parasites includes a brief incubation period and mild, self-manageable gastrointestinal

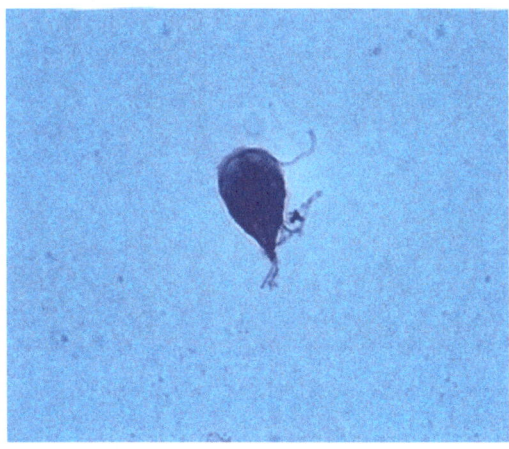

Fig. 19.1 Giemsa stained *Giardia* trophozoite

discomfort. However, it is important to acknowledge that the existing terminology and techniques for identifying *Kudoa* species may undergo significant revisions as more efficient identification methods emerge. This raises the prospect of uncovering additional species with the capacity to infect humans. The symptoms induced by these parasites in humans, such as vomiting and diarrhoea, can easily be mistaken for similar symptoms caused by other agents such as viruses or bacteria. Consequently, instances of misdiagnosis or underdiagnosis are relatively frequent. Furthermore, new hosts and broader distribution patterns continue to be uncovered on a regular basis (Shamsi and Barton 2023).

There is limited knowledge regarding infections caused by *Unicapsula* spp. These parasites might be found in the flesh of edible fish, such as amberjack (*Seriola* spp.). The symptoms closely resemble those of *Kudoa* spp. infections, primarily manifesting as diarrhoea and vomiting, and can occur within a range of 1–12 h following infection (Ohnishi et al. 2018). *Myxobolus* sp. is considered as the probable cause of diarrhoea in immunocompromised patients and has been incidentally found in patients with gastrointestinal symptoms (Boreham et al. 1998; dos Reis et al. 2019). Infection with *Henneguya* spp. causes diarrhoea and is even less commonly reported (McClelland et al. 1997).

19.4 Trematoda

All trematodes use snail or similar invertebrate intermediate hosts and infect humans through either cercariae or metacercariae (see Chap. 5). Cercariae directly penetrate the skin, leading to cercarial dermatitis. Metacercariae, on the other hand, infect humans through oral ingestion, often via contaminated fish and vegetables. Clinical disease in trematode infections is influenced by a number of factors, including worm load, site of infection and morphology of the parasite. An overview of some common zoonotic trematode parasites is provided below.

19.4.1 Lung Flukes

Paragonimus species have been documented worldwide, but their distribution at the species level tends to be localised to specific geographic regions. Over 40 species of *Paragonimus* have been documented to infect both animals and humans (Griffin et al. 2019). The most prevalent/known species is *P. westermani*, which is commonly referred to as the Oriental lung fluke (Fig. 19.2). Human infection occurs through consuming inadequately cooked or raw infected crustaceans such as a crab or crayfish bearing metacercariae of the parasite. More than 50 species of crustaceans, including freshwater crabs, can act as the intermediate host of these parasites. Within the human host, metacercariae undergo excystation in the duodenum. Subsequently, they traverse the intestinal wall into the peritoneal cavity, continue through the abdominal wall and then penetrate various sites in the body such as the brain, liver, intestines, skin, testes, striated muscles and lungs. Once in these organs, they encyst and undergo development into adult parasites. The migratory phase of the infection is usually asymptomatic. The acute phase usually starts with fever and dry cough, and can exhibit as symptoms such as diarrhoea, chest pain, abdominal pain, urticaria, hepatosplenomegaly, pulmonary irregularities and elevated levels of eosinophils in the blood (Yoshida et al. 2019). In

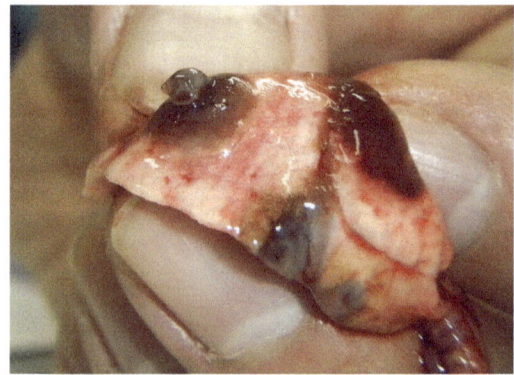

Fig. 19.2 *Paragonimus* fluke in experimentally infected mouse lungs

the chronic phase, pulmonary symptoms may include coughing, the production of discoloured sputum, haemoptysis (coughing up blood) and abnormalities visible in chest X-rays. When adult worms are found in locations outside the lungs, particularly when the brain is affected, more severe manifestations such as epileptic seizures, headaches, visual disturbances and symptoms of meningitis can occur. Risk of human infection also depends on what part of the crayfish is ingested as metacercariae of some species have preferred site in the crustacean host (Coogle et al. 2022).

19.4.2 Liver Flukes

Liver flukes belong to several different families of the order Plagiorchiida. This order includes many significant parasites of humans, many of which are considered zoonotic; these include *Clonorchis sinensis* (the Chinese liver fluke), *Opisthorchis viverrini, O. felineus, Metorchis conjunctus* and *Fasciola* spp. (Miyazaki 1991; Goldsmid et al. 1992). Human infections are a result of the metacercariae present in seafood or attached to plants (in the case of *Fasciola* spp.). A representative example of the flukes that are transmitted to humans by fish is *Clonorchis sinensis*, which is mainly found in South-East Asia. Other than humans, it has many vertebrate reservoir hosts, including members of canine and feline species. Adult parasites typically reside in the bile ducts. While most infections usually involve only a few adult parasites, there have been instances where as many as 6000 adults were present. The first intermediate hosts for these parasites are various snail species such as *Gabbia longicornis, G. fuchsiana* and *Parafossarulus manchouricus*. Secondary intermediate hosts for *Clonorchis sinensis* include over 100 species of freshwater fish and freshwater crustaceans. Individual fish may harbour up to 3000 metacercarial cysts, which become infective approximately 23 days after initial host infection. When raw or undercooked fish are consumed, the metacercarial cysts hatch within the human duodenum and the immature flukes migrate to the bile ducts, maturing within 3–4 weeks. These adult flukes can have a lifespan of up to 26 years, with an average survival of about 10 years. Self-fertilisation is the norm. The majority of infections are asymptomatic; when there are 100 to 1000 flukes in the bile duct (i.e. moderate infections), symptoms may include diarrhoea, abdominal discomfort and mild splenomegaly. Heavier infections can lead to additional symptoms, such as a sudden onset of fever, acute upper right quadrant abdominal pain, liver enlargement, tenderness, oedema and, in some cases, eventuate in carcinoma of the liver.

Opisthorchis viverrini and *O. felineus* are another two major fish-borne liver flukes that can cause serious liver damage to their definitive host. Endemic areas for *O. viverrini* include East Asia including Thailand, Cambodia and Vietnam, as far north as far-eastern Russia and Siberia and Italy for *O. felineus*. The Canadian liver fluke, *Metorchis conjunctus*, infects raw fish-eating mammals in North America such as mink, bears and wolves, occasionally being found in humans (Wobeser et al. 1983; Behr et al. 1998). The biology, pathogenesis and clinical disease of these species are similar to *Clonorchis sinensis*. These liver flukes have been recognised as strong initiators of cancer development and significant contributors to the occurrence of bile duct cancer, also known as cholangiocarcinoma.

Liver flukes belonging to the genus *Fasciola* such as *F. hepatica* and *F. gigantica* have an important relationship with water, as their cercariae hatch from lymnaeid snails and encyst on submerged water plants. Throughout the world, except Antarctica, these parasites infect ruminants and humans as final hosts. People can become infected by eating, for example, poorly-washed raw watercress or other water plants. Due to the durability of the metacercariae, transmission can also be mediated by ingestion of terrestrial plants and crops that were previously submerged in water containing infected snails, for example, due to flood irrigation of fields, as has been reported from Egypt (Grabner et al. 2014).

19.4.3 Intestinal Flukes

These flukes, including the heterophyids *Heterophyes heterophyes* and *Metagonimus yokogawai* and the fasciolid *Fasciolopsis buskii*, have been documented worldwide, with notable prevalence in regions such as Asia, the Middle East and Africa. The largest intestinal fluke of humans is *F. buskii*, which has a similar life cycle to that of other fasciolid species but lives in the intestine rather than the liver. Again, infection of humans occurs by eating raw or undercooked aquatic plants such as water chestnut, water caltrop, lotus, bamboo and other edible plants containing metacercariae. These metacercariae develop from cercariae shed by planorbid snail species of the genera *Segmentina* and *Hippeutis*, in which asexual multiplication occurs. Often, infections in humans are mild; if severe, symptoms include diarrhoea, abdominal pain, fever, ascites, anasarca and intestinal obstruction.

In contrast to *Fasciolopsis buskii*, *H. heterophyes* and *M. yokogawai* are small hermaphroditic flukes typically found attached to the walls of the small intestine, specifically in the jejunum and upper ileum of both humans and various mammals, including dogs, cats and foxes. The life cycles of these parasites involve the release of eggs through faeces into brackish or freshwater environments. Subsequently, these eggs infect snail intermediate hosts, with different flukes using different host species. For *H. heterophyes*, *Cerithidia* spp. serve as intermediate hosts, while for *M. yokogawai*, *Semisulcospira* spp. and *Thiara* spp. are involved. Cercariae emerge from the snail hosts and encyst under the skin of various freshwater animals, including fishes such as cyprinids, salmonids, mullet and tilapia, as well as frogs, tadpoles or other snails. When raw fish is consumed, the metacercariae are released and develop within the small intestine. The flukes reach maturity in approximately 15–20 days, with adult worms having a lifespan of around 2 months.

Light infection of hermaphroditic flukes in humans is usually asymptomatic. Heavy infections can be associated with nausea, intermittent chronic diarrhoea (sometimes with blood passage), abdominal discomfort accompanied by colicky pains and tenderness, mild inflammation of the mucosa at the attachment site and superficial necrosis. Eggs may pass through the intestinal wall into the mesenteric lymphatics or venules or be laid in the peritoneal cavity causing cardiac lesions and heart failure. On rare occasions, eggs have been found in the brain, along with encapsulated adult parasites (Maguire 2015).

19.4.4 Other Flukes

There are other flukes which can be transmitted to humans through the consumption of raw or undercooked infected fish. Among the most prevalent species is *Clinostomum complanatum*, a parasite that infects freshwater fish during its metacercarial stage. The metacercaria of *C. complanatum* can be found in various locations within fish, including under the skin, in the muscle, skull and other organs (Aghlmandi et al. 2018). While fish-eating birds typically serve as the natural final hosts for *Clinostomum*, occasional infections have been reported in other mammals after consuming infected fish. Despite the global distribution of the parasite, documented cases of human infections are limited to a few countries (Chai and Jung 2024). This scarcity in reported cases may be attributed, in part, to a lack of awareness among medical professionals and potential misdiagnosis.

Another group of Trematoda belong to the family Echinostomatidae comprises numerous species of spiny-collared intestinal flukes, many of which are taxonomically poorly resolved. Among these, *Echinostoma revolutum sensu lato* are the most extensively studied and recognized species (Ray et al. 2024). The definitive hosts for members of this family include a wide range of birds and mammals, including humans. These parasites inhabit the intestines of their definitive hosts, where they can cause gastrointestinal symptoms such as abdominal discomfort, diarrhoea, mucosal ulcerations, and mucosal bleeding (Chai 2015). Epidemiological studies have reported varying prevalence rates of human infections. For instance, a study conducted in Taiwan estimated the prevalence of *E. revolutum* infections to range between 2.8% and 6.5% (Yu and Mott 1994). Human cases have been documented in several countries, for example Southeast Asian countries and regions of Russia (Chai and Jung, 2024).

19.5 Cestoda

Tapeworms have indirect life cycles involving crustaceans as first intermediate hosts followed by fish as their second intermediate hosts; the infective stage, which is usually found in the second intermediate host can infect humans after consumption of undercooked infected host (see Chap. 5). The most infamous group of zoonotic aquatic tapeworms includes members of the family Diphyllobothriidae (Hernández-Orts and Scholz 2024). This family encompasses various genera and species, including the well-known *Dibothriocephalus latus* (previously known as *Diphyllobothrium latum*) and other *Dibothriocephalus* species such as *Dibo. dendriticus* and *Dibo. nihonkaiense*, species of *Diphyllobothrium* such as *Diph. balaenopterae* and *Diph. stemmacephalum*, and also *Adenocephalus pacificus* and *Spirometra* spp. (Scholz et al. 2019). Among these, *Dib. latus* is the most well-known and can reach lengths of up to 15 m, with some records indicating lengths of up to 25 m. They have a relatively long lifespan, living for up to 25 years.

Adult parasites inhabit the small intestine of various hosts, including humans, as well as dogs, bears, cats, foxes, martens, minks, seals and other wild mammals. The eggs are excreted in faeces and infect freshwater invertebrates such as copepods and freshwater fish, serving as the first and second intermediate hosts (see Chap. 5). Human infections occur when individuals consume undercooked or raw fish that is infected with these parasites. Most human infections do not display symptoms, but some individuals may experience gastrointestinal issues and vitamin B12 deficiency and in rare cases, extensive infections may result in intestinal obstruction (Schmidt et al. 2012). Members of this parasite family have been reported in various parts of the world. The global trade and consumption of wild-caught fish contribute to human cases occurring regularly outside of their naturally endemic regions.

Another group of tapeworms worthy of a brief mention are those of the order Trypanorhyncha, which are commonly found in seafood. Their larval stages can be found as large cysts embedded in both edible and non-edible parts of fish. The final hosts are usually shark and ray species. They are harmless to humans; however, heavy parasitic loads with larval stages of these parasites make the infected seafood unappealing for consumers.

19.6 Nematoda

Typical seafood-borne nematodes belong to one of the three orders of nematodes, the Ascaridida (e.g. species of the genera *Anisakis*, *Pseudoterranova*, *Contracaecum*, *Hysterothylacium* and *Eustrongylides*), Enoplida (*Capillaria* spp.) and Rhabditida (*Gnathostoma* spp. and *Echinocephalus* spp.). The common trait shared by all these nematodes (except for *Capillaria* spp.) is their life cycle (see Chap. 5) usually involving small invertebrates such as crustaceans as first intermediate hosts, fish or cephalopods as second intermediate or paratenic hosts and vertebrates (mainly mammals, birds, fish and reptiles) as definitive hosts. The critical stage of these nematodes for humans is the third larval stage (L3), which is the definitive infective stage and can be found in marine and freshwater fish and squids (Anderson 2000). Accordingly, humans can become infected with these nematodes by consuming raw or undercooked seafood contaminated with the infective larvae. The larvae can penetrate the gastrointestinal tract and cause inflammation and tissue damage, leading to symptoms such as abdominal pain, nausea, vomiting and diarrhoea. The main differences among the nematodes mentioned here are in their particular life cycle characteristics (i.e. which free-living species are used as intermediate and definitive hosts) as well as their pathogenic effects on humans.

Anisakid nematodes including species of *Anisakis* (Fig. 19.3), *Pseudoterranova* and *Contracaecum* (Fig. 19.4) show a broad range of host specificity, particularly in larval stages and are known to be able to infect other fish-eating animals, such as predatory fish and sea snakes (Shamsi et al. 2017). In humans, they cause anisakidosis, also known as anisakiasis and anisakiosis. In addition to the symptoms described above, anisakid larvae may also migrate to other organs such as liver, pancreas and testis with subsequent symptoms (Shamsi and Barton 2023). In some cases, allergic reactions can occur, manifesting as itching, rash and even anaphylaxis (Mehrdana et al. 2021). Members of the genus *Hysterothylacium* also have the potential to infect humans following consumption of infected seafood. Species of this genus are still commonly considered as anisakids in the medical literature, despite having been reclassified in the family Raphidascarididae (Deardorff and Overstreet 1981; Nadler et al. 2005; Shamsi 2014). The main difference between the life cycle of

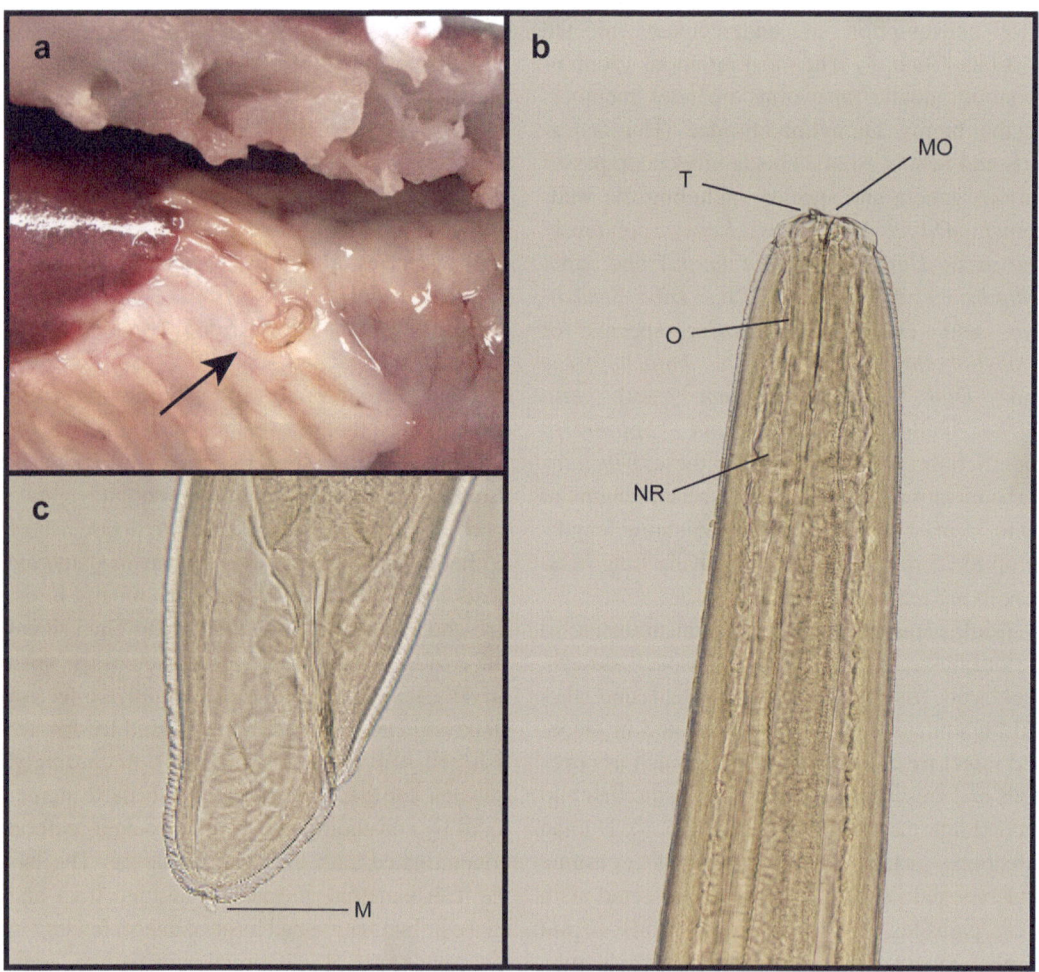

Fig. 19.3 *Anisakis* larva. (**a**) In a dissected fish (arrow points at larva located on the pyloric caecae) and the microscope images of the (**b**) anterior end with the mouth opening (MO), boring tooth (T), nerve ring (NR) and patrial oesophagus (O) and (**c**) posterior end with typical mucron (M)

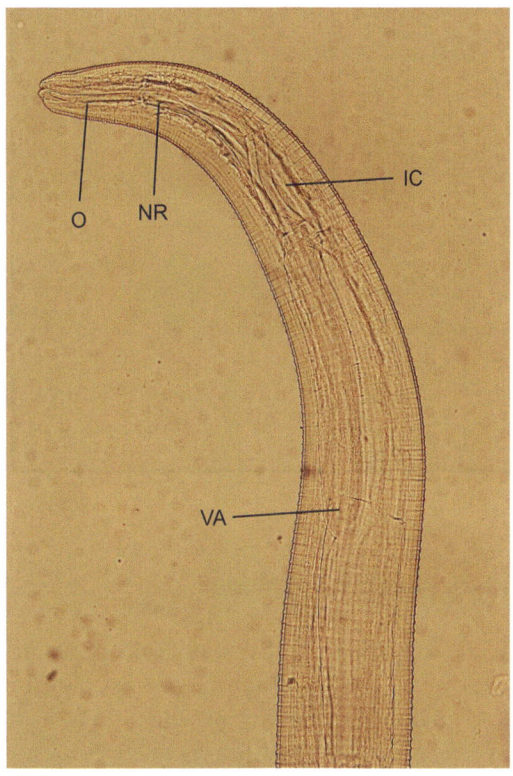

Fig. 19.4 Anterior end of a *Contracaecum* sp. larva collected from a blue mackerel *Scomber australasicus* showing details of the oesophagus (O), nerve ring (NR), intestinal caecum (IC) and ventricular appendix (VA)

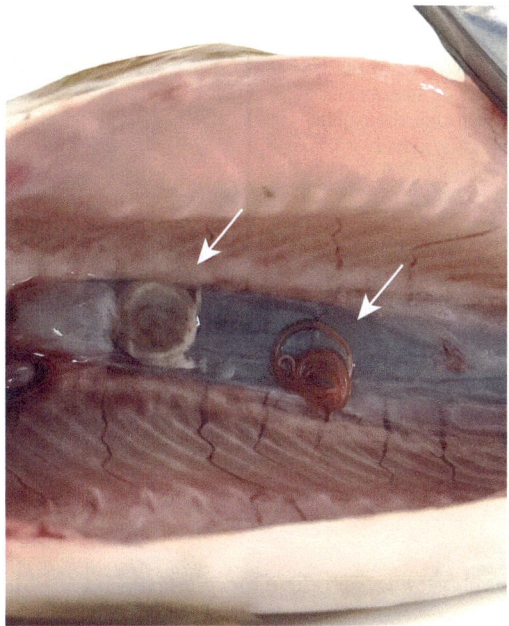

Fig. 19.5 *Eustrongylides* larva (arrows) in the body cavity of brown trout, *Salmo trutta*

Raphidascarididae and Anisakidae is their final hosts, which are large predatory fish for the former and marine mammals and birds for the latter. Despite several reports of human infections and their high prevalence and abundance in edible fish globally, there are controversial debates among authors about the zoonotic significance of species of Raphidascarididae. The definitive hosts of *Eustrongylides* spp. are piscivorous birds, while aquatic oligochaetes and fish serve as the first and second intermediate hosts, respectively (Fig. 19.5). Since Eustrongylides larvae can be found in the muscle tissue of infected fish (Shamsi et al. 2023), this parasite poses a particular zoonotic concern, especially when fish are consumed raw or undercooked. Infections with *Eustrongylides* spp. in humans can lead to symptoms like gastritis and intestinal perforation. Human infection has been reported in Sudan and the USA, after the consumption of raw fish (Eiras et al. 2018). Increased anthropogenic activities seem to lead to increased infection in fish (Lymbery et al. 2010; Shamsi et al. 2023).

Within the order Enoplida, *Paracapillaria* (formerly *Capillaria*) *philippinensis* (causing intestinal capillariasis) is the best-known aquatic species causing human infection (Kasher et al. 2022). The life cycle of Enoplida is not fully understood. Experimental studies indicate that their life cycle includes freshwater fish as intermediate hosts and piscivorous birds as definitive hosts. Embryonated eggs extracted from faeces, when ingested by fish, hatch and mature into larvae within the intestines of the fish and may migrate to the tissues. These infective larvae, when consumed by intended final hosts as well as humans, undergo further development and mature into adult parasites in the small intestine, causing severe protein-losing enteropathy. Autoinfection can occur if eggs become larvated and hatch in the intestine leading to hyperinfection in infected people.

Gnathostomiasis is the disease caused by various species of the family Gnathostomatidae (order Rhabditida), primarily of the genera

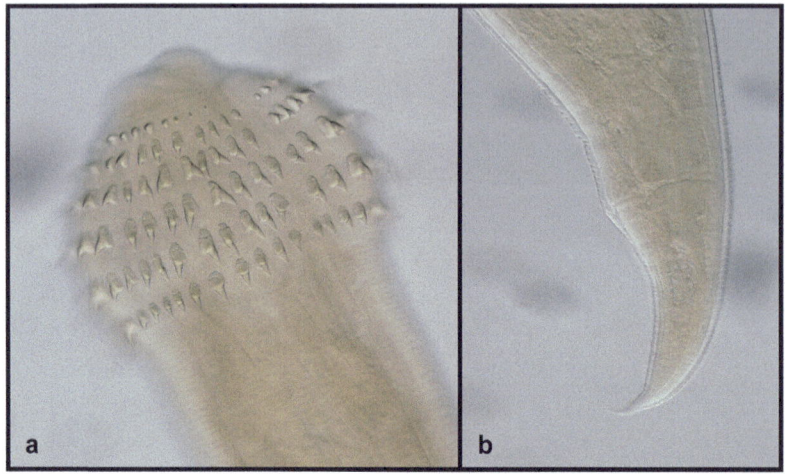

Fig. 19.6 Larval stage of *Echinocephalus* showing the row of spines on anterior end surrounding the mouth opening (**a**) and posterior end (**b**)

Gnathostoma and *Echinocephalus* (Fig. 19.6). These parasites follow an indirect life cycle and human infections typically occur when undercooked fish containing the parasites are consumed. In humans, the infection can manifest as visceral and cutaneous larval migrants, leading to symptoms like fever, urticaria, anorexia, nausea, vomiting, diarrhoea and epigastric or right upper quadrant pain, usually appearing 1–2 days after consuming infected fish.

Additionally, there is a lesser-known parasite within this group, *Echinocephalus* spp. Successful experimental infections of kittens, monkeys and puppies by *Echinocephalus sinensis* obtained from oysters suggest that, under certain conditions, *Echinocephalus* spp. could potentially infect humans. However, doubts have recently arisen regarding the accuracy and sensitivity of available diagnostic tests. The possibility of cross-reactivity between *Echinocephalus* spp. and *Gnathostoma* spp. has been raised, indicating that some cases of gnathostomiasis may actually be due to *Echinocephalus* spp. (Shamsi and Sheorey 2018; Shamsi et al. 2021). The infective stage of these nematodes is found in fish and molluscs and can infect humans when raw or undercooked seafood is consumed. Adult parasites have been discovered in various countries, including Australia and the USA.

19.7 Acanthocephala

Acanthocephalans or thorny-headed worms are parasitic in all classes of vertebrates (Smales 2015). They have heteroxenous life cycles which involve crustaceans and fish as intermediate or paratenic hosts and marine mammals as their definitive hosts (see Chap. 5). Infection of humans by acanthocephalans is rare. Among this group, two aquatic genera, *Bolbosoma* and *Corynosoma* (Fig. 19.7), are known to cause human infection. *Bolbosoma* spp. usually infect whales as adults, whereas seals are the usual final host for *Corynosoma* spp. Crustaceans are the first intermediate host for both species, and various species of marine fish are their paratenic/transport host (Habibi and Shamsi 2018; Costa et al. 2000). Several cases of human *Bolbosoma* and *Corynosoma* infection have been reported (Arizono et al. 2012; Fujita et al. 2016), mostly from Japan.

19.8 Zoonotic Aquatic Parasites and the Changing World

Human activities such as aquaculture, global trade and travel, have experienced significant acceleration in the twenty-first century, leading to

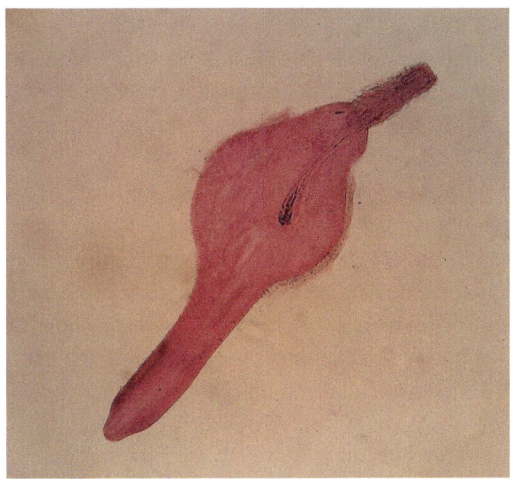

Fig. 19.7 A stained excysted *Corynosoma* larva found in tiger flathead, *Platycephalus richardsoni*

substantial changes in the populations of various parasite species, with some becoming more prominent while others have declined.

Humans share a wide range of parasites with animals, typically those whose natural host is another vertebrate. A notable area of concern in emerging zoonotic parasites is seafood-borne parasites, as well as those which are transmitted by aquatic plants, such as the trematodes mentioned above. Seafood has become a popular and healthy source of protein, with global food fish consumption increasing at a rate twice as high as population growth (FAO 2022). As reviewed above, seafood products can harbour a wide range of parasites transmissible to humans and reports of seafood-borne parasites infecting humans are on the rise.

A meta-analysis of the literature showed a 283-fold increase in *Anisakis* spp. abundance in fish in US coasts (Fiorenza et al. 2020). According to the Food and Agriculture Organisation (FAO) of the United Nations, there was a twofold increase of parasitised seafood products from 2019 to 2020 (https://www.fao.org/in-action/globefish/import-notifications/import-notifications-eu/2020-import-notifications/parasitic-causes/en/). In Japan, nearly 30 years ago, the primary sources of *Anisakis* infection were the spotted chub mackerel (*Scomber japonicus*) and Japanese flying squid (*Todarodes pacificus*) but, just a little more than a decade later, a wide range of 25 cephalopods and 200 fish species were identified as hosts for *Anisakis* spp.

This growing trend can be attributed to various factors, including global warming and changes in human behaviour. It is evident that, as our planet undergoes environmental shifts, the prevalence of these parasites is on the rise, posing significant challenges to global health. One striking aspect of this issue is the increase in zoonotic diseases derived from aquatic animals, often transmitted to humans through the consumption of undercooked seafood. Reports indicate that nearly half of these zoonotic infections can be traced back to seafood consumption (Ziarati et al. 2022). These statistics underscores the importance of ensuring proper food safety measures, especially when it comes to seafood preparation and consumption. Moreover, the global seafood trade adds another layer of complexity. The extensive movement of seafood across international borders increases the risk of spreading parasitic infections to new regions, potentially establishing these parasites in local ecosystems. Illegal, unreported, and unregulated (IUU) fishing practices further exacerbate this issue (Williams et al. 2020), as fish caught through these means often bypass proper health inspections, increasing the likelihood of contaminated products entering the market. Fish substitution and mislabelling also present significant challenges (Williams et al 2020). Consumers and even regulatory bodies may unknowingly handle species that carry higher risks of parasitic infections. This misidentification hampers traceability efforts, complicates outbreak investigations, and undermines public health interventions. Additionally, inconsistencies in fish naming conventions across different regions and markets create further barriers to effective monitoring and control of seafood-borne parasites.

It is particularly noteworthy that these zoonotic threats are not confined to developing countries or rural areas. They have made their presence felt in affluent communities and among well-off populations. The reach of these aquatic parasites serves as a stark reminder that the impact of zoonotic diseases is far-reaching and can affect consumers from all backgrounds. To effectively address this growing concern, it is crucial to adopt a 'One Health' approach, that is, an integrated approach aimed at sustainably optimising the health of plants, animals including humans and ecosystems, by taking into account the interconnectedness of aquatic animals and humans within a shared environment (Antuofermo et al. 2023). By recognising the interplay between these ecosystems, we can better understand and mitigate the potential risks to global health.

In a world marked by environmental changes, globalization, and shifting seafood consumption patterns, addressing these interconnected issues is vital for safeguarding public health and ensuring the sustainability of global fisheries resources.

19.9 Conclusion and Outlook

Beneath the appeal of seafood lies a hidden threat, that is, emerging parasites, that can pose health risks to consumers. A significant challenge persists in terms of educating everyone in the seafood safety chain about the threat of these emerging parasites and diagnosing affected consumers. Cooking seafood at high temperatures is the most efficient method to ensure the safety of the seafood, as heat effectively kills the infective stages of numerous parasites, rendering the seafood safe for consumption. Yet the real challenge lies in preventing these parasites from reaching consumers' plates in the first place.

Education plays a pivotal role in addressing this challenge. It is essential to ensure that everyone involved in the seafood supply chain, from fishermen to cooks and consumers, is well-informed about the potential risks associated with emerging parasites. Understanding how to identify and handle seafood contaminated with these parasites is critical. Moreover, early diagnosis is paramount in cases where consumers do become affected. Symptoms of parasitic infections can often mimic other illnesses, leading to misdiagnosis. Medical professionals must be educated and equipped to recognise the signs of these infections promptly. Research and development efforts should also focus on improving diagnostic tools for detecting emerging parasites in seafood. Swift and accurate testing methods can help identify contaminated batches before they reach consumers, thus preventing potential health crises.

In a world where the demand for seafood is steadily rising, ensuring seafood safety has never been more critical. It is not just about enjoying a delicious meal; it is about safeguarding our health and well-being. By fostering a culture of education and vigilance across the seafood industry and the medical community, we can better protect ourselves from the risks posed by these emerging parasites.

References

Aghlmandi F, Habibi F, Afraii MA, Abdoli A, Shamsi S (2018) Infection with metacercaria of *Clinostomum complanatum* (Trematoda: Clinostomidae) in freshwater fishes from Southern Caspian Sea Basin. Revue de Médecine Vétérinaire 7:147–151

Anderson RC (2000) Nematode parasites of vertebrates: their development and transmission, 2nd edn. CABI Publishing

ANON (2020) First-class seafood: Oh my cod! The tasty gifts of the sea. https://www.visitnorway.com/things-to-do/food-and-drink/seafood/. Accessed 5th May 2020

Antuofermo E, Polinas M, Dessì D, Henriquez FL (2023) Zoonosis associated with parasites and infectious diseases in aquatic animals. Front Vet Sci 10:1227007

Arizono N, Kuramochi T, Kagei N (2012) Molecular and histological identification of the acanthocephalan *Bolbosoma* cf. *capitatum* from the human small intestine. Parasitol Int 61(4):715–718. https://doi.org/10.1016/j.parint.2012.05.011

Behr MA, Gyorkos TW, Kokoskin E, Ward BJ, MacLean JD (1998) North American liver fluke (*Metorchis conjunctus*) in a Canadian Aboriginal population: a submerging human pathogen? Can J Public Health 89(4):258–259. https://doi.org/10.1007/BF03403931

Boreham RE, Hendrick S, O'Donoghue PJ, Stenzel DJ (1998) Incidental finding of *Myxobolus* spores (Protozoa: Myxozoa) in stool samples from patients with gastrointestinal symptoms. J Clin Microbiol 36(12):3728–3730. https://doi.org/10.1128/jcm.36.12.3728-3730.1998

Chai J-Y (2009) Echinostomes in humans. 147-183 In: The biology of echinostomes: from the molecule to the community. editors: Fried B and Toledo R. Springer New York. https://doi.org/10.1007/978-0-387-09577-6

Chai J-Y, Jung, B-K (2024) Epidemiology and geographical distribution of human trematode infections. In *Digenetic Trematodes*. Editors: Toledo R and Fried B 443–505. https://doi.org/10.1007/978-3-031-60121-7_12

Chai J-Y, Darwin Murrell K, Lymbery AJ (2005) Fish-borne parasitic zoonoses: status and issues. Int J Parasitol 35(11–12):1233–1254. https://doi.org/10.1016/j.ijpara.2005.07.013

Coogle J, Sosland S, Bahr NC (2022) A clinical review of human disease due to *Paragonimus kellicotti* in North America. *Parasitol* 149:1327–1333. https://doi.org/10.1017/S0031182021001359

Couso-Pérez S, Ares-Mazás E, Gómez-Couso H (2022) A review of the current status of *Cryptosporidium* in fish. Parasitology 149(4):444–456. https://doi.org/10.1017/S0031182022000099

Deardorff TL, Overstreet RM (1981) Review of *Hysterothylacium* and *Iheringascaris* (both previously = *Thynnascaris*) (Nematoda: Anisakidae) from the northern Gulf of Mexico. Proc Biol Soc Wash 93(4):1035–1079

dos Reis LL, de Jesus LC, Christo Fernandes OC, Barroso DE (2019) First report of *Myxobolus* (Cnidaria: Myxozoa) spores in human feces in Brazil. Acta Amaz 49(2):162–165. https://doi.org/10.1590/1809-4392201802671

Eiras JC, Pavanelli GC, Takemoto RM, Nawa Y (2018) An overview of fish-borne nematodiases among returned travelers [sic] for recent 25 years–unexpected diseases sometimes far away from the origin. Korean J Parasitol 56(3):215–227. https://doi.org/10.3347/kjp.2018.56.3.215

FAO (2022) The state of world fisheries and aquaculture 2022. Towards blue transformation. FAO, Rome. https://doi.org/10.4060/cc0461en

Fiorenza EA, Wendt CA, Dobkowski KA, King TL, Pappaionou M, Rabinowitz P, Samhouri JF, Wood CL (2020) It's a wormy world: Meta-analysis reveals several decades of change in the global abundance of the parasitic nematodes *Anisakis* spp. and *Pseudoterranova* spp. in marine fishes and invertebrates. Glob Change Biol 26(5):2854–2866. https://doi.org/10.1111/gcb.15048

Fujita T, Waga E, Kitaoka K, Imagawa T, Komatsu Y, Takanashi K, Anbo F, Anbo T, Katuki S, Ichihara S, Fujimori S, Yamasaki H, Morishima Y, Sugiyama H, Katahira H (2016) Human infection by acanthocephalan parasites belonging to the genus *Corynosoma* found from small bowel endoscopy. Parasitol Int 65(5, Part A):491–493. https://doi.org/10.1016/j.parint.2016.07.002

Goldsmid JM, Mills A, Kibel M (1992) Helminth infection. In: Campbell AGM, McIntosh N (eds) Forfar and Arneil's textbook of paediatrics, 4th edn. Churchill Livingstone, Edinburgh, pp 1538–1582

Golomazou E, Malandrakis EE, Panagiotaki P, Karanis P (2021) *Cryptosporidium* in fish: implications for aquaculture and beyond. Water Res 201:117357. https://doi.org/10.1016/j.watres.2021.117357

Grabner D, Mohamed F, Nachev M, Méabed E, Sabry AH, Sures B (2014) Invasion biology meets parasitology: a case study of parasite spill-back with Egyptian *Fasciola gigantica* in the invasive snail *Pseudosuccinea columella*. PLoS One 9(2):e88537. https://doi.org/10.1371/journal.pone.0088537

Griffin DO, Gwadz RW, Despommier DD, Hotez PJ, Knirsch CA (2019) Parasitic diseases (7th edn). Amazon Digital Services LLC - KDP Print US

Hernández-Orts JS, Scholz T (2024) Diphyllobothriidae (Broad Tapeworms), in Encyclopedia of Food Safety, Vol. 1-4, Elsevier, pp. V2-582-V582-589

Kasher C, Grossman T, Vainer J, Yanovskay A, Okopnik M, Goldstein LH, Schwartz E, Chazan B (2022) First case of imported *Capillaria philippinensis* in Israel. J Travel Med 29(1):1–2. https://doi.org/10.1093/jtm/taab132

Lymbery AJ, Hassan M, Morgan DL, Beatty SJ, Doupe RG (2010) Parasites of native and exotic freshwater fishes in south-western Australia. *J Fish Biol* 76:1770–1785. https://doi.org/10.1111/j.1095-8649.2010.02615.x

Maguire JH (2015) Trematodes (Schistosomes and liver, intestinal, and lung flukes). In: Bennett JE, Dolin R, Blaser MJ (eds) Mandell, Douglas, and Bennett's principles and practice of infectious diseases, 8th edn. W.B. Saunders, Philadelphia, pp 3216–3226. https://doi.org/10.1016/j.aquaculture.2019.734556

McClelland RS, Murphy DM, Cone DK (1997) Report of spores of *Henneguya salminicola* (Myxozoa) in human stool specimens: possible source of confusion with human spermatozoa. J Clin Microbiol 35(11):2815–2818. https://doi.org/10.1128/jcm.35.11.2815-2818.1997

Mehrdana F, Lavilla M, Kania PW, Pardo MA, Audicana MT, Longo N, Buchmann K (2021) Evidence of IgE-mediated cross-reactions between *Anisakis simplex* and *Contracaecum osculatum* proteins. Pathogens 10(8):950. https://doi.org/10.3390/pathogens10080950

Miyazaki I (1991) An illustrated book of helminthic zoonoses. International Medical Foundation of Japan, SEAMIN publication; no. 62. p 494

Nadler SA, D'Amelio S, Dailey MD, Paggi L, Siu S, Sakanari JA (2005) Molecular phylogenetics and diagnosis of *Anisakis*, *Pseudoterranova*, and *Contracaecum* from northern Pacific marine mammals. J Parasitol 91(6):1413–1429. https://doi.org/10.1645/ge-522r.1

Ohnishi T, Obara T, Arai S, Yoshinari T, Sugita-Konishi Y (2018) Quantitative analysis of *Unicapsula seriolae* in greater amberjack associated with unidentified foodborne disease. Food Hyg Saf Sci 59(1):24–29. https://doi.org/10.3358/shokueishi.59.24

Palomino E (2022) Indigenous Arctic fish skin heritage: sustainability, craft and material innovation, University of the Arts London (United Kingdom), England. PhD Dissertation

Ray M, Trinidad, M, Francis, N, Shamsi, S (2024) Characterization of *Echinostoma* spp. (Trematoda: Echinostomatidae Looss, 1899) infecting ducks in south-eastern Australia. Int J Food Microbiol 110754. https://doi.org/10.1016/j.ijfoodmicro.2024.110754

Rimm EB, Appel LJ, Chiuve SE, Djoussé L, Engler MB, Kris-Etherton PM, Mozaffarian D, Siscovick DS, Lichtenstein AH (2018) Seafood long-chain n-3 polyunsaturated fatty acids and cardiovascular disease: a science advisory from the American Heart Association. Circ 138(1):e35–e47. https://doi.org/10.1161/CIR.0000000000000574

Ryan U, Zahedi A (2019) Molecular epidemiology of giardiasis from a veterinary perspective. *Advances in parasitology* 106:209–254. https://doi.org/10.1016/bs.apar.2019.07.002

Schmidt GD, Roberts LS, Nadler SA (2012) Foundations of parasitology, McGraw Hill, London.

Scholz T, Kuchta R, Brabec J (2019) Broad tapeworms (Diphyllobothriidae), parasites of wildlife and humans: recent progress and future challenges. *International Journal for Parasitology: Parasites and Wildlife* 9:359–369

Shamsi S (2014) Recent advances in our knowledge of Australian anisakid nematodes. Int J Parasitol Parasites Wildl 3(2):178–187. https://doi.org/10.1016/j.ijppaw.2014.04.001

Shamsi S (2019) Seafood-borne parasitic diseases: a "one-health" approach is needed. Aust Fish 4(1):9. https://doi.org/10.3390/fishes4010009

Shamsi S, Barton DP (2023) A critical review of anisakidosis cases occurring globally. Parasitol Res 122:1733–1745. https://doi.org/10.1007/s00436-023-07881-9

Shamsi S, Sheorey H (2018) Seafood-borne parasitic diseases in Australia: are they rare or underdiagnosed? Intern Mede J 48(5):591–596. https://doi.org/10.1111/imj.13786

Shamsi S, Briand MJ, Justine J-L (2017) Occurrence of *Anisakis* (Nematoda: Anisakidae) larvae in unusual hosts in southern hemisphere. Parasitol Int 66(6):837–840. https://doi.org/10.1016/j.parint.2017.08.002

Shamsi S, Williams M, Mansourian Y (2020) An introduction to aboriginal fishing cultures and legacies in seafood sustainability. Sustain For 12(22):9724. https://doi.org/10.3390/su12229724

Shamsi S, Steller E, Zhu X (2021) The occurrence and clinical importance of infectious stage of *Echinocephalus* (Nematoda: Gnathostomidae) larvae in selected Australian edible fish. Parasitol Int 83:102333. https://doi.org/10.1016/j.parint.2021.102333

Shamsi S, Francis N, Masiga J, Barton DP, Zhu X, Pearce L, McLellan M (2023) Occurrence and characterisation of *Eustrongylides* species in Australian native birds and fish. Food Waterborne Parasitol 30:e00189. https://doi.org/10.1016/j.fawpar.2023.e00189

Sponseller JK, Griffiths JK, Tzipori S (2014) The evolution of respiratory cryptosporidiosis: evidence for transmission by inhalation. Clin Microbiol Rev 27(3):575–586. https://doi.org/10.1128/cmr.00115-13

Sung G-H, Park I-J, Koo HS, Park E-H, Lee M-O (2023) Molecular detection and genotype analysis of *Kudoa septempunctata* from food poisoning outbreaks in Korea. Parasites Hosts Dis 61(1):15–23. https://doi.org/10.3347/phd.22034

Tamomh AG, Agena AM, Elamin E, Suliman MA, Elmadani M, Omara AB, Musa SA (2021) Prevalence of cryptosporidiosis among children with diarrhoea under five years admitted to Kosti teaching hospital, Kosti City, Sudan. BMC Infect Dis 21(1):349

Widmer G, Carmena D, Kvac M, Chalmers RM, Kissinger JC, Xiao L, Sateriale A, Striepen B, Laurent F, Lacroix-Lamande S, Gargala G, Favennec L (2020) Update on *Cryptosporidium* spp.: highlights from the Seventh International *Giardia* and *Cryptosporidium* Conference. Parasite 27:14. https://doi.org/10.1051/parasite/2020011

Wobeser G, Runge W, Stewart RR (1983) *Metorchis conjunctus* (Cobbold, 1860) infection in wolves (*Canis lupus*), with pancreatic involvement in two animals. J Wildl Dis 19(4):353–356. https://doi.org/10.7589/0090-3558-19.4.353

Yoshida A, Doanh, PN, Maruyama, H (2019) *Paragonimus* and paragonimiasis in Asia: an update. Acta Trop 199:105074. https://doi.org/10.1016/j.actatropica.2019.105074

Yoshihito T, Tsuyoshi S, Naoki O, Eiki M (2021) A rare case of food poisoning by *Kudoa hexapunctata*. BMJ Case Rep 14(9):e246111. https://doi.org/10.1136/bcr-2021-246111

Yu S-H, Mott KE (1994) Epidemiology and morbidity of food-borne intestinal trematode infections, World Health Organization. Trop Dis Bull 91:R125–R152

Ziarati M, Zorriehzahra MJ, Hassantabar F, Mehrabi Z, Dhawan M, Sharun K, Emran TB, Dhama K, Chaicumpa W, Shamsi S (2022) Zoonotic diseases of fish and their prevention and control. Vet Q 42(1):95–118. https://doi.org/10.1080/01652176.2022.2080298

Open Access This chapter is licensed under the terms of the Creative Commons Attribution-NonCommercial-NoDerivatives 4.0 International License (http://creativecommons.org/licenses/by-nc-nd/4.0/), which permits any non-commercial use, sharing, distribution and reproduction in any medium or format, as long as you give appropriate credit to the original author(s) and the source, provide a link to the Creative Commons license and indicate if you modified the licensed material. You do not have permission under this license to share adapted material derived from this chapter or parts of it.

The images or other third party material in this chapter are included in the chapter's Creative Commons license, unless indicated otherwise in a credit line to the material. If material is not included in the chapter's Creative Commons license and your intended use is not permitted by statutory regulation or exceeds the permitted use, you will need to obtain permission directly from the copyright holder.

Environmental Parasitology: Interactions Between Parasites and Pollutants

20

Bernd Sures, Milen Nachev, Daniel Grabner, Victor Wepener, and Sonja Zimmermann

Abstract

In natural environments, organisms face anthropogenic stressors such as temperature changes, habitat alterations and pollution. Additionally, free-living organisms are often infected by parasites, which can alter their responses to pollutants. Conversely, parasites can also be affected by these pollutants. The interactions between parasites and pollutants are diverse, leading to the development of the field of environmental parasitology over the past 30 years. Today, the field has expanded to investigate a wide range of anthropogenic stressors and their interactions with parasites, including their combined effects on host health. Environmental parasitology also explores the use of parasites as indicators of pollutants and other environmental changes. This chapter focuses on pollutants as major anthropogenic stressors in aquatic ecosystems. Each pollutant has specific properties that determine its toxicity through the phases of exposure, toxicokinetics and toxicodynamics, all of which can be influenced by parasites. The chapter provides an overview of pollutant effects on parasites and their life cycles, combined effects on hosts and subsequent consequences for populations and communities. It also discusses the ecotoxicological perspective, summarising how parasites can modulate toxicity assessments and serve as indicator organisms.

B. Sures (✉)
Aquatic Ecology and Centre for Water and Environmental Research, University of Duisburg-Essen, Essen, Germany

Research Centre One Health Ruhr, Research Alliance Ruhr, University of Duisburg-Essen, Essen, Germany

Water Research Group, Unit for Environmental Sciences and Management, North-West University, Potchefstroom, South Africa
e-mail: bernd.sures@uni-due.de

M. Nachev · D. Grabner
Aquatic Ecology and Centre for Water and Environmental Research, University of Duisburg-Essen, Essen, Germany
e-mail: milen.nachev@uni-due.de; daniel.grabner@uni-due.de

V. Wepener
Water Research Group, Unit for Environmental Sciences and Management, North-West University, Potchefstroom, South Africa
e-mail: Victor.Wepener@nwu.ac.za

S. Zimmermann
Aquatic Ecology and Centre for Water and Environmental Research, University of Duisburg-Essen, Essen, Germany

Water Research Group, Unit for Environmental Sciences and Management, North-West University, Potchefstroom, South Africa
e-mail: sonja.zimmermann@uni-due.de

© The Author(s) 2025
N. J. Smit, B. Sures (eds.), *Aquatic Parasitology: Ecological and Environmental Concepts and Implications of Marine and Freshwater Parasites*, https://doi.org/10.1007/978-3-031-83903-0_20

20.1 Introduction

In the natural environment, organisms are exposed to environmental stressors of anthropogenic origin, such as changes in temperature, habitat alteration and pollution. In addition, free-living organisms can be infected by parasites, which can affect their host in a variety of ways, including alteration of the response to pollutants. Furthermore, parasites themselves can be impacted by pollutants (Fig. 20.1).

Consequently, the interactions between parasites and pollutants can be very diverse. In order to systematically investigate and describe these interactions, the field of environmental parasitology has developed over the past 30 years. The term 'Environmental Parasitology' was first used in 1997 by Lafferty, who discussed the use of parasites to assess environmental quality (Lafferty 1997). Similarly, pioneering work demonstrated a strong metal accumulation potential of various helminths (especially acanthocephalans, see Sures et al. 1994a, b, c; see also Chap. 5). Today, environmental parasitology has a much broader view, not only focusing on pollutants but also investigating the interactions of a wide range of anthropogenic environmental stressors with the occurrence of parasites as well as their combined effects on the health of their hosts (Sures et al. 2023). Furthermore, environmental parasitology focuses on the use of parasites as indicators of pollutants (e.g. accumulation indicators (e.g. Sures et al. 2017a) or other anthropogenic changes in the environment (e.g. habitat alteration, climate change; see Sures et al. 2017b; Marcogliese 2023; Chap. 22). In this chapter, we will focus on pollutants as one of the main anthropogenic stressors in freshwater and marine ecosystems worldwide.

Each pollutant has specific properties that determines its mode of action and ultimately its toxicity to organisms. These properties can be defined by three phases: (a) exposure representing the contact of a pollutant at the interface between environment and organism; (b) toxicokinetics dealing with organisms' processes

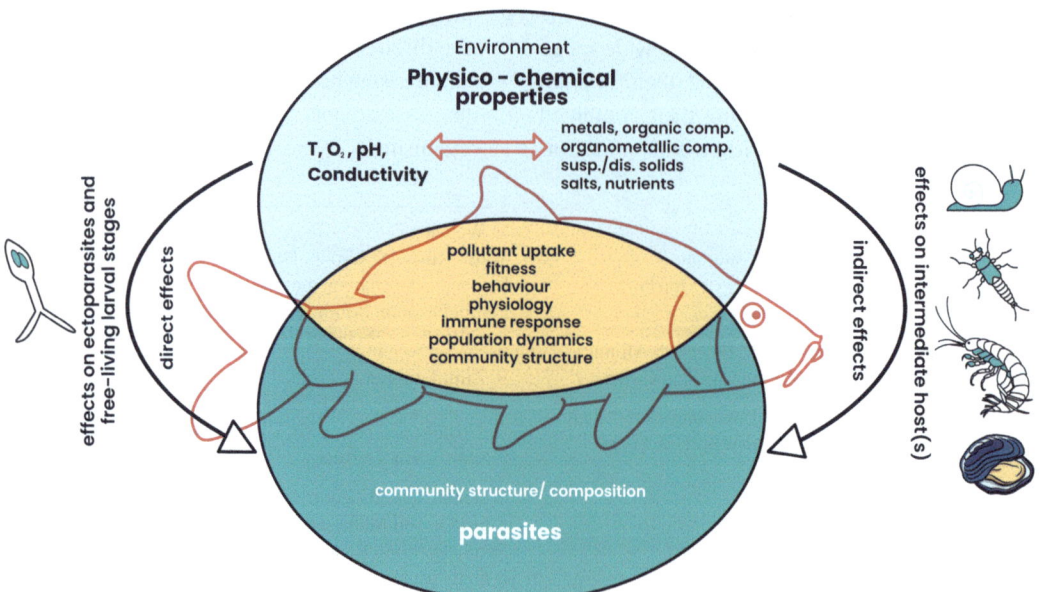

Fig. 20.1 Pollutants in aquatic habitats can directly (adult ectoparasites or free-swimming larval stages) or indirectly (adverse effects on intermediate or definitive hosts) affect the structure and composition of fish parasite communities (adapted from Sures and Nachev 2022)

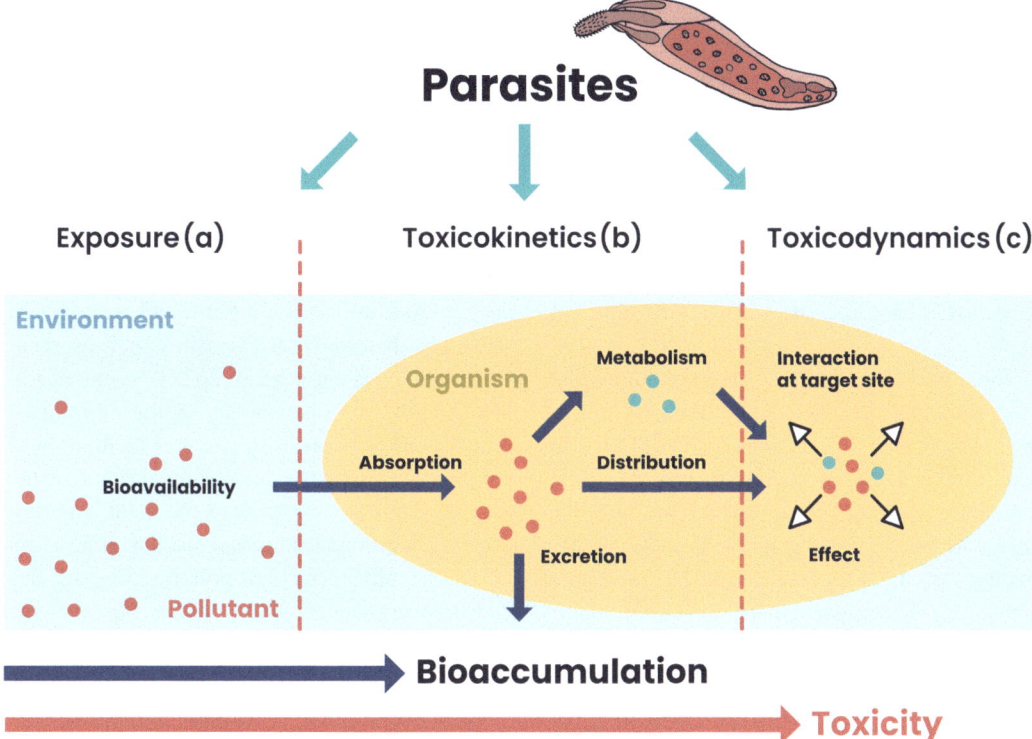

Fig. 20.2 Parasites interfere with all three basic processes in (eco-)toxicology: (a) exposure, (b) toxicokinetics and (c) toxicodynamics (adapted from Zimmermann and Sures 2023)

(absorption, distribution, metabolism, and excretion) of a pollutant over time; and (c) toxicodynamics describing the molecular, biochemical, and physiological effects of pollutants or their metabolites in biological systems. Each of these phases can be modulated by parasites (Fig. 20.2). The aim of this chapter is to provide an overview on the state of knowledge on pollutant effects on parasites and their life cycles, the combined effects of pollutants and parasites on the host organism, as well as subsequent consequences for populations and communities. We relate this to the ecotoxicological perspective (see Zimmermann and Sures 2023), summarise how parasites can modulate the outcome of field or laboratory studies aimed at assessing the toxicity of pollutants and where parasites themselves might be used as indicator organisms (Grabner and Sures 2019; Sures and Nachev 2022; Grabner et al. 2023; Sures et al. 2023).

20.2 Effects of Pollutants on Host–Parasite Systems

The ecological perspective of environmental parasitology focuses on the question of whether and how stressors influence host–parasite interactions (see Fig. 20.1). Pollutants may influence the life cycle of parasites in various ways, depending on their classification in the tree of life (e.g. protists vs. metazoans), host-specificity, life cycle complexity (monoxenous vs. heteroxenous) and their localisation on or within the host (endoparasites vs. ectoparasites). Consequently, direct and indirect effects of environmental conditions on both hosts and their parasites might lead to fluctuations in parasite population and community dynamics (Sures et al. 2023). The composition and the diversity of parasite communities can therefore be used as a measure of ecosystem health and integrity (Hudson et al. 2006; Nachev and Sures 2009; Erasmus et al. 2022a, b). Given the wide variety of adverse effects, the possibility that parasites may interact with pollutant effects leads to a multitude of interactive effects, ranging from synergistic to antagonistic outcomes (Sures et al. 2023), which is also extremely important for ecotoxicological effect studies, as they usually do not consider parasitism of test animals as an important and often co-occurring additional stressor (Grabner and Sures 2019; Grabner et al. 2023).

There are several ways in which pollutants can affect parasites, their associated hosts, as well as the host–parasite interactions (Fig. 20.1). The most direct mode of action results from the immediate contact of the parasite with the external environment (Gheorgiu et al. 2006, 2007; Thieltges et al. 2008; Sures et al. 2017b; Zhong et al. 2022). This is particularly the case for ectoparasites and free-living transmission stages of some endoparasites. For example, pollutants have been shown to negatively affect monogenean communities (Gilbert and Avenant-Oldewage 2021) as well as the transmission efficiency of trematodes, by affecting the longevity, viability and infectivity of cercarial intermediate stages (Pietrock and Marcogliese 2003; Koprivnikar et al. 2006, 2007; Morley and Lewis 2006; Rohr et al. 2008; Raffel et al. 2009; Hua et al. 2016).

Indirect effects on parasites may occur for life stages within their intermediate or definitive hosts. Host organisms can tolerate an optimum range of conditions (e.g. temperature, salinity, pollutant load). If the parameters of one or more factors are outside the optimum range, host populations might decline. As a result, the parasites cannot complete their life cycles which leads to negative effects on parasite abundance and diversity (Fig. 20.1). However, the indirect effects on parasites can be caused by alternations in host physiology and immune system impairment as a result of toxic effects of pollution (see Sect. 20.4). This scenario is in most cases beneficial for the parasites, leading to increased abundance and diversity due to the reduced host defence capabilities (e.g. Sanchez-Ramirez et al. 2007; Pravdová et al. 2021; also see Marcogliese 2004, 2005). For example, Erasmus et al. (2022b) reported a higher parasite diversity in marine fish from polluted habitats in comparison to fish from conservation areas. One underlying mechanism involves the physiological impact of pollution on hosts, which lowers their resistance to parasitic infections, as described by Pérez-del Olmo et al. (2007) for monoxenous parasites. Alternatively, parasites may increase hosts' tolerance to pollution, leading to higher infection rates with heteroxenous parasites such as acanthocephalans (Fanton et al. 2022). In practice, it is difficult to disentangle pollution-induced impairment of parasite stages, and facilitation of parasite development due to impairment of the host, as both are exposed to the same environmental conditions (Sures and Nachev 2022).

The negative effects of pollution on parasites and their associated hosts, and on host-parasite systems in general, might also represent a driving force for microevolutionary processes (adaptations to high pollution levels). Studies already demonstrated that different fish populations (without considering their infection status) can acclimate to different mercury concentrations and thus tolerate high pollution levels (Weis and Weis 1989). Similarly, local adaptation to high copper concentration from mining effluents was found in the genomes of exposed populations of the free-living flatworm *Dugesia gonocephala* (Weigand et al. 2018). But parasites can also induce transgenerational effects on their hosts. In laboratory experiments with three-spined sticklebacks (*Gasterosteus aculeatus*) and the nematode parasite *Camallanus lacustris*, offspring of infected male sticklebacks were more tolerant to infection by the nematode (Kaufmann et al. 2014). Adaptation to both trematode parasites and cadmium (Cd) contamination was found for the Manila clam (*Ruditapes philippinarum*) that showed higher resistance to parasites and pollution if the tested population was exposed to the respective stressor beforehand (Paul-Pont et al. 2010). However, the presence of parasites in a host might support the adaptation process of the host to pollution, as parasites may, on the one hand, serve as pollution sinks (see Sect. 20.3) and, on the other hand, modify the physiological response of the host (see Sects. 20.4 and 20.5) in a way which is beneficial for the host in a polluted environment.

In some instances, parasites have been found to positively affect the host's exposure to pollutants. For example, enhanced tolerance to high metal concentrations was observed in phylogenetically diverse host groups such as crustaceans (Sánchez et al. 2016), amphibians (Akinsanya et al. 2020) and birds (Morrill et al. 2019). Similar trends were also noted in amphipods, the intermediate hosts of acanthocephalans, where infected amphipods exhibited higher LC_{50} values when exposed to the insecticide deltamethrin (Kochmann et al. 2023). This suggests that organic pollutants can also accumulate in the parasite, thereby reducing toxicity for the host. On the community level, parasites can affect the structure and function of food webs, thereby possibly influencing the biomagnification of pollutants by altering trophic interactions like predation rate or activity of prey (Nachev and Sures 2016).

20.3 Parasites Affect Both the Exposure Scenarios and Pollutant Absorption of Their Hosts

Exposure to pollutants always implies that a fraction of the pollutants will be taken up by organisms. The uptake process itself, however, is by definition part of the toxicokinetics. In the following, we will combine exposure scenarios (see Fig. 20.2a) and the subsequent uptake of pollutants (see Fig. 20.2b), whereas the distribution, metabolism and elimination of pollutants within the organisms will be treated in Sect. 20.4 on toxicodynamics. The rationale behind this is that parasites may lower the pollutant exposure concentration for the host, a difference that becomes apparent only when comparing pollutant accumulation patterns in parasites and their hosts.

During the exposure phase, the pollutant present in the environment comes into contact with organisms. There are two general uptake routes of pollutants, that is, through direct contact (e.g. dermal uptake, and uptake via respiratory structures such as lungs and gills) and via food (i.e. uptake via the gastrointestinal tract). In both instances uptake involves the pollutant traversing a biological membrane. The manner and the degree to which the pollutant can move through the membrane are dependent on the nature of the pollutant (e.g. inorganic metal or organic pollutant) and its bioavailability (e.g. ionisation in the case of elements and lipid solubility in case of organic pollutants). The uptake routes and mechanisms are the same for hosts and parasites. However, since parasites are very often present on or in the uptake site of a host (e.g. on gills or in intestines), there is competition between the host and the parasite for the same substance (Sures 2002). The uptake of pollutants by host organisms can therefore be modulated by parasites that compete with their host for the uptake of the same substances and may decrease the exposure of the host to pollutants (Sures and Nachev 2022; Sures et al. 2017a).

There is limited information on the competition for pollutants between fish gills and their gill parasites, such as parasitic isopods and monogeneans, as this is a relatively new field of research. These ectoparasites, which are directly exposed to the surrounding environment and attached to the gill epithelium, might reduce pollutant uptake into the host through the gill membranes. Pérez-del Olmo et al. (2019) conducted a pioneering study on the isopod *Ceratothoa oestroides* parasitising *Boops boops* (Sparidae), focusing on metal contamination in fish ectoparasites. Following the Prestige oil spill in the English Channel in 2002, they observed temporal trends of metal pollutants in the host–parasite system from 2004 to 2006, finding higher concentrations of various toxic metals in the parasites compared to the host muscle. Similarly, Van der Spuy et al. (2023) found that the parasitic isopod *Cinusa tetrodontis*, infecting the buccal cavity of the marine puffer fish (*Amblyrhynchote honckenii*), accumulated higher element levels than the muscle and liver of the host. Although element accumulation in *C. tetrodontis* is likely due to both external exposure and uptake from the host, the infected fish generally had lower element concentrations compared to uninfected fish. This might be interpreted as direct competition for element uptake between gill tissues and parasites. It is also conceivable that reduced metal uptake through fish gills, and therefore lower accumulation rates in infected fish, occurs due to physiological changes imposed by the parasites.

In contrast to these studies, research on metal accumulation patterns in monogeneans has so far focused on freshwater fish species. Nachev et al. (2022) found higher concentrations of various toxic and essential elements in the monogenean *Mazocraes alosae* infecting the Pontic shad (*Alosa immaculata*) from the Danube River in Bulgaria. Similarly, Gilbert et al. (2022) reported higher iron (Fe) and zinc (Zn) accumulation in the monogenean *Paradiplozoon ichthyoxanthon* infecting the gills of two yellowfish species, *Labeobarbus aeneus* and *Labeobarbus kimberleyensis*, from the Vaal Dam in South Africa. These studies suggest that freshwater monogeneans primarily accumulate elements from their hosts rather than the environment, with Zn and Fe being essential elements related to their blood-feeding habits. Accordingly, monogeneans are unlikely to reduce the exposure concentration for their hosts.

Internal parasitic helminths, particularly adult acanthocephalans and cestodes, are believed to compete with their hosts for the intestinal uptake of metals, thereby disrupting the enterohepatic circulation of pollutants (e.g. non-essential metals; see Sures and Siddall 1999; Nachev and Sures 2016). Drawing from experimental studies on chub (*Squalius cephalus*) infected with the acanthocephalan *Pomphorhynchus laevis*, and existing literature, Sures and Siddall (1999) proposed a model explaining the uptake, transport and elimination of metals in parasitised fish (Fig. 20.3).

Freshwater fish experience a constant osmotic inflow of water across their gills, which are the primary site for metal uptake. Metal ions enter the bloodstream through paracellular diffusion (Hodson et al. 1978; Hofer and Lackner 1995). In the liver, most metal ions are removed from the blood and excreted into the intestine via bile (Hofer and Lackner 1995). Bile constituents forms organometallic complexes with metal ions, which can either be reabsorbed by the intestinal wall or excreted with faeces (Fig. 20.3). Host's bile production is crucial for acanthocephalans, as bile salts activate larval cystacanths and aid in cholesterol and fatty acid absorption (Kennedy et al. 1978; Nickol 1985). This suggests that organometallic complexes in the liver are absorbed by acanthocephalans in the intestine along with bile salts. Given that acanthocephalan species typically have toxic element burdens up to 100 times higher than the intestinal wall (summarised in Sures et al. 2017a), they likely compete with the intestinal wall for bioavailable heavy metals, reducing the amount reabsorbed in infected fish (Fig. 20.3). This results in reduced enterohepatic cycling of lead, making acanthocephalans significant metal sinks within the host fish. Supporting this hypothesis, many studies have shown higher element concentrations in uninfected hosts compared to infected ones (summarised in Sures et al. 2017a). Accordingly, acanthocephalans, and most likely cestodes, reduce the intra-intestinal exposure concentrations of metal ions available for intestinal absorption by their fish hosts.

Another example of competition between a parasite and its host in the host's intestine involves essential nutrients. The fish tapeworm *Dibothriocephalus latus* competes with its final host for dietary vitamin B_{12}, as demonstrated in a study where humans infected with the tapeworm were administered ^{60}Co-tagged vitamin B_{12} (Scudamore et al. 1961). Parasite-mediated dissociation of the vitamin B_{12}–intrinsic factor complex within the host's gut lumen makes vitamin B_{12} unavailable to the host, so that heavy or prolonged infection with *D. latus* can lead to vitamin B_{12} deficiency in infected humans (reviewed in Scholz et al. 2009). Furthermore, gastrointestinal parasites may also indirectly affect the pollutant uptake by changing the microbial communities in the host gut that play an important role in the biological availability of nutrients and pollutants to the host (Rooney et al. 2022).

Beyond directly lowering exposure concentrations through their uptake efficiency, parasites

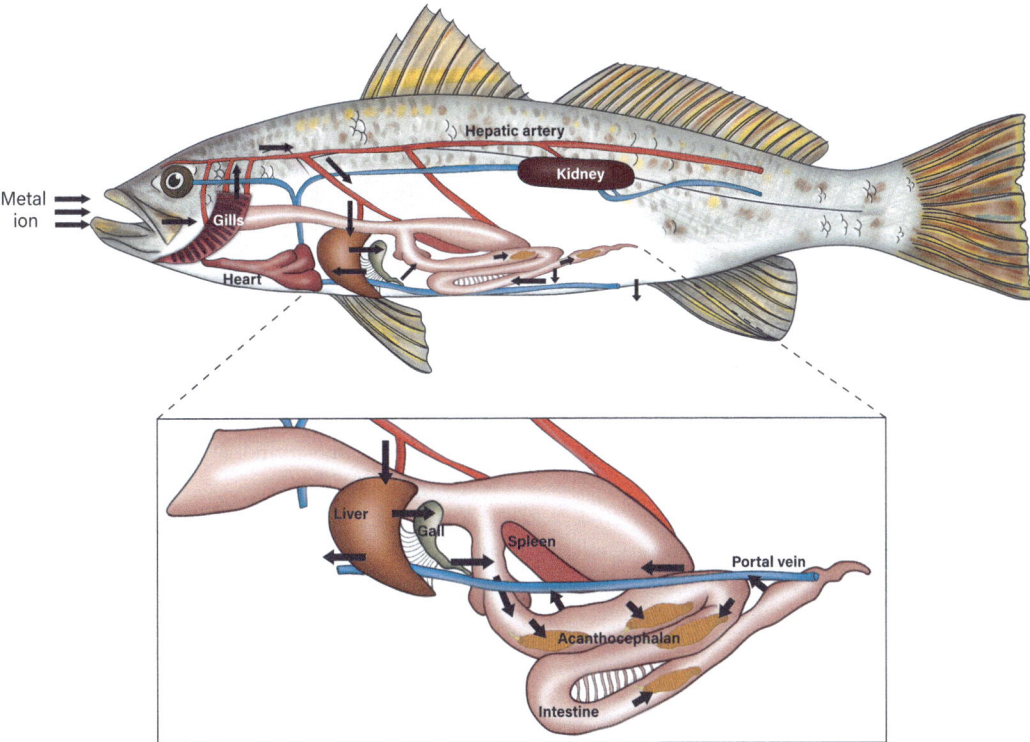

Fig. 20.3 Schematic diagram illustrating the uptake, transport, excretion, and entero-hepatic cycling of metal ions in a fish host, along with the route of metal uptake by intestinal acanthocephalans (redrawn from Sures and Siddall 1999)

can also influence the host's biological structures involved in uptake processes—such as epithelia and their cell membranes—both physically and chemically (including transmembrane channel proteins, passive and active transporters). Salmon lice (Costello 2006) and monogeneans (Gilbert and Avenant-Oldewage 2017) in gills, as well as acanthocephalans and cestodes (e.g. Taraschewski 2000; Dezfuli et al. 2011; de Sales-Ribeiro et al. 2021) in the intestinal wall, can damage the tissues of the main uptake organs in fish. This damage might facilitate the uptake of pollutants due to the partial loss of function of the epithelia that separate the outer environment from the inside of the body. The direct impact of parasite infection on the exposure concentration and uptake of pollutants by the host, that is, on the passage of substances through a host membrane or epithelium, has so far not been studied in detail. However, Nachev et al. (2010) investigated the possible impact of *Pomphorhynchus* sp. infrapopulation size in the intestine of barbel, *Barbus barbus*, on the metal accumulation in the fish host and could not find a significant correlation between these parameters, even though these parasites severely damaged the intestinal wall with their hook-bearing proboscises.

20.4 Effects of Parasites on the Toxicokinetics of Pollutants

In addition to pollutant uptake, the toxicokinetic phase (Fig. 20.2b) comprises distribution, metabolism and elimination of the pollutant by the exposed organism, and considers all processes determining the fate of a pollutant within the organism. Parasites can affect all these processes and thereby increase or decrease the level of accumulated pollutants (Sures and Siddall 2003).

20.4.1 Parasites Affecting the Distribution of Pollutants in the Body of Their Hosts

The presence of adult or larval helminths in the bile ducts might influence further transportation and distribution of the bile liquid and associated pollutants (El-Shazly et al. 2005; Caballero-Mateos et al. 2017; Wu et al. 2020). In addition, liver flukes (*Fasciola hepatica* and *F. gigantica*) that occupy the bile ducts are also known to be able to accumulate different heavy metals directly from their microhabitats and bile liquid (Sures et al. 1998; Lotfy et al. 2013). Physical damage of biological structures of the host will also be addressed later (Sect. 20.4.2), but these aspects have rarely been studied, especially in the context of pollutant transport within hosts.

20.4.2 Parasites Interact with the Metabolism of Pollutants

Within an organism, pollutants can change their chemical speciation or can be metabolised by a process called biotransformation or xenobiotic metabolism. Biotransformation of organic pollutants consists of two main phases (phase I and phase II metabolism) and involves a variety of enzymes and metabolic routes (Zimmermann and Sures 2023). Phase I metabolism consists of reduction, oxidation or hydrolysis reactions of organic substances. These reactions serve to convert lipophilic substances into more polar molecules by adding or exposing a polar functional group such as –OH or –SH. Phase I metabolism is facilitated by inducible cytochrome P450 dependent monooxygenases. In phase II, these metabolites are conjugated with a nontoxic endogenous metabolite that leads to a water-soluble, excretable product. Enzymes involved in these conjugation processes comprise, among others, glutathione S-transferase (GST), N-acetyltransferase (NAT) and sulfotransferase (SULT) (Zimmermann and Sures 2023). Although this biotransformation can take place in all tissues, the main organ is the liver in vertebrates or the hepatopancreas in invertebrates. Parasites are suspected to affect the xenobiotic metabolism of their host as they can affect the host's nutrient metabolism (Hesse et al. 2023), or significantly damage detoxifying tissues such as the liver. Examples of such tissue-dwelling parasites are different species of anisakids (see Chaps. 5 and 9), larval cestodes (e.g. plerocercoids of *Ligula intestinalis* and *Schistocephalus solidus*, see Chap. 5) but also liver flukes (i.e. adult trematodes; see Chaps. 5 and 19). Human schistosomes could also affect the pharmacokinetics of several drugs, mainly those metabolised in the liver (Wilby et al. 2013).

Besides the damaging effects at the cellular and tissue levels, parasites can have significant effects at the molecular level through the inhibition of the enzymes involved in biotransformation. For example, based on non-aquatic mammalian studies, impairment of hepatic microsomal drug metabolising activity was observed during amoebiasis in hamsters (Kumar et al. 1983), murine filariasis (Mostafa et al. 1984; Srivastava et al. 1985), bovine fascioliasis (Facino et al. 1984) and murine leishmaniasis (Coombs et al. 1990; summarised in Samanta et al. 2003). Félix and Silveira (2011) observed an inhibition of cytochrome P450 reductase in *Anopheles gambiae* infected with *Plasmodium berghei*. Furthermore, it is known from human studies that inflammation influences the cytochrome P450 activity (reviewed in Lenoir et al. 2021). The ubiquity of cytochrome P450—the most important xenobiotic-metabolizing enzyme superfamily—across organisms from bacteria to mammals, along with its presence in various parasites, including sporozoans (Wisedpanichkij et al. 2011), flagellates (Zhang et al. 2017), trematodes (Pakharukova et al. 2015), and nematodes (Laing et al. 2015), suggests that parasites themselves have active biotransformation capabilities. As parasites often cause inflammatory processes in their hosts, it stands to reason that parasites would also indirectly affect host cytochrome P450 activity. However, the mechanisms behind this effect are not well understood, with literature available on this topic solely focused on human parasites.

With respect to phase II enzymes, GSTs are very important as they modulate the conjugation of glutathione to hydrophobic ligands and oxidative stress products (Tang et al. 2019). The mechanism is based on the addition of glutathione to hydrophobic substances which leads to a complex that is less reactive and more hydrophilic. GSTs are ubiquitous and have been reported in parasites, for example, in trematodes (reviewed in Mordvinov and Pakharukova 2022). In an experimental study, it was found that laboratory-bred three-spined sticklebacks experimentally infected with *Schistocephalus solidus* (Cestoda) but not exposed to pollutants had significantly lower GST activity when compared to uninfected individuals (Frank et al. 2011). In a study assessing the effect of differently treated wastewaters on amphipods (*Gammarus fossarum*) infected with cystacanths of the acanthocephalan *Polymorphus minutus*, a slight (but non-significant) reduction in GST was found compared to uninfected amphipods (Rothe et al. 2022). Studies on the response of parasite GST to pollutants are lacking. Nevertheless, it was shown that parasites might affect biotransformation by altering GST levels of their hosts.

20.4.3 Do Parasites Influence Pollutant Elimination in Their Hosts?

No specific information is available on the effects of parasites on the direct pollutant elimination capabilities of their hosts. However, it can be assumed that parasites will influence the pollutant elimination rates of their hosts, as they significantly impair host metabolic functions (e.g. Chodkowski and Bernot 2017; Hesse et al. 2023). Additionally, by reducing the reabsorption of substances through the intestinal wall, intestinal parasites likely have a substantial impact on the host's elimination processes. A similar effect may arise from reduced GST activities in the hosts, which would lower elimination rates.

A well-known strategy for removing pollutants from an organism's metabolism involves incorporating them into biologically inert structures, followed by the potential expulsion of these structures. This elimination process is based on isomorphic substitution (e.g. Lingard et al. 1992) wherein a potentially toxic ion is incorporated into crystal structures used as inert structural components such as bones (in the case of vertebrates) or exoskeletons (e.g. molluscs, arthropods). For example, in crustaceans, regular ecdysis allows for the elimination of toxic metals by shedding the outer cuticle (Zauke 1982). To the best of our knowledge, no study has addressed whether parasitic infection might affect isomorphic substitution processes within their hosts. However, parasites themselves may have elimination strategies. Acanthocephalans, known for their high metal accumulation potential, can incorporate metals into inert structures (e.g. eggshells, hooks of the presoma), thereby removing them from their metabolism or excreting them via the cement glands or eggshells, as reported for *Moniliformis moniliformis* (Sures et al. 2000). Similar detoxification strategies exist for monogeneans, which incorporate metals into their clamps and other hard structures (Gilbert and Avenant-Oldewage 2018; Latief et al. 2023).

20.5 Effects of Parasites on Toxicodynamic Processes

Toxicodynamics focuses on the effects of pollutants on organisms, meaning that pollutants can interact at target sites and elicit biological responses, for example, by binding to hormonal receptors or changing enzyme activities or DNA (see Fig. 20.4 for further examples). These toxic effects manifest at various levels of biological organisation, from the molecular level in cells to tissues or organs, and can ultimately affect an organism's survival. Once individual organisms are severely affected, these effects may manifest at the population level, even cascading to the community and ecosystem levels (Fig. 20.4).

There are indications that parasites can modulate the toxic effects of pollutants (Sures 2008;

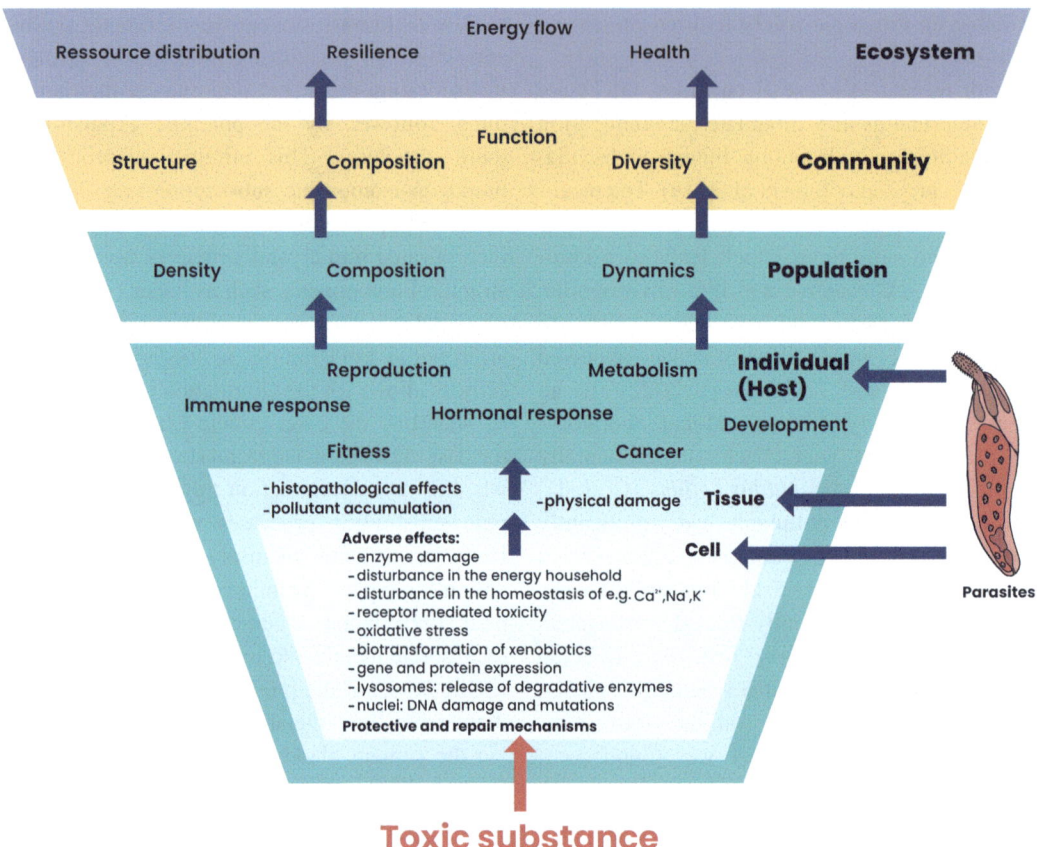

Fig. 20.4 Interactive effects of pollutants and parasites and their outcome on different levels of biological organisation (modified from Sures and Nachev 2022)

Grabner and Sures 2019; Sures and Nachev 2022; Grabner et al. 2023) in different ways, sometimes with unexpected outcomes. Moreover, parasites can also cause adverse effects by directly affecting cells, tissues and/or individuals, for example, by mechanical damage or the simple act of feeding on them. Both exposure to pollutants and parasitic infection can use the same mode of action, can lead to the same adverse effects or can trigger the same protection and repair mechanisms in the host (Fig. 20.4). Following the ecological concept of stressor interaction (Birk et al. 2020), three main types of interactive effects can be distinguished: (1) stressor dominance, that is, the effects of one stressor outweigh those of the other stressor; (2) additive effects, that is parasites and pollutants act independently of each other so that their combined effect is the sum of their individual effects; and (3) one stressor either strengthens (synergistic) or weakens (antagonistic) the effects of the other (Sures and Nachev 2022). Only a fraction of the possible pollutant–parasite combinations has been studied so far, and the mechanisms by which parasites interfere at the various levels of biological organisation are often not well understood.

20.5.1 Parasites Interfere with Pollutant Effects on the Cellular Level

Generally, the main mechanisms of toxicity of chemicals occur in cells at the molecular level, comprising reactions with enzymes, disturbance of energy balance, disturbance in ion homeostasis, overproduction of reactive oxygen species (ROS), receptor-mediated toxicity including endocrine disruption, cellular protective and repair mechanisms and genetic alterations, among others (Zimmermann and Sures 2023). Not all these molecular mechanisms have been studied to show how they are affected by parasites, and the evidence that parasites interfere with certain molecular pathways is mostly indirect. Below, we present examples of studies demonstrating how parasites interfere with specific molecular effects of pollutants.

An example of parasite modulation of ROS production was found in a field study by Molbert et al. (2020). Chub (*Squalius cephalus*) infected with the acanthocephalan parasite *Pomphorhynchus* sp. exhibited lower oxidative damage compared to uninfected fish, regardless of their pollutant load. Molbert et al. (2020) hypothesised that intestinal parasites may mitigate the production of highly reactive molecules generated from parent pollutants in the host intestine, thereby reducing their toxicity. Conversely, the ROS-defence was found to be impaired by microsporidian parasites infecting amphipods. If the parasites were present, the cadmium-induced antioxidant defence was reduced compared to uninfected individuals (Gismondi et al. 2012).

Parasites can disrupt the function of enzymes in host cells (e.g. Sánchez Di Maggio et al. 2020). Additionally, parasites might manipulate ion homeostasis to suit their needs, a phenomenon primarily studied in the malaria parasite *Plasmodium falciparum* (Kirk 2015). However, so far, no studies demonstrate parasite interference with toxicant-related effects on enzymes or subcellular ion concentrations. Similarly, while it is well documented that both parasites and pollutants can affect the host hormone system (endocrine disruption) in vertebrates and invertebrates (Jobling and Tyler 2003; Trubiroha et al. 2010; Grabner and Sures 2019; Kloas et al. 2024), there are no studies on the extent to which parasites modulate the endocrine-disrupting effects of pollutants.

Parasites can also interfere with cellular protective and repair mechanisms following cellular and subcellular damage. These mechanisms include, for example, the induction of heat shock proteins (Hsps) in response to various stressors, and metallothionein (MT) induction following metal exposure. For example, two laboratory studies examined Hsp levels after cadmium (Cd) exposure in larvae of the acanthocephalan *Polymorphus minutus* and its intermediate host *Gammarus fossarum*. In the first study by Frank et al. (2013), combined parasite infection and Cd exposure led to a decrease in Hsp70 induction in the host, whereas either parasite infection or Cd exposure alone increased Hsp70 levels. However, a second study that differentiated between male and female gammarids did not confirm these results (Chen et al. 2015). Instead, this study found that in female gammarids, the combination of Cd exposure and microsporidian (*Dictyocoela duebenum*) infection significantly increased Hsp70 levels, while male gammarids, or those infected with *P. minutus*, did not show this pattern. Additionally, infection with either parasite or Cd exposure alone had no effect on Hsp70 levels (Chen et al. 2015). The discrepancy between the two studies can possibly be explained by the fact that Frank et al. (2013) did not examine the gammarids for microsporidian infections, which can easily be overlooked, potentially resulting in undetected co-infections. In another study with *Gammarus roeselii* as the intermediate host, infection by *P. minutus* completely inhibited Hsp70 induction, regardless of palladium (Pd) exposure, whereas uninfected gammarids showed increased Hsp70 levels after Pd exposure or heat treatment (Sures and Radszuweit 2007). These studies indicate that the impact of parasites on pollution-related Hsp70 induction in their hosts depends highly on the host-parasite system and the type of pollutant. Further systematic investigations considering host age, origin of the host population and exposure conditions (e.g. reconstituted water vs. tap water) are needed to clarify the patterns of the Hsp70 response.

Metallothioneins are metal-binding proteins induced upon exposure to specific metals such as silver (Ag), Cd, copper (Cu), mercury (Hg), Pd, platinum (Pt) and Zn, among others. In laboratory studies, cockles (*Cerastoderma edule*) infected with the digenean *Himasthla elongata* showed a reduced MT response following Cd exposure (Baudrimont and De Montaudouin 2007). However, field studies in metal contaminated areas found that parasites did not affect the MT response (Filipović Marijić et al. 2013; Erasmus et al. 2020). Since parasites can act as a metal sink (Sures and Siddall 1999), they may indirectly alter MT levels in their hosts by decreasing metal accumulation within host tissues (Frank et al. 2013). However, the combined effect of parasite infection and metal exposure on MT induction in the host remains poorly understood.

20.5.2 Combined Effects of Parasites and Pollutants Manifest on Organism Level

Several studies investigated the combined effects of pollutants and parasites on host survival at the organismal level without identifying the underlying mechanisms, often revealing contrasting patterns. It was reported that the simultaneous presence of parasites and pollutants can lead to higher mortality or death at lower pollutant concentrations compared to either stressor alone. Conversely, some studies have found that infected hosts exposed to pollutants may survive longer or only die at higher pollutant concentrations compared to uninfected, pollutant-exposed organisms (Sures 2004). The first reports on the combined effects of parasites and pollutant exposure date back to 1977, when researchers found that fish infected by cestodes were more susceptible to waterborne metals than uninfected conspecifics (Boyce and Yamada 1977; Pascoe and Cram 1977). In both cases, infected fish had shorter survival periods during metal exposure experiments. Guth et al. (1977) demonstrated that patent infections of two trematode species increased the susceptibility of their intermediate host, the mollusc *Lymnaea stagnalis*, to Zn. Such detrimental effects of parasites on host health are often masked by the acute toxicity of pollutants, particularly at higher concentrations. For instance, when the amphipod *Gammarus pulex* was exposed to Cd concentrations of 0.01–1.0 mg/L, the median lethal concentrations (LC) were not significantly different between uninfected crustaceans and those infected by cystacanths of *Pomphorhynchus laevis* (McCahon et al. 1988). However, a subsequent study with the same host–parasite system showed that infected gammarids were significantly more sensitive to Cd than their uninfected counterparts after exposure to a Cd concentration of 2.1 µg/L (Brown and Pascoe 1989). Parasitism is therefore likely more important in determining toxic effects within infected populations during low-level pollutant scenarios than during high-concentration pollution.

Interestingly, antagonistic interactions between pollutants and parasites are also observed in different aquatic hosts. For example, freshwater mussels (*Pisidium amnicum*) that were partially infected with larvae of digenean trematodes exhibited significantly longer survival times when exposed to pentachlorophenol (PCP) compared to uninfected mussels, which had considerably shorter survival periods (Heinonen et al. 2001). Similarly, Sánchez et al. (2016) found that *Artemia parthenogenetica* infected with different larval cestode species displayed increased resistance to rising concentrations of arsenic. The LC_{50} levels (the concentration at which 50% of the organisms die) were significantly higher in infected *A. parthenogenetica* compared to uninfected conspecifics. Such an increased survival of infected organisms under polluted conditions could be seen as a potentially beneficial effect, contrasting with the common perception of parasites as predominantly harmful pathogens (see also Chap. 21). It might therefore be too simplistic to view parasites solely as pathogens, an issue that has also been recently highlighted in the context of the One Health concept (Selbach et al. 2022).

Many effects observed at the organismal level are mediated through reaction cascades as part of the individual stress response. This response manifests by increased stress hormone concentrations, which can trigger a series of physiological reactions such as a depressed immune response or other fundamental physiological changes (Hoole 1997; Wendelaar Bonga 1997). In a laboratory study, Sures et al. (2006) found a

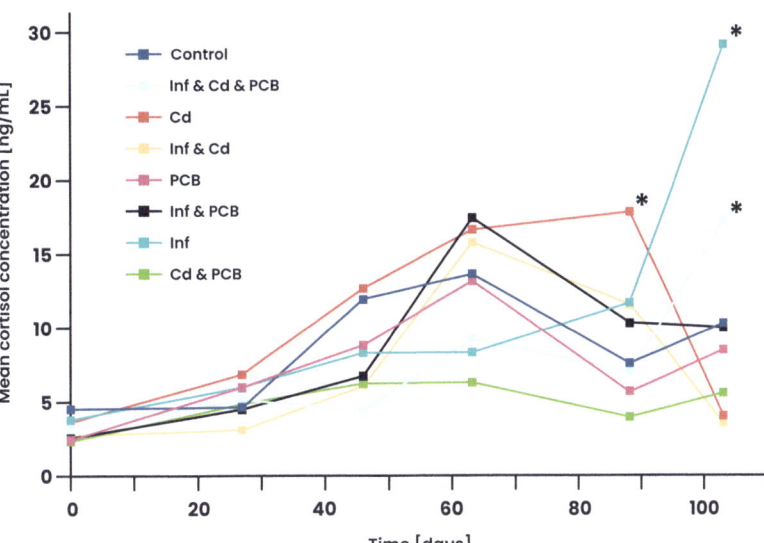

Fig. 20.5 Course of mean cortisol concentrations in eels following different treatments during laboratory exposure, with *: significant difference from controls ($p < 0.05$, U-test) (modified from Sures et al. 2006)

significant impact of parasite infection on the stress response of eels exposed to Cd and/or PCB 126 (3,3′, 4,4′,5-pentachlorobiphenyl). While exposure to the chemicals (alone or in combination) elicited only a slight increase in serum cortisol concentrations in eels, infection with the swim bladder nematode *Anguillicola crassus* resulted in significantly elevated cortisol levels (see Fig. 20.5). The fact that the highest increase in serum cortisol was observed in eels infected solely by *A. crassus* highlights both the overriding role of this parasite as a stressor and the potential for pollution effects on hosts to be exacerbated by parasite infection.

In summary, the examples presented demonstrate that the effects of parasites and pollutants can be antagonistic, additive or synergistic. Depending on their mode of interaction, combined stressor effects can often lead to severe harm, ultimately resulting in the death of the affected organism. However, the opposite can also occur, with infected hosts suffering less during pollutant exposure compared to uninfected conspecifics, leading even to a prolonged lifespan. Various other interactions between pollutants and parasites affecting toxicodynamic processes are conceivable, with potentially surprising outcomes. Accordingly, the study of combined effects between parasites and pollutants or anthropogenic stressors in the environment is an increasingly important field and should be pursued with much broader approaches in the coming years.

20.6 The Importance of Parasites in Ecotoxicology

From the information in the previous sections, it is clear that parasites themselves can take up pollutants to a certain degree (Sect. 20.3), thereby affecting pollutant uptake by their hosts. Reduced pollutant uptake results in lower levels of pollutants accumulated by infected organisms compared to uninfected conspecifics (Fig. 20.6a, b). This reduction in pollutant steady-state concentrations leads to less severe effects from the pollutants (Fig. 20.6c) and potentially beneficial effects of parasites in pollution scenarios. Additionally, both pollutants and parasites can trigger the same metabolic, toxic and protective mechanisms in the host, raising questions about the mutual impact of parasitic infection and pollutant exposure. As exemplified in Sect. 20.2, pollutants might also affect the occurrence of parasites, directly or indirectly.

All these aspects—toxicity of pollutants to parasites, pollutant accumulation in parasites, and the interactive effects of pollutants and parasites—have direct consequences for ecotoxico-

Fig. 20.6 Uptake and accumulation of pollutants in uninfected (**a**) and infected (**b**) host organisms and (**c**) associated intensity of physiological response (modified from Sures 2008 and Gunkel 1994). After pollutants are taken up, they accumulate in the organism until a steady-state concentration is reached (**a**). At this point, the pollutants are excreted if there is no further exposure. (**a**). This accumulation process typically triggers a physiological response (**c**), which may involve specific adverse effects, such as alterations in hormone levels, protein induction, and DNA damage, or more generalized impacts, like changes in heart rate, potentially leading to mortality. However, due to the significant uptake of pollutants by parasites, the steady-state concentration in the host organism may be reduced (**b**), thereby mitigating the adverse effects compared to uninfected hosts

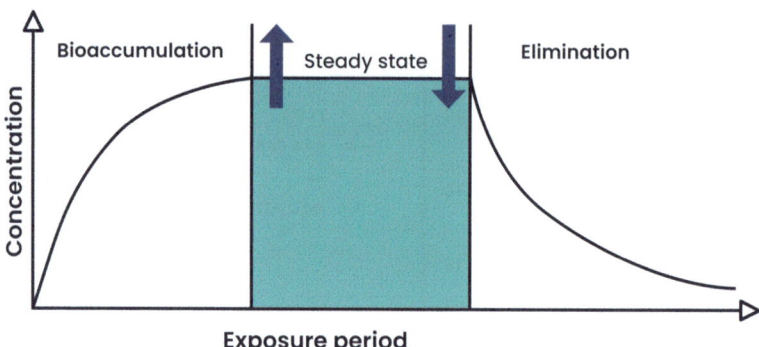

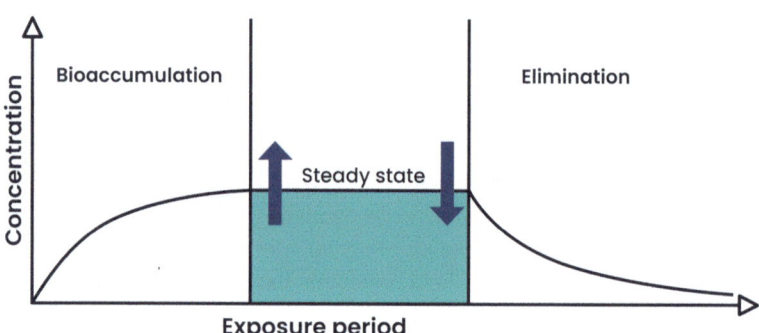

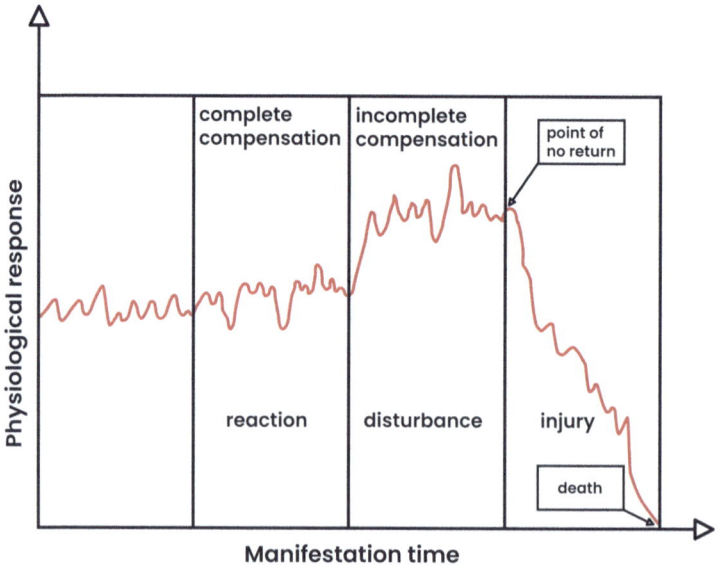

logical research in addition to their parasitological implications. Given the substantial pollutant accumulation in parasites, it has been suggested to use parasites as bioindicators (accumulation indicators) for detecting specific pollutants in different ecosystems. The sensitivity of parasites to pollutants, leading for example to the disappearance of certain species, can be interpreted as a toxic effect of pollutants resulting in changes in parasite diversity, suggesting the use of parasites as effect indicators. Finally, the interactive effects of parasites and pollution have direct implications for traditional ecotoxicological approaches. For instance, the analysis of specific enzyme activity is used as a biomarker to correlate the concentration of a pollutant in the environment with the physiological responses of the organism.

20.6.1 Parasites as Bioindicators

Similarly to free-living organisms, parasites can also be used as bioindicators, that is, organisms that provide valuable information about the quality of the environment, in which their hosts occur. Depending on their mode of response, the bioindicators can be classified as accumulation or effect indicators.

Parasites as Accumulation Bioindicators
The accumulation indication approach relies on the high accumulation potential of some parasitic taxa, which can be used to detect, quantify and study the biological availability of different pollutants. Over the last three decades, many different parasitic taxa such as acanthocephalans, cestodes, nematodes, digeneans (summarised in Sures et al. 2017a), monogeneans (e.g. Nachev et al. 2022; Gilbert et al. 2022) and isopods (Pérez-del Olmo et al. 2019; Van der Spuy et al. 2023) were studied with respect to their pollutant accumulation potential. Acanthocephalans and cestodes appear to be very promising candidates being able to accumulate various essential and non-essential elements (see for example Sures 2004; Sures et al. 2017a therein) as well as different organic pollutants (e.g. Polychlorinated Biphenyls (PCBs), Polycyclic Aromatic Hydrocarbons (PAH) and Organochlorine Pesticides (OCPs); see Brázová et al. 2012, 2021; Molbert et al. 2020, 2021), which were many folds higher than those in the host tissues or the environment. Also, nematodes, digeneans, monogeneans and crustaceans demonstrated promising bioaccumulation potential for selected elements; however, the number of studies on these taxa remain scarce, and the variety of elements they accumulate is considerably lower compared to acanthocephalans and cestodes. There is still a need for additional research addressing marine host–parasite systems, as the predominant studies to date focused mainly on freshwater habitats. The same is true for aspects related to accumulation of organic/lipophilic chemicals by parasites. Considering the vast variety of organic contaminants and their metabolites, the few available studies cannot effectively evaluate the indication potential of parasites, especially for new emerging persisting chemicals (e.g. Per- and Polyfluoroalkyl Substances (PFAS), see Giari et al. 2023) for which information about the bioavailability and distribution in the environment is largely lacking.

One of the reasons why accumulation studies have focused on selected parasitic taxa as mentioned above (acanthocephalans, cestodes and partly nematodes) is due to the criteria that suitable accumulation indicators need to fulfil. Sures (2004) summarised the most important criteria, where in addition to the high accumulation potential and resistance to pollutants, the parasites should be large in size and abundant in the host population. This is needed to obtain enough sample material for the chemical analysis. Metazoan parasites, and particularly acanthocephalans and cestodes, were hence primarily considered in most of the studies, as they are large in size and some species are quite abundant in their definitive host, so they can be easily sampled and identified during monitoring campaigns (see Sures et al. 2005; Nachev and Sures 2009; Nachev et al. 2010). Another important criterion is that the pollutant levels in parasites should correspond to those in the host environment, which can be tested and confirmed in laboratory exposure experiments as has been done with acanthocephalans (e.g. Sures and Siddall 1999; Zimmermann et al. 1999; Sures et al. 2000; Molbert et al. 2021; Le et al. 2022), and also when considering the concentrations in habitats (water, sediment and biota) where the hosts were sampled (e.g. Sures et al. 1999, 2005).

However, given the variety of free-living organisms that qualify as accumulation bioindicators, such as mussels (Wepener and Degger 2020), one might question the need to study pollutant accumulation in parasites. Collecting parasites typically requires sacrificing their hosts, often vertebrates, which complicates the process and raises ethical concerns. Despite this, direct comparisons—primarily of metal accumulation—have shown that parasites can have accumulation levels several times higher than established free-living bioindicators (Sures et al. 1999a). In other words, even when metals are not detected in the host or in traditional free-living bioindicators such as bivalves, evaluating metal occurrence in the environment is still possible through metal accumulation in acanthocephalans within their definitive hosts (Sures et al. 1999a, b). This method has been successfully used to detect trace elements in pristine environments such as Antarctica (Sures and Reimann 2003). Furthermore, the presence of contaminants in acanthocephalans indicates their general biological availability. Because acanthocephalans lack a digestive tract, any substances found within them must have traversed their teguments and membranes, confirming their bioavailability. This unique uptake mechanism, shared by cestodes, positions these gutless taxa as promising indicators for studies on the bioavailability of (nano) particles. For instance, if particulate substances or their constituent elements, such as platinum-group metals, are detected in acanthocephalans (Sures et al. 2005), they must have passed through multiple biological membranes, thereby proving their bioavailability.

Due to the high accumulation potential of some parasites, several parasite taxa might even have a sanitary function for their hosts, reducing the pollutant burdens in their tissues. Sures et al. (2017a) summarised various field and experimental studies demonstrating this, mainly for different essential and non-essential elements. However, infection by parasites can also have the opposite effect by increasing the levels of pollutants in infected host (see Jankovská et al. 2010; Oyoo-Okoth et al. 2010; Carravieri et al. 2020).

There is evidence that parasites can also act as pollution sinks for lipophilic substances (e.g. PCBs, DDT, OCPs and PAH) as has been demonstrated for acanthocephalans (Brázová et al. 2012; Kochmann et al. 2023), cestodes (Brázová et al. 2021), trematodes (Vidal-Martínez et al. 2003) and nematodes (Henríquez-Hernández et al. 2016; Mille et al. 2020). Accordingly, it is conceivable that parasites might also be used as accumulation bioindicators for lipophilic substances (Le et al. 2014).

Parasites as Effect Indicators

A parasite-related effect indicator approach is based on changes in the composition and diversity of parasite communities that reflect the environmental conditions (Hudson et al. 2006; Erasmus et al. 2022a, b). As described in Sect. 20.2, the composition and diversity of parasite communities can be directly and indirectly influenced by environmental factors, in particular by pollution. Thus, changes in the occurrence of a particular parasite species can help to detect and assess changes in the environment, when background information on their host(s) is also available, as well as changes in diversity parameters of parasite communities. For example, the presence and abundance of certain heteroxenous parasites (e.g. trematodes) can serve as a proxy for the presence and abundance of their associated hosts, and provide information on the complexity of ecosystems and the connectivity of different components within it (Sures et al. 2017a, 2023; Selbach et al. 2020; Schwelm et al. 2021). Higher parasite diversity is also an indicator of good environmental quality, low pollution levels (e.g. Nachev and Sures 2009) and intact ecosystems (Hudson et al. 2006; Sures et al. 2017a, 2023). In contrast to the use of parasites as specific accumulation indicators, less research has been conducted on the use of parasites as effect indicators.

20.6.2 Interference of Parasites in Biomarker Analyses in Ecotoxicology

Parasites can modulate molecular and cellular processes involved in toxicokinetic (Sect. 20.4) and toxicodynamic pathways (Sect. 20.5) of pollutants. Many of the underlying mechanisms are commonly used as biomarkers in ecotoxicology to measure the response of test organisms to changes in their environment, for example, the exposure to pollutants (Zimmermann and Sures 2023). Biomarkers are classified as biomarkers of exposure (e.g. MT induction following exposure to metals or acetylcholine esterase inhibition following exposure to organophosphates) and biomarkers of effect (e.g. enzyme activities, DNA damage, oxidative stress marker, histological, behavioural responses) depending on their nature of action (Zimmermann and Sures 2023). The examples given in the previous sections show the wide range of ways in which parasites may influence the effects of chemicals on free-living organisms and thus their biomarker responses. In general, biomarker responses focus on the change rather than the status of the test organism's environment and are therefore measured relative to reference conditions, which might be based on an uncontaminated site, time and sample material (Grabner et al. 2023). The unpredictable impact of parasites on both the reference biomarker response (i.e. the negative control in toxicity tests) and/or the exposed test organism response can significantly affect the outcome of ecotoxicity tests (Grabner and Sures 2019; Grabner et al. 2023; see Fig. 20.7).

For example, in a laboratory study on the effects of the acanthocephalan *P. minutus* on energy reserves of *G. fossarum* exposed to Cd, the parasite infection caused a decrease in the baseline of the lipid content of the unexposed host (Chen et al. 2015). Similarly, gammarids infected with *P. laevis* exhibited significantly reduced concentrations of Hsps, regardless of Pd exposure (Sures and Radszuweit 2007). If ecotoxicologists overlook such physiological effects of parasites in unexposed (reference) organisms, the interpretation of test results becomes nearly impossible.

Moreover, parasite infections often increase the susceptibility of their hosts to various stressors (Combes 1995), sometimes even leading to the death of the test organism (see Sect. 20.5). Sures and Nachev (2022) recently summarised the studies demonstrating that negative acute and chronic effects of chemicals can be intensified by parasites (e.g. lower fish survival, poorer body condition, and various physiological biomarkers), thereby creating an additive effect. Therefore, the toxicity of a substance may be overestimated if the parasite effect is not considered. In turn, the natural condition where parasites are present in the ecosystems will not be reflected by laboratory studies using parasite-free model species for toxicity testing (Grabner et al. 2023).

Synergistic effects of the interaction of parasites and pollutants might lead to unexpected results that cannot be predicted based on the single stressor effects. For example, toxicant effects in parasitised hosts can reach a higher level than expected based on the impact of each single stressor exposure (either pollution or parasite) (Grabner et al. 2023). Furthermore, parasites may increase the stressor tolerance of the host due to an antagonistic effect of the parasite (see Fig. 20.7). For example, acanthocephalan-infected amphipods showed a greater tolerance to elevated salinity levels from 12.6 g/L to 21.5 g/L based on the LC_{50} (Piscart et al. 2007). Antagonistic effects of parasites can also impact the host metabolism, for example, the cellular repair mechanisms (e.g. Hsp70, MTs; see also Sect. 20.5) and there are several examples when parasite infections appear to be beneficial to infected host individuals (see Sect. 20.5.2). In such cases, it is crucial to consider parasitism when assessing pollutant toxicity, as parasites may remove xenobiotics from the host, potentially altering the tolerance levels of exposed and infected hosts compared to uninfected individuals.

In ecotoxicity testing, the biomarker response is recorded with increasing exposure concentration of the test substance, displaying a

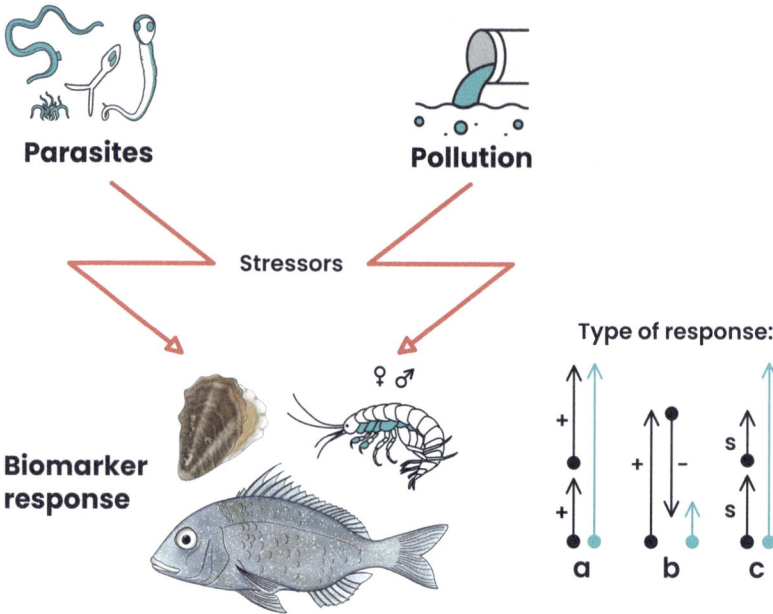

Fig. 20.7 Schematic overview on the combined effects of parasitism and pollution on organisms used in ecotoxicological test set ups. Response types are shown as (**a**) additive effect of both stressors; (**b**) antagonistic effect; (**c**) synergistic effect (black arrows: effect of each of the individual stressors; green arrows: combined effect of both stressors; +/− indicates direction of effect; modified from Grabner et al. 2023)

sigmoid-shaped concentration–response relationship ranging from 0% to 100% at best (see Fig. 20.8a). Parasitic infections can potentially alter concentration–response curves in various ways (e.g. Kochmann et al. 2023). For instance, the curve may shift to the right, indicating a decrease in toxicity (antagonism: a positive effect of the parasite on the negative pollutant effect; Fig. 20.8b). Alternatively, the curve may shift to the left, indicating an increase in toxicity (addition or synergism: a negative effect of both the parasite and the pollutant; Fig. 20.8c). Additionally, the slope of the curve might change, reflecting an alteration in the sensitivity of the test organism to the pollutant (see Fig. 20.8d). However, the modulation of biomarker responses to pollutant exposure in hosts by parasites is not well understood or even studied so far and therefore requires further investigation in future (Grabner et al. 2023; Sures et al. 2017a).

Another aspect of parasite infections in biomarker studies that may have a significant influence on the interpretation of the results is the non-awareness of the presence of parasites in tissue samples used for biomarker analyses. Since parasites themselves may express molecules relevant for detoxification (e.g. xenobiotic metabolising enzymes; see Sect. 20.4.2), biomarker responses analysed in host tissues that are not recognisably infected with parasites would reflect the mixture of host and parasite tissues and may therefore lead to erroneous results. Consequently, toxicity tests are typically designed using non-parasitised free-living organisms to exclude parasites as a secondary stressor and avoid analysing parasitised tissue samples. However, ensuring the test organisms are free from parasites can be challenging and often requires the expertise of parasitologists, particularly for detecting small parasites (<1 mm) such as protists and microsporidians that are easily overlooked. Nevertheless, where possible, parasites should be identified, removed from analysed host tissues and ideally also be analysed separately for their respective responses to pollutants. As outlined above, studies, and therefore knowledge, on the stressor response of parasites, are limited and should be a target for future research.

20.7 Conclusions and Outlook

Host–parasite interactions in polluted environments are highly complex due to the multidimensional nature of the underlying mechanisms. From the host perspective, both parasites and

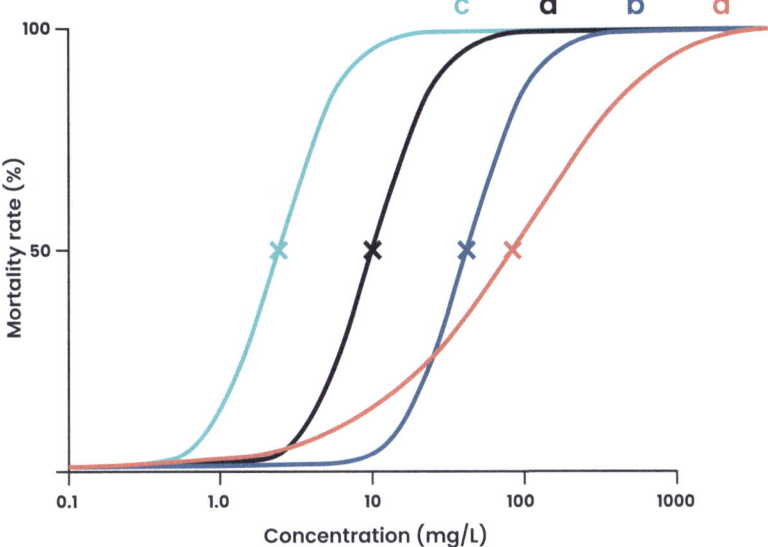

Fig. 20.8 Concentration response-curves of organisms exposed to pollutants with mortality rate on the y-axis x and concentration on the x-axis y (a). The concentration at which 50% of the test organisms die, is the LC_{50}. Parasitic infections may shift concentration-response curves to the right, signifying reduced toxicity (b) or to the left which means a higher toxicity (c). Changes in the slope of the curve may suggest altered sensitivity of the test organism to the pollutant (d)

xenobiotics can influence the physiological status of the free-living host individually or synergistically, sometimes enhancing their combined impact. Conversely, antagonistic interactions are also common, such as parasitism benefiting the host in a polluted environment or pollution reducing infection pressure by decreasing parasite abundance and diversity. These scenarios can have cascading effects across different levels of biological organisation, starting at the molecular level and extending through individual, population and community levels, ultimately leading to significant consequences at ecosystem level. Despite the growing body of scientific literature in the field of environmental parasitology since the early 1990s, there remains a significant knowledge gap due to the vast diversity of parasitic taxa, the multitude of pollutants and their complex interactions with hosts. Consequently, further fundamental research is essential to address the myriad host–parasite–pollutant interactions. Future studies should also account for the increasing number of emerging contaminants and other concurrent anthropogenic impacts on natural ecosystems, such as climate change, habitat degradation and biodiversity loss. Including parasitism in environmental studies, particularly in ecotoxicology, would enhance explanatory power and address more complex scientific questions. Research in environmental parasitology has shown the usefulness of parasites as sentinels, highlighting the need to integrate them into standard monitoring programs and active monitoring initiatives.

References

Akinsanya B, Isibor PO, Onadeko B, Tinuade A-A (2020) Impacts of trace metals on African common toad, *Amietophrynus regularis* (Reuss, 1833) and depuration effects of the toad's enteric parasite, *Amplicaecum africanum* (Taylor, 1924) sampled within Lagos metropolis. Nigeria Heliyon 6:e03570. https://doi.org/10.1016/j.heliyon.2020.e03570

Baudrimont M, De Montaudouin X (2007) Evidence of an altered protective effect of metallothioneins after cadmium exposure in the digenean parasite-infected cockle (*Cerastoderma edule*). Parasitology 134:237–245. https://doi.org/10.1017/S0031182006001375

Birk S, Chapman D, Carvalho L, Spears BM, Andersen HE, Argillier C, Auer S, Baattrup-Pedersen A, Banin L, Beklioğlu M, Bondar-Kunze E, Borja A, Branco P, Bucak T, Buijse AD, Cardoso AC, Couture R-M, Cremona F, de Zwart D, Feld CK, Ferreira MT, Feuchtmayr H, Gessner MO, Gieswein A, Globevnik L, Graeber D, Graf W, Gutiérrez-Cánovas C, Hanganu J, Işkın U, Järvinen M, Jeppesen E, Kotamäki N, Kuijper M, Lemm JU, Lu S, Solheim AL, Mischke U, Moe SJ, Nõges P, Nõges T, Ormerod SJ, Panagopoulos Y, Phillips G, Posthuma L, Pouso S, Prudhomme C, Rankinen K, Rasmussen JJ, Richardson J, Sagouis A, Santos JM, Schäfer RB, Schinegger R, Schmutz S, Schneider SC, Schülting L, Segurado P, Stefanidis K, Sures B, Thackeray SJ, Turunen J, Uyarra MC, Venohr M, von der Ohe PC, Willby N, Hering D (2020) Impacts of multiple stressors on freshwater biota across spatial scales and ecosystems. Nat Ecol Evol 4:1060–1068. https://doi.org/10.1038/s41559-020-1216-4

Boyce NP, Yamada SB (1977) Effects of a parasite, *Eubothrium salvelini* (Cestoda: Pseudophyllidea), on the resistance of juvenile sockeye salmon *Oncorhynchus nerka* to zinc. J Fish Res Board Canada 34:706–709. https://doi.org/10.1139/f77-110

Brázová T, Hanzelová V, Miklisová D (2012) Bioaccumulation of six PCB indicator congeners in a heavily polluted water reservoir in Eastern Slovakia: tissue-specific distribution in fish and their parasites. Parasitol Res 111:779–786. https://doi.org/10.1007/s00436-012-2900-3

Brázová T, Miklisová D, Barčák D, Uhrovič D, Šalamún P, Orosová M, Oros M (2021) Hazardous pollutants in the environment: fish host-parasite interactions and bioaccumulation of polychlorinated biphenyls. Environ Pollut 291:118175. https://doi.org/10.1016/j.envpol.2021.118175

Brown AF, Pascoe D (1989) Parasitism and host sensitivity to cadmium: an acanthocephalan infection of the freshwater amphipod *Gammarus pulex*. J Appl Ecol 26:473–487. https://doi.org/10.2307/2404075

Caballero-Mateos AM, Martínez-Cara JG, Redondo-Cerezo E (2017) Hepatic hydatid cysts causing biliary obstruction. Clin Gastroenterol Hepatol 15:e121–e122. https://doi.org/10.1016/j.cgh.2017.01.001

Carravieri A, Burthe SJ, De La Vega C, Yonehara Y, Daunt F, Newell MA, Jeffreys RM, Lawlor AJ, Hunt A, Shore RF, Pereira MG, Green JA (2020) Interactions between environmental contaminants and gastrointestinal parasites: novel insights from an integrative approach in a marine predator. Environ Sci Technol 54:8938–8948. https://doi.org/10.1021/acs.est.0c03021

Chen H-Y, Grabner DS, Nachev M, Shih H-H, Sures B (2015) Effects of the acanthocephalan *Polymorphus minutus* and the microsporidian *Dictyocoela duebenum* on energy reserves and stress response of cadmium exposed *Gammarus fossarum*. PeerJ 3:e1353. https://doi.org/10.7717/peerj.1353

Chodkowski N, Bernot RJ (2017) Parasite and host elemental content and parasite effects on host nutrient excretion and metabolic rate. Ecol Evol 7:5901–5908. https://doi.org/10.1002/ece3.3129

Combes C (1995) Interactions durables. Ecologie et évolution du parasitisme, Masson, Paris

Coombs GH, Roland Wolf C, Morrison VM, Craft JA (1990) Changes in hepatic xenobiotic-metabolising enzymes in mouse liver following infection with *Leishmania donovani*. Mol Biochem Parasitol 41:17–24. https://doi.org/10.1016/0166-6851(90)90092-Z

Costello MJ (2006) Ecology of sea lice parasitic on farmed and wild fish. Trends Parasitol 22:475–483. https://doi.org/10.1016/j.pt.2006.08.006

de Sales-Ribeiro C, Rivero MA, Fernández A, García-álvarez N, González JF, Quesada-canales O, Caballero MJ (2021) A study on the pathological effects of trypanorhyncha cestodes in dusky groupers epinephelus marginatus from the canary islands. Animals 11:1471. https://doi.org/10.3390/ani11051471

El-Shazly AM, El-Nahas HA, Soliman ME, Abdel-Mageed AA, El-Gharabawy S, Morsy AT, Hamza MM (2005) Cholestasis in human fascioliasis in Dakahlia Governorate, Egypt. J Egypt Soc Parasitol 35:83–94

Erasmus JH, Wepener V, Nachev M, Zimmermann S, Malherbe W, Sures B, Smit NJ (2020) The role of fish helminth parasites in monitoring metal pollution in aquatic ecosystems: a case study in the world's most productive platinum mining region. Parasitol Res 119:2783–2798. https://doi.org/10.1007/s00436-020-06813-1

Erasmus A, Wepener V, Hadfield KA, Sures B, Smit NJ (2022a) Metazoan parasite diversity of the endemic South African intertidal klipfish, *Clinus superciliosus*: factors influencing parasite community composition. Parasitol Int 90:102611. https://doi.org/10.1016/j.parint.2022.102611

Erasmus A, Wepener V, Zimmermann S, Nachev M, Hadfield KA, Smit NJ, Sures B (2022b) High element concentrations are not always equivalent to a stressful environment: differential responses of parasite taxa to natural and anthropogenic stressors. Mar Pollut Bull 184:114110. https://doi.org/10.1016/j.marpolbul.2022.114110

Facino RM, Carini M, Genchi C (1984) Impaired in vitro metabolism of the flukicidal agent nitroxynil by hepatic microsomal cytochrome P-450 in bovine fascioliasis. Toxicol Lett 20:231–236. https://doi.org/10.1016/0378-4274(84)90153-X

Fanton H, Franquet E, Logez M, Cavalli L, Kaldonski N (2022) Acanthocephalan parasites reflect ecological status of freshwater ecosystem. Sci Total Environ 838:156091. https://doi.org/10.1016/j.scitotenv.2022.156091

Félix RC, Silveira H (2011) The interplay between tubulins and P450 cytochromes during plasmodium berghei invasion of *Anopheles gambiae* midgut. PLoS One 6:e24181. https://doi.org/10.1371/journal.pone.0024181

Filipović Marijić V, Vardić Smrzlić I, Raspor B (2013) Effect of acanthocephalan infection on metal, total protein and metallothionein concentrations in European chub from a Sava River section with low metal contamination. Sci Total Environ 463–464:772–780. https://doi.org/10.1016/j.scitotenv.2013.06.041

Frank SN, Faust S, Kalbe M, Trubiroha A, Kloas W, Sures B (2011) Fish hepatic glutathione-S-transferase activity is affected by the cestode parasites *Schistocephalus solidus* and *Ligula intestinalis*: Evidence from field and laboratory studies. Parasitology 138:939–944. https://doi.org/10.1017/S003118201100045X

Frank SN, Godehardt S, Nachev M, Trubiroha A, Kloas W, Sures B (2013) Influence of the cestode *Ligula intestinalis* and the acanthocephalan *Polymorphus minutus* on levels of heat shock proteins (HSP70) and metallothioneins in their fish and crustacean intermediate hosts. Environ Pollut 180:173–179. https://doi.org/10.1016/j.envpol.2013.05.014

Gheorghiu C, Cable J, Marcogliese DJ, Scott ME (2007) Effects of waterborne zinc on reproduction, survival and morphometrics of *Gyrodactylus turnbulli* (Monogenea) on guppies (*Poecilia reticulata*). Int J Parasitol 37:375–381. https://doi.org/10.1016/j.ijpara.2006.09.004

Gheorgiu C, Marcogliese DJ, Scott M (2006) Concentration-dependent effects of waterborne zinc on population dynamics of *Gyrodactylus turnbulli* (Monogenea) on isolated guppies (*Poecilia reticulata*). Parasitology 132:225–232. https://doi.org/10.1017/S003118200500898X

Giari L, Guerranti C, Perra G, Cincinelli A, Gavioli A, Lanzoni M, Castaldelli G (2023) PFAS levels in fish species in the Po River (Italy): new generation PFAS, fish ecological traits and parasitism in the foreground. Sci Total Environ 876:162828. https://doi.org/10.1016/j.scitotenv.2023.162828

Gilbert BM, Avenant-Oldewage A (2017) Trace element and metal sequestration in vitellaria and sclerites, and reactive oxygen intermediates in a freshwater monogenean, *Paradiplozoon ichthyoxanthon*. PLoS One 12:e0177558. https://doi.org/10.1371/journal.pone.0177558

Gilbert BM, Avenant-Oldewage A (2018) Trace element biomineralisation in the carapace in male and female *Argulus japonicus*. PLoS One 13:e0197804. https://doi.org/10.1371/journal.pone.0197804

Gilbert BM, Avenant-Oldewage A (2021) Monogeneans as bioindicators: a meta-analysis of effect size of contaminant exposure toward Monogenea (Platyhelminthes). Ecol Indic 130:108062. https://doi.org/10.1016/j.ecolind.2021.108062

Gilbert BM, Jirsa F, Avenant-Oldewage A (2022) First record of trace element accumulation in a freshwater ectoparasite, *Paradiplozoon ichthyoxanthon* (Monogenea; Diplozoidae), infecting the gills of two yellowfish species, *Labeobarbus aeneus* and *Labeobarbus kimberleyensis*. J Trace Elem Med Biol 74:127053. https://doi.org/10.1016/j.jtemb.2022.127053

Gismondi E, Rigaud T, Beisel J-N, Cossu-Leguille C (2012) Microsporidia parasites disrupt the responses to cadmium exposure in a gammarid. Environ Pollut 160:17–23. https://doi.org/10.1016/j.envpol.2011.09.021

Grabner D, Sures B (2019) Amphipod parasites may bias results of ecotoxicological research. Dis Aquat Org 136:121–132. https://doi.org/10.3354/dao03355

Grabner D, Rothe LE, Sures B (2023) Parasites and pollutants: effects of multiple stressors on aquatic organisms. Environ Toxicol Chem 42:1946–1959. https://doi.org/10.1002/etc.5689

Gunkel G (1994) Bioindikation in aquatischen Ökosystemen — Bioindikation in limnischen und küstennahen Ökosystemen; Grundlagen, Verfahren und Methoden. Reihe: Umweltforschung. Fischer Verlag, Jena

Guth DJ, Blankespoor HD, Cairns J Jr (1977) Potentiation of zinc stress caused by parasitic infection of snails. Hydrobiologia 55:225–229. https://doi.org/10.1007/BF00017554

Heinonen J, Kukkonen JVK, Holapainen JI (2001) Temperature- and parasite-induced changes in toxicity and lethal body burdens of pentachlorophenol in the freshwater clam *Pisidium amnicum*. Environ Toxicol Chem 20:2778–2784. https://doi.org/10.1002/etc.5620201217

Henríquez-Hernández LA, Carretón E, Camacho M, Montoya-Alonso JA, Boada LD, Valerón PF, Cordón YF, Almeida-González M, Zumbado M, Luzardo OP (2016) Influence of parasitism in dogs on their serum levels of persistent organochlorine compounds and polycyclic aromatic hydrocarbons. Sci Total Environ 562:128–135. https://doi.org/10.1016/j.scitotenv.2016.03.204

Hesse T, Nachev M, Khaliq S, Jochmann MA, Franke F, Scharsack JP, Kurtz J, Sures B, Schmidt TC (2023) A new technique to study nutrient flow in host-parasite systems by carbon stable isotope analysis of amino acids and glucose. Sci Rep 13:1054. https://doi.org/10.1038/s41598-022-24933-9

Hodson PV, Blunt BR, Spry DJ (1978) Chronic toxicity of water-borne and dietary lead to rainbow trout (*Salmo gairdneri*) in lake Ontario water. Water Res 12:869–878. https://doi.org/10.1016/0043-1354(78)90039-8

Hofer R, Lackner R (1995) Fischtoxikologie: theorie und praxis. Fischer

Hoole D (1997) The effects of pollutants on the immune response of fish: implications for helminth parasites. Parassitologia 39:219–225

Hua J, Buss N, Kim J, Orlofske SA, Hoverman JT (2016) Population-specific toxicity of six insecticides to the trematode *Echinoparyphium* sp. Parasitology 143:542–550. https://doi.org/10.1017/S0031182015001894

Hudson PJ, Dobson AP, Lafferty KD (2006) Is a healthy ecosystem one that is rich in parasites? Trends Ecol Evol 21:381–385. https://doi.org/10.1016/j.tree.2006.04.007

Jankovská I, Miholová D, Bejček V, Vadlejch J, Šulc M, Száková J, Langrová I (2010) Influence of parasitism on trace element contents in tissues of Red Fox (*Vulpes vulpes*) and its parasites *Mesocestoides* spp. (Cestoda) and *Toxascaris leonina* (Nematoda). Arch Environ Contam Toxicol 58:469–477. https://doi.org/10.1007/s00244-009-9355-2

Jobling S, Tyler CR (2003) Endocrine disruption, parasites and pollutants in wild freshwater fish. Parasitology 126:S103–S107. https://doi.org/10.1017/S0031182003003652

Kaufmann J, Lenz TL, Milinski M, Eizaguirre C (2014) Experimental parasite infection reveals costs and benefits of paternal effects. Ecol Lett 17:1409–1417. https://doi.org/10.1111/ele.12344

Kennedy CR, Broughton PF, Hine PM (1978) The status of brown and rainbow trout, *Salmo trutta* and *S. gairdneri* as hosts of the acanthocephalan, *Pomphorhynchus laevis*. J Fish Biol 13:265–275. https://doi.org/10.1111/j.1095-8649.1978.tb03434.x

Kirk K (2015) Ion regulation in the malaria parasite. Ann Rev Microbiol 69:341–359. https://doi.org/10.1146/annurev-micro-091014-104506

Kloas W, Stöck M, Lutz I, Ziková-Kloas A (2024) Endocrine disruption in teleosts and amphibians is mediated by anthropogenic and natural environmental factors: implications for risk assessment. Philos Trans R Soc B Biol Sci 379:20220505. https://doi.org/10.1098/rstb.2022.0505

Kochmann J, Laier M, Klimpel S, Wick A, Kunkel U, Oehlmann J, Jourdan J (2023) Infection with acanthocephalans increases tolerance of *Gammarus roeselii* (Crustacea: Amphipoda) to pyrethroid insecticide deltamethrin. Environ Sci Pollut Res 30:55582–55595. https://doi.org/10.1007/s11356-023-26193-0

Koprivnikar J, Forbes MR, Baker RL (2006) Effects of atrazine on cercarial longevity, activity, and infectivity. J Parasitol 92:306–311. https://doi.org/10.1645/GE-624R.1

Koprivnikar J, Baker RL, Forbes MR (2007) Environmental factors influencing community composition of gastropods and their trematode parasites in southern Ontario. J Parasitol 93:992–998. https://doi.org/10.1645/GE-1144R.1

Kumar VS, Saxena PN, Tripathi LM, Saxena KC, Mohan Rao VK (1983) Action of antiamoebic drugs on hepatic microsomal drug-metabolizing enzymes of hamster infected with virulent *Entamoeba histolytica*. Indian J Med Res 78:349–353. https://doi.org/10.1016/0014-4800(90)90073-M

Lafferty KD (1997) Environmental parasitology: What can parasites tell us about human impacts on the environment? Parasitol Today 13:251–255. https://doi.org/10.1016/S0169-4758(97)01072-7

Laing R, Bartley DJ, Morrison AA, Rezansoff A, Martinelli A, Laing ST, Gilleard JS (2015) The cytochrome P450 family in the parasitic nematode *Haemonchus contortus*. Int J Parasitol 45:243–251. https://doi.org/10.1016/j.ijpara.2014.12.001

Latief L, Gilbert BM, Avenant-Oldewage A (2023) Biomineralisation and metal sequestration in a crustacean ectoparasite infecting the gills of a freshwater fish. J Comp Physiol B Biochem Syst Environ Physiol 193:271–279. https://doi.org/10.1007/s00360-023-01489-2

Le TTY, Rijsdijk L, Sures B, Jan Hendriks A (2014) Accumulation of persistent organic pollutants in parasites. Chemosphere 108:145–151. https://doi.org/10.1016/j.chemosphere.2014.01.036

Le TTY, Kiwitt G, Nahar N, Nachev M, Grabner D, Sures B (2022) What contributes to the metal-specific partitioning in the chub-acanthocephalan system? Aquat Toxicol 247:106178. https://doi.org/10.1016/j.aquatox.2022.106178

Lenoir C, Rollason V, Desmeules JA, Samer CF (2021) Influence of inflammation on cytochromes P450 activity in adults: a systematic review of the literature. Front Pharmacol 12:733935. https://doi.org/10.3389/fphar.2021.733935

Lingard SM, Evans RD, Bourgoin BP (1992) Method for the estimation of organic-bound and crystal-bound metal concentrations in bivalve shells. Bull Environ Contam Toxicol 48:179–184. https://doi.org/10.1007/BF00194369

Lotfy WM, Ezz AM, Hassan AAM (2013) Bioaccumulation of some heavy metals in the liver flukes *Fasciola hepatica* and *F. gigantica*. Iran J Parasitol 8:552–558

Marcogliese DJ (2004) Parasites: small players with crucial roles in the ecological theater. EcoHealth 1:151–164. https://doi.org/10.1007/s10393-004-0028-3

Marcogliese DJ (2005) Parasites of the superorganism: Are they indicators of ecosystem health? Int J Parasitol 35:705–716. https://doi.org/10.1016/j.ijpara.2005.01.015

Marcogliese DJ (2023) Major drivers of biodiversity loss and their impacts on helminth parasite populations and communities. J Helminthol 97:e34. https://doi.org/10.1017/S0022149X2300010X

McCahon CP, Brown AF, Pascoe D (1988) The effect of the acanthocephalan *Pomphorhynchus laevis* (Muller 1776) on the acute toxicity of cadmium to its intermediate host, the amphipod *Gammarus pulex* (L.). Arch Environ Contam Toxicol 17:239–243. https://doi.org/10.1007/BF01056030

Mille T, Soulier L, Caill-Milly N, Cresson P, Morandeau G, Monperrus M (2020) Differential micropollutants bioaccumulation in European hake and their parasites *Anisakis* sp. Environ Pollut 265:115021. https://doi.org/10.1016/j.envpol.2020.115021

Molbert N, Alliot F, Leroux-Coyau M, Médoc V, Biard C, Meylan S, Jacquin L, Santos R, Goutte A (2020) Potential benefits of acanthocephalan parasites for chub hosts in polluted environments. Environ Sci Technol 54:5540–5549. https://doi.org/10.1021/acs.est.0c00177

Molbert N, Agostini S, Alliot F, Angelier F, Biard C, Decencière B, Leroux-Coyau M, Millot A, Ribout C, Goutte A (2021) Parasitism reduces oxidative stress of fish host experimentally exposed to PAHs. Ecotoxicol Environ Saf 219:112322. https://doi.org/10.1016/j.ecoenv.2021.112322

Mordvinov V, Pakharukova M (2022) Xenobiotic-metabolizing enzymes in trematodes. Biomedicines 10:3039. https://doi.org/10.3390/biomedicines10123039

Morley NJ, Lewis JW (2006) Anthropogenic pressure on a molluscan-trematode community over a long-term period in the Basingstoke Canal, UK, and its implications for ecosystem health. EcoHealth 3:269–280. https://doi.org/10.1007/s10393-006-0058-0

Morrill A, Provencher JF, Gilchrist HG, Mallory ML, Forbes MR (2019) Anti-parasite treatment results in decreased estimated survival with increasing lead (Pb) levels in the common eider Somateria mollissima. Proc R Soc B Biol Sci 286:20191356. https://doi.org/10.1098/rspb.2019.1356

Mostafa MH, El-Bassiouni EA, El-Sewedy SM, Akhnouk S, Tawfic T, Abdel-Rafee A (1984) Hepatic microsomal enzymes in Schistosoma mansoni-infected mice: I. Effect of duration of infection and lindane administration on dimethylnitrosamine demethylases. Environ Res 35:154–159. https://doi.org/10.1016/0013-9351(84)90122-1

Nachev M, Sures B (2009) The endohelminth fauna of barbel (Barbus barbus) correlates with water quality of the Danube River in Bulgaria. Parasitology 136:545–552. https://doi.org/10.1017/s003118200900571x

Nachev M, Sures B (2016) Environmental parasitology: parasites as accumulation bioindicators in the marine environment. J Sea Res 13:45–50. https://doi.org/10.1016/j.seares.2015.06.005

Nachev M, Zimmermann S, Rigaud T, Sures B (2010) Is metal accumulation in Pomphorhynchus laevis dependent on parasite sex or infrapopulation size? Parasitology 137:1239–1248. https://doi.org/10.1017/S0031182010000065

Nachev M, Rozdina D, Michler-Kozma DN, Raikova G, Sures B (2022) Metal accumulation in ecto- and endoparasites from the anadromous fish, the Pontic shad (Alosa immaculata). Parasitology 149:496–502. https://doi.org/10.1017/S0031182021002080

Nickol B (1985) Epizootiology. In: Crompton D, Nickol B (eds) Biology of the Acanthocephala. Cambridge University Press, Cambridge, pp 307–346

Oyoo-Okoth E, Wim A, Osano O, Kraak MH, Ngure V, Makwali J, Orina PS (2010) Use of the fish endoparasite Ligula intestinalis (L., 1758) in an intermediate cyprinid host (Rastreneobola argentea) for biomonitoring heavy metal contamination in Lake Victoria, Kenya. Lakes Reserv Res Manag 15:63–73. https://doi.org/10.1111/j.1440-1770.2010.00423.x

Pakharukova MY, Vavilin VA, Sripa B, Laha T, Brindley PJ, Mordvinov VA (2015) Functional analysis of the unique cytochrome P450 of the liver fluke Opisthorchis felineus. PLoS Negl Trop Dis 9:e0004258. https://doi.org/10.1371/journal.pntd.0004258

Pascoe D, Cram P (1977) The effect of parasitism on the toxicity of cadmium to the three-spined stickleback, Gasterosteus aculeatus L. J Fish Biol 10:467–472. https://doi.org/10.1111/j.1095-8649.1977.tb04079.x

Paul-Pont I, de Montaudouin X, Gonzalez P, Soudant P, Baudrimont M (2010) How life history contributes to stress response in the Manila clam Ruditapes philippinarum. Environ Sci Pollut Res 17:987–998. https://doi.org/10.1007/s11356-009-0283-5

Pérez-del Olmo A, Raga JA, Kostadinova A, Fernández M (2007) Parasite communities in Boops boops (L.) (Sparidae) after the Prestige oil-spill: detectable alterations. Mar Pollut Bull 54:266–276. https://doi.org/10.1016/j.marpolbul.2006.10.003

Pérez-del Olmo A, Nachev M, Zimmermann S, Fernández M, Sures B (2019) Medium-term dynamics of element concentrations in a sparid fish and its isopod parasite after the Prestige oil-spill: shifting baselines? Sci Total Environ 686:648–656. https://doi.org/10.1016/j.scitotenv.2019.05.455

Pietrock M, Marcogliese DJ (2003) Free-living endohelminth stages: at the mercy of environmental conditions. Trends Parasitol 19:293–299

Piscart C, Webb D, Beisel JN (2007) An acanthocephalan parasite increases the salinity tolerance of the freshwater amphipod Gammarus roeseli (Crustacea: Gammaridae). Naturwissenschaften 94:741–747. https://doi.org/10.1007/s00114-007-0252-0

Pravdová M, Kolářová J, Grabicová K, Mikl L, Bláha M, Randák T, Kvach Y, Jurajda P, Ondračková M (2021) Associations between pharmaceutical contaminants, parasite load and health status in brown trout exposed to sewage effluent in a small stream. Ecohydrol Hydrobiol 21:233–243. https://doi.org/10.1016/j.ecohyd.2020.09.001

Raffel TR, Sheingold JL, Rohr JR (2009) Lack of pesticide toxicity to Echinostoma trivolvis eggs and miracidia. J Parasitol 95:1548–1551. https://doi.org/10.1645/GE-2078.1

Rohr JR, Schotthoefer AM, Raffel TR, Carrick HJ, Halstead N, Hoverman JT, Johnson CM, Johnson LB, Lieske C, Piwoni MD, Schoff PK, Beasley VR (2008) Agrochemicals increase trematode infections in a declining amphibian species. Nature 455:1235–1239. https://doi.org/10.1038/nature07281

Rooney J, Northcote HM, Williams TL, Cortés A, Cantacessi C, Morphew RM (2022) Parasitic helminths and the host microbiome – a missing 'extracellular vesicle-sized' link? Trends Parasitol 38:737–747. https://doi.org/10.1016/j.pt.2022.06.003

Rothe LE, Loeffler F, Gerhardt A, Feld CK, Stift R, Weyand M, Grabner D, Sures B (2022) Parasite infection influences the biomarker response and locomotor activity of Gammarus fossarum exposed to conventionally-treated wastewater. Ecotoxicol Environ Saf 236:113474. https://doi.org/10.1016/j.ecoenv.2022.113474

Samanta TB, Das N, Das M, Marik R (2003) Mechanism of impairment of cytochrome P450-dependent metabolism in hamster liver during leishmaniasis. Biochem Biophys Res Commun 312:75–79. https://doi.org/10.1016/j.bbrc.2003.09.227

Sánchez Di Maggio L, Tirloni L, Uhl M, Carmona C, Logullo C, Mulenga A, da Silva VI, Berasain P (2020)

Serpins in *Fasciola hepatica*: insights into host–parasite interactions. Int J Parasitol 50:931–943. https://doi.org/10.1016/j.ijpara.2020.05.010

Sánchez MI, Pons I, Martínez-Haro M, Taggart MA, Lenormand T, Green AJ (2016) When parasites are good for health: Cestode parasitism increases resistance to arsenic in brine shrimps. PLoS Pathog 12:1–19. https://doi.org/10.1371/journal.ppat.1005459

Sanchez-Ramirez C, Vidal-Martinez VM, Aguirre-Macedo ML, Rodriguez-Canul RP, Gold-Bouchot G, Sures B (2007) *Cichlidogyrus sclerosus* (Monogenea: Ancyrocephalinae) and its host, the Nile tilapia (*Oreochromis niloticus*), as bioindicators of chemical pollution. J Parasitol 93:1097–1106. https://doi.org/10.1645/GE-1162R.1

Dezfuli BS, Giari L, Squerzanti S, Lui A, Lorenzoni M, Sakalli S, Shinn AP (2011) Histological damage and inflammatory response elicited by *Monobothrium wageneri* (Cestoda) in the intestine of *Tinca tinca* (Cyprinidae). Parasit Vectors 4:225. https://doi.org/10.1186/1756-3305-4-225

Scholz T, Garcia HH, Kuchta R, Wicht B (2009) Update on the human broad tapeworm (genus *Diphyllobothrium*), including clinical relevance. Clin Microbiol Rev 22:146–160. https://doi.org/10.1128/CMR.00033-08

Schwelm J, Selbach C, Kremers J, Sures B (2021) Rare inventory of trematode diversity in a protected natural reserve. Sci Rep 11:22066. https://doi.org/10.1038/s41598-021-01457-2

Scudamore HH, Thompson JH Jr, Owen CA Jr (1961) Absorption of Co60-labeled vitamin B12 in man and uptake by parasites, including *Diphyllobothrium latum*. J Lab Clin Med 57:240–246

Selbach C, Soldánová M, Feld CK, Kostadinova A, Sures B (2020) Hidden parasite diversity in a European freshwater system. Sci Rep 10:2694. https://doi.org/10.1038/s41598-020-59548-5

Selbach C, Mouritsen KN, Poulin R, Sures B, Smit NJ (2022) Bridging the gap: aquatic parasites in the One Health concept. Trends Parasitol 38:109–111. https://doi.org/10.1016/j.pt.2021.10.007

Srivastava AK, Chatterjee RK, Ghatak S (1985) Hepatic microsomal alterations during *Dipetalonema viteae* infection in *Mastomys natalensis*. Int J Parasitol 15:171–174. https://doi.org/10.1016/0020-7519(85)90083-9

Sures B (2002) Competition for minerals between *Acanthocephalus lucii* and its definitive host perch (*Perca fluviatilis*). Int J Parasitol 32:1117–1122. https://doi.org/10.1016/S0020-7519(02)00083-8

Sures B (2004) Environmental parasitology: Relevancy of parasites in monitoring environmental pollution. Trends Parasitol 20:170–177. https://doi.org/10.1016/j.pt.2004.01.014

Sures B (2008) Host-parasite interactions in polluted environments. J Fish Biol 73:2133–2142. https://doi.org/10.1111/j.1095-8649.2008.02057.x

Sures B, Nachev M (2022) Effects of multiple stressors in fish: how parasites and contaminants interact. Parasitology 149:1822–1828. https://doi.org/10.1017/S0031182022001172

Sures B, Radszuweit H (2007) Pollution-induced heat shock protein expression in the amphipod *Gammarus roeselii* is affected by larvae of *Polymorphus minutus* (Acanthocephala). J Helminthol 81:191–197. https://doi.org/10.1017/S0022149X07751465

Sures B, Reimann N (2003) Analysis of trace metals in the Antarctic host-parasite system *Notothenia coriiceps* and *Aspersentis megarhynchus* (Acanthocephala) caught at King George Island, South Shetland Islands. Polar Biol 26:680–686. https://doi.org/10.1007/s00300-003-0538-4

Sures B, Siddall R (1999) *Pomphorhynchus laevis*: the intestinal acanthocephalan as a lead sink for its fish host, chub (*Leuciscus cephalus*). Exp Parasitol 93:66–72. https://doi.org/10.1006/expr.1999.4437

Sures B, Siddall R (2003) *Pomphorhynchus laevis* (Palaeacanthocephala) in the intestine of chub (*Leuciscus cephalus*) as an indicator of metal pollution. Int J Parasitol 33:65–70. https://doi.org/10.1016/S0020-7519(02)00249-7

Sures B, Taraschewski H, Jackwerth E (1994a) Comparative study of lead accumulation in different organs of perch (*Perca fluviatilis*) and its intestinal parasite *Acanthocephalus lucii*. Bull Environ Contam Toxicol 52:269–273. https://doi.org/10.1007/BF00198498

Sures B, Taraschewski H, Jackwerth E (1994b) Lead accumulation in *Pomphorhynchus laevis* and its host. J Parasitol 80:355–357. https://doi.org/10.2307/3283403

Sures B, Taraschewski H, Jackwerth E (1994c) Lead content of *Paratenuisentis ambiguus* (Acanthocephala), *Anguillicola crassus* (Nematodes) and their host *Anguilla anguilla*. Dis Aquat Org 19:105–107. https://doi.org/10.3354/dao019105

Sures B, Jürges G, Taraschewski H (1998) Relative concentrations of heavy metals in the parasites *Ascaris suum* (Nematoda) and *Fasciola hepatica* (Digenea) and their respective porcine and bovine definitive hosts. Int J Parasitol 28:1173–1178. https://doi.org/10.1016/S0020-7519(98)00105-2

Sures B, Siddall R, Taraschewski H (1999a) Parasites as accumulation indicators of heavy metal pollution. Parasitol Today 15:16–21. https://doi.org/10.1016/S0169-4758(98)01358-1

Sures B, Steiner W, Rydlo M, Taraschewski H (1999b) Concentrations of 17 elements in the zebra mussel (*Dreissena polymorpha*), in different tissues of perch (*Perca fluviatilis*), and in perch intestinal parasites (*Acanthocephalus lucii*) from the subalpine Lake Mondsee, Austria. Environ Toxicol Chem 18:2574–2579. https://doi.org/10.1002/etc.5620181126

Sures B, Jürges G, Taraschewski H (2000) Accumulation and distribution of lead in the archiacanthocephalan *Moniliformis moniliformis* from experimentally infected rats. Parasitology 121:427–433. https://doi.org/10.1017/S003118209900654X

Sures B, Thielen F, Baska F, Messerschmidt J, Von Bohlen A (2005) The intestinal parasite *Pomphorhynchus laevis* as a sensitive accumulation indicator for the platinum group metals Pt, Pd, and Rh. Environ Res 98:83–88. https://doi.org/10.1016/j.envres.2004.05.010

Sures B, Lutz I, Kloas W (2006) Effects of infection with *Anguillicola crassus* and simultaneous exposure with Cd and 3,3′,4,4′,5-pentachlorobiphenyl (PCB 126) on the levels of cortisol and glucose in European eel (*Anguilla anguilla*). Parasitology 132:281–288. https://doi.org/10.1017/S0031182005009017

Sures B, Nachev M, Selbach C, Marcogliese DJ (2017a) Parasite responses to pollution: what we know and where we go in 'Environmental Parasitology'. Parasit Vectors 10:1–19. https://doi.org/10.1186/s13071-017-2001-3

Sures B, Nachev M, Pahl M, Grabner D, Selbach C (2017b) Parasites as drivers of key processes in aquatic ecosystems: facts and future directions. Exp Parasitol 180:141–147. https://doi.org/10.1016/j.exppara.2017.03.011

Sures B, Nachev M, Schwelm J, Grabner D, Selbach C (2023) Environmental parasitology: stressor effects on aquatic parasites. Trends Parasitol 39:461–474. https://doi.org/10.1016/j.pt.2023.03.005

Tang C-L, Zhou H-H, Zhu Y-W, Huang J, Wang G-B (2019) Glutathione S-transferase influences the fecundity of *Schistosoma japonicum*. Acta Trop 191:8–12. https://doi.org/10.1016/j.actatropica.2018.12.027

Taraschewski H (2000) Host-parasite interactions in acanthocephala: a morphological approach. Adv Parasitol 46:1–179. https://doi.org/10.1016/S0065-308X(00)46008-2

Thieltges DW, Jensen KT, Poulin R (2008) The role of biotic factors in the transmission of free-living endohelminth stages. Parasitology 135:407–426. https://doi.org/10.1017/S0031182007000248

Trubiroha A, Kroupova H, Wuertz S, Frank SN, Sures B, Kloas W (2010) Naturally-induced endocrine disruption by the parasite *Ligula intestinalis* (Cestoda) in roach (*Rutilus rutilus*). Gen Comp Endocrinol 166:234–240. https://doi.org/10.1016/j.ygcen.2009.08.010

Van Der Spuy L, Erasmus JH, Nachev M, Schaeffner BC, Sures B, Wepener V, Smit NJ (2023) The use of fish parasitic isopods as element accumulation indicators in marine pollution monitoring. Mar Pollut Bull 194:115385. https://doi.org/10.1016/j.marpolbul.2023.115385

Vidal-Martínez VM, Aguirre-Macedol ML, Noreña-Barroso E, Gold-Bouchot G, Caballero-Pinzón PI (2003) Potential interactions between metazoan parasites of the Mayan catfish *Ariopsis assimilis* and chemical pollution in Chetumal Bay, Mexico. J Helminthol 77:173–184. https://doi.org/10.1079/JOH2002158

Weigand H, Weiss M, Cai H, Li Y, Yu L, Zhang C, Leese F (2018) Fishing in troubled waters: revealing genomic signatures of local adaptation in response to freshwater pollutants in two macroinvertebrates. Sci Total Environ 633:875–891. https://doi.org/10.1016/j.scitotenv.2018.03.109

Weis JS, Weis P (1989) Tolerance and stress in a polluted environment. Bioscience 39:89–95. https://doi.org/10.2307/1310907

Wendelaar Bonga SE (1997) The stress response in fish. Physiol Rev 77:591–625. https://doi.org/10.1152/physrev.1997.77.3.591

Wepener V, Degger N (2020) Monitoring metals in South African harbours between 2008 and 2009, using resident mussels as indicator organisms. Afr Zool 55:267–277. https://doi.org/10.1080/15627020.2020.1799720

Wilby KJ, Gilchrist SE, Ensom MHH (2013) A review of the pharmacokinetic implications of schistosomiasis. Clin Pharmacokinet 52:647–656. https://doi.org/10.1007/s40262-013-0055-8

Wisedpanichkij R, Grams R, Chaijaroenkul W, Na-Bangchang K (2011) Confutation of the existence of sequence-conserved cytochrome P450 enzymes in *Plasmodium falciparum*. Acta Trop 119:19–22. https://doi.org/10.1016/j.actatropica.2011.03.006

Wu X, Wang W, Li Q, Xue Q, Li Y, Li S (2020) Case report: Surgical intervention for fasciolopsis buski infection: a literature review. Am J Trop Med Hyg 103:2282–2287. https://doi.org/10.4269/ajtmh.20-0572

Zauke G-P (1982) Cadmium in Gammaridae (Amphipoda: Crustacea) of the rivers Werra and Weser-II. Seasonal variation and correlation to temperature and other environmental variables. Water Res 16:785–792. https://doi.org/10.1016/0043-1354(82)90005-7

Zhang X, Zhang T, Liu J, Li M, Fu Y, Xu J, Liu Q (2017) Functional characterization of a unique cytochrome P450 in *Toxoplasma gondii*. Oncotarget 8:115079–115088. https://doi.org/10.18632/oncotarget.23023

Zhong Z-H, Li Z-C, Li H, Guo Q-K, Wang C-X, Cao J-Z, Li A-X (2022) Glutathione metabolism in *Cryptocaryon irritans* involved in defense against oxidative stress induced by zinc ions. Parasit Vectors 15:318. https://doi.org/10.1186/s13071-022-05390-9

Zimmermann S, Sures B (2023) Environmental toxicology. In: Hock FJ, Gralinski MR, Pugsley MK (eds) Drug discovery and evaluation: safety and pharmacokinetic assays. Springer-Verlag GmbH, Berlin

Zimmermann S, Sures B, Taraschewski H (1999) Experimental studies on lead accumulation in the eel-specific endoparasites *Anguillicola crassus* (Nematoda) and *Paratenuisentis ambiguus* (Acanthocephala) as compared with their host, *Anguilla anguilla*. Arch Environ Contam Toxicol 37:190–195. https://doi.org/10.1007/s002449900505

Open Access This chapter is licensed under the terms of the Creative Commons Attribution-NonCommercial-NoDerivatives 4.0 International License (http://creativecommons.org/licenses/by-nc-nd/4.0/), which permits any non-commercial use, sharing, distribution and reproduction in any medium or format, as long as you give appropriate credit to the original author(s) and the source, provide a link to the Creative Commons license and indicate if you modified the licensed material. You do not have permission under this license to share adapted material derived from this chapter or parts of it.

The images or other third party material in this chapter are included in the chapter's Creative Commons license, unless indicated otherwise in a credit line to the material. If material is not included in the chapter's Creative Commons license and your intended use is not permitted by statutory regulation or exceeds the permitted use, you will need to obtain permission directly from the copyright holder.

Parasites and Pollutants: Allometric Toxicology for Parasitologists

21

Andy Dobson

Abstract

This chapter discusses the interactions between parasites and pollutants in the environment and their implications for animal health. A mathematical framework is presented to examine these dynamics, focusing on parasitic helminths, their hosts and pollutants. The model is designed to provide general insights into these interactions, without specific calibration for any particular system. The potential of parasites to reduce the burden of pollutants in hosts is highlighted, qualifying it as an ecosystem service. The model incorporates the dynamics of hosts, parasites and pollutants, considering factors such as host survival, parasite distribution, pollutant absorption and excretion rates. The results suggest that parasites can significantly reduce pollutant concentrations in hosts, particularly in larger-bodied hosts. However, the presence of parasites also leads to lower host abundance. The study emphasises the importance of further research to understand these complex interactions and their implications for ecosystems. It also raises questions about the role of parasites in mitigating the impact of pollutants and the potential consequences of parasite loss.

21.1 Introduction

Parasites and pollutants are both prevalent in the environment, leading to significant and increasing levels of human stressors and loss of health. How do they interact? A series of loosely connected research groups have been addressing these huge and important problems over the last few decades and some clarity is beginning to emerge (Sures 2003, 2004; Sures et al. 2023; see also Chap. 20). One thing is plain: there is a lot more we need to know to understand interactions between parasites and pollutants, but everything we are beginning to glimpse has huge implications for human and animal health. In this chapter, I develop an initial mathematical framework for examining the dynamics of interactions between parasitic helminths, their hosts and pollutants in their environment.

In this chapter, I present a very preliminary modelling framework for examining interactions between parasitic worms, their hosts and pollutants. The model is designed to provide very broad, general insights into these problems. I will

A. Dobson (✉)
EEB, Princeton University, Princeton, NJ, USA

Santa Fe Institute, Santa Fe, NM, USA

Smithsonian Tropical Research Institute, Balboa, Panama
e-mail: dobson@princeton.edu

make no attempt to tune its parameters for any particular or specific system. Nevertheless, I have developed the model within a structure that allows most of its parameters to be calibrated using underlying body-size dependent allometric relationships for host and parasite. Similar relationships have been widely used in the literature on pollutants and toxicity. My ultimate aim has been to create a framework that illustrates how interactions between pollutants and parasites operate at population levels and, as is often the case with models, to suggest different ways of interpreting data collected in the field, or additional experiments that could more fully unravel a hugely unexplored set of biological interactions.

One way to frame discussion of the impact of parasites on pollutants is as a previously ignored 'ecosystem service'. A large variety of empirical studies suggest parasitic helminths are able to reduce the burden of pollutants in their hosts (Sures et al. 2017, 2023; see Chap. 20). This would qualify as a formidable ecosystem service. But what are the conditions under which these activities are undertaken? More specifically, what are the relative effects of the parasite and the toxicant on the hosts when acting independently of each other? Crucially, if both parasites and pollutants have detrimental effects on host abundance, how much is the impact of the pollutant reduced by the presence of parasite, or is this only achieved when the parasite reduces the host population to a lower level than is achieved when only the pollutant is present?

21.2 Modelling These Interactions

21.2.1 Macroparasites and a New Ecosystem Service

The classic Anderson and May model (Anderson and May 1978; May and Anderson 1978) for the dynamics of parasitic helminths can readily be modified to include the dynamics of an environmental pollutant. Here, I am going to be spectacularly casual and assume I am considering an unspecified inert pollutant that is present in the environment at some concentration, T_e, and this is taken up or absorbed by the hosts at some rate βH. The pollutant leads to increases in the host mortality rate which scale linearly with its concentration in the host. We could also modify this so that toxicity is described as an LC50 or LD50 (lethal concentration or dose at which 50% of the test organisms die, see Chap. 20 for details).

In the absence of the parasites, the system can be expressed by a single differential equation:

Hosts with pollutant:

$$\frac{dH}{dt} = (b - d - \Delta H)H - \tau_H T_H \quad (21.1)$$

In the absence of the pollutant, the host populations settle to an equilibrium:

$$H^* = (b - d)/\Delta.$$

Where b is host birth rates, d is death rate and Δ is a density-dependent parameter that increases host death rates. We can then consider the pollutant and again readily determine the equilibrium host density:

$$H_T^* = (b - d - \tau_H (T_H / H_T))/\Delta.$$

This expression essentially tells us that host abundance declines linearly with increases in the concentration of the pollutant in the environment, and in the host. When the combined deaths due to the pollutant and 'natural' host mortality equal host birth rate, the population will be driven to extinction. Here it is important to notice that, if the pollutant also has a negative impact on host fecundity, then decline to extinction will occur at lower doses of the pollutant.

We can now add in the parasite dynamics; we will do this by ignoring eventually free-living (i.e. intermediate) parasite infective stages, by assuming these operate on a much faster time scale than the life expectancy of adult parasites or their hosts. Thus, the proportion of infective stages that actually pass from a live adult to infect a new host is given by $H/(H + H_0)$, where H_0 = ratio of mortality rate of free-living larval stages (usually 12–350), to their rate of uptake by their hosts, which is much harder to quantify! This we tend to calibrate H_0 by assuming $H/(H + H_0) \sim 0.0001 \sim 0.01$ and substituting to find β.

Four equations then describe the rate of change of hosts, parasites, pollutants in hosts and pollutants in parasites. A key difference to notice here that is in sharp contrast with the pollutant model is the dynamic nature of the interaction between host and parasite. Parasites can significantly reduce hosts' abundance through their impact on host survival (or fecundity), but they are also entirely dependent upon their hosts for their own persistence. This is not the case for the pollutant; its persistence is independent of the host and the parasite. All the parasites are doing is absorbing the pollutant from the hosts' tissues and storing it in their own tissue.

Hosts with pollutant and parasites

$$\frac{dH}{dt} = (b - d - \Delta H)H - \alpha P - \tau_H T_H \quad (1.2)$$

Parasites

$$\frac{dP}{dt} = \frac{\lambda H}{H + H_0} - \left(\alpha + d + \mu + \tau_H \frac{T_H}{H}\right)P$$
$$-\alpha \frac{P^2}{H} \frac{k+1}{k} \quad (1.3)$$

Pollutant *in hosts (mean concentration per host)*

$$\frac{dT_H}{dt} = \beta_H T_E H$$
$$-\left(d(1+e) + \alpha \frac{P}{H} + \tau_H \frac{T_H}{H}\right)T_H - \beta_P P T_H \quad (1.4)$$

Pollutant *in parasites (mean concentration/parasite/host)*

$$\frac{dT_P}{dt} = \beta_P T_H P - \left(d + \mu + \alpha \frac{P}{H} + \tau_H \frac{T_H}{H} + \tau_P\right)T_P$$
$$(1.5)$$

The model assumes that the parasites have a *per capita* impact on host survival, α, and the pollutant also reduces host survival as its concentration in the host increases, τ_H/H. As both of these terms enter as means, host abundance cancels out in 1.2 above (e.g. $H\alpha P/H = \alpha P$ and $\tau_H T_H H/H = \tau_H T_H$). The parasites are assumed to be aggregated in their distribution, best characterised by the negative binomial distribution with mean P/H and exponent k. Hosts excrete the pollutant at a rate that scales with their metabolic and death rates, whence e, the expulsion rate is a simple multiple of death rate (if $e = 2$, the pollutant is in the host for half its life expectancy). The parasites absorb the pollutant from the host at a rate determined by the mean number of parasites in each host and the concentration of pollutant in host tissue, $(\beta_P T_H/H)(P/H)$. Note that the equations for the pollutant track the total mass of pollutant in the host and parasite population; these figures can then be re-expressed as mean weight per gram of host and parasite tissue in the final output from the calculations.

We can derive expressions for the equilibrium number of parasites per host (in the absence of the pollutant ($HP^* = (b - d)/\alpha$) and the equilibrium concentration of the pollutant in the host in the absence of the parasite ($HT^* = (b - d)/\tau$). These can be used to compare the mean burden and pollutant concentration when both parasite and pollutant are in the system. This can be achieved by solving the four coupled equations numerically and letting the dynamics of the system run to equilibrium (Fig. 21.1). The key result to observe here is that while both parasites and pollutants can reduce host abundance, the parasites tend to dominate this effect and their presence significantly reduces the concentration of pollutants in the hosts. If the parasites are lost from the system, concentration of pollutants in the hosts increases, accompanied by a concomitant reduction in host density.

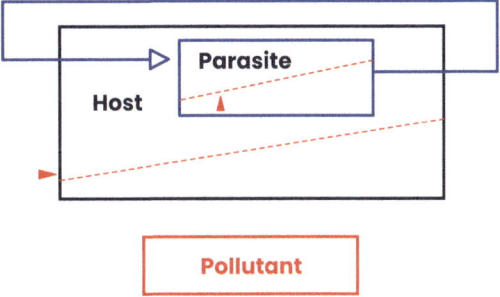

Fig. 21.1 Flow diagram of host-macroparasite model with an environmental pollutant. The pollutant is ingested by the host and the parasite can then absorb it from the host's gut when bound to bile in the alimentary canal. The parasite competes with the host to absorb the pollutant

The relative magnitude of these effects is likely to be highly dependent on host body size. I have tried to capture this by scaling the parameters of the model using known allometric relationships, as these are well characterised for host birth and death rates and average population density. We are only beginning to develop an understanding for these scaling rules in parasitology, where most work has focused on nematodes, rather than cestodes, trematodes or acanthocephalans. Nonetheless, scaling the models using nematodes of terrestrial mammals may provide some broad general insights and generate the stimulus for others to collate data which would allow these relationships to be quantified for other aquatic parasite and host groups. Similarly, although allometric scaling has been a considerable area of interest regarding mammals and birds, there is less work on fish, reptiles, amphibia or any invertebrates. The body size scaling relationships I have used here are collated in Table 21.1; they are based on work with Giulio de Leo, Marino Gatto and others on nematode parasites of mammals (Gatto and de Leo 1998; de Leo et al. 2016). The reviews by Hechinger et al. (2012) and Molnaret et al. (2017) provide excellent introductions into the underlying logic. As mentioned above, these rescaling by body size allow us to compare pollutant concentration per gram of host and parasite tissue.

It is not possible to find any simple analytical insights into the model's behaviour, other than those mentioned briefly above. Instead, I will simply present the results as numerical solutions to the above system of equations which are relatively stable when $k < 0.1$ and run quickly to equilibrium. Results are presented for hosts of four different body sizes, a sequence that quadruples in mass from 0.025 kg up to 1.25 kg. In the absence of the pollutant, the parasites reduce the hosts below their parasite free equilibrium, with larger hosts being proportionally more affected than smaller hosts (Fig. 21.2). In the presence of pollutants, but without parasites, the hosts are again reduced below their carrying capacity (Fig. 21.3). The effect is again more severe for the larger hosts which both acquire pollutants at a faster rate (through their need to

Table 21.1 Main model parameters and their allometric relationship with body size, unit of measure, formula used to calculate them, and corresponding reference. W (kg) is host body size and Ω (mm3) is parasite body size (after De Leo et al. 2016)

Host			
Parameter	*Units*	*Function/value*	*Reference*
K Parasite free carrying capacity for herbivorous mammals	No/km2	=103 W–0.93	Peters (1986, p 167)
r per-capita population growth rate		=0.9 W–0.27	Damuth (1981), Peters (1986)
d per-capita mortality rate	Years–1	=0.4 W–0.26	Peters (1986), Calder (1984)
Parasite			
Parameter	*Units*	*Relationship*	*Reference*
λ per capita fecundity	Eggs/day	=103.5 Ω0.79	Hechinger et al. (2012)
T time to maturity (per-patent period)	Days	=101.4 Ω0.24	Hechinger et al. (2012)
Pat Patent period	Days	=1.15 *T*1.498 = 144 Ω0.359	De Leo et al. (2016)
μ Mortality rate adult parasites	Days–1	=1/Pat = 0.00693 Ω–0.359	De Leo et al. (2016)
σ Maturation rate	Days–1	1/*T* = 0.0398 Ω–0.24	De Leo et al. (2016)
S Fraction of worms surviving to maturity		$\sigma(\Omega)$/ ($\sigma(\Omega) + \mu(\Omega) + d(W)$)	As in Morand and Poulin (2002)
k Clumping parameter of negative binomial distribution	a dimensional	Between 0.01 and 1.0	Shaw and Dobson (1995)
H0 Saturation constant of parasite transmission	No/km2	Constrained so that R0 < 10	De Leo et al. (2016)
α per-capita, per-worm, parasite induced mortality	Hosts/parasite/time	=ε Ωq/W0.75	De Leo et al. (2016)

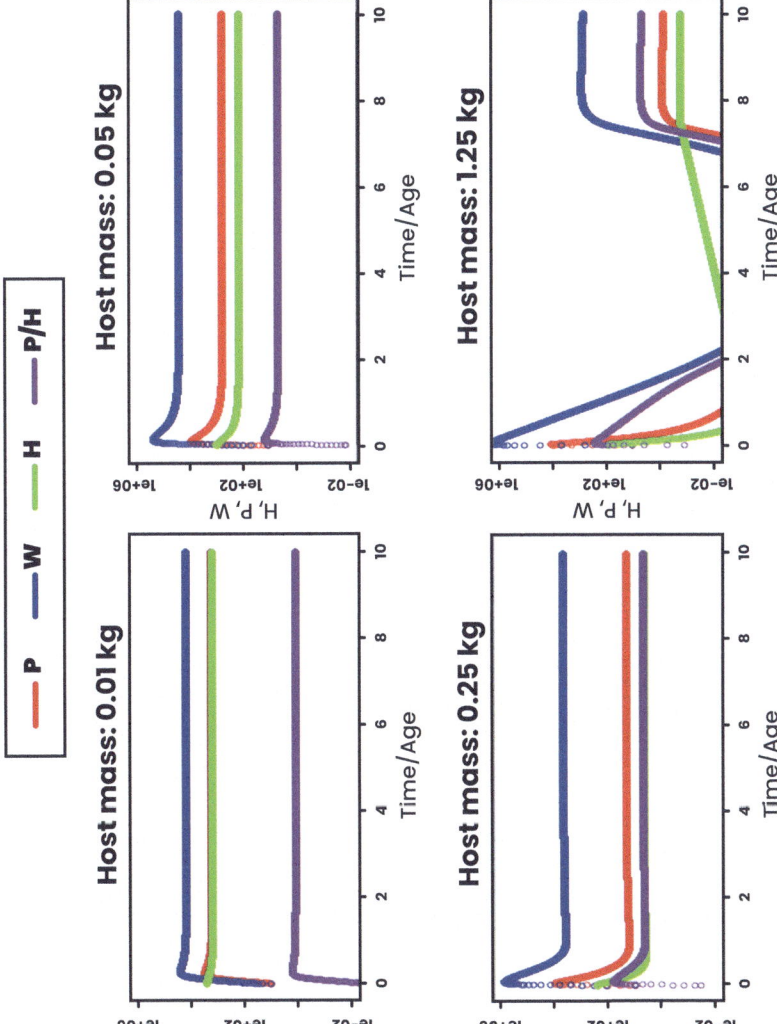

Fig. 21.2 Host–parasite dynamics in the absence of any pollutant, the figure is drawn for hosts of four different body sizes and the model's parameters are calculated from the body size allometries in Table 21.1. The lines in each figure illustrate parasite abundance (P-red), parasite free-living infective stages (W-blue), host abundance (H-green) and mean parasites per host (P/H-purple). Notice the parasites have a larger impact on the host as the hosts increase in body size. The dynamics for the largest host are least stable and the system overshoots before settling to a steady state

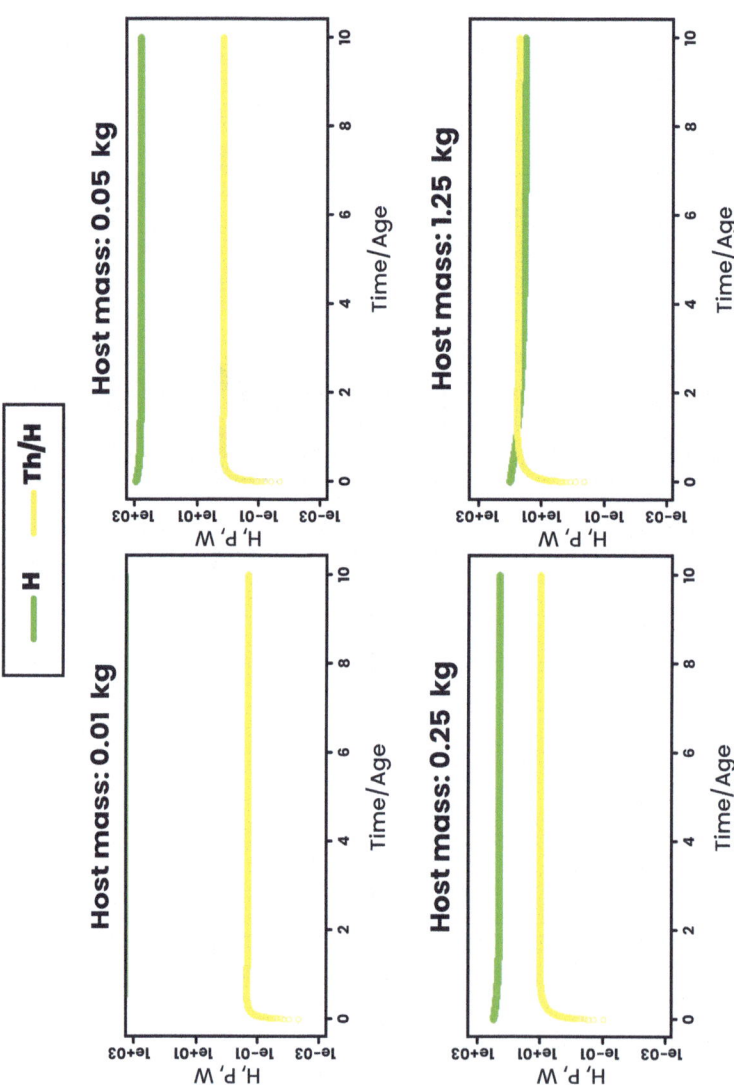

Fig. 21.3 Interaction between host abundance and pollutant in the absence of parasites. Hosts have the same body sizes as in Fig. 21.2. The green line is host abundance (H), and the yellow line is concentration of pollutant in the host (Th/H). Larger hosts accumulate a higher concentration of pollutants and this causes a larger reduction in their equilibrium density. In each case the initial density of the host is its abundance in the absence of the pollutant

eat more) and accumulate it for longer. The latter occurs as pollutant excretion rate is assumed to be a simple multiple of life expectancy and larger hosts have lower death rates. This effect has been commonly observed in other multispecies studies of pollutants, where larger hosts accumulate higher concentrations of pollutants.

The plot begins to thicken when we examine the system with both pollutants and parasites (Fig. 21.4). The host and parasite populations now settle to similar levels of abundance as they did when only the parasites were present, but the levels of pollutants in the hosts decrease as these are now taken up by the parasites. Crucially, the importance of parasites in removing pollutants from the host is more pronounced in the larger hosts, than in the smaller hosts. This is because the worms in the larger hosts live longer and are removing the pollutant from a larger pool accumulated over a longer life span. This would suggest that parasites are much more important in reducing the impact of pollutants on larger-bodied hosts, than in smaller-bodied hosts. It will be intriguing to see if similar effects apply in other groups of hosts, particularly fish and birds. The logic of allometric scaling suggests this should be the case, but the relative magnitude of the effects will certainly be different.

We can examine the broad trends more clearly if we run our model across a more finely parsed range of pollutant concentrations and express our output as concentration of pollutant in host and parasite tissue (Fig. 21.5). When we do this for hosts with relatively small body sizes, we essentially recapture the key results presented in Fig. 21.4 across a wide range of pollutant concentrations. Parasites can reduce the concentration of pollutants in their hosts, but this occurs at the expense of host abundance, which always decreases to lower levels in the presence of the parasite than in the presence of the pollutant.

A more subtle and complex result emerges when we examine the impact of parasites and pollutants on larger-bodied hosts (Fig. 21.6a, b). In the absence of the parasite, increased concentrations of pollutants in the environment lead to increased uptake by the hosts and a reduction in their abundance. When parasites are introduced into the system, they take up significant amounts of pollutants at low pollutant concentrations, but again at the expense of host abundance. As pollutant concentration increases, this eventually drives the parasite to near-extinction and, while this allows the hosts to increase, they are also accumulating increasing concentrations of pollutants. These last two figures suggest there is much more to be explored within this set of models.

21.3 Discussion and Conclusions

A central theme running through many of these results is that both pollutants and parasites have the potential to reduce host populations below the abundance to which they would settle in a world without parasites and pollution. But there is no such thing as a world without parasites, and it will be a long time before we return to a world without significant levels of human-mediated pollutants, which have been an escalating feature of the natural world for around 2 to 5 hundred years. The work by Rachel Carson (Carson 2015) and Ellen Brockovich (Rosskam 2005) only led to transient and local constraints on levels of toxic pollution. It is non-trivially ironic that the parasites that have always been with us have the potential to significantly reduce levels of pollution, though disconcertingly this is aligned with their own ability to reduce host population abundances. Although most ecosystems harbour a healthy and diverse community of parasites and pathogens, the abundances we have quantified in the wild over the last century have always been impacted, and most likely reduced, by the presence of parasites.

The omnipotent presence of parasitic helminths in natural systems may partly explain why we have such difficulty detecting the impact of environmental pollutants on host populations (Dobson et al. 2008). It seems highly likely that the full impact of pollutants is massively buffered by parasites. The best way to test this experimentally would be to find a way of removing or reducing parasites, but this could lead to a significant increase in pollutant levels in the surviving

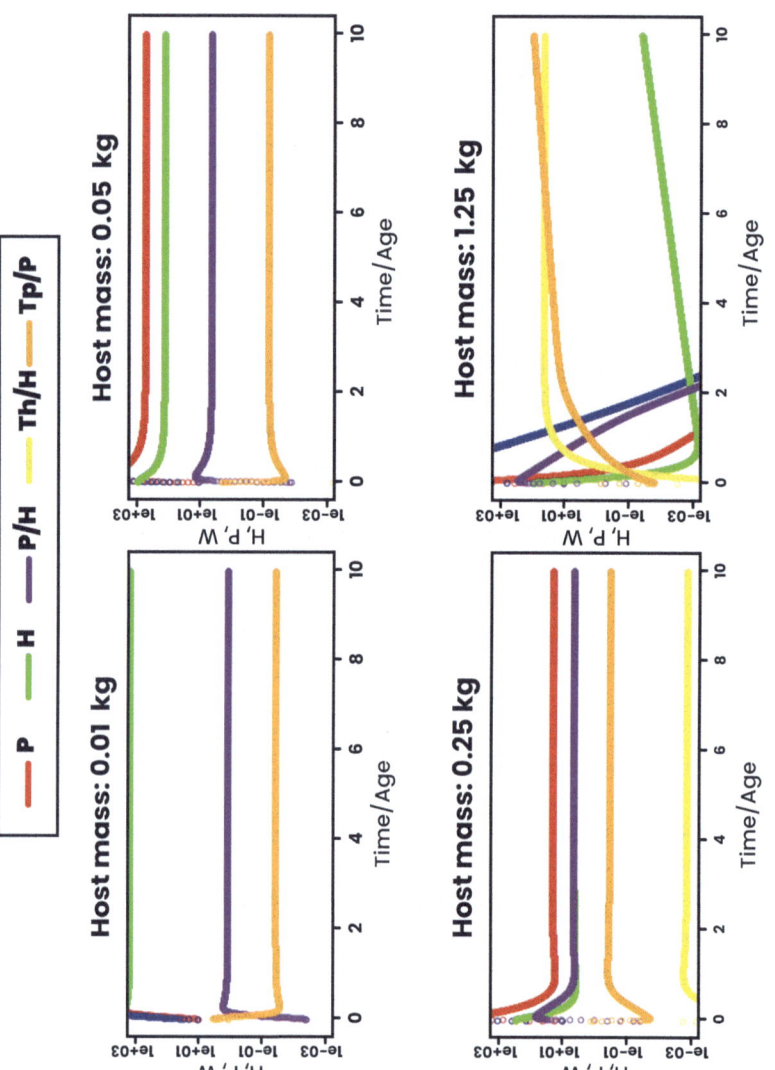

Fig. 21.4 Abundance of host, parasites and pollutant when all are present in the system. Parasite abundance is illustrated in red (P), host abundance in green (H). Mean parasites per host are in purple (P/H), pollutant concentration in host is in yellow (Th/H) and in parasites in orange (Tp/P). In hosts with small body sizes, the parasites reduce the concentration of pollutants and take up most of the pollutant in their own bodies. But, the presence of the parasites always reduces the host abundance to lower levels than occur when only the pollutant is present. In the case of the host with the largest body size the parasites take up so much pollutant that their abundance declines which allows the hosts to increase, but with very high levels of pollutant

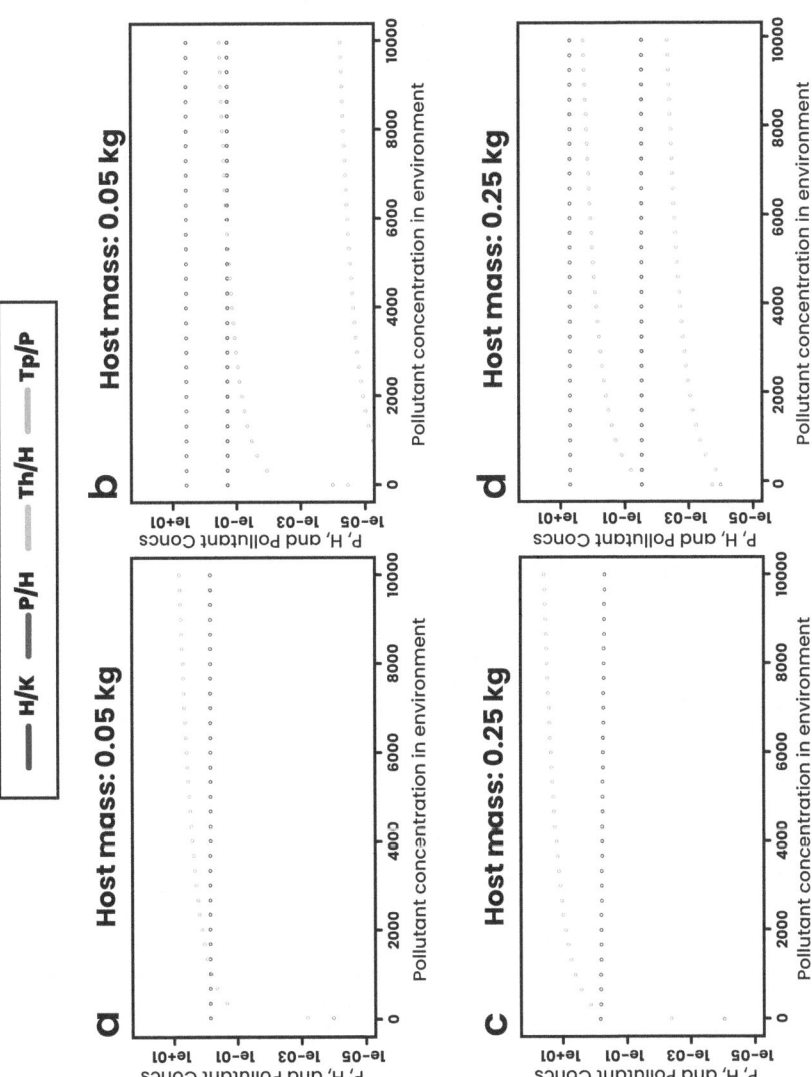

Fig. 21.5 In each of these figures the model for a pollutant concentration illustrated on the x-axis were run. The points are the abundances of host–parasite and pollutant concentration at the end of each model run. The upper figure is for a 0.05 kg host with no parasites (**a**). Host abundance is in blue (H/K) and pollutant concentration in the host in green (Th/H), this rises to a relatively constant level and then saturates. When the parasites are added (**b**) the hosts settle to a lower abundance, but the pollutant concentration in the hosts is much reduced and most pollutant accumulates in the parasites. Similar things happen in (**c**) and (**d**) for a 0.25 kg host

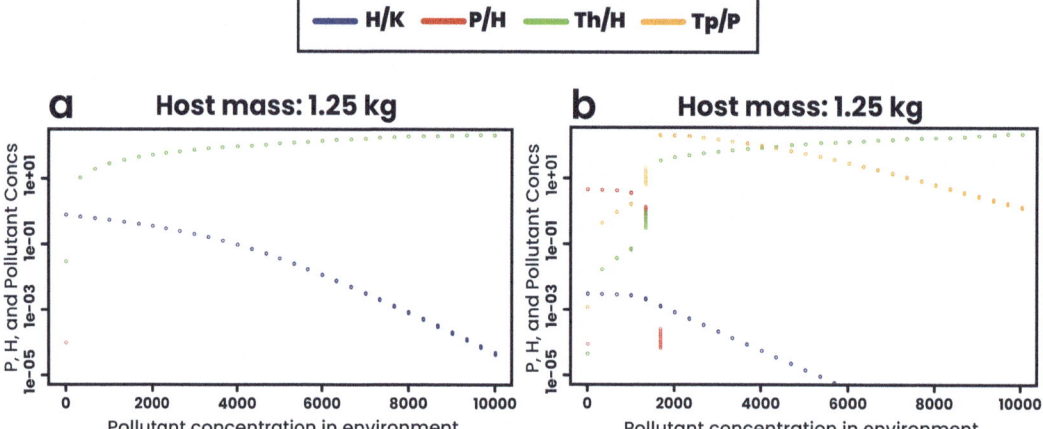

Fig. 21.6 The figure in (**a**) illustrates the concentration of a pollutant in individual hosts (green) and the impact of this on host abundance in the absence of parasites. The figure in (**b**) illustrates how the situation changes when parasites are added. Mean parasite abundance is in red (P/H) and concentration of the pollutant in the parasite in orange (Tp/P). Host abundance and mean parasite burden are still in blue and red, respectively. When pollutant concentrations in the environment are low the parasites reduce the pollutant concentration in the hosts, but still reduce them to a lower abundance than in the absence of parasites. As pollutant concentration in the environment increases this kills the parasites and the hosts concentration of pollutant increases and significantly reduces their abundance

hosts. All of which suggests that a healthy ecosystem is a heavily parasitised ecosystem! Correspondingly, an ecosystem that has lost its parasite diversity may well exhibit extreme levels of adverse effects due to pollution. Underlying this is a deep and tragic irony. Pharmacology has made huge strides over the last century in developing drugs to treat parasitic worms of humans and domestic livestock. All this has occurred over a period when we have been steadily increasing the levels of pollutants in the environment. The huge success of pharmacological parasitology means we may never know the full role that parasites might have played in protecting us from pollution.

There are likely to be important differences in the role played by pollutants and parasites between aquatic and terrestrial ecosystems. Most of the biomass in aquatic ecosystems is in the consumers, particularly fish, most of the parasite biomass in the ecosystem will be in these species and thus parasites of consumers in aquatic systems can potentially sequester a large amount of the pollutants in the system. In terrestrial systems, most of the biomass is in the plants at the base of the food chain, primary and secondary consumers contain a much lower proportion of the biomass and, although they contain a significant diversity and abundance of parasites, the biomass is likely much lower than in aquatic systems. Although most of the evidence we have for the beneficial role of parasites in sequestering pollutants comes from aquatic systems, the models I have presented are parameterised for mammals in terrestrial systems. I have no doubt they could be re-parameterised for fish in aquatic systems, but there is insufficient space to undertake that here. I would thus urge caution in directly extrapolating from the above to aquatic systems, even though I suspect the broad details will be similar.

An area I have not discussed is the role of immunity. Zinc, copper and selenium are vital for immune function in vertebrates, and silver, gold, cadmium and mercury are particularly important from the perspective of pollutants as immune disruptors. These metals replace the element above them in the periodic table and disrupt the enzyme pathways in which they operate (Wood 1974). Cadmium and mercury act as toxicants by replacing zinc in chemical pathways that are vital to innate immunity, such as phagocytosis and cytokine production. They are more common as toxic pollutants than silver and gold, which replace copper and impact the abundance and efficiency of T-cells. Selenium is essential for thyroid function, but less readily replaced by the elements above and below it in the periodic table (sulphur, tellurium and polonium). All of which suggests that parasites that remove cadmium and mercury from host tissue likely have a beneficial effect on host immunity from the perspective of microparasite infections (viruses and bacteria). This is intriguing, as the immune system functions in a state of tension between Type I and Type II immunity. The former protects against viruses and cancers and tends to generate long-term immunological memory, while Type II immunity focuses on parasitic helminths. If parasites are actively reducing host burdens of cadmium and mercury, then they are assisting Type I immunity, although it is unclear if this in turn reduces the efficiency of Type II immunity in inhibiting the fitness of the parasites providing this service.

The potential for metals, particularly cadmium and zinc, to disrupt the immune system suggests there will be interesting and potentially more antagonistic interactions between environmental pollutants and microparasites (viruses and bacteria), as well as cancers. There is insufficient space to explore this here, but the framework described could readily be modified for the simpler life cycles of the classic standardised infection ratio (SIR) pathogens.

There is plainly much more that could be done within the model framework described above. Part of this would focus on parameterisation; there is a considerable amount known about body size and toxicity and I confess myself woefully inadequate in my knowledge of this literature. Similarly, the host–parasite models need to be extended to consider the impact of pollutants on the fecundity of the host and its parasites and the potential impact of the pollutant on the host's immune system and how this modifies its ability to modulate (or amplify!) the impact of parasites.

There are a range of things that could be done to improve the model; many of these would add additional parameters which will make things consistently harder to solve analytically. Nonetheless, it may be relatively straightforward to make uptake rates saturating functions. Similarly, comparisons with the sparse current literature might be improved by calibrating parasite and host deaths in relation to LD50s.

It would take most probably another 20 years to begin to understand this additional material. However, it is plainly an important and understudied synergy that potentially plays a huge role in the functioning of increasingly assaulted natural ecosystems. It is thus very important that we should work towards fully understanding the dynamics of these systems before it is too late to reverse the damage they are causing. Nonetheless, as a parasitologist it totally delights me that the parasitic worms that have entranced me for nearly 50 years may play a vital role in making the world a healthier place!

Acknowledgements I started writing this paper over 20 years ago when I was at a parasite meeting in Germany and Bernd Sures gave a talk about his work on parasites and pollutants. I remember being 'completely blown away' with excitement and on the plane back to Princeton writing down the first iteration of the equations that appear above, essentially adding toxicant dynamics into the classic Anderson and May macroparasite framework. I subsequently gave a talk about these models at a Cary Conference organised by Ric Ostfeld and Felicia Keesing. 'Sloth was my undoing' and I found myself hog-tied between writing a chapter for their book or taking much more time to convert these into a paper for a scientific journal. Inevitably, life intervened and neither got done. The work sat on a computer drive until about 6 months ago when Bernd asked me if I'd like to write a chapter for this book. I dug out my old notes, a very early generation PowerPoint presentation, and decided there might be something worth rescuing. Thanks are therefore due to Bernd for inspiration, patience and perseverance in persuading me to write this all up. Equally inevitably, I received no funding to support this work, but many helpful conversations with Kevin Lafferty, Mercedes Pascual, Peter Hudson, Ian Hatton, Peter Molnar and colleagues at Princeton, NCEAS and the Santa Fe Institute helped the ideas to coalesce.

References

Anderson RM, May RM (1978) Regulation and stability of host-parasite population interactions. I. Regulatory processes. J Anim Ecol 47:219–247

Calder WA (1984) Size, function, and life history. Harvard University Press

Carson R (2015) Silent spring. In: Thinking about the environment. Routledge, pp 150–155

Damuth J (1981) Population density and body size in mammals. Nature, 290(5808), 699–700

De Leo GA, Dobson AP, Gatto M (2016) Body size and meta-community structure: the allometric scaling of parasitic worm communities in their mammalian hosts. Parasitology 143:880

Dobson AP, Lafferty KD, Kuris AM, Hechinger RF, Jetz W (2008) Homage to Linnaeus: How many parasites? How many hosts? PNAS 105:11482–11489

Gatto M, de Leo GA (1998) Interspecific competition among macroparasites in a density-dependent host population. J Math Biol 37:467–490

Hechinger RF, Lafferty KD, Kuris AM (2012) Parasites. In: Sibly RM, Brown JH, Kodric-Brown A (eds) Metabolic ecology: a scaling approach. Wiley, pp 234–247

May RM, Anderson RM (1978) Regulation and stability of host-parasite population interactions. II. Destabilizing processes. J Anim Ecol 47:249–267

Molnár PK, Sckrabulis JP, Altman KA, Raffel TR (2017) Thermal performance curves and the metabolic theory of ecology—a practical guide to models and experiments for parasitologists. J Parasitol 103:423–439

Morand S, Poulin R (2002) Body size–density relationships and species diversity in parasitic nematodes: patterns and likely processes. Evolutionary Ecology Research, 4(7), 951–961

Peters RH (1986) The ecological implications of body size (Vol. 2). Cambridge University Press

Rosskam E (2005) Technical assistance to the grassroots, part ii: Erin Brockovich revisited—a lesson for social policy-making and agency. New Solut 15:107–112

Shaw DJ, Dobson AP (1995) Patterns of macroparasite abundance and aggregation in wildlife populations: a quantitative review. Parasitology, 111(S1), S111–S133

Sures B (2003) Accumulation of heavy metals by intestinal helminths in fish: an overview and perspective. Parasitology 126:S53–S60

Sures B (2004) Environmental parasitology: relevancy of parasites in monitoring environmental pollution. Trends Parasitol 20:170–177

Sures B, Nachev M, Schwelm J, Grabner D, Selbach C (2023) Environmental parasitology: stressor effects on aquatic parasites. Trends Parasitol 39:461–474

Sures B, Nachev SC, Marcogliese DJ (2017) Parasite responses to pollution: what we know and where we go in 'Environmental Parasitology'. Parasit Vectors 10:65

Wood J (1974) Biological cycles for toxic elements in the environment. Science 183:1049–1052

Open Access This chapter is licensed under the terms of the Creative Commons Attribution-NonCommercial-NoDerivatives 4.0 International License (http://creativecommons.org/licenses/by-nc-nd/4.0/), which permits any non-commercial use, sharing, distribution and reproduction in any medium or format, as long as you give appropriate credit to the original author(s) and the source, provide a link to the Creative Commons license and indicate if you modified the licensed material. You do not have permission under this license to share adapted material derived from this chapter or parts of it.

The images or other third party material in this chapter are included in the chapter's Creative Commons license, unless indicated otherwise in a credit line to the material. If material is not included in the chapter's Creative Commons license and your intended use is not permitted by statutory regulation or exceeds the permitted use, you will need to obtain permission directly from the copyright holder.

Climate Change and Parasitism in Aquatic Ecosystems

22

David J. Marcogliese

Abstract

Climate change has become a huge threat to terrestrial and aquatic ecosystems globally. While most attention has been focused on potential effects on free-living organisms, there also has been widespread interest on effects on pathogens and disease. Parasites will no doubt be impacted by climate change, but whether they increase or decrease in abundance, or increase or decrease their distribution, is not obvious. Because transmission of parasites is linked to abundance and diversity of potential hosts, climate change effects on free-living organisms will also affect their parasites. Herein, advances in effects of increasing temperature on parasites and host–parasite interactions are explored, with emphasis on recent advances in acclimation effects and the application of metabolic models to particular host–parasite systems. Climate change will also affect the immune function and physiology of aquatic organisms. Furthermore, climate change does not operate in a vacuum, and there are numerous associated abiotic effects that will impact parasites and their hosts. Effects of precipitation and drought, hydrological changes, eutrophication, acidification, salinity and contaminants in relation to a warming climate are examined for parasites of freshwater and marine ecosystems. Nor do these effects operate independently, and combined effects of multiple stressors on host–parasite systems are discussed. This chapter concludes with an examination of higher order ecological effects and ecosystem consequences of parasitism in a warming climate, and finally, a series of case studies on effects of climate change on particularly well-studied aquatic parasites.

22.1 Introduction

The global biosphere is in the midst of experiencing profound changes to its climate that are unequivocally human induced, according to the latest findings of the Intergovernmental Panel on Climate Change (IPCC) (Arias et al. 2021). Global surface temperatures have increased over 1 °C since the mid-1850s. Various models predict temperature increases from 1.0–1.8 °C to 3.3–5.7 °C in low CO_2 and high CO_2 emission scenarios respectively for the remainder of the twenty-first century (Arias et al. 2021). This warming is considered irreversible for at least centuries. In addition, the frequency of extreme warm conditions will increase, while that of cold

D. J. Marcogliese (✉)
St. Andrews Biological Station, Fisheries and Oceans Canada, St. Andrews, NB, Canada
e-mail: David.Marcogliese@dfo-mpo.gc.ca

© The Authors(s) 2025
N. J. Smit, B. Sures (eds.), *Aquatic Parasitology: Ecological and Environmental Concepts and Implications of Marine and Freshwater Parasites*, https://doi.org/10.1007/978-3-031-83903-0_22

extremes will decrease over that period, with increasing extremes in precipitation in most regions (Arias et al. 2021). Indeed (and not to be alarmist), as I write this chapter during the summer of 2023, the BBC News reported that (i) July 2023 was the hottest day on record since 1940; (ii) June was the hottest month on record globally since 1850–1900; (iii) the entire North Atlantic Ocean was experiencing an intense heatwave from April to July 2023 compared to 1985–1993 averages; (iv) the daily average sea surface temperatures were the highest since 1979 between 50° and 60° N in the Atlantic Ocean, being 5 °C higher than average in 2023; and (v) the extent of Antarctic sea ice was the lowest in July since 1979 (https://www.bbc.com/news/science-environment-66229065; accessed 3 November 2023).

It must be remembered that climate warming does not happen in a vacuum. There are myriad knock-on environmental effects in all ecosystems, including aquatic ones (Marcogliese 2001, 2008, 2016, 2023). In oceans, numerous unprecedented changes have occurred and will continue to occur in the decades to come (Table 22.1). All these perturbations are likely to affect parasitism in marine organisms. Many of these, including deep-ocean warming, acidification and sea level rise, are considered irreversible for millennia (Arias et al. 2021).

Environmental changes in freshwaters are similar in relative magnitude (Table 22.2). Among the most important are an increase in temperature, more intense precipitation, altered hydrology, a decrease in oxygen concentration, eutrophication and an increase in contaminant levels, largely due to prolonged low-flow periods and runoff from heavy precipitation (Jiménez Cisneros et al. 2014).

The Intergovernmental Science-Policy Platform on Biodiversity and Ecosystem Services (IPBES) and the World Wildlife Fund (WWF) have identified climate change as one of the major drivers of biodiversity loss from marine and freshwater ecosystems (IPBES 2019; WWF 2020). Marine and freshwater ecosystems have been altered globally and effects on free-living species are extensive. For those species where data are available, approximately half have shifted ranges poleward and two-thirds have shifted phenology earlier in the season (Pörtner et al. 2022). Ecosystems have experienced loss of local species, increases in disease, and mass mortality events. Irreversible shifts in ecosystems in the oceans have resulted from loss of corals, kelps and seagrasses (Pörtner et al. 2022). Biodiversity and abundance of organisms have decreased in riverine ecosystems (Jiménez Cisneros et al. 2014). Effects on freshwater, coastal and marine ecosystems are, and will be, exacerbated by pollution, habitat fragmentation, land use changes and invasive species, leading to further irreversible changes (Jiménez Cisneros et al. 2014; Settele et al. 2014; Pörtner et al. 2022).

Being poikilothermic organisms, and in aquatic ecosystems mainly parasitic on poikilothermic hosts, parasites are expected to be heavily influenced by temperature. A meta-analysis of experimental studies using a variety of invertebrates and vertebrates found that exposure to temperature as a stressor increased pathogen intensity (Vicente-Santos et al. 2023). Furthermore, due to their complex life cycles and reliance on trophic transmission, parasite biodiversity is linked to the diversity of free-living animals (Hechinger and Lafferty 2005; Hechinger et al. 2007; Kamiya et al. 2014; Johnson et al. 2016). As a consequence, parasites in aquatic ecosystems are expected to be impacted extensively by climate change (reviewed by Dobson and Carper 1992; Patz et al. 1996, 2000; Harvell et al. 1999, 2002; Marcogliese 2001, 2008, 2016, 2023; Altizer et al. 2013; Burge et al. 2014; Barber et al. 2016; Lõhmus and Björklund 2015; Byers 2020, 2021; Rohr and Cohen 2020; Woo et al. 2020). Indeed, range shifts among parasites could lead to 5–10% extinction of species due to habitat loss induced by climate change, while extinction of hosts from climate change and other associated pressures could result in a 30% loss of helminth species alone (Carlson et al. 2017). In general, climate change should affect species with complex life cycles that use multiple hosts to complete their development and reproduction more than those with direct life cycles (Harvell et al. 2002; Rohr et al. 2011). Specialist parasites should also be more at risk than generalists (Rohr et al. 2011), as should those that infect poikilothermic hosts at some point in their life cycle (Harvell et al. 2002). It should be noted that the latter includes the vast majority of parasites in aquatic ecosystems. In their global analysis of wildlife parasites, Carlson et al. (2017) also found that ectoparasites were more likely to experience range reductions than endoparasites in a changing climate.

Numerous research gaps and confounding factors in climate change research into parasitism and disease have been identified (Marcogliese 2001, 2016; Rohr et al. 2011; Altizer et al. 2013). Some outstanding questions include (i) what is the role of temperature variability and acclimation on parasites, hosts and parasite–host interactions; (ii) what are the net effects of temperature changes on parasite fitness; (iii) what are the ecophysiological effects of climate change on parasites and hosts, including on host immunity; (iv) will there be range shifts, contractions or expansions of parasites, and will phenological mismatches occur between parasites and hosts; (v) what are the impacts of the various abiotic changes resulting from climate change on parasitism, including precipitation and water flows, eutrophication, acidification, sea level and salinity and contaminant levels (see Tables 22.1 and 22.2); and (vi) what are the cascading effects of climate-induced changes in parasitism and disease at a community level? These issues will be the focus of attention in the remainder of this chapter.

22.2 Temperature and Parasites

Temperature is undoubtedly the best studied environmental factor affecting parasites. Classic, extensive reviews of its effects on helminths date back over 40 years (Chubb 1977, 1979, 1980, 1982). Higher temperatures will accelerate parasite growth rates, development and maturation (Marcogliese 2001; Barber et al. 2016), but also increase mortality (Dobson and Carper 1992; Harvell et al. 2002) (Fig. 22.1), rendering predictions of net effects difficult (Dobson and Carper 1992; Harvell et al. 2002; Altizer et al. 2013). Not only do effects of temperature on parasites vary among species, even closely-related ones, but they can vary among stages within the same life cycle (Marcogliese 2001; Morley 2011; Altizer et al. 2013; Morley and Lewis 2015; Barber et al. 2016; Scott 2023). In addition, effects on parasites

Table 22.1 Effects of climate change on major environmental conditions in marine waters globally according to the Sixth Assessment Report of the Intergovernmental Panel on Climate Change (from Arias et al. 2021). Trends will continue throughout the remainder of the twenty-first century. The time frame provides reference conditions. Effects on host–parasite systems have been discussed in the various reviews cited

Environmental condition	Direction of effect	Time frame	Review of effects on marine parasites
Temperature	Increase	1850–1900	Harvell et al. (2002), Marcogliese (2008, 2016, 2023), Byers (2020, 2021)
Salinity contrasts	Increase	1950s	Marcogliese (2001, 2008, 2016), Byers (2021)
Upper ocean stratification	Increase	1970s	Byers (2021)
Marine heatwaves	Increase	1900s	Marcogliese (2008, 2016, 2023)
Oxygen levels	Decrease	1950s	Marcogliese (2001), Byers (2021)
Oxygen minimum zones	Expand	1950s	
Acidification	Increase	Late 1900s	Marcogliese (2016), Byers (2021)
Sea level	Increase	1901	Marcogliese (2001, 2016)
Oceanic circulation	Change	1993	Marcogliese (2001)
Ice cover	Decrease	Late 1970s	Marcogliese (2001)

Table 22.2 Effects of climate change on major environmental conditions in freshwaters globally, in part from the Fifth Assessment Report of the Intergovernmental Panel on Climate Change (from Marcogliese 2001, 2008, 2016; Jiménez Cisneros et al. 2014). Effects on host–parasite systems have been discussed in the various reviews cited. The time frame for most of these changes is the same as those listed in Table 23.1 (Arias et al. 2021)

Environmental condition	Direction of effect	Review of effects on freshwater parasites
Temperature	Increase	Marcogliese (2001, 2008, 2016)
Precipitation	Increase	Marcogliese (2001, 2008, 2016)
Extreme weather	Increase	Marcogliese (2016)
Hydrology	Change	Marcogliese (2001, 2016)
Eutrophication	Increase	Marcogliese (2001, 2008, 2016)
Water quality	Decrease	This study
Stratification	Increase and time shift	Marcogliese (2001)
Ice cover	Decrease	Marcogliese (2001)
Acidification	Change	Marcogliese (2001)
Sea level rise	Increase	Marcogliese (2001, 2016)
UV radiation	Increase	Marcogliese (2001)

are further complicated by effects of increasing temperature on their hosts, particularly poikilothermic ones (Marcogliese 2001; Altizer et al. 2013). Predictions are further complicated still by nonlinear effects of climate change on parasite or host fitness, which should decrease at temperatures both lower and higher than optimal (Rohr et al. 2011; Altizer et al. 2013; Rohr and Cohen 2020). Essentially, in order to accurately predict effects of climate change on parasites with complex life cycles, effects of temperature on all parasite stages and on all hosts in the life cycle, as well as the various host–parasite interactions, must be known (Fig. 22.1; Barber et al. 2016). Temperature can also moderate habitat selection and behaviour of both hosts and parasites, with implications for host encounter rates with and resistance to parasites (Fig. 22.1; Barber et al. 2016). For example, swimming speed of both miracidia and cercariae of the human trematode pathogen *Schistosoma mansoni* was positively related to temperature, implying that warmer temperatures may lead to increased transmission (Nguyen et al. 2020). However, these behavioural modifications are not necessarily predictable. Increases in temperature affect cercarial activity levels in separate species of trematodes at different magnitudes and sometimes opposite directions (Koprivnikar and Poulin 2009; Koprivnikar et al. 2010; Selbach and Poulin 2020). Both cercarial emergence and activity of cercariae of the intertidal trematode *Maritrema novaezealandense* increased on average with temperature, but were also affected by parasite genotype, rendering a complex situation even more complicated (Berkhout et al. 2014). Indeed, the direction of the temperature response varied with genotype.

Free-living stages of parasites are directly affected by environmental conditions, including temperature (Pietrock and Marcogliese 2003). As a group, where effects of temperature on free-living stages have been synthesised in large-scale analyses, trematodes are among the best studied. In one meta-analysis of published studies, cercarial emergence from the molluscan intermediate host increased up to 200-fold following a 10 °C increase, with a mean of an eightfold increase (Poulin 2006). These increases in cercarial productivity were higher than would be predicted from the sensitivity of most physiological functions, where two- to threefold increases would be expected. In contrast, in another meta-analysis of low and mid-latitude species that incorporated both a minimum emergence temperature threshold and acclimation, cercarial emergence was not affected between 20 and 25 °C, while declining at higher temperatures (Morley and Lewis 2013). Temperature also had no effect on cer-

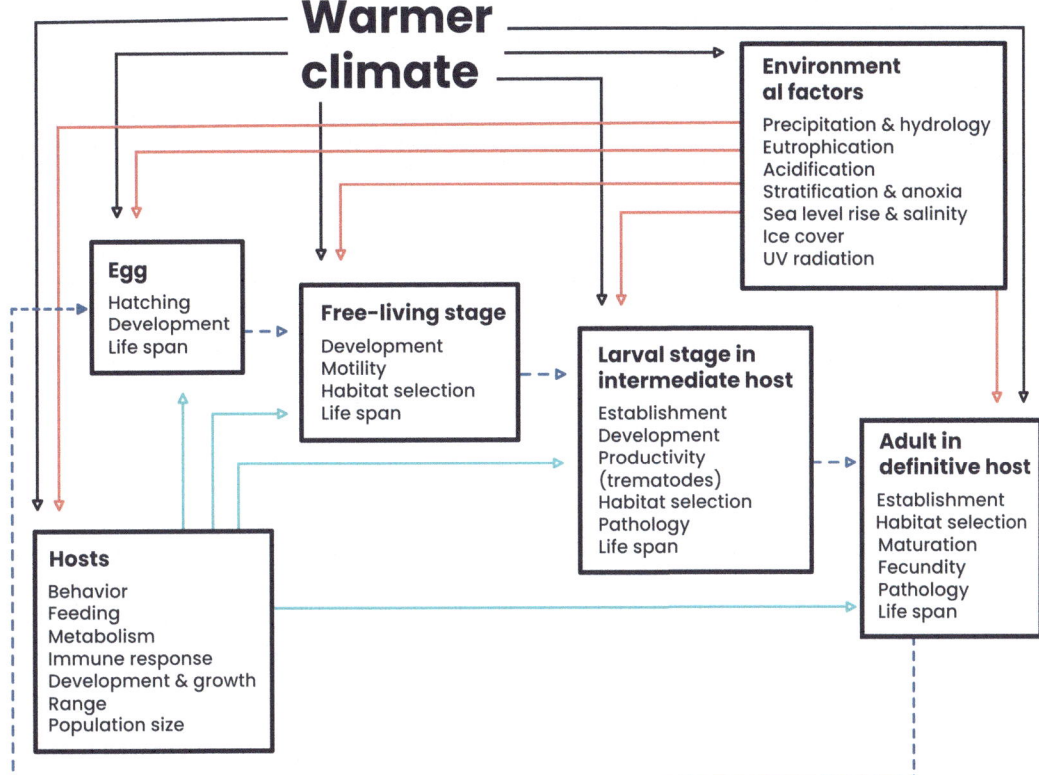

Fig. 22.1 Conceptual model depicting the parasite life-cycle traits susceptible to modification by a warming climate for a hypothetical helminth parasite possessing a free-living egg, a free-living infective stage, one poikilothermic intermediate host and one poikilothermic definitive host. The host traits influenced by temperature pertain to both the intermediate and the definitive host. Solid black lines: direct effects of temperature on parasite stages and hosts; dashed dark blue lines: sequential progression of parasite life history stages; solid light blue lines: effects of modification of host traits due to a warming climate on parasite life cycle stages; solid red lines: effects of environmental perturbations associated with a warmer climate on parasite life cycle stages and hosts. It should be noted that effects will vary not only among parasite species, but among stages within species. Furthermore, effects will vary depending on the combination of stressors experienced in a particular ecosystem. Modified and redrawn from Marcogliese (2001)

carial development with the molluscan host. Together, these results suggest some degree of thermostability (Morley and Lewis 2013). The two studies yielded vastly different predictions for the effects of climate change on trematode transmission. Trematode egg development increases with temperature, but slows at high temperatures, a typical invertebrate response (Morley and Lewis 2017). In contrast, egg hatching shows thermostability across trematode species at midrange temperatures, then declines at high temperatures (Morley and Lewis 2017). Thermostability was also observed in trematode miracidia, which hatch from eggs and infect the molluscan host. Indeed, both survival and metabolism of miracidia were not affected to a great degree by temperature, and even less so than cercariae, reflecting thermostability at this life cycle stage as well (Morley 2012). Lastly, miracidial and cercarial infectivity to the next host increased with temperature within optimal thermal ranges, subsequently declining at higher temperatures (Morley and Lewis 2015). Metacercarial infectivity, in contrast, is highest at low temperatures, and declines as temperature increases.

For fish pathogens with direct life cycles, such as sea lice (*Lepeophtheirus salmonis*) parasitic on salmonids, one effect of warmer temperatures is to increase development but also mortality of larval stages and possibly limit dispersal (Costello 2006; Fast and Dalvin 2020). However, prolonged warmer temperatures into late autumn extend the infectious period, allowing infection prevalence and abundance to increase. Reduced generation time and prolonged exposure periods probably will result in greater overall reproduction as well (Fast and Dalvin 2020). However, warmer temperatures above a particular threshold (18 °C) during summer could limit parasite development and even halt reproduction (Fast and Dalvin 2020). Intertidal eastern oysters (*Crassostrea virginica*) exposed to warmer air temperatures have higher intensities of the protist parasite *Perkinsus marinus* compared to subtidal, submerged oysters that are exposed to cooler temperatures (Malek and Byers 2018). While not detrimental to the oysters themselves, experimental exposure of oysters to warmer-than-average air temperatures of 35 °C resulted in a peak in parasite intensity, suggesting climate warming will have a negative effect on eastern oysters through increased infection levels of *P. marinus* (Malek and Byers 2018).

One difficulty in evaluating effects of climate change on parasites is the absence of long-term data. Recently, Wood et al. (2023) examined long-term changes between 1880 and 2019 in parasites of marine fishes in Puget Sound, off the northwestern coast of the United States, using specimen collections from museums to obtain the historical data, in addition to historical records of temperature and other environmental factors. They found a decline in 52% of certain species over time. These declines occurred in helminth parasites with three or more hosts in their life cycles, but not those with one or two hosts, suggesting that life cycle complexity contributes to parasite vulnerability to climate change (Wood et al. 2023). Although temperature only explained a limited amount of the variability in parasite abundance over time, they found no effect of contaminant levels or host density on abundance.

22.2.1 Thermal Acclimation and Temperature Variability

Temperature variability and thermal extremes are expected to increase with climate change (Arias et al. 2021). Thermal acclimation capacity reflects an organism's ability to respond to varying temperatures (Rohr et al. 2013, 2018). Because of their small size and faster metabolic rates compared to their hosts, parasites may be expected to acclimate faster to varying temperatures, and thus proliferate under conditions of climate change (Raffel et al. 2013; Rohr and Cohen 2020). Indeed, a large-scale analysis of 500+ species of ectotherms suggests that small organisms acclimate faster than large ones (Rohr et al. 2018). While effects of temperature on parasites have been studied for decades (Marcogliese 2001), earlier experimental work failed to account for acclimation (Morley and Lewis 2013; Marcogliese 2016) yet, in a study of trematode cercarial development and emergence from their molluscan intermediate hosts, consideration of acclimation effects led to conclusions that differed from previous studies (Morley and Lewis 2013). In a set of experiments that incorporated acclimation effects, cercarial release of the freshwater trematode *Ribeiroia ondatrae* from its snail intermediate host [*Planorbella* (=*Helisoma*) *trivolvis*] was affected by acclimation temperature and its duration (Paull et al. 2015). Acclimation effects also varied with temperature and differed between *R. ondatrae* and its green frog (*Lithobates clamitans*) tadpole hosts (Altman et al. 2016). Cercarial establishment in tadpoles, as well as elimination of metacercarial cysts, also varied with acclimation temperatures, but not to the same degree (Altman et al. 2016).

Host–parasite interactions are affected not only by an increase in temperature but also by variations in temperature (Marcogliese 2016). Exposure to a shifting temperature regime led to higher intensity of the chytrid fungus *Batrachochytrium dendrobatidis* (*Bd*) on Cuban treefrogs (*Osteopilus septentrionalis*) and greater host mortality compared to a constant temperature regime (Raffel et al. 2013; see also Chap. 4). Furthermore, a comprehensive analysis of *Bd* infecting anurans in the genera *Atelopus* and *Telmatobius* in Central and South America clearly linked host population declines to variability in temperature associated with global El Niño climatic events (Rohr and Raffel 2010). Cercarial production of *M. novaezealandense* from infected New Zealand mud snails (*Zeacumantus subcarinatus*) and infectivity to the next host (the amphipod *Paracalliope novizealandiae*) in the life cycle were affected differently by exposure to variable temperatures compared with constant ones in an intertidal trematode (Studer and Poulin 2013). It is worth noting that transmission success was highest in a heat wave treatment (Studer and Poulin 2013). In another heat wave experiment, exposure of the snail intermediate hosts (*Lymnaea stagnalis*) to the trematode *Echinoparyphium aconiatum* resulted in increased infection, whether exposure was 3 or 7 days. However, in a second experiment, increased infection success was only observed after 7 days (Leicht and Seppälä 2014). Temperature shifts also affected the magnitude of cercarial release in the *R. ondatrae*-*P. trivolvis* parasite–host system (Paull et al. 2015). Overall, these studies demonstrate that acclimation effects and temperature variability have different effects on parasites and their life cycle stages, their hosts and host–parasite interactions, rendering predictions problematic (Marcogliese 2016).

22.2.2 Net Effects of Climate Change on Parasite and Host Traits

As stated, temperature has different effects on parasites and their life cycle stages, their hosts and host–parasite interactions (Marcogliese 2001, 2016). Consequently, it is imperative for our understanding of parasitism and disease in a changing climate that temperature effects on host–parasite systems be examined across a parasite's entire life cycle (Barber et al. 2016; Marcogliese 2016, 2023). A few studies have been able to examine effects of warmer temperatures simulating climate change on transmission to and infections in multiple intermediate hosts within a parasite's life cycle (Table 22.3). These studies clearly highlight the nonlinearities that can result when considering temperature effects on both hosts and parasites (see Rohr et al. 2011). Furthermore, they reinforce the notion that, despite the common patterns described in Sect. 22.2, net effects are variable among species and host–parasite systems, rendering predictions difficult.

In the *R. ondatrae* system mentioned above, exposure to high temperatures resulted in an increased rate of development of parasite eggs and cercariae, while increasing cercarial mortality (Paull and Johnson 2011). It also enhanced growth, fecundity and mortality of *P. trivolvis*, the snail intermediate host, while in infected snails, growth rate increased but fecundity was reduced (Paull and Johnson 2011). In the second intermediate host, in this case the Pacific chorus frog (*Pseudacris regilla*), cercarial penetration increased but their establishment decreased with increasing temperature (Paull et al. 2012). Metacercarial intensity also decreased at the highest temperature (26 °C), but pathology in the form of malformations was highest at the intermediate temperature (20 °C) and lowest at the highest temperature (Paull et al. 2012).

Cercarial productivity from the snail *Z. subcarinatus*, infection of the amphipod second intermediate host (*P. novizealandiae*) and metacercarial development all peaked at intermediate temperatures (20–25 °C) in *M. novaezealandense* (Studer et al. 2010). Amphipod survival decreased with temperature, but even though parasite transmission was reduced, parasite-induced mortality was greatest at the highest temperatures (≥30 °C) (Studer et al. 2010).

Another study on a marine trematode, *Himasthla elongata,* provided comparable information. At warm temperatures (22 °C), cercarial productivity from the first intermediate host snail, the periwinkle *Littorina littorea* and infectivity to the second intermediate host, the blue mussel (*Mytilus edulis sensu lato*), peaked while cercarial survival was reduced (Díaz-Morales et al. 2022), similar to the studies above. Snail mortality also increased at warm temperatures. The authors concluded that mortality of cercariae and snails at warmer temperatures would more than

Table 22.3 Net effects of increases in temperature on trematode parasites in intermediate hosts from diverse ecosystems determined from experimental studies on the parasite and its hosts, and host–parasite interactions. Asterisk denotes heatwave

Parasite	Ecosystem	Temperatures (°C)	First intermediate host	Second intermediate host	Net effect	References
Ribeiroia ondatrae	Freshwater	13, 20, 26	Snail (*Planorbella trivolvis*)	Amphibian tadpole (*Pseudacris regilla*)	Metacercarial numbers lowest at 26 °C; Tadpole malformations highest at 20 °C	Paull and Johnson (2011), Paull et al. (2012)
Maritrema novaezealandense	Intertidal	16, 20, 25, 30, 34	Snail (*Zeacumantus subcarinatus*)	Amphipod (*Paracalliope novizealandiae*)	Metacercarial development highest at 25 °C	Studer et al. (2010)
Maritrema novaezealandense (temperature variability study)	Intertidal	15, 15 + 5, 15 + 10, 15 + 15, 15 + hw*	Snail (*Zeacumantus subcarinatus*)	Amphipod (*Paracalliope novizealandiae*)	Amphipod infections highest at 15 + 5 °C and 15 °C + hw	Studer and Poulin (2013)
Himasthla elongata	Intertidal	10, 22, 28	Snail (*Littorina littorea*)	Mussel (*Mytilus edulis sensu lato*)	Net infectivity in mussels highest at 22 °C	Díaz-Morales et al. (2022)

offset any increases in cercarial productivity and infectivity, thus reducing transmission (Díaz-Morales et al. 2022). Curiously, in a separate experiment on the same host–parasite system, moderate infections of *H. elongata* in blue mussels (*M. edulis sensu lato*), in combination with increasing temperatures, had additive negative effects on mussel survival, but high infection levels appeared to protect mussels from the negative impacts of temperature stress on those hosts (Selbach et al. 2020).

22.2.3 Ecophysiological Approaches

Accurate models are required to predict effects of climate changes on host–parasite interactions (Rohr et al. 2013; Molnár et al. 2017). Two approaches are possible. The first, standard susceptible-infected-recovered (SIR) models are data-intensive and unique to specific host–parasite systems. The second, more general strategic models are based on general physiological responses to temperature change and can account for nonlinearities in the data (Altizer et al. 2013; Rohr et al. 2013; Rohr and Cohen 2020). In that regard, the Metabolic Theory of Ecology holds great promise (Rohr et al. 2011, 2013; Altizer et al. 2013; Rohr and Cohen 2020). This application develops relationships between temperature and parasite and host body size and metabolism to predict the effects of climate change on host–parasite interactions (Rohr et al. 2011; Altizer et al. 2013; Molnár et al. 2017). The model was successfully applied to an Arctic terrestrial nematode-caribou system (Molnár et al. 2013) but requires testing on other parasites (Molnár et al. 2017), especially in aquatic environments.

Parasites and their hosts may respond differently to the same temperature conditions (Marcogliese 2001; Altizer et al. 2013) and possess different optimal peaks in performance. The responses of host and parasite vital rates can be described by thermal performance curves (TPCs), which are typically nonlinear and diminish either gradually or abruptly at the edges of tolerable temperature ranges (Molnár et al. 2017; Byers, 2020, 2021). TPCs can be compared between hosts and parasites to determine which will benefit with increases in temperature. More specifically, TPCs can be calculated for parasite growth or replication and the host ability to reduce or limit parasite growth (Raffel et al. 2013; Rohr et al. 2013). TPCs can also be used to compare survival of infected and uninfected hosts over a range of temperatures (Byers 2020). Gehman et al. (2018) experimentally demonstrated a shift in thermal optima for survival of the crab *Eurypanopeus depressus* infected (or not) with the rhizocephalan *Loxothylacus panopaei*, such that parasitised hosts had a lower thermal optimum and lethal maximum for survival than uninfected crabs. Essentially, the infected crabs had lower optimal temperatures for survival compared to uninfected ones (Gehman et al. 2018). Consequently, parasite prevalence was predicted to decline with increases in temperature, and extinction would occur with 2 °C warming. Thus, these authors suggested that *L. panopaei* is unable to expand northward from its current range along the eastern coast of the United States without adaptation (Gehman et al. 2018). A comparison of TPCs for the parasitic dinoflagellate *Haematodinium* spp. in two different crab species suggested that the parasites would fare better than their hosts at warmer temperatures (Shields 2019). In contrast, if a host possesses a broader thermal range than a parasite, perhaps due to a greater acclimatisation capacity owing to its larger body size, it may be able to find thermal refuge from its parasite(s) (Gsell et al. 2023). Another complicating factor is that infection can alter a host's thermal limits to that host's detriment (Hector et al. 2021).

Parasites are predicted to outperform their hosts as temperatures shift, and parasites should attain their highest abundance in cases where they respond better to temperature shifts than their hosts, rather than optimal temperatures determined in isolation (Rohr and Cohen 2020). This is termed the Thermal Mismatch Hypothesis (TMH) developed for the chytrid fungus *Bd* in amphibians across almost 600 populations (Cohen et al. 2017). It suggests that hosts from cold climates are more at risk of infection under warm conditions, while those from warm climates are more at risk under cold conditions (Cohen et al. 2017; Rohr and Cohen 2020). Indeed, an analysis of >7000 terrestrial and freshwater host populations observed just that, with ectothermic hosts displaying the strongest results (Cohen et al. 2020). Furthermore, parasites with direct life cycles showed stronger pathological effects than those with complex life cycles (Cohen et al. 2020). Models predicted that ectotherms from the temperate zone will experience increases in infection risk, while infection risk will be reduced in those from tropical regions (Cohen et al. 2020). Temperate-zone helminths fared best in temperate areas, while fungal parasite prevalence decreased in the tropics. The TMH best explained the timing and location of declines in a large number of populations of *Atelopus* spp. due to *Bd* compared to alternative models and was supported by laboratory experiments (Cohen et al. 2019; Rohr and Cohen 2020).

22.2.4 Climate Change and the Immune Response

The immune system in poikilotherms, like all physiological processes, is temperature-dependent (Bowden 2008; Scharsack and Franke 2022). Immune function and parasite prevalence both vary seasonally in fish due to variations in temperature (Magnadottir 2010), but depend on the adaptive temperature range of the host, which is species specific (Scharsack and Franke 2022). Lysozyme activity, plasma immunoglobulin M (IgM) concentration and gene expression of the major histocompatibility complex (MHC) and cytokines are all temperature dependent in fish, with lysozyme and IgM levels positively correlated with temperature (Bowden 2008). The innate immune response, however, is relatively temperature independent (Magnadóttir 2006). Indeed, within the tolerable temperature range in a species of fish, the innate immune response functions at relatively low temperatures and across a wide temperature range, while the acquired immune response functions better at higher temperatures (Magnadóttir 2006, 2010; Scharsack and Franke 2022). Both systems, and overall immune function, work best at intermediate temperatures (Scharsack and Franke 2022). Typically, acute temperature changes reduce phagocytosis activity, respiratory burst performance and antibody production (Martin et al. 2010). Indeed, temperature extremes, especially heatwaves, may result in lower immune function and increased disease (Martin et al. 2010). Rapid and large temperature shifts lead to stress and immune dysfunction, and thus variability in temperature can also lower disease resistance (Harvell et al. 1999; Scharsack and Franke 2022). Nevertheless, given that warm temperatures increase immune enzyme activity (Altizer et al. 2013), global warming may have positive effects on ectotherm immunity to disease (Martin et al. 2010; Rohr et al. 2011), as long as temperatures are not near the host's upper permissive range (Scharsack and Franke 2022). Alternatively, long-term warm temperatures may result in chronic stress, impeding the immune response and resistance to disease (Marcogliese 2001, 2008; Harvell et al. 2002; Scharsack and Franke 2022). In addition, infections by pathogens and parasites can also lower upper thermal limits (Marcogliese 2008; Hector et al. 2021). A fundamental question that arises is 'Who benefits most from climate change, the parasite or the host?' (Scharsack et al. 2016). Although increased temperatures accelerate parasite growth and development, host immune activity may also increase. Furthermore, if the immune response modulates host–parasite interactions, climate change may be expected to have less of an influence on parasite population dynamics (Harvell et al. 2007).

Raffel et al. (2006) explored how changes in temperature can alter optimal immune function. By examining immune function in red-spotted newts (*Notophthalmus viridescens*) collected seasonally, they tested two predictions operating at different time scales. First, over the short term, the lag hypothesis states that temperature-dependent immune responses lag behind temperature changes. Second, the seasonal acclimation hypothesis states that immune cell production declines as temperature drops over the long term until the host acclimates to the lower temperatures. Results backed both hypotheses: A lag of a few days was observed on lymphocyte levels in newts in response to changes in temperature; in contrast, seasonal acclimation effects were observed in newts, lowering levels of lymphocytes, neutrophils and eosinophils in the blood in the autumn (Raffel et al. 2006). The authors concluded that temperature variability may increase susceptibility to infection (Raffel et al. 2006).

Climate change may affect parasites differently, depending on the processes regulating their populations. A study on two monoxenous nematodes in European rabbits (*Oryctolagus cuniculus*) over two decades deserves mention, as no comparable studies exist for aquatic ecosystems. Populations of *Graphidium strigosum* are unregulated in the rabbit, while those of *Trichostrongylus retortaeformis* are regulated by the immune response (Mignatti et al. 2016). Climate warming was associated with an increase in larval infective stages in the soil for both nematodes. Yet, intensity of *G. strigosum* increased in the rabbits over time, while no long-term trend was observed for *T. retortaeformis* (Mignatti et al. 2016). Of course, these results may not apply to ectotherms, where the immune response varies with temperature.

In general, in those fishes that have been studied (Fig. 22.2), the acquired immune response and immune gene expression are increased, while the innate immune response is reduced at the highest temperature tested (Dittmar et al. 2014). Effects of temperature on immune function in aquatic ectotherms are perhaps best studied in the three-spined stickleback (*Gasterosteus aculeatus*) (see also Sect. 22.6). Surprisingly, measuring a number of immune parameters in sticklebacks from two different populations, both the innate and acquired immune responses were found to be highest at suboptimal temperatures (13 °C). Innate and acquired immune responses were lower following an increase from optimal (18 °C) to warmer (24 °C) temperatures (see also Franke et al. 2017, 2019). These include respiratory burst as an indicator of the innate immune response and lymphocyte proliferation, reflecting the acquired immune response (Dittmar et al. 2014; Franke et al. 2017, 2019; Scharsack et al. 2021). Immune function was disrupted after exposure to a simulated heatwave at 28 °C and subsequent return to optimal temperatures (Dittmar et al. 2014). Interestingly, cold-adapted fish from a stream environment responded more strongly and were more susceptible to the heatwave effect than those from a more variable habitat. Following exposure of three-spined sticklebacks held in experimental mesocosms to a heatwave, survival of those with an intermediate number of major histocompatibility complex (MHC) IIb alleles survived best and had the lowest number of parasites compared to those with low and high numbers of MHC IIb alleles (Wegner et al. 2008). Thus, MHC diversity may play a role in effects of extreme climatic events on fish health and resistance to parasitism. A laboratory study examining gene expression in 14 immune-associated genes found that exposure to increasing temperatures affected gene expression in three-spined sticklebacks. Prevailing short-term thermal effects, where temperatures were altered post-parasite exposure, had a stronger impact on gene expression than lagged long-term thermal effects where the temperature regime was altered pre-parasite exposure, although the latter effects were not negligible (Stewart et al. 2018). Twelve of the 14 genes displayed changes in gene expression between 7 and 23 °C, while 6 of 14 were affected between 7 and 15 °C, in the prevailing temperature effects experiment (Stewart et al. 2018). However, in a long-term mesocosm experiment allowing differentiation of thermal effects and nonthermal seasonal effects on immune gene expression, latent seasonal effects proved as important as thermal effects, with implications for parasitic disease severity (Stewart et al. 2018). In general, in three-spined sticklebacks, there is a host stress response and immune activity is lower following acute temperature increases. In chronic exposures to higher temperatures, the immune response shows some signs of recovery but there are still indications of stress to the host (Scharsack et al. 2016).

Like other poikilotherms, amphibians display temperature-dependent immunity (Rollins-Smith and Woodhams 2012; Rollins-Smith 2017). It is

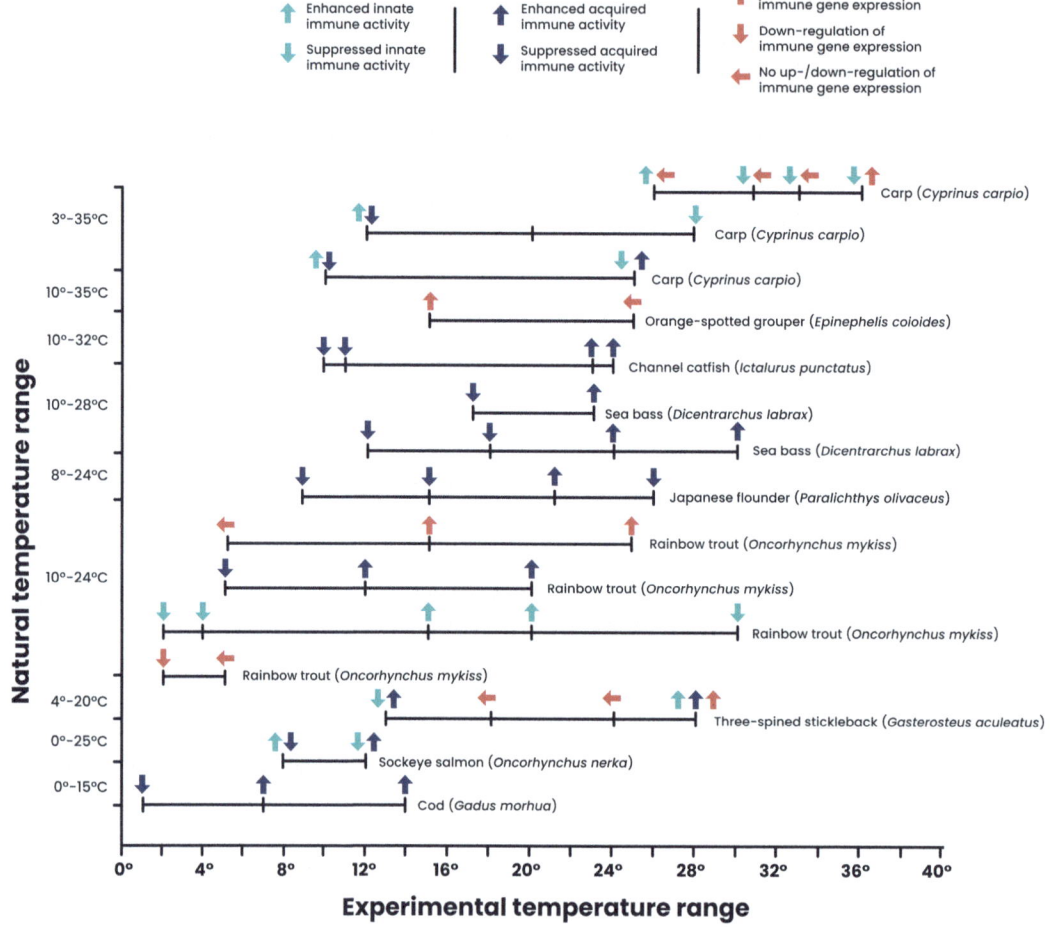

Fig. 22.2 Schematic of temperature effects on immune activity in several freshwater and marine fishes derived from published experimental studies. Horizontal bars on the *x*-axis show the temperature range studied. Arrows indicate the temperature at which a particular effect on the immune system occurred for that species. The *y*-axis displays the natural temperature range of the different fishes (Redrawn with permission from Dittmar et al. 2014). Original references can be found in Dittmar et al. (2014)

well-known that amphibian immunity is impaired at low temperatures (Rollins-Smith 2017), but relatively little is known about the effects of exposure to warm temperatures (Rollins-Smith and Le Sage 2023). However, the amphibian immune system is similar to that of other vertebrates, although it is completely re-organised at metamorphosis (Rollins-Smith and Woodhams 2012). It should be expected that amphibian immunity may be enhanced at high temperatures, as is in fish. The Pacific chorus frog (*P. regilla*) displayed the highest clearance rate of trematode metacercariae in experimental infections carried out between 17 and 26 °C at the highest temperature tested (Paull et al. 2012). Evidence suggests amphibians may be better able to resist infection by chytrid fungi (*Batrachochytrium* spp.) at warmer temperatures (Rollins-Smith 2017). The microbiome of larval amphibians has received some recent attention and is believed to function in disease resistance. These microbial communities are altered at elevated temperatures, possibly reducing resistance to pathogens (Rollins-Smith and Le Sage 2023).

Much less is known about the effects of temperature on the immune system in invertebrates. The invertebrate immune system is non-adaptive and consists of both innate and humoral components (Mydlarz et al. 2006; Ellis et al. 2011). Compared to vertebrates, there are relatively few studies on invertebrates (Ellis et al. 2011). In freshwater, phenoloxidase and antibacterial activities were lower at 30 °C compared to 15 °C in the pond snail, *Lymnaea stagnalis* (Seppälä and Jokela 2011). A simulated heat wave had no effect on the immune response in *L. stagnalis* over 1–5 days, but haemocyte concentration and phenoloxidase-like activity were reduced during simulated heatwaves maintained over longer periods (Leicht et al. 2013, 2017; Salo et al. 2017). In a marine isopod, *Idotea balthica*, a simulated heatwave resulted in a reduction in immunocompetence, in particular in phagocytosis (Roth et al. 2010). In contrast, phenoloxidase activity was reduced after the simulated heatwave in males, but not females (Roth et al. 2010). Haemocyte numbers were not affected by simulated heatwaves in either the isopod or *L. stagnalis* (Roth et al. 2010; Seppälä and Jokela 2011). A meta-analysis revealed that prolonged heatwaves have a strong effect on the oxidative status of aquatic vertebrate ectotherms (Messina et al. 2023). Numerous other examples of effects of an increase in temperature on innate and humoral components of the immune system in marine invertebrates are provided in Table 22.4 and in Mydlarz et al. (2006). In general, the duration of exposure is important, and effects depend on whether stressors such as temperature are tested in combination or in isolation (Ellis et al. 2011). Overall, temperature increases will impact components of both the innate and humoral immune response, including antimicrobial proteins, the phenoloxidase system, antibacterial proteins and antioxidant enzyme activity, among others (Mydlarz et al. 2006; Ellis et al. 2011). Corals and other organisms that harbour obligate algal symbionts are especially impacted by temperature stress (Mydlarz et al. 2006). Indeed, disease outbreaks in immunocompromised hosts associated with climate change are considered best known in corals. Thermal stress, including warming, leads to the expulsion of symbiotic algae from the corals, resulting in bleaching, which is believed to increase susceptibility to opportunistic pathogens, such as the fungus *Aspergillus sydowii* (Mydlarz et al. 2006).

22.3 Range Shifts, Seasonality and Phenology

Global warming has led to host range shifts in both freshwaters and the oceans (Pörtner et al. 2022). These shifts will bring parasites into new habitats and into contact with new, naïve hosts (Marcogliese 2001; Altizer et al. 2013). For example, Marcogliese (2001) extrapolated a total of 83 potential parasitic introductions into the North American Great Lakes that could arrive with invading fishes expanding their ranges northwards in a changing climate. Nevertheless, it is unclear if latitudinal shifts in parasite species distributions due to climate change will result in net increases or decrease in disease (Lafferty 2009). Furthermore, climate change will likely provide conditions favourable to the invasion and establishment of alien species outside their native ranges (Marcogliese 2001; Walther et al. 2009; Byers 2021).

Perhaps the best recognised association of a range change in parasitism associated with temperature and climate is that of the eastern oyster *Crassostrea virginica*, and its protozoan parasite, *Perkinsus marinus*, the causative agent of Dermo disease. On the east coast of the United States, this oyster expanded its northern distribution from Long Island to Maine during a warm winter period and disease outbreaks are now observed in areas much further north of its previous range (see Harvell et al. 2002; Marcogliese 2008; Burge et al. 2014; Byers 2021; Okon et al. 2023). Epizootics of another disease of eastern oysters, MSX disease, caused by the protist *Haplosporidium elson*, have also spread northward along the eastern seaboard, following warm winter conditions (see Marcogliese 2008; Burge

Table 22.4 Effects of increased temperature on innate and humoral components of the immune response in marine invertebrates. Results derived from review by Ellis et al. (2011). Original sources may be found in Ellis et al. (2011)

Species	Immune function	Temperature treatment	Response
Crustaceans			
Pacific white-leg shrimp (*Penaeus vannamei*)	Total haemocyte count (THC)	Increase	Decrease in first 3 d, then recovery
	Phenoloxidase activity	Increase	Decrease
	Serine protease activity	Increase	Decrease
	Proteinase inhibitor activity	Increase	Decrease
Indian spiny lobster (*Panulirus homarus*)	Superoxide dismutase (SOD) activity	Increase	Increase
Molluscs			
Blue mussel (*Mytilus edulis*)	MGD2 (antimicrobial protein)	Heat shock	Increase
	Phagocytic activity	Increase	Increase
	THC	Increase	Increase
	Differential cell counts	Increase	No change
Mediterranean mussel (*Mytilus galloprovincialis*)	MGD2	Heat shock	Increase
	AMP gene expression	Heat shock	Decrease
	Myticin mRNA	Heat shock	Increase
	Defensin regulation	Increase	Increase
	Myticin B regulation	Increase	Increase
	Mytilin B regulation	Increase	No effect
Eastern oyster (*Crassostrea virginica*)	Reactive oxygen intermediate (ROI) production	Sudden increase	No effect
Zhinhong scallop (*Chlamys farreri*)	SOD activity	Increase	No effect
	Acid phosphatase activity	Increase	Decrease
Manila clam (*Ruditapes philippinarum*)	Leucine aminopeptidase concentration	Temperature & pathogen challenge	Altered
Echinoderms			
Caribbean sea fan coral (*Gorgonia ventalina*)	Amoebocyte number	Increase	Increase
	Antifungal activity	Warm temperature + *Aspergillus sydowii*	Increase
Common starfish (*Asterias rubens*)	ROI production	Increase	Decrease
	Coelomocyte number	Increase	No change
Japanese common sea cucumber (*Apostichopus japonicus*)	Lysozyme activity	Acute change	Altered
	SOD activity	Acute change	Altered
	Catalase activity	Acute change	Altered
	Myeloperoxidase activity	Acute change	Altered

et al. 2014). Other pathogens such as sea lice may also be expected to expand their distribution northward (Fast and Dalvin 2020).

Shifts in the phenology of hosts and seasonal transmission patterns will also impact host–parasite interactions (Marcogliese 2001; Altizer et al. 2013). Many parasites have adapted their life cycles to the seasonal occurrence of their hosts to promote transmission between intermediate and definitive hosts (Marcogliese 2001). However, there is little information on phenological changes in parasite abundance with temperature. Paterson et al. (2023) searched for general patterns in the abundance of trematodes in both intermediate and definitive hosts from freshwater ecosystems during the transition from spring to summer, as seasonal temperatures warm up. Aside from a weak tendency for prevalence to increase between spring and summer in first intermediate snail hosts, they were unable to detect any universal trends in prevalence or abundance between the seasons in second intermediate or definitive hosts, regardless of host taxa or habitat, among representatives from 39 different trematode families. Indeed, population declines from spring to summer were as common as increases (Paterson et al. 2023). Clearly, these results indicate varying species-specific responses and once again render predictions of effects of climate change on parasites problematic.

Exposure to high temperatures accelerated intermediate snail host growth, egg production and mortality, as well as tadpole development. They also promoted both egg and cercarial development of the trematode *R. ondatrae* (Paull and Johnson 2011; Paull et al. 2012). This parasite causes malformations in anurans, but there is a critical window in amphibian development when the tadpoles are most sensitive to limb deformities caused by the parasite (Bowerman and Johnson 2003; Paull et al. 2012). This critical window explains in part why some amphibian species have higher rates of deformities than others. Such shifts in the timing of host–parasite interactions could move parasite transmission to tadpoles outside the critical window (see Paull and Johnson 2018). If this happens, then the malformation rate in the amphibian would decline. Although it is not known if malformations increase the vulnerability of metamorphosed anurans to predation (Johnson et al. 2004), it is not unreasonable to presume so. Any reductions in predation on infected amphibians could lower transmission to the definitive hosts and limit subsequent parasite reproduction.

Severe limb malformations caused by *R. ondatrae* were 50% higher in Pacific chorus frogs (*Pseudacris regilla*) from warmer low-altitude sites compared to cooler high-elevation areas, after controlling for parasite intensity (McDevitt-Galles et al. 2020). The phenological synchrony between host and parasite was considered to be greatest at the warmer temperatures, leading the authors to conclude that climate warming could increase malformation rates (McDevitt-Galles et al. 2020).

Timing of cercarial release of two trematodes from snails and infection of mosquitofish (*Gambusia affinis*) was altered in a reservoir exposed to thermal effluents in South Carolina, United States. The period of transmission of *Tylodelphys scheuringi* from *Planorbella trivolvis* to mosquitofish was extended in the reservoir, although cercarial release ceased during the warmest months, compared to ambient conditions (Aho et al. 1982). Similarly, timing of cercarial release of *Posthodiplostomum* (= *Ornithodiplostomum*) *ptychocheilus* from *Physa* sp. was also prolonged and also ceased during the warmest months, compared to reference waters. However, while the period of recruitment of this trematode by mosquitofish was extended, it was also interrupted during the warmest months (Camp et al. 1982). Recruitment of the cestode *Schyzocotyle* (= *Bothriocephalus*) *acheilognathi* continued over an entire year in mosquitofish in another thermally altered reservoir in North Carolina, whereas there was no recruitment during colder months in ambient waters (Granath and Esch 1983). As an aside, these studies also illustrated the value of using thermally altered cooling reservoirs to simulate effects of climate change on parasite population dynamics (Dobson and Carper 1992; Marcogliese 2001).

22.4 Associated Environmental Effects

As noted above, environmental effects on climate change are not limited to temperature (Tables 23.1 and 23.2). Knock-on effects permeate through both freshwater and marine ecosystems. It should be pointed out here that the tolerance of parasites and hosts for any of the environmental conditions described below may be quite different. If the parasite's tolerance is broader than that of its host, it may fare better with environmental change. However, if the host's tolerance is broader, it may be able to find an environmental refuge from its parasites, similar to the thermal refuge effect previously described (Gsell et al. 2023).

22.4.1 Precipitation and Hydrology

Variability in precipitation is expected to increase, with the frequency and intensity of heavy precipitation increasing in most regions globally (Arias et al. 2021). Droughts are also expected to increase (Arias et al. 2021). These changes could have profound effects on freshwater and coastal ecosystems (Fig. 22.3). Effects on parasites in freshwaters have been reviewed by Marcogliese (2001, 2016). With heavy precipitation comes increased runoff and higher streamflow. Parasite free-living stages can be removed from habitats in strong currents, as well as experience low infectivity rates (Marcogliese 2001, 2016). These include those of myxozoans such as *Myxobolus cerebralis*, the pathogen causing whirling disease in salmonids and trematodes such as *Schistosoma mansoni* and *Diplostomum spathaceum* (see Marcogliese 2001, 2016). Similarly, strong current flow can disrupt transmission of ectoparasites with direct life cycles (Marcogliese 2001).

Marcogliese (2001) predicted that high streamflow would reduce parasite diversity, in part based on the experimental results of studies on myxozoans and trematodes. Furthermore, parasite diversity was negatively correlated with streamflow in the plains killifish (*Fundulus zebrinus*) and the fathead minnow (*Pimephales promelas*) in the Platte River system, Nebraska, United States, with a decline in prevalence and abundance of larval trematodes (Janovy and Hardin 1988; McDowell et al. 1992; Janovy et al. 1997). In the Richelieu River, Quebec, Canada, parasite species richness was actually higher in spottail shiners (*Notropis hudsonius*) during a high-water year compared to a low-water year, but overall parasite abundance was greater in the low-water year (Marcogliese et al. 2016). In addition, mean myxozoan infracommunity species richness in spottail shiners was negatively correlated with water flow in the St. Lawrence River, Quebec. Rainfall was shown to have seasonal effects on parasites of snails and fish in Celestun coastal lagoon, in Yucatan, Mexico. The proportion of infected hosts for several parasite species predominantly displayed lag-correlated increases associated with the prevailing rainfall patterns over long-term periods, although immediate increases in prevalence also were observed (Pech et al. 2010). Rainfall also influences the prevalence of the chytrid fungus *Bd* in amphibians, being higher during the rainy months in the Brazilian Atlantic Forest (Ruggeri et al. 2018). In contrast, prevalence of *Bd* was negatively associated with annual precipitation in seven species of anuran across sites in Costa Rica (Whitfield et al. 2017).

Water levels are expected to drop in some freshwater ecosystems with climate change, including the Great Lakes and St. Lawrence River, because evaporation is predicted to exceed precipitation (Marcogliese 2001). Consequently, infections with trematodes such as *Diplostomum* spp., which causes cataracts and blindness in fish, and those causing blackspot have been predicted to increase (Marcogliese 2001). In the Pacific Northwest, climatic changes in temperature and precipitation are expected to lead to earlier spring runoff, lower spring floods and extended periods of lower and warmer summer discharge, all of which could result in enhanced transmission of the myxozoan *M. cerebralis* (Alexander and Bartholomew 2020).

While precipitation is expected to increase in many parts of the world, it is decreasing in others, resulting in decline in stream flow globally

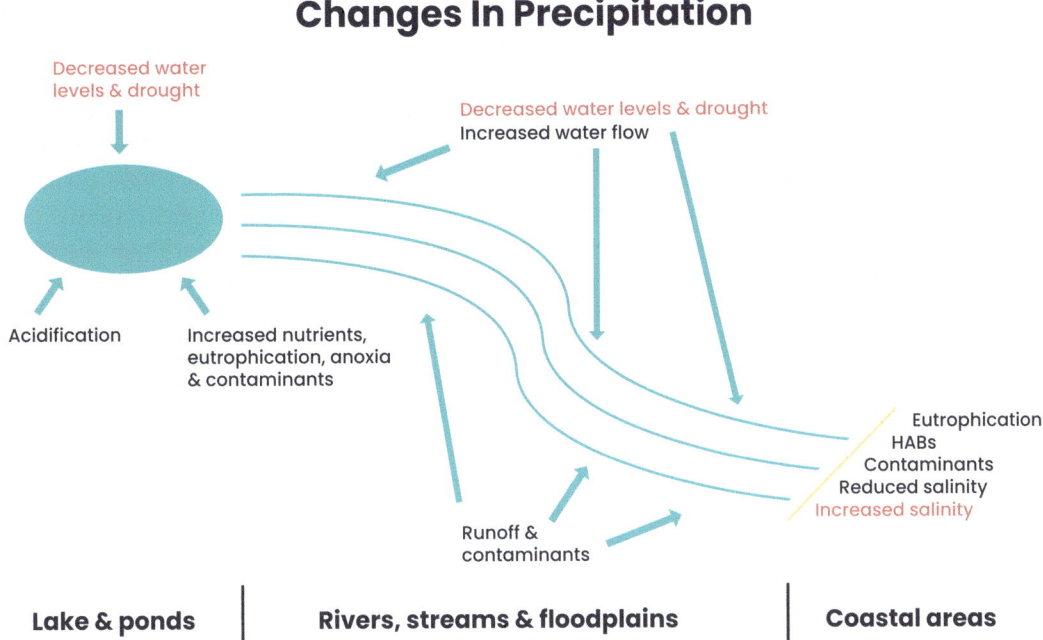

Fig. 22.3 General schematic of abiotic and biotic effects of changes in precipitation on lakes, rivers and coastal areas. Effects of increased precipitation are shown in black text. Effects of decreased precipitation are shown in red text. Changes will affect parasites and their communities in different ways, depending on their sensitivity to the various abiotic parameters, their mode of transmission, and their intermediate and definitive hosts. Many of the abiotic and biotic effects shown would be exacerbated by increasing temperature. *HABs* harmful algal blooms

(Zhang et al. 2023), loss of wetlands and floodplain area (Marcogliese 2001; Schindler 2001), extended dry seasons (Fu et al. 2013), and more frequent regional droughts (Feyen and Dankers 2009; Espinoza et al. 2019; Stokstad 2021; Bochow and Boers 2023). An extensive 16-month drought in the United Kingdom in 1976–77 had huge effects on animal and plant life. Freshwater ecosystems experienced higher temperatures, lower water levels, reductions in dissolved oxygen and eutrophication from nutrient inputs (Morley and Lewis 2014a). Fish farms were subjected to numerous outbreaks of disease, including costiasis, caused by the protist *Ichthyobodo necator*, and saprolegniosis, caused by the fungus *Saprolegnia parasitica*. Low stream flows, increasing temperatures and eutrophication were associated with the emergence of various parasites in aquatic wildlife, while others declined. However, parasites, including protists, trematodes and fungi, responded positively to the drought conditions overall (Morley and Lewis 2014a). Decreasing water levels and increasing drought affect the seasonality and phenology of parasite transmission processes (Marcogliese 2001). For example, prevalence of metacercariae of *R. ondatrae* in metamorphs of the Pacific chorus frog dropped significantly with reductions in pond area. With ponds drying more rapidly, tadpoles metamorphosed faster, and consequently suffered reduced exposure to infective cercariae (Paull and Johnson 2018). These effects predominated over those of eutrophication on infection intensity and degree of malformations in frogs (Paull and Johnson 2018; see Sect. 22.4.2 Eutrophication, below). Differences in infections were observed between temporary and permanent ponds in another frog-trematode system. Mean abundance of metacercariae of the trematode *Telorchis* sp. was four times higher in grey treefrog (*Dryophytes versicolor*) tadpoles in temporary ponds compared to permanent ponds (Kiesecker and Skelly 2001). Tadpoles exposed to infected snails (*Pseudosuccinea columella*) in temporary ponds experienced significantly higher mortality and lower metamorphosis success, compared to no effect in permanent ponds, yet the presence of uninfected snails had no effect on anuran growth and development in either type of pond (Kiesecker and Skelly 2001). The higher infection rate, with subsequent effects on the anuran host, was attributed to increasing water temperature and declining water volume as ponds dried (Kiesecker and Skelly 2001). Similarly, outbreaks of Bd in the foothill yellow-legged frog (*Rana boylii*) from the San Francisco Bay Area watershed was observed during extremely low stream flows in autumn and following absence of peak flows in winter (Adams et al. 2017). The low autumnal streamflow forced the frogs to concentrate their numbers in drying pools, while the lack of peak winter flow permitted American bullfrogs [*Lithobates catesbeianus* (=*Rana catesbeiana*)] that can distribute the infection, to expand their local range. Furthermore, intensity of Bd was negatively correlated with low streamflow (Adams et al. 2017).

22.4.2 Eutrophication

Eutrophication is predicted to increase in freshwater ecosystems with climate change (Table 22.2). Coastal marine ecosystems have experienced an overall increase in phytoplankton blooms between 2003 and 2020 (Dai et al. 2023) and eutrophication has increased in numerous estuarine ecosystems (Byers 2021). Intense phytoplankton blooms have spiked in freshwater lakes globally between 1984 and 2012 (Ho et al. 2019). Oceans, coastal marine waters and lakes have all experienced declines in dissolved oxygen, associated at least in part to climate change (Jane et al. 2021). Exposure to hypoxia, even if only temporary, can inhibit immune function in fishes and crustaceans (Bowden 2008; Uribe et al. 2011; Abdel-Tawwab et al. 2019; Shields 2019; Byers 2021). The effects of nutrient input and eutrophication on parasites have been reviewed and summarised numerous times (Marcogliese 2004, 2005; Johnson and Carpenter 2008; McKenzie and Townsend 2007; Blanar et al. 2009; Johnson et al. 2010a; Budria 2017; Byers 2021). Indeed, the subject has also been reviewed specifically in the direct context of climate change (Marcogliese 2001, 2016; Byers 2021). Some common patterns and trends regarding parasites of freshwater fishes have emerged associated with eutrophication (reviewed in Marcogliese 2001). Compared with oligotrophic systems, eutrophic ones are characterised by parasites of cyprinid fishes and birds, that are more generalist, allogenic, possess direct or short life cycles with oligochaetes and zooplanktonic intermediate hosts, and tend to be digeneans and cestodes. In contrast, oligotrophic systems tend to have parasites of salmonid fishes, specialists, autogenic, possess complex life cycles with benthic intermediate hosts, and tend to be nematodes and acanthocephalans. Thus, a shift can be expected from the latter group to the former group as eutrophication proceeds (Marcogliese 2001).

Meta-analyses have shed further light on how eutrophication affects parasites in aquatic ecosystems. Lafferty (1997) found that in general eutrophication had a positive effect on parasites. A general analysis of overall patterns led to the same conclusion by Blanar et al. (2009). However, when accounting for effect sizes, Blanar et al. (2009) revealed that eutrophication had negative effects on trematode populations and those parasites with complex life cycles. Somewhat different results were generated by another meta-analysis that accounted for both effect size and natural variation in parasite population parameters (Vidal-Martínez et al. 2010). In this analysis, eutrophication had positive effects on parasites and their communities. Furthermore, positive effects were observed on most parasite taxa (Vidal-Martínez et al. 2010).

However, none of the analyses accounted for the level of eutrophication. Overall effects have been shown to vary with the intensity of eutrophication, which may explain some of the contrasting results in the analyses described above (Fig. 22.4). Nutrient input tends to promote parasite populations and diversity, most likely because increased primary production promotes the populations of free-living invertebrates that serve as

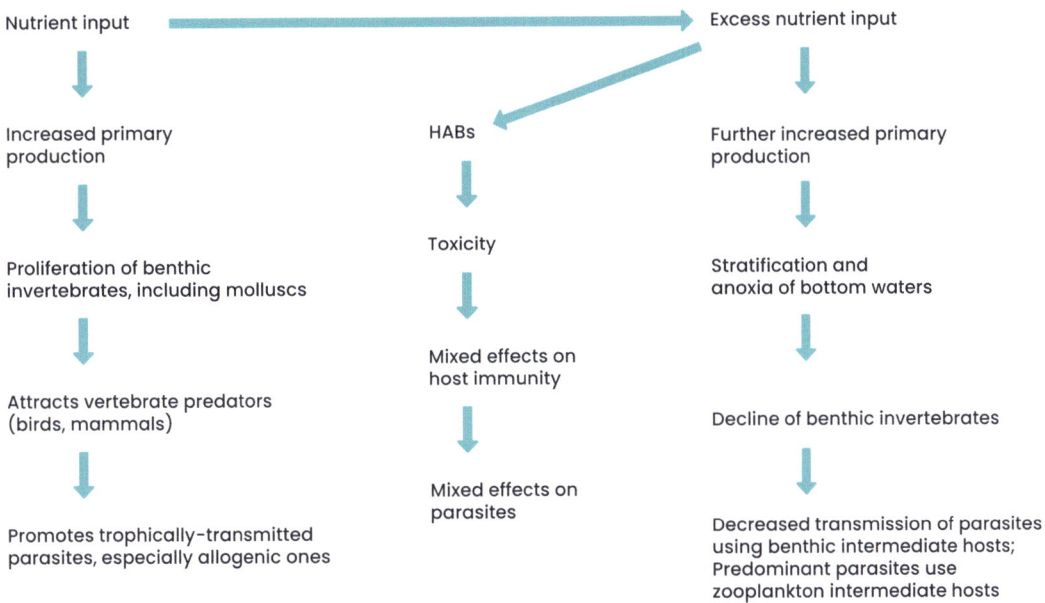

Fig. 22.4 Conceptual model of chain of events associated with eutrophication and subsequent effects on free-living biota and parasites. After a threshold is reached, profound changes occur that affect species composition of parasites. Changes will affect individual parasite species in different ways, depending on their sensitivity to the various abiotic parameters, their mode of transmission, and their intermediate and definitive hosts. While parasite species richness declines with excessive eutrophication, certain parasite species will proliferate. *HABs* harmful algal blooms

intermediate hosts (Marcogliese 2001). As eutrophication increases, water quality decreases and bottom waters become anoxic. Not only does hypoxia occur in lakes and the oceans, but it is increasing in riverine ecosystems (Zhi et al. 2023). Consequently, an ecosystem will see a decline in the populations and species richness of free-living animals, with the concomitant reduction in parasites that rely on these hosts for transmission (Marcogliese 2001). Parasites of small-bodied fishes in the Baltic Sea illustrate this phenomenon quite nicely. As eutrophication increases and bottom waters become anoxic, fish parasite communities experience a transition from allogenic parasites with long, complex life cycles involving benthic intermediate hosts, to those of low diversity dominated by autogenic parasites with shorter life cycles utilising zooplankton as intermediate hosts (Zander 1998; Zander and Reimer 2002). The long-term studies in the Baltic over wide spatial scales of varying degrees of eutrophication also illustrate why digeneans respond positively to eutrophication, at least initially. As productivity rises, the snail intermediate hosts flourish, and the local system with abundant benthic prey attracts birds to feed there, providing all the components necessary for allogenic trematodes to complete their life cycles. Thus, trematodes are favoured but, once eutrophication proceeds to a level where bottom waters become anoxic, the snail hosts and other bottom dwellers can no longer survive, fewer birds frequent the area and both free-living and parasite diversity decline (Zander and Reimer 2002). Another large-scale study across an island chain in the Central Pacific found that productivity was correlated with overall abundance of trophically-transmitted parasites in five species of fish, but not in parasites with direct life cycles (Wood et al. 2015). Among parasite taxa, positive associations with productivity were seen for trematodes and cestodes, but not other helminths. The authors attributed results to an increase in zooplankton and molluscan intermediate hosts (Wood et al. 2015). Moreover, the association of parasites with productivity held for specialists, but not generalists. Again, the authors attributed this to the availability of their respective hosts, which was more limited for the specialist parasites (Wood et al. 2015).

Eutrophication may contribute to higher levels of disease in aquatic ecosystems, especially generalist parasites with direct or simple life cycles (McKenzie and Townsend 2007; Johnson et al. 2010a). The trematode *R. ondatrae* causes malformations in their anuran intermediate hosts. Increasing nutrient input results in greater numbers of the molluscan first intermediate host, *Planorbella* spp. (Johnson and Chase 2004). Mesocosm experiments demonstrated that snails proliferated due to high algal productivity (Johnson et al. 2007). Malformations increased in amphibian hosts, in part due to higher densities of infected snail hosts, but also due to enhanced parasite cercarial production in snails that experienced greater growth and condition at higher nutrient levels (Johnson et al. 2007). Indeed, a field study of *R. ondatrae* and *Echinostoma* spp. in snail (*Planorbella trivolvis*) and frog (*P. regilla*) intermediate hosts suggested that eutrophication effects can dominate temperature effects in their study system (Paull and Johnson 2018). Abundance of metacercariae of the trematode *Posthodiplostomum* spp. in pumpkinseed (*Lepomis gibbosus*) increased across a series of streams in southern Ontario, Canada, concordantly reflecting the degree of eutrophication (Chapman et al. 2015). The authors attributed this to the positive influence of eutrophication on the snail first intermediate hosts. Other parasites known to proliferate with eutrophication are those transmitted by oligochaetes, including the nematode *Eustrongylides* spp., a known pathogen of waterfowl, and various myxozoans, which can be pathogenic in fish (Marcogliese 2001, 2005, 2016; Alexander and Bartholomew 2020).

Eutrophication promotes outbreaks of the myxozoan *Tetracapsuloides bryosalmonae*, the causative agent of proliferative kidney disease in salmonids (Okamura et al. 2011). It is transmitted by bryozoan alternate hosts that experience population increases with eutrophication (see also Chap. 3). However, nutrient levels also promote parasite growth and development (Okamura et al. 2011, Okamura 2016). Mean myxozoan prevalence and infracommunity species richness in spottail shiners (*Notropis hudsonius*) in the St. Lawrence River were higher downstream of the City of Montreal, Quebec, Canada, between 1999 and 2004 (Marcogliese and Cone 2001; Marcogliese et al. 2009). This effect was attributed to enhanced nutrient input from the city's municipal sewage treatment plant, leading to higher abundance of oligochaete alternate hosts downstream (Marcogliese et al. 2009). Parasite species richness was also associated with eutrophication in spottail shiners from the Richelieu River and common shiners (*Luxilus cornutus*) from the Bras d'Henri watershed in southwestern Quebec, Canada (Marcogliese and Cone 2021). For a marine example, aspergillosis, a disease caused by the generalist fungus *Aspergillus sydowii* in the common gorgonian sea fan *Gorgonia ventalina*, was more severe in experiments with nutrient enrichment (Bruno et al. 2003).

Another consequence of eutrophication is the generation of harmful algal blooms (HABs) caused by algae and cyanobacteria (Figs. 22.3 and 22.4) that may have profound effects on marine and freshwater ecosystems (Patz et al. 1996; Johnson et al. 2010a). Cyanobacterial biomass has increased in north temperate and subarctic lakes since approximately 1800, an effect associated with increased nutrient concentration (Taranu et al. 2015). There has also been a global increase in coastal phytoplankton blooms, many of which are associated with HABs, linked to sea surface temperatures between 2003 and 2020 (Dai et al. 2023). Exposure to HABs and their associated cyanotoxins is known to negatively impact immune function in both invertebrates and fish but enhances host resistance in other studies (see Budria 2017). Furthermore, exposure to HABs may be detrimental to parasites, including the pathogen *Perkinsus olseni* in Manila clams, *Ruditapes philippinarum* (see references in Budria 2017). Exposure of leopard frogs (*Lithobates pipiens*) to a low dose of the cyanotoxin microcystin-LR (MC-LR) resulted in higher infection levels of metacercariae of *Echinostoma* sp. (Milotic et al. 2018). However, exposure of three species of trematodes to relatively low concentrations of MC-LR did not reduce cercarial survival (Milotic et al. 2019). In fact, survival of cercariae of one trematode species, an unidentified strigeid, actually increased (Milotic et al. 2019). In another study, exposure of unidentified echinostome cercariae to a wide range of concentrations of MC-LR increased mortality in a dose-dependent manner, but low concentrations did not affect their infectivity (Buss et al. 2019). Curiously, exposure of green frog (*Lithobates clamitans*) tadpoles to the lowest MC-LR concentration increased their susceptibility to infection, while a slightly higher concentration did not (Buss et al. 2019). Consequently, these authors concluded that high levels of MC-LR should limit parasite transmission, while infections at lower levels are determined by non-linear effects on tadpoles, but not parasites. In planktonic water fleas (*Daphnia* sp.) from a Swiss lake, cyanobacterial density was correlated with outbreaks of the ichthyosporean gut parasite, *Caullerya mesnili* (Tellenbach et al. 2016). Furthermore, experiments demonstrated that exposure of *Daphnia* sp. to cyanobacteria increased their susceptibility to the parasite (Tellenbach et al. 2016). However, overall, we have little information in general on effects of HABs on host–parasite interactions. In addition, given the widespread ecological effects of HABs on ecosystems, we can expect a plethora of indirect effects on parasites and their hosts, but there is little known in this regard.

22.4.3 Acidification

Both fresh and marine waters are expected to undergo, or are already undergoing, acidification as a result of climate change. In freshwater, increased precipitation and runoff will lead to acidification of poorly-buffered headwater streams and lakes (Marcogliese 2001; Schindler 2001). In the oceans, increased water absorption of CO_2 generated by the burning of fossil fuels will cause a decline in pH (reviewed in Byers 2020, 2021). Numerous organisms are sensitive to pH disturbances, including certain fish and invertebrates (reviewed in Marcogliese 2001; Byers 2021). While studies of effects of pH on immune response in fish provide conflicting results (Bowden 2008; Uribe et al. 2011), acidification is known to cause oxidative stress in both aquatic invertebrates and vertebrates, an energetically costly physiological condition that may trade off against immune function (Baag and Mandal 2022; Thomas et al. 2022; Messina et al. 2023; Okon et al. 2023). Furthermore, some molluscs and crustaceans become more susceptible to disease when exposed to acidification, including *Perkinsus marinus*, a serious pathogen of oysters (Shields 2019; Okon et al. 2023). However, relatively little work exists exploring the effects of acidification on other parasites and host–parasite relationships (see Marcogliese 2001, 2005; Byers 2020, 2021), although free-living stages of parasites are known to tolerate a wide range of pH (Pietrock and Marcogliese 2003).

Parasite species richness was reduced in a number of freshwater systems exposed to low pH, including in perch (*Perca fluviatilis*) in Finnish acidified reservoirs (Halmetoja et al. 2000) and American eels (*Anguilla rostrata*) from Nova Scotian rivers in eastern Canada (Cone et al. 1993; Marcogliese and Cone 1996; Marcogliese and Cone 1997). In both these ecosystems, parasite species decline was principally due to the loss of trematodes, which resulted from the decline of acid-sensitive molluscan intermediate hosts (Marcogliese 2005). Thus, any acidification resulting from climate change is expected to particularly impact the trematode fauna. Exposure to acidified conditions also significantly increased the mortality of amphipod intermediate hosts (*Gammarus pulex*) infected with larval acanthocephalans (*Pomphorhynchus laevis*) compared to uninfected amphipods (McCahon and Poulton 1991).

In the oceans, the most sensitive species are those that undergo calcification of their shells and external structures, such as corals and molluscs (Byers 2021). Most work concerning acidification in marine waters to date has been experimental, with results varying with parasites and their hosts. In one study, pH had no effect on cercarial longevity of two trematode species released from the intertidal horn snail, *Cerithideopsis (= Cerithidea) californica*, the first intermediate host (Koprivnikar et al. 2010). However, in another study on four trematode species released from either the New Zealand mud snail *Zeacumantus subcarinatus* or the littorinid *Austrolittorina cincta*, the first intermediate hosts, all four species experienced a reduced period of activity with decreasing pH (MacLeod and Poulin 2015a). For one of these species, *Maritrema novaezealandense*, susceptibility of the second intermediate host, the intertidal amphipod *Paracalliope novizealandiae*, increased at the lowest pH tested (Harland et al. 2015). Thus, for this parasite, parasite transmission was highest at pH 7.4, despite the reduced longevity of the cercariae at this level of acidification (Harland et al. 2015). Two of the other species, *Philophthalmus* sp. and *Parorchis* sp., use *Z. subcarinatus* and *A. cincta* as first intermediate hosts, and encyst as metacercariae on hard substrates in the external environment. Cercarial production of *Philophthalmus* sp. was highest, while metacercarial survival was lowest, at the lowest pH tested (Guilloteau et al. 2016). In contrast, survival of *Parorchis* sp. was unaffected by pH (Guilloteau et al. 2016). In yet another study, the absolute cercarial life span of *Himasthla* sp. cercariae released from the marine snail *Littorina scutulata*, the first intermediate host, varied non-linearly with pCO_2, which is related to acidification (Franzova et al. 2019). In another experiment, cercarial transmission of *Himasthla elongata* to the second intermediate host cockle *Cerastoderma edule* was more successful at a lower pH (Magalhães et al. 2018). Cercarial release of an echinostome trematode from the intertidal snail *Echinolittorina peruviana*, the first intermediate host, was unaffected by pCO_2, while that of a philophthalmid was highest at elevated concentrations (Leiva et al. 2019). Clearly, patterns of cercarial release and longevity vary greatly among trematode parasites with acidification, preventing any generalisations. Non-linear and contrasting interactions with temperature further complicate matters. In Franzova et al. (2019), different temperatures caused even further variations in response to pCO_2, whereas in Leiva et al. (2019), pCO_2 increased emergence of the philopthalmid, but not the echinostome, at the highest temperatures tested.

Parasites also have different effects on hosts, depending on pH. While pH negatively affected shell growth, length and tensile strength in *Z. subcarinatus*, infection with any of three different trematode species further reduced shell growth tensile strength at reduced pH (MacLeod and Poulin 2015b). While acidified waters reduced oxygen consumption rates and tissue glucose content in *Z. subcarinatus*, infections with each of three trematode species affected snail metabolism differently at the various levels of acidity (MacLeod and Poulin 2016). Thus, by affecting host metabolism, parasites may further affect host survival and fitness with acidification.

22.4.4 Salinity

Sea level is increasing and will continue to increase with climate change (Arias et al. 2021). Sea level rise will be accompanied by intrusion of salt water into freshwaters in coastal areas (Fig. 22.3) causing changes in salinity in these areas, especially where water levels and flow rates have decreased (Marcogliese 2001). Subsequent effects on parasites have been discussed in Marcogliese (2001, 2016). In contrast, where precipitation and subsequent runoff and water flows increase, salinity in estuaries and other coastal environments may be expected to decrease, with subsequent effects on parasitism and disease (Byers 2020, 2021). Salinity is known to impact the parasite fauna of fishes that migrate between areas of varying salinity, or inhabit areas of different salinities, including flounders and salmonids (MacKenzie et al. 1995;

El-Darsh and Whitfield 1999; Marcogliese and Jacobson 2015). Storm surges of saline waters into low saline areas can also kill stenohaline (i.e. low-salinity variation tolerance) intermediate hosts and free-living parasite stages (Overstreet 1997, 2007). However, effects of salinity on a parasite and its free-living stages depend on the freshwater or marine habitat of the species in question (Pietrock and Marcogliese 2003). Furthermore, salinity affects each stage of a parasite, as well as its host, differently (Overstreet 1993). A meta-analysis on a variety of hosts including both invertebrates and vertebrates suggested that exposure to salinity as a stressor in experimental studies resulted in an increase in pathogen intensity (Vicente-Santos et al. 2023). Exposure of freshwater and anadromous fishes to elevated salinity leads to alterations in numerous components of both the innate and adaptive immune response (Bowden 2008; Uribe et al. 2011).

Studies of effects of salinity on parasites may be best accomplished by examining euryhaline hosts from different ecosystems. A circumglobal analysis of parasites of three-spined sticklebacks (*Gasterosteus aculeatus*) from freshwater, brackish and marine habitats revealed that within each of three separate geographical regions (Eurasia, eastern North America, and western North America) similarities among parasite communities were greater in areas of similar salinity (Poulin et al. 2011). Among numerous invertebrate and fish hosts in the Baltic Sea, where salinity decreases from west to east, marine and brackish water parasites occur throughout the whole area (Zander and Reimer 2002). Freshwater parasites, with their limited tolerance of saline conditions, are primarily limited to the eastern portion of the Baltic Sea (Zander and Reimer 2002). Even low salinity levels appear to limit the parasite fauna of freshwater fishes. Parasites of bream (*Abramis brama*) and roach (*Rutilus rutilus*) were compared between a freshwater lake and the Kiel Canal, Germany, where salinity varies from 2.5 to 15 ppt (Rückert et al. 2007). Parasite species richness was lower in the canal for both hosts. The freshwater parasites appeared to have limited salinity tolerance, being negatively affected even at low salinities (Rückert et al. 2007). Byers (2021) suggests that many marine ectoparasites are sensitive to low salinities, while endoparasites may be buffered against salinity changes. Nevertheless, others have observed that parasite species richness, including endoparasites, is higher in waters of lower salinity (Blanar et al. 2011). The Bothnian Bay, Finland, a low-salinity region (0.5–3.5 ppt) of the Baltic Sea, is inhabited by primarily freshwater fishes, but includes some marine ones. Here, marine parasites are found in both marine and freshwater fishes, while freshwater parasites also occur in the marine fishes (Valtonen et al. 2001). However, only eight of the 63 parasite species found are of marine origin. Parasite occurrence appears to be dictated by the invertebrate species present. Marine molluscs are absent and marine copepods are limited, explaining the absence of marine trematodes and the impoverished marine cestode fauna. However, freshwater and brackish water copepods are abundant and, consequently, freshwater cestodes are common (Valtonen et al. 2001).

Studies on specific hosts also yield interesting patterns. Parasites were examined in the euryhaline mummichog (*Fundulus heteroclitus*) from an upper low-salinity site and a lower high-salinity site in each of two river estuaries in New Brunswick, Canada (Blanar et al. 2011). There was a distinct difference in parasite community structure between the upstream and downstream sites in each river. Notably, salinity had a stronger influence on parasite community structure than did eutrophication and other contaminants (Blanar et al. 2011).

A number of important diseases are affected by salinity. Decreases in rainfall and runoff into coastal waters will result in increased salinity in many coastal areas. For example, salinity has increased over time in Chesapeake Bay on the eastern United States, although it is predicted to vary due to increases in precipitation (Shields 2019). Increasing salinity is expected to lead to greater infections of the parasitic dinoflagellate *Haematodinium perezi* in crabs. Although blue crabs (*Callinectes sapidus*) can tolerate low salinities, viability of the parasite's dinospores declines at lower salinities (Shields 2019; Byers 2021). The oyster pathogen *Perkinsus marinus* also fares poorly in low saline conditions (Burge et al. 2014; Shields 2019; Okon et al. 2023). Indeed, in the Gulf of Mexico, where temperatures are always suitable for *P. marinus*, infections are limited by low salinities dictated by cyclical climatic conditions (Burge et al. 2014). Planktonic larval stages of sea lice (*L. salmonis*) are susceptible to variations in salinity. Larval development is inconsistent below 30 ppt (Brooks 2005). Hatching and larval activity increases above 25 ppt, while the adult stages on fish are more tolerant of low salinity (Fast and Dalvin 2020). An increased epizootic of sea lice on farmed salmon (*Salmo salar*) in Scotland was associated with a decreased river discharge and subsequent increase in salinity in coastal areas during an intense drought in the United Kingdom (Morley and Lewis 2014a). Hypersaline conditions may also increase an oyster's susceptibility to disease (Okon et al. 2023). In freshwater, increasing salinity from 0.5 ppt to 3.5–4.5 ppt in mesocosm experiments reduced transmission of the chytrid fungus and lowered mortality in the green-and-golden bell frog, *Litoria aurea* (Clulow et al. 2018).

Salinity is known to affect various life cycle stages of marine parasites. Survival and infectivity of parasite free-living stages are generally reduced after exposure to freshwater (Pietrock and Marcogliese 2003). For example, eggs of the sealworm [*Phocanema* (=*Pseudoterranova) decipiens*] hatched in fresh, brackish and sea water, but survival of the hatched larvae was much lower in freshwater compared to the other treatments (Measures 1996). Cercariae of intertidal parasites, such as *Maritrema subdolum*, are known to be tolerant of a wide range of salinities (Mouritsen 2002), but this is not always the case. Transmission of the trematode *Philophthalmus* sp. is strongly affected by relatively small decreases in salinity (Lei and Poulin 2011). *Philophthalmus* cercariae are released from the intertidal New Zealand mud snail and encyst as metacercariae on hard external substrates. Transmission is completed through accidental ingestion by shorebirds. At lower salinities of 25 and 30 ppt, cercarial output from the snail is reduced, encystment time prolonged, encystment success reduced and metacercarial survival lower, compared to at 35 ppt (Lei and Poulin 2011). Transmission of *M. novaezealandense* from its first intermediate snail host (*Z. subcarinatus*) to a second intermediate amphipod host (*P. novizealandiae*) was also impacted by salinity (Studer and Poulin 2012). However, metacercarial development proceeded faster at lower salinity (Studer and Poulin 2012). Low salinity reduced cercarial output and survival but did not affect cercarial infectivity or amphipod susceptibility. Low salinity also reduced cercarial activity of the trematode *H. elongata* released from first intermediate host periwinkles (Littorinidae) (Bonmarito et al. 2020). Infectivity of the second intermediate host, the blue mussel (*Mytilus edulis sensu lato*), was also reduced in low salinity, as was mussel susceptibility to infection (Bonmarito et al. 2020). Thus, any reductions in salinity in coastal areas due to climate change would likely reduce transmission of this parasite. Curiously, however, Koprivnikar and Poulin (2009) obtained contrasting results, with equal or higher cercarial emergence at 30 ppt compared to 35 ppt, although the mud snails (*Z. subcarinatus*) originated from the same area as in Lei and Poulin (2011).

Freshwater parasites and their hosts may be particularly sensitive to salinity changes. In general, free-living stages of freshwater parasites can survive slight increases in salinity, but trematode cercarial infectivity is impeded after a short time (Pietrock and Marcogliese 2003). Survival of free-living stages encased in eggs or a cuticle is usually prolonged, but protection is not complete (Pietrock and Marcogliese 2003). However, hosts may vary in their sensitivity. For example, larval anurans vary in their susceptibility to salinity. Wood frogs [*Lithobates* (= *Rana*) *sylvaticus*] and spring peepers (*Pseudacris crucifer*) are much more susceptible to sodium chloride (NaCl) than American toads (*Anaxyrus americanus*). Experimental pre-exposure of tadpoles of these species to a salinity of 1 ppt followed by exposure to cercariae of echinostome trematodes singly or in combination led to mixed results (Buss and Hua 2023). Following parasite exposure, wood frogs and spring peepers pre-exposed to NaCl had higher intensities of echinostomes than American toads, while no differences were observed in the control environment. This suggests that a saline environment negatively impacts resistance to these trematodes in wood frogs and spring peepers, but not American toads (Buss and Hua 2023). Similarly, exposure of wood frogs to salinities as low as 0.6 ppt results in higher infection intensities of *R. ondatrae*, but not in northern leopard frogs exposed to *Echinostoma* sp. (Milotic et al. 2017). A possible explanation is that wood frogs exposed to saline conditions displayed reduced activity, which is an anti-cercarial behaviour (Milotic et al. 2017). Prevalence of the human trematode parasite *S. mansoni* in its snail intermediate host (*Biomphalaria alexandrina*) declined when exposure to increasing salinities between 0.38 and 5.7 ppt, while cercarial survival peaked at 3.8 ppt (Yu et al. 2022). Curiously, infections of the freshwater larval acanthocephalan *Polymorphus minutus* in its intermediate amphipod host (*Gammarus roeselii*) increased the amphipod's tolerance to salinity variations (Piscart et al. 2007). Such physiological changes could allow freshwater invertebrates to survive saltwater intrusions.

Hypersaline conditions may detrimentally affect parasites. For example, trematodes were absent in an endemic threatened fish in New Mexico, United States, the White Sands pupfish (*Cyprinodon tularosa*), at a hypersaline site, likely due to the absence of intermediate host springsnails (*Juturnia tularosae*) (Rogowski and Stockwell 2006). Fewer species of trematodes and cestodes occurred in black-striped pipefish (*Syngnathus abaster*) and marbled goby (*Pomatoschistus marmoratus*) from a high-salinity coastal lagoon in the Iberian Peninsula, compared to other parasitological surveys in those hosts elsewhere (Almeida et al. 2023). Hypersaline conditions resulting from a severe drought in southern Africa virtually eliminated an invasive fish pathogen. In the Phongolo River floodplain, South Africa, the Mozambique tilapia (*Oreochromis mossambicus*) was infected with a parasitic copepod (*Lernaea cyprinacea*) at a mean intensity of 22 per fish. During the drought, not a single parasite was found on these fish (Welicky et al. 2017).

22.4.5 Contaminants

Climate change may have important effects on the availability and metabolism of contaminants by biota (Table 22.5). Increasing temperature enhances the toxicity of contaminants and their rate of uptake (Schiedek et al. 2007; Noyes et al. 2009; Hooper et al. 2013). Furthermore, salinity and pH affect bioavailability and toxicity of certain contaminants such as metals (Pietrock and Marcogliese 2003; Schiedek et al. 2007; Noyes et al. 2009; Hooper et al. 2013). However, climate change could also moderate effects of contaminants by reducing exposure (Landis et al. 2014). Stress associated with climate change may reduce tolerance to contaminants and an organism's ability to recover from exposure to toxicants (Moe et al. 2013). Furthermore, exposure to contaminants may pose an additional risk to organisms at the limits of their physiological tolerance range to temperature, salinity, pH and other natural stressors (Noyes et al. 2009; Hooper et al. 2013). In addition, other changes associated with climate change may affect contaminants (Fig. 22.3). For example, low pH results in increased solubility and toxicity, and thus bioavailability, of metals to biota (Overstreet and Howse 1977; Sures et al. 2023). Increased frequency of extreme weather events such as storms can also release high concentrations of contaminants from sediments, resulting in acute exposure to invertebrates and fish (Overstreet 2007).

Contaminants are known to affect parasites and host–parasite interactions (see also Chap. 20). Numerous reviews and syntheses have addressed pollution, contaminants and parasites in the last three decades, including some meta-analyses (e.g. Khan and Thulin 1991; Poulin 1992; Overstreet 1993; MacKenzie et al. 1995; Lafferty 1997; MacKenzie 1999; Williams and MacKenzie 2003; Sures 2004, 2008; Marcogliese 2005, 2023; Blanar et al. 2009; Vidal-Martínez

Table 22.5 Potential effects of climate change and associated abiotic changes on the toxicity of contaminants

Climate change effect	Effect on contaminants	References
Increased temperature	Enhance toxicity	Noyes et al. (2009)
	Alter biotransformation to metabolites	
	Affect absorption, distribution, metabolism, excretion	Hooper et al. (2013)
	Impair homeostasis	Noyes et al. (2009)
Increased precipitation & storms	Increase runoff and contaminant concentration	Noyes et al. (2009)
Salinity changes	Alter bioavailability	Noyes et al. (2009)
	Increase toxicity	
	Affect absorption, distribution, metabolism, excretion	Hooper et al. (2013)
Acidity	Affect absorption, distribution, metabolism, excretion	Hooper et al. (2013)

et al. 2010; Sures et al. 2017, 2023; Gilbert and Avenant-Oldewage 2021; Sures and Nachev 2022). In summary, the effects of contaminants on parasites are often negative and can be direct, as on ectoparasites and free-living stages, or indirect, acting on the intermediate and definitive hosts of endoparasites (Lafferty 1997; Sures et al. 2017; Sures and Nachev 2022). Overall, parasite diversity tends to decrease in polluted ecosystems (MacKenzie 1999; Marcogliese 2004). Nevertheless, some parasites may increase in prevalence or abundance, due to a compromised immune response on the part of the host (MacKenzie et al. 1995; MacKenzie 1999; Marcogliese 2005). These parasites tend to be those with direct life cycles, including protists and monogeneans. For example, prevalence of the chytrid fungus *Bd* in amphibians from over 4000 waterbodies in Mexico was positively associated with poor water quality from urban and industrial wastes, likely due to immunocompromising effects of the contaminants on the hosts (Jacinto-Maldonado et al. 2023).

In terms of direct exposure, contaminants tend to reduce the longevity and infectivity of a parasite's free-living infective stages, depending on the concentration and exposure time (Pietrock and Marcogliese 2003). Effects on infectivity of free-living stages usually manifest themselves sooner than those affecting survival (Pietrock and Marcogliese 2003). For trematodes, contaminants have been shown to affect development and hatching of miracidia, survival and infectivity of cercariae, and infectivity of metacercariae (Morley et al. 2003). Contaminants are also known to impair a host's immune response in both invertebrates and vertebrates (Fournier et al. 2005; Mydlarz et al. 2006; Martin et al. 2010; Rehberger et al. 2017). For chytrid fungi, exposure to different pesticides affects zoospore production and zoosporangia growth to varying degrees (Hanlon and Parris 2012). Application of the agricultural antifungal pesticide triazole at concentrations 10 times lower than that required to inhibit *Bd* growth prevented chytrid infections in Alpine newts, *Ichthyosaura alpestris* (Barbi et al. 2023). Triazole bioaccumulated on the newts' skin tenfold, effectively preventing infection (Barbi et al. 2023). Survival of animals such as invertebrates and fish exposed to contaminants is often reduced when infected, although in some instances infection may actually increase the host's tolerance (Morley et al. 2003; Marcogliese and Pietrock 2011).

However, while contaminant concentrations and their toxicity are expected to increase with climate change, it is not known if the magnitude of these changes will be sufficient to affect parasite and host–parasite interactions at this point in time. Parasite communities are not necessarily affected by low-to-moderate degrees of pollution and, in fact, may only show changes when contaminant concentrations are high (Marcogliese et al. 2006).

22.4.6 Combined Effects of Multiple Stressors

It clearly is important to understand the effects and interactions of multiple stressors on organisms (Sih et al. 2004; Holmstrup et al. 2010). The combined effects of multiple stressors may be additive, synergistic, antagonistic or reversed, the latter being when the net combined effect is in the opposite direction to the sum of their individual effects (Jackson et al. 2016). However, interactions between stressors can be complex, rendering predictions of effects on parasitism and disease are fraught with uncertainty (Cable et al. 2017). Importantly for climate change, a meta-analysis of freshwater studies revealed that increased temperature combined with nutrification resulted in additive net effects, but the overall mean net combined effect of increased temperature and a second stressor was antagonistic. Indeed, in freshwater systems, the most common interaction between effects among free-living organisms was antagonistic (Jackson et al. 2016). In contrast, in coastal and marine ecosystems, effects of paired stressors were variably antagonistic and synergistic, while a combination of three stressors was mainly synergistic (Crain et al. 2008). Synergistic effects were more common when acidification and increased temperature were specifically considered in another meta-analysis of marine ecosystems (Harvey et al. 2013).

A number of synthetic studies have examined the combined effects of parasites and other stressors on host organisms (Sures 2008; Marcogliese and Pietrock 2011; Sures et al. 2017; Grabner and Sures 2019; Grabner et al. 2023). Not only can infection with parasites modify a host's growth and development, it can also alter their biochemical responses to other stressors, including biomarkers used in environmental monitoring (Sures 2008; Marcogliese and Pietrock 2011; Sures et al. 2017; Grabner and Sures 2019; Grabner et al. 2023). Again, the combination may have additive, synergistic, antagonistic or neutral effects on a host's health and survival (Pietrock and Marcogliese 2003; Sures 2008; Marcogliese and Pietrock 2011; Cable et al. 2017; Sures et al. 2017, 2023; Sures and Nachev 2022; Grabner et al. 2023). In general, the combined effects of parasitism and another abiotic stressor are considered to be either additive or synergistic (Marcogliese and Pietrock 2011; Sures and Nachev 2022). Many of these abiotic stressors are those associated with climate change, including temperature, acidification, salinity, reduced oxygen content, HABs, drought, UV-B radiation and a variety of contaminants (Marcogliese and Pietrock 2011). Some recent studies demonstrate a variety of nonlethal interactions between parasites and stressors associated with climate change. While reductions in pH reduced shell growth, shell length and tensile strength in the New Zealand mud snail *Z. subcarinatus*, infections with three separate trematode species each differently modified these measures of shell integrity (MacLeod and Poulin 2015b). Infections of cockles with the trematode *Himasthla elongata* virtually altered all oxidative stress biomarkers tested compared to unparasitised cockles at all temperatures, salinities and pH levels incorporated into the experiments (Magalhães et al. 2018). A combination of trematode infection with low salinity reduced condition in the second intermediate host, the blue mussel *M. edulis sensu lato* (Bommarito et al. 2022). As a final example, drought-stressed snail intermediate hosts (*Biomphalaria alexandrina*) infected with the trematode *Schistosoma mansoni* failed to exhibit a significant prepatent reproductive period, termed fecundity compensation, unlike snails in an unstressed environment (Gleichsner et al. 2016).

Fewer studies examine the combined effects of multiple abiotic stressors on parasites. Early results showed variable responses, with no consistent patterns (Pietrock and Marcogliese 2003). However, in recent years, a number of studies demonstrate that a combination of higher temperature and ocean acidification affects the

Table 22.6 Selected studies demonstrating the effects of multiple stressors associated with climate change on parasites and host–parasite interactions. Unless otherwise indicated, all parasites are trematodes

Parasite	Host	Stressors	Interactions	References
Acanthoparyphium spinulosum	*Cerithideopsis californica* (Mollusca)	Temperature × salinity	Increased mortality, decreased activity of cercariae	Koprivnikar et al. (2010)
		Salinity × pH	Increased mortality, decreased activity of cercariae	
Euhaplorchis californiensis		Temperature × salinity	No combined effect	
		Salinity × pH	No combined effect	
Maritrema novaezealandense	*Zeacumantus subcarinatus* (Mollusca)	Temperature × salinity	No combined effect on numbers of emerging cercariae	Koprivnikar and Poulin (2009)
Philophthalmus sp.			No combined effect in one trial. Cercarial output decreased with increased temperature but more so at lowest salinity (30 ppt)	
Himasthla sp.	*Littorina scutulata* (Mollusca)	Temperature × pCO_2	No combined effect on cercarial active swimming period; decrease in total cercarial life span	Franzova et al. (2019)
Maritrema subdolum	*Peringia* (= *Hydrobia*) *ulvae* (Mollusca)	Temperature × salinity	No combined effect on cercarial survival; temperature-dependent cercarial emergence, with lower emergence with increasing salinity at 15 °C, higher emergence with increasing salinity at 20 °C and 25 °C	Mouritsen (2002)
Himasthla elongata	*Littorina littorea* (first intermediate host, Mollusca); *Mytilus edulis* (second intermediate host, Mollusca)	Temperature × salinity	No combined effect on cercarial activity, infectivity or mussel susceptibility	Bonmarito et al. (2020)
Metschnikowia bicuspidata (Fungi)	*Daphnia longispina* species complex	Temperature × pesticide	Greater effect of tebuconazole on parasite at higher temperature	Cuco et al. (2018)
Metschnikowia bicuspidata (Fungi)	*Daphnia longispina* species complex	Temperature × cyanobacteria	Decrease fungal epidemics	Manzi et al. (2020)
Batrachochytrium dendrobatidis (Fungi)	Green and golden bell frog (*Litoria aurea*)	Temperature × salinity	Increasing temperature 13 to 25 °C and raising salinity 0.5 ppt to 3.5–4.5 ppt cleared frogs of infection	Clulow et al. (2018)

immune response in crustaceans and molluscs, and also increases parasite transmission (Baag and Mandal 2022). Most of the studies examining the combined effects of stressors associated with climate change on parasites and host–parasite interactions are on marine trematodes, involving temperature, salinity and/or acidification (Table 22.6). Two-way interactions of temperature and salinity increased cercarial mortality and decreased cercarial activity in *Acanthoparyphium spinulosum*, but not in *Euhaplorchis californiensis* (Koprivnikar et al. 2010). Similarly, two-way interactions of temperature and pH increased cercarial mortality and decreased cercarial activity in the former trematode species, but not in the latter (Koprivnikar et al. 2010). Salinity–temperature interactions did not affect cercarial output of *M. novaezea-*

landense from the New Zealand mud snail *Z. subcarinatus* but did so inconsistently for *Philophthalmus* sp. from the same host (Koprivnikar and Poulin 2009). The interactive effects of temperature and pCO_2 did not affect the active swimming period, but did reduce the cercarial life span, in *Himasthla* sp. emerging from *Littorina scutulata* (Franzova et al. (2019). Perhaps the most intriguing results can be found in Mouritsen (2002), who found interactive effects of salinity and temperature on cercarial numbers, but not longevity, of *Maritrema subdolum* emerging from *Peringia ulvae*. Cercarial emergence was positively temperature dependent. Lower numbers of cercariae emerged with increasing salinity at 15 °C, while higher numbers were released with increasing salinity at 20 °C and 25 °C (Mouritsen 2002). Clearly, interactive effects between stressors such as temperature, salinity and pH vary unpredictably among both host and parasite species.

One last aspect of parasite-contaminant interactions deserves mention. It is well established that a variety of intestinal parasites, especially acanthocephalans and cestodes, can bioaccumulate different contaminants (for details, see Chap. 20), most notably heavy metals, at much higher concentrations than their hosts (Sures 2004, 2008; Sures et al. 2017). To date, this capacity has been demonstrated for over 50 helminth species (Sures et al. 2017). Furthermore, it appears that they can offer some protection to the host by diverting the contaminants away from them (Sures 2008; Sures et al. 2017). Conceivably, with the increasing bioavailability of contaminants with climate change, those parasites will absorb even higher levels of contaminants. However, as we do not know the degree to which contaminants may become more bioavailable, at this point it is mere speculation as to whether climate change will impact the bioaccumulation capacity of parasites.

22.5 Ecological Interactions and Ecosystems

The importance of parasites in ecosystems is without doubt (Hatcher et al. 2012). Aside from overall effects on fish health and mortality, and, as such, are capable of regulating host populations, they also impact other host traits such as behaviour and interactions with other species (Marcogliese 2004; Buck 2019; see Chap. 7). Furthermore, they comprise a substantial proportion of the biomass within a number of ecosystems (Kuris et al. 2008; Lagrue and Poulin 2016; Preston et al. 2021) and affect food web structure (Lafferty et al. 2008). Aside from direct interactions between two species, organisms may affect other species through indirect effects. These effects can propagate through a food web through a chain of interacting species. They may be density-mediated indirect effects (DMIEs) via chains of interactions between species within a food web that depend on the density of the initiator organism (Werner and Peacor 2003). Alternatively, one organism can affect another by moderating the phenotype (e.g. behaviour) of the second species, known as a trait-mediated indirect effect (TMIE), that then propagates through the food web through further interactions (Werner and Peacor 2003). Both DMIEs and TMIEs may occur within host–parasite interactions through a parasite's consumptive effects on the host, or its non-consumptive effects (Buck and Ripple 2017). For example, a common TMIE among parasites would be host manipulation that increases its vulnerability to predation (Buck and Ripple 2017). Of course, these ecological interactions between parasites, hosts and other species are likely affected by temperature and other abiotic changes associated with climate change. However, to what degree and in what direction remains pure speculation at this point in time.

However, a couple of studies have examined the effects of parasites on grazing by gastropods in relation to temperature. For example, while infected periwinkles consumed less than those infected by trematodes at 18 °C, no difference was observed at 21 °C. Results imply that parasitism could counteract any negative impacts of periwinkle grazing (a TMIE) with climate change (Larsen and Mouritsen 2009). A more complex experiment involved the gastropod *Planorbella trivolvis* and various trematodes (Resetarits et al. 2023). Infected snails suffered greater mortality, reducing consumption, but ate more than uninfected snails at 18 °C, with a net positive effect on resource consumption. Thus, there were negative lethal effects (a DMIE) and positive nonlethal effects (a TMIE) of parasites on resource consumption (Resetarits et al. 2023). However, these effects varied with temperature. At 24 °C, lethal and nonlethal effects on ingestion rates were equal. At 30 °C, parasite-induced host mortality was high but nonlethal effects of the parasite on snail consumption again outweighed the lethal effects, a net positive effect of parasitism, similar to results at 18 °C (Resetarits et al. 2023). These results demonstrate that the lethal and nonlethal effects of parasitism may not co-vary with temperature.

An important functional role of parasites in a food web is the consumption of free-living stages by predators (Johnson et al. 2010b; Thieltges et al. 2013), of which cercariae are the best studied (Koprivnikar et al. 2023). A tremendously wide array of freshwater and marine predators, including fish, insects, crustaceans, molluscs, oligochaetes, rotifers and bryozoans feed on trematode cercariae (Koprivnikar et al. 2023). Temperature increases will affect all animals' metabolism and vital rates. These effects should scale up to behavioural processes, including feeding rates. Thus, predation on free-living stages of parasites should increase with temperature, up to a maximum. Conceivably, such increased consumption under warmer conditions could reduce infectivity to the next host. Goedknegt et al. (2015) experimentally examined this question with three potential cercarial predators of the trematode *Renicola parvicaudatus*. They found that predation by the eastern oyster *Magallana gigas* and the barnacle *Austrominius modestus* reduced the number of cercariae, shed by the first-intermediate periwinkle, that were able to subsequently infect the second intermediate host, mussels (*Mytilus edulis*). Furthermore, the predators reduced infections in the mussels in a temperature dependent manner (Goedknegt et al. 2016). The authors suggested that increased cercarial production and infectivity may be offset by increased predation in a warming climate (Goedknegt et al. 2016).

In the 1960s, Robert Paine developed the concept of keystone species which, at the time, referred to a key predator in an ecosystem whose feeding activities limited the spread of a competitively dominant prey species, such as grazers and filter feeders (Paine 1966, 1969). The concept was applied to parasitology by Minchella and Scott (1991), who defined a keystone parasite as one that limits a competitively dominant species through negative effects on its abundance. Some examples are provided in Minchella and Scott (1991), Marcogliese (2004, 2008) and Sures et al. (2017). Mouritsen and Poulin (2002) actually linked the mass mortalities of key marine invertebrates caused by keystone parasites to the North Atlantic climate oscillation. In related work, Thomas et al. (1998) applied the concept of ecosystem engineering to parasites. These are parasites that hinder a host's engineering properties and capacity, or modify the host in such a way that alters the habitat of other free-living species.

The amphipod *Corophium volutator* is extremely abundant on intertidal sediments and an important prey item for shorebirds on both sides of the North Atlantic Ocean (Mouritsen et al. 2005). Previous population collapses of this amphipod, linked to heat waves, had huge effects on the intertidal ecosystem of the Danish Wadden Sea, due to the loss of the amphipod's burrowing behaviour and astronomically high numbers. The amphipod thus is considered an ecosystem engineer (reviewed in Mouritsen and Poulin 2002; Poulin and Mouritsen 2006; Marcogliese 2008). Transmission rates of cercariae of two species of microphallid trematodes shed from their first intermediate host, *Peringia ulva*, to *C. volutator*, the second intermediate host, is temperature-sensitive, as is parasite-induced host mortality. Mouritsen et al. (2005) developed a model simulating the effects of temperature on the impact of microphallid trematodes on their second intermediate host in the Danish Wadden Sea. The simulation demonstrated that a 3.8 °C increase in temperature would result in collapse of the amphipod population, which would have cascading effects on the local macroinvertebrate fauna (Mouritsen et al. 2005; Poulin and Mouritsen 2006). The latter authors predicted that climate-mediated change on cercarial production could have major impacts on intertidal ecosystems. Warm temperatures and high microphallid parasite levels in *P. ulva* almost eradicated *C. volutator* in an outdoor mesocosm experiment. While elevated temperatures alone negatively affected macrofaunal diversity, the high parasite treatment with high temperature increased macrofaunal diversity (Larsen and Mouritsen 2014). This experiment illustrated the confounding, but key, role parasites may play with climate change.

A related experiment examined the effects of the microphallid trematode *Maritrema novaezealandense* on invertebrate community structure under different temperature scenarios Mouritsen et al. 2018). The parasite uses the New Zealand mud snail *Z. subcarinatus* as its first intermediate host and a variety of amphipods as second intermediate hosts. At current levels of temperature and parasitism, the parasite had little effect on the amphipod community. In addition, temperature alone had little effect. With parasites present, increasing the temperature from 17 °C to 21 °C decreased the abundance of epibenthic amphipods, but had little effect on infaunal species (Mouritsen et al. 2018). Simulating a heat-wave by increasing the temperature from 19 °C to 25 °C together with high parasite pressure almost completely eliminated the amphipod population from the experimental enclosures, while exposure to the heat wave alone reduced epibenthic amphipods but not infaunal ones (Mouritsen et al. 2018). Friesen et al. (2021) examined the effects of temperature on parasite–host interactions in a community of crustacean invertebrates and two trematode species in a follow-up mesocosm experiment. Abundance of one amphipod [*Indocalliope indica* (=*Paracalliope fluviatilis*)] declined significantly when exposed to elevated temperature and parasites, while higher temperatures resulted in parasite-induced mortality of another amphipod (*Paracorophium excavatum*) (Friesen et al. 2021). In contrast, two species of isopod host were affected to a much lesser degree. The authors suggested that, in the presence of rising temperatures, parasites may have large effects on invertebrate communities and ecosystems (Friesen et al. 2021), in agreement with the previous studies cited above.

Claar and Wood (2020) point out that care must be taken to differentiate the effects of pulse heat stress caused by events such as heatwaves versus gradual warming, as the former acts on physiological processes over short time scales. This, of course, related to effects of acclimation, as discussed previously (see Sect. 22.2.1). Pulse heat affects parasites, their hosts and host–parasite interactions that are not predictable from simple correlations of parasites or their hosts with temperature (Claar and Wood 2020). Adding further uncertainty to the question, in terms of disease impact, parasite intensity is not always correlated with virulence during heat stress (Hector et al. 2023).

22.6 Case Studies

There are a number of host–parasite systems that are worthy of singling out due to an extensive array of studies on effects of climate change on these particular parasites. While some have received attention because they are pathogens of humans or aquatic resources, others have been amenable to experimental study. These studies have been aided by the suitability of the parasite for experimentation, as in the case of trematodes in their molluscan hosts, or the establishment of other good laboratory models. In addition, the application of experimental laboratory and field mesocosm approaches has greatly advanced the field (Marcogliese 2023).

22.6.1 Ribeiroia ondatrae (Trematoda)

This trematode has been the subject of extensive study due to the fact that it causes malformation in amphibians (Johnson et al. 2004). The parasite

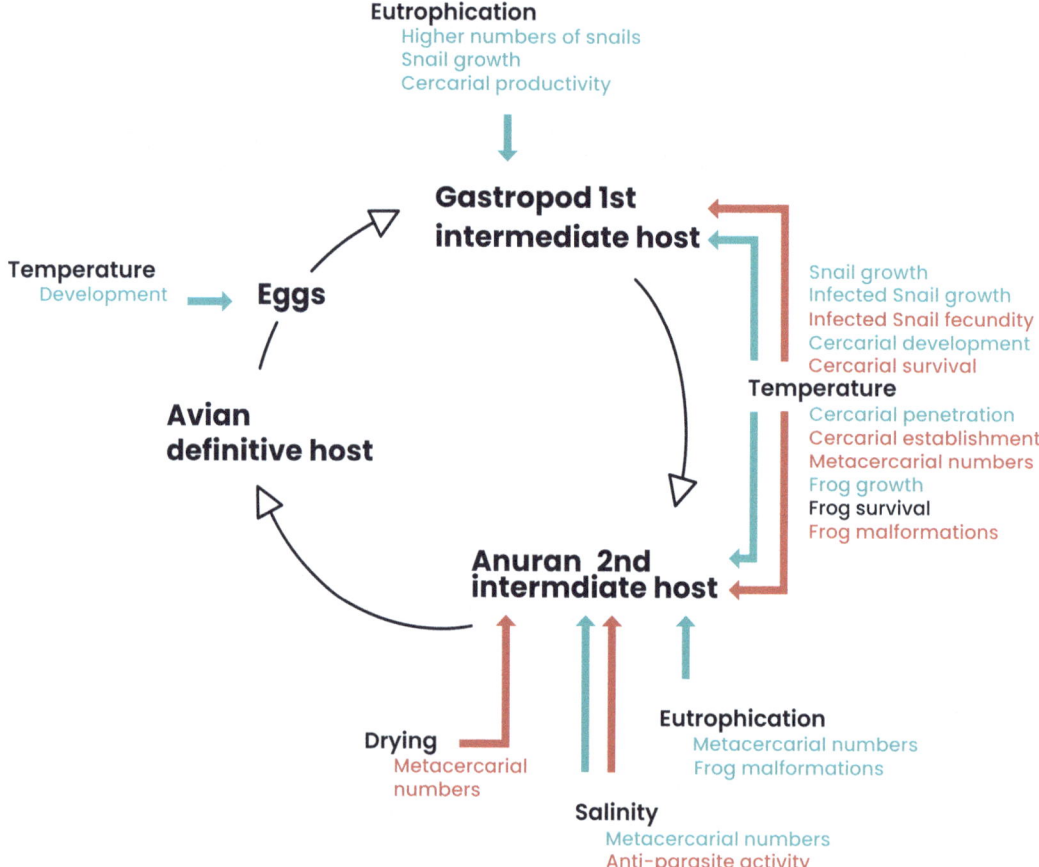

Fig. 22.5 Effects of temperature and other abiotic stressors associated with climate change (drought, eutrophication, salinity) on various life cycle stages and hosts of the trematode *Ribeiroia ondatrae*. Temperature effects reflect increases from 20 °C to 26 °C. Positive effects on parasite life cycle stages and intermediate hosts are shown in teal text and arrows; negative effects in red text and arrows; neutral effects in black text. Information from Johnson and Chase (2004), Johnson et al. (2007), Paull and Johnson (2011, 2018), Paull et al. (2012), and Milotic et al. (2017)

has become an excellent laboratory model, at least for the phases of infection in the first and second intermediate hosts. Studies on the trematode have been referred to repeatedly already, with relation to effects of temperature, desiccation, eutrophication and salinity (Fig. 22.5). However, it is appropriate and worthwhile at this time to synthesise the results.

There are profound effects of temperature on early life cycle stages, including egg and cercarial development, cercarial longevity and infectivity and metacercarial establishment (Marcogliese 2016). Furthermore, host–parasite interactions including infected snail fecundity, infected frog growth and survival, survival and prevalence of malformations have also been examined with relation to temperature (Paull and Johnson 2011; Paull et al. 2012; reviewed in Marcogliese 2016). In general, effects of temperature vary for both parasite life cycle stages and host–parasite interactions, with maximum effects occurring at lowest, middle or highest temperatures tested. Field work suggests that higher temperatures promote synchrony between the occurrence of the infective parasites and susceptible frogs, at least for the Pacific chorus frog (*P. regilla*) (McDevitt-Galles et al. 2020).

Eutrophication increases infections in frogs, as increased nutrients and algal growth result in a higher abundance of snail intermediate hosts (Johnson and Chase 2004; Johnson et al. 2007). Furthermore, snails grow larger under these conditions and release more cercariae. Consequently, the rate of malformations is higher in frogs under eutrophic conditions (Johnson and Chase 2004; Johnson et al. 2007). Importantly, effects of eutrophication can overwhelm temperature effects (Paull and Johnson 2018). The parasite is also susceptible to low levels of desiccation (Paull and Johnson 2018), but intensities increase in anurans exposed to salinity (Milotic et al. 2017).

22.6.2 Schistosoma spp. (Trematoda)

Schistosoma spp. are serious pathogens of humans. Early work suggested that schistosomiasis would spread into temperate areas with climate change, but that view has been called into question (see Lafferty 2009). Indeed, temperature has a complex impact on the prevalence and abundance of *Schistosoma mansoni*, the main cause of disease in humans (Mangal et al. 2008). Modelling suggests that disease increases in humans up to 30 °C, but then crashes at 35 °C due to higher mortality of the snail intermediate host (Mangal et al. 2008), resulting in unstable epidemic cycles. In terms of distribution, a recent modelling effort incorporating temperature sensitivity of parasite life cycle stages with that of the snail intermediate host, demonstrated that infection risk of *S. mansoni* was expected to increase in 24–36% of eastern Africa over the next 50 years as a result of increasing temperatures (Stensgaard et al. 2016). However, they also predicted a general decrease in risk in 30–37% of the region (Stensgaard et al. 2016; see also Lafferty 2009).

Parasite abundance has been linked to the size of snail populations that serve as first intermediate hosts (Patz et al. 2000). The distribution of disease is affected not only by temperature, but by the type of water body, precipitation, stream flow and altitude, all of which affect the distribution and abundance of snail hosts (Mas-Coma et al. 2009). Changes in precipitation leading to increases in flooding are expected to increase transmission in sub-Saharan Africa (Adekiya et al. 2020).

In addition, habitat alterations can affect the spread of disease (Marcogliese 2023). For example, the formation of reservoirs due to dam construction increases habitat of snail intermediate hosts for *S. mansoni* (Steinmann et al., 2006; Mas-Coma et al. 2009). However, the actual situation is more complex than merely the creation of a good snail habitat. Dams have been shown to block the migration of predatory river prawns (*Macrobrachium* spp.) that consume and control snails (Sokolow et al. 2017). This is noteworthy because, in areas of increasing drought, including areas where schistosomiasis may be endemic, dam construction is one possible remediation technique to alleviate water shortages and low water levels (Marcogliese 2001; Jiménez Cisneros et al. 2014). In the simulations by Stensgaard et al. (2016), companion models of snail habitat suitability for eastern Africa suggested that modified habitat plays an important role in distribution of the intermediate host.

Free-living stages of *Schistosoma* spp., like other parasites, are influenced not only by temperature, but by pH, salinity and contaminants (Pietrock and Marcogliese 2003; Adekiya et al. 2020; Douchet et al. 2023). These factors also impact the snail intermediate hosts and host–parasite interactions (Adekiya et al. 2020). As discussed above, increased precipitation with climate change will increase runoff, and therefore contaminants, particularly from agricultural areas. This is in addition to potential direct effects on free-living stages of the parasite (Pietrock and Marcogliese 2003). A number of commonly-used agrochemicals have been shown to indirectly increase infection risk by *Schistosoma* spp. (Halstead et al. 2018). Certain herbicides and fertilisers increase snail biomass by promoting periphyton growth, which serves as snail food. Insecticides may also eliminate snail predators (Halstead et al. 2018; Haggerty et al. 2022).

Fig. 22.6 Plerocecoids of *Schistocephalus solidus*, removed from the body cavity of threespine stickleback (*Gasterosteus aculeatus*), the second intermediate host. The parasite has a circumpolar distribution. Infections with the cestode have serious fitness consequences for both female stickleback reproduction, but these effects vary from extreme to rather mild in comparison depending on the host population. Much less pronounced effects have been observed in populations from Alaska (see Barber 2013). Photograph credit: David J. Marcogliese

These studies are supported by mathematical simulations on transmission of *Schistosoma haematobium* that demonstrate that herbicides increase parasite transmission through bottom-up effects on the snail intermediate hosts, but decrease transmission through direct effects on miracidial survival, snail reproduction and snail survival (Hoover et al. 2020). Similarly, insecticides increase transmission through top-down effects on snail predators, but decrease infection via effects on survival of miracidia, cercariae and snails, as well as snail reproduction. Although results vary between pesticides, they show that agrochemicals increase human mortality due to proliferating *S. haematobium*.

22.6.3 Schistocephalus solidus (Cestoda)

Climate change studies on transmission of this cestode have been aided greatly by the development of an *in vitro* model that rears worms from the plerocercoid stage to adulthood, a process that normally occurs in birds (see Barber 2013). The parasite infects copepods as first intermediate hosts, and three-spined sticklebacks as second intermediate hosts. Both these hosts are also relatively easy to maintain in captivity, creating opportunities for detailed experimental studies. Sticklebacks infected with *S. solidus* (Fig. 22.6) have become especially useful to examine host–parasite interactions at the physiological, behavioural and ecological level (Barber 2013).

Macnab and Barber (2012) performed some fascinating experiments clearly demonstrating the effects of warmer temperatures on growth of *S. solidus* in three-spined sticklebacks. Plerocercoids from fish reared at 20 °C were almost four times as heavy as those from fish reared at 15 °C. This is especially notable, as even minor increases in plerocercoid mass result in increases in worm fecundity, and thus these large increments in mass would result in huge increases in parasite egg production (Macnab and Barber 2012). In addition, sticklebacks naturally infected with *S. solidus* demonstrated a predilection for warmer waters, a behavioural change that would serve to increase parasite growth and fecundity (Macnab and Barber 2012).

Sticklebacks reared at cold temperatures grew faster than those reared at warm temperatures, while cestodes grew faster at warm temperatures (Franke et al. 2017). Warmer temperatures appeared to benefit the parasite, while fish fitness and immunity were stronger at cold temperatures (Franke et al. 2017). Cestodes and fish from three different habitats varying widely in temperature were used as source populations of both fish and parasites for experiments examining fish growth and immune function as well as cestode maturation and fecundity (Franke et al. 2019). At low temperatures (13 °C), sticklebacks grew faster and had higher immune activity, and the parasites took much longer time to reach infectivity (64 days). At higher temperatures (18 °C and 24 °C), parasite growth and development were much faster, reaching infectivity in 36 days (Franke et al. 2019; Scharsack et al. 2021). Furthermore, egg production from cultured worms and hatching increased strongly while host immunity and growth were impaired when infected sticklebacks were maintained at the higher temperatures (Franke et al. 2019). Results again suggest that global warming will benefit *S.*

solidus in this host–parasite system due to enhanced parasite growth and impaired host immune function (Franke et al. 2019; Scharsack et al. 2021). These findings are important because they demonstrate that exposure of larval parasites in ectothermic hosts to warmer temperatures can affect parasite reproduction and fitness in an endothermic host (see Morley and Lewis 2014b).

22.6.4 Batrachochytrium dendrobatidis (Fungi)

The chytrid fungus *Bd* is a devastating disease of amphibians, known as chytridiomycosis, that is responsible for population declines, extirpations and even extinctions globally (Lips et al. 2006; Skerratt et al. 2007; Lips 2016; Scheele et al. 2019; see also Chap. 4). These losses are now recognised as the greatest loss of biodiversity as a result of disease ever recorded (Scheele et al. 2019). Consequently, the parasite has received a great deal of attention, especially in relation to climate change. Results of these studies have been somewhat contentious. Unlike many other pathogens, chytrids fare better and are more pathogenic at cooler temperatures. For example, infection intensities in seven species of frog experiencing huge declines across Costa Rica were negatively related to mean annual temperature (Whitfield et al. 2017). Amphibian extinctions due to chytridiomycosis in Costa Rica were linked to sea surface and air temperatures (Pounds et al. 2006). The authors proposed that temperatures in cold mountainous locations were warming due to climate change, approaching those optimal for chytrid growth and promoting disease epidemics (Pounds et al. 2006). A subsequent study refuted this idea and linked the spread of disease outbreaks to introductions into South America and the Andes (Lips et al. 2008).

A laboratory study demonstrated that lower temperatures decreased chytrid fecundity but increased maturation rates; they also increased infectivity while slowing growth (Woodhams et al. 2008). Modelling various population parameters, they determined that populations of *Bd* can grow across a broad temperature range up to 25 °C, and maintain high growth rates at the lower end of that range. The disease was prevalent at low temperatures, presumably in part because amphibian immunity is suppressed and thus host factors may contribute to amphibian declines at cooler temperatures (Woodhams et al. 2008).

Using climatic predictions from the IPCC at the time and published data on *Bd*'s thermal tolerance and temperature-dependent growth rates, Rödder et al. (2010) predicted that future climate change would restrict the parasite's global distribution. Again, using global climate predictions, seasonality was demonstrated to be important in the outbreak of disease. Outbreaks were associated with the onset of spring thaw and transmission from year-round resident amphibian species in the French Pyrenees. Earlier spring thaws with climate change were predicted to augment epidemics in local amphibians (Clare et al. 2016). A more complex modelling effort involved 13 climatic variables, an extensive global chytrid distribution database, and based findings on IPCC projections. Results suggested that outbreaks of *Bd* vary regionally depending on different climatic conditions, but that risks of expansion were high at higher altitudes and latitudes, especially in temperate North America (Xie et al. 2016).

Yet another model making use of thermal tolerances for both *Bd* and amphibian hosts mapped global infection patterns across regional temperatures and canopy cover (REF). Thermal tolerances were experimentally determined for common frogs from Costa Rica from a number of taxonomic families. Results suggested that infection prevalence will decrease as the gap widens between thermal tolerances of host and parasite (TMH), and varied with amphibian phylogeny (Nowakowski et al. 2016). However, another set of experiments showed that infection with *Bd* lowered a host's critical thermal maximum temperature (Greenspan et al. 2017).

As in other studies of disease in ectotherms, temperature variation and host acclimation may be important. In experiments with *Bd* in red-spotted newts, infections were greater in those experiencing a shift in temperature compared to newts that were acclimated at that temperature, whether low or high (Raffel et al. 2015). The acclimation effects were stronger in downward temperature shifts compared to an upward change in temperature. However, experiments on green frog (*Lithobates clamitans*) tadpoles, while indicating that infections decreased with increasing temperature as expected, acclimation had little effect on infection intensity (Altman and Raffel 2019). In another set of experiments, acclimation enhanced thermal tolerance, but its effects could be counteracted by high pathogen intensities (Greenspan et al. 2017).

22.7 Concluding Remarks

Clearly, the effects of climate change on parasites and host–parasite systems are complex. This is because effects vary even among closely related species, as well as their hosts. Also, the various stages of a parasite's life cycle and their interaction with the particular host in that portion of the life cycle respond differently to temperature. Our understanding has been aided by the application of laboratory and mesocosm experiments to particular host–parasite systems that have clearly demonstrated the complexity of temperature effects on different components of a parasite's life cycle. Significant advances have been made in the application of the Metabolic Theory of Ecology (MTE) to temperature effects on selected host–parasite systems. Additionally, advances in molecular techniques have permitted a more thorough examination of effects of temperature on the host immune response. However, not only are effects of temperature alone difficult to predict, but they are modified by other climate-associated abiotic stressors, rendering predictions even more complicated. These include precipitation and hydrology, eutrophication, acidification, salinity and contaminants, as well as the combined effects of multiple stressors. Other abiotic and biotic factors associated with climate change not covered herein include ice cover, oceanic circulation, UV radiation changes and the spread of invasive species (Marcogliese 2001). Furthermore, many environmental factors, especially anthropogenic ones, may mask direct effects of climate change (Marcogliese 2008; Lafferty 2009). Ultimately, thermal effects and other climate-related abiotic changes have effects on parasites of keystone species and ecosystem engineers that in turn have profound impacts on entire ecosystems.

Byers (2021) suggests that constructing TPCs for both hosts and parasites is a good starting point to evaluate effects of global warming on host–parasite systems. More examples are required from a variety of phylogenetically diverse taxa. It should also be noted that for parasites with complex life cycles, TPCs may vary among life cycle stages, a fact that should be taken into account. Furthermore, not only should TPCs be constructed for uninfected hosts, they should also be made for infected hosts, especially for nonlethal endpoints, as these may change once the host becomes infected. A mathematical modelling approach that combines parasite population models with their physiological tolerances (MTE) further enhances our predictive capacity for effects of climate change on parasite dynamics and transmission (see Molnár et al. 2013, 2017). In terms of experimentation, more host–parasite model systems are required that are amenable to experimental laboratory and mesocosm approaches in order to test effects not only of temperature but other stressors associated with climate change. For example, the fish monogenean *Gyrodactylus* spp. model system, which has been extensively in parasite epidemiology (e.g. Tadiri et al. 2019), could be applied more broadly to questions related to climate change. In terms of field studies, it has already been pointed out that detailed studies of parasite population dynamics and transmission in thermal effluents provide a good proxy for effects of climate change on host–parasite systems (Marcogliese 2001). In addition, comprehensive sampling of host–parasite systems over relatively large spatial scales that encompass natural and anthropogenic stressors across environmental gradients with adequate replication is required to disentangle the relative importance of various environmental drivers associated with climate change in determining parasite abundance and diversity (see, for example, McDevitt-Galles et al. 2020).

References

Abdel-Tawwab M, Monier MN, Hoseinifar SH, Faggio C (2019) Fish response to hypoxia stress: growth, physiological, and immunological biomarkers. Fish Physiol Biochem 45:997–1013. https://doi.org/10.1007/s10695-019-00614-9

Adams AJ, Kupferberg SJ, Wilber MQ, Pessier AP, Grefsrud M, Bobzien S, Vredenburg VT, Briggs CJ (2017) Extreme drought, host density, sex, and bullfrogs influence fungal pathogen infection in a declining lotic amphibian. Ecosphere 8:e01740. https://doi.org/10.1002/ecs2.1740

Adekiya TA, Aruleba RT, Oyinloye BE, Okosun KO, Kappo AP (2020) The effect of climate change and the snail-schistosome cycle in transmission and biocontrol of schistosomiasis in Sub-Saharan Africa. Int J Environ Res Public Health 17:181. https://doi.org/10.3390/ijerph17010181

Aho JM, Camp JW, Esch GW (1982) Long-term studies on the population biology of *Diplostomulum scheuringi* in a thermally altered reservoir. J Parasitol 68:695–708

Alexander JD, Bartholomew JL (2020) Myxoboliosis (*Myxobolus cerebralis*). In: Woo PTK, Leong J-A, Buchmann K (eds) Climate change and infectious fish diseases. CAB International, Wallingford, pp 381–403

Almeida D, Cruz A, Llinares C, Torralva M, Lantero E, Fletcher DH, Oliva-Paterna FJ (2023) Fish morphological and parasitological traits as ecological indicators of habitat quality in a Mediterranean coastal lagoon. Aquat Conserv Mar Freshw Ecosyst 33:1229–1244. https://doi.org/10.1002/aqc.3996

Altizer S, Ostfeld RS, Johnson PTJ, Kutz S, Harvell CD (2013) Climate change and infectious diseases: from evidence to a predictive framework. Science 341:514–519. https://doi.org/10.1126/science.1239401

Altman KA, Raffel TR (2019) Thermal acclimation has little effect on tadpole resistance to *Batrachochytrium dendrobatidis*. Dis Aquat Org 133:207–216. https://doi.org/10.3354/dao03347

Altman KA, Paull SH, Johnson PTJ, Golembieski MN, Stephens JP, LaFonte BE, Raffel TR (2016) Host and parasite thermal acclimation responses depend on the stage of infection. J Anim Ecol 85:1014–1024. https://doi.org/10.1111/1365-2656.12510

Arias PA, Bellouin N, Coppola E, Jones RG, Krinner G, Marotzke J et al (2021) Technical summary. In: Masson-Delmotte V, Zhai P, Pirani A, Connors SL, Péan C, Berger S et al (eds) Climate change 2021: the physical science basis. Contribution of working group I to the sixth assessment report of the intergovernmental panel on climate change. Cambridge University Press, Cambridge, pp 33–144. https://doi.org/10.1017/9781009157896.002

Baag S, Mandal S (2022) Combined effects of ocean warming and acidification on marine fish and shellfish: a molecule to ecosystem perspective. Sci Tot Environ 802:149807. https://doi.org/10.1016/j.scitotenv.2021.149807

Barber I (2013) Sticklebacks as model hosts in ecological and evolutionary parasitology. Trends Parasitol 29:556–566. https://doi.org/10.1016/j.pt.2013.09.004

Barber I, Berkhout BW, Ismail Z (2016) Thermal change and the dynamics of multi-host parasite life cycles in aquatic ecosystems. Integr Comp Biol 56:561–572. https://doi.org/10.1093/icb/icw025

Barbi A, Goessens T, Strubbe D, Deknock A, Van Leeuwenberg R, De Troyer N et al (2023) Widespread triazole pesticide use affects infection dynamics of a global amphibian pathogen. Ecol Lett 26:313–322. https://doi.org/10.1111/ele.14154

Berkhout BW, Lloyd MM, Poulin R, Studer A (2014) Variation among genotypes in responses to increasing temperature in a marine parasite: evolutionary potential in the face of global warming? Int J Parasitol 44:1019–1027. https://doi.org/10.1016/j.ijpara.2014.07.002

Blanar CA, Munkittrick KR, Houlahan J, MacLatchy DL, Marcogliese DJ (2009) Pollution and parasitism in aquatic animals: a meta-analysis of effect size. Aquat Toxicol 93:18–28. https://doi.org/10.1016/j.aquatox.2009.03.002

Blanar CA, Marcogliese DJ, Couillard CM (2011) Natural and anthropogenic factors shape metazoan parasite community structure in mummichog (*Fundulus heteroclitus*) from two estuaries in New Brunswick, Canada. Folia Parasitol 58:240–248

Bochow N, Boer N (2023) The South American monsoon approaches a critical transition in response to deforestation. Sci Adv 9:eadd9973

Bommarito C, Khosravi M, Thieltges DW, Pansch C, Hamm T, Pranovi F, Vajedsamiei J (2022) Combined effects of salinity and trematode infections on the filtration capacity, growth and condition of mussels. Mar Ecol Prog Ser 699:33–44. https://doi.org/10.3354/meps14179

Bommarito C, Pansch C, Khosravi M, Pranovi F, Wahl M, Thieltges DW (2020) Freshening rather than warming drives trematode transmission from periwinkles to mussels. Mar Biol 167:46. https://doi.org/10.1007/s00227-020-3657-3

Bowden TJ (2008) Modulation of the immune system of fish by their environment. Fish Shellfish Immunol 25:373–383. https://doi.org/10.1016/j.fsi.2008.03.017

Bowerman J, Johnson PTJ (2003) Timing of trematode-related malformations in Oregon spotted frogs and Pacific treefrogs. Northwest Nat 84:142–145

Brooks KM (2005) The effects of water temperature, salinity, and currents on the survival and distribution of the infective copepodid stage of sea lice (*Lepeophtheirus salmonis*) originating on Atlantic salmon farms in the Broughton Archipelago of British Columbia, Canada. Rev Fish Sci 13:177–204. https://doi.org/10.1080/10641260500207109

Bruno JF, Petes LE, Harvell CD, Hettinger A (2003) Nutrient enrichment can increase the severity of coral diseases. Ecol Lett 6:1056–1061. https://doi.org/10.1046/j.1461-0248.2003.00544.x

Buck JC (2019) Indirect effects explain the role of parasites in ecosystems. Trends Parasitol 35:835–847. https://doi.org/10.1016/j.pt.2019.07.007

Buck JC, Ripple RJ (2017) Infectious agents trigger trophic cascades. Trends Ecol Evol 32:681–694. https://doi.org/10.1016/j.tree.2017.06.009

Budria A (2017) Beyond troubled waters: the influence of eutrophication on host–parasite interactions. Funct Ecol 31:1348–1358. https://doi.org/10.1111/1365-2435.12880

Burge CA, Eakin CM, Friedman CS, Froelich B, Hershberger PK, Hofmann EE et al (2014) Climate change influences on marine infectious diseases: implications for management and society. Annu Rev Mar Sci 6:249–277. https://doi.org/10.1146/annurev-marine-010213-135029

Buss N, Hua J (2023) Host exposure to a common pollutant can influence diversity–disease relationships. J Anim Ecol 92:2151–2162. https://doi.org/10.1111/1365-2656.13988

Buss N, Wersebe M, Hua J (2019) Direct and indirect effects of a common cyanobacterial toxin on amphibian-trematode dynamics. Chemosphere 220:731–737. https://doi.org/10.1016/j.chemosphere.2018.12.160

Byers JE (2020) Effects of climate change on parasites and disease in estuarine and nearshore environments. PLoS Biol 18:e3000743. https://doi.org/10.1371/journal.pbio.3000743

Byers JE (2021) Marine parasites and disease in the era of global climate change. Annu Rev Mar Sci 13:397–420. https://doi.org/10.1146/annurev-marine-031920-100429

Cable J, Barber I, Boag B, Ellison AR, Morgan ER, Murray K, Pascoe EL, Sait SM, Wilson AJ, Booth M (2017) Global change, parasite transmission and disease control: lessons from ecology. Philos Trans R Soc B 372:20160088. https://doi.org/10.1098/rstb.2016.0088

Camp JW, Aho JM, Esch GW (1982) A long-term study on various aspects of the population biology of *Ornithodiplostomum ptychocheilus* in a South Carolina cooling reservoir. J Parasitol 68:709–718

Carlson CJ, Burgio KR, Dougherty ER, Phillips AJ, Bueno VM, Clements CF et al (2017) Parasite biodiversity faces extinction and redistribution in a changing climate. Sci Adv 3:e1602422. https://doi.org/10.1126/sciadv.1602422

Chapman J, Marcogliese DJ, Suskic CD, Cooke SJ (2015) Variation in parasite communities and health indices of juvenile *Lepomis gibbosus* across a gradient of watershed land-use and habitat quality. Ecol Indic 57:564–572. https://doi.org/10.1016/j.ecolind.2015.05.013

Chubb JC (1977) Seasonal occurrence of helminths in freshwater fishes. Part I. Monogenea. Adv Parasitol 15:133–199

Chubb JC (1979) Seasonal occurrences of helminths in freshwater fishes. Part II. Trematoda. Adv Parasitol 17:141–313

Chubb JC (1980) Seasonal occurrence of helminths in freshwater fishes. Part III. Larval Cestoda and Nematoda. Adv Parasitol 18:1–120

Chubb JC (1982) Seasonal occurrence of helminths in freshwater fishes. Part IV. Adult Cestoda, Nematoda and Acanthocephala. Adv Parasitol 20:1–292

Claar DC, Wood CL (2020) Pulse heat stress and parasitism in a warming world. Trends Ecol Evol 35:704–715. https://doi.org/10.1016/j.tree.2020.04.002

Clare FC, Halder JB, Daniel O, Bielby J, Semenov MA, Jombart T et al (2016) Climate forcing of an emerging pathogenic fungus across a montane multi-host community. Philos Trans R Soc B 371:20150454. https://doi.org/10.1098/rstb.2015.0454

Clulow S, Gould J, James H, Stockwell M, Clulow J, Mahony M (2018) Elevated salinity blocks pathogen transmission and improves host survival from the global amphibian chytrid pandemic: implications for translocations. J Appl Ecol 55:830–840. https://doi.org/10.1111/1365-2664.13030

Cohen JM, Venesky MD, Sauer EL, Civitello DJ, McMahon TA, Roznik EA, Rohr JR (2017) The thermal mismatch hypothesis explains host susceptibility to an emerging infectious disease. Ecol Lett 20:184–193. https://doi.org/10.1111/ele.12720

Cohen JM, Civitello DJ, Venesky MD, McMahon TA, Rohr JR (2019) An interaction between climate change and infectious disease drove widespread amphibian declines. Glob Change Biol 25:927–937. https://doi.org/10.1111/gcb.14489

Cohen JM, Sauer EL, Santiago O, Spencer S, Rohr JR (2020) Divergent impacts of warming weather on wildlife disease risk across climates. Science 370:eabb1702. https://doi.org/10.1126/science.abb1702

Cone DK, Marcogliese DJ, Watt WD (1993) Metazoan parasite communities of yellow eels (*Anguilla rostrata*) in acidic and limed rivers of Nova Scotia. Can J Zool 71:177–184

Costello MJ (2006) Ecology of sea lice parasitic on farmed and wild fish. Trends Parasitol 22:475–482. https://doi.org/10.1016/j.pt.2006.08.006

Crain CM, Kroeker C, Halpern BS (2008) Interactive and cumulative effects of multiple human stressors in marine systems. Ecol Lett 11:1304–1315. https://doi.org/10.1111/j.1461-0248.2008.01253.x

Cuco AP, Castro BB, Gonçalves F, Wolinska J, Abrantes N (2018) Temperature modulates the interaction between fungicide pollution and disease: evidence from a *Daphnia*-microparasitic yeast model. Parasitology 145:939–947. https://doi.org/10.1017/S0031182017002062

Dai Y, Yang S, Zhao D, Hu C, Xu W, Anderson DM et al (2023) Coastal phytoplankton blooms expand and intensify in the 21st century. Nature 615:280–284. https://doi.org/10.1038/s41586-023-05760-y

Díaz-Morales DM, Bommarito C, Vajedsamiei J, Grabner DS, Rilov G, Wahl M, Sures B (2022) Heat sensitivity of first host and cercariae may restrict parasite transmission in a warming sea. Sci Rep 12:1174. https://doi.org/10.1038/s41598-022-05139-5

Dittmar J, Janssen H, Kuske A, Kurtz J, Scharsack JP (2014) Heat and immunity: an experimental heat wave alters immune functions in three-spined sticklebacks (*Gasterosteus aculeatus*). J Anim Ecol 83:744–757. https://doi.org/10.1111/1365-2656.12175

Dobson A, Carper R (1992) Global warming and potential changes in host-parasite and disease– vector relationships. In: Peters RL, Lovejoy TE (eds) Global warming and biological diversity. Yale University Press, New Haven, CT, pp 201–217

Douchet P, Gourbal B, Loker ES, Rey O (2023) *Schistosoma* transmission: scaling-up competence from hosts to ecosystems. Trends Parasitol 39:563–574. https://doi.org/10.1016/j.pt.2023.04.001

El-Darsh HEM, Whitfield PJ (1999) Digenean metacercariae (*Timoniella* spp., *Labratrema minimus* and *Cryptocotyle concava*) from the flounder, *Platichthys flesus*, in the tidal Thames. J Helminthol 73:103–113

Ellis RP, Parry H, Spicer JI, Hutchinson TH, Pipe RK, Widdicombe S (2011) Immunological function in marine invertebrates: responses to environmental perturbation. Fish Shellfish Immunol 30:1209–1222. https://doi.org/10.1016/j.fsi.2011.03.017

Espinoza JC, Ronchail J, Marengov JA, Segura H (2019) Contrasting north–south changes in Amazon wet-day and dry-day frequency and related atmospheric features (1981–2017). Clim Dynam 52:5413–5430. https://doi.org/10.1007/s00382-018-4462-2

Fast MD, Dalvin S (2020) Lepeophtheirosis (*Lepeophtheirus salmonis*). In: Woo PTK, Leong J-A, Buchmann K (eds) Climate change and infectious fish diseases. CAB International, Wallingford, pp 471–498

Feyen L, Dankers R (2009) Impact of global warming on streamflow drought in Europe. J Geophys Res 114:D17116. https://doi.org/10.1029/2008JD011438

Fournier M, Robert J, Salo HM, Dautremepuits C, Brousseau P (2005) Immunotoxicology of amphibians. Appl Herpetol 2:297–309

Franke F, Armitage SAO, Kutzer MAM, Kurtz J, Scharsack JP (2017) Environmental temperature variation influences fitness trade-offs and tolerance in a fish-tapeworm association. Parasites Vectors 10:252. https://doi.org/10.1186/s13071-017-2192-7

Franke F, Raifarth N, Kurtz J, Scharsack JP (2019) Consequences of divergent temperature optima in a host–parasite system. Oikos 128:869–880. https://doi.org/10.1111/oik.05864

Franzova VA, MacLeod CD, Wang T, Harley CDG (2019) Complex and interactive effects of ocean acidification and warming on the life span of a marine trematode parasite. Int J Parasitol 49:1015–1021. https://doi.org/10.1016/j.ijpara.2019.07.005

Friesen O, Poulin R, Lagrue C (2021) Temperature and multiple parasites combine to alter host community structure. Oikos 130:1500–1511. https://doi.org/10.1111/oik.07813

Fu R, Yin L, Li W, Arias PA, Dickinson RE, Huang L et al (2013) Increased dry-season length over southern Amazonia in recent decades and its implication for future climate projection. Proc Natl Acad Sci USA 110:18110–18115. https://doi.org/10.1073/pnas.1302584110

Gehman A-LM, Hall RJ, Byers JE (2018) Host and parasite thermal ecology jointly determine the effect of climate warming on epidemic dynamics. Proc Natl Acad Sci USA 115:744–749. https://doi.org/10.1073/pnas.1705067115

Gilbert BM, Avenant-Oldewage A (2021) Monogeneans as bioindicators: a meta-analysis of effect size of contaminant exposure toward Monogenea (Platyhelminthes). Ecol Indic 130:108062. https://doi.org/10.1016/j.ecolind.2021.108062

Gleichsner AM, Cleveland JA, Minchella DJ (2016) One stimulus—two responses: host and parasite life-history variation in response to environmental stress. Evolution 70:2640–2646. https://doi.org/10.1111/evo.13061

Goedknegt MA, Welsh JE, Drent J, Thieltges DW (2015) Climate change and parasite transmission: how temperature affects parasite infectivity via predation on infective stages. Ecosphere 6:96. https://doi.org/10.1890/ES15-00016.1

Grabner D, Sures B (2019) Amphipod parasites may bias results of ecotoxicological research. Dis Aquat Org 136:121–132. https://doi.org/10.3354/dao03355

Grabner D, Rothe LE, Sures B (2023) Parasites and pollutants: effects of multiple stressors on aquatic organisms. Environ Toxicol Chem 42:1946–1959. https://doi.org/10.1002/etc.5689

Granath WO Jr, Esch GW (1983) Seasonal dynamics of *Bothriocephalus acheilognathi* in ambient and thermally altered areas of a North Carolina cooling reservoir. Proc Helminthol Soc Wash 50:205–218

Greenspan SE, Bower DS, Roznik EA, Pike DA, Marantelli G, Alford RA, Schwarzkopf L, Scheffers BR (2017) Infection increases vulnerability to climate change via effects on host thermal tolerance. Sci Rep 7:9349. https://doi.org/10.1038/s41598-017-09950-3

Gsell AS, Biere A, de Boer W, de Bruijn I, Eichhorn G, Frenken T et al (2023) Environmental refuges from disease in host–parasite interactions under global change. Ecology 104:e4001. https://doi.org/10.1002/ecy.4001

Guilloteau P, Poulin R, MacLeod CD (2016) Impacts of ocean acidification on multiplication and caste organisation of parasitic trematodes in their gastropod host. Mar Biol 163:96. https://doi.org/10.1007/s00227-016-2871-5

Haggerty CJE, Halstead NT, Civitello DJ, Rohr JR (2022) Reducing disease and producing food: effects of 13 agrochemicals on snail biomass and human schistosomes. J Appl Ecol 59:729–741. https://doi.org/10.1111/1365-2664.14087

Halmetoja A, Valtonen ET, Koskenniemi E (2000) Perch (*Perca fluviatilis* L.) parasites reflect ecosystem conditions: a comparison of a natural lake and two acidic reservoirs in Finland. Int J Parasitol 30:1437–1444

Halstead NT, Hoover CM, Arakala A, Civitello DJ, De Leo GA, Gambhir M et al (2018) Agrochemicals increase risk of human schistosomiasis by supporting higher densities of intermediate hosts. Nat Commun 9:837. https://doi.org/10.1038/s41467-018-03189-w

Hanlon SM, Parris MJ (2012) The impact of pesticides on the pathogen *Batrachochytrium dendrobatidis* independent of potential hosts. Arch Environ Contam Toxicol 63:137–143. https://doi.org/10.1007/s00244-011-9744-1

Harvell CD, Kim K, Burkholder JM, Colwell RR, Epstein PR, Grimes DJ et al (1999) Emerging marine diseases-climate links and anthropogenic factors. Science 285:1505–1510. https://doi.org/10.1126/science.285.5433.1505

Harvell CD, Mitchell CE, Ward JR, Altizer S, Dobson AP, Ostfeld RS, Samuel MD (2002) Climate warming and disease risks for terrestrial and marine biota. Science 296:2158–2162. https://doi.org/10.1126/science.1063699

Harvell D, Jordán-Dahlgren E, Merkel S, Rosenberg E, Raymundo L, Smith G, Weil E, Willis B (2007) Coral disease, environmental drivers, and the balance between coral and microbial associates. Oceanography 20:172–195

Harvey BP, Gwynn-Jones D, Moore PJ (2013) Meta-analysis reveals complex marine biological responses to the interactive effects of ocean acidification and warming. Ecol Evol 3:1016–1030. https://doi.org/10.1002/ece3.516

Hatcher MJ, Dick JTA, Dunn AM (2012) Diverse effects of parasites in ecosystems: linking interdependent processes. Front Ecol Environ 10:186–194. https://doi.org/10.1890/110016

Hechinger RF, Lafferty KD (2005) Host diversity begets parasite diversity: bird final hosts and trematodes in snail intermediate hosts. Proc R Soc B 272:1059–1066. https://doi.org/10.1098/rspb.2005.3070

Hechinger RF, Lafferty KD, Huspeni TC, Brooks AJ, Kuris AM (2007) Can parasites be indicators of free-living diversity? Relationships between species richness and the abundance of larval trematodes and of local benthos and fishes. Oecologia 151:82–92. https://doi.org/10.1007/s00442-006-0568-z

Hector TE, Sgrò CM, Hall MD (2021) Thermal limits in the face of infectious disease: How important are pathogens? Glob Change Biol 27:4469–4480

Hector TE, Gehman A-LM, King KC (2023) Infection burdens and virulence under heat stress: ecological and evolutionary considerations. Philos Trans R Soc B 378:20220018. https://doi.org/10.1098/rstb.2022.0018

Ho JC, Michalak AM, Pahlevan N (2019) Widespread global increase in intense lake phytoplankton blooms since the 1980s. Nature 574:667–670. https://doi.org/10.1038/s41586-019-1648-7

Holmstrup M, Bindesbøl A-M, Oostingh GJ, Duschl A, Scheil V, Köhler H-R et al (2010) Interactions between effects of environmental chemicals and natural stressors: a review. Sci Tot Environ 408:3746–3762. https://doi.org/10.1016/j.scitotenv.2009.10.067

Hooper MJ, Ankley GT, Cristol DA, Maryoung LA, Noyes PD, Pinkerton KE (2013) Interactions between chemical and climate stressors: a role for mechanistic toxicology in assessing climate change risks. Environ Toxicol Chem 32:32–48. https://doi.org/10.1002/etc.2043

Hoover CM, Rumschlag SL, Strgar L, Arakala A, Gambhir M, de Leo GA, Sokolow SH, Rohr JR, Remais JV (2020) Effects of agrochemical pollution on schistosomiasis transmission: a systematic review and modelling analysis. Lancet Planet Health 4:e280–e291

IPBES (2019) Summary for policymakers of the global assessment report on biodiversity and ecosystem services of the Intergovernmental Science-Policy Platform on Biodiversity and Ecosystem Services. S. Díaz, J. Settele, E. S. Brondízio, H. T. Ngo, M. Guèze, J. Agard et al., editors. IPBES secretariat, Bonn

Jacinto-Maldonado M, González-Salazar C, Basanta MD, García-Peña GE, Saucedo B, Lesbarrères D, Meza-Figueroa D, Stephens CR (2023) Water pollution increases the risk of chytridiomycosis in Mexican amphibians. EcoHealth 20:74–83. https://doi.org/10.1007/s10393-023-01631-0

Jane SF, Hansen GJA, Kraemer BM, Leavitt PR, Mincer JL, North RL et al (2021) Widespread deoxygenation of temperate lakes. Nature 594:66–70. https://doi.org/10.1038/s41586-021-03550-y

Janovy J Jr, Hardin EL (1988) Diversity of the parasite assemblage of *Fundulus zebrinus* in the Platte River of Nebraska. J Parasitol 74:207–213

Janovy J Jr, Snyder SD, Clopton RE (1997) Evolutionary constraints on population structure: the parasites of *Fundulus zebrinus* (Pisces: Cyprinodontidae) in the South Platte River of Nebraska. J Parasitol 83(83):584–592. https://doi.org/10.2307/3284228

Jiménez Cisneros BE, Oki T, Arnell NW, Benito G, Cogley JG, Döll P et al (2014) Freshwater resources. In: Field CB, Barros VR, Dokken DJ, Mach KJ, Mastrandrea MD, Bilir TE et al (eds) Climate change 2014: impacts, adaptation, and vulnerability. Part A: global and sectoral aspects. Contribution of Working Group II to the fifth assessment report of the intergovernmental panel on climate change. Cambridge University Press, Cambridge, pp 229–269

Johnson PTJ, Carpenter SR (2008) Influence of eutrophication on disease in aquatic ecosystems: patterns, processes, and predictions. In: Ostfeld RS, Keesing F, Eviner VT (eds) Infectious disease ecology. Princeton University Press, Princeton, pp 71–99. https://doi.org/10.1515/9781400837885.71

Johnson PTJ, Chase JM (2004) Parasites in the food web: linking amphibian malformations and aquatic eutrophication. Ecol Lett 7:521–526. https://doi.org/10.1111/j.1461-0248.2004.00610.x

Johnson PTJ, Sutherland DR, Kiinsella JM, Lunde KB (2004) Review of the trematode genus *Ribeiroia* (Psilostomidae): ecology, life history and pathogenesis with special emphasis on the amphibian malformation problem. Adv Parasitol 57:191–253

Johnson PTJ, Chase JM, Dosch KL, Hartson RB, Gross JA, Larson DJ, Sutherland DR, Carpenter SR (2007) Aquatic eutrophication promotes pathogenic infection in amphibians. Proc Nat Acad Sci 104:15781-15786. 10.1073/pnas.0707763104

Johnson PTJ, Chase JM, Dosch KL, Hartson RB, Gross JA, Larson DJ, Sutherland DR, Carpenter SR (2010a) Aquatic eutrophication promotes pathogenic infection in amphibians. Ecol Appl 20:16–29

Johnson PTJ, Dobson A, Lafferty KD, Marcogliese DJ, Memmott J, Orlofske SA, Poulin R, Thieltges DW (2010b) When parasites become prey: ecological and epidemiological significance of eating parasites. Trends Ecol Evol 25:362–371. https://doi.org/10.1016/j.tree.2010.01.005

Johnson PTJ, Wood CL, Joseph MB, Preston DL, Haas SE, Springer YP (2016) Habitat heterogeneity drives the host-diversity-begets-parasite-diversity relationship: evidence from experimental and field studies. Ecol Lett 19:752–761. https://doi.org/10.1111/ele.12609

Kamiya T, O'Dwyer K, Nakagawa S, Poulin R (2014) Host diversity drives parasite diversity: meta-analytical insights into patterns and causal mechanisms. Ecography 37:689–697. https://doi.org/10.1111/j.1600-0587.2013.00571.x

Khan RA, Thulin J (1991) Influence of pollution on parasites of aquatic animals. Adv Parasitol 30:201–238

Kiesecker JM, Skelly DK (2001) Effects of disease and pond drying on gray tree frog growth, development, and survival. Ecology 82:1956–1963

Koprivnikar J, Poulin R (2009) Effects of temperature, salinity, and water level on the emergence of marine cercariae. Parasitol Res 105:957–965. https://doi.org/10.1007/s00436-009-1477-y

Koprivnikar J, Lim D, Fu C, Brack SHM (2010) Effects of temperature, salinity, and pH on the survival and activity of marine cercariae. Parasitol Res 106:1167–1177. https://doi.org/10.1007/s00436-010-1779-0

Koprivnikar J, Thieltges DW, Johnson PTJ (2023) Consumption of trematode parasite infectious stages: from conceptual synthesis to future research agenda. J Helminthol 97:e33. https://doi.org/10.1017/S0022149X23000111

Kuris AM, Hechinger RF, Shaw JC, Whitney KL, Aguirre-Macedo L, Boch CA et al (2008) Ecosystem energetic implications of parasite and free-living biomass in three estuaries. Nature 2008(454):515–518. https://doi.org/10.1038/nature06970

Lafferty KD (1997) Environmental parasitology: what can parasites tell us about human impacts on the environment? Parasitol Today 13:251–255

Lafferty KD (2009) The ecology of climate change and infectious diseases. Ecology 90:888–900. https://doi.org/10.1890/08-0079.1

Lafferty KD, Arim M, Briggs J, Leo GD, Dobson P, Dunne JA et al (2008) Parasites in food webs: the ultimate missing links. Ecol Lett 11:533–546. https://doi.org/10.1111/j.1461-248.2008.01174.x

Lagrue C, Poulin R (2016) The scaling of parasite biomass with host biomass in lake ecosystems: are parasites limited by host resources? Ecography 39:507–514. https://doi.org/10.1111/ecog.01720

Landis WG, Rohr JR, Moe SJ, Balbus JM, Clements W, Fritz A (2014) Global climate change and contaminants, a call to arms not yet heard? Integrat Environ Assess Manage 10:483–484

Larsen MH, Mouritsen KN (2009) Increasing temperature counteracts the impact of parasitism on periwinkle consumption. Mar Ecol Prog Ser 383:141–149. https://doi.org/10.3354/meps08021

Larsen MH, Mouritsen KN (2014) Temperature-parasitism synergy alters intertidal soft-bottom community structure. J Exp Mar Biol Ecol 460:109–119

Lei F, Poulin R (2011) Effects of salinity on multiplication and transmission of an intertidal trematode parasite. Mar Biol 158:995-1003. 10.1007/s00227-011-1625-7

Leicht K, Seppälä O (2014) Infection success of *Echinoparyphium aconiatum* (Trematoda) in its snail host under high temperature: role of host resistance. Parasit Vectors 7:192

Leicht K, Jokela J, Seppälä O (2013) An experimental heat wave changes immune defense and life history traits in a freshwater snail. Ecol Evol 3:4861–4871. https://doi.org/10.1002/ece3.874

Leicht K, Seppälä K, Seppälä O (2017) Potential for adaptation to climate change: family-level variation in fitness-related traits and their responses to heat waves in a snail population. BMC Evol Biol 17:140. https://doi.org/10.1186/s12862-017-0988-x

Leiva NV, Manríquez PH, Aguilera VM, González MT (2019) Temperature and pCO2 jointly affect the emergence and survival of cercariae from a snail host: implications for future parasitic infections in the Humboldt Current system. Int J Parasitol 49:49–61. https://doi.org/10.1016/j.ijpara.2018.08.006

Lips KR (2016) Overview of chytrid emergence and impacts on amphibians. Phil Trans R Soc B 371:20150465. 10.1098/rstb.2015.0465

Lips KR, Brem F, Brenes R, Reeve JD, Alford RA, Voyles J, Carey, Livo L, Pessier AP, Collins JP (2006) Emerging infectious disease and the loss of biodiversity in a Neotropical amphibian community. Proc Nat Acad Sci 103:3165-3170. 10.1073/pnas.0506889103

Lips KR, Diffendorfer J, Mendelson JR, Sears MW (2008) Riding the wave: reconciling the roles of disease and climate change in amphibian declines. PLoS Biol 6:e72. 10.1371/journal.pbio.0060072

Lõhmus M, Björklund M (2015) Climate change: what will it do to fish-parasite interactions? Biol J Linn Soc 116:397–411. https://doi.org/10.1111/bij.12584

MacKenzie K (1999) Parasites as pollution indicators in marine ecosystems: a proposed early warning system. Mar Poll Bull 38:955–959

MacKenzie K, Williams HH, Williams B, McVicar AH, Siddall R (1995) Parasites as indicators of water quality and the potential use of helminth transmission in marine pollution studies. Adv Parasitol 35:85–144

MacLeod CD, Poulin R (2015a) Differential tolerances to ocean acidification by parasites that share the same host. Int J Parasitol 45:485–493. https://doi.org/10.1016/j.ijpara.2015.02.007

MacLeod CD, Poulin R (2015b) Interactive effects of parasitic infection and ocean acidification on the calcification of a marine gastropod. Mar Ecol Prog Ser 537:137–150. https://doi.org/10.3354/meps11459

Macleod CD, Poulin R (2016) Parasitic infection alters the physiological response of a marine gastropod to ocean acidification. Parasitology 143:1397–1408. https://doi.org/10.1017/S0031182016000913

Macnab V, Barber I (2012) Some (worms) like it hot: fish parasites grow faster in warmer water, and alter host thermal preferences. Glob Change Biol 18:1540–1548. https://doi.org/10.1111/j.1365-2486.2011.02595.x

Magalhães L, de Montaudouin X, Figueira E, Freitas R (2018) Trematode infection modulates cockles biochemical response to climate change. Sci Tot Environ 637–638:30–40. https://doi.org/10.1016/j.scitotenv.2018.04.432

Magnadóttir B (2006) Innate immunity of fish (overview). Fish Shellfish Immunol 20:137–151. https://doi.org/10.1016/j.fsi.2004.09.006

Magnadottir B (2010) Immunological control of fish diseases. Mar Biotechnol 12:361–379. https://doi.org/10.1007/s10126-010-9279-x

Malek JC, Byers JE (2018) Responses of an oyster host (*Crassostrea virginica*) and its protozoan parasite (*Perkinsus marinus*) to increasing air temperature. PeerJ 6:e5046. https://doi.org/10.7717/peerj.5046

Mangal TD, Paterson S, Fenton A (2008) Predicting the impact of long-term temperature changes on the epidemiology and control of schistosomiasis: a mechanistic model. PLoS One 3:e1438. https://doi.org/10.1371/journal.pone.0001438

Manzi F, Agha R, Lu Y, Ben-Ami F, Wolinska J (2020) Temperature and host diet jointly influence the outcome of infection in a *Daphnia*-fungal parasite system. Freshw Biol 65:757–767. https://doi.org/10.1111/fwb.13464

Marcogliese DJ (2001) Implications of climate change for parasitism of animals in the aquatic environment. Can J Zool 79:1331–1352. https://doi.org/10.1139/z01-067

Marcogliese DJ (2004) Parasites: small players with crucial roles in the ecological theater. EcoHealth 1:151–164

Marcogliese DJ (2005) Parasites of the superorganism: are they indicators of ecosystem health? Int J Parasitol 35:705–716. https://doi.org/10.1016/j.ijpara.2005.01.015

Marcogliese (2008) The impact of climate change on the parasites and diseases of aquatic animals. OIE Rev Sci Tech 27:467-484

Marcogliese DJ (2016) The distribution and abundance of parasites in aquatic ecosystems in a changing climate: more than just temperature. Integr Comp Biol 56:611–619. https://doi.org/10.1093/icb/icw036

Marcogliese DJ (2023) Major drivers of biodiversity loss and their impacts on helminth parasite populations and communities. J Helminthol 97:e34. https://doi.org/10.1017/S0022149X2300010X

Marcogliese DJ, Cone DK (1996) On the distribution and abundance of eel parasites in Nova Scotia: influence of pH. J Parasitol 82:389-399. 10.2307/3284074

Marcogliese DJ, Cone DK (1997) Parasite communities as indicators of ecosystem stress. Parassitologia 39:227–232

Marcogliese DJ, Cone DK (2001) Myxozoan communities parasitizing *Notropis hudsonius* (Cyprinidae) at selected localities of the St. Lawrence River, Quebec: possible effects of urban effluents. J Parasitol 87:951–956

Marcogliese DJ, Cone DK (2021) Myxozoan communities in two cyprinid fishes from mesotrophic and eutrophic rivers. J Parasitol 107:39–47. https://doi.org/10.1645/20-76

Marcogliese DJ, Jacobson KC (2015) Parasites as biological tags of marine, freshwater and anadromous fishes in North America from the tropics to the Arctic. Parasitology 142:68–89. https://doi.org/10.1017/S0031182014000110

Marcogliese DJ, Pietrock M (2011) Combined effects of parasites and contaminants on animal health: parasites do matter. Trends Parasitol 27:123–130. https://doi.org/10.1016/j.pt.2010.11.002

Marcogliese DJ, Gendron AD, Plante C, Fournier M, Cyr D (2006) Parasites of spottail shiners (*Notropis hudsonius*) in the St. Lawrence River: effects of municipal effluents and habitat. Can J Zool 84:1461–1481

Marcogliese DJ, Gendron AD, Cone DK (2009) Impact of municipal effluents and hydrological regime on myxozoan parasite communities of fish. Int J Parasitol 39:1345–1351. https://doi.org/10.1016/j.ijpara.2009.04.007

Marcogliese DJ, Locke SA, Gélinas M, Gendron AD (2016) Variation in parasite communities in spottail shiners (*Notropis hudsonius*) linked with precipitation. J Parasitol 102:27–36. https://doi.org/10.1645/12-31

Martin LB, Hopkins WA, Mydlarz LD, Rohr JR (2010) The effects of anthropogenic global changes on immune functions and disease resistance: ecoimmunology and global change. Ann N Y Acad Sci 1195:129–148. https://doi.org/10.1111/j.1749-6632.2010.05454.x

Mas-Coma S, Valero MA, Bargues MD (2009) Climate change effects on trematodiases, with emphasis on zoonotic fascioliasis and schistosomiasis. Vet Parasitol 163:264–280. https://doi.org/10.1016/j.vetpar.2009.03.024

McCahon CP, Poulton MJ (1991) Lethal and sub-lethal effects of acid, aluminium and lime on *Gammarus pulex* during repeated simulated episodes in a Welsh stream. Freshw Biol 25:169–178

McDevitt-Galles T, Moss WE, Calhoun DM, Johnson PTJ (2020) Phenological synchrony shapes pathology in host–parasite systems. Proc R Soc B 287:20192597. https://doi.org/10.1098/rspb.2019.2597

McDowell MA, Ferdig MT, Janovy J Jr (1992) Dynamics of the parasite assemblage of *Pimephales promelas* in Nebraska. J Parasitol 78:830–836

McKenzie VJ, Townsend AR (2007) Parasitic and infectious disease responses to changing global nutrient cycles. EcoHealth 4:384–396. https://doi.org/10.1007/s10393-007-0131-3

Measures LN (1996) Effect of temperature and salinity on development and survival of eggs and free-living larvae of sealworm (*Pseudoterranova decipiens*) Can J Fish Aquat Sci 53:2804-2807. 10.1139/f96-241

Messina S, Costantini D, Eens M (2023) Impacts of rising temperatures and water acidification on the oxidative status and immune system of aquatic ectothermic vertebrates: a meta-analysis. Sci Tot Environ 868:161580. https://doi.org/10.1016/j.scitotenv.2023.161580

Michelle C., Jackson Charlie J. G., Loewen Rolf D., Vinebrooke Christian T., Chimimba (2016) Net effects of multiple stressors in freshwater ecosystems: a meta-analysis Abstract Global Change Biology 22(1) 180-189 10.1111/gcb.2016.22.issue-1 10.1111/gcb.13028

Mignatti A, Boag B, Cattadori IM (2016) Host immunity shapes the impact of climate changes on the dynamics of parasite infections. Proc Natl Acad Sci USA 113:2970–2975. https://doi.org/10.1073/pnas.1501193113

Milotic D, Milotic M, Koprivnikar J (2017) Effects of road salt on larval amphibian susceptibility to parasitism through behavior and immunocompetence. Aquat Toxicol 189:42–49. https://doi.org/10.1016/j.aquatox.2017.05.015

Milotic M, Milotic D, Koprivnikar J (2018) Exposure to a cyanobacterial toxin increases larval amphibian susceptibility to parasitism. Parasitol Res 117:513–520. https://doi.org/10.1007/s00436-017-5727-0

Milotic M, Milotic D, Koprivnikar J (2019) Effects of a cyanobacterial toxin on trematode cercariae. J Parasitol 105:598–605. https://doi.org/10.1645/18-170

Minchella DJ, Scott MS (1991) Parasitism: a cryptic determinant of animal community structure. Trends Ecol Evol 6:250–254

Moe SJ, de Schamphelaere K, Clements WH, Sorensen MT, Van den Brin PJ, Liess M (2013) Combined and interactive effects of global climate change and toxicants on populations and communities. Environ Toxicol Chem 32:49-61. 10.1002/etc.2045

Molnár PK, Kutz SJ, Hoar BM, Dobson AP (2013) Metabolic approaches to understanding climate change impacts on seasonal host-macroparasite dynamics. Ecol Lett 16:9–21. https://doi.org/10.1111/ele.12022

Molnár PK, Sckrabulis JP, Altman KA, Raffel TR (2017) Thermal performance curves and the Metabolic Theory of Ecology—a practical guide to models and experiments for parasitologists. J Parasitol 103:423–439. https://doi.org/10.1645/16-148

Morley NJ (2011) Thermodynamics of cercarial survival and metabolism in a changing climate. Parasitology 138:1442–1452. https://doi.org/10.1017/S0031182011001272

Morley NJ (2012) Thermodynamics of miracidial survival and metabolism. Parasitology 139:1640–1651. https://doi.org/10.1017/S0031182012000960

Morley NJ, Lewis JW (2013) Thermodynamics of cercarial development and emergence in trematodes. Parasitology 140:1211–1224. https://doi.org/10.1017/S0031182012001783

Morley NJ, Lewis JW (2014a) Extreme climatic events and host–pathogen interactions: the impact of the 1976 drought in the UK. Ecol Complex 17:1–19. https://doi.org/10.1016/j.ecocom.2013.12.001

Morley NJ, Lewis JW (2014b) Temperature stress and parasitism of endothermic hosts under climate change. Trends Parasitol 30:221–227. https://doi.org/10.1016/j.pt.2014.01.007

Morley NJ, Lewis JW (2015) Thermodynamics of trematode infectivity. Parasitology 142:585–597. https://doi.org/10.1017/S0031182014001632

Morley NJ, Lewis JW (2017) Thermodynamics of egg production, development and hatching in trematodes. J Helminthol 91:284–294. https://doi.org/10.1017/S0022149X16000249

Morley NJ, Irwin SWB, Lewis JW (2003) Pollution toxicity to the transmission of larval digeneans through their molluscan hosts. Parasitology 126:S5–S26. https://doi.org/10.1017/S0031182003003755

Mouritsen KN (2002) The *Hydrobia ulvae–Maritrema subdolum* association: influence of temperature, salinity, light, water-pressure and secondary host exudates on cercarial emergence and longevity. J Helminthol 76:341–247. https://doi.org/10.1079/JOH2002136

Mouritsen KN, Poulin R (2002) Parasitism, climate oscillations and the structure of natural communities. Oikos 97:462–468

Mouritsen KN, Tompkins DM, Poulin R (2005) Climate warming may cause a parasite-induced collapse in coastal amphipod populations. Oecologia 146:476–483. https://doi.org/10.1007/s00442-005-0223-0

Mouritsen KN, Sørensen MM, Poulin R, Fredensborg BL (2018) Coastal ecosystems on a tipping point: global warming and parasitism combine to alter community structure and function. Glob Change Biol 24:4340–4356. https://doi.org/10.1111/gcb.14312

Mydlarz LD, Jones LE, Harvell CD (2006) Innate immunity, environmental drivers, and disease ecology of marine and freshwater invertebrates. Annu Rev Ecol Evol Syst 37:251–288. https://doi.org/10.1146/annurev.ecolsys.37.091305.110103

Nguyen KH, Gemmell BJ, Rohr JR (2020) Effects of temperature and viscosity on miracidial and cercarial movement of *Schistosoma mansoni*: ramifications for disease transmission. Int J Parasitol 50:153–159. https://doi.org/10.1016/j.ijpara.2019.12.003

Nowakowski AJ, Whitfield M, Eskew EA, Thompson ME, Rose P, Caraballo BL, Kerby JL, Todd BD (2016) Infection risk decreases with increasing mismatch in host and pathogen environmental tolerances. Ecol Lett 19:1051–1061. https://doi.org/10.1111/ele.12641

Noyes PD, McElwee MK, Miller HD, Clark BW, Van Tiem LA, Walcott KC, Erwin KN, Levin ED (2009) The toxicology of climate change: environmental contaminants in a warming world. Environ Int 35:971–986. https://doi.org/10.1016/j.envint.2009.02.00

Olivia Roth Joachim, Kurtz Thorsten B. H., Reusch (2010) A summer heat wave decreases the immunocompetence of the mesograzer Idotea baltica Marine Biology 157(7) 1605–1611 https://doi.org/10.1007/s00227-010-1433-5

Okamura B (2016) Hidden infections and changing environments. Integr Comp Biol 56:620–629. https://doi.org/10.1093/icb/icw008

Okamura B, Hartikainen H, Schmidt-Posthaus H, Wahli T (2011) Life cycle complexity, environmental change and the emerging status of salmonid proliferative kidney disease. Freshw Biol 56:735–753. https://doi.org/10.1111/j.1365-2427.2010.02465.x

Okon EM, Birikorang HN, Munir MB, Kari ZA, Téllez-Isaías G, Khalifa NE et al (2023) Global analysis of climate change and the impacts on oyster diseases. Sustainability 15:12775. https://doi.org/10.3390/su151712775

Overstreet RM (1993) Parasitic diseases of fishes and their relationship with toxicants and other environmental factors. In: Couch JA, Fournie JW (eds) Pathobiology of marine and estuarine organisms. CRC Press, Boca Raton, FL, pp 111–156

Overstreet RM (1997) Parasitological data as monitors of environmental health. Parassitologia 39:167–175

Overstreet RM (2007) Effect of a hurricane on fish parasites. Parassitologia 49:161–168

Overstreet RM, Howse HD (1977) Some parasites and diseases of estuarine fishes in polluted habitats of Mississippi. Ann N Y Acad Sci 298:427–462

Paine RT (1966) Food web complexity and species diversity. Am Nat 100:65–75

Paine RT (1969) A note on trophic complexity and community stability. Am Nat 103:91–93

Paterson RA, Poulin R, Selbach C (2023) Global analysis of seasonal changes in trematode infection levels reveals weak and variable link to temperature. Oecologia. https://doi.org/10.1007/s00442-023-05408-8

Patz JA, Epstein PR, Burke TA, Balbus JM (1996) Global climate change and emerging infectious diseases. JAMA 275:217–223. https://doi.org/10.1001/jama.1996.03530270057032

Patz JA, Graczyk TK, Geller N, Vittor AY (2000) Effects of environmental change on emerging parasitic diseases. Int J Parasitol 30:1395–1405. https://doi.org/10.1016/S0020-7519(00)00141-7

Paull SH, Johnson PTJ (2011) High temperature enhances host pathology in a snail-trematode system: possible consequences of climate change for the emergence of disease: climate change and a snail-trematode interaction. Freshw Biol 56:767–778. https://doi.org/10.1111/j.1365-2427.2010.02547.x

Paull SH, Johnson PTJ (2018) How temperature, pond-drying, and nutrients influence parasite infection and pathology. EcoHealth 15:396–408. https://doi.org/10.1007/s10393-018-1320-y

Paull SH, LaFonte BE, Johnson PTJ (2012) Temperature-driven shifts in a host-parasite interaction drive nonlinear changes in disease risk. Glob Change Biol 18:3558–3567. https://doi.org/10.1111/gcb.12018

Paull SH, Raffel TR, LaFonte BE, Johnson PTJ (2015) How temperature shifts affect parasite production: testing the roles of thermal stress and acclimation. Funct Ecol 29:941–950. https://doi.org/10.1111/1365-2435.12401

Pech D, MaL A-M, Lewis JW, Vidal-Martínez VM (2010) Rainfall induces time-lagged changes in the proportion of tropical aquatic hosts infected with metazoan parasites. Int J Parasitol 40:937–944. https://doi.org/10.1016/j.ijpara.2010.01.009

Pietrock M, Marcogliese DJ (2003) Free-living endohelminth stages: at the mercy of environmental conditions. Trends Parasitol 19:293–299. https://doi.org/10.1016/S1471-4922(03)00117-X

Piscart C, Webb D, Beisel JN (2007) An acanthocephalan parasite increases the salinity tolerance of the freshwater amphipod Gammarus roeseli (Crustacea: Gammaridae). Naturwiss 94:741–747. https://doi.org/10.1007/s00114-007-0252-0

Pörtner H-O, Roberts DC, Adams H, Adelekan I, Adler C, Adrian R et al (2022) Technical summary. In: Pörtner H-O, Roberts DC, Poloczanska ES, Mintenbeck K, Tignor M, Alegría A et al (eds) Climate change: impacts, adaptation and vulnerability. Contribution of Working Group II to the Sixth Assessment Report of the Intergovernmental Panel on Climate Change. Cambridge University Press, Cambridge, pp 37–118. https://doi.org/10.1017/9781009325844.002

Poulin R (1992) Toxic pollution and parasitism in freshwater fish. Parasitol Today 8:58–61

Poulin R (2006) Global warming and temperature-mediated increases in cercarial emergence in trematode parasites. Parasitology 132:143–151. https://doi.org/10.1017/S0031182005008693

Poulin R, Mouritsen KN (2006) Climate change, parasitism and the structure of intertidal ecosystems. J Helminthol 80:183–191

Poulin R, Blanar CA, Thieltges DW, Marcogliese DJ (2011) The biogeography of parasitism in sticklebacks: distance, habitat differences and the similarity in parasite occurrence and abundance. Ecography 34:540–551. https://doi.org/10.1111/j.1600-0587.2010.06826.x

Pounds JA, Bustamante MR, Coloma LA, Consuegra JA, Fogden MPL, Foster PN, Ron SR, Sanchez-Azofeifa GA, Still CJ, Young BE (2006) Widespread amphibian extinctions from epidemic disease driven by global warming. Nature 439:161–167. https://doi.org/10.1038/nature04246

Preston DL, Layden TJ, Segui LM, Falke LP, Brant SV, Novak M (2021) Trematode parasites exceed aquatic insect biomass in Oregon stream food webs. J Anim Ecol 90:766–775. https://doi.org/10.1111/1365-2656.13409

Raffel TR, Rohr JR, Kiesecker JM, Hudson PJ (2006) Negative effects of changing temperature on amphibian immunity under field conditions. Funct Ecol 20:819–828. https://doi.org/10.1111/j.1365-2435.2006.01159.x

Raffel TR, Romansic JM, Halstead NT, McMahon TA, Venesky MD, Rohr JR (2013) Disease and thermal acclimation in a more variable and unpredictable climate. Nat Clim Change 3:146–151. https://doi.org/10.1038/nclimate1659

Raffel TR, Halstead NT, McMahon TA, Davis AK, Rohr JR (2015) Temperature variability and moisture synergistically interact to exacerbate an epizootic disease. Proc R Soc B: 282:20142039. 10.1098/rspb.2014.2039

Rehberger K, Werner I, Hitzfeld B, Segner H, Baumann L (2017) 20 Years of fish immunotoxicology – what we know and where we are. Crit Rev Toxicol 47:509–535. https://doi.org/10.1080/10408444.2017.1288024

Resetarits EJ, Ellis WT, Byers JE (2023) The opposing roles of lethal and nonlethal effects of parasites on host resource consumption. Ecol Evol 13:e9973. https://doi.org/10.1002/ece3.9973

Rödder D, Kielgast J, Lötters S (2010) Future potential distribution of the emerging amphibian chytrid fungus under anthropogenic climate change. Dis Aquat Org 92:201–207. https://doi.org/10.3354/dao02197

Rogowski DL, Stockwell CA (2006) Parasites and salinity: costly tradeoffs in a threatened species. Oecologia 146:615–622. https://doi.org/10.1007/s00442-005-0218-x

Rohr JR, Cohen JM (2020) Understanding how temperature shifts could impact infectious disease. PLoS Biol 18:e3000938. https://doi.org/10.1371/journal.pbio.3000938

Rohr JR, Raffel TR (2010) Linking global climate and temperature variability to widespread amphibian declines putatively caused by disease. Proc Natl Acad Sci USA 107:8269–8274. https://doi.org/10.1073/pnas.0912883107

Rohr JR, Dobson AP, Johnson PTJ, Kilpatrick AM, Paull SH, Raffel TR, Ruiz-Moreno D, Thomas MB (2011) Frontiers in climate change–disease research. Trends Ecol Evol 26:270–277. https://doi.org/10.1016/j.tree.2011.03.002

Rohr JR, Raffel TR, Blaustein AR, Johnson PTJ, Paull SH, Young S (2013) Using physiology to understand climate-driven changes in disease and their implications for conservation. Conserv Physiol 1:1–15. https://doi.org/10.1093/conphys/cot022

Rohr JR, Civitello DJ, Cohen JM, Roznik EA, Sinervo B, Dell AI (2018) The complex drivers of thermal acclimation and breadth in ectotherms. Ecol Lett 21:1425–1439. https://doi.org/10.1111/ele.13107

Rollins-Smith LA (2017) Amphibian immunity–stress, disease, and climate change. Dev Comp Immunol 66:111–119. https://doi.org/10.1016/j.dci.2016.07.002

Rollins-Smith LA, Le Sage EH (2023) Heat stress and amphibian immunity in a time of climate change. Philos Trans R Soc B 378:20220132. https://doi.org/10.1098/rstb.2022.0132

Rollins-Smith LA, Woodhams DC (2012) Amphibian immunity: staying in tune with the environment. In: Demas GE, Nelson RJ (eds) Eco-immunology. Oxford University Press, Oxford, pp 92–143

Ruggeri J, de Carvalho-e-Silva SP, James TY, Toledo LF (2018) Amphibian chytrid infection is influenced by rainfall seasonality and water availability. Dis Aquat Org 127:107–115. https://doi.org/10.3354/dao03191

Rückert S, Klimpel S, Palm HW (2007) Parasite fauna of bream *Abramis brama* and roach *Rutilus rutilus* from a man-made waterway and a freshwater habitat in northern Germany. Dis Aquat Org 74:225-233. 10.3354/dao074225

Salo T, Stamm C, Burdon FJ, Räsänen K, Seppälä O (2017) Resilience to heat waves in the aquatic snail *Lymnaea stagnalis*: additive and interactive effects with micropollutants. Freshw Biol 2:1831–1846. https://doi.org/10.1111/fwb.12999

Scharsack JP, Franke F (2022) Temperature effects on teleost immunity in the light of climate change. J Fish Biol 101:780–796. https://doi.org/10.1111/jfb.15163

Scharsack JP, Franke F, Erin NI, Kuske A, Büscher J, Stolz H, Samonte IE, Kurtz J, Kalbe M (2016) Effects of environmental variation on host–parasite interaction in three-spined sticklebacks (*Gasterosteus aculeatus*). Zoology 119:375–383. https://doi.org/10.1016/j.zool.2016.05.008

Scharsack JP, Wieczorek B, Schmidt-Drewello A, Büscher J, Franke F, Moore A et al (2021) Climate change facilitates a parasite's host exploitation via temperature-mediated immunometabolic processes. Glob Change Biol 27:94–107. https://doi.org/10.1111/gcb.15402

Scheele BC, Pasmans F, Skerratt LF, Berger L, Martel A, Beukema W et al (2019) Amphibian fungal panzootic causes catastrophic and ongoing loss of biodiversity. Science 363:1459–1463

Schiedek D, Sundelin B, Readman JW, Macdonald RW (2007) Interactions between climate change and contaminants. Mar Poll Bull 54:1845–1856. https://doi.org/10.1016/j.marpolbul.2007.09.020

Schindler DW (2001) The cumulative effects of climate warming and other human stresses on Canadian freshwaters in the new millenium. Can J Fish Aquat Sci 58:18–29

Scott ME (2023) Helminth-host-environment interactions: Looking down from the tip of the iceberg. J Helminthol 97:e59. https://doi.org/10.1017/S0022149X23000433

Selbach C, Poulin R (2020) Some like it hotter: trematode transmission under changing temperature conditions. Oecologia 194:745–755. https://doi.org/10.1007/s00442-020-04800-y

Selbach C, Barsøe M, Vogensen TK, Samsing AB, Mouritsen KN (2020) Temperature–parasite interaction: Do trematode infections protect against heat stress? Int J Parasitol 50:1189–1194. https://doi.org/10.1016/j.ijpara.2020.07.006

Seppälä O, Jokela J (2011) Immune defence under extreme ambient temperature. Biol Lett 7:119–122. https://doi.org/10.1098/rsbl.2010.0459

Settele J, Scholes R, Betts R, Bunn S, Leadley P, Nepstad D et al (2014) Terrestrial and inland water systems. In: Field CB, Barros VR, Dokken DJ, Mach KJ, Mastrandrea MD, Bilir TE et al (eds) Climate change: impacts, adaptation, and vulnerability. Part A: global and sectoral aspects. Contribution of working group II to the Fifth Assessment Report of the Intergovernmental Panel on Climate Change. Cambridge University Press, Cambridge, pp 271–359

Shields JD (2019) Climate change enhances disease processes in crustaceans: case studies in lobsters, crabs, and shrimps. J Crust Biol 39:673–683. https://doi.org/10.1093/jcbiol/ruz072

Sih A, Bell AM, Kerby JL (2004) Two stressors are far deadlier than one. Trends Ecol Evol 19:274–276

Skerratt LF, Berger L, Speare R, Cashins S, McDonald KR, Phillott AD, Hines HB, Kenyon N (2007) Spread of chytridiomycosis has caused the rapid global decline and extinction of frogs. EcoHealth 4:125–134. https://doi.org/10.1007/s10393-007-0093-5

Sokolow SH, Jones IJ, Jocque M, La D, Cords O, Knight A et al (2017) Nearly 400 million people are at higher risk of schistosomiasis because dams block the migration of snail-eating river prawns. Philos Trans R Soc B 372:20160127. https://doi.org/10.1098/rstb.2016.0127

Steinmann P, Keiser J, Bos R, Tanner M, Utzinger J (2006) Schistosomiasis and water resources development: systematic review, meta-analysis, and estimates of people at risk. Lancet Infect Dis 6:411–425

Stensgaard A-S, Booth M, Nikulin G, McCreesh N (2016) Combining process-based and correlative models improves predictions of climate change effects on *Schistosoma mansoni* transmission in eastern Africa. Geo Health 11:406. 10.4081/gh.2016.406

Stewart A, Hablützel PI, Brown M, Watson HV, Parker-Norman S, Tober AV, Thomason AG, Friberg IM, Cable J, Jackson JA (2018) Half the story: thermal effects on within-host infectious disease progression in a warming climate. Glob Change Biol 24:371–386. https://doi.org/10.1111/gcb.13842

Stokstad E (2021) Streams that flow only part of the year are getting even drier. Science 724:373

Studer A, Poulin R (2012) Effects of salinity on an intertidal host–parasite system: Is the parasite more sensitive than its host? J Exp Mar Biol Ecol 412:110–116. https://doi.org/10.1016/j.jembe.2011.11.008

Studer A, Poulin R (2013) Differential effects of temperature variability on the transmission of a marine parasite. Mar Biol 160:2763–2773. https://doi.org/10.1007/s00227-013-2269-6

Studer A, Thieltges D, Poulin R (2010) Parasites and global warming: net effects of temperature on an intertidal host–parasite system. Mar Ecol Prog Ser 415:11–22. https://doi.org/10.3354/meps08742

Sures B (2004) Environmental parasitology: relevancy of parasites in monitoring environmental pollution. Trends Parasitol 20:170–177

Sures B (2008) Host–parasite interactions in polluted environments. J Fish Biol 73:2133–2142

Sures B, Nachev M (2022) Effects of multiple stressors in fish: how parasites and contaminants interact. Parasitology 149:1822–1828. https://doi.org/10.1017/S0031182022001172

Sures B, Nachev M, Selbach C, Marcogliese DJ (2017) Parasite responses to pollution: what we know and where we go in 'Environmental Parasitology'. Parasit Vectors 10:65. https://doi.org/10.1186/s13071-017-2001-3

Sures B, Nachev M, Schwelm J, Grabner D, Selbach C (2023) Environmental parasitology: stressor effects on aquatic parasites. Trends Parasitol 39:461–474. https://doi.org/10.1016/j.pt.2023.03.005

Tadiri CP, Kong JD, Fussmann GF, Scott ME, Wang H (2019) A data-validated host-parasite model for infectious disease outbreaks. Front Ecol Evol 7:307. https://doi.org/10.3389/fevo.2019.00307

Taranu Z, Gregory-Eaves I, Leavitt PR, Bunting L, Buchaca T, Catalan J et al (2015) Acceleration of cyanobacterial dominance in north temperate subarctic lakes during the Anthropocene. Ecol Lett 18:375–384. https://doi.org/10.1111/ele.12420

Tellenbach C, Tardent N, Pomati F, Keller B, Hairston NG Jr, Wolinska J, Spaak P (2016) Cyanobacteria facilitate parasite epidemics in *Daphnia*. Ecology 97:3422–3432

Thieltges DW, Amundsen P-A, Hechinger RF, Johnson PTJ, Lafferty KD, Mouritsen KN, Preston DL, Reise K, Zander CD, Poulin R (2013) Parasites as prey in aquatic food webs: implications for predator infection and parasite transmission. Oikos 122:1473–1482. https://doi.org/10.1111/j.1600-0706.2013.00243.x

Thomas F, Poulin R, Guegan J-F, Renaud F (1999) Parasites and ecosystem engineering: what roles could they play? Oikos 84:167-171

Thomas A, Ramkumar A, Shanmugam A (2022) CO_2 acidification and its differential responses on aquatic biota – a review. Environ Adv 8:100219. https://doi.org/10.1016/j.envadv.2022.100219

Uribe C, Folch H, Enriquez R, Moran G (2011) Innate and adaptive immunity in teleost fish: a review. Vet Med 56:486–503

Valtonen ET, Pulkkinen K, Poulin R, Julkunen M (2001) The structure of parasite component communities in brackish water fishes of the northeastern Baltic Sea. Parasitology 122:471-481. 10.1017/S0031182001007491

Vicente-Santos A, Willink B, Nowak K, Civitello DJ, Gillespie TR (2023) Host–pathogen interactions under pressure: a review and meta-analysis of stress-mediated effects on disease dynamics. Ecol Lett 26:2003–2020. https://doi.org/10.1111/ele.14319

Vidal-Martínez VM, Pech D, Sures B, Purucker ST, Poulin R (2010) Can parasites really reveal environmental impact? Trends Parasitol 26:44–51

Walther G-R, Roques A, Hulme PE, Sykes MT, Pyšek P, Kühn I et al (2009) Alien species in a warmer world:

risks and opportunities. Trends Ecol Evol 24:686–693. https://doi.org/10.1016/j.tree.2009.06.008

Wegner KM, Kalbe M, Milinski M, Reusch TB (2008) Mortality selection during the 2003 European heat wave in three-spined sticklebacks: effects of parasites and MHC genotype. BMC Evol Biol 8:124. https://doi.org/10.1186/1471-2148-8-124

Welicky RL, De Swardt J, Gerber R, Netherlands EC, Smit NJ (2017) Drought-associated absence of alien invasive anchorworm, *Lernaea cyprinacea* (Copepoda: Lernaeidae), is related to changes in fish health. Int J Parasitol Parasit Wildl 6:430–438. https://doi.org/10.1016/j.ijppaw.2017.01.004

Werner EE, Peacor SD (2003) A review of trait-mediated indirect interactions in ecological communities. Ecology 84:1083–1100

Whitfield SM, Alvarado G, Abarca J, Zumbado H, Zuñiga I, Wainwright M, Kerby J (2017) Differential patterns of *Batrachochytrium dendrobatidis* infection in relict amphibian populations following severe disease-associated declines. Dis Aquat Org 126:33–41. https://doi.org/10.3354/dao03154

Williams HH, MacKenzie K (2003) Marine parasites as pollution indicators: an update. Parasitology 126:S27-S41. 10.1017/S0031182003003640

Woo PTK, Leong J-A, Buchmann K (eds) (2020) Climate change and infectious fish diseases. CAB International, Wallingford

Wood CL, Baum JK, Reddy SMW, Trebilco R, Sandin SA, Zgliczynski BJ, Briggs AA, Micheli F (2015) Productivity and fishing pressure drive variability in fish parasite assemblages of the Line Islands, equatorial Pacific. Ecology 96:1383–1398

Wood CL, Welicky RL, Preisser WC, Leslie KL, Mastick N, Greene C, Maslenikov KP, Tornabene L, Kinsella JM, Essington TE (2023) A reconstruction of parasite burden reveals one century of climate-associated parasite decline. Proc Natl Acad Sci USA 120:e2211903120. https://doi.org/10.1073/pnas.2211903120

Woodhams DC, Alford RA, Briggs CJ, Johnson M, Rollins-Smith LA (2008) Life-history trade-offs influence disease in changing climates: strategies of an amphibian pathogen. Ecology 89:1627–1639

WWF (2020) Living planet report 2020 - bending the curve of biodiversity loss. Almond REA, Grooten M, Petersen T editors. WWF, Gland

Xie GY, Olson DH, Blaustein AR (2016) Projecting the global distribution of the emerging amphibian fungal pathogen *Batrachochytrium dendrobatidis* based on IPCC climate futures. PLoS One 11:e0160746. 10.1371/journal.pone.0160746

Yu IA, Trevor J, Vannatta JT, Gutierrez SO, Minchella DJ (2022) Opportunity or catastrophe? Effect of sea salt on host-parasite survival and reproduction. PLoS Negl Trop Dis 16:e0009524. https://doi.org/10.1371/journal.pntd.0009524

Zander CS (1998) Ecology of host-parasite relationships in the Baltic Sea. Naturwiss 85:426–436

Zander CD, Reimer LW (2002) Parasitism at the ecosystem level in the Baltic Sea. Parasitology 124:S119–S135

Zhang Y, Zheng H, Zhang X, Leung LR, Liu C, Zheng C et al (2023) Future global streamflow declines are probably more severe than previously estimated. Nat Water 1:261–271. https://doi.org/10.1038/s44221-023-00030-7

Zhi W, Klingler C, Liu J, Li L (2023) Widespread deoxygenation in warming rivers. Nat Clim Change 13:1105–1113. https://doi.org/10.1038/s41558-023-01793-3

Open Access This chapter is licensed under the terms of the Creative Commons Attribution-NonCommercial-NoDerivatives 4.0 International License (http://creativecommons.org/licenses/by-nc-nd/4.0/), which permits any non-commercial use, sharing, distribution and reproduction in any medium or format, as long as you give appropriate credit to the original author(s) and the source, provide a link to the Creative Commons license and indicate if you modified the licensed material. You do not have permission under this license to share adapted material derived from this chapter or parts of it.

The images or other third party material in this chapter are included in the chapter's Creative Commons license, unless indicated otherwise in a credit line to the material. If material is not included in the chapter's Creative Commons license and your intended use is not permitted by statutory regulation or exceeds the permitted use, you will need to obtain permission directly from the copyright holder.

Parasites in Aquaculture

Cecilia Power, Sho Shirakashi, Nathan J. Bott, and Barbara F. Nowak

Abstract

Seafood contributes a significant proportion of protein to human diet. Due to overfishing, seafood is increasingly supplied by aquaculture. Farmed fish can be affected by a range of diseases, including parasitic diseases, which can have economic impact and cause adverse effects on health of consumers. Risks of outbreaks of parasitic diseases are affected by farming systems. Parasitic infections in aquaculture are particularly common in farming systems with reduced biosecurity, for example, sea cages. As only one commercial vaccine is available against fish parasitic disease and there is a high risk of development of resistance of parasites to medications, selective breeding and other preventative methods are increasingly adopted by aquaculture industry. Current issues and future challenges include improvements in diagnostic methods, co-infections with other parasites and development of novel seafood production systems such as integrated aquaculture, recirculated systems and cellular aquaculture.

23.1 Introduction

On 15 November 2022, the global human population reached 8 billion and it is projected to reach 10 billion in the mid-2080s (UN 2023). As the human population continues to increase, one of the most important challenges we face is food security. While Zero Hunger is one of 17 United Nations Sustainable Development Goals, moderate or severe food insecurity affected more than 29% of the global population in 2021 (FAO et al. 2022). Based on its environmental footprint and nutritional value, seafood is an important component of sustainable and healthy diet (Farmery et al. 2022). Globally, seafood contributes around 17% of animal source protein to human diets; however, seafood provides 50% or more of animal protein for some countries, including Small Island Developing States (Farmery et al. 2022). As wild fish stocks become overfished, seafood increasingly is supplied by aquaculture industry.

C. Power · N. J. Bott
School of Science, RMIT University,
Bundoora, VIC, Australia
e-mail: cecilia.power@rmit.edu.au;
nathan.bott@rmit.edu.au

S. Shirakashi
Aquaculture Research Institute, Kindai University,
Wakayama, Japan
e-mail: shirakashi@kindai.ac.jp

B. F. Nowak (✉)
School of Science, RMIT University,
Bundoora, VIC, Australia

Institute of Marine and Antarctic Studies, University of Tasmania, Launceston, TAS, Australia
e-mail: b.nowak@utas.edu.au

This is expected to continue with an increasing per capita seafood consumption requiring doubling in the aquaculture production volume by 2050 (FAO 2020; Stentiford and Holt 2022). Globally, more than 50% of seafood comes from farmed animals and this contribution has been increasing over time, but there is a high variability between geographical regions and countries (FAO 2020; Stentiford and Holt 2022). While the highest per capita producer of seafood from aquaculture is Norway (270.1 kg pp.$^{-1}$ pa^{-1}), most countries in Europe produce less than 25% of seafood they consume from aquaculture and most countries producing at least 50% of national seafood demand from aquaculture are in Asia (Stentiford and Holt 2022).

Fish farming has a lot of benefits; however, it faces multiple challenges including environmental sustainability and biosecurity issues. A diverse range of freshwater and marine fish species can be farmed in different geographical locations (Table 23.1). Life cycles have been closed for many fish species, so they are produced in commercial hatcheries before going to grow-out production stage. However, some fish production still relies on catching wild fish and then fattening them in sea cages. Grow-out can occur in different farming systems. Based on biosecurity and control over water, fish farming systems are classified as semi-open (for example, sea cages), semi-closed (for example, flow-through ponds) and closed (recirculating aquaculture systems [RAS]) (Fig. 23.1). In land-based semi-closed facilities, biosecurity is related to speed and volume of water flow, which determines the system's potential for removal and inactivation of parasite free-living stages through filtration, ultraviolet (UV) or ozone treatment, or other means. If all water is recirculated, full biosecurity should be ensured as the entry of pathogens is limited to initial introduction of fish and initial water source. Parasitic infections in aquaculture are particularly common in farming systems with reduced biosecurity. Sea cage farming is common for marine fish and salmonid grow-out as it requires less investment in the farm infrastructure and has lower running costs than land facilities. However, other than through the location of sea cages, environmental conditions and biosecurity cannot be controlled in sea cage farming. The biosecurity issues include parasitic infections, including both ectoparasites and endoparasites (Table 23.1). As fish are farmed at high stocking density and in confinement, risk of parasite transmission is greater in culture than in wild fish populations (Murray and Peeler 2005). Parasites with simple life cycles, requiring only one host species, can be highly successful in aquaculture. While most farmed fish are fed manufactured diets, reducing the risk of orally transmitted parasites, some parasites with complex life cycles infect through free-living life stages and those can still become a significant issue in fish farming.

Owing to their simple direct life cycle and rapid proliferation rates (see Chap. 5), monogeneans are among the most prevalent and problematic parasites in both freshwater and marine finfish aquaculture (Hoai 2020). *Gyrodactylus* spp. and *Dactylogyrus* spp. are frequently observed in freshwater aquaculture, while in marine aquaculture, parasites belonging to Mazocraeidea, commonly referred to as 'gill flukes', and Capsalidae, known as 'skin flukes', particularly *Benedenia* spp. and *Neobenedenia girellae*, represent the most problematic Monogenea groups. Infections caused by these parasites tend to be chronic, requiring periodic and repeated treatments to prevent retarded growth and mortality in fish. Presently, the primary strategy to controlling monogenean parasites in fish farms focuses on eliminating worms from the fish through bath treatment or, less frequently, administering oral antiparasitic drugs (Buchmann 2022).

Caligids (Copepoda), another group of parasites with direct life cycle (see Chap. 6), can affect a range of fish species farmed in the marine environment and can cause significant losses, in particular in Atlantic salmon, *Salmo salar* L., marine grow-out. In Europe and Atlantic Canada, *Lepeophtheirus salmonis* is the main sea lice species, whereas, in Pacific Canada, *Caligus elongatus* and, in Chile, *Caligus rogercresseyi* affect farmed Atlantic salmon. *L. salmonis* has two subspecies in the Northern Hemisphere: *L. sal-*

Table 23.1 Examples of the main farmed fish species and their parasites: Share of global production by live weight, separate for marine (M) and freshwater (F)

Species	Share of global production in 2020 (%)	Main producers (country)	Farming systems (grow-out)	Main parasites of concern
Atlantic salmon *Salmo salar*	23.6 (M)	Norway, Chile, UK, Canada, Faroe Islands, Australia	Sea cages	*Lepeophtheirus salmonis* (c), *Neoparamoeba perurans* (p)
Milkfish *Chanos chanos*	14 (M)	Philippines, Indonesia, Taiwan	Sea cages, ponds	*Amyloodinium ocellatum* (d), *Lernaea cyprinacea* (c), *Caligus epidemicus* (c), *Haplorchis yokogawai* (t)[a], *Procerovum calderoni* (t)[a], *Rocinella typicus* (i), *Capillaria* sp. (n)
Grass carp *Ctenopharyngodon idellus*	11.8 (F)	China, Malaysia, Singapore, Borneo, Indonesia, Thailand, Taiwan, Philippines	Ponds	*Ichthyophthirius multifiliis* (p), *Clonorchis sinensis* (t)[a], *Bothricephalus gowkongensis* (ce), *Dactylogyrus lamellatus* (m), *Sanguinicola armatus* (t)
Silver carp *Hypophthalmichthys molitrix*	10 (F)	China, India, Bangladesh	Ponds	*Clonorchis sinensis* (t)[a], *Dactylogyrus hypophthalmichthys* (m), *Dactylogyrus aristichthys* (m), *Sanguinicola lungensis* (t), *Lernaea* sp. (c)
Nile tilapia *Oreochromis niloticus*	9 (F)	China, Indonesia, Mexico, Honduras, Colombia, Brazil	Cages	*Cichlidogyrus halli* (m), *Cichlidogyrus mbirizei* (m), *Cichlidogyrus thurstonae* (m), *Cichlidogyrus tilapiae* (m), *Scutogyrus longicornis* (m), *Prohemistomum vivax* (t)[a]
Common carp *Cyprinus carpio*	8.6 (F)	China, India, Bangladesh, Indonesia, Israel, Czechia, Poland	Ponds	*Ichthyophthirius multifiliis* (p), *Clonorchis sinensis* (t)[a], *Sanguinicola inermis* (d), *Khawia japonensis* (ce)
Catla *Catla catla*	7.2 (F)	India, Bangladesh, Myanmar Laos Pakistan Thailand	Ponds, carp polyculture	*Lernaea* spp. (c), *Dactylogyrus kalyanensis* (m), *Dactylogyrus labei* (m)
Gilthead bream *Sparus aurata*	3.4 (M)	Turkey, Greece, Egypt, Spain, Tunisia, Croatia	Sea cages	*Amyloodinium ocellatum* (d), *Sparicotyle chrysophrii* (m), *Ceratothoa parallela* (i)
Japanese amberjack *Seriola quinqueradiata*	1.6 (M)	Japan, Republic of Korea	Sea cages	*Benedenia seriolae* (m), *Neobenedenia girellae* (m), *Heteraxine heterocerca* (m), *Microsporidium seriolae* (mi), *Caligus spinosus* (c), *Paradeontacylix buri* (t), *Myxobolus acanthogobii* (my)
Barramundi *Lates calcarifer*	1.3 (M)	Australia, Indonesia, Thailand, Israel, Papua New Guinea, Vanuatu, Malaysia, Singapore, Taiwan, Vietnam	Sea cages	*Neobenedenia girellae* (m), *Cruoricola lates* (t), *Parasanguinicola vastispina* (t), *Lernanthropus* sp. (c)

Parasites: *c* copepod, *ce* cestode, *d* dinoflagellate, *i* isopod, *m* monogenean, *mi* microsporidia, *my* myxosporea, *n* nematode, *p* protist, *t* trematode
Source: FAO (2022)
[a]Parasites of concern to human health

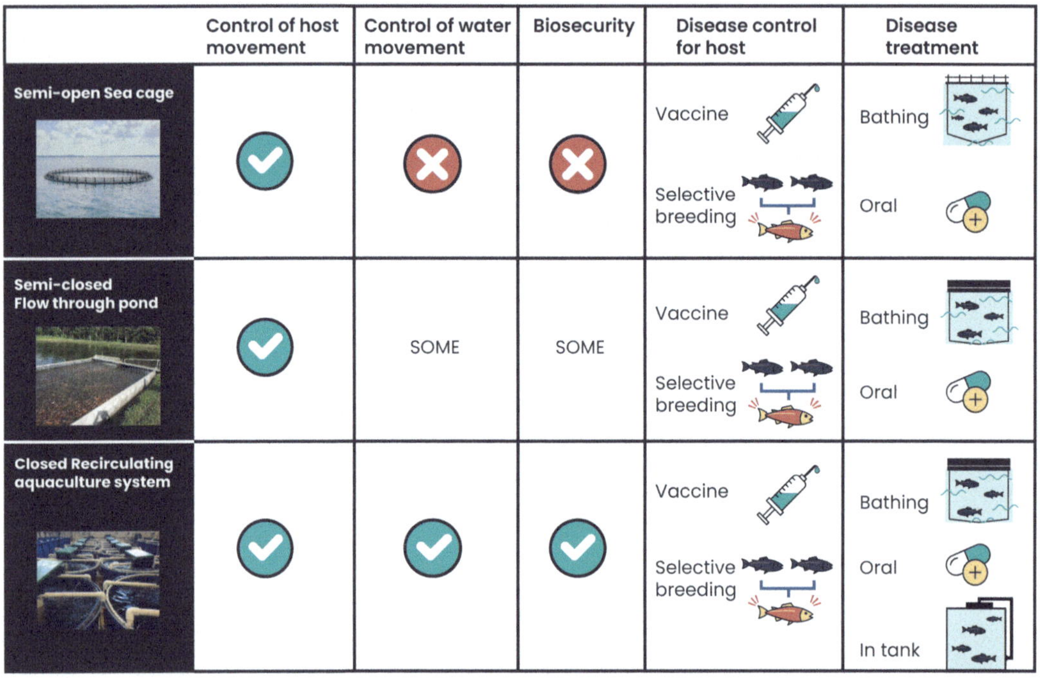

	Control of host movement	Control of water movement	Biosecurity	Disease control for host	Disease treatment
Semi-open Sea cage	✓	✗	✗	Vaccine / Selective breeding	Bathing / Oral
Semi-closed Flow through pond	✓	SOME	SOME	Vaccine / Selective breeding	Bathing / Oral
Closed Recirculating aquaculture system	✓	✓	✓	Vaccine / Selective breeding	Bathing / Oral / In tank

Fig. 23.1 Effect of farming system on control and treatment of diseases

monis oncorhynchi in the Pacific and *L. salmonis salmonis* in the Atlantic Oceans (Skern-Mauritzen et al. 2014). Management of sea lice infections includes a combination of preventative treatments and continuous on-demand delousing (Barrett et al. 2020a, b). Despite improved fish health management reducing the infections, physical or chemical removal of the parasites is still required on most of the affected farms, including the use of cleaner fish (Barrett et al. 2020a, b). The rapid evolutionary capacity of sea lice to develop resistance to treatments has been a continuing challenge (Barrett et al. 2020a, b; Hamre et al. 2021).

Blood flukes (Digenea: Aporocotylidae) are known to parasitise the circulatory system of fishes, and severe infections have been associated with mass mortality events in cultured amberjacks (*Seriola* spp.), sea breams (Sparidae) and bluefin tunas (*Thunnus* spp.) (Power et al. 2020). Fish blood flukes have a two-host life cycle, which involves an invertebrate intermediate host (typically a terebellid polychaete) and a fish definitive host (see Chap. 5). The most speciose genus, *Cardicola* spp., includes three species that affect farmed and ranched bluefin tuna in the Southern, Pacific and Atlantic Oceans (Balli Garza et al. 2016). Management strategies to control *Cardicola* spp. infections of bluefin tuna include site selection, separation or removal of intermediate host and praziquantel treatment (Huston et al. 2020; Norbury et al. 2022).

Free-living marine amoeba, *Neoparamoeba perurans* (see Chap. 2), can colonise gills of fish cultured in the marine environment, becoming parasitic and causing amoebic gill disease (AGD; Young et al. 2007; Crosbie et al. 2012). White, mucoid gill lesions are characteristic gross signs for this disease (Zilberg and Munday 2000; Adams and Nowak 2001; Adams et al. 2004; Taylor et al. 2009). Histologically, the disease manifests as epithelial hyperplasia leading to lamellar fusion and formation of interlamellar vesicles (Adams and Nowak 2003). This disease affects farmed Atlantic salmon in almost all geographical locations as well as other fish species. Elevated water temperature and high salinity are the main risk factors for the outbreak of this disease (Clark and Nowak 1999). Freshwater bathing is the most common treatment against AGD.

This chapter focuses on parasitic diseases of fish farmed in sea cages, the economic impact of these diseases and potential human health impact of the parasites. Prevention and treatments used in mariculture against parasitic diseases are reviewed based on examples including monogeneans, blood fluke, sea lice and amoebae. Finally, current issues and potential future challenges are discussed. Please refer to other chapters of this book for detailed information on biology of the parasites, including their life cycles (see Chaps. 2–6).

23.2 Economic Impact

Parasitic diseases have a significant economic impact on fish culture. For example, globally, the direct and indirect costs of sea lice *L. salmonis* infections range from USD 500 million to 1 billion (Barker et al. 2019), including, in 2018, USD 100 million of cleaner fish (Barrett et al. 2020a), while management of amoebic gill disease costs Atlantic salmon producers in Australia AUD 40 million/year (Kube et al. 2012). The economic impact can be due to production losses (fish mortality or increased feed conversion ratio) or increased husbandry costs due to treatment or other measures. Costs of parasitic diseases to aquaculture industry can be either sporadic and unpredictable or regular and predictable (Shinn et al. 2015). Sporadic or unpredictable costs are associated with outbreaks of uncommon parasitic diseases and usually include cost of mortalities followed by other production loss and costs of management, for example treatment. For the known and common parasitic infections, the cost is regular or predictable and usually it is more focused on the costs of prevention and management. For example, AGD-related mortalities causing economic impact when the disease first affected a new geographical area are then reduced by regular treatment (for example, freshwater bathing) and selective breeding, but while the disease management significantly reduces the impact of AGD on fish, the production costs are still increased due to the costs of the treatment and monitoring (Kube et al. 2012). An example of sporadic costs are mortalities caused by first outbreaks of AGD, which resulted in USD 12.5 million loss in Norway in 2006 and USD 80 million in Scotland in 2013 (Shinn et al. 2015). Monogenean infections cause significant costs to Japanese fish farmers; for example, Japanese amberjack, *Seriola quinqueradiata*, producers lost USD 201 million in 2001 due to infections by *Benedenia seriolae*, while *Neobenedenia girellae* currently can cause similar annual losses (Shinn et al. 2015). Salmon producers in British Columbia lost CAD 6 million in 2015 because of infections by the myxozoan *Kudoa thyrsites* (Braden et al. 2018). Infections by the ciliate *Cryptocaryon irritans* resulted in loss of USD 856 to the Spanish producers of greater amberjack, *Seriola dumerili* (Shinn et al. 2015). Parasitic diseases have a potential to increase costs of fish production in aquaculture during the current intensification and expansion (Shinn et al. 2015).

23.3 Human Health Impact

While the majority of parasites found in farmed fish are considered safe for human consumption, there are a few that can transmit zoonotic diseases or pose public health concerns. Farmed fish products are generally considered safer than wild-caught fish, which carry a high risk of various parasite infections (see Chap. 19). Nonetheless, while some zoonotic parasites can still be found in farmed fish, there is limited parasite occurrence in farmed fish especially when the fish are raised in certain environments facilitating infections of specific parasite groups.

The consumption of raw or undercooked freshwater fish farmed in open fields carries a zoonotic risk due to infections from larval parasites. Metacercariae of opisthorchiid, heterophyid and echinostomatid trematodes are primary zoonotic parasites found in freshwater farmed fish (dos Santos and Howgate 2011; Chai and

Jung 2022). These parasites are particularly problematic in Asian countries, where freshwater aquaculture is expanding rapidly, and undercooked fish consumption is common (Phan et al. 2011). The risk of zoonotic transmission is higher in rural traditional fish farm ponds where the water is contaminated by parasite eggs from reservoir hosts such as humans, cats, dogs and pigs, as well as birds (Nguyen et al. 2010; Bhar 2022). Such contamination combined with the high density of snail intermediate host allows the establishment of the life cycles of liver flukes such as *Opisthorchis* spp. and *Clonorchis sinensis* as well as the intestinal flukes *Haplorchis* spp. and *Echinochasmus* spp. Infections of these zoonotic trematodes among farmed fish, such as cyprinids, catfish and tilapia, can be high, over 50%, in certain regions (Chi et al. 2008; Saleh et al. 2009; Thien et al. 2009; Phan et al. 2010; Clausen et al. 2012, Thien and Murrell 2020; Acosta-Pérez et al. 2022). The zoonotic diphyllobothriid tapeworm could also infect fishes farmed in infested lakes. *Dibothriocephalus* spp. (formerly known as *Diphyllobothrium* spp.) infections have been reported in farmed rainbow trout, *Oncorhynchus mykiss*, in Chile and North African catfish, *Clarias gariepinus*, in Nigeria, but extensive surveys of freshwater farmed salmonids in Japan showed no risk of infection (Torres et al. 2002, 2010; Cabello 2007; Watanabe et al. 2014; Rufai and Adetona 2017).

Anisakid nematodes are the most significant zoonotic parasites in marine fish (Buchmann and Mehrdana 2016; Bao et al. 2019; Rahmati et al. 2020) (also see Chap. 19). The ingestion of live third-stage larvae of *Anisakis simplex* complex, *Pseudoterranova decipiens* complex, as well as *Contracecum osculatum* complex, can result in severe acute abdominal pain, nausea and vomiting, caused by the larval invasion of the gastric mucosa (Anisakidosis) (Hochberg et al. 2010). Moreover, sensitised patients may also develop allergic reactions, such as rash, oedema, itching and even anaphylaxis, triggered by the ingestion of even dead, frozen, or heat-treated worm materials (Audicana et al. 2002; Ivanović et al. 2017). The risk of anisakid infection is almost negligible in farmed fish raised on formulated or frozen feed (Skov et al. 2009). In fact, *Anisakis* infection is notably absent in salmonid aquaculture, where fish are raised from eggs and fed formulated feed (Angot and Brasseur 1993; Inoue et al. 2000; Levsen and Maage 2016; Fioravanti et al. 2021; Karami et al. 2022). However, there still exists a minor risk of anisakid infection in farmed fish that inadvertently ingest infected crustacean or wild fish while raised in open sea cages. A few such cases have been reported in farmed European sea bass, *Dicentrarchus labrax*, in Italy and Atlantic salmon in Canada (Marty 2008; Cammilleri et al. 2018). The risk is further amplified when wild-caught fish are utilised for farming or fattening. There was an incidence of high *Anisakis* infection among farmed greater amberjack raised from wild-caught juveniles that had been imported to Japan for farming purposes (Yoshinaga et al. 2006). These fish were either discarded or required to be frozen prior to being sold. In recent years, chub mackerel, *Scomber japonicus*, has become a popular aquaculture species in Japan, and *Anisakis* infections have been reported in farmed mackerel originating from the wild. Given the emerging concern regarding *Anisakis* allergy (Daschner and Cuéllar 2020; Rahmati et al. 2020; Rama and Silva 2022), and the need to ensure the safety of aquaculture products, proper management and monitoring measures are crucial to prevent anisakid infections in mariculture (Polimeno et al. 2021; Ziarati et al. 2022).

Certain species of Myxosporea (Cnidaria: Myxozoa) have the potential to pose a public health issue. Mass food poisoning incidents associated with the consumption of farmed olive flounder, *Paralichthys olivaceus*, have been reported in Japan and Korea. Investigations identified *Kudoa septempunctata* as the causative agent (Kawai et al. 2012; Iwashita et al. 2013; Kim et al. 2018). While allergic reactions to *Kudoa* spp. have previously been reported in Spain (Martínez de Velasco et al. 2007, 2008), this is the first recognition of a myxosporean parasite affecting human health. The Japanese government has since listed *K. septempunctata* as a food poisoning agent in 2013. *K. septempunctata* infections are subclinical as the parasite forms pseudocysts within fish myofibers. Consumers unknowingly ingest numerous spores when eating raw infected flounder sashimi and sushi. Freezing and cooking can prevent food poisoning; however, olive flounder is primarily distributed, often alive, for raw consumption and thus, pose a particular risk (Yokoyama et al. 2016). Ingesting *K. septempunctata* myxospores in quantities exceeding 10^7 can cause acute transient vomiting, diarrhoea and abdominal pain due to the loss of intestinal cell integrity by sporoplasm, which is non-infective to humans (Ohnishi et al. 2013, 2016; Sugita-Konishi et al. 2014). Although symptoms are typically mild and self-healing, over 40 annual cases with more than 400 patients have been reported in Japan. While *Kudoa* food poisonings have decreased in Japan due to extensive management and infection monitoring in flounder farms, cases from wild and imported farmed flounder still occur. Certain species of other multivalvulid myxosporeans in marine fish, particularly those in genera *Kudoa* and *Unicapsula*, are also suspected of being related to food poisoning (Suzuki et al. 2015; Ohnishi et al. 2018; Tachibana and Watari 2021). Infections of these parasites in farmed fish intended for raw consumption require specific attention, as no control measures currently exist for myxosporean infections in open water mariculture.

Zoonotic protists, namely *Cryptosporidium* spp., *Giardia duodenalis*, and possibly *Toxoplasma gondii*, also pose a potential transmission risk from farmed fish (Moratal et al. 2020). Various freshwater and marine fish species harbour zoonotic *Cryptosporidium* including *C. parvum, C. hominis, C. ubiquitum* and *C. scrofarum* (Golomazou et al. 2021; Couso-Pérez et al. 2022; Moratal et al. 2022). Fish can serve as both host and physical carrier, and contamination during fish processing and preparation can lead to human infection (Robertson et al. 2020). Although evidence of transmission of these zoonotic protists from farmed fish and shellfish to humans remains limited, it warrants further attention, especially given the global increase in fish consumption and aquaculture industries. Continued research and monitoring can help identify potential risks and ensure the safety of aquaculture products for human consumption.

23.4 Control

23.4.1 Physical and Geographical Distancing

Parasitic diseases can be avoided if the host and parasite do not share the same environment (Barrett et al. 2020a, b; Huston et al. 2020). Locating fish farms in the areas where the parasites are uncommon or where oceanographic conditions, such as currents or depth, make the contact between free-living stages of parasites and the fish difficult, can reduce the risk of parasitic infection (Samsing et al. 2017). Parasitic infections, in particular blood fluke, *Cardicola* spp. and sea lice, *Caligus chiastos*, were significantly reduced when southern bluefin tuna, *T. maccoyii*, were ranched offshore (Kirchhoff et al. 2011). Greater distance between the bottom of the pen and ocean floor, increased water flow due to currents and fewer encounters with wild fish likely contributed to this reduction (Kirchhoff et al. 2011). Another way of avoiding parasitic infection is fallowing farming sites when the infective stages of parasite are present; this can decrease the risk of parasitic disease outbreaks, as reported for sea lice (Bron et al. 1993). Fallowing is a temporary stop of production at a farming site, usually after harvest or transfer of the fish to another site, so the site is without any farmed fish. It is used to restore the natural environment and to reduce the potential for re-infections. Fallowing periods are rarely mandatory; their duration differ and is based not just on scientific evidence but also logistics and commercial priorities.

Light traps, filtering and trapping were tried experimentally to remove sea lice from the farming environment before they encounter salmon without much success (Barrett et al. 2020a, b). A wide range of physical barriers including skirts or snorkel sea cages have been tested as a prevention of parasitic diseases with varying success, potentially depending on the parasite and environmental conditions (Wright et al. 2017). Sea lice infections were reduced by 84% in snorkel sea cages on a commercial farm in Norway (Wright et al. 2017). While these systems show promising results for reducing AGD gill scores during one outbreak in autumn/winter, the scores were higher in snorkel sea cages than control sea cages in spring/summer (Wright et al. 2017). In another study, both prevalence and intensity of infection of Atlantic salmon by tapeworm *Eubothrium* sp. were reduced in snorkel pens (Geitung et al. 2021).

Behavioural manipulations can be used to separate fish and infectious stages of parasites in the water column. For example, deep feeding and use of artificial lights were successfully used to encourage farmed Atlantic salmon to swim deeper than the depth where sea lice are most common (for review, see Barrett et al. 2020b). Submerged pens are another way to move the fish deeper than where the free-living stages of parasites usually occur.

Implementing prophylactic measures is crucial for the effective management of monogenean infections in fish farms, thereby reducing the associated cost and labour-intensive efforts required for curative treatments. A classical approach is mechanical removal of monogenean eggs entangled and accumulated in culture pen nets through routine cleaning and net replacement (Fig. 23.2) (Huston et al. 2020). Employing biocontrol agents such as cleaner shrimp and fish for monogenean egg removal has demonstrated promising results in reducing *Neobenedenia girellae* infection and may offer a noble management strategy to mitigate the labour and time associated with net cleaning (Vaughan et al. 2018, Shirakashi and Takagi 2021) (Fig. 23.3). Another effective approach is to exploit the innate behaviour of monogenean larvae. Oncomiracidia of many problematic monogeneans exhibit positive phototaxis (Shirakashi et al. 2021a). The use of submersible sea cages at some mariculture sites for protection against tidal waves has resulted in considerably fewer monogenean infection issues. This can be attributed to the reduced chance of encounters between fish and monogenean larvae, which predominantly occur in the upper water column due to their phototaxis (Shirakashi et al. 2013). Modifying the lighting conditions of culture site, for instance, by shading, could also alter larval distribution and help to prevent infection (Yamamoto et al. 2014). Research into developing traps to attract monogenean larvae is in progress, and this may evolve into a sophisticated prophylactic measure in the future (Skilton et al. 2020).

Fig. 23.2 Eggs of the skin fluke *Benedenia seriolae* entangled in a net

23.4.2 Vaccines and Immunostimulants/Functional Diets

Commercial vaccines have been used against bacterial and viral diseases in aquaculture. These are delivered mostly by injection or immersion and are based on a wide range of antigen preparations. Traditionally, most fish vaccines are based on killed whole pathogens or live attenuated pathogens; however, more recently subunit vaccines have been used. Despite a lot of research and a large number of patents filed for antigens and vaccines against fish parasites in the last 20 years (patents search using Google search engine on 6 June 2023), to the best of our knowledge there is only one

Fig. 23.3 Siganid fish (**a**) and shrimp (**b**) as potential biocontrol agents for removing skin fluke eggs

commercially available vaccine against sea lice, Providean AquaTec Sea Lice (Shivam et al. 2021). While humoral antibody responses and acquired immunity were observed in certain fish-monogenean systems, no vaccine candidate antigen has been identified to date (Ilgová et al. 2021; Kar et al. 2022). Many experimental vaccines, ranging from live attenuated or killed pathogen to subunit and nucleic acid preparations, have been tested without much success against other parasitic diseases, for example AGD (Zilberg and Munday 2001; Valdenegro-Vega et al. 2015).

Immunostimulants improve innate immune response and can be delivered orally as part of functional feeds. Functional feeds provide additional health and physiological benefits and are becoming more commonly used by aquaculture industry (Tacchi et al. 2011). Atlantic salmon survival was improved when the fish were fed a commercial functional diet (Protec Gill Skretting) during an experimental challenge (Mullins et al. 2020). Mucus viscosity, polysaccharide concentration and lysozyme activity were significantly increased in the fish fed Protec Gill. The survival of *N. perurans* was significantly reduced when exposed to some of the functional ingredients in Protec Gill. The patent for this diet has already been granted in Norway, Denmark and Australia. Similarly, functional feeds promising improved resistance to sea lice infections have been available on the market (for example, Shield, Skretting, or Robust EWOS).

23.4.3 Selective Breeding/Hybrids/Gene Editing

Given the challenges in developing a vaccine against fish parasites and limited impact of immunostimulants, alternative preventive strategies are being explored. Selective breeding is an essential tool in management of many parasitic diseases for a number of farmed fish species. Recent advancements in genomic studies have identified alleles responsible for resistance in certain monogenean-fish systems. For example, quantitative trait loci associated with resistance against *Benedenia seriolae* in *Seriola quinqueradiata* have been reported, and polymorphic alleles in the major histocompatibility complex II found to be linked to the susceptibility of *Oreochromis niloticus* to *Gyrodactylus cichlidarum* (Ozaki et al. 2013; Uchino et al. 2020; Chen et al. 2021). These discoveries have led to the selective breeding of resistant fish strains, and, currently, *B. seriolae*-resistant *S. quinqueradiata* are commercially produced by aquaculture companies (Akita et al. 2023). Similarly, selective breeding has been used for increasing resistance of Atlantic salmon against sea lice infections with the two major breeding companies in Norway (AquaGen and SalmoBreed) advertising family lines for sea lice resistance (Barrett et al. 2020a, b). Two generations of genomic selection only for salmon lice resistance resulted in 40–45% reduction in infections (Øvergard et al. 2018). The Atlantic salmon breeding programme in

Tasmania has been selecting for AGD resistance since 2004 (Elliott and Kube 2009; Kube et al. 2012). While initially based on phenotype selection, genomic selection has been successfully used by the Atlantic salmon industry in Tasmania to improve the resistance of Atlantic salmon to AGD (Elliott and Kube 2009; Kube et al. 2012). Other countries farming Atlantic salmon have included gill damage and amoebic load as traits for genomic selection in their breeding programmes.

Additionally, certain hybrid fish have been shown to exhibit greater resistance to Monogenea. For instance, a hybrid grouper, *Epinephelus bruneus* × *E. lanceolatus*, has demonstrated increased resistance to *Benedenia epinepheli* compared to its one of the parent species *E. bruneus* (Chuda et al. 2018). Similarly, hybrid leuciscids between *Rutilus rutilus* and *Abramis brama* harboured significantly fewer *Dactylogyrus* and *Gyrodactylus* species compared to the parent fish (Dedić et al. 2023). Farming hybrids was suggested as an AGD management option for salmon industry with brown trout ♀ × Atlantic salmon ♂ (TS) population exhibiting the highest levels of resistance to this disease (Maynard et al. 2016; Adams et al. 2023). Similarly, hybridisation of Atlantic salmon with other more resistant salmonid species was proposed for sea lice (Fleming et al. 2014). The culture of families and hybrid fish resistant to parasites represents a realistic and promising preventive approach for managing infections in aquaculture.

A new approach, which has a future in programming of disease resistance in farmed fish, is gene editing (Gratacap et al. 2019; Houston et al. 2020; Blix et al. 2021). With a genetically modified (GM) salmon strain approved for human consumption by the USA and Canada (Waltz 2017), there is a potential for genome edited fish to be available on the market. However, genes responsible for the disease resistance would need to be identified before gene editing could be used to increase fish resistance to parasitic diseases.

23.5 Treatment

23.5.1 Chemical

Traditionally, parasitic diseases were treated using chemicals. For the treatment to be successful, it not only has to be delivered to the parasite at the lethal dose, but also the safety margin has to be large enough, so that the treatment has minimal side effects on the host. The impacts of environmental factors, in particular temperature and host factors, including fish species and its physiological and health state on the efficacy of the treatment must be considered. Furthermore, continuous and sustained use of the same treatment can result in the development of resistance in the parasite. For example, sea lice developed resistance to most chemical treatments in all salmon producing countries affected by this parasite (Aaen et al. 2015; Godwin et al. 2022). Similarly, the tapeworm, *Eubothrium* spp., affecting Atlantic salmon farmed in parts of Norway, was reported to develop resistance to praziquantel (Geitung et al. 2021).

Chemical treatments can be delivered as a bath (mostly effective for ectoparasites) or orally; this method has been used successfully for both ecto- and endoparasites. In chemical bath treatment, the culture sea cages can be encased by a skirt, and chemicals are introduced to the water, eliminating the need for fish handling. This approach, though, is labour-intensive, time-consuming, stressful for fish and demands significant costs for chemicals, while also posing environmental concerns related to the discharge of chemicals into the environment. Additionally, hydrogen peroxide can be toxic to fish at high water temperature, potentially leading to accidental losses of the treated fish.

Bath treatments have been used against a range of ectoparasites. For example, immersing infected fish in chemical solutions such as hydrogen peroxide and formaldehyde, for durations ranging from several minutes to an hour, has proven effective to remove monogeneans from skin and fins. However, these bath treatments might occasionally be less effective against those infecting gills, due to the host's excessive mucus secretion, which could potentially protect the parasites from the treatment solutions.

Several oral anthelminthic drugs are available for aquaculture applications, offering more convenient alternatives to bath treatments. Praziquantel is a well-known and effective drug against various flatworms, including monogeneans and blood flukes (Bader et al. 2019; Norbury et al. 2022). Although limited studies have reported praziquantel's toxicity to fish, the effective doses for monogeneans and blood flukes are generally much lower than the toxic dose for fish, rendering it a relatively safe drug option. Nevertheless, its usage in aquaculture is hindered by its low palatability, which causes fish to reject medicated feeds. Recently, in Japan, a more palatable form of praziquantel medication has become commercially available, potentially facilitating increased utilisation. No palatability issues are noted for praziquantel treatment in southern bluefin tuna, where locally caught sardines are injected with praziquantel and fed to fish with great results (Norbury et al. 2022). Control of blood fluke infection in southern bluefin tuna is currently highly dependent on the continued efficacy of praziquantel, so ongoing monitoring is important to evaluate its overall effectiveness as a control measure over time (Power et al. 2023).

Benzimidazoles, such as albendazole, mebendazole, fenbendazole and its pro-drug febantel, are also employed in aquaculture (Buchmann 2022). These medications exhibit fewer palatability issues compared to praziquantel and are particularly effective against hematophagous gill flukes, such as heteraxinids and diclidophirids. However, benzimidazoles may be toxic to certain fish species, necessitating caution regarding administration doses (Shirakashi et al. 2021b). Praziquantel and other chemicals can also be used for bath treatments, but the high drug costs constrain their frequent usage in aquaculture industries. Another issue with long-term use of praziquantel is the development of resistance, for example as reported for the salmon infecting tapeworm in Norway (Geitung et al. 2021).

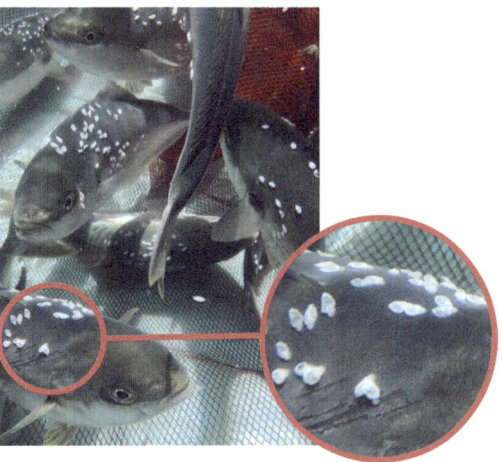

Fig. 23.4 *Seriola* undergoing freshwater bath treatment. Skin fluke, *Benedenia seriolae*, become opaque and detach within a few minutes

23.5.2 Non-Chemical (Herbal, Biocontrol, Use of Life Cycle and Parasite Ecology)

Freshwater Bathing

Modifying the salinity of the environmental water presents an effective non-chemical treatment for ectoparasites. Freshwater bath treatment was first developed in the 1950s to manage *Benedenia seriolae* infections in Japanese amberjack aquaculture, representing the first major disease in mariculture. Upon immersion in freshwater, worms on the fish's body readily become opaque and detach within a span of 1–5 min (Fig. 23.4). Freshwater baths, where the affected fish are exposed for 3–4 h to oxygenated fresh water, remain the main commercial treatment for AGD (Parsons et al. 2001; Powell et al. 2015). While the technology has progressed and many companies moved from sea cage-based bath to bathing in wellboats, the principle remains unchanged for more than 30 years and has been adopted at all sites affected by AGD where freshwater is available (Powell et al. 2015). Freshwater baths have also been used to manage sea lice infections (Powell et al. 2015). This method remains popular for treating various ectoparasites due to its simplicity, cost-effectiveness, and eco-friendliness. Nevertheless, it is labour-intensive and highly stressful for fish, as they must be transferred between a sea cage and a freshwater tank, where they are confined at an extremely high density. Fish must be fasted before, and sometimes even after, the bath treatment to cope with the stress. Consequently, frequent treatment hinders fish growth. Furthermore, transporting large quantities of freshwater to a farming site can be difficult, particularly in offshore mariculture. The cost and the logistics, including the need to access large volumes of fresh water, are the main disadvantages of freshwater bathing.

Herbal Treatment
Herbal medicines and plant-derived substances are anticipated to become safe and environmentally friendly therapeutic alternatives to conventional antiparasitic drugs (Zhu 2020, Wunderlich et al. 2017). A variety of plant-derived substances has been tested against a wide range of monogenean species, with *Dactylogyrus* spp. in freshwater fish being the most common target (Doan et al. 2020). Many botanical substances have demonstrated lethal efficacy against monogeneans in *in vitro* tests, with some also yielding positive results when used for infected fish. For example, bath treatments with garlic-derived substances exhibited high effectiveness in removing *Gyrodactylus* spp. from fish (Abd El-Galil and Aboelhadid 2012, Schelkle et al. 2013, Fridman et al. 2014). Oral administration appeared to be less effective; however, feeding supplementation containing garlic-derived substances over several weeks led to a significant reduction of *Anacanthorus penilabiatus*, *Gyrodactylus turnbulli* and *Dactylogyrus* sp. (Martins et al. 2002; Fridman et al. 2014). Such herbal supplemented diets enhance immune responses and have shown preventive efficacy against important monogeneans in mariculture, such as *Neobenedenia girellae* and *Zeuxapta seriolae* (Militz et al. 2013; Ingelbrecht et al. 2020; Fernández-Montero et al. 2021). These plant-derived dietary supplementations are now commercialised and gaining popularity within the aquaculture industry.

Physical Removal of Parasites from Fish
Physical removal of ectoparasites from fish has been used on commercial farms to manage sea lice infections on Atlantic salmon. Cleaner fish, including lumpfish, *Cyclopterus lumpus* and ballan wrasse, *Labrus bergylta*, have been used, particularly in countries significantly affected by sea lice infections, such as Norway and Scotland (Fig. 23.5). Physical delousing can also be done using mechanical or thermal commercial methods (for review, see Barrett et al. 2020a). Mechanical delousing applies high-pressure water, brushes, or negative pressure and turbulence. Thermal delousing involves pumping farmed salmon to a wellboat and expose them to 28–34°C for 30 s. While these methods can reduce sea lice loads, it can have adverse effects on salmon and cleaner fish health. Based on Norwegian salmon industry data from 2012 to 2017, thermal delousing caused greatest mortality increases (elevated mortality for 31% of treatments), followed by mechanical (25%) (Overton et al. 2019).

Fig. 23.5 Farmed lumpfish are effective cleaner fish used by Atlantic salmon industry against sea lice. (Photo ©Jon Bryan)

23.6 Interactions Between Parasites of Wild and Farmed Fish

The interactions between parasites of wild and farmed fish can be complex and varied, depending on the specific parasite species, its host-specificity and its life cycle, environmental conditions and the management practices of fish farms. Most parasites affecting farmed fish have a simple life cycle and require only one host; those are more likely to be shared with wild fish. Parasites with a broader host range could infect more species of fish. If wild populations of the same species or closely related species to the one farmed are present in the farming area, then parasites with narrow host range can infect both wild and farmed fish.

Parasites can be transmitted between wild and farmed fish through various means, including waterborne transmission, farmed fish escapees, and interactions with wild fish in the surrounding area. Semi-open aquaculture systems (e.g. sea cages) pose the highest risk for parasite exchange between farmed and wild fish. Uncontrolled exchange of water in sea cage systems permits parasites with free-living life stages to infect wild and farmed fish (e.g. sea lice, blood flukes, monogeneans) and there is a higher risk of fish escapees that if infected may spread the parasites to wild populations further away from the farming area. Sometimes, wild fish co-habit sea cages with farmed fish or freely move into and out of the sea cage, resulting in closer interactions and increasing the risk of sharing parasites.

Parasite transmission can occur from farmed populations to wild or wild fish can be the source of infection. Naive hosts or host populations that do not have a co-evolutionary history with the transmitted parasite can be particularly vulnerable, leading to negative effects on the host species, communities and ecosystems. For example, the monogenean *Gyrodactylus salaris* was introduced with Swedish Atlantic salmon resistant to this parasite into a Norwegian river where the local Atlantic salmon had no co-evolutionary history with *G. salaris*. The Norwegian Atlantic salmon population was highly susceptible resulting in up to 99% mortality of infected fish (Adolfsen et al. 2021). Sea lice *L. salmonis salmonis* and *C. clemensi* originating from farmed salmon in North America have been shown to infect wild juvenile pink salmon, *Oncorhynchus gorbuscha* and chum salmon, *Oncorhynchus keta*, when passing salmon farms during their migration, proposed as the cause of the population decline and local risk of extinction of the wild host species (Krkošek et al. 2005). However, the causal relationship has not been proven and this issue remains debated. For example, it has been shown that pink salmon and chum salmon are relatively resistant to sea lice infections (Jones and Hargreaves 2007, 2009) and that a single *L. salmonis* affected swimming performance and post-swim whole body ions only for the smallest pink salmon and with a sea louse stage of chalimus 3 or greater (Nendick et al. 2011). Furthermore, productivity of wild salmon was not negatively associated with either farm lice numbers or farmed fish production based on long-term data (Marty et al. 2010). All countries where sea lice are a problem in farmed Atlantic salmon have sea lice regulations, requiring regular monitoring of farmed salmon and treatment with the trigger for treatment ranging from 0.2 female (Norway) to 3.0 mobile (Canada) (Vormedal 2023).

The monogenean *Gyrodactylus salaris* was introduced to river Vefsna in Northern Norway during stocking with hatchery reared Baltic salmon from Sweden in the 1970s and many wild salmon populations became locally extinct (Johnsen and Jensen 1991). This introduction was a result of farmed fish being used to stock a river without appropriate health checks and had a significant impact on wild salmon populations in Norway, including average mortality of 86% after spreading to 51 rivers (Adolfsen et al. 2021). The piscicide rotenone has been used to eradicate the parasite together with the infected host (Adolfsen et al. 2021).

Wild fish can act as reservoirs for certain parasites, serving as a source of infection for farmed fish. For example, reef fish act as a reservoir of myxosporeans (*Kudoa*) for several commercial species, including Mahi mahi, *Coryphaena hippurus* L., and Japanese amberjack (Egusa and Nakajima 1980; Langdon et al. 1992; Burger et al. 2008). Infection with *Kudoa islandica*, which gradually increased in brood-stock and the first-generation juvenile fish, stopped land-based experimental aquaculture of spotted wolffish, *Anarhichas minor*, in Iceland (Kristmundsson and Freeman 2014). The brood-stock water supply was unfiltered whereas the water for juvenile fish was sand filtered. This parasite is common in wild Atlantic lumpfish, *Cyclopterus lumpus* and wild Atlantic wolffish, *Anarhichas lupus*, from Icelandic waters (Kristmundsson and Freeman 2014). Sand filters were not effective, while either ultraviolet irradiation or ozonation was suitable for control of another myxosporean, *Kudoa neurophila* (see Cobcroft and Battaglene, 2013), suggesting that any susceptible stages of the species affected by myxosporeans present in local wild fish should be farmed in land-based facilities where ultraviolet irradiation or ozonation of all incoming water is possible.

Sea lice *Caligus chiastos* infect ranched Southern bluefin tuna and are associated with eye damage and reduced condition factor (Hayward et al. 2008). These parasites are introduced to ranching sites at chalimus or adult stage by Degen's leatherjacket, *Thamnocanis degni*, attracted to the tuna cages by leftover feed (Hayward et al. 2011). Locating farm sites near rocky reefs, where the leatherjacket is common, can increase the risk of infection while moving tuna farms offshore eliminated it (Kirchhoff et al. 2011).

To mitigate the risks of introducing parasites to farmed or wild fish populations, it is crucial to reduce the interactions between farmed and wild fish, and maintain good management practices and biosecurity measures on the farm, including quarantine protocols where appropriate, regular health monitoring, rapid diagnostics and multiple control and treatment options (Fig. 23.6).

23.7 Current Issues and Future Challenges

23.7.1 Diagnosis

There are several diagnostic techniques utilised by industry and government personnel, veterinarians and researchers to detect parasites in aquaculture. Techniques generally utilised (but not limited to) include microscope-based and molecular-based methods, each with their own benefits and limitations. Microscope-based methods involve screening fish organs, body fluids, or environmental samples for detection of parasites. Fish organs can also be stained or fixed and processed for histology to improve detection. Microscopy is cheap and accessible, usually only requiring basic equipment, and some applications can be performed on-site or in the field. While successful, microscope-based methods are limited as they are time-consuming, require taxonomic expertise, and can only detect certain life stages in a targeted organ, for example, adults or eggs of blood flukes in heart or gill of fish, thus overlooking migrating stages such as infecting cercariae or emerging miracidia (Power et al. 2020). Traditional microscopy methods have been used to complement other diagnostic techniques, helping to overcome the limitations and improve the accuracy of parasite detection in aquaculture.

Molecular-based methods have advanced rapidly in the last 30 years, reducing reliance on microscopy and histology. Molecular-based methods primarily involve the use of polymerase chain reaction (PCR), which amplifies specific regions of parasite DNA or RNA from fish or environmental samples (eDNA) through a series of repeated cycles of heating and cooling (see Chap. 14 for more on eDNA). Parasite DNA or RNA can be quantified using quantitative PCR (qPCR) or droplet digital PCR (ddPCR). PCR-based methods offer higher sensitivity and specificity than traditional methods, but their use requires specialised laboratory equipment and a level of expertise in molecular biology. Decreasing costs associated with DNA sequencing have facilitated its increasing use in detection of parasites and allowed for

MANAGEMENT OPTION	Amoebic Gill Disease (Atlantic Salmon)	Sea lice infection (Atlantic Salmon)	Blood fluke infection (bluefin tuna)	Monogenean infection (amberjack)
Functional Diet	Commercial	Commercial	NA	Commercial
Biological Control		Cleaner fish		
Physical and mechanical controls		Barriers: Skirts; Snorkel Sea Cages; Bubble Curtains	Offshore farming; physical removal of intermediate host	Cage alignment/placement; physical removal of eggs
Chemical treatments	Hydrogen peroxide	Emamectin, benzoate, azamethiphos, deltamethrin, hydrogen peroxide	Praziquantel	Praziquantel, fenbendazole/ febantel, formalin, hydrogen peroxide
Non-chemical treatments	Fresh water	Thermal Water; Fresh Water; Mechanical Delousing	NA	Fresh water
Vaccination	Experimental	Commercial	NA	NA
Selective breeding	Commercial	Commercial	NA	Commercial

Fig. 23.6 Examples of management for some parasitic diseases affecting farmed fish

precise identification and differentiation of closely related species or genotypes. Third-generation nanopore sequencing on the MinION platform (Oxford Nanopore Technologies, UK) offers a simple low-cost portable device for real-time sequencing data, making it attractive for use on-site or in remote settings. The MinION has been demonstrated for rapid multiplex identification (15–30 min) of several nematode species, including fish parasite *Anisakis simplex* (Knot et al. 2020), and human protozoan parasites *Plasmodium* spp. (Imai et al. 2017).

Recent work improving molecular diagnostics has focused on the development of field-based tools using isothermal amplification of nucleic acids. Techniques including loop-mediated isothermal amplification (LAMP) and recombinase polymerase amplification (RPA) are rapid

(10–30 min), sensitive, specific and operate at constant low temperatures (60–65 °C for LAMP, 35–42 °C for RPA), so can be performed using simple heating devices or body heat for RPA (Notomi et al. 2000; Piepenburg et al. 2006). End-point analysis can be performed using a lateral flow assay, providing quick results and enabling detection at point-of-care (POC). POC detection using isothermal amplification has been demonstrated for blood flukes *Cardicola* spp. (Power 2022), and human parasites *Fasciola hepatica* and *Schistosoma japonicum* (Cabada et al. 2017; Sun et al. 2016). Isothermal amplification can be paired with CRISPR enzymes (Cas 13a) to further increase sensitivity and reduce non-specific amplification (SHERLOCK method, see Gootenberg et al. 2017), and utilised for detection of human parasites *Leishmania* spp. and *Plasmodium* spp. (Lee et al. 2020; Dueñas et al. 2022).

Rapid sample preparation using dipstick technology enables DNA extraction within 60 s without the use of specialised equipment or reliance on trained personnel and has been demonstrated for use on human parasite *S. japonicum* in combination with LAMP for POC detection (Mason and Botella 2020; Aula et al. 2021). There is a need to improve POC diagnostics for aquaculture parasites by incorporating sample preparation steps (cell lysis, DNA extraction) with current POC steps (DNA amplification, visualisation) to increase its ease of use in aquaculture settings.

Collecting samples for diagnosis in aquaculture can be challenging. The availability and accessibility of affected fish, especially in large-scale facilities, can make sampling representative of the entire population difficult. Additionally, fish have complex anatomical structures, such as gills or scales, which require specialised sampling techniques for accurate diagnosis. The lack of standardised diagnostic methods can make it challenging to compare results between laboratories, impeding effective disease monitoring and control efforts. There has been a recent effort to develop standard diagnostic procedures for aquatic animal health in Australia and New Zealand (ANZSDP), but so far has only been developed for the parasitic disease Bonamiasis affecting flat oyster *Ostrea angasi* (Corbeil et al. 2009). Disease surveillance is important to assess the risk of the introduction and spreading of parasites, however a balanced cost–benefit relationship is required. Reducing costs and labour associated with parasite diagnostics will improve uptake of routine surveillance by industry personnel, allowing for early detection and minimising risk of disease outbreaks.

Artificial intelligence (AI) technologies are emerging as a method of parasite detection and management in aquaculture. Large ectoparasites can be detected by gross examination (e.g. sea lice *Caligus* spp., *Lepeophtheirus* spp.); however, visual diagnosis can be time-consuming, so image-based machine-learning techniques have been developed to detect wounds and the presence of lice in salmon fish farms (Gupta et al. 2022). Real-time mortality tracking using AI can provide automated mortality monitoring and alert aquaculture personnel, which may help to identify root causes faster and address disease outbreaks as quickly as possible (Ranjan et al. 2023). AI methods can reduce labour costs in certain applications of parasite surveillance, but they require significant capital and infrastructure investment, so use of AI in aquaculture will likely increase as AI technologies become more affordable over time.

The choice of diagnostic technique depends on factors such as the target parasite, the resources available, the desired sensitivity and specificity, and the laboratory expertise. Often, a combination of techniques is used to enhance diagnostic accuracy and efficiency in aquaculture settings. Addressing the challenges of diagnosis requires a multifaceted approach, including improved diagnostic techniques (particularly at POC), standardised protocols, enhanced surveillance systems and collaboration among stakeholders to ensure a healthy and sustainable aquaculture industry.

23.7.2 Co-Infections

Parasitic diseases are often part of a co-infection that may involve other parasites or bacterial, viral, or fungal pathogens. For example, Complex Gill Disease (CGD) is a condition affecting farmed Atlantic salmon. CGD involves multiple pathogens, including microsporidian parasite *Desmozoon lepeophtherii*, salmon gill poxvirus and *Candidatus* Branchiomonas cysticola (Herrero et al. 2018), and can co-occur with AGD (Herrero et al. 2018).

Co-infections with two species from the same genus of blood flukes can occur in greater amberjack farmed in Mediterranean (Rigos et al. 2021) and ranched Southern bluefin tuna (Polinski et al. 2013) while co-infections with two species of monogeneans affect farmed yellowtail kingfish, *Seriola lalandi* (Williams et al. 2007). Even two closely related species of parasites have different susceptibility to treatments, so health management of fish affected by co-infections can be complex. As most experimental infections use single parasite species, our understanding of co-infections is limited. However, it is clear that co-infections can have adverse effects on disease control. For example, experimental co-infection of Atlantic salmon with sea lice *Caligus rogercresseyi* and the bacterial pathogen *Piscirickettsia salmonis*, of vaccinated Atlantic salmon, decreased performance of the fish and increased their bacterial load, explaining poorer performance of vaccines in the field, where they are often co-infected, than in experimental challenges using single pathogen (Figueroa et al. 2017). Performance of an experimental recombinant protein vaccine against AGD was most likely adversely affected by a concurrent infection with *Yersinia ruckeri*, which was detected post-trial (Valdenegro-Vega et al. 2015). Co-infections make mortality investigations particularly challenging as often it is hard to attribute the cause and cost of the fish loss to one of the parasites or pathogens. For example, Atlantic salmon mortalities in Chile in 2007 were co-infected by *Neoparamoeba perurans*, *C. rogercresseyi* and Infectious Salmon Anaemia Virus (Bustos et al. 2011; Valdes-Donoso et al. 2013). The interactions between different pathogens, host and management methods in co-infections are currently poorly understood and require further research.

23.7.3 Integrated Multitrophic Aquaculture

Integrated multitrophic aquaculture (ITMA) is based on farming together species from different trophic levels, for example seaweed, shellfish and fish (Chopin et al. 2012). Shellfish rafts have been proposed as a biosecurity barrier by filtering pathogens (Chopin et al. 2012). Blue mussel, *Mytilus edulis*, ingested free-living stages of sea lice, *Lepeophtheirus salmonis* (see Molloy et al. 2011). Similarly, *Mytilus chilensis* filtered larval stages of *Caligus rogercresseyi*, sea lice affecting salmon farming in Chile (Montory et al. 2020). Research on *N. perurans* suggested that blue mussel, *Mytilus edulis*, could reduce concentration of the amoebae in water (Rolin et al. 2016). However, this study provided no evidence for mussels filtering the amoebae out of water with no detection of *N. perurans* DNA in mussels from fish farms. The results of this study should be interpreted with caution as it was based on a tank experiment and field sampling and the sample storage was not optimal (Rolin et al. 2016). Furthermore, while the pathogenicity of the isolate used was confirmed 2 years after isolation, this experiment was run 2 years later, and it is possible that the characteristics of the amoeba changed in the culture (Bridle et al. 2015).

23.7.4 Recirculating Aquaculture Systems and Cellular Aquaculture

At the other extreme, recirculating aquaculture systems (RAS) and cellular aquaculture promise not only control over parasitic life cycles but also ensure biosecurity. RAS are promoted as solutions for environmental sustainability and climate change adaptation for aquaculture (Ahmed and Turchini 2021). However, higher salinity can present a challenge and pathogens can establish in RAS, so fish can be affected by parasitic infections. Even if the intake water is effectively disinfected, parasites can be introduced by fish entering the facility without adequate quarantine. Parasites, including the genera *Gyrodactylus*, *Chilodonella*, *Trichodina*, *Epistylis*, *Trichophrya*, *Ichthyophthirius* and *Ichtyobodo*, were detected in rainbow trout farmed in freshwater RAS (Noble and Summerfelt 1996). There is little published information on parasites in marine RAS. However, RAS for marine fish species often runs at reduced salinity, which would limit the range of potential ectoparasites to those that can survive that salinity.

There is a growing interest in cellular aquaculture based on the development of cell-based seafood across multiple species (Rubio et al. 2019). Marine cell cultures have particular physiological characteristics making them particularly attractive for scaled cellular production, including the tolerance to hypoxia, high buffering capacity and low-temperature growth conditions (Rubio et al. 2019). While contamination of cell lines with pathogens has been reported (Mirjalilia et al. 2005), among all aquaculture platforms cellular aquaculture has the biggest potential to eliminate parasitic infections in seafood. However, cellular aquaculture is still in a developmental stage and commercial production is still years away and its sustainability has already been questioned (Telesetsky 2023). In particular, the median global warming potential (GWP) of cell-based meat produced with conventional energy sources appears significantly higher for cellular aquaculture than for fish from capture fishing or aquaculture (Telesetsky 2023). Furthermore, due to the current high energy needs associated with using existing bioreactor technologies, energy required for cellular production exceeds other fish production methods (Telesetsky 2023).

While research works on marine RAS and cellular aquaculture are promising avenues to deal with parasitic infections during seafood production, more research and successful upscaling is required to take them to commercial level. As it is likely that they will not replace but complement sea cage farming, there is a need to continue research on parasitic diseases affecting farmed fish, in particular novel control and treatment methods.

23.8 Conclusions

While farmed fish can be a sustainable option for food security, fish parasites affect the farming industry. Currently, parasitic infections of fish increase production costs and cause animal welfare issues. Limited management options are available for most fish parasites. Some of the parasites could potentially impact the health of consumers. In the future, the risk of parasitic diseases will continue in semi-open or semi-closed production systems. Farming fish in RAS will reduce the risk of parasitic infections. Cellular aquaculture will further lower this risk. However, as most fish will be farmed in semi-open or semi-closed systems, the problems with parasitic diseases will continue. This means that there is a need for fish parasitology training as well as interdisciplinary research on fish parasites.

References

Aaen SM, Helgesen KO, Bakke MJ, Kaur K, Horsberg TE (2015) Drug resistance in sea lice: a threat to salmonid aquaculture. Trends Parasitol 31(2):72–81. https://doi.org/10.1016/j.pt.2014.12.006

Abd El-Galil MA, Aboelhadid SM (2012) Trials for the control of trichodinosis and gyrodactylosis in hatchery reared *Oreochromis niloticus* fries by using garlic. Vet Parasitol 185(2–4):57–63. https://doi.org/10.1016/j.vetpar.2011.10.035

Acosta-Pérez VJ, Ángeles-Hernández JC, Vega-Sánchez V, Zepeda-Velázquez AP, Añorve-Morga J, Ponce-Noguez JB, Reyes-Rodríguez NE, De-La-Rosa-Arana JL, Ramírez-Paredes JG, Gómez-De-Anda FR (2022) Prevalence of parasitic infections with zoonotic potential in Tilapia: a systematic review and meta-analysis. Anim 12(20):2800. https://doi.org/10.3390/ani12202800

Adams MB, Nowak BF (2001) Distribution and structure of lesions in the gills of Atlantic salmon, *Salmo salar* L., affected with amoebic gill disease. J Fish Dis 24(9):535–542. https://doi.org/10.1046/j.1365-2761.2001.00330.x

Adams MB, Nowak BF (2003) Amoebic gill disease: sequential pathology in cultured Atlantic salmon, *Salmo salar* L. J Fish Dis 26(10):601–614. https://doi.org/10.1046/j.1365-2761.2003.00496.x

Adams MB, Ellard K, Nowak BF (2004) Gross pathology and its relationship with histopathology of amoebic gill disease (AGD) in farmed Atlantic salmon, *Salmo salar* L. J Fish Dis 27(3):151–161. https://doi.org/10.1111/j.1365-2761.2004.00526.x

Adams MB, Maynard BT, Rigby M, Wynne JW, Taylor RS (2023) Reciprocal hybrids of Atlantic salmon (*Salmo salar*) x brown trout (*S. trutta*) confirm a heterotic response to experimentally induced amoebic gill disease (AGD). Aquac 572:739535. https://doi.org/10.1016/j.aquaculture.2023.739535

Adolfsen P, Bardal H, Aune S (2021) Fighting an invasive fish parasite in subarctic Norwegian rivers – the end of a long story? Manag Biol Invasions 12(1):49–65. https://doi.org/10.3391/mbi.2021.12.1.04

Ahmed N, Turchini GM (2021) Recirculating aquaculture systems (RAS): environmental solution and climate change adaptations. J Clean Prod 297:126604. https://doi.org/10.1016/j.jclepro.2021.126604

Akita K, Yoshida K, Noda T, Suzuki T, Hotta T, Shinoda R, Chujo T, Ogawa H, Fujinami Y, Ozaki A (2023) Heritability of resistance to benedeniosis in Japanese yellowtail (*Seriola quinqueradiata*) estimated based on long term repeated measurements in field trials. Aquac 562:738856. https://doi.org/10.1016/j.aquaculture.2022.738856

Angot V, Brasseur P (1993) European farmed Atlantic salmon (*Salmo salar* L.) are safe from anisakid larvae. Aquac 118(3–4):339–344. https://doi.org/10.1016/0044-8486(93)90468-E

Audicana MAT, Ansotegui IJ, de Corres LF, Kennedy MW (2002) *Anisakis simplex*: dangerous—dead and alive? Trends Parasitol 18(1):20–25. https://doi.org/10.1016/s1471-4922(01)02152-3

Aula OP, McManus DP, Mason MG, Botella JR, Gordon CA (2021) Rapid parasite detection utilizing a DNA dipstick. Exp Parasitol 224:108098. https://doi.org/10.1016/j.exppara.2021.108098

Bader C, Starling DE, Jones DE, Brewer MT (2019) Use of praziquantel to control platyhelminth parasites of fish. J Vet Pharmacol Ther 42(2):139–153. https://doi.org/10.1111/jvp.12735

Balli Garza J, Mladineo I, Shirakashi S, Nowak BF (2016) Diseases in tuna aquaculture. In: Advances in tuna aquaculture. Elsevier, pp 253–272. https://doi.org/10.1016/B978-0-12-411459-3.00008-4

Bao M, Pierce GJ, Strachan NJ, Pascual S, González-Muñoz M, Levsen A (2019) Human health, legislative and socioeconomic issues caused by the fish-borne zoonotic parasite *Anisakis*: challenges in risk assessment. Trends Food Sci Technol 86:298–310. https://doi.org/10.1016/j.tifs.2019.02.013

Barker SE, Bricknell IR, Covello J, Purcell S, Fast MD, Wolters W, Bouchard DA (2019) Sea lice, *Lepeophtheirus salmonis* (Krøyer 1837), infected Atlantic salmon (*Salmo salar* L.) are more susceptible to infectious salmon anemia virus. PLoS One 14(1):e0209178. https://doi.org/10.1371/journal.pone.0209178

Barrett LT, Overton K, Stien LH, Oppedal F, Dempster T (2020a) Effect of cleaner fish on sea lice in Norwegian salmon aquaculture: a national scale data analysis. Int J Parasitol 50(10–11):787–796. https://doi.org/10.1016/j.ijpara.2019.12.005

Barrett LT, Oppedal F, Robinson N, Dempster T (2020b) Prevention not cure: a review of methods to avoid sea lice infestations in salmon aquaculture. Rev Aquac 12(4):2527–2543. https://doi.org/10.1111/raq.12456

Bhar R (2022) Opisthorchiids, heterophyids and aquaculture: a brief review. MCAES 2:10–18, 22. https://doi.org/10.55162/MCAES.02.040

Blix TB, Dalmo RA, Wargelius A, Myhr AI (2021) Genome editing on finfish: current status and implications for sustainability. Rev Aquac 13(4):2344–2363. https://doi.org/10.1111/raq.12571

Braden LM, Rasmussen KJ, Purcell SL, Ellis L, Mahony A, Cho S, Whyte SK, Jones SRM, Fast MD (2018) Acquired protective immunity in Atlantic salmon *Salmo salar* against the myxozoan *Kudoa thyrsites* involves induction of MHII+ CD83+ antigen-presenting cells. Infect Immun 86(1):e00556–e00517. https://doi.org/10.1128/IAI.00556-17

Bridle AR, Davenport DL, Crosbie PB, Polinski M, Nowak BF (2015) *Neoparamoeba perurans* loses virulence during clonal culture. Int J Parasitol 45(9–10):575–578. https://doi.org/10.1016/j.ijpara.2015.04.005

Bron JE, Sommerville C, Wootten R, Rae GH (1993) Fallowing of marine Atlantic salmon, *Salmo salar* L., farms as a method for the control of sea lice, *Lepeophtheirus salmonis* (Kroyer, 1837). J Fish Dis 16(5):487–493. https://doi.org/10.1111/j.1365-2761.1993.tb00882.x

Buchmann K (2022) Control of parasitic diseases in aquaculture. Parasitology 149(14):1985–1997. https://doi.org/10.1017/S0031182022001093

Buchmann K, Mehrdana F (2016) Effects of anisakid nematodes *Anisakis simplex* (sl), *Pseudoterranova decipiens* (sl) and *Contracaecum osculatum* (sl) on fish and consumer health. Food Waterb Parasit 4:13–22. https://doi.org/10.1016/j.fawpar.2016.07.003

Burger MAA, Barnes AC, Adlard RD (2008) Wildlife as reservoirs for parasites infecting commercial species: host specificity and a redescription of *Kudoa amamiensis* from teleost fish in Australia. J Fish Dis 31(11):835–844. https://doi.org/10.1111/J.1365-2761.2008.00958.X

Bustos PA, Young ND, Rozas MA, Bohle HM, Ildefonso RS, Morrison RN, Nowak BF (2011) Amoebic gill disease (AGD) in Atlantic salmon (*Salmo salar*) farmed in Chile. Aquac 310(3-4):281–288. https://doi.org/10.1016/j.aquaculture.2010.11.001

Cabada MM, Malaga JL, Castellanos-Gonzalez A, Bagwell KA, Naeger PA, Rogers HK, Maharsi S, Mbaka M, White AC (2017) Recombinase polymerase amplification compared to real-time polymerase chain reaction test for the detection of *Fasciola hepatica* in human stool. Am J Trop Med Hyg 96(2):341–346. https://doi.org/10.4269/ajtmh.16-0601

Cabello FC (2007) Aquaculture and public health. The emergence of diphyllobothriasis in Chile and the world. Rev Méd Chile 135(8):1064–1071. https://doi.org/10.4067/s0034-98872007000800016

Cammilleri G, Costa A, Graci S, Buscemi MD, Collura R, Vella A, Pulvirenti A, Cicero A, Giangrosso G, Schembri P (2018) Presence of *Anisakis pegreffii* in farmed sea bass (*Dicentrarchus labrax* L.) commercialized in Southern Italy: a first report. Vet Parasitol 259:13–16. https://doi.org/10.1016/j.vetpar.2018.06.021

Chai JY, Jung BK (2022) General overview of the current status of human foodborne trematodiasis. Parasitology 149(10):1262–1285. https://doi.org/10.1017/S0031182022000725

Chen J, Zheng Y, Zhi T, Xu X, Zhang S, Brown CL, Yang T (2021) MHC II α polymorphism of Nile tilapia, *Oreochromis niloticus*, and its association with the susceptibility to *Gyrodactylus cichlidarum* (Monogenea) infection. Aquac 539:736637. https://doi.org/10.1016/j.aquaculture.2021.736637

Chi TT, Dalsgaard A, Turnbull JF, Tuan PA, Darwin Murrell K (2008) Prevalence of zoonotic trematodes in fish from a Vietnamese fish-farming community. J Parasitol 94(2):423–428. https://doi.org/10.1645/GE-1389.1

Chopin T, Cooper JA, Reid G, Cross S, Moore C (2012) Open water integrated multi-trophic aquaculture: environmental biomitigation and economic diversification of fed aquaculture by extractive aquaculture. Rev Aquac 4(4):209–220. https://doi.org/10.1111/j.1753-5131.2012.01074.x

Chuda H, Ieda K, Shirakashi S, Masuma S (2018) Differences in susceptibility to the skin fluke *Benedenia epinepheli* between *Epinephelus bruneus*, *E. septemfasciatus*, and a new hybrid grouper Kue-Tama, *E. bruneus* × *E. lanceolatus*. Aquac 491:346–350. https://doi.org/10.1016/j.aquaculture.2018.03.027

Clark A, Nowak BF (1999) Field investigations of amoebic gill disease in Atlantic salmon, *Salmo salar* L., in Tasmania. J Fish Dis 22(6):433–443. https://doi.org/10.1046/j.1365-2761.1999.00175.x

Clausen JH, Madsen H, Murrell KD, Van PT, Thu HNT, Do DT, Thi LAN, Manh HN, Dalsgaard A (2012) Prevention and control of fish-borne zoonotic trematodes in fish nurseries, Vietnam. Emerg Infect Dis 18(9):1438. https://doi.org/10.3201/eid1809.111076

Cobcroft JM, Battaglene SC (2013) Ultraviolet irradiation is an effective alternative to ozonation as a sea water treatment to prevent *Kudoa neurophila* (Myxozoa: Myxosporea) infection of striped trumpeter, *Latris lineata* (Forster). J Fish Biol 36(1):57–65. https://doi.org/10.1111/j.1365-2761.2012.01413.x

Corbeil S, Handlinger J, StJ Crane M (2009) Bonamiasis in Australian *Ostrea angasi*. Department of Agriculture, Fisheries and Forestry, Canberra

Couso-Pérez S, Ares-Mazás E, Gómez-Couso H (2022) A review of the current status of *Cryptosporidium* in fish. Parasitology 149(4):444–456. https://doi.org/10.1017/S0031182022000099

Crosbie PBB, Bridle AR, Cadoret K, Nowak BF (2012) In vitro cultured *Neoparamoeba perurans* causes amoebic gill disease in Atlantic salmon and fulfils Koch's postulates. Int J Parasitol 42(5):511–515. https://doi.org/10.1016/j.ijpara.2012.04.002

Daschner A, Cuéllar C (2020) Progress in *Anisakis* allergy research: milestones and reversals. Curr Treat Options Allergy 7:457–470. https://doi.org/10.1007/s40521-020-00273-9

Dedić N, Vetešník L, Šimková A (2023) Monogeneans in intergeneric hybrids of leuciscid fish: Is parasite infection driven by hybrid heterosis, genetic incompatibilities, or host-parasite coevolutionary interactions? Front Zool 20(1):5. https://doi.org/10.1186/s12983-022-00481-w

Doan HV, Soltani E, Ingelbrecht J, Soltani M (2020) Medicinal herbs and plants: Potential treatment of monogenean infections in fish. Rev Fish Sci Aquac 28(2):260–282. https://doi.org/10.1080/23308249.2020.1712325

dos Santos CAL, Howgate P (2011) Fishborne zoonotic parasites and aquaculture: a review. Aquac 318(3–4):253–261. https://doi.org/10.1016/j.aquaculture.2011.05.046

Dueñas E, Nakamoto JA, Cabrera-Sosa L, Huaihua P, Cruz M, Arévalo J, Milón P, Adaui V (2022) Novel CRISPR-based detection of *Leishmania* species. Front Microbiol 13:958693. https://doi.org/10.3389/fmicb.2022.958693

Egusa S, Nakajima K (1980) *Kudoa amamiensis* n. sp. (Myxosporea: Multivalvulida) found in cultured yellowtails and wild damselfishes from Amami-Ohshima and Okinawa, Japan. Bull Jpn Soc Sci Fish 46(10):1193–1198

Elliott N, Kube PD (2009) Development and early results of the Tasmanian Atlantic salmon breeding program. Proc Assoc Advmt Anim Breed Genet 18:362–365

FAO (2020) Fisheries and aquaculture statistics. Global aquaculture and fisheries production 1950–2018 (Fishstat). Fisheries and Aquaculture Department, Rome

FAO (2022) The State of World Fisheries and Aquaculture 2022. Towards Blue Transformation, Rome

FAO, IFAD, UNICEF, WFP, WHO (2022) The state of food security and nutrition in the world 2022. Repurposing food and agricultural policies to make healthy diets more affordable. FAO, Rome. https://doi.org/10.4060/cc0639en

Farmery AK, Alexander K, Anderson K, Blanchard JL, Carter CG, Evans K, Fischer MA, Fleming Frusher S, Fulton EA, Haas B, MacLeod CA, Murray L, Nash KL, Pecl GT, Rousseaun Y, Trebilco R, van Putten IE, Mauli S, Dutra L, Greeno D, Kaltavara J, Watson R, Nowak B (2022) Food for all: designing sustainable and secure future seafood systems. Rev Fish Biol Fish 32(1):101–121. https://doi.org/10.1007/s11160-021-09663-x

Fernández-Montero Á, Torrecillas S, Acosta F, Kalinowski T, Bravo J, Sweetman J, Roo J, Makol A, Docando J, Carvalho M (2021) Improving greater amberjack (*Seriola dumerili*) defenses against monogenean parasite *Neobenedenia girellae* infection through functional dietary additives. Aquac 534:736317. https://doi.org/10.1016/j.aquaculture.2020.736317

Figueroa C, Bustos P, Torrealba D, Dixon B, Soto C, Conejeros P, Gallardo JA (2017) Coinfection takes its toll: sea lice override the protective effects of vaccination against a bacterial pathogen in Atlantic salmon. Sci Rep 7(1):17817. https://doi.org/10.1038/s41598-017-18180-6

Fioravanti ML, Gustinelli A, Rigos G, Buchmann K, Caffara M, Pascual S, Pardo MÁ (2021) Negligible risk of zoonotic anisakid nematodes in farmed fish from European mariculture, 2016 to 2018. Eur Secur 26(2):1900717. https://doi.org/10.2807/1560-7917.ES.2021.26.2.1900717

Fleming M, Hansen T, Skulstad OF, Glover KA, Morton C, Vøllestad LA, Fjelldal PG (2014) Hybrid salmonids: ploidy effect on skeletal meristic characteristics and sea lice infection susceptibility. J Appl Ichthyol 30(4):746–752. https://doi.org/10.1111/jai.12530

Fridman S, Sinai T, Zilberg D (2014) Efficacy of garlic-based treatments against monogenean parasites infecting the guppy (*Poecilia reticulata* (Peters)). Vet Parasitol 203(1–2):51–58. https://doi.org/10.1016/j.vetpar.2014.02.002

Geitung L, Wright DW, Stien LH, Oppedal F, Karlsbakk E (2021) Tapeworm (*Eubothrium* sp.) infestation in sea caged Atlantic salmon decreased by lice barrier snorkels during a commercial-scale study. Aquac 541:736774. https://doi.org/10.1016/j.aquaculture.2021.736774

Godwin SC, Bateman AW, Kuparinen A, Johnson R, Powell J, Speck K, Hutchings JA (2022) Salmon lice in the Pacific Ocean show evidence of evolved resistance to parasiticide treatment. Sci Rep 12(1):4775. https://doi.org/10.1038/s41598-022-07464-1

Golomazou E, Malandrakis E, Panagiotaki P, Karanis P (2021) *Cryptosporidium* in fish: implications for aquaculture and beyond. Water Res 201:117357. https://doi.org/10.1016/j.watres.2021.117357

Gootenberg JS, Abudayyeh OO, Lee JW, Essletzbichler P, Dy AJ, Joung J, Verdine V, Donghia N, Daringer NM, Freije CA, Myhrvold C, Bhattacharyya RP, Livny J, Regev A, Koonin EV, Hung DT, Sabeti PC, Collins JJ, Zhang F (2017) Nucleic acid detection with CRISPR-Cas13a/C2c2. Science 356(6336):438–442. https://doi.org/10.1126/science.aam9321

Gratacap RL, Wargelius A, Edvardsen RB, Houston RD (2019) Potential of genome editing to improve aquaculture breeding and production. Trends Genet 35(9):672–684. https://doi.org/10.1016/j.tig.2019.06.006

Gupta A, Bringsdal E, Knausgård KM, Goodwin M (2022) Accurate wound and lice detection in Atlantic salmon fish using a convolutional neural network. Aust Fish 7(6). https://doi.org/10.3390/fishes7060345

Hamre LA, Oldham T, Oppedal F, Nilsen F, Glover KA (2021) The potential for cleaner fish-driven evolution in the salmon louse *Lepeophtheirus salmonis*: genetic or environmental control of pigmentation? Ecol Evol 11(12):7865–7878. https://doi.org/10.1002/ece3.7618

Hayward CJ, Aiken HM, Nowak BF (2008) An epizootic of *Caligus chiastos* on farmed southern bluefin tuna *Thunnus maccoyii* off South Australia. Dis Aquat Org 79(1):57–63. https://doi.org/10.3354/dao01890

Hayward CJ, Svane I, Lachimpadi SK, Itoh N, Bott NJ, Nowak BF (2011) Sea lice infections of wild fishes near ranched Southern bluefin tuna (*Thunnus maccoyii*) in South Australia. Aquaculture 320(3–4):178–182. https://doi.org/10.1016/j.aquaculture.2010.10.039

Herrero A, Thompson KD, Ashby A, Rodger HD, Dagleish MP (2018) Complex gill disease: an emerging syndrome in farmed Atlantic salmon (*Salmo salar* L.). J Comp Pathol 163:23–28. https://doi.org/10.1016/j.jcpa.2018.07.004

Hoai TD (2020) Reproductive strategies of parasitic flatworms (Platyhelminthes, Monogenea): the impact on parasite management in aquaculture. Aquac Int 28:421–447. https://doi.org/10.1007/s10499-019-00471-6

Hochberg NS, Hamer DH, Hughes JM, Wilson ME (2010) Anisakidosis: perils of the deep. Clin Infect Dis 51(7):806–812. https://doi.org/10.1086/656238

Houston RD, Bean TP, Macqueen DJ, Gundappa MK, Jin YH, Jenkins TL, Selly SLC, Martin SAM, Stevens JR, Santos EM, Davie A, Robledo D (2020) Harnessing genomics to fast-track genetic improvement in aquaculture. Nat Rev Genet 21(7):389–409. https://doi.org/10.1038/s41576-020-0227-y

Huston DC, Ogawa K, Shirakashi S, Nowak BF (2020) Metazoan parasite life cycles: significance for fish mariculture. Trends Parasitol 36(12):1002–1012. https://doi.org/10.1016/j.pt.2020.07.011

Ilgová J, Salát J, Kašný M (2021) Molecular communication between the monogenea and fish immune system. Fish Shellfish Immunol 112:179–190. https://doi.org/10.1016/j.fsi.2020.08.023

Imai K, Tarumoto N, Misawa K, Runtuwene LR, Sakai J, Hayashida K, Eshita Y, Maeda R, Tuda J, Murakami T, Maesaki S, Suzuki Y, Yamagishi J, Maeda T (2017) A novel diagnostic method for malaria using loop-mediated isothermal amplification (LAMP) and MinION™ nanopore sequencer. BMC Infect Dis 17(1):1–9. https://doi.org/10.1186/S12879-017-2718-9/TABLES/3

Ingelbrecht J, Miller TL, Lymbery AJ, Maita M, Torikai S, Partridge G (2020) Anthelmintic herbal extracts as potential prophylactics or treatments for monogenean infections in cultured yellowtail kingfish (*Seriola lalandi*). Aquac 520:734776. https://doi.org/10.1016/j.aquaculture.2019.734776

Inoue K, Oshima SI, Hirata T, Kimura I (2000) Possibility of anisakid larvae infection in farmed salmon. Fish Sci 66(6):1049–1052. https://doi.org/10.1046/j.1444-2906.2000.00167.x

Ivanović J, Baltić MŽ, Bošković M, Kilibarda N, Dokmanović M, Marković R, Janjić J, Baltić B (2017) Anisakis allergy in human. Trends Food Sci Technol 59:25–29. https://doi.org/10.1016/j.tifs.2016.11.006

Iwashita Y, Kamijo Y, Nakahashi S, Shindo A, Yokoyama K, Yamamoto A, Omori Y, Ishikura K, Fujioka M,

Hatada T (2013) Food poisoning associated with *Kudoa septempunctata*. J Emerg Med 44(5):943–945. https://doi.org/10.1016/j.jemermed.2012.11.026

Johnsen BO, Jensen AJ (1991) The *Gyrodactylus* story in Norway. Aquaculture 98(1–3):289–302. https://doi.org/10.1016/0044-8486(91)90393-L

Jones SRM, Hargreaves NB (2007) The abundance and distribution of *Lepeophtheirus salmonis* (Copepoda: Caligidae) on pink (*Oncorhynchus gorbuscha*) and chum (*O. keta*) salmon in coastal British Columbia. J Parasitol 93(6):1324–1331. https://doi.org/10.1645/GE-1252.1

Jones SRM, Hargreaves NB (2009) Infection threshold to estimate *Lepeophtheirus salmonis*-associated mortality among juvenile pink salmon. Dis Aquat Org 84(2):131–137. https://doi.org/10.3354/dao02043

Kar B, Mohapatra A, Parida S, Sahoo P (2022) Vaccines for parasitic diseases of fish. In: Fish immune system and vaccines. Springer, pp 125–157

Karami AM, Marnis H, Korbut R, Zuo S, Jaafar R, Duan Y, Mathiessen H, Al-Jubury A, Kania PW, Buchmann K (2022) Absence of zoonotic parasites in salmonid aquaculture in Denmark: causes and consequences. Aquaculture 549:737793. https://doi.org/10.1016/j.aquaculture.2021.737793

Kawai T, Sekizuka T, Yahata Y, Kuroda M, Kumeda Y, Iijima Y, Kamata Y, Sugita-Konishi Y, Ohnishi T (2012) Identification of *Kudoa septempunctata* as the causative agent of novel food poisoning outbreaks in Japan by consumption of *Paralichthys olivaceus* in raw fish. Clin Infect Dis 54(8):1046–1052. https://doi.org/10.1093/cid/cir1040

Kim JJ, Ryu S, Lee H (2018) Foodborne illness outbreaks in Gyeonggi province, Korea, following seafood consumption potentially caused by *Kudoa septempunctata* between 2015 and 2016. Osong Public Health Res Perspect 9(2):66–72. https://doi.org/10.24171/j.phrp.2018.9.2.05

Kirchhoff NT, Rough K, Nowak BF (2011) Moving cages further offshore: effects on southern bluefin tuna, *T. maccoyii*, parasites, health and performance. PLoS One 6(8):e23705. https://doi.org/10.1371/journal.pone.0023705

Knot IE, Zouganelis GD, Weedall GD, Wich SA, Rae R (2020) DNA barcoding of nematodes using the MinION. Front Ecol Evol 8:100. https://doi.org/10.3389/fevo.2020.00100

Kristmundsson A, Freeman MA (2014) Negative effects of *Kudoa islandica* n. sp. (Myxosporea: Kudoidae) on aquaculture and wild fisheries in Iceland. Int J Parasitol Parasites Wildl 3(2):135–146. https://doi.org/10.1016/j.ijppaw.2014.06.001

Krkošek M, Lewis MA, Volpe JP (2005) Transmission dynamics of parasitic sea lice from farm to wild salmon. P Roy Soc B-Biol Sci 272(1564):689–696. https://doi.org/10.1098/rspb.2004.3027

Kube PD, Taylor RS, Elliott NG (2012) Genetic variation in parasite resistance of Atlantic salmon to amoebic gill disease over multiple infections. Aquaculture 364–365:165–172. https://doi.org/10.1016/j.aquaculture.2012.08.026

Langdon JS, Thorne T, Fletcher WJ (1992) Reservoir hosts and new clupeoid host records for the myoliquefactive myxosporean parasite *Kudoa thyrsites* (Gilchrist). J Fish Dis 15(6):459–471. https://doi.org/10.1111/j.1365-2761.1992.tb00678.x

Lee RA, de Puig H, Nguyen PQ, Angenent-Mari NM, Donghia NM, Mcgee JP, Dvorin JD, Klapperich CM, Pollock NR, Collins JJ (2020) Ultrasensitive CRISPR-based diagnostic for field-applicable detection of *Plasmodium* species in symptomatic and asymptomatic malaria. PNAS 17(41):25722–25731. https://doi.org/10.1073/pnas.2010196117

Levsen A, Maage A (2016) Absence of parasitic nematodes in farmed, harvest quality Atlantic salmon (*Salmo salar*) in Norway – results from a large scale survey. Food Control 68:25–29. https://doi.org/10.1016/j.foodcont.2016.03.020

Martínez de Velasco G, Rodero M, Chivato T, Cuéllar C (2007) Seroprevalence of anti-*Kudoa* sp. (Myxosporea: Multivalvulida) antibodies in a Spanish population. Parasitol Res 100:1205–1211. https://doi.org/10.1007/s00436-006-0390-x

Martínez de Velasco G, Rodero M, Cuéllar C, Chivato T, Mateos JM, Laguna R (2008) Skin prick test of *Kudoa* sp. antigens in patients with gastrointestinal and/or allergic symptoms related to fish ingestion. Parasitol Res 103:713–715. https://doi.org/10.1007/s00436-008-1017-1

Martins ML, Moraes F, Miyazaki D, Brum C, Onaka E, Fenerick JJ, Bozzo F (2002) Alternative treatment for *Anacanthorus penilabiatus* (Monogenea: Dactylogyridae) infection in cultivated pacu, *Piaractus mesopotamicus* (Osteichthyes: Characidae) in Brazil and its haematological effects. Parasite 9(2):175–180. https://doi.org/10.1051/parasite/2002092175

Marty GD (2008) Anisakid larva in the viscera of a farmed Atlantic salmon (*Salmo salar*). Aquaculture 279(1–4):209–210. https://doi.org/10.1016/j.aquaculture.2008.04.006

Marty GD, Saksida SM, Quinn TJ (2010) Relationship of farm salmon, sea lice, and wild salmon populations. PNAS 107(52):22599–22604. https://doi.org/10.1073/pnas.1009573108

Mason MG, Botella JR (2020) Rapid (30-second), equipment-free purification of nucleic acids using easy-to-make dipsticks. Nat Protoc 15(11):3663–3677. https://doi.org/10.1038/s41596-020-0392-7

Maynard BT, Taylor RS, Kube PD, Cook MT, Elliott NG (2016) Salmonid heterosis for resistance to amoebic gill disease (AGD). Aquaculture 451:106–112. https://doi.org/10.1016/j.aquaculture.2015.09.004

Militz TA, Southgate PC, Carton AG, Hutson KS (2013) Dietary supplementation of garlic (*Allium sativum*) to prevent monogenean infection in aquaculture. Aquaculture 408:95–99. https://doi.org/10.1016/j.aquaculture.2013.05.027

Mirjalilia A, Parmoora E, Moradi Bidhendib S, Sarkaric B (2005) Microbial contamination of cell cultures:

a 2-year study. Biologicals 33(2):81–85. https://doi.org/10.1016/j.biologicals.2005.01.004

Molloy SD, Pietrak MR, Bouchard DA, Bricknell I (2011) Ingestion of *Lepeophtheirus salmonis* by the blue mussel *Mytilus edulis*. Aquaculture 311(1–4):61–64. https://doi.org/10.1016/j.aquaculture.2010.11.038

Montory JA, Chaparro OR, Averbuj A, Salas-Yanquin LP, Büchner-Miranda JA, Gebauer P, Cumillaf JP, Cruces E (2020) The filter-feeding bivalve *Mytilus chilensis* capture pelagic stages of *Caligus rogercresseyi*: a potential controller of the sea lice fish parasites. J Fish Dis 43(4):475–484. https://doi.org/10.1111/jfd.13141

Moratal S, Dea-Ayuela MA, Cardells J, Marco-Hirs NM, Puigcercós S, Lizana V, López-Ramon J (2020) Potential risk of three zoonotic protozoa (*Cryptosporidium* spp., *Giardia duodenalis*, and *Toxoplasma gondii*) transmission from fish consumption. Food Secur 9(12):1913. https://doi.org/10.3390/foods9121913

Moratal S, Dea-Ayuela MA, Martí-Marco A, Puigcercós S, Marco-Hirs NM, Doménech C, Corcuera E, Cardells J, Lizana V, López-Ramon J (2022) Molecular characterization of *Cryptosporidium* spp. in cultivated and wild marine fishes from Western Mediterranean with the first detection of zoonotic *Cryptosporidium ubiquitum*. Animals 12(9):1052. https://doi.org/10.3390/ani12091052

Mullins J, Nowak B, Leef M, Røn Ø, Berger Eriksen T, McGurk C (2020) Functional diets improve survival and physiological response of Atlantic salmon (*Salmo salar*) to amoebic gill disease. J World Aquacult Soc 51(3):634–648. https://doi.org/10.1111/jwas.1269

Murray AG, Peeler EJ (2005) A framework for understanding the potential for emerging diseases in aquaculture. Prev Vet Med 67(2–3):223–235. https://doi.org/10.1016/j.prevetmed.2004.10.012

Nendick L, Sackville M, Tang S, Brauner CJ, Farrell AP (2011) Sea lice infection of juvenile pink salmon (*Oncorhynchus gorbuscha*): effects on swimming performance and postexercise ion balance. Can J Fish Aquat Sci 68(2):241–249. https://doi.org/10.1139/F10-150

Nguyen TLA, Madsen H, Dalsgaard A, Nguyen TP, Dao THT, Murrell KD (2010) Poultry as reservoir hosts for fishborne zoonotic trematodes in Vietnamese fish farms. Vet Parasitol 169(3–4):391–394. https://doi.org/10.1016/j.vetpar.2010.01.010

Noble AC, Summerfelt ST (1996) Diseases encountered in trout farmed in recirculation systems. Annu Rev Fish Dis 6:65–92. https://doi.org/10.1016/S0959-8030(96)90006-X

Norbury LJ, Shirakashi S, Power C, Nowak BF, Bott NJ (2022) Praziquantel use in aquaculture – current status and emerging issues. Int J Parasitol Drugs Drug Resist 18:87–102. https://doi.org/10.1016/j.ijpddr.2022.02.001

Notomi T, Okayama H, Masubuchi H, Yonekawa T, Watanabe K, Amino N, Hase T (2000) Loop-mediated isothermal amplification of DNA. Nucleic Acids Res 28(12):e63. https://doi.org/10.1093/nar/28.12.e63

Ohnishi T, Kikuchi Y, Furusawa H, Kamata Y, Sugita-Konishi Y (2013) *Kudoa septempunctata* invasion increases the permeability of human intestinal epithelial monolayer. Foodborne Pathog Dis 10(2):137–142. https://doi.org/10.1089/fpd.2012.1294

Ohnishi T, Fujiwara M, Tomaru A, Yoshinari T, Sugita-Konishi Y (2016) Survivability of *Kudoa septempunctata* in human intestinal conditions. Parasitol Res 115:2519–2522. https://doi.org/10.1007/s00436-016-5036-z

Ohnishi T, Obara T, Arai S, Yoshinari T, Sugita-Konishi Y (2018) Quantitative analysis of *Unicapsula seriolae* in greater amberjack associated with unidentified foodborne disease. Shokuhin Eiseigaku zasshi. J Food Hyg Soc Jpn 59(1):24–29. https://doi.org/10.3358/shokueishi.59.24

Øvergard A-C, Hamre LA, Grotmol S, Nilsen F (2018) Salmon louse rhabdoviruses: impact on louse development and transcription of selected Atlantic salmon immune genes. Dev Comp Immunol 86:86–95. https://doi.org/10.1016/j.dci.2018.04.023

Overton K, Dempster T, Oppedal F, Kristiansen TS, Gismervik K, Stien LH (2019) Salmon lice treatments and salmon mortality in Norwegian aquaculture: a review. Rev Aquac 11(4):1398–1417. https://doi.org/10.1111/raq.12299

Ozaki A, Yoshida K, Fuji K, Kubota S, Kai W, Aoki JY, Kawabata Y, Suzuki J, Akita K, Koyama T (2013) Quantitative trait loci (QTL) associated with resistance to a monogenean parasite (*Benedenia seriolae*) in yellowtail (*Seriola quinqueradiata*) through genome wide analysis. PLoS One 8(6):e64987. https://doi.org/10.1371/journal.pone.0064987

Parsons H, Nowak B, Fisk D, Powell M (2001) Effectiveness of commercial freshwater bathing as a treatment against amoebic gill disease in Atlantic salmon. Aquaculture 195(3–4):205–210. https://doi.org/10.1016/S0044-8486(00)00567-6

Phan VT, Ersbøll AK, Nguyen TT, Nguyen KV, Nguyen HT, Murrell D, Dalsgaard A (2010) Freshwater aquaculture nurseries and infection of fish with zoonotic trematodes, Vietnam. Emerg Infect Dis 16(12):1905. https://doi.org/10.3201/eid1612.100422

Phan VT, Ersbøll AK, Do DT, Dalsgaard A (2011) Raw-fish-eating behavior and fishborne zoonotic trematode infection in people of northern Vietnam. Foodborne Pathog Dis 8(2):255–260. https://doi.org/10.1089/fpd.2010.0670

Piepenburg O, Williams CH, Stemple DL, Armes NA (2006) DNA detection using recombination proteins. PLoS Biol 4(7):e204. https://doi.org/10.1371/journal.pbio.0040204

Polimeno L, Lisanti MT, Rossini M, Giacovazzo E, Polimeno L, Debellis L, Ballini A, Topi S, Santacroce L (2021) *Anisakis* allergy: is aquacultured fish a safe and alternative food to wild-capture fisheries for *Anisakis simplex*-sensitized patients? Biology 10(2):106. https://doi.org/10.3390/biology10020106

Polinski M, Belworthy Hamilton D, Nowak BF, Bridle AR (2013) SYBR, TaqMan, or both: highly sensitive, non-invasive detection of *Cardicola* blood fluke species in Southern bluefin buna (*Thunnus maccoyii*). Mol Biochem Parasitol 191(1):7–15. https://doi.org/10.1016/j.molbiopara.2013.07.002

Powell MD, Reynolds P, Kristensen T (2015) Freshwater treatment of amoebic gill disease and sea-lice in seawater salmon production: considerations of water chemistry and fish welfare in Norway. Aquaculture 448:18–28. https://doi.org/10.1016/j.aquaculture.2015.05.027

Power C (2022) Southern bluefin tuna and their blood flukes. PhD thesis. Royal Melbourne Institute of Technology, Melbourne

Power C, Nowak BF, Cribb TH, Bott NJ (2020) Bloody flukes: a review of aporocotylids as parasites of cultured marine fishes. Int J Parasitol 50(10–11):743–753. https://doi.org/10.1016/j.ijpara.2020.04.008

Power C, Carabott M, Widdicombe M, Coff L, Rough K, Nowak BF, Bott NJ (2023) Effects of company and season on blood fluke (*Cardicola* spp.) infection in ranched Southern bluefin tuna: preliminary evidence infection has a negative effect on fish growth. PeerJ 11:e15763. https://doi.org/10.7717/peerj.15763

Rahmati AR, Kiani B, Afshari A, Moghaddas E, Williams M, Shamsi S (2020) World-wide prevalence of *Anisakis* larvae in fish and its relationship to human allergic anisakiasis: a systematic review. Parasitol Res 119:3585–3594. https://doi.org/10.1007/s00436-020-06892-0

Rama TA, Silva D (2022) *Anisakis* allergy: raising awareness. Acta Méd Port 35(7–8):578–583. https://doi.org/10.20344/amp.15908

Ranjan R, Sharrer K, Tsukuda S, Good C (2023) MortCam: an artificial intelligence-aided fish mortality detection and alert system for recirculating aquaculture. Aquac Eng 102:102341. https://doi.org/10.1016/j.aquaeng.2023.102341

Rigos G, Katharios P, Kogiannou D, Cascarano CM (2021) Infectious diseases and treatment solutions of farmed greater amberjack *Seriola dumerili* with particular emphasis in Mediterranean region. Rev Aquac 13(1):301–323. https://doi.org/10.1111/raq.12476

Robertson LJ, Lalle M, Paulsen P (2020) Why we need a European focus on foodborne parasites. Exp Parasitol 214:107900. https://doi.org/10.1016/j.exppara.2020.107900

Rolin C, Graham J, McCarthy U, Martin SAM, Matejusova I (2016) Interactions between *Paramoeba perurans*, the causative agent of amoebic gill disease, and the blue mussel, *Mytilus edulis*. Aquaculture 456:1–8. https://doi.org/10.1016/j.aquaculture.2016.01.019

Rubio N, Datar I, Stachura D, Kaplan D, Krueger K (2019) Cell-based fish: a novel approach to seafood production and an opportunity for cellular agriculture. Front Sustain Food Syst 3:43. https://doi.org/10.3389/fsufs.2019.00043

Rufai MA, Adetona AJ (2017) Parasites fauna of farmed African catfish (*Clarias gariepinus*) in Osogbo, Osun State, Southwestern Nigeria. Annales of West University of Timisoara. Ser Biol 20:123–130. https://doi.org/10.1007/s00436-011-2491-4

Saleh R, Abou-Eisha A, Fadel H, Helmy Y (2009) Occurrence of encysted metacercariae of some zoonotic trematodes in freshwater fishes and their public health significance in Port Said province, Abbassa. Int J Aquac, Special Issue for Global Fisheries & Aquaculture Research Conference, Cairo International Convention Center, pp 24–26

Samsing F, Johnsen I, Dempster T, Oppedal F, Treml EA (2017) Network analysis reveals strong seasonality in the dispersal of a marine parasite and identifies areas for coordinated management. Landsc Ecol 32:1953–1967. https://doi.org/10.1007/s10980-017-0557-0

Schelkle B, Snellgrove D, Cable J (2013) In vitro and in vivo efficacy of garlic compounds against *Gyrodactylus turnbulli* infecting the guppy (*Poecilia reticulata*). Vet Parasitol 198(1–2):96–101. https://doi.org/10.1016/j.vetpar.2013.08.027

Shinn AP, Pratoomyot J, Bron JE, Paladini G, Brooker EE, Brooker AJ (2015) Economic costs of protistan and metazoan parasites to global mariculture. Parasitology 142(1):196–270. https://doi.org/10.1017/S0031182014001437

Shirakashi S, Takagi T (2021) Consumption of monogenean eggs by crustaceans and fish. Aquac Res 52(5):1915–1924. https://doi.org/10.1111/are.15040

Shirakashi S, Hirano C, Ishitani H, Ishimaru K (2013) Diurnal pattern of skin fluke infection in cultured amberjack, *Seriola dumerili*, at different water depths. Aquaculture 402:19–23. https://doi.org/10.1016/j.aquaculture.2013.03.014

Shirakashi S, Asai N, Miura M (2021a) Phototactic responses in four monogenean oncomiracidia. Parasitol Res 120:3173–3180. https://doi.org/10.1007/s00436-021-07280-y

Shirakashi S, Miwa S, Katsuki T, Harakawa S, Kawakami H, Nakayasu C, Mori KI (2021b) Evaluations of lethal and sub-lethal toxicity of febantel in the juvenile Japanese amberjack *Seriola quinqueradiata*. Fish Pathol 56(2):79–88. https://doi.org/10.3147/jsfp.56.79

Shivam S, El-Matbouli M, Kumar G (2021) Development of fish parasite vaccines in the OMICs era: progress and opportunities. Vaccine 9(2):179. https://doi.org/10.3390/vaccines9020179

Skern-Mauritzen R, Torrissen O, Glover KA (2014) Pacific and Atlantic *Lepeophtheirus salmonis* (Krøyer, 1838) are allopatric subspecies: *Lepeophtheirus salmonis salmonis* and *L. salmonis oncorhynchi* subspecies novo. BMC Genet 15:109. https://doi.org/10.1186/1471-2156-15-32

Skilton DC, Saunders RJ, Hutson KS (2020) Parasite attractants: identifying trap baits for parasite management in aquaculture. Aquaculture 516:734557. https://doi.org/10.1016/j.aquaculture.2019.734557

Skov J, Kania PW, Olsen MM, Lauridsen JH, Buchmann K (2009) Nematode infections of maricultured and wild fishes in Danish waters: a comparative study. Aquaculture 298(1–2):24–28. https://doi.org/10.1016/j.aquaculture.2009.09.024

Stentiford GD, Holt CC (2022) Global adoption of aquaculture to supply seafood. Environ Res Lett 17(4):041003. https://doi.org/10.1088/1748-9326/ac5c9f

Sugita-Konishi Y, Sato H, Ohnishi T (2014) Novel foodborne disease associated with consumption of raw fish, olive flounder (*Paralichthys olivaceus*). Food Saf 2(4):141–150. https://doi.org/10.14252/foodsafetyfscj.2014026

Sun K, Xing W, Yu X, Fu W, Wang Y, Zou M, Luo Z, Xu D (2016) Recombinase polymerase amplification combined with a lateral flow dipstick for rapid and visual detection of *Schistosoma japonicum*. Parasit Vectors 9(1):476. https://doi.org/10.1186/s13071-016-1745-5

Suzuki J, Murata R, Yokoyama H, Sadamasu K, Kai A (2015) Detection rate of diarrhoea-causing *Kudoa hexapunctata* in Pacific bluefin tuna *Thunnus orientalis* from Japanese waters. Int J Food Microbiol 194:1–6. https://doi.org/10.1016/j.ijfoodmicro.2014.11.001

Tacchi L, Bickerdike R, Douglas A, Secombes CJ, Martin SAMM (2011) Transcriptomic responses to functional feeds in Atlantic salmon (*Salmo salar*). Fish Shellfish Immunol 31(5):704–715. https://doi.org/10.1016/j.fsi.2011.02.023

Tachibana T, Watari T (2021) A novel case of food poisoning caused by the consumption of Pacific bluefin tuna infected with *Kudoa hexapunctata*. Clin Case Rep 9(6):e04222. https://doi.org/10.1002/ccr3.4222

Taylor RS, Muller WJ, Cook MT, Kube PD, Elliott NG (2009) Gill observations in Atlantic salmon (*Salmo salar*, L.) during repeated amoebic gill disease (AGD) field exposure and survival challenge. Aquaculture 290(1-2):1–8. https://doi.org/10.1016/j.aquaculture.2009.01.030

Telesetsky A (2023) Cellular mariculture: challenges of delivering sustainable protein security. Mar Policy 147:105400. https://doi.org/10.1016/j.marpol.2022.105400

Thien PC, Murrell KD (2020) Seasonal prevalence of zoonotic trematode parasites in commercial catfish species in the Ho Chi Minh City area. Aquac Rep 17:100315. https://doi.org/10.1016/j.aqrep.2020.100315

Thien CP, Dalsgaard A, Nhan NT, Olsen A, Murrell KD (2009) Prevalence of zoonotic trematode parasites in fish fry and juveniles in fish farms of the Mekong Delta, Vietnam. Aquaculture 295(1–2):1–5. https://doi.org/10.1016/j.aquaculture.2009.06.033

Torres P, Lopez JC, Cubillos V, Lobos C, Silva R (2002) Visceral diphyllobothriosis in a cultured rainbow trout, *Oncorhynchus mykiss* (Walbaum), in Chile. J Fish Dis 25(6):375–379. https://doi.org/10.1046/j.1365-2761.2002.00381.x

Torres P, Quintanilla J, Rozas M, Miranda P, Ibarra R, San Martín M, Raddatz B, Wolter M, Villegas A, Canobra C (2010) Endohelminth parasites from salmonids in intensive culture from southern Chile. J Parasitol 96(3):669–670. https://doi.org/10.1645/GE-2211.1

Uchino T, Tabata J, Yoshida K, Suzuki T, Noda T, Fujinami Y, Ozaki A (2020) Novel *Benedenia* disease resistance QTLs in five F1 families of yellowtail (*Seriola quinqueradiata*). Aquaculture 529:735622. https://doi.org/10.1016/j.aquaculture.2020.735622

UN (2023) Global Issues: Population. Available at https://www.un.org/en/global-issues/population; accessed on 7 June 2023

Valdenegro-Vega VA, Cook M, Crosbie P, Bridle AR, Nowak BF (2015) Vaccination with recombinant protein (r22C03), a putative attachment factor of *Neoparamoeba perurans*, against AGD in Atlantic salmon (*Salmo salar*) and implications of a co-infection with *Yersinia ruckeri*. Fish Shellfish Immunol 44(2):592–602. https://doi.org/10.1016/j.fsi.2015.03.016

Valdes-Donoso P, Mardones FO, Jarpa M, Ulloa M, Carpenter TE, Perez AM (2013) Co-infection patterns of infectious salmon anaemia and sea lice in farmed Atlantic salmon, *Salmo salar* L., in southern Chile (2007–2009). J Fish Dis 36(3):353–360. https://doi.org/10.1111/jfd.12070

Vaughan DB, Grutter AS, Hutson KS (2018) Cleaner shrimp are a sustainable option to treat parasitic disease in farmed fish. Sci Rep 8(1):13959. https://doi.org/10.1038/s41598-018-32293-6

Vormedal I (2023) Sea-lice regulation in salmon-farming countries: how science shape policies for protecting wild salmon. Aquac Int:1–17. https://doi.org/10.1007/s10499-023-01270-w

Waltz E (2017) First genetically engineered salmon sold in Canada. Nature 548(7666):148. https://doi.org/10.1038/nature.2017.22116

Watanabe T, Sawada M, Yanagida T, Ogawa K (2014) Investigation of the infection with *Diphyllobothrium nihonkaiense* plerocercoids and *Metagonimus* metacercariae in freshwater salmonids cultured in Japan. Fish Pathol 49(4):198–201. https://doi.org/10.3147/jsfp.49.198

Williams RE, Ernst I, Chambers CB, Whittington ID (2007) Efficacy of orally administered praziquantel against *Zeuxapta seriolae* and *Benedenia seriolae* (Monogenea) in yellowtail kingfish *Seriola lalandi*. Dis Aquat Org 77(3):199–205. https://doi.org/10.3354/dao01824

Wright DW, Stien LH, Dempster T, Vågseth T, Nola V, Fosseidengen JE, Oppedal F (2017) 'Snorkel' lice barrier technology reduced two co-occurring parasites, the salmon louse (*Lepeophtheirus salmonis*) and the amoebic gill disease causing agent (*Neoparamoeba perurans*), in commercial salmon sea-cages. Prev Vet Med 140:97–105. https://doi.org/10.1016/j.prevetmed.2017.03.002

Wunderlich AC, Zica E, Ayres V, Guimarães AC, Takeara R (2017) Plant-derived compounds as an alternative treatment against parasites in fish farming: a review. IntechOpen, London, pp 115–135. https://doi.org/10.5772/67668

Yamamoto S, Fukushima A, Ishimaru K, Shirakashi S (2014) Shading of net cage is an effective control measure against skin fluke *Neobenedenia girellae* infection in chub mackerel *Scomber japonicus*.

Fish Sci 80:1021–1026. https://doi.org/10.1007/s12562-014-0781-3

Yokoyama H, Funaguma N, Kobayashi S (2016) In vitro inactivation of *Kudoa septempunctata* spores infecting the muscle of olive flounder *Paralichthys olivaceus*. Foodborne Pathog Dis 13(1):21–27. https://doi.org/10.1089/fpd.2015.2003

Yoshinaga T, Kinami R, Hall KA, Ogawa K (2006) A preliminary study on the infection of anisakid larvae in juvenile greater amberjack *Seriola dumerili* imported from China to Japan as mariculture seedlings. Fish Pathol 41(3):123–126. https://doi.org/10.3147/jsfp.41.123

Young ND, Crosbie PBB, Adams MB, Nowak BF, Morrison RN (2007) *Neoparamoeba perurans* n. sp., an agent of amoebic gill disease of Atlantic salmon (*Salmo salar*). Int J Parasitol 37(13):1469–1481. https://doi.org/10.1016/j.ijpara.2007.04.018

Zhu F (2020) A review on the application of herbal medicines in the disease control of aquatic animals. Aquaculture 526:735422. https://doi.org/10.1016/j.aquaculture.2020.735422

Ziarati M, Zorriehzahra MJ, Hassantabar F, Mehrabi Z, Dhawan M, Sharun K, Emran TB, Dhama K, Chaicumpa W, Shamsi S (2022) Zoonotic diseases of fish and their prevention and control. Vet Q 42(1):95–118. https://doi.org/10.1080/01652176.2022.2080298

Zilberg D, Munday BL (2000) Pathology of experimental amoebic gill disease in Atlantic salmon, *Salmo salar L.*, and the effect of pre-maintenance of fish in sea water on the infection. J Fish Dis 23(6):401–407. https://doi.org/10.1046/j.1365-2761.2000.00252.x

Zilberg D, Munday BL (2001) Responses of Atlantic salmon, *Salmo salar L.*, to *Paramoeba* antigens administered by a variety of routes. J Fish Dis 24(3):181–183. https://doi.org/10.1046/j.1365-2761.2001.00280.x

Open Access This chapter is licensed under the terms of the Creative Commons Attribution-NonCommercial-NoDerivatives 4.0 International License (http://creativecommons.org/licenses/by-nc-nd/4.0/), which permits any non-commercial use, sharing, distribution and reproduction in any medium or format, as long as you give appropriate credit to the original author(s) and the source, provide a link to the Creative Commons license and indicate if you modified the licensed material. You do not have permission under this license to share adapted material derived from this chapter or parts of it.

The images or other third party material in this chapter are included in the chapter's Creative Commons license, unless indicated otherwise in a credit line to the material. If material is not included in the chapter's Creative Commons license and your intended use is not permitted by statutory regulation or exceeds the permitted use, you will need to obtain permission directly from the copyright holder.

Species List of Parasites, Hosts and Vectors of Aquatic Parasitology: Ecological and Environmental Concepts and Implications of Marine and Freshwater Parasites

Russell Q. -Y. Yong

Introduction

In the following, we provide lists of all species and viruses mentioned in the book. The tables are listed in a phylogenetically arranged order, starting with viruses (Table A1) and followed by bacteria (Table A2), protists (Table A3), fungi including microsporidians (Table A4), all other parasites (Table A5) and all hosts, vectors or otherwise free-living species (Table A6). Nomenclature and authorities were primarily sourced from the World Register of Marine Species (WoRMS) and supporting/subsidiary databases, e.g. MolluscaBase, FishBase and World of Copepods, AlgaeBase, AmphibiaWeb and GBIF, and cross-checked at all times against primary literature that was most current at the time of this list's publication.

R. Q.-Y. Yong
Water Research Group, Unit for Environmental Sciences and Management, North-West University, Potchefstroom, South Africa
e-mail: 49933884@mynwu.ac.za

Table A1 Mentioned viruses listed alphabetically by family, with the species arranged in alphabetical order within each family

Family	Species	Chapter
Orthomyxoviridae	Infectious Salmon Anaemia Virus (*Isavirus*)	23
Paramyxoviridae	Rinderpest morbillivirus (*Morbillivirus*)	8
Poxviridae	Salmon Gill Pox Virus (SGPV)	23
Poxviridae	Squirrelpox Virus (SQPV)	8
Rhabdoviridae	Carp spirivirus [Spring Viraemia of Carp (SVC)] Fijan, 1976	6

Table A2 Mentioned bacteria listed alphabetically by family, with the species arranged in alphabetical order within each family

Family	Species	Chapter
Anaplasmataceae	*Neorickettsia* Philip, Hadlow & Hughes, 1953	15
Borreliaceae	*Borrelia garinii* Baranton, Postic, Saint Girons, Boerlin, Piffaretti, Assous & Grimont, 1992	6
Burkholderiales *incertae sedis*	*Candidatus* Branchiomonas cysticola Toenshoff, Kvellestad, Mitchell, Steinum, Falk, Colquhoun & Horn, 2012	23
Chlamydiaceae	*Chlamydia* Jones, Rake & Stearns 1945	2
Enterobacteriaceae	*Escherichia coli* (Migula, 1895)	2, 14
Enterobacteriaceae	*Photorhabdus asymbiotica* Fischer-Le Saux, Viallard, Brunel, Normand & Boemare, 1999	7
Enterobacteriaceae	*Xenorhabdus nematophila* (Poinar & Thomas, 1965)	7
Flavobacteriaceae	*Flavobacterium columnare* (Bernardet & Grimont, 1989)	7
Holosporaceae	*Holospora undulata* (Haffkine, 1890)	7
Legionellaceae	*Legionella* Brenner, Steigerwalt & McDade, 1979	2, 14
Legionellaceae	*Legionella pneumophila* Brenner, Steigerwalt & McDade, 1979	2
Mycobacteriaceae	*Mycobacterium* Lehmann & Neumann 1896	2
Piscirickettsiaceae	*Piscirickettsia salmonis* Fryer, Lannan, Giovannoni & Wood, 1992	23
Pseudomonadaceae	*Pseudomonas* Migula, 1894	2
Pseudomonadaceae	*Pseudomonas aeruginosa* (Schröter, 1872)	2
Pyrodictiaceae	*Pyrolobus fumarii* Blochl, Rachel, Burggraf, Hafenbradl, Jannasch & Stetter, 1997	12
Staphylococcaceae	*Staphylococcus aureus* Rosenbach, 1884	2
Sulfolobaceae	*Sulfolobus acidocaldarius* Brock, Brock, Belly & Weiss, 1972	12
Vibrionaceae	*Vibrio* Pacini, 1854	2
Vibrionaceae	*Vibrio penaeicida* Ishimaru, Akagawa-Matsushita & Muroga, 1995	14
Yersiniaceae	*Yersinia ruckeri* Ewing, Ross, Brenner & Fanning, 1978	23

Table A3 Mentioned protists listed alphabetically by family, with the species arranged in alphabetical order within each family

Family	Species	Chapter
Acanthamoebidae	*Acanthamoeba* Volkonsky, 1931	2
Adeleidae	*Adelina tribolii* Bhatia, 1937	8
Amphileptidae	*Amphileptus* Ehrenberg, 1830	2
Amphileptidae	*Amphileptus balticus* (Fenchel, 1965)	2
Anabaenaceae	*Dolichospermum (Anabaena) flosaquae* (Brébisson in Bornet & Flauhault, 1886)	4
Ancistrocomidae	*Stegotricha enterikos* Bower & Meyer, 1993	2
Babesiidae	*Babesia* Starcovici, 1893	2
Balantidiidae	*Balantidium coli* (Malmsten, 1857)	2
Blastodinidae	*Blastodinium* Chatton, 1906	2
Calyptosporidae	*Calyptospora* Overstreet, Hawkins & Fournie, 1984	2
Chilodonellidae	*Chilodonella* Strand, 1928	23
Chilodonellidae	*Chilodonella hexasticha* (Kiernik, 1909)	2
Chilodonellidae	*Chilodonella piscicola* (Zacharias, 1894)	2
Claustrosporidiidae	*Claustrosporidium aselli* (Plugfelder, 1948)	2
Claustrosporidiidae	*Claustrosporidium gammari* Larsson, 1987	2
Cryptobiaceae	*Cryptobia salmositica* Katz, 1951	2
Cryptosporidiidae	*Cryptosporidium* Tyzzer, 1907	2, 15, 19, 23
Cryptosporidiidae	*Cryptosporidium hominis* Morgan-Ryan, Fall, Ward, Hijjawi, Sulaiman, Fayer, Thompson, Olson, Lal & Xiao, 2002	23
Cryptosporidiidae	*Cryptosporidium parvum* Tyzzer, 1912	23
Cryptosporidiidae	*Cryptosporidium scrofarum* Kváč, Kestřánová, Pinková, Květoňová, Kalinová, Wagnerová, Kotková, Vítovec, Ditrich, McEvoy, Stenger & Sak, 2013	23
Cryptosporidiidae	*Cryptosporidium ubiquitum* Fayer, Santín & Macarisin, 2010	23
Didiniidae	*Didinium nasutum* (Müller, 1773)	7
Dinoflagellata *incertae sedis*	*Atelodinium* Chatton, 1920	2
Eimeriidae	*Eimeria* Schneider, 1875	2
Eimeriidae	*Goussia* Labbe, 1896	2
Entamoebidae	*Entamoeba histolytica* Schaudinn, 1903	2
Ephelotidae	*Ephelota gigantea* Noble, 1929	2
Epistylididae	*Epistylis* Ehrenberg, 1830	2, 23
Foettingeriidae	*Vampyrophrya pelagica* (Chatton & Lwoff, 1930)	2
Folliculinidae	*Halofolliculina corallasia* Antonius & Lipscomb, 2001	2
Folliculinidae	*Mirofolliculina limnoriae* (Giard, 1883)	2
Fragilariaceae	*Synedra* Ehrenberg, 1830	16
Gomphosphaeriaceae	*Gomphosphaeria* Kützing, 1836	4
Haemogregarinidae	*Haemogregarina bigemina* Laveran & Mesnil, 1901	2
Haemogregarinidae	*Haemogregarina bigemina* Laveran & Mesnil, 1901	6
Haemogregarinidae	*Haemogregarina stepanowi* (Danilewsky, 1885)	2
Haemohormidiidae	*Cardiosporidium* Gaver & Stephan, 1907	2
Haemohormidiidae	*Haemohormidium* Henry, 1910	2
Haemohormidiidae	*Nephromyces* Giard, 1888	2
Haemoproteidae	*Haemoproteus* Kruse, 1890	2
Haplosporidiidae	*Bonamia* Pichot, Comps, Tigé, Grizel & Rabouin, 1980	2
Haplosporidiidae	*Haplosporidium nelsoni* Haskin, Stauber & Mackin, 1966	2, 22

(continued)

Table A3 (continued)

Family	Species	Chapter
Haplozoonidae	*Haplozoon* Dogiel, 1906	2
Hexamitidae	*Giardia* Künstler, 1882	15
Hexamitidae	*Giardia duodenalis* Stiles, 1902	2, 19, 23
Hexamitidae	*Spironucleus salmonicida* (Jørgensen, 2006)	2
Holophryidae	*Cryptocaryon irritans* Brown, 1951	2, 23
Ichthyophthiriidae	*Ichthyophthirius* Fouquet, 1876	23
Ichthyophthiriidae	*Ichthyophthirius multifiliis* Fouquet, 1876	2
Kyaroikeidae	*Kyaroikeus paracetarius* Jin, Qu, Wei, Montagnes, Fan & Chen, 2020	2
Kyaroikeidae	*Planilamina ovata* Ma, Overstreet, Sniezek, Solangi & Wayne Coats, 2006	2
Labyrinthulaceae	*Labyrinthula* Cienkowski, 1864	8
Lecudinidae	*Lankesteria ascidiae* (Lankester, 1872)	2
Leucocytozoidae	*Leucocytozoon* Berestneff, 1904	2
Licnophoridae	*Licnophora auerbachii* (Cohn, 1866)	2
Marteiliidae	*Marteilia refringens* Grizel, Comps, Bonami, Cousserans, Duthoit & Le Pennec, 1974	2, 14
Mayorellidae	*Mayorella* Schaeffer, 1926	2
Microcystaceae	*Microcystis* Lemmermann, 1907	4
Mikrocytiidae	*Paramikrocytos canceri* Hartikainen, Stentiford, Bateman, Berney, Feist, Longshaw, Okamura, Stone, Ward, Wood & Bass, 2014	2
N/A	*Ichthyobodo* Pinto, 1928	23
N/A	*Ichthyobodo necator* (Henneguy, 1884)	2, 22
N/A	*Leishmania* Ross, 1903	23
Oodiniaceae	*Amyloodinium ocellatum* (Brown, 1946)	2, 23
Orchitophryidae	*Orchitophrya stellarum* Cépède, 1907	2
Paradinophycidae	*Paradinium* Chatton, 1910	2
Parameciidae	*Paramecium caudatum* Ehrenberg, 1833	7
Paramoebidae	*Paramoeba invadens* Jones, 1985	2, 8
Perkinsidae	*Perkinsus marinus* (Mackin, Owen & Collier, 1950)	2, 22
Perkinsidae	*Perkinsus olseni* Lester & Davis, 1981	2, 22
Phytomyxea *incertae sedis*	*Phagomyxa algarum* Karling, 1944	2
Phytomyxea *incertae sedis*	*Phagomyxa odontellae* Schnepf, Kühn & Bulman, 2000	2
Piridae	*Caullerya mesnili* (Chatton, 1907)	22
Plasmodiidae	*Plasmodium* Marchiafava & Celli, 1885	23
Plasmodiidae	*Plasmodium berghei* Vincke & Lips, 1948	20
Plasmodiidae	*Plasmodium chabaudi* Landau, 1965	7
Plasmodiidae	*Plasmodium falciparum* (Welch, 1897)	20
Plasmodiidae	*Plasmodium relictum* (Grassi & Feletti, 1891)	2
Plasmodiophoridae	*Maullinia ectocarpii* Maier, Parodi, Westermeier & Müller, 2000	2
Plasmodiophoridae	*Plasmodiophora bicaudata* J. Feldmann, 1941	2
Pseudocolliniidae	*Fusiforma themisticola* Chantangsi, Lynn, Rückert, Prokopowicz, Panha & Leander, 2013	2
Sarcocystidae	*Sarcocystis neurona* Dubey, Davis, Speer, Bowman, de Lahunta, Granstrom, Topper, Hamir, Cummings & Suter, 1991	15
Sarcocystidae	*Toxoplasma gondii* (Nicolle & Manceaux, 1908)	2, 15, 23
Selenidiidae	*Selenidium pendula* Giard, 1884	2
Syndiniaceae	*Haematodinium* Chatton & Poisson, 1930	22
Syndiniaceae	*Haematodinium perezi* Chatton & Poisson, 1931	22

(continued)

Table A3 (continued)

Family	Species	Chapter
Tetrahymenidae	*Lambornella clarki* Corliss & Coats, 1976	2
Tetrahymenidae	*Tetrahymena corlissi* Thompson, 1955	2
Thecamoebidae	*Thecamoeba* Fromentel, 1874	2
Trichodinidae	*Trichodina* Ehrenberg, 1830	23
Trichophryidae	*Trichophrya* Claparède & Lachmann, 1859	23
Trypanosomatidae	*Trypanosoma murmanense* Nikitin, 1927	2
Trypanosomatidae	*Trypanosoma otospermophili* (Wellman & Wherry, 1910)	7
Trypanosomatidae	*Trypanosoma pleuronectidium* Robertson, 1906	2
Vahlkampfiidae	*Naegleria fowleri* Carter, 1970	2
Vannellidae	*Vannella* Bovee, 1965	2
Vermamoebidae	*Vermamoeba vermiformis* (Page, 1967)	2
Vexilliferidae	*Neoparamoeba perurans* Young, Crosbie, Adams, Nowak & Morrison, 2007	2, 23
Vexilliferidae	*Vexillifera* Schaeffer, 1926	2
Zoothamniidae	*Zoothamnium* Bory de St. Vincent, 1824	2

Table A4 Mentioned fungi listed alphabetically according to higher classification. Within each higher classification, families are arranged alphabetically and, within each family, species are listed in alphabetical order

Higher classification	Family	Species	Chapter
Chytridiomycetes	Batrachochytriaceae	*Batrachochytrium dendrobatidis* Longcore, Pessier & Nichols, 1999	4, 7, 13, 14, 18, 22
Chytridiomycetes	Batrachochytriaceae	*Batrachochytrium salamandrivorans* Martel, Blooi, Bossuyt & Pasmans, 2013	4, 18
Chytridiomycetes	Chytridiaceae	*Rhizophydium planktonicum* Canter, 1948	4
Chytridiomycetes	Chytridiaceae	*Rhizophydium sphaerocarpum* (Zopf, 1884)	4
Eurotiomycetes	Aspergillaceae	*Aspergillus* Micheli, 1729	4
Eurotiomycetes	Aspergillaceae	*Aspergillus sydowii* (Bainier & Sartory, 1913)	4, 22
Fungi	Saccharomycetaceae	*Candida albicans* (Robin, 1923)	4
Microsporidia	Amblyosporidae	*Amblyospora* Hazard & Oldacre, 1975	3
Microsporidia	Culicosporidae	*Edhazardia aedis* (Kudo, 1930)	3
Microsporidia	Encephalitozoonidae	*Endoreticulatus* Brooks, Becnel & Kennedy, 1988	3
Microsporidia	Enterocytozoonidae	*Desmozoon lepeophtherii* Freeman & Sommerville, 2009	23
Microsporidia	Enterocytozoonidae	*Ecytonucleospora hepatopenaei* (Tourtip, Wongtripop, Situnyalucksana, Stentiford, Bateman, Sriuratana, Chayaburakul & Withyachumnarkul, 2009)	3
Microsporidia	Enterocytozoonidae	*Enterocytozoon* Desportes, Lecharpentier, Galian, Bernard, Cochand-Priollet, Lavergne, Ravisse & Modigliani, 1985	3
Microsporidia	Glugeidae	*Glugea anomala* (Moniez, 1887)	3
Microsporidia	Glugeidae	*Pleistophora mulleri* (Pfeiffer, 1895)	4, 7
Microsporidia	Glugeidae	*Pseudoloma neurophilia* Matthews, Brown, Larison, Bishop-Stewart, Rogers & Kent, 2001	3
Microsporidia	Metchnikovellidae	*Amphiamblys* Caullery & Mesnil, 1914	3
Microsporidia	N/A	*Cucumispora dikerogammari* (Ovcharenko & Kurandina, 1987)	3

(continued)

Table A4 (continued)

Higher classification	Family	Species	Chapter
Microsporidia	N/A	*Cucumispora ornata* Bojko, Dunn, Stebbing, Ross, Kerr & Stentiford, 2015	3
Microsporidia	N/A	*Dictyocoela duebenum* Terry, Smith, Sharpe, Rigaud, Littlewood, Ironside, Rollinson, Bouchon, MacNeil, Dick & Dunn, 2004	3, 20
Microsporidia	N/A	*Fibrillanosema crangonycis* Galbreath, Smith, Terry, Becnel & Dunn, 2004	3
Microsporidia	N/A	*Microsporidium seriolae* Egusa, 1982	23
Microsporidia	N/A	*Mitosporidium* Haag, James, Pombert, Larsson, Schaer, Refardt & Ebert, 2014	3
Microsporidia	N/A	*Neoflabelliforma magnivora* (Larsson, Ebert, Mangin & Vávra, 1998)	3
Microsporidia	N/A	*Nucleophaga* Dangeard, 1895	3
Microsporidia	N/A	*Paramicrosporidium* Corsaro, Walochnik, Venditti, Steinmann, Müller & Michel, 2014	3
Microsporidia	N/A	*Vavraia culicis* (Weiser, 1947)	3
Microsporidia	Nosematidae	*Nosema apis* (Zander, 1909)	3
Microsporidia	Nosematidae	*Nosema bombycis* Nägeli, 1857	3
Microsporidia	Nosematidae	*Nosema ceranae* Fries, Feng, da Silva, Slemenda & Pieniazek, 1996	3
Microsporidia	Nosematidae	*Nosema granulosis* Terry, Smith, Bouchon, Rigaud, Duncanson, Sharpe & Dunn, 1999	3
Microsporidia	Nosematidae	*Paranosema locustae* (Canning, 1953)	3
Microsporidia	Nosematidae	*Vairimorpha* Pilley, 1976	3
Microsporidia	Pleistophoridae	*Ovipleistophora diplostomuri* Lovy & Friend, 2017	3
Microsporidia	Pleistophoridae	*Ovipleistophora ovariae* (Summerfelt, 1964)	3
Microsporidia	Pleistophoridae	*Trachipleistophora hominis* Hollister, Canning, Weidner, Field, Kench & Marriott, 1996	3
Microsporidia	Thelohaniidae	*Thelohania* Henneguy, 1892	3
Microsporidia	Thelohaniidae	*Thelohania contejeani* Henneguy, 1892	4, 8
Microsporidia	Unikaryonidae	*Unikaryon* Canning, Foon & Joe, 1974	3, 7
Oomycetes	Saprolegniaceae	*Saprolegnia* Nees, 1823	4
Oomycetes	Saprolegniaceae	*Saprolegnia ferax* (Gruithuisen, 1821)	8
Oomycetes	Saprolegniaceae	*Saprolegnia parasitica* (Coker, 1923)	14, 22
Oomycetes	Leptolegniaceae	*Aphanomyces astaci* Schikora, 1906	4, 8, 14, 18
Oomycetes	Leptolegniaceae	*Aphanomyces invadans* David & Kirk, 1997	4
Saccharomycetes	Metschnikowiaceae	*Metschnikowia bicuspidata* (Metschnikoff, 1884)	22
Sordariomycetes	Nectriaceae	*Fusarium* Link, 1809	4
Sordariomycetes	Nectriaceae	*Fusarium crassum* (Sandoval-Denis & Crous, 2019)	4
Sordariomycetes	Nectriaceae	*Fusarium falciforme* (Carrión, 1951)	4
Sordariomycetes	Nectriaceae	*Fusarium keratoplasticum* Short, O'Donnell, Thrane, Nielsen, Zhang, Juba & Geiser, 2013	4
Sordariomycetes	Nectriaceae	*Fusarium solani* (Martius, 1842) species complex	4

Table A5 Mentioned parasites listed alphabetically according to higher taxonomic classification. Within each higher classification, families are arranged alphabetically and, within each family, species are listed in alphabetical order

Higher classification	Family	Species	Chapter
Acanthocephala	Echinorhynchidae	*Echinorhynchus gadi* Zoega in Müller, 1776	15
Acanthocephala	Echinorhynchidae	*Echinorhynchus truttae* Schrank, 1788	18
Acanthocephala	Moniliformidae	*Moniliformis moniliformis* (Bremser, 1811)	5, 20
Acanthocephala	Neoechinorhynchidae	*Neoechinorhynchus emyditoides* Fisher, 1960	11
Acanthocephala	Paracanthocephalidae	*Acanthocephalus rhinensis* Amin, Thielen, Münderl, Taraschewski & Sures, 2008	5
Acanthocephala	Paracanthocephalidae	*Acanthocephalus* Koelreuter, 1771	5
Acanthocephala	Paracanthocephalidae	*Acanthocephalus tumescens* (von Linstow, 1896)	18
Acanthocephala	Polymorphidae	*Andracantha* Schmidt, 1975	15
Acanthocephala	Polymorphidae	*Bolbosoma* Porta, 1908	15, 19
Acanthocephala	Polymorphidae	*Bolbosoma capitatum* (von Linstow, 1880)	15
Acanthocephala	Polymorphidae	*Corynosoma* Lühe, 1904	15, 19
Acanthocephala	Polymorphidae	*Corynosoma australe* Johnston, 1937	15, 17
Acanthocephala	Polymorphidae	*Corynosoma cetaceum* Johnston & Best, 1942	15
Acanthocephala	Polymorphidae	*Corynosoma enhydri* Morozov, 1940	15
Acanthocephala	Polymorphidae	*Corynosoma hamanni* (Linstow, 1892)	15
Acanthocephala	Polymorphidae	*Corynosoma magdaleni* Montreuil, 1958	15
Acanthocephala	Polymorphidae	*Corynosoma semerme* (Forssell, 1904)	5, 15
Acanthocephala	Polymorphidae	*Corynosoma strumosum* (Rudolphi, 1802)	15
Acanthocephala	Polymorphidae	*Hexaglandula corynosoma* (Travassos, 1915)	9
Acanthocephala	Polymorphidae	*Polymorphus* Lühe, 1911	5
Acanthocephala	Polymorphidae	*Polymorphus minutus* (Zeder, 1800)	5, 7, 20, 22
Acanthocephala	Polymorphidae	*Profilicollis altmani* (Perry, 1942)	11, 15
Acanthocephala	Polymorphidae	*Profilicollis major* (Lundström, 1942)	15
Acanthocephala	Polymorphidae	*Profilicollis novaezelandensis* Brockerhoff & Smales, 2002	11
Acanthocephala	Polymorphidae	*Southwellina hispida* (Van Cleave, 1925)	9
Acanthocephala	Pomphorhynchidae	*Pomphorhynchus* Monticelli, 1905	5
Acanthocephala	Pomphorhynchidae	*Pomphorhynchus bosniacus* Kiskároly & Čanković, 1967	18
Acanthocephala	Pomphorhynchidae	*Pomphorhynchus laevis* (Zoega in Müller, 1776)	5, 7, 11, 18, 20, 22
Acanthocephala	Pomphorhynchidae	*Pomphorhynchus tereticollis* (Rudolphi, 1809)	5, 11
Acanthocephala	Quadrigyridae	*Acanthogyrus* Thapar, 1927	13
Acanthocephala	Tenuisentidae	*Paratenuisentis ambiguus* (van Cleave, 1921)	5
Amphipoda	Cyamidae	*Balaenocyamus balaenopterae* (K.H. Barnard, 1931)	15
Amphipoda	Cyamidae	*Cyamus boopis* Lütken, 1870	15
Amphipoda	Cyamidae	*Cyamus scammoni* Dall, 1872	6
Amphipoda	Cyamidae	*Isocyamus deltobranchium* Sedlak-Weinstein, 1992	15

(continued)

Table A5 (continued)

Higher classification	Family	Species	Chapter
Annelida	Erpobdellidae	*Croatobranchus mestrovi* Kerovec, Kucinic & Jalzic, 1999	12
Annelida	Piscicolidae	*Calliobdella nodulifera* (Malm, 1863)	2
Annelida	Piscicolidae	*Johanssonia arctica* (Johansson, 1898)	2
Annelida	Piscicolidae	*Piscicola salmositica* Meyer, 1946	2
Arachnida	Argasidae	*Ornithodoros capensis* Neumann, 1901	2
Arachnida	Demodecidae	*Demodex folliculorum* (Simon, 1842)	14
Arachnida	Eylaidae	*Eylais* Latreille, 1796	8
Arachnida	Halarachnidae	*Halarachne halichoeri* (Allman, 1847)	13, 15
Arachnida	Halarachnidae	*Orthohalarachne* Newell, 1947	15
Arachnida	Hydrachnidae	*Hydrachna* Müller, 1776	8
Arachnida	Ixodidae	*Ixodes ornithorhynchi* Lucas, 1846	13
Arachnida	Ixodidae	*Ixodes ricinus* (Linnaeus, 1758)	6
Arachnida	Ixodidae	*Ixodes uriae* White, 1852	6
Ascothoracida	Dendrogastridae	*Ulophysema oeresundense* Brattström, 1936	6
Aspidogastrea	Aspidogastridae	*Aspidogaster conchicola* von Baer, 1827	5
Aspidogastrea	Aspidogastridae	*Cotylaspis insignis* Leidy, 1857	5
Aspidogastrea	Aspidogastridae	*Lobatostoma manteri* Rohde, 1973	5
Aspidogastrea	Aspidogastridae	*Multicotyle purvisi* Dawes, 1941	5
Bivalvia	Margaritiferidae	*Pseudunio auricularius* (Spengler, 1793)	13
Branchiura	Argulidae	*Argulus* Müller, 1785	6
Branchiura	Argulidae	*Argulus coregoni* Thorell, 1865	6, 7, 10
Branchiura	Argulidae	*Argulus foliaceus* (Linnaeus, 1758)	6
Branchiura	Argulidae	*Argulus japonicus* Thiele, 1900	6, 10, 16
Branchiura	Argulidae	*Argulus mexicanus* Pineda, Páramo & Del Rio, 1995	6
Branchiura	Argulidae	*Argulus stizostethii* Kellicott, 1880	8
Branchiura	Argulidae	*Argulus yucatanus* Poly, 2005	6
Branchiura	Argulidae	*Chonopeltis* Thiele, 1900	6
Branchiura	Argulidae	*Chonopeltis minutus* Fryer, 1977	13
Branchiura	Argulidae	*Dipteropeltis* Calman, 1912	6
Branchiura	Argulidae	*Dolops* Audouin, 1837	6
Cestoda	Bothriocephalidae	*Bothriocephalus gowkongensis* Yeh, 1955	23
Cestoda	Bothriocephalidae	*Schyzocotyle acheilognathi* (Yamaguti, 1934)	13, 22
Cestoda	Caryophyllaeidae	*Wenyonia virilis* Woodland, 1923	11
Cestoda	Diphyllobothriidae	*Adenocephalus pacificus* Nybelin, 1931	15, 19
Cestoda	Diphyllobothriidae	*Baylisia* Markowski, 1952	15
Cestoda	Diphyllobothriidae	*Baylisiella* Markowski, 1952	15
Cestoda	Diphyllobothriidae	*Baylisiella tecta* (Linstow, 1892)	15
Cestoda	Diphyllobothriidae	*Dibothriocephalus* Lühe, 1899	12
Cestoda	Diphyllobothriidae	*Dibothriocephalus dalliae* (Rausch, 1956)	19
Cestoda	Diphyllobothriidae	*Dibothriocephalus dendriticus* (Nitzsch, 1824)	19
Cestoda	Diphyllobothriidae	*Dibothriocephalus latus* (Linnaeus, 1758)	5, 15, 19, 20
Cestoda	Diphyllobothriidae	*Dibothriocephalus nihonkaiensis* (Yamane, Kamo, Bylund & Wikgren, 1986)	19
Cestoda	Diphyllobothriidae	*Dibothriocephalus ursi* (Rausch, 1954)	19
Cestoda	Diphyllobothriidae	*Diphyllobothrium* Cobbold, 1858	23

(continued)

Table A5 (continued)

Higher classification	Family	Species	Chapter
Cestoda	Diphyllobothriidae	*Diphyllobothrium balaenopterae* (Lönnberg, 1892)	19
Cestoda	Diphyllobothriidae	*Diphyllobothrium cordatum* (Leuckart, 1863)	19
Cestoda	Diphyllobothriidae	*Diphyllobothrium lanceolatum* (Krabbe, 1865)	19
Cestoda	Diphyllobothriidae	*Diphyllobothrium stemmacephalum* Cobbold, 1858	15, 19
Cestoda	Diphyllobothriidae	*Flexobothrium* Yurakhno, 1989	15
Cestoda	Diphyllobothriidae	*Glandicephalus perfoliatus* (Railliet & Henry, 1912)	15
Cestoda	Diphyllobothriidae	*Ligula intestinalis* (Linnaeus, 1758)	11, 15, 20
Cestoda	Diphyllobothriidae	*Plicobothrium globicephalae* Rausch & Margolis, 1969	15
Cestoda	Diphyllobothriidae	*Pyramicocephalus phocarum* (Fabricius, 1780)	15
Cestoda	Diphyllobothriidae	*Schistocephalus* Creplin, 1829	7
Cestoda	Diphyllobothriidae	*Schistocephalus solidus* (Müller, 1776)	5, 7, 9, 11, 15, 16, 20, 22
Cestoda	Diphyllobothriidae	*Spirometra* Faust, Campbell & Kellogg, 1929	5
Cestoda	Diphyllobothriidae	*Tetragonoporus calyptocephalus* Skryabin, 1961	15
Cestoda	Eutetrarhynchidae	*Prochristianella clarkeae* Beveridge, 1990	13
Cestoda	Lacistorhynchidae	*Callitetrarhynchus gracilis* (Rudolphi, 1819)	9
Cestoda	Lacistorhynchidae	*Grillotia adenoplusius* (Pintner, 1903) *taxon inquirendum*	9
Cestoda	Lacistorhynchidae	*Grillotia carvajalregorum* Menoret & Ivanov, 2009	17
Cestoda	Lecanicephalidae	*Floriparicapitus* Cielocha, Jensen & Caira, 2014	13
Cestoda	Lytocestidae	*Khawia japonensis* (Yamaguti, 1934)	23
Cestoda	Onchobothriidae	*Acanthobothrium* Blanchard, 1848	13
Cestoda	Otobothriidae	*Fossobothrium perplexum* Beveridge & Campbell, 2005	13
Cestoda	Otobothriidae	*Pristiorhynchus palmi* Schaeffner & Beveridge, 2013	13
Cestoda	Otobothriidae	*Proemotobothrium linstowi* (Southwell, 1912)	13
Cestoda	Phyllobothriidae	*Clistobothrium* Dailey & Vogelbein, 1990	15
Cestoda	Phyllobothriidae	*Clistobothrium delphini* (Bosc, 1802)	15
Cestoda	Phyllobothriidae	*Clistobothrium grimaldii* (Moniez, 1899)	15
Cestoda	Proteocephalidae	*Proteocephalus* Weinland, 1858	12
Cestoda	Pterobothriidae	*Pterobothrium australiense* Campbell & Beveridge, 1996	13
Cestoda	Rhinoptericolidae	*Rhinoptericola megacantha* Carvajal & Campbell, 1975	9
Cestoda	Tentaculariidae	*Tentacularia coryphaenae* Bosc, 1802	9
Cestoda	Tetrabothriidae	*Anophryocephalus* Baylis, 1922	15
Cestoda	Tetrabothriidae	*Anophryocephalus anophrys* Baylis, 1922	15
Cestoda	Tetrabothriidae	*Priapocephalus* Nybelin, 1922	15
Cestoda	Tetrabothriidae	*Strobilocephalus* Baer, 1932	15

(continued)

Table A5 (continued)

Higher classification	Family	Species	Chapter
Cestoda	Tetrabothriidae	*Tetrabothrius* Rudolphi, 1819	15
Cestoda	Tetrabothriidae	*Trigonocotyle* Baer, 1932	15
Cestoda	Tetrabothriidae	*Trigonocotyle prudhoei* Markowski, 1955	15
Cestoda	Tetraphyllidea *incertae sedis*	*Anthobothrium/Phyllobothrium pristis* (see *Anthobothrium pristis* Woodland, 1934)	13
Cestoda	Triaenophoridae	*Eubothrium* Nybelin, 1922	23
Cestoda	Triaenophoridae	*Eubothrium salvelini* (Schrank, 1790)	12
Chromista	Bachelotiaceae	*Bachelotia antillarum* (Grunow, 1868)	2
Chromista	Ectocarpaceae	*Ectocarpus siliculosus* (Dillwyn, 1809)	2
Chromista	Laminariaceae	*Macrocystis pyrifera* (Linnaeus, 1771)	2
Chromista	Triceratiaceae	*Odontella* Agardh, 1832	2
Cirripedia	Coronulidae	*Xenobalanus globicipitis* Steenstrup, 1852	15
Cirripedia	Peltogastridae	*Briarosaccus callosus* Boschma, 1930	7
Cirripedia	Pollicipedidae	*Anelasma squalicola* (Lovén, 1844)	6
Cirripedia	Sacculinidae	*Loxothylacus panopaei* (Gissler, 1884)	7, 18, 22
Cirripedia	Sacculinidae	*Loxothylacus texanus* Boschma, 1933	10
Cirripedia	Sacculinidae	*Sacculina* Thompson, 1836	6, 7
Copepoda	Caligidae	*Caligus* Müller, 1785	7, 10, 13
Copepoda	Caligidae	*Caligus chiastos* Lin & Ho, 2003	23
Copepoda	Caligidae	*Caligus elongatus* von Nordmann, 1832	23
Copepoda	Caligidae	*Caligus epidemicus* Hewitt, 1971	23
Copepoda	Caligidae	*Caligus furcisetifer* Redkar, Rangnekar & Murti, 1949	13
Copepoda	Caligidae	*Caligus rogercresseyi* Boxshall & Bravo, 2000	10, 23
Copepoda	Caligidae	*Caligus spinosus* Yamaguti, 1939	23
Copepoda	Caligidae	*Lepeophtheirus salmonis* (Krøyer, 1837)	7, 9, 10, 11, 14, 22, 23
Copepoda	Caligidae	*Lepeophtheirus salmonis oncorhynchi* Skern-Mauritzen, Torrissen & Glover, 2014	23
Copepoda	Caligidae	*Lepeophtheirus salmonis salmonis* (Krøyer, 1837)	23
Copepoda	Chondracanthidae	*Chondracanthodes deflexus* Wilson, 1932	12
Copepoda	Cyclopoida *incertae sedis*	*Ophelicola kurambia* Conradi, Bandera, Marin & Martin, 2015	12
Copepoda	Lernaeidae	*Lernaea* Linnaeus, 1758	6, 23
Copepoda	Lernaeidae	*Lernaea cyprinacea* Linnaeus, 1758	12, 22, 23
Copepoda	Lernaeopodidae	*Salmincola* Wilson, 1915	12
Copepoda	Lernanthropidae	*Lernanthropus* de Blainville, 1822	23
Copepoda	Mytilicolidae	*Mytilicola intestinalis* Steuer, 1902	18
Copepoda	Mytilicolidae	*Mytilicola orientalis* Mori, 1935	18
Copepoda	Nicothoidae	*Choniomyzon inflatus* Wakabayashi, Otake, Tanaka & Nagasawa, 2013	6
Copepoda	Pandaridae	*Pandarus rhincodonicus* Norman, Newbound & Knott, 2000	13
Copepoda	Pandaridae	*Perissopus dentatus* Steenstrup & Lütken, 1861	13
Copepoda	Pennellidae	*Lernaeocera branchialis* (Linnaeus, 1767)	10
Copepoda	Pennellidae	*Peniculus minuticaudae* Shiino, 1956	6

(continued)

Table A5 (continued)

Higher classification	Family	Species	Chapter
Copepoda	Pennellidae	*Pennella balaenoptera* Koren & Danielssen, 1877	15
Copepoda	Serpulidicolidae	*Abyssotaurus vermiambatus* Brenke, Fanenbruck & George, 2018	12
Decapoda	Pinnotheridae	*Nepinnotheres novaezelandiae* (Filhol, 1885)	10
Insecta	Echinophthiriidae	*Echinophthirius horridus* (von Olfers, 1816)	6, 15
Insecta	Echinophthiriidae	*Lepidophthirus macrorhini* Enderlein, 1904	6
Insecta	Figitidae	*Leptopilina boulardi* (Barbotin, Carton & Kelner-Pillault, 1979)	8
Insecta	Ichneumonidae	*Venturia canescens* (Gravenhorst, 1829)	7
Insecta	Menoponidae	*Colpocephalum californici* Price & Beer, 1963	13
Insecta	Trichodectidae	*Lutridia exilis* Giebel, 1861	13
Isopoda	Aegidae	*Alitropus typus* (as "*Rocinella typicus*") Milne-Edwards, 1840	23
Isopoda	Cryptoniscidae	*Liriopsis pygmaea* (Rathke, 1843)	7
Isopoda	Cymothoidae	*Ceratothoa oestroides* (Risso, 1827)	20
Isopoda	Cymothoidae	*Ceratothoa parallela* (Otto, 1828)	23
Isopoda	Cymothoidae	*Cinusa tetrodontis* Schioedte & Meinert, 1884	20
Isopoda	Cymothoidae	*Cymothoa excisa* Perty, 1833	10
Isopoda	Cymothoidae	*Riggia paranensis* Szidat, 1948	7
Isopoda	Gnathiidae	*Caecognathia* Dollfus, 1901	10
Isopoda	Gnathiidae	*Gnathia africana* (Barnard, 1914)	6
Isopoda	Gnathiidae	*Gnathia marleyi* Farquharson, Smit & Sikkel, 2012	10
Isopoda	Gnathiidae	*Paragnathia formica* (Hesse, 1864)	6
Isopoda	Hemioniscidae	*Hemioniscus balani* Buchholz, 1866	7
Monogenea	Acanthocotylidae	*Acanthocotyle lobianchi* Monticelli, 1888	10
Monogenea	Ancyrocephalidae	*Anacanthorus penilabiatus* Boeger, Husak & Martins, 1995	23
Monogenea	Ancyrocephalidae	*Cichlidogyrus halli* (Price & Kirk, 1967)	23
Monogenea	Ancyrocephalidae	*Cichlidogyrus mbirizei* Muterezi Bukinga, Vanhove, Van Steenberge & Pariselle, 2012	23
Monogenea	Ancyrocephalidae	*Cichlidogyrus thurstonae* Ergens, 1981	23
Monogenea	Ancyrocephalidae	*Cichlidogyrus tilapiae* Paperna, 1960	23
Monogenea	Ancyrocephalidae	*Enterogyrus* Paperna, 1963	5
Monogenea	Ancyrocephalidae	*Scutogyrus longicornis* (Paperna & Thurston, 1969)	23
Monogenea	Capsalidae	*Benedenia* Diesing, 1858	23
Monogenea	Capsalidae	*Benedenia epinepheli* (Yamaguti, 1937)	23
Monogenea	Capsalidae	*Benedenia seriolae* (Yamaguti, 1934)	23
Monogenea	Capsalidae	*Entobdella* Blainville in Lamarck, 1818	5
Monogenea	Capsalidae	*Entobdella soleae* (Van Beneden & Hesse, 1863)	10
Monogenea	Capsalidae	*Neobenedenia* Yamaguti, 1963	10
Monogenea	Capsalidae	*Neobenedenia girellae* (Hargis, 1955)	23
Monogenea	Capsalidae	*Neobenedenia melleni* (MacCallum, 1927)	23

(continued)

Table A5 (continued)

Higher classification	Family	Species	Chapter
Monogenea	Dactylogyridae	*Dactylogyrus* Diesing, 1850	5, 13, 14, 23
Monogenea	Dactylogyridae	*Dactylogyrus aristichthys* Long & Yu, 1958	23
Monogenea	Dactylogyridae	*Dactylogyrus hypophthalmichthys* Akhmerov, 1952	23
Monogenea	Dactylogyridae	*Dactylogyrus kalyanensis* Musselius & Gusev, in Gusev, 1976	23
Monogenea	Dactylogyridae	*Dactylogyrus labei* Musselius & Gusev, in Gusev, 1976	23
Monogenea	Dactylogyridae	*Dactylogyrus lamellatus* Akhmerov, 1952	23
Monogenea	Diclidophoridae	*Diclidophora merlangi* (Kuhn, 1829)	17
Monogenea	Diclidophoridae	*Neoheterobothrium affine* (Linton, 1898)	5
Monogenea	Diplectanidae	*Lamellodiscus* Johnston & Tiegs, 1922	9, 11
Monogenea	Diplozoidae	*Diplozoon* von Nordmann, 1832	5
Monogenea	Diplozoidae	*Paradiplozoon* Akhmerov, 1974	13
Monogenea	Diplozoidae	*Paradiplozoon ichthyoxanthon* Avenant-Oldewage, le Roux, Mashego & van Vuuren, 2013	20
Monogenea	Discocotylidae	*Discocotyle sagittata* (Leuckart, 1842)	10
Monogenea	Enoplocotylidae	*Enoplocotyle kidokoi* Kearn, 1993	10
Monogenea	Gyrodactylidae	*Gyrodactylus* von Nordmann, 1832	9, 13, 22, 23
Monogenea	Gyrodactylidae	*Gyrodactylus cichlidarum* Paperna, 1968	23
Monogenea	Gyrodactylidae	*Gyrodactylus conei* Leis, Bailey, Katona, Standish, Dziki, McCann, Perkins, Eckert & Baumgartner, 2023	13
Monogenea	Gyrodactylidae	*Gyrodactylus salaris* Malmberg, 1957	5, 14, 23
Monogenea	Gyrodactylidae	*Gyrodactylus salinae* Paladini, Huyse & Shinn, 2011	12
Monogenea	Gyrodactylidae	*Gyrodactylus turnbulli* Harris, 1986	23
Monogenea	Gyrodactylidae	*Ieredactylus rivuli* Schelkle, Paladini, Shinn, King, Johnson, van Oosterhout, Mohammed & Cable, 2011	12
Monogenea	Gyrodactylidae	*Macrogyrodactylus* Malmberg, 1957	5
Monogenea	Heteraxinidae	*Heteraxine heterocerca* (Goto, 1894)	23
Monogenea	Heteraxinidae	*Zeuxapta seriolae* (Meserve, 1938)	23
Monogenea	Hexabothriidae	*Epicotyle torpedinis* (Price, 1942)	10
Monogenea	Hexabothriidae	*Hexabothrium appendiculatum* (Kuhn, 1829)	10
Monogenea	Mazocraeidae	*Mazocraes alosae* Hermann, 1782	20
Monogenea	Microbothriidae	*Dermopristis pterophila* Ingelbrecht, Morgan & Martin, 2022	13
Monogenea	Microbothriidae	*Leptocotyle minor* (Monticelli, 1888)	10
Monogenea	Microcotylidae	*Atrispinum salpae* (Parona & Perugia, 1890)	10
Monogenea	Microcotylidae	*Sparicotyle chrysophrii* (Van Beneden & Hesse, 1863)	23
Monogenea	Monocotylidae	*Calicotyle* Diesing, 1850	13
Monogenea	Monocotylidae	*Heterocotyle* Scott, 1904	13
Monogenea	Monocotylidae	*Merizocotyle* Cerfontaine, 1894	13
Monogenea	Monocotylidae	*Neoheterocotyle* Hargis, 1955	13
Monogenea	Polystomatidae	*Metapolystoma ohlerianum* Landman, Verneau, Vences & du Preez, 2023	13

(continued)

Table A5 (continued)

Higher classification	Family	Species	Chapter
Monogenea	Polystomatidae	*Nanopolystoma brayi* du Preez, Wilkinson & Huyse, 2008	13
Monogenea	Polystomatidae	*Nanopolystoma lynchi* du Preez, Wilkinson & Huyse, 2008	13
Monogenea	Polystomatidae	*Oculotrema hippopotami* Stunkard, 1924	5
Monogenea	Polystomatidae	*Polystomum integerrimum* (Frölich, 1791)	5
Monogenea	Polystomatidae	*Polystoma* Zeder, 1800	5
Monogenea	Polystomatidae	*Pseudodiplorchis americanus* (Rodgers & Kuntz, 1940)	12
Monogenea	Pseudodactylogyridae	*Pseudodactylogyrus anguillae* (Yin & Sproston, 1948)	10
Monogenea	Pseudodactylogyridae	*Pseudodactylogyrus bini* (Kikuchi, 1929)	7, 10
Monogenea	Urogyridae	*Urogyrus* Bilong Bilong, Birgi & Euzet, 1994	5
Myxozoa	Ceratomyxidae	*Ceratomyxa* Thélohan, 1892	3
Myxozoa	Ceratomyxidae	*Ceratomyxa puntazzi* Alama-Bermejo, Raga & Holzer, 2011	14
Myxozoa	Ceratomyxidae	*Ceratonova shasta* (Noble, 1950)	3
Myxozoa	Chloromyxidae	*Chloromyxum* Mingazzini, 1890	3
Myxozoa	Enteromyxidae	*Enteromyxum leei* (Diamant, Lom & Dyková, 1994)	3
Myxozoa	Kudoidae	*Kudoa* Meglitsch, 1947	3, 19
Myxozoa	Kudoidae	*Kudoa hexapunctata* Yokoyama, Suzuki & Shirakashi, 2014	19
Myxozoa	Kudoidae	*Kudoa islandica* Kristmundsson & Freeman, 2014	23
Myxozoa	Kudoidae	*Kudoa iwatai* Egusa & Shiomitsu, 1983	3
Myxozoa	Kudoidae	*Kudoa neurophila* (Grossel, Dyková, Handlinger & Munday, 2003)	23
Myxozoa	Kudoidae	*Kudoa septempunctata* Matsukane, Sato, Tanaka, Kamata & Sugita-Konishi, 2010	19, 23
Myxozoa	Kudoidae	*Kudoa thyrsites* (Gilchrist, 1923)	3, 23
Myxozoa	Myxidiidae	*Cystodiscus anoxis* Hartigan, Fiala, Dyková, Rose, Phalen & Šlapeta, 2012	13
Myxozoa	Myxidiidae	*Myxidium coryphaenoideum* Noble, 1966	3
Myxozoa	Myxidiidae	*Myxidium giardi* Cépède, 1906	7
Myxozoa	Myxidiidae	*Myxidium lieberkuehni* Bütschli, 1882	3
Myxozoa	Myxidiidae	*Myxidium rhodei* Léger, 1905	3
Myxozoa	Myxidiidae	*Zschokkella nova* Klokacewa, 1914	3
Myxozoa	Myxobolidae	*Henneguya psorospermica* Thélohan, 1892	3
Myxozoa	Myxobolidae	*Henneguya salminicola* Ward, 1919	3
Myxozoa	Myxobolidae	*Henneguya* Thélohan, 1892	3, 19
Myxozoa	Myxobolidae	*Myxobolus* Bütschli, 1882	3, 13, 19
Myxozoa	Myxobolidae	*Myxobolus acanthogobii* Hoshina, 1952	23
Myxozoa	Myxobolidae	*Myxobolus cerebralis* Hofer, 1903	3, 12, 18, 22
Myxozoa	Myxobolidae	*Myxobolus intimus* Zaika, 1965	3
Myxozoa	Myxobolidae	*Myxobolus iquitoensis* Mathews, Mertins, Milanin, Espinoza, Flores-Gonzales, Audebert & Morandini, 2020	13
Myxozoa	Myxobolidae	*Thelohanellus kitauei* Egusa & Nakajima, 1981	3

(continued)

Table A5 (continued)

Higher classification	Family	Species	Chapter
Myxozoa	Parvicapsulidae	*Parvicapsula minibicornis* Kent, Whitaker & Dawe, 1997	9
Myxozoa	Saccosporidae	*Buddenbrockia* Schröder, 1910	3
Myxozoa	Saccosporidae	*Tetracapsuloides bryosalmonae* Canning, Tops, Curry, Wood & Okamura, 2002	3, 14, 22
Myxozoa	Sinuolineidae	*Sinuolinea lophii* (Freeman, Yokoyama & Ogawa, 2008)	3
Myxozoa	Sphaerosporidae	*Sphaerospora molnari* Lom, Dyková, Pavlásková & Grupcheva, 1983	3
Myxozoa	Trilosporidae	*Unicapsula* Davis, 1924	19, 23
Nematoda	Acuariidae	*Acuaria* Bremser, 1811	12
Nematoda	Acuariidae	*Skrjabinoclava morrisoni* Wong & Anderson, 1987	5
Nematoda	Ancylostomatidae	*Uncinaria* Frölich, 1789	5
Nematoda	Ancylostomatidae	*Uncinaria hamiltoni* Baylis, 1933	15
Nematoda	Ancylostomatidae	*Uncinaria lucasi* Stiles, 1901	9, 15
Nematoda	Ancylostomatidae	*Uncinaria lyonsi* Kuzmina & Kuzmin, 2015	15
Nematoda	Ancylostomatidae	*Uncinaria sanguinis* Marcus, Higgins, Slapeta & Gray, 2014	15
Nematoda	Anguillicolidae	*Anguillicola crassus* Kuwahara, Niimi & Itagaki, 1974	5, 18, 20
Nematoda	Anisakidae	*Anisakis* Dujardin, 1845	5, 15, 17, 19
Nematoda	Anisakidae	*Anisakis berlandi* Mattiucci, Cipriani, Webb, Paoletti, Marcer, Bellisario, Gibson & Nascetti, 2014	9
Nematoda	Anisakidae	*Anisakis pegreffii* Campana-Rouget & Biocca, 1955	9, 11
Nematoda	Anisakidae	*Anisakis simplex* (Rudolphi, 1809)	11, 15, 17
Nematoda	Anisakidae	*Anisakis simplex* (Rudolphi, 1809) complex	9, 15, 23
Nematoda	Anisakidae	*Contracaecum* Railliet & Henry, 1912	13, 15, 17, 19
Nematoda	Anisakidae	*Contracaecum ogmorhini* Johnston & Mawson, 1941	15
Nematoda	Anisakidae	*Contracaecum osculatum* (Rudolphi, 1802) complex	9, 15, 23
Nematoda	Anisakidae	*Contracaecum rudolphii* Hartwich, 1964	9
Nematoda	Anisakidae	*Phocanema* Myers, 1959	15
Nematoda	Anisakidae	*Phocanema bulbosum* (Cobb, 1889)	9
Nematoda	Anisakidae	*Phocanema cattani* (George-Nascimento & Urrutia, 2000)	15
Nematoda	Anisakidae	*Phocanema decipiens* (Krabbe, 1878) complex	5, 9, 15, 22, 23
Nematoda	Anisakidae	*Phocascaris* Höst, 1932	15
Nematoda	Anisakidae	*Pseudoterranova* Mozgovoi, 1951	13, 15, 17, 19
Nematoda	Anisakidae	*Skrjabinisakis* Mozgovoi, 1951	15
Nematoda	Anisakidae	*Skrjabinisakis paggiae* (Mattiucci, Nascetti, Dailey, Webb, Barros, Cianchi & Bullini, 2005)	12
Nematoda	Anisakidae	*Skrjabinisakis physeteris* (Baylis, 1923)	11
Nematoda	Anisakidae	*Sulcascaris sulcata* (Rudolphi, 1819)	5, 9
Nematoda	Ascarididae	*Ascaris lumbricoides* Linnaeus, 1758	5

(continued)

Table A5 (continued)

Higher classification	Family	Species	Chapter
Nematoda	Camallanidae	*Camallanus lacustris* (Zoega in Müller, 1776)	20
Nematoda	Camallanidae	*Procamallanus neocaballeroi* (Caballero-Deloya, 1977)	11
Nematoda	Capillariidae	*Capillaria* Zeder, 1800	19, 23
Nematoda	Capillariidae	*Paracapillaria philippinensis* (Chitwood, Valesquez & Salazar, 1968)	19
Nematoda	Cystidicolidae	*Cystidicola stigmatura* (Leidy, 1886)	7
Nematoda	Dioctophymatidae	*Eustrongylides* Jägerskiöld, 1909	19, 22
Nematoda	Dracunculidae	*Dracunculus medinensis* (Linnaeus, 1758)	5, 13
Nematoda	Echinomermellidae	*Echinomermella matsi* Jones & Hagen, 1987	8
Nematoda	Gnathostomatidae	*Echinocephalus* Molin, 1858	19
Nematoda	Gnathostomatidae	*Echinocephalus sinensis* Ko, 1975	19
Nematoda	Gnathostomatidae	*Gnathostoma* Owen, 1836	19
Nematoda	Heterakidae	*Heterakis gallinarum* (Schrank, 1788)	8
Nematoda	Onchocercidae	*Acanthocheilonema spirocauda* (Leidy, 1858)	6, 15
Nematoda	Panagrolaimidae	*Panagrolaimus davidi* Timm, 1971	12
Nematoda	Parafilaridae	*Parafilaroides* Dougherty, 1946	15
Nematoda	Philometridae	*Philometra* Costa, 1845	9
Nematoda	Pseudaliidae	*Halocercus* Baylis & Daubney, 1925	15
Nematoda	Pseudaliidae	*Halocercus delphini* Baylis & Daubney, 1925	15
Nematoda	Pseudaliidae	*Stenuroides herpestis* Gerichter, 1951	15
Nematoda	Pseudaliidae	*Stenurus* Dujardin, 1845	5
Nematoda	Pseudaliidae	*Stenurus minor* (Kuhn, 1829)	15
Nematoda	Raphidascarididae	*Hysterothylacium* Ward & Magath, 1917	17, 19
Nematoda	Rhabdiasidae	*Rhabdias hylae* Johnston & Mawson, 1942	18
Nematoda	Rhabdiasidae	*Rhabdias pseudosphaerocephala* Kuzmin, Tkach & Brooks, 2007	18
Nematoda	Rhabdochonidae	*Rhabdochona* Railliet, 1916	13
Nematoda	Rhabdochonidae	*Rhabdochona lichtenfelsi* Sánchez-Álvarez, García-Prieto & Pérez-Ponce de León, 1998	11
Nematoda	Rhabdochonidae	*Rhabdochona longleyi* Moravec & Huffman, 1988	12
Nematoda	Tetrameridae	*Placentonema gigantissima* (Gubanov, 1951)	5, 15
Nematoda	Trichinellidae	*Trichinella* Railliet, 1895	15
Nematoda	Trichinellidae	*Trichinella nativa* Britov & Boev, 1972	15
Nematoda	Trichinellidae	*Trichinella spiralis* (Owen, 1835)	15
Nematoda	Trichinellidae	*Trichinella zimbabwensis* Pozio, Foggin, Marucci, la Rosa, Sacchi, Corona, Rossi & Mukaratirwa, 2002	5
Nematoda	Trichostrongylidae	*Graphidium strigosum* (Dujardin, 1846)	22
Nematoda	Trichostrongylidae	*Trichostrongylus retortaeformis* (Zeder, 1800)	22
Nematoda	Trichostrongylidae	*Trichostrongylus tenuis* (Mehlis in Creplin, 1846)	7
Nematomorpha	Chordodidae	*Chordodes formosanus* Chiu, Huang, Wu & Shiao, 2011	13

(continued)

Table A5 (continued)

Higher classification	Family	Species	Chapter
Nematomorpha	Chordodidae	*Paragordius tricuspidatus* (Dufour, 1828)	7
Nematomorpha	Gordiidae	*Acutogordius taiwanensis* Heinze, 1952	13
Nematomorpha	Gordiidae	*Gordius chiashanus* Chiu, Huang, Wu, Lin, Shen & Shiao, 2020	13
Pentastomida	Linguatulidae	*Linguatula* Frölich, 1789	6
Pentastomida	Reighardiidae	*Reighardia sternae* (Diesing, 1864)	6
Pentastomida	Sebekidae	*Sambonia* Noc & Giglioli, 1922	6
Pentastomida	Sebekidae	*Sebekia* Sambon, 1922	6
Pentastomida	Subtriquetridae	*Subtriquetra* Sambon, 1922	6
Petromyzonti	Petromyzontidae	*Petromyzon marinus* Linnaeus, 1758	10
Tantulocarida	Basipodellidae	*Serratotantulus chertoprudae* Savchenko & Kolbasov, 2009	6
Trematoda	Aephnidiogenidae	*Stegodexamene anguillae* MacFarlane, 1951	11
Trematoda	Allocreadiidae	*Bunodera luciopercae* (Müller, 1776)	7
Trematoda	Allocreadiidae	*Crepidostomum* Braun, 1900	12
Trematoda	Allocreadiidae	*Crepidostomum farionis* (Müller, 1780)	12
Trematoda	Aporocotylidae	*Cardicola* Short, 1953	23
Trematoda	Aporocotylidae	*Cardicola dhangali* Hutson, Vaughan & Blair, 2019	15
Trematoda	Aporocotylidae	*Cruoricola lates* Herbert, Shaharom-Harrison & Overstreet, 1994	23
Trematoda	Aporocotylidae	*Paradeontacylix buri* Ogawa, Akiyama & Grabner, 2015	23
Trematoda	Aporocotylidae	*Parasanguinicola vastispina* Herbert & Shaharom, 1995	23
Trematoda	Aporocotylidae	*Sanguinicola armatus* Plehn, 1905	23
Trematoda	Aporocotylidae	*Sanguinicola inermis* Plehn, 1905	23
Trematoda	Aporocotylidae	*Sanguinicola lungensis* Tang & Ling, 1975	23
Trematoda	Brachycladiidae	*Brachycladium* Looss, 1899	15
Trematoda	Brachycladiidae	*Campula* Cobbold, 1858	15
Trematoda	Brachycladiidae	*Cetitrema* Skrjabin, 1970	15
Trematoda	Brachycladiidae	*Hunterotrema* McIntosh, 1960	15
Trematoda	Brachycladiidae	*Nasitrema* Ozaki, 1935	15
Trematoda	Brachycladiidae	*Odhneriella* Skrjabin, 1915	15
Trematoda	Brachycladiidae	*Orthosplanchnus* Odhner, 1905	15
Trematoda	Brachycladiidae	*Orthosplanchnus arcticus* Kurochkin & Nikol'skii, 1972	15
Trematoda	Brachycladiidae	*Orthosplanchnus fraterculus* Odhner, 1905	15
Trematoda	Brachycladiidae	*Oschmarinella* Skrjabin, 1947	15
Trematoda	Brachycladiidae	*Oschmarinella rochebruni* (Poirier, 1886)	15
Trematoda	Brachycladiidae	*Synthesium* Stunkard & Alvey, 1930	15
Trematoda	Brachycladiidae	*Synthesium pontoporiae* (Raga, Aznar, Balbuena & Dailey, 1994)	15
Trematoda	Brachycladiidae	*Zalophotrema* Stunkard & Alvey, 1929	15
Trematoda	Brauninidae	*Braunina cordiformis* Wolf, 1903	15
Trematoda	Bucephalidae	*Bucephalus minimus* (Stossich, 1887)	11
Trematoda	Bucephalidae	*Prosorhynchus squamatus* Odhner, 1905	8
Trematoda	Cladorchiidae	*Chiorchis* Fischoeder, 1901	15

(continued)

Table A5 (continued)

Higher classification	Family	Species	Chapter
Trematoda	Clinostomidae	*Clinostomum complanatum* (Rudolphi, 1814)	19
Trematoda	Cryptogonimidae	*Retrovarium* Miller & Cribb, 2007	9
Trematoda	Cyathocotylidae	*Prohemistomum vivax* (Sonsino, 1892)	23
Trematoda	Derogenidae	*Deropegus aspina* (Ingles, 1936)	9, 11
Trematoda	Dicrocoeliidae	*Dicrocoelium dendriticum* (Rudolphi, 1819)	7
Trematoda	Diplostomidae	*Diplostomum* von Nordmann, 1832	5, 7, 12
Trematoda	Diplostomidae	*Diplostomum pseudospathaceum* Niewiadomska, 1984	11
Trematoda	Diplostomidae	*Diplostomum spathaceum* (Rudolphi, 1819)	22
Trematoda	Diplostomidae	*Posthodiplostomum* Dubois, 1936	11
Trematoda	Diplostomidae	*Posthodiplostomum minimum* (MacCallum, 1921)	3
Trematoda	Diplostomidae	*Posthodiplostomum ptychocheilus* (Faust, 1917)	7, 22
Trematoda	Diplostomidae	*Tylodelphys scheuringi* (Hughes, 1929)	22
Trematoda	Echinochasmidae	*Echinochasmus* Dietz, 1909	23
Trematoda	Echinochasmidae	*Stephanoprora* Odhner, 1902	15
Trematoda	Echinostomatidae	*Echinoparyphium aconiatum* Dietz, 1909	22
Trematoda	Echinostomatidae	*Echinostoma* Rudolphi, 1809	15, 22
Trematoda	Echinostomatidae	*Echinostoma caproni* Richard, 1964	10
Trematoda	Echinostomatidae	*Echinostoma revolutum* (Fröhlich, 1802)	10
Trematoda	Echinostomatidae	*Echinostoma trivolvis* (Cort, 1914)	7
Trematoda	Echinostomatidae	*Hypoderaeum conoideum* (Bloch, 1782)	10
Trematoda	Echinostomatidae	*Pseudechinoparyphium echinatum* (von Siebold, 1837) *taxon inq.*	10
Trematoda	Fasciolidae	*Fasciola* Linnaeus, 1758	5, 19
Trematoda	Fasciolidae	*Fasciola gigantica* Cobbold, 1855	19, 20
Trematoda	Fasciolidae	*Fasciola hepatica* Linnaeus, 1758	10, 19, 20, 23
Trematoda	Fasciolidae	*Fasciolopsis* Looss, 1899	5
Trematoda	Fasciolidae	*Fasciolopsis buskii* (Lankester, 1857)	19
Trematoda	Fellodistomidae	*Proctoeces humboldti* George-Nascimento & Quiroga, 1983	7
Trematoda	Fellodistomidae	*Steringophorus thulini* Bray & Gibson, 1980	12
Trematoda	Gonocercidae	*Gonocerca phycidis* Manter, 1925	12
Trematoda	Gorgoderidae	*Phyllodistomum umblae* (Fabricius, 1780)	12
Trematoda	Gymnophallidae	*Gymnophallus choledochus* Odhner, 1900	11
Trematoda	Hemiuridae	*Lecithochirium grandiporum* (Rudolphi, 1819)	9
Trematoda	Heterophyidae	*Ascocotyle* Looss, 1899	15
Trematoda	Heterophyidae	*Ascocotyle longa* Ransom, 1920	15
Trematoda	Heterophyidae	*Ascocotyle patagoniensis* Hernández-Orts, Montero, Crespo, García, Raga & Aznar, 2012	15
Trematoda	Heterophyidae	*Cercaria batillariae* Shimura & Ito, 1980	18
Trematoda	Heterophyidae	*Euhaplorchis californiensis* Martin, 1950	22
Trematoda	Heterophyidae	*Galactosomum* Looss, 1899	15
Trematoda	Heterophyidae	*Haplorchis* Looss, 1899	23
Trematoda	Heterophyidae	*Haplorchis yokogawai* (Katsuta, 1932)	23

(continued)

Table A5 (continued)

Higher classification	Family	Species	Chapter
Trematoda	Heterophyidae	*Heterophyes* Cobbold, 1866	15
Trematoda	Heterophyidae	*Heterophyes heterophyes* (von Siebold, 1852)	19
Trematoda	Heterophyidae	*Heterophyopsis* Tubangui & Africa, 1938	15
Trematoda	Heterophyidae	*Metagonimus yokogawai* (Katsurada, 1912)	19
Trematoda	Heterophyidae	*Phocitrema* Goto & Ozaki, 1930	15
Trematoda	Heterophyidae	*Pholeter gastrophilus* (Kossack, 1910)	15
Trematoda	Heterophyidae	*Pholeter* Odhner, 1914	15
Trematoda	Heterophyidae	*Procerovum calderoni* (Africa & Garcia, 1935)	23
Trematoda	Himasthlidae	*Acanthoparyphium spinulosum* Johnston, 1917	22
Trematoda	Himasthlidae	*Curtuteria australis* Allison, 1979	7
Trematoda	Himasthlidae	*Himasthla elongata* (Mehlis, 1831)	5, 7, 18, 20, 22
Trematoda	Himasthlidae	*Himasthla leptosoma* (Creplin, 1829)	11
Trematoda	Himasthlidae	*Himasthla littorinae* Stunkard, 1966	12
Trematoda	Labicolidae	*Labicola elongata* Blair, 1979	15
Trematoda	Lepidapedidae	*Lepidapedon zubchenkoi* Campbell & Bray, 1993	12
Trematoda	Leucochloridiidae	*Leucochloridium paradoxum* (Carus, 1835)	5
Trematoda	Microphallidae	*Atriophallophorus winterbourni* Blasco-Costa, Seppälä, Feijen, Zajac, Klappert & Jokela, 2019	11
Trematoda	Microphallidae	*Maritrema* Nicoll, 1907	15
Trematoda	Microphallidae	*Maritrema novaezealandense* Martorelli, Fredensborg, Mouritsen & Poulin, 2004	8, 9, 11, 12, 22
Trematoda	Microphallidae	*Maritrema subdolum* Jägerskiöld, 1909	7, 8, 22
Trematoda	Microphallidae	*Microphallus* Ward, 1901	15, 18
Trematoda	Microphallidae	*Microphallus claviformis* (Brandes, 1889)	8
Trematoda	Microphallidae	*Microphallus papillorobustus* (Rankin, 1940)	7
Trematoda	Microphallidae	*Microphallus pygmaeus* (Levinsen, 1881)	11
Trematoda	Microphallidae	*Plenosoma* Ching, 1960	15
Trematoda	Notocotylidae	*Ogmogaster antarctica* Johnston, 1931	15
Trematoda	Notocotylidae	*Ogmogaster heptalineata* Carvajal, Duran & George-Nascimento, 1983	15
Trematoda	Notocotylidae	*Paramonostomum antarcticum* Graefe, 1968	12
Trematoda	Notocotylidae	*Tristriara anatis* Belopolskaja in Skrjabin, 1953	11
Trematoda	Nudacotylidae	*Nudacotyle* Barker, 1916	15
Trematoda	Opecoelidae	*Coitocaecum parvum* Crowcroft, 1945	11
Trematoda	Opecoelidae	*Plagioporus shawi* (McIntosh, 1939)	9, 11, 17
Trematoda	Opisthorchiidae	*Amphimerus* Barker, 1911	15
Trematoda	Opisthorchiidae	*Apophallus* Lühe, 1909	15
Trematoda	Opisthorchiidae	*Clonorchis sinensis* (Cobbold, 1875)	19, 23
Trematoda	Opisthorchiidae	*Cryptocotyle* Lühe, 1899	9, 15
Trematoda	Opisthorchiidae	*Cryptocotyle lingua* (Creplin, 1825)	9
Trematoda	Opisthorchiidae	*Delphinicola* Yamaguti, 1933	15
Trematoda	Opisthorchiidae	*Metorchis* Looss, 1899	15
Trematoda	Opisthorchiidae	*Metorchis bilis* (Braun, 1790)	15

(continued)

Table A5 (continued)

Higher classification	Family	Species	Chapter
Trematoda	Opisthorchiidae	*Metorchis conjunctus* (Cobbold, 1860)	19
Trematoda	Opisthorchiidae	*Opisthorchis* Blanchard, 1895	15, 19, 23
Trematoda	Opisthorchiidae	*Opisthorchis felineus* (Rivolta, 1884)	19
Trematoda	Opisthorchiidae	*Opisthorchis viverrini* (Poirier, 1886)	14, 19
Trematoda	Opisthorchiidae	*Pricetrema* Ciurea, 1933	15
Trematoda	Opisthorchiidae	*Pseudamphistomum* Lühe, 1908	15
Trematoda	Opisthorchiidae	*Pseudamphistomum truncatum* (Rudolphi, 1819)	15
Trematoda	Opisthotrematidae	*Folitrema* Blair, 1981	15
Trematoda	Opisthotrematidae	*Lankatrema* Crusz & Fernand, 1954	15
Trematoda	Opisthotrematidae	*Lankatrematoides* Blair, 1981	15
Trematoda	Opisthotrematidae	*Lankatrematoides gardneri* Blair, 1981	15
Trematoda	Opisthotrematidae	*Moniligerum* Dailey, Vogelbein & Forrester, 1988	15
Trematoda	Opisthotrematidae	*Opisthotrema* Fischer, 1884	15
Trematoda	Opisthotrematidae	*Opisthotrema australe* Blair, 1981	15
Trematoda	Opisthotrematidae	*Pulmonicola pulmonalis* (von Linstow, 1904)	15
Trematoda	Opisthotrematidae	*Pulmonicola* Poche, 1926	15
Trematoda	Paragonimidae	*Paragonimus* Braun, 1899	19
Trematoda	Paragonimidae	*Paragonimus westermani* (Kerbert, 1878)	19
Trematoda	Paramphistomidae	*Calicophoron sukari* (Dinnik, 1954)	13
Trematoda	Philophthalmidae	*Parorchis* Nicoll, 1907	22
Trematoda	Philophthalmidae	*Philophthalmus* Looss, 1899	22
Trematoda	Philophthalmidae	*Philophthalmus attenuatus* Bennett & Presswell, 2019	11
Trematoda	Philophthalmidae	*Philophthalmus zalophi* Dailey, Perrin & Parás, 2005	15
Trematoda	Phocidae	*Halichoerus grypus* (Fabricius, 1791)	15
Trematoda	Plagiorchiidae	*Plagiorchis* Lühe, 1899	8, 16
Trematoda	Psilostomidae	*Ribeiroia ondatrae* (Price, 1931)	5, 7, 11, 18, 22
Trematoda	Renicolidae	*Renicola parvicaudatus* (Stunkard & Shaw, 1931)	5, 11, 12, 22
Trematoda	Reniferidae	*Paralechriorchis syntomentera* (Sumwalt, 1926)	11
Trematoda	Rhabdiopoeidae	*Faredifex* Blair, 1981	15
Trematoda	Rhabdiopoeidae	*Haerator* Blair, 1981	15
Trematoda	Rhabdiopoeidae	*Rhabdiopoeus* Johnston, 1913	15
Trematoda	Rhabdiopoeidae	*Taprobanella* Crusz & Fernand, 1954	15
Trematoda	Schistosomatidae	*Schistosoma* Weinland, 1858	14, 15
Trematoda	Schistosomatidae	*Schistosoma haematobium* (Bilharz, 1852)	10, 22
Trematoda	Schistosomatidae	*Schistosoma japonicum* Katsurada, 1904	10, 23
Trematoda	Schistosomatidae	*Schistosoma mansoni* Sambon, 1907	7, 10, 13, 15, 22
Trematoda	Schistosomatidae	*Trichobilharzia* Skrjabin & Zakharow, 1920	14
Trematoda	Schistosomatidae	*Trichobilharzia ocellata* (La Valette St. George, 1855)	10
Trematoda	Stichocotylidae	*Stichocotyle nephropis* Cunningham, 1884	17
Trematoda	Strigeidae	*Cotylurus* Szidat, 1928	11
Trematoda	Telorchiidae	*Telorchis* Lühe, 1899 nec Looss, 1899	22
Trematoda	Troglotrematidae	*Nanophyetus salmincola* (Chapin, 1926)	9, 11, 15

Table A6 Species mentioned as hosts, vectors or otherwise free-living taxa, listed alphabetically according to higher taxonomic classification. Within each higher classification, families are arranged alphabetically and, within each family, species are listed in alphabetical order

Taxon higher classification	Family	Species	Chapter
Amphibia	Bufonidae	*Anaxyrus americanus* (Holbrook, 1836)	22
Amphibia	Bufonidae	*Atelopus* Duméril and Bibron, 1841	22
Amphibia	Bufonidae	*Rhinella marina* (Linnaeus, 1758)	18
Amphibia	Caeciliidae	*Caecilia* cf. *pachynema* Günther, 1859	13
Amphibia	Caeciliidae	*Caecilia gracilis* Shaw, 1802	13
Amphibia	Ceratophryidae	*Lepidobatrachus llanensis* Reig & Cei, 1963	12
Amphibia	Hylidae	*Hyla versicolor* LeConte, 1825	22
Amphibia	Hylidae	*Litoria aurea* (Günther, 1864)	22
Amphibia	Hylidae	*Osteopilus septentrionalis* (Duméril & Bibron, 1841)	22
Amphibia	Hylidae	*Pseudacris crucifer* (Wied-Neuwied, 1838)	22
Amphibia	Hylidae	*Pseudacris regilla* (Baird & Girard, 1852)	7, 8, 22
Amphibia	Mantellidae	*Aglyptodactylus madagascariensis* (Duméril, 1853)	13
Amphibia	Ranidae	*Lithobates catesbeianus* (Shaw, 1802)	22
Amphibia	Ranidae	*Lithobates clamitans* (Latreille, 1801)	22
Amphibia	Ranidae	*Lithobates pipiens* (Schreber, 1782)	22
Amphibia	Ranidae	*Lithobates sylvatica* (LeConte, 1825)	22
Amphibia	Ranidae	*Rana boylii* Baird, 1854	22
Amphibia	Ranidae	*Rana cascadae* Slater, 1939	8
Amphibia	Salamandridae	*Ichthyosaura alpestris* (Laurenti, 1768)	22
Amphibia	Salamandridae	*Notophthalmus viridescens* (Rafinesque, 1820)	22
Amphibia	Salamandridae	*Salamandra salamandra* (Linnaeus, 1758)	18
Amphibia	Salamandridae	*Triturus cristatus* (Laurenti, 1768)	14
Amphibia	Scaphiopodidae	*Scaphiopus couchii* Baird, 1854	12
Amphibia	Telmatobiidae	*Telmatobius* Wiegmann, 1834	22
Amphibia	Typhlonectidae	*Typhlonectes compressicaudata* (Duméril & Bibron, 1841)	13
Amphibia	Typhlonectidae	*Typhlonectes natans* (Fischer, 1880)	13
Amphibia	Xenopidae	*Xenopus laevis* (Daudin, 1802)	4
Amphipoda	Asellidae	*Asellus aquaticus* (Linnaeus, 1758)	4
Amphipoda	Corophiidae	*Chelicorophium curvispinum* (Sars, 1895)	16
Amphipoda	Corophiidae	*Corophium arenarium* Crawford, 1937	8
Amphipoda	Corophiidae	*Corophium volutator* (Pallas, 1766)	8, 22
Amphipoda	Corophiidae	*Paracorophium excavatum* (Thomson, 1884)	22
Amphipoda	Corophiidae	*Paracorophium lucasi* Hurley, 1954	8
Amphipoda	Crangonyctidae	*Crangonyx pseudogracilis* Bousfield, 1958	3, 7
Amphipoda	Gammaridae	*Dikerogammarus haemobaphes* (Eichwald, 1841)	3
Amphipoda	Gammaridae	*Dikerogammarus villosus* (Sowinsky, 1894)	3, 16
Amphipoda	Gammaridae	*Gammarus* Fabricius, 1775	2
Amphipoda	Gammaridae	*Gammarus duebeni* Liljeborg, 1852	3, 4, 7
Amphipoda	Gammaridae	*Gammarus fossarum* Koch, 1836	3, 20
Amphipoda	Gammaridae	*Gammarus insensibilis* Stock, 1966	7
Amphipoda	Gammaridae	*Gammarus pulex* (Linnaeus, 1758)	2, 7, 18, 20, 22
Amphipoda	Gammaridae	*Gammarus roeselii* Gervais, 1835	3, 7, 20, 22
Amphipoda	Gammaridae	*Gammarus tigrinus* Sexton, 1939	4
Amphipoda	Hyperiidae	*Themisto libellula* (Lichtenstein in Mandt, 1822)	2
Amphipoda	Paracalliopiidae	*Indocalliope indica* (Barnard, 1935)	22

(continued)

Table A6 (continued)

Taxon higher classification	Family	Species	Chapter
Amphipoda	Paracalliopiidae	*Paracalliope novizealandiae* (Dana, 1853)	22
Amphipoda	Phoxocephalidae	*Proharpinia stephenseni* (Schellenberg, 1931)	8
Amphipoda	Pontogeneiidae	*Paramoera chevreuxi* (Stephensen, 1927)	8
Amphipoda	Talitridae	*Orchestia* Leach, 1814	2
Amphipoda	Talitridae	*Talorchestia* Dana, 1852	2
Annelida	Alvinellidae	*Alvinella pompejana* Desbruyères & Laubier, 1980	12
Annelida	Naididae	*Tubifex* Lamarck, 1816	12
Annelida	Nereididae	*Hediste diversicolor* (Müller, 1776)	8
Annelida	Spionidae	*Polydora* Bosc, 1802	18
Annelida	Spionidae	*Pygospio elegans* Claparède, 1863	8
Annelida	Teredinidae	*Teredo navalis* Linnaeus, 1758	18
Aves	Accipitridae	*Circus cyaneus* (Linnaeus, 1766)	7
Aves	Chionidae	*Chionis albus* (Gmelin, 1789)	12
Aves	Haematopodidae	*Haematopus ostralegus* Linnaeus, 1758	7
Aves	Laridae	*Larus delawarensis* Ord, 1815	7
Aves	Phalacrocoracidae	*Gulosus aristotelis* (Linnaeus, 1761)	9
Aves	Phalacrocoracidae	*Phalacrocorax carbo* (Linnaeus, 1758)	9
Aves	Phasianidae	*Lagopus lagopus* (Linnaeus, 1758)	7
Aves	Phasianidae	*Perdix perdix* (Linnaeus, 1758)	8
Aves	Phasianidae	*Phasianus colchicus* Linnaeus, 1758	8
Aves	Scolopacidae	*Calidris alpina* (Linnaeus, 1758)	8
Aves	Scolopacidae	*Calidris pusilla* (Linnaeus, 1766)	5
Aves	Spheniscidae	*Spheniscus demersus* (Linnaeus, 1758)	2
Bivalvia	Cardiidae	*Cerastoderma edule* (Linnaeus, 1758)	7, 8, 20, 22
Bivalvia	Dreissenidae	*Dreissena* Van Beneden, 1835	18
Bivalvia	Mytilidae	*Brachidontes exustus* (Linnaeus, 1758)	7
Bivalvia	Mytilidae	*Mytilus chilensis* Hupé, 1854	23
Bivalvia	Mytilidae	*Mytilus edulis* Linnaeus, 1758	8, 18, 22, 23
Bivalvia	Mytilidae	*Mytilus galloprovincialis* Lamarck, 1819	7, 8, 22
Bivalvia	Mytilidae	*Perna canaliculus* (Gmelin, 1791)	10
Bivalvia	Mytilidae	*Perna perna* (Linnaeus, 1758)	7
Bivalvia	Ostreidae	*Crassostrea virginica* (Gmelin, 1791)	2, 22
Bivalvia	Ostreidae	*Magallana gigas* (Thunberg, 1793)	2, 18, 22
Bivalvia	Ostreidae	*Ostrea angasi* Sowerby II, 1871	23
Bivalvia	Ostreidae	*Ostrea edulis* Linnaeus, 1758	2
Bivalvia	Pectinidae	*Aequipecten opercularis* (Linnaeus, 1758)	2
Bivalvia	Pectinidae	*Argopecten irradians* (Lamarck, 1819)	8
Bivalvia	Pectinidae	*Chlamys farreri* (Jones & Preston, 1904)	22
Bivalvia	Sphaeriidae	*Pisidium amnicum* (Müller, 1774)	7, 20
Bivalvia	Tellinidae	*Macoma balthica* (Linnaeus, 1758)	8
Bivalvia	Veneridae	*Austrovenus stutchburyi* (Wood, 1828)	7, 8
Bivalvia	Veneridae	*Ruditapes philippinarum* (Adams & Reeve, 1850)	20, 22
Cephalopoda	Ommastrephidae	*Illex illecebrosus* (Lesueur, 1821)	17
Cephalopoda	Ommastrephidae	*Ommastrephes bartramii* (Lesueur, 1821)	17
Cephalopoda	Ommastrephidae	*Todarodes pacificus* (Steenstrup, 1880)	19
Chlorophyta	Ulvaceae	*Ulva lactuca* Linnaeus, 1753	8
Cirripedia	Chthamalidae	*Chthamalus fissus* Darwin, 1854	7
Cirripedia	Elminiidae	*Austrominius modestus* (Darwin, 1854)	22

(continued)

Table A6 (continued)

Taxon higher classification	Family	Species	Chapter
Cnidaria	Acroporidae	*Acropora* Oken, 1815	2
Cnidaria	Actiniidae	*Anthopleura hermaphroditica* (Carlgren, 1899)	8
Cnidaria	Gorgoniidae	*Gorgonia ventalina* Linnaeus, 1758	22
Cnidaria	Hydridae	*Hydra vulgaris* Pallas, 1776	3
Copepoda	Cyclopidae	*Cyclops* Müller, 1785	2
Copepoda	Cyclopidae	*Macrocyclops albidus* (Jurine, 1820)	7
Crustacea	Artemiidae	*Artemia* Leach, 1819	12
Crustacea	Artemiidae	"*Artemia parthenogenetica*"	18, 20
Crustacea	Artemiidae	*Artemia franciscana* Kellogg, 1906	18
Crustacea	Artemiidae	*Artemia salina* (Linnaeus, 1758)	18
Crustacea	Astacidae	*Austropotamobius pallipes* (Lereboullet, 1858)	8, 18
Crustacea	Astacidae	*Pacifastacus leniusculus* (Dana, 1852)	8, 18
Crustacea	Daphniidae	*Daphnia* Müller, 1785	8, 22
Crustacea	Daphniidae	*Daphnia longispina* Müller, 1776 complex	22
Decapoda	Cambaridae	*Faxonius limosus* (Rafinesque, 1817)	7
Decapoda	Carcinidae	*Carcinus maenas* (Linnaeus, 1758)	18
Decapoda	Hippolytidae	*Lysmata* Risso, 1816	7
Decapoda	Lithodidae	*Paralithodes camtschaticus* (Tilesius, 1815)	2
Decapoda	Lithodidae	*Paralomis granulosa* (Hombron & Jacquinot, 1846)	7
Decapoda	Palaemonidae	*Macrobrachium* Spence Bate, 1868	22
Decapoda	Palinuridae	*Panulirus homarus* (Linnaeus, 1758)	22
Decapoda	Panopeidae	*Eurypanopeus depressus* (SI Smith, 1869)	7, 18, 22
Decapoda	Panopeidae	*Panopeus herbstii* Milne Edwards, 1834	7
Decapoda	Penaeidae	*Penaeus vannamei* Boone, 1931	22
Decapoda	Portunidae	*Callinectes sapidus* Rathbun, 1896	2, 10, 22
Echinodermata	Asteriidae	*Asterias rubens* Linnaeus, 1758	22
Echinodermata	Asteriidae	*Pisaster ochraceus* (Brandt, 1835)	2
Echinodermata	Loveniidae	*Echinocardium cordatum* (Pennant, 1777)	6
Echinodermata	Stichopodidae	*Apostichopus japonicus* (Selenka, 1867)	22
Echinodermata	Strongylocentrotidae	*Strongylocentrotus* Brandt, 1835	8
Fungi	Agaricaceae	*Agaricus bisporus* (J. E. Lange, 1926)	4
Fungi	Saccharomycetaceae	*Saccharomyces cerevisiae* Meyen in E.C. Hansen, 1883	4
Fungi	Trichocomaceae	*Penicillium chrysogenum* Thom, 1910	4
Gastropoda	Batillariidae	*Batillaria attramentaria* (Sowerby II, 1855)	18
Gastropoda	Batillariidae	*Batillaria cumingii* (Crosse, 1862)	8
Gastropoda	Batillariidae	*Zeacumantus subcarinatus* (Sowerby, 1855)	8, 22
Gastropoda	Bythiniidae	*Gabbia fuchsiana* (Möllendorff, 1888)	19
Gastropoda	Bythiniidae	*Gabbia longicornis* (W. H. Benson, 1842)	19
Gastropoda	Bythiniidae	*Parafossarulus manchouricus* (Bourguignat, 1860)	19
Gastropoda	Calyptraeidae	*Crepidula fornicata* (Linnaeus, 1758)	18
Gastropoda	Cerithiidae	*Cerithidea* Swainson, 1840	19
Gastropoda	Cochliopidae	*Heleobia australis* (d'Orbigny, 1835)	15
Gastropoda	Cochliopidae	*Juturnia tularosae* Hershler, Liu & Stockwell, 2002	22
Gastropoda	Cominellidae	*Cominella glandiformis* (Reeve, 1847)	8
Gastropoda	Fissurellidae	*Fissurella crassa* Lamarck, 1822	7
Gastropoda	Hydrobiidae	*Peringia ulvae* (Pennant, 1777)	8, 22
Gastropoda	Littorinidae	*Austrolittorina cincta* (Quoy & Gaimard, 1833)	22

(continued)

Table A6 (continued)

Taxon higher classification	Family	Species	Chapter
Gastropoda	Littorinidae	*Echinolittorina peruviana* (Lamarck, 1822)	22
Gastropoda	Littorinidae	*Laevilitorina caliginosa* (A. Gould, 1849)	12
Gastropoda	Littorinidae	*Littorina littorea* (Linnaeus, 1758)	8, 22
Gastropoda	Littorinidae	*Littorina obtusata* (Linnaeus, 1758)	12
Gastropoda	Littorinidae	*Littorina saxatilis* (Olivi, 1792)	12
Gastropoda	Littorinidae	*Littorina scutulata* A. Gould, 1849	22
Gastropoda	Lottiidae	*Lottia alveus* (Conrad, 1831)	8
Gastropoda	Lottiidae	*Notoacmea elongata* (Quoy & Gaimard, 1834)	8
Gastropoda	Lymnaeidae	*Biomphalaria alexandrina* (Ehrenberg, 1831)	22
Gastropoda	Lymnaeidae	*Biomphalaria glabrata* (Say, 1818)	7, 10
Gastropoda	Lymnaeidae	*Ladislavella elodes* (Say, 1821)	7
Gastropoda	Lymnaeidae	*Lymnaea stagnalis* (Linnaeus, 1758)	10, 18, 20, 22
Gastropoda	Lymnaeidae	*Pseudosuccinea columella* (Say, 1817)	22
Gastropoda	Muricidae	*Nucella lapillus* (Linnaeus, 1758)	18
Gastropoda	Nassariidae	*Ilyanassa obsoleta* (Say, 1822)	8
Gastropoda	Physidae	*Physa* Draparnaud, 1801	22
Gastropoda	Planorbidae	*Hippeutis* Charpentier, 1837	19
Gastropoda	Planorbidae	*Planorbella trivolvis* (Say, 1817)	7, 22
Gastropoda	Planorbidae	*Segmentina* Fleming, 1818	19
Gastropoda	Pomatiopsidae	*Oncomelania hupensis* Gredler, 1881	10
Gastropoda	Potamididae	*Cerithideopsis californica* (Haldeman, 1840)	7, 8, 17, 18, 22
Gastropoda	Semisulcospiridae	*Semisulcospira* Boettger, 1886	19
Gastropoda	Tateidae	*Potamopyrgus antipodarum* (Gray, 1843)	18
Gastropoda	Thiaridae	*Thiara* Röding, 1798	19
Gastropoda	Trochidae	*Diloma subrostratum* (Gray, 1835)	8
Insecta	Brachycentridae	*Brachycentrus americanus* (Banks, 1899)	4
Insecta	Corixidae	*Cenocorixa bifida* (Hungerford, 1926)	8
Insecta	Corixidae	*Cenocorixa expleta* (Uhler, 1895)	8
Insecta	Culicidae	*Aedes* Meigen, 1818	18
Insecta	Culicidae	*Aedes sierrensis* (Ludlow, 1905)	2
Insecta	Culicidae	*Anopheles gambiae* Giles, 1902	20
Insecta	Libelullidae	*Leucorrhinia intacta* (Hagen, 1861)	8
Insecta	Nepidae	*Nepa cinerea* Linnaeus, 1758	7
Insecta	Pyralidae	*Ephestia kuehniella* (Zeller, 1879)	7
Insecta	Pyralidae	*Galleria mellonella* (Linnaeus, 1758)	7
Insecta	Pyralidae	*Plodia interpunctella* (Hübner, 1813)	7
Insecta	Tenebrionidae	*Tribolium castaneum* (Herbst, 1797)	8
Insecta	Tenebrionidae	*Tribolium confusum* Jacquelin du Val, 1863	8
Insecta	Trigonidiidae	*Nemobius sylvestris* (Bosc, 1792)	7
Isopoda	Asellidae	*Asellus* Geoffroy, 1762	2
Isopoda	Idoteidae	*Idotea baltica* (Pallas, 1772)	22
Isopoda	Limnoriidae	*Limnoria* Leach, 1814	2
Mammalia	Balaenidae	*Balaena mysticetus* Linnaeus, 1758	2
Mammalia	Balaenidae	*Eubalaena* Gray, 1864	2
Mammalia	Balaenopteridae	*Balaenoptera bonaerensis* Burmeister, 1867	15
Mammalia	Balaenopteridae	*Megaptera novaeangliae* (Borowski, 1781)	15
Mammalia	Cervidae	*Alces alces* Peterson, 1952	7
Mammalia	Delphinidae	*Delphinus delphis* Linnaeus, 1758	9, 15
Mammalia	Delphinidae	*Globicephala* Lesson, 1828	15
Mammalia	Delphinidae	*Orcinus orca* (Linnaeus, 1758)	15

(continued)

Table A6 (continued)

Taxon higher classification	Family	Species	Chapter
Mammalia	Delphinidae	*Pseudorca crassidens* (Owen, 1846)	15
Mammalia	Delphinidae	*Stenella coeruleoalba* (Meyen, 1833)	15
Mammalia	Delphinidae	*Steno bredanensis* (Cuvier in Lesson, 1828)	15
Mammalia	Delphinidae	*Tursiops truncatus* (Montagu, 1821)	15
Mammalia	Dugongidae	*Dugong dugon* (Müller, 1776)	2, 15
Mammalia	Eschrichtiidae	*Eschrichtius robustus* (Lilljeborg, 1861)	6
Mammalia	Herpestidae	*Herpestes ichneumon* (Linnaeus, 1758)	15
Mammalia	Leporidae	*Oryctolagus cuniculus* (Linnaeus, 1758)	22
Mammalia	Monodontidae	*Delphinapterus leucas* (Pallas, 1776)	2, 15
Mammalia	Mustelidae	*Enhydra lutris* (Linnaeus, 1758)	15
Mammalia	Mustelidae	*Enhydra lutris nereis* (Merriam, 1904)	13
Mammalia	Mustelidae	*Lutra lutra* (Linnaeus, 1758)	13
Mammalia	Odobenidae	*Odobenus rosmarus* (Linnaeus, 1758)	15
Mammalia	Ornithorhynchidae	*Ornithorhynchus anatinus* (Shaw, 1799)	13
Mammalia	Otariidae	*Arctocephalus australis* (Zimmermann, 1783)	15
Mammalia	Otariidae	*Arctocephalus forsteri* (Lesson, 1828)	15
Mammalia	Otariidae	*Arctocephalus pusillus* (Schreber, 1775)	15
Mammalia	Otariidae	*Callorhinus ursinus* (Linnaeus, 1758)	9, 15
Mammalia	Otariidae	*Eumetopias jubatus* (Schreber, 1776)	9, 15
Mammalia	Otariidae	*Neophoca cinerea* (Péron, 1816)	15
Mammalia	Otariidae	*Otaria flavescens* Shaw, 1800	15
Mammalia	Otariidae	*Zalophus californianus* (Lesson, 1828)	2, 15
Mammalia	Otariidae	*Zalophus wollebaeki* Sivertsen, 1953	15
Mammalia	Phocidae	*Erignathus barbatus* (Erxleben, 1777)	2, 9, 15
Mammalia	Phocidae	*Halichoerus grypus* (Fabricius, 1791)	2, 13
Mammalia	Phocidae	*Leptonychotes weddellii* (Lesson, 1826)	15
Mammalia	Phocidae	*Mirounga leonina* (Linnaeus, 1758)	15
Mammalia	Phocidae	*Monachus monachus* (Hermann, 1779)	13, 15
Mammalia	Phocidae	*Pagophilus groelandicus* (Erxleben, 1777)	2
Mammalia	Phocidae	*Phoca vitulina* Linnaeus, 1758	15
Mammalia	Phocidae	*Pusa hispida* (Schreber, 1775)	2, 15
Mammalia	Phocoenidae	*Phocoena phocoena* (Linnaeus, 1758)	2, 15
Mammalia	Physeteridae	*Physeter macrocephalus* Linnaeus, 1758	15
Mammalia	Pontoporiidae	*Pontoporia blainvillei* (Gervais & d'Orbigny, 1844)	15
Mammalia	Sciuridae	*Sciurus carolinensis* Gmelin, 1788	8
Mammalia	Sciuridae	*Sciurus vulgaris* Linnaeus, 1758	8
Mammalia	Sciuridae	*Urocitellus richardsoni* (Sabine, 1822)	7
Mammalia	Ursidae	*Ursus maritimus* Phipps, 1774	15
Mammalia	Ziphiidae	*Mesoplodon mirus* True, 1913	15
Nemertea	Lineidae	*Lineus ruber* (Müller, 1774)	8
Pisces	Acipenseridae	*Scaphirhynchus albus* (Forbes & Richardson, 1905)	13
Pisces	Acipenseridae	*Acipenser baerii* Brandt, 1869	13
Pisces	Acipenseridae	*Acipenser sturio* Linnaeus, 1758	13
Pisces	Alosidae	*Alosa immaculata* Bennett, 1835	20
Pisces	Alosidae	*Sardinops sagax* (Jenyns, 1842)	9
Pisces	Anarhichadidae	*Anarhichas lupus* Linnaeus, 1758	23
Pisces	Anarhichadidae	*Anarhichas minor* Olafsen, 1772	23
Pisces	Anguillidae	*Anguilla anguilla* (Linnaeus, 1758)	7, 18

(continued)

Table A6 (continued)

Taxon higher classification	Family	Species	Chapter
Pisces	Anguillidae	*Anguilla rostrata* (Lesueur, 1817)	22
Pisces	Arhynchobatidae	*Sympterygia bonapartii* Müller & Henle, 1841	17
Pisces	Atherinopsidae	*Odontesthes* Evermann & Kendall, 1906	15
Pisces	Blenniidae	*Hypsoblennius sordidus* (Bennett, 1828)	10
Pisces	Blenniidae	*Salariopsis fluviatilis* (Asso, 1801)	13
Pisces	Carangidae	*Seriola* Cuvier, 1816	23
Pisces	Carangidae	*Seriola dumerili* (Risso, 1810)	23
Pisces	Carangidae	*Seriola lalandi* Valenciennes, 1833	23
Pisces	Carangidae	*Seriola quinqueradiata* Temminck & Schlegel, 1845	23
Pisces	Carangidae	*Trachinotus blochii* (Lacepède, 1801)	5
Pisces	Carangidae	*Trachurus lathami* Nichols, 1920	17
Pisces	Carangidae	*Trachurus trachurus* (Linnaeus, 1758)	17
Pisces	Carapidae	*Encheliophis* Müller, 1842	7
Pisces	Carcharhinidae	*Carcharhinus hemiodon* (Valenciennes, 1839)	13
Pisces	Carcharhinidae	*Carcharhinus obsoletus* White, Kyne & Harris, 2019	13
Pisces	Centrarchidae	*Lepomis gibbosus* (Linnaeus, 1758)	22
Pisces	Centrarchidae	*Lepomis macrochirus* Rafinesque, 1810	3
Pisces	Characidae	*Astyanax mexicanus* (De Filippi, 1853)	12
Pisces	Cichlidae	*Oreochromis grahami* (Boulenger, 1912)	12
Pisces	Cichlidae	*Oreochromis mossambicus* (Peters, 1852)	22
Pisces	Cichlidae	*Oreochromis niloticus* (Linnaeus, 1758)	2, 23
Pisces	Clariidae	*Clarias* Scopoli, 1777	2
Pisces	Clariidae	*Clarias gariepinus* (Burchell, 1822)	23
Pisces	Clariidae	*Heteroclarias* hybrid	2
Pisces	Clupeidae	*Clupea harengus* Linnaeus, 1758	9
Pisces	Congridae	*Conger conger* (Linnaeus, 1758)	9
Pisces	Coryphaenidae	*Coryphaena hippurus* Linnaeus, 1758	23
Pisces	Cottidae	*Myoxocephalus quadricornis* (Linnaeus, 1758)	15
Pisces	Cyclopteridae	*Cyclopterus lumpus* Linnaeus, 1758	23
Pisces	Cyprinidae	*Abramis brama* (Linnaeus, 1758)	22
Pisces	Cyprinidae	*Barbus barbus* (Linnacus, 1758)	5, 20
Pisces	Cyprinidae	*Carassius auratus* (Linnaeus, 1758)	16
Pisces	Cyprinidae	*Ctenopharyngodon idella* (Valenciennes, 1844)	3
Pisces	Cyprinidae	*Cyprinus carpio* Linnaeus, 1758	3, 10, 23
Pisces	Cyprinidae	*Danio rerio* (Hamilton, 1822)	3
Pisces	Cyprinidae	*Hypophthalmichthys molitrix* (Valenciennes, 1844)	23
Pisces	Cyprinidae	*Labeobarbus aeneus* (Burchell, 1822)	20
Pisces	Cyprinidae	*Labeobarbus kimberleyensis* (Gilchrist & Thompson, 1913)	20
Pisces	Cyprinidae	*Leuciscus leuciscus* (Linnaeus, 1758)	3
Pisces	Cyprinidae	*Luxilus cornutus* (Mitchill, 1817)	22
Pisces	Cyprinidae	*Notemigonus crysoleucas* (Mitchill, 1814)	3
Pisces	Cyprinidae	*Notropis hudsonius* (Clinton, 1824)	22
Pisces	Cyprinidae	*Pimephales promelas* Rafinesque, 1820	7, 22
Pisces	Cyprinidae	*Rhodeus amarus* (Bloch, 1782)	3
Pisces	Cyprinidae	*Rutilus rutilus* (Linnaeus, 1758)	3, 15, 22
Pisces	Cyprinidae	*Rutilus rutilus* × *Abramis brama*	23
Pisces	Cyprinidae	*Saurogobio* Bleeker, 1870	7

(continued)

Table A6 (continued)

Taxon higher classification	Family	Species	Chapter
Pisces	Cyprinidae	*Squalius cephalus* (Linnaeus, 1758)	18, 20
Pisces	Cyprinodontidae	*Cyprinodon tularosa* Miller & Echelle, 1975	12, 22
Pisces	Eleotridae	*Milyeringa veritas* Whitley, 1945	12
Pisces	Esocidae	*Esox lucius* Linnaeus, 1758	3
Pisces	Etmopteridae	*Etmopterus spinax* (Linnaeus, 1758)	9
Pisces	Fundulidae	*Fundulus heteroclitus* (Linnaeus, 1766)	22
Pisces	Fundulidae	*Fundulus parvipinnis* Girard, 1854	8
Pisces	Fundulidae	*Fundulus zebrinus* Jordan & Gilbert, 1883	22
Pisces	Gadidae	*Gadus morhua* Linnaeus, 1758	2, 9, 10, 15
Pisces	Gadidae	*Merlangius merlangus* (Linnaeus, 1758)	10, 17
Pisces	Gadidae	*Pollachius pollachius* (Linnaeus, 1758)	10
Pisces	Galaxiidae	*Galaxias maculatus* (Jenyns, 1842)	13
Pisces	Gasterosteidae	*Gasterosteus aculeatus* Linnaeus, 1758	3, 5, 7, 8, 12, 20, 22
Pisces	Gasterosteidae	*Gasterosteus wheatlandi* Putnam, 1867	8
Pisces	Gasterosteidae	*Pungitius pungitius* (Linnaeus, 1758)	8, 12
Pisces	Gobiidae	*Elacatinus evelynae* (Böhlke & Robins, 1968)	7
Pisces	Gobiidae	*Neogobius melanostomus* (Pallas, 1814)	7, 18
Pisces	Gobiidae	*Pomatoschistus flavescens* (Fabricius, 1779)	3
Pisces	Gobiidae	*Pomatoschistus marmoratus* (Risso, 1810)	22
Pisces	Gonorynchidae	*Chanos chanos* (Forsskål 1775)	23
Pisces	Grammatidae	*Gramma* Poey, 1868	7
Pisces	Haemulidae	*Haemulon flavolineatum* (Desmarest, 1823)	10
Pisces	Holocentridae	*Holocentrus* Scopoli, 1777	7
Pisces	Holocentridae	*Holocentrus rufus* (Walbaum, 1792)	10
Pisces	Holocentridae	*Myripristis* Cuvier, 1829	7
Pisces	Holocentridae	*Sargocentron* Fowler, 1904	7
Pisces	Labridae	*Diproctacanthus* Bleeker, 1862	7
Pisces	Labridae	*Labroides* Bleeker, 1851	7
Pisces	Labridae	*Labrus bergylta* Ascanius, 1767	23
Pisces	Labridae	*Notolabrus celidotus* (Bloch & Schneider, 1801)	8
Pisces	Labridae	*Thalassoma* Swainson, 1839	7
Pisces	Latidae	*Lates calcarifer* (Bloch, 1790)	23
Pisces	Lophiidae	*Lophius piscatorius* Linnaeus, 1758	3
Pisces	Loricariidae	*Otocinclus cocama* Reis, 2004	13
Pisces	Lutjanidae	*Lutjanus synagris* (Linnaeus, 1758)	10
Pisces	Macrouridae	*Coryphaenoides rupestris* Gunnerus, 1765	3
Pisces	Merlucciidae	*Merluccius hubbsi* Marini, 1933	17
Pisces	Merlucciidae	*Merluccius merluccius* (Linnaeus, 1758)	9, 17
Pisces	Monacanthidae	*Thamnaconus degeni* (Regan, 1903)	23
Pisces	Moronidae	*Dicentrarchus labrax* (Linnaeus, 1758)	23
Pisces	Mugilidae	*Mugil liza* Valenciennes, 1836	15
Pisces	Ophidiidae	*Raneya brasiliensis* (Kaup, 1856)	15
Pisces	Paralichthyidae	*Paralichthys isosceles* Jordan, 1891	17
Pisces	Paralichthyidae	*Paralichthys olivaceus* (Temminck & Schlegel, 1846)	5, 23
Pisces	Percidae	*Gymnocephalus* Bloch, 1793	18
Pisces	Percidae	*Perca fluviatilis* Linnaeus, 1758	16, 22
Pisces	Percophidae	*Percophis brasiliensis* Quoy & Gaimard, 1825	17
Pisces	Phycidae	*Urophycis brasiliensis* (Kaup, 1858)	17

(continued)

Table A6 (continued)

Taxon higher classification	Family	Species	Chapter
Pisces	Platycephalidae	*Platycephalus richardsoni* Castelnau, 1872	19
Pisces	Pleuronectidae	*Hippoglossoides platessoides* (Fabricius, 1780)	9
Pisces	Pleuronectidae	*Microstomus kitt* (Walbaum, 1792)	10
Pisces	Pleuronectidae	*Parophrys vetulus* Girard, 1854	13
Pisces	Pleuronectidae	*Platichthys flesus* (Linnaeus, 1758)	10
Pisces	Poeciliidae	*Gambusia affinis* (Baird & Girard, 1853)	22
Pisces	Poeciliidae	*Poecilia mexicana* Steindachner, 1863	12
Pisces	Poeciliidae	*Poecilia reticulata* Peters, 1859	12
Pisces	Pomacentridae	*Stegastes diencaeus* (Jordan & Rutter, 1897)	10
Pisces	Pristidae	*Anoxypristis cuspidata* (Latham, 1794)	13
Pisces	Pristidae	*Pristis clavata* Garman, 1906	13
Pisces	Pristidae	*Pristis pectinata* Latham, 1794	13
Pisces	Pristidae	*Pristis pristis* (Linnaeus, 1758)	13
Pisces	Pristidae	*Pristis zijsron* Bleeker, 1851	13
Pisces	Rajidae	*Raja clavata* Linnaeus, 1758	17
Pisces	Rajidae	*Rostroraja alba* (Lacepède, 1803)	13
Pisces	Rhincodontidae	*Rhincodon typus* Smith, 1828	13
Pisces	Rivulidae	*Anablepsoides hartii* (Boulenger, 1890)	12
Pisces	Salmonidae	*Oncorhynchus gorbuscha* (Walbaum, 1792)	23
Pisces	Salmonidae	*Oncorhynchus keta* (Walbaum, 1792)	23
Pisces	Salmonidae	*Oncorhynchus mykiss* (Walbaum, 1792)	7, 9, 10, 17, 18, 23
Pisces	Salmonidae	*Salmo salar* Linnaeus, 1758	2, 5, 10, 11, 17, 22, 23
Pisces	Salmonidae	*Salmo trutta* Linnaeus, 1758	3, 18
Pisces	Salmonidae	*Salvelinus alpinus* (Linnaeus, 1758)	12
Pisces	Salmonidae	*Salvelinus fontinalis* (Mitchill, 1814)	18
Pisces	Salmonidae	*Salvelinus leucomaenis* (Pallas, 1814)	7
Pisces	Salmonidae	*Salvelinus namaycush* (Walbaum, 1792)	7
Pisces	Sciaenidae	*Cynoscion guatucupa* (Cuvier, 1830)	17
Pisces	Sciaenidae	*Micropogonias furnieri* (Desmarest, 1823)	17
Pisces	Sciaenidae	*Micropogonias undulatus* (Linnaeus, 1766)	10
Pisces	Scombridae	*Katsuwonus pelamis* (Linnaeus, 1758)	17
Pisces	Scombridae	*Scomber australasicus* Cuvier, 1832	19
Pisces	Scombridae	*Scomber japonicus* Houttuyn, 1782	19, 23
Pisces	Scombridae	*Thunnus* South, 1845	23
Pisces	Scombridae	*Thunnus maccoyii* (Castelnau, 1872)	23
Pisces	Scombridae	*Thunnus orientalis* (Temminck & Schlegel, 1844)	9
Pisces	Scombridae	*Thunnus thynnus* (Linnaeus, 1758)	9
Pisces	Scophthalmidae	*Scophthalmus maximus* (Linnaeus, 1758)	10
Pisces	Serranidae	*Epinephelus bruneus* × *Epinephelus lanceolatus*	23
Pisces	Somniosidae	*Scymnodon macracanthus* (Regan, 1906)	13
Pisces	Sparidae	*Boops boops* (Linnaeus, 1758)	20
Pisces	Sparidae	*Lagodon rhomboides* (Linnaeus, 1766)	10
Pisces	Sparidae	*Pagellus* Valenciennes, 1830	9
Pisces	Sparidae	*Pagrus pagrus* (Linnaeus, 1758)	17
Pisces	Sparidae	*Sparus aurata* Linnaeus, 1758	23
Pisces	Syngnathidae	*Syngnathus abaster* Risso, 1827	22
Pisces	Terapontidae	*Bidyanus bidyanus* (Mitchell, 1838)	13
Pisces	Tetraodontidae	*Amblyrhynchote honckenii* (Bloch, 1785)	20

(continued)

Table A6 (continued)

Taxon higher classification	Family	Species	Chapter
Pisces	Torpedinidae	*Torpedo suessii* Steindachner, 1898	13
Pisces	Trichiuridae	*Trichiurus japonicus* Temminck & Schlegel, 1844	9
Pisces	Urolophidae	*Urolophus javanicus* (Martens, 1864)	13
Pisces	Zeidae	*Zenopsis conchifer* (Lowe, 1852)	17
Plantae	Zosteraceae	*Zostera* Linnaeus, 1758	2
Plantae	Zosteraceae	*Zostera marina* Linnaeus, 1758	8
Reptilia	Cheloniidae	*Caretta caretta* (Linnaeus, 1758)	9
Reptilia	Emydidae	*Emys orbicularis* (Linnaeus, 1758)	2
Reptilia	Iguanidae	*Amblyrhynchus cristatus* Bell, 1825	6
Rhodophyta	Bangiaceae	*Porphyra* Agardh, 1824	8
Rhodophyta	Corallinaceae	*Corallina officinalis* Linnaeus, 1758	8
Rhodophyta	Gigartinaceae	*Chondrus crispus* Stackhouse, 1797	8
Rhodophyta	Phyllophoraceae	*Mastocarpus stellatus* (Stackhouse, 1796)	8
Rhodophyta	Rhodomelaceae	*Melanothamnus harveyi* (Bailey, 1848)	8
Rhodophyta	Wrangeliaceae	*Spermothamnion* Areschoug, 1847	8
Rotifera	Brachionidae	*Keratella* Bory de St. Vincent, 1822	16
Thermoprotei	Pyrodictiaceae	*Pyrolobus fumarii* Blochl, Rachel, Burggraf, Hafenbradl, Jannasch & Stetter, 1997	12
Trematoda	Aporocotylidae	*Paradeontacylix buri* Ogawa, Akiyama & Grabner, 2015	23
Tricladida	Dugesiidae	*Dugesia gonocephala* (Duges, 1830)	20
Tunicata	Cionidae	*Ciona intestinalis* (Linnaeus, 1767)	2

Glossary of Terms Used in Aquatic Parasitology: Ecological and Environmental Concepts and Implications of Marine and Freshwater Parasites

Glossary

Abyssal Referring to regions of the ocean 3000–6000 m below the surface.

Acanthella Larval stage in the life cycle of acanthocephalans that occurs after the acanthor stage and before the cystacanth stage. Once the acanthor is ingested by the intermediate host, it hatches and penetrates the host's gut wall. Inside the host's tissues, the acanthor develops into the acanthella.

Acanthocephala A group of exclusively endoparasitic species that attach as adults in the intestines of vertebrates of all classes with the help of an eversible proboscis equipped with spines. They are also referred to as thorny-headed worms or spiny-headed worms.

Acanthor First larval stage in the life cycle of acanthocephalans. It develops from the fertilised egg within the egg capsule and is characterised by having a fully formed proboscis, hooks, and other structures necessary for infecting the intermediate host.

Acclimation Pre-exposure to new thermal conditions to minimise disturbance to organisms when subjected to the new experimental conditions. Important to consider when examining effects of temperature variability on host susceptibility and parasite transmission.

Acetabulum (pl. acetabula) Specialised structure of eucestodes for attachment to the intestinal wall of the definitive host. Acetabula are characterised by their sucker or cup-shaped structures.

Acidification Reduction in pH in aquatic systems. In oceans, due to increased absorption of CO_2 from the atmosphere. In freshwaters, due to acid precipitation from airborne pollutants.

A-cyprid Larval stage in the Ascothoracida.

Accumulation bioindicator Organism that accumulates substances in its tissues above ambient levels.

Adaptation The process by which parasites develop physical features or mechanisms that enhance their ability to infect and survive novel hosts, contributing to successful establishment and colonisation.

Adoral zone of membranelles (AZM) An area near the mouth of a ciliate, with many membranelles to move food in the direction of the mouth.

Alien species New species that are intentionally or unintentionally introduced into an ecosystem assisted by direct or indirect human intervention; also called non-indigenous or non-native species.

Allogenic life cycle Complex parasite life cycle involving freshwater intermediate hosts and a terrestrial definitive host that can disperse the parasite among distinct aquatic habitats.

Allopatric (noun Allopatry) Two or more related species which have separate non-overlapping geographic distributions.

Amoebic dysentery Is caused by *Entamoeba histolytica* and associated with symptoms such as abdominal pain, fever, bloody or mucous diarrhoea.

Ancestral polymorphism Presence of alleles (or haplotypes) at a polymorphic nuclear or mitochondrial gene, which have not been fixed as alternative between two taxa during the speciation process. Consequently, shared alleles (or haplotypes) may be found in both species.

Anthelminthic Medication or substance used to treat parasitic worm infections.

Anthropocene Current geological epoch, begun in the mid-twentieth century when human activities became the most significant impact on the Earth's geology and ecosystems.

Apical complex Characteristic organelle of species in the phylum Apicomplexa, made up of rhoptries, micronemes, polar rings, and, if present, the conoid (spirally arranged microtubules), used for cell penetration and feeding.

Aquaculture The culture of aquatic organisms, such as fish, shellfish, and aquatic plants, under controlled conditions.

Asphalt lake A surface deposit of natural asphalt consisting of hydrocarbons, sulphur, metals, and volcanic ash oils.

Aspidogastrea Subclass of trematodes, with simple life cycles, usually including only one host and often lacking asexually produced free-living larval stages.

Autogenic life cycle Complex parasite life cycle that involves only freshwater hosts, i.e. that is completed entirely in the freshwater environment.

Benthic zone The bottom of a waterbody.

Bioaccumulation Accumulation of substances, e.g. pollutants within organisms.

Binary fission Is a mode of asexual reproduction. After duplication and separation of the genetic material of the parent cell, it divides into two identical daughter cells.

Bioindicator An organism or community of organisms that respond to environmental changes by altering their vital functions (**effect indicator**) and/or chemical composition (**accumulation indicator**), thus providing insights into the condition of their environment.

Biomagnification Increasing pollutant concentrations along food webs with higher concentrations in organisms of higher trophic levels.

Biomarker Measurable biological characteristic as response of test organisms to changes in their environment. Biomarkers can be classified as biomarkers of exposure and biomarkers of effect. In general, biomarker responses focus on the change rather than the status of the organism's environment and are therefore measured relative to reference conditions.

Bioreactor A device or vessel used for the cultivation of microorganisms, e.g. animal cells in cellular aquaculture.

Biosecurity Measures and protocols put in place to prevent the introduction and spread of diseases, pests, and other harmful biological agents.

Biotransformation (also called xenobiotic metabolism) Metabolisation of (organic) pollutants in organisms with the goal to detoxify them. In phase I lipophilic substances are converted into polar molecules, which are conjugated to nontoxic endogenous metabolite leading to water-soluble, excretable products. These will be conjugated to a nontoxic endogenous metabolite in phase II that leads to a water-soluble, excretable product.

Biramous Usually used in reference to a crustacean appendage (limb) in having two branches (rami) coming off of the limb basis: the outer branch, or exopodite (exopod), and the inner branch, or endopodite (endopod).

Blastocyst Fleshy capsule or vesicle that encases the plerocercoid of trypanorhynch cestodes.

Bopyridium Third life cycle stage of epicaridian crustacean isopods (Bopyroidea), which develops into a male or female adult.

Bothria Specialised structure of eucestodes for attachment to the intestinal wall of the definitive host. Bothria are typically elongated, slit-like grooves or depressions.

Brine lake/pool An underwater pool, in which highly concentrated saltwater collects in a depression on the ocean floor (basically a hypersaline lake at the bottom of the sea).

Canonical microsporidians Group of derived microsporidians that contains most of the microsporidian diversity known today. In a molecular phylogeny, the canonical microsporidians (also called 'long-branch microsporidians') are separated from the more basal groups (called 'short-branch microsporidians') by a long branch in the phylogenetic tree.

Casual/introduced species Alien species that occurs only occasionally and sparsely but is unable to build self-reproducing populations.

Cellular Aquaculture Development of cell-based seafood products using in vitro cultures of fish tissues.

Cercaria (pl. cercariae) Free-swimming larval stage of digeneans produced by rediae or sporocysts within the first intermediate host. Cercariae leave the first intermediate host and often infect a second intermediate host by penetrating its tissues. However, in some species, cercariae can directly infect the final host by penetrating its skin.

Cercomer Bladder-like structure with six larval hooks, characteristic feature of the procercoid of eucestodes.

Cestoda A group of Platyhelminthes, also known as tapeworms. They are dorsoventrally flattened parasites usually segmented, lacking a digestive tract and typically consist of a differentiated scolex and a chain of proglottides, each including a set of reproductive organs. They occur as adults in the intestines of vertebrates of all classes.

Cestodaria Basal subgroup of cestodes, whose monophyly is not supported. Cestodaria possess 10 hooks (decanth) in the larval stage and show unsegmented bodies without a scolex as adults.

Cetacea Infraorder of marine mammals that includes whales, dolphins, and porpoises.

Chalimus Third larval stage of siphonostomatoid copepods, between the infective copepodid stage and the adult.

Chemoautotroph An organism that can obtain energy from chemical reactions to produce organic compounds.

Chitin Nitrogenous polysaccharide that makes up the main component of fungal cell walls and exoskeletons of arthropods.

Cholangiocarcinoma Bile duct cancer.

Coevolution Evolutionary change driven by the reciprocal selective pressures that two or more interacting species exert on each other.

Co-infection Simultaneous presence of multiple parasitic, bacterial, viral, or fungal pathogens in a host.

Colonisation Phenomenon of a parasite establishing a population where none was present at the time.

Complex life cycle Life cycle of species (typically a parasite), which involves multiple distinct stages or forms in its development, often includes various morphological and ecological changes, and it may alternate between different hosts or environments.

Compound-specific Isotope Analysis (CSIA) Measurement of the stable isotope ratios of e.g. carbon and nitrogen of individual compounds (e.g. amino acids or fatty acids) extracted from complex matrices to derive additional information e.g. on metabolic pathways.

Congeneric Species that belong to the same genus.

Consumption Process of seizing the energy contained in a particular resource. In parasitism (unlike in predation), the parasite consumes the energy contained in its host.

Copepodid Free-swimming larval stage of copepods. This is the second larval stage, between naupliar stages and adult.

Coracidium (pl. coracidia) Free-swimming, ciliated larval stage found in the life cycle of cestodes. Its main function is to find and infect the first intermediate host.

Corollarium To be statistically consistent, Nm must be estimated on Fst values inferred over a large and similar number of parasites individuals from each population collected from host species and/or geographic locality.

Cospeciation Joint and approximately simultaneous speciation of two closely associated and interacting lineages, such as a host and its parasite.

Cotylocidium (pl. cotylocidia) Sexually produced larvae belonging to some species of aspidogastreans.

Cryptic species Species that are genetically distinct from each other but morphologically indistinguishable.

Cuticle Major part of the Crustacea exoskeleton made up of a chitinous, layered matrix.

Cyprid Second larval stage in many crustaceans. It is an infective stage, with usually one or two instars.

Cystacanth Encysted and infective larval stage in the life cycle of acanthocephalans. It occurs within the intermediate host, typically an arthropod such as an insect or crustacean.

Cytostome Feeding groove or mouth of ciliates.

Darwinian shortfall Lack of knowledge about the tree of life and evolution of species and their traits, the phylogenetic relationships of species and edge lengths.

Definitive host Host that harbours the adult form of a parasite, in which the parasite sexually multiplies and completes its life cycle.

Density-mediated indirect effect (DMIE) Density of a species (in this case, a host) is altered by the presence of a second species (in this case, a parasite), such that the changes affect a third species.

Deoxyribonucleic acid (DNA) A molecule that carries the genetic information which determines an organism's traits and characteristics.

Diagnostic loci A locus/genetic marker can be designated as 'diagnostic', when fixed alleles or fixed nucleotide site differences are present between two species. In other words, those fixed differences are not shared between the two species, thus indicating a lack of gene flow between them.

Digenea Most diverse subclass of trematodes, with complex life cycles including several hosts and sexual and asexual reproduction.

Dilution effect An invasive host acts as an unsuitable host for a parasite, in which no further development takes place preventing further transmission and resulting in a net lower infection prevalence in native hosts.

Dinospore Biflagellated, motile, and infective stage of parasitic dinoflagellates.

Dioecious Individual organisms are either male or female, i.e. species have different sexes.

Diplozoic Bilaterally symmetrical cells with two sets of organelles, including nucleus and cytoskeleton as a result of an incomplete cell division.

Diporpa Characteristic larval stage from monogeneans belonging to the family Diplozoidae.

Direct effect Effect exerted directly by a species on the traits or population density of another; there are four types of direct effects.

Dissemination pathway In the context of diseases, it is the route along which a pathogen is spread from the point of release (source) to the location and host where the disease is expressed.

DNA barcoding Identification of a species by using a unique genetic marker that distinguishes the species from all other species.

DNA metabarcoding Process of simultaneous identification of species in a sample using DNA barcoding. Sequencing is typically performed on a high-throughput sequencer such as Oxford Nanopore or Illumina NovaSeq.

Ecosystem engineer An organism that physically affects the availability of abiotic or biotic resources to other organisms, essentially affecting local habitats structure. May be a parasite.

Ecosystem Community of species interacting together within a habitat.

Ecotoxicology A branch of toxicology dedicated to examining the adverse effects of natural or synthetic contaminants on all organisms in ecosystems, considering the broader environmental context.

Ectoparasite Parasitic organism that lives on the external surface of a host.

Effect indicator Organism that changes its vital functions in response to pollutant exposure.

Eltonian shortfall Lack of knowledge on the interactions of species and the effect of these interactions' effect on individual survival and fitness.

Endobiont An organism living inside another organism.

Endoparasite Parasitic organism that lives inside the body of its host. Endoparasites can reside within tissues, organs, or body cavities.

Enemy Release Hypothesis (ERH) Invasive species can succeed due to the lack or release of enemies (e.g. predators, pathogens, and parasites) in the recipient environment.

Environmental DNA (eDNA) DNA that can be extracted from environmental samples (such as soil, water, faeces, or air) without isolating target organisms first. eDNA can include free DNA, cells, or whole organisms.

Epibiont An organism living on the surface of another organism.

Epidemic A greater number of disease outbreaks than expected in a short time and often over a wide geographical area.

Equilibrium isotope effects Result from thermodynamic processes usually between two substances in chemical equilibrium exchange reactions that simultaneously proceed both forward and backward and will eventually come to an equilibrium with the heavier isotopes concentrating where the bond strengths are strongest.

Established species Alien species that has persisted over several generations without human intervention.

Eukaryote (adj. eukaryotic) Organisms having cells that contain a nucleus surrounded by a membrane, and whose DNA is bound together by proteins into chromosomes.

Euryxenous Parasites with several, not related host species (i.e. more than one family of hosts), syn. generalists, polyxenous, polyxenic.

Eutrophication High nutrient levels (especially phosphorus and nitrogen) in aquatic systems that are associated with increased primary production. May lead to oxygen depletion of bottom waters.

Exophthalmia Abnormal protrusion of the eyeball.

Extrasporogonic stages Phases in the life cycle of some parasitic organisms that occur outside the final site of infection. These stages involve processes such as replication and development that are essential for the parasite's survival and transmission.

Extremophile An organism living and thriving under extreme abiotic conditions that most organisms would not be able to withstand, such as extreme temperatures, radiation, or pH.

Extremotolerant An organism able to tolerate (at least for some time) the stress of extreme abiotic conditions that most organisms would not be able to withstand, such as extremely high or low temperatures, radiation, or pH.

Extrusome Membrane-bound organelle that can release its content/structures, used for protection or prey capture.

Fallowing Period of inactivity or rest; often used in aquaculture where a farming site is temporarily stopped from production to restore the natural environment and control diseases.

Fluke Parasitic platyhelminth usually with suckers for attachment to the host.

Founder effect Loss of genetic variation occurring when a new population is established by a very small number of colonising individuals from the larger source population.

F-statistic (Fst) also known as fixation index. This statistic is commonly used as a measure of population subdivision and it is also useful to estimate the interpopulational gene flow (see below), under the neutrality theory. The parameter was introduced by Wright (1951) to describe the genetic population structure of diploid organisms. Fst can be interpreted as the variance of allele frequencies among parasite populations from different hosts and geographic areas. It can be estimated by using dedicate software programmes, such as DNASP and Arlequin.

Gall Locally restricted, abnormal growth of plant cells/tissue caused by infection.

Gene flow Transfer of genetic material between parasite populations resulting from the movement of its individuals, allowed by the movement of its hosts and/or free-living stages. Gene flow can be inferred from the spatial distribution of the alleles/haplotypes estimated at the intraspecific level, by genetic/molecular markers of the parasite populations over a spatial scale, by a statistical approach. This approach is based on the 'island model' or 'stepping stone' theories of population genetic structure. This means a species is assumed to be subdivided in populations (islands) of equal size N, where all the individuals have the same probability to exchange alleles of equal probability or the model wherein only adjacent demes exchange alleles. However, the alleles frequencies of a population are under the effect of the natural genetic drift phenomenon, which is correlated to the population size (N) of the parasite, which in turn can lead to the genetic variation of a single populations. Thus, both the gene flow and the genetic drift are acting on a population. As a consequence, the statistical estimate of gene flow can be inferred indirectly, taking into consideration the measure of the genetic differentiation (Fst) of each single population, and the absolute number of individuals exchanged between different populations per generations, under the neutrality theory. It would be the scenario of N close or equal to zero, which is generally correlated to an Fst value close or equal to 1 in this case reproductive isolation between the parasite populations would exist.

Gene Segment of DNA that serves as the basic unit of heredity in living organisms.

Generalist See Euryxenous.

Genetic drift Unpredictable changes over time in the relative frequencies of different alleles in a population, owing to chance events.

Genetic marker DNA fragment that allows to distinguish a species from others due to mutations in this gene. Typical genetic markers for barcoding are the mitochondrial cytochrome c oxidase subunit 1 (COI) gene for invertebrates, the 12S mitochondrial ribosomal gene for vertebrates, the 18S nuclear gene for unicellular eukaryotes, the ITS (internal transcribed spacer) fragment for fungi, and the 16S ribosomal gene for prokaryotes.

Genetic variability The genetic variability (in some instances, also reported as genetic diversity) is intended as the genetic polymorphism of a population/species. It can be estimated at the following parameters: observed and expected heterozygosity (Ho and/or He), mean number of alleles (A), mean number of private alleles (Ap) and haplotypes (Nh).

Genomic Selection Breeding technique that uses genomic information to select individuals with specific genetic traits.

Global warming potential (GWP) Measure of the total greenhouse gas emissions associated with a particular product or activity.

Gonocyte Is a life stage in some dinoflagellate life cycles that produces many sporocytes through sequential mitotic division.

Gonosphere Reproductive cyst stage where flagellated spores are formed.

Habitat Environment in which a community of species live; its biotic and abiotic characteristics determine which species occur in it, and where they occur. In the case of internal parasites, the host body itself is the habitat.

Hadal Referring to regions of the ocean > 6,000 m below the surface.

Haemocoel The body cavity of arthropods and molluscs.

Haemocyte Morphologically distinct cell type of the invertebrate (e.g. Arthropoda) haemolymph.

Harmful algal bloom (HAB) Proliferation of algae and/or cyanobacteria that cause harm, through either production of toxins (cyanotoxins) that pass up the food chain or depletion of oxygen in bottom waters, leading to dead zones.

Heterotrophic Organisms that are dependent on feeding on organic compounds produced by other organisms for biomass production.

Heteroxenous A parasite requires at least two hosts, a definitive host and at least one intermediate host species to complete the life cycle.

Hidden diversity Presence of a greater number of distinct species within a group of organisms initially classified as a single taxon (species) based on traditional morphological or genetic characteristics. This concept highlights the importance of molecular techniques in revealing previously unrecognised diversity.

Histology Study of the microscopic structure of tissues and organs in living organisms; involves the microscopic examination of fixed, sectioned, and stained samples attached to glass slides.

Homeobox proteins Are a large family of transcription factors (TFs) that contain a highly conserved DNA-binding domain of 60 amino acids known as the homeodomain and are involved in the patterns of anatomical development.

Horizontal transmission Transmission within one host generation, e.g. in microsporidians by spores released from one host that are taken up by another host individual (see also 'Vertical transmission').

Host Organism that gives food and shelter to another organism (often a parasite).

Host-specific See Monoxenous.

Host-switching Evolutionary event through which a parasite colonises a new host species, leading to a reduction in its host specificity.

Hutchinsonian shortfall Absence of knowledge on species' ecology, tolerances, and sensitivity to changes in habitat.

Hybridisation Mating between individuals from genetically distinct species that produces offspring, which represents hybrid of first (F1) generation.

Hydrothermal vents Underwater geysers or hot springs that typically form around active volcanic regions and release hot water of up to 400°C that remains fluid under the intense pressure of the deep sea.

Hyperparasitism A parasite infecting another parasite species.

Hyperplasia Increased cell production in a tissue or organ.

Hypersaline lake A waterbody with concentrations of dissolved salt exceeding 35 g/l.

Hypertrophy Excessive growth of host cells that can lead to enlargement of host organs and tissue.

Hyphae Long and slender fungal tubes that develop from a spore of fungal body and may or may not be divided into sections by cross-walled septa.

Hypnospore Thick-walled, non-motile resting cyst.

Immunocompromised Animals are immunocompromised when they have a weakened immune system due to a particular condition that suppresses their immune system.

Immunostimulant A substance or agent that enhances the natural immune response in organisms.

Indirect effect Impact on the traits or population density of a species resulting from the direct interaction between two other species.

Integrated multitrophic aquaculture (ITMA) Farming approach that involves cultivating species from different trophic levels, such as seaweed, shellfish, and fish together.

Intermediate host An essential host for parasites with indirect (heteroxenous) life cycle that harbours a larval stage of the parasite in which the parasite grows, and/or asexually reproduces, but does not reach its sexual maturity.

Internode Horizontal stems or rhizomes of seagrass are composed of segments. At the nodes of the rhizome, stems or leaves form and the sections in between these nodes are called internodes.

Interspecies competition Direct or indirect antagonistic interaction between species requiring access to limited habitat resources; in parasitism, the host itself is considered as a limited resource.

Intertidal zone Shoreline region that is covered by water during high tide and exposed to air during low tide.

Introgression Incorporation of alleles from a species into the gene pool of a second divergent species, through hybridisation and backcrossing of F1 hybrids with individuals from one of the two parental species. The resultant offspring hybrid categories are 'backcrosses' individuals.

Invasional meltdown hypothesis Scenario in which a group of alien species can aid one another through the process of invasion by increasing their chances of survival and overall invasion success, including the respective ecological repercussions.

Invasive species (IS) Alien species (free-living or parasitic) that has been able to reproduce and spread at a significant distance from the original point of introduction. Its introduction is likely to cause environmental, human, or economic harm.

Isospace x–y range of isotopic values typical of a given species and representing the specific ecological niche of this species in terms of the $\delta^{13}C$ and $\delta^{15}N$ values.

Isotope fractionation Stepwise differences (called discrimination factor, Δ) between the isotopic ratios of naturally occurring stable isotopes of carbon ($\delta^{13}C$) and nitrogen ($\delta^{15}N$) between consumers of different trophic levels.

Isotope Labelling ^{15}N or ^{13}C-labelled resources are experimentally added in field or laboratory experiments and then traced through the food chain, often referred to as a pulse-chase study.

Isotopic Mixing Models Allow translation of isotopic data into estimates for food source contribution to consumers.

Keartonian shortfall Concerns the lack of availability of visual representations of what species look like, to promote which traits to look for and create an understanding of a species' appearance, behaviour, and habitat.

Kentrogon Non-segmented larvae of Rhizocephala. Injects the vermigon into the host.

Keystone parasite Parasite infecting a competitively dominant host species, reducing its density so that species diversity increases.

Keystone species A predator of a competitively dominant species, reducing its density so that species diversity increases.

Kinetic isotope effects Have their origin in the energy differences between isotopologues during the formation of an active transitional complex in unidirectional, rate-dependent reactions such as diffusion and chemical reactions.

Kinetid Elementary repeating organellar complex of the typical ciliate cortex. The term describes the flagellar and ciliary apparatus, including basal bodies (kinetosome), ancillary fibres, and microtubular roots.

Kinety Single structurally and functionally integrated somatic file or typically longitudinally oriented row of kinetids.

LC_{50}/LD_{50} Concentration or dose of a substance (e.g. a pollutant) at which 50 percent of the organisms die.

Life cycle A series of stages in the life of a parasite, through which the parasite goes through physical changes allowing it to reach adulthood and produce new organisms. Parasites can have direct or indirect life cycles. During direct life cycle, they require only one host type (i.e. intermediate or definitive), during indirect cycle, both intermediate and definitive hosts are essential.

Linnean shortfall Discrepancy between the number of taxonomically described species and the actual number of existing species.

Local adaptation Phenomenon in which different populations of the same species each possess traits that confer them higher fitness in their particular local environment, resulting from natural selection driven by local biotic and abiotic factors.

Long-branch microsporidians See canonical microsporidians.

Lorica A secreted and/or assembled test, envelope, case, shell, or theca, usually open at one end.

Lungworm (marine mammals) Parasitic nematode members of the families Crenosomatidae, Parafilaroididae, and Pseudaliidae (superfamily Metastrongyloidea) specific to the respiratory tract, cranial, and auditory sinuses as well as the circulatory system of cetaceans and pinnipeds.

Macroevolution Evolution above the species level, occurring on geological time scales and involving the origins of major traits and new lineages.

Major Histocompatibility Complex (MHC) Polymorphic immunoglobulin receptors responsible for antigen presentation to T-cells in the innate and adaptive immune response. Class I genes are involved in the immune response to intracellular pathogens; class II genes in the response to extracellular pathogens.

Marsupium Brood pouch of some crustaceans. Formed with lamellar oostegites on the ventral surface of the female.

Membranelle Several serially arranged oral polykinetids.

Meront This life cycle stage is also called schizont and describes an enlarged trophozoite that has undergone repeated replication of its nucleus and organelles.

Merozoite A small, specialised, infective stage produced through multiple fission of the meront, mainly in Apicomplexa.

Metabolic Theory of Ecology (MTE) Modelling approach that relates an organism's (parasite or host) metabolic rate to its body size and temperature that can be used to predict responses to climate warming.

Metacercaria (pl. metacercariae) Infective larval stage for the definitive host in the life cycle of digenean trematodes. After invasion of a cercaria into the second intermediate host, the cercaria encysts itself or in combination with host responses and develops into a metacercaria. These cysts provide protection to the larvae and help them survive until they are ingested by the definitive host in which they excyst and mature into adults. However, there are also species that encyst on plants, in the water or do not encyst at all.

Metacestode Generic term for larval stages of cestodes, e.g. coracidium, plerocercoid, procercoid.

Metastenoxenous Parasites with more than one host but restricted to one family (also referred to as generalists).

Methane seep An area of the ocean floor where methane diffuses into the water column above it.

Microbiome Refers to the community of microorganisms that live in (e.g. inside the gastrointestinal tract) or on (e.g. the epidermis of a plant) a host organism.

Microevolution Changes in allele frequencies in a population over time scales of a few to many generations, resulting from genetic drift or natural selection.

Micropredator Small predator-like organisms that feed entirely on other organisms but do not typically kill their hosts.

Microtriches Hair-like structures on the external surface of tapeworms, increasing the surface area and aiding in the absorption of nutrients from the host's digestive system.

Miracidium (pl. miracidia) Free-swimming, ciliated larval stage found in the life cycle of digenean trematodes. Its main function is to find and infect the first intermediate host, which is usually a mollusc.

Mitosome Reduced mitochondria in the Microsporidia that lost their function for aerobic respiration but are responsible for parts of the sulphur and iron metabolism.

Mixotrophic Organisms that can photosynthesise and feed on other carbon sources.

Mono-/Multipodial With only one single pseudopodium or multiple pseudopodia.

Monoecious Male and female reproductive organs are present. Monoecious organisms are also referred to as hermaphroditic.

Monogenean Simple life cycle parasites infecting the gills and skin of fish, amphibians, reptiles, cetaceans, or cephalopods, and in rare cases mammals. Only a few species have developed into endoparasites.

Monoxenous Parasites requiring one host species to complete the life cycle, syn. host-specific, monoxenic, oioxenous, specialists, species-specific.

Monozoic Continuous, unsegmented body plan.

Nauplius First larval stage in many crustaceans, usually with three pairs of appendages, a median eye, and little or no segmentation.

Necrosis Premature death of body tissue due to external factors such as injury, radiation, or chemicals.

Nematocysts Specialised stinging cells found in the tentacles of cnidarians, such as jellyfish and sea anemones. These cells contain a coiled, thread-like structure that can be rapidly discharged to capture prey or defend.

Nematoda Free-living or parasitic unsegmented worm-like organisms that inhabit both terrestrial and aquatic habitats. In their parasitic form, they infect crustaceans, fish, molluscs, reptiles, and mammals. They are also referred to as roundworms.

Neodermis Syncytial cytoplasmic type of body cover of the Neodermata (Trematoda, Monogenea, and Cestoda) and Acanthocephala.

Nephrocytes Excretory cells located in the gills of Crustacea.

Non-targeted eDNA analysis A molecular workflow that aims to broadly assess the diversity of organisms present in an environmental sample, e.g., all eukaryotic species, all fish species or selected groups of benthic invertebrate species. This approach typically uses the so-called universal primers that bind well to a broad range of phylogenetically distantly related organisms.

North Atlantic climate oscillation Interannual and decadal fluctuations in winter temperatures, precipitation, wind conditions, and oceanic currents on both sides of the North Atlantic Ocean.

Nutrient cycle Is the movement and switching of inorganic and organic matter for the construction of new matter.

Oioxenous See Monoxenous.

Oncomiracidium (pl. oncomiracidia) Ciliated larvae in the life cycle of monogeneans.

Oncosphere Initial stage in the life cycle of eucestodes, also known as six-hook larva. It usually possesses a protective outer layer called the oncosphere membrane, which can show specific adaptations to the environment, e.g., cilia for purposeful movements in water (see coracidium).

Oogonium Reproductive structure in e.g. brown algae producing female gametes.

Oostegite Large, flexible plate-like expansion. It is the basal segment of a thoracic appendage in many female crustaceans. Forms a brood pouch (marsupium).

Opisthaptor Specialised attachment organ found in the monogeneans. This organ is located at the posterior end of the worm's body and is used for attachment to the host. The opisthaptor typically consists of various attachment organs such as hooks, clamps, suckers, or a combination of these structures.

Opportunistic pathogen Potentially infectious disease-causing agent; rarely causes disease in organisms with healthy immune system.

Orifice An opening, e.g. mouth or hole in a cell, spore, or cyst.

Pansporoblast Group of sporoblasts, clusters of sporogonic cells, from which spores are formed. They are typically developed in plasmodia or pansporocysts.

Pansporocyst Spore-forming stage developing in annelid hosts.

Pantochelis Initial post-embryonic (larval) phase in hyperiid amphipods. This stage occurs within the marsupium of the adult female and the pleon remains undifferentiated and lacks pleopods.

Parapatric distribution Adjacent geographical distribution.

Parasite spillback An invasive host may acquire native parasites in the novel environment and transmit them forward to a native host resulting in disease amplification.

Parasite spillover If parasites, co-introduced with their hosts into the new environment, become established and infect native species.

Parasite An organism that lives in or on an organism of another species (its host) and benefits by deriving nutrients at the other's expense.

Parasitoid An invertebrate whose offspring develop and live as parasites either externally or internally on a host, ultimately killing it.

Paratenic host Host in a heteroxenous life cycle, in which a parasite can survive for a limited period without undergoing any development or significant changes. The presence of a paratenic host can be advantageous for the parasite in terms of transportation and survival but is not essential for the parasite's development or reproduction.

Parthenita (pl. parthenitae) Generic term for larval stages of digenean trematodes, e.g. miracidia, rediae, cercariae.

Probably bette Asexual reproduction in which an embryo develops from an unfertilized egg, without the involvement of a male gamete (sperm).

Pathogen Macro- or microorganism able to cause disease in another organism.

Pedigerous Having legs or feet.

Pelagic zone The open water zone of a waterbody.

Pereopod Thoracic appendage (leg on the person) in crustaceans, used for locomotion.

Phoront Stage in a polymorphic life cycle of a ciliate during which the organism is carried about on or in the integument of another organism (generally a metazoan).

Photoautotrophs Organisms that can use light energy for the production of organic compounds (via photosynthesis).

Pinnipedia Suborder of aquatic carnivores that includes the families Odobenidae (walrus), Otariidae (sea lions and fur seals), and Phocidae (true seals).

Plasmodium Mass of multinucleated protoplasm in the life cycles of certain protists.

Platyhelminthes Phylum with a few free-living, predatory species, a few ectoparasitic and many endoparasitic worms, including a number of human pathogenic parasites.

Pleopod A biramous abdominal (pleon) appendage in crustaceans.

Plerocercoid Larval stage of cestodes in the second intermediate host. It is usually larger than the procercoid and already possesses a well-defined scolex and a relatively short, subtly segmented strobila.

Polar filament/Polar tube Ejectable filament in the spores of both microsporidians and myxozoans. In microsporidians the polar filament is a hollow tube that injects the sporoplasm in the host cell and in myxozoans, the polar filament is used for the attachment of the spore to the host.

Polaroplast Stacks of flattened membranes at the anterior end of microsporidian spores. Osmotic swelling of the polaroplast induces the ejection of the polar filament.

Pollutant exposure Behaviour of the pollutant within the environment, including its entry in organisms.

Polykinetid A kinetid composed of three or more kinetosomes and their fibrillar associates, forming a cirrus or membranelle.

Polymerase chain reaction (PCR) A laboratory technique in molecular biology used to amplify or replicate a specific segment of DNA.

Poly-paraphyletic relationships Phylogenetic relationships of group of organisms classified together based on shared characteristics, but this grouping is not based on a common evolutionary ancestor.

Population density Number of individuals of a particular species per unit area or, in aquatic habitats, per unit volume.

Praniza Larval phase in gnathiid isopods (Gnathiidae). It is the fed, haematophagous phase that will drop off from its host once it has completed its meal and find a suitable benthic substrate to reside in until it moults.

Prestonian shortfall Lack of knowledge on any given species' abundance and changes thereof through space and time.

Primer A short, typically 18–25 base-pair long single-stranded DNA sequence that is used in a PCR reaction to anneal to a binding site or a template molecule. For the PCR two primers are needed that define the boundaries of the amplified fragment, i.e. the amplicon.

Private alleles/haplotypes Private alleles are those observed by nuclear gene loci, and haplotypes from mitochondrial ones, found at the intraspecific level, only in a specific one population of a parasite species. It means that private/unique haplotypes are those found as not shared with other populations of the species.

Proboscis Spiny, retractable organ used by acanthocephalans for attaching to the host's intestinal wall. It is equipped with hooks or spines to anchor the parasite securely.

Procercoid Larval stage of cestodes in the first intermediate host, characterised by its elongated and cylindrical shape.

Proglottids Individual, segmented units that make up the body of tapeworms.

Prohaptor Specialised attachment organ found in the monogeneans. The prohaptor is located at the anterior end of the worm's body, opposite to the posterior attachment organ known as the opisthaptor. Similar to the opisthaptor, the prohaptor consists of various attachment structures.

Prophylactic Preventive measure or action taken to reduce the risk of disease.

Protostomous The mouth of the ciliate sits at the apical end of the cell.

Protandry Maturation of male reproductive organs before female reproductive organs.

Protelean An organism that has parasitic juveniles and free-living adult stages.

Protist Eukaryotes with unicellular, colonial, filamentous, or parenchymatous organisation, which lack vegetative tissue differentiation except for reproduction.

Protogyny Maturation of female reproductive organs before male reproductive organs.

Protomont Brief stage in the polymorphic life cycle of some ciliates, recognised as separate stage between the feeding trophont and the often encysted true tomont, based on kinetome features.

Protoplast The living content of a cell.

Protopleon The second developmental (larval) phase in hyperiid amphipods (up to five stages). This phase is post-marsupial and initially unable to swim. The pleopods progressively develop in different stages.

Pseudopodium Temporary plasma membrane-enclosed extension of the cytoplasm, used for locomotion and food uptake.

Quarantine Isolation procedures used to monitor and prevent the spread of diseases in new or potentially infected organisms.

Raunkiæran shortfall Lack of knowledge about species' traits and their ecological functions.

Recirculating aquaculture system (RAS) Aquaculture system designed to use recycled water, purified using biofilters.

Redia (pl. rediae) Larval stage in the life cycle of digenean trematodes, which occurs within the first intermediate host. Rediae are elongated, sac-like structures that contain a primitive digestive system, including a mouth, a simple gut, and sometimes rudimentary organs like a birth pore. They actively move within the host's tissue and feed on it or other larvae. Rediae are responsible for asexual reproduction within the intermediate host. They produce additional generations of rediae and eventually produce the next larval stage, cercariae.

Reference database Public or private database where reference sequences are stored. Commonly used public DNA reference databases include NCBI GenBank (universal), BOLD (mostly metazoans), Silva (microbes), diat.barcode (diatoms), and UNITE (fungi).

Reference sequence Sequence of a genetic marker analysed for a taxonomically well-identified species that is deposited in a database. With reference sequence comparisons, eDNA sequences can taxonomically be assigned to the lowest taxonomic level currently represented in the reference database.

Ribonucleic acid (RNA) Molecule that plays a vital role in the transmission of genetic information and the synthesis of proteins in cells.

Salinity Concentration of salt in a given body of water.

Saprophyte Organism, such as a fungus or a bacterium, that resides and nourishes itself by consuming deceased and decomposing plant and animal material.

Schizogony Asexual, multiple-fission stage in the life cycle of Apicomplexa.

Scolex Head-like anterior end of a cestode.

Seal hookworm Parasitic nematodes belonging to the family Ancylostomatidae and genus *Uncinaria* that infect pinnipeds.

Seal louse (pl. lice) Amphibious insects of the family Echinophthiriidae that live attached to the hair of pinnipeds and the river otter.

Sealworm or codworm Anisakid nematodes belonging to the *Phocanema decipiens* (syn. *Pseudoterranova decipiens* species) complex that infect marine and euryhaline invertebrates and fishes as intermediate/paratenic hosts and pinnipeds as definitive hosts.

Sequencing Process of determining the precise order of nucleotide bases in a DNA or RNA molecule; can be done for entire genomes or specific regions of DNA or RNA.

Short-branch microsporidians See canonical microsporidians.

Sirenia An order of aquatic mammals, composed by three species of manatees (family Trichechidae) and one species of dugong (family Dugongidae).

Soda lake A highly alkaline waterbody, often with pH values of 9 or above.

Somatic kinetid Kinetids confined to the somatic region (all of a ciliate's body, except for the oral region) used for locomotion.

Somite Segment or subdivision of the body.

Specialist species See Monoxenous.

Species-specific See Monoxenous.

Sporangium Sealed capsule housing spores generated in fungi and algae.

Sporoblast A sporozoan reproductive stage that develops spores and sporozoites.

Sporocyst Larval stage in the life cycle of digenean trematodes, which occurs within the first intermediate host. Sporocysts are simple, elongated, sac-like structures without a digestive system. They absorb nutrients directly from the host's tissues. Sporocysts are responsible for asexual reproduction within the intermediate host. Within the sporocyst, further asexual reproduction occurs, leading to the development of multiple rediae or cercariae, which will eventually leave the sporocyst to infect the next host.

Sporoplasm Infective cell or group of cells in the spores of microsporidians and myxozoans.

Stable isotopes Non-radioactive forms of elements that have the same number of protons but a different number of neutrons, giving them slightly different atomic masses. Unlike radioactive isotopes, they do not decay over time.

Stranding Event where marine mammals, whether alive or deceased, become stranded or washed up onto beaches or shorelines.

Strobila Chain of repeating segments, which form the body of cestodes. The strobila is formed by a series of proglottids, which are individual, reproductive, and connected segments.

Subunit vaccine Contains only specific components (subunits) of a pathogen, rather than the whole organism, to trigger an immune response in recipient.

Sympatric distribution Overlapping geographical distribution.

Sympatric (noun sympatry) Two or more related species which have the same or overlapping geographic distributions.

Synhospitalic When two or more related parasitic species occur together in or on the same host, regardless of their microhabitat in or on the host.

Syzygy Association of two gamonts (male, female) at the start of gregarine sexual reproduction.

Tantulus The only larval stage of Tantulocarida.

Targeted eDNA analysis A molecular workflow that aims to specifically test for the presence of a single target species. Sometimes this approach is also termed '**active surveillance**'.

Taxonomically restricted genes Also known as orphan genes, are genes that are unique to a particular taxonomic group or species, and do not have homologues (similar genes) in other organisms. These genes may play important roles in species-specific adaptations and traits.

Tens Rule Hypothesis The assumption is that upon transport, only 10% of introduced species manage to establish, from which again 10% will be able to build permanent, self-reproducing populations (without human intervention) and of which in turn only 10% become invasive.

Thermal mismatch hypothesis (TMH) Differences in thermal tolerances or thermal optima between small-bodied parasites and their larger hosts. It predicts that hosts should be more susceptible to parasites when environmental conditions deviate from their thermal optima.

Thermal performance curves (TPCs) These describe the ability of an organism (parasite or host) to perform a physiological function across a range of temperatures.

Theront Dispersal stage in the polymorphic life cycle of a number of parasitic or histophagous ciliates.

Tomite A stage in the polymorphic life cycle of a number of parasitic or histophagous ciliates. Numerous tomites emerge from a cyst, in which the tomont divided.

Tomont Prefission or dividing stage in the polymorphic life cycle of a number of parasitic or histophagous ciliates; large, typically encysted, may divide a number of times in quick succession.

Toxicodynamics Study of the interactions of a chemical at the site of action with a biological target and its adverse effects.

Toxicokinetics Study of the absorption, distribution, metabolism, and elimination (ADME) of chemicals in an organism.

Trait Phenotype; relates to physical, physiological, immunological, and behavioural characteristics of an individual in a species.

Trait-mediated indirect effect (TMIE) When the phenotype of a species (in this case, a host) is altered by the presence of a second species (in this case, a parasite), such that the changes affect a third species.

Translocation Bringing a species to a new territory; the phases of transport and introduction often cannot be separated.

Trematode Also referred to as flukes, are a diverse group of endoparasitic flatworms that parasitise both invertebrates and vertebrates.

Trichogon Simple amoeboid structure in Rhizocephala. It is inserted into the virgin externa and becomes a dwarf male.

Trophic transfer Process through which elements, including contaminants move from one trophic level to another.

Trophont Feeding or growing life stage in general or specifically between the tomite and tomont in the polymorphic life cycle of a number of parasitic or histophagous ciliates.

Trophozoite Feeding and growing stage of some protists (especially Sporozoa).

Turnover rate Time that is needed for a tissue to completely incorporate and turn over nutrients from the nutritional pool.

Uniramous Usually used in reference to a crustacean appendage (limb) having one branch (ramus) coming off of the limb basis; undivided.

Urosome The posterior tagma of the body in certain crustaceans, positioned behind the main articulation and typically comprising the fifth pedigerous somite and the abdominal somites.

Vermigon A migratory, multicellular, vermiform larval stage (in Rhizocephala) injected by a kentrogon larva into the host.

Vertical transmission Transmission from one generation to the next. Found in microsporidians that infect the gonadal tissue of female hosts and can thereby enter the oocytes. The developing offspring will then carry the infection.

Virulence Reduction in fitness incurred by the host following infection by a parasite, measured as reduced reproductive output, decreased survivorship, etc.

Wallacean shortfall Incomplete knowledge about the geographic distribution of species.

Whale louse (pl. lice) Ectoparasitic amphipods of the family Cyamidae that live on the body surface of cetaceans.

Whale worm Larvae of anisakid nematodes belonging to genus *Anisakis* that infect marine and euryhaline invertebrates and fishes with cetaceans as final hosts.

Zoonosis An infectious agent that may be transmitted from animals (e.g. a fish) to humans.

Zoonotic organism Is causing a disease that is naturally transmissible between vertebrate animals and humans.

Zoospore A motile, asexual spore, with one or two flagella in certain algae, fungi, and protists.

Zoosporulation Asexual multiplication within a zoosporangium producing zoospores.

Zuphea A larval phase in gnathiid isopods (Gnathiidae). It is the unfed, clearly segmented phase which will search for a suitable host on which to feed.

The manufacturer's authorised representative in the EU is Springer Nature Customer Service Centre GmbH, Europaplatz 3, 69115 Heidelberg, Germany. If you have any concerns regarding our products, please contact ProductSafety@springernature.com

Printed and bound by CPI Group (UK) Ltd, Croydon, CR0 4YY